ELEMENTARY SCHOOL MATHEMATICS

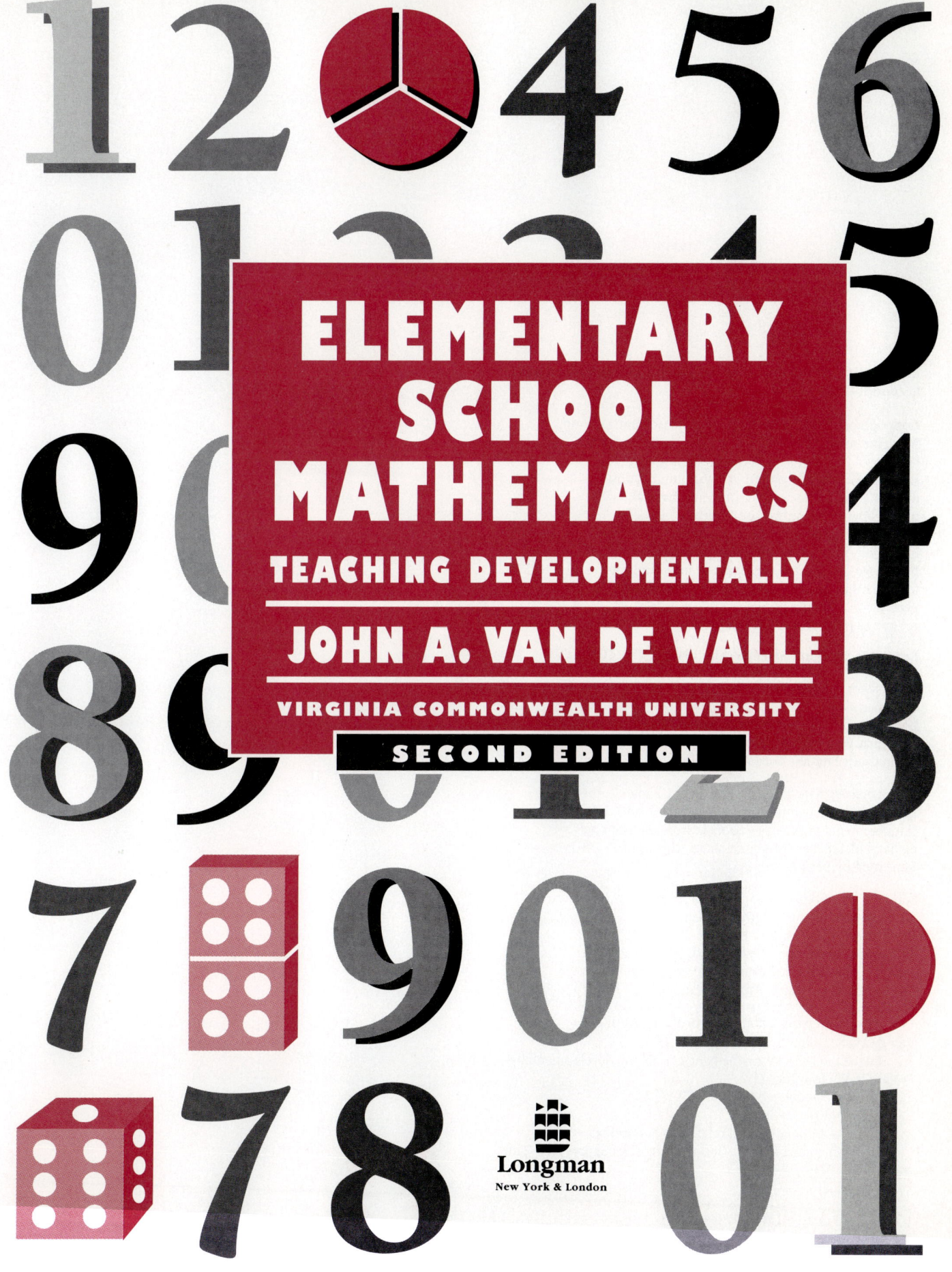

**Elementary School Mathematics: Teaching Developmentally,
Second Edition**

Copyright © 1994, 1990 by Longman Publishing Group.
All rights reserved.
No part of this publication may be reproduced,
stored in a retrieval system, or transmitted in any
form or by any means, electronic, mechanical,
photocopying, recording, or otherwise, without
the prior permission of the publisher.
Classroom instructors and teachers who have
purchased the text for personal use may duplicate
the black-line masters for classroom use without
permission from the publisher.
All others must secure written permission.

Longman, 10 Bank Street, White Plains, N.Y. 10606

Associated companies:
Longman Group Ltd., London
Longman Cheshire Pty., Melbourne
Longman Paul Pty., Auckland
Copp Clark Pitman, Toronto

The *Standards* statements at the end of Chapter 2, the
position statements in Chapters 21 and 23, and the
full statements in the Appendix are from the National
Council of Teachers of Mathematics, copyright © 1991.
Reprinted by permission.

Senior acquisitions editor: Laura McKenna
Development editor: Virginia L. Blanford
Production editor: Ann P. Kearns
Text and cover design: Seventeenth Street Studios
Text art: The Wheetley Company
Production supervisor: Anne P. Armeny

Library of Congress Cataloging-in-Publication Data
Van de Walle, John A.
    Elementary school mathematics: teaching developmentally / John A.
    Van de Walle—2nd ed.
        p.   cm.
    ISBN 0-8013-1184-5
    Includes bibliographical references and index.
    1. Mathematics—Study and teaching (Elementary) I. Title.
QA 135.5.V34   1993
372.7—dc20   93-6099
                                                          CIP
1 2 3 4 5 6 7 8 9 10-CRW-9796959493

CONTENTS

PREFACE ix

CHAPTER 1
TEACHING MATHEMATICS: REFLECTIONS AND DIRECTIONS 1

Thinking about Teaching Mathematics 1
The Revolution in School Mathematics 1
Forces Driving the Revolution 2
The NCTM *Curriculum and Evaluation Standards* 2
The *Professional Standards for Teaching Mathematics* 4
Teaching Mathematics 5
   Reflections on Chapter 1: Writing to Learn 5
   For Discussion and Exploration 6
   Suggested Readings 6

CHAPTER 2
DOING MATHEMATICS: LEARNING IN A MATHEMATICAL ENVIRONMENT 7

Mathematics and Children 7
Doing Mathematics in the Classroom 9
Examples of Problem-solving Explorations 11
The Teacher and the Mathematical Environment 17
   Reflections on Chapter 2: Writing to Learn 19
   For Discussion and Exploration 19
   Suggested Readings 19

CHAPTER 3
DEVELOPING UNDERSTANDING IN MATHEMATICS 21

Knowledge of Mathematics 21
Understanding Mathematics 23
Helping Children Develop Relational Understanding 28
Connecting Conceptual and Procedural Knowledge 32
Teaching Developmentally 35
   Reflections on Chapter 3: Writing to Learn 37
   For Discussion and Exploration 37
   Suggested Readings 37

CHAPTER 4
DEVELOPING PROBLEM-SOLVING PROCESSES 39

Problem Solving in the Curriculum 39
Strategies and Problems 45
Developing Strategies 57
Developing Metacognitive Habits 59
Attending to Affective Goals 60
Assessing Problem Solving 62
   Reflections on Chapter 4: Writing to Learn 63
   For Discussion and Exploration 63
   Suggested Readings 64

CHAPTER 5
ASSESSMENT IN THE CLASSROOM 65

Assessment and the NCTM *Standards* 65
What to Look For 66
Performance Assessments 68
Other Assessment Options 74
Portfolios in the Assessment Plan 79
Diagnostic Interviews 81
Grading 83
    Reflections on Chapter 5: Writing to Learn 84
    For Discussion and Exploration 84
    Suggested Readings 85

CHAPTER 6
THE DEVELOPMENT OF NUMBER CONCEPTS AND NUMBER SENSE 87

Early Number Sense 87
The Beginnings of Number Concepts 87
Procedural Knowledge of Numbers 90
Development of Number Relationships: Numbers through 10 93
Relationships for Numbers 10 to 20 102
Expanding Early Number Sense 104
Summary 106
    Reflections on Chapter 6: Writing to Learn 107
    For Discussion and Exploration 108
    Suggested Readings 108

CHAPTER 7
DEVELOPING MEANINGS FOR THE OPERATIONS 109

Two Sources of Operations Meanings 109
Addition and Subtraction Concepts 110
Word Problems for Addition and Subtraction 115
Multiplication Concepts 118
Word Problems for Multiplication 122
Division Concepts 124
Word Problems for Division 128
Translation Problems in the Upper Grades 129
    Reflections on Chapter 7: Writing to Learn 131
    For Discussion and Exploration 131
    Suggested Readings 132

CHAPTER 8
HELPING CHILDREN MASTER THE BASIC FACTS 133

A Three-Step Approach to Fact Mastery 133
Strategies for Addition Facts 136
Strategies for Subtraction Facts 141
Strategies for Multiplication Facts 146
Division Facts and "Near Facts" 150
Making It Work 150
Fact Remediation with Upper-Grade Students 151
    Reflections on Chapter 8: Writing to Learn 152
    For Discussion and Exploration 152
    Suggested Readings 153

CHAPTER 9
WHOLE NUMBER PLACE-VALUE DEVELOPMENT 154

Early Development of Place-Value Ideas 155
Models for Place Value 157
Developing Place-Value Concepts and Procedures 161
Number Sense Development 169
Numbers beyond 1000 173
Diagnosis of Place-Value Concepts 176
    Reflections on Chapter 9: Writing to Learn 177
    For Discussion and Exploration 177
    Suggested Readings 178

CHAPTER 10
PENCIL-AND-PAPER COMPUTATION WITH WHOLE NUMBERS 179

Algorithms: A New Perspective 179
Developing an Addition Algorithm 181
Developing a Subtraction Algorithm 184
Developing a Multiplication Algorithm 185
Developing a Division Algorithm 193
    Reflections on Chapter 10: Writing to Learn 199
    For Discussion and Exploration 200
    Suggested Readings 200

CHAPTER 11
MENTAL COMPUTATION AND ESTIMATION 201

Alternative Forms of Computation 201
Mental Methods in the Curriculum 202
Mental Addition and Subtraction 203
Mental Multiplication 206
Mental Division 208
Computational Estimation 209
Computational Estimation Strategies 211
Estimation Exercises 214
Estimating with Fractions, Decimals, and Percents 217
Evaluating Mental Computation and Estimation 217
   Reflections on Chapter 11: Writing to Learn 218
   For Discussion and Exploration 219
   Suggested Readings 219

CHAPTER 12
THE DEVELOPMENT OF FRACTION CONCEPTS 221

Children and Fraction Concepts 221
Three Categories of Fraction Models 222
Developing the Concept of Fractional Parts 223
Fraction Number Sense 231
Equivalent Fraction Concepts 234
Other Meanings of Fractions 239
   Reflections on Chapter 12: Writing to Learn 239
   For Discussion and Exploration 240
   Suggested Readings 240

CHAPTER 13
COMPUTATION WITH FRACTIONS 242

Number Sense and Algorithms 242
Addition and Subtraction 243
Multiplication 247
Division 252
   Reflections on Chapter 13: Writing to Learn 255
   For Discussion and Exploration 256
   Suggested Readings 256

CHAPTER 14
DECIMAL AND PERCENT CONCEPTS AND DECIMAL COMPUTATION 257

Connecting Fraction and Decimal Concepts 257
Developing Decimal Number Sense 263
Introducing Percents 267
Computation with Decimals 270
   Reflections on Chapter 14: Writing to Learn 272
   For Discussion and Exploration 273
   Suggested Readings 273

CHAPTER 15
DEVELOPING THE CONCEPTS OF RATIO AND PROPORTION 274

Proportional Reasoning 274
Informal Activities to Develop Proportional Reasoning 278
Solving Proportions 285
   Reflections on Chapter 15: Writing to Learn 289
   For Discussion and Exploration 289
   Suggested Readings 289

CHAPTER 16
DEVELOPING MEASUREMENT CONCEPTS 291

The Meaning and Process of Measuring 291
Developing Measurement Concepts and Skills 292
Measuring Activities 295
Introducing Standard Units 307
Estimating Measures 310
Developing Formulas 312
Time and Clock Reading 317
   Reflections on Chapter 16: Writing to Learn 319
   For Discussion and Exploration 319
   Suggested Readings 320

CHAPTER 17
GEOMETRIC THINKING AND GEOMETRIC CONCEPTS 321

Explorations for Your Reflection 321
Informal Geometry: What and Why 324
The Development of Geometric Thinking 325
Informal Geometry Activities: Level 0 328
Informal Geometry Activities: Level 1 340
Informal Geometry Activities: Level 2 356
Informal Geometry on the Computer 360
   Reflections on Chapter 17: Writing to Learn 364
   For Discussion and Exploration 364
   Suggested Readings 365

CHAPTER 18
LOGICAL REASONING: ATTRIBUTE AND PATTERN ACTIVITIES 367

Objectives 367
Attribute Materials and Activities 368
Working with Patterns 373
    Reflections on Chapter 18: Writing to Learn 380
    For Discussion and Exploration 381
    Suggested Readings 381

CHAPTER 19
EXPLORING BEGINNING CONCEPTS OF PROBABILITY AND STATISTICS 382

Probability and Statistics in Elementary Schools 382
An Introduction to Probability 382
Developing Concepts of Probability 384
Simulations 390
Gathering and Making Sense of Data 391
Collecting Data 391
Graphical Representations 392
Descriptive Statistics 397
    Reflections on Chapter 19: Writing to Learn 401
    For Discussion and Exploration 402
    Suggested Readings 402

CHAPTER 20
PREPARING FOR ALGEBRA 404

What Is Prealgebra? 404
Exploring Topics in Number Theory 404
Roots and Irrationals 409
Integer Concepts 411
Operations with the Integers 412
Developing Concepts of Variable 417
Graphs and Relationships 421
More about Functions 423
    Reflections on Chapter 20: Writing to Learn 426
    For Discussion and Exploration 426
    Suggested Readings 427

CHAPTER 21
TECHNOLOGY AND ELEMENTARY SCHOOL MATHEMATICS 428

Calculators in the Classroom 428
Reasons for Using Calculators 430
When and Where to Use Calculators 431
Practical Considerations Concerning Calculators 432
Computers and Mathematics Education 434
Instructional Software 434
Computer Tools 437
Learning through Programming 438
    Reflections on Chapter 21: Writing to Learn 442
    For Discussion and Exploration 443
    Suggested Readings 443

CHAPTER 22
PLANNING FOR DEVELOPMENTAL INSTRUCTION 445

Conceptual and Procedural Knowledge: Balance and Pace 445
Guidelines for Developmental Lessons 446
Developing Student Responsibility for Concepts 452
Cooperative Learning Groups 453
Homework 454
The Basal Textbook: An Overview 456
Suggestions for Textbook Use 457
    Reflections on Chapter 22: Writing to Learn 458
    For Discussion and Exploration 458
    Suggested Readings 458

CHAPTER 23
MATHEMATICS AND CHILDREN WITH SPECIAL NEEDS 460

Exceptional Children 460
Children with Perceptual and Cognitive Processing Deficits 460
Slow Learners and the Mildly Mentally Handicapped 463
Mathematics for the Gifted and Talented 464
    Reflections on Chapter 23: Writing to Learn 466
    For Discussion and Exploration 466
    Suggested Readings 467

APPENDIX: SUMMARIES OF CHANGES IN CONTENT AND EMPHASIS A–1

REFERENCES R–1

INDEX I–1

BLACK-LINE MASTERS AND MATERIALS CONSTRUCTION TIPS

PREFACE

Education in any discipline helps students learn to think, but education also must help students take responsibility for their thoughts. While this objective applies to all subjects, it is particularly apt in mathematics education because mathematics is an area in which even young children can solve a problem and have confidence that the solution is correct—not because the teacher says it is, but because its inner logic is so clear.

Everybody Counts: A Report to the Nation on the Future of Mathematics Education (National Research Council, 1989)

This book is designed as a guide and a resource to help you with the challenging and rewarding task of helping children develop ideas and relationships that make up mathematics. Children (and adults) do not learn by remembering rules or mastering mechanical skills. They use ideas that they have to develop new ideas and modify old ones. The challenge for the teacher is to engage students in activities where they will create that clear inner logic, not to simply master mindless rules.

The first edition of this text was completed about the same time that real and significant change in school mathematics began in earnest. The National Council of Teachers of Mathematics had just published its seminal document, the *Curriculum and Evaluation Standards for School Mathematics*. The reaction to the *Standards* in the years that have followed has been little short of phenomenal. As you read this book, mathematics in schools is experiencing the most profound and productive changes that have ever taken place. It is an exciting time for mathematics education. This edition of *Elementary School Mathematics* reflects many of the changes that have already occurred and is designed to prepare you to be a part of the continuation of that change.

AN OVERVIEW: WHAT TO EXPECT IN THIS BOOK

Let me give you a brief guide to the way the book is written and some of the things you may expect to find here.

CHAPTERS 1–5: IDEAS ON WHICH TO BUILD

The first five chapters provide a framework for thinking about mathematics and children learning mathematics. Chapter 1, "Teaching Mathematics: Reflections and Directions," describes the revolution referred to above. It introduces you to the NCTM *Standards* and the broad directions of change that we are currently experiencing.

Completely new in this edition, Chapter 2, "Doing Mathematics: Learning in a Mathematical Environment," will help you understand what it means to *know* and *do* mathematics. When you were in elementary school, mathematics probably meant getting answers, memorizing rules, and learning how to do computations. Today, even the youngest children are being challenged to think and solve problems—to "do" mathematics. Chapter 2 will get you in involved in doing mathematics as children should.

ix

Chapter 3, "Developing Understanding in Mathematics," describes mathematical knowledge and what it means to understand mathematics. A constructivist approach to learning is developed. Most importantly, these fundamental principles provide us with general guidelines for teaching mathematics—teaching *developmentally*.

Chapter 4, "Developing Problem-solving Processes," turns to the processes of mathematics as problem-solving. It includes suggestions for helping you and your students become mathematical thinkers—problem solvers.

Chapter 5, "Assessment in the Classroom," reflects the most notable *Standards*-oriented change in this edition of the text. It is an introduction to new assessment strategies for the classroom and how they affect and interact with instruction. Assessment is viewed as an integral part of instruction and so this chapter belongs with the foundational ideas of teaching mathematics. While assessment strategies could consume an entire course, Chapter 5 will provide you with some initial insights into alternative, performance-based assessment in mathematics.

These first five chapters describe a foundation on which you can develop your own personal approach to the excitement of helping children learn mathematics:

- The directions of change
- What it means to do mathematics
- Helping children construct mathematics concepts
- Helping children become problem solvers
- Assessing what children know

CHAPTERS 6–20: ACTIVITIES, LEARNING, AND CHILDREN

Chapters 6 through 20 build on this foundation, as each examines a specific area of the curriculum. *Teaching developmentally* is a child-oriented activity, not a teacher-telling activity. Rather than mindlessly following the superior knowledge of the teacher, children must be mentally active and engaged in the construction of new ideas and relationships. *Teaching developmentally* means that we must constantly think of the mathematics we teach from the vantage point of the child who is learning it rather than from our own. I try in Chapters 6 through 20 to help you see mathematics as might a child who is trying to construct these ideas.

These chapters contain an abundance of activities for children, each boxed in red and identified by a number and a title. While providing you with a resource for your teaching, these activities are also there for *you*. You, after all, are also constructing new knowledge—knowledge about teaching mathematics to children. Just as children must be mentally active and engaged in learning mathematics, you must be actively engaged in learning about children *learning* mathematics. By actually doing the activities as you read through the book, you can get an idea of how children might react to or learn from the activity. It is a good idea to at least try one or two per section.

Consider reading this text not just with a highlighter or pencil to take notes, but with some simple materials at hand—counters, grid papers, a calculator, blocks, and so on. They need not be as nice as those you might prepare for your students. Some beans or pennies make good counters.

The Black-line Masters section has masters that can provide many manipulatives and also assorted grids and mats. These pages are perforated for easier photocopying, and there are instructions to help you make things for your students as well. Add a simple $5.00 calculator, and these materials will be sufficient to get you into most activities.

Try not only doing some of the activities as you read along, but work hard at doing them from the perspective of a child. Avoid using your adult knowledge, which makes many elementary-level tasks seem trivial. The goal of each activity is not to be able to do the activity, or to get the answer, but to *construct* ideas. Activities are designed to encourage children to think. Try to figure out how the activity might affect a child's thinking. Children and adults do not think alike. Reflecting on how children learn from activities is the best way to grow as a teacher.

CHAPTERS 21–23: SPECIAL CONSIDERATIONS

The last three chapters of the text may be read at any time. While calculator activities are found in nearly every chapter and computer activities in many, Chapter 21, "Technology and Elementary School Mathematics," provides additional perspective on the role of calculators and computers in mathematics teaching.

Chapter 22, "Planning for Developmental Instruction," discusses planning lessons, suggestions for homework, and the role of the basal textbook.

My philosophy is that the principles of quality mathematics instruction are essentially the same for all children. There are, however, some additional considerations that should be kept in mind when working with special-needs children, and Chapter 23, "Mathematics and Children with Special Needs," explores these ideas. Discussed are children with learning disabilities, mentally handicapped children, and the mathematically gifted.

SPECIAL FEATURES OF THIS TEXT

The new edition of *Elementary School Mathematics* retains all of the features that I believe helped make the first edition so successful. These include the following.

ACTIVITIES

This book is a text first and a resource second. I have tried to make it informative and complete as a text, but I have also included as many activities as the size of the book would allow. Each activity now has a short, descriptive title, written with you in mind, rather than children, to assist you in recalling or referring back to activities as needed.

DRAWINGS

This text contains no decorative or nonfunctional art. Every drawing is an integral part of the text information and should not be overlooked. In this edition, we have added color in a way that makes the drawings even more functional. Drawings are included whenever a picture seemed to be worth more than the proverbial thousand words and when the idea is especially important. (I frequently tell my own students to "read" the pictures.) Often activities or teaching ideas are embedded in these drawings, even though they may not be labeled as such in the text.

EXPLORATIONS

Each chapter ends with a few questions "For Discussion and Exploration." These are suggestions for going beyond what the text itself can offer. Consider these as food for thought, even if you do not pursue them. Some of these suggestions may be useful if you are in a practicum setting, or for discussion with an experienced teacher or your colleagues.

BLACK-LINE MASTERS

The Black-line Masters section at the end of the book offers an extensive collection of black-line masters and some brief directions for making important materials. Suggestions for their use are found throughout the book. These pages are perforated, and you are encouraged to tear them out and duplicate them to make materials for use in activities. (Permission to copy these pages can be found on the copyright page.)

COMPUTER AND CALCULATOR ACTIVITIES

This book, along with the NCTM *Standards*, strongly advocates the use of calculators as a tool for learning. Calculator-based activities are included in nearly every chapter and are identified with an icon. Computer-based activities are included in those instances when the computer can have a unique impact on learning. In addition, an entire chapter, Chapter 21, provides an overview of the role that technology can play in mathematics learning.

NEW TO THIS EDITION

EMPHASIS ON DOING MATHEMATICS

In the years since the *Curriculum Standards* was released, I have found it extremely important for teachers, both pre-service and in-service, to reconceptualize what it means to *do mathematics*. For this reason, I added a completely new chapter, Chapter 2. Included there are activities like the ones with which I usually begin my classes. My hope is to get teachers involved in really doing mathematics in the same spirit as children, actively engaged in a problem-solving format. These activities are also a vehicle for introducing calculators as learning tools and a few important manipulatives.

A CONSTRUCTIVIST APPROACH

Chapter 3, in which the constructivist approach to learning mathematics is developed, was entirely rewritten from the first edition. Although no fundamental idea has changed, I believe that the presentation of these ideas is now clearer. I would like to acknowledge Jim Hiebert and Tom Carpenter, whose chapter "Learning and Teaching with Understanding" (1992) was a significant influence in rewriting this chapter.

MORE EXPLICIT REFLECTION OF THE *STANDARDS*

The second edition reflects the *Standards* more explicitly than the first edition. In that regard, the first chapter provides considerable information about the directions articulated in the *Standards* documents. References to the *Standards* are also more frequent throughout the text.

ASSESSMENT

Perhaps the most notable *Standards*-oriented change is found in Chapter 5. Chapter 5 provides an extensive introduction to new assessment strategies—including performance-based assessment, portfolios, group assessment, and observations—and how these strategies can be an integral part of instruction that should at least open students' eyes about alternative or performance-based assessment in mathematics. Although no single chapter can adequately address all of these new ideas, I believe you will find this a good beginning.

REFLECTIVE THINKING

This is a very densely packed text, with more detail than you may want on first reading. To help you focus on the "big" ideas, I have added questions at the end of each chapter under the heading "Reflections on Chapter *N*: Writing to Learn." The notion of helping children reflect on a new idea in order to learn and understand it is a guiding principle of this book. *Your* reflection on ideas you have read or discussed in class is also the best way for you to learn. As I have written this book I have been amazed at how my own understanding has grown as a result. I am deeply convinced that writing and/or discussing an idea is the very best way there is to develop an understanding of that idea. I hope these questions will help you in that constructive endeavor.

As you have probably noticed, a brochure from the Cuisenaire® Company of America accompanies each copy of this edition of *Elementary School Mathematics*. This brochure provides you with access to a wide array of math manipulatives for your own use in learning how to help children learn mathematics, or for use in your elementary school classroom.

NOTES TO THE INSTRUCTOR

I chose to write a book that talks to teachers. *Elementary School Mathematics*, Second Edition, provides teachers and prospective teachers with enough information about how children learn, and enough conceptual activities for helping them learn, that the book will serve them as a resource as well as a text. As a result, this book is long. I hope that you look on that length as a luxury and not a burden. It allows you to make choices and provide emphasis on topics or activities that are most important to you. Those topics that you choose not to cover in class, and those activities that you elect not to explore, are described in sufficient detail that they can be studied independently. Your students will carry those ideas and discussions with them beyond your course if they see this book as a resource rather than just a text. There is plenty from which to choose.

As I have talked with friends around the country who have used the first edition, I have been struck by the many different styles of methods courses to which the book has been adapted. These conversations have convinced me that there are many ways to teach teachers and just as many ways that this book may serve to help. I wish you and your teachers much success and excitement.

ACKNOWLEDGMENTS

Much of the credit for the success of the first edition belongs to the mathematics educators who gave time from their own professional endeavors and took great care in offering comments on the original draft manuscript. Each provided many, many helpful suggestions and insights that served to substantially improve the quality of the book. Regardless of how many subsequent editions this text may see, I will always be most sincerely indebted to John Dossey (Illinois State University), Bob Gilbert (Florida International University), Warren Crown (Rutgers), and Steven Willoughby (University of Arizona), each of whom reviewed and commented on the entire first-edition manuscript. I am also deeply indebted to Arthur Baroody (University of Illinois–Champaign) and James Bruni (Herbert H. Lehman College, CUNY), who reviewed significant portions of that manuscript.

In preparing this second edition, I have received thoughtful input from the following educators who offered comments on the first edition and/or the manuscript for the revision:

Nadine S. Bezuk, San Diego State University

Sandra L. Canter, Ball State University

Lynn Columba, Lehigh University

Warren D. Crown, Rutgers University

Clarence J. Dockweiler, Texas A & M University

Lowell Gadberry, Southwestern Oklahoma State University

Thomas Gibney, University of Toledo

Bruce Godsave, State University of New York, Geneseo

Thomas Kandl, Slippery Rock University

Rochell Kaplan, William Paterson College, NJ

Gerald Kulm, Texas A & M University

Ann L. Madsen, University of Texas at Austin

Bruce Mitchell, Michigan State University

Joanne S. Rankin, Eastern Michigan University

James E. Riley, Western Michigan University

Lois Silvernail, Spring Hill College, AL

Martin A. Simon, Pennsylvania State University

Mary Beth Ulrich, Pikeville College, KY

Janet J. Woerner, California State University, San Bernardino

Each of these educators challenged me to think through many issues and rewarded me with their helpful input. I am especially grateful to Martin Simon, who provided many insightful suggestions for the improvement of Chapters 2 and 3. My students at Virginia Commonwealth University have given me some of the most important feedback possible. I am fortunate to learn from them as they work at learning in my classes. I thank them for their sincerity and help.

Finally, and most importantly, I repeat the thought that ended the Preface to the first edition. This book would never have reached completion, much less a second edition, if it were not for the constant love, support, encouragement, and patience far beyond the ordinary that my wife continues to give without reservation or complaint. She has been my support throughout our lives together. A professional in her own right, she frequently takes a greater share of daily tasks and endures long nights and weekends alone while I work on "the book."

With all my love, thank you, Sharon.

1 TEACHING MATHEMATICS: REFLECTIONS AND DIRECTIONS

THINKING ABOUT TEACHING MATHEMATICS

Most of the book is about teaching mathematics, or, perhaps in better words, helping children learn mathematics. So let's just start there: *teaching mathematics*.

What kinds of images and emotions does that simple phrase bring to your mind? Consider first the *mathematics* part. What do you think mathematics is all about? What is mathematics in the elementary school? Pause right now and reflect on your own ideas about the topic of mathematics. What is it? How does it make you feel? What does it mean to "do mathematics"? Where do calculators and computers fit in? What parts of the subject seem to you to be most important? Write down three or four of your strongest thoughts about mathematics. Compare your thoughts with those of others.

Next focus on the *teaching* part. Someday soon you will find yourself in front of a class of students, or perhaps you are already teaching. Your goal is for children to learn mathematics. What general ideas will guide the way you will teach mathematics? Do you think your ideas are influenced by your view of what mathematics is? Do children learn mathematics differently than they do other topics? How can you make it interesting and enjoyable? If mathematics is not exactly your favorite subject, do you think it had anything to do with the way you were taught? How can you help children like the subject more than you do?

These are hard questions. They do not have simple, unique answers with which everyone will agree.

THE REVOLUTION IN SCHOOL MATHEMATICS

It is reasonable to say that the United States is in the middle of a revolution in school mathematics, a revolution that is more positive, more pervasive, and more widely accepted than any change that has preceded it. From kindergarten to college, changes are occurring in what mathematics is taught and the manner in which it is taught. While the momentum for change was building for some time before, a reasonable date for the beginning of this revolution is 1989, the year that the National Council of Teachers of Mathematics (NCTM) published the *Curriculum and Evaluation Standards for School Mathematics*. This landmark document provided standards and direction for the mathematics that should be taught in our schools. The publication of the *Standards* has received unanimous support and praise from nearly every sector of the education, business, and political communities of our society. It has provided the philosophy and direction for curriculum reform in virtually every state and local school district throughout the nation.

In the same year as the release of the *Standards* by NCTM, the Mathematical Sciences Education Board released *Everybody Counts: A Report to the Nation on the*

Future of School Mathematics (National Research Council, 1989). This very readable document outlined the nature of mathematics, the needs of our changing society, and the problems with our past efforts in mathematics, and echoed the directions for mathematics suggested by the *Curriculum Standards*. Certainly these two documents were not the actual causes of change. The need and the knowledge required to even produce these documents had been building for years. However, anyone wishing to understand what is happening in mathematics education in the nineties cannot go wrong by beginning with these two books.

FORCES DRIVING THE REVOLUTION

A growing body of educational research has given us real insights into just how children learn about numbers, fractions, geometry, or other aspects of mathematics. This increased understanding of the learning process significantly influences teaching methods. As for the curriculum in school mathematics, two factors have provided a significant impetus for the change we are now experiencing: (1) the needs of society in a highly technological and global economy, and (2) advances in technology, most specifically calculators and computers.

THE DEMANDS OF SOCIETY

Years ago, school mathematics was focused almost entirely on the skills of pencil-and-paper computation. This was, at the time, appropriate. The vast majority of jobs in industry and agriculture demanded little more. Few of the nation's students were ever expected to study mathematics in college and subsequently to contribute to the research efforts of the mathematics and science community. It became fashionable (and unfortunately remains fashionable) to proclaim one's personal incompetence in the areas of mathematics and science. For no other subject in the curriculum are Americans so quick to proclaim their inadequacies.

In a world that is increasingly complex and dominated by quantitative information in every facet of its economy, mathematical thinking is not just more important but essential for even the most ordinary of jobs. Mathematical thinking is not at all the same as the computational skills of yesterday's school mathematics. It involves the ability and the habits of reasoning and solving problems. It includes having number sense—an intuition about numbers, their magnitudes, their effects in operations, and their relationships to real quantities and phenomena. It implies the ability to meaningfully interpret charts and graphs and to understand basic concepts of probability and data interpretation. It includes spatial sense—a familiarity with shapes and relationships among them. These are the basic skills of today's society. Higher-order thinking skills remain entirely human. These skills of the mind are expected of everyone in the modern workplace.

It is not that the United States is doing a worse job of teaching mathematics than in the past. Evidence suggests we do about as well today as we ever did (Willoughby, 1990). The problem is that we have not moved as quickly as other nations in changing what we teach and how we teach it.

THE INFLUENCE OF TECHNOLOGY

Technology pervades our everyday life from our microwave ovens and videodisc players to electronic banking and supermarket scanners. Computerization is replacing nearly every mundane facet of the workplace. In mathematics education, the calculator and the computer in particular have had a significant effect on school mathematics. This effect manifests itself in three significant ways.

First, the calculator and the computer have drastically reduced the importance of low-level computational skills. It remains critically important that all students master basic facts such as $12-7$ or 6×8. These tools assist in mental computations, estimations, and many aspects of numeric reasoning. But long and tedious computations are simply obsolete.

Second, the calculator and computer provide new instructional approaches to significant ideas. Activities have been designed with simple calculators that assist children in developing basic ideas about number such as place value, relationships between fractions and decimals, and the relative magnitudes of numbers. Estimation skills, mental mathematics, and even drill of basic facts can be enhanced by the calculator. Contrary to persistent beliefs held by parents and some teachers, there is, after hundreds of studies over 15 years, *no evidence* that calculators will have a negative effect on basic skills or concepts. Quite the opposite is true. The calculator is a powerful teaching tool.

Third, technology has changed what we are able to teach. This is especially obvious in the upper grades. The graphing calculator and computer software now easily perform calculations, do tedious statistical procedures, and accurately draw, measure, and manipulate all manner of geometric shapes and constructions. These technologies are readily available, opening new worlds that were never before accessible to students. As an example, the graphing calculator permits students to quickly display the graph of almost any equation, then make changes in the equation and instantly observe the results. Many graphs can be sketched and compared in minutes. Students are able to focus on relationships that the various graphs exhibit and on real applications of graphs.

THE NCTM *CURRICULUM AND EVALUATION STANDARDS*

As noted earlier, the release of the *Curriculum and Evaluation Standards for School Mathematics* by NCTM in 1989 was a significant event in school mathematics. With its virtually universal acceptance, it gave momentum and articulated direction to a reform movement that, for ten years before 1989, could only be described as one of incremental change. As a result of the *Standards*, objectives are being rewritten, textbooks are changing, teaching methods are different, and assessment practices are being completely revised.

WHAT ARE THE *STANDARDS*?

The *Standards* establishes a "vision of what it means to be mathematically literate" in today's society. The effect has been to provide impetus and initial directions for the revolution in school mathematics. While not a curriculum for mathematics, the *Standards* provides an over-arching philosophy for mathematics education as well as direction and focus for each specific content area and for assessment. Woven throughout are examples of appropriate learning activities that suggest the intended spirit of instruction. The effect of the document will be felt throughout the 1990s. It will continue to have far-reaching implications for mathematics education and society well into the next century.

The *Standards* is divided into four sections: K–4, 5–8, 9–12, and Evaluation. Within each section are 13 or 14 standards or statements about a particular area of mathematics. As noted in its introduction, "A standard is a statement that can be used to judge the quality of a mathematics curriculum or methods of evaluation. Thus, standards are statements about what is valued" (p. 2).

THE VISION OF THE *STANDARDS*

The *Curriculum Standards* outlines five goals for students: (1) to value mathematics, (2) to become confident in their ability to do mathematics, (3) to become mathematical problem solvers, (5) to learn to communicate mathematically, and (5) to learn to reason mathematically. These goals (discussed more fully in the next chapter) are aimed at the development of *mathematical power*. This term refers to students' abilities to "explore, conjecture, and reason logically, as well as the ability to use a variety of mathematical methods effectively to solve nonroutine problems" (p. 5).

The Appendix contains two charts taken from the *Curriculum Standards* that outline the relative changes in content emphasis that the document suggests. Even a casual review of these charts can provide a snapshot of suggested changes from the curriculum of the 1980s. It would be a good idea to spend some time with this material, both now and as you go from chapter to chapter through the book. You will very likely recognize the descriptions under the heading "Less Emphasis" as being very similar to what you recall from your own school experiences. The "More Emphasis" column indicates the direction of the 1990s.

FOUR THEMES OF THE *STANDARDS*

In each of the three grade-level sections of the *Curriculum Standards*, the first four standards have the same labels:

1. Mathematics as Problem Solving
2. Mathematics as Communication
3. Mathematics as Reasoning
4. Mathematical Connections

These four standards represent over-arching themes for the mathematics curriculum; they can be applied to nearly every area and every lesson. To teach with these four standards clearly in mind is what it means to teach in a "*Standards*-oriented" manner.

Mathematics as Problem Solving

According to the *Curriculum Standards*, "Problem solving should be the central focus of the mathematics curriculum." This means much more than learning to solve word problems. Rather, mathematics *as* problem solving means that problem solving is a part of all real mathematical activity. The "Mathematics as Problem Solving" standard speaks to learning a variety of general problem-solving strategies such as a guess-and-check approach or looking for a pattern. It talks about being able to formulate problems and assess results. It speaks about confidence in solving problems. These processes and attitudes apply to all of mathematics. Problem solving is a way of thinking and reasoning used in the learning and the doing of all mathematics.

Mathematics as Communication

The communication standards at each level point to the importance of being able to talk about, write about, describe, and explain mathematical ideas. Symbolism in mathematics, along with visual aids such as charts and graphs, should become ways of expressing mathematical ideas to others. This means that students should learn not only to interpret the language of mathematics but to use that language themselves. Learning to communicate in mathematics makes accessible the world of mathematics beyond the classroom. It also fosters interaction and exploration of ideas within the classroom as students learn in an active, verbal environment.

Mathematics as Reasoning

To reason is as integral to mathematics as problem solving. The *Standards* tells us that reasoning should be a part of mathematical activity from kindergarten on. To observe and extend a pattern, defend a result, or decide if an answer is correct are all activities that involve logical reasoning. When reasoning is part of all mathematics, students learn that mathematics is not a collection of arbitrary rules but a system that makes sense and can be figured out.

Mathematical Connections

The theme of connections is really threefold. First, the connection standards refers to connections within and among mathematical ideas. Addition and subtraction are intimately related. Fractional parts of a whole are connected to concepts of decimals and percents.

Second, the symbols and procedures of mathematics should be clearly connected to the conceptual knowledge that the symbolism represents. Rules such as "invert the divisor and multiply" should never be learned in the absence of well-developed supporting concepts.

Third, mathematics should be connected to the real world and to other disciplines. Children should see that mathematics plays a significant role in art, science, and social studies. This suggests that mathematics should frequently be integrated with other discipline areas and also that real applications of mathematics in the real world should be explored. Mathematics should be viewed as a meaningful and relevant discipline, in terms of both how it is done and how it is used.

THE EVALUATION STANDARDS

The evaluation section reflects the basic themes of the *Standards* while presenting one of the most significant challenges in the document. It is clear that if the mathematics curriculum is going to focus on problem solving, reasoning, communication, and connections, then the assessment program must also focus on these themes. A valid assessment program should move the curriculum in mathematics away from a focus on answers and answer-getting to a focus on the things that are most important. A main theme of the evaluation standards is finding out what students *do* know instead of what they *do not* know. In that spirit, assessment is a much more open-ended activity. Performance tasks may pose questions with no singular correct answer. The questions may require students to explain their approach, defend their solutions, or even to make two different arguments based on the same data. The evaluation standards suggest that the line between instruction and assessment should be blurred. As children are actively doing mathematics in the spirit of the *Standards*, teachers can make observations and incorporate their findings into their total assessment program. Products of student activities (solved problems, projects, explorations) can be collected into portfolios to demonstrate student growth. When students are taught using manipulatives or calculators, assessment should permit students to use these same materials. Alternative forms of assessment including performance-based tasks, portfolios, group assessments, interviews, projects, and reports are all methods that are gaining prominence in other disciplines as well. In mathematics, these methods represent a significant departure from the past. Without change in the area of assessment, however, the reform in mathematics will never be complete.

THE PROFESSIONAL STANDARDS FOR TEACHING MATHEMATICS

Two years after the *Curriculum Standards* was published, NCTM issued a second standards document, the *Professional Standards for Teaching Mathematics*. The *Professional Standards* focuses on instruction and teachers while the *Curriculum Standards* was aimed at content and assessment. The *Professional Standards* asserts that teachers are the key agents of change in the classroom. If the revolution begun by the *Curriculum Standards* is to come to fruition, teachers must learn to shift from a teacher-centered to a child-centered approach to instruction. Research in cognitive psychology and mathematics education suggests that students actively construct their own meanings through assimilation of new information and engaging experiences. They do not simply absorb knowledge placed before them by the teacher. The *Professional Standards* describes teaching that supports the *Curriculum Standards* and incorporates this constructivist view of learning in its suggestions for instruction.

FIVE SHIFTS IN CLASSROOM ENVIRONMENT

The introduction to the *Professional Standards* lists five major shifts in the environment of the mathematics classroom. The authors of the *Professional Standards* see these shifts as necessary to allow students to develop mathematical power. According to the document, teachers need to shift:

- toward classrooms as mathematics communities—away from classrooms as simply a collection of individuals;
- toward logic and mathematical evidence as verification—away from the teacher as the sole authority for right answers;
- toward mathematical reasoning—away from merely memorizing procedures;
- toward conjecturing, inventing, and problem solving—away from an emphasis on mechanistic answer finding;
- toward connecting mathematics, its ideas, and its applications—away from treating mathematics as a body of isolated concepts and procedures.

MATHEMATICS FOR ALL STUDENTS

In both standards documents, but highlighted in the *Professional Standards*, is the phrase "mathematics for *all* students." Both new teachers and those who have been in the classroom for some time should be keenly aware that the standards documents are not elitist ideals reserved for gifted students. The NCTM has endorsed the following statements:

> As a professional organization and as individuals within that organization, the Board of Directors sees the comprehensive mathematics education of every child as its most compelling goal. By "every child" we mean specifically—
>
> - students who have been denied access in any way to educational opportunities as well as those who have not;
> - students who are African American, Hispanic,

- American Indian, and other minorities as well as those who are considered to be a part of the majority;
- students who are female as well as those who are male; and
- students who have not been successful in school and mathematics as well as those who have been successful.

(*Professional Standards for Teaching Mathematics*, p.4)

The issue of *all* children is very important in considering the over-all message of the standards documents. It articulates clearly that the mathematics of the *Curriculum Standards* and the instructional methods of the *Professional Standards* are desired and appropriate for all students. No students should receive a second-level curriculum or second-best instruction.

TEACHING STANDARDS

The first section of the *Professional Standards* addresses standards for the teaching of mathematics. The six standards are arranged in four categories: providing worthwhile mathematical tasks, encouraging discourse among students and between students and teacher, providing for an environment in which learning will be enhanced, and ongoing analysis of both teaching and learning.*

TEACHING MATHEMATICS

At the outset of this chapter you were asked to consider your thoughts about the phrase *teaching mathematics*. It is quite possible that much of what you have read about so far is different from your own school experiences with school mathematics. Or, you may have read this outline of the revolution in school mathematics with a nodding familiarity. In either case, living up to the standards documents is a significant yet exciting challenge.

AN INVITATION TO LEARN

As a teacher or prospective teacher, among the first things you may need to confront in facing the challenge of the standards documents are some of your personal beliefs and ideas about what it means to *do* mathematics, how one goes about *learning* mathematics, and what it means to *evaluate* mathematics. The next four chapters of this book are designed to help you. They lay the foundation for teaching developmentally.

The remainder of the book examines specific topics in mathematics and offers suggestions that follow up on these basic ideas in keeping with a standards approach to teaching mathematics.

New directions in mathematics education have opened the world of exciting investigations in mathematics to all students. Elementary mathematics can no longer be equated with mundane computational skills.

Teaching mathematics is an exciting adventure. Perhaps the most exciting part is that you can, and will, grow along with your students.

◆◆◆◆◆

REFLECTIONS ON CHAPTER 1: WRITING TO LEARN

At the end of each chapter of this book you will find a series of questions under this same heading. The questions are designed to help you reflect on the most important or "big" ideas of the chapter. Writing (or talking aloud with a peer) is an excellent way to organize new ideas and incorporate them into your own knowledge base. The writing (or discussion) will help make the ideas "yours." After you have written your responses in your own words, return to the text to compare what you have written with the book. Make changes if necessary or discuss differences with your instructor.

1. Describe the societal factors that have contributed to the need for more thinking, reasoning, and problem solving in school mathematics.
2. Of the three ways that technology has influenced mathematics in schools, which two do you see as the most significant? Explain your choices.
3. Describe briefly each of the first four thematic standards that appear in each of the three grade-level sections of the *Curriculum Standards*.
4. What is the main message in your mind of the evaluation section of the *Curriculum Standards*?
5. This chapter has only outlined a few of the ideas found in the *Professional Standards*. Among these are five shifts in the classroom environment from what has been traditional to what must be in order to support the *Standards*. Examine these five shifts and describe in a few sentences what seems to be most significant to you about them.

Note: The most important thing that you could do to understand this chapter would be to read one or more of the following:

a. The introduction to the *Curriculum Standards* and the first four standards for at least one grade level.
b. *Everybody Counts*. While the entire booklet is an evening's read, the chapters entitled "Opportunity," "Curriculum," "Teaching," and "Change" may be most appropriate at this juncture.
c. *Reshaping School Mathematics: A Philosophy and Framework for Curriculum*.

These and other readings are listed in the Suggested Readings that follow.

---

*A complete listing of the six teaching standards can be found in Chapter 2, pages 18–19.

♦♦♦♦♦
FOR DISCUSSION AND EXPLORATION

1. Get a copy of the NCTM publication *Curriculum and Evaluation Standards for School Mathematics* and select one recommendation that you find especially important. Compare the content of the recommendation with a current basal textbook for a particular grade level. How do the two compare? If there are differences, why do you think they exist?

2. What proportion (percentage) of time is currently being spent in schools on pencil-and-paper computation? Make a guess for several different grade levels. How much time to you think is spent on other forms of computation (estimation, mental computation, calculators)? How much time is spent on problem solving? Compare your estimates with a textbook series and/or discuss these areas with a classroom teacher or curriculum specialist. Are the emphases you observe appropriate? What changes need to be made?

3. How does curriculum get changed? Select any of the following roles: teacher, district supervisor, state supervisor. Select a specific change that you believe should be made in curriculum or instruction in mathematics. How could you implement that change? What factors would make change difficult? How do you think change happens?

♦♦♦♦♦
SUGGESTED READINGS

Brownell, W. A. (1987). Arithmetic teacher classic: Meaning and skill—Maintaining the balance. *Arithmetic Teacher*, 34(8), 18–25.

Enochs, L. G. (Ed.). (1990). NCTM Standards [Special issue] *School Science and Mathematics*, 90(6).

Lindquist, M. M. (1989). It's time to change. In P. R. Trafton (Ed.), *New directions for elementary school mathematics*. Reston, VA: National Council of Teachers of Mathematics.

Mathematical Sciences Education Board, National Research Council. (1990). *Reshaping school mathematics: A philosophy and framework for curriculum*. Washington, DC: National Academy Press.

Mathematical Sciences Education Board, National Research Council. (1991). *Counting on you: Actions supporting mathematics teaching standards*. Washington, DC: National Academy Press.

National Council of Supervisors of Mathematics. (1989). Essential mathematics for the twenty-first century: Position of the NCSM. *Arithmetic Teacher*, 37(1), 44–46.

National Council of Teachers of Mathematics: Commission on Standards for School Mathematics. (1989). *Curriculum and evaluation standards for school mathematics*. Reston, VA: The Council.

National Council of Teachers of Mathematics: Commission on Teaching Standards for School Mathematics. (1991). *Professional standards for teaching mathematics*. Reston, VA: The Council.

National Research Council. (1989). *Everybody counts: A report to the nation on the future of mathematics education*. Washington, DC: National Academy of Sciences.

National Research Council. (1990). *Reshaping school mathematics: A philosophy and framework for curriculum*. Washington, DC: National Academy of Sciences.

Payne, J. N. (1990). New directions in mathematics education. In J. N. Payne (Ed.), *Mathematics for the young child*. Reston, VA: National Council of Teachers of Mathematics.

Steen, L. A. (1990). Mathematics for all Americas. In T. J. Cooney (Ed.), *Teaching and learning mathematics in the 1990s*. Reston, VA: National Council of Teachers of Mathematics.

Willoughby, S. S. (1990). *Mathematics education for a changing world*. Alexandria, VA: Association for Supervision and Curriculum Development.

2 DOING MATHEMATICS: LEARNING IN A MATHEMATICAL ENVIRONMENT

◆ ACCORDING TO THE NCTM PUBLICATION *Curriculum Standards*, "Knowing mathematics is doing mathematics." Exactly what does that mean? How would you describe what you are doing when you are doing mathematics? Mathematics is more accurately described as "a science of pattern and order" rather than a collection of rules. Even young children can be immersed in an environment where they themselves are actually doing mathematics rather than practicing rules provided by others. Once we have a better idea of what this business called mathematics really is, we can look at the classroom environment we want to create to help children actually do mathematics.

MATHEMATICS AND CHILDREN

It may be that the description of mathematics you will read about here is not the same as your own personal experiences. That's OK! It is OK to come to this point with whatever beliefs were developed in your previous mathematics experiences. However, it is not OK to accept these outdated ideas about mathematics and expect to be a quality teacher. Your obligation and challenge as you read this chapter and this book is to reconceptualize your own understanding of what it means to know and do mathematics so that the children with whom you work will have an exciting and accurate vision of mathematics.

MATHEMATICS AS A SCIENCE OF PATTERNS AND ORDER

Mathematics is much more than computation with pencil and a paper and getting answers to routine exercises. In fact, it can easily be argued that computation, such as doing a long division, is not mathematics at all. Calculators can do the same thing and calculators can only calculate—they cannot do mathematics. At the same time, the *invention* of a method of doing long division is certainly mathematics. That involves a search for some order in our number system coupled with ideas about what division means. Knowing how to graph the equation of a parabola is simply following rules. Figuring out why certain forms of equations always produce parabolic graphs involves a search for patterns in the way numbers behave. Even the youngest school children can and should be involved in significant searches for patterns and order. Have you ever noticed that 6 + 7 is the same as 5 + 8 and 4 + 9? What is the pattern? What are the relationships? When two odd numbers are multiplied the result is also odd, but if the same numbers are added or subtracted, the result is even. There is a logic behind simple results such as these, an order and a pattern. And pattern is not just in numbers but in everything around us. The world is full of order and pattern: in nature, in art, in buildings, and in music. Pattern and order are found in commerce, science, medicine, manufacturing, and sociology. Mathematics discovers this order and uses it in a multitude of fascinating ways, improving our lives and expanding our knowledge. School must begin to help children with this process of discovery.

The formulation of mathematics as a *science of patterns and order* was articulated elegantly in the book *Everybody Counts* (1989). It is not an idea that is obvious to most of us and you may be wondering, "Who cares?" However, as you reflect on activities that involve problem solving, searching for solutions before precise methods have been laid down, or examining the environment in terms of its structure and beauty, the significance of the statement will become more evident. The mechanistic activities of computing sums or finding common denominators pale in comparison.

NEW GOALS FOR STUDENTS

In the introduction to the *Curriculum Standards*, five important educational goals are described for all students, not just those few who traditionally gravitate toward mathematics. These goals state that all students should

1. learn to value mathematics,
2. become confident in their ability to do mathematics,
3. become mathematical problem solvers,
4. learn to communicate mathematics, and
5. learn to reason mathematically.

As you will see, these goals are intricately bound together in the process of doing mathematics. All can and will be addressed within a classroom reflecting a true mathematical environment.

Valuing Mathematics

"Why do I have to know this stuff?" This refrain is heard all too often by teachers of mathematics, especially in the upper grades. The mathematics of computation for the sake of computation is certainly open to criticism. But the mathematics of reasoning, problem solving, and patterns is intimately connected with the very fabric of our society. Virtually every job role in today's society requires mathematics and, more importantly, mathematical thinking. Today's employers are searching for the ability to solve problems that have never been encountered before. Children need to see themselves learning to reason and learning to solve problems, not just learning skills. Furthermore, whenever these are realistic and connected with real data from real-life situations, the inescapable conclusion is that this is "important stuff."

Being Confident

As the need for mathematical thinking pervades our society, the need to feel ownership of that thinking power is all-important. We are entering an era when it can no longer be fashionable to announce, "I never was any good at math." Children must, from the earliest grade and continually throughout their school experiences, be made to feel a personal success with solving problems, figuring things out, making sense of the world. It is difficult to develop confidence in your mathematical abilities if all you do in that domain is follow unfathomable rules that seem to come down from above. In a passive, rule-oriented curriculum, the best you might expect is confidence in following rules and remembering procedures. But only by making your own sense of things, discovering a pattern, figuring out a relationship, solving a problem, or inventing a new procedure can you begin to develop a sense of "I can do mathematics!" As teachers of children, it is imperative that we provide an atmosphere in which such self-confidence can develop in every child.

Being a Mathematical Problem Solver

Problem solving has been a focus of school mathematics for nearly two decades, and yet we still have not made it the essence of mathematics. Problem solving is much more than finding answers to those exercises labeled "problem solving." Problem solving and the process of searching for pattern and order are virtually synonymous. Mathematics *is* problem solving. So our task is to investigate mathematical ideas in a problem-solving environment, to help children experience mathematics the way real mathematics is actually done.

Communicating Mathematics

Mathematics is thinking, solving problems, and searching for order. The symbolism so closely associated with mathematics is only a means of recording and expressing mathematics and conveying these ideas to others. Children need to view written work in mathematics with this perspective. In the same sense, children need to learn other modes of mathematical expression, including oral and written reports, drawings, graphs, and charts. Each day should include discussion and/or writing about the mathematical thinking that is going on in the classroom. No better way exists for wrestling with an idea than to attempt to articulate it to others. Mathematical expression, therefore, is part of the process and not an end in itself.

Reasoning in Mathematics

In much of the work of Marilyn Burns (Burns, 1987, 1990a, 1990b; Burns & Tank, 1988) with children, a recurrent phrase appears in the many writing tasks she has her children do: *We (or I) think the answer is _____. We think this because . . . .* Children can and do learn that the reasoning behind their answers is at least as important as the answer itself. Our goal as teachers is to help our students develop the understanding that an argument or a rationale for a response is always part of that response. No longer should we hear children say, "Do we divide in this one?" but rather, "I think we should use division because . . . ." The habit is best started in kindergarten, but seventh and eighth graders can also learn the habit (and the satisfaction and value) of defending their own solutions.

MATHEMATICS FROM THE CHILD'S PERSPECTIVE

For most children, an understanding of the nature of mathematics is primarily acquired through classroom experiences with mathematics.

Mathematics Does Not Come from the "Math God"

In the school mathematics most of us are used to, the curriculum, consisting mostly of computation, has been broken into hundreds of little bite-sized chunks. Each chunk (e.g., "Invert the divisor and multiply") is explained briefly and practiced extensively. Children emerge from these experiences with a view that mathematics is an endless series of meaningless rules, handed down by the teacher, who in turn must get them from some very smart source. The students' role in this exercise is largely passive; accept what you are told, and try to master each new rule. The very fact that children are neither asked nor required to understand these rules says loudly to children, "You can't possibly comprehend these ideas."

So where do these rules come from? Is there a "math god," someone who has all the rules and figures them out and to whom the teacher is somehow connected? In such an environment what else could children come to believe?

A major premise of this text (one which is also implicit in the *Standards*) is that the ideas to which children are exposed in mathematics can and should be completely understood by them. There are no exceptions! In general, children are capable of learning all of the mathematics we want them to learn and they can learn it in a meaningful manner. *There is no Math God!* Children must believe this, not because we tell them, but because the experiences we provide make it abundantly clear. For some the struggle to learn may take more time; but all can learn.

Problems and Rules

By fifth or sixth grade there are many children who simply refuse to attempt a problem that has not been first explained: "You haven't shown us how to do these." This is a natural consequence of the bits-and-pieces, rules-without-reasons approach to mathematics. Children come to accept that every problem must have a method of solution already determined, that there is only one way to solve any problem, and that there is no expectation that they could solve a problem unless someone gave them the solution method ahead of time. Mathematics viewed this way is certainly not a science of patterns and order. In fact, it is not mathematics at all.

In reality the *search* for the solution method and the subsequent *defense* of that method are what we want children to learn. When the focus is placed on the reasoning and the search for solutions, the answer to the question, "Why do I need to know this?" is simple: because you are learning to think. Mindless procedures are routinely carried out by computers or those on the lowest rung of the pay scale. Virtually every productively employed person in our society needs to know how to think and to solve problems they have never before seen. Tom Romberg (1992), chair of the NCTM Commission on Standards for School Mathematics, says that, "Only if the emphasis is put on the process of *doing* is mathematics likely to make sense to students (p. 61)."

DOING MATHEMATICS IN THE CLASSROOM

Romberg makes the distinction between the knowledge of mathematics and the record of that knowledge by use of a powerful analogy to music:

> Like mathematics, music has many branches categorized in a variety of ways (classical, jazz, rock; instrumental, vocal); it has a sparse notational system for preserving information (notes, time-signatures, clefs) and theories that describe the structure of compositions (scales, patterns). However, no matter how many of the artifacts of music one has learned, it is not the same as *doing* music. It is only when one performs that one knows music. Similarly, in mathematics one can learn the concepts about numbers, how to solve equations, and so on, but that is not doing mathematics. Doing mathematics involves solving problems, abstracting, inventing, proving, and so forth. (Romberg, 1992, p. 61)

Certainly it is true that the notes and scales of mathematics are important. But we will never teach real mathematics and entice students into the discipline if we do not continually immerse them in "playing the music" of mathematical explorations.

THE VERBS OF DOING MATHEMATICS

Envision for a moment an elementary mathematics class where students are "doing" mathematics. What verbs would you use to describe the activity in this classroom? Stop for a moment and make a short list before reading further. Many children in traditional mathematics classes tend to think of mathematics as "work" or "getting answers." They talk about "plussing" and "doing times" (multiplication). The following collection of verbs is taken from throughout the NCTM *Standards* document and reflects a more appropriate view of doing mathematics:

- explore
- investigate
- conjecture
- solve
- construct
- discover
- represent
- formulate
- conjecture
- justify
- verify
- explain
- predict
- develop
- describe
- use

When children are engaged in the kinds of activities suggested by this list, they have to be actively involved with the ideas under consideration. It is virtually impossible for them to be passive observers. These verbs also can be associated with the five goals for students set by the *Curriculum Standards*. Take any of them and reflect back on the goals. Which goal or goals would you associate with the verb you selected? For example, *discovering* is an activity that can directly affect a child's personal confidence in his or her ability to do mathematics: "If I discovered this on my own (or in my group) I must be pretty good."

A major shift is called for from an environment that focuses on answers and answer getting to one that focuses on the thinking process itself; teaching how to solve problems is much different from teaching how to get answers.

A PROBLEM-SOLVING ENVIRONMENT FOR LEARNING MATHEMATICS

Real mathematics rarely involves cranking through a series of procedures that have already been formulated. That type of activity is simply answer getting. Outside the traditional mathematics classroom, in the real world, once procedures are well in place, computers or calculators are frequently employed to do the laborious repetitive work. Good mathematics classroom activities avoid routine answer getting and instead focus on the process of problem solving.

A classroom environment that encourages active mental involvement and problem solving will generally include most of the following features:

1. a problem or task for exploration
2. a spirit of inquiry
3. a frequent use of models (physical materials)
4. verbal expression and group work
5. encouragement of self-validation

In the next section you will see some examples that will encourage you to engage in such an environment. First, let's consider each of these features.

◆ Problems or Tasks for Exploration

A lesson that begins with "How can we figure out . . . ?" or "How many different ways do you think you can find to . . . ?" or other similar phrases sets a goal for students to achieve without simply telling them how to do something. A good task can provide an intellectual challenge, stimulate discussion, and encourage cooperation. But most of all, it can promote thinking and problem solving. Students can be asked to work on these questions individually, in pairs, or in groups. The tasks may be short and simple with a two-minute discussion, or they may be quite involved and require a full period or even several days of exploration.

◆ A Spirit of Inquiry

When tasks without solutions are presented regularly, we can help students realize that being perplexed or confused is natural and OK. At the same time, we want to help students learn that they themselves can work their way out of this confusion. To promote this spirit of inquiry, teachers can incorporate the following approaches in their instruction:

1. provide hints and direction rather than solutions;
2. encourage risk taking; and
3. listen to and accept ideas—avoid censorship and evaluation of solutions.

Hints. If students, blocked in an approach to a problem or task, are simply shown how to proceed, a sense of lacking self-worth can easily begin to develop. When hints and encouragement are provided instead of solutions, students find their own paths to solutions, connect them to their own ideas, and develop a sense of confidence in solving problems on their own. Keep students focused on doing something productive. Learning will take place in the process of exploration, and your hints should be aimed at getting students actively working. One can learn without getting an answer. One can get an answer without learning.

Risk Taking. Most students have some ideas concerning a problem, but many lack the confidence to share them with peers or teachers for fear of being wrong. Teachers can help overcome such insecurities by praising the efforts of *all* students who volunteer an idea, not just those who are correct or exceptionally creative. Students need to learn that doing or saying something is rewarded more than doing nothing. As this lesson is learned they will venture their own ideas more willingly and frequently be surprised to find out how much they are able to contribute.

Listening. When ideas are presented by students, immediate teacher evaluation can be stifling. It demonstrates to students that it is the teacher who really has the answers and not the students. By being an active listener and accepting the students' comments, the teacher becomes part of the inquiry and not the judge or evaluator. Responses such as, "That's a good idea. Who else has an idea?" or "That's interesting. What do the rest of you think?" can easily become a profitable habit.

◆ Frequent Use of Models

As you will learn throughout this book, physical materials give students something to think with, something to do. They encourage an exploratory approach. Materials can encourage students to reflect on new or evolving ideas with the feedback coming from the materials. For example, figuring out how to sort a collection of 7 hundreds blocks, 2 tens, and 3 ones into five equal piles (Figure 2.1) allows children to experiment with different possible strategies. (You may wish to reflect on how you might do this task.) By contrast, to do the long division on paper $5\overline{)723}$, is very answer-oriented and requires little reflection. Children's attention here is focused on getting finished. Simple errors only cause frustration or anxiety. When children do not have command of a procedure, there is no way for them to tell if what they are doing makes sense.

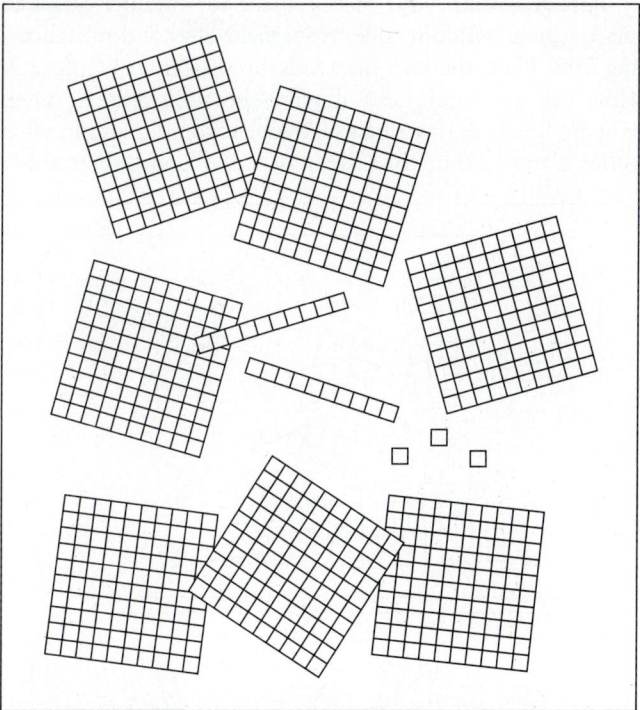

FIGURE 2.1: *How can you divide this amount into five equal piles?*

◆ Verbal Expression and Group Work

A problem-solving environment is not a silent one in which students diligently work quietly at their seats. In order to explore, test ideas, and create relationships, young children need to express their thoughts in words. Explaining, questioning, showing, and making observations and logical arguments verbally forces us to formulate thoughts and connect ideas. Even hearing themselves talk is frequently a way that students discover errors or observe new relationships in their own thoughts. Responding to classmates involves evaluation of other students' ideas in relationship to their own. Persuading others that your ideas are correct is a significant process in the development of new ideas.

When two or more children work together, they must frequently decide on which of their various approaches should be pursued. This leads to an examination and selection of strategies that rarely takes place when a child works alone. Negotiation about the merits of alternative approaches is a valuable learning experience.

When six or seven small groups are each working on a task, there is six to seven times as much opportunity for individuals to express their ideas. Within a small group the personal risk factor is diminished considerably. Children who would not dream of offering an idea in front of the whole class will readily test their thoughts with a few classmates within a small group. Most importantly, the very spirit of group interaction and trying new ideas to solve a problem is at the very heart of what it means to do mathematics. (For more about group work, see Chapter 22.)

◆ Encouragement of Self-Validation

Mathematics classes should not be seen by students as a game of testing answers against the teacher's or those in the back of the book. Teachers should regularly turn the question "Is this right?" directly back to the student: "Why do you think it's right? How could we tell?" Reasoning to the validity of their own answers conveys to students that mathematics makes sense and is not mysterious (does not come from the math god). When students learn that the teacher thinks they are capable of appropriate judgments, they begin to conclude that it must be possible to make sense of these ideas. Older children, who may have had several years of teachers consistently serving as the arbiter of correctness, have a difficult time with this approach. It never occurs to them that they themselves could possibly tell on their own if an answer is correct or not. In contrast, in classrooms where students have learned to justify their results, answers tend to be followed with reasons as a matter of course. The result is increased confidence, a sense of satisfaction, and interest in further exploration.

It should be noted that the validation of results as described here refers to logical explanations, reasoned arguments, or a conceptual defense of one's ideas. In contrast, checking one relatively mindless procedure with another (e.g., checking subtraction with addition) is not quite the same thing. To know that a result is right means that it makes sense, you can explain it, it logically has to be that way.

EXAMPLES OF PROBLEM-SOLVING EXPLORATIONS

A problem-solving environment is encouraged throughout this book. However, five examples of tasks that particularly exemplify such an environment are presented here. The purpose is threefold: first, to provide you with firsthand examples of the type of explorations that have just been discussed; second, to present some commonly used materials (models) that you will find in most good mathematics programs; and third, to illustrate how a problem-solving approach can be used for different instructional purposes.

FIVE EXPLORATIONS

Each of these five explorations is described in enough detail that you can pursue it on your own or preferably in a small group. (Your instructor may decide to use one or more of these in class.) However, further discussion and possible extensions are found in the section that follows, entitled "Notes and Observations on the Explorations." You may want to spend some time with one or more of the explorations before reading that section.

The heading for each task includes a title and also suggests an instructional purpose that posing the task might serve in the classroom.

◆ Start and Jump: Searching for Patterns

This is a good activity for young children to do with a calculator. It takes advantage of the automatic constant feature found on most (but not all) simple calculators.* If you are not familiar with this feature it might be good to learn about that first.

The task is to search for patterns in a long series of numbers that kids can generate with their calculator using a "start" number and a "jump" number.

Part A. Begin with a start number of 3 and a jump number of 5. Write down the start number at the top of your list. On your calculator press the start number (3), then +, then the jump number (5). Next press = and record the result (8) as the second number in the list. Continue to press = and record each new result until your list goes beyond 130. Now, within the list, look for as many patterns as you can find. Write each idea down or describe it to a partner. No idea is too simple or too complex.

[*Note:* Part B should not be started until this much has been thoroughly explored. If several groups are working on the activity, a sharing of patterns found is important. It is rare that any one group will find all the patterns that the other groups found.]

Part B. Repeat Part A, but change the start number. Keep the jump number the same. You may want to try several different start numbers. What about a start number larger than 9 or larger than 99? What stays the same and what is different?

Part C. Repeat the activity but use a different jump number. You will likely find that there is a lot more variation when jump numbers are changed than when start numbers are changed. A few hints: Look at the length of the repeating patterns. Try smaller jump numbers first. Do even jump numbers work differently than odd? Does it matter if they are prime? Are the numbers 2 and 5 important?

◆ Combining Piles: Computation

This addition activity is most appropriate for second graders or early third-grade children who have not yet been taught traditional ways to add large numbers with pencil and paper. Older children can do the same activity but must be told that the "usual" way of adding with a pencil is not permitted. Activities similar to this one can be done with each of the other three operations.

---

*The automatic constant feature automatically stores the last operation of addition and subtraction. For example, if you press 3 + 2 =, the display shows 5. A subsequent press of the = will add 2 to the 5 so that the display now shows 7. Each press of = will add 2 to the current display. If you press another number, say 27, and then =, the calculator will add 2 to the 27, displaying 29. Subtraction works in a similar manner.

To Begin. You will need some ones, tens, and hundreds pieces, also referred to as *base ten* pieces. (These materials are discussed in Chapter 9 and are available in a variety of forms, including a poster-board version you can make.)

Part A. With your pieces, make the number 367 and place it in a "random" pile. Now make a second pile showing 298. Place the two piles side by side as in Figure 2.2. How can we decide how much is here altogether? When you are finished, describe exactly what you did and in what order. Can you think of a different way of going about this?

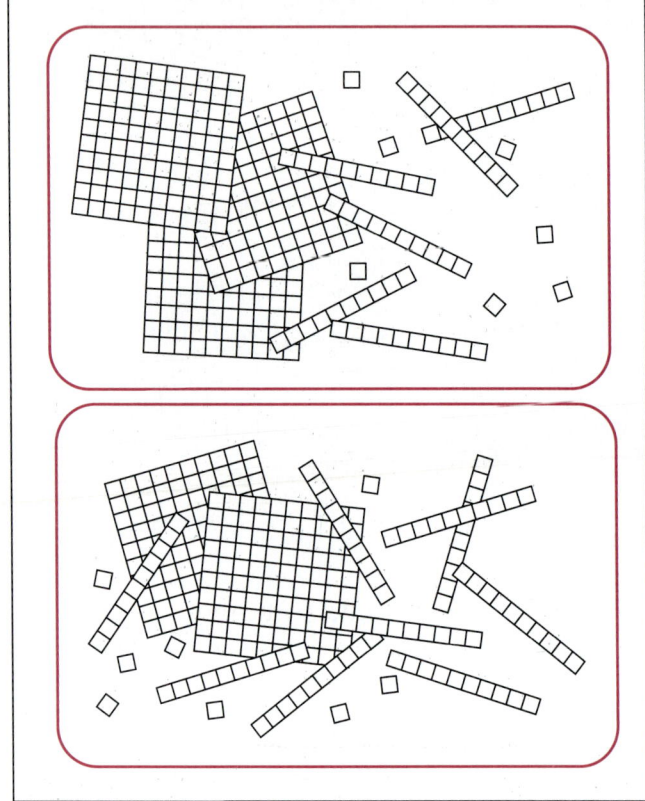

FIGURE 2.2: *How much in the two piles altogether?*

Part B. How would you record what you did? If you had to do the same thing again but were just doing it mentally (without having the pieces), what would you do?

◆ In-Between Numbers: Development of a Concept

In this activity a calculator is again used as a thinking device. The main idea here is to begin to think of decimal numbers as numbers in between whole numbers and to lay groundwork for further conceptual development of fractions.

Part A. Use your calculator to determine the product of 7 and 23 and the product of 7 and 24. Write these down. By using only the multiplication function on the calculator, try to find a number □ so that 7 × □ is 165.*

---

*Most simple calculators have an automatic constant for multiplication. If you press 7 × 2 the display shows 14. If you now press 5 = the display will show 35. The calculator retains the first factor, in this example "5 x" and multiplies that by whatever is currently in the display each time = is pressed. Try 2 × = = ...

Part B. Describe how you would convince a friend of yours that you have found a number between 23 and 24.

Part C. How many numbers do you think there are between 23 and 24? Explain

◆ Finding Areas: Making Connections within Mathematics

This activity is best done with a geoboard and rubber bands. A geoboard is a board with pins or nails regularly spaced in a square array as in Figure 2.3. Rubber bands are stretched around the nails permitting children to easily "draw" figures composed of straight line segments.

An alternative to a geoboard is to use dot paper such as that found in black-line master 30.

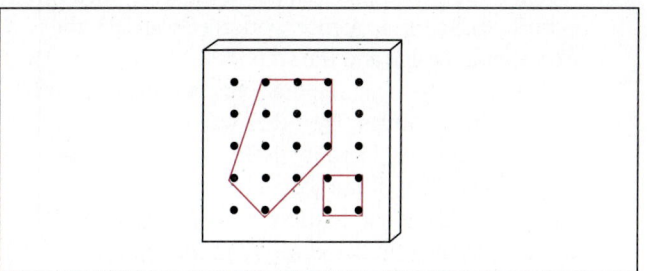

FIGURE 2.3: *A geoboard is an easy way to draw shapes without using a pencil or a ruler. Changes are easily made, and the board has built-in units for length and area.*

Background. Use the smallest square on the geoboard as one unit of area and the side of that square as one unit of length. The geoboard is one model that can be used to develop the area formula for rectangles, namely, that the *area of a rectangle is the length of one side (the base) times the height.* Following that discussion, students can make parallelograms that have two parallel sides going the same way as the edges of the board. As shown in Figure 2.4, a triangular piece on one end of the parallelogram can be shown to be the same as a missing piece on the other end. The result is a rectangle that has the same area as the original parallelogram. Therefore, a simple connection can be made between the area of a rectangle and the area of a parallelogram. In fact, if you are careful to distinguish *height* from the slanted side of a parallelogram, the formulas for both parallelograms and rectangles are the same. (Can you demonstrate this for any parallelogram that you can draw?)

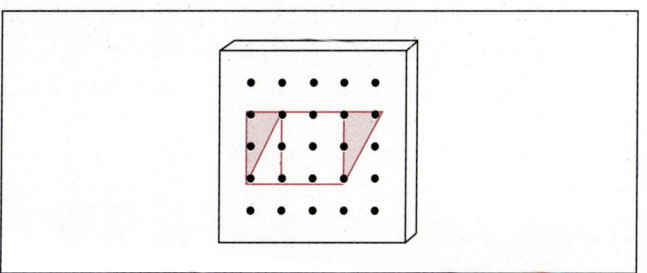

FIGURE 2.4: *Make some parallelograms like this, and transform them into rectangles that have the same area as well as the same base and height as the original parallelogram.*

Part A. On your geoboard make any triangle you wish as long as one side is parallel to (goes the same way as) one side of the geoboard. Call this side the base of the triangle. Record the length of the base. Now figure out (any way you wish) the area of the triangle, and record that as well. Repeat this for at least four triangles. Can you find a pattern? If you want a hint for finding the area of your triangles, try making two that are exactly the same size and putting them together to form a parallelogram. (There are other ways to approach this task and this is only one possible hint. Students should be permitted to explore other approaches as well.)

Part B. Devise a formula for the area of a triangle, and explain why you think it will always work.

Part C. Find the area of a trapezoid (two sides parallel and two sides not parallel). One idea is to use a method similar to that used with the triangles.

◆ One Equation: Consolidation of Simple Concepts

Children need ample opportunity to play with an idea in mathematics. Most mathematical concepts are not learned by children in one or two lessons. Rather, they require numerous experiences over time to reflect on an idea and to bring what may be related but superficially different ideas together. In the United States, we often neglect to provide opportunities for students to look at an idea from different perspectives or to think about it in different contexts and in different ways. Lessons tend to have short, quick explanations of an idea followed immediately by a series of drill exercises. We are so focused on the production of answers that we fail to focus on the integration of concepts. In Japan, a country frequently contrasted with the United States, a glaring difference is found in the time that a class may spend with one simple idea. A single equation might very well carry the discussion for an entire period as the following example suggests.

Part A. Consider the equation $4 + 5 = 9$ (or at an even lower grade level: "Four and five is the same as nine"). Why do you believe this is true? Find three different ways to convince someone else that $4 + 5$ is 9. *Hint:* Some arguments might include some kind of counters, others a drawing, a number line, or perhaps references to other numbers or facts.

Part B. Make up three different word stories that go with this equation. Here is an example: "Mario and Lamond were playing a game. Mario had scored only 4 points, which was 5 points behind Lamond. Lamond had 9 points."

Part C. What subtraction equation(s) can you find that mean the same as $4 + 5 = 9$? Explain why you think your new equation means the same thing as the original one. You might want to use counters or a drawing in your explanation.

14   2 / DOING MATHEMATICS: LEARNING IN A MATHEMATICAL ENVIRONMENT

NOTES AND OBSERVATIONS ON THE EXPLORATIONS

In the spirit of this chapter it would be inappropriate to simply list "answers" to the five tasks in the previous section. In this section a few notes are provided about the kinds of things that children working on these tasks may well discover or observe. For each of the tasks, the notes are followed by a few observations concerning the mathematics children might be doing in these tasks that extends beyond the immediate content involved.

If you want to really explore one or more of these tasks yourself, skip this section for now. You can come back later and compare your ideas with those described here. Do not be concerned if your own discoveries or approaches are not the same as those offered here. Children will approach these ideas in lots of different ways as well.

Notes are listed with the titles of the corresponding tasks.

◆ Start and Jump

The first sequence in Figure 2.5 illustrates a list with a jump number of 5 starting with 3. Students have made the following observations:

- The ones digit alternates—3, 8, 3, 8, 3, 8, 3, ...
- There is an odd/even pattern.
- The second digit goes in pairs—1, 1, 2, 2, 3, 3, ... They count, sort of.

Here you might ask about the first two numbers in the list, 3 and 8. Also, what happens after 98 in the tens place? Some say the pattern starts over again.

Start = **3**, Jump = **5**	Start = **2**, Jump = **4**	Start = **3**, Jump = **6**
3	2	3
8	6	9
13	10	15
18	14	21
23	18	27
28	22	33
33	26	39
38	30	45
43	34	51
48	38	57
53	42	63
58	46	69
63	50	75
68	54	81
...	...	...
98	90	93
103	94	99
108	98	105
113	102	111
...	...	...

FIGURE 2.5: *A few start-and-jump sequences: What patterns can you find?*

Others argue that it continues: 10, 10, 11, 11, ... There are two possible ways to think about the numbers 103 and 108. They can have 10 tens or 1 hundred and no tens. A possibility here is to suggest that each successive number in the list be modeled with base ten pieces. Each time ▣ is pressed, add five ones, and make trades when reasonable. What happens after nine tens and eight ones? It depends on what you do with the tens pieces.

- If you add the numbers in pairs you get the sequence 11, 31, 51, 71, ...
- If you add the digits in each number you get a really interesting result: 3, 8, 4, 9, 5, 10, 6, 11, 7, 12, 8, 13, ... Every other number is in a sequence.

Again, it is interesting to ask about the three-digit numbers. To make the sequence work nicely at 103, the sum should be 13, and the sum for 108 should be 18. One way of looking at the numbers permits this, and the pattern continues. The other way is to declare that the sum for 103 is 4 and for 108 is 9. The pattern begins again. If each number is being represented by base ten pieces, what do these sums of digits correspond to? (The number of pieces in the representation.)

Once children begin to look at digit sums, someone is likely to try subtraction or multiplication. Interesting results and more questions pop up again. If you always subtract the tens digit from the ones digit, negative numbers result. If you subtract smaller from larger, the patterns go up and back. Multiplication patterns are a bit more obscure, but they are also there. What about division? When you change the start number but keep the jump number at 5, the numbers change; but every observation made earlier is still there.

When the jump number is changed, a lot of new ideas are likely to appear. Many depend on how much children have been exposed to looking for patterns and what they know about factors of numbers, primes, common factors, and the like. The reason it might be good to begin this exercise with a jump number of 5 is that 5 + 5 is 10, and so a lot of patterns based on place-value concepts are likely to pop up in pairs as we have just seen. With a jump number of 3, for example, the pattern in the ones digit is 10 numbers long before it begins to repeat. No matter what the start number is, the numbers in the pattern can be traced clockwise around this circle:

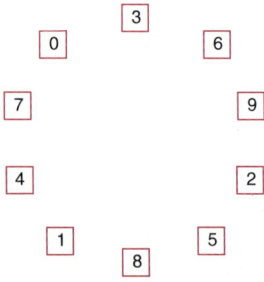

Now the question might well be, what other jump numbers have patterns like this? Do they all have ten numbers in them? Clearly not the jump number 5 (which happens to be a prime). Maybe it's because 3 is odd? Well, 5 is also odd. Eventually you will get to numbers that have a factor of either 2 or 5. For example, the jump number 6 has a pattern that is 5 numbers long. And, of course, you can still explore those other patterns that were observed before.

While pattern was the main idea here, the structure of place value in terms of ones and tens pops up several times as when you need to decide whether 103 is 10 tens and 3 ones or is it 1 hundred, no tens, and 3 ones. Adding up the digits is the same as counting how many pieces are being used. The concept of a common divisor is implicit in the discussion of different jump numbers and how long the sequences will be.

◆ Combining Piles

The interesting feature of this activity is that almost everyone will do it quite differently from the way we usually add two numbers with pencil and paper. Virtually all children will first pick up the big pieces or hundreds and put them together, then the tens, and finally the ones. In this example, it is probable that they will then put 10 tens together and place them with the hundreds and do a similar thing with the ones. To children, there is no clear reason why they should actually exchange the 10 tens for a hundred piece or the 10 ones for a ten piece.

As long as children know a little about ones, tens, and hundreds (or in first grade, just ones and tens), it is interesting that they can actually combine large numbers such as these without having mastered the traditional (and laborious) pencil-and-paper methods. Nor is there any clear "right" way to proceed. If some children choose to make exchanges, that's fine. If some would prefer starting with the ones pieces, that also is fine. Children are in control and are not being placed in a follow-the-rules mode.

The Combining Piles activity has the potential of helping with a variety of important skills. Children who explore putting numbers together or taking them apart in this way eventually are able to perform the same operations mentally. Mental computation always begins with the left-most digits, while pencil-and-paper computation for addition and subtraction begins with the ones. Children are gaining background for mental computation, a skill that is much more important to mathematical thinking than pencil-and-paper computation.

If students are encouraged to find a way to record or write down what they do when they combine the pieces, it is not unreasonable to get them to invent their own pencil-and-paper procedures. The methods they may actually invent for doing these computations with pencil and paper are not nearly as important as the larger, general idea that it is possible to invent ways to record what you have done and thought about. It is not necessary to rely on procedures that come from the teacher.

◆ In-Between Numbers

Assuming that children have had some exposure to decimals, say late fourth grade or above, this activity helps to develop the notion that a number such as 23.6 is between 23 and 24. When they find that $(7 \times 23.5) < 165 < (7 \times 23.6)$, many children will be stumped as to what to do next. Suggest that they are looking for a number that is larger than 23.5 but not as large as 23.6. This may be a good place to try modeling these numbers with base ten pieces. (These ideas are covered more completely in Chapter 14.) If the 10×10 square is used to stand for one whole, then 3.5 is represented by 3 squares and 5 strips. (As you will see later, it is difficult to have desktop models for a number like 23.56, so we compromise by not showing the 2 tens.) Something a bit more than 23.5 can be seen to be 3 squares, 5 strips, and some small squares. These amounts are then tried in the calculator accordingly. It is not at all unreasonable for children to continue on a trial-and-error basis to extend their decimal estimates to three or four places in search of a "best" answer.

Once students begin to add more digits to the right of 2.5, they find that while they get obviously larger numbers (why is this obvious?), they do not get a number as large as 2.6. This helps with the notion that there are many, many numbers between 25 and 26. The use of the base ten strips and squares helps them express their ideas about these numbers in terms of physical models.

Children who have never even heard of decimals can actually do this activity if the suggestion to try numbers such as 23.1, 23.2, . . . , 23.9 is made. Careful lists of the resulting products of these numbers with 7 illustrate that these are successively larger numbers. It may be reasonable to move to a second or even a third decimal place by suggesting the existence of another string of numbers between 23.5 and 23.6. While not the best introduction to decimals, it certainly is an interesting one that allows children to explore a new idea without rules.*

The use of a try-and-adjust or guess-and-check approach is actually firsthand experience with a powerful problem-solving strategy that will be discussed in Chapter 5. This helps kids understand that you do not always have a clear-cut procedure for doing a task; often you make guesses and try things out. Here, then, process or thinking strategies are values of the task that go beyond the content of place value and decimal meanings.

◆ Finding Areas

If two identical triangles are put together along a matching side, a parallelogram is formed (see Figure 2.6a, p. 16). The restriction to make one side parallel to the edge of the geoboard was done so that the heights of the triangles could be measured in geoboard units. The result is the same for any triangle, of course; it is just that the heights

---

*This activity is a modification of one described by Goldenberg (1991) in an article about fourth graders thinking about decimals.

are not nice whole numbers as they will be with this restriction on the geoboard. Assuming the parallelogram formula (area = length of base × height) is established, children can discover the areas of the triangles as being half of the parallelogram made with two triangles. However, on the geoboard, it is also possible to figure the areas of triangles without using a formula, and many children will do just that. This is actually preferable since it is more believable to count squares and parts of squares than it is to put numbers into the formulas. Then when the formula is discovered as it can be using this approach, it is confirmed by the children's own data.

Figure 2.6b illustrates how two trapezoids can be put together in a similar manner. Here the base of the parallelogram is the sum of the top and bottom sides of the trapezoid, but the height is the same.

If this activity served to convince children that mathematics formulas were not invented by such super-smart people that they could never understand them on their own, then it would be serving a useful purpose. Beyond that, mathematics is a highly integrated and connected discipline. The more connections we see and understand, the more meaningful the ideas will be. Here, not only are the formulas for the areas of rectangles, parallelograms, triangles, and trapezoids all connected, but the method of developing some of these is the same: Put two like pieces together to see if you can make a shape for which we already know the formula. The same procedures can also be applied to three-dimensional models and volume formulas. (As noted earlier, the method of putting two triangles together may not occur to all children, and many children will find other ways to find these areas. For most approaches, however, there are similar, unifying ideas that can be capitalized on. It is not a requirement that students "discover" the approach that is suggested here.) In Chapter 16 you will see that formulas for volumes can be connected in similar ways as the formulas for the shapes in this exercise and that the same process and thinking is used to derive them. Mathematics is not only figuring things out but figuring out *how* we figure things out.

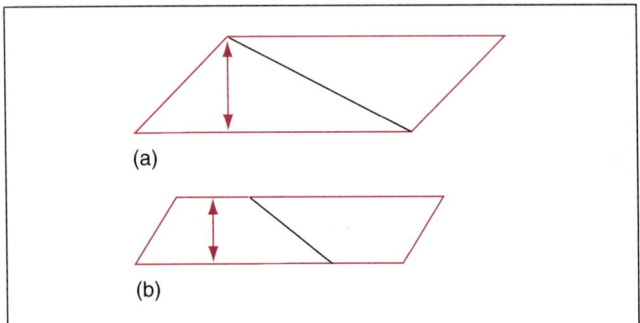

FIGURE 2.6: *(a) Two triangles form a parallelogram. The parallelogram has the same base and same height as the triangle. (b) Two trapezoids can be put together to form a parallelogram. What is the base and height of the parallelogram?*

◆ One Equation

How young children respond to this challenge is very much a function of their experiences with illustrating number concepts and equations in different ways. In addition to the use of models shown in the figure, children might also use number relationships. For example, "Double four is eight, and so four and five must be one more than eight." A similar argument is made around double five. Or a child might think of two sets of five fingers and note that four and five is one less than this. Also good is simply counting up orally or counting on from four or counting on from five. (See Figure 2.7.)

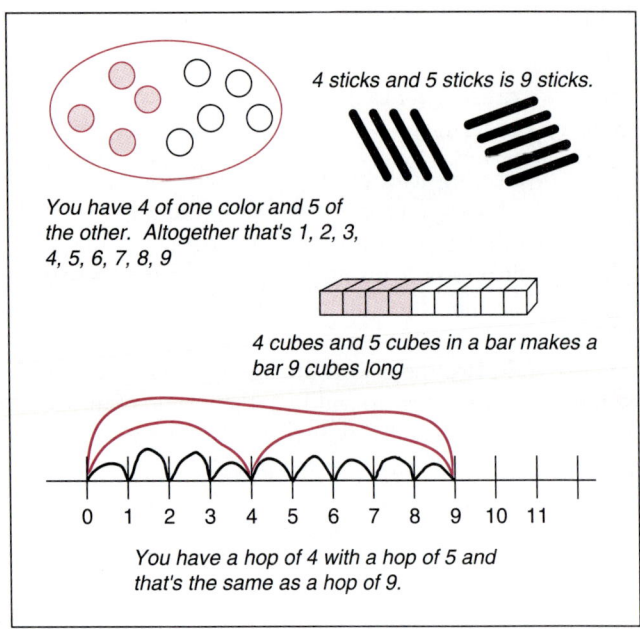

FIGURE 2.7: *There are several possible ways to show why 4 + 5 = 9. In the eyes of children, these are each quite different.*

Word stories offered by children will also vary with their experience. It may be useful to provide a context in which all of their stories must fit. This will help you to get truly different stories and not just the same structure with different contexts. The extent of the variations that are possible is a broader topic than can be discussed here. However, it would be interesting to see how many stories you and your peers can come up with that are really different, not just different contexts. Notice that these are stories, not "problems" with questions. How do you think the effect would be different if you asked for problems instead of stories?

Even at a very early age, children can learn that mathematics is a well-connected interrelated collection of ideas. This exercise helps in that regard. In a traditional program, children in the first grade are asked to complete page after page of simple addition and subtraction facts such as 4 + 5 = 9. Generally, they have to count to get the answers. The focus of the activity turns to counting and answer getting. There is no reflection on the relationship between addition and subtraction, on how 4 + 5 is related to 5 + 5 or to double 4, or that 1 less than 10 is 9. Word stories likewise are problems requesting answers: "Teacher, do we add on this one?" In the present task, children make relationships to a variety of real problem relationships by making up the stories themselves.

Helping young children come to the realization that $9 - 5 = 4$ is equivalent to $4 + 5 = 9$ is not easy. With an extensive use of various models and encouragement to produce and describe explanations, children begin to construct relationships such as this on their own.

THE TEACHER AND THE MATHEMATICAL ENVIRONMENT

So far we have discussed the nature of mathematics as a science of searching for patterns and relationships. In that discussion we explored what it means to *do* mathematics. The previous activities were suggested as tasks students might work on in order to be *doing* mathematics. But the task alone is not sufficient for making mathematics happen. In this section we will examine a bit further the total cultural environment of the mathematics classroom and how the teacher plays a significant role in creating that environment.

THE CLASSROOM ENVIRONMENT

The climate in which relevant and engaging tasks are presented will ultimately structure how children approach these tasks and what benefit they derive from them. Lauren Resnick suggests that

> Becoming a good mathematical problem solver—becoming a good thinker in any domain—may be as much a matter of acquiring the habits and dispositions of interpretation and sense-making as of acquiring any particular set of skills, strategies, or knowledge. If this is so, we may do well to conceive of mathematics education less as an instructional process (in the traditional sense of teaching specific, well-defined skills or items of knowledge), than as a socialization process. (1988, p. 58)

We cannot as teachers simply tell children how to think or what habits to acquire. Processes and habits of thought are acquired over time within a community where such thinking and habits are the norm. Lampert (1990) talks of a community of mathematical discourse in which students evaluate their own assumptions and those of others and argue about what is mathematically true. Her goal is to let students believe that they are the authors of ideas and arguments. Reasoning and mathematical argument, in this environment, are the sources of an idea's legitimacy—not the teacher. As Schoenfeld put it, "'Figuring it out' is what mathematics is all about" (1988, p. 87). The classroom environment must be about "figuring it out" as well.

In the traditional view of mathematics as consisting of a fixed set of rules and procedures, "figuring it out" has very little place other than deciding what rule or procedure to use. Mathematics is commonly known as the epitome of the "well-structured discipline." Resnick (1988) makes a compelling argument that perhaps we should begin to envision mathematics as an ill-structured discipline:

> We need to take seriously, with and for young learners, the propositions that mathematical statements can have more than one interpretation, that interpretation is the responsibility of every individual using mathematical expressions, and that argument and debate about interpretations and their implications are as natural in mathematics as they are in politics or literature. Such teaching would aim to develop both capability and disposition for finding relationships among mathematical entities and between mathematical statements and situations involving quantities, relationships, and patterns. (p. 33)

Return to Romberg's analogy of learning about music. Classrooms must be social environments in which children learn more than just the notes—the rules and the procedures. Classrooms need to be environments where students wrestle with how the notes might be played and what ideas seem to make the most sense. In short, mathematics classrooms must be cultural environments or communities in which the music of mathematics is played and the notes and scales are learned along the way.

THE TEACHER'S ROLE IN CREATING THE ENVIRONMENT

Being the teacher responsible for creating this environment of mathematical discourse may sound overwhelming. You may have viewed mathematics as relatively easy to teach since it involved only dispensing rules and managing practice. The preceding pages have painted a different view. Let's be honest. Creating a classroom culture and environment in which children are *doing* mathematics is *not easy*. When you first begin you will be filled with misgivings and apprehensions. There is no reason to believe that you will be an expert from the start. You need help and experience. Experience will come if you just work at it. A significant amount of help is also available in the NCTM's *Professional Standards for Teaching Mathematics*.

In Chapter 1 it was noted that the *Professional Standards* suggests five major shifts in classroom environments. The first of these was "toward classrooms as mathematical communities—away from classrooms as simply a collection of individuals." We have included a listing of each of the six standards for the teaching of mathematics. For each of these, the NCTM document includes six or seven pages of discussion and excellent examples that go well beyond the statement of the standard. This section of the *Professional Standards* is important reading for all teachers.

STANDARD 1: WORTHWHILE MATHEMATICAL TASKS
The teacher of mathematics should pose tasks that are based on—
- sound and significant mathematics;
- knowledge of students' understandings, interests, and experiences;
- knowledge of the range of ways that diverse students learn mathematics;

and that
- engage students' intellect;
- develop students' mathematical understandings and skills;
- stimulate students to make connections and develop a coherent framework for mathematical ideas;
- call for problem formulation, problem solving, and mathematical reasoning;
- promote communication about mathematics;
- represent mathematics as an ongoing human activity;
- display sensitivity to, and draw on, students' diverse background experiences and dispositions;
- promote the development of all students' dispositions to do mathematics.

STANDARD 2: THE TEACHER'S ROLE IN DISCOURSE
The teacher of mathematics should orchestrate discourse by—
- posing questions and tasks that elicit, engage, and challenge each student's thinking;
- listening carefully to students' ideas;
- asking students to clarify and justify their ideas orally and in writing;
- deciding what to pursue in depth from among the ideas that students bring up during a discussion;
- deciding when and how to attach mathematical notation and language to students' ideas;
- deciding when to provide information, when to clarify an issue, when to model, when to lead, and when to let a student struggle with difficulty;
- monitoring students' participation in discussions and deciding when and how to encourage each student to participate.

STANDARD 3: STUDENTS' ROLE IN DISCOURSE
The teacher of mathematics should promote classroom discourse in which students—
- listen to, respond to, and question the teacher and one another;
- use a variety of tools to reason, make connections, solve problems, and communicate;
- initiate problems and questions;
- make conjectures and present solutions;
- explore examples and counterexamples to investigate a conjecture;
- try to convince themselves and one another of the validity of particular representations, solutions, conjectures, and answers;
- rely on mathematical evidence and argument to determine validity.

STANDARD 4: TOOLS FOR ENHANCING DISCOURSE
The teacher of mathematics, in order to enhance discourse, should encourage and accept the use of—
- computers, calculators, and other technology;
- concrete materials used as models;
- pictures, diagrams, tables, and graphs;
- invented and conventional terms and symbols;
- metaphors, analogies, and stories;
- written hypotheses, explanations, and arguments;
- oral presentations and dramatizations.

STANDARD 5: LEARNING ENVIRONMENT
The teacher of mathematics should create a learning environment that fosters the development of each student's mathematical power by—
- providing and structuring the time necessary to explore sound mathematics and grapple with significant ideas and problems;

- using the physical space and materials in ways that facilitate students' learning of mathematics;
- providing a context that encourages the development of mathematical skill and proficiency;
- respecting and valuing students' ideas, ways of thinking, and mathematical dispositions;

and by consistently expecting and encouraging students to—
- work independently or collaboratively to make sense of mathematics;
- take intellectual risks by raising questions and formulating conjectures;
- display a sense of mathematical competence by validating and supporting ideas with mathematical argument.

STANDARD 6: ANALYSIS OF TEACHING AND LEARNING

The teacher of mathematics should engage in ongoing analysis of teaching and learning by—
- observing, listening to, and gathering information about students to assess what they are learning;
- examining effects of the tasks, discourse, and learning environment on students' mathematical knowledge, skills, and dispositions;

in order to—
- ensure that every student is learning sound and significant mathematics and is developing a positive disposition toward mathematics;
- challenge and extend students' ideas;
- adapt or change activities while teaching;
- make plans, both short- and long-range;
- describe and comment on each student's learning to parents and administrators, as well as to the students themselves.

REFLECTIONS ON CHAPTER 2: WRITING TO LEARN

1. Explain what is meant by "Mathematics is a science of patterns and order." Contrast this view with traditional school mathematics that is largely a collection of rules and procedures.
2. Pick one of the *Curriculum Standards'* five goals for students that you think is most important. Explain the goal itself and why you selected it as important.
3. What do you think is meant by the suggestion that elementary students might begin to think that mathematics comes from a "math god"? What is one instructional idea that teachers can use to prevent this concept from developing?
4. Why is doing pencil-and-paper computation not "doing mathematics"?
5. What is a "problem-solving environment"?
6. Why does Lauren Resnick think mathematics might be better characterized as an ill-structured discipline rather than a well-structured one?
7. Explain what you think the NCTM's *Professional Standards* means by "discourse" and what the teacher can do to promote the quality of discourse in the classroom.

FOR DISCUSSION AND EXPLORATION

1. What factors do today's teachers face that get in the way of implementing real *doing* of mathematics as described in this chapter and in books such as *Everybody Counts*? What should teachers do to deal with these factors?
2. Explore the teacher's edition of any current basal textbook series for any grade level of your interest. Pick one chapter, and identify lessons or activities that promote *doing* mathematics and/or a problem-solving environment.
3. Select one of the six Teaching Standards in the *Professional Standards*, and discuss the implications for instruction as you see them and any other reactions you may have to the standard.
4. Read in the *Professional Standards* the first vignette (pp. 11–15) in which a sixth-grade teacher of five years confronts her own realization that she needs to change. In reaction to this vignette you might

   - try her lesson with fifth- to seventh-grade children;
   - design a lesson on a topic of your own choice that would serve the same purpose of getting children actively involved in real mathematics; and/or
   - take a lesson out of a fifth- to seventh-grade book, and discuss how this teacher might now decide to teach the lesson.

SUGGESTED READINGS

Backhouse, J., Haggarty, L., Pirie S., & Stratton, J. (1992). *Improving the learning of mathematics.* Portsmouth, NH: Heinemann.

Baker, D., Semple, C., & Stead, T. (1990). *How big is the moon?* Portsmouth, NH: Heinemann.

Baker, J., & Baker, A. (1990). *Mathematics in process.* Portsmouth, NH: Heinemann.

Ball, D. L. (1991). Improving, not standardizing, teaching. *Arithmetic Teacher, 39*(1), 18–22.

Ball, D. L. (1991). What's all this talk about discourse? *Arithmetic Teacher, 39*(2), 44–48.

Baroody, A. J. (1989). Manipulatives don't come with guarantees. *Arithmetic Teacher, 37*(2), 4-5.

Baroody, A. J. (1993). *Problem solving, reasoning, and communicating (K–8): Helping children think mathematically.* New York: Merrill.

Borasi, R. (1990). The invisible hand operating in mathematics instruction: Students' conceptions and expectations. In T. J. Cooney (Ed.), *Teaching and learning mathematics in the 1990s.* Reston, VA: National Council of Teachers of Mathematics.

Burns, M. (1992). *About teaching mathematics: A K–8 resource.* Sausalito, CA: Marilyn Burns Education Associates.

Campbell, P. F. (1990). The vision of problem solving in the *Standards. Arithmetic Teacher, 37*(9), 14–17.

Cobb, P., & Merkel, B. (1989). Thinking strategies: Teaching arithmetic through problem solving. In P. R. Trafton (Ed.), *New directions for elementary school mathematics.* Reston, VA: National Council of Teachers of Mathematics.

Cobb, P., Yackel, E., Wood, T., & Wheatley, G. (1988). Creating a problem-solving atmosphere. *Arithmetic Teacher, 36*(1), 46–47.

Corwin, R. B. (1993). Doing mathematics together: Creating a mathematical culture. *Arithmetic Teacher, 40,* 338–341.

Garofalo, J. (1989). Beliefs and their influence on mathematical performance. *Mathematics Teacher, 82,* 502–505.

Garofalo, J., & Durant, K. (1991). Where did that come from? A frequent response to mathematics instruction. *School Science and Mathematics, 91,* 318–321.

Hyde, A. A., & Hyde, P. R. (1991). *Mathwise: Teaching mathematical thinking and problem solving.* Portsmouth, NH: Heinemann.

Koehler, M. S., & Prior, M. (1993). Classroom interactions: The heartbeat of the teaching/learning process. In D. T. Owens (Ed.), *Research ideas for the classroom: Middle grades mathematics.* New York: Macmillan Publishing Co.

Lampert, M. (1990). When the problem is not the question and the solution is not the answer: Mathematical knowing and teaching. *American Educational Research Journal, 27,* 29–63.

Lappan, G., & Schram, P. W. (1989). Communication and reasoning: Critical dimensions of sense making in mathematics. In P. R. Trafton (Ed.), *New directions for elementary school mathematics.* Reston, VA: National Council of Teachers of Mathematics.

National Council of Teachers of Mathematics. (1990). *Reaching higher: A problem-solving approach to elementary school mathematics* [Video and teacher's guide]. Reston, VA: The Council.

Schoenfeld, A. H. (1989). Problem solving in context(s). In R. I. Charles & E. A. Silver (Eds.), *The teaching and assessing of mathematical problem solving.* Reston, VA: National Council of Teachers of Mathematics.

Schroeder, T. L., & Lester, F. K., Jr. (1989). Developing understanding in mathematics via problem solving. In P. R. Trafton (Ed.), *New directions for elementary school mathematics.* Reston, VA: National Council of Teachers of Mathematics.

Silver, E. A., & Smith, M. S. (1990). Teaching mathematics and thinking. *Arithmetic Teacher, 37*(8), 34–37.

Small, M. S. (1989). Do you speak math? *Arithmetic Teacher, 37*(5), 26–29.

Stenmark, J. K., Thompson, V., & Cossey, R. (1986). *Family math.* Berkeley, CA: Lawrence Hall of Science, University of California.

Stoessiger, R., & Edmunds, J. (1992). *Natural learning and mathematics.* Portsmouth, NH: Heinemann.

Sullivan, P., & Clarke, D. (1991). Catering to all abilities through "good questions." *Arithmetic Teacher, 39*(2), 14–18.

Trowell, J. M. (1990). *Projects to enrich school mathematics: Level 1.* Reston, VA: National Council of Teachers of Mathematics.

Whitin, D. J. (1989). The power of mathematical investigations. In P. R. Trafton (Ed.), *New directions for elementary school mathematics.* Reston, VA: National Council of Teachers of Mathematics.

Whitin, D. J., Mills, H., & O'Keefe, T. (1990). *Living and learning mathematics: Stories and strategies for supporting mathematical literacy.* Portsmouth, NH: Heinemann.

Yackel, E., Cobb, P., Wood, T., Wheatley, G., & Merkel, G. (1990). The importance of social interaction in children's construction of mathematical knowledge. In T. J. Cooney (Ed.), *Teaching and learning mathematics in the 1990s.* Reston, VA: National Council of Teachers of Mathematics.

DEVELOPING UNDERSTANDING IN MATHEMATICS

◆ IN THIS CHAPTER THE COGNITIVE DIMENSIONS OF KNOWING mathematics are considered. In the first two sections we will consider the nature of mathematical knowledge and what it means to understand mathematics. The next two sections turn to the corresponding instructional implications for helping children develop or construct mathematical knowledge and understanding.

KNOWLEDGE OF MATHEMATICS

Knowledge consists of internal or mental representations of ideas that our mind has constructed. Not all knowledge is of the same type, and it can be categorized and described in different ways. It will be useful here to look at a general categorization of knowledge into three types: physical knowledge, logico-mathematical knowledge, and social knowledge. These distinctions will then help us understand the two types of knowledge found in the domain of mathematics: conceptual knowledge and procedural knowledge.

PIAGET'S THREE TYPES OF KNOWLEDGE

Jean Piaget, on whose theories of learning much of the current tenets of constructivism are based, made distinctions among physical knowledge, logico-mathematical knowledge, and conventional or social knowledge (Kamii, 1985, 1989; Labinowicz, 1985).

◆ Physical Knowledge

Physical knowledge is knowledge for which there are real, physical examples out there in the physical world. It is knowledge of things we can see, touch, hear, and the like. While the knowledge of these physical things is a representation in our minds, the source of that knowledge is physical reality. For example, we all have a concept of shoe, an idea that is somehow represented in each of our minds. Over time we constructed that concept by abstracting the essential features of shoes from our experiences with the hundreds or thousands of shoes we have seen and reflected on. Concepts of objects, colors, sounds, and smells are all examples of physical knowledge.

◆ Logico-mathematical Knowledge

Logico-mathematical knowledge is knowledge of relationships (logical associations or logical connections). There are no physical examples of the concepts and ideas in this category.

Consider the block and the stick in Figure 3.1 (p. 22). We probably have a lot of physical knowledge of objects such as these: They are hard, made of wood, and have a color; one rolls and one does not; they make a noise when dropped; and so on. These ideas are physical knowledge, and the block and stick are exemplars of that physical knowledge. Now notice that in the figure the block is *above* the stick. The stick is *longer than* the block. The two objects are *different*. We typically would say that we can "see" these ideas of above, longer, and different. However, if we could see "above," for example, then we could touch it, put our finger on it, point to it, or in some way tell where it is. You see the block and the stick, but you do not see "above." What kind of things are these ideas of above, longer, and different? They are *relationships*. They are logical constructs of the mind. The source of these ideas is the action of our own minds.

The physical objects can help create relationships because they give us something to see and reflect on. But seeing objects does not mean "seeing" the relationship(s). You can look at the block and stick and not construct or "see" aboveness.

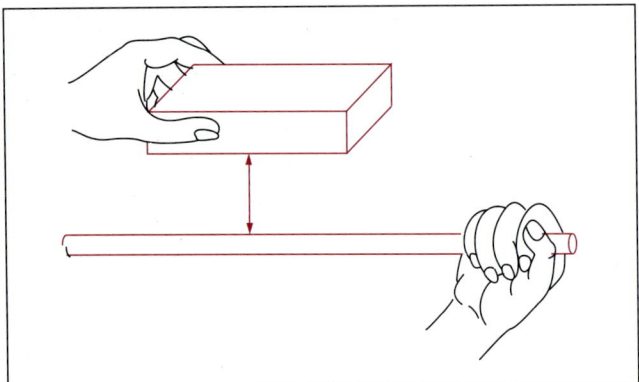

FIGURE 3.1: *The block is "above" the stick. The stick is "longer than" the block.*

Conventional Knowledge

Conventional knowledge is knowledge of things that have been arbitrarily agreed upon by society. Social interaction is the principal source. We learn the name of a friend by being told the name. The name is arbitrarily assigned by parents. Social or conventional knowledge includes the names and symbols used to stand for things (such as numerals, letters of the alphabet, mathematical symbols, commercial logos, etc.). The fact that Thanksgiving is celebrated on the last Thursday in November is an example of conventional knowledge.

CONCEPTUAL AND PROCEDURAL KNOWLEDGE OF MATHEMATICS

When thinking about learning and teaching mathematics it has been found useful to distinguish between *conceptual knowledge* and *procedural knowledge* (Hiebert & Carpenter, 1992; Hiebert & Lefevre, 1986; Hiebert & Lindquist, 1990). *Conceptual knowledge* consists of relationships constructed internally and connected to already existing ideas. It is the type of knowledge Piaget referred to as logico-mathematical knowledge. *Procedural knowledge* of mathematics is knowledge of the symbolism that is used to represent mathematics and of the rules and the procedures that one uses in carrying out routine mathematical tasks.

Conceptual Knowledge of Mathematics

Ideas such as seven, rectangle, ones/tens/hundreds (as in place value), sum, product, equivalent, ratio, and negative are all examples of relationships.

Figure 3.2 shows three blocks commonly used to represent ones, tens, and hundreds. By the middle of second grade most children have seen pictures of these or have used the actual blocks. It is quite common for these children to

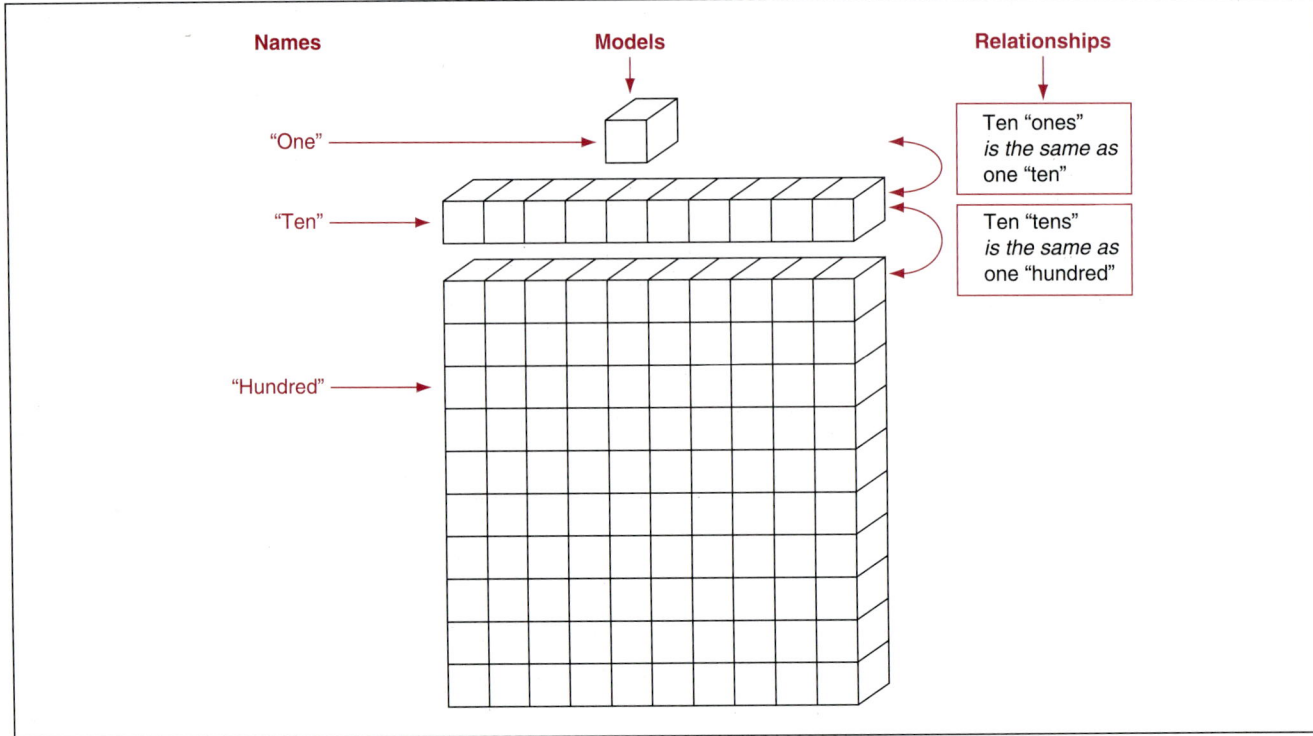

FIGURE 3.2: *Objects and names of objects are not the same as relationships between the objects.*

be able to identify the rod as the "ten" piece and the large square block as the "hundred" piece. Does this mean that they have the concept of ten and hundred? All that is known for sure is that they have learned the names for these objects; they have the common conventional knowledge of the blocks. The mathematical concept of ten is that it *is the same as ten ones*, not a rod. The concept is the relationship between the rod and the small cube. It is not the rod or a bundle of ten sticks or any other model of a ten. This relationship called "ten" must be created by children within their own minds.

In Figure 3.3 the shape labeled A is a rectangle. But, if we call shape B "one" or a "whole" then we might refer to shape A as "one half." The idea of "half" is the *relationship* between shapes A and B, a relationship that must be constructed in our minds. It is not in either rectangle. In fact, if we decide to call shape C the whole, then shape A becomes "one fourth." The physical rectangle did not change in any way. The concepts of "half" and "fourth" are not in the rectangle A; we can neither see them nor touch them. We construct these logical entities in our minds. The rectangles help us "see" the relationships, but what we see are rectangles, not fraction concepts.

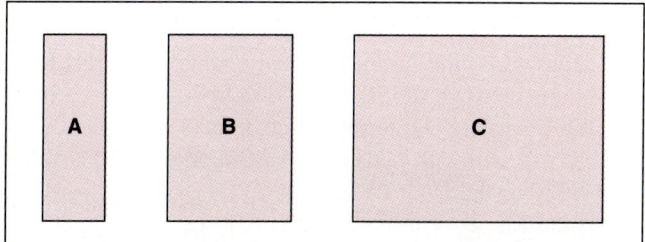

FIGURE 3.3: *Three shapes—different relationships*

Relationships are also involved in the concept of rectangle. What we actually see in Figure 3.3 is a collection of line segments. That these segments form rectangles is due to the relationships between them: Each segment *is at a right angle with* its *adjacent* segment and is *parallel* to the *opposite* segment, opposite segments are the *same length*, and the rectangles consist of *a collection of four* segments. Notice that each italicized word or phrase is a relationship.

For teachers of mathematics the fact that mathematical concepts are relationships is extremely significant. It means that it is essentially impossible to show children an example of a mathematical concept. We can show things or objects to children where the concept consists of relationships within or among the objects, but we can only be sure that the children are seeing objects. They must create the relationships. It is critical that we get children to be active mentally, to reflect on the ideas we present. That is the only way that the mind can construct a relationship. A passive learner will only see objects, not relationships.

◆ Procedural Knowledge of Mathematics

Knowledge of mathematics consists of more than concepts. Many concepts are represented by special words and symbols. Procedures have been developed for performing certain tasks. While this procedural knowledge can be supported by concepts, it is essentially made up of conventions—conventional knowledge.

Symbolism includes expressions such as $(9 - 5) \times 2 = 8$, $\pi, <, \neq$, and so on. What meaning is attached to this symbolic knowledge depends on how it is understood; what connections to concepts and other ideas the individual connects to the symbols.

Procedures are the step-by-step routines that are learned in order to accomplish some task. "To add two three-digit numbers you first add the numbers in the right-hand column. If the answer is 10 or more, put the one above the second column, and write the other digit under the first column. Proceed in a similar manner for the next two columns in order." We can say that one who can accomplish a task such as this has knowledge of that procedure. Again, what understanding may or may not accompany the procedural knowledge can vary considerably and depends on what connections are made to conceptual knowledge that supports the procedure.

Some procedures are very simple and may even by confused with conceptual knowledge. For example, seventh-grade children may be shown how to add the integers -7 and $+4$ by combining 7 red "negative" checkers with 4 yellow "positive" checkers. Pairs consisting of one red and one yellow checker are removed and the result noted. In this example there would be three red negative checkers remaining, and the students would record -3 as the sum. This might be called a manipulative procedure. Notice that it is possible to master a procedure such as this with or without understanding.

A computer program is a suitable analogy to a mathematical procedure (also known as an *algorithm*). Computers do exactly as they have been programmed, one step at a time. Once programmed, the computer "knows" that piece of procedural knowledge. Clearly the computer has no other knowledge connected with the program, so it is also accurate to say that the computer has no understanding of the procedure.

It is important to point out here that procedural knowledge of mathematics is important knowledge to have. Procedures allow us to do routine tasks easily. Symbolism is a powerful mechanism for conveying mathematical ideas to others and for "doodling around" with an idea as we *do* mathematics. To the extent that procedural knowledge is intimately connected with conceptual knowledge, procedures and symbolism become powerful tools in the construction of new knowledge.

UNDERSTANDING MATHEMATICS

All mathematical knowledge, conceptual and procedural, is potentially integrated or connected with other ideas that we have. We say that a mathematical concept or procedure is *understood* if there is some connection or integration with existing ideas. The manner in which a particular piece of knowledge is connected in our own minds modifies substantially how well it is understood (Backhouse, Haggarty, Pirie, & Stratton, 1992; Davis, 1986; Hiebert & Carpenter, 1992; Janvier, 1987; Schroeder & Lester, 1989).

RELATIONAL UNDERSTANDING VERSUS INSTRUMENTAL UNDERSTANDING

Understanding is not an all-or-nothing proposition. You have certainly heard or used expressions such as "I understand that *better* now," or "I have a *more complete* understanding than before." These phrases suggest that understanding can grow, that it can be weak (few connections) or strong (highly connected or integrated). One way that we can think about an individual's understanding, then, is that it exists along a continuum. At one extreme is a very rich set of connections. The understood idea is associated with many other existing ideas in a meaningful network of concepts and procedures. Hiebert and Carpenter (1992) refer to "webs" of interrelated ideas. Borrowing a term made popular by Richard Skemp (1978, 1979), this richly connected end of the continuum of understanding will be referred to in this text as *relational understanding*. At the other end of the continuum, ideas are completely isolated or nearly so. (Conceptual knowledge must be at least minimally connected since existing ideas were required to construct the relationships involved. Procedural knowledge, however, exists all too often without any support from existing concepts.) Again borrowing terminology from Skemp, understanding at the end of the continuum where knowledge is isolated and not integrated with other ideas will be referred to as *instrumental understanding*. Knowledge that is learned instrumentally is learned by rote through drill and practice.

Understanding Conceptual Knowledge

As new relationships (mathematical concepts) are constructed, they are almost certainly connected to the ideas that were used to construct them. By way of example, the preschool child begins with very primitive concepts of number such as *one*, *two*, and *more than*. He or she learns the procedural knowledge of counting through imitation at first (probably instrumental knowledge, no connections). Counting activities, answering "how many?" questions, playing games with score keeping, and in general a long series of successive reflections on quantity and counting procedures eventually help the child develop concepts of numbers such as *seven*. The concept seven is already connected to the counting procedure, to the construct of more than, and probably to other number concepts as well. It would be hard for a child to learn *seven* without some connections, without some rudimentary understanding. There are many more relationships that can be connected to the number seven. Seven is 1 more than 6, it is 2 less than 9, it is the combination of 3 and 4 or 2 and 5, it is odd, it is small compared to 73 and large compared to one-tenth, it is the number of days in the week, it is prime, and on and on. The web of ideas connected to a number can grow large and involved.

This web of integrated ideas around seven can itself be connected to other ideas that are being developed and understood. Addition and subtraction fact mastery can be strongly related to number concepts. Mental computation procedures are highly dependent on number concepts. Continuing with the seven example, one way to think about 26×7 is to think first of 25×7: 4 twenty-fives (100) and 3 twenty-fives (75), so that's 175, and then there is 7 more—175 . . . then 5 more—180 . . . and the last 2—182.

As another example of a rich understanding, consider the many ideas that a learner could potentially associate with the concept of *ratio* (Figure 3.4). Unfortunately, many children learn only procedural knowledge connected with ratio, such as "given one ratio, how do you find an equivalent ratio?" If one sees the connections between all of the ideas of Figure 3.4 (and there are more), then that person's understanding of ratio and of the various connected concepts is very good—very *relational*.

A few more examples of the potential for understanding follow:

- Decimal concepts are very much tied to the ideas of whole-number place value (the relationships between ones, tens, and hundreds are the same as the relationships between thousandths, hundredths, and tenths). Decimals are connected to fractions (4.75 is the same as $4\frac{3}{4}$) and with percentages (25% off of $120 is $\frac{1}{4}$ of $120 or $0.25 \times \$120$).
- Multiplication concepts are closely related to division concepts.
- The algebraic concept of variable is related to solving equations, constructing formulas from patterns, visualizing graphs, expressing changes in a phenomenon, and generalizing most any numeric relationship.

Understanding Procedural Knowledge

When talking about understanding procedural knowledge, understanding the knowledge of symbols, rules, and procedures, the most important connection we want is to meaningful concepts. It was "rules without reasons" that Skemp was referring to when he coined the term *instrumental understanding*. To understand a procedure in mathematics means that when we go through the steps they make sense and we understand why we are doing them. This does not mean we will necessarily think about the concepts involved as we do the procedures.

Once you have mastered the procedure for multiplying fractions, for example, you multiply tops with tops and bottoms with bottoms mechanically. However, research strongly suggests that without an understanding of why a procedure makes sense, that procedures tend to get confused ("Do I multiply the bottom numbers or do I get a common denominator?"), are easily forgotten, and lead to a very negative view of what mathematics is all about.

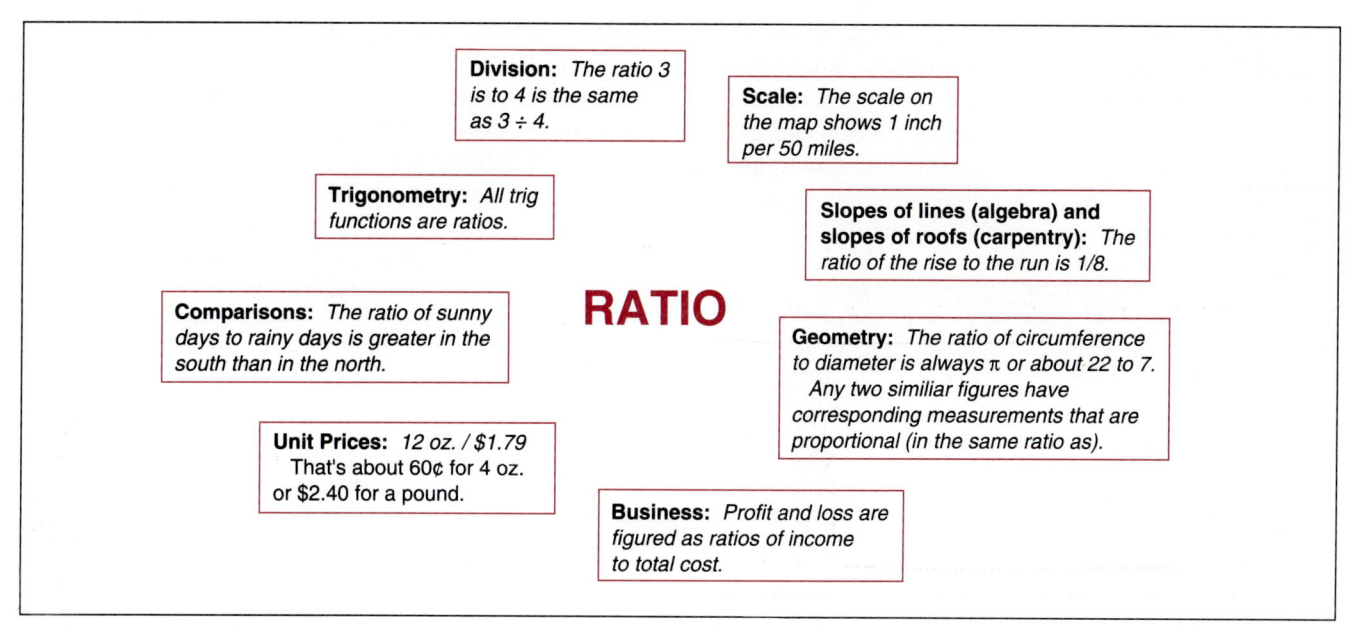

FIGURE 3.4: Potential web of associations with the concept of ratio

Division: The ratio 3 is to 4 is the same as 3 ÷ 4.

Scale: The scale on the map shows 1 inch per 50 miles.

Trigonometry: All trig functions are ratios.

Slopes of lines (algebra) and slopes of roofs (carpentry): The ratio of the rise to the run is 1/8.

Comparisons: The ratio of sunny days to rainy days is greater in the south than in the north.

RATIO

Geometry: The ratio of circumference to diameter is always π or about 22 to 7. Any two similiar figures have corresponding measurements that are proportional (in the same ratio as).

Unit Prices: 12 oz. / $1.79 That's about 60¢ for 4 oz. or $2.40 for a pound.

Business: Profit and loss are figured as ratios of income to total cost.

No one who is competent in the discipline of mathematics uses procedures that he or she cannot explain. Every single mathematical procedure is potentially connected with large networks of information. In many instances two or more different procedures are connected through a conceptual linkage. For example, to get half of a number you can divide by 2 or multiply by 0.5. Conceptual relationships give meaning to both procedures and indicate that they are equivalent while on the surface they appear quite distinct.

Procedural knowledge is most susceptible to instrumental understanding or rote rule learning. Many of us were unfortunately taught instrumentally; that is, taught to master rules that we learned by rote memorization with no hint of why they worked. You may know the rule "invert the divisor and multiply" but not be able to make up a simple word problem to go with $\frac{3}{4} \div \frac{1}{2}$. What does it mean to divide $\frac{3}{4}$ by $\frac{1}{2}$? On the Fourth National Assessment of Educational Progress (NAEP), roughly 80% of seventh-grade students were able to correctly express $5\frac{1}{4}$ as $\frac{21}{4}$. However, when asked to select the meaning of $5\frac{1}{4}$ ($5 + \frac{1}{4}$, $5 - \frac{1}{4}$, $5 \times \frac{1}{4}$, or $5 \div \frac{1}{4}$), fewer than half of the children chose the correct expression. Too many children are using procedures with fractions without an understanding of the concepts behind them (Kouba et al., 1988a).

At the second- and third-grade levels, many children are able to subtract with pencil and paper but are unable to explain the meaning of the little numbers that they write when they "borrow." They also do not understand that the number written after regrouping is the same quantity as before (Figure 3.5). These same children are able to use sticks and bundles of 10 sticks to do the same subtraction (Cauley, 1988). They seem to have conceptual knowledge of place value and regrouping and also procedural knowledge of regrouping but have failed to connect the two ideas. No matter how rich or useful their understanding of place value may be, the regrouping process is left disconnected from these concepts; it is known only instrumentally.

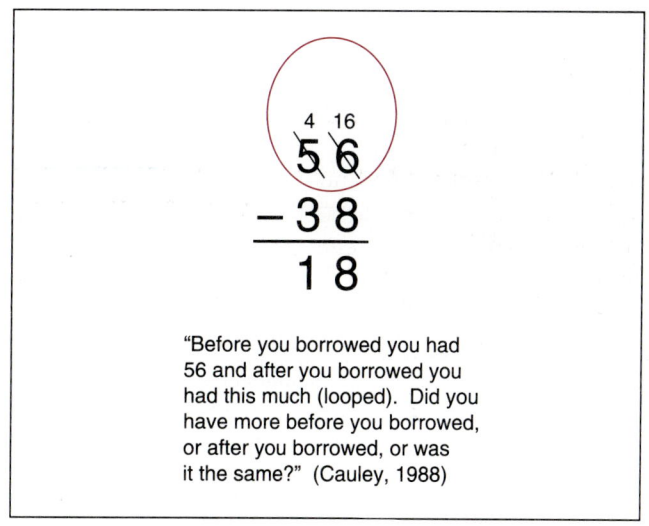

$$\begin{array}{r}{\scriptstyle 4\ \ 16}\\ \cancel{5}\cancel{6}\\ -3\ 8\\ \hline 1\ 8\end{array}$$

"Before you borrowed you had 56 and after you borrowed you had this much (looped). Did you have more before you borrowed, or after you borrowed, or was it the same?" (Cauley, 1988)

FIGURE 3.5: *To the question in the figure, roughly one third of second- and third-grade children responded less, another third thought there was more, and one third knew they were the same.*

The last example illustrates that it is possible to have taught children the corresponding conceptual knowledge related to a procedure but fail in helping them connect the concepts and procedure. Unfortunately, it is much more common to find children who simply do not possess the conceptual knowledge that supports a procedure.

Many children do not have even a belief or expectation that rules in mathematics should make any sense. A classic example of failed expectations that mathematics makes sense (Figure 3.6, p. 26) is provided by Erlwanger (1975) and is one that has been observed in various forms by many other teachers. In an interview, a fifth-grade child was asked to add $\frac{3}{4} + \frac{1}{4}$. The child wrote $\frac{4}{8}$ and confidently reduced it to $\frac{1}{2}$. The student explained that you add the tops and the bottoms, noting that was how he had been taught (likely a confusion with multiplication). The child also demonstrated, without prompting, that you could alternately do it with a drawing. Quickly drawing and shading parts of a circle he explained that the result was one whole and he wrote $\frac{3}{4} + \frac{1}{4} = 1$. The child was not bothered by the discrepancy in the two answers to the same problem. "It depends on which way the teacher tells you to do it." In his mind, the context (symbols vs. pie sections) was partly responsible for the meaning given to the problem. He was unaware of the error in the symbolic version.

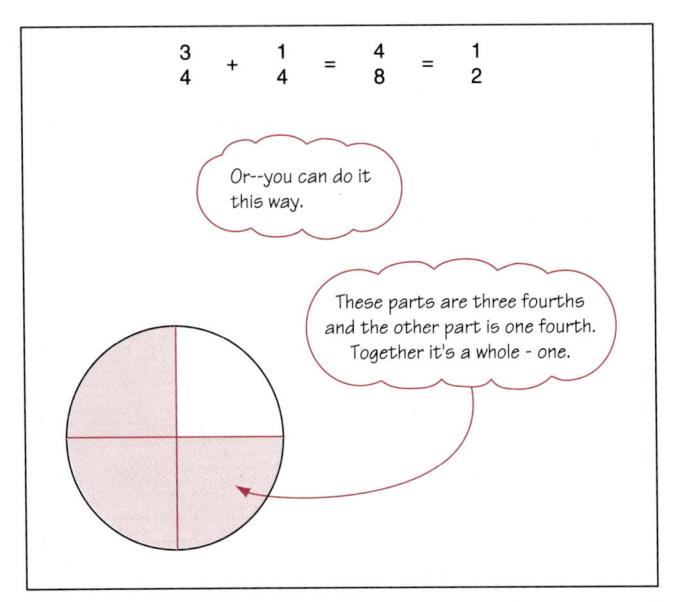

FIGURE 3.6: *Fifth-grade child gives two explanations and answers for the same problem. (Erlwanger, 1975)*

The following are examples of rules or procedures that are frequently learned instrumentally, without connection to a conceptual basis:

Turn the second fraction upside down and multiply.

In division, after you subtract, "bring down" the next number.

$(7 + \square = 12)$ Subtract the two numbers and put the answer in the box.

Area is length $\times$ width.

To change decimals to percents, move the decimal point two places to the right.

Divide by the "% number" to get the "of number" (20 is 35% of what?).

Line the ruler up with one end of the object, and read the number on the ruler at the other end.

The "key words" *in all* in a word problem mean you should add.

◆ The Individual Nature of Understanding

Since understanding is measured by the quantity and quality of the connections or associations that an individual is able to make with other already-formed networks of ideas, it follows that understanding depends in a very large degree to what networks and ideas the child brings to the task of understanding. Consider the idea of ratio with its many potential connections as noted in Figure 3.4. Do you, for example, currently possess all of the ideas surrounding this concept of ratio? Do you think the average seventh- or eighth-grade student does? This is a somewhat extreme example, selected earlier to indicate the potential for growth of understanding. But the point to be made remains valid: *Understanding is a highly subjective aspect of knowledge.* This feature of understanding should be ever-present in the mind of teachers dealing with a class of children. "Do you understand?" Clearly, different children could answer this question in the affirmative and all have different degrees and types of understanding.

The individual, subjective nature of understanding has three immediate implications for quality developmental instruction:

1. For each new concept or procedure you wish your students to understand, consider ahead of time the nature of that understanding. That is, to what other ideas can and should the new ideas be anchored or connected? After answering this question you will be able to select or design activities that will be aimed at making these specific connections.

2. Make every effort to find out what ideas potentially related to the new knowledge your students have to begin with. Children cannot connect ideas to things that are not there.

3. As instruction proceeds, make continual efforts to listen to your children in order to ascertain that the connections you planned on being made are in fact being constructed by the students.

There are many ways to listen. They all require that we provide for active involvement of our students so that we can watch what they do, hear how they explain, observe what they draw, see how they react. Passive students in a teacher-directed classroom provide us with very little to listen to.

BENEFITS OF RELATIONAL UNDERSTANDING

As you may already have guessed, to teach for relational understanding requires a lot of work and effort. Concepts and connections develop over time, not in a day. Instructional materials must be made. The classroom must be organized for group work and maximum interaction with and among the children. There are important reasons why this effort is not only worthwhile but essential for quality instruction.

Relational understanding

1. is intrinsically rewarding;
2. enhances memory;
3. requires that less be remembered;

4. helps with learning new concepts and procedures;
5. improves problem-solving abilities;
6. can be self-generative; and
7. has a positive effect on attitudes and beliefs.

◆ Intrinsically Rewarding

Nearly all people, and certainly children, enjoy learning. This is especially true when new information connects with ideas already possessed. The new knowledge makes sense, it fits, it feels good. Children who learn by rote must be motivated by external means: for the sake of a test, to please a parent, fear of failure, or to receive some reward. Such learning is distasteful. Rewards of an extra recess or a star on a chart may be effective in the short run but do nothing to encourage a love of the subject when the rewards are removed. Mathematics learned relationally is frequently just plain fun.

◆ Enhances Memory

Memory is a process of retrieving information. First, when concepts and procedures are learned relationally there is much less chance that the information will deteriorate; connected information is simply more likely to be retained over time than disconnected information. Second, retrieval of the information is easier. Connected information is like a filing system; each piece is in a related folder, and related folders are in related files, and files in related drawers, and so on. Finding a complete file of related ideas assures finding the particular piece of information. Retrieving disconnected information is more like finding a needle in a haystack.

A large portion of instructional time in American schools is devoted to reteaching and review. If ideas were learned relationally instead of instrumentally, it is likely that much less time would be spent on review.

◆ Less to Remember

A negative effect of the behaviorist influence on education was the fragmenting of mathematics into seemingly endless lists of isolated skills, concepts, rules, and symbols. Each was to be mastered before moving on. The lists grow so large that teachers and students become overwhelmed. When ideas are learned relationally, they become part of a larger web of information. Frequently, the network is so well constructed that whole chunks of information are stored and retrieved as single entities rather than isolated bits. For example, knowledge of place value subsumes rules about lining up decimal points, ordering decimal numbers, whether to move decimal points to the right or left in decimal/percent conversions, rules concerning rounding and estimating, and a host of other ideas. Similarly, knowledge of equivalent fractions ties together rules concerning common denominators, reducing fractions, and changing between mixed numbers and whole numbers.

◆ Helps Learning New Concepts and Procedures

An idea fully understood in mathematics is more easily extended to learn a new idea. Several examples of this have already been noted: Number concepts and relationships help in the mastery of basic facts, fraction knowledge and place-value knowledge come together to make decimal learning easier, and decimal concepts directly enhance an understanding of percentage concepts and procedures. In Chapter 2 we saw how area formulas for rectangles can be expanded to area formulas for parallelograms, triangles, and trapezoids. Many of the ideas of elementary arithmetic become the model for understanding ideas in algebra. Reducing fractions by finding common prime factors is the same thing as canceling out common factors.

Without these connections, each new piece of information children encounter would need to be wholly learned in isolation. The result would be fragmented, rote learning.

◆ Enhances Problem-solving Abilities

Between 1973 and 1986, the NAEP gathered data on the mathematics proficiency of the nation's 9-, 13-, and 17-year-olds. A consistent trend was a significantly lower level of performance in both problem solving and concepts than in traditional computational skills (Dossey et al., 1988). While the results are largely a reflection of the emphasis on basic skills in the U.S. curriculum, they also point out that skills developed in isolation are not very useful when it comes to solving problems and thinking. Problem solving requires both procedural and conceptual knowledge. Both are much more useful to the problem solver when intertwined and connected (Silver, 1986).

◆ Self-generative

The term *organic* is used by Skemp (1978) to denote this searching and growth quality of relational understanding. Skemp notes that when knowledge or gaining knowledge is found to be pleasurable, people who have had that experience of pleasure are likely to seek or invent new ideas on their own, especially when confronted with problematic situations. "Inventions that operate on understandings can generate new understandings, suggesting a kind of snowball effect. As networks grow and become more structured they increase the potential for invention" (Hiebert & Carpenter, 1992, p. 74).

◆ Attitudes and Beliefs

Relational understanding has an affective effect as well as a cognitive effect. When learning relationally, the learner tends to develop a positive self-concept about his or her ability to learn and understand mathematics. There is a definite sense of "I can do this! I understand!" A sense of self-worth is developed. Knowledge learned relationally is not foreign or strange. There is no reason to fear it or to be in awe of it. Mathematics then makes sense. It is not some mysterious world in which only "smart people" dare enter.

At the other end of the continuum, instrumental understanding has the potential of producing real mathematics anxiety. Math anxiety is a real phenomenon that involves definite fear and avoidance behavior. It is self-destructive. The more one fears and avoids mathematics, the more one is reinforced in beliefs of inadequacy.

Relational understanding also promotes a positive view about mathematics itself. Sensing the connectedness of mathematics and the way that it makes sense, students are more likely to gravitate toward it or to describe the discipline in positive terms. Students positively disposed to the discipline of mathematics are much more likely to pursue its study. Encouraging a much larger percentage of children to study mathematics and related fields is a serious national agenda.

COMMITMENT TO A GOAL

The importance of helping our students develop a relational understanding of mathematics cannot possibly be overstated. The benefits, as noted in the previous section, are enormous, and the negative effects of instrumental learning of mathematics are all too obvious and well known. If we as teachers wish to teach mathematics in the spirit of the NCTM's *Standards*, to have students who like and pursue mathematics, we must make a firm commitment to always teach for relational understanding. There is absolutely no exception. All mathematics taught to children can be understood. Our students should never, never get the idea that mathematics comes from the math god. Instrumental understanding has no place in school mathematics at any level!

HELPING CHILDREN DEVELOP RELATIONAL UNDERSTANDING

You want relational understanding for your students, but how do you go about it? This section turns to the subject of how children develop the knowledge and understanding based on a constructivist approach to learning. Based on this approach, we will then develop some general principles and guidelines that can inform our teaching regardless of the grade level or the specific content involved.

A CONSTRUCTIVIST APPROACH TO LEARNING

Since the mid-1980s, constructivism has played a major role in mathematics education, and constructivist approaches to learning are beginning to influence the teaching of mathematics as well as curriculum development. Even a casual look at the two *Standards* documents will support this statement. While possibly a contemporary position, constructivism is firmly rooted in the cognitive school of psychology and the theories of Piaget dating back at least as far as 1960. The constructivist approach is also in complete harmony with the work of Zoltan Dienes (1960) and Jerome Bruner (1960), two psychologists who have had a profound effect on mathematics education.

◆ Two Examples of Constructing Knowledge

Listed below is a string of numbers. Before reading further, try spending about one minute memorizing it so that you can repeat it orally or in writing.

2581114172023

How did you approach the task? Many separate the list into smaller chunks: for example, 258-111-417-2023. The four or five chunks are easier to remember than the entire string of 13 separate numbers. If you tried this method or one similar, how long do you think you will remember the list? an hour? a day? If you practice, especially if you say it aloud, invent a sing-song cadence, and perhaps write it down 40 or 50 times, your memory would probably be improved. Most people know several number strings, such as phone numbers and Social Security numbers, that they have learned in just that way.

If you think your mastery of the number string is weak and that you will likely forget it soon, look again at the numbers. However, this time look for some kind of pattern or rule. Try it now!

In a group of adults given this same memory task, one woman had the string "mastered" in less than 20 seconds and was quite confident that she would recall the string two or three weeks later with no practice. She pointed out that the list starts with 2, and then 3 is added to each successive number: 2, 5, 8, 11, 14, 17, 20, 23. She commented that it was because she was in a mathematics class that convinced her there must be some logic or pattern involved, so she looked for one from the outset.

What can be learned from this example? First, the idea of adding 3 each time is not visible in that string of numbers. It is a relationship of "3-more-than" between certain numbers. You had to construct that relationship.

It is significant that a disposition to look for a pattern or relationship played a key role. Recall from Chapter 2 the idea of creating a community of discourse in the mathematics classroom. In such an environment students expect to find relationships, and they will be more actively involved in searching for them.

Once the +3 relationship is observed, the string is very easy to recall. This has little to do with the quantity of material to be mastered. Relationships within the new material are integrated with your existing ideas of pattern, addition, number, and 3-more-than. These connections provide stability.

Finally, there is a positive feeling of satisfaction at having accomplished the task so easily when it seemed a bit formidable at the outset. Even if you had to be told about the pattern, by being able to integrate it into your framework of ideas it becomes a clever, albeit not very profound, tidbit of knowledge. And it's yours!

Children do not always construct the same knowledge that we intend for them to construct. Consider a third-grade child who has made a quite common error in subtraction, as shown in Figure 3.7. The child was presented with a situation that was partly familiar and partly not. What was familiar was that the problem appeared on a mathematics worksheet, it was subtraction, and the class had been doing subtraction with borrowing. This context narrowed the choice of ways to give (construct) meaning to the situation. But this problem was a little different from the child's existing ideas. She knew she should borrow from the next column but the next column contained a 0. She could not take 1 from the 0. That part was different. The child decided that "the next column" must mean the next one that has something in it. She therefore borrowed from the 6 and ignored the 0. This child gave her own meaning to the rule "borrow from the next column."

```
     5  13
     ↘  ↘
     6 0 3
  -  2 5 7
  ─────────
           6
```

There is nothing in this next column, so I'll borrow from the 6.

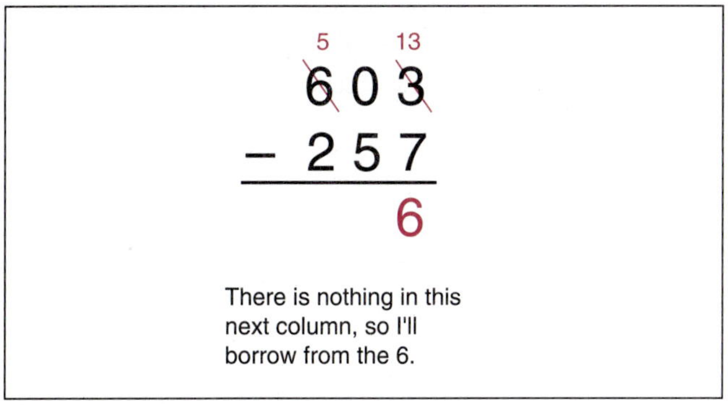

FIGURE 3.7: *Children sometimes invent incorrect meanings by extending poorly understood rules.*

In this example the existing ideas that were brought to the problem-solving situation were unfortunately quite limited and consisted mostly of procedural information that was not connected to a firm conceptual base of place-value knowledge. What we bring to the task of learning largely influences how we construct our knowledge. When a difficulty arises (what Piaget called *disequilibrium*), a child searches through the local network of ideas (those that are immediately related to the task) and uses what is available. Few third-grade children consciously think about place-value concepts while doing routine computation. However, had the procedural knowledge of borrowing been more tightly connected to place-value conceptual knowledge, it is much more likely that this child's understanding of the task would have been modified correctly. Had this quite common example occurred in your classroom, it would not be clear if knowledge of ones, tens, and hundreds was incomplete, was present but not being used (due to weak connections to the procedure), or simply was not present at all.

Children rarely give random responses (Ginsburg, 1977; Labinowicz, 1985). Their answers tend to make sense in terms of their personal perspective or in terms of the knowledge they are using to understand the situation. The connection or integration of new ideas with existing knowledge is a key principle of learning. In many instances children's existing knowledge is incomplete or inaccurate, or perhaps the knowledge we assume is there simply is not. In such situations, as in the present example, new knowledge may be constructed inaccurately.

◆ Constructing Knowledge

Two hallmarks of the constructivist position will help guide our teaching of mathematics. First, constructing knowledge is a highly active endeavor on the part of the learner (Baroody, 1987). To construct and understand a new idea involves making connections between old ideas and new ones. "How does this fit with what I already know?" "How can I understand this in the face of my current understanding of this idea?" The learner must play an active role rather than a passive one in the learning process. In each of the preceding examples, reflective thinking was going on. The learner searched for what was known in order to make sense of the situation. In classrooms, children must be encouraged to wrestle with new ideas, work at fitting them into existing networks, and challenge their own ideas and those of others. Put simply, *constructing knowledge requires reflective thought*.

Second, networks of ideas that presently exist in the learner's mind are the principal determining factors in how an idea will be constructed. These networks, frequently referred to as *cognitive schemas*, are both the product of constructing knowledge and the tools with which new knowledge is constructed. The schemas or integrated networks are used to give meaning to new ideas. The more connections with the existing schema, the better the ideas are understood. Fitting an idea into the existing schema is what Piaget referred to as *assimilation*. As learning occurs, the networks change, are rearranged, are added to, or are otherwise modified. Through reflective thought, schemas are constantly being modified or changed so that ideas fit better with what we know. This modification is the *accommodation* process referred to by Piaget.

REFLECTIVE THOUGHT

An adherent to the constructivist theory cannot possibly view children as "empty vessels" or "blank slates." Learners are not passive receptors of knowledge. If it were so, teaching would simply be a matter of carefully sequencing the content, communicating it to the students, and providing for adequate practice. Such an "absorptionist" view is founded in the behaviorist theories of Thorndike and Skinner and has proven completely inadequate in helping children learn to think, reason, and solve unfamiliar problems.

How, then, should we teach mathematics? One response might be that we don't teach at all; we get children to learn.

A principal feature of the constructivist position is that learning is a mentally active process, not a passive one. The corresponding key to effective teaching (or getting children to learn mathematics) is to get children to be active thinkers so that their minds will be working at forming relationships, making connections, and integrating concepts and procedures. Active thinking about or mentally working on an idea can be called *reflective thought*.

SIX WAYS TO PROMOTE REFLECTIVE THOUGHT

It is a major premise of this book that the single most important question we need to be continually asking about our teaching is, "*How can we structure lessons to promote reflective thought?*" Here are six ideas. Perhaps you will be able to add to the list.

1. Create a problem-solving environment.
2. Use models: manipulatives, drawings, calculators.
3. Encourage interaction and discussion.
4. Use cooperative learning groups.
5. Require self-validation of responses.
6. Listen actively.

Create a Problem-solving Environment

A problem-solving environment was a major topic of discussion in Chapter 2. In fact, the remaining five ideas for promoting reflective thought are contained in the description of a problem-solving environment (see p. 10). There is a point to be made. Since the late 1970s, problem solving has been promoted as the focus of school mathematics with the implication that we learn mathematics so that we can solve problems. A different view, one more in keeping with the NCTM's *Standards*, is that understanding should be the focus of school mathematics and problem solving seen as a means of acquiring that understanding (Schroeder & Lester, 1989). In Chapter 2 we said, "Mathematics *is* problem solving." Here we say, "Problem solving is a *vehicle* for learning mathematics." Together, the two statements correctly suggest that we help our students *learn* mathematics by having them *do* mathematics.

Use Models: Manipulatives, Drawings, Calculators

A *model* for a mathematical concept refers to any objects, pictures, drawings, and in many cases even calculators that embody or in some way illustrate the relationship(s) that make up the concept and can thereby help students construct or understand that concept. Recall that mathematical concepts are relationships and as such have no physical exemplars "out there" in the physical world. To model a concept, the materials should help the student reflect on the situation and thereby form the appropriate relationship.

Models of various types make it much easier for children at all grade levels to do the type of thinking that is necessary to create mathematical ideas. Models are things to reason with, concrete materials to talk about when referring to abstract relationships, a means of testing conjectures, props for articulating explanations.

Encourage Interaction and Discussion

To explain an idea orally or in written form forces us to wrestle with that idea until it is really ours and we personally understand it. The more we try to explain something or argue reasonably about something, the more connections we will search for and utilize in our explanation or in our argument. Talking gets the talker involved.

When children are asked to respond to and critique others they are similarly forced to attend, to assimilate what is being said into personal mental schemes. Frequently, when we get involved verbally with an idea we find ourselves changing or modifying the idea in midstream. The reflective thought required to make an explanation or argue a point is a true learning experience in itself (Yackel et al., 1990). There has long been a "writing to learn" movement in education that is slowly but surely growing in mathematics education (Azzolino, 1990; Countryman, 1992). Writing includes journals, formal essays, and reports on problem solutions or methods and is also a significant tool in assessment. Countryman states, "The writer reflects on, returns to, and builds upon what has gone before" (p. 59). Writing is an important form of interaction.

Use Cooperative Learning Groups

Placing children in groups of three or four to work on a problem is perhaps the best single classroom strategy a teacher can employ to implement all of the other strategies on this list of six ideas. A classroom arranged in groups of three or four children has many times the amount of interaction and discussion going on as can be accomplished in a full-class setting. Children are much more willing and able to speak out, explore ideas, explain things to their group, question and learn from one another, pose arguments, and have their own ideas challenged in a friendly atmosphere of learning. Children will take risks within a small group that they would never dream of doing in front of an entire class.

While the groups are at work, the teacher has the opportunity to be an active listener to six or more different discussions. Teachers can take part in groups and promote appropriate interaction. As an added bonus, the group structure means it is easier to manage the use of manipulatives.

Require Self-validation of Responses

As noted in Chapter 2, when children are required to explain or defend their responses there is a positive effect on how children view mathematics and their own abilities within that discipline. Confidence and self-worth is clearly promoted. What was not mentioned is the obvious requirement for reflective thinking that self-validation places on the student. To defend and/or explain eliminates guessing or responses based on rote learning. A habit of having children explain their answers is, then, another excellent mechanism for getting the same benefits that were discussed under interaction and discussion.

Listen Actively

To promote reflective thinking requires that teaching be very child-centered, not teacher-centered. By placing attention on the children's thoughts instead of ours, we encourage children to do more thinking and hence to search for and strengthen more internal connections—that is, understanding. When children respond to questions or make an observation in class, an interested but very nonevaluative response is a way to ask for an elaboration. "Tell me more about that, Karen," or, "I see. Why do you think that?" Even a simple "Um-hmm," followed by silence is very effective, permitting the child and others to continue their thinking.

USING MODELS TO TEACH MATHEMATICS

Models play such an important role in teaching developmentally that it is worth discussing them a bit further.

Examples of Models

To be a model, the material must embody or exhibit the relationship that we want children to construct. In Figure 3.8 common examples of models are illustrated for a variety of mathematical concepts. Consider each of the concepts and the corresponding model. Try to separate the physical model from the relationship that the model exhibits.

For the examples in Figure 3.8:

The concept of *six* is a relationship between sets that match the words *one, two, three, four, five, six*. Changing a set of counters by adding one changes the relationship. The difference between the set of 6 and the set of 7 is the relationship "one more than."

The concept of *length* could not be developed without making comparisons of the length attribute ("longness") of different objects. The length measure of an object is a relationship of the length of the object to the length of the unit.

The concept of *rectangle* is a combination of spatial and length relationships. The dot paper can illustrate the relationship of the opposite sides being of equal length and the sides meeting at right angles.

The concept of *hundred* is not in the larger square but in the relationship of that square to the strip (ten) and the little square (one).

Chance is a relationship between the frequency of an event happening compared with all possible outcomes.

Countable objects can be used to model *number* and related ideas such as *one more than*.

Base 10 concepts (ones, tens, hundreds) are frequently modeled with strips and squares. Sticks and bundles of sticks are also common.

Length involves a comparison of the length attribute of different objects. Rods can be used to measure length.

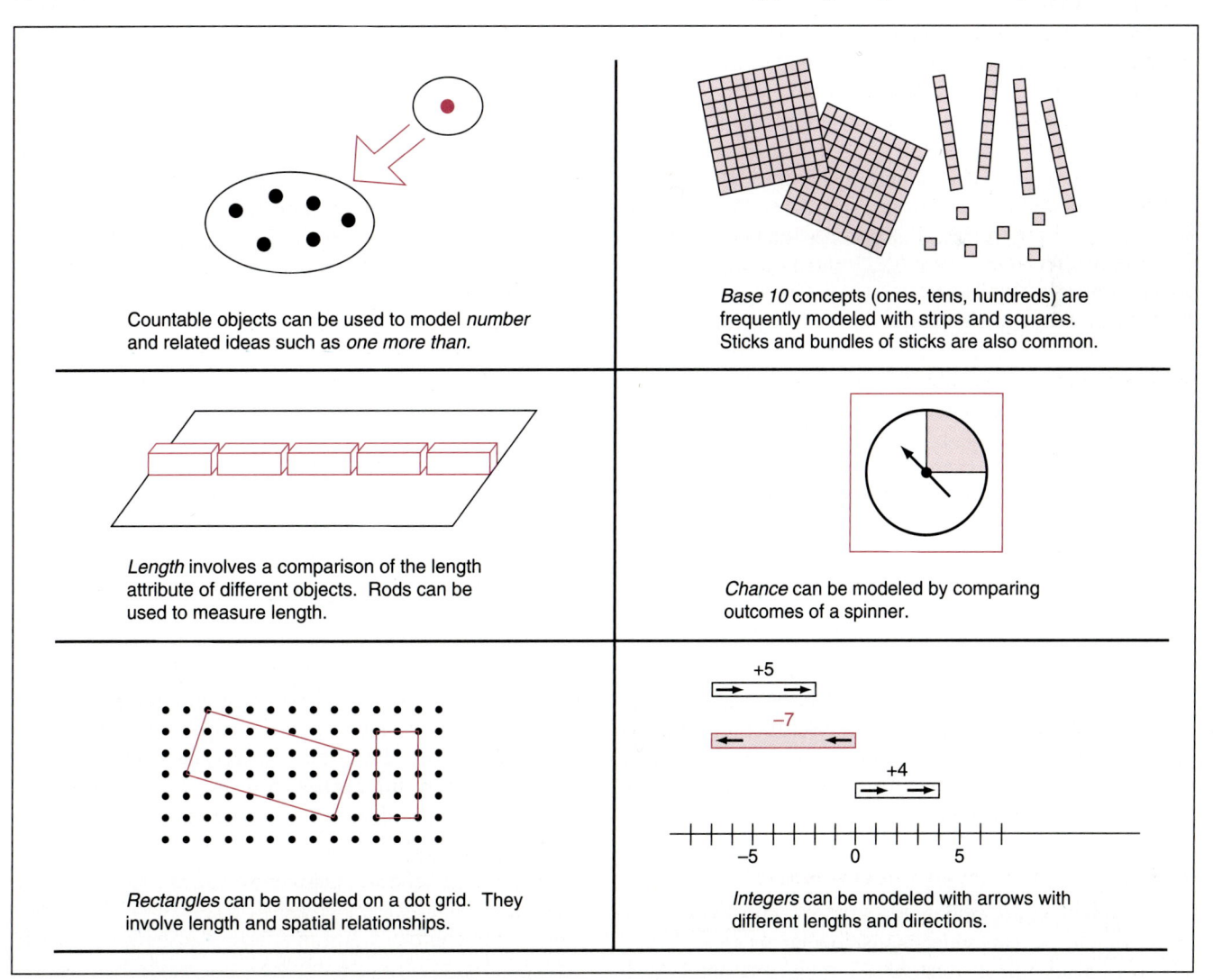

Chance can be modeled by comparing outcomes of a spinner.

Rectangles can be modeled on a dot grid. They involve length and spatial relationships.

Integers can be modeled with arrows with different lengths and directions.

FIGURE 3.8: *Examples of models to illustrate mathematics concepts*

The spinners can be used to create relative frequencies. These can be predicted by observing relationships of sectors of the spinner. Note how chance and probability are integrated with ideas of fractions and ratio.

The concept of a *negative integer* is based on the relationship "is the opposite of." Negative quantities only exist in relationship to positive quantities. Arrows on the number line are not themselves negative quantities but model the "opposite of" relationship in terms of length and direction.

A model is only effective if the students actually construct the desired relationship. That construction cannot be forced. If a model consists of physical materials that can actually be moved (relationships changed) or drawings that students make, the relationships exhibited have a better chance of being "seen." But the teacher should never take that seeing for granted. When children manipulate something (change, move, count, compare, draw, measure), there is a better chance that they will have to reflect on how and why they are doing that particular action.

Firsthand physical interaction with something is simply a better thinking tool than passively observing it. Thinking with the model refines and confirms the ideas we are constructing.

It is important to include calculators in any list of common models. At every grade level, K–16, the calculator models a wide variety of numeric relationships by quickly and easily demonstrating the effects of these ideas. A few examples will help make the point. If the calculator is made to count by increments of 0.01 (press $+$ 0.01 $=$), the meaning of 0.01 as one-hundredth is illustrated. For example, press 3 $+$ 0.01. How many presses of $=$ are required to get from 3 to 4? Doing the required 100 presses and observing how the display changes along the way is quite impressive. Especially note what happens after 3.19, 3.29, and so on.

As a second example, how would you use the calculator to divide 348 by 26, giving both quotient and remainder, without pressing the $\div$ key? There are at least three solutions to this task. (Can you find one or two of them?) Solving the problem helps students develop relationships between multiplication and division.

Attributes of models such as color, texture, or other attractive features of certain materials may have some motivational value. However, these features alone do not make the model valuable. If the desired relationship is not embodied by the materials, they will probably not help the child construct it.

◆ Three Uses of Models

There are three appropriate uses of models:

1. **To help children develop new concepts or relationships**. This suggests that models are generally most appropriate during the introductory phases of a new unit.

2. **To help children make connections between concepts and symbols**. Once a concept has been formed and students are learning how to record the idea on paper, the same models that were used in the initial development now serve as a connecting link. This is covered further in the next section.

3. **To assess children's understanding**. The first Evaluation Standard in the NCTM's *Standards* calls for alignment of the way we assess student learning with the corresponding instructional approaches and activities that were used in instruction. "If students' understandings are closely related to the use of physical materials, they should be allowed to use these materials to demonstrate their knowledge during assessment" (p. 195). (Refer to Chapter 5 for a full discussion.)

A good way to think about a model is as a "Thinker Toy"—something to promote thought, to think with.* Models should never be used as answer-getting devices. The distinction between "thinker toy" and "answer getter" is not always clear. The difference is in the focus or the purpose of the activity as it is perceived by the children.

When children are still in the process of wrestling with new concepts or figuring out a new procedure, models serve as aids to thinking. In contrast, when children are using materials such as blocks, counters, or number lines in a well-practiced, established manner with the focus being on cranking out answers to a series of problems, the use of the materials is mindless and inappropriate.

The model is no longer serving to promote thinking. Even the use of manipulatives can be learned instrumentally.

CONNECTING CONCEPTUAL AND PROCEDURAL KNOWLEDGE

Early in the chapter the distinction between conceptual and procedural knowledge was made. Both kinds of knowledge are important and need to be understood; both need to be connected to a broader network of ideas. For procedural knowledge, perhaps the most important connection to be made is to the conceptual knowledge that supports it. Without this critical connection, procedural knowledge is learned instrumentally. An instrumentally learned procedure is terribly rigid and not likely to be appropriately used outside of the narrow setting of practice pages in which it was developed. With a connection to the conceptual basis, not only is the procedure understood, but the user has ready access to all of the ideas that are related to the concept. The

---

*The term "Thinker Toy" is taken from Seymour Papert's popular book, *Mindstorms: Children, Computers, and Powerful Ideas* (1980). In the book the inventor of the Logo computer language describes his vision of the computer as a powerful and flexible device that encourages learners of any age to play with ideas and work through problems in virtually all disciplines.

result is a much more useful procedure both in applications and in the development of new information (Hiebert & Carpenter, 1992).

Three broad guidelines for helping children connect conceptual and procedural knowledge can guide your instructional strategies.

1. Develop conceptual knowledge and understanding first.
2. Incorporate models and language into the conceptual development to be used as future linkages with procedures.
3. Use translation activities where students are required to make and explain concepts in two or more modes: models, language, and symbols.

CONCEPTUAL KNOWLEDGE FIRST

The vast majority of the research suggests that the concept–procedure connection is best made when conceptual knowledge is developed first. When procedures have been taught and practiced instrumentally, there is heavy resistance on the part of students to want to find out why the rules make sense. Furthermore, it is generally more efficient to begin with concepts since students are not trying to master a procedure that they do not understand.

A Unit View, Not a Lesson View

As a rough guideline, at least the first 50% to 60% of the total time spent on a topic should be devoted to concept development and making connections with procedural knowledge. Here "total time" refers to the entire unit or chapter—not single lessons.

Most basal textbooks are written with nearly every lesson being a combination of concepts and procedures or of procedural knowledge only. However, development of a new concept is usually a matter of days or even weeks. That means that the developmental teacher should make modifications in the use of textbook materials. Textbooks, especially the teacher's editions, almost always include excellent suggestions for teaching concepts. These activities should be considered for possible instructional purposes along with those found in other resources. Just because there are procedural exercises in the text lesson does not mean they must be done that day. The textbook should always be viewed as a resource rather than a series of lesson plans.

Symbolism Records Ideas Already Learned

Children should generally view symbolism as simply a way to record or represent the ideas that they have already experienced, discussed, and understood. Written procedures should evolve as methods of doing or recording meaningful processes and ideas. In this manner, the ideas remain the focus, and there is a reasonable rationale for the symbolism. In many instances, students will find that the procedural version of an idea is much more efficient than working through the conceptual ideas that are initially learned. In such instances, instead of the procedure being viewed as meaningless, it is seen as providing power. Symbols are very efficient tools for representing ideas. Even excluding the obvious example of pencil-and-paper computations, equations, symbols, charts, and graphs are all instances of a picture being worth a thousand words. Approached as a record of ideas, students can begin to appreciate that power as well.

The following lines from Roach Van Allen provide food for thought:

What I can do, I can think about.

What I can think about I can talk about.

What I can say, I can write.

What I can write, I can read. The words remind me of what I did, thought, and said.

I can read what I can write and what other people can write for me to read.

(Cited in Labinowicz, 1980, p. 176)

Although written about a language approach to reading, these words make compelling common sense when applied to mathematics. Consider what it would mean if a child could not do a procedure conceptually. Could he or she think about it? talk about it? write about it? Children should only be asked to use symbolism for ideas they have explored, reflected on, and discussed—what they have done, thought about, and talked about.

DEVELOP LINKAGES: MODELS AND LANGUAGE

During the period in which conceptual knowledge is being developed, models and language will play a major role. These same external representations of the concept will then be available when you are helping children connect symbolic representations of the concept.

Models as Links

When some form of external materials (models) is used to develop a concept, keep in mind that the concept is the representation the child constructs mentally. There will be, however, a significant association between the concept represented internally in children's minds and the concept represented externally by the model. It is then appropriate to use the same models to make connections to the symbols. The idea is summarized in Figure 3.9. The first connection is between concept and model. Next, activities are designed explicitly to link models and symbolism. If both of these associations are clearly constructed by children, the chance that the connection between concepts and symbolism will be made is significantly enhanced (Carpenter, 1986).

♦ Language as Links

Language, both oral and written, has already been pointed to as a significant tool in promoting reflective thought and thus aids in constructing concepts and understanding. It is useful to think of language in much the same way as the models in Figure 3.9. However, language is more flexible and personal and can be used with or without the presence of models. When children are encouraged to talk about an idea in the process of developing a concept, they can then "talk through" the same conceptual ideas when working with symbols. Frequently, the exact same words can be used when working with models as when working with symbols (Figure 3.10).

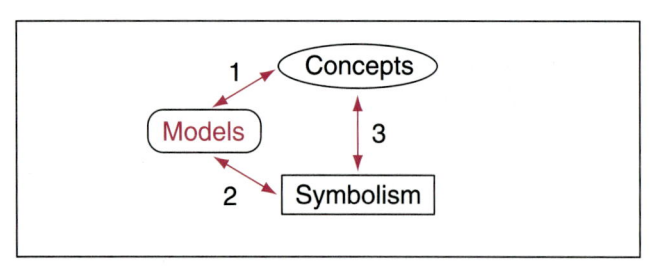

FIGURE 3.9: *Using models to develop concepts creates the first connection. Activities are then designed to make the second connection so that the third and most important connection will be developed.*

When language is established as a linkage, it is always available to children in helping them recall important conceptual ideas. When stumped with a procedural or problem-solving situation, the child who can talk about the ideas involved has a good chance of retrieving previously established connections to make sense of the task.

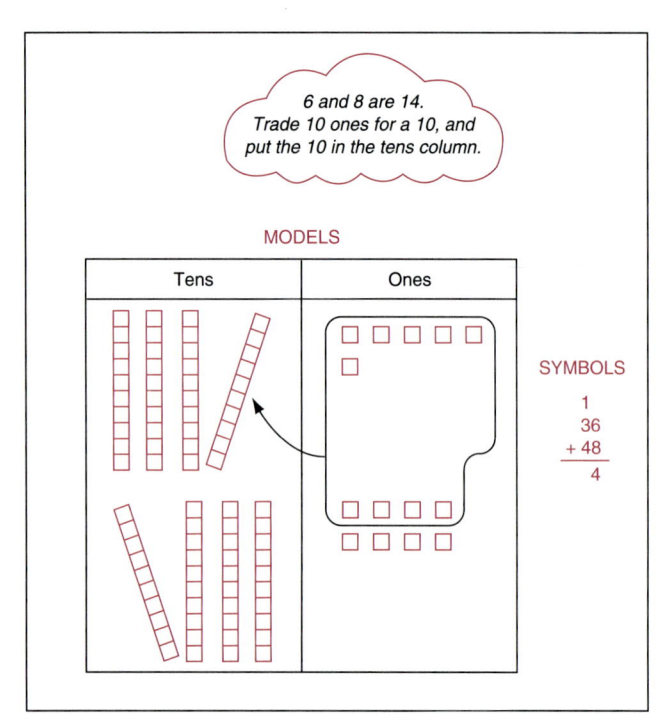

FIGURE 3.10: *The same oral language is frequently used in both manipulative and symbolic modes.*

USE TRANSLATION ACTIVITIES

We can never be certain what representation of a concept children have constructed internally. What we can do is observe how they interact with external representations. Figure 3.11 illustrates the three broad categories of external representations we have been discussing: models, language, and symbolism. The triangle provides a framework for designing activities frequently referred to as *translation activities*.* In a translation activity, children are given an idea in any one of the three external representational modes and are required to give or explain the idea in either or both of the other modes. Translations between models and language are conducted early on during conceptual development and serve to establish models and language as linkages. Later, as symbolism is introduced to represent the concepts, a variety of activities can emphasize translations along any of the three sides of the triangle. In their work with fractions and proportion concepts, Lesh, Post, and Behr (1987) found that "these 'translation (dis)abilities' are significant factors influencing both mathematical learning and problem-solving performance, and that strengthening or remediating these abilities facilitates the acquisition and use of elementary mathematical ideas" (p. 36).

Of the six possible translations, the four that involve models seem to be the most significant. Translations from one model to a different model that embodies the same concept are also quite profitable.

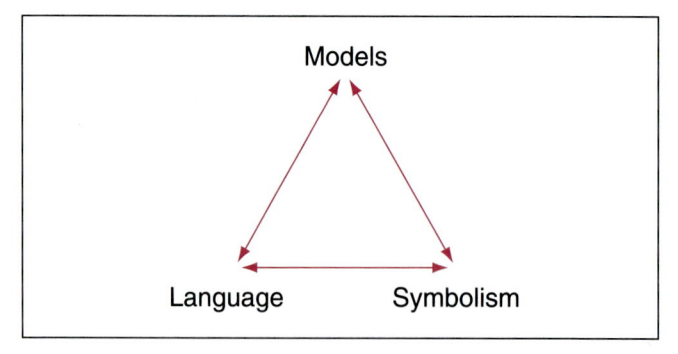

FIGURE 3.11: *Translation activities require children to make explicit connections between one of three external representations of an idea and either or both of the remaining representations.*

♦ Models to Symbols and/or Language

Translations that begin with models may have children record and/or talk about an idea as they work through it using manipulatives. An increasingly common approach in the primary grades is to provide children with some form of a work mat on which counters are manipulated to illustrate a concept. For each example, they then write an equation or tell about the concept orally. Figure 3.12 shows a simple two-part mat on which children separate nine counters into two sets. For each they write a corresponding addi-

---

*The triangle framework of models, language, and symbolism was originally suggested by Payne and Rathmell (1975).

tion equation. In a class discussion the children might read the mats as follows: "Two and seven is the same as nine."

In Figure 3.13 children slice various shaded fractional parts of a square into smaller fractional parts by drawing lines in the opposite direction. For each drawing, the equivalent fraction equation is written with an emphasis on the two products displayed in the pictures.

Translation activities involving step-by-step procedures might have two children working together. One child uses manipulative materials to model the procedure, and the other child writes down each step as it is being done. Numerous examples of recording a step at a time can be found in Chapter 10 on computation.

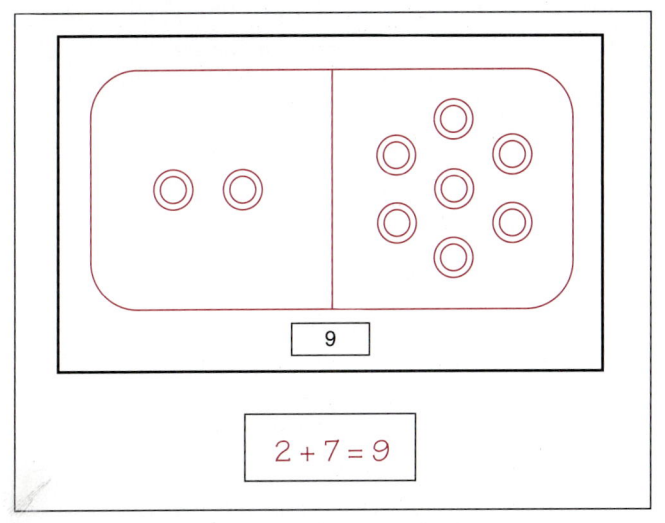

FIGURE 3.12: *Models translated to symbols in a simple addition exercise.*

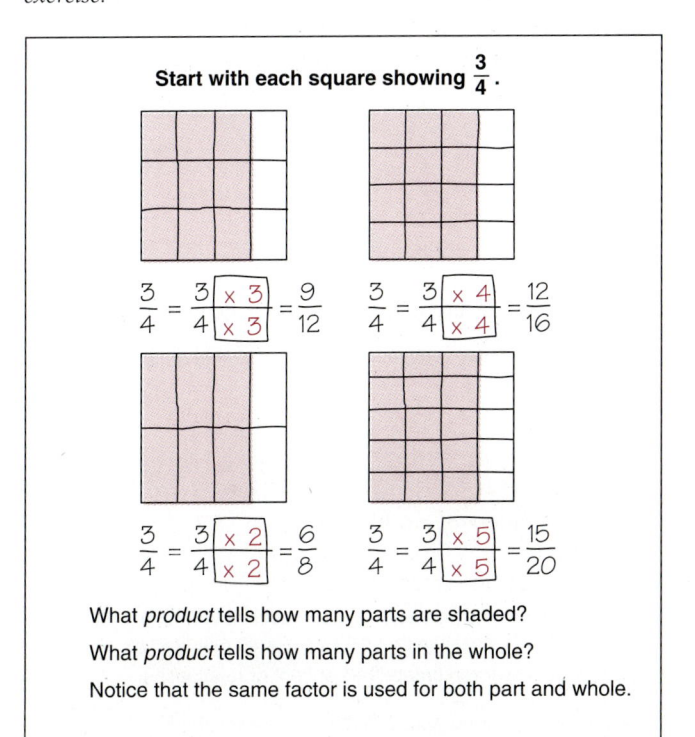

FIGURE 3.13: *Students work with a drawing model of equivalent fractions and write the corresponding equations.*

Symbols to Models

Translations that begin with symbolism generally have students use models to explain an idea. The teacher might work a symbolic procedure through at the board with the class and then have children provide explanations using models. For example, in subtraction with carrying, "Use your ones, tens, and hundreds pieces to explain what the little one stands for at the top of this column." A 10×10 square grid is a common model for percent concepts. Children can be asked to make a drawing on the grid to illustrate why 75% and $\frac{3}{4}$ of a whole are the same amounts. These would be accompanied by verbal explanations as well.

TEACHING DEVELOPMENTALLY

This chapter has presented you with a full plate of nontrivial ideas about understanding mathematics, how children learn mathematics, and implications for instruction. Quite possibly some of these ideas are a bit fuzzy at this point. As reflective thinking is important for children in learning mathematics, so should you reflect on the ideas in this chapter as you read through other sections of this book and as you select and plan lessons for your children.

KEY IDEAS IN A DEVELOPMENTAL APPROACH

The listing below is simply a summary listing of the major points that have been made. A teacher who keeps these ideas in mind can be said to be basing his or her instruction on a constructivist view of learning or, in the terminology of this book, a developmental approach.

1. **Knowledge is constructed by each individual; it is not absorbed passively.** The understanding of an idea is determined by the number and quality of the connections that a learner makes with other existing ideas. Well-understood ideas are imbedded in a network of related concepts, ideas, and procedures.

2. **Knowledge and understanding is a personal matter for each learner.** The quantity and quality of the learner's existing ideas is a major determining factor in how new ideas will be learned and understood. Ideas are not poured into an empty vessel. The learner must use whatever ideas are available in his or her cognitive makeup to construct new ones.

3. **Reflective thought is the single most important ingredient in an effective learning activity.** Learners must be mentally active in the construction of new ideas. If ideas are to be integrated into a larger web of ideas, those existing networks must be actively engaged in the learning process.

4. **Teaching is a child-centered activity.** Before a lesson, the effective teacher designs activities by consid-

ering, "What will the children be doing and thinking as they do this?" During a lesson, the effective teacher listens to the children to try to determine how ideas are being constructed.

REFLECTING ON STUDENT ACTIVITIES

Throughout this book, in every student textbook, in every article you read or in-service workshop you attend, you will hear and read about suggestions for activities that someone believes are effective in helping children learn some aspect of mathematics. As a teacher, you will constantly be selecting from these activities and perhaps designing your own as you plan lessons. Here is a four-step guide you can use when considering a new activity for your lessons.

The first two steps of this guide are routine, practical considerations. The third step requires you to focus on what the activity is trying to accomplish. The third step asks you to be clear about why you want to do this. You must know what relationships you want your children to construct by virtue of doing this activity.

The fourth question is the most important point in understanding how an activity will or will not accomplish its purpose. Try to put yourself in the children's position as you think about the activity. What is it about the activity that will force or at least improve the chances that the children will reflect on the desired relationship? There are many activities that appear on the surface to be fantastic, and yet children are so involved with following the directions, or managing materials, or worrying about losing a game, and so on, that the critical feature of the activity is overshadowed, and very little reflective thinking actually takes place. Remember it is not the activity you want children to learn but rather the relationships that the activity causes or encourages them to construct.

Recall the importance of listening to children as you teach. Being very conscious of how an activity will cause the desired reflective thought will guide your interaction during a lesson. "Are my children thinking about this activity the way they need to or the way I thought they would? How can I direct their thinking so that the activity is effective?" Two teachers can conduct the same activity with two similar classrooms of children and yet have very different levels of effectiveness. The effective teacher is the one who has considered how the activity will cause the desired thought and who actively listens to the children to see that the desired thinking is taking place.

Practice using this little guide as you read this book. To help you better understand an activity, try answering the fourth question. Work toward thinking about activities from the view of what happens inside children's minds, not just what they are doing with their hands. Good activities are *minds-on* activities, not just *hands-on* activities.

TEACHER FOUR-STEP REFLECTIVE THOUGHT GUIDE FOR MATHEMATICS LEARNING ACTIVITIES

STEP 1: How Is the Activity Done?
Actually do the activity. Learn as much about the actual doing aspect of the activity as possible.
How do *you* do it?
How would *children* do it? (They don't know what you do!)
What materials are needed?
What are the steps?
What is written down or recorded?

STEP 2: What Does the Teacher Have to Do?
Focus on what you need to conduct the activity in your class.
What directions would you need to give to students so that they could do the activity? How should they be given? Oral? Written? Demonstration?
How would you have to prepare students for the activity? What would they need to know beforehand to do it?

STEP 3: What Is the Purpose of the Activity?
What is the activity designed to develop?
Concepts and relationships?
Connections between concepts and procedures?
Procedural knowledge (practice, skill, drill)?

STEP 4: How Does the Activity Accomplish the Purpose?
How does the activity promote the required reflective thought for its specific purpose? What must children reflect on to do the activity? Will it necessarily promote reflective thought or is it possible that it may be done mindlessly?
For conceptual and connecting activities: How will the activity help children create the concept or relationship in their own minds?
For procedural activities: Is the activity effective in providing the necessary practice? Does it capitalize on or contribute to conceptual development?

REFLECTIONS ON CHAPTER 3: WRITING TO LEARN

1. How would Piaget classify conceptual knowledge of mathematics? What is the significance of this for teaching mathematics?
2. Contrast conceptual and procedural knowledge of mathematics.
3. How is it that understanding can exist on a continuum from a very strong to a very weak understanding? What are the ends of this continuum called?
4. Why do you think it is important to understand mathematics? (There are seven ideas discussed in the text. Put a few of these that you feel are most important together and make a good argument for relational understanding.)
5. Explain in your own words what it means to construct knowledge instead of absorbing knowledge.
6. Think up some synonyms or phrases that mean the same thing as reflective thought. Explain.
7. The six ways the text lists to promote reflective thought are so important that you should probably be able to list them and discuss each briefly.
8. What is a *model* for teaching mathematics? Select one, and explain why it meets the definition.
9. Models should be used in some assessments. What are the other two times in instruction when they should be used? Explain briefly.
10. What is a translation activity? Give an example. What is the purpose of a translation activity?
11. Tie together the ideas of reflective thinking and the individual nature of learning to explain why developmental instruction must be child-centered and not teacher-centered.

FOR DISCUSSION AND EXPLORATION

1. Discuss the meaning and validity of the following statement: Children see what they understand rather than understand what they see.
2. Read the first chapter of Labinowicz's *Learning from Children* (1985) or the first chapter of *Children's Mathematical Thinking* (Baroody, 1987). Relate the ideas of these authors to those in this chapter.
3. Visit an elementary school classroom to observe several mathematics lessons over a period of a few days. Discuss how it appears that children are learning. Is the general approach based more on a constructivist theory of learning or on an absorption theory? Take special note of the use of physical or even picture models. How and for what apparent purpose are these being used? How are students encouraged to discuss and articulate ideas?
4. Examine the teacher's edition of a current popular basal textbook. Select any chapter, and explore the development within that chapter in terms of concept development and the introduction and use of symbolism. Consider the use of models or other means of connecting concepts and procedures. Select activities from throughout the chapter that could be used as good developmental or concept activities. Use the text to outline a concept development lesson and a different lesson connecting concepts and procedures.

SUGGESTED READINGS

Baroody, A. J. (1987). *Children's mathematical thinking: A developmental framework for preschool, primary, and special education teachers.* New York: Teachers College Press.

Davis, R. B. (1986). *Learning mathematics: The cognitive science approach to mathematics education.* Norwood, NJ: Ablex.

Eisenhart, J., Borko, H., Underhill, R., Brown, C., Jones, D., & Agard, P. (1993). Conceptual knowledge falls through the cracks: Complexities of learning to teach mathematics for understanding. *Journal for Research in Mathematics Education, 24,* 8–40.

Erlwanger, S. H. (1975). Case studies of children's conceptions of mathematics—Part I. *Journal of Children's Mathematical Behavior, 1*(3), 157–183.

Ginsburg, H. P., & Baron, J. (1993). Cognition: Young children's construction of mathematics. In R. J. Jensen (Ed.), *Research ideas for the classroom: Early childhood mathematics.* New York: Macmillan Publishing Co.

Greenes, C., Schulman, L., & Spungin, R. (1992). Stimulating communication in mathematics. *Arithmetic Teacher, 40,* 78–82.

Hart, L. C., Schultz, K., Najee-ullah, D., & Nash, L. (1992). The role of reflection in teaching. *Arithmetic Teacher, 40,* 40–42.

Hiebert, J. (1984). Children's mathematics learning: The struggle to link form and understanding. *The Elementary School Journal, 84,* 497–513.

Hiebert, J. (1987). The struggle to link written symbols with understandings: An update. *Arithmetic Teacher, 36*(7), 38–44.

Hiebert, J., & Lindquist, M. M. (1990). Developing mathematical knowledge in the young child. In J. N. Payne (Ed.), *Mathematics for the young child.* Reston, VA: National Council of Teachers of Mathematics.

Juraschek, W. (1983). Piaget and middle school mathematics. *School Science and Mathematics, 83,* 5–13.

Kamii, C. (1990). Constructivism and beginning arithmetic (K–2). In T. J. Cooney (Ed.), *Teaching and learning mathematics in the 1990s.* Reston, VA: National Council of Teachers of Mathematics.

Kloosterman, P., & Gainey, P. H. (1993). Students' thinking: Middle grades mathematics. In D. T. Owens (Ed.), *Research ideas for the classroom: Middle grades mathematics*. New York: Macmillan Publishing Co.

Labinowicz, E. (1980). *The Piaget primer: Thinking, learning, teaching*. Menlo Park, CA: Addison-Wesley.

Labinowicz, E. (1985). *Learning from children: New beginnings for teaching numerical thinking*. Menlo Park, CA: Addison-Wesley.

Post, T. R. (1988). Some notes on the nature of mathematics learning. In T. R. Post (Ed.), *Teaching mathematics in grades K–8: Research-based methods,* 2nd ed. Boston: Allyn & Bacon.

Schwartz, J. E. (1992). "Silent teacher" and mathematics as reasoning. *Arithmetic Teacher, 40,* 122–124.

Skemp, R. (1978). Relational understanding and instrumental understanding. *Arithmetic Teacher, 26*(3), 9–15.

Van de Walle, J. A. (1983). Focus on the connections between concepts and symbolism. *Focus on Learning Problems in Mathematics, 5*(1), 5–13.

4 DEVELOPING PROBLEM-SOLVING PROCESSES

◆ PROBLEM SOLVING IS THE PROCESS OF THINKING, OF SEARCHING, for patterns and regularities. In simpler terms, problem solving is "figuring it out," making sense of puzzling or difficult situations. In the last chapter we saw how reflective thought could be enhanced through a problem-solving approach; learning mathematics *via* problem solving. In this chapter we look at teaching *about* problem solving, how we can help children become better problem solvers. Thus, we have come full circle, returning to the themes of Chapter 2: Knowing mathematics is doing mathematics, and doing mathematics is solving problems.

PROBLEM SOLVING IN THE CURRICULUM

It should also be reemphasized that problem solving is very much a factor in constructing mathematics. We do not learn mathematics first and then figure out afterward how to use that knowledge to solve problems. At the same time, the mathematical knowledge possessed does modify one's problem-solving ability. But problem solving also involves its own strategies or processes, the ability to monitor and regulate these processes, and a collection of attitudes and beliefs about problem solving. These aspects of problem solving require explicit attention in an instructional program that intends to develop students' abilities to solve problems—and to learn and do mathematics.

WHAT IS A PROBLEM?

If we are going to talk about problem solving, it would be good to have a clear understanding of what a problem is. Two definitions are offered here:

1. A *problem* is a doubtful or difficult question; a matter of inquiry, discussion, or thought; a question that exercises the mind (*Oxford English Dictionary*).

2. A *problem* is a situation or task for which:
   (a) The person confronting that task wants or needs to find a solution.
   (b) The person has no readily available procedure for finding the solution.
   (c) The person makes an attempt to find the solution.
   (Charles & Lester, 1982, p. 5)

Both definitions point to the fact that problems are not routine exercises. Traditionally in mathematics we have referred to most any answer-getting exercise as a problem. Routine activities such as finding the percentage that one number is of another or adding two three-digit numbers were referred to as problems. In reality, such exercises were and are rarely assigned before the teacher explains how they are to be done, models how they are to be done, and has students then practice doing the "problems" in the text. These exercises do not include the significant element required for a problem, namely *blockage:* the need to confront, figure out, find order in, reason, conjecture, test, and so on. The Charles and Lester definition points to two other factors: *desire* and *effort*. All three of these ingredients of a problem, *desire, blockage,* and *effort,* point to the fact that what may be a problem for one person may not be a problem for another. Mathematical knowledge available to an older student may remove the blockage component that makes the situation a real problem for a younger child. The desire and effort aspects relate to affective considerations: Do I really care enough about this problem to work on it?

COMPONENTS OF PROBLEM-SOLVING ABILITY

Problem-solving ability does not develop over a few weeks or months. Nor is it a topic that is taught at a particular grade level. Growth in the ability to solve problems is slow and continuous. We need to address problem solving virtually every day, in every lesson, beginning in kindergarten and continuing through high school, because problem solving and learning mathematics are so intimately connected. However, as we involve our students in problem-solving activities, it is beneficial to be clearly aware of what is involved in good problem-solving ability. In addition to the knowledge base, three other aspects of students' cognitive makeup need to be developed: (1) problem-solving *strategies,* problem-solving approaches that can be generalized; (2) *metacognitive* processes, monitoring and regulating our own thought processes; and (3) *beliefs and attitudes,* including self-confidence, willingness, and perseverance in the area of problem solving.

Strategies

Strategies for solving problems are identifiable methods of approaching a task that are completely independent of the specific topic or subject matter. George Polya, a virtual legend in the area of mathematical problem solving, used the term *heuristic* to refer to these thought patterns or approaches to problems. They include such things as working backwards from the conclusion, searching for a pattern, making a sketch or picture, and many others.

Problem-solving heuristics also include habits and schemes appropriate for virtually all problems. These include such things as stopping to fully understand the problem at the outset, identifying relevant and irrelevant information, selecting an appropriate strategy, assessing the reasonableness of the answer, and others. These general heuristics will be discussed a bit later.

To have a better feel for what a strategy or heuristic is, try to solve the following problem. Its one that has become a classic in the problem-solving literature. Feel free to work with someone else on the problem.

> **PIGS AND CHICKENS**
>
> One day Farmer Brown was counting his pigs and chickens. He noticed that they had 60 legs and that there were 22 animals in all. How many of each kind of animal (pigs and chickens) did he have?

If you are still reading and have not stopped to solve the pigs-and-chickens problem, try to do so now. (Go ahead. Do it!) When you get finished and are confident of your solution, examine how you solved the problem. How would you describe your approach? At least three different general approaches or strategies are possible for this problem. One commonly used approach is to guess-and-check or preferably, try-and-adjust. Making a drawing and creating a chart or table are two other possibilities. Figure 4.1 illustrates these three methods. You may have used yet a different approach. Even if you used one of these approaches, it may look substantially different on paper from what you see here.

Third-graders with no instruction in the use of strategies will almost all "solve" this problem by adding 22 and 60. (Note that the problem includes the words "in all" that many believe tell them they should add.) Many fifth-graders will attempt to divide 60 by 22. It seems the fifth-graders are at least sensitive to adding legs and animals (Lester, 1985). These students are products of traditional programs where "problems" are more like exercises. You find the two numbers that are in the problem and select an operation. Students use key words and information about the numbers. Fifth-graders tend to stop because 22 "won't go into" 60. Many straightforward, routine problems found in textbooks tend to reinforce this behavior since, sadly, for many of these problems, such superficial approaches frequently work.

Research evidence strongly suggests that more general and powerful strategies such as try-and-adjust are teachable, that students who are taught strategies do in fact use them, and that, in general, strategy instruction improves problem-solving performance (Campione, Brown, & Connell, 1989; Lester, 1985; Schoenfeld, 1992; Suydam, 1987). It is generally accepted that some overt instruction in strategies is appropriate. As students learn strategies they must also learn to self-select and use them in a wide variety of contexts and different areas of mathematics. That is, we do not want to teach strategies for the sake of strategies but rather to recognize and encourage the use of a wide variety and combination of strategies in all areas of mathematics instruction.

Metacognition

Metacognition refers to conscious monitoring (being aware of how and why you are doing something) and regulation (choosing to do something or deciding to make changes) of one's own thought process.

Guess-and-Check method

22 animals, so try 11 each.

11 pigs   = 44 legs
11 chickens = 22 legs
              —————
              66 legs

That's too many--need fewer pigs.
Try 9 pigs. That leaves 13 chickens.

9 × 4 = 36
13 × 2 = 26
         ————
         62 legs

Still too many.
Try 8 pigs and 14 chickens.

8 × 4 = 32
14 × 2 = 28
         ————
         60   That's it!

Draw a Picture Method

22 animals. Each has at least 2 legs.

That's 44 legs. I'll add 2 more legs until I get 60 legs.

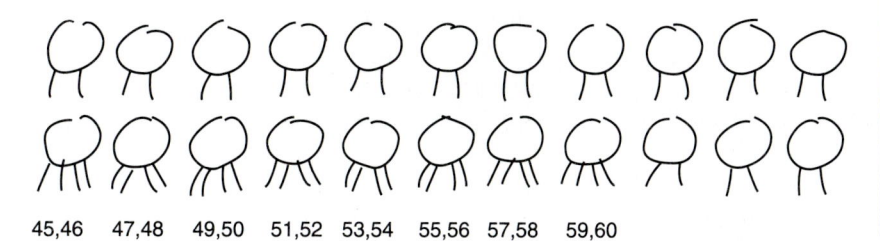

45,46 47,48 49,50 51,52 53,54 55,56 57,58 59,60

I added 2 more legs to 8 animals.
So, 8 pigs and 14 chickens.

Pigs	Chickens	Pig Legs	Chicken Legs	Total
1	21	4	42	46
2	20	8	40	48
3	19	12	38	50
4	18	16	36	52
5	17	20	34	54	← Hey. These go up by 2. I can extend this column to 60.
6	16			56
7	15			58
8	14			60	← Ans: 8 pigs & 14 chickens
9	13			
10	12			

FIGURE 4.1: *Three ways to solve the "pigs-and-chickens" problem*

There seems to be a strong connection between problem-solving success and instruction that has integrated cognitive monitoring and regulatory practices with strategy instruction. Good problem solvers clearly monitor their thinking regularly and automatically. They are deliberate about their problem-solving actions. They recognize when they are stuck or do not fully understand and make conscious decisions to switch strategies, rethink the problem, search for related content knowledge that may help, and so forth.

Poor problem solvers apparently have not learned this behavior or are unaware that it is at all useful. They are usually impulsive, spending very little time reflecting on a novel problem. The tendency is to select a plan of attack very quickly and then stick with this initial approach regardless of lack of progress. Nor are poor problem solvers able to explain why they used the selected strategy or if they even believe it should work (Schoenfeld, 1992).

As with strategies, there is evidence that metacognitive behavior can be learned (Campione, Brown, & Connell, 1989; Garofalo, 1987; Lester, 1989). Further, students who learn to monitor and regulate their own problem-solving behaviors do show improvement in problem solving.

Beliefs and Attitudes

How students feel about problem solving and the subject of mathematics in general has a significant effect on how they approach problems and ultimately on how well they succeed in mathematics. Suppose that your students had the following negative ideas about mathematics:

> Mathematics problems have one and only one right answer and there is only one way to go about solving them.
>
> The average student has no chance unless the teacher has told you how to solve the problem.
>
> If you know the mathematics that means you will know how to solve the problems quickly and easily.
>
> If you can't figure out how to do the problem right away you must be stupid or the problem is impossible.
>
> I am no good at solving problems. I must be stupid.

It does not take a lot of imagination to figure out that students harboring such ideas are not going to do very well. They will not persevere, they will be unwilling to try, and they will not be very receptive to a nonprescriptive approach.

Attitudes (likes, dislikes, preferences) are nearly as important as beliefs. Children who enjoy solving problems and feel satisfaction or pleasure at finally conquering a perplexing problem are much more likely to persevere, make second and third attempts, and even search out new problems. Negative attitudes have just the opposite effect.

A GENERAL PROBLEM-SOLVING SCHEME

The components of problem solving (knowledge, strategies, metacognition, and beliefs and attitudes) all come to play throughout the entire process of solving a problem. George Polya (1957) is very well known for his general four-step approach to solving problems:

1. *Understand the problem*
2. *Devise a plan or decide on an approach for attacking the problem*
3. *Carry out the plan*
4. *Look back at the problem, the answer, and what you have done to get there.*

Variations of George Polya's four-step approach have become standard models for problem-solving behavior. Polya's four steps do not direct or describe how to teach problem solving, but they identify the goals of problem-solving behavior that students should construct. In Figure 4.2, a problem is solved, with the solver's thought processes provided at each step. As you read the following discussion of Polya's four steps, refer to the solution of this problem as an example. Even better, start with an unfamiliar problem yourself, and go through it following the general scheme described. Have a "metacognitive experience."

Understanding the Problem

Many students (and adults) are overwhelmed when they first see an unfamiliar situation or problem. A first step in problem solving is to calmly examine the information in a problem. Articulate or write down all relevant information. Decide what information is important and what seems unimportant. Examine the conditions of the problem that will have an impact on the problem. Be very clear about what is being asked for in the problem. It may be helpful to reformulate some of the information, perhaps make lists of knowns and unknowns, to draw pictures, charts, or diagrams. Begin to think of what similar situations you have experienced that may be like this problem or contribute to its solution. Write these down, or test to see if they really are similar situations.

Besides the obvious benefit of being clear about what a problem is asking, the very act of going through this understanding phase is calming. It gets the problem solver doing something productive without having to decide what to do. Thinking about a problem is a very active and involved process. The amount of time and effort spent in this first phase is a major difference between good problem solvers and poor problem solvers.

Devising or Selecting a Plan

Once the problem is well understood, the experienced problem solver begins to change his or her focus of thought. Rather than getting information and ideas *from* the problem, thoughts turn to what can be *brought to* the problem. "Does this problem lend itself to special cases? Is a drawing or chart a useful approach? Have I ever done anything similar to this before? How was that solved? Could I make a guess, and if so, would that help? Could I test a solution? Perhaps the problem could be restated in such a way to give me a better clue to the solution, or perhaps there is a special case of the problem that can be solved quite easily." And so on.

For a novice problem solver who has not developed a collection of heuristics, the selection of a plan may be reaching into a nearly empty bag of tricks. Even experienced problem solvers may try several unproductive approaches and begin to wonder if in fact they have a method of solving the problem. Students in school must be helped to develop a set of solution strategies as part of their repertoire. In addition, they must learn to recognize when a particular situation calls for this or that approach.

Carrying Out the Plan

Carrying out the plan is partly a matter of following through with the approach selected, being careful of each step along the way. The more sophisticated aspect of carrying out the plan is the self-monitoring of your own progress and the methods you are using, a metacognitive

4 / DEVELOPING PROBLEM-SOLVING PROCESSES 43

Problem: How many squares on an 8 x 8 checkerboard?

Understanding the Problem

Seems simple. There are 64 little squares. There must be something else. What other squares could there be? The board itself is a square. What about a 2 x 2 square? There could be all sorts of sizes.

There are 1-squares, 2-squares, 3-squares,...,8-squares (biggest). But they could also overlap. Wow! I need to find out how many there are of each size and add them up. I probably could tackle each of those separately. Need to remember that they overlap. There really is no way I can guess what the answer is.

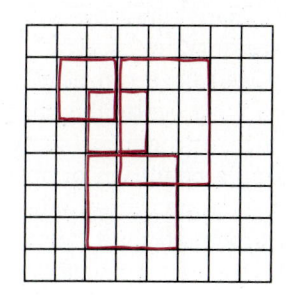

Devising or Selecting a Plan

I think this seems like a very visual problem, so I'll try to draw a picture. In my picture I can draw all of the squares of one type and then count them, and then I can draw another to count the next size, etc. Maybe I will see some sort of pattern to this as I go and I really won't have to draw them all. But to see a pattern I'd better draw them in some sort of order.

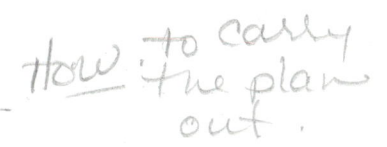

Carrying Out the Plan

This is getting very messy. I can't really see what I have drawn (sense blockage). There must be some system that I can use to help keep track. Instead of drawing pictures, maybe it would help to cut out a little square and sort of slide it around the board.

There are clearly 64 little squares. The 2 x 2 paper square can slide across the top of the board in seven different places. But then I can move it down one row and get seven different squares. And slide down again, and again. There are seven rows that I can slide that one on; each has seven squares. That's 64 + 49. Try the 3 x 3 square. It slides across in six places. Go down. Six rows. 6 times 6 squares. Now 64 and 49 and 36. Hey, looks like 8 x 8 then 7 x 7 then 6 x 6. Check out a 4 x 4 square. Yes! Five places, five rows. 5 x 5.

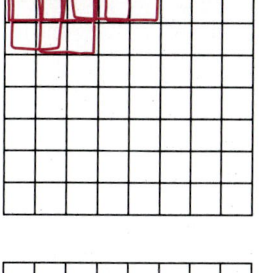

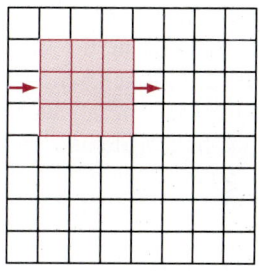

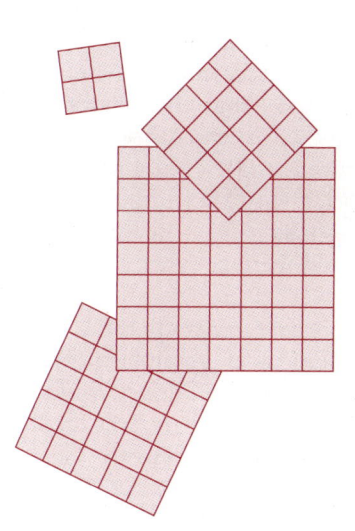

That's the pattern. Get the calculator and add these up.
64 + 49 + 36 + 25 + 16 + 9 + 4 + 1 = 204

Looking Back

Does 204 squares sound right? I never would have guessed that so I can't tell. Did I miss any? I counted squares from 1 x 1 to 8 x 8. That is all the sizes. There is no other way to draw squares. Any square I can draw would be counted in my scheme. Yes, I believe I can trust that.

This was sort of a draw a picture, systematic count, look for a pattern approach. The picture gave me an idea for how to count them. But the pattern was so clear I didn't need to count them all. I could have started with a smaller checkerboard. A 6 x 6 board wouldn't have been much easier. But I could sure solve it for a 2 x 2 board almost without trying. Five squares—4 and 1. Wonder if that would have helped.

(In fact, the smaller-problem approach can be pursued, and another similar pattern discovered.)

Hey, I can do this for any size board. Just add up all the square numbers to $n \times n$. (There is a formula for this, and discovery or use of it depends on the background of the solver.) What else? Rectangles! Could I count them? They come in different sizes and can be positioned in different ways. (This, too, is solvable in a similar manner.) What if the checkerboard wasn't square? Maybe 6 x 12 or something. What if I used a grid of equilateral triangles? Could I count triangles using a similar approach? What other grids and shapes are there?

(Each of the conjectures above leads to a solvable problem, perhaps a bit more difficult, but approached using similar techniques.)

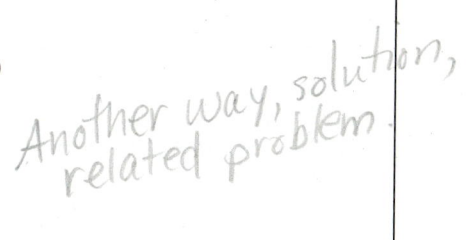

FIGURE 4.2: *An illustration of Polya's four steps of problem solving*

activity. You may have been working feverishly on an approach for a long time, producing lots of pencil markings, charts, drawings, calculations, and so on. But are you making any progress? Or a particular approach may very well have run head on into a dead end. You must learn to recognize this and decide if the blockage is due to the approach or some other factor, such as overlooking a condition or an incorrect assumption or incorrect data recording. When blockage midway in a solution process occurs, some recycling of the general scheme is called for. Sometimes it requires returning to the original problem and the understanding stage. At other times it is the approach to the problem that requires rethinking. In such cases a return to the general process of devising a plan is called for to search for a new way of approaching the problem.

◆ Looking Back

When a solution to a problem is found, the problem-solving process is not over. Three significant looking-back activities should always be considered:

1. Look at the answer.
2. Look at the solution process.
3. Look at the problem itself.

In the real world, problems are never solved when an answer is found. There is no answer book or teacher to verify that the answer is correct. Consequently some effort must be made to be sure that the answer arrived at is indeed a solution. The method of doing this varies with the problem. Does the answer seem to be reasonable? Is it in the ballpark of what was expected? Could there be other answers? Is there a way to verify the answer by checking it against all of the conditions? Are there any contradictions between the answer and the conditions of the problem? Sometimes the answer can only be accepted or rejected by looking at the process used to arrive at the answer. Was the logic used appropriate? Were the calculations correct?

How was the problem solved? This question sometimes helps validate the solution as just discussed. However, it is an important step in itself. Could it have been solved a different way? an easier way? What was the method used to solve it? The last question helps identify or label the strategy to make it more readily used in future problems.

Finally, consider in retrospect the problem itself. Now that the problem is solved, are there other questions similar to or related to this problem that can be answered? If the number were bigger or different somehow, could you still solve the problem? Is there a general case that can be solved because of having solved this special case? What features of the problem might be changed to create a similar problem that may be solvable, interesting, challenging, or more useful?

INSTRUCTIONAL GOALS FOR PROBLEM SOLVING

The components of problem solving and the general scheme of solving problems suggest goals for instruction that are useful to keep in mind.

◆ Strategy Goals

1. To help students develop a collection of problem-solving strategies or heuristics that are useful in a variety of problem-solving settings.
2. To improve students' ability to analyze an unfamiliar problem, identify wanted and needed information, ignore nonessential information, and to state clearly what the problem is asking.
3. To improve students' ability to select and use a strategy or combinations of strategies that is appropriate for the problem at hand.
4. To improve students' ability to self-assess the solutions to problems in light of the information in the problem and the approaches that are used.
5. To help students learn to go beyond the solution and to extend, modify, or generalize problems, so that the total result is more than the initial answer.

◆ Metacognitive Goal

To help students develop the habit and ability profitably to monitor and regulate their thinking processes at each stage of the problem-solving process.

For example, students might think, "Have I checked everything this problem is saying? Do I have a clear understanding of what the goal is? Is this strategy I am using helping make progress? Have I checked my answer against all conditions in the problem?"

◆ Affective Goals

1. To develop in students a view of mathematics as a discipline concerned with thinking, sense making, and a search for patterns and regularities.
2. To develop students' self-confidence in their ability to *do* mathematics and to confront unfamiliar tasks without being given a ready-made prescription for a solution.
3. To improve students' willingness to attempt unfamiliar problems.
4. To improve students' perseverance in their attempts at solving problems and not to be easily discouraged with initial setbacks.
5. To help students learn to enjoy and sense personal reward in the process of thinking, pattern searching and solving problems—to enjoy doing mathematics.

STRATEGIES AND PROBLEMS

PROBLEM TYPES AND PROBLEM USES

Problems in the classroom are a bit like teaching tools that you can use for various purposes. Different problems will help our students learn different things. Therefore, you can select different problems to serve different agendas in the curriculum. One overly simplified way to classify problems is to use the terms "routine" and "nonroutine." Most of the problems we will look at in this chapter fall into the nonroutine category.

Routine Problems or Translation Problems

A *routine* problem, also called a *translation* problem, presents a "real" situation that can be solved by the application of one or more of the four arithmetic operations. These simple story-problems are usually broken down further into one-step or multistep problems. The following are examples of one-step and two-step translation problems respectively:

> Each crate of oranges weighs approximately 38 pounds. Vito's Van Service is to pick up 24 crates of oranges. How many pounds are in the shipment? 38 × 24 = 912 lbs
>
> The local candy store purchased candy in cartons of 12 boxes per carton. The price paid for one carton was $42.50. Each box contained 8 candy bars, which the store was going to sell individually. What was the candy store's cost for each candy bar? 12 ÷ 42.50 = 3.54 ÷ 8 = .44

These problems are called translation problems, because we solve them by making translations between three different languages. The problem situation is presented in oral or written words, a *word language*. This must be translated to a *symbolic language* or computational form. Frequently a third language, that of *models*, is used as an intermediary to assist in the translation to symbols. The model, which may be a drawing or actual materials such as counters, helps clarify which operation is required. When the computation is complete, the result is translated back to the context of the original problem. Figure 4.3 illustrates these translations for the first problem.

Students' ability to solve simple and complex translation problems is closely related to their understanding of the meanings of the operations. In fact, for years, word problems were found at the ends of the chapters where an operation was taught. They were also found in chapters on computation, so that students could see some application for the skills they were learning.

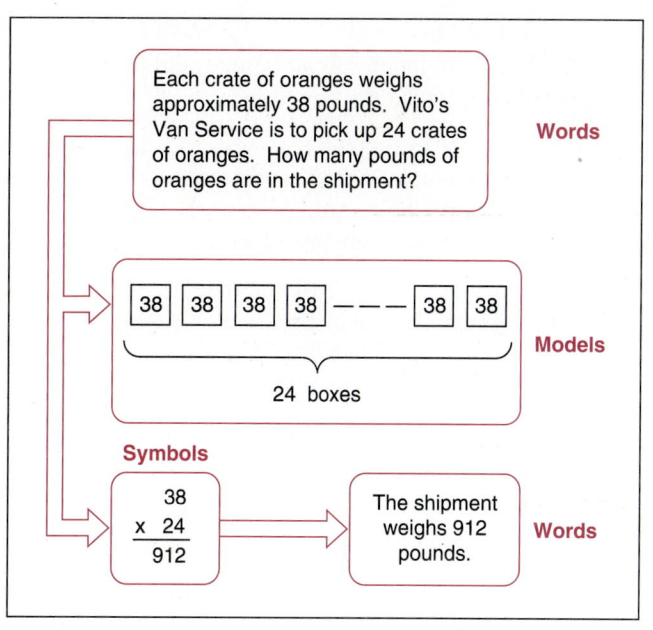

FIGURE 4.3: *Translations in routine word problems*

Translation problems (one-step and multistep word problems) should now be seen as only one small part of a total program of problem solving. They should be encountered every week of the year regardless of what chapter the class is working on.

Translation problems are addressed again in Chapter 7 on the meanings of the operations.

Nonroutine Problems

Any problem we might pose to students for which no routine method of solution has been established can be called *nonroutine*. It is with these problems that students must use one or more of the general heuristics or strategies discussed earlier. Some problems may have several different solution approaches. Occasionally there is either too much and/or not enough data in the problem as presented. Some nonroutine problems consist of projects requiring data gathering or constructions or measurements to be performed. Some are quite realistic and others (e.g., the pigs-and-chickens problem) offer somewhat absurd contexts. What they have in common and what makes them all valuable is that solution approaches are never routine. Each problem presents its own unique set of circumstances with which the solver has to deal.

It may be useful to look briefly at some subcategories of nonroutine problems, although these categories are far from being clearly defined and are certainly not mutually exclusive.

Modified Translation Problems. As already mentioned, translation problems can have too much information, insufficient information, or both.

> George and Bernie each bought 3 bags of marbles. The marbles cost 69¢ a bag. Each bag has 25

marbles including 5 special cat-eye marbles. How many marbles did the boys buy?

Mrs. Saunders gave Annette $10.00 to spend at the circus. She bought cotton candy that cost $1.25 and went on 3 rides. Admission to the circus was $1.50. How much money did she have left over?

Brenda is reading a new storybook that has 327 pages. Today she stopped reading on page 120. She reads 30 pages in a day.

The last problem is an example of one without a question. This one requires students to formulate relationships among the given bits of information so that they can decide what questions they could answer with the given data or could answer if they had more data. Children can make up questions for each other to answer.

These variations of translation problems are an excellent way to encourage students to pay more attention to problem information and analyze what is given and asked for. Too many children simply grab two numbers in a problem and select an operation that seems to go with the numbers. (Add or subtract if they are big numbers. Divide a little number into a big number, and then try multiplication if it doesn't go evenly.)

Process Problems. Process problems, also called nonstandard problems, require the use of one or more general strategies for solving problems. The pigs-and-chickens problem and the squares-on-the-checkerboard problem were good examples of process problems. Many of these require no computation at all but are more involved with logic or geometry. For illustration purposes, here are two more process problems:

HANDSHAKES

At Mrs. Brown's party there were 10 guests, including Mrs. Brown. If everyone at the party shook hands with everyone else, how many handshakes would that be?

STOP-SIGN DIAGONALS

A stop sign has eight sides. It is an octagon. How many diagonals (lines from one corner to another corner that is not adjacent) can be drawn?

These two problems do not have the same answer, but knowing how to solve one of them can be a big help in solving the other—if, of course, one recognizes how the two problems are similar. The heuristic or strategy of drawing a picture is one way to see how the problems are alike. The strategy of using a simpler problem is also useful

for both problems. What if there were only three guests (a triangle), then four guests (a square), five guests (pentagon), and so on? Can you discover a pattern?

There is also a basic difference between these two problems. The handshake problem is a bit silly. No one would ever ask such a question in the real world. The stop-sign problem is directly related to a geometric discussion of shapes and diagonals. It would fit well within a unit on geometry. In thinking about learning mathematics through a problem-solving approach, many process problems are simply problems about the content that is being studied. This is an ideal blending of teaching problem-solving strategies and teaching content. The next problem introduces another aspect of problems that some feel is very important.

REMODELING

Suppose you are remodeling a room that is 21 feet by 28 feet. At the lumber company you find that molding strips to go around the baseboard come in 10- and 16-foot lengths. What should you buy in order to have:
(a) the fewest seams?
(b) the least waste?
(c) the lowest cost?
(16-ft. strips are $1.25 per foot, and 10-ft. strips are $1.10 per foot.)*

The remodeling problem is very realistic or a real "real-world" problem. It is still a process problem and strategies such as draw-a-picture and try-and-adjust are quite applicable. When problems are very realistic they give a message to students that real people actually use the mathematics they themselves are learning.

The position taken in this book is that we should highly value those problems with a mathematics content connection and those with real-world contexts. However, to exclude problems like the pigs-and-chickens problem because it is not realistic closes off a large collection of rich problems that have proven potential to help children develop good strategies, practice mental monitoring, and improve attitudes concerning problem solving and mathematics.

Open-ended and Project Problems. Many excellent nonroutine problems can involve students or groups of students in extended explorations. Consider the following problems:

What path patterns can you make by flipping a triangle over and over on its edges?†

---

*Adapted from Lesh & Zawajewski (1992).
†Adapted from Baker & Baker (1990).

How would you explain what a negative number is?

What patterns or regularities can you find on a 0 to 99 chart (Figure 4.4)?

The church is building a new parking lot. Where should the entrances and exits be, and how should the lines be painted on the lot?

What is the best price to charge for tickets to the school play? What should the refreshments be, and what should be charged for them?

Which of these five brands of paper towels is the "best"?

The examples include three that are connected to content areas and three that are more real "real-world" types. In each there are many "right" answers, different ways of thinking about what the problem is requesting, and a requirement to seek additional data or make measurements. Students might work in groups for several periods on these problems, or they might be out-of-class assignments. Results should be written up and presented to the class for discussion. These problems are good opportunities for students to pull ideas together from more than one content area and to see that mathematics is not so narrowly rule-driven. The presentation of results is an opportunity for students to take pride in their accomplishments as they defend ideas and solutions in front of peers.

0	1	2	3	4	5	6	7	8	9
10	11	12	13	14	15	16	17	18	19
20	21	22	23	24	25	26	27	28	29
30	31	32	33	34	35	36	37	38	39
40	41	42	43	44	45	46	47	48	49
50	51	52	53	54	55	56	57	58	59
60	61	62	63	64	65	66	67	68	69
70	71	72	73	74	75	76	77	78	79
80	81	82	83	84	85	86	87	88	89
90	91	92	93	94	95	96	97	98	99

FIGURE 4.4: *What patterns can you find on the 0 to 99 chart?*

STRATEGIES FOR SOLVING PROBLEMS

This section will acquaint you with those problem-solving strategies or heuristics that are commonly taught in the elementary school or are found frequently in resource materials. Each of the nine strategies is explained briefly, followed by a problem solved using the strategy. When looking at these solutions, realize that different people may work these problems differently, even if they were using the same general strategy. The solution is there to illustrate the particular strategy. Other problems are then suggested that can be solved with a similar approach. If you are just learning these strategies, it would be profitable for you to try to work the unsolved problems using the suggested strategy. Sometimes you may feel that a different approach may well be possible or even easier for you. That's fine. However, in order to learn the strategies, you should make an attempt to use the given approach. Frequently you will find yourself using more than one strategy. That is certainly acceptable. Strategies are often used in combination. These problems represent a variety of difficulty levels, but most have been chosen from those suitable for grades 3 to 8. Problems that lend themselves to the same strategies exist for both lower and higher grades.

◆ Try-and-Adjust

The try-and-adjust strategy involves making a guess at the solution based on an estimation of a reasonable answer or even just a blind "shot in the dark." However, the strategy is only useful in problems where you can tell if an answer is correct by checking against the conditions of the problem. You can make a guess at any problem, but with many you would not be able to tell if your guess was correct. Further, in problems where try-and-adjust or guess-and check is most useful, a check of an incorrect guess should give you information about how to make an adjustment. Therefore, the strategy is most likely profitable if you sense that successive trials are moving you closer and closer to the correct solution.

MARBLES

Larry and Pete play marbles almost daily. Since Larry is the better player, he agrees that, when he wins, Pete pays him 5 marbles, but if Pete wins, Larry will pay Pete 8 marbles. In one month the boys played 26 games, and they each ended with just as many marbles as when they began. How many games did Larry win? (Solution given in Figure 4.5, p.48.)

Try using a guess-and-check method with these problems:

LIZARDS

Ross collects lizards, beetles, and worms. He has more worms than lizards and beetles together. There are 12 creatures in all, with a total of 26 legs. If lizards have 4 legs and beetles have 6, how many lizards does Ross have?

1st try

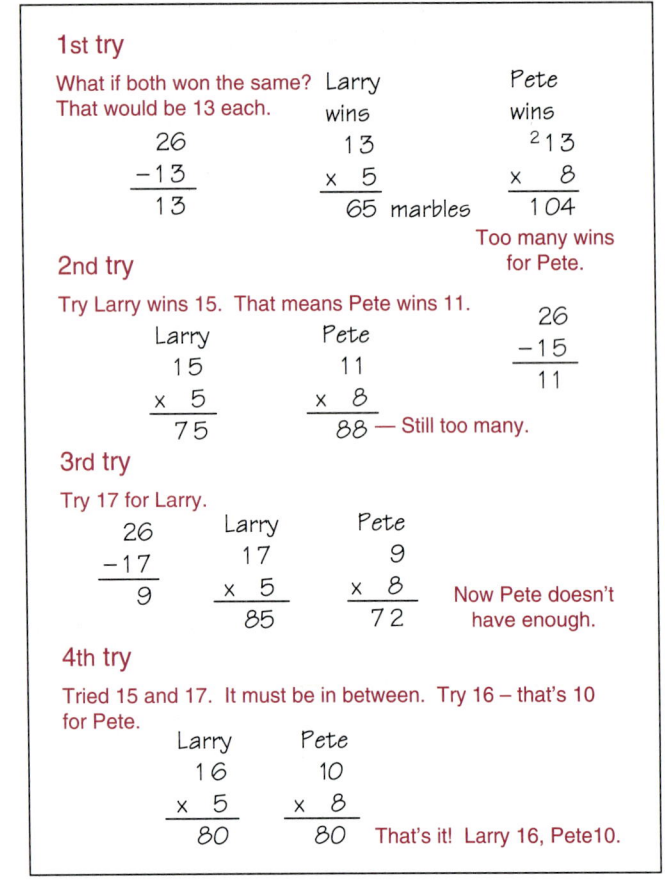

What if both won the same?
That would be 13 each.

```
    26
  − 13
  ----
    13
```

Larry wins	Pete wins
13 × 5 ---- 65 marbles	²1 3 × 8 ---- 1 0 4

Too many wins for Pete.

2nd try

Try Larry wins 15. That means Pete wins 11.

Larry	Pete	
15 × 5 ---- 7 5	11 × 8 ---- 8 8	26 −15 ---- 11

— Still too many.

3rd try

Try 17 for Larry.

```
    26
  − 17
  ----
     9
```

Larry	Pete
17 × 5 ---- 8 5	9 × 8 ---- 7 2

Now Pete doesn't have enough.

4th try

Tried 15 and 17. It must be in between. Try 16 – that's 10 for Pete.

Larry	Pete
1 6 × 5 ---- 8 0	10 × 8 ---- 8 0

That's it! Larry 16, Pete 10.

FIGURE 4.5

NINE DIGITS

Use the digits 1 through 9, each exactly one time, to fill in this addition problem:

```
    □ □ □
    □ □ □
  + _____
    □ □ □
```

◆ Draw a Picture, Act it Out, Use Models

Even the best mathematicians and thinkers draw pictures to help them think about the relationships in a problem. Pictures here refer to any sort of sketch that helps organize the data and relationships involved in the problem. Just because a problem is about elephants does not mean we should draw pictures of elephants. Simple dots or *X*'s will do fine. Lines, loops (for sets), dots (for things), or geometric drawings (circles, rectangles, triangles, and the like) are a few of the kinds of things we can draw easily to help solve problems. Think about sketching the ideas and relationships, and avoid nonessential elements of the problem.

In some problems the action in the problem can actually be carried out by children doing what the problem describes and gathering data as they go. Other problems lend themselves to using counters, blocks, or geometric models to help in thinking about the problem. Using models and acting out a problem can be looked on as variations of the draw-a-picture strategy, although many will list these as separate approaches. With drawing, models, and actions, the idea is to get something visual in front of us to help determine the necessary relationships.

Drawing a picture is a strategy that permeates doing mathematics, not just the solution of process problems. It is frequently used in connection with other strategies.

CUTS

Jane wanted a board cut into 8 equal pieces. The lumber company charges 60 cents for cutting a board into 4 equal pieces. How much will it charge for cutting Jane's board? (Solution given in Figure 4.6.)

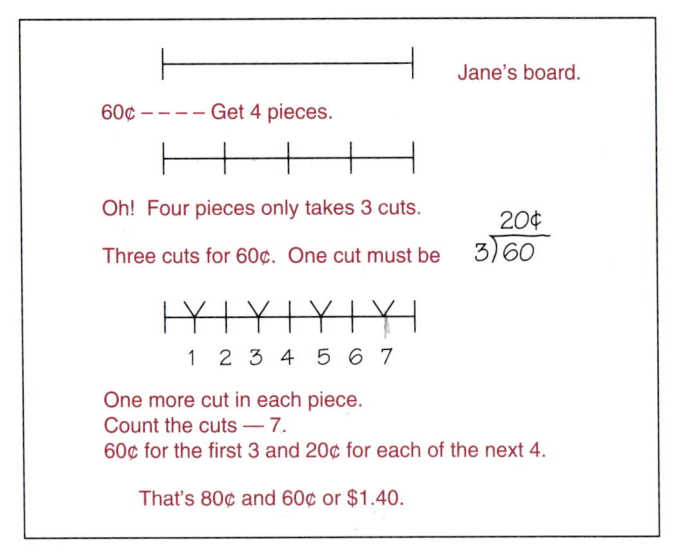

```
 ├─────────────────┤    Jane's board.
60¢ – – – – Get 4 pieces.

 ├─────┼─────┼─────┼─────┤
Oh!  Four pieces only takes 3 cuts.
                                          20¢
Three cuts for 60¢.  One cut must be   3)60

 ├─┼─┼─┼─┼─┼─┼─┼─┤
   1  2  3  4  5  6  7
One more cut in each piece.
Count the cuts — 7.
60¢ for the first 3 and 20¢ for each of the next 4.

That's 80¢ and 60¢ or $1.40.
```

FIGURE 4.6

Use a drawing to help with these problems:

COMPUTER CENTER

Terri's new computer center is in 3 sections that sit side by side: a desk, a printer stand, and a set of drawers. The desk is as long as the stand and drawers together. The stand, which is 16 inches wide, is $\frac{1}{3}$ the width of the drawers. How wide is the entire unit?

SPIDER AND FLY

In the drawing, a spider is in the center position and is trying to catch the fly on the corner to the right.

4 / DEVELOPING PROBLEM-SOLVING PROCESSES   49

They take turns moving one position each turn, always staying on the paths. If the spider moves first, how many moves before the fly is captured by the spider?

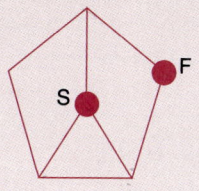

blocks will he need to have a stairs with 20 steps? (Solution given in Figure 4.7.)

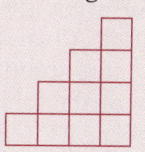

Find a pattern to help with these:

HUNDREDTH POWER

Two to the 100th power (that is $2 \times 2 \times 2 \times \ldots \times 2$ with 100 twos) is a very, very large number. If you could and did calculate it, what would the digit in the ones place be?

CLUB MEMBERSHIP

Chip and Patrick decided to start a club. They agreed that each would get one new member by the end of the month. In order to join, each new member has to agree that at the end of their first month in the club they will bring in one new member who will likewise agree to recruit a new member after being in the club one month. Recruiting new members is a one-time-only responsibility of new members, and no other members are added except by this method. How many members will the club have at the end of one year?

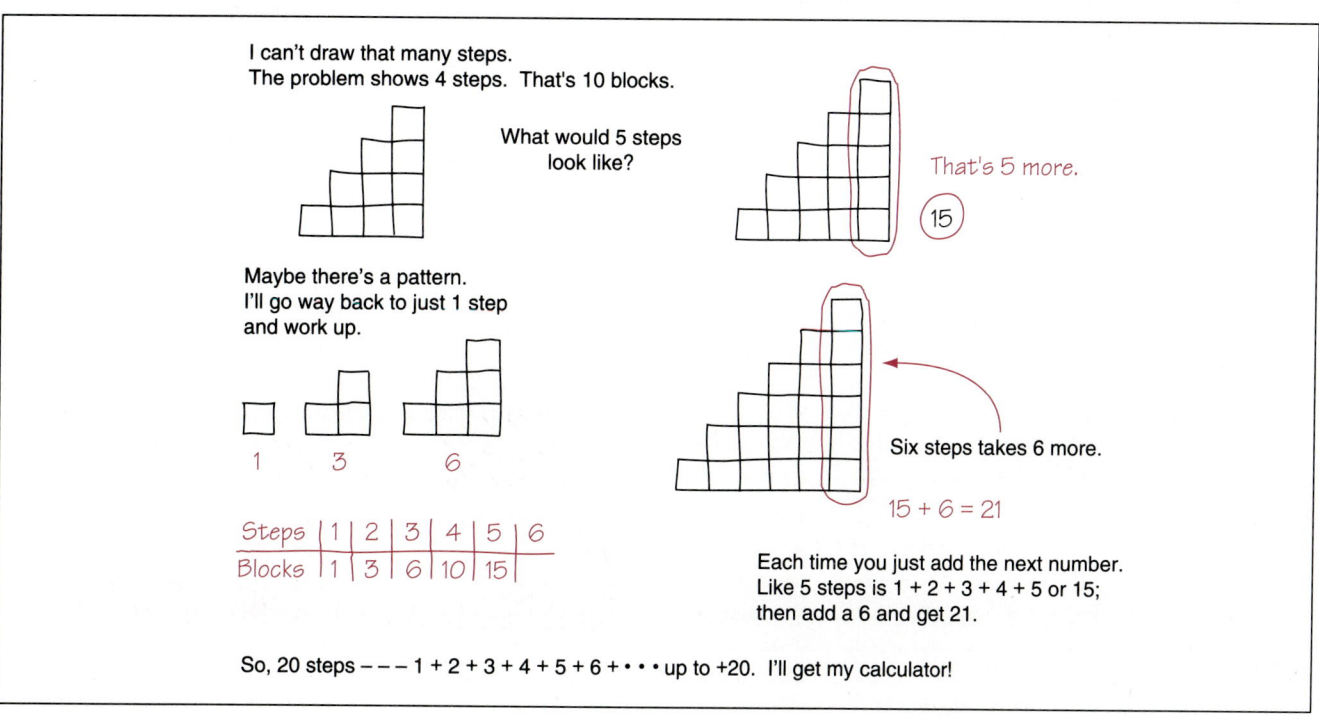

◆ Look for a Pattern

Mathematics is filled with patterns. With practice we can learn to expect a pattern to exist in certain situations. Patterns may occur in problems where there is a progression of circumstances or in geometric problems in which there are successive variations or elements added to a figure in a regular manner. Sometimes the pattern is not apparent to problem solvers because they do not have (or do not think of) the appropriate relationships that form the pattern. Looking for a pattern involves arranging the elements of the problem in such a way that the pattern will emerge and then bringing to bear other ideas that will make the pattern "visible."

STAIR STEPS

Carlos is building stair steps with toy blocks as shown. This set of stairs has 4 steps. How many

---

I can't draw that many steps.
The problem shows 4 steps. That's 10 blocks.

What would 5 steps look like?

That's 5 more.
15

Maybe there's a pattern.
I'll go way back to just 1 step and work up.

1   3   6

Steps	1	2	3	4	5	6
Blocks	1	3	6	10	15	

Six steps takes 6 more.

15 + 6 = 21

Each time you just add the next number.
Like 5 steps is 1 + 2 + 3 + 4 + 5 or 15;
then add a 6 and get 21.

So, 20 steps — — — 1 + 2 + 3 + 4 + 5 + 6 + • • • up to +20. I'll get my calculator!

FIGURE 4.7

Make a Table or Chart

The make-a-table strategy is useful when there is a series of numbers in a problem or where the answer could be found in a list if it were available. Charts or tables consist of rows or columns that list important variables in the problem. Usually one row of the table begins at some natural starting place and progresses in an orderly, numeric manner, for example: 1, 2, 3, 4, . . . or 5, 10, 15, ... The other parts of the table are filled in accordingly. At some point along the way the table should yield the answer. To use this strategy, you need to learn what helpful things might go in the chart or table; what should the headings be?

The make-a-table strategy is frequently used in conjunction with another strategy. Drawing a picture can be used to generate the elements in a table. As the table is constructed, we can look for a pattern. In the stairstep problem just discussed, a table was constructed showing the number of blocks in each set of stairs. The table helped in seeing the pattern.

TEMPERATURE

Mark turned on the radio at 6:00 A.M. and heard the temperature was −13 degrees. That night he heard that the temperature had risen 3 degrees each hour until 3:00 P.M., when it was at its highest for the day. What was the high temperature that day? (Solution given in Figure 4.8.)

I bet I could list each hour and then figure out what the temperature is at that time.

Time	6	7	8	9	10	11	12	1	2	3
Temp	-13	-10	-7	-4	-1	2				

Three degrees up from minus 13 is minus 10,... and then minus 7,...

FIGURE 4.8

Make a table to solve these problems:

BLUE EYES

Carlotta read that 2 of every 5 people have blue eyes. She decided to use this idea to predict how many students in her class of 32 would have blue eyes. What would she predict?

DAYS OFF

Bert and Ernie work part-time at the local supermarket. Bert works one day and then has three straight days off. Ernie works a day and then does not get a chance to work again for 4 days. Bert and Ernie are both working on Wednesday. How many days will it be before they are both working together again? Assume their schedules continue and that the store is open every day.

Make an Organized List

Sometimes referred to as systematic counting, the strategy of making an organized list is useful in problems that require some method of counting things or making certain that every possibility has been covered. It may appear on the surface that simple counting of the different ways or situations would be impossible. To make the situation manageable for counting, we can try to devise an organizational scheme that permits listing all of the elements in some way. If the scheme or organized list is adequate, not only is it usually easier to count, but we will be able to tell that we have included all things that need counting, and none were counted twice.

An organized list is also useful when you are looking for one or more of something from an unruly collection of things, and you are not sure which combination you want or how many acceptable combinations there may be. If you can construct an organized list, all of the cases for which you are searching will show up.

POSTAGE

Pablo has 4 three-cent stamps and 3 four-cent stamps. What different amounts of postage can he make with these stamps? (Solution given in Figure 4.9.)

Now you try these:

LICENSE PLATES

Suppose that you are making bike license plates, using 4 letters and numbers on each. The first of the 4 positions must be a letter, and the last 3 places must be numbers. The only letters to be used are A, B, and C, and the only numbers are 1 through 3. Numbers may be used twice but not three times on one plate. How many license plates can be made?

COMMITTEE SIZE

You are trying to decide how many members each of four committees, A, B, C, and D, should have. Each committee is to have a different number of members. Committee A will be the smallest, B the next largest, C next, and D the largest. You have 18 total committee positions. How many different ways can you distribute the 18 positions among the four committees?

If he uses all of the stamps that's the most. Now if I start taking away 3¢ stamps one at a time, — — — —

Fours	Threes	$
3	4	24
3	3	21
3	2	18
3	1	15
3	0	(12)

−3
−3
−3
−3

4   3
×3   ×4
12   12 = 24

That's all the ways to do it with 3 fours. I can do the same thing with 2 fours and 1 four and no fours.

2	4	20		1	4	16		0	4	(12) X
2	3	17		1	3	13		0	3	9
2	2	14		1	2	10		0	2	6
2	1	11		1	1	7		0	1	3
2	0	8		1	0	4		0	0	(0) ?

That's 20 ways altogether. WHOOPS! Twelve cents happened twice! Should I count that? Problem says how many *different amounts*. Those aren't different. When he gets zero, he isn't using any stamps. Don't count zero. That leaves 18.

FIGURE 4.9

◆ Work Backward

The work-backward strategy is sometimes useful when a series of events takes place and we know the result but need to determine the start condition. If the events are a matter of arithmetic operations such as earning or spending money, the task is to reverse these operations. In other problems a series of unknown events results in some condition, and the task is to find out what that sequence of events is. If not all steps are obvious, occasionally the last one is relatively clear. From the last step you can then get to the next to last, and so on.

NIM

Two players play a game of Nim in which they take turns playing 1, 2, or 3 square tiles on this strip.

The play starts at one end, and players add their tiles adjacent to those already played. The person who plays the last tile wins. If you start, how can you be assured of winning? (Solution given in Figure 4.10.)

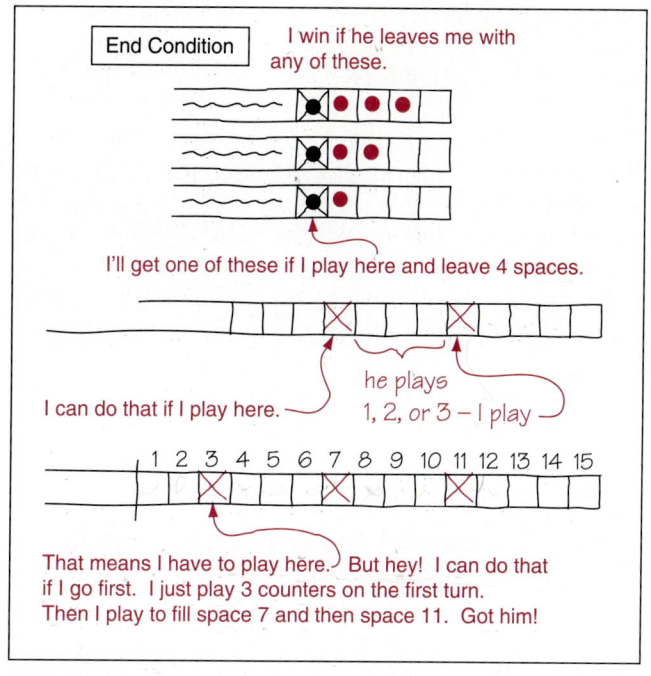

End Condition — I win if he leaves me with any of these.

I'll get one of these if I play here and leave 4 spaces.

I can do that if I play here. — he plays 1, 2, or 3 — I play

1 2 3 4 5 6 7 8 9 10 11 12 13 14 15

That means I have to play here. But hey! I can do that if I go first. I just play 3 counters on the first turn. Then I play to fill space 7 and then space 11. Got him!

FIGURE 4.10

Working backward is useful in each of these problems:

RAKING LEAVES

Molly, Max, and Buz earned $150 raking leaves, but they each earned a different amount. They agreed to share equally. Since Molly had the most, she took half of her money and shared that part equally with Max and Buz. But then Buz had too much, so he gave $10 each to Molly and Max. Finally, Max gave Molly $2, and they all had the same amount. How much did each earn originally?

SHARING INHERITANCE

Terika and Wanda inherited a collection of figurines from their great-grandmother. They agreed on a plan to decide who would get which figures. Since Terika was older, she would go first and select $\frac{1}{4}$ of the figurines for herself. Then Wanda would get to choose half of the remaining figures. That would leave 6 figurines that they would keep as "common property" until a later date. How many figurines did they inherit to begin with?

◆ Logical Reasoning

The title of this strategy has been applied to a variety of problem types that are quite a bit different from each other. The one thing they all have in common is that they use some form of "if-then" reasoning. In some cases the if-then strategy is used to consider different possible cases or scenarios. For example, "The number must be either even or odd. *If* it is even *then* it must . . . . And *if* it is odd *then* . . . ."

The example that is worked for you here is typical of logic problems that employ the use of a matrix (rows and columns of cells). Each cell of the matrix stands for a combination of two or more possible pairings. *If* you can determine the contents of one cell, *then* that provides information concerning other cells in the matrix. One of the problems left for you to do is clearly of this variety.

The remaining unsolved problem also uses if-then thinking, but the matrix approach is not appropriate. See if you can identify your use of if-then reasoning as you work it.

TOM, DICK, AND HARRY

Tom, Dick, and Harry work in a bank. One is the manager, one is the cashier, and one is the teller. The teller, who was an only child, earns the least. Harry, who married Tom's sister, earns more than the manager. What job does each one have? (Solution given in Figure 4.11.)

They could each be one of these three things—I'll put that in a chart. Now I can X out a space if the man can't have that job.

	Manager	Cashier	Teller
Tom		X	X
Dick		X	
Harry	X	yes	X

The teller is an only child—therefore he isn't Tom—Tom has a sister.

If Harry earns more than the manager, then he isn't the manager. Put an X in there.

And Harry can't be the teller because the teller earns the least, and Harry earns more than someone— that means he isn't least. That's only one spot left for Harry. He has to be the cashier.

If Harry is the cashier, then Dick and Tom get X's there.

I can finish the rest of this easily.

FIGURE 4.11

Here are two logic problems for you to try:

COOKING OUT

Four families are planning their monthly cookout together. One family will bring the main dish, one dessert, one the beverages, and one the cups and plates. This month the Browns said they would buy drinks or paper products but did not have time to cook. Since the Smiths fixed steaks last month, they did not want to fix the main dish. By popular demand, Mrs. Framar agreed to make her prize-winning cheesecake. The Goldings said they were wiling to do whatever was necessary. Who should do what?*

MIXED BAG

Chris's grandfather has put 15 pieces of candy in a bag. There are 5 flavors, 3 pieces of each. Chris gets to reach in and take one at a time, but he has to quit when he has 3 different flavors. What are the most pieces Chris could expect to get?

◆ Try a Simpler Problem

At times we encounter problems that seem to be overwhelming due to their complexity and/or large numbers. When this happens, it may be useful to try the same problem with much smaller numbers of fewer conditions. By working the easier problem, we hope that one of two things will happen: (1) Solving the easier problem may help us see a way of solving the original problem; or (2) there is a chance that working a series of simpler problems, starting with the easiest possible and systematically increasing the difficulty, may lead to a pattern or generalization that will solve the original problem.

CHECKERS TOURNAMENT

In a club checkers tournament, each player played every other player exactly once. If there were 20 players, how many games were played? (Solution in Figure 4.12.)

In addition to trying a simpler problem, drawing a picture or using physical models will help with this one:

TILES AND RECTANGLES

Suppose you have 36 identical rectangular tiles. If you are to put them together to make larger rectangles in such a way that short sides are

---

*Adapted from Hyde & Hyde (1991).

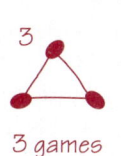

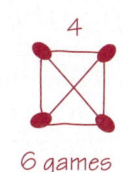

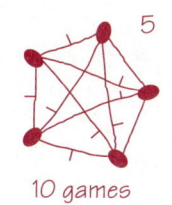

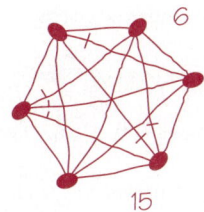

> Wow! I'll never figure this out for 20 players.
>
> Maybe I could do it for 3 or even 4 players. I can draw a little picture for that.
>
> 3 — 3 games
> 4 — 6 games
> 5 — 10 games
> 6 — 15
>
> Let's see—7 is going to really be a mess! Maybe I can look at these numbers—
>
> 3 → 6 → 10 → 15 → Add 3, then 4, then 5...
> That's just like the stairsteps!
>
> $1 + 2 + 3 + 4 + 5 + 6 + \sim\sim + 20$
>
> Do I add up to 20? No, I think only 19. It's always 1 less. Two players is 1, then three is 1 + 2, and four is 1 + 2 + 3. Yes—just add the numbers up to 19.

FIGURE 4.12

matched only with short sides and long sides only with long sides, in how many different ways can you do it?

LOCKERS

Each locker in a row of 100 on a long hallway has the door closed. One hundred students line up with the following directions. Student one is to open every locker. Student two goes to every second locker beginning with locker number two and closes it. The third student begins with locker three and goes to every third locker in the row. If the locker is open, she closes it. If it is closed, she opens it. The fourth student begins with the fourth locker and changes the door status (open or closed) of that door and every fourth door. The fifth student does likewise for every fifth locker starting with number five and so on. When all 100 students have passed the row of lockers, which ones will be open and which will be closed? *

As another exercise you might also return to the problem concerning how many squares are on a checkerboard that is solved in detail for you in Figure 4.2. That problem can be worked using the simpler-problem strategy. Begin with the smallest possible board which is 1 × 1. Then go to a 2 × 2 board, a 3 × 3 board, and so on. You want to see if there is a pattern from one size board to the next or a relationship between the size of the board and the number of squares.

◆ Write an Equation or Open Sentence

Sometimes the numeric relationships involved in a problem can be written down in terms of equations or inequalities. Young children can use boxes or triangles to represent unknown numbers in their equations. In either case, such equations or inequalities are referred to as *open sentences,* since until a value replaces the boxes or variables they are neither true nor false. They are "open." If there is only one equation required, usually straightforward computation will solve the problem. If there are two or more conditions in the problem, several equations or sentences involving two or more unknowns may be required. Even children in the fourth grade can then use a try-and-adjust procedure to find a solution, while older children can use simple algebraic techniques. Writing down an equation to represent numeric relationships in a problem is an algebraic skill that is practiced in the seventh and eighth grades.

GIFT SHOP

Harvey's Gift Shop had a sale on figurines and posters. A figurine and a poster together cost $8. Arthur spent $51 dollars buying 7 figurines and 2 posters. How much did each cost? (Solution given in Figure 4.13, p. 54.)

---

*If you want a hint for the locker problem, consider numbering the lockers, and then look at the factors of each locker number.

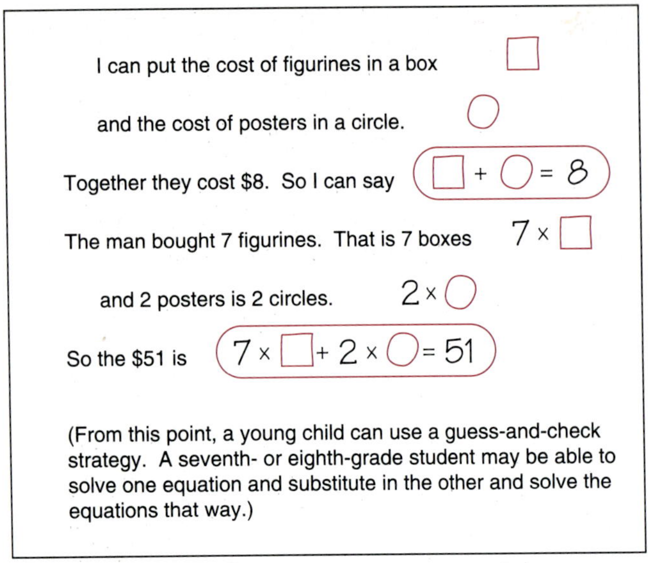

54 4 / DEVELOPING PROBLEM-SOLVING PROCESSES

I can put the cost of figurines in a box □

and the cost of posters in a circle. ○

Together they cost $8. So I can say $\boxed{□ + ○ = 8}$

The man bought 7 figurines. That is 7 boxes $7 \times □$

and 2 posters is 2 circles. $2 \times ○$

So the $51 is $\boxed{7 \times □ + 2 \times ○ = 51}$

(From this point, a young child can use a guess-and-check strategy. A seventh- or eighth-grade student may be able to solve one equation and substitute in the other and solve the equations that way.)

FIGURE 4.13

It is also possible that in *your* approach to these problems, there will be some strategies you do not use at all. A good idea is to write down all of your thinking and first attempts. Even write down the name of the strategy or strategies you think you want to try. If you change your approach midstream, make a note of that too. Then compare your notes, ideas, and methods with others.

No answers are given. Part of good problem solving is to decide if your answer makes sense and to convince yourself that you have actually solved the problem. In the real world there are no answer books.

Now try an equation approach on this one.

ABE AND ZACK

Abe and Zack live in opposite directions from the town. The distance from Abe's house to Zack's house is 29 miles. Abe is 5 miles closer to town than Zack is. How far does each live from town?

TEST AVERAGE

Tina has already taken 3 of the 4 tests to be given in her history class. Her grades are 79, 95, and 89. She wants to know if it's possible to get a 90 average, and if so, what grade she will need on the last test to do that.

◆ Mixed Problems for You to Work On

Recall that part of what makes a problem a problem is not having a well-defined method of getting at the answer—blockage. In the problems you just worked through, a strategy was strongly suggested, so in some sense the blockage was reduced or perhaps even eliminated. The purpose of being so directed was to expose you to specific strategies and to label or name them for you.

Now that you have seen some strategies for solving problems and had a chance to try them a bit, here are some more problems on which you can work. By removing the strong clue concerning which strategy to use, the feature of problem solving that requires the solver to search for a method of solution is now included. It is possible to use each of the nine strategies at least once while solving these problems, and you are likely to use many in combination.

RECTANGLES

How many rectangles are in this drawing?

BOXES

You have 50 boxes to deliver to a series of stops. At the first stop you are to leave 1 box. At the second stop you are to leave 2 more boxes than the first. At the third stop you are to leave 2 more than at the last stop, and so on. At what stop will you not have enough boxes to make a delivery?

TWENTY-SEVEN CENTS

How many different ways can you make 27 cents, using pennies, nickels, dimes, or quarters?

SWIMMING

Robyn and Lois swam toward each other at the same speed from opposite ends of the pool. They passed each other after 12 seconds. They continued swimming at the same speed and lost no time in turning. In how many seconds will they pass each other again?

SOCKS

(This is an oldie.) A sock drawer has 20 blue socks and only 10 brown socks. They are all mixed up. You reach into the drawer in the dark and pull out socks one at a time. How many should you take out before you are certain to have 2 socks of the same color? (What if the drawer also had 15 green socks?)

4 / DEVELOPING PROBLEM-SOLVING PROCESSES

MAGIC DOUGH

A baker has perfected a magic bread dough that, when placed in the oven, doubles in size every minute. He has further determined a special measure of this dough that will exactly fill the oven in 30 minutes. If the baker puts 2 special measures in the oven, when will the oven be full?

BOX TOPS

Diana and Debbie are saving cereal box tops to get a free record album. Diana has 9 more box tops than Debbie. Together they have 35. How many do they each have?

BUY AND SELL

A man buys a necklace for $6, sells it for $7, buys it back for $8, and sells it again for $9. How much does the man make or lose in all of this buying and selling?

DART SCORES

The dart board has 3 rings. The bull's-eye is worth 11 points, the middle ring is worth 7 points, and the outer ring is worth 3 points. Marshal threw 5 darts and all hit the target. His score was 43. Where might the darts have landed?

EGGBERT

Eggbert had a basket of eggs that he was taking to market. On the way he met a poor lady in need of food. He gave her half of all of his eggs, plus half an egg, and traveled on. Later he met a poor man in need of food. He gave the man half of his eggs, plus half an egg, and traveled on to market. There he sold the 10 eggs he had remaining. No eggs were broken at any time. How many eggs did he have to begin with?

BASEBALL CARDS

Stewart has a collection of baseball cards. If he puts them in piles of 2, he has 1 left over. He also has 1 left over if he puts them in piles of 3 or 4. In piles of 7 he has none left over. What is the smallest number of cards he could have?

CONSECUTIVE FACTORS

The product of three consecutive numbers is 32,736. What are the numbers? (Use a calculator!)

EMPEROR'S BANQUET

You have been invited to the emperor's banquet. The emperor is a rather strange host. Instead of sitting with his guests at his large round dining table, he walks around the table pouring oatmeal on the head of every other person. He continues this process, pouring oats on the head of everyone who has not had oats until there is only one person left. You will know when you arrive where the first seat is to get the oats and that the emperor always rotates to the left with his oats. You do not know in advance how many people are there. Devise a system that will tell you in what seat to sit in order to not get "oatmealed" once you know how many people are at the banquet.*

CHANGE FOR A DOLLAR

You have exactly $1.60 in change. It consists of 19 coins of 3 kinds, nickels, dimes, and quarters. How many of each do you have?

PHONE CABLES

Some small towns set up an independent phone system. Between each pair of towns they stretched a main cable connecting the two. In all there were 28 cables installed before all towns were connected. Can you discover how many towns were there?

RECTANGLE DIAGONAL

Suppose that you draw a rectangle on squared grid paper. If a straight line is drawn diagonally from corner to corner, how many squares will this diagonal pass through? Devise a system that will work for any size rectangle.

---

*Adapted from Countryman (1992). The problem also appears in other places in different forms.

FIVE BUCKS A HOLE

Sam and Ben have been playing golf on their vacation. They bet $5.00 on each hole. Over two weeks they played 45 holes that did not end in a tie, and Ben was ahead $125. How many holes did Ben win?

MY AGE

Five more than my age times 3 is the same as 77 less my age. How old am I?

STRANGE CHANGE

If you spend $1.85 and give the clerk $10.00, you will receive $8.15 in change. Notice that the digits 1, 5, and 8 are the same in both the amount spent and in the change. What other amounts could you spend for which the change out of $10.00 uses the same digits as the amount spent?*

Hint: Why must the amounts all end in a 5? Why not in a 0? Why not in some other digit?

WIDGETS

Widgets can be purchased wholesale at 3 for $5.00 and sold at 2 for $5.00. In order to make a profit of $50.00, how many widgets must be purchased?

CANTEENS

Three men stranded in the center of the desert have 15 canteens, all the same size. Five are full of water, 5 are exactly half full, and 5 are empty. Each man plans to take a different route out of the desert. How can they share the water and the canteens equally so that if they come to an oasis they will have an equal ability to take on more water.†

COOKIE CONTEST

Ann, Betty, Connie, and Denise had a contest to see who could sell the most Girl Scout cookies. Betty sold more than Denise. Ann sold the fewest. Connie sold more than Betty. In what order did the contest come out?

TWO TEAM HANDSHAKE

At the end of each game the two Little League teams line up, and each player shakes the hand of every player on the other team. If each team has 25 players, how many handshakes is that?

CAN'T MAKE CHANGE

What is the most change you can have and still not be able to change a nickel, a dime, a quarter, a half dollar, or a dollar?

DARTS

A dart board scores 9 points for a bull's-eye with 7, 5, 3, and 1 point respectively for each successive ring inside to outside. Six darts are thrown, and all score something. With 6 darts, how can each of these scores be made: 8, 12, 18, 27, 28, 56?

EGG TIMERS

With just a 7-minute and an 11-minute "hourglass" how can you time a 15-minute hard-boiled egg?

PRIMARY-LEVEL PROBLEMS

The problems in the previous section are suitable for children in grades 4 to 8. Children in the primary grades can also solve nonroutine problems and benefit from the experience in the same way. Here are a few problems taken from the first- and second-grade books of a popular problem-solving program (Charles et al., 1985).

Paul Penguin is learning how to fish. When he first started fishing he didn't catch many fish, but he improved every day. On day one he caught only one fish, but on day two he caught four fish. Complete the table to decide how many fish he caught on day five (Grade 1, p. 88; see Figure 4.14).

Pete went to Winnie's Toy Store to buy some toy cars and trucks to add to his collection. He spent 9¢ in all. Which toy cars or trucks did he buy? Draw a ring around each toy he bought. *(Children have a page with five toys priced 1¢, 4¢, 7¢, 3¢, and 5¢)* (Grade 1, p. 59.)

---

*Adapted from Burns (1992).
†Adapted from Hyde & Hyde (1991).

Freddie Frog, Sadie Snail, and Reggie Robin asked Bruno Bear how old he is. Bruno said, "You will have to figure it out for yourself, but I will tell you three things.

1. I am less than 12 years old.
2. I am more than 9 years old.
3. I am not 11 years old."

How old is Bruno Bear? (Grade 2, p. 71).

In Chapter 18, pattern activities and activities with attribute materials are explored. These activities help to develop strategies similar to "Look for a Pattern" and "Logical Reasoning." The easier versions of patterning and attribute explorations are often used as part of the problem-solving program for grades K–2 even though there are extensions for both that reach as high as grade 8.

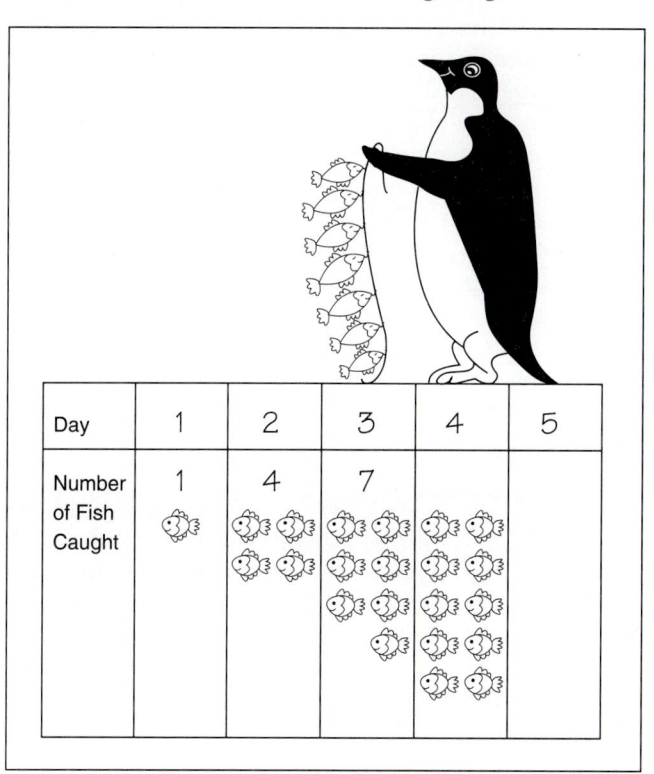

FIGURE 4.14: *First-grade problem. How many fish caught on the fifth day?* SOURCE: *From Charles and Lester,* Problem-Solving Experiences in Mathematics: Grade 1. *Copyright 1985 by Addison-Wesley Publishing Company. Reprinted by permission.*

DEVELOPING STRATEGIES

Now we turn to a discussion of *teaching* children *about* problem solving. Here the focus of instruction is on helping children develop a general scheme similar to Polya's four-step approach including strategies for solving problems.

The teaching plan described here is modeled after the three-part approach described, developed, and tested by Charles and Lester (1982). It is also very nicely carried out in the resource books, *Problem-solving Experiences in Mathematics* (Charles et al., 1985). The plan focuses on teaching actions *before, during,* and *after* the time students are solving a problem. The three-part approach is an excellent scheme for a teacher at any level to prepare a lesson plan for problem solving.

Problems can be presented in any appropriate manner: read orally, prepared on an overhead or chalkboard, distributed on paper, or taken from books. It is generally a good idea to have students work in cooperative groups of three or four to maximize interaction and discussion.

For the purpose of the discussion, consider the following problem:

> Lisa has 12 coins in her purse that have a total value of 45 cents. What coins might she have?

TEACHER ACTIONS BEFORE STUDENTS BEGIN

There are some standard things you can expect to do with all problems. These include reading the problem, having students restate the problem in their own words, asking for identification of what is known and what is being asked for. Besides these things, good planning includes preparation of some specific questions related to the problem.

How many coins does Lisa have?
How much are they worth in all?
What different kinds of coins are there?
How much is a quarter (and other coins) worth?
Could she have a half dollar?
What is one way to have 45 cents? How many coins?

With each problem the first questions should cover all of the important data in the problem, and the questions can be answered directly. Any vocabulary that may be confusing should be clarified. Make few assumptions about students' knowledge. Help students clarify any implicit assumptions in the problem. For example, in the checkerboard problem in Figure 4.2, the squares can be of all sizes and can overlap, but that is not clearly in the problem.

Some questions can begin to give hints and are appropriate for beginning problem solvers at this time. They can also be held back as hints for individual groups later on. Could there be pennies? Could there be less than five? Could there be six?

The purpose in this *before* stage is to model the type of thinking and problem-understanding behavior that some day they will do on their own. There is no reason to hide assumptions, use vague wording, double meanings, or in any way to be tricky. That would discourage students and defeat the purpose.

TEACHER ACTIONS WHILE STUDENTS ARE WORKING

The next thing is to get your students working. Exactly what you plan to do at this time depends on the prior experiences of your students, the strategy that you want to draw out of the lesson, and whether they have experienced that strategy before. For a totally novice class, it may be wise simply to set them to work and prepare a series of hints and suggestions to give to individual groups. Another option is to guide a full class discussion toward a particular solution strategy and then set the groups to work with the hope that many will follow up on your lead.

Whatever your specific approach, it should never communicate to students that there is some predetermined "right" way to do the problem. Accept ahead of time that some students will solve problems in ways you had not planned and that had not even occurred to you. If your goal was to use the problem to teach the organized list strategy, then plan your hints and suggestions to steer students toward that strategy. However, do not be upset that several groups found the guess-and-check heuristic to be more useful to them. In the final analysis, if your class solves one problem using three different methods or strategies, that will be more valuable to their development than 30 problems all solved exactly as you prescribe.

Suppose your intention in using Lisa's coin problem is to work on the organized list strategy. The first thing you should do in preparation is to solve the problem yourself according to this strategy and reflect on the kinds of decisions you made and in what order. Then think about things that could go astray or cause difficulty. You might prepare the following hints and questions:

> Would it help to find *all* of the ways to make 45 cents? Could you make a list of them?
>
> What is the fewest number of coins that could possibly be used?
>
> What is the greatest number of quarters you could have?
>
> What would you have left to worry about after one quarter? (20 cents)
>
> Perhaps it would help to organize the different ways in some kind of chart. What would you need to keep track of?

Notice that these questions aim at the strategy of finding all the ways to make 45 cents, starting with the least number of coins. If a group decided to start with all pennies and work from that direction, the strategy is exactly the same but with different specifics. Both should be encouraged. If you have thought about that ahead of time, you will be prepared to help that group without letting them think they are on the wrong track. Never force your method on students, especially if theirs is just as effective.

You should try to work the problem ahead of time with as many different approaches as you can that seem to be probable. If you have only one approach in mind, you may communicate to students that your approach is "the correct" one. Students with that perception will have difficulty building self-confidence in their own abilities to solve problems.

In planning, it is a good idea to list as many hints, questions, and suggestions as you can. Write them down as they come to you, and then examine the list. Put the hints and questions in an order that goes from least guided to most guided. You may not need them all for all groups. You want the students to do as much thinking on their own as possible. The more help you give them, the more you are solving the problem for them. On the other hand, the kind of reflective thinking required to produce just the right hint or guidance is difficult to do while walking from group to group if you have not thought through some possible guidance ahead of time.

During the actual lesson be an active listener. If a group is searching for a place to begin, some of the first hints in your list may be appropriate. If a group has started on a strategy but needs some help in using it effectively, you need to decide if intervention and assistance is more helpful or if this is a time to let students learn from their struggles. When students are working hard but clearly going in the wrong direction, try to praise their work and ask questions that will help them discover their own misdirection. You will frequently encounter approaches you have not thought of. They may be more inventive and valuable than what you had planned. Do not be too quick to guide students in another direction.

Besides helping those groups that need guidance, you must also be prepared for those groups or individuals that zoom directly to the solution ahead of the rest of the class. First, require that they have carefully checked their reasoning and their answer. Encourage them to be ready to explain their approach. But then what? Be prepared to offer an extension or a related challenge. Simply giving the group another problem can be construed as punishment for being effective and clever. Rather, your praise of their excellent work can include a challenge to go beyond the original problem:

> That is super work! Are you sure there is only one solution? (There are two solutions to the coin problem, but the class need not know that.)
>
> I wonder if this group could figure out what other amounts you could have with exactly 12 coins. Could you change the amount from 45 cents to something else, so that there is only one answer to this problem?

TEACHER ACTIONS AFTER THE PROBLEM IS SOLVED

After a sufficient time, return to a full class discussion. Ask several groups to give their answers, and then discuss with the class how they can decide if the answer is correct. You want the students to learn to validate or assess their own answers. Do not be an answer book!

In the example of 45 cents and 12 coins, it is relatively straightforward to check that the solution meets both conditions. But is it the only solution? That question can turn the discussion back to the processes used.

While students were working, you will have been able to determine the approaches used by different groups. Select one or two groups to explain how they did the problem. (Each group should always have both a recorder and a spokesperson.) As different solutions are described, help the class identify the strategies used. As noted earlier, there are at least two approaches to this example problem that use the organized list strategy. The same problem can also be done with a try-and-adjust approach. If a group has done it that way and you have time to discuss it, then there is an opportunity to both see another strategy and communicate that there is no mystical "right" way to solve a problem. The try-and-adjust method is less likely to turn up two solutions, but there is no way anyone could know that before doing the problem. At the very least, a quick acknowledgment of a group's different approach will keep them from thinking they were wrong in what they did.

If a strategy used is a new one for the class, then it should be labeled and added to your classroom list of strategies. A chart or bulletin board with a growing list of useful approaches will be valuable as your problem-solving program continues.

Finally, discuss the problem itself. What else is known, or what else can be done now that we have solved the problem? Encourage students to look for generalizations if appropriate. The coin example does not lend itself to a generalization in the same way as the checkerboard problem.

Here are some ideas that you might consider when looking for a new but related problem (Charles & Lester, 1982):

Can the wanted and given information be reversed?

What if the context were changed? (Suppose that instead of money, Lisa had boxes of candy of different sizes.)

Can other conditions be added, changed, or removed? (What if Lisa had "more than 12 coins" and/or "less than 75 cents"?)

If the numbers were just changed, would the problem be any easier or harder? Could we make one up and solve it easily now?

Are there any problems that seemed like this one somehow? (The coin problem has the same structure as the pigs-and-chickens problem.)

Can the problem be generalized? (Sometimes you can get to this by working toward larger and larger numbers.)

DEVELOPING METACOGNITIVE HABITS

The lesson plan just described focuses on important problem-solving strategies. However, if children are to become responsible for their own actions in problem solving, they must also learn to develop the metacognitive habits of monitoring and regulating their own thought processes. Ultimately, we want students to solve problems without our guidance. Therefore, all teaching actions are designed to develop the habits that will make this happen. Children will not spontaneously develop them.

MONITORING AND REGULATING DURING THE PROBLEM-SOLVING PROCESS

During all three phases of the problem-solving lesson you want to include questioning aimed at students' thinking. A simple formula consists of three questions: *What* are you doing? *Why* are you doing it? *How* does it help you? In some form or another, these are questions we want children to begin asking themselves.

Before, During, and After

While the exact form of the question does not have to be in these same words, the idea is to be persistent with this reflective questioning as you go through a lesson. In the understanding stage, students need to be aware that they are (or are not) looking for and clarifying the information given and required. When they notice some feature of the problem or a piece of information, they need to learn to ask, "How can this help me? What does that tell me about what I should be doing?" A general goal is to delay acting on the problem until students are confident that they fully understand the problem and have given it some thought.

As students are working on the problem, the "What?" "Why?" and "How does it help me?" questions begin to focus on strategies being employed. "What information is in the chart we made? Why did we draw that picture? Could a different picture help more? Are we making any progress? Is there any pattern developing? If we finish this list of all the ways to do it, how will that help? What will we know when we do this?" These and similar questions are the type students need to learn to ask themselves. As they are learning, you are assisting by doing the asking.

In the after or looking-back stage, the answer becomes the focus. "Have we checked our solution? Why will that tell us if it is right or not? How else could this be checked?" Reflection on the strategy is actually summarizing what went on and reflecting on the decisions already made. "What method did we use? Why did we do it that way and not some other way? Could we have done it differently?" The third part of looking back is looking at the problem itself. "Have we checked to see if we can extend or modify this problem? What does the answer tell us? Is this like other problems we've worked before?"

The student answers to the "What?" "Why?" and "How does it help?" questions will be very problem-specific. In fact, students have a very difficult time answering these questions at first, and you will need to help them understand that this is a good way to help them solve problems. There is no prescription or formula for solving real problems, and so the only way we can possibly know if we are doing it "right" is to keep asking ourselves.

Group Reflection

Some of this questioning can go on with the full class, especially during the before and after stages. In the during phase of problem solving, you will be more likely to focus your reflective questioning on individual groups. Whenever possible, actually join a group and work with it as an active member. Here you can model the reflective monitoring questions, forming them in terms of "we" instead of as an outsider. "What are *we* doing here? We need to know why *we* are doing this. Can *we* tell where this is leading? How does this get *us* closer to an answer?"

By joining the group you model questioning that you want the group to eventually do on their own. After you feel that students are beginning to understand the process of monitoring, each group can designate (or you can assign) a member to be the monitor. The monitor's job is to be the reflective questioner the same way you have modeled when working with them.

POST-PROBLEM REFLECTION

Another approach is to have students report on their problem-solving activity after it is over. This can be done orally, in a written format, and through self-rating forms. A five-minute discussion after a problem is over can focus on what types of things were done in solving the problem. "What did you do that helped you understand the problem? Did you find any numbers or information you didn't need? How did you decide what to do? Did you think about your answer after you got it? How did you decide your answer was right?"

Questioning similar to this tells students that all of these things are important. If they know you are going to be asking them, there is an increased chance they will think about the questions ahead of time. Oral discussions with the class will be most helpful with younger children for whom written formats are difficult.

An excellent reflective technique that you can begin as early as second or third grade is to have students make written reports of their problem solutions after the problem has been solved. Even when a problem has been done in a group setting, individuals can still write up the problem and the solution. The report should include a statement of the problem (if not already printed on the paper), a full explanation of the solution, and some discussion of the answer. After getting comfortable with this much, ask students to add in what they did to understand the problem, why they solved the problem the way they did, and what decisions they may have made along the way.

The open-ended reporting on a problem is not only a good way to get students to reflect on their own thinking but also a good item to periodically include in your assessment program. Students can select reports that they feel especially good about each month to be included in a portfolio illustrating their accomplishments and progress.

Many students will have trouble putting their solutions on paper and even more difficulty reflecting on their thought processes. Another approach to consider is occasionally to have students respond to a checklist shortly after solving a problem. For each of the statements shown in Table 4.1 students circle one of these responses: NO, MAYBE, or YES (NO—I didn't do this; MAYBE—I may have done this; YES—I did do this).

One suggestion is to use the results of a checklist such as this one as a basis for class discussion. Students can talk about why they responded as they did and hear what other students were thinking about while solving the same problem. The responses also provide information that you can use to help the class or individuals improve their monitoring habits.

Neither the written problem reports nor the checklist approach need be done for every problem. The idea is periodically (once every week or two) to get students to think about their monitoring and control processes. Just doing the reports provides feedback to the students and makes them aware that this self-reflection is valuable.

Finally, the ability to monitor one's own thinking is something that develops very, very slowly. If your students have had no prior metacognitive training, it may take the entire year before you see any real progress. Do not expect students even to think about self-monitoring without your input for a long time. Be persistent, but be patient! The evidence that exists at this time suggests that the effort is well worthwhile.

ATTENDING TO AFFECTIVE GOALS

Look back for a moment at the affective goals of a problem-solving program. You will see that these are not things that are taught but rather are instilled in children by virtue of the classroom climate and by the attitudes and values that you, the teacher, demonstrate on a daily basis.

DEVELOPING POSITIVE STUDENT ATTITUDES

If students do not demonstrate a willingness and interest in solving problems, it is very difficult to teach them problem-solving processes. Willingness to solve problems, even enjoyment, should be one of the highest priorities in your problem-solving program. What can be done to develop that type of positive attitude?

> **Before you began to solve the problem—what did you do?**
>
> 1. I read the problem more than once.
> 2. I tried to find everything out about the problem that I could.
> 3. I asked myself, "Do I really understand what the problem is asking me?"
> 4. I thought about what information I needed to solve this problem.
> 5. I asked myself, "Have I ever worked a problem like this before?"
> 6. I asked myself, "Is there information in this problem I do not need?"
>
> **As you worked the problem—what did you do?**
>
> 7. I kept looking back at the problem as I worked.
> 8. I had to stop and rethink what I was doing and why.
> 9. I checked my work as I went along step by step.
> 10. I had to start over and do it differently.
> 11. I asked myself, "Is what I am doing getting me closer to the answer?"
>
> **After you finished working the problem—what did you do?**
>
> 12. I checked to see if all my calculations were correct.
> 13. I went over my work to see if it still seemed like a good way to do the problem.
> 14. I looked at the problem to see if my answer made sense.
> 15. I thought about a different way to solve the problem.
> 16. I tried to see if I could tell more than what the problem asked for.
>
> ———
> SOURCE: Table 4.1 is a modification of one suggested by Fortunato, Hecht, Tittle, & Alvarez (1991).

TABLE 4.1: *Suggestions for a Metacognitive Checklist*

◆ Build in Success

Nothing succeeds like success. This is a cliché but very true in this context. In the beginning of the year, plan problems that you are confident your students can solve. The success should clearly be your students' and not due to your careful guidance. Avoid creating a false success that depends on you showing the way at every step and curve. Students will quickly see that you are the one that solves the problems, and without you, they will believe, they are helpless. Rather, if problems are solved through their own efforts, you can correctly heap praise on their outstanding good work. Even as much as two months of relatively simple problems that did nothing to move the students forward in their confidence would be more worthwhile than frustrating students with weak self-concepts about problem solving.

◆ Praise Efforts and Risk Taking

Students need to hear frequently that they are "good thinkers" capable of good, productive thought. When students volunteer ideas, listen carefully and actively to the idea, and give credit for the thinking and the risk that children take by venturing to speak out. The more that students are encouraged, the more effort they will expend to receive that praise. Conversely, if weak or incorrect ideas are put down or ignored, the children who ventured forth with those ideas will think twice before ever trying again. After some time the reluctance turns into a belief that says "I'm no good! I can't think of good ideas."

◆ Use Nonevaluative Responses

Develop a response form that is nonevaluative, that can be used all of the time for both good and weak ideas. For example, "That's a good idea. Who else has an idea?" As another example, "That's good thinking. I can tell you are really thinking about this situation." Even the child who spurts out without thinking can be rewarded: "I always want to hear your good ideas, Cindy. Be sure you've thought through what you want to say first." Be sure to return to this child and reward the more reflective thought.

◆ Listen to Many Students

Avoid ending a discussion with the first correct answer. As you make nonevaluative responses, you will find many children repeating the same idea. Were they just copying a known leader? Perhaps, but more likely they were busy thinking and did not even hear what had already been said by those who were a bit faster. If 10 hands go up, 10 children may have been doing good thinking. If you forget to listen to lots of children, lots of children do not hear your praise.

◆ Provide Special Successes for Special Children

Not all children will develop the same problem-solving abilities. Those who are slower or not as strong need success also. One way to provide their success is to involve them in groups with strong and supportive children. In group settings all children can be made to feel the success of the group work. Slower students can also be quietly given special hints that move them toward success.

Better students need special successes also. Be especially prepared with a super challenge for these students. Be careful, however, to be discreet.

THE TEACHER AND THE PROBLEM-SOLVING CLIMATE

There is no more important ingredient in developing student problem-solving abilities than the teacher. The teacher sets the tone and the spirit—the climate for the classroom.

◆ Enthusiasm

A teacher's enthusiasm for a topic is easily detected by students. If the teacher is "into" problem solving, is clearly enjoying the activity, and is spirited in discussions with students, then students will begin to share in that enthusiasm. This can be as simple as excitement over finding a new problem to solve or a bit of exuberant behavior when someone has demonstrated a good idea.

◆ Values

If problem solving is to be the most important part of the mathematics curriculum, then your students should have every indication of that value system. Your value system around problem solving is reflected in the amount of time devoted to it, compared with other topics, and its position relative to other work in the day and week.

Perhaps more important than time and emphasis is evaluation. Students need to know that the problem-solving work is being evaluated, monitored, and made a significant part of the regular grading scheme. Parents should also be kept informed about your problem-solving program. Let them know of the part it has in grades, the time devoted to it in class, and the importance that you believe thinking skills have for their children.

◆ Challenge

Problem solving should be fun and serious business at the same time. One way to bridge this gap is to approach problem solving as a challenge. It is important and fun to conquer challenges. Monotonous work is rarely important and never fun.

◆ Model Problem-Solving Behavior

Let your students know that you, too, are still growing as a problem solver. Occasionally do problems with them that you have never seen before. With older children, share problems that you have found in magazines or books that you find challenging and interesting. Become a problem solver yourself, and let your students know that it is important to you.

ASSESSING PROBLEM SOLVING

Today, assessment in the mathematics classroom has broadened well beyond the end-of-the-chapter test. It is an ongoing process, integrated with instruction, and is designed to look at students' mental processes, understanding of concepts and procedures, and their attitudes and beliefs. Assessment of problem solving is no different in this respect than assessment of the rest of the mathematics program. In this section we will focus briefly on assessment of the specific problem-solving components discussed in this chapter. In Chapter 5, assessment is discussed more fully, and the ideas touched on here are included in the broader scheme of assessment in mathematics.

WHAT TO ASSESS

In days long past, students' problem-solving abilities were evaluated by providing problems on tests. Because tests are threatening, problems were generally restricted to routine one- and two-step problems that could be solved quickly and routinely. Scoring was based solely on the answers which were either right or wrong. Many times students got correct results using incorrect reasoning while others with excellent reasoning and problem-solving skills got incorrect answers through careless errors. This outdated approach simply did not assess students' problem-solving abilities.

The goals of problem solving were listed earlier under the headings Strategy Goals, Metacognitive Goals, and Affective Goals. A good assessment strategy will reflect these goals. Each can be rephrased in terms of student behaviors. Here are examples for each area.

◆ Strategy Goals

- Is able to determine the relevant information in a problem
- Can clearly state what is being asked for
- Given a strategy can use it appropriately
- Uses different strategies (which are used, which not)
- Makes appropriate selections of strategies
- Is able to self-assess answers in terms of the problem and process used
- Goes beyond the solution when possible

◆ Metacognitive Goals

- Shows evidence of monitoring thinking processes (before, during, after)
- Makes appropriate strategy adjustments while working on a problem
- Is able to reflect on and explain choices made in the solution process

◆ Affective Goals

- Demonstrates willingness to try to solve problems
- Demonstrates self-confidence in problem-solving abilities

- Enjoys solving problems
- Perseveres in attempting to solve problems even in the face of slow progress
- Believes in personal ability to solve problems
- Believes that methods of solving problems can be discovered and do not consist of predetermined rules to be memorized

You may wish to modify this list, add to it, or delete some items. What is important is that attention is focused on all aspects of problem solving. These processes, attitudes, and beliefs are important for children to develop, and they are therefore important enough to be assessed.

WHEN TO ASSESS

Any time that students are engaged in a problem-solving activity is an opportunity to assess problem-solving behaviors. It is while students are engaged in problem solving that you can find out how they are doing. Assessment does not need to be separate from instruction. It is a myth that we have to teach first and test afterward. Remember also, that while this chapter focused on teaching *about* problem solving, a problem-solving approach (teaching *via* problem solving) is recommended for learning activities in all content areas. Following that approach expands significantly the opportunities you will have to "catch" your students in problem-solving activities.

You cannot and should not try to evaluate every child at every opportunity. Consider problem-solving assessment an ongoing activity, and gather information from targeted students or groups of students during a particular lesson. Nor should you feel that all objectives need be assessed with each observation. Select those objectives for which you currently need data, and make it a point to gather that information across all students over a reasonable period of time. As you rotate objectives, gathering data regularly, you will be able to demonstrate growth over time on all objectives for each individual in the class.

PROBLEM-SOLVING PORTFOLIOS

A portfolio is a selection of students' work collected over a period of several weeks to as much as a year. A problem-solving portfolio will have selected items illustrating growth or achievement in various aspects of the entire problem-solving process, including metacognitive and affective objectives. Included in the portfolio might be student solutions to problems, group solutions (duplicated for all group members), products from problem-solving projects, observations made by you (either anecdotal or on a prepared checklist), attitude inventories, or journal entries.

Chapter 5 includes a more complete discussion of portfolios. A portfolio component to problem-solving assessment is highly recommended, since it is one of the best ways to illustrate growth over lengthy periods of time and allows the student the opportunity to demonstrate his or her own abilities.

REFLECTIONS ON CHAPTER 4: WRITING TO LEARN

1. Why is a problem for one person not necessarily a problem for another? The computational exercises that we all used to call problems are not problems at all. Why is that?
2. Explain briefly the essential features of each step of the four-step plan devised by George Polya. In the looking-back stage, what three things do we want students to look at?
3. The goals of problem-solving instruction have been categorized into three areas: Strategy, Metacognition, and Affect. Pick one or two goals from each area that seem most important, and describe them briefly.
4. What makes a nonroutine problem nonroutine? Contrast these problems with routine problems.
5. There are nine problem-solving strategies described in this chapter. Describe each briefly, including when they are most useful.
6. A problem-solving lesson structure has been suggested consisting of three parts: Before, During, and After. Describe the teaching actions in each part or phase.
7. Describe one or two ways to help children develop metacognitive habits.
8. Describe one or two ways to help children develop positive attitudes toward problem solving.
9. When should you assess problem solving? Explain briefly.
10. Why would you want to use portfolios in your problem-solving assessment scheme?

FOR DISCUSSION AND EXPLORATION

1. What would you do if you had a class do a process problem, and no group used the strategy that was your objective for the day?
2. Review the affective objectives for problem solving. How do you stack up on these objectives? What major factors in your experiences caused these attitudes toward problem solving?
3. Pick any process problem that you and your study group have not worked. Assign one member of the group the task of being the metacognitive monitor. As you solve the problem, make a special effort to attend to the monitoring and control agendas

discussed in the chapter. How much of this is part of your natural problem-solving approach?

4. Examine a commercially available problem-solving program. Two possibilities are *Problem-solving Experiences in Mathematics* (Charles et al., 1985) and *The Problem Solver: Activities for Learning Problem Solving Strategies* (Goodnow, Hoogeboom, Moretti, Stephens, & Scanlin, 1987). How do these programs fit your understanding of the needs of problem-solving instruction?

5. Try some problem-solving instruction in a classroom. Remember that there will be big differences in how children work with problems, depending on the amount of such experiences they have had.

6. Examine a basal textbook series for problem solving. What types of problems are provided? How frequently are process problems provided? What strategies are introduced? What form of problem-solving evaluation is provided? It will very likely be necessary to examine the teacher editions of the books to find everything you want to know.

7. How much effort should be given to problem-solving evaluation by the teacher in the classroom? What if no evaluation were done? Could there be too much time, effort, or pressure placed on evaluation? What are the positive benefits of evaluation? What are the negative features?

8. Find out if your state or local school district has a performance-assessment program or portfolio-assessment program. If so, get some information about the program, and examine the way problem solving is assessed.

◆◆◆◆◆

SUGGESTED READINGS

Brown, S. L., & Walter, M. I. (1983). *The art of problem posing*. Hillsdale, NJ: Lawrence Erlbaum.

Cemen, P. B. (1989). Developing a problem-solving lesson. *Arithmetic Teacher, 37*(2), 14–19.

Charles, R., et al. (1985). *Problem-solving experiences in mathematics* (grades 1 to 8). Menlo Park, CA: Addison-Wesley.

Charles, R., & Lester, F. (1982). *Teaching problem solving: What, why & how*. Palo Alto, CA: Dale Seymour.

Charles, R., Lester, F., & O'Daffer, P. (1987). *How to evaluate progress in problem solving*. Reston, VA: National Council of Teachers of Mathematics.

Fortunato, I., Hecht, D., Tittle, C. K., & Alvarez, L. (1991). Metacognition and problem solving. *Arithmetic Teacher, 39*(4), 38–40.

Goodnow, J., Hoogeboom, S., Moretti, G., Stephens, M., & Scanlin, A. (1987). *The problem solver series* (grades 1 to 8). Palo Alto, CA: Creative Publications.

Hembre, R., & Marsh, H. (1993). Problem solving in early childhood: Building foundations. In R. J. Jensen (Ed.), *Research ideas for the classroom: Early childhood mathematics*. New York: Macmillan Publishing Co.

Kersh, M.E., & McDonald, J. (1991). How do I solve thee? Let me count the ways! *Arithmetic Teacher, 39*(2), 38–41.

Kroll, D. L., & Miller, T. (1993). Insights from research on mathematical problem solving in the middle grades. In D. T. Owens (Ed.), *Research ideas for the classroom: Middle grades mathematics*. New York: Macmillan Publishing Co.

Mason, J., Burton, L., & Stacey, K. (1982). *Thinking mathematically*. London: Addison-Wesley.

Meyer, C., & Sallee, T. (1983). *Make it simpler: A practical guide to problem solving in mathematics*. Menlo Park, CA: Addison-Wesley.

Ohio Department of Education. (1980). *Problem solving . . . a basic mathematics goal: A resource for problem solving*. Columbus, OH: The Department.

Ohio Department of Education. (1980). *Problem solving . . . a basic mathematics goal: Becoming a better problem solver*. Columbus, OH: The Department.

Otis, M. J., & Offerman, T. R. (1988). How *do* you assess problem solving? *Arithmetic Teacher, 35*(8), 49–51.

Szetela, W. (1986). The checkerboard problem extended, extended, extended, ... *School Science and Mathematics, 86*, 205–222.

Worth, J. (1990). Developing problem-solving abilities and attitudes. In J. N. Payne (Ed.), *Mathematics for the young child*. Reston, VA: National Council of Teachers of Mathematics.

5 ASSESSMENT IN THE CLASSROOM

◆ ONE OF THE MOST SIGNIFICANT FACTORS FUELING THE REVolution in school mathematics is a nationwide shift in assessment practices. No longer are teachers able to say "Yes, but...what about all of those skills that are on the standardized test?" While some traditional skills remain important, testing from the classroom level to the district and state level is shifting away from multiple-choice, computationally oriented tests that have had a stranglehold on what is taught in schools. Traditional testing programs have been designed to find out what students do *not* know. They have been aimed at the lowest level of skill development. Today, teachers and school districts alike are beginning to move toward performance-based assessments that encourage students to demonstrate the thinking and the conceptual knowledge that is the heart of knowing mathematics.

Alternative assessments focusing on students' performance, frequently referred to as *performance-based assessments*, take on a variety of forms including observations, open-ended tasks, group assessments, interviews, and portfolios. Calculators and manipulative materials are being used on tests and other assessments. Attitudes and dispositions concerning mathematics are also important in the total assessment scheme.

The challenge of the new directions in assessment is significant. While it is taking time for school districts and teachers to rethink and restructure their thinking about assessment, it is clear that the nation is changing its long-standing habit of testing only the lowest-level mathematics skills.

ASSESSMENT AND THE NCTM *STANDARDS*

The NCTM *Standards* documents are among the best places to look in order to understand just how student assessment fits into the curriculum and instruction picture.

CHANGES IN EMPHASIS

The *Standards* calls for a number of changes in the nature of assessment and assessment practices that suggest a classroom evaluation program considerably different from the one you may have personally experienced. In the past, teachers presented information, students practiced procedures, and the testing came last. A change from this approach is directed by the *Standards*. The following shifts are suggested (p. 191):

From	To
◆ Testing that focuses on finding out what students do not know	◆ Assessing to find out what students do know, how they think and reason, and how they view mathematics
◆ Simply counting correct answers to determine a numeric grade	◆ Making assessment an integral part of every-day instruction
◆ A focus on specific and isolated skills	◆ A focus on a broad range of tasks that take into account the true nature of mathematics
◆ Using exercises or word problems requiring only limited skills	◆ The use of problem-solving situations requiring a collection of mathematical ideas brought together as the student wishes
◆ Using only written tests to gather information from students	◆ Using a variety of assessment techniques including student writing and self-assessments, demonstrations, and observations
◆ Excluding the use of calculators, computers, and manipulative materials from any testing situation	◆ Using calculators, computers, and manipulatives in assessment

PURPOSES

In the past, the main reason for testing was to assign grades. In the alternative view of assessment, the main purpose of assessment—perhaps the only reason—is to improve learning. With that in mind, a good assessment program in the classroom should

- inform the teaching decisions that are made on a day-by-day basis by focusing on students' thinking;
- inform students about what is important as thinking, concepts, and communication are valued over low-level skills;
- communicate progress to students on an ongoing basis as they take an active role in the evaluation process; and
- communicate to parents the mathematics their child is doing, including the breadth of the ideas and the quality of his or her thinking and problem-solving skills.

ADVANTAGES

When the focus of assessment is on children doing mathematics and related thought processes and attitudes, the entire purpose of teaching tends to shift as well. Teachers are provided with real access to students' thinking, not just their performance level on a predetermined set of skills. How students think, how their ideas are connected and accessed, how they select and use skills in applied settings, how they monitor and control their thinking, and how they feel about mathematics and themselves as doers of mathematics are all critical information when teaching from a developmental or constructivist point of view. Teaching itself tends to become a more interesting activity. Your ability to listen and question is improved as these same skills are used daily in the assessment process. Respect for your students will be greater. The monotony of "teaching as telling" is set aside as students begin to take on more responsibility for their learning.

Students probably gain the most. As the thinking process becomes important due to the assessment program, students focus more on, and consequently learn better, the skills of thinking and problem solving. They learn that there are many ways to approach problems and that most real problems do not have a single right answer. They begin to take a greater responsibility for their thinking, improve their listening skills, and ask better questions. Self-confidence improves because they are allowed to show what they know. Conceptual knowledge is tied to a broader web of ideas as students work together to solve problems and communicate ideas.

Tests given at the end of an instructional unit or chapter were generally seen by both teachers and students as necessary, yet painful, things to endure. They added little or nothing to the learning process. Students learned that the only things that really counted were the routine skills that appear on the tests. The tests themselves were threatening. Alternative, performance-based strategies make assessment a plus to learning rather than a burden.

WHAT TO LOOK FOR

Following up on the premise that assessment should focus on what children do know, as opposed to what they do not know, requires us to broaden our vision of knowing well beyond a list of skills. To know and to do mathematics involves mental processes (reasoning, using problem-solving strategies, and metacognitive activity); a knowledge base (relational understanding of concepts and procedures and methods of communicating ideas); and affective factors (attitudes, values, beliefs, self-confidence).

MENTAL PROCESSES

Problem solving and reasoning are such integral parts of what it means to know and do mathematics that it is imperative that teachers know what sorts of mental processes their students are developing and using. Listed below are the types of things you should be looking for.

You will recognize some of the strategy and metacognitive processes from the discussion of problem solving in Chapter 4.

Recall that in that chapter the discussion was teaching children *about* problem solving. The processes listed here are applicable to any mathematical activity in any content area where students are faced with a perplexing situation and are trying to figure it out or make sense of it. Therefore, the opportunity to observe children using these processes occurs at any time students are using a problem-solving approach to learning (see Chapters 2 and 3).

Reasoning

- draws logical conclusions about mathematical ideas
- uses models, symbolism, properties, and other relationships to explain ideas
- justifies answers and defends processes
- uses spatial reasoning
- uses proportional reasoning*
- uses both inductive and deductive arguments
- assesses arguments made by others

Strategies

- determines the relevant information in a problem
- clearly states what is being asked for
- uses different strategies
- makes appropriate selections of strategies
- self-assesses answers in terms of the problem and process used
- goes beyond the solution or makes generalizations when appropriate
- formulates problems from given situations

Metacognition

- shows evidence of monitoring thinking processes (before, during, after solving problems)
- makes appropriate strategy adjustments while working on a problem
- reflects on and explains choices made in the solution process

KNOWLEDGE BASE

As you consider a particular content area, you want to look at the knowledge that students possess and the level of their understanding. Recall from Chapter 3 that understanding of an idea is determined by the number and complexity of the connections that the student has made with

---

*Appropriate for middle-grade students only.

other ideas. Relational understanding is the richly connected end of the continuum and should always be our main content goal. This means that for any particular concept or procedure, what you should be looking for is evidence of the quantity and quality of connections that students are able to make among concepts, among procedures, and between procedures and concepts.

Conceptual Understanding

- uses or explains concepts
- uses a variety of models to explain concepts; explains how two or more models illustrate the same concept
- applies previously learned ideas to new concept areas
- communicates ideas meaningfully using words, models, drawings, charts, and graphs

Procedural Understanding

- provides explanations of procedures and symbolism using models and/or other procedures
- invents procedures using a conceptual basis
- uses procedures correctly
- uses alternative procedures appropriately

AFFECTIVE FACTORS

Affective considerations are intimately connected to understanding and processes. Attitudes and beliefs also affect how eagerly students will pursue mathematics.

- demonstrates willingness to try to solve problems or approach unfamiliar situations
- demonstrates self-confidence in problem-solving abilities
- enjoys solving problems and believes in personal ability to solve problems
- perseveres in attempting to solve problems even in the face of slow progress
- believes that mathematics makes sense
- believes that methods of solving problems can be discovered and do not consist of predetermined rules to be memorized

ALIGNMENT WITH INSTRUCTION

It makes common sense to assess students' knowledge using the same modes and materials that were used to develop that knowledge, even though such has rarely been the case in the past. This includes the use of manipulative materials and calculators and assessment in group formats.

Students' understanding will, in a developmental classroom, be closely connected with a wide variety of models or materials. These were used as "thinker toys" to help them develop ideas and connections. Students should be allowed to use these materials to demonstrate their understanding. It is not reasonable to always expect children to show you what they understand about a mathematical idea without the use of materials. After all, you needed to use those materials to show these ideas and relationships to them.

Calculators and computers should likewise not be excluded from assessment activities without good reason. If the focus of the activity is on concepts, or any aspect of problem solving, the calculator should be there to avoid having tedious, time-consuming computations detract from the focus of the activity. A possible exception to the ready availability of calculators are tests that focus on a routinized pencil-and-paper computational skill. However, a heavy emphasis on such skills is not appropriate in today's curriculum.

When students learn in groups, they are communicating and using reasoning and problem-solving skills that are important components of the curriculum. You have a much better chance of catching a child exhibiting proficiency with these skills in group settings than working alone.

Resistance to these ideas may be the result of a lifelong exposure to assessments consisting entirely of tests viewed essentially as hurdles for students. Narrowly defined skills were prescribed at the outset and tested at the end. Evaluation was a matter of finding what percentage of these skills were mastered. In this scheme of things the use of manipulatives and calculators and working with fellow students during an evaluation are seen as cheating. When assessment is viewed as a window to students' minds, then the use of materials, calculators, and group evaluations simply broadens the window and clarifies the view.

PERFORMANCE ASSESSMENTS

A *performance assessment* consists of two components: (1) a task, open-ended problem, project, or investigation that is presented to students; and (2) observations of students as they work and/or examination of products produced. A problem-solving approach to instruction (Chapters 2 and 3) involves the same kind of activities. Problem situations and explorations should be commonplace in your classroom as a means of promoting reflective thought and engaging students in doing mathematics. Every time such an activity is taking place there is an opportunity (not a requirement) for assessing conceptual understanding, problem-solving strategies, procedural knowledge, attitudes, and beliefs.

By systematically or even informally observing students as they work on a problematic situation you can assess processes used as well as the final products or answers. You can see how students test and try out ideas that you may never even see on the final paper. Since the tasks in performance situations are frequently open-ended, you also have an opportunity to observe the extent of knowledge that could never be tapped through single-answer test exercises.

Performance assessments provide students with the opportunity to show their creativity, abilities, and understanding. There is much greater motivation to do well on open-ended, engaging tasks than on narrow skill exams.

EXAMPLES OF PERFORMANCE TASKS

There is no significant value in classifying different types of performance assessment tasks although they are certainly not all alike. Some are relatively short questions or situations that ask students to explain, solve, analyze, or react to something. At the other extreme students may be engaged in projects or investigations requiring several days, gathering data, making displays or models, or doing extensive writing. What all performance tasks have in common is that they are engaging, allowing students the chance to do something meaningful that you can observe.

Consider these two examples:

Explain two different ways to multiply 4×276 in your head. Which way is the easiest to use? Would you use a different way to multiply 5×398? Explain why you would use the same or different methods.

Enrollment data for the school provides information about the students and their families as shown here:

	School	Class
Siblings		
none	36	5
one	89	4
two	134	17
more than two	93	3
Race		
African-American	49	11
Asian-American	12	0
Hispanic	72	3
White	219	15
Travel-to-school method		
walk	157	10
bus	182	19
other	13	0

If someone asked you how typical our class was of the rest of the school, how would you answer? Write an explanation of your answer. Include one or more charts or graphs that you think would support your conclusion.

Both questions are open-ended and have a variety of ways that they could be answered. One is context-free, and the other is more realistic. Each provides the opportunity to include a variety of concepts and to use different processes. In the mental multiplication example, ideas about rounding, breaking a number into parts, adjusting an answer, and use of tens and hundreds are all potentially included. The second example permits a wide range of possibilities in the area of ratios, proportion, percentages, decimal and fraction connections, and the use of a variety of charts and graphs. In neither case is there a preconceived right answer, and students at different levels of understanding would exhibit more or less evidence of connections with and among concepts and procedures.

Another important aspect of performance assessments is that they need not be separate from instruction. The tasks are as much a learning experience for children as they are an opportunity to assess. *Assessment need not stop instruction or be separate from instruction.* Here are some additional performance tasks that offer opportunities for assessment:*

GRAPHS

Here is a graph. What does it tell you?

Hours Watching Television

Marcia
Joel
Terry
Paulo
Shana

DIVISION

Dave says $13 \div 4$ is $3\frac{1}{4}$. Martha says it's 3 R 1. Zach says they are both wrong. He thinks it's 3.25. Why are all three of them correct? Describe a situation where Dave's answer makes the most sense. Describe one where Martha's answer is reasonable and one where Zach's answer is reasonable.

BOXES

If 4 squares all the same size are cut from the 4 corners of a regular sheet of typing paper, the paper can then be folded up to make an open box. (a) What size should the squares be so the box has the largest volume? (b) What happens to your answer if you start with a different size piece of paper? Try a square piece and a half sheet of paper cut lengthwise. (c) For one of your pieces of paper describe a relationship between the size of the square and the volume of the box. What is the relationship between the size of the square and the surface area of the box?

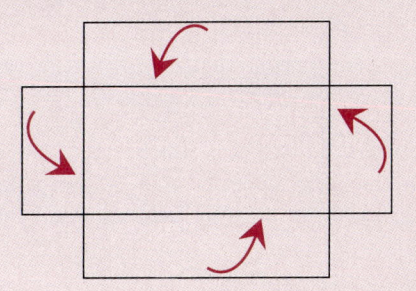

BALLPOINTS

At the Dollar Store, one dollar buys two ballpoint pens. The Pencil Point stationery store offers three of the same pens for two dollars.

(a) Which store has the better buy?
(b) Can you explain your answer in two different ways?

MYSTERY BAG

We reached in the bag of colored blocks 10 times. We got 3 reds, 1 yellow, and 6 green blocks. Each time we put the block back in the bag. If there are 20 blocks in the bag, what colors and how many of each do you think are in the bag? How sure can you be? Explain.

SHARES

Leila had 6 gumdrops, Darlene had 2, and Melissa had 4. They wanted to share them equally. How will they do it? Draw a picture to help explain your answer.

TOP AND BOTTOM

In the fraction $\frac{3}{4}$, how would you explain the meaning of the top number to a third-grader? Do the

---

*Many of these tasks are adaptations of tasks found in the *Vermont Mathematics Portfolio Project: Resource Book* (Vermont Department of Education, 1991); *Assessment Alternatives in Mathematics: An Overview of Assessment Techniques that Promote Learning* (Stenmark, 1989); *A Sampler of Mathematics Assessment* (Pandey, 1991); and *Mathematics Assessment: Myths, Models, Good Questions, and Practical Suggestions* (Stenmark, 1991).

same for the bottom number. Use two different ways to explain each.

FOUR LITERS

You have an 8-liter pail and a 3-liter pail, but you want exactly 4 liters of water. If you can fill them as often as you like, how can you measure exactly 4 liters?

SIX

Think about the number 6 broken into 2 different amounts. Draw a picture to show a way that 6 things can be in 2 parts. Think up a story to go with your picture.

TWO TRIANGLES

Tell everything you can about these two triangles.

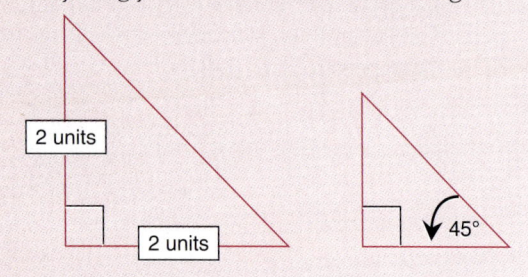

HAMBURGERS

You and a friend have been reading about McDonald's. The article says that 7% of Americans eat at McDonald's each day. Your friend thinks this is impossible. You know that there are about 250,000,000 Americans and 9,000 McDonald's restaurants. Is the article's statistic reasonable, or is your friend correct?

ALPHABET STATS

(Task given orally.)

(a) Predict on your own what you think are the five most used letters in the alphabet. Which do you think is used the most?

(b) Now work as a group, and answer the same questions.

(c) In a story book, find a sentence that is not too short. Write it down, and find out how many times each letter of the alphabet appears. Do this on your own, and then compile your results with the other members of the group.

(d) Here is a list that gives the accepted order of usage. How do your results compare?

E T A O N I S R H L D C U P F M W Y B G V K Q X J Z

(e) Who, where, and when would a person need to know this information?

TRIANGLES

List all the things you can think of that are alike and all the things that are different about these two triangles.

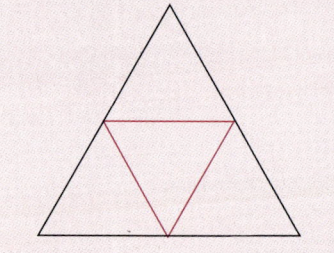

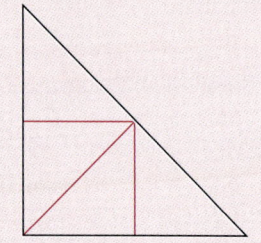

QUARG

As a visitor from another planet, Quarg is confused about our number system. He asks you if 7 + 24 is 724. How would you answer Quarg?

◆ Thoughts on the Nature and Use of Performance Tasks

The examples offered here were selected to illustrate the wide range of possibilities in performance-assessment tasks. Notice that some are process problems you might select to focus especially on strategies and other problem-solving processes. Some are directly related to content knowledge such as number concepts, place value, fractions, geometry, or measurement. Even here the students are left with a variety of approaches and/or answers. Some tasks are quite realistic or real "real world," and others are a bit more contrived. While realism is important, a more important feature is that students find the tasks engaging and thought-provoking. Some tasks will require a relatively short time for solution, while others are extensive explorations. Many tasks are ideally suited for group work, but certainly that is not a requirement.

More Than Just the Answer

Students will soon learn that you are interested in much more than the answers to these problems. Discussions should often focus on some of these features of the solution:

How did different students think about the task?

How did you approach it?

What was hard or easy?

What ideas did you try and abandon?

What pictures or models helped them?

What other ideas did the problem cause you to think about?

What other problems have you solved that were similar in some way?

How confident were you with your methods, with the results?

What did you like (dislike) about the task?

What did you learn while working on the task that you had not thought about before?

These are just a few ideas to begin to help you get information relative to the earlier lists of mental processes, knowledge base, and affective considerations. A class discussion can provide feedback. That is probably the way to begin getting students used to thinking about these ideas. Soon you want to have individuals or groups (or both) write down their solutions and to include some of these additional points. You may want to select one or two questions similar to those just listed for students to specifically respond to. Even students in the K–1 range should be encouraged to draw pictures to go with their solutions and to use these pictures in oral interactions. In the upper grades, students may be asked to write about their solutions and reflections as part of class or as a homework assignment.

CREATING PERFORMANCE TASKS

Where will you get all of the tasks needed for your grade level and your particular purposes? First, keep your eyes open, and steal every idea you can get your hands on. Good ideas can be found in articles in the *Arithmetic Teacher* and *Mathematics Teacher*, in the *Standards*, and in the *Addenda Series*.* File ideas by content area or write them in your plan book for future reference. When an idea looks interesting but is a bit too hard or too easy for your class, modify it. Almost every task in the collection provided here could be adjusted both up or down in difficulty level. (Try a few right now to get a feel for how this can be done.)

As textbooks begin to adapt more completely to a *Standards*-oriented curriculum, more and more of these tasks will appear in your books. In the meantime, it is not at all impossible to devise some open-ended challenges of your own. Here are just a few suggestions.

Change from Products to Explanations

When the student text is asking students for specific answers or skills, try selecting one example, and ask for explanations instead. For example, from a page of subtraction facts, pick two that are very different, and ask for different ways that could be used to think of the answers. Or, pick two related facts, such as 6 + 6 and 7 + 6, and ask how knowing one helps with the other. Instead of a series of adding fractions with common denominators, you might select $\frac{1}{4} + \frac{3}{2}$ and ask, "Why is a common denominator necessary?" or "Is there a way to get the answer without using common denominators?" or "Explain why the answer is the same if you use 8 as a common denominator." Given a routine word problem, supply the answer, and ask for a justification.

Use Models

Models (both drawings and manipulative materials) open up a wide range of problem situations. If you have used two or more different models to teach an idea, have students show one idea in two models and explain how they are alike and different. This approach works well with counters and number lines, with assorted materials for place value, with assorted fraction and percent models, or with different models for integers.

Another approach with models is to use "How many ways . . . ?" questions. "How many ways can you show what 3.25 means?" "How many ways can you separate 12 counters into 2 or more parts?" "How many ways can you find to measure the height of the ceiling in this room?"

You should certainly use models when asking for explanations. "Why is 2.4 more than 2.389?" You may want to specify a model, let the students select one, or ask for different presentations using different models.

Collect and Use a Variety of Contexts

To add a bit of real "real world" to your tasks, try starting with a situation, a job role, or data, and brainstorm ways to meld that with the content on which you are working.

Situations to think about might include mysteries; design competitions; treasure hunts; banks, stores, or warehouses; merit badges; and so on. At times the results may be contrived, but keep trying. Share initial thoughts with colleagues, and then share the results as well.

Job roles provide another way to tickle your creativity. Consider how stories and situations might be woven around some of these: museum curator (grant money,

---

*The *Curriculum and Evaluation Standards for School Mathematics Addenda Series* is a collection of 21 booklets at all grade levels, published by NCTM, to provide classroom activities that will assist teachers in the implementation of the *Standards*.

design of museum), engineer (build or design a bridge, a swimming pool, etc.), product designer (package design, research needs and uses), community planner (predict or describe future trends, amounts, ratios, appearances of buildings, space travel, etc.).

Charts, graphs, world records (sports, *Guinness Book of Records*, natural phenomena such as weather or earthquakes), census data (local and national), school and school district data, prices in ads, catalogs, or menus—all of these are data sets that open up all sorts of possibilities. You may want to use an idea from a real data collection and modify the numbers or the amount of information to better suit your grade level. This can be done with imaginary or futuristic names and places. Data can be used to have students make up their own problems, or you can pose your own open-ended questions. For example, given world high-jump records over the last 20 years, students could be asked to look for trends, to predict the record that will be set in the next Olympics, to make graphs or charts to illustrate the information, to plot a graph showing changes, and so on.

SCORING PERFORMANCE ASSESSMENT TASKS

It may be possible to get some feel for how your students are doing relative to mental process, knowledge base, and affective factors, but it is important to be specific about what we want to look for and what we see as students engage in these tasks. Specificity is important for you, the teacher, to be able to adjust and improve your instruction. It is also important for communicating to students (and parents) about how they are doing and what is expected of them. This process can be a quite different experience for you than simply grading papers and determining the percent correct. This section will offer some suggestions for your consideration. There is, however, no single right way or method, and you will need to adapt the ideas found here to specific situations.

Holistic Scoring

Holistic scoring is an approach to scoring students' performance by looking at the total effort and assigning a score, such as 0, 1, or 2, to the performance based on some predetermined set of criteria or rubric. A holistic approach can be used to assess any particular aspect of the process, especially mental processes and students' knowledge base. An *analytic scale* uses the holistic approach on several particular aspects that can be observed in one situation. The scoring system shown in Figure 5.1 can be used to analyze the three different parts of the problem-solving process. The descriptive phrases for each score constitute the rubric for these three scales. Notice that focusing on particular aspects of students' work provides more information than a single score. Using this particular scale, a score of 3, 3, 6 would mean something quite different from a score of 6, 6, 0, although both scores total 12 points. Note that this scale

Analytic Scoring Scale

Understanding the Problem	0 – Complete misunderstanding of the problem
	3 – Part of the problem misunderstood or misinterpreted
	6 – Complete understanding of the problem
Planning a Solution	0 – No attempt, or totally inappropriate plan
	3 – Partially correct plan based on correct interpretation of part of the problem
	6 – Plan could lead to a correct solution if implemented properly
Getting an Answer	0 – No answer, or wrong answer based on an inappropriate plan
	1 – Copying error, computational error, or partial answer for a problem with multiple answers
	2 – Incorrect answer although this answer follows logically from an incorrect plan
	3 – Correct answers and correct label for the answer

FIGURE 5.1. SOURCE: Adapted from Charles, Lester, & O'Daffer (1987) by Kroll, Masingila, & Mau (1992, p. 18).

Rubric for Scoring Understanding

6 Points In-depth	Demonstrates full understanding. Uses concepts and procedures correctly and goes beyond the required minimal explanations. Explains with appropriate models in a flexible manner.
5 Points Thorough	Shows basic understanding but does not go beyond minimal explanations. Can use models in explanations but is rigid in interpretation.
4 Points Satisfactory	Essentially correct understanding but not completely integrated with other ideas. Some minor points are missing or unclear. Not sure in use of models in explanations.
3 Points Gaps	Gaps in concepts or understanding are apparent. Ideas are not well connected to concepts or to other procedures.
2 Point Fragmented	Partial understanding. Arguments are very weak or are incomplete.
1 Point Little	Shows little evidence of understanding. Fails to use appropriate reasoning in explanations.

FIGURE 5.2: *A scale for scoring understanding of conceptual or procedural knowledge in a performance task.*

uses higher point values for understanding and planning, indicating that the teacher places a higher premium on those aspects of the process.

Understanding of Task
Source
- Explanation of task
- Reasonableness of approach
- Correctness of response leading to inference of understanding

Final Rating
1. Totally misunderstood
2. Partially understood
3. Understood
4. Generalized, applied, extended

Quality of Approaches/Procedures
Source
- Demonstrations
- Descriptions (oral or written)
- Drafts, scratch work, etc.

Final Rating
1. Inappropriate or unworkable approach/procedure
2. Appropriate approach/procedure some of the time
3. Workable approach/procedure
4. Efficient or sophisticated approach/procedure

Decisions along the Way
Source
- Changes in approach
- Explanations (oral or written)
- Validation of final solution
- Demonstration

Final Rating
1. No evidence of reasoned decision-making
2. Reasoned decision-making posssible
3. Reasoned decisions/adjustments inferred with certainty
4. Reasoned decisions/adjustments shown/explained

Outcomes of Activities
Source
- Solutions
- Extensions— observations, connections, applications, synthesis, generalizations, abstractions

Final Rating
1. Solution without extensions
2. Solution with observations
3. Solution with connections or application(s)
4. Solution with synthesis, generalization, or abstraction

Language of Mathematics
Source
- Terminology
- Notation/symbols

Final Rating
1. No or inappropriate use of mathematical language
2. Appropriate use of mathematical language some of the time
3. Appropriate use of mathematical language most of the time
4. Use of rich, precise, elegant, appropriate mathematical language

Mathematical Representations
Source
- Graphs, tables, charts
- Models
- Diagrams
- Manipulatives

Final Rating
1. No use of mathematical representation(s)
2. Use of mathematical representation(s)
3. Accurate and appropriate use of mathematical representation(s)
4. Perceptive use of mathematical representation(s)

Clarity of Presentation
Source
- Audio/video tapes (or transcripts)
- Written work
- Teacher interview/observations
- Journal entries
- Student comments on cover sheet
- Student self-assessment

Final Rating
1. Unclear (e.g., disorganized, incomplete, lacking detail)
2. Some clear parts
3. Mostly clear
4. Clear (e.g., well organized, complete, detailed)

FIGURE 5.3. SOURCE: Adapted from *Looking Beyond "The Answer."* The report of Vermont's Mathematics Portfolio Assessment Program: Pilot Year Report 1990–91.

Similar scales and rubrics can be devised to focus on other aspects of a performance task. For example, the scale in Figure 5.2 could be used for most concepts. By using general statements for the rubric, a student's understanding of various concepts can be compared over time, or growth within one concept area can be tracked.

The seven scales in Figure 5.3 were used by the state of Vermont in rating performance tasks. It is quite reasonable to select and assemble different scales for different tasks. During a particular period you may want to focus especially on mental processes. At other times you may wish to focus on concepts or procedures or some combination of factors. Select different scales at different times to obtain scores on those aspects of instruction that are currently important to you or for which you need information.

◆ Observations

Scales like those discussed so far are generally used when there is written work that can be scored out of class. Students' written work provides opportunity for your comments, more time for your analysis, and a record that can be used in grading if desired. This approach does take time and requires you to make certain inferences about students' thought processes.

You may want to modify the approach so that you can gather information as students are working. Make a chart with a space by each student's name for short comments. Be sure to date entries so that you will be able to tell later what problems were being worked on at the time. It is important to realize that not every student's work need be scored on every performance task, nor does every performance task need to be scored.

Another useful approach is simply to keep some index cards handy on which you make comments based on your observations. This is especially useful for reasoning and metacognitive processes and affective considerations. (See Figure 5.4.) By selecting a few things to be looking for and writing them down on your lesson plans, you will be reminded to look for these things and make notes. The card approach allows you to put the comments into individual student folders later for use in conferences with the student or the parents or for use in grading.

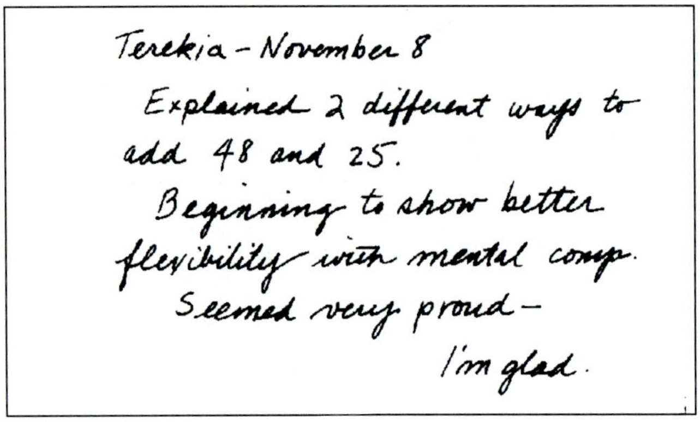

> *Terekia – November 8*
> *Explained 2 different ways to add 48 and 25.*
> *Beginning to show better flexibility with mental comp.*
> *Seemed very proud –*
> *I'm glad.*

FIGURE 5.4: *Comment cards are a good way to record anecdotal information about most any aspect of a student's performance.*

OTHER ASSESSMENT OPTIONS

Performance assessments are not the only way to gather information about your students. In this section a variety of other options are suggested. A good assessment plan will probably include a mix of approaches selected to serve different needs.

OBSERVING AND LISTENING

As your instruction becomes more student-centered, you will realize that students are constantly providing you with information about their knowledge and the processes they are using. The more actively we involve students, the more information they provide us.

◆ **Questioning and Responding**

We have a tendency in mathematics classes to ask questions that have very specific right/wrong answers. A further tendency is to explore or follow up only on incorrect responses. Students whose answers are challenged are more likely to change them than to defend them. A questioning technique that looks only for correct answers will fail to help you see how students are constructing ideas. It will intimidate less able students and will be dominated by the top students in your class.

In order to get more information during classroom interchanges we need to (1) develop an open-ended questioning technique and (2) use a more inquiring form of response, encouraging students to defend or explain both correct and incorrect answers.

Here are some examples of closed and open questioning for the same situations:

SECOND-GRADE SUBTRACTION ALGORITHM

$$72 \\ -45$$

Closed: You can't take 5 from 2, so what should we do with the 7?

Open: Is there anything special about this problem? If we are going to begin in the ones column, what do you think we should do first?

FOURTH-GRADE ONE-STEP TRANSLATION PROBLEM (OR SIMPLE WORD PROBLEM)

Closed: What kind of a problem is this: add, subtract, multiply, or divide?

Open: What could we do to help us understand this problem? Could you draw a picture, perhaps using a number line? Explain how your picture is the same as the problem. What operation goes with your picture?

SIXTH-GRADE MEASUREMENT: SELECTING AN APPROPRIATE UNIT

Closed: What unit should be used to measure the length of the room?

Open: How would you measure the length of this room? What choices of units do you have? Why would some units seem more appropriate than others?

An open question will encourage students to think about several related ideas. It will implicitly or explicitly indicate that there may be more than one way to

5 / ASSESSMENT IN THE CLASSROOM 75

respond. It may indicate that the answer can be figured out from what is known and does not require rote memorization.

The second part of good questioning is to respond to students in a manner that helps them think and lets you see what they are thinking. You may like some of the following techniques or develop some similar ones of your own:

Waiting or Nonjudgment. An immediate judgment of a response stops any further pondering or reflection on the part of the students. If, after a student answers, you simply nod thoughtfully, or say, "Uhuh," or "Hmm," the student is encouraged to elaborate on the response. Other students can join in, offer alternative ideas, or continue to think. A similar response is, "That's a good idea. Who else has an idea about this?" When used consistently, this response tells students that the time for thinking is not yet over.

Request for Rationale. Most students have learned to expect judgmental responses from teachers and to have teachers tell them if their answer is correct. These students are initially shocked when teachers begin to respond to their answers with phrases such as the following:

"That's a good idea. Why do you think that might be right?"

"Are you sure? How do you know that?"

"I see. Can you explain your answer to the rest of us?"

"Show us how you figured that out. Maybe you could use these base-ten blocks to help you."

These responses should be used frequently and with both correct and incorrect answers. Soon the request for self-validation will become the expected norm, and students even in the lower grades will begin to offer unsolicited explanations along with their answers.

Search for Alternative Ideas. A similar approach is to accept the first response without judgment and then ask if other students may have a different idea. "I see. That's one possible idea. Who else has an idea about this?" Alternative responses by different classmates can then be evaluated in a discussion that looks at the rationale for each. Allow plenty of wait time for the second and third responses, because it is during this time that slower or even average students are given an opportunity to think through the idea. Accept repeat answers, since they are quite likely original thoughts and not copies. Students frequently pay little attention to the response of others because they are busy working through the idea on their own.

Use of Models

One of the most startling effects of using models in a classroom is the amount of information it provides the teacher almost at a glance. When students are engaged in a task at their desks with a model, casually walking about the classroom permits the teacher to see how each student is dealing with the concept at hand. When students are comfortable with an idea, the use of the model is very likely to be self-assured and direct. When, on the other hand, they are experiencing some difficulty, it will become very evident. It is hard to cover up or disguise lack of understanding when a model is being used.

Written Work

Written work provides a ready source of information about your students. Most of the time this information is much more than just the number of items correct.

In Figure 5.5(a), two exercises from the same worksheet are incorrect. However, the first is a simple counting error, while the second indicates some misconnection between base ten concepts and the written numerals. Figure 5.5(b) shows several errors that might be made on a worksheet involving integer computation. Some of these may be due to carelessness, basic fact difficulties, incorrect application of rules for integers, or a variety of conceptual problems. A look at similar exercises would help decide what sort of difficulty the student may be having. In either example, to simply count each exercise as wrong and to ignore the distinction between the errors is not taking advantage of feedback that should be used to guide instruction or even to assign grades.

Recording Observations

Much of what you hear and observe will immediately guide your instruction or help you plan more effective tasks for subsequent lessons. However, if you occasionally use some form of recording along with your observations, you will be more focused in what you are looking for, and you will have data that you can later use in discussion with parents or long-range planning. By making some sort of check-

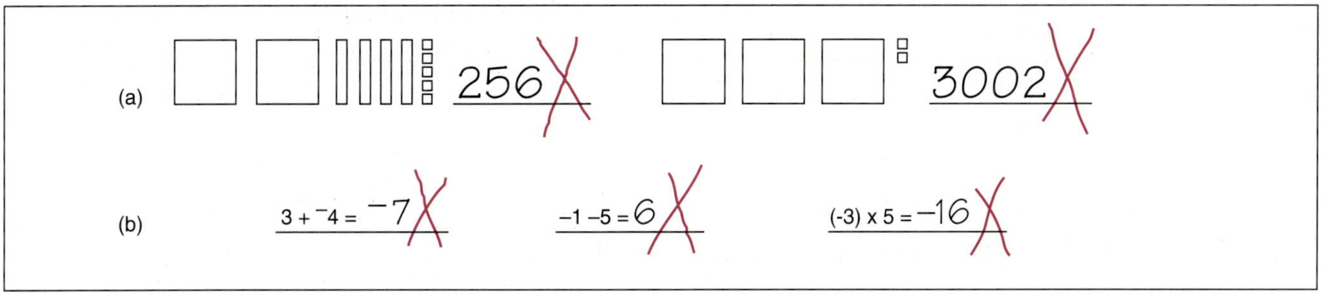

(a) [base-ten blocks image] 256 ✗   [base-ten blocks image] 3002 ✗

(b) $3 + {}^-4 = {}^-7$ ✗ ${}^-1 - 5 = 6$ ✗ $(-3) \times 5 = -16$ ✗

FIGURE 5.5: *Worksheet errors*

list that you keep handy, you can record information without interfering with your instruction. In Figure 5.6, two approaches are suggested. The first is designed to record observations about a number of broadly based objectives and would require several pages to include names for all students. Since these are the types of objectives that tend to cross specific content, you might use this form of recording over a period of several weeks or more. The objectives listed in the figure are simply suggestions and should be modified to suit your particular purposes.

The second form is more focused. All students are listed on one page, and only one objective per page is addressed. What you are looking for is more likely to be tied to specific content. Here are other examples of things for which you might use a form like this:

Uses sets to explain equivalent fraction concepts

Can use missing part thinking for subtraction facts

Selects appropriate metric unit for measuring lengths

BROAD CHECKLIST	Group 4			
	Dustin	Felando	Angelo	Shannon
Processes				
Justifies answers				
Listens carefully to others and evaluates				
Reflects on and explains choices				
Appears to monitor personal progress				
Knowledge				
Explains concepts easily				
Uses models appropriately				
Can explain procedures				
Affect				
Willing to tackle hard tasks				
Perseveres				
Enjoys solving problems and/or making explanations				
Shows confidence				

FOCUSED CHECKLIST

Names	Shows evidence of good number sense with fractions Comments-Dates
Demetria	
Thomas	
Sheena	
Calvin	

FIGURE 5.6: *Two forms of observation checklists*

Explains mental computation methods

Can relate two different models for place value (*or other topic*)

Solves problems involving proportions and explains reasoning

Clearly affective and process objectives could be included here as well.

STUDENT SELF-ASSESSMENT

Stenmark (1989) notes that "the capability and willingness to assess their own progress and learning is one of the greatest gifts students can develop. . . . Mathematical power comes with knowing how much we know and what to do to learn more" (p. 26). A quick and simple self-assessment is a good way to have students reflect on a unit of study or a particular lesson or activity.

Self-assessment can focus on most any aspect of learning mathematics. Usually some form of checklist or a few short open-ended questions are useful.

Looking at the Knowledge Base

- What did you learn this week about fractions?
- What did you find difficult (easy) about the work you did on multiplication?
- How does this activity relate to things we have done before?
- What new questions did this activity make you think about?
- What one thing about division do you think you may need more work on?

Looking at Processes and Group Activity

- Did you spend time trying to understanding this problem before you began to work on it?
- Which of these strategies seems hard for you: working backward, try-and-adjust, looking for a pattern? Why is it hard?
- What did you do to be sure your answer was correct?
- What worked well in your group today?
- How could you help the group work better?
- How did you help the group?

Looking at Affective Issues

- When we do problems in mathematics I usually feel...
- I like doing mathematics (or I do not like doing mathematics) because . . .

Attitudes and beliefs may be difficult for children to write about. An inventory where they can respond *Yes, No,* or *Maybe* to a series of statements is another approach. Encourage students to add comments under an item if they wish. Here are some items you could use to build such an inventory:

_____ I feel sure of myself when I get an answer to a problem.

_____ I sometimes just put down anything so I can get it over with.

_____ I like to work on really hard math problems.

_____ Math class makes me feel nervous.

_____ If I get stuck, I usually just quit or go to another problem.

_____ I am not as good in math as most of the other students in this class.

_____ Mathematics is my favorite subject.

_____ I do not like to work at problems that are hard to understand.

_____ Memorizing rules is the only way I know to learn mathematics.

_____ I will work a long time at a problem until I think I've solved it.

A simple technique is to ask students to write a sentence at the end of any work they do in mathematics class saying how that activity made them feel. Young children can draw a face on each page to tell you about their feelings.

In Chapter 4, Table 4.1, there is a self-reporting checklist to help students reflect on their metacognitive activities. You may want to use some form of that checklist in your assessment scheme for problem solving.

WRITING AND REPORTING

Having students write in mathematics is a natural way to help them reflect on what they are learning as well as a good way for you to see what they know or feel about a subject. Writing (or oral reporting) is a form of communication, a major objective for mathematics instruction. It helps students make connections and helps teachers see what connections exist.

Student writing can take on many forms.

- At the end of a problem, students can give an explanation of their solution. Marilyn Burns' books include many examples where students write down answers and then follow with, "We think this because . . ." The explanations include drawings, computations, graphs, . . . anything that helps explain the process. These are frequently followed with comments by the students about how they liked the task they just worked on.

- Students can be asked to write an explanation for others or for students in an earlier grade. For example,

"Write an explanation of why 4 × 7 is the same as 7 × 4 and why this works for 6 × 49 and 49 × 6."

- In language education, the notion of publishing a story is often used. Published stories are displayed prominently for all to see. Students are required to edit and revise stories before publication. Similar ideas can be used in mathematics. Children can publish stories about the solution to a problem, a pattern they have discovered on the hundreds board or during a geometry lesson, why two ideas are related, how a particular idea is used in a story found in the newspaper, and so on. The story or idea is turned in or shared with peers for feedback. A revision is made based on the feedback. Final editing, including the addition of necessary drawings, is completed before actual publishing.

- Students can write short papers or write in their journals about how they feel. For example, "Write to [a teacher, friend, grandparent] describing how you feel about the work we did today in mathematics." Other starters include: "What I am most confused about is . . ." "Today in mathematics I learned . . ." "When we work in groups in mathematics I feel . . ."

Within the writing process it is hard to separate learning from assessment, which is how assessment should be. Writing requires careful reflection and organization of ideas. It includes many forms of communication and linkages between ideas and symbols. It allows children to use their own creativity. All of these things and more contribute to your own understanding of the children in your class.

TESTS

A test is nothing more than an assessment that is prepared in common for all students in the class. With that definition it need not be a collection of low-level skill exercises that simply require students to master a procedure with no requirement for understanding. While simple tests of computational skills may have some role in your classroom, the use of such tests should certainly be limited. Like all other forms of assessment, tests should reflect the goals of your instruction. With relational understanding as your goal, tests should be designed to find out what concepts students have and how their ideas are connected. Tests of procedural knowledge should go beyond just knowing how to perform an algorithm and should allow and require the student to demonstrate a conceptual basis for the process.

Examples

1. Write a multiplication problem that has an answer between the answers to these two problems:

   $$\begin{array}{cc} 49 & 45 \\ \times 25 & \times 30 \end{array}$$

2. (a) In this division exercise, what number tells how many tens were shared among the 6 sets?

   (b) Instead of writing the remainder as "R 2," Elaine writes "$\frac{2}{3}$". Explain the difference between these two ways of handling the leftover part.

   $$\begin{array}{r} 49\ R2 \\ 6\overline{)296} \\ 24 \\ \overline{56} \\ 54 \\ \overline{2} \end{array}$$

3. On the square grid draw two figures with the same area but different perimeters. List the area and perimeter of each.

4. For each subtraction fact, write an addition fact that helps you think of the answer to the subtraction.

   $$\begin{array}{cccc} 12 & 9 & 9 & 14 \\ -3 & +3 & -4 & -7 \\ \overline{9} & \overline{12} & & \end{array}$$

5. Draw pictures of arrows to show why $^-3 + {}^-4$ is the same as $^-3 - {}^+4$.

Some Testing Options

As indicated by the previous examples, tests can include more than simple computations, and there is much more information to be found in a good test than simply correct or incorrect answer getting. Here are some things to think about when constructing a test.

1. Use calculators all the time. Except for the simplest computation tests, the calculator allows students to focus on what you really want to test. It also communicates a positive attitude about calculator use to your students.

2. Use manipulatives and/or drawings. Students can use appropriate models to work on test questions when those same models have been used to develop concepts. Note the use of drawings in Example 5 above. Simple drawings can be used to represent counters, base ten pieces, fraction pieces, and the like. Teach students how to draw the models before asking them to draw on a test. (See drawings on p. 79.)

3. Include opportunities for explanations.

4. Assess affective factors on the test. There are many ways to do this. One idea is to have students rate their personal confidence level on each question as A, B, or C. Another idea is to have students draw a face showing how they feel about the question or the test in general. Ask students to complete a sentence that begins "I liked this test because . . ." or "I did not like this test because . . ."

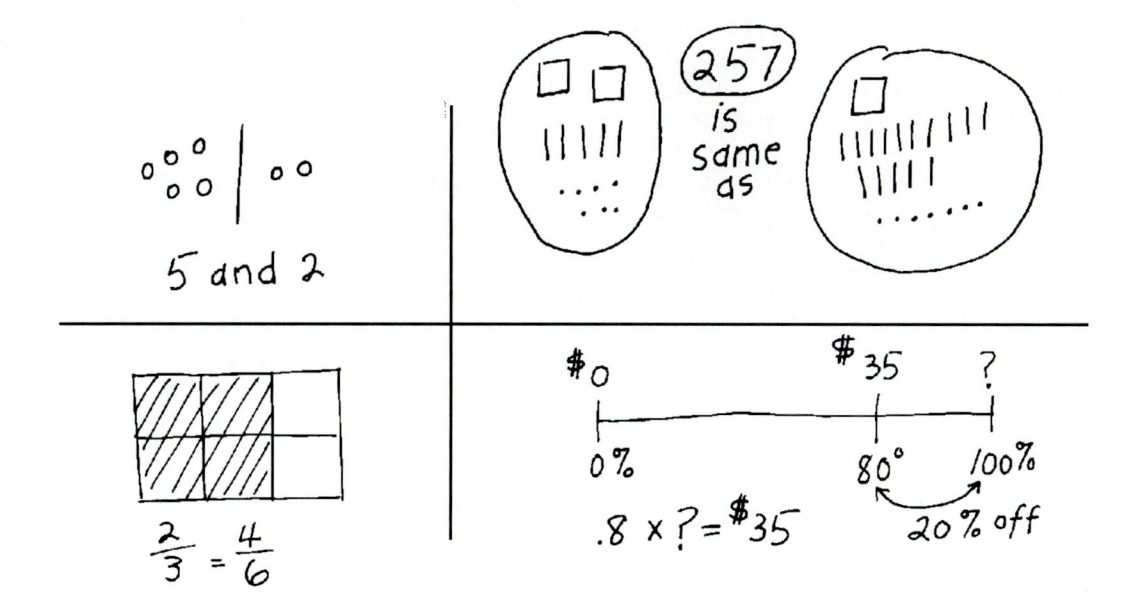

5. Try to avoid "preanswered" tests. These are tests with only one correct answer, whether it is a calculation, a multiple choice, or a fill-in-the-blank. Tests of this type tend to fragment what children have learned and hide most of what they know. Rather, construct tests that allow students the opportunity to show what they know.

◆ What about Timed Tests?

"Speed with arithmetic skills has little, if anything, to do with mathematical power. The more important measure of children's mathematical prowess is their ability to use numbers to solve problems, confidently analyze situations that call for the use of numerical calculations, and be able to arrive at reasonable numerical decisions they can explain and justify" (Burns, 1989, p. 5). Reflection on this quotation should give us pause whenever we are tempted to give a timed test. Reasoning and pattern searching is never facilitated by restricting time. Some children simply cannot work well under pressure or situations that invoke stress. For those children a timed test is certainly not one that is attempting to find out what they know.

Speed tests have been most popular for tests of basic facts. While speed encourages children to memorize facts, speed in a testing situation is debilitating for many and provides you, the teacher, with no positive benefits. Very short (one row of eight facts) speed drills, if presented in a lighthearted manner, may be useful. Drills of this nature should not, however, be used as assessments and certainly not for grading purposes.

In every instance, timed tests reward few and punish many. They can have a lasting negative impact on student attitudes. They should be avoided whenever possible. Speed is much less important in real-life situations than reasoning and a willingness to attack difficult situations. Mathematics *as problem solving, as communication,* and *as reasoning* cannot be assessed in a timed situation.

PORTFOLIOS IN THE ASSESSMENT PLAN

A *portfolio* is an assembly of many types of materials selected with both student and teacher input, designed to demonstrate progress and growth in student attitudes, understandings, and problem-solving processes. The materials in the portfolio may include assignments, projects, reports, student writings, worksheets from texts or other sources, comments by teachers, observations from interviews, self-evaluations of group and individual efforts, and so on. The portfolio captures the essence of the assessment agenda by providing a comprehensive and open-ended forum for students to demonstrate what they know as opposed to what they do not know. It is an excellent form of communication between student and teacher, student and parent, and parent and teacher.

PORTFOLIO CONTENT SUGGESTIONS

There are many different ways to design a portfolio assessment program for your classroom, and you should not feel confined by rigid prescriptions. Many states are developing portfolio assessment programs to be used as a main component of their state assessment plans. State models provide teachers with good ideas for their own portfolio projects. However, external assessments may appear overly prescriptive or rigid in designating portfolio contents or other aspects of portfolios and consequently may not serve your particular needs or agendas. Your own portfolio program should be carried out to serve your needs and to communicate with your students and parents. It should be designed with particular goals in mind. These plans and goals should be discussed with your students from the outset so they are fully involved in the process.

Table 5.1 is taken from the superb NCTM booklet, *Mathematics Assessment: Myths, Models, Good Questions, and*

GOAL	EVIDENCE, EXAMPLES, AND COMMENTS
Positive mathematical disposition 　Motivation 　Curiosity 　Perseverance 　Risk taking 　Flexibility 　Self-responsibility 　Self-confidence 　. . .	◆ Journal entries depicting enthusiasm for mathematics ◆ Photographs of large, colorful mathematical graphics by students ◆ Problem solution with an added paragraph beginning with "On the other hand . . ." or "What if . . . ." ◆ Log of a week's or month's work showing a single important problem or investigation worked on over a period of time ◆ Homework paper with a description of several approaches to a problem ◆ Student-written planning calendar outlining work to be done ◆ Mathematics autobiography . . .
Growth in mathematical understanding 　Concept development 　Problem-solving skills 　Communication skills 　Construction of mathematics 　Reflection on approaches and solutions to problems or tasks 　. . .	◆ Similar items collected at regular intervals from the beginning of the year ◆ Written explanation of why an algorithm works ◆ Diagram, table, or similar organized representation that clarifies a problem situation ◆ Solution that defines assumptions, includes counter-examples ◆ Photographs of a mathematics project ◆ Journal entries delineating solution justifications and variations in strategies ◆ Student identification of papers that need more work, with reasons ◆ A paper that starts with "Today in math class I learned . . ." ◆ Inclusion of drafts as well as finished work . . .
Mathematical reasoning in a variety of mathematical topics 　Estimation, number sense, number operations, and computation 　Measurement, geometry, and spatial sense 　Statistics and probability 　Fractions and decimals 　Patterns and relationships 　. . .	◆ Report on an investigation (e.g., number patterns in sums of sequential numbers) ◆ Student-planned statistical survey, with accompanying graphic displays ◆ Written report of a probability experiment and accompanying theoretical design ◆ Response to an open-ended question regarding measurement of geometric shapes ◆ Student explanation of what $\frac{1}{2}$ minus $\frac{1}{3}$ means ◆ Diagram examples of multiplication using a number line, a rectangular array, and repeated groups of physical objects ◆ Annotated drawing illustrating the Pythagorean theorem ◆ Representation of an area model solution for a statistical problem . . .
Mathematical connections 　Connecting a mathematical idea to other mathematical topics, to other subject areas, and to real-world situations 　. . .	◆ Papers that show authentic use of mathematics in other curricular areas such as science or social studies ◆ Student reflections on how mathematics is meaningful as it is used in the adult world ◆ Examples from nature of occurrences of Pascal's triangle number patterns ◆ A report on the relationship among arithmetic, algebra, and geometry with demonstrative examples on a coordinate grid ◆ Student-constructed table of equivalent fractions, decimal numbers, and percents, with examples of where each type of number is used ◆ Mathematical art project ◆ Report about a person in history or personally known who contributed to mathematics . . .
Group problem-solving 　Development of skills in working with others 　Communication 　. . .	◆ A task design or plan ◆ Group paper that includes the names of the members of the group and the tasks each did ◆ Group self-assessment sheet ◆ Videotape or audiotape of group working on problems or making oral reports ◆ Group-work report of trying a second or third strategy applied when the first one didn't work
Use of tools 　Integration of technology—use of calculators and computers and so on 　Use of manipulatives 　. . .	◆ Computer-generated statistical analysis of a problem ◆ Frequent mention of use of calculators on open-ended problems ◆ Diagrams representing use of manipulative material . . .
Teacher and parent involvement 　Communication between teacher and parent and between parent and student 　Parental understanding of educational objectives and values 　. . .	◆ Consistency of program demonstrated by items from every grading period ◆ Anecdotal report ◆ Informal assessment sheet ◆ Interview of student ◆ Teacher- or parent-written comments ◆ Assessment by teacher of student work ◆ Student presentation of portfolio to parent during parent-teacher conference . . .

TABLE 5.1: *Goals and Suggested Portfolio Contents.* SOURCE: From the *National Council of Teachers of Mathematics,* copyright 1991. Reprinted by permission.

Practical Suggestions (Stenmark, 1991). It provides ideas for the type of entries that would be included to assess a variety of important goals. Any given portfolio will include only a sampling of these ideas and may contain only 8 to 10 entries. You may decide that for a particular portfolio you want to stress only a few goals or a broad range of goals.

PORTFOLIO MANAGEMENT

At first, the idea of keeping a portfolio for every child may be a bit daunting. Included in this section are some ideas found in the Stenmark booklet.

Use a Working Portfolio

Set up folders for each student, and keep them in a place where students can easily get to them. This working portfolio will hold all of the student's work over the period as well as journal entries, writings, tests, self-assessments, or project materials. When students revise or correct work, the original and the revisions should be kept to indicate progress. The working portfolio is also the place where your observation notes or other items that originate with you can be kept. When groups turn in reports or projects, make copies for each group member's work portfolio.

At the end of the unit or grading period, or whenever it is time to assemble the final portfolio, students select and clip together the items from the working portfolio that they wish to be included. You will want to have a class discussion about how to select items and what the items should show. Encourage children to pick things that tell about themselves in mathematics; their strengths, weaknesses, points of pride, places they need help, and progress made over the period. Students should include a paragraph about each of their selections and why they were included.

After the students have made their selections, you should select additional items that you want included. Mark these in some way to note that you selected them and not the student. These selected works go into the assessment or final portfolio. Students then review their portfolios and write a cover letter that discusses the contents as a whole and includes their personal perspective on things such as problem-solving skills, confidence, attitudes, growth in understanding, and so on. This portfolio is now ready for evaluation and subsequent sharing with parents. Materials remaining in the work portfolio should be sent home.

Portfolio Evaluation

You will probably have made comments on most or all of the materials as they were completed. Now you are looking at the portfolio as a whole. Whatever your approach, it is very important to provide feedback. The portfolio may or may not be a component of your grading scheme. It is an excellent way to encourage students and to build confidence. Your comments should be made with this in mind. If you plan to include the portfolio in your grading scheme, some type of rating scale will be useful.

Regardless of the purpose, you will find it useful to establish some focus or criteria for your evaluation before you begin. A form can be devised with several criteria specified and a place for written comments under or next to each. Stenmark's book suggests the following criteria but cautions that you select only a few criteria to review at one time (p. 43):

- Formulates and understands the problem or task
- Chooses a variety of strategies
- Shows development of mathematical concepts
- Constructs mathematical ideas—inventing, discovering, extending, integrating, connecting, critiquing ideas and procedures
- Shows thinking and reflection involved in mathematical reasoning, conjecturing, exploring, and processing
- Uses appropriate mathematical language and notation
- Interprets results—verifying, summarizing, applying to new cases
- Shows development of group problem-solving skills
- Relates mathematics to other subject areas and to the real world
- Shows development of positive attitudes—confidence, flexibility, willingness to persevere, appreciation of the value and beauty of mathematics
- Shows evidence of self-assessment and self-correction of work

The criteria should be discussed with the class so that they understand your values and what the criteria actually mean. The comments you make should be included in the portfolio and can be used in your holistic scoring scheme if you intend to use the portfolio for grades.

DIAGNOSTIC INTERVIEWS

An interview is simply a one-on-one discussion with a child to help you see how children are thinking about a particular subject, what processes they use in solving problems, or what attitudes and beliefs they may have. A structured interview may be as short as five to ten minutes. More open-ended or exploratory interviews may last as long as a half hour.

REASONS TO CONSIDER AN INTERVIEW

There are several reasons why it may occasionally be well worth the time and effort to conduct interviews. The most obvious reason is that you need more information concerning a particular child and how he or she is constructing concepts or using a procedure. Remediation will almost always be more successful if you can pinpoint *why* a student is having difficulty before you try to fix the problem.

A second reason is to get information either to plan your instruction or to assess the effectiveness of your instruction.

For example, are you sure that your students have a good understanding of equivalent fractions, or are they just doing the exercises according to rote rules? At the end of Chapter 10, a collection of tasks is suggested to get at students' understanding of place value. Concept tasks such as these might be used in interviews conducted at the beginning of a unit to see what ideas you need to work on. They can be used later in the unit or at the end to assess growth or retention of ideas. Be sure not to include diagnostic tasks as instructional exercises. Practice with the tasks may make them routine and thereby mask understanding.

Another reason is to grow as a teacher, to learn more about the way children learn and think. You might decide to conduct at least one interview a week, changing the topic and the children you interview. In addition to growth in your knowledge of children, you will also grow in your ability to conduct good interviews.

PLANNING AN INTERVIEW

There is no magic right way to plan or structure an interview. In fact, flexibility is a key ingredient. You should, however, have some overall game plan before you begin, and be prepared with key questions and materials. Figure 5.7 is a basic skeleton of one interview strategy. Depending on the topic and situation, some elements may be omitted or others added.

Begin an interview with questions that are easy or closest to what the child is likely to be able to do, usually some form of procedural exercise. For numeration or computation topics, begin with a pencil-and-paper task such as doing a computation, writing or comparing numerals, or solving a simple translation problem. When the opening task has been completed, ask the child to explain what was done. "How would you explain this to a second-grader (or your younger sister)?" "What does this (point to something on the paper) stand for?" "Tell me about why you do it that way." At this point you may try a similar task but with a different feature; for example, after doing 372 − 54 try 403 − 37. The second problem has a zero in the tens place, a possible source of difficulty.

The next phase of the interview involves the use of models or drawings that the child can use to demonstrate understanding of the earlier procedural task. Computations can be done with base-ten materials, blocks or counters can be used, number lines explored, grid paper used for drawing, and so on. Be careful not to interject or teach. The temptation to do so is sometimes overwhelming. Watch and listen. Next, explore connections between what was done with models and what was done with pencil and paper. Many children will do the very same task and get two different answers. Does it matter to the child? How do they explain the discrepancy? Can they connect actions using models to what they wrote or explained earlier?

Alternative beginnings to an interview include making an estimate of the answer to either a computation or a word problem, doing a computation mentally, or trying to predict the solution to a given task. Notice that the interview does not generally proceed the way instruction does. That is, in an interview, the conceptual explanations and discussions in general come after the procedural activity. Your goal is not to use the interview to teach but to find out where the child is in terms of concepts and procedures at this time.

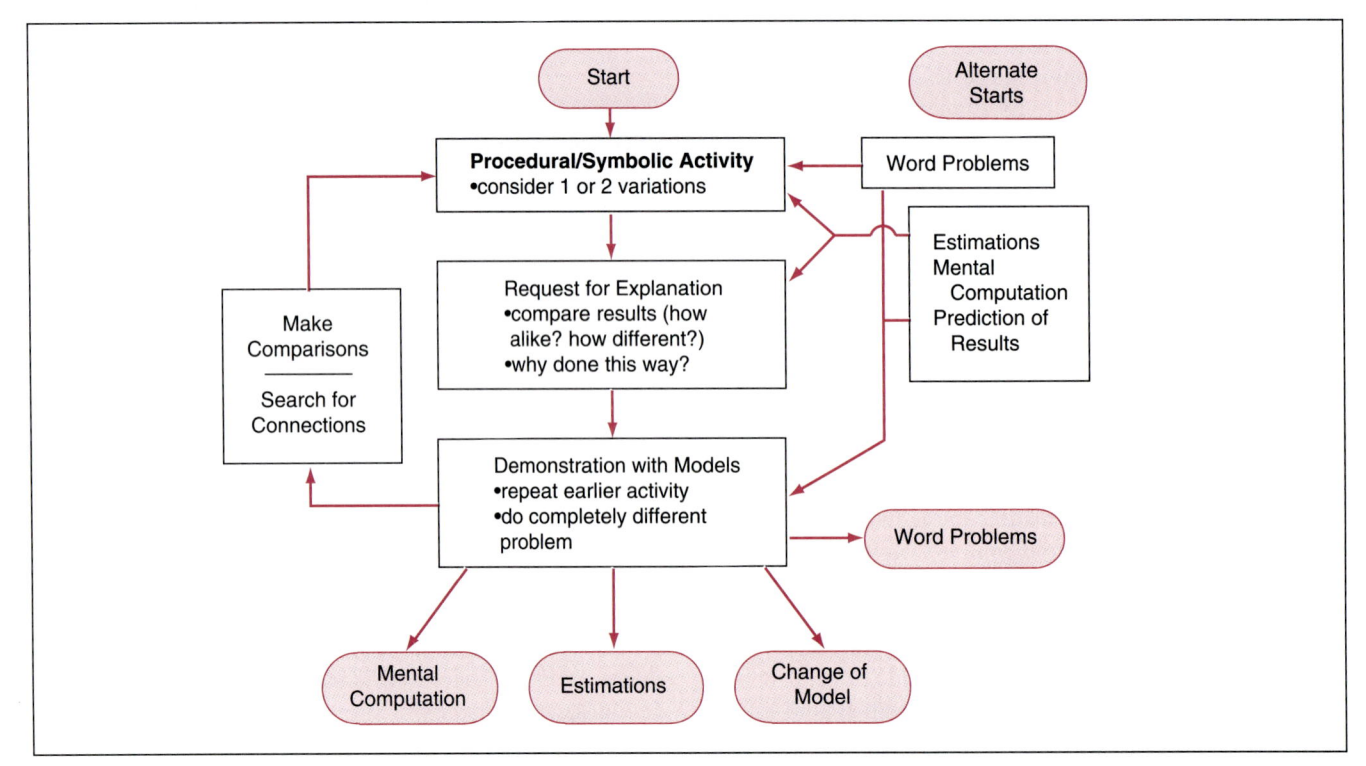

FIGURE 5.7: *A skeleton plan for a diagnostic interview*

SUGGESTIONS FOR EFFECTIVE INTERVIEWS

The following suggestions have been adapted from excellent discussions of interviewing children by Labinowicz (1985, 1987), Scheer (1980), and Liedtke (1988).

1. **Be accepting and neutral as you listen to the child.** Smiles, frowns, and other body language can make a child think that the answer he or she gave is right or wrong. Develop neutral responses such as, "Uhuh," "I see," or even a silent nod of the head.

2. **Avoid cuing or leading the child.** "Are you sure about that?" "Now look again closely at what you just did." "Wait. Is that really what you mean?" These responses will indicate to children that they have made some mistake and cause them to change their responses. This can mask what they really think and understand. A similar form of leading is a series of easily answered questions that direct the student to a correct response. That is teaching, not interviewing.

3. **Do wait silently.** Wait after you ask a question. Give the student plenty of time before you try a different question or probe. After the child makes a response, wait again! This second wait time is even more important because it allows and encourages the child to elaborate on the initial thought and provide you with more information. Wait even when the response is correct. Waiting also can be relaxing and give you a bit more time to think about the direction you want the interview to take. Your wait time will almost never be as long as you imagine it is.

4. **Do not interrupt.** Let children's thoughts flow freely. Encourage children to use their own words and ways of writing things down. Interviewing with questions or by correcting language can be distracting to the child's thinking.

5. **Phrase questions in a directive manner to prevent avoidance.** Use "Show me . . . ," "Do . . . ," or "Try . . . ," rather than "Can you . . . ," or "Will you . . ." In the latter form, the child can simply say "no," leaving you in a vacuum.

6. **Avoid confirming a request for validation.** Students frequently follow answers or actions with, "Is that right?" This query can easily be answered with a neutral, "That's fine," or "You're doing OK," regardless of whether the answer is right or wrong.

Interviewing is not an easy thing to do well. Many teachers are timid about it and fail to take the time. But not much damage is possible, and the rewards of listening to children, both for you and your students, are so great you really do not want to pass it up.

GRADING

Myth: A grade is an average of a series of scores on tests and quizzes. The accuracy of the grade is dependent primarily on the accuracy of the computational technique used to calculate the final numeric grade.

Reality: A grade is a statistic that is used to communicate to others the achievement level that a student has attained in a particular area of study. The accuracy or validity of the grade is dependent on the information that is used in preparing the grade, the judgment of the grader, and the alignment of the assessments with the true goals and objectives of the course.

CONFRONTING THE MYTH

Most experienced teachers will tell you that they know a great deal about their students in terms of what they know, how they perform in different situations, their attitudes and beliefs, and their various levels of skill attainment. Good, experienced teachers have always been engaged in ongoing performance assessment, albeit informal and usually nondocumented. In the past, however, even these good teachers believed in the test score as the determinant of grades.

Frequently, they found themselves hoping that the tests would produce scores that agreed with what they already knew about their students. A student's low test score might be met with laments such as, "If he hadn't made that careless error," or "Johnny really understands this material, he just gets nervous on tests," or "She's my best problem solver. This test just didn't ask the things she is good at." An unusually high score for a student can cause similar difficulties. "He got the right answer, but he clearly did the problem wrong. The answer just happened to be right," or "He only memorized the rules. He never is able to explain things in class." For these teachers, adherence to the myth of test scores and grades places them in a moral dilemma of having to assign grades they know are not true indicators of student achievement. When the tests tend to agree with what is known, the teacher has to wonder what was gained by giving the test other than to produce numbers in the grade book.

The myth of grading by statistical number crunching is so firmly ingrained in schooling at all levels you may find it hard to abandon. If one thing is clear from the discussions in this chapter, it should be that it is quite possible to gather a wide variety of rich information about students' understanding, about their attitudes and beliefs, and about their mental processes. To ignore all of this information in favor of a handful of numbers on tests, tests that usually focus on low-level skills and that almost inevitably ignore cognitive processes, understanding, attitudes, and beliefs, is certainly not fair to students, to parents, or to ourselves as teachers.

This chapter has been about assessment, the gathering of

information to find out what our students know. A subjective grading decision made on the basis of daily observations, student projects, group reports, student self-evaluations, and portfolio evidence of growth will undoubtedly be much more valid than any test average computed to three decimal places. Remember that there is always enormous subjectivity in construction of the tests and the determination of the numeric scale. Let's admit that grading is subjective and arm ourselves with the best information available.

GRADING ISSUES

When we use the information of alternative assessments in assigning grades we still must make some hard decisions. Some are philosophical, some require school or district agreements about grades, and all require us to look at our objectives.

Alignment

The grades you assign should reflect all of your objectives. If affective considerations have been important this grading period, they should be reflected in the grades. The same is true for metacognitive skills, for perseverance, for appropriate use of models, for going beyond the required, for creativity, and so on. If you are restricted to a single grade for mathematics, different factors probably have different weights or values in making up the grade. Student X may be fantastic at reasoning and truly love mathematics, yet very weak in traditional skills and maybe careless. Student Y may be mediocre in problem solving but possess good communication skills. How much weight should you give to cooperation in groups, or for written versus oral reports, or for computational skills? There are no simple answers to these questions. However, they should be addressed at the beginning of the grading period and not the night you set out to assign grades.

A multidimensional reporting system is a big help. If you can assign several grades for mathematics and not just one, then your report to parents is more meaningful. Even if the school's report card does not permit multiple grades, you can devise a supplement indicating several grades for different objectives. A place for comments is also helpful. This form can be shared with students periodically during a grading period and can easily accompany a report card.

Grading Scales

Many school systems have "standard" grading scales: "Above 94 is an A, above 87 a B, ..." These scales tend to perpetuate the myth that numbers are needed to produce grades. Such a scale need not deter you from using alternative assessment strategies. Simply transfer your subjective grade to a corresponding number. For example, a "high B" might be a 92.

Such games with numbers do not, however, solve the more important issues surrounding grading scales:

- Is progress to be considered or only final achievement?
- Is ability a factor, or must all students be evaluated on the same scale?
- When different students study different material, are the criteria the same? That is, is excellent performance on material that is perhaps below grade level deserving of the same grade as excellent performance on advanced work?
- Should special-needs children mainstreamed into your classroom be graded the same as all of the rest? To what degree are disabilities to be taken into account?

These and similar issues should be addressed by the faculty in your school and should be clarified in the reports that you send beyond the classroom. As grading goes beyond the average of test scores, the need for more information than can be delivered by a single number or letter becomes apparent.

REFLECTIONS ON CHAPTER 5: WRITING TO LEARN

1. Summarize the changes in emphasis in the area of assessment suggested by the NCTM *Standards*.
2. What does this mean: "Assessment should attempt to find out what students know instead of what they do not know"?
3. What are some of the kinds of things we should be looking for in our assessment plan under each of the categories of Mental Processes, Knowledge Base, and Affective Factors?
4. What is a performance assessment?
5. Describe *holistic scoring,* and explain how this can be used with performance assessments.
6. Besides performance assessments, what are other ways that a teacher can gather information about students?
7. What is a portfolio, and how can a portfolio program be set up and used in a mathematics class? (Do you think this could be done at any grade level?)
8. Why do you not need to average a set of scores to assign grades?
9. What are some real issues surrounding grades?

FOR DISCUSSION AND EXPLORATION

1. Select a chapter from the teacher's edition of any standard textbook at the grade level of interest. What assessment alternatives and suggestions are

offered in the form of tests, performance tasks, or other ideas that would help you with assessment? How adequate would the assessment of students on this chapter be if only these ideas are followed? Note that most series now offer additional assessment guides in booklets that teachers are given when the text is adopted. Your instructor may be able to help you find some of these if you do not have access to them.

2. Pick a topic other than problem solving for a particular grade level. Design several test items that you think could help you assess understanding of concepts in that topic. If you selected a procedural topic, what assessment items would you want to consider other than a student's ability to carry out the procedure?

3. Read any of the following, and make a short report on the aspect of the reading that you find most relevant or important.
   a. The assessment section of the *Standards*.
   b. In the *Professional Standards*, under "Standards for Teaching" read Standard 6: Analysis of Teaching and Learning, and in the section on "Evaluation of Teaching," Standard 7: Assessing Student's Understanding of Mathematics.
   c. *Mathematics Assessment: Myths, Models, Good Questions, and Practical Suggestions* (Stenmark, 1991).
   d. *Assessment Alternatives in Mathematics: An Overview of Assessment Techniques that Promote Learning* (Stenmark, 1989).
   e. Any booklet published by your state department of education (or some other state) that addresses alternative assessment, portfolio assessment, or some related topic.

4. Read "Assessing for Learning: The Interview Method," by Labinowicz in the November 1987 *Arithmetic Teacher*. Try conducting all or part of the diagrammed interview that accompanies that short article with a second- or third-grade child. Alternatively, select a topic and grade level, and design a short interview for that topic using the basic guidelines provided by Labinowicz. As another resource, consult "Diagnosis in Mathematics: The Advantages of an Interview," by Liedtke in the November 1988 *Arithmetic Teacher*.

5. What do you think is the single most important issue surrounding grading? How will you confront this issue in the classroom?

◆◆◆◆◆

SUGGESTED READINGS

Badger, E. (1992). More than testing. *Arithmetic Teacher, 39*(9), 7–11.

Clarke, D. J., Clarke, D. M., & Lovitt, C. J. (1990). Changes in mathematics teaching call for assessment alternatives. In T. J. Cooney (Ed.), *Teaching and learning mathematics in the 1990s*. Reston, VA: National Council of Teachers of Mathematics.

Harvey, J. G. (1991). Using calculators in mathematics changes testing. *Arithmetic Teacher, 38*(7), 52–54.

Hebert, E. A. (1992). Portfolios invite reflection—from students *and* staff. *Educational Leadership, 49*(2), 58–61.

Hopkins, M. H. (1992). The use of calculators in assessment of mathematics achievement. In J. T. Fey (Ed.), *Calculators in mathematics education*. Reston, VA: National Council of Teachers of Mathematics.

Knight, P. (1992). How I use portfolios in mathematics. *Educational Leadership, 49*(2), 71–72.

Labinowicz, E. (1987). Assessing for learning. *Arithmetic Teacher, 35*(3), 22–25.

Liedtke, W. (1988). Diagnosis in mathematics: The advantages of an interview. *Arithmetic Teacher, 36*(3), 26–29.

Lindquist, M. M. (1988). Assessing through questioning. *Arithmetic Teacher, 35*(5), 16–19.

Mathematical Sciences Education Board, National Research Council. (1993). *Measuring up: Prototypes for mathematics assessment*. Washington, DC: National Academy Press.

National Council of Teachers of Mathematics. (1992) Assessment [Focus Issue]. *Arithmetic Teacher, 39*(6).

National Council of Teachers of Mathematics. (1992). Alternative assessment [Theme Issue]. *Mathematics Teacher, 85*(8).

Pandey, T. (1991). *A sampler of mathematics assessment*. Sacramento: California Department of Education.

Richardson, K. (1988). Assessing understanding. *Arithmetic Teacher, 35*(6), 39–41.

Robinson, G. E., & Bartlett, K. T. (1993). Assessment and the evaluation of learning. In R. J. Jensen (Ed.), *Research ideas for the classroom: Early childhood mathematics*. New York: Macmillan Publishing Co.

South Carolina Educational Television, Kuhs, T. (Dir.), (1992). *Mathematics assessment: Alternative approaches* [video and guide book]. Reston, VA: National Council of Teachers of Mathematics.

Spangler, D. A. (1992). Assessing students' beliefs about mathematics. *Arithmetic Teacher, 40*, 148–152.

Stenmark, J. K. (Ed.). (1991). *Mathematics assessment: Myths, models, good questions, and practical suggestions*. Reston, VA: National Council of Teachers of Mathematics.

Vermont Department of Education. (1991). *Looking beyond "the answer:" The report of Vermont's mathematics portfolio assessment program*. Montpelier, VT: Author.

Webb, N. (Ed.). (1993). *Assessment in the mathematics classroom*. Reston, VA: National Council of Teachers of Mathematics.

Webb, N., & Briars, D. (1990). Assessment in mathematics classrooms, K–8. In T. J. Cooney (Ed.), *Teaching and learning mathematics in the 1990s*. Reston, VA: National Council of Teachers of Mathematics.

Webb, N., & Romberg, T. A. (1992). Implications of the NCTM *Standards* for mathematics assessment. In T. A. Romberg (Ed.), *Mathematics assessment and evaluation: Imperatives for mathematics educators*. Albany, NY: State University of New York Press.

Webb, N. L., & Welsch, C. (1993). Assessment and evaluation for middle grades. In D. T. Owens (Ed.), *Research ideas for the classroom: Middle grades mathematics*. New York: Macmillan Publishing Co.

Wiggins, G. (1992). Creating tests worth taking. *Educational Leadership, 49*(2), 26–33.

6 THE DEVELOPMENT OF NUMBER CONCEPTS AND NUMBER SENSE

◆ NUMBER IS A COMPLEX AND MULTIFACETED CONCEPT. A RICH understanding of number, a relational understanding, involves many different ideas, relationships, and skills. To be able only to count accurately and to match a numeral with an appropriate set is minimal understanding at the instrumental end of the understanding continuum. A relational understanding of number includes not only counting and numeral recognition but also an intricate web of more and less relationships, part-part-whole ideas, anchoring to special numbers such as five and ten, connections with real quantities and measures in the environment, and much more. Experience suggests that these relationships do not develop automatically, and it is our job to provide children with a rich assortment of activities that will help them construct these many ideas of number.

EARLY NUMBER SENSE

Number sense was described by Howden (1989), as a "good intuition about numbers and their relationships. It develops gradually as a result of exploring numbers, visualizing them in a variety of contexts, and relating them in ways that are not limited by traditional rules and procedures" (p.11). The NCTM *Standards* defines number sense as involving five components:

1. well-understood number meanings,
2. multiple relationships on numbers,
3. recognition of the relative magnitude of numbers,
4. knowledge of the effect of operations on numbers, and
5. referents to measure of things in the real world.

These definitions, and many others that have appeared in the literature, speak to a knowledge of numbers that pervades all numbers including both large and small whole numbers, decimals, fractions, percentages, and so on. Whenever we think of helping children develop an understanding of any type of numbers, that should include the development of an intuition for these numbers in the broadest sense.

In this chapter we will focus on only a small yet very important piece of number sense, that which involves numbers up to about 20. The focus is on helping children construct a variety of important relationships on these numbers and connections to the real world. These are the relationships that form the very foundation of much of the numeric thinking and number sense development that will come later.

THE BEGINNINGS OF NUMBER CONCEPTS

Parents help children count their fingers, toys, people at the table, and other small sets of objects. Questions concerning "who has more?" or "are there enough?" are part of the daily life of children as early as two to three years of age.

Considerable evidence has been gathered indicating that these very young children have some form of understanding of number and counting (Gelman & Gallistel, 1978; Gelman & Meck, 1986; Baroody, 1987; Fuson & Hall, 1983; Ginsburg, 1977).

A developmental approach to number will capitalize on the simple beginning ideas about counting and number that children have before entering kindergarten. It will help them construct new ideas about number to expand, refine, and enhance the concepts that have been developing in the preschool years.

COUNTING

The Development of Counting Skills

Counting involves at least two separate skills. First, one must be able to produce the standard list of counting words in order: "One, two, three, four, ..." Second, one must be able to connect this sequence in a one-to-one manner with the items in the set being counted. Each item must get one and only one count.

Experience and guidance are the major factors in the development of these counting skills. Many children come to kindergarten able to count sets of 10 or beyond. At the same time, children from impoverished backgrounds may require considerable practice to make up their experience deficit. The size of the set is also a factor related to success in counting. Obviously longer number strings require more practice to learn. The first 12 counts involve no pattern or repetition, and many do not recognize a pattern in the teens.

Counting a set of objects that can be moved as they are counted is easier than counting objects that cannot be moved or touched. Counting a set that is ordered in some way, such as in a string of dots or other pattern, is easier than counting a randomly displayed set. There seems to be little advantage in making counting tasks difficult. Therefore, children still learning the skills of counting, that is, matching oral number words with objects, should be given sets of blocks or counters that they can move or pictures of sets that are arranged for easy counting.

Meaning Attached to Counting

There is a difference between being able to count (as just described) and knowing what counting tells. When we count a set, the last number word used is the name of the "manyness" of the set or the cardinality of the set. While very young children seem to have some concept of quantity or manyness, they do not immediately connect this concept with the act of counting. When children understand that the last count word names the quantity of the set, they are said to have the cardinality principle. While experience is a major factor, most children by age $4\frac{1}{2}$ have made this connection (Fuson & Hall, 1983).

To determine if children have the cardinality rule, they can be given a set of objects and asked, "How many?" If they repeat or emphasize the last count, then it can be inferred that they have switched meaning from a count use of the last word to a cardinality use of the last word. For example, "One, two, three, four, five, *six*. (There are six.)" If this cardinal usage is not clear, the "How many?" question can be repeated. If the child then announces the total without counting, it is clear that he or she is using the cardinal meaning of the counting word. However, a recount of the entire set would indicate that the question "How many?" was interpreted by the child as a command to count rather than a request for the quantity in the set.

In the classroom you can help students develop a deeper sense of cardinality through counting activities such as the following.

6.1 COUNTING SETS

Have children count several sets where the number of objects is the same, but the objects are very different in size. Discuss how they are alike and different. "Were you surprised that they were the same amount? Why or why not?"

6.2 COUNT AND REARRANGE

Have students count a set. Then rearrange the set and ask "How many now?" If they see no need to count over, you can infer that they have connected the cardinality to the set regardless of its arrangement. If they choose to count again, discuss why they think the answer is the same.

6.3 FIND THE SAME AMOUNT

Give children a collection of cards with sets on them. The dot cards (Black-line Masters) are one possibility. Have the children pick up any card in the collection and then find another card with the same amount to form a pair. Continue to find other pairs. Watch how children are doing this task. Children whose number ideas are completely tied to counting and nothing more will select cards at random and count each dot. Others will be able to find a card that appears to have about the same number of dots. This is a significantly higher level of understanding. Also observe how the dots are counted. Are the counts made accurately? Is each counted only once?

An important variation of **Find the Same Amount** is to have children make a set using simple counters that matches the set they see on the card (Figure 6.1). During the construction of the set the children have many opportunities to stop and compare. In this way, they are actually doing more reflection than when searching for a matching card.

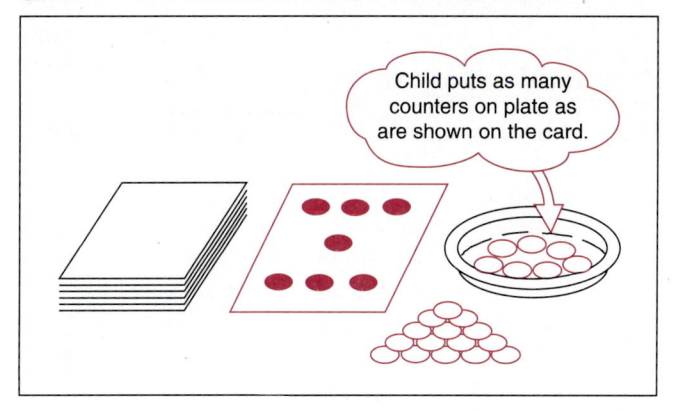

FIGURE 6.1: *An early number activity. Make a set that has as many.*

Once counting becomes a meaningful activity, it becomes a major tool in the development of other relationships on number. Meaningful counting is only the beginning of number concept development, not the end goal.

THE RELATIONSHIPS OF MORE, LESS, AND SAME

The concepts of *more*, *less*, and *same* are basic relationships contributing to the overall concept of *number*. The beginnings of these ideas are developed before children enter school. An entering kindergarten child can almost always choose the set that is *more* if presented with two sets that are quite obviously different in number. In fact, Baroody states "A child unable to use 'more' in this intuitive manner is at considerable educational risk" (1987, p. 29). Classroom activities should help children build on this basic notion and refine it.

Researchers have found that almost all children use counting rather than matching as the basis of comparison. Children can learn to use matching, but it does not seem to be the natural method that they use intuitively (Fuson, Secada, & Hall, 1983; Baroody, 1987). In fact, children who are not taught to use matching are in no way less able to make comparisons than those who are. There seems to be no real reason to force children to engage in one-to-one matching activities if they can successfully use counting to make comparisons.

◆ More, Less, and Same

While the concept of *less* is very similar to the concept of *more*, the word *less* proves to be more difficult for children than does *more*. A possible explanation is that children have many opportunities to use the word *more* and have very limited exposure to the word *less*. Given two unequal sets, selecting the set with more is logically the same as not selecting the set that has less. To help children with the concept of *less*, frequently pair it with the word *more* and make a conscious effort to ask "which is less" questions as well as "which is more" questions. For example, suppose that your class has correctly selected the set that has more from two that are given. Immediately follow with the question, "Which is less?" The unfamiliar term and concept can be connected with the better-known idea.

For all three concepts (more, less, and same) children should construct sets using counters as well as make comparisons or choices between two given sets. The activities described here include both types.

6.4 MAKE SETS OF MORE/LESS/SAME

At a work station or table provide about eight cards with sets of 4 to 12 objects, a set of small counters or blocks, and some word cards labeled More, Less, and Same. Next to each card students make three collections of counters: a set that is more, one that is less, and one that is the same. The appropriate labels are placed on the sets (Figure 6.2).

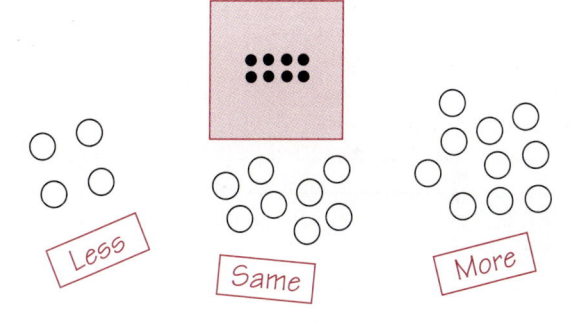

FIGURE 6.2: *Making sets that are more, less, and same*

As a seatwork variation, a worksheet such as that shown in Figure 6.3 (p. 90), can be provided. On a given day students can make sets that are less than the given set (or more, or the same). After making the set with small counters, they can use a crayon to make a dot for each counter.

6.5 MORE AND LESS WAR

The familiar game of war can be played with any cards that have sets on them, or you can use standard playing cards with the face cards removed. To stress the idea of less, periodically play the game where the winner is the person with less rather than more. Another variation is to spin a more/less spinner on each turn. The spinner face is divided in two parts labeled MORE and LESS. If the spinner shows LESS, the player with less wins, and vice versa (Figure 6.4).

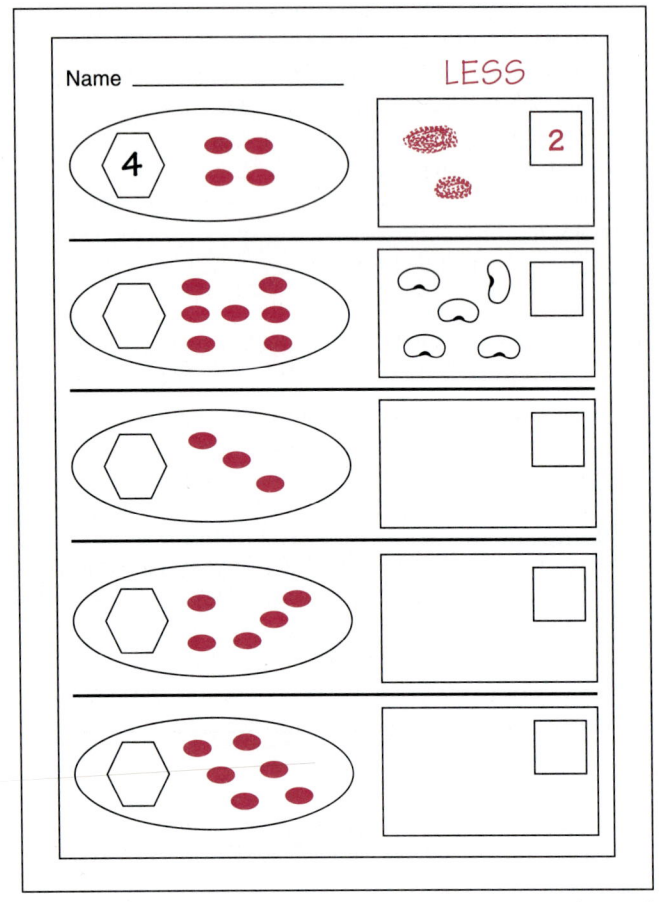

FIGURE 6.3: *A more, less, same worksheet that can be used with counters.*

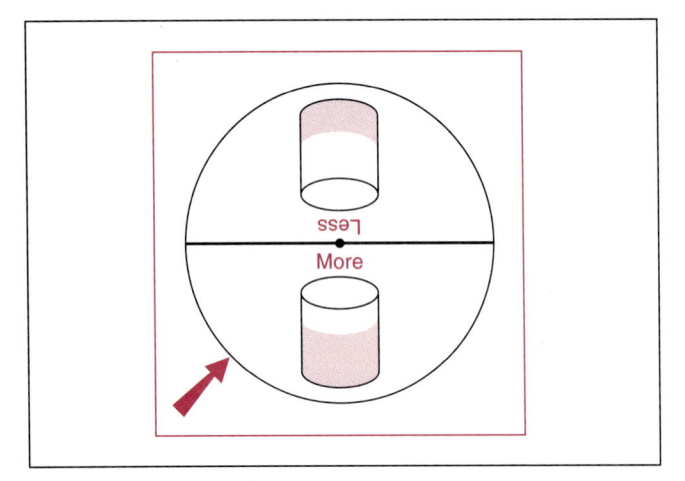

FIGURE 6.4: *A More/Less spinner*

PROCEDURAL KNOWLEDGE OF NUMBERS

Some procedural knowledge of numbers has already been discussed, namely knowledge of the standard counting words in order (oral counting) and the skill of using this counting sequence to accurately count a set. Other procedural skills related to number concepts are the ability to count on and count back, the ability to read and write numerals, and the use of symbols to represent the relationships of more, less, and same (>, <, =).

NUMERAL WRITING AND RECOGNITION

Helping children to read and write single-digit numerals is similar to teaching them to read and write letters of the alphabet. Traditionally, instruction has involved various forms of repetitious practice.

Children trace over pages of numerals, repeatedly write the numbers from 0 to 10, make the numerals from clay, trace them in sand, write them on the chalkboard or in the air, and so on. The principle has been that numeral form is a rote activity that can best be learned through practice.

Baroody (1987) argues convincingly that while some numeral writing practice is clearly necessary, much time can be saved by taking a meaningful approach. Such an approach involves focusing directly on the defining characteristics of each numeral and on a motor plan for forming the numerals. The characteristics of the numerals should be articulated and discussed. For example, 1, 4, and 7 are made up of straight lines, while 2 and 5 have a straight and a curve. Similar numerals should be taught together in order to focus on the properties that are necessary to distinguish them from each other. The numerals 6 and 9, for example, are distinguished from all other numerals by a closed loop on a stick. The loop for the 6 is at the bottom, and the loop for the 9 is at the top. These features distinguish one from the other.

The calculator is a good instructional tool for numeral recognition. In addition to helping children with numerals, early activities can help develop familiarity with the calculator so that more complex activities are possible.

6.6 FIND AND PRESS

Provide each child with a calculator. (A calculator should be standard equipment in the desks of all children at all grade levels.) Be sure that children begin by pressing the clear key [C]. Then you say a number, and the children press that number on the calculator. If you have an overhead calculator you can then show the children the correct key so that they can confirm their responses. As an alternative you can write the number on the board for children to check. For this activity begin with single-digit numbers. Then progress to two or three numbers called in succession. For example, call "Three, seven, one" (not three hundred seventy-one). Children press the complete string of numbers as called.

6 / THE DEVELOPMENT OF NUMBER CONCEPTS AND NUMBER SENSE

As a variation to the **Find and Press** activity, have children practice pressing their addresses, phone numbers, bus numbers, ages, the number of brothers and sisters, the number of windows in the room, the number of children in their group, and so on. That is, use numeral recognition exercises to connect numbers to real things in children's lives. This begins their number sense development.

ORAL COUNTING

Oral counting that does not attempt to enumerate a set is simply an exercise to practice the number sequence. While the forward sequence is relatively familiar to most young children, counting on and counting back are difficult skills for many. Frequent short practice drills are recommended.

6.7 EXERCISE COUNTING

Correlate rhythmic counting with some form of movement. For example, in a count-to-7 exercise, children might stretch their hands above their heads and clap on each count up to 7. Then in the same rhythm and without slowing down, lean over and clap each count, again from 1 to 7, down near their knees. A rhythmic count is useful to keep everyone together. Use a xylophone, triangle chime, or small bell to strike out a rhythm. Count to different target numbers on different days.

6.8 UP AND BACK COUNTING

Counting up to and back from a target number in a rhythmic fashion is an important counting exercise. For example, line up five children and five chairs in front of the class. As the whole class counts from 1 to 5, the children sit down one at a time. When the target number, 5, is reached, it is repeated; the child who sat on 5 now stands; and the count goes back to 1. As the count goes back, the children stand up one at a time, and so on, "1, 2, 3, 4, 5, 5, 4, 3, 2, 1, 1, 2, ..." Kindergarten and first-grade children find exercises such as this quite fun and challenging. Any movement (clapping, turning around, jumping jacks, and the like) can be used as the count goes up and back in a rhythmic manner.

A variation of the last activity is to count up and back *between* two numbers. For example, start with 4 and count to 11 and back to 4, and so on Keep a rhythm as in the other exercises.

6.9 CALCULATOR UP AND BACK

The calculator provides an excellent counting exercise for young children because it includes seeing the numerals as the count is made. Have each child press ＋1 ＝＝＝＝＝. The display will go from 1 to 5 with each ＝ press. The count should also be made out loud in a rhythm as in the other exercises. To start over, press the clear key and repeat. Counting up and back is also possible, but the end numbers will not be repeated. The following illustrates the key presses and what the children would say in rhythm:*

3 ＋ 1 ＝＝＝＝ － 1 ＝＝＝＝ ＋ 1＝＝…
"3 plus 1, 4, 5, 6, 7, minus 1, 6, 5, 4, 3, plus 1, 4, 5,…"

COUNTING SETS

Children need practice applying their counting skills to counting sets. Many useful games exist in which children count a set and then make another set of the same amount. Usually a die is rolled or a card drawn that shows a set of objects. The player then counts the corresponding number of counters or moves a marker around a track on a board.

6.10 FILL-UP

Fill-Up can be played by two or three players. Each player has a board with spaces marked and a collection of counters to place in the spaces. In turn each child draws a dot card. He or she then counts out that many counters. When the correct amount has been counted (to the side of the board), the counters are placed on the board, one in each space. The first to fill up their board is the winner. The number of spaces on the board, the arrangement, and the number of dots on the cards can all be varied to suit the ability level of the students. Figure 6.5 (p. 92) shows a board and some cards that could be used.

---

*To do activity 6.9 requires that the calculator have an automatic constant feature for addition and subtraction. This feature automatically stores the last operation of addition and subtraction. For example, if you press 3 ＋ 2 ＝, the display shows 5. A subsequent press of the ＝ will add 2 to the display and result in a display of 7. If you press 4 ＝, the 2 will be added to the 4. Subtraction works in a similar manner.

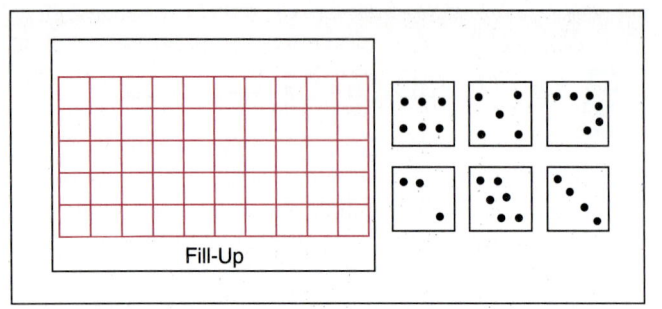

FIGURE 6.5: *The game "Fill-Up": Take turns. Draw a card. Put that many counters on your board*

6.11 CALCULATOR COUNTING

The calculator can also be useful in counting objects. Children first clear the calculator and press ⊞ 1. At this point the calculator is poised to help them count objects. Decide on what to count. For example count a set of papers on the bulletin board or the number of plants on your window sill. As you or a child point to each object, the children press ⊟ and say the count aloud. This method can easily be used to count larger sets such as the number of letters in the alphabet, the number of ears in the room, the stripes or stars on the flag, and so on.

good homework activity

Send the calculator counting activity home, and encourage parents to use this method of counting objects around the home. Prepare a list of counts to bring in to compare: number of plates on the dinner table, the number of windows on the first floor, doors, pets, pictures on the walls, clocks, coats in the closet, stair steps, and so on. The calculator helps motivate the activity and shows parents that it can be used as a learning tool.

Perhaps the most common textbook exercises have children match sets with numerals. Children are given pictured sets and asked to write or match the number that tells how many. Alternatively, they may be given a number and told to make or draw a set with that many objects. Many teacher resource books describe cute learning-center activities where children put a numeral with the correct-sized set—frogs (with dots) on numbered lily pads, for example. It is important to note that these frequently over-worked activities involve only the skills of counting sets and numeral recognition or writing. When children are successful with these activities, little is gained by continuing to do them.

Counting on and counting back from a number are skills that have already been discussed under oral counting. The following activities connect these skills to situations involving objects.

6.12 COUNTING ON WITH COUNTERS

Give each child a collection of 10 or 12 small counters that they line up left to right on their desks. Tell them to count four counters and push them under their left hands (Figure 6.6). Then say, "Point to your hand. How many are there?" (4) "So let's count like this." F-o-u-r (pointing to their hand), five, six, ..." Repeat with other numbers under the hand. The same activity can be done on the overhead projector. Display the counters, count off some, and cover them. Ask how many are covered, and then collectively count starting with the amount covered.

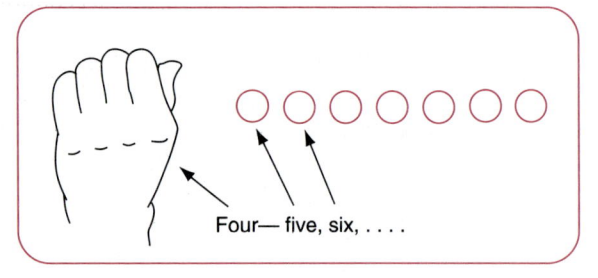

Four— five, six, ...

FIGURE 6.6: *Counting on: Hide four. Count, starting with those hidden*

6.13 STAND/SIT COUNTING ON

Line up some children in front of the room. Count off some of them from one end and have them sit. Then have the class count all of the children beginning with the seated group as a single count and count on from there.

In each of these counting-on exercises, the children first count the group. You stop them, ask how many are there, and then count using a count-on procedure. The children can understand that they are counting on to enumerate the set because they in fact counted the first group. A more abstract version has the first set previously counted.

6.14 REAL COUNTING ON

On the overhead, cover some counters with a card before the light is turned on and place additional counters to the side of the card, as in Figure 6.7.

6 / THE DEVELOPMENT OF NUMBER CONCEPTS AND NUMBER SENSE 93

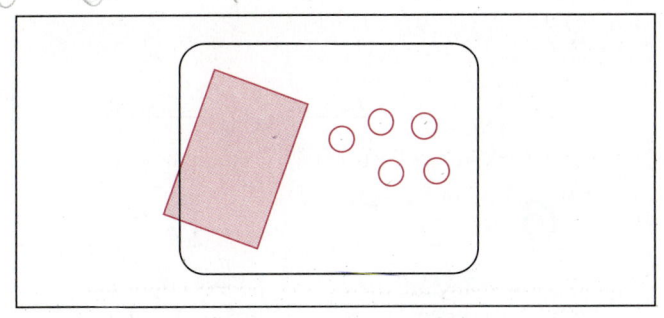

FIGURE 6.7: *Four under card. How many altogether? Count.*

Worksheets can be made with examples such as those shown in Figure 6.8. Notice the progression from an illustration of the above activities to the use of a numeral and dots to two numerals.

Applying counting-back skills is more difficult for children but can be developed in a similar manner as counting on. To learn about counting-back, the children first count out a set as they did in the counting-on activities. These are covered. The covered set is pointed to, saying the number covered. Then, as counters are removed one at a time, the backward count is made.

DEVELOPMENT OF NUMBER RELATIONSHIPS: NUMBERS THROUGH 10

Set-to-numeral matchings and other counting activities are vitally important to number development. However, they only focus on the basic meaning of number and on accurate counting skill. Once children have acquired a concept of manyness or cardinality and can adequately use their counting skills in meaningful ways, little more is to be gained from the kinds of counting activities that have been described so far. More relationships must be created in order for children to have what is called "number sense" or a flexible concept of number not completely tied to counting.

A COLLECTION OF NUMBER RELATIONSHIPS

Figure 6.9 (p. 94) illustrates the four different types of relationships that children can and should develop on numbers:

- **Spatial relationships:** Children can learn to recognize sets of objects in patterned arrangements and tell how many without counting. For most numbers there are several common patterns. Patterns can also be made up of two or more easier patterns for smaller numbers.
- **One and two more, one and two less:** The two-more-than and two-less-than relationships are more than just the ability to count on two or count back two. Children should know that 7, for example, is *1 more than* 6 and also *2 less than* 9.
- **Anchors or "benchmarks" of 5 and 10:** Since 10 plays such a large role in our numeration system and because two fives make up 10, it is very useful to develop relationships for the numbers 1 to 10 to the important anchors of 5 and 10.
- **Part-part-whole relations:** To conceptualize a number as being made up of two or more parts is the most important relationship that can be developed about numbers. For example, 7 can be thought of as a set of 3 and a set of 4 or a set of 2 with a set of 5.

The standard number curriculum has placed almost no emphasis on these relationships. When children have learned to count and to read and write numerals, the next step has been, unfortunately, to begin addition and subtraction. The development of these relationships is not something that occurs overnight but rather requires many months at the kindergarten and first grade level. In the sections that follow, activities that can be used to help children develop these relationships are discussed. The principal tool that children will use in this development is the one number tool they possess: counting. Initially, then, you will notice a lot of counting, and you may wonder if you are making progress. Have patience! As children construct these new relationships, counting will become less and less necessary as children begin to use the more powerful ideas they have constructed.

SPATIAL RELATIONSHIPS—PATTERNED SET RECOGNITION

Many children learn to recognize the dot arrangements on standard dice due to the many games involving dice. Similar

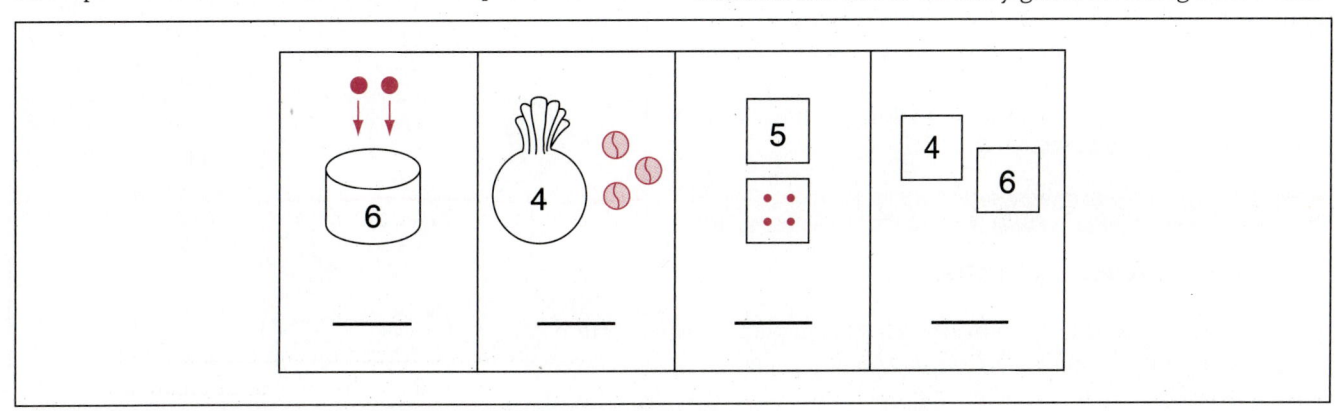

FIGURE 6.8: *A progression of counting-on activities for worksheets*

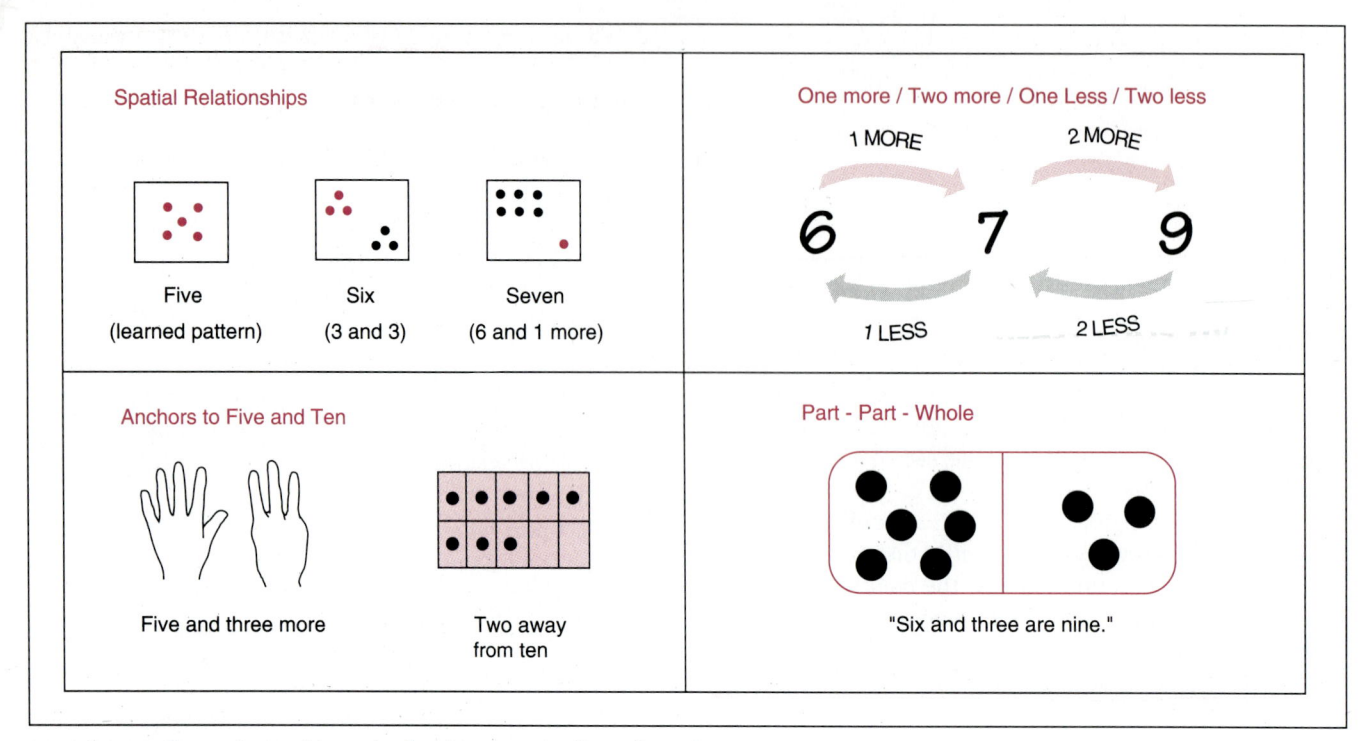

94 6 / THE DEVELOPMENT OF NUMBER CONCEPTS AND NUMBER SENSE

FIGURE 6.9: *Four relationships to be developed on small numbers*

instant recognition of patterned arrangements can be developed for other patterns as well. The activities suggested here are designed to encourage reflective thinking about the patterns so that the relationships will be constructed. The patterns represent more than an abstract symbolism for the number of dots. Since much repetitive counting goes into learning to recognize the patterns, they are intimately connected to number concepts. Recognition of the pattern enhances knowledge of the particular number concept. Quantities up to 10 can be known and named without the routine of counting. This then can aid in counting on (from a known patterned set) or learning combinations of numbers (see a pattern of two known smaller patterns). In general, the idea is to help children access real quantities and yet be free of the tedium of counting.

A good set of materials to use in pattern-recognition activities is a set of dot plates. These can be made using luncheon-size paper plates and the peel-off dots commonly available in stationery stores. A reasonable collection of patterns is shown in Figure 6.10. Note that some patterns are combinations of two smaller patterns or a pattern with one or two additional dots. These should be made in two colors. Keep the patterns compact. If the dots are spread out, the patterns are hard to see.

6.15 LEARNING PATTERNS

To introduce the patterns, provide each student with about 10 counters and a piece of construction paper as a mat. Hold up a dot plate for about 3 seconds.

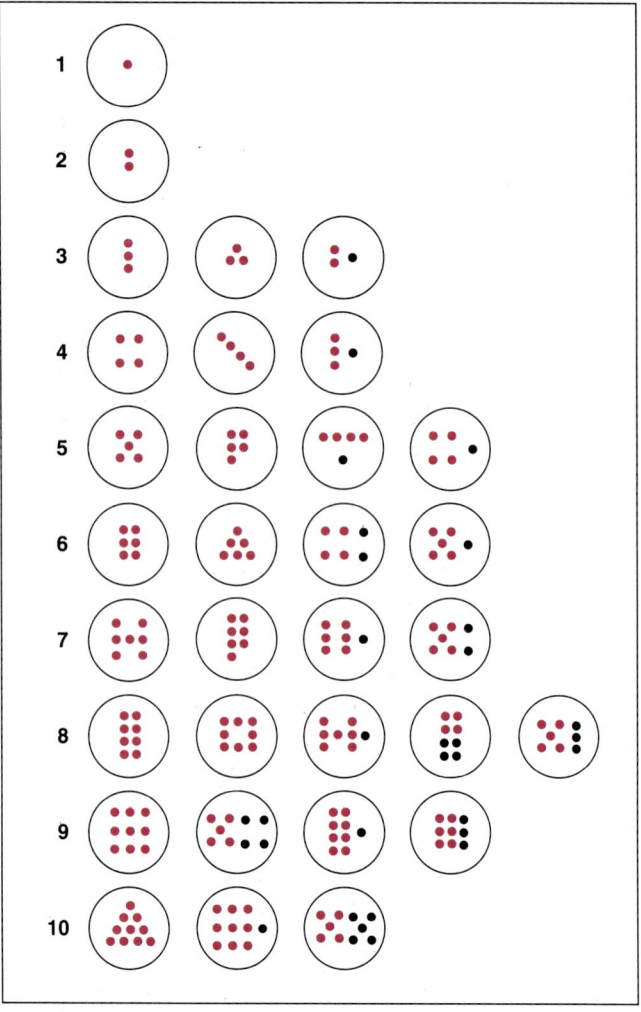

FIGURE 6.10: *A useful collection of dot patterns for "dot plates"*

6 / THE DEVELOPMENT OF NUMBER CONCEPTS AND NUMBER SENSE

"How many dots did you see? How did you see it? Make the pattern you saw using the counters on the mat." Spend some time discussing the configuration of the pattern and how many dots. Do this with a few new patterns each day.

6.16 DOT PLATE FLASH

Hold up a dot plate for only 1 to 3 seconds. "How many? How did you see it?" Children like to see how quickly they can recognize and say how many dots. Include lots of easy patterns and a few of those with more dots as you build their confidence. Students can also flash the dot plates to each other as a workstation activity.

6.17 DOMINOES

Make a set of dominoes out of poster board and put a dot pattern on each end. The dominoes can be about 5 cm by 10 cm. The same patterns can appear on lots of dominoes with different pairs of patterns making up each one. Let the children play dominoes in the regular way, matching up the ends. As a speed activity, spread out all of the dominoes, and see how fast they can play all of the dominoes or play until no more can be played. Regular dominoes could also be used, but there are not as many patterns.

The dot plates and patterned sets in general can easily be used in many other activities, as we will see. However, the instant recognition activities with the plates are exciting and can be done in five minutes at any time of day or between lessons. There is value in using them at any primary grade level and at any time of year.

ONE AND TWO MORE/ ONE AND TWO LESS

When children count they have no reason to reflect on the way one number is related to another. The goal is only to match number words with objects until they reach the end of the count. To learn that 6 and 8 are related by the twin relationships to two-more-than and two-less-than requires reflection on these ideas within tasks that permit counting. Counting on (or back) one or two counts is a useful tool in constructing these ideas.

6.18 ONE-LESS-THAN DOMINOES

Use the dot pattern dominoes or a standard set and play "one-less-than" dominoes. Play in the usual way, but instead of matching ends a new domino can be added if it has an end that is one less than the end on the board. A similar game can be played for two less or one (two) more.

(Subsequent activities in this section can be done with any of the four relationships, but each will be described for only one.)

6.19 ONE-MORE-THAN DOT PLATES

Using the dot plates, flash the plates and have the students say 1 more than the amount on the plate.

6.20 MAKE A TWO-MORE-THAN SET

With the dot plates or any set of dot cards, have students construct a set of counters that is 2 more than the set. Similarly, spread out 8 to 10 dot cards, and find another card for each that is two less than the card shown. (Omit the one and two card for two less than, and so on.)

In activities where children find a set or make a set, they can add a numeral card (a small card with a number written on it) to all of the sets involved. They can also be encouraged to take turns reading a *number sentence* to their partner. If, for example, a set has been made that is 2 more than a set of 4, the child can read this by saying the number sentence "Two more than four is six."

The following activities involve numerals and sets together or just numerals. These are especially important in late first grade through the third grade.

6.21 ONE-MORE-THAN RESPONSE CARDS

Provide each child with six to eight number cards about the size of index cards. (Children can cut up paper and make their own.) On the cards put the numbers you will need for the day (e.g., 5 through 10). Now flash a dot plate and have students hold up the card that is 1 more than the plate. Nothing is said out loud, and all students respond. Similarly, you can hold up a numeral card or say a number orally, and they respond with their numeral cards.

more-than result. For example, press 5. Two more is seven. Press = to confirm. Since the +2 is stored, the process can be repeated over and over.

Many activities that have simple student responses to simple input, such as those that have been described, can be made into machine games. Draw a funny-looking "machine" on the board. It requires an input hopper and an output chute, as shown in Figure 6.11. Tell the students what the machine does. For example, "This is a magic one-more-than machine. It takes in a number up here and spits out a number that is one more."

ANCHORING NUMBERS TO FIVE AND TEN

Here again we want to help children relate a given number to other numbers, specifically to 5 and 10. These relationships are especially useful in thinking about various combinations of numbers. For example, in each of the following, consider how the knowledge of 8 as "5 and 3 more" and as "2 away from 10" can play a role: $5+3$, $8+6$, $8-2$, $8-3$, $8-4$, $13-8$. (It may be worth stopping here to consider the role of 5 and 10 in each of these examples.) Later, similar relationships can be used in development of mental computation skills on larger numbers such as $68+7$.

The most common and perhaps most important model for this relationship is the ten-frame. Probably developed by Robert Wirtz (1974), the ten-frame is simply a 2×5 array in which counters or dots are placed to illustrate numbers. Ten-frames can be simply drawn on a piece of construction paper. Nothing fancy is required, and each child can have one. The ten-frame has been incorporated into a variety of activities in this book and is becoming more popular in standard textbooks for children.

Before doing any activities, show children how to "show numbers" on the ten-frame. It is only by adhering to the notion of filling up the top row first and then adding more to the second row that the relationships to 5 and 10 become apparent. (See Figure 6.12.)

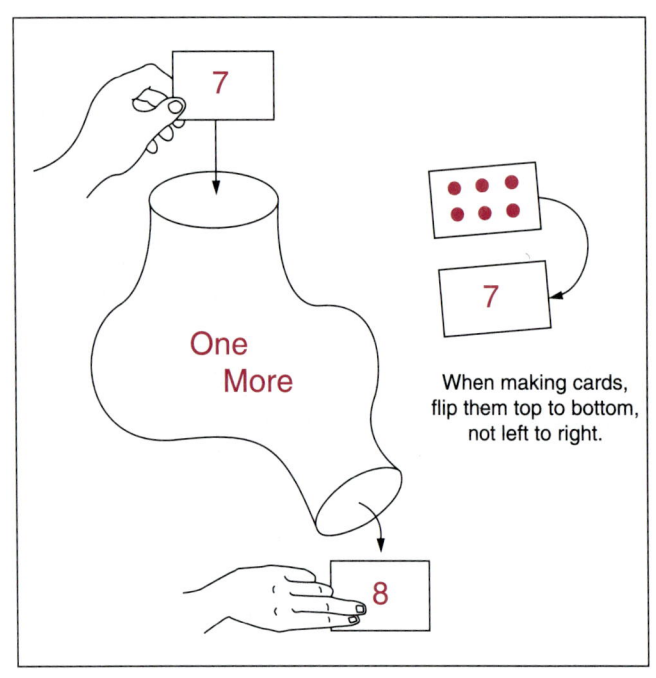

FIGURE 6.11: *A one-more-than machine drawn on the chalkboard. Machines like this one can be used for a variety of teacher-directed drill*

One way to operate a machine is to prepare some cards with the input on one side and the output on the reverse. Hold the card over the input hopper, and as you slide it down the board toward the output the children call out what will come out. Just as the card gets to the chute, flip it over to confirm. The input side of the cards can be dot patterns or numerals. Machine activities can also be done without cards by just announcing the input numbers and thereby making a quick and easy drill.

6.22 CALCULATOR TWO-MORE-THAN

The calculator is every child's personal input-output machine. For example, press + 2 = to store plus two. Now press any number, and predict the two-

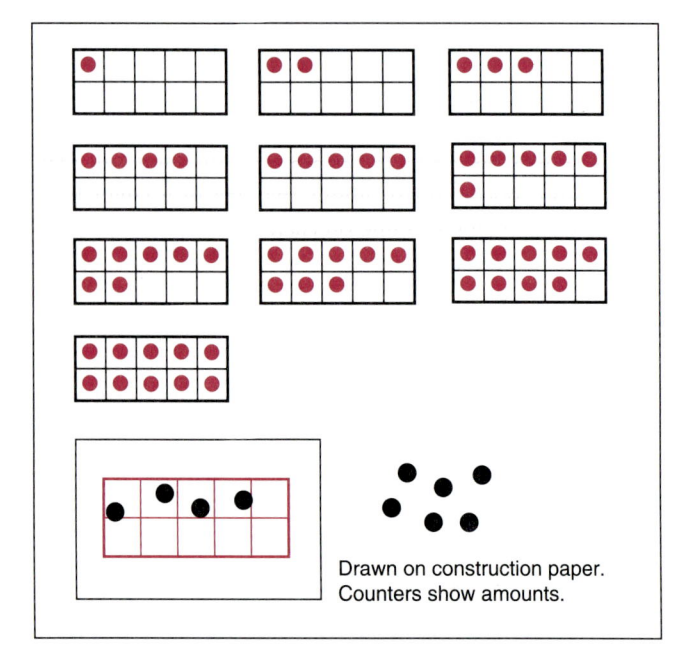

FIGURE 6.12: *Ten-frames*

6.23 CRAZY MIXED-UP NUMBERS

Crazy Mixed-Up Numbers is a great introductory game adapted from *Mathematics Their Way* (Baratta-Lorton, 1976). All children make their ten-frame show the same number. The teacher then calls out numbers between 0 and 10. After each number the children examine their ten-frames and decide how many more counters need to be added ("plus") or removed ("minus"). They then call out plus (or minus) whatever amount is appropriate. If, for example, the frames showed six, and the teacher called out "four," the children would respond, "Minus two!" and then change their ten-frames accordingly. The activity continues with a random list of numbers. Children can play this game independently by preparing strips of about 15 "crazy mixed-up numbers." One child plays "teacher," and the rest use the ten-frames. Children like to make up their own number strips.

6.24 TEN-FRAME TRANSLATION

Give students a number in some form other than a ten-frame, and have them show that number on their ten-frames. Show them a dot card, hold up fingers, show some counters on the overhead, hold up a numeral card, or simply say a number out loud. These activities could also be done independently at a workstation by providing the number cards or sets along with the ten-frames and counters. Worksheets could be prepared for independent seatwork, as shown in Figure 6.13. The children would use an actual ten-frame and then record the result on the paper.

Ten-frame cards provide another useful variation. Make cards from poster board about the size of a small index card, and draw a ten-frame on each. Dots are drawn in the frames. A set of 20 cards consists of a 0 card, a 10 card, and two each of the numbers 1 to 9.

Important variations of the previous activity include

Saying the number of spaces on the card instead of the number of dots.

Saying one more than the number of dots (or two more, and also less than).

Saying the "ten fact." For example: "Six and four make ten."

After students have become familiar with the ten-frame, simply having a large blank ten-frame drawn on the board can profitably influence children's thinking about five and ten as they do number activities. Try looking at a ten-frame while doing the following two activities.

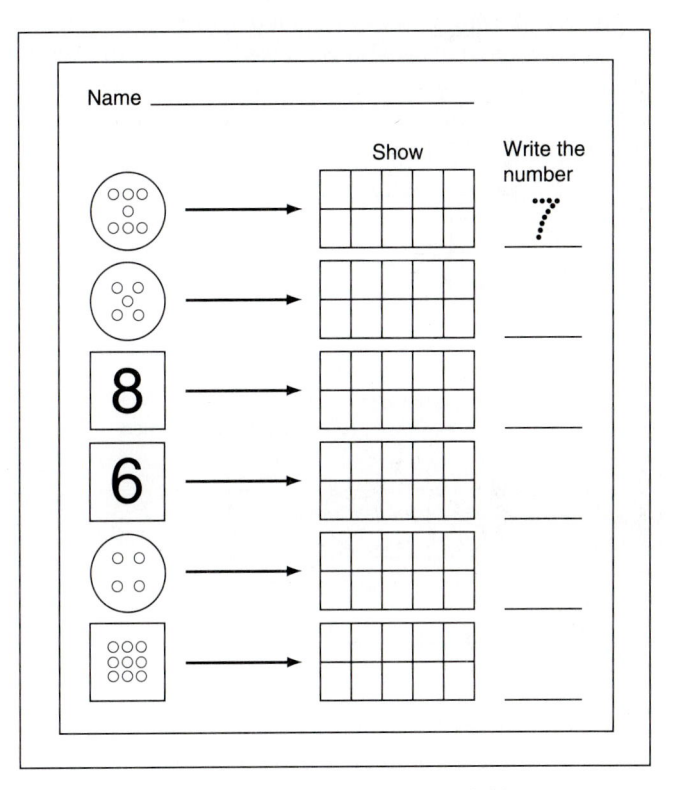

FIGURE 6.13: *A ten-frame record sheet. Children can use a large ten-frame with counters and then draw dots in the ten-frames.*

6.25 TEN-FRAME FLASH

Flash ten-frame cards to the class or group, and see how fast they can tell how many dots are shown. This activity is fast-paced, takes only a few minutes, can be done at any time, and is a lot of fun if you encourage speed.

6.26 FIVE-AND

In **Five-And** the teacher calls out numbers between 5 and 10. The children respond "Five and _____" using the appropriate number. For example, if you say, "Eight!" the children respond, "Five and three."

6.27 MAKE-TEN

In **Make-Ten** the children respond to a number called by calling how many more are needed to make 10. This is most effective with numbers between 5 and 10.

◆ **Ten-Frame Alternatives**

In Japan, number tiles are used almost exclusively to represent numbers (Hatano, 1982). These consist of small squares and unmarked strips that are as long as five squares. It may be, as some have suggested, that the five tiles should be marked to show the five squares. Both ideas are shown in Figure 6.14(a). Tiles can easily be made from poster board using squares of about one inch.

Another variation is also borrowed from Robert Wirtz. He glued beans onto popsicle sticks or tongue depressors, as shown in Figure 6.14(b). These could also be made by drawing dots on strips of poster board.

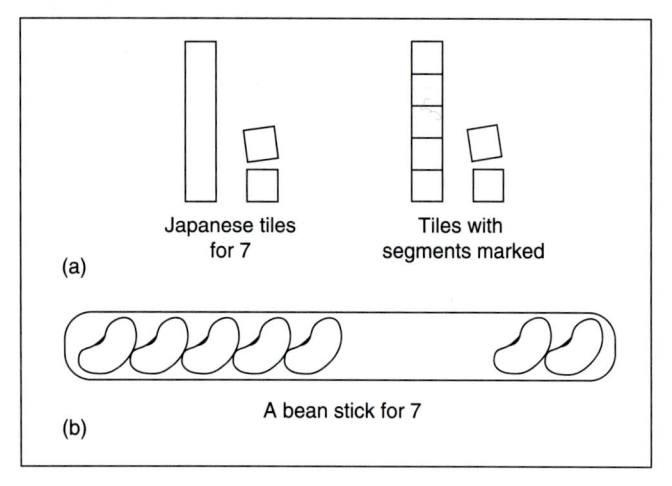

FIGURE 6.14: *Alternatives to ten-frames*

The basic idea with ten-frames, tiles, and bean sticks is to encourage children to think about numbers in relation to 5 and 10. Most of the activities described for ten-frames can be modified to be done with tiles or bean sticks, and certainly other activities can be devised in the same spirit.

Once again the calculator provides a symbolic approach. If [+]5 is stored, then a press of any key from 0 to 5 can be thought of as putting that many counters in the second row. As a variation, if the minus sign on the display is ignored, the children can store [−]10, and then a press of any number followed by [=] shows how much more to get to 10. Children should try to predict the result each time before pressing [=].

PART-PART-WHOLE RELATIONSHIPS

Consider what you think about when you count a set of seven objects. What is significant is what you do *not* think about. Specifically, counting a set will never cause you to focus on the fact that it could be made up of two parts and what the size of those parts might be. Without activities that focus on quantities in terms of two or more parts, many children simply continue to use counting as their principal means of accessing quantity. A noted researcher in children's number concepts, Lauren Resnick, states:

> Probably the major conceptual achievement of the early school years is the interpretation of numbers in terms of part and whole relationships. With the application of a Part-Whole schema to quantity, it becomes possible for children to think about numbers as compositions of other numbers. This enrichment of number understanding permits forms of mathematical problem solving and interpretation that are not available to younger children. (1983, p. 114)

A study of kindergarten children examined the effects of part-part-whole activities on number concepts (Fischer, 1990). With only 20 days of instruction to develop the part-whole structure, children showed significantly higher achievement than the control group on number concepts, word problems, and place-value concepts.

◆ **The Basic Ingredients of Part-Part-Whole Activities**

Most part-part-whole activities focus on a single number for the entire activity. Thus, a child or group of children working together might work on the number 7 throughout the activity. Children use a wide variety of materials and formats to either build the designated quantity in two or more parts, or else they start with the full amount and separate it into two or more parts. A group of two or three children may work on one number within one activity for 5 to 20 minutes. Kindergarten children will usually begin these activities working on either 4 or 5. As concepts develop, the children can extend their work to numbers from 6 to 12. It is not unusual to find children in the second grade who have not developed firm part-part-whole constructs for numbers in the 7 to 12 range.

It is important that as they do these activities children say or "read" the parts aloud and/or write them down on some form of recording sheet. Reading or writing the combinations serves as a means of encouraging reflective thought focused on the part-whole relationship. Writing can be in the form of drawings, numbers written in blanks (_____ and _____) or in the form of addition equations if these have been introduced (3 + 5 = 8). There is a clear connection between part-part-whole concepts and addition and subtraction ideas.

A special and most important variation of part-part-whole activities is referred to as *missing-part* activities. In a missing-part activity children have one of two parts hidden. They know the whole amount and use their already developed knowledge of the parts of that whole to try to tell what the covered or hidden part is. If they do not know or are unsure, they simply uncover the unknown part and say the full combination as they would normally. These activities should never be conducted as "problems" or "tests" with right and wrong answers. They are simply learning activities where children try to think of a part they cannot see. Missing-part activities provide maximum reflection on the combinations for a number. They also are the forerunner to subtraction concepts. With a whole of 8 but with only 3 showing, the child can later learn to write "8 − 3 = 5."

In most of the part-part-whole activities that involve materials, children will set out materials in two or more parts to illustrate their designated whole. It is reasonable to have children make at least eight sets of materials. Each can be placed on a small mat or piece of construction paper. Two or three children working together may have quite a large number of displays, all using the same materials and all representing the same quantity in two or more parts. As you come around to observe, ask individuals to "read a number sentence" to go with the designs. Encourage children to read their designs to each other.

◆ **Part-Part-Whole Activities**

6.28 BUILD IT IN PARTS

The ideas here are illustrated in Figure 6.15.

FIGURE 6.15: *Two-color materials for building parts of six*

Make sets with "two-color counters" such as lima beans spray-painted on one side or picture mat board cut into small squares.

Make bars of Unifix cubes or other plastic cubes that interlock. Make each bar with two colors.

Color rows of squares on one-inch grid paper.

Make combinations with "dot strips," which are simply strips of poster board with dots on them. Make lots of strips with from one to four dots and some strips with five to 10 dots. The strips are only as long as the row of dots that is on them. (Punch holes in the dots to make a set for the overhead projector.)

Make combinations with two-column strips. These are cut from tagboard ruled in one-inch squares. Except for the single square, all pieces are cut from two columns of squares. Odd numbers will have an odd square on one end. (Punch holes in center of the squares to make a set for the overhead projector.)

6.29 PART-PART-WHOLE DESIGNS

Have students make designs with different materials. Each design has the same prescribed number of objects. When talking with the children about their designs, ask how they can see their design in two parts or in three parts.

Make arrangements of wooden cubes.

Make designs with pattern blocks. It is a good idea to use only one or two shapes at a time.

Make designs with flat toothpicks. These can be dipped in white glue for a permanent record on small squares of construction paper.

Make designs with touching squares or triangles. Cut a large supply of small squares or triangles out of construction paper. These can also be pasted down.

It is both fun and useful to challenge children to see their designs in different ways producing different number combinations. In Figure 6.16 (p.100), decide how children look at the designs to get the combinations listed under each.

6.30 TWO OUT OF THREE

Make lists of three numbers, two of which total the whole that children are focusing on. For the number 5 here is an example list:

2 — 3 — 4
5 — 0 — 2
1 — 3 — 2
3 — 1 — 4
2 — 2 — 3
4 — 3 — 1

100 6 / THE DEVELOPMENT OF NUMBER CONCEPTS AND NUMBER SENSE

With the list on the board or overhead, children can take turns selecting the two numbers that make the whole. All children should have a set of five counters to use to confirm the choice. The same activity clearly can be made into a worksheet format.

◆ Missing-Part Activities

Missing-part activities require some way for a part to be hidden or unknown. Usually this is done with two children working together or else in a teacher-directed manner with the class. Again, the focus of the activity remains on a single designated quantity as the whole.

6.31 COVERED PARTS

A set of counters equal to the target amount is counted out and the rest put aside. One child places the counters under a margarine tub or piece of tagboard. He or she then pulls some out into view. (This amount could be none, all, or any amount in between.) For example, if 6 is the whole and 4 are showing, the other child says, "Four and *two* are six." If there is hesitation or if the hidden part is unknown, the hidden part is immediately shown. The focus is on learning and thinking, not on testing and anxiety.

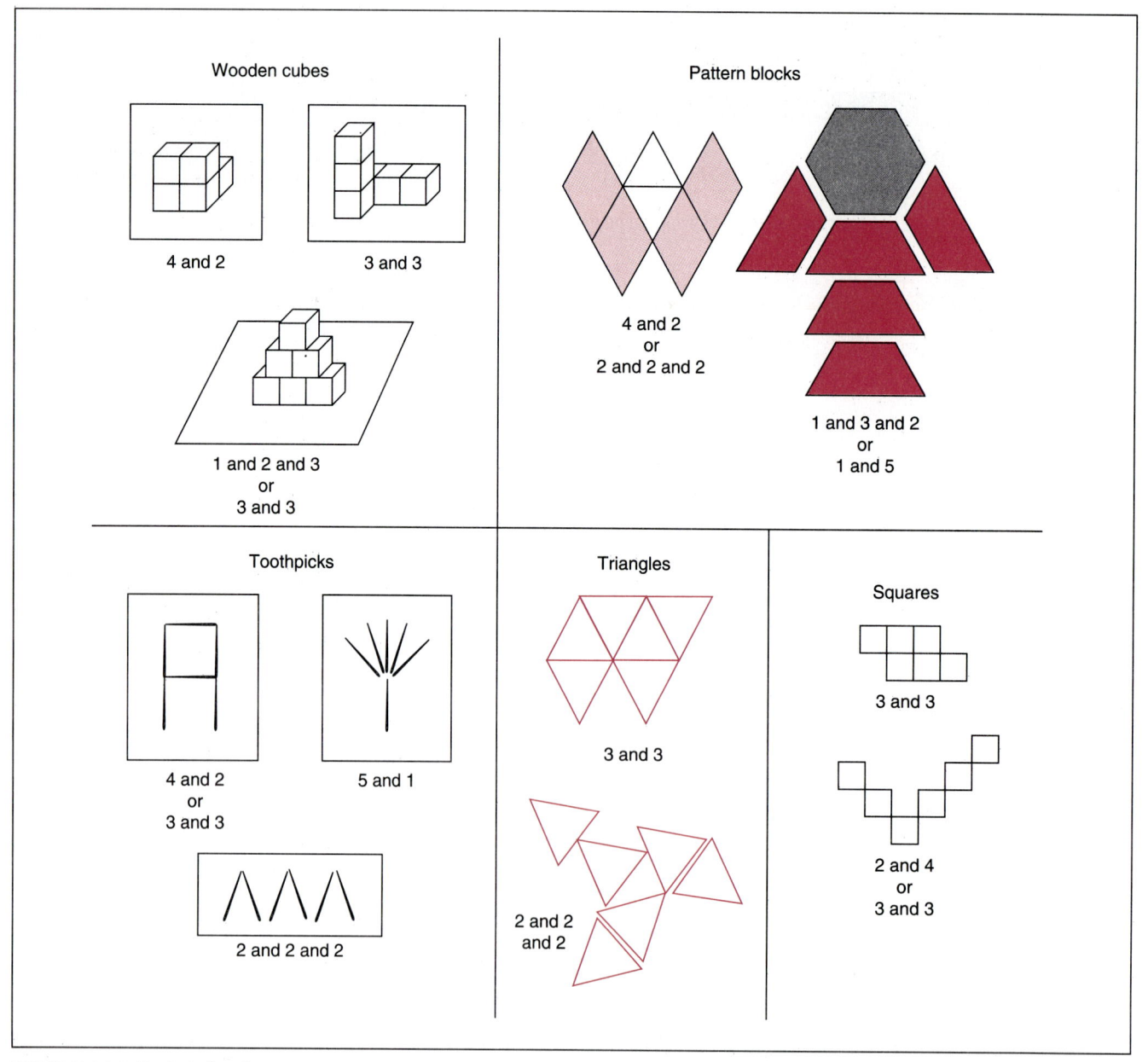

FIGURE 6.16: *Designs for six*

6 / THE DEVELOPMENT OF NUMBER CONCEPTS AND NUMBER SENSE   101

6.32 MISSING-PART CARDS

For each number 4 to 10, make missing-part cards on strips of 3-in. by 9-in. tagboard. Each card has a numeral for the whole and two dot sets with one set covered by a flap. For the number 8 you need nine cards with the visible part ranging from 0 to 8 dots. Students use the cards as in **Covered Parts**, saying "Four and two is six" for a card showing four dots and hiding two. (See Figure 6.17.)

6.33 I WISH I HAD

Hold out a bar of Unifix cubes, a dot strip, a two-column strip, or a dot plate showing 7 or less. Say "I wish I had seven." The children respond with the part that is needed to make 7. Counting on can be used to check. The game can either focus on a single whole, or the "I wish I had" number can change each time.

The following are also missing-part activities but are completely symbolic.

6.34 MISSING PART MACHINE

Draw a "machine" on the board similar to the one in Figure 6.11. This time the machine is a "parts-of-8" machine (or any number). A number "goes in" and the students say the other part. If 3 goes in a parts-of-8 machine, a 5 comes out.

6.35 CALCULATOR MISSING PART

On the calculator, store the whole number by pressing ⊟ 8 ⊟, for example. (Ignore the minus sign.) Now if any number from 0 to 8 is pressed followed by ⊟, the display shows the other part. Children should try to say the other part before they press ⊟.

DOT CARD ACTIVITIES

Many good number development activities involve more than one of the relationships discussed so far. As children learn about ten-frames, patterned sets, and other relationships, the dot cards found in Black-line Masters provide a wealth of activities. The cards contain dot patterns, patterns that require counting, combinations of two and three simple patterns, and ten-frames with "standard" as well as unusual placements of dots. When children use these cards for almost any activity that involves number concepts, the cards make them think about numbers in many different ways. The dot cards add another dimension to many of the activities already described and can be used effectively in the following activities. (See Figure 6.18, p. 102.)

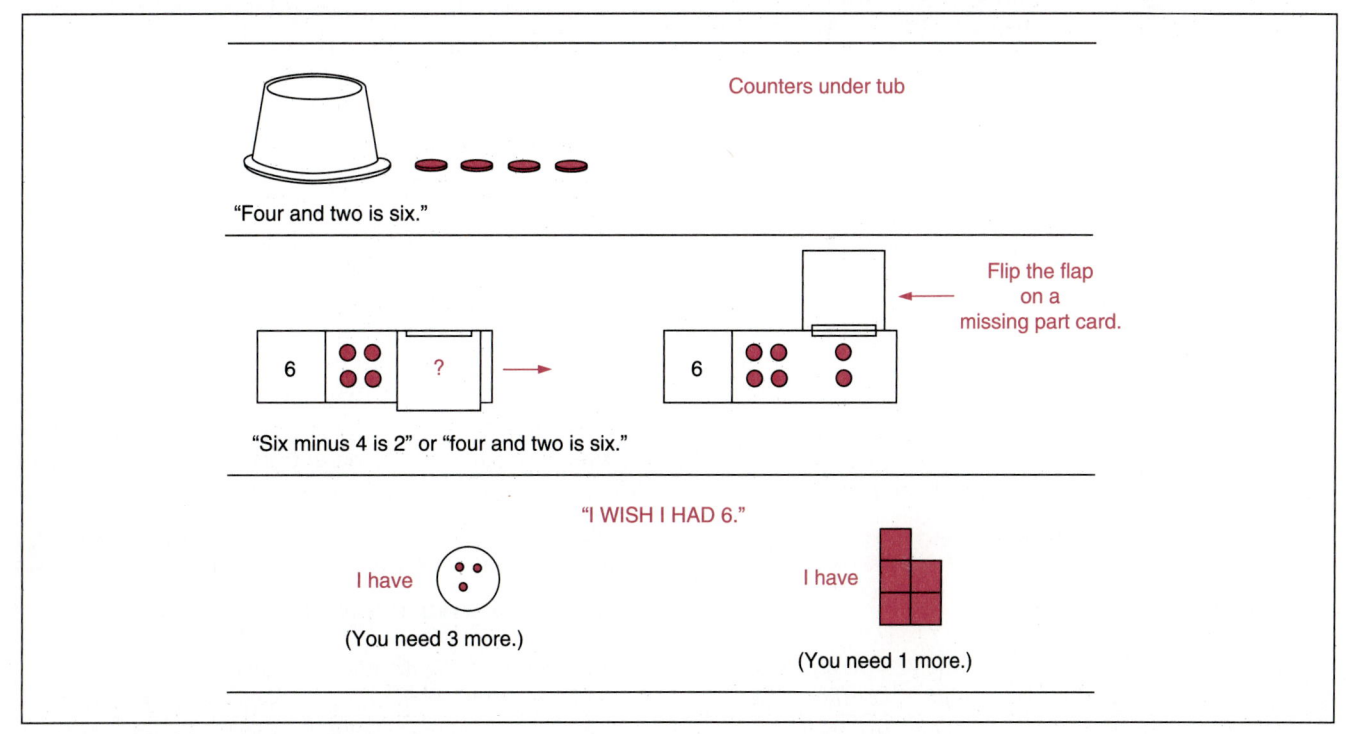

FIGURE 6.17: *Missing-part activities*

FIGURE 6.18: *Dot cards*

◆◆◆◆◆◆◆◆◆◆◆◆◆◆◆◆◆◆◆◆◆

6.36 DOUBLE WAR

The game of **Double War** (Kamii, 1985) is played like war, but on each play both players turn up two cards instead of one. The winner is the one with the larger total number. Children playing the game can use many different number relationships to determine the winner without actually finding the total number of dots.

◆◆◆◆◆◆◆◆◆◆◆◆◆◆◆◆◆◆◆◆◆

6.37 DOT-CARD TRAINS

Make a long row of dot cards from 0 up to 9 and then start back again to 1 and then up, and so on. Alternatively, begin with 0 or 1, and make a two-more/two-less train.

◆◆◆◆◆◆◆◆◆◆◆◆◆◆◆◆◆◆◆◆◆

6.38 DIFFERENCE WAR

Besides dealing out the cards to the two players as in regular war, prepare a pile of about 50 counters. On each play the players turn over their cards as usual. The player with the greater number of dots wins as many counters from the pile as the difference between the two cards. The players keep their cards. The game is over when the counter pile runs out. The player with the most counters wins the game.

◆◆◆◆◆◆◆◆◆◆◆◆◆◆◆◆◆◆◆◆◆

6.39 MISSING PART COMBOS

Select a number between 5 and 12 and find combinations of two cards that total that number. When students have found at least 10 combinations, one student can then turn face down one card in each group. The next challenge is to name the card that was turned down.

RELATIONSHIPS FOR NUMBERS 10 TO 20

Even though young children daily experience numbers up to 20 and beyond, it should not be assumed that they will automatically extend the rich set of relationships they have developed on smaller numbers to the numbers beyond 10. And yet these numbers play a big part in many simple counting activities, in basic facts, and in much of what we do with mental arithmetic. Relationships on these numbers are just as important as relationships on the numbers through 10.

A PRE-PLACE-VALUE RELATIONSHIP WITH 10

A set of 10 should play a major role in children's initial understanding of numbers between 10 and 20. When children see a set of 6 with a set of 10, they should know without counting that the total is 16. However, the numbers between 10 and 20 are *not* an appropriate place to discuss place-value concepts. That is, proir to a much more complete development of place-value concepts (appropriate for second grade and beyond), children should not be asked to explain the "1" in "16" as representing "one ten." (Stop for a moment and say to yourself: "one ten." Think about it! What would this mean to a five-year-old? Ten is a lot. How can it be one? This is initially a strange idea.) The inappropriateness of discussing "one ten and six ones" (what's a one?) does not mean that a set of 10 should not figure prominently in the discussion of the teen numbers. The following activities illustrate this idea.

◆◆◆◆◆◆◆◆◆◆◆◆◆◆◆◆◆◆◆◆◆

6.40 TEN AND SOME MORE

Use a simple two-part mat, and have children count out 10 counters onto one side. Next have them put five counters on the other side. Together count all of the counters by ones. Chorus the combination: "Ten and five is fifteen." Turn the mat around: "Five and ten is fifteen." Repeat with other numbers in a random order but without changing the 10 side of the mat.

The **Crazy Mixed-Up Numbers** activity (6.23) can be extended to include numbers to 20. Provide each child with two ten-frames drawn on a construction paper mat. One ten-frame is placed under the other. For numbers greater than 10, the new rule is that one ten-frame must be completely filled. Periodically say a number sentence for a number represented. For numbers less than 10, one part is always 5: "Five and three make eight." For numbers greater than 10, one part is always 10: "Ten and seven make seventeen."

Frequently represent numbers between 10 and 20 using one ten-frame filled and additional counters or dots not in a ten-frame. It is not necessary to always have the frame to the left of the single counters. The oral description is simply "Ten and _____ " or "_____ and ten." The place-value notion of one ten is avoided until place-value concepts are firmly established.

EXTENDING MORE AND LESS RELATIONSHIPS

The relationships of one-more-than, two-more-than, one-less-than, and two-less-than are important for all numbers. However, these ideas are built on or connected to the same concepts for numbers less than 10. The fact that 17 is 1 less than 18 is connected to the idea that 7 is 1 less than 8. Children may need help in making this connection.

6.41 MORE AND LESS EXTENDED

On the overhead show 7 counters, and ask what is 2 more, or 1 less, and so on. Now add a filled ten-frame to the display (or 10 in any pattern), and repeat the questions. Pair up questions by covering and uncovering the ten-frame as illustrated in Figure 6.19.

• DOUBLE AND NEAR-DOUBLE RELATIONSHIPS

The use of doubles (double 6 is 12) and near-doubles (13 is double 6 and 1 more) is generally considered a strategy for memorizing basic addition facts. There is no reason why children should not begin to develop these relationships long before they are concerned with memorizing basic facts. Doubles and near-doubles are simply special cases of the general part-part-whole construct.

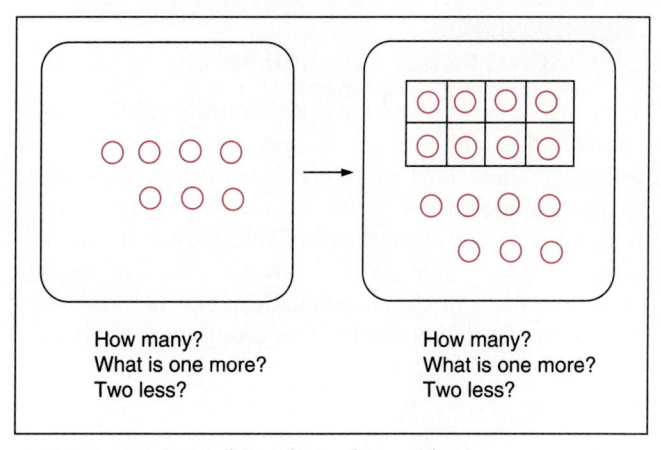

FIGURE 6.19: *Extending relationships to the teens*

6.42 DOUBLE IMAGES

Relate the doubles to special images. Thornton (1982) helped first graders connect doubles to these visual ideas:

Double 3 is the bug double: three legs on each side.

Double 4 is the spider double: four legs on each side.

Double 5 is the hand double: two hands.

Double 6 is the egg carton double: two rows of six eggs.

Double 7 is the two-week double: two weeks on the calendar.

Double 8 is the crayon double: two rows of eight crayons in a box.

Double 9 is the 18-wheeler double: two sides, nine wheels on each side.

Children should draw pictures or make posters for each double. (See Figure 8.3.)

6.43 WHAT'S THE DOUBLE NUMBER ?

Give a number orally, and ask students to tell what double it is. "What is fourteen?" (Double 7!) When students can do this well, use any number up to 20. "What is seventeen?" (Double 8 and 1 more.)

6.44 THE DOUBLE MAKER

On the calculator, store the "double maker" (2 ⊠ ⊟). Now a press of any digit followed by ⊟ will produce the double. Children can work in pairs or individually to try to beat the calculator.

EXPANDING EARLY NUMBER SENSE

So far we have discussed the development of number meanings and some very specific relationships on numbers. These ideas are very much a part of what is meant by number sense. In this section we examine ways to broaden the early knowledge of numbers even further. Relationships of numbers to real-world measures and the use of numbers in simple estimations can help children develop the flexible, intuitive ideas about numbers that are most desired.

NUMBER AND THE REAL WORLD

Activities in this section are designed to help children connect ideas about numbers to real things in their world. These activities should be interspersed with those already discussed.

Estimation and Measuring

One of the best ways for children to think of real quantities is to associate numbers with measures of things. In the early grades, measures of length, weight, and time are good places to begin. Just measuring and recording results will not be very effective, however, since there is no reason for children to be interested in or to think about the result. To help children think or reflect a bit on what number might tell how long the desk is or how heavy the book is, it would be good if they could first write down or tell you an estimate. To produce an estimate is, however, a very difficult task for young children. They do not understand the concept of "estimate" or "about." For example, suppose that you have cut out of poster board an ample supply of very large foot prints, say about 18 inches long. All are exactly the same size. You would like to ask the class, "About how many footprints will it take to measure across the rug in our reading corner?" The key word here is *about,* and it is one that you will need to spend a lot of time helping children understand. To this end, the request of an estimate can be made in ways that help with the concept of about and still permit children to respond. Here are some suggestions that can be applied to this example and to most other early estimation activities:

- **More or less than** _____? Will it be more or less than 10 footprints? Will the apple weigh more or less than 20 wooden blocks? Are there more or less than 15 Unifix cubes in this long bar?
- **Closer to** _____ **or to** _____? Will it be closer to 5 footprints or closer to 20 footprints? Will the apple weigh close to 10 cubes or closer to 30 cubes? Does this bar of Unifix have closer to 10 cubes or closer to 50 cubes?
- **Less than** _____, **between** _____ **and** _____, **or more than** _____? Will it take less than 10, between 10 and 20, or more than 20 footprints? Will the apple weigh less than 5 cubes, between 5 and 15 cubes, or more than 15 cubes? Are there less than 20, between 20 and 50, or more than 50 cubes in this Unifix bar?
- **About** _____. Use one of these numbers: 5, 10, 15, 20, 25, 30, 35, 40, ... About how many footprints? About how many cubes will the apple weigh? About how many cubes are in this Unifix bar?

This list of estimation-question formats is ordered from the easiest to most difficult. However, notice that each clearly indicates to the children that they need not come up with an exact amount or number. Asking for estimates using these formats has several advantages. It helps children learn what you mean by "about." Every child can select a response without having to pull a number out of the air. Response cards can be used to quickly poll the class. For example, if each child has a green and a yellow card, these can be labeled "more" and "less," or "closer to 10" and "closer to 30," and so on. Then, when the measurement estimation question is posed, all can respond quickly and at the same time. One child's response will not influence the others.

For almost all measurement estimation activities at the K–2 level, it is good to use an informal measuring unit instead of standard units such as feet, centimeters, pounds, kilograms. This avoids the problem of children's lack of familiarity with the unit. If, however, you want them to learn about a unit of measure, say the meter, then by all means include that unit in these exercises.

Another suggestion is to estimate several things in succession using the same unit. This will help children develop an understanding of relative measures. For example, suppose you are estimating and measuring "around things" using a string. To measure, the string is wrapped around the object and then measured in some unit such as craft sticks. After measuring the distance around Demetria's head, estimate the distance around the waste basket or around the globe, or around George's wrist. Each successive measure helps children with the new estimates.

An alternative approach is to estimate and measure the same item successively with different sized units. You will be surprised when young children guess that the measure is smaller when the unit gets smaller. The ideas that smaller units produce larger measurements, and large units small measurements, are difficult relationships for children to construct. (See Chapter 16 for a more complete discussion of measurement.)

More Connections

Here are some additional activities that can help children connect numbers to real situations.

6.45 ADD A UNIT TO YOUR NUMBER

Write a number on the board. Now suggest some units to go with it, and ask the children what they can think of that fits. For example, suppose the number is 9. "What do you think of when I say 9 *dollars*? 9 *hours*? 9 *cars*? 9 *kids*? 9 *meters*? 9 *o'clock*? 9 *hand spans*? 9 *gallons*?" Spend some time with each unit. Let children suggest units as well. Be prepared to explore some of the ideas either immediately or as projects or homework tasks.

6.46 IS IT REASONABLE?

As with **Add a Unit**, pick a number and a unit. For example, 15 feet. Could the teacher be 15 feet tall? Could your living room be 15 feet wide? Can a man jump 15 feet high? Could three children stretch their arms 15 feet? Pick any number, large or small, and a unit with which children are familiar. Then make up a series of these questions.

6.47 THINGS THAT ARE SEVEN

Pick any number (seven is used here as an example) and have groups of children find ways to tell about that number. Seven might be the days in a week, the number of people in Mandy's family, or the number of kittens in Inky's litter. Other children might want to show seven in some measurement way such as a stack of seven books, seven glasses of water in a jug, seven long giant steps, the length of seven children lying down head to toe, or how many hops they can hop in seven seconds. Groups can present their ideas orally, make a written report, contribute a page to put on the bulletin board, or present a demonstration.

The last three activities will require a lot of help from you in the beginning. Children need many opportunities to connect number with their environment, and they will not have these connections when you begin. A first-grade teacher of children from very impoverished backgrounds noted: "They all have fingers, the school grounds are strewn with lots of pebbles and leaves, and pinto beans are cheap. So we count, sort, compare, and talk about such objects. We've measured and weighed almost everything in this room and almost everything the children can drag in" (in Howden, 1989, p. 6). This teacher's children had a rich and long list of responses to the question: "What comes to your mind when I say twenty-four?" In another school in a professional community where test scores are high, the same question brought almost no response from a class of third graders. It can be a very rewarding effort to help children connect their number ideas to the real world.

◆ Graphs

Graphing activities are another good way to connect children's worlds with number. Chapter 19 discusses ways to make graphs with children in grades K–2. Graphs can be quickly made of most any data that can be gathered from the students: favorite ice cream, color, sports team, pet; number of sisters and brothers, kids who ride different buses, types of shoes, number of pets, and so on. Graphs can be connected to content in other areas. A unit on sea life might lead to a graph of favorite sea animals.

Once a simple bar graph is made, it is very important to take a few minutes to ask as many number questions as is appropriate for the graph. In the early stages of number development (grades K–1), the use of graphs for number relationships and for connecting numbers to real quantities in the children's environment is a more important reason for building graphs than the graphs themselves. The graphs focus attention on counts of realistic things, an important connection. Equally important, bar graphs clearly exhibit comparisons between and among numbers that are rarely made when only one number or quantity is considered at a time. See Figure 6.20 (p. 106) for an example of a graph and questions that can be asked. At first children will have trouble with the questions involving differences, but repeated exposure to these ideas in a bar-graph format will improve their understanding. These comparison concepts add considerably to children's understanding of number.

EXTENSIONS TO EARLY MENTAL MATH

Teachers in about the second and third grade can capitalize on some of the early number relationships and extend them to numbers up to a hundred. A useful set of materials to help with these relationships is the little base-ten-frames found in the Black-line Masters section. Each child can have a set of 10 tens and a set of frames for each number one to nine with an extra five.

The one-more-than/one-less-than relationships can be extended to one more *ten* (or one less *ten*) as shown in Figure 6.21(a)(p. 107). If you later include a part-filled frame, children can almost as easily think about 10 (or 20) more or less than 63.

The relationship of numbers to the ten anchor can be used for adding numbers "through ten." That is, to add 6 and 9, think of building from 9: one more to get to ten (leaving 5 from the 6) and 5 more is 15. In Chapter 8, this idea is developed in more detail. With the little base-ten-frames, children can easily extend the same approach to adding a number to any two-digit number, as illustrated in Figure 6.21(b). Subtracting mentally can be done similarly.

Part-part-whole ideas provide another powerful method of giving children a mental control of numbers up to 100. For example, the parts-of-8 ideas can be extended to parts of 80. First work with 8 counters as before. Children separate the counters into two parts and say the combination: "Two and six, seven and one, and so on." Now substitute 8 of the full little base-ten-frames for the 8 counters and repeat: "Twenty and sixty, seventy and ten, and so on." Talk with the students about the similarities between the parts-of-8 activity and the parts-of-80. Use the same materials in a missing-part activity. "Thirty and what makes 80?" Any missing-part activity can be adapted to parts of 80 or 90 or 50, and so on.

After spending some time using only tens, substitute two five-frames for one of the tens. Repeat the part-part-whole and missing-part activities, but now make the rule that one of the five-frames must be in each part. The children can begin to think about separating 80 into two parts ending in 5. What they will soon discover is that this really involves thinking about parts of 7 instead of parts of 8. (Why?) You can see this progression in Figure 6.21(c). The two fives can also be exchanged for any two cards with a sum of ten.

By extending the ideas of parts of small numbers to parts of large numbers by using sets of ten, children are actually preparing for mental addition and subtraction of two-digit numbers. Twenty and 40 is an extension of 2 and 4. There are a variety of ways to think about 35 and 45 or even 26 or 44. Missing-part is actually subtraction: 80 − 35 is the same as 35 and what makes 80.

More will be said about early mental computation in Chapter 11. The point to be made here is that early relationships on number have a greater impact on what children know than may be apparent at first. Even teachers in the upper grades may profitably consider the use of ten-frames and part-part-whole activities.

SUMMARY

Number sense is not a topic we teach and that children master. It is a way of thinking about number and quantity that is flexible, intuitive, and very individualistic. It grows slowly over years of exposure to all sorts of activities that

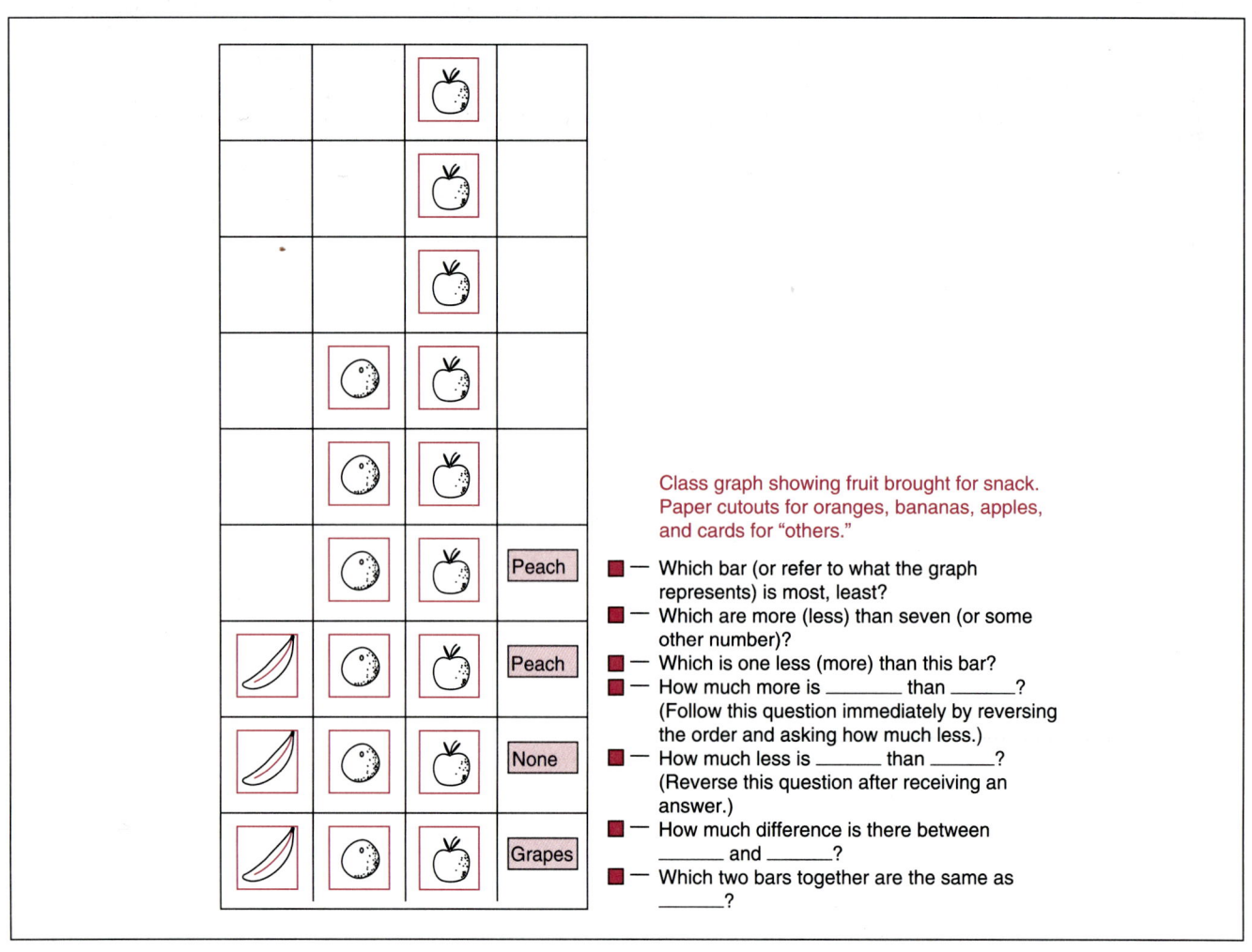

Class graph showing fruit brought for snack. Paper cutouts for oranges, bananas, apples, and cards for "others."

■ — Which bar (or refer to what the graph represents) is most, least?
■ — Which are more (less) than seven (or some other number)?
■ — Which is one less (more) than this bar?
■ — How much more is _____ than _____? (Follow this question immediately by reversing the order and asking how much less.)
■ — How much less is _____ than _____? (Reverse this question after receiving an answer.)
■ — How much difference is there between _____ and _____?
■ — Which two bars together are the same as _____?

FIGURE 6.20: *Comparisons in a bar graph*

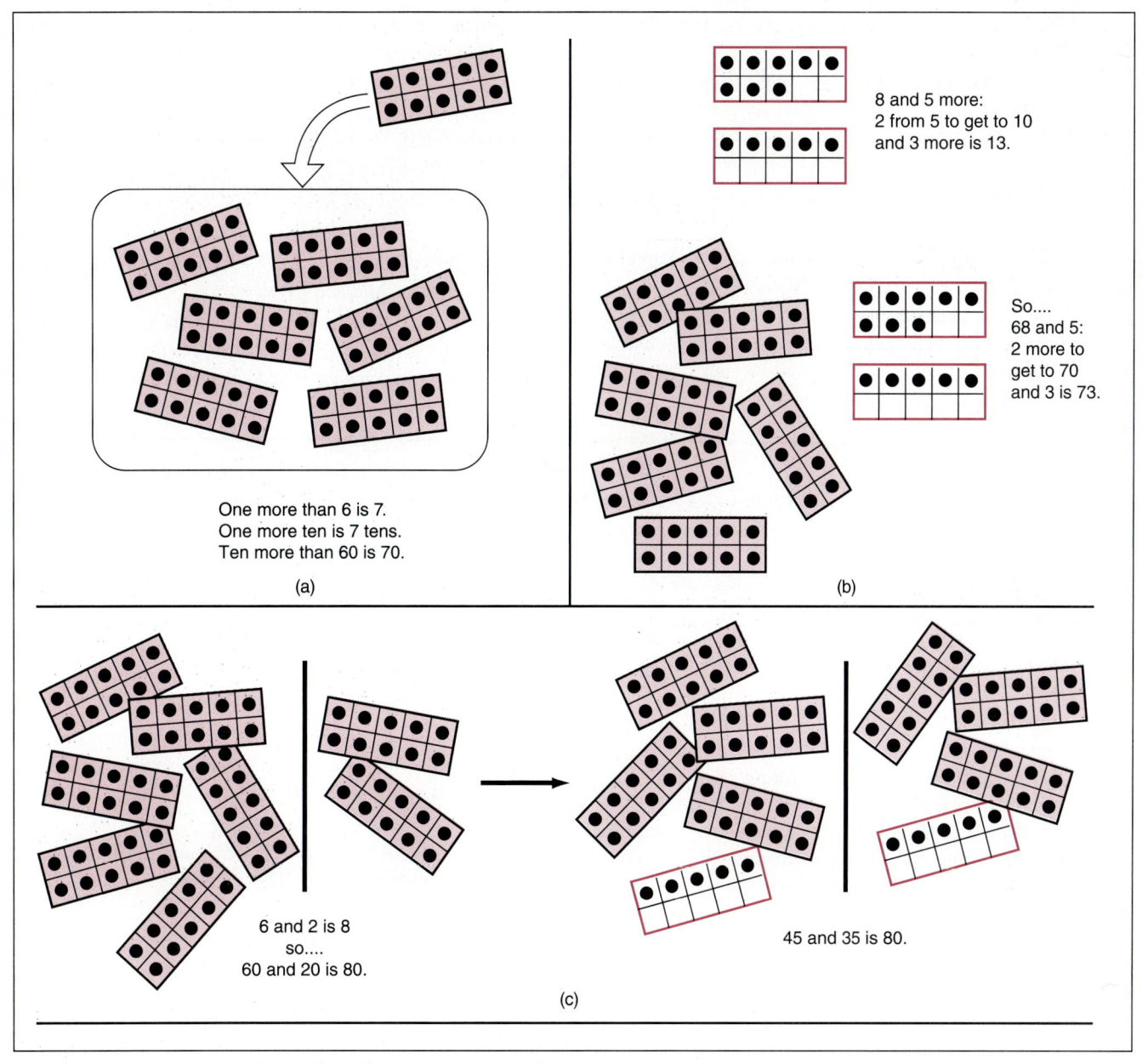

FIGURE 6.21: *Extending early number relationships to mental computation activities*

cause children to think about number in many ways and in many different contexts. Number sense is also a way of teaching that can and should pervade all of our instruction that has to do with number in any way. In this chapter we looked at number sense at the very beginning of the number curriculum. You have also seen a brief glimpse of how number sense is a factor in mastery of basic facts, in mental mathematics, and in measurement estimation. Throughout this book you will find the number sense agenda whenever numbers are involved.

◆◆◆◆◆
REFLECTIONS ON CHAPTER 6: WRITING TO LEARN

1. How can you tell if children are attaching any meaning to their counting?

2. Describe an activity that is a "set to numeral match" activity, and describe what ideas the child must have to do these activities meaningfully and correctly.

3. What are the four types of relationships that have been described for small numbers? Explain briefly what each of these means, and give at least one activity for each.

4. Describe a missing-part activity, and explain what should happen if the child trying to give the missing part does not know it.

5. Describe in detail how to do three different calculator activities for developing number ideas, and briefly explain what ideas each activity is designed to develop. (Check your key presses out on a calculator to be sure you have these correct.)

6. For numbers between 10 and 20, describe how to develop each of these ideas:
   - The idea of the teens as a set of ten and some more
   - Extension of the one-more/less concept to the teens
7. Describe in two or three sentences your own idea of what number sense is.
8. What are three ways that children can be helped to connect numbers to real-world ideas?
9. Describe two examples of how early number relationships can be used to develop some early mental computation skills.

◆◆◆◆◆
FOR DISCUSSION AND EXPLORATION

1. Examine a textbook series for Grades K–2. Compare the treatment of counting and number concept development with that presented in this chapter. What ideas are stressed? What ideas are missed altogether? If you were teaching in one of these grades, how would you plan your number concept development program? What part would the text play?
2. Explore number concepts of some children in the first or second grade. Consider checking their understanding of counting on, counting back, and their use of the relationships of one (two) more and less. Try using dot plates and ten-frame cards to see how children relate to these representations of numbers.
3. Discuss how much time out of the year can and should be spent on the development of number relationships in grades K, 1, or 2.
4. Many teachers in grades above the second find that their children do not possess the number relationships discussed in this chapter but rely heavily on counting. Given the pressures of other content at these grades, how much effort should be made to remediate these number concept deficiencies?
5. Discuss the importance of number sense in the K–4 curriculum. You might want to read "Standard 6: Number Sense and Numeration" in the K–4 section of the *Standards*. Do you find that teachers and/or current textbooks are adequately addressing the issue of number sense? Is it really important? Why?

◆◆◆◆◆
SUGGESTED READINGS

Baratta-Lorton, M. (1976). *Mathematics their way*. Menlo Park, CA: Addison-Wesley.

Baratta-Lorton, M. (1979). *Workjobs II*. Menlo Park, CA: Addison-Wesley.

●Burk, D., Snider, A., & Symonds, P. (1988). *Box it or bag it mathematics: Teachers resource guide* [Kindergarten, First–Second]. Salem, OR: The Math Learning Center.

●Burton, G. (1993). *Number sense and operations: Addenda series, grades K–6*. Reston, VA: National Council of Teachers of Mathematics.

Coombs, B., & Harcourt, L. (1986). *Explorations 1*. Don Mills, Ontario: Addison-Wesley.

Fischer, F. E. (1990). A part-part-whole curriculum for teaching number in the kindergarten. *Journal for Research in Mathematics Education, 21*, 207–215.

Fuson, K. C. (1989). *Children's counting and concepts of number*. New York: Springer-Verlag.

●Greenes, C., Schulman, L., & Spungin, R. (1993). Developing sense about numbers. *Arithmetic Teacher, 40*, 279–284.

Kelly, B., Wortzman, R., Cornwall, J., Kennedy, N., Maher, A., & Nimigon, B. *Mathquest one: Teacher's edition*. Don Mills, Ontario: Addison-Wesley.

Kroll, D. L., & Yabe, T. (1987). A Japanese educator's perspective on teaching mathematics in the elementary school. *Arithmetic Teacher, 35*(2), 36–43.

Leutzinger, L. P., & Bertheau, M. (1989). Making sense of numbers. In P. R. Trafton (Ed.), *New directions for elementary school mathematics*. Reston, VA: National Council of Teachers of Mathematics.

Liedtke, W. (1983). Young children—small numbers: Making numbers come alive. *Arithmetic Teacher, 31*(1), 34–36.

National Council of Teachers of Mathematics. (1990). *Number sense now! Reaching the NCTM Standards* [Video and teacher's guide]. Reston, VA: The Council.

Payne, J. N., & Huinker, D. M. (1993). Early number and numeration. In R. J. Jensen (Ed.), *Research ideas for the classroom: Early childhood mathematics*. New York: Macmillan Publishing Co.

●Thompson, C. S., & Rathmell, E. C. (Eds.). (1989). Number sense. Special issue of *Arithmetic Teacher, 36*(6).

●Van de Walle, J. A. (1988). The early development of number relations. *Arithmetic Teacher, 35*(6), 15–21, 32.

Van de Walle, J. A. (1990). Concepts of number. In J. N. Payne (Ed.), *Mathematics for the young child*. Reston, VA: National Council of Teachers of Mathematics.

●Van de Walle, J. A., & Watkins, K. B., (1993). Early development of number sense. In R. J. Jensen (Ed.), *Research ideas for the classroom: Early childhood mathematics*. New York: Macmillan Publishing Co.

7 DEVELOPING MEANINGS FOR THE OPERATIONS

◆ THIS CHAPTER IS ABOUT HELPING CHILDREN CONNECT DIFFERent meanings, interpretations, and relationships to the four operations so that they can effectively use these operations in other settings.

It should be clear that the activities of this chapter are not designed to produce basic fact mastery (being able to respond quickly 4 × 9 = ☐ or 12 − 8 = ☐) or computational skill. Conceptual knowledge of the operations is certainly important in mastering basic facts but, as will be seen in Chapter 8, it takes more than an understanding of the operations to develop mastery.

The main thrust of this chapter is on helping children develop what might be termed "operation sense," a highly integrated understanding of the four operations and the many different but related meanings these operations take on in real contexts.

TWO SOURCES OF OPERATIONS MEANINGS

For all four of the operations, models (usually sets of counters and number lines) and word stories or word problems are the two basic tools the teacher has to help students develop operation concepts.

BASIC MEANINGS DEVELOPED WITH MODELS

Counter or set models for the operations include moveable objects, pictures of sets in various arrangements, arrays (things arranged in rows and columns) and variations of these basic ideas. Number lines are also a good model at the upper grade level but can cause more difficulty than help at the primary level.

The various relationships that constitute the meanings for the operations come from arranging or seeing the models in different configurations and connecting these models with written symbolism. Models, then, help children construct meanings for the operations in the following way: Children *do something with the model* (arrange, separate, join, draw arrows on the number line, count rows and lengths of rows, and so on) and then *write number sentences* (equations) *that mean the same thing as the model.*

At the early stages for addition, the "do-then-write" approach is almost identical to many of the number concept activities discussed in Chapter 6. The difference is the association with symbolism. Part-part-whole ideas and addition are closely related. For subtraction, the writing includes both + and − equations since there is an added objective of connecting the ideas of addition and subtraction. Similarly, when first learning about multiplication, both + and × equations are written for the same models illustrating, for example, that 3 × 5 and 5 + 5 + 5 are related. Division concepts are not only developed through models but are connected to multiplication ideas.

Note that in all of the do-then-write activities, children are to write complete equations. The activities begin with models, and what is written is a full equation that tells about the model. An equation such as 12 − 8 = 4 tells something about the model and how it relates the quantities 12, 8, and 4. The child is doing much more than simply filling in the answer to 12 − 8 = ☐, an activity that is more of a drill than a concept-development activity.

OPERATION MEANINGS FROM WORD PROBLEMS

Another very important way to help children construct operation meanings is from word problems or word stories. The word problem provides an opportunity for examining a much more diverse set of meanings for each operation. Models remain a critical part of the development.

With a word problem or word story, the relationship that is the operation is in the semantics of the story. Children use models to help them analyze the structure of the word problem and to eventually connect that structure or relationship to symbolic equations. Activities with word problems typically begin with a reading of the word story, and then *children use models to "solve" the problem, then write equations to tell what they did.*

As you will see, word stories inject many more variables into the whole notion of understanding an operation. While removing 4 counters from 11 counters may be a model for 11 − 4 = 7, word problems for this situation may be quite different. One story might be about earning $4 less than the $11 earned yesterday (11 − 4 = 7). Another might be about having 11 apples, eating some, and having 7 left (11 − 4 = 7). The action with counters may be the same, but it is quite likely that children will write different equations. Children will need to relate one equation to another that means the same thing. In this case, 11 − 4 = 7 and 11 − 7 = 4.

TRANSLATIONS: MODELS, WORDS, AND SYMBOLS

It is useful to think of models, word problems, and symbolic equations as three separate languages. Each language can be used to express the relationships involved in one of the operations. The model and word problem languages more clearly illustrate the relationships involved, while the equation is a convention used to stand for ideas. Given these three languages, a powerful approach to helping children develop operation meaning is to have them make translations from one language to another. The translations from words to models to symbols and also the translations between models and symbols have already been discussed. Once children develop a familiarity with an assortment of models and drawings and they gain exposure to a wide variety of meanings through word problems, they can begin to make translations from any one language to the other two (Figure 7.1). These translation activities can be done in short 10-minute periods every week, all year long.

In a translation exercise students are provided with an expression in one of the three languages: a model (usually in the form of a drawing), a word problem, or an equation. They are then asked to come up with expressions in each of the other two "languages" that represent the same relationships. Since different cooperative groups or individuals will devise different ideas, the results provide a wonderful source of discussion.

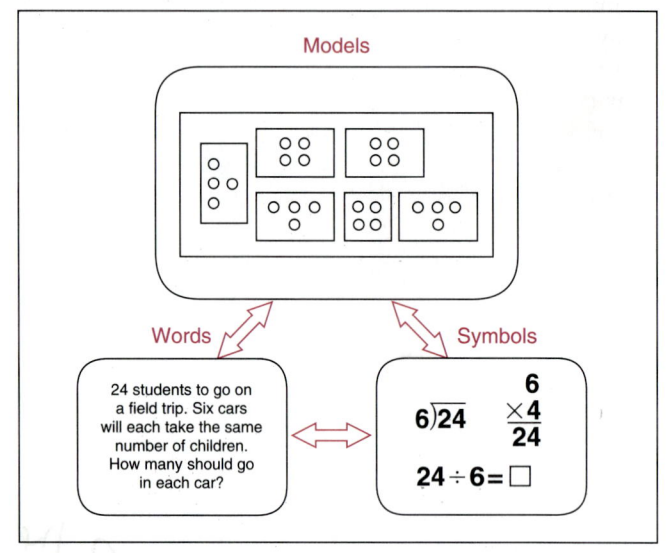

FIGURE 7.1: *Translations can be made from any one of three languages to the other two.*

ADDITION AND SUBTRACTION CONCEPTS

Addition and subtraction concepts are very closely related. Both can be derived from the same basic relationships between sets: either a part-part-whole relationship or a comparison relationship. For each of these there are several models that children can effectively use in the classroom to learn about addition and subtraction.

PART-PART-WHOLE MEANINGS

The part-part-whole concept as discussed in Chapter 6 is a way of understanding that a quantity or *whole* can be composed of two or more separate quantities or *parts.* All materials that can be used to model a part-part-whole meaning of addition or subtraction are simply variations of the ideas in Chapter 6.

Part-Part-Whole Addition Concepts

When the parts of a set are known, *addition* is used to name the whole in terms of the parts. Some researchers have made a distinction between addition situations that involve a joining or "put-together" action and those that are static, having no action. For example, if you place five red counters on a mat and then add or join three blue counters with them, the result is a set with two parts, five and three. If you have a mat with five counters on one side and three on the other, then that also is a set made of two parts, five and three, but there was no action or joining. Children will probably not make the distinction between action and nonaction, but it does lead to different types of word story problems, as will be seen.

Examine the different variations of part-part-whole models for addition shown in Figure 7.2. Each can be written as 5 + 3 = 8. Some of these are the result of a definite put-together or joining action, some are not. It really makes no difference. What is important is to notice that in every example both of the parts are distinct—even after two parts are joined. If counters are used, then the two parts should be kept in separate piles, kept in separate sections of a mat, or be two distinct colors. To illustrate the value of keeping the parts distinct, try this. Set out 5 counters. Now add 3 more of the same type to the pile. Now there is only one set remaining, the whole set of 8. In order for children to see a relationship between the two parts and the whole, the image of the 5 and 3 must be kept in their minds. By keeping the two sets distinguishable, it is possible to reflect on the action after it takes place. "These red chips are the ones I started with. Then I added these five blue ones, and now I have eight altogether."

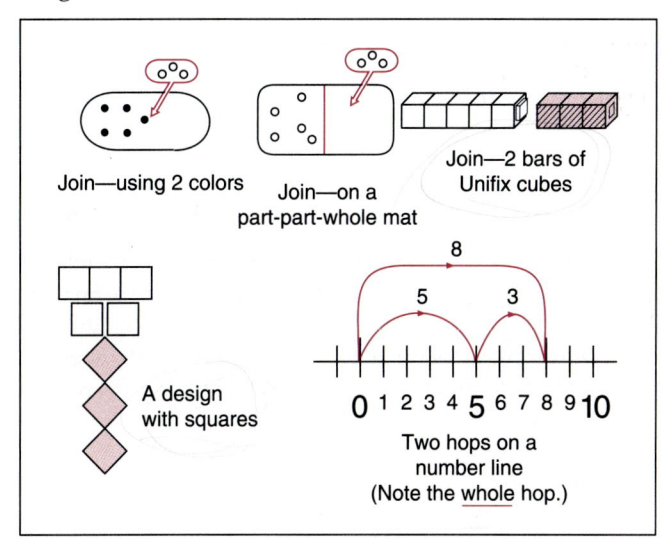

FIGURE 7.2: *Part-part-whole models for 5 + 3*

A number line presents some real conceptual difficulties for first- and second-grade children, and its use as a model at that level is generally not recommended. A number line measures distances from zero the same way a ruler does. The concepts of length measures are poorly developed in early grades. Children focus on the dots or numerals on a number line instead of the spaces (distances or unit lengths). They think of numbers in terms of sets and objects, not lengths. However, if arrows (hops) are drawn for each number in an exercise, the length concept is more clearly illustrated. To model the part-part-whole concept of three plus five, start by drawing an arrow from 0 to 3, indicating, "This much is three." Do not point to the dot for 3, saying "This is three."

To connect these part-part-whole ideas to addition symbolism, recall that the idea is to have children do something with a model and then write what they do in the form of an equation.

7.1 EQUATIONS WITH NUMBER PATTERNS

Recall the two activities from Chapter 6 where children made designs or built sets for a specific number (6.28, **Build It in Parts**, 6.29, **Part-Part-Whole Designs**; and Figures 6.15 and 6.16). Convert these activities to addition concept activities by simply having the children write a plus or addition equation (number sentence) for each of their designs. Remember that in those early activities, the children said a number combination, such as "Four and five is nine." Now they also write the equation "4 + 5 = 9."

There are many ways that children can "write" an equation for the designs in the last activity. Small pieces of paper about the size of an index card can be cut up, and children can write on these. The designs can be copied with crayon or pasted onto the sheet and a little addition booklet made. Remember that in these activities, all of the sums or wholes will be the same as children focus on a single number. Children can use small cards with prewritten numerals and signs on them to avoid having to use pencils and make erasures. They also can write equations in a list on a blank sheet of paper. (See Figure 7.3.)

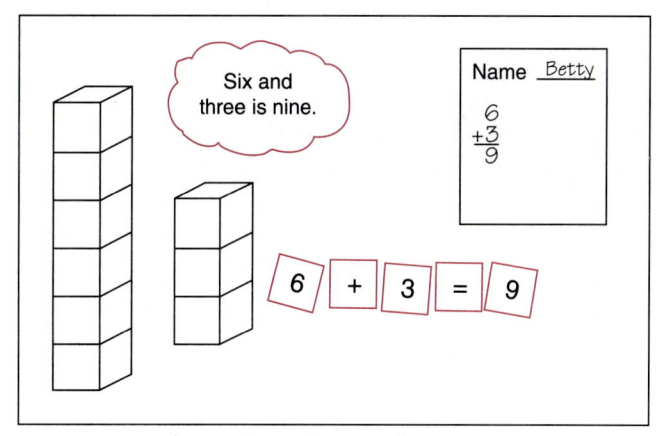

FIGURE 7.3: *Writing about what you do*

Another variation is to have two children work together, do something with a model, say the combination, and together write the equation. With this approach the children can vary the total amounts they construct each time—within appropriate limits, of course.

The popular games of *Mathematics Their Way* (Baratta-Lorton, 1976), *Workjobs II* (Baratta-Lorton, 1979), and many excellent activities in *Box It or Bag It* (Burk, Snider, & Symonds, 1988), and the *Explorations* books (Coombs & Harcourt, 1986) are similar to the activity just described. These books can provide you with a wealth of creative variations on the basic approach.

So far the examples have not involved any action or joining. In joining activities, the amounts you begin with and the amount added or joined will change with each example.

7.2 JOIN ACTION FOR ADDITION

Provide children with a mat depicting some action as in Figure 7.4. (Other ideas are a sliding board, a dump truck, or any motif illustrating action.) With the mat, provide a worksheet that indicates how many to start with and how many to join. Note that in Figure 7.4, children write the complete equation for each example. They also record the numbers in the drawing of the action. It is important to have them use different colors of counters for the two parts so that after the joining action, the two parts remain distinct and the equation can be written after the fact. That is, the equation represents a record of the action. The action does not "solve" the equation.

FIGURE 7.4: *Worksheet and "mat" for join addition*

◆ Part-Part-Whole Subtraction Concepts

In the part-part-whole model, when the whole and one of the parts are known, *subtraction* names the other part. This definition is in agreement with the drastically overused language of "take-away." If you start with a whole set of 9 and remove a set of 4, the two sets that you know are the sets of 9 and 4. The expression $9 - 4$, read "nine *minus* four," names the five remaining. Therefore, nine minus four is five. Compare the models in Figure 7.5(a), showing a remove action, with those in Figure 7.5(b), showing only a missing part but no action. All are models for $9 - 4 = 5$. Again, there is no need to make this distinction with children.

Two points should be made about the models in Figure 7.5. First, in Figure 7.5(b), you may have a tendency to say those are models for $9 - 5$ rather than $9 - 4$, because it looks as if 5 were removed. But the *known* part is what you see. The expression $9 - 4$ tells what the unknown part is. When you take 4 away from a set of 9, you also know the whole (9) and the part, which is 4 (the four you counted). The expression $9 - 4$ names those that remain.

The second point is about keeping the parts distinct, as was discussed with addition. In take-away situations, there is a tendency to focus only on those left. They become the "answer." Those that were removed are often just shoved aside or returned to the supply box. As with addition, this prevents children from reflecting on the situation and reversing it. If the part removed in a $9 - 4 = 5$ situation is then returned, that is exactly an addition action. Both parts and the total are present, again helping students connect the ideas of addition and subtraction.

Activities for part-part-whole subtraction concepts are similar to those for addition with two important differences:

1. The amount left (for action situations) or the unknown part (nonaction situations) should initially be covered. The child is encouraged to "think addition," or "What goes with this part to make the whole amount?"

2. Two equations should almost always be written for the model. That is, if the whole amount is 9, and 3 are showing, the child first writes "$9 - 3 = 6$." Then the unknown part is uncovered, and the corresponding addition equation is written as well. The single model for both equations serves as a means of helping the child construct the relationship between addition and subtraction.

7.3 MISSING PART SUBTRACTION

A fixed number of counters is placed on a part-part-whole mat. One child separates the counters into two parts while the other child hides his or her eyes. The first child covers one of the two parts with a sheet of tagboard revealing the other part. The second child says the subtraction sentence. For example, "Eight minus three (the visible part) is five (the covered part)." The covered part can be revealed if necessary for the child to say how many are there. Both the subtraction equation and the addition equation can then be written.

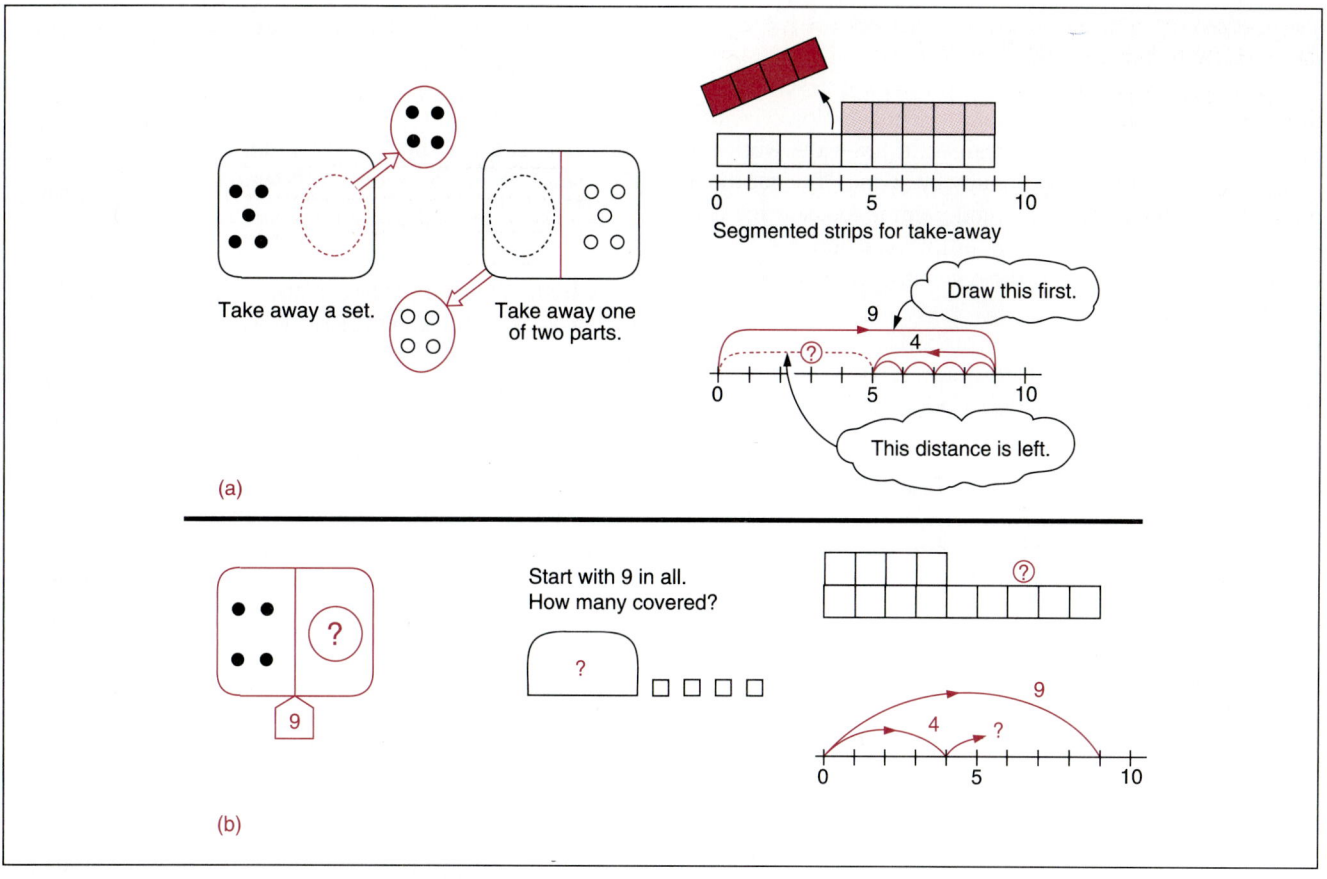

FIGURE 7.5: *Models for 9 − 4 as missing part*

7.4 THE TAKE-APART GAME

The **Take-Apart Game** (Thompson & Van de Walle, 1984a) is played by two children using a fixed number of counters. One child places all of the counters under his or her left hand and then removes some leaving them just to the right. The second child says, "You started with eight, you took away three, and there are five left." The amount under the hand can be shown as the child gets to that point of the sentence if needed. Both children can then record two equations for what they did and said: 8 − 3 = 5 and 5 + 3 = 8.

As an alternative to the **Take-Apart Game**, a mat showing action such as the one in Figure 7.4 can be used. Now the whole amount is placed in the circle to be poured into the dish. This can be specified on a worksheet, a spinner can be used, cards can be drawn, or whatever. Then the amount to be poured into the dish is removed from the circle and placed in the dish part of the mat. This number again comes from the worksheet or from a die or spinner. The subtraction equation should reflect the action.

Note that in both of these remove-action activities, the situation ends with two parts clearly distinct. The removed part remains in the activity or on the mat. This then allows the child to also write the addition equation after having written the subtraction equation. If the whole amount is covered as it is in the **Take-Apart Game**, and the counters are removed from under the cover, children will be encouraged to think about the remaining hidden part in terms of "What goes with this part I see to make the whole?" The mental activity is think-addition instead of a "count what's left" approach. The part-part-whole addition/subtraction worksheet in Figure 7.6 (p. 114) is a way to connect two related subtraction equations to the corresponding addition equation. This relationship between addition and subtraction must be constructed by the children. We cannot make them learn this by simply having them write equations for number families, as will be seen in many standard texts.

COMPARISON MEANINGS

The comparison relationship involves two distinct sets or quantities and the difference between them. We can refer to these as the smaller set, the larger set, and the *difference* amount. Several ways of modeling the difference relationship are shown in Figure 7.7 (p. 114).

Comparison Addition Concepts

If the smaller of two sets and the difference between them are known, then *addition* tells how many are in the larger set. For example, if you know that your stack of cubes is 3 cubes shorter than the one behind the wall and your stack has 5, then 5 + 3 tells how big the larger stack is.

For some reason, this relationship is almost never used to illustrate a meaning for addition in schools. However, the relationship does exist in the real world, even the world of children. If Suzy is taunting Tommy by saying, "I've got three more pennies than you have," and Tommy looks at his five pennies, then Tommy is exactly in this situation. He knows the smaller of two sets and the difference amount. Addition tells how many Suzy has.

While you may not find the comparison model for addition in your second- or third-grade texts, you may find story problems that involve that relationship.

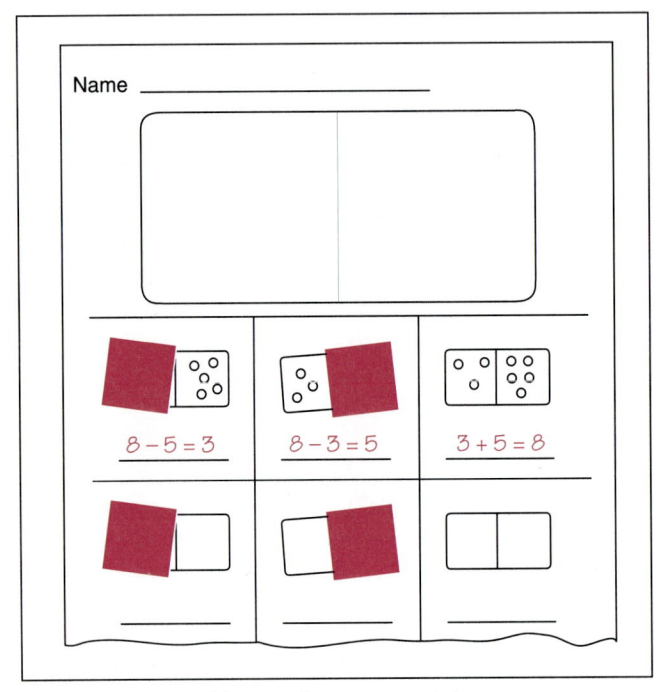

FIGURE 7.6: *An addition/subtraction worksheet*

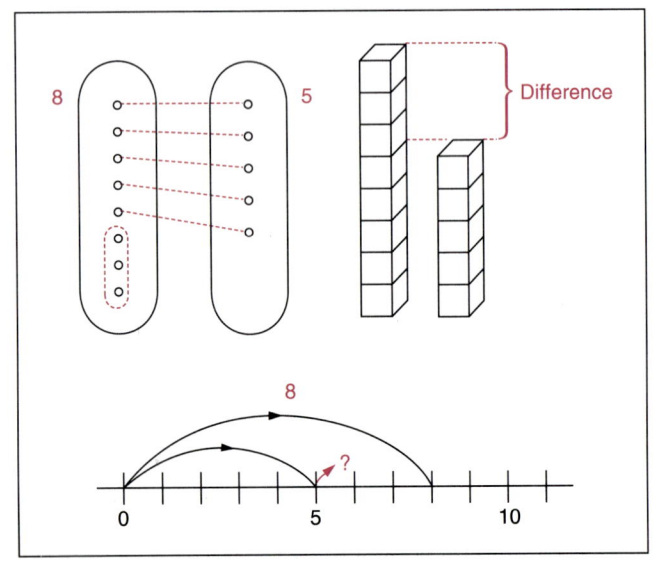

FIGURE 7.7: *The difference between 8 and 5 is 3.*

Comparison Subtraction Concepts

A comparison meaning for subtraction occurs in two situations:

1. The larger of the two sets and the difference is known and the smaller set is unknown.
2. Both of the sets are known, and the difference is unknown.

In both cases subtraction names the unknown amount. The "what is the difference" situation is fairly common in the curriculum but is not given anywhere near the same emphasis as the take-away concept for subtraction. As a result, when students drilled in take-away confront a word story such as the following, they experience difficulty. "Yesterday the mailman delivered 8 pieces of mail and today only 5 pieces. How many more did he bring yesterday than today?" The difficulty is involved with the fact that there are 13 things in the problem. If they model this with counters, they will need a set of 8 and another set of 5. Now they have two sets, and for many it makes sense at this point to add rather than subtract.

◆◆◆◆◆◆◆◆◆◆◆◆◆◆◆◆◆◆◆◆◆◆◆◆

7.5 THE COMPARE GAME

The **Compare Game** is played by two players as a way to connect the comparison relationship to subtraction symbolism (Thompson & Van de Walle, 1984a). A fixed number of blocks or bar of Unifix cubes is lined up on the left side of a mat. One child places a second bar, no longer than the first, on the right side of the mat, as in Figure 7.8. The other child says, "The difference between eight and three is five." If necessary the bars are placed side by side to make the comparison. A challenge version is played by placing the first bar under the mat out of sight. Players know how many are hidden and try to make the comparison mentally. A subtraction equation is written on paper or with the aid of numeral cards.

◆◆◆◆◆◆◆◆◆◆◆◆◆◆◆◆◆◆◆◆◆◆◆◆

7.6 COMPARISON WORKSHEET

Make a worksheet where children are given two of the three numbers in the comparison relationship. They model the relationship, and write two corresponding equations. Figure 7.9 shows one possibility, where children color in the large and small amounts on strips of squares. Alternatively, children could use connecting cubes or square tiles to do the same thing. Notice that in this format any of the three numbers (large, small, or difference) can be omitted and the model used to determine it. Also note that two children may each correctly write different subtraction equations for the same model.

7 / DEVELOPING MEANINGS FOR THE OPERATIONS   **115**

This is not the complete list of addition and subtraction properties you may have learned but only those that have a direct impact on children's learning.

◆ The Order Property in Addition

Put simply, the *order* (or *commutative*) property says that it makes no difference in which order two numbers are added. Most children find little difficulty with this idea. Since it is quite useful in problem solving and for mastering basic facts, there is value in spending some time helping children construct the order relationship.

7.7 THE TURN-AROUND PROPERTY

Show a set on a part-part-whole mat on the overhead projector. Ask the children to say the addition name for how much is shown, for example, "Eight plus four." Now, draw careful attention to the mat as you turn it completely around. Again request the addition name, "Four plus eight." Discuss which is more, eight plus four or four plus eight. Repeat for a few other examples, and eventually have your students discuss the rule that the order makes no difference. Notice that the rule is not arbitrary, but comes from the students and the model.

◆ Zero in Addition and Subtraction

Many children have difficulty with addition and subtraction when zero is involved. There may be many reasons for this. Zero as a number is probably not understood as early as other numbers. There is an intuitive idea that addition "makes numbers bigger" and subtraction "makes numbers smaller." To help with this problem, some effort should be made to model addition and subtraction involving zero without simply providing arbitrary sounding "rules." Figure 7.10 (p. 116) illustrates several ideas.

WORD PROBLEMS FOR ADDITION AND SUBTRACTION

While models provide straightforward, basic concepts of addition and subtraction, word problems involve a wider variety of relationships and semantic differences. Researchers have separated problems into categories based on the kind of relationships involved: *join and separate problems, part-part-whole problems*, and *compare problems* (Carpenter & Moser, 1983; Thomspon & Hendrickson, 1986).

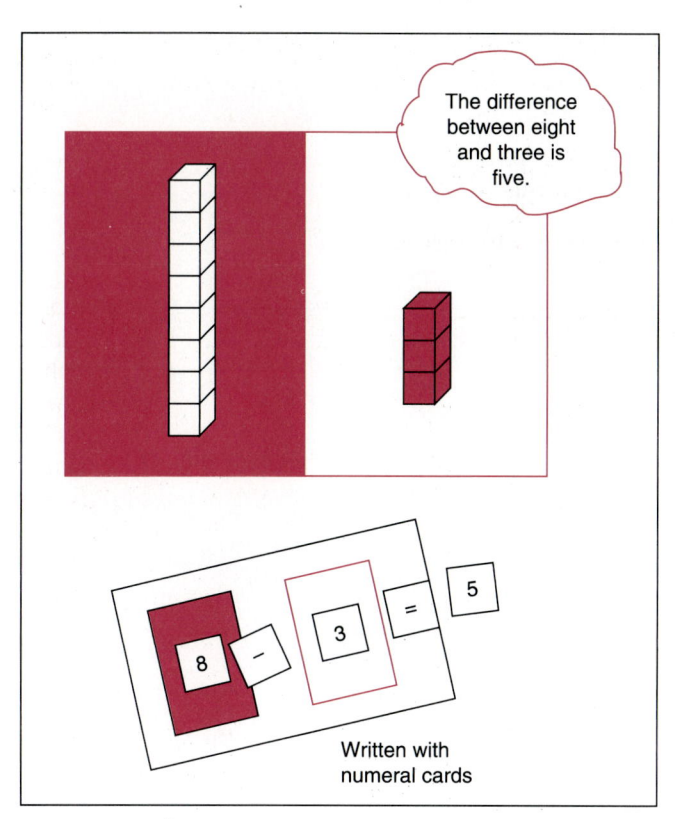

FIGURE 7.8: *The compare game*

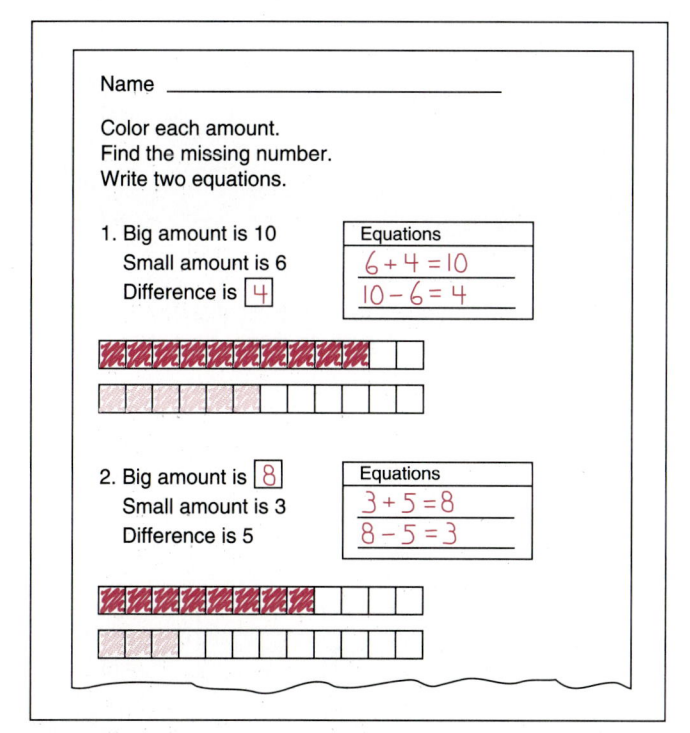

FIGURE 7.9: *A comparison activity worksheet*

PROPERTIES THAT ARE IMPORTANT FOR CHILDREN

There are a few properties or ideas that deserve some attention because they are helpful to children. The names of the properties are not important for children, but the ideas are.

JOIN AND SEPARATE PROBLEMS

For the action of joining, there are three quantities involved: an *initial* amount, a *change* amount (the part being added or joined), and the *resulting* amount (the

116   7 / DEVELOPING MEANINGS FOR THE OPERATIONS

Zero and six is six.

$\begin{array}{r}0\\+6\\\hline 6\end{array}$

Eight minus zero is eight.
Eight plus zero is eight.

$\begin{array}{r}8\\-0\\\hline 8\end{array}$ $\begin{array}{r}8\\+0\\\hline 8\end{array}$

8
3 and 5 is 8 $3 + 5 = 8$
6 and 2 is 8 $6 + 2 = 8$
8 and 0 is 8 $8 + 0 = 8$

"The difference between seven and zero is seven."

$\begin{array}{r}7\\-0\\\hline 7\end{array}$

"The difference between six and six is zero."
$6 - 6 = 0$

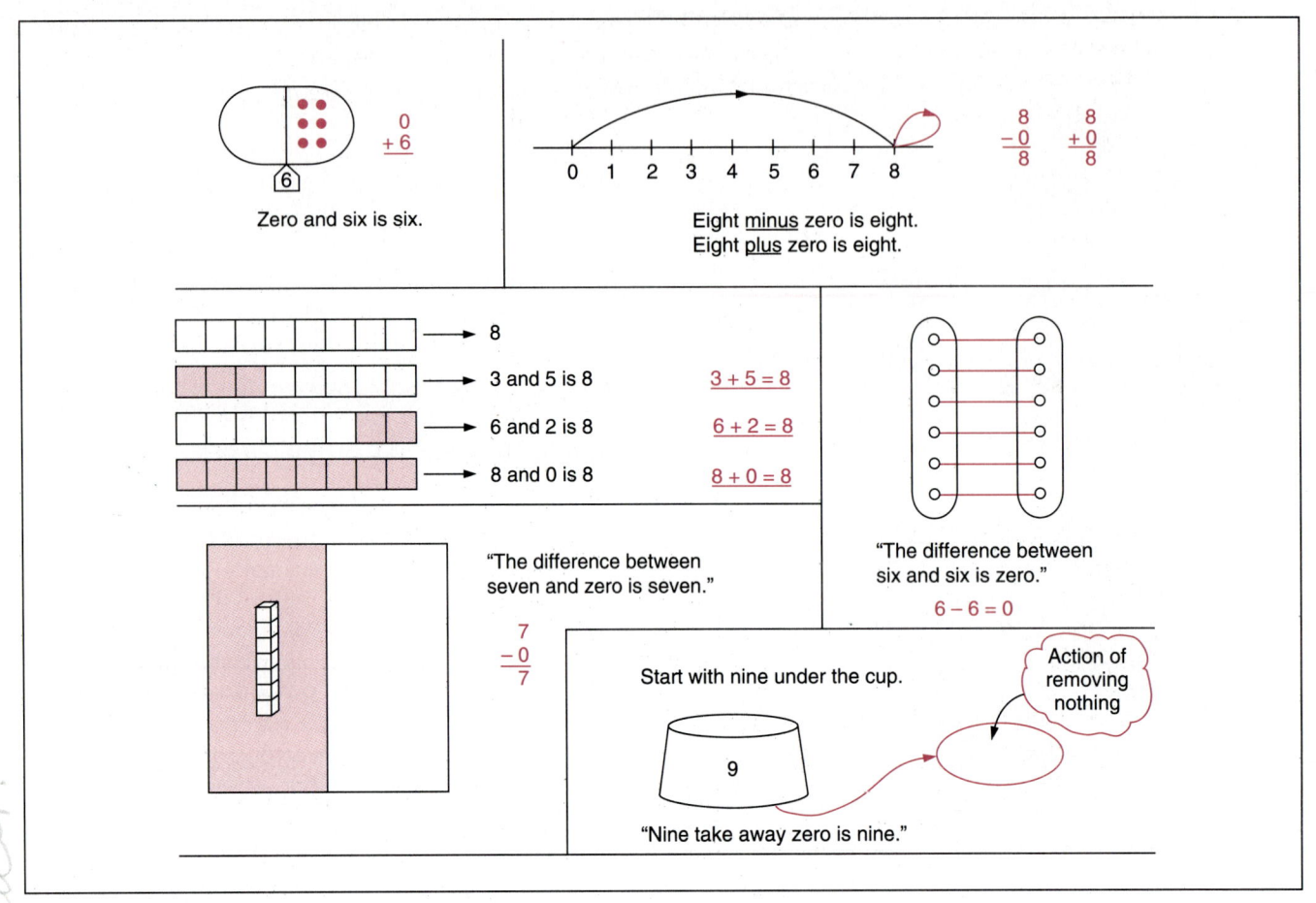

Start with nine under the cup.

Action of removing nothing

"Nine take away zero is nine."

FIGURE 7.10: *Zero in addition and subtraction*

amount there after the action is over). Notice that the whole in *join* situations is the result amount, and the initial and change amounts are parts. Since there are three numbers involved, any one of them can be unknown and a word problem devised for that situation. Here are examples for the three possibilities of join problems.

JOIN: RESULT UNKNOWN

Sandra had 8 pennies. George gave her 4 more. How many pennies does Sandra have altogether? (Initial = 8, Change = 4, Result = ?)

JOIN: CHANGE UNKNOWN

Sandra had 8 pennies. George gave her some more. Now Sandra has 12 pennies. How many did George give her? (Initial = 8, Change = ? Result = 12.)

JOIN: INITIAL UNKNOWN

Sandra had some pennies. George gave her 4 more. Now Sandra has 12 pennies. How many pennies did Sandra have to begin with? (Initial = ? Change = 4, Result = 12.)

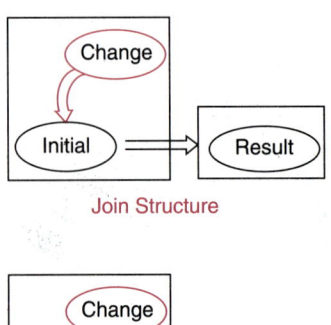

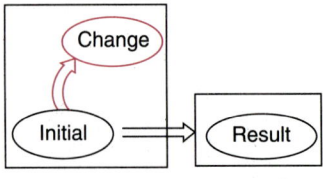

Join Structure

Separate Structure

For the action of separate, there are three similar possibilities for word problems depending on which quantity is unknown. Notice that in the *separate* problems, the initial amount is the whole, and the change and result amounts are the parts. As with the join problems, here are examples of the three possibilities using a simple context and the same numbers in the three positions. Think about the drawing as you examine the problems.

SEPARATE: RESULT UNKNOWN

Sandra had 12 pennies. She gave 4 pennies to George. How many pennies does Sandra have now? (Initial = 12, Change = 4, Result = ?)

SEPARATE: CHANGE UNKNOWN

Sandra had 12 pennies. She gave some to George. Now she has 8 pennies. How many did she give to George? (Initial = 12, Change = ? Result = 8)

SEPARATE: INITIAL UNKNOWN

Sandra had some pennies. She gave 4 to George. Now Sandra has 8 pennies left. How many pennies did Sandra have to begin with? (Initial = ? Change = 4, Result = 8)

When making up your own problems, a good idea is to first think of a context (Martians, shopping, a magic tree that grows doughnuts and a child who comes to eat them, a car going down the road and how many miles has it gone, etc.). With a context to provide a story line, put some numbers into the join and separate structures to create some of these six different change problems.

PART-PART-WHOLE PROBLEMS

While the difference between action and no action is not necessarily obvious to children when using models, there is a difference in story problems. Here the absence of action is more significant. Remember, the meaning of addition and subtraction is a relationship, not an action.

PART-PART-WHOLE: WHOLE UNKNOWN

George has 4 pennies and 8 nickels. How many coins does he have?

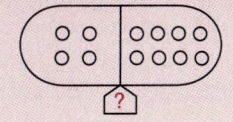

PART-PART-WHOLE: PART UNKNOWN

George has 12 coins. Eight of his coins are pennies, and the rest are nickels. How many nickels does George have?

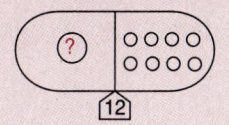

COMPARE PROBLEMS

There are three types of compare problems corresponding to which quantity is unknown (smaller, larger, or difference). For each of these, two examples are given: one in terms of "more," the other in terms of "less."

COMPARE: DIFFERENCE UNKNOWN

George has 12 pennies and Sandra has 8 pennies. How many more pennies does George have than Sandra?

George has 12 pennies. Sandra has 8 pennies. How many fewer pennies does Sandra have than George?

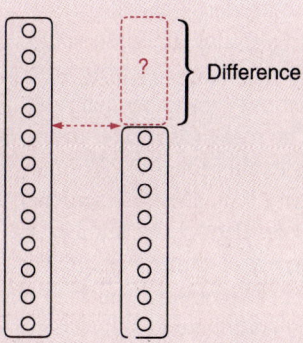

COMPARE: LARGER UNKNOWN

George has 4 more pennies than Sandra. Sandra has 8 pennies. How many pennies does George have?

Sandra has 4 fewer pennies than George. Sandra has 8 pennies. How many pennies does George have?

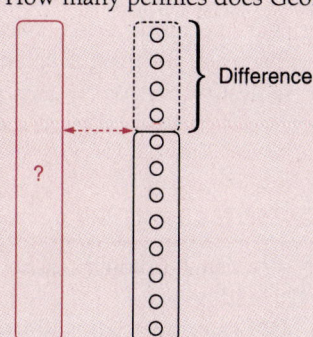

COMPARE: SMALLER UNKNOWN

Sandra has 4 fewer pennies than George. George has 12 pennies. How many pennies does Sandra have?

George has 4 more pennies than Sandra. George has 12 pennies. How many pennies does Sandra have?

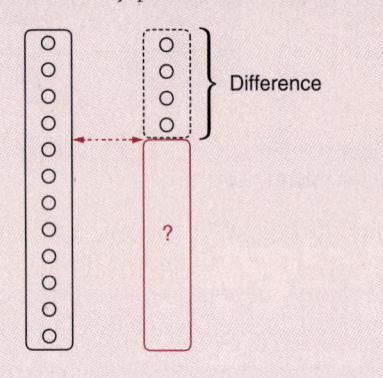

USING ADDITION AND SUBTRACTION WORD PROBLEMS IN THE CLASSROOM

The important goal here is not just getting answers, but to analyze relationships. Provide children with counters, strips, Unifix, or other blocks. Present the problems to children orally. If children work in pairs or small groups, they can discuss their ideas out loud.

In the early grades, children solve world problems with their models and afterward write an equation that *tells what they did.* Equations can be "written" with numeral cards so that changes can be made easily. The answer to a problem may be any number in the equation. Thompson and Hendrickson (1986) suggest that after the sentence is written, children should circle or identify the number that tells the answer. In this way a variety of correct equations is possible for the same word problem. Discussions will help children make connections between addition and subtraction in different forms. For example, $8 + \boxed{4} = 12$ and $12 - 8 = \boxed{4}$ are both likely equations for the first compare problem on page 117.

Not all of the various addition and subtraction word problems are equally easy. The join and separate problems with the initial set unknown and compare problems with the reference set unknown are generally the most difficult (Thompson & Hendrickson, 1986). However, all problems should be explored, even in the first grade. As children are exposed to a richer and more varied collection of problem types, they can begin to use these different problem structures as they make up their own word problems in the translation exercises described earlier in the chapter.

MULTIPLICATION CONCEPTS

Multiplication also can be thought of in several different ways. Traditional textbooks have focused almost exclusively on the notion of repeated addition, and that can be thought of as the elementary school "definition" of multiplication. Other concepts (combinations and area concepts) also are important. In textbooks, these tend to surface in problem-solving pages rather than in the section that develops meaning for the operations.

REPEATED ADDITION

By far the most common concept of multiplication is that of *repeated addition*. When a whole is represented in two or more equal parts, *multiplication* can be used to name the whole amount. It is important for children to connect the new multiplication concept with their already existing knowledge of addition. For example, the addition expression $4 + 4 + 4 + 4 + 4 + 4$ and the expression 6×4 represent the same relationship. They are 2 ways to express the quantity 24: one with addition, the other with multiplication. The 6 and 4 are referred to as the first and second *factors*, and the total, 24, is the *product*. Notice that each factor represents a different idea. The first factor tells *how many sets*, while the second factor represents the *number in each set*.

Sets and number lines are models for multiplication, since they are also the models for addition. A model not generally used for addition, but extremely important and widely used for multiplication, is the *array*. An *array* is any arrangement of things in rows and columns, such as a rectangle of square tiles or blocks.

In order to make clear the connection to addition, early activities should also include writing an addition sentence that represents the same model. A variety of models is shown in Figure 7.11 with both addition and multiplication expressions. Notice that the products are not included; only addition and multiplication "names" are written. This is to avoid the tedious counting of large sets. The purpose of these activities is to associate the symbolic knowledge of multiplication with the concept. Mastering the basic facts is a later objective. Another approach is to write one sentence that expresses both concepts at once; for example $9 + 9 + 9 + 9 = 4 \times 9$.

Multiplication is usually first introduced in second grade and developed with more emphasis in the third and fourth grades. The general approach of writing expressions for models is the same as with addition and subtraction. As noted earlier, multiplication and addition "names" can be used to avoid tedious counting and still explore large products. Fact mastery is a later objective.

7.8 FINDING MULTIPLICATION NAMES

Start with a number that has several factors; for example, 12, 18, 24, 30, or 36. Have students use a model to find multiplication expressions for this number. With counters, students attempt to find ways to separate the counters into equal subsets. With arrays, students try to build rectangles with blocks or draw rectangles on grid paper that have the given number of squares. For each such arrangement of sets or appropriate rectangles, both an addition and a multiplication equation should be written. Students could be given two or three numbers to explore for an independent assignment.

7.9 MODELS FOR MULTIPLICATION

A more directive exercise is to specify the number of sets and the size of each. For example, "Make six rows of four squares in each." Addition and multiplication names are written for each instance. When physical models are used, students can draw pictures of their sets or arrays as a record of their activity. Examples of these exercises are shown in Figure 7.12 (p. 120).

7.10 MULTIPLICATION ON A 10 × 10 ARRAY

A transparency of a 10 × 10 array (Black-line Masters) can be used with a large L-shaped piece of poster board to outline arrays of various sizes. As you change the array shown with the L, students say the corresponding multiplication name for that array (Figure 7.13, p. 120). The array has four 5 × 5 sections, making it easy to see how many rows and columns are being shown. The array can be duplicated for students to use at their seats to give them an instant model for an multiplication combination up to 10 × 10.

7.11 FOLD-UPS

Student-made "fold-ups" are another variation of the array. Cut paper lengthwise into strips about 5 or 6 cm wide. Have students fold up the strips from one end in folds about 1 cm wide. Fold back one fold and write 0 × 6 on the roll. Roll down another fold. Write 1 × 6 on the fold, and make a row of six evenly spaced dots on the paper, as shown in Figure 7.14 (p. 121). Unroll another fold, write 2 × 6, and make a second row of dots exactly under the first. Continue to unfold and make rows of dots until there are nine rows for 9 × 6. The fold-up helps to emphasize the repeated-addition concept and can be used later to help with basic fact memory. Turn it sideways to see that 4 × 6, for example, is the same as 6 × 4.

6 × 4 = 4 + 4 + 4 + 4 + 4 + 4

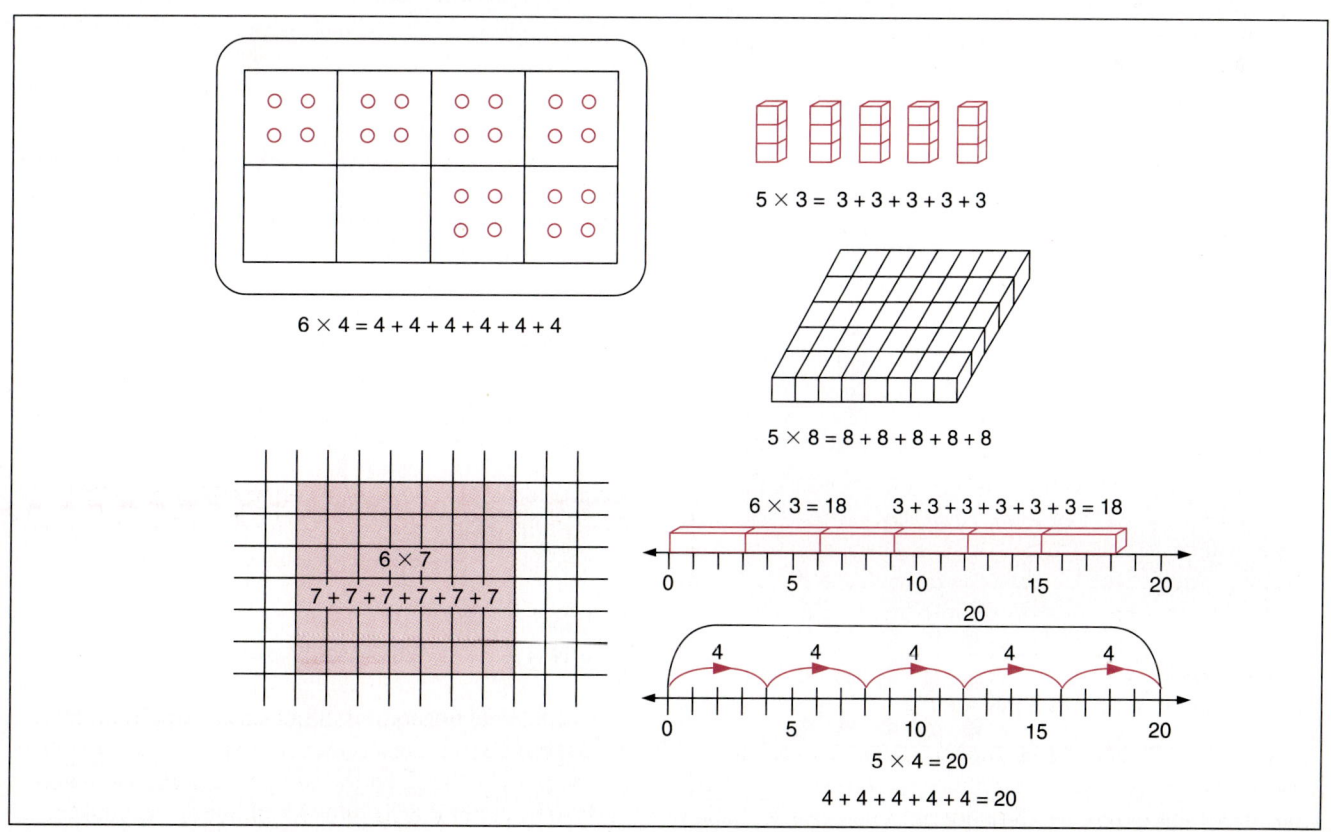

6 × 7
7 + 7 + 7 + 7 + 7 + 7

5 × 3 = 3 + 3 + 3 + 3 + 3

5 × 8 = 8 + 8 + 8 + 8 + 8

6 × 3 = 18   3 + 3 + 3 + 3 + 3 + 3 = 18

5 × 4 = 20
4 + 4 + 4 + 4 + 4 = 20

FIGURE 7.11: *Models for repeated addition*

120 7 / DEVELOPING MEANINGS FOR THE OPERATIONS

Draw		Draw	
3 rows 8 in each	Write × and + names	6 rows 4 in each	Write × and + names

○ ○ ○ ○ ○ ○ ○ ○
○ ○ ○ ○ ○ ○ ○ ○
○ ○ ○ ○ ○ ○ ○ ○

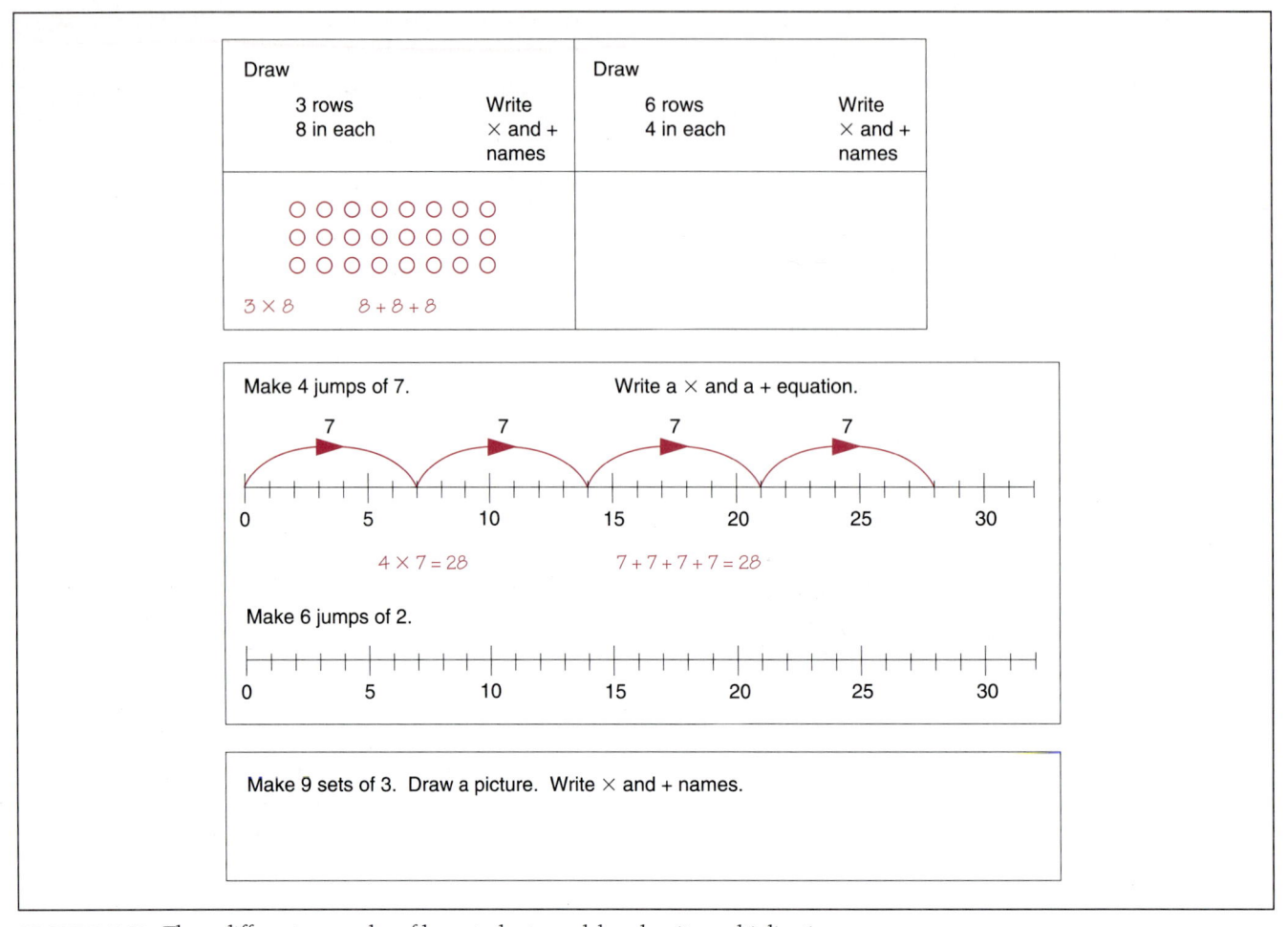

3 × 8   8 + 8 + 8

Make 4 jumps of 7.   Write a × and a + equation.

0   5   10   15   20   25   30

4 × 7 = 28   7 + 7 + 7 + 7 = 28

Make 6 jumps of 2.

0   5   10   15   20   25   30

Make 9 sets of 3. Draw a picture. Write × and + names.

FIGURE 7.12: *Three different examples of how students model and write multiplication names*

Four times eight

FIGURE 7.13: *A 10 × 10 array can be used to quickly show models for 1 × 1 to 10 × 10.*

7.12 MULTIPLICATION ON THE CALCULATOR

The calculator is a good way to relate multiplication to addition. Students can be told to find various products on the calculator without using the ⊠ key. For example, 6 × 4 can be found by pressing ⊞ 4 ═ ═ ═ ═ ═ ═. Let students use the same technique to add up sets of counters they have made. Students will want to confirm these results with the use of the ⊠ key.

COMBINATIONS (CARTESIAN PRODUCTS)

A very different concept of multiplication involves only two sets. Multiplication can be used to express the number of possible pairings between the elements of two sets. (This is also known as the *Cartesian product* concept of multiplication.) In elementary and middle school, this concept is usu-

7 / DEVELOPING MEANINGS FOR THE OPERATIONS   121

ally modeled in terms of some real context. Figure 7.15 illustrates several ways to model the combination concept. The array model for the combination concept helps illustrate that the two concepts of multiplication are not as different as they first appear.

The need to count possible combinations or outcomes, such as in simple probability experiments, suggests that an increased emphasis on this concept may be appropriate. At present, this concept is given only limited attention, mostly in the upper grades.

◆◆◆◆◆◆◆◆◆◆◆◆◆◆◆◆◆◆◆◆◆◆

7.13 HOW MANY PAIRS?

Provide students with small squares of paper, four red and six yellow. Label these A through D and A through F, respectively. Pose the task of finding out how many different pairs are possible, with each pair having one red square and one yellow one. Encourage students to work together, record results, and look for a pattern. This is a good example of a problem-solving approach to the combination concept of multiplication.

▸ AREA

Consider the rectangle in Figure 7.16 (p. 122) filled in with square units. The multiplication 4 × 7 tells how many squares there are. Compare that with the task of using the length-times-width formula for areas of a rectangle. Now we are multiplying two *lengths* to get an *area*. Neither factor is the size of a set or a number of sets but a length. Furthermore, the product is an area, a different unit from either factor. While there is a clear connection between the

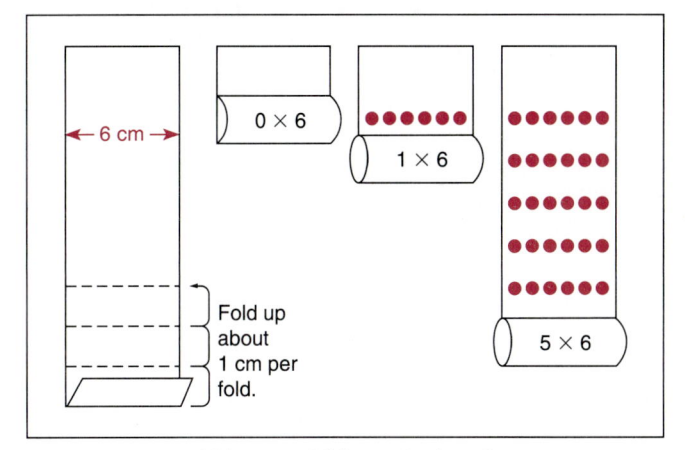

FIGURE 7.14: *A fold-up model for multiples of six*

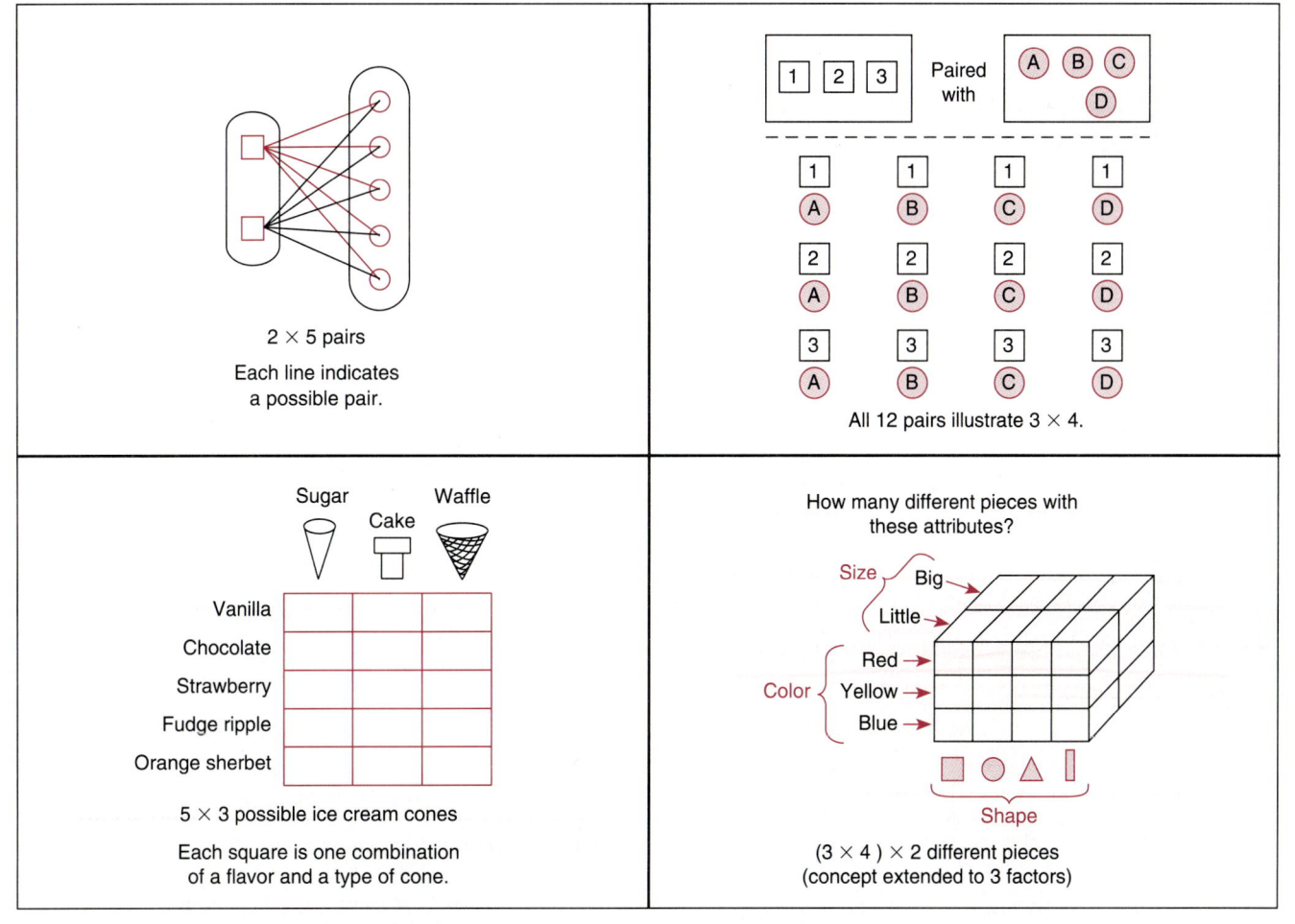

FIGURE 7.15: *Models for the combination concept of multiplication*

array of squares and the length × length = area relationship, the latter is frequently considered as a separate meaning of multiplication. In fact, the area concept for multiplication will be used extensively in Chapter 10 to develop computation procedures for multiplication.

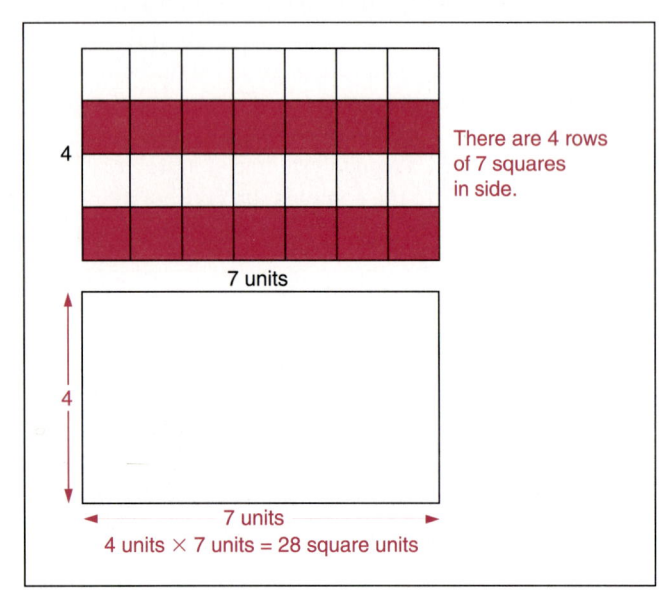

FIGURE 7.16: *Length × Length = Area*

MULTIPLICATION PROPERTIES IMPORTANT TO CHILDREN

The Order Property in Multiplication

Since the two factors in a multiplication expression carry different meanings, it is not intuitively obvious that 3 × 8 is the same as 8 × 3 or that, in general, the order of the numbers makes no difference (the commutative property). A picture of three sets of eight objects cannot immediately be seen as eight piles of three objects. Eight hops of 3 land at 24, but it is not clear from the model that three hops of 8 will land at the same point.

The array, on the other hand, is quite powerful in illustrating the order property, as is shown in Figure 7.17. Children should draw or build arrays and use them to demonstrate why each array represents two different multiplications with the same product.

The Role of Zero and One in Multiplication

Factors of 0 and, to a lesser extent, 1 can cause difficulty for children. They are somewhat extreme cases. In one second-grade textbook (Bolster et al., 1988) a train pulling four cars of six people models 4 × 6. The train pulling six empty cars models 6 × 0, while a train with no cars of 6 models 0 × 6. Factors of 1 can be interpreted similarly. On the number line, five hops of 0 land at 0. No hops of 5 is the same result. It is fun to discuss factors of 0 and 1 in terms of an array.

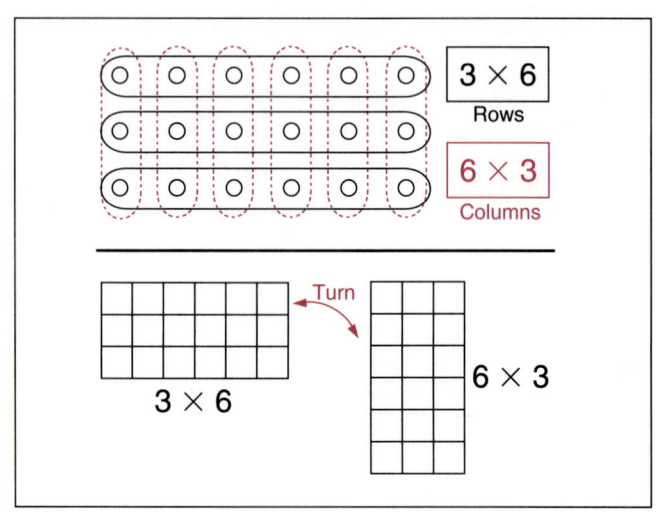

FIGURE 7.17: *Two ways an array can be used to illustrate the order property for multiplication.*

The Distributive Property

It may be argued that the distributive property is not essential for young children to know. In the formal form, $a \times (b + c) = a \times b + a \times c$, that may be true. But the concept involved may prove useful in relating one basic fact to another, and it is also involved in the development of two-digit computation. Figure 7.18 illustrates how the array model can be used to illustrate that a product can be broken up into two parts. Children could be asked to draw specific rectangles, slice them into two parts, and write the corresponding equation.

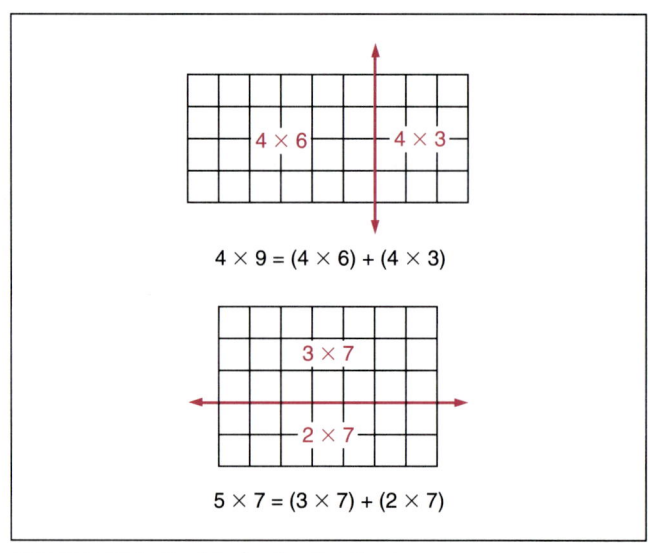

FIGURE 7.18: *Models for the distributive property*

WORD PROBLEMS FOR MULTIPLICATION

A scheme for categorizing both multiplication and division problems together (as was done for addition and subtraction) is possible (Hendrickson, 1986). For clarity, however, types of multiplication problems are discussed here, and related division problems are discussed later in the chapter.

Three types of multiplication word problems can be identified. The first two, rate times a quantity and multiples of a quantity, are ultimately related to or modeled by the repeated-addition concept. The third involves combinations or Cartesian products. Word problems for the area concept are not included, since those problems are essentially measurement tasks.

RATE TIMES A QUANTITY

In each of the following, one of the numbers is a *rate* or comparison of some amount to one set or one item. Rate examples include the "number in each set," the "cost per unit item," or the "speed per unit time." The second number is an actual *quantity*. It is the number of sets or the number of units to which the rate applies.

> Mark has 4 bags of apples. There are 6 apples in each bag. How many apples does Mark have altogether?
>
> Apples cost 7 ¢ each. Jill bought 4 apples. How much did they cost in all?
>
> Peter can walk 4 miles per hour. If he walks at that rate for 3 hours, how far will he have walked?

In Figure 7.19, each problem is modeled with sets, an array, and a number line. An examination of the models will show how these three problems are alike. In the modeled form, the rate can be seen as the number in each set or the size of the set, and the quantity is the number of the sets. It is difficult for children to deal with price and time rates in problems. Helping them translate these ideas to models can help them relate these ideas to the more basic equal-addition concept they have developed for multiplication. According to Quintero (1986), children's difficulty with multiplication comes from understanding the situations in the problems and the various number uses, and not in selecting the correct operation.

MULTIPLES OF A QUANTITY

In these problems only one number is an actual quantity and stands as a reference set. The other number is a *multiplier* that indicates how many copies of the reference are in the total or product. These problems can be interpreted using any of the repeated-addition models. Examples are shown in Figure 7.20 (p. 124).

> Jill picked 6 apples. Mark picked 4 times as many apples as Jill. How many apples did Mark pick?
>
> Mark now has 4 times as many dollars saved as he had last year. Last year he had $6. How many dollars does he have now?

COMBINATION PROBLEMS (A QUANTITY TIMES A QUANTITY)

> Sam Slick is really excited about his new clothing purchases. He bought 4 pairs of pants and 3 jackets, and they all can be mixed or matched. For how many days can Sam wear a different outfit if he wears one new pair of pants and one new jacket each day?

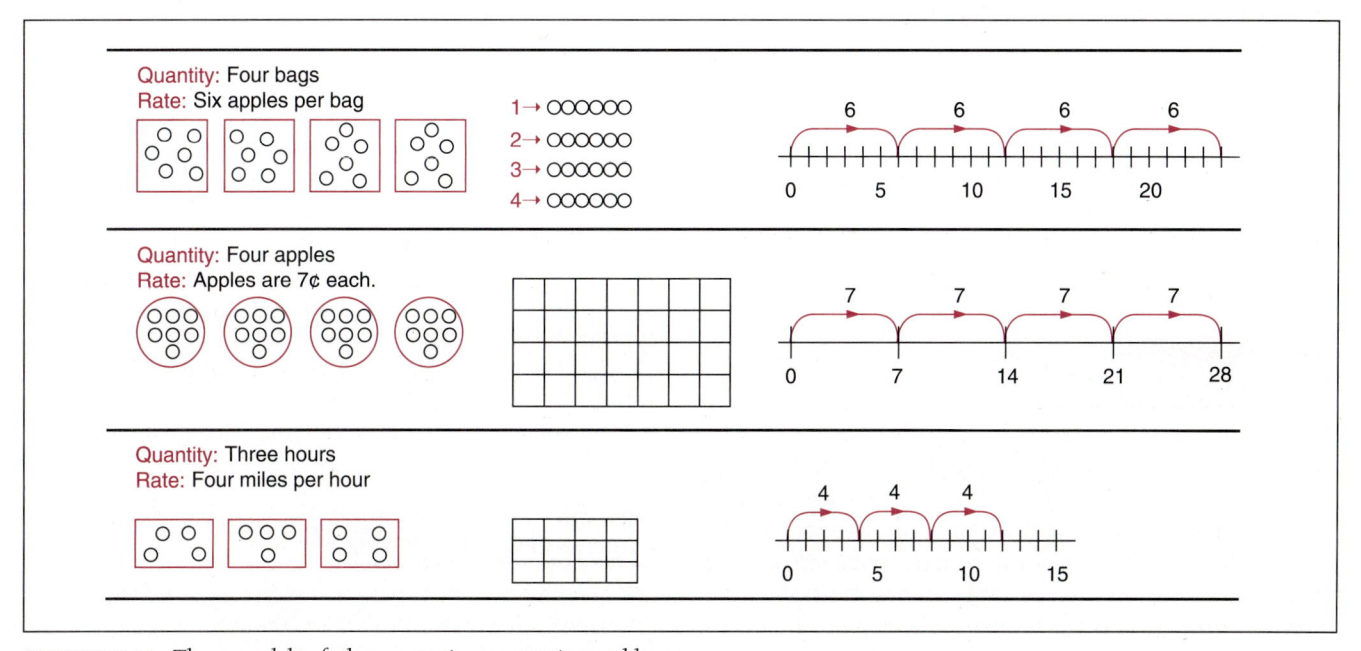

FIGURE 7.19: *Three models of three rate times quantity problems*

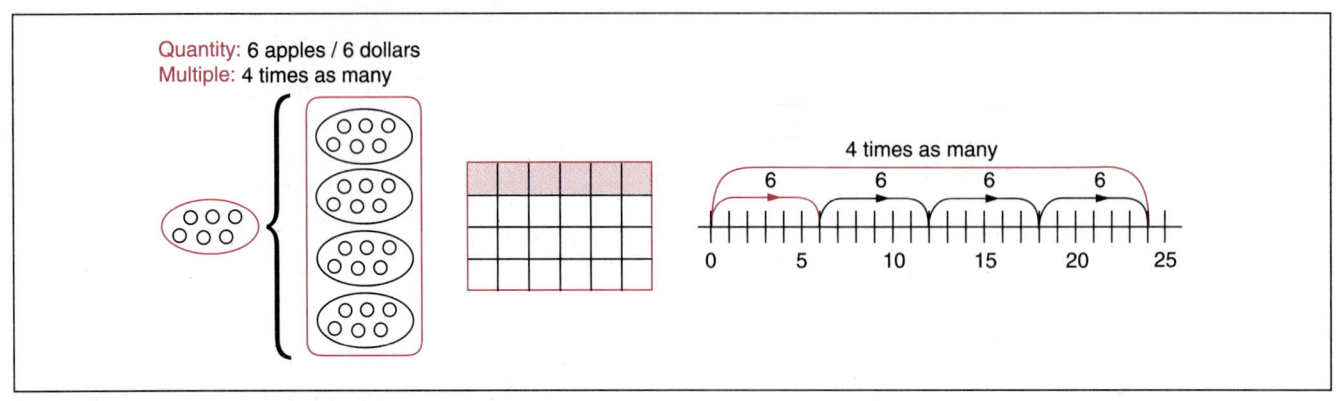

FIGURE 7.20: *Three models of multiples of quantity problems*

> You want to make a set of attribute pieces that have 3 colors and 6 different shapes. If you want your set to have exactly one piece for every possible combination of shapes and colors, how many pieces will you need to make?

> An experiment involves tossing a coin and rolling a die. How many different possible results or outcomes can this experiment have?

These problems reflect directly the combination concept of multiplication. There are different ways to model these problems: indicate pairings using lines, model all pairs directly, or use an array. Figure 7.21(a) shows how the coin-and-die experiment could be modeled all three ways.

Combination problems can have more than two quantities. For example, Sam may also have six different ties (4 × 3 × 6); the attribute pieces could include two sizes and have one, two, or three holes in each piece (3 × 6 × 2 × 3); and the experiment may include two coins instead of one (2 × 2 × 6). Modeling combinations from three or more sets is possible but quickly becomes tedious [Figure 7.21(b)]. However, in the upper grades it is worthwhile exploring such combinations with small numbers to see that this concept of multiplication can be extended to any number of sets. As probability and finite mathematics become more and more common in high school, early experiences with combination problems become more important.

DIVISION CONCEPTS

Corresponding to the repeated-addition concept of multiplication, there are two different concepts of division, depending on which factor is unknown. The models that we can use to illustrate division concepts are exactly the same as those for multiplication. In fact, when a division is modeled, the result always looks like a multiplication model. A division concept related to the combination concept is possible but unusual and is not discussed here.

Before going on, it might be useful to do the following little activity as a readiness for what you will be reading.

Get a set of 18 counters, or make a bar of 18 connecting cubes.

> A: Separate your pile or bar into 3 equal-sized sets or bars. How many *in each set* or bar? Make a drawing of what you have, and write next to it the multiplication equation that corresponds to the result. In your equation, circle the number that tells the answer. Write a full statement of both the question and the answer.
>
> B: Put your counters or blocks back into one set of 18 (or get 18 new counters). Use the materials to find out how many *sets of 3* there are in a set of 18. Make a drawing of what you have, and write next to it the multiplication equation that corresponds to the result. In your equation, circle the number that tells the answer. Write a full statement of both the question and the answer.

Reflect for a moment on what you did in each case. How were the two situations different? Develop for yourself a way of talking about these two situations in a way that you could translate to any similar situation regardless of the numbers. In both of the above situations, the division equation 18 ÷ 3 = 6 and the computational form $3\overline{)18}^{6}$ are each correct expressions to represent the situation. Notice that the number you circled in one situation was the first factor, and in the other it was the second factor. (If you did not do it this way, go back, and decide now which way seems appropriate given the convention that the first factor tells the number of sets.) In one case you *partitioned* the set of 18 into 3 equal parts. In the other, you *measured off* sets of 3. (Can you apply these terms to your two situations?)

FAIR SHARING OR PARTITION

If a quantity is to be separated evenly into a given number of subsets, then *division* expresses the number in each subset. For example, if 24 counters are to be separated into 6

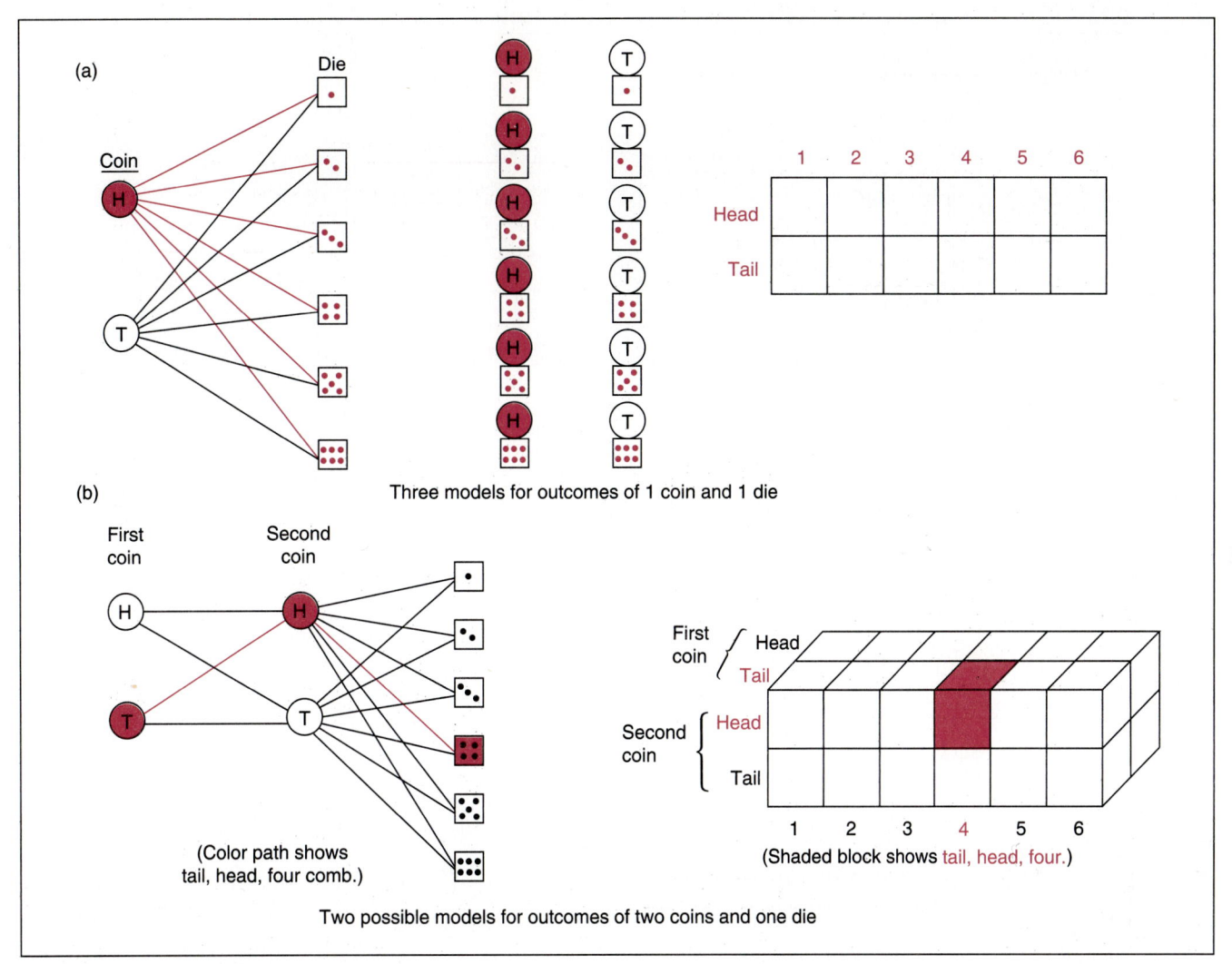

FIGURE 7.21: *Models of combination situations*

equal piles, the expression 24 ÷ 6 is used to tell *how many are in each subset*. This meaning of division is referred to as the *partition* concept or *fair sharing*.

Besides counters, partitioning can be modeled with a number line or an array. On a number line, the distance from 0 to 24 could be partitioned or separated into six equal parts. The size of each hop or part corresponds to the number of things in each set. If 24 items are put in 6 rows, 24 ÷ 6 refers to the length or number of things in each row (Figure 7.22, p. 126).

In Figure 7.22, notice that after the partitioning is completed for each model, a multiplication relationship is apparent. The equation 24 ÷ 6 = ☐ is clearly related to 6 × ☐ = 24. The *size of the set* is what is unknown.

REPEATED SUBTRACTION OR MEASUREMENT

If a quantity is to be measured out into sets of a specified size, then *division* expresses the number of such sets that can be made. For example, if the 24 counters in the previous example were to be arranged in as many sets of 6 as possible, then the expression 24 ÷ 6 tells *how many sets of 6 can be made*. This meaning of division is referred to as *measurement* or *repeated subtraction*. (The equal-size sets are "subtracted" from the total.)

To model the measurement concept of division on a number line, the total distance from 0 to 24 is measured out in jumps or lengths of 6. Alternatively, the lengths could be subtracted by starting at 24 and working bacward. If the 24 counters were to be arranged in rows of 6, division tells how many such rows are required. All three models for measurement division are shown in Figure 7.23 (p. 126).

As before, after the sets of 6 are made, the models look like a multiplication situation. For the measurement concept of division, each resulting model illustrates 4 × 6 = 24. That is, the measurement concept for 24 ÷ 6 = ☐ is related to ☐ × 6 = 24. The *number of sets* is unknown, with the size of the sets and the total amount given.

NOTATION, LANGUAGE, AND REMAINDERS

Unfortunately, there are two commonly used symbols for division: 24 ÷ 6 and 6)̄24̄. The second form is the computational form. It would probably not exist if there were no pencil-and-paper computational procedure that made use of it.

126   7 / DEVELOPING MEANINGS FOR THE OPERATIONS

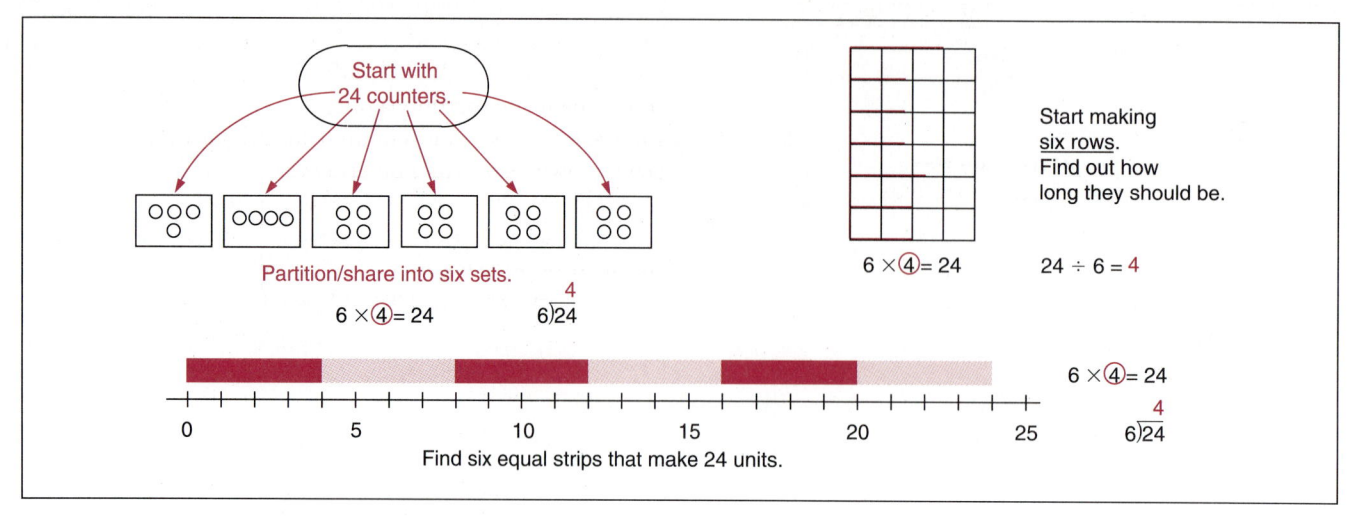

FIGURE 7.22: *Three models of the partition concept for 24 ÷ 6*

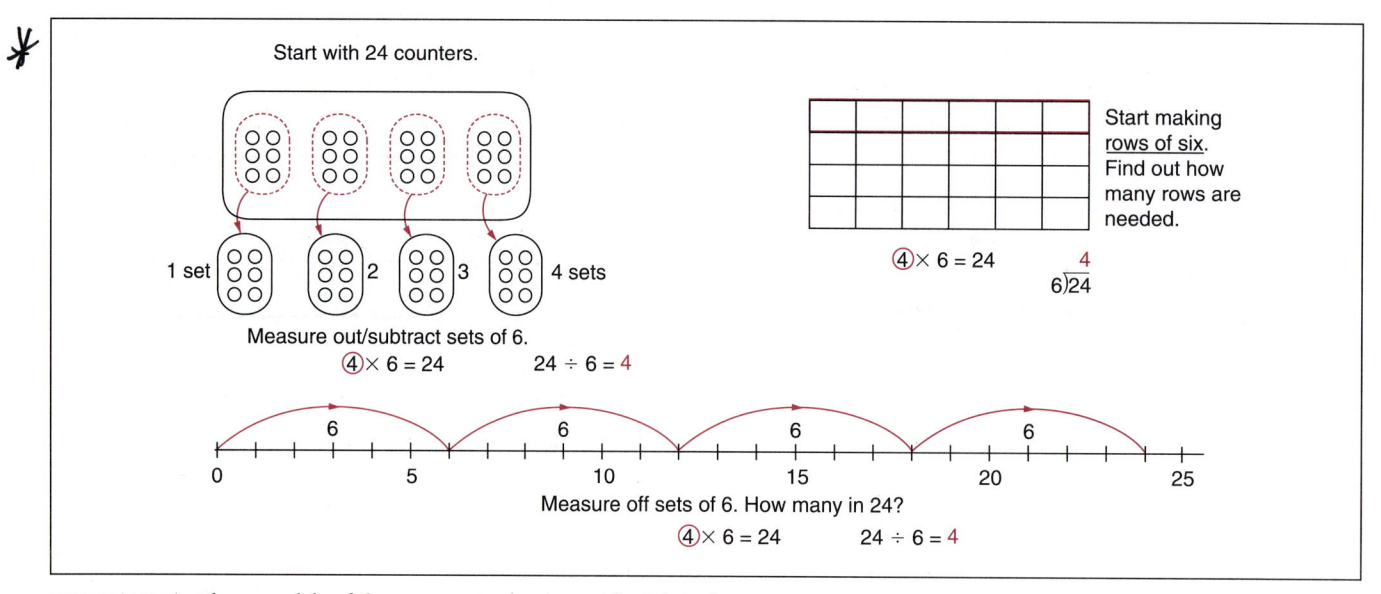

FIGURE 7.23: *Three models of the measurement concept for 24 ÷ 6*

(The other three operations use a vertical form for computation.)

These two forms cause some troubles worth noting. First, the order of the numbers is reversed. Therefore, to read them both as "twenty-four divided by six," one is read left to right and other right to left. To compound this difficulty is the meaningless but traditional phrase, "six *goes into* twenty-four." This expression probably originated with the computational form in conjunction with the question, "How many sixes are in twenty-four?" The "goes into" terminology is so ingrained in our society that most adults and teachers continue to use it as if it carried some clear meaning. The phrase has not been in student textbooks for many years. Perhaps it would help if we realize that "goes into" is language that is connected to *our* tradition and understanding, not children's. If "goes into" is in your vernacular, and it probably is, try not to use it in the classroom.

While the terms *multiplier* and *multiplicand* are rarely used anymore, the terms *divisor, dividend,* and *quotient* are still around. However, the words *factor* and *product* more clearly connect the concept of division to multiplication and can be used instead, as shown here.

$$\text{factor} \overline{)\text{product}}^{\text{factor}} \qquad \text{divisor} \overline{)\text{dividend}}^{\text{quotient}}$$

More often than not, division does not result in a simple whole-number result: problems such as 33 ÷ 6 = ☐ are more common that 30 ÷ 5 = ☐. In the absence of a word-problem context, a *remainder* can be dealt with in only two ways: it can either remain a quantity left over or be partitioned into fractions. In Figure 7.24, 11 ÷ 4 = ☐ is modeled to show fractions.

DIVISION ACTIVITIES

The partition and measurement concepts of division are the most important. When modeling and recording these concepts, children should write both division and multiplication equations to emphasize the connection between the operations.

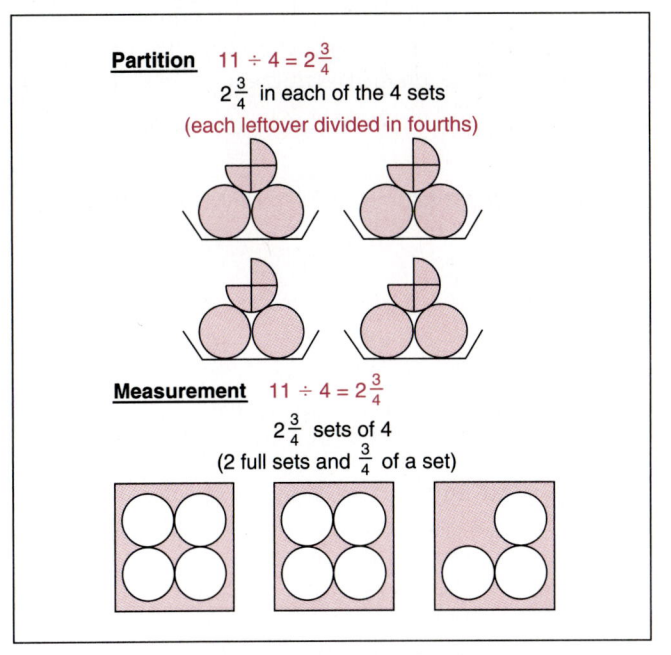

FIGURE 7.24: *Remainders expressed as fractions*

Partition $11 \div 4 = 2\frac{3}{4}$
$2\frac{3}{4}$ in each of the 4 sets
(each leftover divided in fourths)

Measurement $11 \div 4 = 2\frac{3}{4}$
$2\frac{3}{4}$ sets of 4
(2 full sets and $\frac{3}{4}$ of a set)

having them draw arrays on centimeter grid paper. Present the exercises by specifying how many squares are to be in the array. You can then specify the number of rows that should be made (partition) or the length of each row (measurement). How could children model fractional answers using drawings of arrays on grid paper?

Other materials that can be useful include long trains of Unifix or connecting cubes as the whole or lengths specified on a meter stick. With a meter stick, Cuisenaire rods can be used to make subsets along the edge of the meter stick.

Children in the fourth to eighth grades can do the same type of activity using various materials with the idea of exploring different ways of thinking about remainders. For example, if you had 32 ounces of water, how many 6-ounce cups can be made? ($5\frac{1}{3}$) How many ounces in each glass if you have 5 glasses you wish to fill equally? Use quantities such as hours, miles, seconds, dollars, feet, liters, and so on. Include situations where it is reasonable to have fractional parts (hours, yards of material) and those where it is not (counts of things such as marbles, trucks). The idea is to reflect on how the "left-over" part should be dealt with: sometimes as a fractional part and sometimes as a remainder.

7.14 LEARNING ABOUT DIVISION

Provide children with an ample supply of counters and some way to place them into small groups. Small paper cups work well. Either orally or by way of a worksheet, have children count out a number of counters to be the whole or total set. They record this number: "Start with *thirty-one*." Next specify either *the number of equal sets to be made* or the *size of the sets to be made*: "Separate your counters into *four* equal-sized sets," or "Make as many sets of *four* as is possible." Next, have the children write the corresponding multiplication equation for what their materials show. Under the multiplication equation, they write the division equation.

7.15 THE BROKEN DIVISION KEY

Have children work in groups to find different methods of using the calculator to solve division exercises *without using the divide key*. For partition situations, a trial-and-error approach can be used. For $45 \div 6$, press 6 ⊠ and a trial quotient and then ⊟. If the trial is too large or too small, just enter a different try and press ⊟. The factor of 6 is retained. For measurement, the method of counting backwards from the total or forward from zero (storing a ⊟ constant and pressing ⊟) can be used. These calculator methods are just as conceptual as using physical models, since they clearly model the meanings of the operation. (To clearly understand how this is done, try it out.)

WHY NOT DIVISION BY ZERO?

Some children are simply told "Division by zero is not allowed." While this fact is quite important, it is not always understood by children. To avoid an arbitrary rule, pose problems to be modeled that involve zero. "Take thirty counters. How many sets of zero can be made?" Or, "Put twelve blocks in zero equal groups. How many in each group?" Even if they think they may have a result that makes sense, suggest that they write the corresponding multiplication equations as they have done for other exercises.

This activity can be used as a first introduction to division. Be sure to include both types of exercises: number of equal sets and size of sets. Discuss with the class how these two are different, yet each is related to multiplication, and each is written as a division equation. You can show both ways to write division equations at this time. Do the **Learning about Division** activity several times. At first use whole quantities that are multiples of the divisor (no remainders) but soon include situations with remainders. (Note that it is technically incorrect to write $31 \div 4 = 7$ R 3. However, in the beginning, that form may be the best to use.)

The activity can be varied by changing the model. Have children build arrays using square tiles or blocks, or by

WORD PROBLEMS FOR DIVISION

For the purpose of simplicity, division problems can be reduced to two categories corresponding to the two most prevalent meanings of division: partition and measurement. However, recall that for multiplication word problems, there were two types that related to the basic model of repeated addition: rate times a quantity and multiples of a quantity. Therefore, within the measurement and partition categories discussed here, you will note problems that correspond to each of these multiplication types. The difference is worth noting and can help us provide our students with more variation in the types of word problems we give them.

Each problem given as an example involves the same numbers and contexts as the multiplication problems on page 123 to which they correspond. It may be helpful for you to make these comparisons so that you can strengthen your own relationship between multiplication and division. Another suggestion is to use counters, arrays, or number lines to model these problems. Then match the model with the corresponding multiplication example problem.

MEASUREMENT PROBLEMS

These first examples correspond to the rate-times-quantity multiplication problems. The quantity (number of sets) is the unknown, while the total and the rate (size of set) are given.

> Mark has 24 apples. He put them into bags of 6 apples each. How many bags did Mark use?
>
> Jill bought apples at 7¢ apiece. The total cost of her apples was 35¢. How many apples did Jill buy?
>
> Peter walked 12 miles at a rate of 4 miles per hour. How many hours did it take Peter to walk the 12 miles?

The next two examples are modeled in a similar manner but correspond to the multiple-times-a-quantity problems. Here the multiplier is unknown. See if you can see how these are the same and how they are a little different from the examples just given.

> Mark picked 24 apples, and Jill picked only 6. How many times as many apples did Mark pick than Jill did?
>
> This year Mark saved $24. Last year he saved $6. How many times as much money did he save this year over last year?

PARTITION PROBLEMS

These first partition problems correspond to the rate-times-quantity problems with the rate (size of set) the unknown and the quantity (number of sets) and total given.

> Mark has 24 apples. He wants to share them equally among his 4 friends. How many apples will each friend receive?
>
> Jill paid 35¢ for 5 apples. What was the cost of 1 apple?
>
> Peter walked 12 miles in 3 hours. How many miles per hour (how fast) did Peter walk?

The second and third examples are more difficult because price ratios and rates of speed are more difficult for children to understand and model. At the upper grades, progress can be made with rate problems like these by using models to help children see how they are like the easier sharing and repeated-subtraction problems.

The next two problems involve an unknown quantity (set size) with a given multiplier of the set and a total. Again, you are encouraged to see how these problems are alike and different from the examples just given and to relate them to the multiplication examples.

> Mark picked 24 apples. He picked 4 times as many apples as Jill. How many apples did Jill pick?
>
> This year Mark saved 4 times as much money as he did last year. If he saved $24 this year, how much did he save last year?

USING DIVISION WORD PROBLEMS IN THE CLASSROOM

It is not important for students to be able to label these problems as measurement or partition. A good class activity is to have students represent the same problem using different models. The more flexibility that students develop modeling these one-step translation problems, the better will be their concepts of the operations.

By the time that division is introduced, regular weekly exercises with simple translation problems should involve all four operations. Do not announce, "Today we will do a division problem." Simply present a problem, have students model the problem, and write one or more equations to go with the model. The more general translation activities from model to word problem can now become quite varied and interesting. One model can easily provoke word problems for two or more different operations, and an interesting discussion can follow.

REMAINDERS IN WORD PROBLEMS

Earlier (Figure 7.24) it was noted that when division problems do not come out evenly, the remainder is either "left over" or can be partitioned to form a fraction. In real contexts, remainders sometimes have three additional effects on answers:

1. The remainder is discarded, leaving a smaller whole number answer.
2. The remainder can "force" the answer to the next highest whole number.
3. The answer is rounded to the nearest whole number for an approximate result.

The following problems illustrate all five possibilities.

> You have 30 pieces of candy to share fairly with 7 children. How many will each receive? Answer: 4 and 2 left over. (left over)
>
> Each jar holds 8 ounces of liquid. If there are 46 ounces in the pitcher, how many jars will that be? Answer: 5 and $\frac{6}{8}$ jars. (partitioned as a fraction)
>
> The rope is 25 feet long. How many 7-foot jump ropes can be made? Answer: 3. (discarded)
>
> The ferry can hold 8 cars. How many trips will it have to make to carry 25 cars across the river? Answer: 4. (forced to next whole number)
>
> Six children are planning to share a bag of 50 pieces of bubble gum. About how many will each get? Answer: about 8 (rounded, approximate result)

Students should not just think of remainders as "R 3" or left over. They should be put in context and dealt with accordingly.

TRANSLATION PROBLEMS IN THE UPPER GRADES

So far in this chapter we have been discussing the meanings of the operations and simple one-step problems that amplify the meanings of those operations. To this end, the word problems have been relatively simple, and the numbers involved have been relatively small. The value of these problems is in allowing children to use simple models to analyze the problems. By fourth grade and beyond, most children have already developed meanings for the operations. The problems that are encountered include large numbers with decimals and fractions, the data is frequently imbedded in charts or tables, problems involve more data than is necessary, and many problems have two or sometimes even three steps. In this section we will discuss how the methods that have been used in this chapter can be applied to help children analyze these more complex situations.

DEALING WITH LARGE NUMBERS IN PROBLEMS

Consider the following problem:

> In building the road through the subdivision, a low section in the land was filled in with dirt that was hauled in by trucks. The complete fill required 638 truckloads of dirt. The average truck carried $6\frac{1}{4}$ cubic yards of dirt, which weighed 17.3 tons. How many tons of dirt were used in the fill?

Typically, in a sixth- to eighth-grade book, problems of this type are found in a series of problems revolving around the same context or theme. Data may be found in a graph or chart or perhaps a short news item or story. Most likely the problems will include all four of the operations. Students have difficulty deciding on the correct operation and even in finding the appropriate data for the problem. Many students will find two numbers in the problem and literally guess at the correct operation. These children simply do not have any tools for analyzing problems. At least two strategies can be taught that are very helpful:

1. Think about the answer before solving the problem.
2. Solve a simpler problem that is just like this one.

● Think about the Answer before Solving the Problem

Poor problem solvers fail to spend adequate time thinking about the problem and what it is about before they rush in and begin doing calculations. Their focus is on "number crunching" with the belief that computation is what solves problems. This is clearly not the case. Rather, children should spend some time talking about (later thinking about) what the answer might look like. For our sample problem above this might go as follows:

What is happening in this problem?— Some trucks were bringing dirt in to fill up a hole.

What will the answer tell us?— How many tons of dirt were needed in the fill.

Will that be a small number of tons or a large number of tons?— Well, there are 17.3 tons on a truck. Oh, but there were a lot of trucks, not just one. It's probably going to be a whole lot of tons.

About how many do you think it will be?— Wow! It's going to be really big. If there were just 100 trucks it would be 1730 tons. It might be close to ten thousand tons. That's a lot of tons!

In this type of discussion, two things are happening. First, the students are asked to focus on the problem and the meaning of the answer instead of on numbers. The numbers are not important in thinking about the structure of the problem. Second, there is a rough estimate of the

130   7 / DEVELOPING MEANINGS FOR THE OPERATIONS

answer. Sometimes this can simply be based on common sense. If the problem were to ask for the price of a box of cereal, for instance, then a ballpark answer in the range of $1 to $3 might be suggested without even looking at the numbers in the problem. In any event, thinking about what the answer tells and about how large it might be is a very useful step in solving the problem. When an answer is finally arrived at, it should be given in a full sentence and compared with this early estimate.

◆ Work a Simpler Problem

Clearly the reason that models are rarely used with problems such as the dirt problem is that the numbers are impossible to model easily. Dollars and cents, distances in thousands of miles, time in minutes and seconds are all examples of data likely to be found in the upper grades, and all are difficult to model. The general problem-solving strategy of "try a simpler problem" can almost always be applied to problems with unwieldy numbers.

For translation problems, a simple strategy has these steps:

1. Substitute small whole numbers for all necessary numbers in the problem.
2. Model the problem using the new numbers (counters, drawing, number line, array.)
3. Write an equation that solves the small-number version of the problem.
4. Write the corresponding equation with the original numbers used where the small number substitutes were.
5. Use a calculator to do the computation.
6. Write the answer in a complete sentence and decide if it makes sense.

A good rule of thumb is to substitute numbers between 3 and 25. Avoid the use of 1 and 2 as relationships might be obscured. For the dirt problem posed at the start of this section, Figure 7.25 shows what might be done. Also in Figure 7.25, an alternative is shown in which only one of the numbers is made smaller, and the other number is illustrated symbolically. That method is just as effective.

The idea is to provide students with a tool they can use to analyze the structure of a problem and not just guess at what computation to do. It is much more useful to have students do a few problems where they must use a model or drawing to justify their solution than to give them a lot of problems where they can guess at a solution and not know if their guess is correct.

◆ TWO-STEP PROBLEMS

If your students are going to work with multistep problems, be sure they can analyze one-step problems in the way that we have discussed. For multistep problems you may need some additional ideas. Consider the following problem:

> Willard Sales decides to add widgets to its line of sale items. To begin with, Willard bought 275 widgets

---

Change all numbers:

638 (7) trucks of dirt — don't need
6 1/4 yards of dirt in a truck — this
weighs 17.3 tons (4)

11,165 tons of dirt were used.

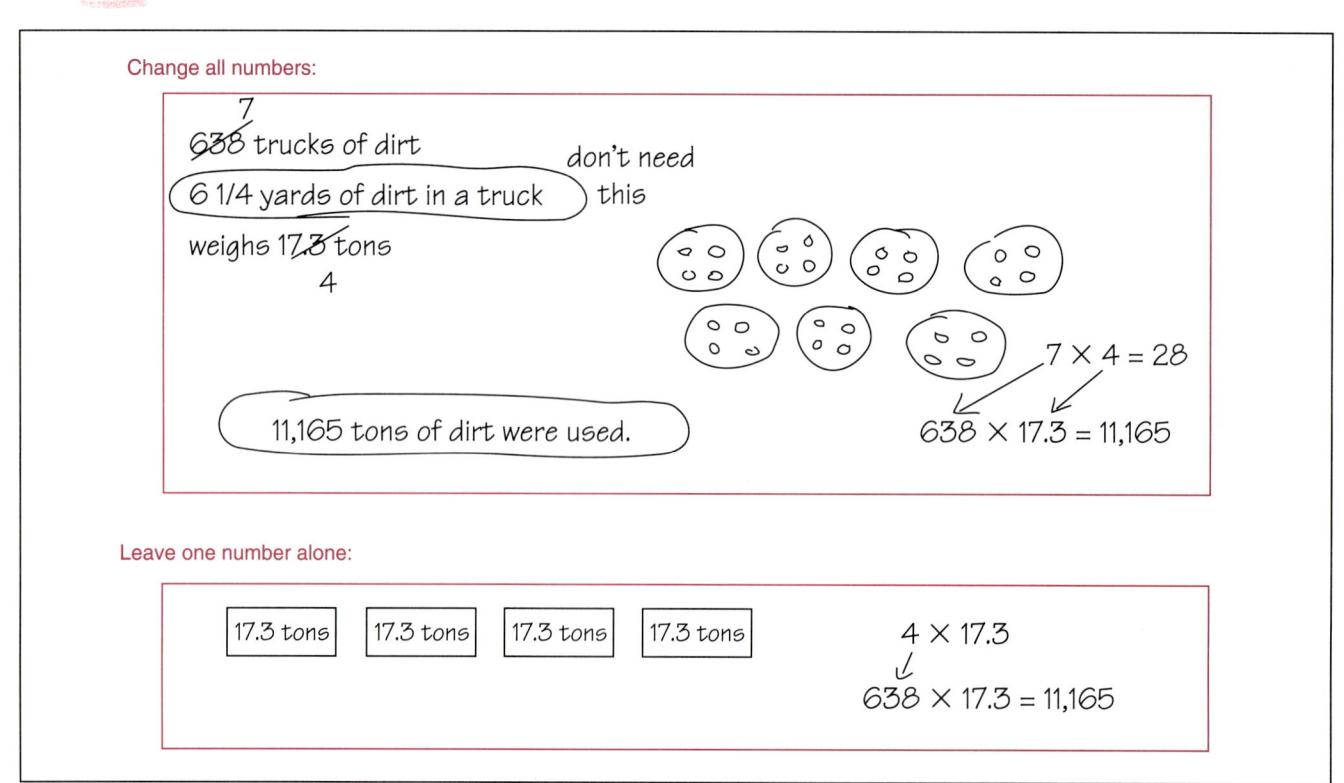

7 × 4 = 28
638 × 17.3 = 11,165

Leave one number alone:

| 17.3 tons | 17.3 tons | 17.3 tons | 17.3 tons |

4 × 17.3
638 × 17.3 = 11,165

FIGURE 7.25: *Working a simpler problem—two possibilities*

> wholesale for $3.69 each. In the first month they sold 205 widgets at $4.99 each. How much did Willard Sales make or lose on the widgets? Do you think Willard Sales should continue to sell widgets?

Begin by considering the same questions that were suggested in the previous section: What's happening in this problem? (Essentially something is being bought and sold at two different prices.) What will the answer tell us? (How much profit or loss there was.) As with all multiple-step problems, the important question is "What would we need to know in order to answer the question?" In this case you need to know two things: the amount spent and the amount of income. Neither is given. Take them one at a time. How can we find out how much Willard spent? How can we find out how much income there was? Notice that both of these questions are one-step problems. If students can solve these one-step problems, they can solve the larger question.

Examine the requirements for solving this and any other multi-step problems:

1. Identify what the answer will tell us and thus what the problem is asking.
2. Identify what is needed to answer that question.
3. Solve the problem(s) that will provide the necessary information.
4. Solve the original problem using the newly found information.

Only Step 2 is different from working on one-step problems. That, then, is the step that needs practice. One way to do this is to pose problems and practice only the first two steps. Help children learn to talk about a problem without worrying about computation and the specific numbers involved. *What will the answer to the problem tell us? What do we need to know in order to find that out?*

Notice that in all of the discussion about translation problems, the emphasis has been on analysis of the problems and not on getting answers. It is worth repeating, then, that more profit is gained from having students work with only a few problems where they are required to defend or demonstrate their analysis than it is to have them do a lot of problems with no explanations required. Tests and quizzes should require the same demonstration of analysis. Require explanations and drawings to justify solutions. Never fail to utilize calculators for the computations. Your objective is to get students to analyze a problem and think through its structure. Be sure that is what you test.

◆◆◆◆◆
REFLECTIONS ON CHAPTER 7: WRITING TO LEARN

1. Give three examples of translations. For each example, start with a different "language"—words, models, or symbols—and then make up the other two.
2. Define addition in terms of parts and wholes. How is subtraction defined so that students have to "think addition" and therefore connect addition and subtraction concepts?
3. Make up a context for some story problems, and make up six different join and separate problems; three with a join action and three with a separate action. For all six problems use the same number family: 9, 4, 13.
4. Make up a comparison word problem. Next, change the problem to provide an example of all six different possibilities for comparison problems.
5. Use three different models to illustrate what 3×7 means in terms of repeated addition. What is the meaning of each factor? What is the meaning of 3×7 in the combination concept? Make a drawing to illustrate what you mean.
6. Make up word problems to illustrate the difference between rate times a quantity and multiple of a quantity.
7. Make up two different word problems for $36 \div 9$. For one, the modeling should result in four sets of nine, and for the other the modeling should result in nine sets of four. Which is which? Which of your problems is a measurement problem, and which is a partition problem?
8. Make up realistic measurement and partition division problems where the remainder is dealt with in each of these three ways: (1) discarded, (2) made into a fraction, and (3) forces the answer to the next whole number.
9. Explain how to help students analyze problems when the numbers are not small whole numbers but rather are large or are fractions or decimals that do not lend themselves to using counters.

◆◆◆◆◆
FOR DISCUSSION AND EXPLORATION

1. Examine a basal textbook at one or more grade levels. Identify how, and in what chapters, the meanings for the operations are developed. Discuss the relative focus on meanings of the operations with models, one-step translation problems, and mastery of basic facts.
2. Look up the *Arithmetic Teacher* article, "Verbal Multiplication and Division Problems: Some Difficulties and Some Solutions" (Hendrickson,

1986), and compare his classifications of multiplication and division problems with those presented here. Which problems match? Which were not included in this presentation?

3. Consider these problems that involve a rate times a rate and therefore constitute a type of problem different from those discussed.

Mark put 6 candies in each bag and gave 3 bags to each of his friends. How many pieces of candy did each friend get?

There are 8 ounces of cereal in a box. The store sells 3 boxes of cereal for $2. How many ounces do you get for $2?

Discuss ways that these problems might be modeled.

◆◆◆◆◆

SUGGESTED READINGS

Baroody, A. J., & Standifer, J. D. (1993). Addition and subtraction in the primary grades. In R.J. Jensen (Ed.), *Research ideas for the classroom: Early childhood mathematics.* New York: Macmillan Publishing Co.

Burk, D., Snider, A., & Symonds, P. (1988). *Box it or bag it mathematics: Teachers resource guide* [Kindergarten, First–Second]. Salem, OR: The Math Learning Center.

Burns, M. (1989). Teaching for understanding: A focus on multiplication. In P.R. Trafton (Ed.), *New directions for elementary school mathematics.* Reston, VA: National Council of Teachers of Mathematics.

Burns, M. (1992). *About teaching mathematics: A K–8 resource.* Sausalito, CA: Marilyn Burns Education Associates.

Carey, D. A. (1991). Number sentences: Linking addition and subtraction word problems and symbols. *Journal for Research in Mathematics Education, 22,* 266–280.

Carey, D. A. (1992). Students' use of symbols. *Arithmetic Teacher, 40,* 184–186.

Carpenter, T. P., Carey, D. A., & Kouba, V. L. (1990). A problem-solving approach to the operations. In J. N. Payne (Ed.), *Mathematics for the young child.* Reston, VA: National Council of Teachers of Mathematics.

Huinker, D. M. (1989). Multiplication and division word problems: Improving students' understanding. *Arithmetic Teacher, 37*(2), 8–12.

Katterns, B., & Carr, K. (1986). Talking with young children about multiplication. *Arithmetic Teacher, 33*(8), 18–21.

Kouba, V. L., & Franklin, K. (1993). Multiplication and division: Sense making and meaning. In R. J. Jensen (Ed.), *Research ideas for the classroom: Early childhood mathematics.* New York: Macmillan Publishing Co.

Mahlios, J. (1998). Word problems: Do I add or subtract? *Arithmetic Teacher, 36*(3), 48–52.

Quintero, A. H. (1985). Conceptual understanding of multiplication: Problems involving combination. *Arithmetic Teacher, 33*(3), 36–39.

Rathmell, E. C., & Huinker, D. M. (1989). Using "Part-Whole" language to help children represent and solve word problems. In P. R. Trafton (Ed.), *New directions for elementary school mathematics.* Reston, VA: National Council of Teachers of Mathematics.

Silverman, F. L., Winograd, K., & Strohauer, D. (1992). Student-generated story problems. *Arithmetic Teacher, 39*(8), 6–12.

Sowder, L., Threadgill-Sowder, J., Moyer, M. B., & Moyer, J. C. (1986). Diagnosing a student's understanding of operation. *Arithmetic Teacher, 33*(9), 22–25.

Talton, C. F. (1988). Let's solve the problem before we find the answer. *Arithmetic Teacher, 36*(1), 40–45.

Thompson, C. S., & Hendrickson, A. D. (1986). Verbal addition and subtraction problems: Some difficulties and some solutions. *Arithmetic Teacher, 33* (7), 21–25.

Thompson, C. S. & Van de Walle, J. A. (1984). Modeling subtraction situations. *Arithmetic Teacher, 32* (2), 8–12.

Weiland, L. (1985). Matching instruction to children's thinking about division. *Arithmetic Teacher, 33* (4), 34–35.

8 HELPING CHILDREN MASTER THE BASIC FACTS

◆ BASIC FACTS FOR ADDITION AND MULTIPLICATION REFER TO those combinations where both addends or both factors are less than 10. Subtraction and division facts correspond to addition and multiplication facts. Thus, 15 − 8 = 7 is a subtraction fact, because both parts are less than 10.

Mastery of a basic fact means that a child can give a quick response (less than 3 seconds) without resorting to nonefficient means, such as counting. Work toward mastery of addition and subtraction facts typically begins in the first grade. Most books include all addition and subtraction facts for mastery in the second grade, although much additional drill is usually required in grade 3 and even higher. Multiplication and division facts are generally a target for mastery in the third grade with more practice required in grades 4 and 5. Many children in grade 8 and above do not have a complete command of the basic facts.

The age of readily available calculators in no way diminishes the importance of basic fact mastery; quite the contrary. With the shift in emphasis from pencil-and-paper computation to mental computation and estimation skills, command of basic facts is more important today than ever. Further, it is true that *all* children are able to master basic facts—including those children with learning disabilities and slow learners. Children simply need to construct efficient mental tools that will help them. This chapter is about helping children develop those tools.

A THREE-STEP APPROACH TO FACT MASTERY

Every teacher of grades 4 to 10 knows children who are still counting on their fingers, making marks in the margins to count on, or simply guessing at answers. These children have certainly been drilled enough and have been given more than adequate opportunity to practice their facts over their previous years in school. Why haven't they mastered their facts? The simple answer is that they have not developed any efficient methods of producing a fact answer. The endless practice they have endured has at best made them very fast counters. Drill of inefficient mental methods does not produce mastery!

Fortunately, we do know quite a bit about helping children develop fact mastery and it has little to do with quantity of drill or drill techniques. Three components or steps to this end can be identified:

1. Develop a strong understanding of the operations and of number relationships.

2. Develop efficient strategies for fact retrieval.

3. Provide practice in the use of and in the selection of the strategies.

133

THE ROLE OF NUMBER AND OPERATION CONCEPTS

Number relationships play a significant role in fact mastery. For example, an efficient mental strategy for 8 + 5 is to think *8 and 2 more is 10—that leaves 3—10 and 3 is 13.* This requires the relationship between 8 and 10 (8 is 2 away from 10), the part-part-whole knowledge of 5 (2 and 3 more makes 5), and the fact that 10 and 3 is 13. For 6 × 7 it is efficient to think *5 × 7 and 7 more.* For many children the efficiency of this approach is lost because they need to count on 7 to get from 35 to 42. With an extension of the number relationships just noted it is possible to think *35 and 5 more is 40, and 2 more is 42.* In addition to the role of 10 and part-part-whole relationships, the relationships of one- and two-more-than, one- and two-less-than, doubles, and visualization of numbers in patterned arrangements to enable thinking of a number as a "single unit" rather than a count all play a significant part in the construction of efficient strategies. Always remember that children construct new ideas integrated with existing ones. When these number relationships are not present, the only existing relationships children have are based in counting.

The meanings of the operations also play a role as children attempt to construct efficient strategies. The ability to relate 6 × 7 to 5 × 7 and 7 more is clearly based on an understanding of the meanings of the first and second factors. To relate 13 − 7 to "7 and what makes 13" is dependent on a clear understanding of how addition and subtraction are related. The commutative or "turn-around" property for addition and multiplication is a very powerful idea. This idea reduces the number of addition and multiplication facts from 100 each to 55 each. Turn-arounds are especially important for the multiplication facts where many children do not realize that knowing 3 × 9 means they also know 9 × 3. These are just a few examples of the role that operation sense plays in the mastery of basic facts.

Teachers in the upper grades who are faced with students who have not mastered basic facts will do well to first investigate what command of number relationships and operations the students have. Without these relationships and concepts, the strategies discussed throughout this chapter will necessarily be learned in an instrumental manner. The result will not be nearly as effective as had these basic ideas been established first.

DEVELOPMENT OF EFFICIENT STRATEGIES

An efficient strategy is one that can be done mentally and quickly. The emphasis is on "efficient." Clearly, counting is not an efficient strategy. If practice or drill is undertaken when this is the only strategy available to a child, then at best you are providing practice in counting.

We have already seen some efficient strategies; the use of building up through 10 in adding 8 + 5 and the use of the related fact 5 × 7 to help with 6 × 7. Think for a moment how you think about 6 + 6? What about 9 + 5 or 2 + 7? You may think that you just "know" these. What is more likely is that you used some ideas similar to double six (for 6 + 6), 10 and 4 more (for 9 + 5), and your knowledge of the two-more-than relationships (for 2 + 7). Your response may be so automatic by now that you are not reflecting on the use of these relationships or ideas. That is one of the features of efficient mental processes—they become automatic with use.

Certainly many children have learned their facts without being taught efficient strategies. You, for example, were probably not taught strategies for the facts. Research suggests that we develop or learn many of these methods in spite of the drill we may have endured. Relational understanding of numbers and operations permit many individuals to invent their own strategies, probably without conscious thought. For those children who do not spontaneously develop efficient fact strategies, it is our job to help them do so by engaging them in activities that will encourage construction of these helpful relationships.

DRILL OF EFFICIENT METHODS AND STRATEGY SELECTION

Drill certainly plays a significant role in fact mastery and the use of old fashioned methods such as flash cards and fact games can be effective.

◆ Avoid Premature Drill

The strategies for some facts may not seem completely efficient at first glance. However, if practice is delayed until a strategy has been rather firmly developed, the practice will be *of that strategy* and will not be aimed at rote memory. The repetitious use of the strategy makes its use much quicker. Furthermore, the confidence it engenders in giving a response allows the strategy itself to fall into the background as the fact response becomes more automatic. Fact responses learned via strategies do become quick and automatic with practice. The important key is to delay practice until it is clear that the strategy is well understood. To begin drill activities before strategies are adequately understood can cause children to simply revert to less efficient methods.

◆ Practice Strategy Selection or Strategy Retrieval

Many teachers who have tried teaching fact strategies report that the method works great while the children are focused on whatever strategy they are working on. That is, they acknowledge that children can learn and use strategies. But, they continue, when the facts are all mixed up or the child is not in a fact-practice mode, the strategies tend to "go out the window" and the old counting habits return. The use of a strategy for basic facts is much like using a problem-solving strategy. We not only need to know how to use the strategy, but we need to learn to select the appropriate strategy when it is needed. This selection of a strategy or retrieval of the strategy from our personal repertoire is as important a part of fact-strategy instruction as the strategies themselves.

For example, suppose that your children have been practicing the near-doubles facts for addition; those that have addends that are one apart such as 4 + 5 or 8 + 7. The strategy is to derive the unknown 8 + 7 fact from the more well-known double, 7 + 7. So, children get quite skilled at doubling the smaller addend and adding one. All of the facts they are practicing are selected to fit this model. On other days they have learned and practiced strategies for facts like 9 + 4 (make ten facts) and probably others. Later, on a worksheet or in a mental math exercise, the children are faced with a mixture of facts. In a single exercise a child might see:

```
   7      4      2      8
  +6     +9     +6     +5
```

There is no mind set or reminder to use different processes for each. Especially if the children have previously been habituated to counting to get answers, they will very likely revert to counting and ignore the efficient methods they so recently practiced. Note that in practicing the strategy, there is no need to decide what strategy may be useful. All of the facts during the near-doubles practice were near doubles, and the strategy worked. Later, however, there is no one to suggest the strategy.

Strategy selection or *strategy retrieval* is the process of deciding what strategy is appropriate for a particular fact. If you don't think to use a strategy, you probably won't. A simple activity that is useful is to prepare a list of facts selected from two or more strategies and then, one fact at a time, ask children to name a strategy that would work for that fact. Further, they should explain why they picked the strategy and demonstrate its use. This type of activity turns the attention to the features of a fact that lend it to this or that strategy. The children are reflecting first on matching strategies with facts and secondly on use of the strategy.

AN OVERVIEW OF THE APPROACH

The use of strategies is not at all a new idea. Brownell and Chazal (1935) recognized that children use different thought processes with different facts. Since the mid-1970s there has been a strong interest among mathematics educators in the idea of directly teaching strategies to children (for example, Rathmell, 1978; Thornton & Noxon, 1977; Thornton & Toohey, 1984; Steinberg, 1985; Baroody, 1985; Fuson, 1984, 1992; Bley & Thornton, 1981). Many of the ideas that appear in this chapter have been adapted from the work of these researchers.

Regardless of the strategy or the operation, the general approach is roughly the same and is outlined here. Most of the remainder of the chapter is devoted to explaining the strategies.

Introduction of a Strategy

Initial work with a strategy begins with teacher-directed discussions. Models or drawings are used to help explain how a strategy works. Children can use materials at their desks to help construct the new idea. Discussions about the strategy and how it works are important. Note that the use of materials is to develop the strategy. The models are not there as answer-getting devices. Student explanations are very important, as they require children to reflect on the process and thereby make it their own. The models will help in that effort.

At times, the strategy requires the use of concepts or ideas developed earlier. You need to check to see if your students have these ideas. These prerequisites may be special number relationships, such as the notion that 10 and 3 is 13. Prerequisites may include mastery of a set of facts to which the new set of facts is to be related. One can hardly use the strategy of "double and one more" for 6 + 7 if double 6 is not a known fact. These prerequisites will be pointed out as each strategy is discussed.

Practice the Strategy

When you are comfortable that children are able to use the strategy without recourse to physical models and that they are beginning to use it mentally, it is time to practice it. It is a good idea to have as many as 10 different activities for each strategy or group of facts. File-folder or boxed activities can be used by children individually, in pairs or even in small groups. With a large number of activities, children can work on the facts that they need the most.

Flash cards are among the most useful approaches to fact-strategy practice. For each strategy make several sets of flash cards using all of the facts that fit that strategy. On the cards, label the strategy and/or include other drawings or cues to remind the children of the strategy. Examples appear throughout the chapter.

Other activities involve the use of special dice made from wooden cubes, teacher-made spinners, matching activities where a helping fact or a relationship is matched with the new fact being learned, and games of all sorts. Most any existing game involving fact drill can be modified to drill only one strategy or one collection of facts.

Add New Strategies

As children move into the practice phase with one strategy, the prerequisites and strategy for a new group of facts can be introduced and developed. With each new strategy, more activities and flash cards are prepared, and children work on those strategies for which they need the most help.

Practice Strategy Selection

After children have worked on two or three strategies, strategy selection drills are very important. These can be conducted quickly with the full class or a group, or independent games and activities can be prepared. Examples are described toward the end of the chapter.

STRATEGIES FOR ADDITION FACTS

The strategies for addition facts are directly related to one or more number relationships. In Chapter 6, numerous activities were suggested to develop these relationships. When the class is working on addition facts, the number relationship activities can and should be included with those described here. The teaching task is to help children connect these number relationships to the basic facts.

ONE-MORE-THAN, TWO-MORE-THAN FACTS

Each of the 36 facts highlighted in the chart has at least one addend of 1 or 2. These facts are a direct application of the one-more-than and two-more-than relationships.

+	0	1	2	3	4	5	6	7	8	9
0		1	2							
1	1	2	3	4	5	6	7	8	9	10
2	2	3	4	5	6	7	8	9	10	11
3		4	5							
4		5	6							
5		6	7							
6		7	8							
7		8	9							
8		9	10							
9		10	11							

8.1 ONE MORE/TWO MORE IS +1/+2

Ask "What is one more than seven?" As soon as you get a response, ask "What is one plus seven?" or hold up a 1 + 7 flash card. Be sure to connect the one-more-than-seven relationship to both 1 + 7 and 7 + 1.

8.2 ONE/TWO MORE THAN DICE

Make a die labeled +1, +2, +1, +2 one more, and two more. Use with another die labeled 4, 5, 6, 7, 7, and 9. After each roll of the dice, children should say the complete fact: "Four and two is six."

8.3 ONE/TWO MORE THAN MATCH

In a matching activity, children can begin with a number, match that with the one that is two more, and then connect that with the corresponding basic fact.

8.4 LOTTO FOR +1/+2

A lotto-type board can be made on a file folder. Small fact cards can be matched to the numbers on the board. The back of each fact card can have a small answer number to use as a check.

Figure 8.1 illustrates some of these activities and shows several possibilities for flash cards. Notice as you read through the chapter that activities such as the dice or spinner games and the lotto type activity can be modified for almost all of the various strategies in the chapter. These are not repeated for each strategy. Examples of flash cards are included for every strategy.

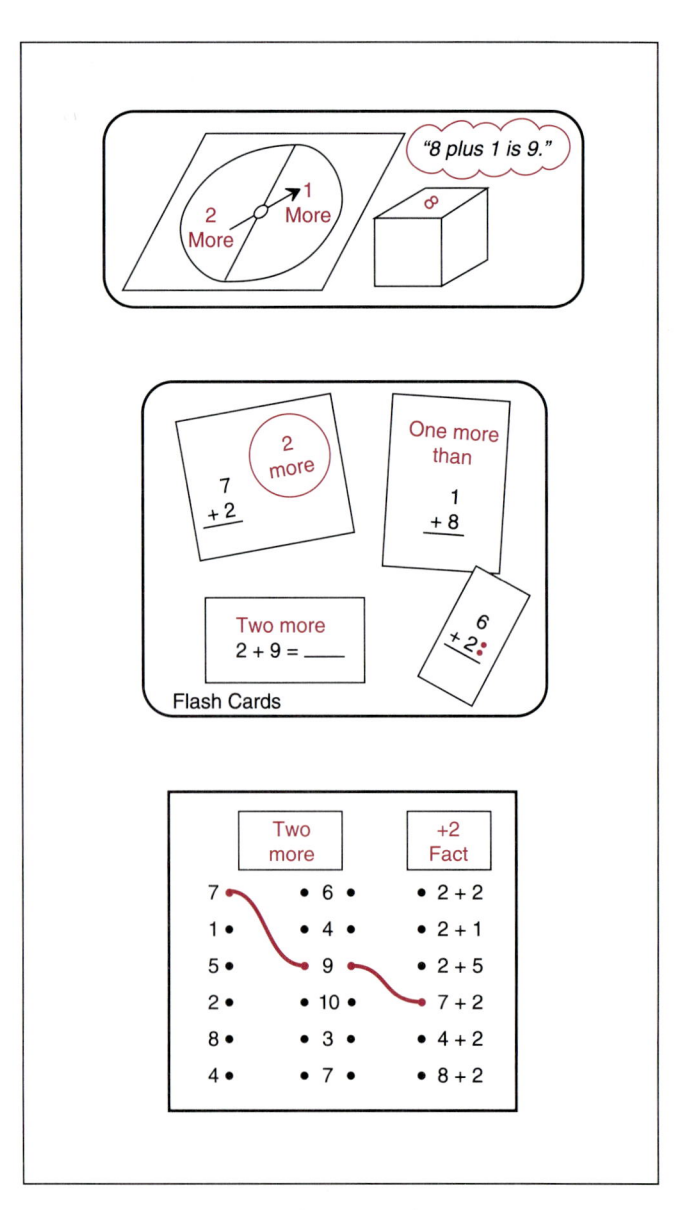

FIGURE 8.1: *One-more and two-more facts*

FACTS WITH ZERO

Nineteen facts have a zero as one of the addends. While such problems are generally easy, some children overgeneralize the idea that answers to addition are bigger. Flash card and dice games should stress the concept of zero (Figure 8.2).

+	0	1	2	3	4	5	6	7	8	9
0	0	1	2	3	4	5	6	7	8	9
1	1									
2	2									
3	3									
4	4									
5	5									
6	6									
7	7									
8	8									
9	9									

Flash Cards

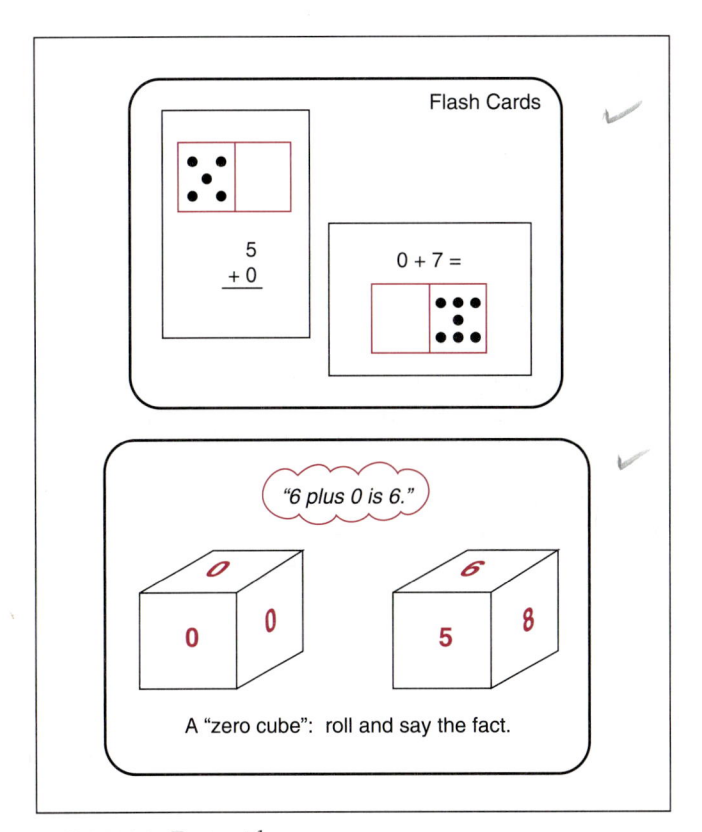

5 + 0

0 + 7 =

"6 plus 0 is 6."

A "zero cube": roll and say the fact.

FIGURE 8.2: *Facts with zero*

8.5 WHAT'S ALIKE?—ZERO FACTS

Write about 10 zero facts on the board, some with the zero first and some with the zero second. Discuss how all of these facts are alike. Have children use counters and a part-part-whole mat to model the facts at their seats.

DOUBLES

There are only 10 doubles facts from 0 + 0 to 9 + 9, as shown here. These 10 facts are relatively easy to learn and become a powerful way to learn the near-doubles (addends one apart). Some children use them as anchors for other facts as well.

+	0	1	2	3	4	5	6	7	8	9
0	0									
1		2								
2			4							
3				6						
4					8					
5						10				
6							12			
7								14		
8									16	
9										18

8.6 DOUBLE IMAGES

Make picture cards for each of the doubles and include the basic fact on the card as shown in Figure 8.3 (p. 138).

8.7 CALCULATOR DOUBLES

Use the calculator and enter the "double maker" (2 × =). Let one child say, for example, "Seven plus seven." The child with the calculator should press 7, try to give the double (14) and then press 7 = to see the correct double on the display. (Note that the calculator is also a good way to practice 11 and 12 facts.)

NEAR-DOUBLES

+	0	1	2	3	4	5	6	7	8	9
0		1								
1	1		3							
2		3		5						
3			5		7					
4				7		9				
5					9		11			
6						11		13		
7							13		15	
8								15		17
9									17	

These facts are also called the *doubles plus 1* facts and include all combinations where one addend is one more than the other. The strategy is to double the smaller number and add 1. Be sure students know the doubles before you start this strategy. (See Figure 8.4, p. 138.)

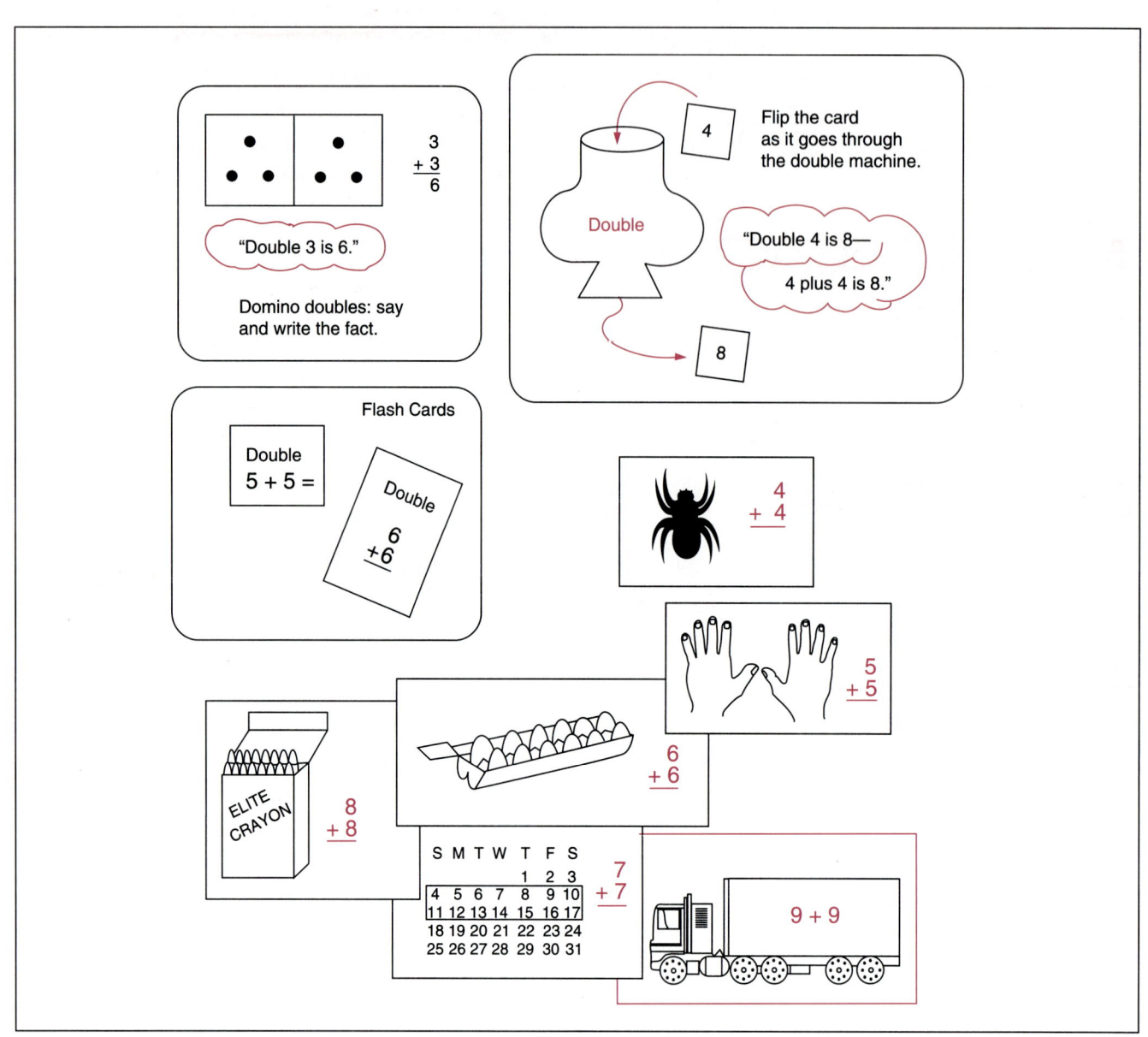

FIGURE 8.3: *Doubles facts*

FIGURE 8.4: *Near-double facts*

After discussing the strategy with the class, write 10 or 15 near-doubles facts on the board. Use vertical and horizontal formats, and vary which addend is the smaller. Quickly go through the facts. First have students only identify which number to double. The next time through, have students name the double that will be used. The third time, have students say the double and then the near-double.

8.8 MATCH WITH A DOUBLE

Use two sets of doubles cards and a complete set of near-doubles cards. Mix both sets and have students match them up, giving the answer for both facts.

8.9 DOUBLE DICE +1

Roll a single die with numerals or dot sets and say the complete double-plus-one fact. That is, for seven, students should say "seven plus eight is 15."

MAKE-TEN FACTS

+	0	1	2	3	4	5	6	7	8	9
0										
1										10
2									10	11
3									11	12
4									12	13
5									13	14
6									14	15
7									15	16
8		10	11	12	13	14	15	16	17	
9	10	11	12	13	14	15	16	17	18	

These facts all have at least one addend of 8 or 9. The strategy is to build onto the 8 or 9 up to 10 and then add on the rest. For 6 + 8: start with 8, then 2 more makes 10 and that leaves 4 more of 14.

Before using this strategy, be sure that children have learned to think of the numbers 11 to 18 as 10 and some more. Many second- and third-grade children have not constructed this relationship. (Refer to the section of Chapter 6 on "Relationships for Numbers 10 to 20.")

Provide a lot of time with the last activity. When children seem to have this idea, try the same activity, but without counters. Make ten-frame cards on acetate. Show an 8 (or 9) card on the overhead. Place other cards beneath it one at a time. Suggest mentally "moving" two dots into the 8 ten-frame. Have students say the complete fact: "Eight and four is twelve." The activity can be done independently with the little base-ten-frame cards (Black-line Masters).

Make flash cards with either one or two ten-frames, with reminders to "make ten" out of the 8 or the 9.

8.10 MAKE TEN ON THE TEN-FRAME

Give students a mat with two ten-frames (Figure 8.5). Flash cards are placed next to the ten-frames, or a fact can be given orally. The students should first model each number in the two ten-frames, and then decide on the easiest way to show (without counting) what the total is. The obvious choice is to move counters into the frame showing either 8 or 9. Get students to explain what they did. Have them explain that 9 + 6 is the same as 10 + 5.

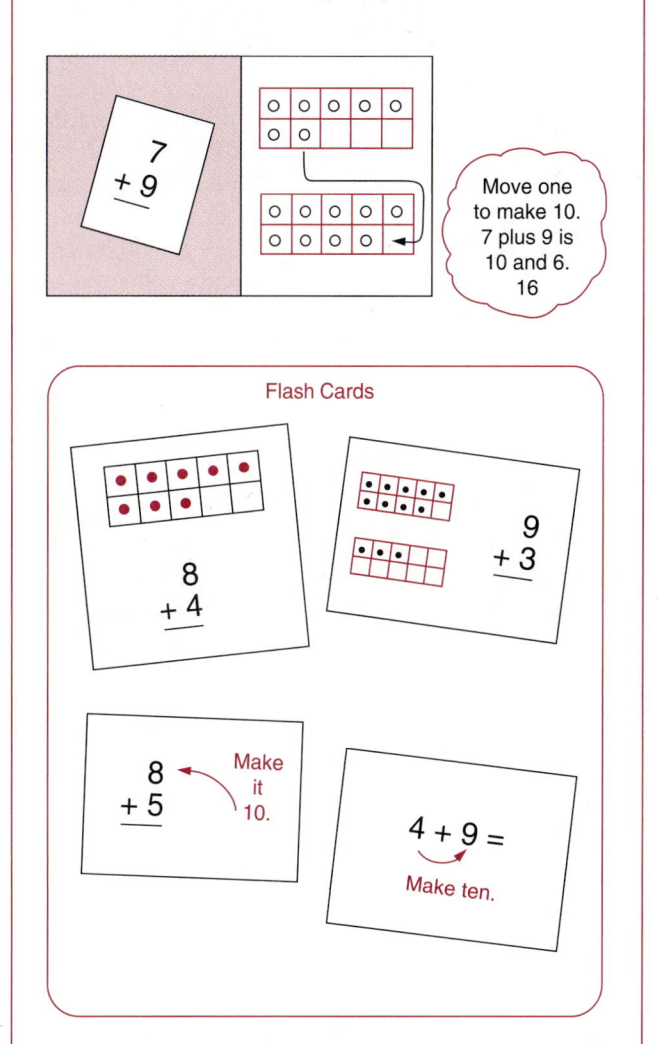

FIGURE 8.5: *Make-ten facts*

OTHER STRATEGIES AND THE LAST SIX FACTS

To appreciate the power of strategies on fact learning, consider the following. We have discussed only five ideas or strategies (one- or two-more-than, zeros, doubles, near-doubles, and make-ten) and yet these ideas have covered 88 of the 100 addition facts! Further, these ideas are really

140 8 / HELPING CHILDREN MASTER THE BASIC FACTS

not at all new but rather the application of important relationships. The 12 remaining facts are really only six facts, and their respective turn-arounds as shown on the chart.

+	0	1	2	3	4	5	6	7	8	9
0										
1										
2										
3						8	9	10		
4							10	11		
5				8				12		
6				9	10					
7				10	11	12				
8										
9										

To help children with these and perhaps other facts, there are a variety of additional ideas that might be employed. These ideas as well as those already discussed should be considered as a flexible set of possible tools for children's thought processes. Strategies should never be seen as "rules" or required procedures that children must follow. You will undoubtedly find children who use relationships and ideas not discussed here. That is fine! In fact, as your children come up with novel approaches, share them with the class. It is difficult to predict what is best or what will work since different children bring different relationships to the task.

◆ Doubles Plus Two or Two-Apart Facts

Of the six remaining facts, three have addends that differ by 2: 3 + 5, 4 + 6, and 5 + 7. There are two possible relationships that might be useful here, each depending on knowledge of doubles. Some children find the idea of extending the near-doubles to double-plus-*two*. In that sense, 4 + 6 (or 6 + 4) is double 4 and 2 more. A different idea is to take one from the larger addend and give it to the smaller. Using this idea, the 5 + 3 fact is transformed into the double 4 fact —*double the number in between.*

◆ Make Ten Extended

Three of the six facts have a 7 as one of the addends. The same make-ten strategy is frequently extended to these facts as well. For 7 + 4 the idea is *7 and 3 more makes 10 and 1 left is 11.* You may decide to introduce this idea at the same time that you initially introduce the make-ten strategy. It is interesting to note that Japan, mainland China, Korea, and Taiwan all teach an addition strategy of building through ten and do so in the first grade. The ideas may be easier in these countries because of the languages. The numbers beyond ten are all named in a regular manner, putting ten first. Twelve is said "ten two" and fifteen is "ten five." Many U.S. second-graders do not know what ten plus any number is (Fuson, 1992).

◆ Counting On

Counting on is the most widely promoted of the strategies not included here. It is generally taught as a strategy for all facts that have a 1, 2, or a 3 as one of the addends and thus includes the One- and Two-More-Than facts. For the fact 3 + 8 the child starts with 8 and counts three counts: *9, 10, 11.* There are several reasons this approach is down-played in this text. First, it is frequently applied to facts where it is not efficient, such as 8 + 5. It is difficult to explain to young children that they should count for some facts but not others. Secondly, it is much more procedural than conceptual. Finally, if other strategies are used, it is not necessary.

◆ Ten-Frame Facts

If you have been keeping track, all of the remaining six facts have been covered by the discussion thus far with a few being touched by two different thought patterns. The ten-frame and five-bar models (see Chapter 6) are so valuable in seeing certain number relationships that these ideas cannot be passed by in thinking about facts. The ten-frame helps children learn the combinations that make ten. Both five-bars and ten-frames immediately model all of the facts from 5 + 1 to 5 + 5 and the respective turn-arounds. Even 5 + 6, 5 + 7, and 5 + 8 are quickly seen as 2 fives and some more when depicted with these powerful models (Figure 8.6).

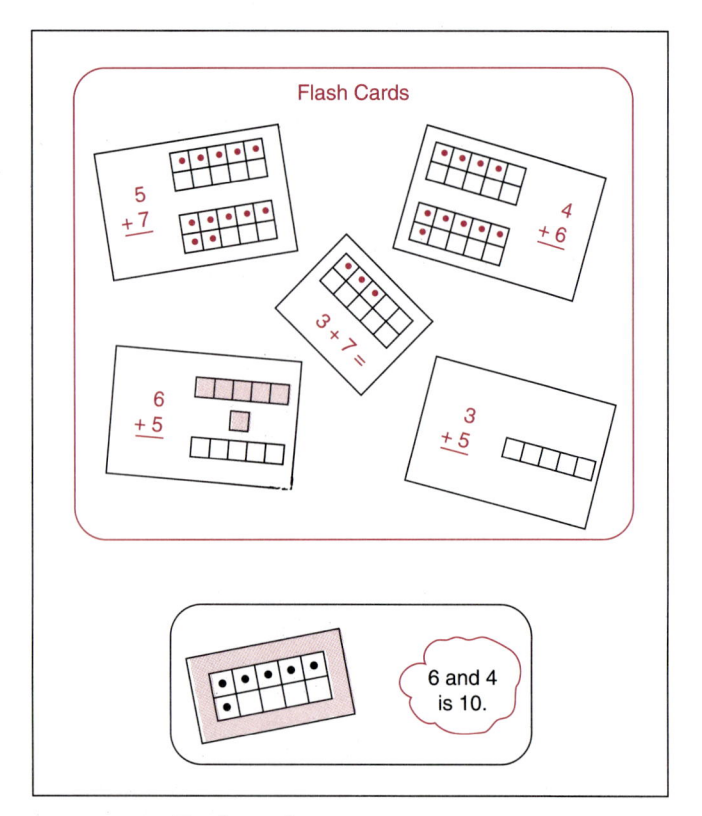

FIGURE 8.6 : *Ten-frame facts*

A good idea might be to group the facts shown in the chart here and practice them using one or two ten-frames as a cue to the thought process.

	0	1	2	3	4	5	6	7	8	9
0						5				
1						6				10
2						7			10	
3						8		10		
4						9	10			
5	5	6	7	8	9	10	11	12	13	14
6					10	11				
7				10		12				
8			10			13				
9		10				14				

The next two activities are suggestive of the type of relationships that can be developed.

◆◆◆◆◆◆◆◆◆◆◆◆◆◆◆◆◆◆◆◆◆◆

8.11 CALCULATOR PLUS-FIVES

Use the calculator to practice adding five. Enter ⊞ 5 ⊟. Next enter any number, and say the sum of that number plus five before pressing ⊟. Continue with other numbers. (The ⊞ 5 ⊟ need not be repeated.) If a set of five-bar models or a ten-frame is present, the potential for strengthening the five and ten relationships is heightened.

◆◆◆◆◆◆◆◆◆◆◆◆◆◆◆◆◆◆◆◆◆◆

8.12 SAY THE TEN FACT

Hold up a ten-frame card and have children say the "ten fact." For a card with 7 dots, the response is "seven and three is ten." Later, with a blank ten-frame drawn on the board, say a number less than 10. Children start with that number and complete the "ten fact." If you say, "four," they say, "four plus six is ten." Use the same activities in independent or small group modes.

STRATEGIES FOR SUBTRACTION FACTS

Subtraction facts prove to be more difficult than addition. This is especially true when children have been taught addition through a count-count-count approach; for 13 − 5: *count* 13, *count* off 5, *count* what's left. There is little evidence to suggest that anyone who has mastered subtraction facts has used this approach in helping them learn the facts. Unfortunately, many sixth, seventh, and eighth graders are still counting.

SUBTRACTION AS THINK-ADDITION

In Figure 8.7, both the take-away and missing-part concepts of subtraction are shown modeled in such a way that students are encouraged to think "What goes with this part to make the total?" When done in this *think-addition* man-

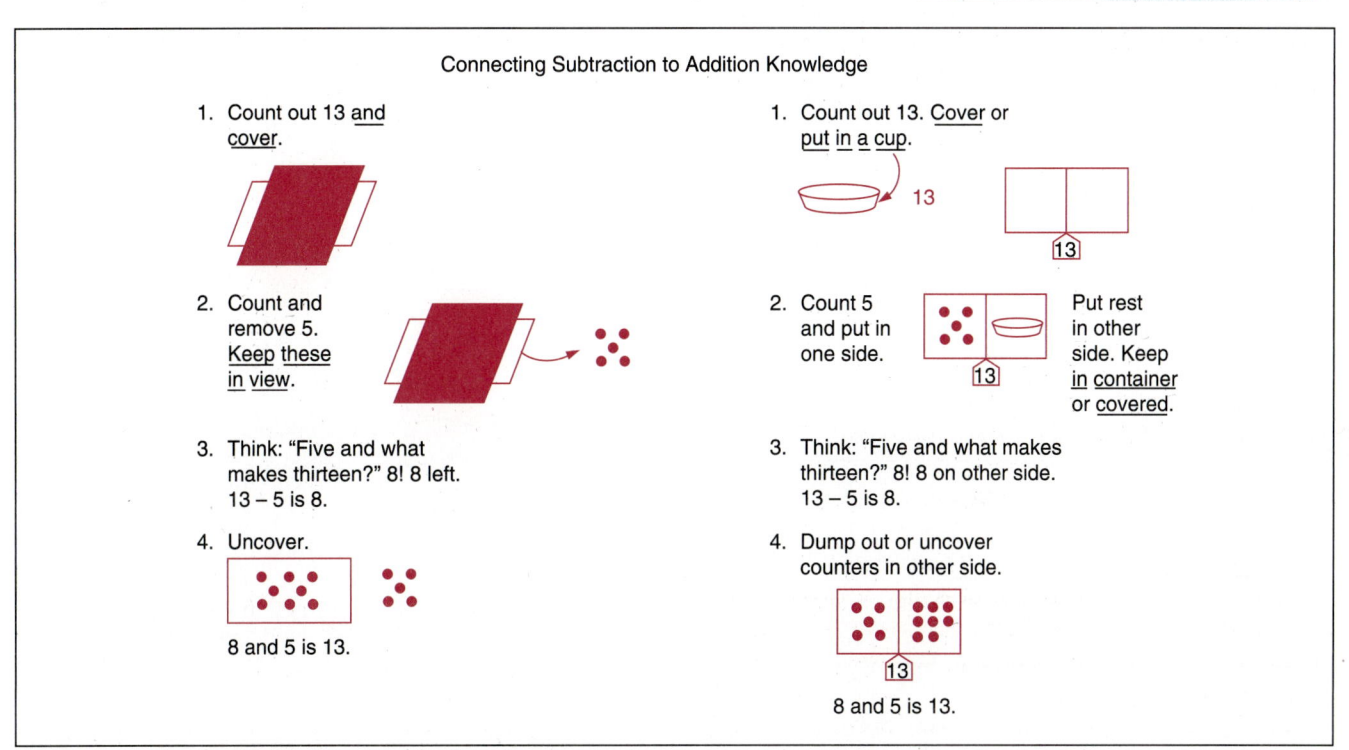

Connecting Subtraction to Addition Knowledge

1. Count out 13 and cover.

2. Count and remove 5. Keep these in view.

3. Think: "Five and what makes thirteen?" 8! 8 left. 13 − 5 is 8.

4. Uncover.
   8 and 5 is 13.

1. Count out 13. Cover or put in a cup.

2. Count 5 and put in one side.
   Put rest in other side. Keep in container or covered.

3. Think: "Five and what makes thirteen?" 8! 8 on other side. 13 − 5 is 8.

4. Dump out or uncover counters in other side.
   8 and 5 is 13.

FIGURE 8.7: *Using a think-addition model for subtraction*

ner, the child uses known addition facts to produce the unknown quantity or part. (You might want to revisit missing-part activities in Chapter 6 and part-part-whole subtraction concepts in Chapter 7). If this important relationship between parts and wholes—between addition and subtraction—can be made, subtraction facts can be much easier. When children see 9 − 4, you want them to spontaneously think, "Four and *what* make nine." By contrast, observe a third-grade child who struggles with this fact. The idea of thinking addition never occurs. Instead, the child will begin to count either back from 9 or up from 4. The value of think-addition cannot be overstated.

SUBTRACTION FACTS WITH SUMS TO TEN

Think-addition is most immediately applicable to those subtraction facts with sums of 10 or less. These are generally introduced with a goal of mastery in the first grade. Sixty-four of the 100 subtraction facts fall into this category. If think-addition is to be used effectively, it is essential that addition facts be mastered first. There is evidence that suggests children learn very few if any subtraction facts without first mastering the corresponding addition facts. In other words, mastery of 3 + 5 can be thought of as prerequisite knowledge for learning the facts 8 − 3 and 8 − 5.

Rather than lump all of these 64 facts together in one rather formidable bundle, a good idea is to group them in the same way that the corresponding addition facts were grouped.

◆ Facts with Zero

This set of facts includes those involving minus zero and those with a difference of zero (e.g., 7 − 0, 7 − 7).

Using the think-addition approach, model these facts on the overhead. You may be surprised to find some children confused by them, especially those involving subtracting zero.

Make flash cards as shown in Figure 8.8.

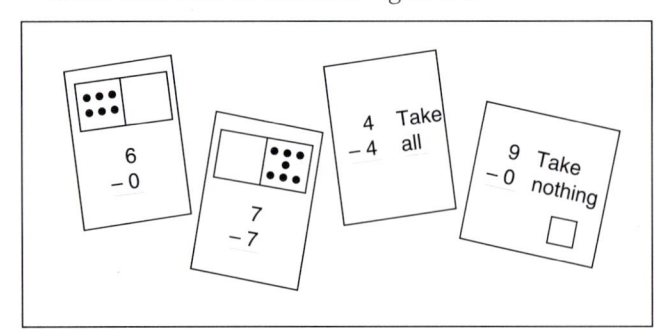

FIGURE 8.8: *Flash cards for zero facts*

◆ One-Less-Than, Two-Less-Than Facts

This group includes all facts with differences of one or two as well as those that involve −1 or −2 (e.g., 8 − 7, 8 − 6, 8 − 1, 8 − 2). The relationships of more than and less than must be connected.

Review the one- and two-less-than relationships using any of the ideas in Chapter 6. With these relationships in students' minds, model facts in this group on the overhead projector using the think-addition approach. Frequently follow one fact with its partner. That is, follow 9 − 2 with 9 − 7 and discuss how the two facts are alike. Then write the 7 + 2 = 9 fact. Help children see how the two-more-than and two-less-than relationships are involved in both facts. Use the calculator to practice "minus two" and "minus one" facts. Press ⊟ 2 ⊟ to make a "minus two" machine. Make flash cards for all 36 facts in this group. Use the words "one less" or "two less" on all cards, as shown in Figure 8.9.

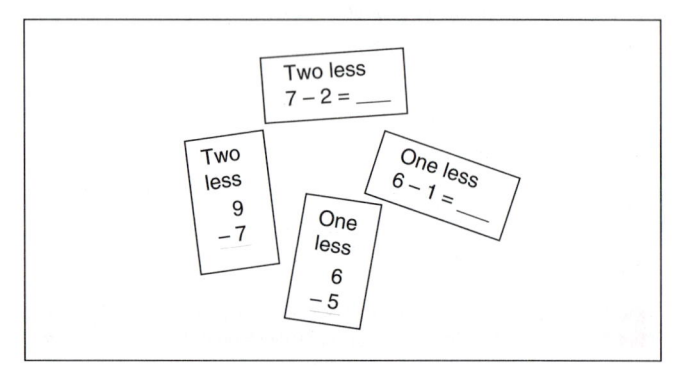

FIGURE 8.9: *Flash cards for one-less and two-less*

◆ Ten-Frame Facts

This group includes all facts with the first number of 10, those involving −5, and those with a difference of 5. Therefore, 10 − 7, 8 − 5, and 8 − 3 are all in this group. Make flash cards for all three types of facts in this group. All are shown in Figure 8.10. Use all the cards together. Similar drawings can be incorporated into other activities.

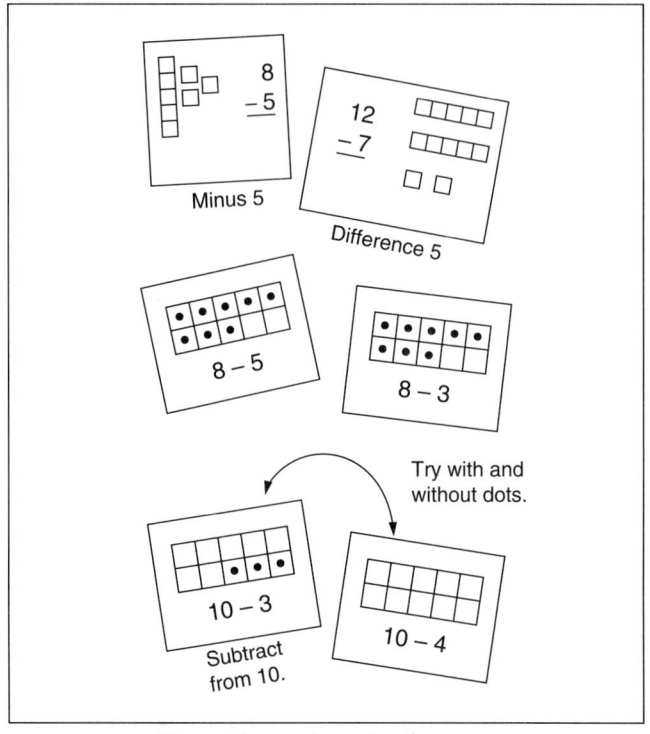

FIGURE 8.10: *Five and ten subtraction facts*

8.13 PLUS AND MINUS TEN-FACTS

Show a ten-frame card, and have students say a subtraction fact and then an addition fact. For a frame showing 6 dots, students would say, "Ten minus six is four," and then, "Six and four is ten."

◆ Doubles and Near-Doubles

This group of facts corresponds to the addition facts of the same name. It is useful to mix the doubles and near-doubles.

Review the addition doubles and the near-doubles. Then model the double and near-double subtraction facts on the overhead projector, using the think-addition approach. For each subtraction fact, ask if it was a double or a near-double, and have children say both facts. ("Near-double! Nine minus four is five. Four plus five is nine.")

8.14 DOUBLE AND NEAR-DOUBLE SEARCH

Write a large number of subtraction facts on a page. Make roughly a third of them doubles, a third near-doubles, and the other third other facts. Feel free to repeat facts several times. Have students circle with a crayon all of the doubles facts and then answer them. Next have them circle the facts that are near-doubles with a different crayon and answer them. The remaining facts can be left blank. Next to each subtraction fact, have them write the addition double or near-double that goes with it.

THE 36 "HARD" SUBTRACTION FACTS: SUMS GREATER THAN 10

Before reading further, look at the three subtraction facts shown here, and try to reflect on what thought process you use to get the answers. Even if you "just know them," think about what a likely process might be.

$$\begin{array}{ccc} 14 & 12 & 15 \\ -9 & -6 & -6 \end{array}$$

Many people will use a different strategy for each of these facts. For 14 − 9, it is easy to start with 9 and work up through 10: 9 and 1 more to 10 and 4 more is 5. For the 12 − 6 fact it is quite common to hear "double six," a think-addition approach. For the last fact, 15 − 6, ten is used again but probably by working backward from 15— a take-away process: Take away 5 to get 10 and one more leaves 9. We could call these three approaches respectively, build-up-through-ten, think-addition, and back-down-through-ten. Each of the remaining 36 facts with sums of 11 or more can be learned using one or more of these strategies. Figure 8.11 shows how these facts, in three overlapping groups, correspond to the strategy that is most likely to be used.

11 − 2	11 − 3	11 − 4	11 − 5	11 − 6	11 − 7	11 − 8	11 − 9
12 − 3	12 − 4	12 − 5	12 − 6	12 − 7	12 − 8	12 − 9	
13 − 4	13 − 5	13 − 6	13 − 7	13 − 8	13 − 9		
14 − 5	14 − 6	14 − 7	14 − 8	14 − 9			
15 − 6	15 − 7	15 − 8	15 − 9				
16 − 7	16 − 8	16 − 9					
17 − 8	17 − 9						
18 − 9							

Build up through ten

Back down through ten

Think-addition (any fact)

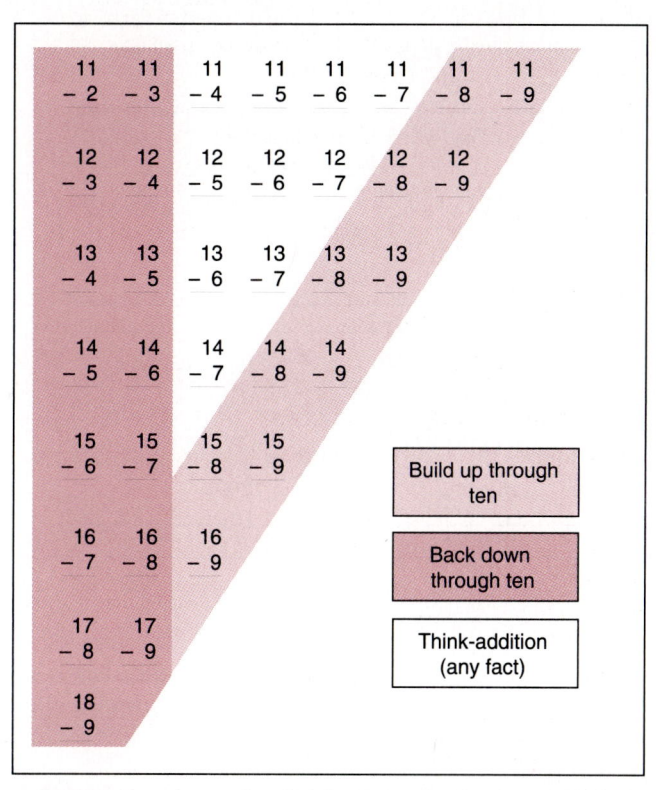

FIGURE 8.11: *The 36 "hard" subtraction facts*

◆ Build-Up through Ten

This group includes those facts where the part or subtracted number is either 8 or 9. Examples are 13 − 9 and 15 − 8.

8.15 BUILD UP THROUGH THE TEN-FRAME

On the board or overhead draw a ten-frame with 9 dots. Discuss how you can build numbers between 11 and 18, starting with 9 in the ten-frame. Stress the idea of *one more to get to* 10 and then the rest of the number. Repeat for a ten-frame showing 8. Next, with either the 8 or 9 ten-frame in view, call out numbers from 11 to 18 and have students respond with the difference between that number and the one on the ten-frame. Later, use the same approach, but show fact cards to connect this idea with the symbolic subtraction fact (Figure 8.12, p. 144).

◆ Back Down through Ten

Here is one strategy that is really take-away and not think-addition. It is useful for all of those facts that have a difference of 8 or 9, such as 15 − 6 or 13 − 5. For example, with 15 − 6 you start with the total of 15 and

8 / HELPING CHILDREN MASTER THE BASIC FACTS

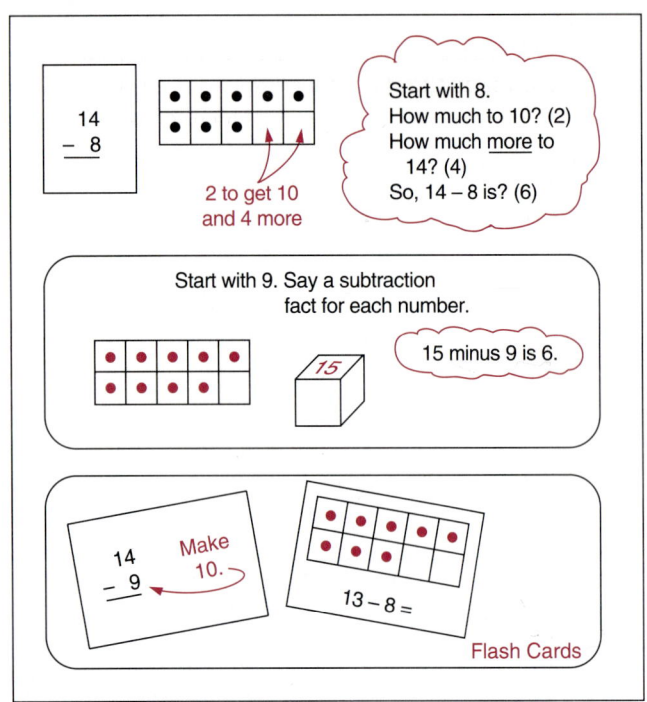

FIGURE 8.12: *Build up through ten*

take off five. That gets you down to 10. Then take off 1 more to get 9. For 14 − 6, just take off 4 and then take off 2 more to get 8. Here we are working backward with 10 as a "bridge."

8.16 DOWN THROUGH THE TEN-FRAME

Start with two ten-frames on the overhead, one filled completely and the other partially filled as in Figure 8.13. For 13, for example, discuss what is the easiest way to think about taking off 4 counters or 5 counters. Repeat with other numbers between 11 and 18. Have students write or say the corresponding fact.

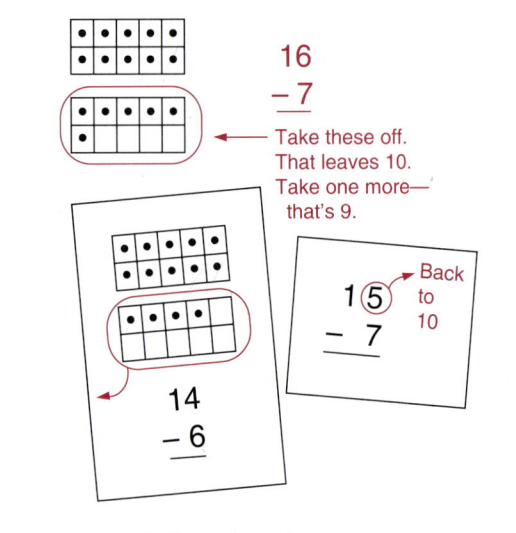

FIGURE 8.13: *Back down through ten*

Extend Think-Addition

Think-addition remains one of the most powerful ways to think about subtraction facts. When the think-addition concept of subtraction is well developed, many will use that approach for all facts. (Notice that virtually everyone uses a think-multiplication approach for division. Why?)

It does not seem particularly useful to separate the 36 "hard facts" according to the corresponding addition strategies. What may be more important is to listen to children's thinking as they attempt to answer subtraction facts that they have not yet mastered. If they are not using one of the three ideas suggested here it is a good bet that they are counting—an inefficient method. Work hard at the think-addition concept. Show a fact such as 12 − 5 and say, "Five and what makes twelve?" Continue using this phrase until it becomes habit for the child as well. Also return to missing-part activities as found in Chapter 6. Use wholes greater than 10 and have the students write both an addition and subtraction fact for each.

The activities that follow are all of the think-addition variety. There is, of course, no reason why these activities could not be used for all of the subtraction facts. They need not be limited to the "hard facts."

8.17 MISSING-NUMBER CARDS

Show children, without explanation, families of numbers with the sum circled as in Figure 8.14(a). Ask why they think the numbers go together and why one number is circled. When this number family idea is fairly well understood, show some families with one number replaced by a question mark [Figure 8.14(b)] and ask what number is missing. When students understand this activity, explain that you have made some missing-number cards based on this idea. Each card has two of three numbers that go together in the same way. Sometimes the circled number is missing (the sum), and sometimes one of the other numbers is missing (a part). The cards can be made both vertically and horizontally with the sum appearing in different positions. The object is to name the missing number.

8.18 MISSING-NUMBER WORKSHEETS

Make copies of the blank form found in Appendix B to make a wide variety of drill exercises. In a row of 13 "cards," for example, put all of the combinations from two families with different numbers missing and with numbers and blanks in different positions. An example is shown in Figure 8.15. After filling in numbers, run the sheet off, and have students fill in the missing numbers. Another idea is to group together facts from one strategy or number relation or perhaps mix facts from two strategies on one page. Actual flash cards can be made this way and put in packets for individual practice. Have students write an addition fact and a subtraction fact to go with each missing-number card. This is an important step because many children are able to give the missing part in a family but do not connect this knowledge with subtraction.

facts on the board. Rather than call out answers, students find the addition fact that helps with the subtraction fact. On your signal, each student holds up the appropriate fact. For 12 − 4 or 12 − 8 the students would select 4 + 8. The same idea of matching a subtraction fact with a helping-addition fact can be made into a matching card game or a matching worksheet.

FIGURE 8.14: Introducing "missing-number" cards

8.19 FIND A PLUS FACT TO HELP

Select a group of subtraction facts that you wish to practice. Divide a sheet of paper into small cards, about 10 or 12 to a sheet. For each subtraction fact, write the corresponding addition fact on one of the cards. Two subtraction facts can be related to each addition fact. Duplicate the sheet, and have students cut the cards apart. Now write one of the subtraction

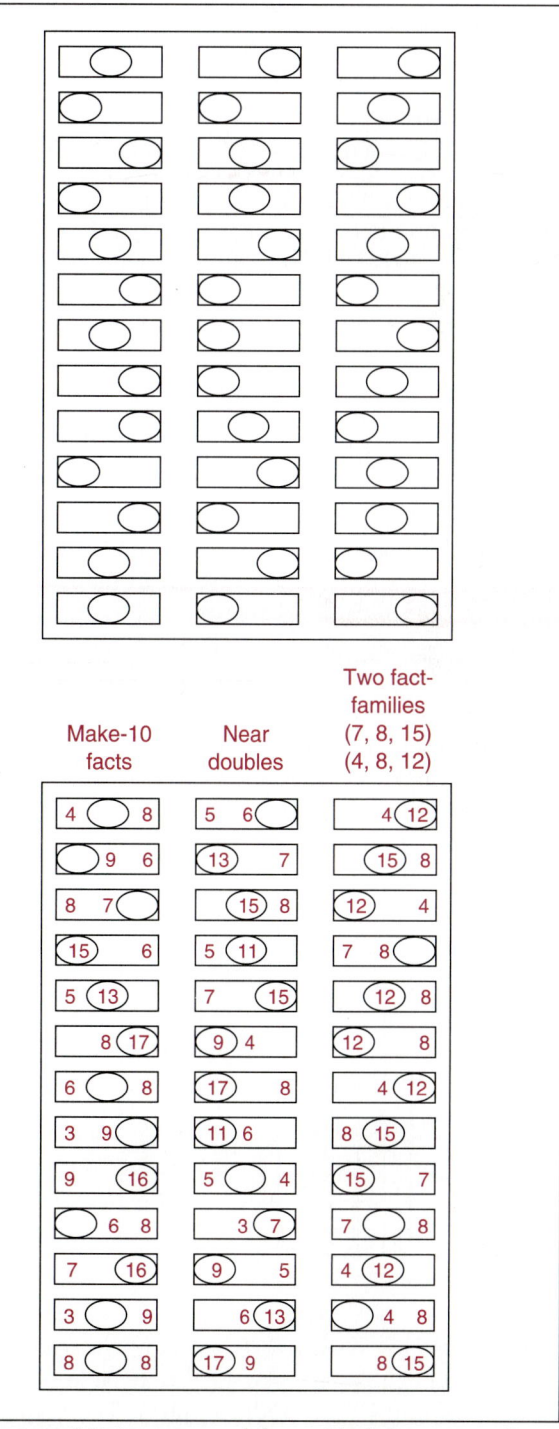

FIGURE 8.15: Missing-part worksheets. Black-line master 8 can be used to fill in any sets of facts you wish to emphasize.

8.20 WRITE THE PLUS HELPER

Supply a list of facts, and have students first write the helping fact and then answer both. Begin by giving a list of addition facts, and have students write either subtraction fact, they wish. Later, give them the subtraction fact and have them write the addition fact.

8.21 CALCULATOR 2× FACTS

Review the concept of doubles from addition. Play "Say the Double." You say a number and the children say the double of that number. Use the calculator to practice doubles (press 2 × =).

Make and use flash cards with the related addition fact or word "double" as a cue (Figure 8.16).

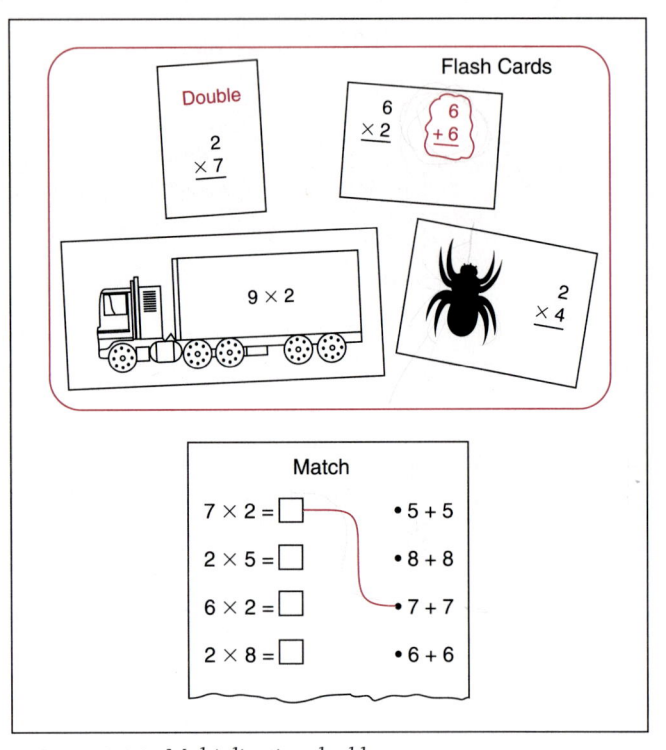

FIGURE 8.16: *Multiplication doubles*

STRATEGIES FOR MULTIPLICATION FACTS

Multiplication facts can also be mastered by relating new facts to existing knowledge. For example, the facts with a factor of 2 are related to the addition doubles. The fact 4 × 7 can be found from double 7 and then double again. While models are sometimes used, they are there for constructing relationships, not for getting answers. Counting the elements in six rows of eight will seldom help a child master the 6 × 8 fact.

Since the first and second factors in multiplication stand for different things (7 × 3 is 7 threes and 3 × 7 is 3 sevens), it is imperative that students completely understand the commutative property (go back and review Figure 7.17). For example, 2 × 8 is related to the addition fact, double 8. But the same relationship also is applied to 8 × 2. Most of the fact strategies are more obvious with the factors in one order than in the other, but turnaround facts should always be learned together.

Of the five groups or strategies discussed next, the first four strategies are generally easier and cover 75 of the 100 multiplication facts.

DOUBLES

Those facts that have a 2 as a factor are equivalent to the addition doubles and should already be known by students who know their addition facts. The major problem is to realize that not only is 2 × 7 double 7, but so is 7 × 2.

FIVES FACTS

This group consists of all facts with a 5 as first or second factor, as shown here.

Practice counting by fives to at least 45. Connect counting by fives with rows of five dots. Point out that six rows is a model for 6 × 5, eight rows is 8 × 5, and so on.

8 / HELPING CHILDREN MASTER THE BASIC FACTS 147

8.22 CLOCK FACTS

Focus on the minute hand of the clock. When it points to a number, how many minutes after the hour is it? Draw a large clock, and point to numbers 1 to 9 in random order. Students respond with the minutes after. Now connect this idea to the multiplication facts with 5. Hold up a flash card and then point to the number on the clock corresponding to the other factor. In this way, the five facts become the "clock facts."

ZEROS AND ONES

×	0	1	2	3	4	5	6	7	8	9
0	0	0	0	0	0	0	0	0	0	0
1	0	1	2	3	4	5	6	7	8	9
2	0	2								
3	0	3								
4	0	4								
5	0	5								
6	0	6								
7	0	7								
8	0	8								
9	0	9								

Include the clock idea on flash cards or to make matching activities (Figure 8.17).

Thirty-six facts have at least one factor that is either a 0 or a 1. These facts, while apparently easy, tend to get confused with "rules" that some children learned for addition. The fact 6 + 0 stays the same, but 6 × 0 is always zero. The 1 + 4 fact is a one-more idea, but 1 × 4 stays the same. Make flash cards and games that reflect a conceptual approach to these facts. For zero, stress sets of nothing. Zero sets of 6 is difficult to conceptualize. One set of 6 for 1 × 6 is just as easy as 6 ones for 6 × 1 (Figure 8.18).

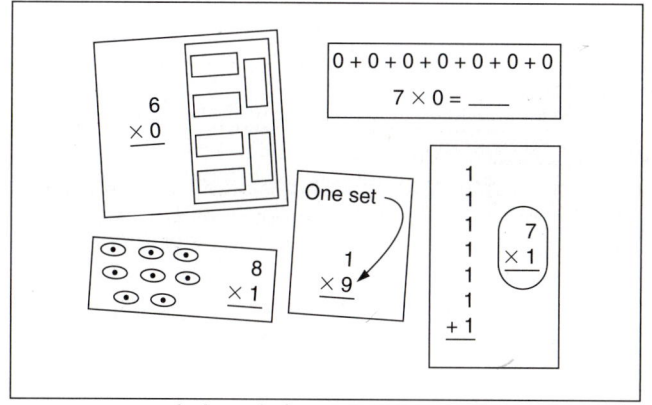

FIGURE 8.18: *Flash cards for zeros and ones facts*

NIFTY NINES

×	0	1	2	3	4	5	6	7	8	9
0										0
1										9
2										18
3										27
4										36
5										45
6										54
7										63
8										72
9	0	9	18	27	36	45	54	63	72	81

Those facts with a factor of 9 include the largest products, but can be among the easiest to learn. The table of nines facts includes some nice patterns that are fun to discover. Two of

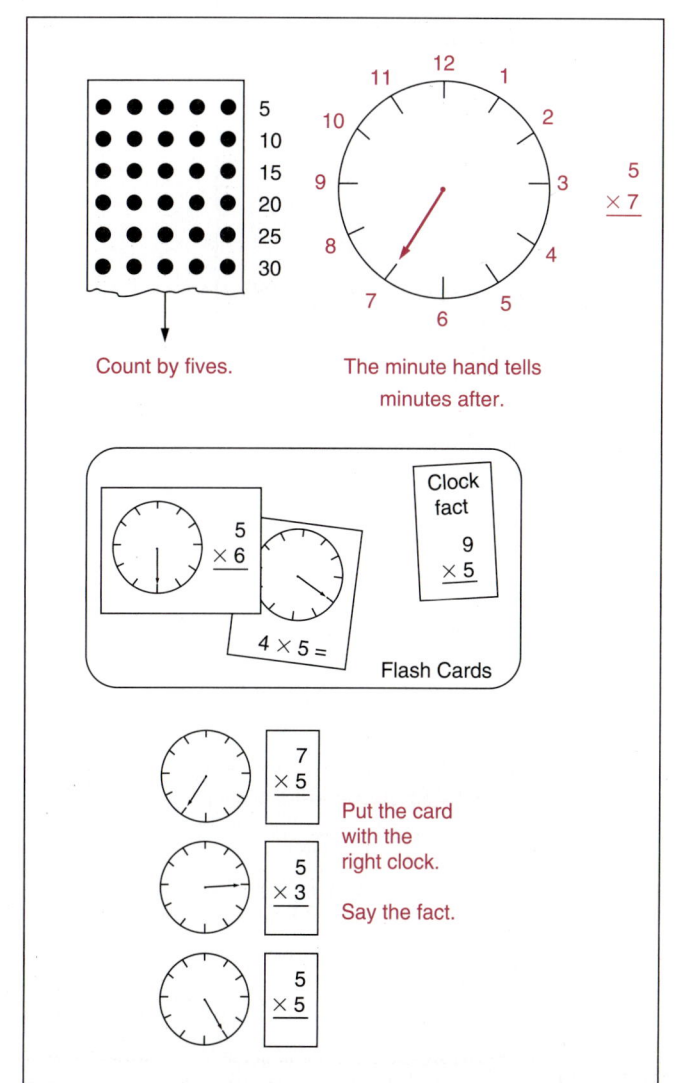

FIGURE 8.17: *Fives facts*

these patterns are useful for mastering the nines: (1) the tens digit of the product is always one less than the "other" factor (the one other than 9), and (2) the sum of the two digits in the product is always 9. So these two ideas can be used together to quickly get any nine fact. For 7 × 9: 1 less than 7 is 6, 6 and 3 make 9, so 63 (Figure 8.19).

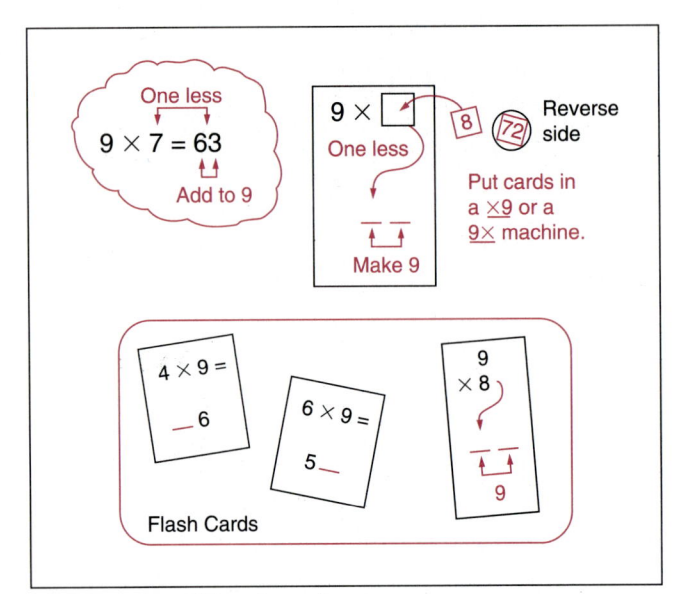

FIGURE 8.19: *Nifty nines rule*

It may be necessary to practice parts of nine. If you say "six," the class answers "three."

Make and use flash cards like those shown in Figure 8.19. Notice that some cards practice only the one-less part of the rule that determines the tens digit, and others practice the sum-to-9 rule to determine the ones digit.

Either part of the nifty nines rule can be incorporated into a matching activity: 4 × 9 can be matched with either ____6 or 3____.

8.23 A NIFTY NINE MACHINE

A "machine" like that shown in Figure 8.19 can be an independent activity or could be drawn on the board for full class practice.

A warning: While the nines strategy can be quite successful, it also can cause confusion. Because two separate rules are involved, and a conceptual basis is not apparent, children may confuse the two rules or attempt to apply the idea to other facts.

An alternative strategy for the nines is almost as easy to use. Notice that 7 × 9 is the same as 7 × 10 less one set of 7, or 70 − 7. This can be easily modeled by displaying rows of 10 cubes, with the last one a different color, as in Figure 8.20. For students who can easily subtract 4 from 40, 5 from 50, and so on, this strategy may be preferable.

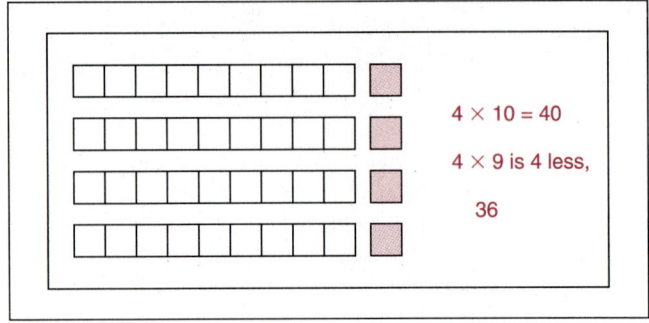

FIGURE 8.20: *Another way to think of the nines*

USE A HELPING FACT

The chart shows the remaining 25 multiplication facts. It is worth pointing out to children that there are actually only 15 facts remaining to master because 20 of them consist of 10 pairs of turnarounds.

×	0	1	2	3	4	5	6	7	8	9
0										
1										
2										
3					9	12		18	21	24
4					12	16		24	28	32
5										
6					18	24		36	42	48
7					21	28		42	49	56
8					24	32		48	56	64
9										

These 25 facts can be learned by relating each to an already known fact or *helping* fact. For example, 3 × 8 is connected to 2 × 8 (double 8 and 8 more). The 6 × 7 fact can be related to either 5 × 7 (5 sevens and 7 more) or to 3 × 7 (double 3 × 7). The helping fact must be known, and the ability to do the mental addition must also be there.

How to find a helping fact that is useful varies with different facts and sometimes depends on which factor you focus. Figure 8.21 illustrates models for four overlapping groups of facts and the thought process associated with each.

The *double and double again* approach is applicable to all facts with a 4 as one of the factors. Remind children that the idea works when the 4 is the second factor as well as the first. For 4 × 8, double 16 is also a difficult fact. Help children with this by noting, for example, that 15 + 15 is 30, 16 + 16 is two more, or 32. Adding 16 + 16 on paper defeats the purpose.

Double and one more set is a way to think of facts with one factor of 3. With an array or a set picture, the double part can be circled, and it is clear that there is one more set. Two facts in this group involve difficult mental additions.

8 / HELPING CHILDREN MASTER THE BASIC FACTS   **149**

admittedly difficult, this approach is used by many children, and it becomes the best way to think of one or two particularly difficult facts. "What is seven times eight—Oh, that's forty-nine and seven more—fifty-six." The process can become almost automatic.

Model the relationships between hard facts and helping facts with the use of arrays and set models as shown in Figure 8.21. Spend time having different students think out loud as they use a helping fact. Students describing these relationships is probably the most effective tool you have to help students with these facts.

Use flash cards and matching activities (Figure 8.22) to help children remember to use a helper. Be aware that one student's preferred approach to a fact may not be the same as another student's.

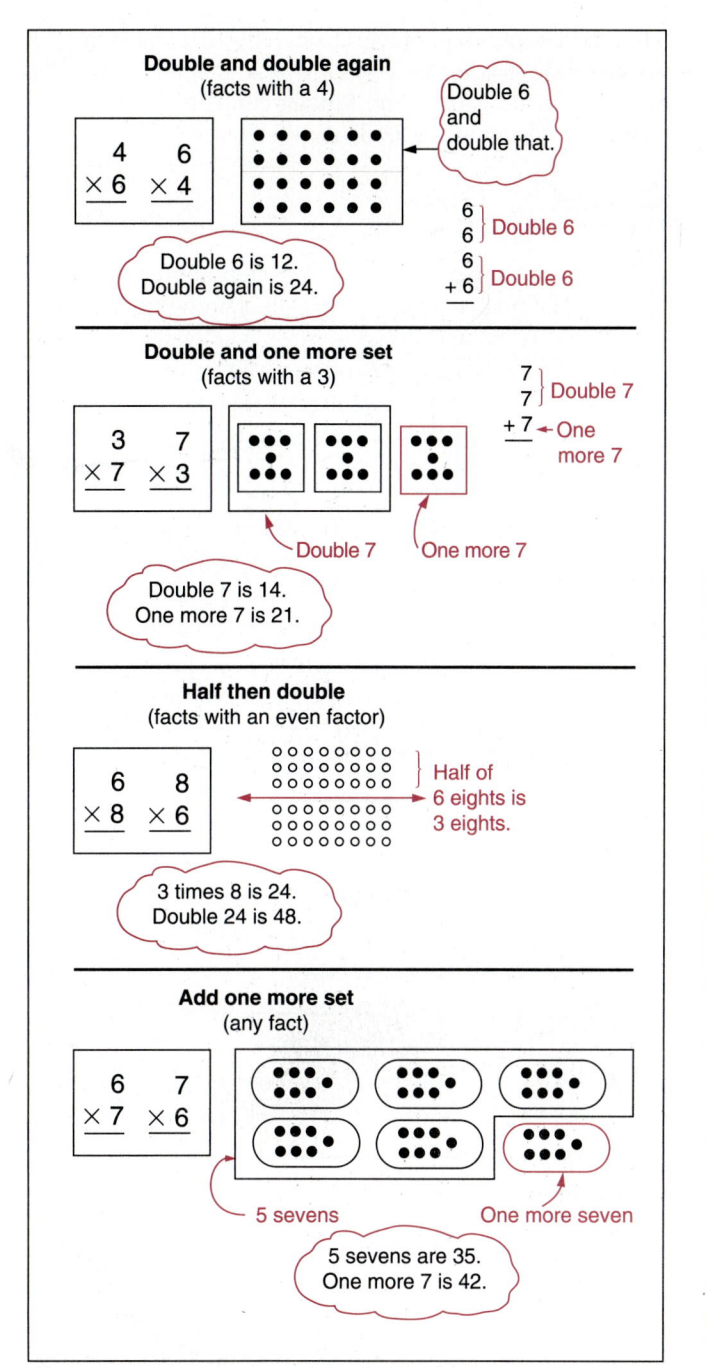

FIGURE 8.21: *Finding a helping fact*

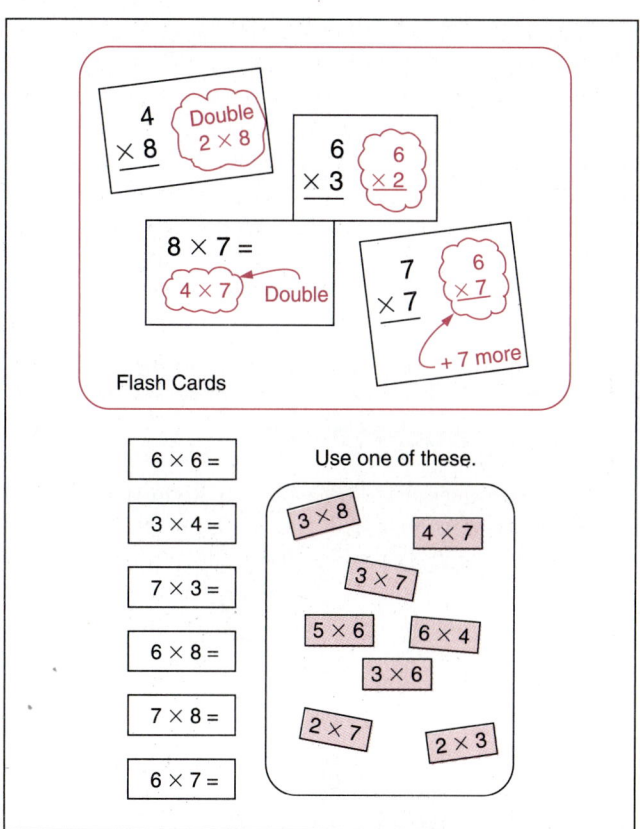

FIGURE 8.22: *Practice thinking of a helping fact*

If either factor is even, a *half then double* approach can be used. Select the even factor and cut it in half. If the smaller fact is known, that product is doubled to get the new fact. For 6 × 7, half of 6 is 3. Three times 7 is 21. Double 21 is 42. For 8 × 7, the double of 28 may be hard, but it remains an effective approach to that traditionally hard fact. (Double 25 is 50 + 2 times 3 is 56).

Many children prefer to go to a fact that is "close" and then *add one more set* to this known fact. For example, think of 6 × 7 as 6 sevens. Five sevens is close: That's 35. Six sevens is one more seven or 42. When using 5 × 8 to help with 6 × 8, the set language "six eights" is very helpful in remembering to add 8 more and not 6 more. While

SOME SPECIAL FACTS

Many fact programs isolate the *squares* or those facts with both factors the same. Children can draw square arrays on grid paper to see why a fact such as 7 × 7 is called a square. This isolation and labeling of the square may be helpful to some children.

Thornton and Toohey (1984) use special images for certain facts. The 3 × 3 fact is the "tic-tac-toe fact" associated with a tic-tac-toe board. The 3 × 7 and 7 × 3 facts are the "three-week facts."

DIVISION FACTS AND "NEAR FACTS"

Children are likely to use the corresponding multiplication fact to think of a division fact. If we are trying to think of 36 ÷ 9, we tend to think, "Nine times what is thirty-six." Most of us have memorized 42 ÷ 6, not as a separate and unrelated fact, but one closely tied to 6 × 7. (Would it not be wonderful if subtraction were so closely related to addition?)

It is an interesting question to ask, "When children are working on a page of division facts, are they practicing division or multiplication?"

DIVISION "NEAR FACTS"

Exercises such as 50 ÷ 6 are much more prevalent in computations and even in real situations than division "facts" or division without remainders. To determine the answer to 50 ÷ 6, most people run through a short sequence of the multiplication facts, comparing each product to 50: "Six times seven (low), six times eight (close), six times nine (too high). Must be eight. That's forty-eight and two left over." This process can and should be practiced. That is, children should be able to do problems with one-digit divisors and one-digit answers plus remainders and do them mentally.

8.24 THE PRICE IS RIGHT

To practice "near facts," try a **Price Is Right** type of exercise. As illustrated in Figure 8.23, the idea is to find the one-digit factor that makes the product as close to the target without going over. Help children develop the process of going through the multiplication facts as described. This can be a drill with the full class by preparing a list for the overhead, or it can be a worksheet or flash card activity.

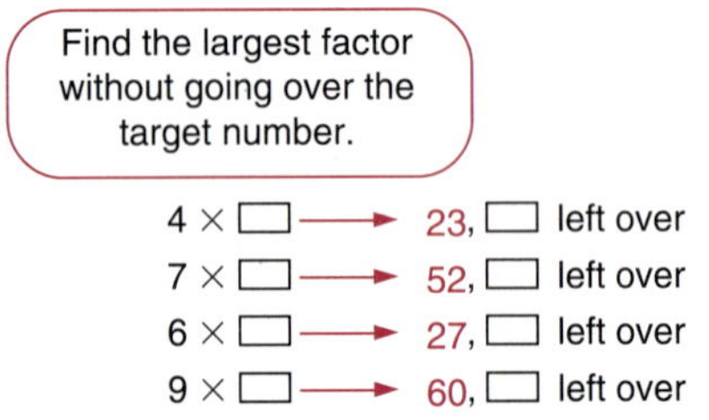

> Find the largest factor without going over the target number.
>
> 4 × ☐ ⟶ 23, ☐ left over
> 7 × ☐ ⟶ 52, ☐ left over
> 6 × ☐ ⟶ 27, ☐ left over
> 9 × ☐ ⟶ 60, ☐ left over

FIGURE 8.23: *A* **Price Is Right** *exercise for division*

MAKING IT WORK

Basic facts have been a major concern of teachers as long as mathematics has been taught in schools. Mastery of facts is as important today as ever. Today, however, there are methods that research has proven to be effective. After having waded through a lot of details in this chapter, it may be useful to look again at the fundamentals of a successful approach to helping our students.

THREE STEPS TO SUCCESS

We began this chapter with a three-step approach to fact mastery. The first step was to develop a strong understanding of the operations and of number relationships. At this point you should better appreciate the value of number concepts and operations meanings in mastering facts. It is never too late to develop some of the really valuable relationships of number and operation sense as described in Chapters 6 and 7.

The second step was to develop efficient strategies for fact retrieval. Remember to check to see if the required prerequisite concepts or ideas that a strategy may depend on are actually present. Developing a strategy involves discussion of the strategy and why it works especially for that particular set of facts. As you have seen, models and discussions are not a means of getting answers but devices to promote reflective thinking about the strategies.

The third step was to practice the strategy and to practice selection of the strategy. There are two important points to be made here:

1. Do not begin a lot of practice before the strategy seems well developed. That is, practice *efficient* strategies, not inefficient counting.
2. It is absolutely necessary to practice selection or retrieval of appropriate strategies.

The following activities are aimed at strategy selection and can be used with strategies for any operation. These activities should begin soon after you have two or more strategies in place.

8.25 CIRCLE THE STRATEGY

On a worksheet, have students circle the facts that belong to a specified strategy they have been working on and then answer only those facts. The same approach can be used for two or even three strategies on one sheet.

8.26 SORT THEM AS YOU DO THEM

Mix ordinary flash cards from two or more strategies into a single packet. Prepare simple little pictures or labels for the strategies that are in the packet. For each card, students first decide which of the labeled strategies they will use for that fact, place it in the appropriate pile, and then answer the fact.

◆◆◆◆◆◆◆◆◆◆◆◆◆◆◆◆◆

8.27 NAME IT, THEN USE IT

Orally give facts to the class or group, and first have them select and name a strategy they will use to answer the fact.

Strategy sorting is a crucial part of your fact strategy program. Children will generally use a strategy consistently when they know that all facts belong to one approach. When taken out of this mindset, as in a mixed-drill exercise, they forget to use the strategies and revert to inefficient counting.

FEATURES OF EFFECTIVE DRILLS

Having said that drill should not begin until an efficient strategy is in place, there are still features of any fact drill that will make them more effective.

1. Keep fact drill periods relatively short.
2. Encourage speed.
3. Provide immediate feedback.

These three ideas should be taken as rules of thumb. Not every drill can adhere to all three maxims, but each has its merits.

◆ Keep Drill Periods Short

Excessive drill can easily become tedious and boring. Drills with flash cards or oral drills can be conducted in 5- to 10-minute segments. They can be done at any time of the day—between subjects, in free time at a learning station, before or after lunch, or as a warmup exercise during morning calendar activities. Be certain that it is strategies and/or strategy sorting that is being drilled, but do not overwhelm or cause stress with lengthy drills.

◆ Encourage Speed

Earlier it was noted that if children are pressured too early to give fact answers in speed drills, they will revert to inefficient counting methods. Under pressure, children forget new ideas and fall back on the secure and familiar. Counting is secure and familiar. Strategies are new. Give strategy development a lot of time so that the strategies become a more secure part of your children's thought patterns.

At the same time, speed is a significant factor in getting children to memorize instead of using less efficient counting procedures (Davis, 1978). Therefore, listen to your children. As they become proficient with a particular strategy, try to increase their speed. For example, work faster and faster through a small deck of flash cards all for a single strategy. After successful strategy sorting, increase pressure to go faster with more varied sets of facts.

A short speed drill is much better than a speed drill of 100 facts. Consider rows of 8 to 10 facts. See how fast students can write the facts for one row instead of the whole page. Check answers after each row to provide feedback. Subsequent rows can have many repetitions of facts that are being worked on. After 10 speed rows, children will not only get faster and feel better, they will have practiced a few new facts over and over.

◆ Provide Immediate Feedback

The frequent suggestion of flash cards in this chapter is partly because this method provides individual students with immediate feedback on every fact. Not only do they know if they are right or wrong, but they see the correct answer after each response. The calculator is another method of providing quick feedback in those situations that lend themselves to calculator drill (adding or multiplying by the same number). There are also many computer software programs that drill facts. Virtually all use some form of instant feedback. Certainly not every activity can provide immediate feedback, but it is a feature to look for in good fact drills. Note, for example, that a long worksheet graded overnight provides no feedback. The same worksheet broken into short segments can have a separate answer key for each section.

FACT REMEDIATION WITH UPPER-GRADE STUDENTS

Children who have not mastered their basic facts by the sixth or seventh grade are in need of something other than more practice. They have certainly seen and practiced those facts endless times over the past several school years. They need a new approach. In addition to the ideas that have already been suggested, the following three ideas are very important when working with children in the upper grades:

1. Do an individual diagnosis. Each child brings different ideas and knowledge to the task. Each child has different needs. Identify strengths and weaknesses before you start.
2. Provide hope. These students know they have failed. It is no fun to be a failure. The prospect of more drill cannot be hopeful. You must provide it from the start.
3. Build in success. As you begin a well-designed fact program for a child who has experienced failure, be sure that successes come quickly and easily.

DIAGNOSE

Find out exactly which facts are mastered (less than three-second response) and which need practice. For a fact not mastered, ask the student how he or she would figure it out. What thought processes are they using? Sometimes you can select strategies that build on ideas they already have. Find out about their concepts of the operations and their command of number relationships. It may be more important to work on number relationships or a think-addition concept of subtraction than to begin directly with subtraction facts. Check on their use of turnaround facts. If they know 3 × 7, do they also know 7 × 3? Many do not even know these facts are related or do not take advantage of that relationship.

Looking for answers to questions like these will permit you to tailor an effective drill program for the remedial student, whereas simply providing more pages of practice is likely to contribute only to the child's ability to count.

PROVIDE HOPE

Children who have experienced difficulty with fact mastery can begin to believe that they cannot learn facts or that they are doomed to finger counting forever. These children have either not seen or have not mastered efficient techniques such as the strategies described in this chapter. Let these children know that you will provide them with some new ideas to help them and *you will help them!* Take that burden on yourself and spare them the prospect of more defeat.

BUILD IN SUCCESS

Begin with easy strategies and introduce only a few new facts at a time. Even with pure rote drill, repetitive practice with five facts in three days will provide more success than introducing 15 facts in a week. Success builds success! With strategies as an added assist, success comes even more quickly. Point out to children how one idea, one strategy, is all that is required to learn many facts. Use fact charts to show what set of facts you are working on. It is surprising how the chart quickly fills up with mastered facts.

Keep reviewing newly learned facts and those that were already known. This is success. It feels good, and failures are not as apparent.

◆◆◆◆◆

REFLECTIONS ON CHAPTER 8: WRITING TO LEARN

1. Explain in about one paragraph each part of the three-step approach to fact mastery.
2. For each addition fact strategy:
   a) List at least three facts for which the strategy can be used.
   b) Explain the thinking process and/or concepts that are involved in using the strategy. Use a specific fact as an illustration.
   c) Describe at least one activity that is designed to help children construct the strategy.
3. What is meant by subtraction as "think-addition?" How can you help children develop a think-addition thought pattern for subtraction?
4. For subtraction facts with sums greater than 10, it is reasonable that as many as three different thought patterns or strategies might be used. Describe each of these.
5. Why is the turn-around property (commutative property) so important in multiplication fact mastery?
6. For each multiplication strategy except for "use a helping fact," answer questions a, b, and c as in 2 above.
7. The "last 25" multiplication facts involve using a fact that has already been learned and working from that fact to the new or harder fact. Four different ways to make this connection with a helping fact were described. Some are only applicable to certain facts. Describe each of these approaches and list the facts for which the approach is applicable.
8. How do you help children who have been drilling their basic facts for years and still have not mastered them?

◆◆◆◆◆

FOR DISCUSSION AND EXPLORATION

1. Read the short chapter, "Suggestions for Teaching the Basic Facts of Arithmetic," by Ed Davis in the 1978 NCTM Yearbook, *Developing Computational Skills*. Davis's 10 principles of teaching fact mastery are worthy of consideration even though he is not talking about a strategy approach. Which ones are most important? In view of the current chapter, are there any of the 10 you disagree with?
2. Try using a calculator as a basic fact drill device. One idea is to have someone say a fact and let the child try to press the keys and say the answer before pressing ⊟. Soon, the calculator actually is seen as slow. Another approach uses the automatic constant features. To practice multiplication facts with a factor of 8, press 8 ⊠ ⊟. Then press any number followed by ⊟ to get the product. Again try to beat the calculator. The same approach works for all four operations. (Different calculators may operate differently.)
3. Explore a computer software program that drills basic facts. There are perhaps more of these pro-

grams than any other area of drill-and-practice software. Many utilize a variation of an arcade format in order to encourage speed. Very few, if any, have organized the facts around thinking strategies. Do you think these programs are effective? How would you utilize such a piece of software in a classroom with only one or two available computers?

4. One view of thinking strategies is that they are little more than a collection of tricks for kids to memorize. This view suggests that direct drill may be more effective and less confusing. Discuss the question, "Is teaching children thinking strategies for basic fact mastery in keeping with a developmental view of teaching mathematics?" For some perspective on the issue, you might want to consult *Learning from Children* (Labinowicz, 1985). In chapter 5, Labinowicz presents a discussion of this issue, although it is highly biased in favor of the view in this book.

5. Examine a recently published second-, third-, or fourth-grade textbook and determine how thinking strategies for the basic facts have been developed. Compare what you find with the groupings of facts in this chapter. How would you use the text effectively in your program? (Note: Before 1988, only two textbook series, Harper and Row and Addison-Wesley, had an explicit program of fact mastery for the basic facts. Other series used many of these same ideas but were less overt in their approach.)

◆◆◆◆◆
SUGGESTED READINGS

Baroody, A. J. (1984). Children's difficulties in subtraction: Some causes and cures. *Arithmetic Teacher, 32*(3), 14–19.

Baroody, A. J. (1985). Mastery of the basic number combinations: Internalization of relationships or facts? *Journal for Research in Mathematics Education, 16,* 83–98.

Brownell, W. A., & Chazal, C.B. (1935). The effects of premature drill in third-grade arithmetic. *Journal of Educational Research, 29,* 17–28.

Davis, E. J. (1978). Suggestions for teaching the basic facts of arithmetic. In M. N. Suydam (Ed.), *Developing computational skills.* Reston, VA: National Council of Teachers of Mathematics.

Flexer, R. J. (1986). The power of five: The step before the power of ten. *Arithmetic Teacher 34*(3), 5–9.

Grove, J. (1978). A pocket multiplier. *Arithmetic Teacher, 25*(6), 25.

Labinowicz, E. (1985). *Learning from children: New beginnings for teaching numerical thinking.* Menlo Park, CA: Addison-Wesley.

Lessen, E. I., & Cumblad, C. L. (1984). Alternatives for teaching multiplication facts. *Arithmetic Teacher, 31*(5), 46–48.

Rathmell, E. C. (1978). Using thinking strategies to teach the basic facts. In M. N. Suydam (Ed.), *Developing computational skills.* Reston, VA: National Council of Teachers of Mathematics.

Rightsel, P. S., & Thornton, C. A. (1985). 72 addition facts can be mastered by mid-grade 1. *Arithmetic Teacher, 33*(3), 8–10.

Steinberg, R. M. (1985). Instruction on derived facts strategies in addition and subtraction. *Journal for Research in Mathematics Education, 16,* 337–355.

Thornton, C. A. (1990). Strategies for the basic facts. In J. N. Payne (Ed.), *Mathematics for the young child.* Reston, VA: National Council of Teachers of Mathematics.

Thornton, C. A., & Smith, P. (1988). Action research: Strategies for learning subtraction facts. *Arithmetic Teacher, 35*(8), 8–12.

Thornton, C. A., & Toohey, M. A. (1984). *A matter of facts: Addition, subtraction, multiplication, division.* Palo Alto, CA: Creative Publications.

9 WHOLE NUMBER PLACE-VALUE DEVELOPMENT

◆ A FULL UNDERSTANDING OF PLACE VALUE INCLUDES A COMPLEX array of ideas and relationships that develop over the K to 6 grade span. In kindergarten, children begin learning to count to 100. By second grade they are talking about tens and ones and are using these ideas in many ways. By fourth grade, students are working with numbers involving four or more digits. Numbers are experienced in computations, on calculators, in mental computations and estimations, and in connection with real-world quantities and measures. In fourth and fifth grades, the ideas of whole-number place value are extended to decimals.

In addition to number meanings, students should begin in the early grades to develop some number sense for large numbers. Number sense must continue to be developed throughout the elementary years. Number sense for the whole numbers refers to:

- a sense of the relative size of numbers (185 is large compared to 15, small compared to 1219, and about the same as 179);
- a connection to real-world concepts (estimation of quantities, knowing what would be reasonable numbers for the capacity of a football stadium or school cafeteria, the dollars required to purchase a sweatshirt or a TV or a car, or the weight of an adult;
- a flexible use of numbers in estimation (using 250 instead of 243 in a computation because "four 250s" is easy to work with; thinking about 1296 as "about 13 hundred");
- a knowledge of the effect of operating with large numbers (adding 1000 increases 3472 by less than a third; 1000 times 3472 will be over 3 million, while division by 1000 will produce a very small result).

Number sense is a general theme in today's curriculum and deserves much greater attention than it has in the past. While pencil-and-paper computation has diminished in importance, the *Standards* view of computation is much broader. As you will see in Chapter 10, not only are children expected to have some facility with pencil-and-paper computation, there is a real value attached to students inventing ways of doing computations as well as understanding why the usual methods work. Estimation and mental computation skills round out the computational curriculum. Neither number sense nor computational understanding can possibly be developed without a firm understanding of place value.

The first section of this chapter focuses on early development of place-value concepts. This is important even for teachers of the upper grades because it highlights the complexity of the concepts that are often assumed in grades 4 to 8. Unfortunately many older students have not constructed these ideas. In later sections, activities are suggested that build on this early development. The chapter concludes with a section on assessing place-value concepts so that you can have a better feel for the ideas your students may have about these complex and important concepts.

EARLY DEVELOPMENT OF PLACE-VALUE IDEAS

It is important for teachers to understand what ideas about numbers children have and to reflect on how new and complex ideas are built on these ideas.

NUMBER IDEAS BEFORE PLACE VALUE

It is tempting to think that children know quite a lot about numbers with two digits (from 10 to 100) even as early as kindergarten. After all, most kindergarten children can and should learn to count to 100 and even to count out sets of things that may have 20 or 30 objects in them. They do daily calendar activities, count children in the room, turn to specified page numbers in their books, and so on. While these students do understand these numbers, that understanding is quite different from yours. It is based on a "one-more-than" or "count-by-ones" approach to quantity.

A Pre–Place-Value Snapshot of Number

Ask first- or even second-grade children to count out 53 tiles, and most will be able to do that with the possible exception of careless errors. It remains a tedious if not formidable task. Watch as children count the tiles, and you will note that they are being counted out one at a time and put into the pile with no use of any type of grouping. Next, have them write the number that tells how many tiles were just counted. If they remember the number, most children will be able to write it. Some may write "35" instead of "53," a simple reversal.

So far, so good. Now ask the children to write the number that is 10 more than the number they just wrote. Most will begin to count in some manner, either starting from 1 or starting with 53. Those counting on from 53 will also find it necessary to keep track of the counts, probably on their fingers. Many if not most children in the first and early second grade will not be successful at this task, and almost none will know immediately that 10 more is 63. Asking for the number that is 10 less is even more problematic. Third-grade children have similar difficulties writing the number that is either 10 more or 100 more than 376 (Labinowicz, 1985).

Finally, show a large collection of cards, each with a ten-frame drawn on it. Discuss how the cards each have ten spaces and that each will hold ten tiles. Demonstrate putting tiles on the cards by filling up one of the ten frames with tiles. Now ask, "How many cards like this do you think it will take if we want to put all of these tiles (the 53 counted out) on the cards?" A not unusual response is "53." Others will say they do not know and a few will try to put the tiles on the cards to figure it out.

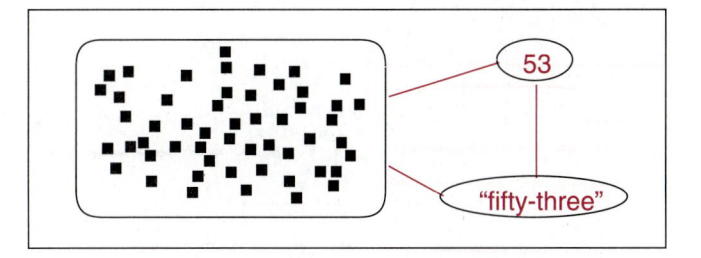

Pre–Place-Value Quantity Is Tied to Counts by Ones

The children just described know there are 53 tiles "because I counted them." Writing the number and saying the number are usually done correctly, but this procedural knowledge is connected to the count by ones. With minimal instruction, children can tell you that the 5 is in the tens place or that there "are 3 ones." It is likely that this is simply a naming of the positions with little understanding. If children have been exposed to base ten materials, it is reasonable to expect them to name a rod of ten as a "ten" and a small cube as a "one." These same children, however, may not be readily able to tell how many ones are required to make a ten. It is quite easy to attach words to both materials and groups without realizing what the materials or symbols are supposed to represent.

Children do know that 53 is a "lot," and that it's more than 47 (because you count past 47 to get to 53). They think of the "53" that they write as a single number. They do not know that the 5 represents five groups of 10 things and the 3, three singles (Ross, 1986, 1989).

This is where children are before they have constructed place-value ideas. It is on this knowledge that place-value concepts must be built. It is important to realize that children do many things that may suggest they understand these numbers, but that understanding may be rather superficial.

THE BASIC IDEAS OF PLACE VALUE

Place-value understanding requires an integration of new and difficult-to-construct concepts of grouping by tens (*base ten* concepts) with procedural knowledge of how groups are recorded in our place-value scheme, how numbers are written, and how they are spoken.

Integration of Groupings by Ten with Counts by Ones

Recognizing that children can count out a set of 53, we want to help them see that making groupings of ten and leftovers is a way of counting or showing that same quantity. Each of the groups in Figure 9.1 (p. 156) has 53 tiles. We want children to construct the idea that all of these are the same and that the sameness is clearly evident by virtue of the groupings of tens.

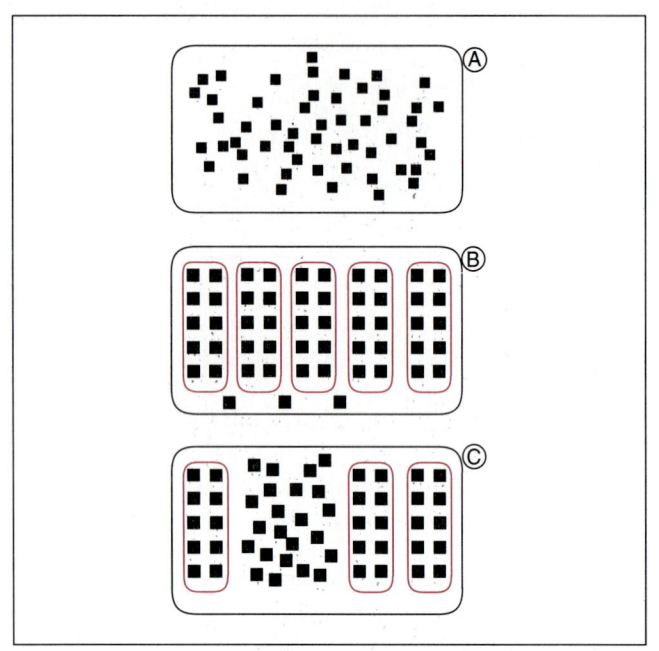

FIGURE 9.1: *Three equivalent groupings of 53 objects. Group A is 53 because "I counted them (by ones)." Group B has 5 tens and 3 more. Group C is the same, but now some groups are broken into singles.*

There is a subtle yet profound difference between two groups of children: those who know that group B is really 53 because they understand the idea that 5 groups of 10 and 3 more is the same amount as 53 counts by one, and those children who simply say "it's 53," because they have been *told* that when things are grouped this way, it's called 53. The latter children may not be sure how many they will get if they count the tiles in set B by ones or if the groups were "ungrouped" how many there would then be. The children who understand will see no need to count set B by ones. They understand the "fifty-threeness" of sets A and B to be the same.

The ideas in the last paragraph are important for you to understand so that the activities discussed later will make sense. Spend some time with these ideas before reading further.

The recognition of the equivalence of groups B and C is another step in the conceptual development that we hope to obtain. Groupings with fewer than the maximum number of tens can be referred to as *equivalent groupings* or *equivalent representations*. To understand that these representations are equivalent indicates that groupings by tens is not just a rule that is followed, but that any grouping by tens, including all or some of the singles, can help to tell how many.

The Role of Counting in Constructing Base Ten Ideas

Just as counting is the vehicle with which children construct various relations on small numbers up to 10, counting plays a key role in constructing base-ten ideas about quantity and connecting these concepts to symbols and oral names for numbers.

Children can count sets such as those in Figure 9.1 in three different ways. Each way helps children think about the quantities in a different way (Thompson, 1990).

1. **Counting by ones.** This is the method children have to begin with. Initially a count by ones is the only way they are able to name a quantity or "tell how many." All three of the sets in Figure 9.1 can be counted by ones. Before base ten ideas develop, this is the only way children could be convinced that all three sets are the same.

2. **Counting by groups and singles.** In group B of Figure 9.1, counting by groups and singles would go like this: "One, two, three, four, five *bunches of ten*, and one, two, three *singles*." Consider how novel this method would be for a child who had never thought about counting a group of things as a single item. Also notice how this counting does not tell directly how many items there are. This counting must be coordinated with a count by ones before it can be a means of telling "how many."

3. **Counting by tens and ones.** This is the way adults would probably count group B and perhaps group C: "Ten, twenty, thirty, forty, fifty, ... fifty-one, fifty-two, fifty-three." While this count ends by saying the number that is there, it is not as explicit as the second method in counting the number of groups. Nor will it convey a personal understanding of "how many" unless it is coordinated with the more meaningful count by ones.

Regardless of the specific activity that you may be doing with children, helping them integrate the grouping-by-tens concept with what they know about number from counting by ones should be your foremost objective. That means children should frequently have the opportunity to count sets of objects in several ways. If first counted by ones, the question might be, "What will happen if we count these by groups and singles (or by tens and ones)?" If a set has been grouped into tens and singles, and counted accordingly, "How can we be really certain that there are 53 things here?" or "What do you think we will get if we count by ones?" It is inadequate to tell children that these counts will all be the same. That is a relationship they must construct themselves through reflective thought, not because the teacher says it works that way.

Integration of Groupings with Words

The way we say a number such as "fifty-three" must also be connected with the grouping-by-tens concept. Again, the counting methods provide a connecting mechanism. The explicit count by tens and ones results in saying the number of groups and singles separately, "five tens and three." This is an acceptable, albeit nonstandard, method of naming this quantity. Saying the number of tens and singles

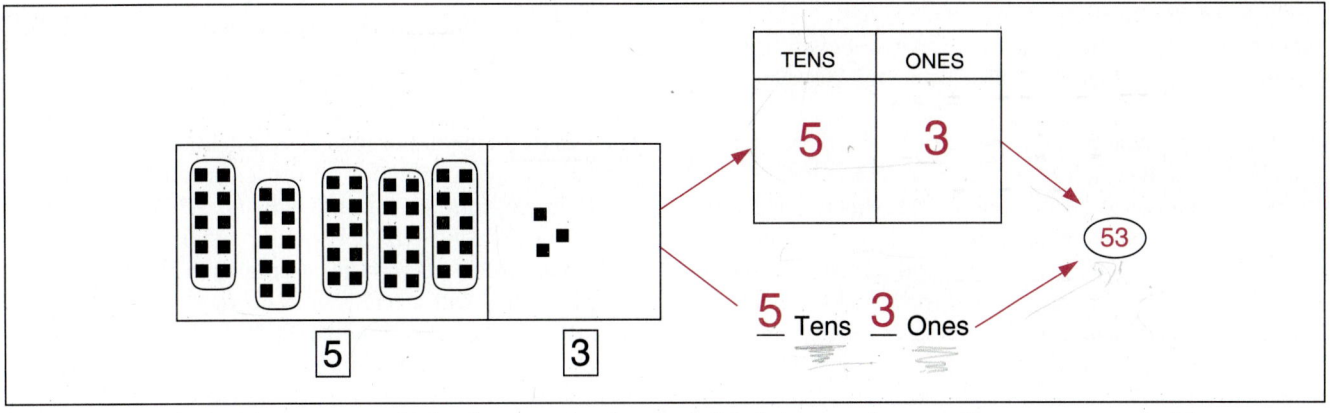

FIGURE 9.2: *Groupings by 10 are matched with numerals, placed in labeled places, and eventually written in standard form.*

separately in this fashion can be called *place-value language* for a number. Children can associate the place-value language with the usual language: "five tens and three—fifty-three."

Notice that there are several variations of the place-value language for fifty-three: five tens and three, five tens and three ones, five groups of ten and three leftovers, five tens and three singles, and so on. Each may be used interchangeably with the standard name, fifty-three. For three-digit numbers the same flexibility is available: Two hundred thirty can be twenty-three tens or two hundreds and three tens. For a number with singles there are more options.

It can easily be argued that place-value language should be used throughout the second grade, even in preference to standard oral names.

◆ Integration of Groupings with Place-Value Notation

In a like manner, the symbolic scheme that we use for writing numbers (ones on the right, tens to the left of ones, and so on) must be coordinated with the grouping scheme. Activities can be designed so that children physically associate a tens and ones grouping with the correct recording of the individual digits, as is indicated in Figure 9.2.

Language again plays a key role in making these connections. The explicit count by groups and singles matches the individual digits as the number is written in the usual left-to-right manner. Counting can help with both ways of interpreting the two digits: as a designation of tens and ones and as a single number representing a full amount.

A similar coordination is necessary for hundreds.

A RELATIONAL UNDERSTANDING

Figure 9.3 (p.158) summarizes the ideas that have been discussed so far.

- Before place-value concepts develop, children's knowledge of number is based on counts by ones.
- The conceptual knowledge of place value consists of the base ten grouping ideas:

- A collection of objects can be grouped in sets of 10 and some leftover singles.
- There can be equivalent representations with fewer than the maximum groupings.
- The base-ten grouping ideas must be integrated with oral and written names for numbers.
- In addition to counting by ones, children use two other ways of counting: by groups and singles separately and by tens and ones. All three methods of counting are coordinated as the principal method of integrating the concepts, the written names, and the oral names.

The base ten or grouping ideas are the conceptual knowledge of place value while counting, oral names, and written names fall under the category of procedural knowledge. A relational understanding of place value integrates all of these ideas.

MODELS FOR PLACE VALUE

The key instructional tool for developing the conceptual knowledge of place value and also for connecting these concepts to symbolism is the use of base ten models.

BASE TEN MODELS SHOW A TEN-MAKES-ONE RELATIONSHIP

A good base ten model for ones, tens, and hundreds is *proportional*. That is, a *ten* model is physically 10 times larger than the model for a *one*, and a *hundred* model is 10 times larger than the ten model. Base ten models can be categorized as *groupable* and *pregrouped*.

◆ Groupable Models

Models that most clearly illustrate the relationships between ones, tens, and hundreds are those for which the ten can actually be made or grouped from the singles. When you bundle 10 popsicle sticks, the bundle of 10 literally *is the same as* the 10 ones from which it was made. Examples of these "groupable" models are shown in Figure 9.4(a). These could also be called "put-together-take-apart" models.

Of the groupable models, beans or counters in cups are the cheapest and easiest for children to use. (Paper or plastic portion cups can be purchased from restaurant supply houses.) The plastic Unifix cubes (or the equivalent) are highly attractive and provide a good transition to pregrouped tens sticks. Bundles of Popsicle sticks are perhaps the best known base ten model, but small hands have trouble with rubber bands and actually making the bundles. Hundreds are possible with most groupable materials, but are generally not practical for most activities in the classroom.

◆ Pregrouped or Trading Models

At some point there is a need to easily represent hundreds. Therefore, models that are pregrouped must be introduced. As with all base-ten models, the ten piece is physically equivalent to 10 ones, and a hundred piece equivalent to 10 tens [Figure 9.4(b)]. However, children cannot actually take them apart or put them together. When 10 single pieces are accumulated, they must be exchanged for or *traded* for a ten, and, likewise, tens must be traded for hundreds.

The chief advantage of these models is the ease of use and the efficient way they model numbers as large as 999. The disadvantage is the increased potential for children to use them without reflecting on the ten-to-one relationships. For example, in a ten-for-ones activity, children are told to trade 10 ones for a ten. It is quite possible for children to make this exchange without attending to the "tenness" of the piece they call a ten. While no model, including the groupable models, will guarantee that children are reflecting on the ten-to-one relationships in the materials, with trading pieces, we need to make an extra effort to see that children understand that a ten piece really is the same as 10 ones.

(See Black-line Masters and Materials Construction Tips for making base ten strips and squares.)

◆ NONPROPORTIONAL MATERIALS

Consider Figure 9.5. In this text, colored counters, abacuses, and money are not considered to model base ten ideas since the relationships must first exist in the mind of the

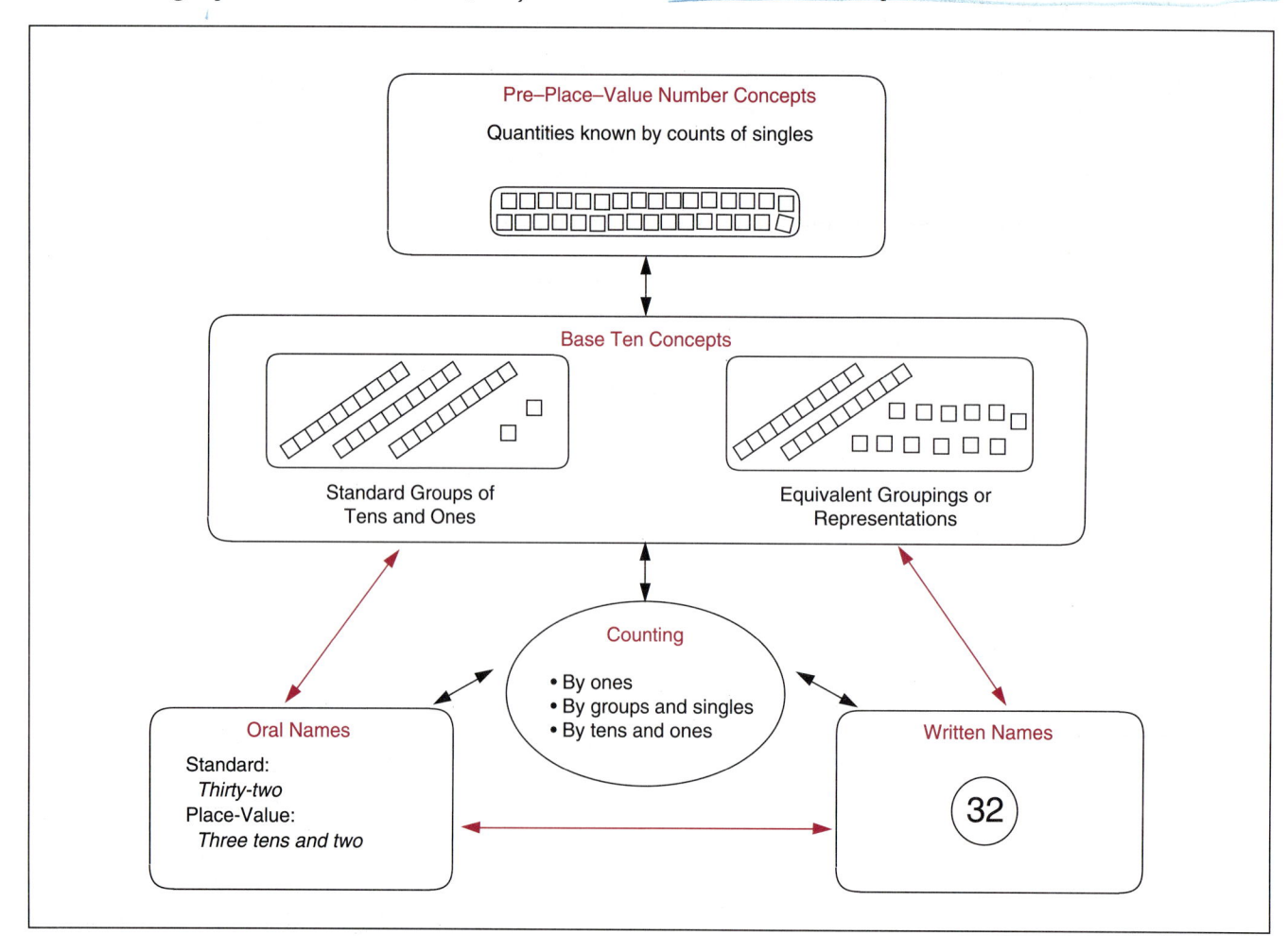

FIGURE 9.3: *Relational understanding of place value integrates three components: concepts, oral names, and numerals. Ideas are built on earlier pre–place-value ideas of number. Counting is a key feature in development of ideas and in the integration of new ideas with old.*

9 / WHOLE NUMBER PLACE-VALUE DEVELOPMENT   159

(a) Groupable Base Ten Models

Counters and cups.
Ten single counters are placed in a cup.
Hundred: ten cups in a margarine tub.

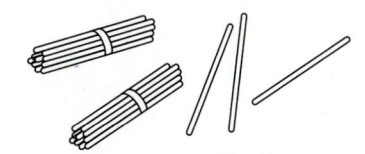

Bundles of sticks (wooden craft sticks, coffee stirrers)
(if bundles are left intact, these are a pregrouped model)
Hundreds: Ten bundles in a big bundle.

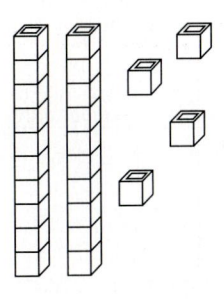

Unifix (or any interlocking cubes).
Ten single cubes form a bar of 10.
Hundreds: ten bars on cardboard backing.

(b) Pregrouped Base Ten Models

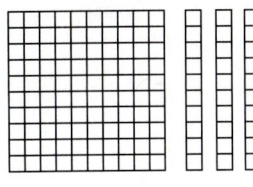

Teacher-made "strips and squares."
Made from mount board and
poster board. See Black-line Masters
and Materials Construction Tips.

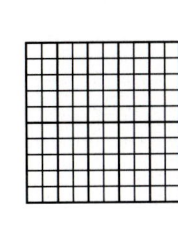

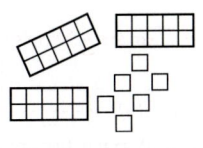

Ten-frame version of strips and
squares. Made the same way but
ten is in arrangement of ten-frame
and hundred shows the tens.

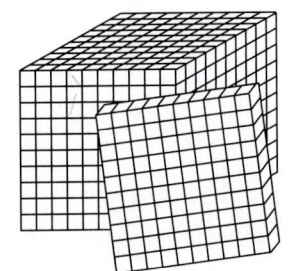

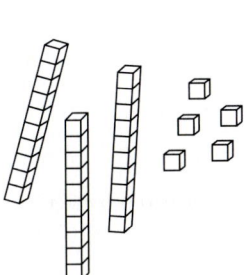

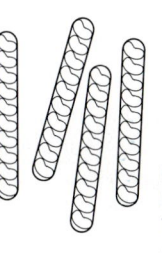

Wooden or plastic units, longs, flats, and block.
Also known as Dienes blocks or base ten
blocks. Expensive, durable, easily handled,
only model with thousand.

"Raft"

Bean sticks.
Beans glued to craft sticks.
Ten sticks in a raft is also
made from cardboard.

See directions and alternatives in Black-line
Masters and Materials Construction Tips.

FIGURE 9.4: *(a) Groupable base ten models (b) Pregrouped base ten models*

Chip Trading Materials

Green	Blue	Yellow	Red

10 red chips are traded for
a yellow, 10 yellows for a blue.

(a)

Abacus Models

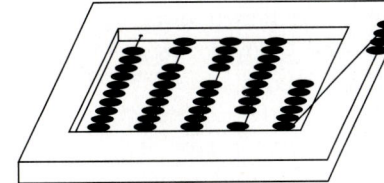

Ten-bead abacus

(b)

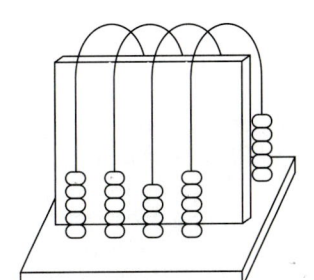

Beads on wires slide
over back to front

FIGURE 9.5: *Nonproportional materials*

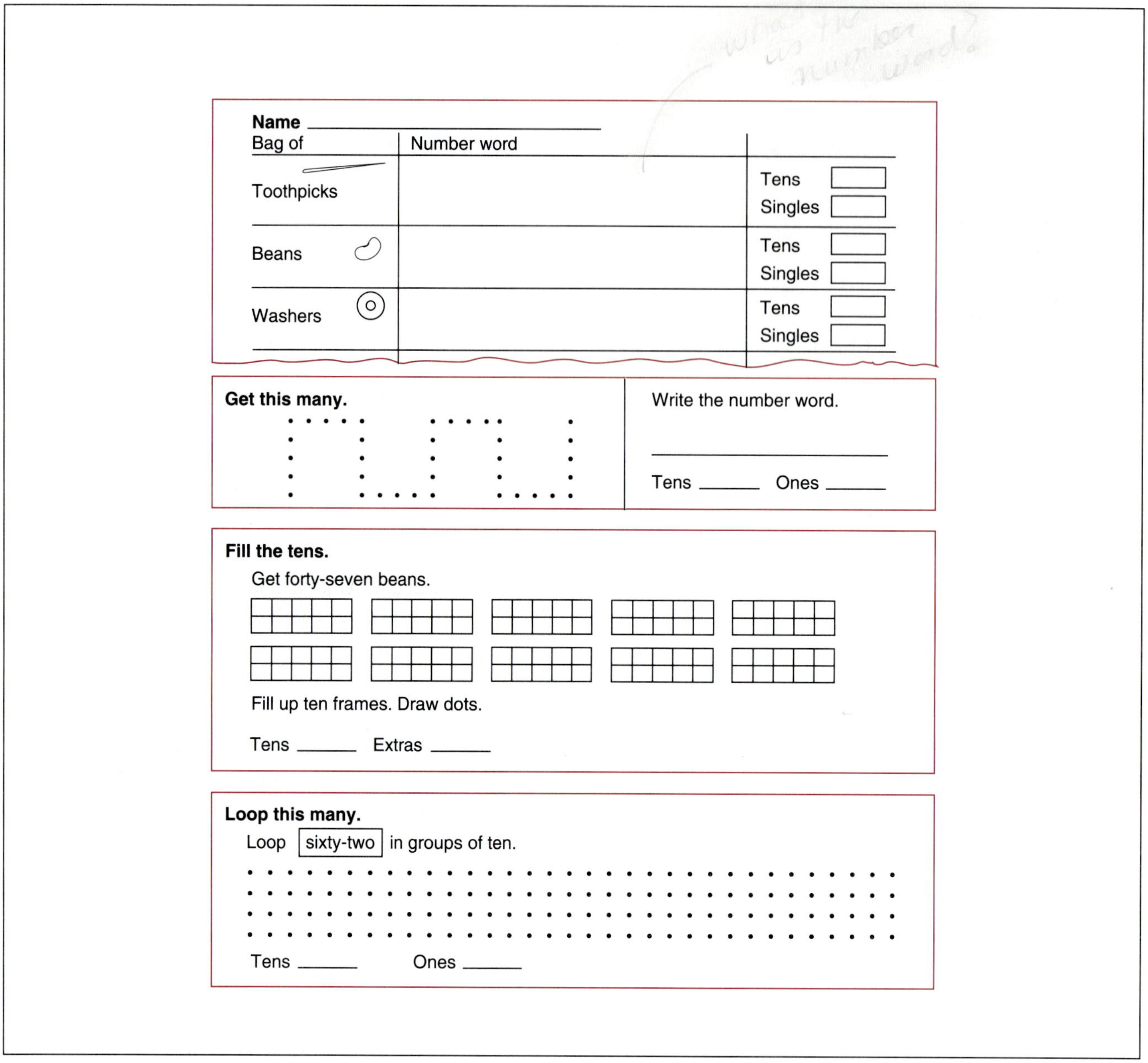

FIGURE 9.6: *Number words and making groups of 10*

child, and the materials play no part in helping to develop that relationship. With *Chip Trading Materials* (Davidson, 1975), for example, different colored chips are used to represent different place values; red for ones, yellow for tens, and so on, as shown in Figure 9.5(a). Trades are made between pieces in a variety of excellent activities, all of which can be done with proportional base ten models. With the chips, the base can be changed arbitrarily by simply designating different exchange rates.

Somewhat similar to the colored-chip model are the various forms of the abacus, some of which are shown in Figure 9.5(b).

The use of money as a model involves the same issue as colored chips. Pennies, dimes, and dollars are frequently used by teachers and by textbooks as a place-value model. If the relationship that a dime is worth 10 pennies and that a dollar is worth 10 dimes is part of the children's existing understanding, then money will model base ten place value. On the other hand, if the teacher must explain or impose this relationship, then the trade rate between pennies and dimes is simply a "rule of the game" from the children's view. They can learn to obey the rule, but the model is not helping them see why 10 ones is the same as 1 ten. The relationship should be in the model.

The pivotal question may be, "Why not just use proportional models?"

DEVELOPING PLACE-VALUE CONCEPTS AND PROCEDURES

Now that you have some idea of what we want to accomplish, we can look at some activities designed to help. This is one area of mathematics where the conceptual and procedural knowledge are both developed in an integrated or coordinated manner. While a particular activity may focus on grouping or on oral or written names, all activities involve models and add to the complete coordination of concepts, oral names, and written symbols as depicted in Figure 9.3.

Number Words		
eleven	ten-	one
twelve	twenty-	two
thirteen	thirty-	three
fourteen	forty-	four
fifteen	fifty-	five
sixteen	sixty-	six
seventeen	seventy-	seven
eighteen	eighty-	eight
nineteen	ninety-	nine

FIGURE 9.7: A chart to help children write number words

GROUPING ACTIVITIES

In these activities, children make groupings of 10 and record or say the amounts. If the activity is done as seatwork without teacher guidance, the number word is frequently included in the worksheet. The idea is to have students count the quantity using whatever current understanding they may have. If we use the number word instead of the numeral, children will not be artificially matching tens and ones with individual digits without confronting the actual quantity in a manner meaningful to them.

9.1 GROUPS OF 10

Prepare bags of counters of different types. For example, bags may have toothpicks, buttons, beans, plastic chips, Unifix cubes, craft sticks, and so on. Each bag should have a label that the children can easily copy. Children have a record sheet similar to the top example in Figure 9.6 (refer to p. 160). The bags can be placed at stations around the room or each pair of children can be given one. After recording the bag label, the children dump out and count the contents. The amount is recorded as a number word. Then the counters are grouped in as many tens as possible. The groupings are recorded on the form. Bags are traded or children move to another station after returning all counters to the bag.

If children have difficulty writing the number words, a chart can be displayed for students to copy from (Figure 9.7).

Variations of the **Groups of 10** activity are suggested by the other record sheets in Figure 9.6. In **Get This Many**, the children count the dots and then count out the corresponding number of counters. Small cups in which to put the groups of 10 should be provided. Notice that the activity requires students to address quantities in a way they understand, to record the amount in words, and then make the groupings.

The activity starts where the students are and develops the idea of groups. Make various countable designs with the dots. **Fill the Tens** and **Loop This Many** each begin with an oral name (number word), and students must count the indicated amount and then make groups.

The following activity is another variation but includes an estimation component that not only adds interest but can contribute to number sense.

9.2 ESTIMATING GROUPS OF TENS AND ONES

Show students a length that they are going to measure. For example, the length of the chalkboard, the length of a student lying down, or the distance around a sheet of newspaper. At one end of the length, line up 10 units (e.g., 10 Unifix in a bar, 10 toothpicks, rods, blocks). On a recording sheet (Figure 9.8, p. 162) students write down a guess of how many groups of ten and leftovers they think will fit into the length. Next they find the actual measure, placing units along the full length. These are counted by ones and also grouped in tens. Both results are recorded.

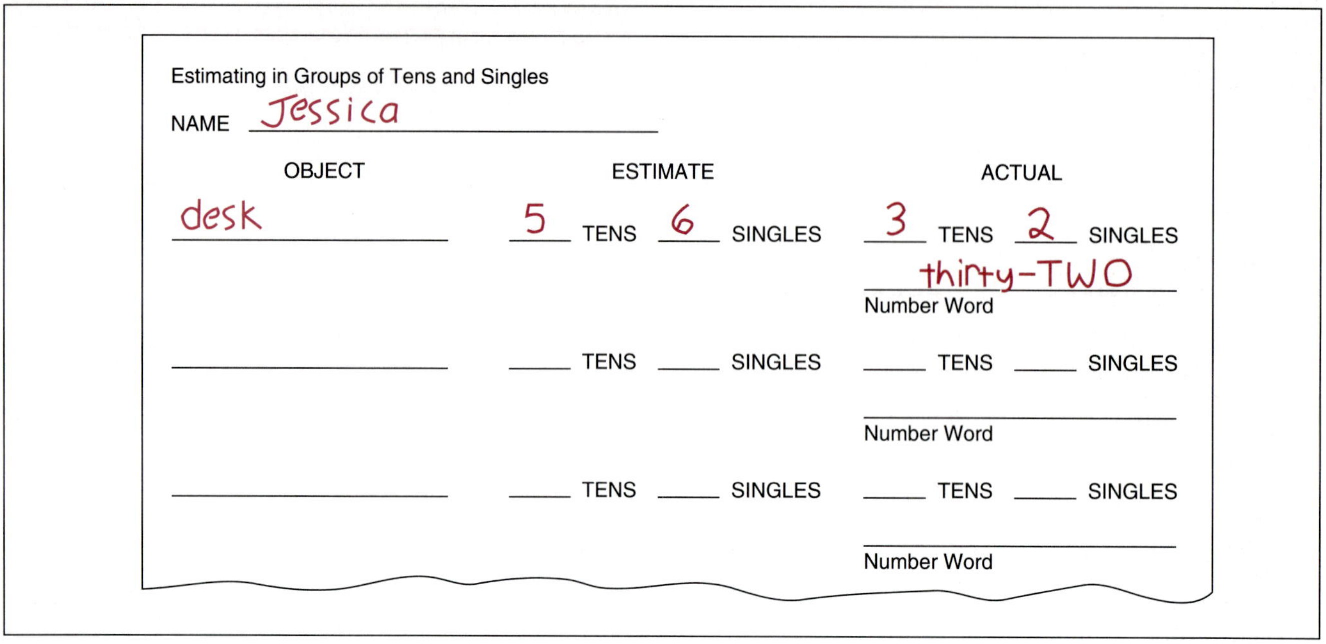

```
┌─────────────────────────────────────────────────────────────────────┐
│   Estimating in Groups of Tens and Singles                          │
│   NAME  Jessica                                                     │
│         OBJECT              ESTIMATE                 ACTUAL         │
│    desk              5 TENS  6 SINGLES       3 TENS  2 SINGLES      │
│                                                  thirty-TWO         │
│                                                  Number Word        │
│                                                                     │
│   _____        ___ TENS ___ SINGLES    ___ TENS ___ SINGLES     │
│                                                  _____        │
│                                                  Number Word        │
│                                                                     │
│   _____        ___ TENS ___ SINGLES    ___ TENS ___ SINGLES     │
│                                                  _____        │
│                                                  Number Word        │
└─────────────────────────────────────────────────────────────────────┘
```

FIGURE 9.8: *Record sheet for* **Estimating Groups of Tens and Ones**

Notice that all place-value components are included in the last activity. Children can work in pairs to measure a series of lengths around the room, or this can be a teacher-directed activity focused on a single measure. A similar estimation approach could be added to **Groups of 10**, where students first estimate the quantity in the bags. Estimation adds reflective thought concerning quantities expressed in groups.

◆ "Ones," "Tens," and "Hundreds" Are Strange Words

As students begin to make groupings of ten, the language of these groupings must also be introduced. At the very start of grouping, language such as "groups of ten and leftovers" or "bunches of tens and singles" is most meaningful. For tens, use whatever terminology fits: bars of 10, cups of 10, bundles of 10. Eventually you can abbreviate this simply to "ten." Surprisingly, there is no hurry to use the word "ones" for the leftover counters. Language such as "four tens and seven" works very well.

Reflect for a moment on how strange it must sound to say "seven *ones*." Certainly children have never said they were seven ones years old. The use of the word "ten" as a singular group name is even more mysterious. Consider the phrase *ten ones makes one ten*. The first "ten" carries the usual meaning of 10 things; the amount that is 1 more than 9 things. But the other "ten" is a singular noun, a thing. How can something the child has known for years as the name for a lot of things suddenly become one thing? Bunches, bundles, cups, and groups of 10 make more sense in the beginning than "a ten."

The word "hundred" is equally strange and yet usually gets less attention. It must be understood in three ways: as 100 single objects, as 10 tens, and as a singular thing. These word names are not as simple as they seem on the surface.

◆ Equivalent Representations

An important variation of the grouping activities is aimed at the equivalent representations of numbers. (Refer back to Figure 9.3.) For example, if working with the children who have just completed the **Groups of 10** activity for a bag of counters, ask "What is another way you can show your forty-two besides four groups and two singles? Let's see how many ways you can find." Interestingly, most children will go next to 42 singles. The following activities are also directed to the idea of equivalent representations.

◆◆◆◆◆◆◆◆◆◆◆◆◆◆◆◆◆◆◆◆◆◆◆

9.3 MAKE OR BREAK

Prepare bags with groupable materials (bundles of sticks or cups with snap-on lids or Unifix). In each bag place from 25 to 75 or 80 counters. About half of the bags should have the maximum numbers of tens already made. The rest should have from 10 to 19 singles. Label each bag. Students have a supply of **Make or Break** record slips (Figure 9.9). They dump out the materials and record the number of tens and ones. They circle MAKE if a ten can be made, and proceed to make a ten from the singles. If no tens can be made they circle BREAK and unbundle one of the tens. The items are all counted, and the number word is written that tells how many. The materials are returned to the bag. (Notice that the next student to use the bag will have the reverse MAKE or BREAK decision.)

9 / WHOLE NUMBER PLACE-VALUE DEVELOPMENT 163

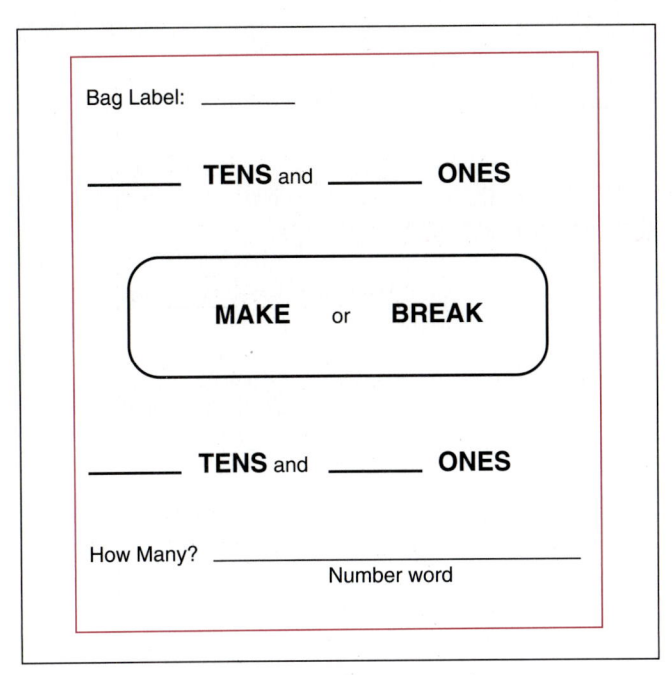

FIGURE 9.9: *Record sheet for Make or Break activity*

The next activity is similar but is done using pregrouped materials.

9.4 THREE OTHER WAYS

Students work in groups or in pairs. First they show "four hundred sixty-three" on their desks with strips and squares in the standard representation. Next they find and record at least three other ways of showing this number.

A variation of **Three Other Ways** is to challenge students to find a way to show an amount with a specific number of pieces. "Can you show 463 with 31 pieces?" (There is more than one way to do this.) Students in grades 4 or 5 can get quite involved with finding all the ways to show a three-digit number.

After children have had sufficient experiences with pregrouped materials, a "dot-stick-and-square" notation can be used for recording ones, tens, and hundreds. By third grade, children can use small squares for hundreds, as shown in Figure 9.10. Use the drawings as a means of telling the children what pieces to get out of their own place-value kits and as a way for children to record results.

The next activity begins to incorporate oral language with equivalent representation ideas.

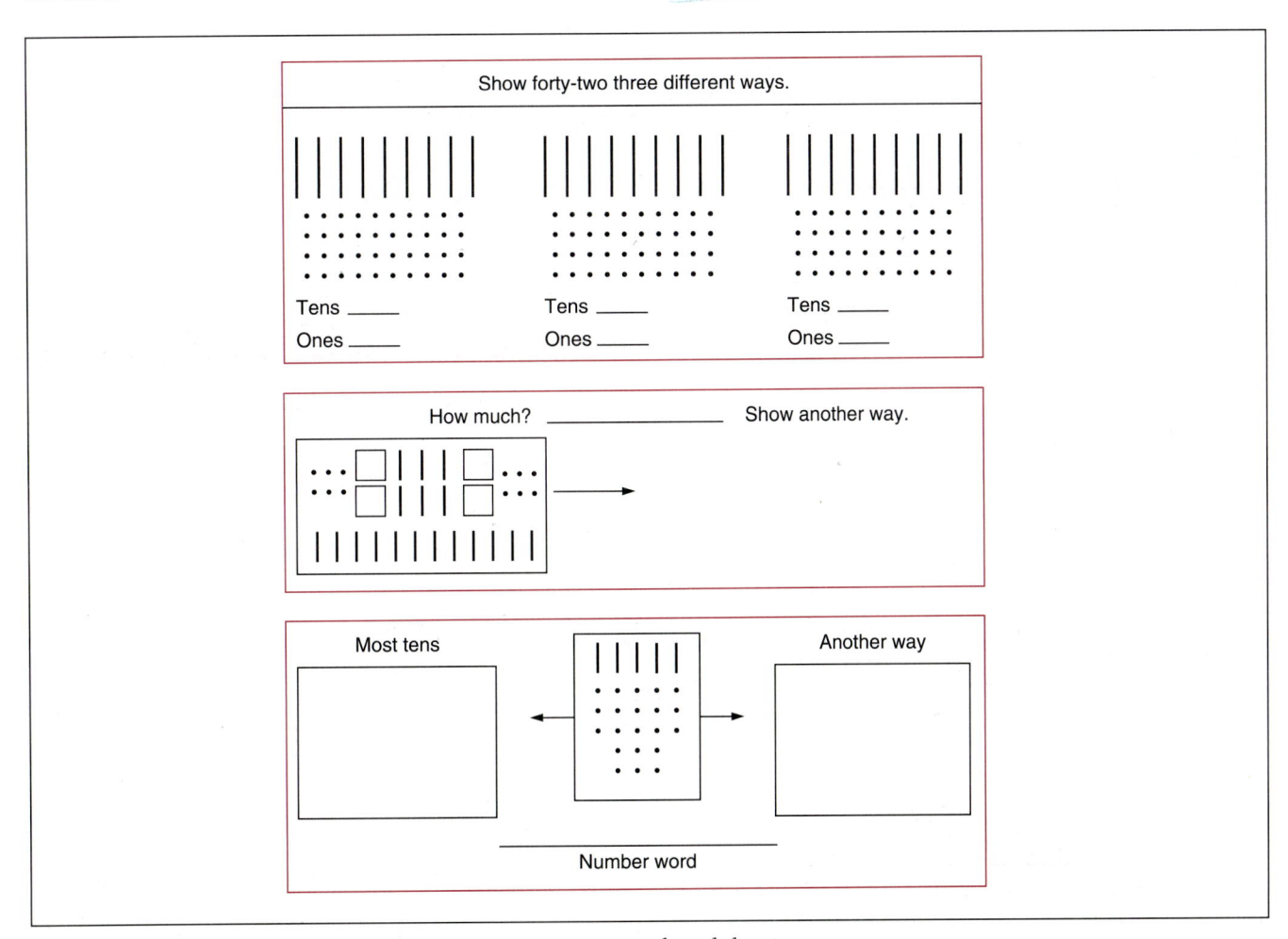

FIGURE 9.10: *Equivalent representation exercises using square-stick-and-dot pictures*

9.5 BASE TEN RIDDLES

Base ten riddles can be presented orally or in written form. In either case, children should use base-ten materials to help solve them. The examples here illustrate a variety of possibilities with different levels of difficulty.

I have 23 ones and 4 tens. Who am I?

I have 4 hundreds, 12 tens, and 6 ones. Who am I?

I have 30 ones and 3 hundreds. Who am I?

I am 45. I have 25 ones. How many tens do I have?

I am 341. I have 22 tens. How many hundreds do I have?

I have 13 tens, 2 hundreds, and 21 ones. Who am I?

If you put 3 more tens with me, I would be 115. Who am I?

I have 17 ones. I am between 40 and 50. Who am I?

I have 17 ones. I am between 40 and 50. How many tens do I have?

ORAL NAMES FOR NUMBERS

The standard name of the collection in Figure 9.11 is "forty-seven." A more explicit terminology is "four tens and seven ones." This latter form will be referred to as *place-value language*.

The more explicit place-value language is rarely misunderstood by children working with base ten materials and encourages thinking in terms of groups instead of a large pile of singles.

"Four tens and seven ones—forty-seven"

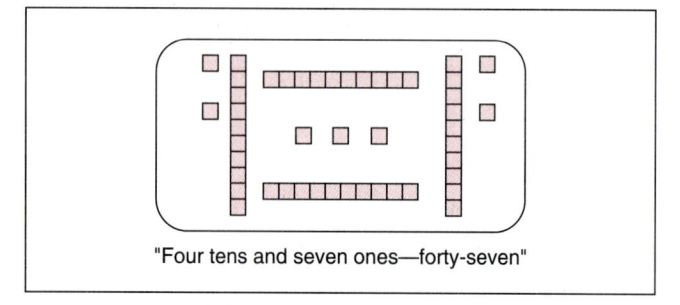

FIGURE 9.11: *Mixed model of 47*

Two-Digit Number Names

In first and second grade, children need to connect the base ten concepts with the oral number names they have used many times. They know the words but have not thought of them in terms of tens and ones. The following sequence is suggested:

Start with the names "twenty," "thirty," "forty," . . . , "ninety."

Next do all names "twenty" through "ninety-nine."

Emphasize the teens as exceptions. Acknowledge that they are backwards and do not fit the patterns.

Almost always use base ten models while learning oral names. Use place-value language paired with standard language.

9.6 COUNTING ROWS OF 10

Use a 10 × 10 array of dots on the overhead projector. Cover up all but two rows [Figure 9.12(a)]. "How many tens? (2) Two tens is called *twenty*. Have the class repeat. Sounds a little like twin. Show another row. "Three tens is called *thirty*. Four tens *forty*. Five tens should have been *fivety* rather than *fifty*. The names *sixty*, *seventy*, *eighty*, and *ninety* all fit the pattern. Slide the cover up and down the array, asking how many tens and the name for that many.

Use the same 10 3 10 array to work on names for tens and ones. Show, for example, four full lines, "forty." Next, expose one dot in the fifth row. "Four tens and one. Forty-one." Add more dots one at a time. "Four tens and two. Forty-two." "Four tens and three. Forty-three" [Figure 9.12(b)]. When that pattern is established, repeat with other decades from twenty through ninety.

The basic approach described with the 10 × 10 dot picture should be repeated with other base ten models.

9.7 COUNTING WITH BASE TEN MODELS

Show some tens pieces on the overhead. Ask how many tens. Ask for the usual name. Add a ten or remove a ten and repeat the questions. Next add some ones. Always have children give the place-value name and the standard name. Continue to make changes in the materials displayed by adding or removing 1 or 2 tens and by adding and removing ones. For this activity show the tens and ones pieces in different arrangements rather than the standard left-to-right order for tens and ones. The idea is to connect the names to the materials, not the order they are in.

Reverse the activity by having children use place-value pieces at their desks. For example, you say, "make sixty-three." The children make the number with the models and then give the place-value name.

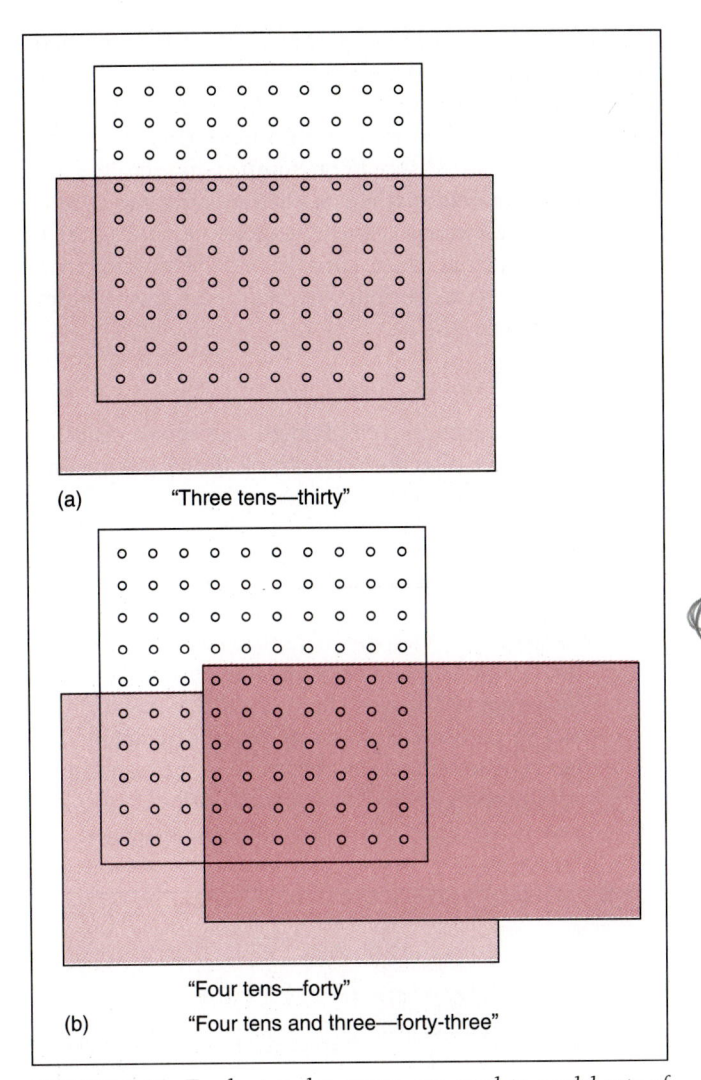

FIGURE 9.12: *Ten-by-ten dot arrays are used to model sets of ten and singles.*

(a) "Three tens—thirty"

(b) "Four tens—forty"
"Four tens and three—forty-three"

◆◆◆◆◆◆◆◆◆◆◆◆◆◆◆◆◆◆◆◆◆◆

9.8 TENS, ONES, AND FINGERS

Ask your class: "How can you show thirty-seven fingers?" (This question is really fun if preceded by a series of questions asking for different ways to show 6 fingers, 8 fingers, and other amounts less than 10.) Soon children will figure out that four children are required. Line up four children and have three hold up 10 fingers and the last child, 7 fingers. Have the class count the fingers by tens and ones. Ask for other children to show different numbers of fingers. Emphasize the number of sets of 10 fingers and the single fingers (place-value language) and pair this with the standard language.

◆◆◆◆◆◆◆◆◆◆◆◆◆◆◆◆◆◆◆◆◆◆

In all of the preceding activities, it is important to occasionally count an entire representation by ones. Remember that the count by ones is the young child's principal linkage with the concept of quantity. For example, suppose you have just had children use Unifix cubes to make 42. Try asking, "Do you think there really are forty-two blocks there?" Many children are not convinced, and the count by ones is very significant.

The language pattern for two-digit numbers is best developed and connected with models using numbers 20 and higher. That is where the emphasis should be. That is *not* to imply that we should hide the teens from children. Teens can and should appear in any of the previous activities but should be noted as the exceptions to the verbal rules you are developing.

One approach to the teen numbers is to "back into them." Show, for example, 6 tens and 5 ones. Get the standard and place-value names from the children as before. Remove tens one at a time, each time asking for both names. The switch from 2 tens and 5 ("twenty-five") to 1 ten and 5 ("fifteen") is a dramatic demonstration of the backward names for the teens. Take that opportunity to point them out as exceptions. Count them by ones. Say them in both languages. Add tens to get to those numbers that follow the rules. Return to a teen number. Continue to contrast the teens with the numbers 20 and above.

◆ Three-Digit Number Names

The approach to three-digit number names is essentially the same as for two-digit names. Show mixed arrangements of base ten materials. Have children give the place-value name and the standard names. Vary the arrangement from one example to the next by changing only one type of piece. That is, add or remove only ones, or only tens, or only hundreds.

Similarly, at their desks have children model numbers that you give to them orally using the standard names. By the time that children are ready for three-digit numbers, the two-digit number names, including the difficulties with the teens, are usually mastered. The major difficulty is with numbers involving no tens, such as 702. As noted earlier, the use of place-value language is quite helpful here. The zero-tens difficulty is more pronounced when writing numerals. Children frequently write 7002 for "seven hundred two." The emphasis on the meaning in the oral place-value language form will be a significant help.

◆ TRADING ACTIVITIES AND WRITTEN NAMES

In trading activities, students either make or break groups as they add or remove counters from the mat. For example, if you have 4 tens and 8 ones and then add 3 more ones, a *trade* must be made. The word *trade* is also used when a ten or hundred is broken or unbundled as when you want to remove 5 ones from 7 tens and 2 ones.

In these activities, students use a place-value mat for their materials. The mat provides a second connecting link between the groups of counters and the written numerals. (The first link is counting by tens and ones.) The activities also lay the groundwork for *regrouping* in computation ("carrying" and "borrowing").

Place-value mats are simple mats divided into two or three sections, as shown in Figure 9.13. While there is no requirement to have anything printed on the mats, it is strongly recommended that two ten-frames be drawn in the ones place as shown. (See Black-line Masters and Materials Construction Tips for directions for making the mats.) As children accumulate counters on the ten-frames, the number of counters there and the number needed to make a set of 10 is always clearly evident, eliminating the need for frequent and tedious counting (Thompson & Van de Walle, 1984b). Most illustrations of place-value mats in this book will show two ten-frames, even though that feature is strictly an option and is not one commonly seen in standard texts.

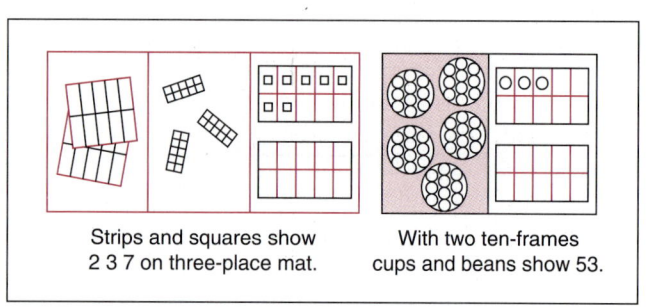

FIGURE 9.13: *Place-value mats with two ten-frames in the ones place to organize the counters and promote the groups-of-ten concept.*

◆ The Forward Game

The game is described first as a first- or second-grade teacher-directed group activity using counters and cups. Any groupable base-ten material can be substituted. Variations are described later.

Each child has an empty place-value mat, a supply of counters, and small cups. Explain that each time you signal (snap fingers, ring a bell), they are to place one more counter on the ones or singles side of the mat. Whenever a ten-frame is filled (or there are 10 counters on the ones side), the class should call out in unison, "ten!" They then take their counters off the ten-frame, put them in a cup, and place the cup of 10 on the left side or tens place of their mats. Periodically pause, and have the class read their mats in unison. When you say "read your mat," children point to the tens side and then the ones side, saying what is there as they point: "Two tens and six." It is also important to occasionally ask, "How many counters are on the mat?" or "How many is that?" Continue in this manner for 10 or 15 minutes or as long as seems reasonable for your class. If you stop at, say, 4 tens and 5, you can write this on the board and begin there the next day. Go to 9 tens and 9 on a two-place mat.

◆ The Backward Game

The "Backward game" is simply the reverse process from the forward game.

Have children place, say, 4 cups of 10 and 6 singles on their mats. Explain that each time you give the signal, they are to remove one counter from their boards instead of putting one on. When there are no more counters on the ones side, a cup of 10 should be dumped out onto the ones side and a ten-frame filled with the counters. Then a counter can be removed. Single counters should never be removed from cups, since cups are to always contain 10 counters. As before, periodically stop, and have children read their mats.

◆ Variations of the Forward and Backward Games

For first- or second-grade children, the variations below should only come after the games have been played as just described. Older children can begin with one of these variations.

1. **Change the amount to put on or take off.** The first change should be to two at a time. In this way, the change to or from tens will always come out even. Then try putting on or taking off three at each signal. Always put all three on the board before making a ten. Use the second ten-frame when there are more than 10 counters on the ones side. In the backward game, nothing should be removed until all three can be removed. Figure 9.14 illustrates the sequence.

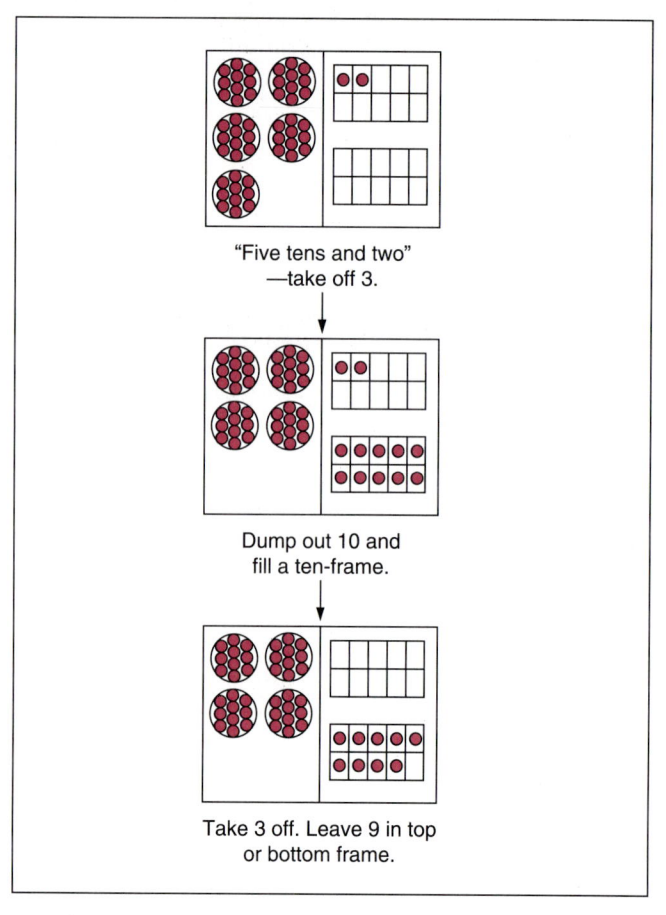

FIGURE 9.14: *Backward sequential grouping. Taking 3 off at each signal.*

2. **Use pregrouped materials.** This is a significant switch for young children because now they *trade* 10 ones for a ten piece instead of grouping the singles into a ten.

3. **Play as a game.** Each of two players has a mat. In turn, players roll a die to determine how many singles should be put on their mat (forward) or taken off of their mat (backward). In the forward version, the mats begin empty, and the first to reach a designated goal is the winner. In the backward game, both players start with a designated amount on the mats, and the winner is the first to clear his or her mat. In the forward game, the entire amount rolled should be placed on the mat before any groups or trades are made. In the backward game, nothing should be removed from the mat until the entire amount can be removed at once. That is, if a ten must be dumped (broken, traded), it should be done before any singles are removed.

4. **Use hundreds on a three-place mat.** Trading can start at any point such as 3 hundreds, 5 tens, and 6. Ones can be added or removed in fixed amounts such as four at a time, or a die or spinner can be used to determine the amounts.

5. **Add or remove both tens and ones on each move.** In a teacher-directed format, simply announce how many ones and tens to put on or remove. Vary the amounts each time. For independent activities and for the two-player game version, use two different dice, perhaps designating a red die for tens and a white die for ones. Cubes can be marked "tens" and "ones", and numerals can be written on them to make dice. You may wish to use only the numbers 0, 1, 2, and 3 on the dice instead of 1 through 6. In forward games, both tens and ones should be added to the mats before any trades are made, as shown in

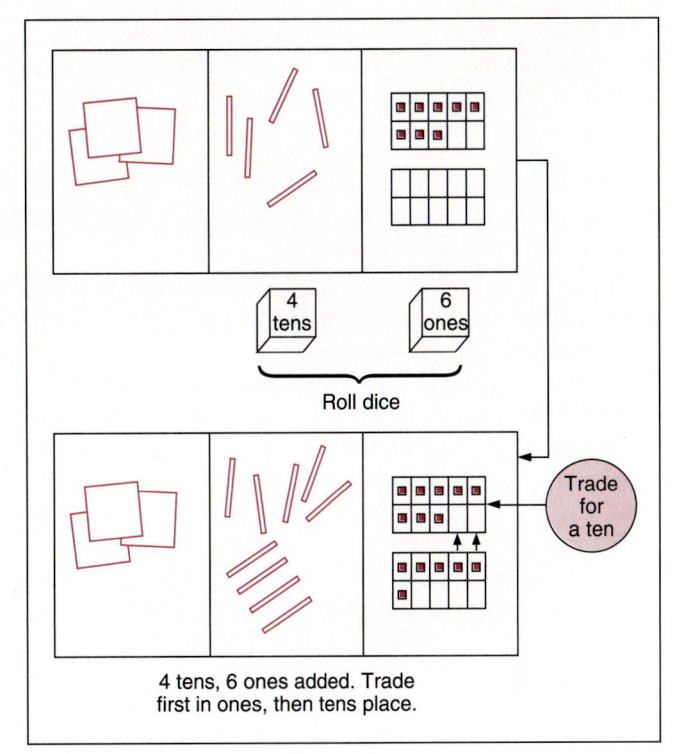

FIGURE 9.15: *The forward trading game using two dice*

Figure 9.15. In backward games, no partial amounts of ones or tens should be removed. You may wish to add the rule of always working first with the ones column and then the tens. This latter rule is totally arbitrary but does match the standard procedure for adding and subtracting with pencil and paper.

6. **Worksheet version.** Worksheets can be made as in Figure 9.16 to designate the amounts to be added or removed. Children use materials as directed and record the results with the square, stick, and dot notation. This is especially useful to increase trades

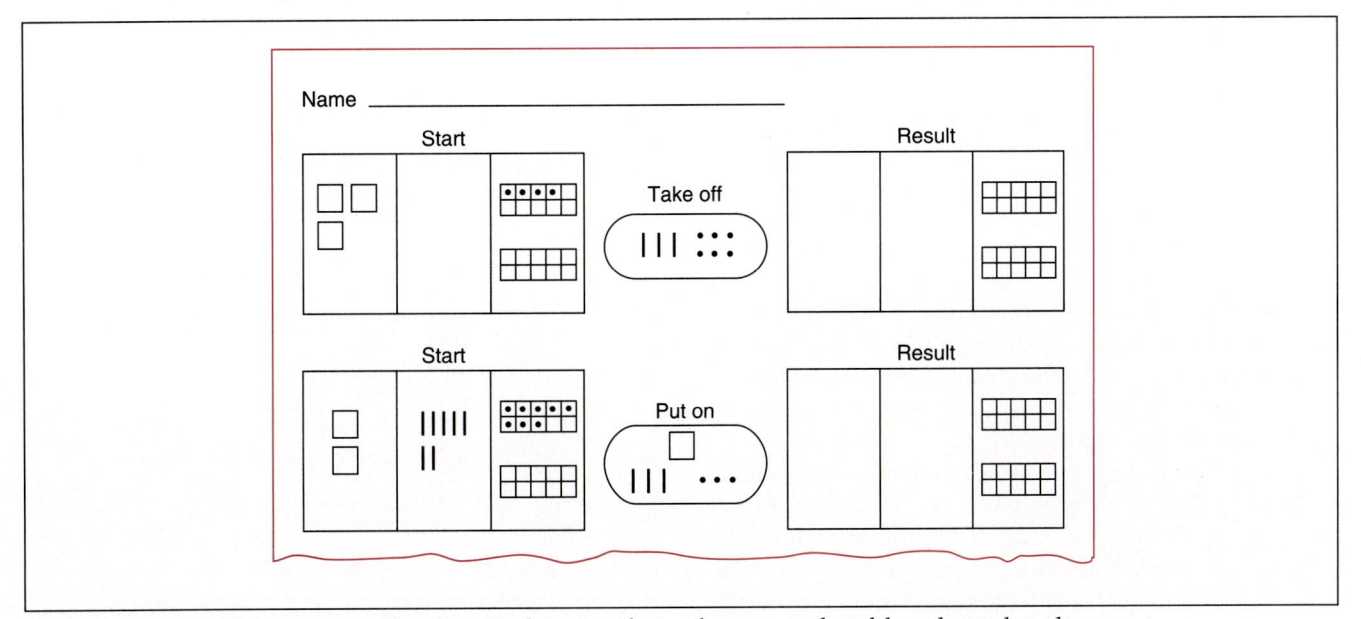

FIGURE 9.16: *A worksheet version of trading. Students use place-value mats and models and record results.*

involving hundreds or to practice the double trade required in the backward game when there are no tens. The games in this form are simply a concept level of addition and subtraction with regrouping.

◆ Connections with Symbols

The trading games are a good way to focus on how numbers with two or three digits are written. This goal can be addressed in a variety of ways.

An idea borrowed from *Mathematics their way* (Baratta-Lorton, 1976) is to put numerals directly on the place-value mats. Make "numeral flips" by stacking small tagboard numeral cards 0 to 9 (0 on top) and connecting them with loops of string through punched holes. One flip is placed on each side of the mat, as shown in Figure 9.17. As materials change, the children change the flips accordingly.

Also from *Mathematics their way* is the idea of using place-value recording strips. These are simply strips of paper ruled in two columns (or three columns). As children make changes on their mats, they also record on the strips. In those versions where the same amount is either put on or taken off each time, children only record each new amount. When the end of a strip is reached, another blank strip is simply taped on at the bottom. Children can make long strips showing the count by ones or twos or threes from "00" to "99" and back [Figure 9.18(a)]. Children can work independently with their mats and their recording strips.

In versions of trading activities where varying amounts are put on or removed, the strips can be used to indicate these change amounts as well. For example, if children were playing a forward-trading game in which they rolled a die for tens and a die for ones, they would record the amounts rolled and the result on the strips [Figure 9.18(b)]. This looks exactly like symbolic addition or subtraction, but is approached as recording the turns in the game. This activity is an essential readiness for computation.

Another way to connect symbols to the sequential grouping and trading activities is to use a calculator. For example, if students are putting 3 more on at each signal, they can store ⊞ 3 in the calculator and press ⊟ after each move. The calculator can then serve as a check to what they have done. A written record is still appropriate.

At first, the trading games with numerals are a means of teaching how to write numbers meaningfully. Later, the place-value mats simply offer a connection between groups (materials) and symbols that many older children must renew or make clear again.

Not all connecting activities should involve materials arranged on a mat. If materials for a two- or three-digit number are presented in a mixed form as in Figure 9.11, students must be able to coordinate the illustrated concepts with both the oral name and the correct written name and not be dependent on a place-value mat.

The next activities are designed to help children make connections between all three representations: groups, oral language, and written forms. They can be done with two- or three-digit numbers in grades 2 to 4.

9.9 SAY IT/PRESS IT

Display some ones and tens (and hundreds) so the class can see. (Use the overhead projector or magnetized pieces on the board, or simply draw using the square-stick-dot method.) Arrange the materials in a mixed design, not in the standard left-to-right format. Students first say the amount shown in place-value language ("four hundreds, one ten, and five"), then in standard language ("four hundred fifteen"), and finally they enter it on their calculators. Have someone share his or her display. Make a change in the materials and repeat.

Say It/Press It is especially good for helping with tens (note the example in the activity description) and for three-digit numbers with zero tens. If you show 7 hundreds and 4 ones, the class says "seven hundreds, *zero* tens, and four—seven hundred (slight pause) four." The pause and the place-value language suggest the correct three-digit number to press or write. Many students have trouble with this example and write "7004," writing exactly what they hear in the standard name.

The next two activities simply change the representation that is presented first to the students.

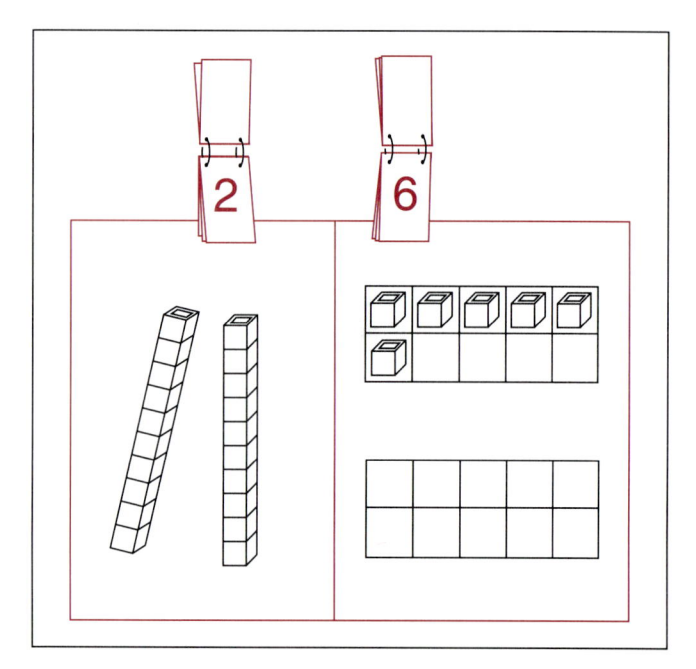

FIGURE 9.17: *Flip cards connect model and numeral.*

9 / WHOLE NUMBER PLACE-VALUE DEVELOPMENT

Students having difficulties should put materials on a place-value mat to help with the proper form.

Independent worksheet activities can combine all three forms for numbers. Figure 9.19 (p. 170) shows two variations. A third variation would give the numeral and have children supply the number and draw the model.

NUMBER SENSE DEVELOPMENT

The discussion so far has addressed the three main components of place-value understanding: the integration of base ten groupings, oral, and written names. These are the foundation ideas of place value. But students need to expand these ideas beyond basic numeration concepts and reading and writing numerals. Even though this section highlights number-sense activities, number sense is not a new or separate curriculum topic. These activities should be integrated along with the development of concepts and should be an ongoing feature of your instruction.

RELATIVE MAGNITUDE

Relative magnitude refers to the size relationship one number has with another—is it much larger, much smaller, close to, or about the same? There are several quick activities that can be done with a number line sketched on the board. The number line can help children see how one number is related to another.

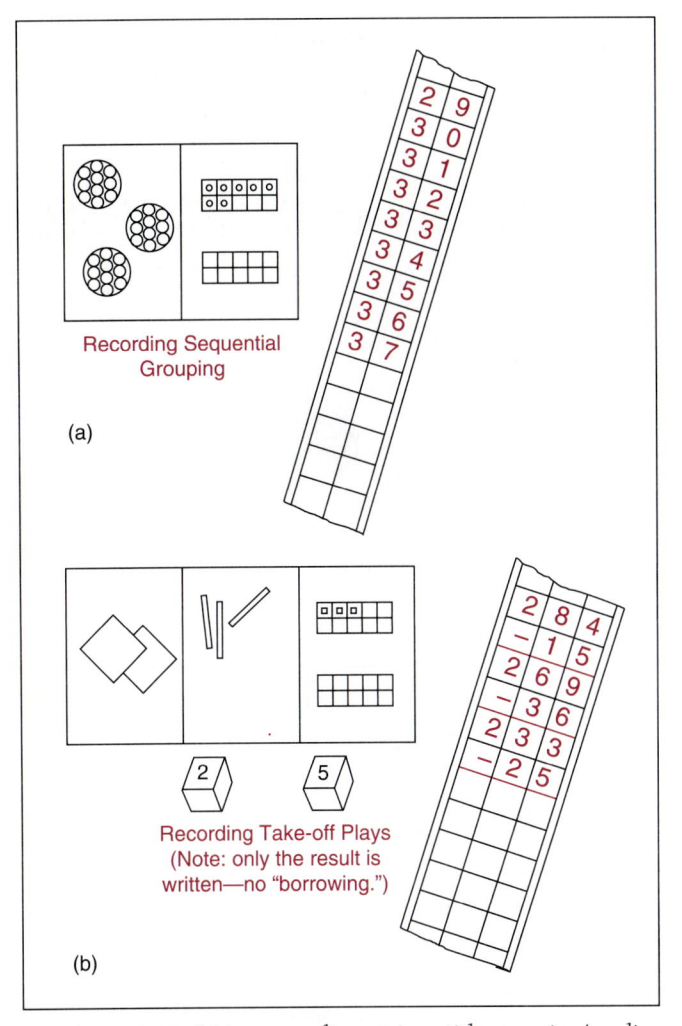

FIGURES 9.18: *Using recording strips with grouping/trading activities.*

9.10 SHOW IT/PRESS IT

Say the standard name for a number (either two or three digits). At their desks students use their own base ten models to show that number and press it on their calculators (or write it). Again, pay special attention to the teens (they sound backwards) and the case of zero tens.

9.11 SHOW IT/SAY IT

In this variation you silently write a number (or press it on your overhead calculator), and the students show it at their desk with their models. On your cue, all say the amount in unison.

In the last two activities you could have students record each example on paper using the square-stick-dot notation.

9.12 WHO AM I?

Sketch a line labeled 0 and 100 at the ends. Mark a point that corresponds to your secret number. (Estimate the position the best you can.) Students try to guess your secret number. For each guess, place and label a mark on the line.

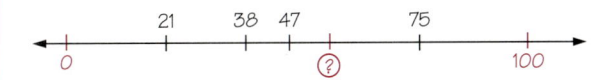

Continue marking each guess until your secret number is discovered. As a variation, the endpoints can be other than 0 and 100. For example, try 0 and 1000, or 200 and 300, or 500 and 800.

9.13 SQUEEZE

Label only the endpoints of the number line, but subdivide the line into 10 equal segments. Have one stu-

170 9 / WHOLE NUMBER PLACE-VALUE DEVELOPMENT

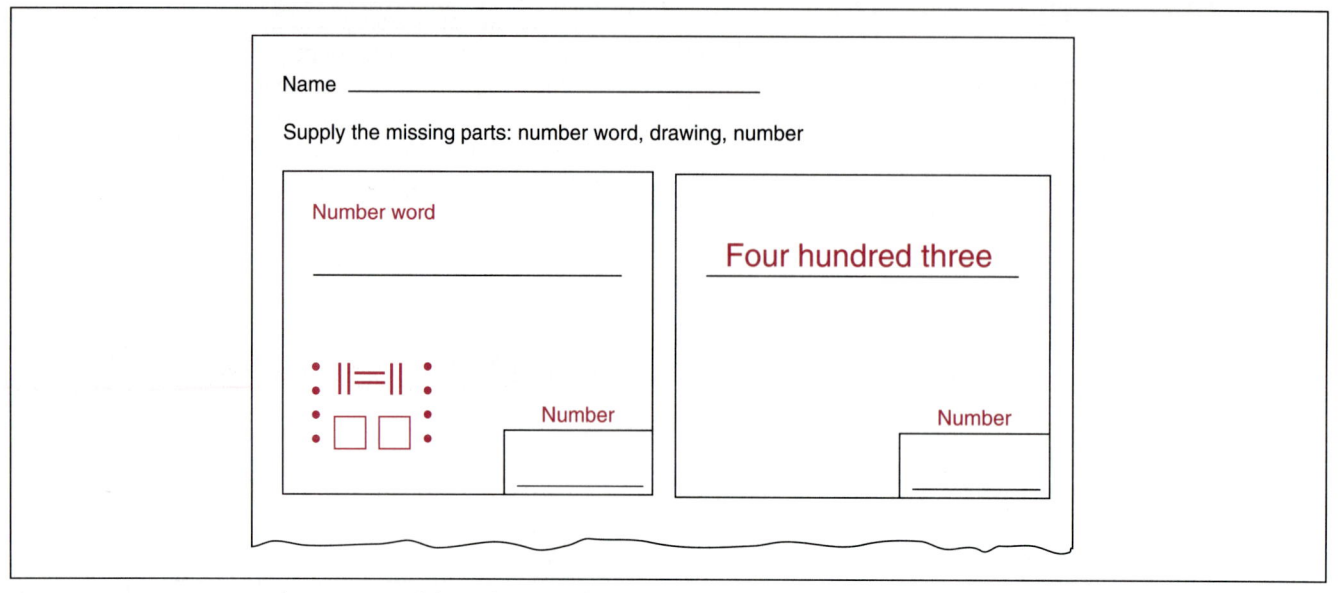

Name _____

Supply the missing parts: number word, drawing, number

Number word	Four hundred three
_____	_____
:||=||:	
: ☐ ☐ :	
Number	Number
_____	_____

FIGURE 9.19: *Connecting language, models, and numerals*

---

dent stand at each end with a marker card. As students guess your secret number, announce that the guess is either too high or too low. If the guess is too high, the student on the high end moves in to mark the guess by holding the marker card at the appropriate place. When the guess is too low, the left-end student moves in and marks that number in a similar manner. Guessing continues as the ends squeeze in on the secret number.

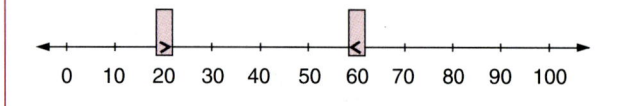

```
  ┌─────┐           ┌─────┐
──┼──┬──┼──┬──┬──┬──┼──┬──┬──┬──→
  0  10 20 30 40 50 60 70 80 90 100
```

9.14 WHO COULD THEY BE?

Label two points of a number line (not necessarily the ends).

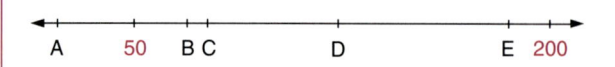

```
←──┬──┬──┬────┬────────┬──→
   A  50 B C  D        E  200
```

Ask students what numbers they think different points labeled with letters might be and why they think that. In the example shown here, B and C are less than 100 but probably more than 70. E could be about 180. You can also ask where 75 might live or where is 400? About how far apart are A and D? Why do you think D is more than 100?

In the next activity, some of the same ideas are discussed without benefit of a number line.

9.15 CLOSE, FAR, AND IN BETWEEN

Put any three numbers on the board.

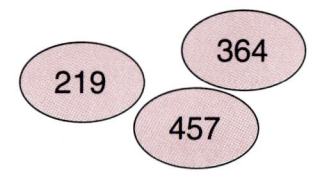

(219) (364)
   (457)

With these three numbers as referents, ask questions such as the following, and encourage discussion of all responses:

Which two are closest? Why?

Which is closest to 300? To 250?

Name a number between 457 and 364.

Name a multiple of 25 between 219 and 364.

Name a number that is more than all of these.

About how far apart are 219 and 500? 219 and 5000?

If these are "big numbers," what are some small numbers? about the same numbers? numbers that make these seem small?

APPROXIMATE NUMBERS AND ROUNDING

In our number system, some numbers are "nice" in that they are easy to think about and work with. The idea of what makes a nice number is sort of fuzzy. However, numbers such as 100, 500, and 750 are easier to use than 94, 517, and 762. Multiples of 100 are very nice, and multiples of 10 are not too bad either. Multiples of 25 (50, 75, 425, 675, . . .) are nice because they combine into 100s and 50s rather easily, and we can mentally place those between multiples of 100s. Multiples of 5 are a little easier to work with than those that are not.

Flexible thought with numbers and many estimation skills are related to the ability to substitute a nice number for one that is not so nice. The substitution may be to make a mental computation easier, to compare it to a familiar reference, or simply to store the number in memory more easily.

The choice of a nice number to substitute for a less manageable one is never completely clear-cut. There is not a "correct" or "best" substitute for $327.99 or 57 pounds. The choices depend on the need for clarity, how much accuracy is needed, and how the substitute will be used. For example, nice substitutes for $327.99 might be $300, $320, $325, $328, $330, or even $350. In a given situation there may be more than one good choice of substitutes. (For each of these substitutes can you think of a situation where using that number might be more useful than the others?)

In the past, rounding numbers was the principal method of selecting a substitute nice number. Students were taught rules for rounding numbers to the nearest 10 or nearest 100. Unfortunately, the emphasis was placed on correctly using the rule. (If the next digit is 5 or more, round up; otherwise leave the number alone. Put zeros on the end.) A context to suggest why they may want to round numbers was usually a lesser consideration.

The activities here are designed to help students recognize what nice numbers are and to identify a nice-number substitute.*

9.16 NICE NUMBER SKIP COUNTS

Count by 5s, 10s, 25s, and 50s with your students. The 5s and 10s are fairly easy, but the skill is certainly worth practicing. Counts by 25s or 50s may be hesitant at first. Students can use a calculator to assist with their counting and connect the counts with numerals. (Press ⊞ 25 ⊟ ⊟ ⊟ . . .) At first, start all counts at zero. Later start at some multiple of your skip amount. For example, begin at 275 and count by 25s. Counts by 10s should also begin at numbers ending in 5 as well as multiples of 10. Also count backwards by these same amounts. (Press 650 ⊟ 50 ⊟ ⊟ ⊟ . . .)

---

*The term "nice number" is not one found in standard usage. It is strictly an invention of the author. It has no commonly accepted definition.

For children in grades 1 and 2, a 100s chart can be used in connection with skip-counting 5s and 10s. Older children should count to at least 1000 and should discuss the patterns that they see in these counts.

Too often children are expected to count coins without any preparation or background. It is not that money is hard for children but, rather, skip-counting by different amounts is difficult. The next activity extends **Nice Number Skip Counts** in preparation for counting money.

9.17 MONEY COUNTS

The goal of this activity is to practice shifting from one skip count to another. Explain to the class that they will start counting by one number and at your signal they will shift to a count by a different number. Begin with only two different amounts, say 25 and 10. Write these numbers on the board. Point to the larger number (25) and have students begin to count. After 3 or more counts, raise your hand to indicate a pause in the counting. Then lower your hand, pointing to the smaller number (10). Children continue the count from where they left off but now count by 10s. Use any two of these numbers: 100, 50, 25, 10, 5, and 1. Always start with the larger. Later, try three numbers, still in descending order.

Note that the counts in Money Counts are the same that are used when counting coins or money. These skills can be applied to bills and coins. Plastic "money" is available for use on the overhead and is quite an effective substitute.

◆ Rounding

To "round" a number simply means to substitute a nice number that is close. The close number can be any nice number and need not be a multiple of 10 or 100 as has been traditional.

9.18 NEAR AND NICE

The idea is to say or write a number and have students select a close nice-number substitute. Begin by requesting a close multiple of 50. Explain that these are the numbers that you get when you count by 50s. It is a good idea to include a real-world measurement as well. For example, "Instead of 243 feet, let's use _____ feet." Pause at the blank, and students fill in an appropriate number, here 250 feet. The nearest 25 is sometimes difficult. For example, is 463 miles closer to 450 or 475 miles? Actually either is quite a good substitute. In fact, place your emphasis on selecting a *close* nice number rather than the *closest* or *best* nice number.

A number line with nice numbers highlighted can be useful in helping children select near-nice numbers. An unlabeled number line like the one shown in Figure 9.20 can be made using three strips of poster board taped end to end. Labels are written above the line on the chalk board. The ends can be labeled 0 and 100, 100 and 200, . . . , 900, and 1000. The other markings then show multiples of 25, 10, and 5. Indicate a number above the line that you want to round. Discuss the marks (nice numbers) that are close.

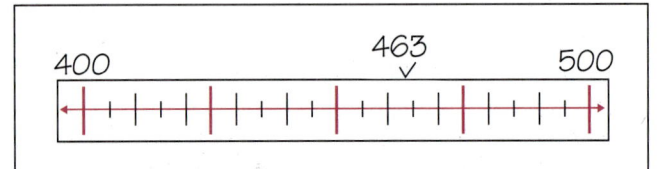

FIGURE 9.20: *A blank number line can be labeled in different ways to help students with near and nice numbers.*

USING TENS AND HUNDREDS

Mental computation is an important part of having number sense. While developing place-value concepts, students can begin to add and subtract tens and hundreds and to combine numbers mentally.

9.19 HUNDREDS BOARD THINKING

For these activities use a transparent 10 × 10 grid on the overhead projector or a large laminated 10 × 10 grid. Explain that this is a hundreds board with the numbers hidden.

Neighbors. On a blank hundreds board, point to a square and ask students to name the number. Write it in. Next, point to squares that surround this number, its neighbors (Figure 9.21).

Diagonally Around. Start on a number anywhere on the border of the hundreds board, and have students name the number. Then point to successive squares on a diagonal from this starting square as children name the numbers you point to. When you get to the next edge, "bounce" by continuing on a diagonal in the reflected direction. After three such bounces, you will be headed directly toward the number you began with (Figure 9.21).

9.20 ARROW MATH

With the hundreds board in view, write a number on the board followed by one or more arrows. The arrow can point up, down, left, right, or even diagonally. Each arrow represents a move of one square on the board.

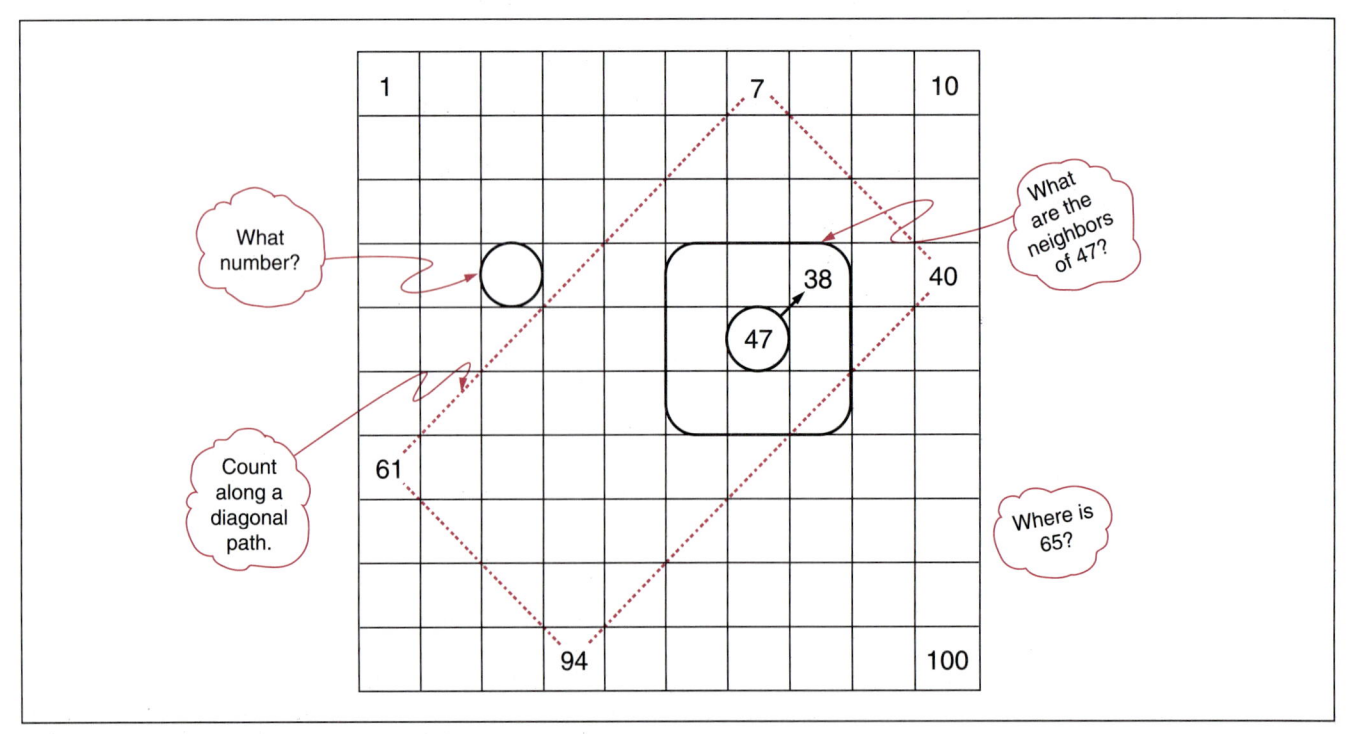

FIGURE 9.21: *Number names and concepts on a hundreds board*

After students become adept at **Arrow Math,** talk about what each arrow means. The left and right arrows are one-less and one-more arrows. The up and down arrows are the same as ten less and ten more. (What do each of the four possible diagonal arrows represent?)*

Arrow Math can easily be extended to numbers to 1000. For example:

793 ⟶ ↓↓   (814)

The next activity is a good follow-up to **Arrow Math.**

◆◆◆◆◆◆◆◆◆◆◆◆◆◆◆◆◆◆◆

9.21 CALCULATOR TENS AND HUNDREDS

Two students work together with one calculator. They press ➕ 10 🟰 to make a "plus 10 machine." One child enters any number. The other student says and/or writes the number that is ten more. The 🟰 is pressed for confirmation. The roles are then reversed. The same activity can be done with plus any multiple of ten or multiple of 100. On their turn to challenge, students can select what kind of machine to use. For example, they may first press 0 ➕ 300 🟰 and then press 572. The other student would say or write 872 and then press 🟰 for confirmation. This game can also be played using ➖ instead of ➕ to practice mentally subtracting tens or hundreds.

Many of the skills and concepts developed so far also appear in the next activity, which combines symbolism with base ten representation.

◆◆◆◆◆◆◆◆◆◆◆◆◆◆◆◆◆◆◆

9.22 NUMBERS, SQUARES, STICKS, AND DOTS

As illustrated in Figure 9.22, prepare a worksheet where a numeral and some base ten pieces are shown. Students write the total. Note how **Money Counts,** place-value concepts, and symbolism are all included.

---

*Adapted from activities found in *Mental Math for the Primary Grades* (Hope, Leutzinger, Reys, & Reys, 1988).

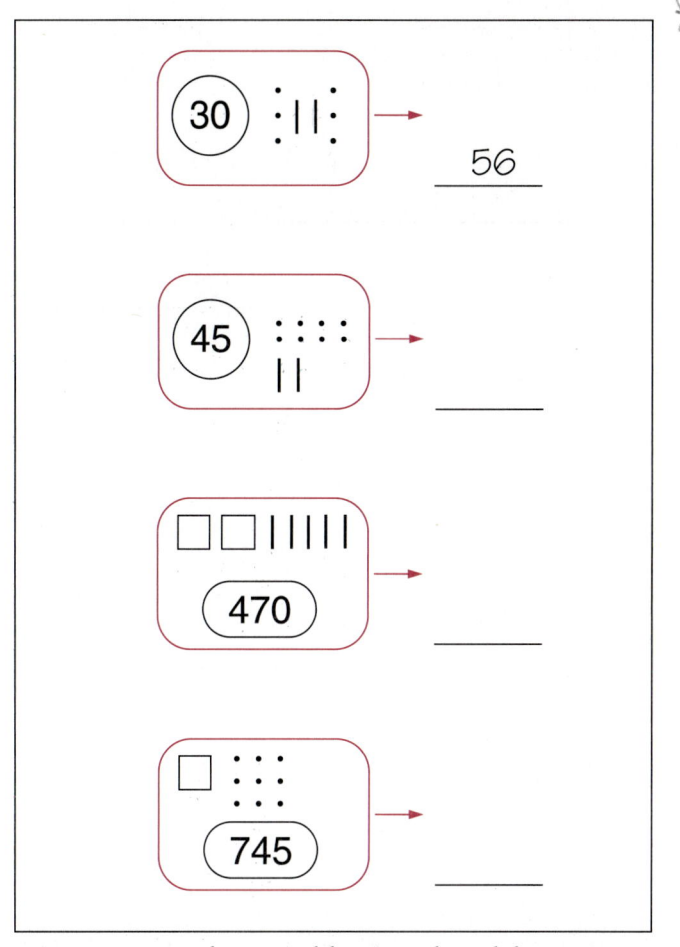

30 :||: ⟶ 56

45 :::: ⟶ ___
    ||

□□ ||||| ⟶ ___
470

□ ::: ⟶ ___
745

FIGURE 9.22: *Combining models, numerals, and skip-counting*

NUMBERS BEYOND 1000

For children to have good concepts of numbers beyond 1000, the conceptual ideas that have been carefully developed must be extended. This is sometimes difficult to do because physical models for thousands are not commonly available. At the same time, number-sense ideas must also be developed. In many ways, it is these informal ideas about very large numbers that are the most important.

EXTENDING THE PLACE-VALUE SYSTEM

Two important ideas developed for three-digit numbers should be carefully extended to larger numbers. First, the grouping idea should be generalized. That is, ten in *any position* makes a single thing in the next position, and vice-versa. Second, the oral and written patterns for numbers in three digits is duplicated in a clever way for every three digits to the left. These two related ideas are not as easy for children to understand as adults seem to believe. Because models for large numbers are so difficult to have or picture, textbooks must deal with these ideas in a predominantly symbolic manner. That is not sufficient!

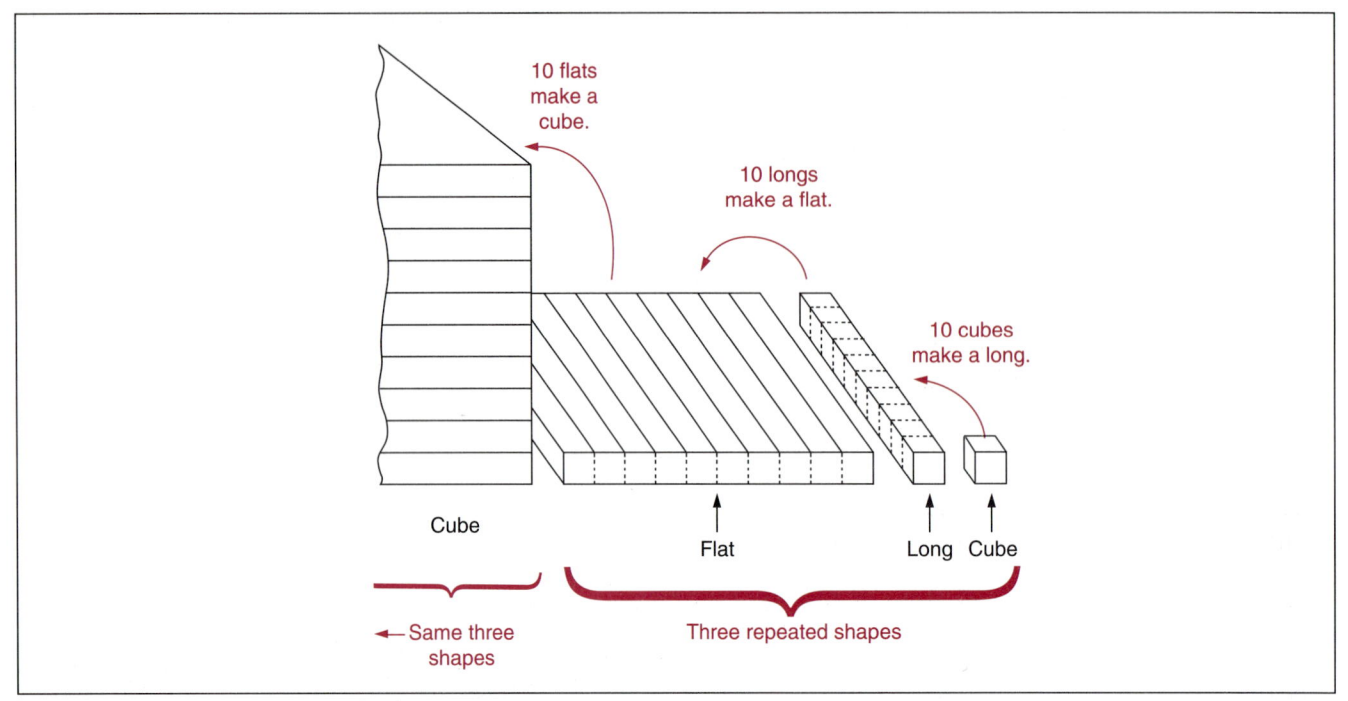

FIGURE 9.23: *Every three places the shapes repeat.*

9.23 WHAT COMES NEXT?

Have a **What Comes Next?** discussion with the base ten strips and squares. The unit or ones piece is a 1-cm square. The tens piece is a 10 × 1 strip. The hundreds piece is a square, 10 cm by 10 cm. What is next? Ten hundreds is called a *thousand*. What shape? It could be a strip made of 10 hundreds squares. Tape 10 hundreds together. What is next? (Reinforce the idea of "ten makes one" that has progressed to this point.) Ten one-thousand strips would make a square 1 meter (m) on a side. Draw one on butcher paper and rule off the 10 strips inside to illustrate the ten-makes-one idea. Continue. What is next? Ten ten-thousand squares would go together to make a strip. Draw this 10-cm by 1-m strip on a long sheet of butcher paper and mark off the 10 squares that make it up. You may have to go out in the hall.

How far you want to extend this square, strip, square, strip sequence depends on your class and your needs. The idea that 10 in one place makes one in the next can be brought home dramatically. There is no need to stop. It is quite possible with older children to make the next 10-m by 10-m square using masking tape on the cafeteria floor or with chalk lines on the playground. The next strip is 100-m by 10-m. This can be made on a large playground using kite string for the lines. By this point the payoff includes an appreciation of the increase in size of each successive amount as well as the *ten-makes-one* progression. The 100-m by 10-m strip is the model for 10 million, and the 10-m by 10-m square models 1 million. The difference between 1 and 10 million is dramatic. Even the concept of 1 million tiny centimeter squares is dramatic.

The three-dimensional wooden or plastic base ten materials are all available with a model for thousands, which is a 10-cm cube. These models are expensive, but having at least one large cube to show and talk about is a good idea.

Try the "What comes next?" discussion in the context of these three-dimensional models. The first three shapes are distinct: a *cube*, a *long*, and a *flat*. What comes next? Stack 10 flats and they make a cube, same shape as the first one only 1000 times larger. What comes next? (See Figure 9.23.) Ten cubes makes another long. What comes next? Ten big longs makes a big flat. The first three shapes have now repeated. Ten big flats will make an even bigger cube, and the triplet of shapes begins again.

A good discussion revolves around the metric dimensions of each successive cube. The million cube is 1 meter on an edge. The billion cube is 10 meters on an edge or about the size of a three-story building.

Each cube has a name. The first one is the unit cube, the next is a thousand, the next a million, then a billion, and so on. Each long is 10 cubes; 10 units, 10 thousands, 10 millions. Similarly, each flat shape is 100 cubes.

To read a number, first mark it off in triples from the right. The triples are then read, stopping at the end of each to name the unit (or cube shape) for that triple (Figure 9.24). Leading zeros are ignored. If students can learn to read numbers like 059 (fifty-nine) or 009 (nine), they should be able to read any number. To write a number, use the same scheme. If first mastered orally, the system is quite easy.

It is important for children to realize that the system

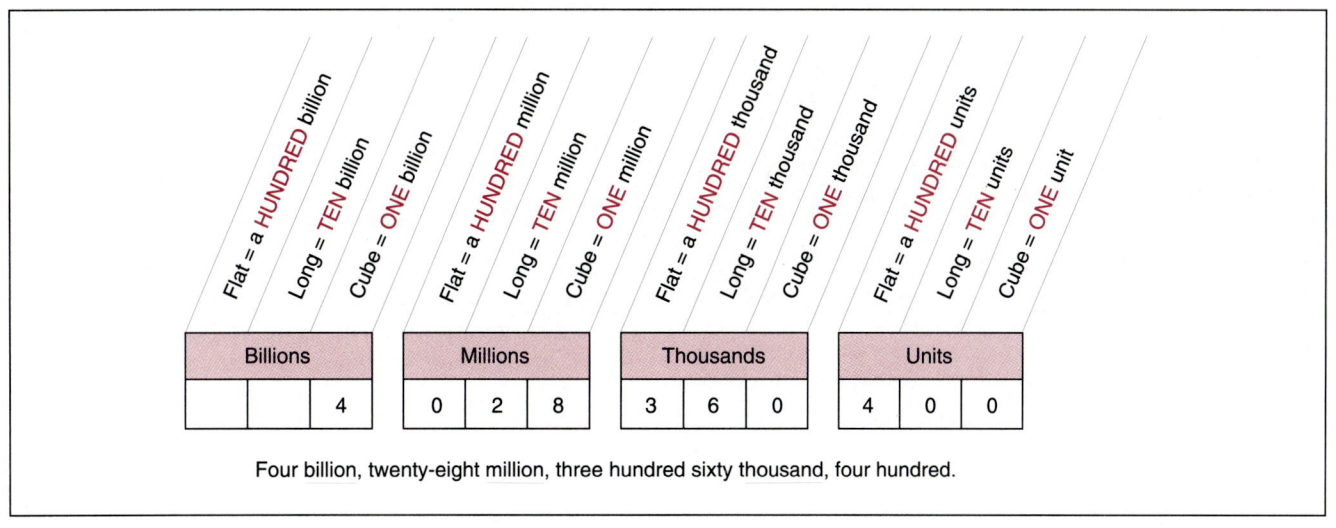

FIGURE 9.24: *The triples system for naming large numbers*

Four billion, twenty-eight million, three hundred sixty thousand, four hundred.

(Labels on the chart, from left to right: Flat = a HUNDRED billion, Long = TEN billion, Cube = ONE billion, Flat = a HUNDRED million, Long = TEN million, Cube = ONE million, Flat = a HUNDRED thousand, Long = TEN thousand, Cube = ONE thousand, Flat = a HUNDRED units, Long = TEN units, Cube = ONE unit. Digits: Billions 4, Millions 028, Thousands 360, Units 400.)

does have a logical structure, is not totally arbitrary, and can be understood.

CONCEPTUALIZING LARGE NUMBERS

The ideas discussed in the previous section are only partially helpful in thinking about the actual quantities involved in very large numbers. For example, in extending the square, strip, square, strip sequence, some appreciation for the quantities of a thousand or of a hundred thousand is included. But it is hard for anyone to translate quantities of small squares into quantities of other items, distances, or time.

Creating References for Special Big Numbers

In these activities numbers like 1000, 10,000, or even 1,000,000 are translated literally or imaginatively into something that is easy or fun to think about. Interesting quantities become lasting reference points or benchmarks for large numbers and thereby add meaning to numbers encountered in real life.

9.24 COLLECTING TEN THOUSAND

Collections: As a class or grade-level project, collect some type of object with the objective of reaching some specific quantity. Some examples: 1000 or 10,000 buttons, walnuts, old pencils, jar lids, pieces of junk mail. If you begin aiming for 100,000 or a million, be sure to think it through. One teacher spent nearly 10 years with her classes before amassing a million bottle caps. It takes a small dump truck to hold that many!

*Idea: collect pop can tabs. After so many have been collected these can be donated to make a wheel chair.

9.25 SHOWING TEN THOUSAND

Illustrations: Sometimes it is easier to create large amounts. For example, start a project where students draw 100 or 200 or even 500 dots on a sheet of paper. Each week different students contribute a specified number. Another idea is to cut up newspaper into pieces the same size as dollar bills to see what a large quantity would look like. Paper chain links can be constructed over time and hung down the hallways with special numbers marked. Let the school be aware of the ultimate goal.

9.26 HOW LONG?/HOW FAR?

Real and imagined distances: How long is a million baby steps? Other ideas that address length: toothpicks, dollar bills, or candy bars end to end; children holding hands in a line; blocks or bricks stacked up; children lying down head to toe. Real measures can also be used: feet, centimeters, meters.

9.27 A LONG TIME

Thinking of time: How long is 1000 seconds? How long is 1,000,000 seconds? a billion? How long would it take to count to 10,000 or 1,000,000? (To make the counts all the same, use your calculator to do the counting. Just press the ⊟.) How long would it take to do some task like buttoning a button 1000 times?

◆ Estimating Large Quantities

The foregoing activities aim at a specific number. The reverse idea is to select a large quantity and find some way to measure, count, or estimate it.

◆◆◆◆◆◆◆◆◆◆◆◆◆◆◆◆◆◆◆◆◆

9.28 REALLY LARGE QUANTITIES

How many:

Candy bars would cover the floor of your room?

Steps would an ant take to walk around the school building?

Grains of rice would fill a cup? a gallon?

Quarters could be stacked in one stack floor to ceiling?

Pennies can be laid side by side down an entire block?

Pieces of notebook paper would cover the gym floor?

Seconds have you (or the teacher) lived?

Big-number projects need not take up large amounts of class time. They can be explored over several weeks as take-home projects, group projects, or, perhaps best of all, translated into great schoolwide estimation contests.

DIAGNOSIS OF PLACE-VALUE CONCEPTS

Assuming that the age level is appropriate, almost all children can be taught *to do* the activities described in this chapter. Even the most *meaningful* activities with place-value pieces can be performed without the child constructing the concepts involved. Going through the motions will not guarantee that concepts are formed. This is, of course, true of any mathematics activity. However, for the concept of place value, the distinction between conceptual thinking and surface-level performance of activities is frequently very difficult to discern.

The diagnostic tasks presented here are designed to help teachers look more closely at children's understanding of place value. They are not suggested as definitive tests but as means of obtaining information for the thoughtful teacher. These tasks have been used by several researchers and are adapted primarily from Labinowicz (1985) and Ross (1986). The tasks are designed for one-on-one settings. They should not be used as instructional activities.

◆ Task: Counting Skills

A variety of oral counting tasks provide insight into the counting sequence.

Count forward for me, starting at 77.

Count backwards, starting at 55.

Count by tens.

Count by tens starting at 34.

Count backwards by tens starting at 130.

In the tasks that follow, the manner in which the child responds is at least as important as the answers to the questions. For example, counting individual squares on tens pieces that are known to the child as "tens" will produce correct answers, but indicates the structure of tens is not being utilized.

◆ Task: One More/Ten More (Also "Less")

Write the number 342. Have the child read the number. Next have him or her write the number that is *1 more than* the number. Next ask for the number that is *10 more than* the number. Following the responses, you may wish to explore further with models. One less and 10 less can be checked the same way.

◆ Task: Digit Correspondence

Dump out 36 blocks. Ask the child to count the blocks, and then have him or her write the number that tells how many are there. Circle the 6 in 36 and ask, "Does this part of your thirty-six have anything to do with how many blocks there are?" Then circle the 3 and repeat the question exactly. Do not give clues. Based on responses to the task, Ross (1989) has identified five distinct levels of understanding of place value.

◆ Task: Using Tens

Dump out 47 counters and have the child count them. Next show the child at least 10 cards, each with a ten-frame drawn on it. (Spaces should be large enough to hold the counters used.) Ask, "If we wanted to put these counters in the spaces on these cards, how many cards could we fill up?" (If the ten-frame has been used in class to model sets of 10, use a different "frame" such as a 10-pin arrangement of circles. Be sure the child knows there are 10 spaces on each card.)

◆ Task: Using Groups of 10

Prepare cards with 10 beans or other counters glued to the cards in an obvious arrangement of 10. Supply at least 10 cards and a large supply of the beans. After you are sure that the child has counted several cards of beans and knows there are 10 on each, say, "Show me thirty-four beans." (Does the child count individual beans or use the cards of 10?) The activity can also be done with hundreds.

An understanding of equivalent representations, critical for understanding computation, can be assessed with the following activity.

◆ Task: Understanding Trades

Have the child represent a number using any base ten materials. Next trade one or more tens for ones (or hundreds for tens). Ask "What number is this now?" or "Is there more now or less, or is it just the same?"

For the next two tasks use a board made from a half sheet of posterboard, a box lid or some other cover, and pregrouped base ten materials.

◆ Task: Base Ten Ideas with Symbols

1. Show a board with some base ten pieces covered and some showing. Tell the child how many pieces are hidden under the cover, and ask him or her to figure out how much is on the board altogether [Figure 9.25(a)].

2. Show a board partially covered as before. Tell the child how many pieces are on the board altogether, and ask how many are hidden [Figure 9.25(b)].

(a) "How many on the board altogether?"

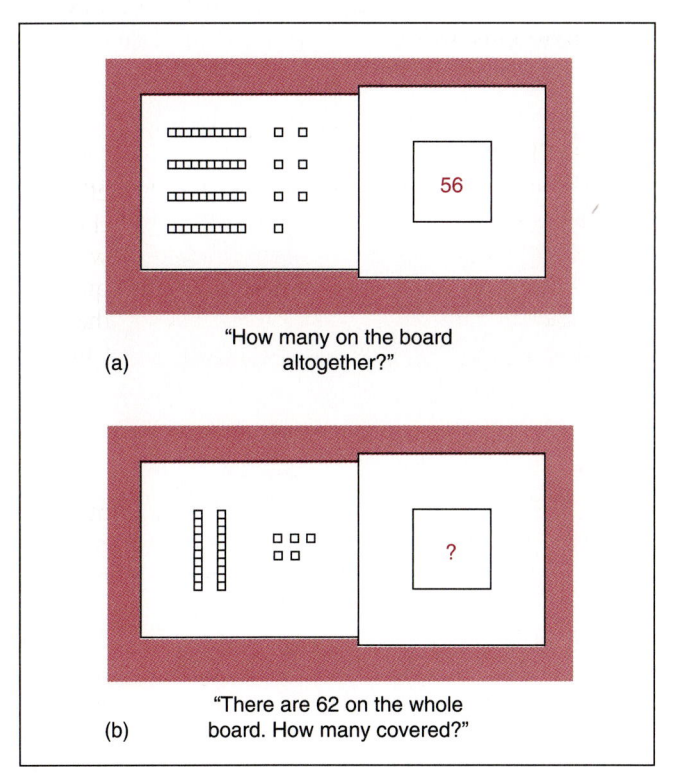

(b) "There are 62 on the whole board. How many covered?"

FIGURES 9.25: *Two diagnostic activities*

The last two tasks could also involve hundreds. The amounts that you tell the child could be given in written form instead of orally.

◆◆◆◆◆ REFLECTIONS ON CHAPTER 9: WRITING TO LEARN

1. Explain how a child who has not yet developed base ten concepts understands quantities as large as, say, 85. Contrast this with a child who understands these same quantities in terms of base ten groupings.

2. What is meant by *equivalent representations*?

3. Explain the three ways one can count a set of objects and how these methods of counting can be used to coordinate concepts and oral and written names for numbers.

4. Describe the two types of physical models for base ten concepts. What is the significance of the difference between these two types of models? Why is an abacus not considered a model for place value?

5. Describe an activity for developing base ten grouping concepts, and reflect on how the activity encourages children to construct base ten concepts.

6. Describe trading activities including some variations and methods of connecting symbolism with these activities.

7. Describe an activity that does not use a place-value mat that has as an objective connecting base ten concepts with place-value written symbolism.

8. This chapter suggests activities around three aspects of number sense. Describe each of these components of number sense and at least one corresponding activity.

9. What are two different ideas that you would want children to know about very large numbers (beyond 1000)? Describe one or two activities for each.

10. Look at each of the tasks suggested for diagnosing place-value understanding. What specific idea of place value does each task look at?

◆◆◆◆◆ FOR DISCUSSION AND EXPLORATION

1. Examine the excellent activities in the resource book *Picturing Numeration* (Madell & Larkin, 1977). Pay special attention to those in which models and symbols are used in the same exercise.

2. Based on the suggestions in the last section and on the content found in the basal textbook, design a diagnostic interview for a child at a particular grade level and conduct the interview. (See Chapter 5 for interview techniques.) It is a good idea to take a friend to act as an observer, and/or use a tape recorder or video recorder to keep track of how the interview went. For a related discussion of children's place-value thinking and suggestions for preparing your interview, read Chapters 10 and 11 of *Learning from Children* (Labinowicz, 1985).

3. The developments of place value in the *Mathematics Their Way* program (Baratta-Lorton, 1976) and *Box It or Bag It* (Burk, Snider, & Symonds, 1988) include several weeks of "readiness" in which the trading activities are conducted in bases 4, 5, and 6. For example, children make sets of five cubes into a Unifix bar or put groups of four beans in a cup. The

groups are given humorous names such as "Bozos," so place-value mats are read "three Bozos and two." This means that children can make more groups in less time, avoid tedious counts to 10, and learn to use a word (e.g., "Bozos") to designate groups without the confusion of using a number word (ten) for a group name. Read the place-value chapter in these books or preferably find a first- or second-grade teacher who uses this approach. Discuss the pros and cons of this readiness activity.

SUGGESTED READINGS

Baratta-Lorton, M. (1976) *Mathematics their way.* Menlo Park, CA: Addison-Wesley.

Bickerton-Ross, L. (1988). A practical experience in problem solving: A "10,000" display. *Arithmetic Teacher, 36*(4), 14–15.

Creative Publications. (1986). *Hands on base ten blocks.* Palo Alto, CA: Creative Publications.

Joslyn, R. E. (1990). Using concrete models to teach large-number concepts. *Arithmetic Teacher, 38*(3), 6–9.

Kamii, C. (1986). Place value: An explanation of its difficulty and educational implications for the primary grades. *Journal of Research in Childhood Education, 1,* 75–86.

Kurtz, R. (1983). Teaching place value with the calculator. In L. G. Shufelt (Ed.), *The agenda in action.* Reston, VA: National Council of Teachers of Mathematics.

Labinowicz, E. (1985). *Learning from children: New beginnings for teaching numerical thinking.* Menlo Park, CA: Addison-Wesley.

Madell, R., & Larkin, E. (1977). *Picturing numeration.* Palo Alto, CA: Creative Publications.

Payne, J. N., & Huinker, D. M. (1993). Early number and numeration. In R. J. Jensen (Ed.), *Research ideas for the classroom: Early childhood mathematics.* New York: Macmillan Publishing Co.

Reys, B. J. (1991). *Developing number sense in the middle grades: Addenda series, grades 5–8.* Reston, VA: National Council of Teachers of Mathematics.

Ross, S. H. (1989). Parts, wholes, and place value: A developmental perspective. *Arithmetic Teacher, 36*(6), 47–51.

Thompson, C. S. (1990). Place value and larger numbers. In J. N. Payne (Ed.), *Mathematics for the young child.* Reston, VA: National Council of Teachers of Mathematics.

Thompson, C. S., & Van de Walle, J. A. (1984). The power of 10. *Arithmetic Teacher, 32*(3), 6–11.

Van de Walle, J. A., & Thompson, C. S. (1981). Give bean sticks a new look. *Arithmetic Teacher, 28*(7), 6–12.

10 PENCIL-AND-PAPER COMPUTATION WITH WHOLE NUMBERS

◆ THE WORD *ALGORITHM* REFERS TO A RULE OR PROCEDURE FOR solving a problem. The whole-number computational algorithms are the specific procedures used with pencil and paper to compute exact answers to arithmetic problems such as 489 + 367 or 56 × 39. In this country most schools teach the same four algorithms, although others do exist. Therefore, this chapter concentrates on teaching what have become the standard algorithms for whole numbers.

It is important to remember that there are other algorithms for computation besides pencil-and-paper methods. Mental computation procedures are also algorithms. The procedure for using a calculator is an algorithm in the broadest sense. Within this chapter, the phrase "pencil-and-paper whole-number computational algorithm" is generally shortened to "algorithm" for convenience.

To do something "algorithmically" implies a mindless or mechanical following of rules or procedures. What has been widely accepted is the value of developing algorithmic skills but with a very clear conceptual basis; that is, to connect the algorithms to concepts rather than to teach them as rules without reasons.

ALGORITHMS: A NEW PERSPECTIVE

As we began the 1990s, it was still true that pencil-and-paper computation held a dominant place in the curriculum. There is, however, a long overdue trend toward viewing the algorithms with a different perspective. In the past, computational proficiency was a required skill in our society. We taught computation because it was necessary for everyday living. It was important for both students and adults to be able to compute sums of long columns of four- and five-digit numbers, to find products such as 378 × 2496, and to do long divisions such as 71.8 ⟌5072.63. Today, computations with numbers as large as these are virtually never required of us, thanks to the readily available calculator.

ALGORITHMS OUTSIDE THE CLASSROOM

Outside the classroom, in the real world, when a situation calls for an exact answer to a computation, three or four choices are available. Often the quickest and easiest is to use a mental computation. If, due to the numbers involved, that is not reasonable, most people reach for a calculator. Or, if there is a long series of repetitive computations, a computer spreadsheet may be appropriate. A pen or pencil is usually the last resort. Even then, if several computations are necessary and they are lengthy or tedious, most reasonable adults will look again for a calculator. Virtually all job situations will have a calculator available. Considering the limited need for computational proficiency in the real world, it is quite likely that children spend much, much more time on pencil-and-paper computation *in* school than adults do outside of school. This seems to be a sad use of valuable school time.

STUDENT-INVENTED ALGORITHMS

The tradition in algorithm instruction is to guide children to learn the procedure exactly as we prescribe it. Steps are modeled and practiced in a fairly rigid manner. This guided approach is no less evident when the instruction is completely conceptual. In fact, in this chapter you will learn how best to conduct such guided development of the algorithms.

In contrast to this very guided approach is the increasingly popular idea of letting students invent their own methods of computing. That is, before children have been taught or guided in the usual algorithms, a student-invented approach suggests that students figure out their own methods of arriving at a result. The student invention is done in a problem-solving atmosphere, usually with children working in groups with or without the aid of base ten materials. A computation, preferably in a real context, is presented. Base ten materials or special grid papers may be supplied. Preliminary class discussion might provide some hints or suggestions as to a place to begin without being prescriptive. Students might work for a full period on a single computation, presenting their particular approach to the full class. The focus of the discussion is on the invented methods, not the answer. After all, even kids know that the fastest way to the answer is with a calculator.

Values of Student-invented Procedures

Rarely will children invent the usual or traditional algorithms that they will later be taught. There is little reason to expect that children would invent the same procedures that have evolved and been perfected over many, many years. Students tend to follow their intuitions when given free rein in solving a problem. When inventing a computational method, intuition tends to suggest "left-handed" approaches. If the computation involves three-digit numbers, we (and most children) work first with the big pieces—the hundreds. Except for division, the standard algorithms are all right-handed—they begin with the ones or little pieces.

If students' invented methods are different from the standard methods, why is there any value in student-invented approaches? There are several important answers. (Before reading further, can you think of any?)

First, students who are inventing algorithms are clearly *doing mathematics*—looking for relationships and solving problems. Algorithm invention is an excellent form of problem solving. In the real world, engineers and scientists, shopkeepers and housekeepers, all have occasion to invent algorithms. They search for clever and useful methods of performing routine and repetitive tasks when a standard procedure has not already been invented. From searching for the best pattern and tools to mow the lawn, to the design of a faster computer chip, real life is full of algorithm invention, and it always will be.

Second, when students see that they can actually figure out computations without being told how, and that different students in the same class have done the task in a variety of ways, they also see clearly that mathematics does not come from a math god. It makes sense. It can be invented by ordinary people.

Third, invented algorithms will always be highly conceptual and reliant on base ten ideas. They will provide an excellent basis for developing mental computation methods. Recall that invented algorithms almost always are left-handed. So also are mental procedures.

Finally, the invented procedures are a good background for the standard pencil-and-paper procedures that are likely to come next.

In this chapter, the discussion of each algorithm will begin with a section entitled "Exploration and Invented Algorithms." One or more examples will be provided of tasks that might be used before an algorithm is developed. In a true *Standards*-oriented curriculum, a significant amount of time would be devoted to these explorations or student inventions, and much less time than has been traditional would be spent on the standard algorithms.

STANDARD PENCIL-AND-PENCIL ALGORITHMS

Perhaps we will see a day when pencil-and-paper algorithms are gone from the curriculum. That day is not yet here. Slowly, with the *Standards* as a guide, computational skill is being deemphasized, and tedious computations involving many digits have almost disappeared. Skill with two-digit divisors and three-digit multipliers is certainly a waste of precious school time. A good case can probably be made for addition and subtraction with pencil and paper, and we are likely to be teaching that for quite a while.

Estimation and Computation

As you go through this chapter, the focus is clearly on a meaningful development of the algorithms. In the classroom, as you work with children doing pencil-and-paper computation exercises, it is not only appropriate but important to consider estimation and mental computation at the same time. For example, before children even begin a computation, have them write down a rough estimate of the result. "Is it going to be closer to 500 or 1000? Would 60 or 600 seem about right for this one? Let's try to get this one in our heads first." And then discuss different methods used. These discussions will almost always have a place-value orientation. Point out how estimation and mental computations are not the same as doing them on paper. These discussions can add to better place-value understanding, to an appreciation for different methods and approaches, to increased mental process skills, and to better understanding of the algorithms.

Calculators and Computation

Even the calculator belongs in the pencil-and-paper computation program. For example, do a computation with the class using traditional approaches. Use the calculator and get the same result. "What would happen if we changed the problem from 32 × 8 to 34 × 8?" Try it with the calculator. What is the difference in the results? "Why is it 16 more? Can you explain that with the base ten pieces? What if we changed the 21 to 42?" The calculator is helping with the tedium, but the focus is still on the meaning of the algorithm, and the entire discussion is excellent background for mental computation skills.

Estimation, mental computation, and calculators all contribute to each other in a broader view of computational skill. There is a time to address each topic separately as they are addressed in this text. It is also appropriate to compare, contrast, integrate, and discuss these methods with your students.

PREREQUISITES FOR ALL ALGORITHMS

Concept of the Operation

Each algorithm is based on one meaning of the operation. That meaning should be familiar to children so that attention can focus on the algorithm. Children also must learn that the algorithm result applies to all meanings of the operation. For example, while long division is usually based on the partition concept, it can be used to solve problems involving the measurement concept.

Place-Value Concepts and Trading

Each of the algorithms depends conceptually on the place-value system of numeration to make it work. Most importantly, trading activities with base ten materials should be totally familiar both conceptually and orally to any class being taught any computational algorithm. If your students do not have this knowledge and cannot talk about trading in a meaningful way, it will be very difficult to teach the algorithms in a meaningful manner. With a firm grasp of the concepts involved in trading or grouping, including familiarity with the language and the models, the algorithms can be taught in much less time and with much less remediation required later. The following discussion of each algorithm is based on the assumption that children have these prerequisites.

Textbooks and curriculum guides have come to refer to trading or grouping as *regrouping*, although that term seems to have no conceptual value for children. Language such as "trade," "make a group," or "break a ten," or whatever words were used when those activities were done have the most promise of carrying any meaning for children. The most traditional terms of "borrowing," "carrying," and "bring down" do not at all relate to conceptual activities. It is strongly recommended that this older terminology not be used. It may be familiar to adults but not to children.

Basic Facts

In the initial stages of learning the algorithms, base ten models are used, and it is possible to do activities without fact mastery. The focus of attention during these beginning stages of instruction is on the use of place-value concepts. If necessary, children can use counting methods with the base ten models.

However, the end goal of algorithm development is a reasonably efficient use of the procedure without models. This level requires fact mastery. Any children still counting on their fingers or referring to a multiplication fact chart will have their thoughts diverted from the procedure itself. Your goal should be to have all required basic facts mastered before your students begin computation in a totally symbolic manner.

DEVELOPING AN ADDITION ALGORITHM

Some work with addition algorithms usually begins in late first grade. The greatest emphasis is placed on both the addition and subtraction algorithms in the second and third grades. Especially as early as grade 1, and also in grade 2, children's attention should be primarily on conceptual aspects of the algorithms and not on skill development. Students should be using base ten materials for most of their work with addition and subtraction algorithms.

EXPLORATION AND INVENTED ALGORITHMS

Two possibilities exist for letting students invent their own procedures for addition. One is to use base ten materials, and the other is to focus on more mental aspects.

10.1 COMBINING BIG NUMBERS

Give students a word problem similar to the following: Farmer Brown picked 36 apples from one tree and 48 more from another tree. How many did he pick altogether?

In this activity, once the suggestion of using base ten pieces is made, children will be very likely to gather the tens pieces before the ones and make a trade for a ten last (Figure 10.1). The method chosen

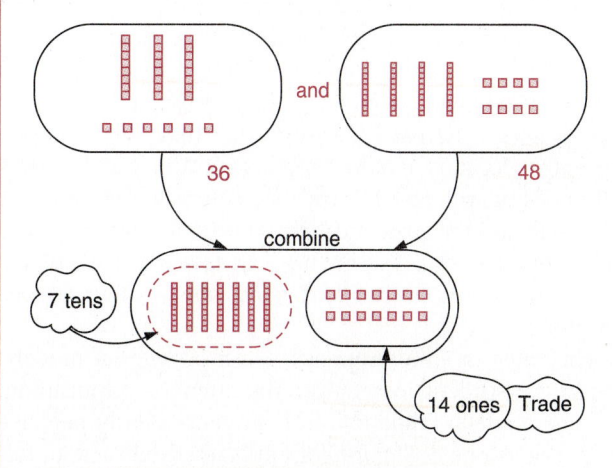

FIGURE 10.1: *Gathering the big pieces first*

is not important. Success with this activity demonstrates the principles of trading and more importantly communicates to the students that they actually know how to add two big numbers. Children with hundreds models can combine numbers such as 348 and 276 in a similar way. Again, it is both natural and acceptable to begin combining the big pieces first. All activities should involve trading from the first example.

Madell (1985) has demonstrated that children who have been given ample experiences with base ten materials will quite readily solve problems with these models, will always begin by working with the largest pieces, and, by third grade, will transfer most of these informal manipulative methods to mental computational skills. Kamii (1989; Kamii and Joseph, 1988) argues that physical models are not necessary for children to invent various methods of computation. With second-grade children she uses activities such as the following that involve only the chalkboard and oral discussions.

10.2 INVENTING MENTAL ADDITION

On the chalkboard, present children who have not yet learned the usual algorithms with problems such as the following (only one at a time):

```
  20      78      39      36
+ 30    +  5    + 29    + 46
```

Instead of using models, simply ask children to come up with an answer mentally. For each problem, accept and record on the board as many answers as the children want to propose. After writing down all responses without judgment, return to those who provided answers and have them explain. The students, not the teacher, decide if the methods make sense.

Kamii has demonstrated that children invariably begin with the tens position instead of the ones position when given tasks such as those in the previous activity. Details of their methods will vary considerably from child to child and also with the particular numbers used. For example, with 36 + 46, most children begin with "thirty and forty is seventy," but will vary in how they complete the problem. Some will add the sixes and "*take ten from the twelve to make eighty, and then that is eighty-two.*" Others will say "*six and four is ten, so that's eighty, and then there are two more to make eighty-two.*"

The value in either approach (with or without models) lies in the student interaction, the attention to meaning, and the student validation. The students clearly perceive their task as one of "figuring it out" rather than one of following rules that they may or may not understand.

DIRECTED DEVELOPMENT OF AN ADDITION ALGORITHM

As you can tell from the previous discussion, the usual algorithm for addition (and also subtraction) will be contrary to children's natural inclination to begin with the left-hand numbers. Explain to children that they are going to learn a method of adding on paper that most "big people" use. It is not the only way to add or even the best way—it is just a way that you want them to learn.

◆ Addition with Models Only

Begin with base ten materials on place-value mats. In the beginning there is no written work except for the possible recording of an answer. The idea is to make the entire algorithm make sense first. Later, children will learn to write or record what they have already developed conceptually.

Have students make two numbers on their place-value mats. Direct attention first to the ones column, and decide if a trade is necessary (or if a group of 10 can be made). If so, have the children make the trade. Repeat the process in the tens column. If you are working with hundreds, go on to the third column. Explain that from now on you want them always to begin working right to left, starting with the ones column. An example is shown in Figure 10.2. Children solve the problems with models and record only

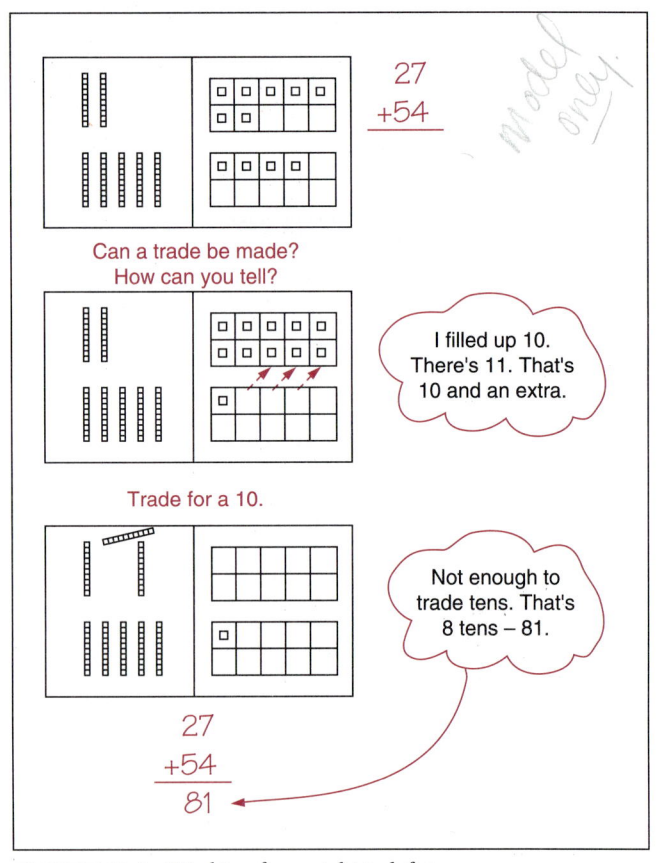

FIGURE 10.2: *Working from right to left*

10 / PENCIL-AND-PAPER COMPUTATION WITH WHOLE NUMBERS   183

the answers. At the concept level, no symbolism is used to represent trades (carrying).

Continue at this level for several days or until children are able to add, using a right-to-left process, without any intervention or teacher direction.

◆ Connections with Symbolism

Reproduce pages with simple place-value charts similar to those shown in Figure 10.3. The charts will help young children record numerals in columns. The general idea is to have children record on these pages each step of the procedure they do with the base ten models *as it is done*. The first few times you do this, guide each step carefully, as is illustrated in Figure 10.4 and the Black-line Masters. A similar approach would be used for three-digit problems.

A variation of the recording sheet that has been used successfully with children permits them to write down the sum in the ones column and thereby see and think about

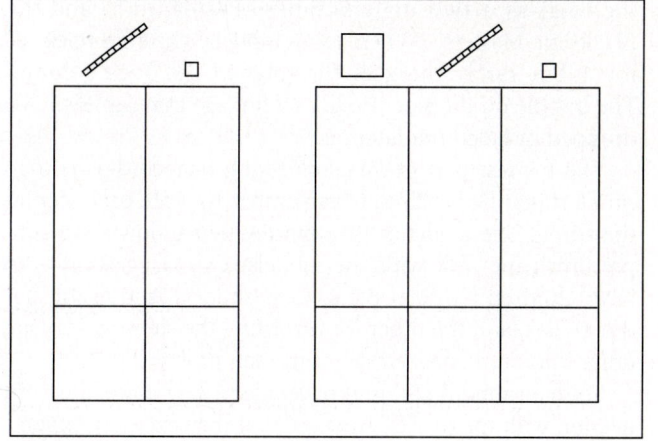

FIGURE 10.3: *Blank recording charts are helpful. These are actual size.*

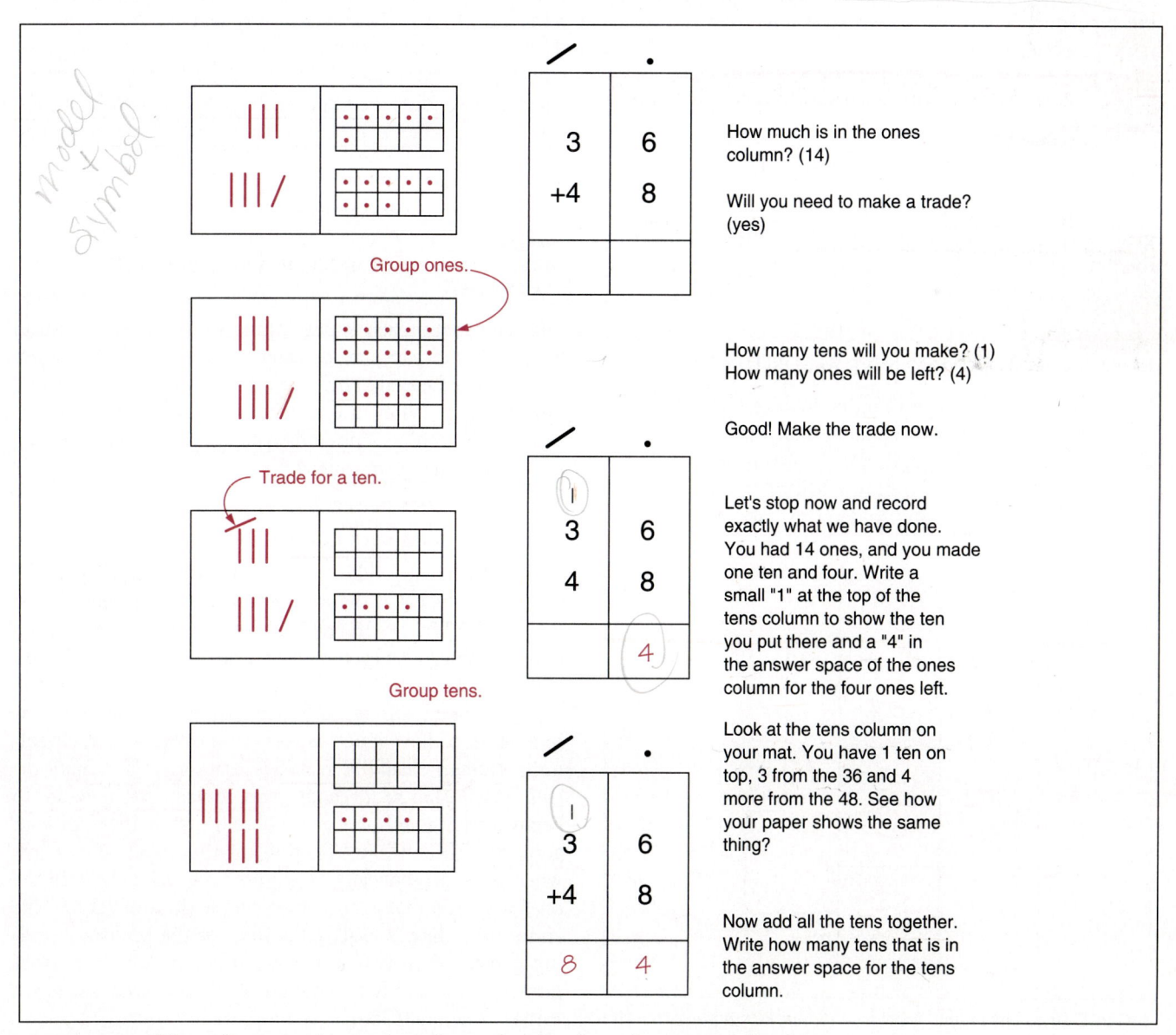

FIGURE 10.4: *Help students record on paper each step that they do on their mats as they do it.*

How much is in the ones column? (14)

Will you need to make a trade? (yes)

How many tens will you make? (1)
How many ones will be left? (4)

Good! Make the trade now.

Let's stop now and record exactly what we have done. You had 14 ones, and you made one ten and four. Write a small "1" at the top of the tens column to show the ten you put there and a "4" in the answer space of the ones column for the four ones left.

Look at the tens column on your mat. You have 1 ten on top, 3 from the 36 and 4 more from the 48. See how your paper shows the same thing?

Now add all the tens together. Write how many tens that is in the answer space for the tens column.

the total as a two-digit number (Figure 10.5, and the Black-line Masters). When using models, children trade 10 for 1 but rarely think of the total in the ones column. The bubble on the side provides a linkage that can easily be dropped or faded out later on.

The hardest part of this connecting or record-as-you-go phase is getting children to remember to write each step as they do it. The tendency is to put the pencil down and simply finish the task with the models. One suggestion is to have children work in pairs. One child is responsible for the models and the other for recording the steps as they are done. Children reverse roles with each problem.

To determine if children have adequately made the connection with the models, first see that they can do problems and record the steps. Then ask them to explain the written form. For example, ask what the little 1 at the top of the column stands for and also why they only wrote a 4 in the ones column when the total was 14. When children demonstrate a connection with concepts and also can do the procedure without any assistance, then they are ready to do problems without models.

DEVELOPING A SUBTRACTION ALGORITHM

The subtraction algorithm is also taught in the second grade with reteaching and more practice with three digits in the third grade.

EXPLORATIONS AND INVENTED ALGORITHMS

While the traditional algorithm that involves "borrowing" is generally more difficult for children than addition, children are just as capable of inventing their own methods of subtraction as they are inventing addition methods. Both approaches for encouraging student-invented methods that were suggested for addition (with and without models) are also appropriate for subtraction. Instead of presenting these ideas as activities, a few comments are all that are necessary here.

If children are given a take-away type of word problem for subtraction and asked to use their base ten pieces to solve it, they will be most likely to model only the initial amount. The change or take-away amount is contained in this initial amount. The student task is to figure out how to remove it.

Regardless of whether the children are using models or a strictly mental approach as suggested by Kamii, children will again work first with the tens or the left-hand numbers. In a problem such as 72 − 38, they will be most likely to begin by removing 3 tens from the 7 tens (or "thirty from seventy leaves forty"). This parallels what happens with addition. Their next step may surprise you. They will rarely trade a ten for ones, which in this example would make 12 ones. Rather, they may take 8 from one of the remaining 4 tens ("*That makes thirty, and two and two more is thirty-four.*") or take away the 2 ones that are there and then take 6 more from one of the tens ("*Six from forty is thirty-four.*"). These ideas are much more in keeping with efficient mental-computation procedures and should never be discouraged. You may even want to practice these ideas.

DIRECTED DEVELOPMENT OF A SUBTRACTION ALGORITHM

The general approach to developing a subtraction algorithm for pencil-and-paper is the same as for addition. Children are instructed exactly how to model the exercises first using only models. When the procedure is completely understood with models, a do-and-write approach to connecting it with a written form is used.

◆ Subtraction with Models Only

Start by having children model the top number in a subtraction problem on the top half of their place-value mats. A good method of dealing with the bottom number is to have children write each digit on a small piece of paper and place these near the bottom of their mats in the respective columns, as in Figure 10.6. The paper numbers serve as reminders of how much is to be removed from each column. *Nothing should be removed from any column until the total amount can be taken off.* Also explain that they are to begin working with the ones column first as they did with addition. The only justification for this latter rule is that when done symbolically, it is a bit easier to work with the ones first, and that is how most people do subtraction. The entire procedure is exactly the same as the backward-trading games when two dice were used. In fact, the paper numerals are simply a variation of the two dice, as shown earlier in Figure 9.15.

Use a set of your own models to follow through the

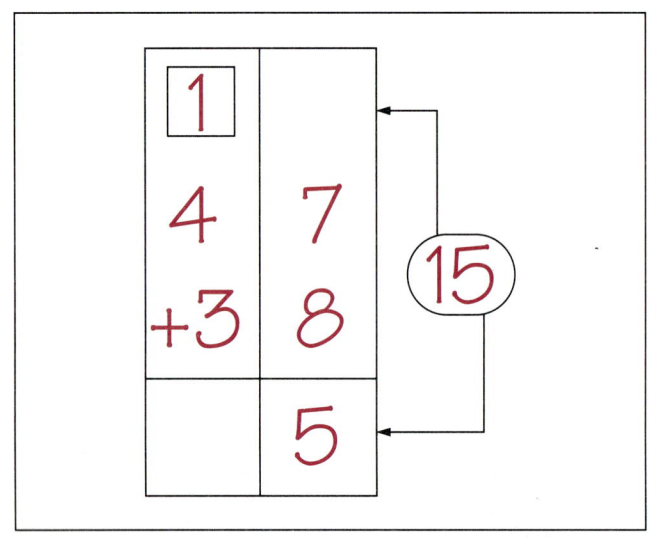

FIGURE 10.5: *The bubble provides children with a place to write the sum in the ones column.*

10 / PENCIL-AND-PAPER COMPUTATION WITH WHOLE NUMBERS   **185**

experience base for working with symbols.

A zero in the ones or tens place of the bottom number means there is nothing to take away and leaves some children wondering what they are supposed to do.

With a zero in the ones place of the top number there are no materials in that column, which presents an unusual situation when done with models. Make a special effort to include problems such as 70 − 36 or 520 − 136 while children are working with models and before they begin learning how to record.

The most difficult zero case is the one with no tens in the top number, as in 403 − 138. With models, children must make a double trade, exchanging a hundreds piece for 10 tens and then one of the tens for 10 ones. With models it is relatively clear that the hundred should not be exchanged directly for ones. Symbolically, that is a very common error: "Take 1 from the 4 and put it with the 3 to make 13." (Notice the lack of logic.)

If you have not done so already, use base ten models to do one of these these subtraction exercises with zeros.

◆ Connections with Symbolism

Children should continue using materials and recording only answers until they can comfortably do problems without assistance and demonstrate an understanding. At that point, the process of recording each step as it is done can be introduced in the same way as was suggested for addition. The same recording sheets (Figure 10.3) can be used. Figure 10.7 (p. 186) shows a sequence for a three-digit subtraction problem, indicating the recording for each step.

When an exercise has been solved and recorded, children should soon be able to explain the meanings of all of the markings at the top of the problem in terms of base ten materials. This ability to explain symbolism is a signal for moving children on to a completely symbolic level.

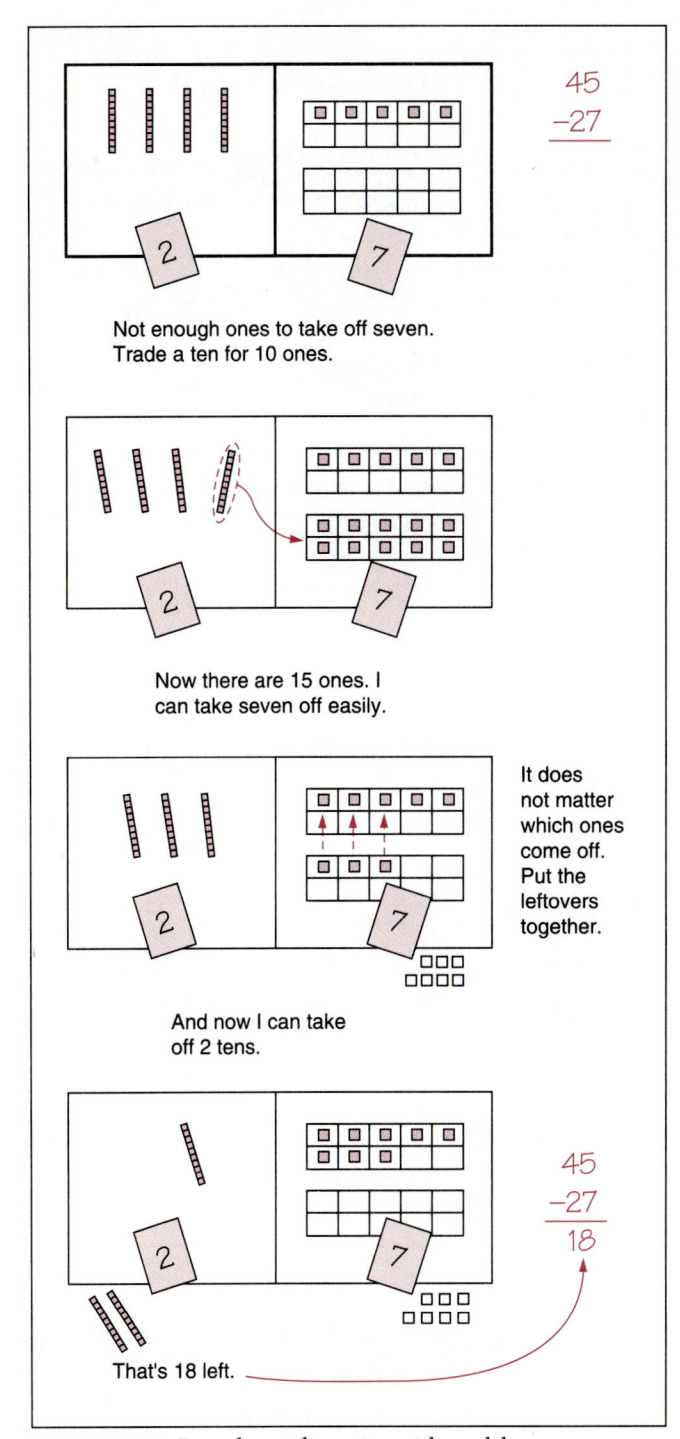

FIGURE 10.6: *Two-place subtraction with models*

steps in Figure 10.6. Notice how the empty ten-frame is filled with 10 ones when the backward trade is made. Without a frame, many children add onto the 5 ones already there and end up with only 10 ones.

◆ Difficulties with Zeros

Exercises in which zeros are involved anywhere in the problem tend to cause special difficulties, especially in symbolic exercises without models. Enough attention should be given to these cases while still using models to provide an

DEVELOPING A MULTIPLICATION ALGORITHM

The multiplication algorithm is probably the most difficult for children to understand in a conceptual manner, especially when the bottom number or multiplier is a two-digit number. The easier case of a one-digit multiplier is generally developed in the third grade, and multipliers with two digits are introduced in fourth and fifth grades. It is difficult to argue that children need to develop skills with larger multipliers.

EXPLORATIONS AND INVENTED ALGORITHMS

There are two distinctly different conceptual approaches to the development of the traditional pencil-and-paper algorithm for multiplication. One is based on an area concept of

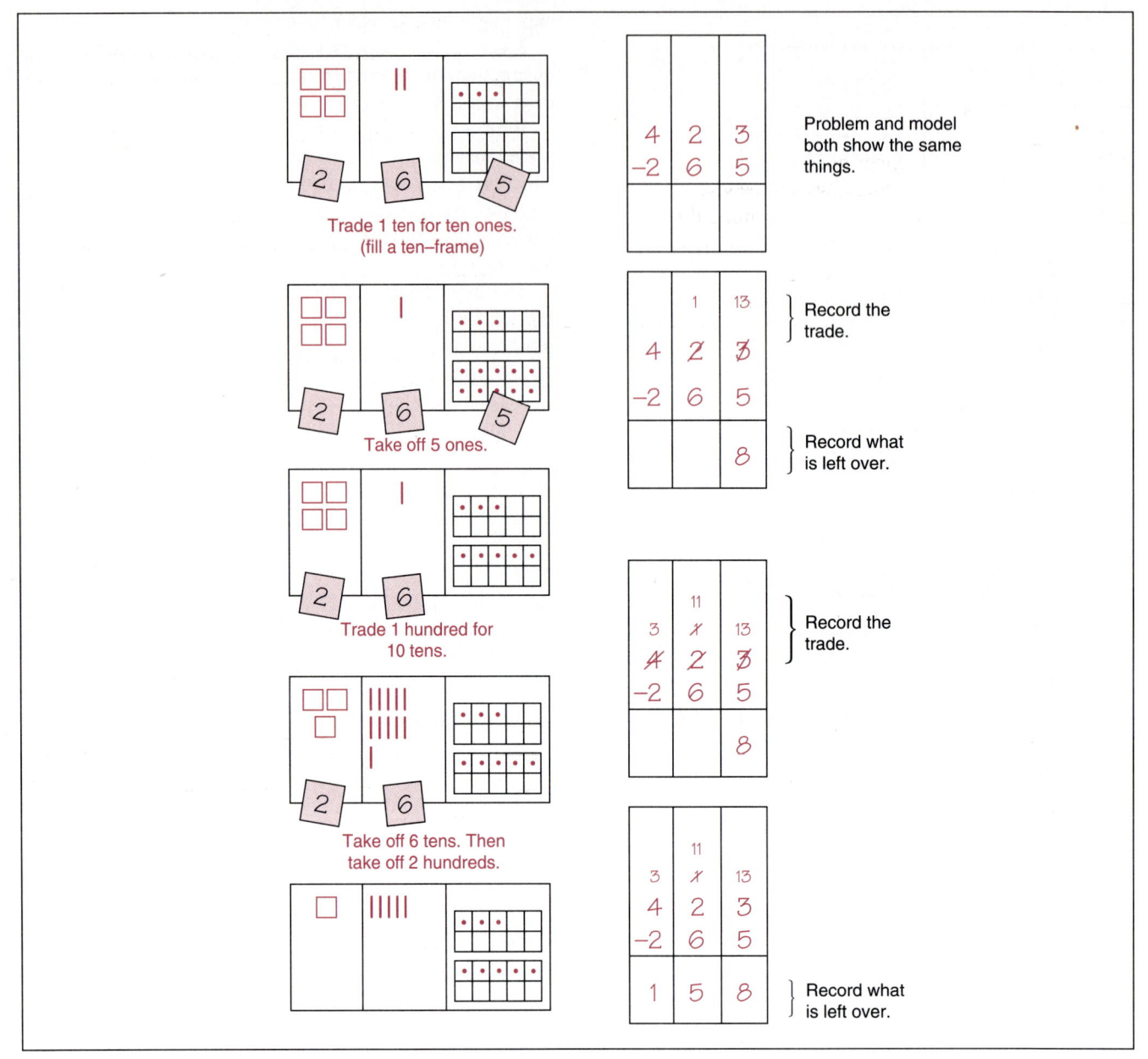

FIGURE 10.7: *Help children record each step as they do it.*

multiplication, where each of the factors is a length of a rectangle and the product is an area—the stuff inside the rectangle. The other approach is based on repeated addition where one factor indicates the number of sets and the other the size of the sets. The area model will be developed in some detail in this chapter with a much shorter discussion of the repeated addition approach. Regardless of the directed development you may teach later, informal explorations of both ideas are certainly worthwhile.

◆ Exploration of an Area Model

In these explorations, children are presented in some manner with a rectangle and its dimensions. Their task is to determine how big the rectangle is or how many *square ones* will fit inside.

10.3 HOW MANY FIT INSIDE?

To prepare for this activity, carefully draw a rectangle on a piece of poster board that measures 47 cm by 36 cm. (Dimensions are only suggestions.) Be sure that each corner is square. Cut out the rectangle to use as a template. On large sheets of paper or poster board, trace around the model so that you will have enough rectangles of the same size for all of your students in groups of three to five. Provide them with base ten materials. (Wooden or plastic base ten pieces are best, but cardboard strips and squares are adequate.) The students are to use the materials to find out how many little ones will fit in the rectangle. Groups should be encouraged to figure out clever ways to do the counting. They should draw pictures to explain how they arrived at their result.

Even children in the upper grades can benefit from the last activity. These older students should be given the artificial constraint of devising a way to determine the number of ones without using any multiplication fact beyond 9 × 9. (*It would be an excellent idea to draw a rectangle and fill it in yourself so that you can better follow this discussion.*) Most children will fill the rectangle first with as many hundreds pieces as possible. For the numbers suggested in the activity, that would be 12 hundreds, probably in a 3 × 4 array (1,200 square ones). While there are many possible ways to fill in the rectangle, one obvious approach is to put the 12 hundreds in one corner. This will leave narrow regions on two sides that can be filled with tens pieces and a final small rectangle that will hold ones. These four regions can each be counted, recorded, and added up using concepts of ones, tens, and hundreds. If you have not actually done this activity yourself, skip forward to Figure 10.17, and see how the pieces would fit into the rectangle.

An alternative to actually filling in a rectangle with base ten models is to provide students with a drawing of a rectangle filled with a *base ten grid* that has heavier lines marking 10 × 10 squares. An example is shown in Figure 10.8. Base ten grid paper can be found in Black-line Masters and Materials Construction Tips. Students can be given a sheet of the grid paper and told to outline a rectangle of specified dimensions. The task becomes one of figuring out how many little squares are inside. Colored pens or pencils are recommended for drawing on the grid paper.

Both of these approaches will be suggested in the directed development of the algorithm. Anything done with base ten grid paper can be done with physical models.

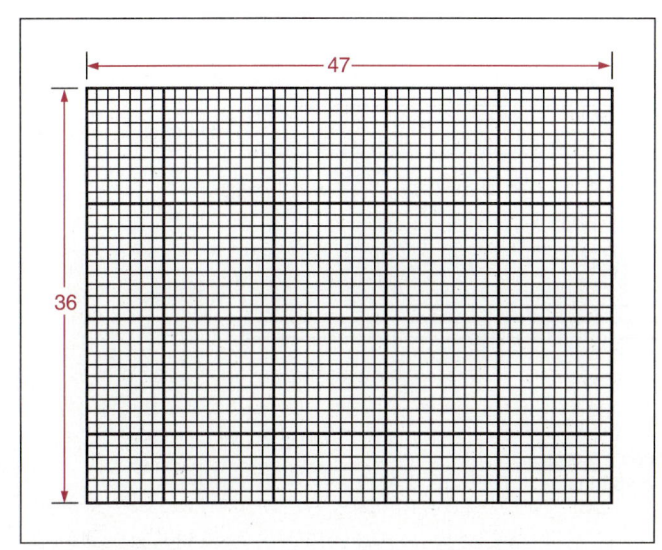

FIGURE 10.8: *How big is this rectangle? (How many squares are inside?)*

◆ Exploring Repeated Addition

When given a multiplication problem and left to invent their own methods (as was done with addition and subtraction), children are most likely to use a repeated addition approach. For example, consider the following problem:

> There were four performances of the school play last weekend, and each one was a sellout. The auditorium holds 368 people. How many people attended the play over all four performances?

If explored with the use of base ten materials, students will probably make four collections of 3 hundreds, 6 tens, and 8 ones and combine them in some way. Some may simply shove all like pieces together and count. Others will look at four groups of 3 hundreds and note (or record) that there are 12 hundreds. Similarly, they will record 24 tens and 32 ones. The exact details of how they combine the 20 tens or the 30 ones will vary from group to group. Encourage children to write down partial answers if they want.

You might want to present problems without models by simply writing the factors on the board:

$$\begin{array}{r} 368 \\ \times 4 \\ \hline \end{array}$$

Again, it is almost certain that children who have not been taught the standard pencil-and-paper algorithm will begin by working with the hundreds. (It would be best to begin with a two-digit example such as 86 × 4.)

When done without models, the details of students' invented procedures will be considerably more varied than for addition and subtraction and will also vary with the numbers involved. For example, 98 × 4 might be done by noting that 98 is almost 100 and then working backward. How would you do 53 × 7 or 115 × 8?

DIRECTED DEVELOPMENT OF A MULTIPLICATION ALGORITHM

With more emphasis being placed on understanding computation, the area model for the standard multiplication algorithm has received increased emphasis in many textbooks. While not necessarily easier to develop than a repeated addition approach, it has some distinct advantages. First, it generalizes much more easily to a two-digit multiplier. (Can you imagine making 57 piles of 86?) Perhaps more importantly, the area model can be used for products of fractions, products of decimals, and products of algebraic expressions. As the *Standards* suggests, it is important to capitalize on connections within mathematics whenever that is possible.

The development presented here has several distinct steps, each of which must be understood completely before going on to the next. These are listed here and examined in depth in the sections that follow:

1. Conceptual development of *base ten products*. Models are used to fill in actual rectangles, or base ten grids are used. A correlated oral language is a critical component.
2. Recording base ten products. The language of ones, tens, and hundreds developed in the first step is used as a guide to determine in which columns to record the numbers.
3. The standard algorithm for multiplying a two-digit by a one-digit number is developed by combining base ten products with the recording scheme.

ONE-DIGIT MULTIPLIERS

The steps just listed are explored first to develop the algorithm for one-digit multipliers and then are repeated for two-digit multipliers.

◆ Developing Base Ten Products: Ones × Ones and Ones × Tens

These exercises involve filling in rectangles with base ten pieces. If the dimensions of the rectangles are in centimeters, then centimeter base ten strips and squares will fit the dimensions the way that you want. Wooden or plastic models all are in metric dimensions and are much easier to work with for these tasks.

It is probably better if you have students work together in small groups of two or three and give them predrawn rectangles. You will need to use large sheets of paper or tagboard. One way to prepare these rectangles for your class is to cut one of each size you want from poster board and quickly trace around with a felt marker.

Provide each group of children two rectangles with these dimensions:

Rectangle A: 3 ones by 5 ones (Note: 3 ones is 3 cm in length.)

Rectangle B: 3 ones by 5 tens

Next have them fill in the rectangles with base ten pieces, using the largest pieces that will fit as shown in Figure 10.9. "How many *pieces* are in the small rectangle?" (15) "What *kind* of pieces are they?" (ones) "How many *pieces* are in the long rectangle?" (15). "And what are they?" (tens) Have children "read" the dimensions of the rectangles and the area (amount filled inside) using base ten language, not centimeters: "3 ones by (times) 5 ones is 15 *ones*. 3 ones by 5 tens is 15 *tens*."

The goal is to generalize two base ten products: *ones × ones are ones, and ones × tens are tens*. Draw other rectangles, and label in terms of tens and ones. Students should measure these with centimeter rulers. Have students read the length, width, and areas using the base ten language. Factors used are always less than 10. For example, "6 *ones* times 7 *tens* is 42 *tens*." The relationship between rectangle dimensions and the type of base-ten pieces that can fill them is the essential concept behind this approach to the algorithm. Students should know why the product is tens or ones (because the rectangle is filled in with that type of piece). The language used in reading the rectangle will provide an important link with symbolism.

Rectangles could also be drawn on base ten grid paper instead of filling them in with strips and squares. Provide base ten grid paper, and give the outside dimensions in terms of ones and tens. Students draw the corresponding rectangles on the grid. Each rectangle is filled with *all ones* or with *all tens*. Always read the rectangle with the class, as in Figure 10.10. Another good idea is to have children shade in one or two pieces in each rectangle. Talk about how many base ten pieces would be inside if you had a full-size rectangle filled with physical models.

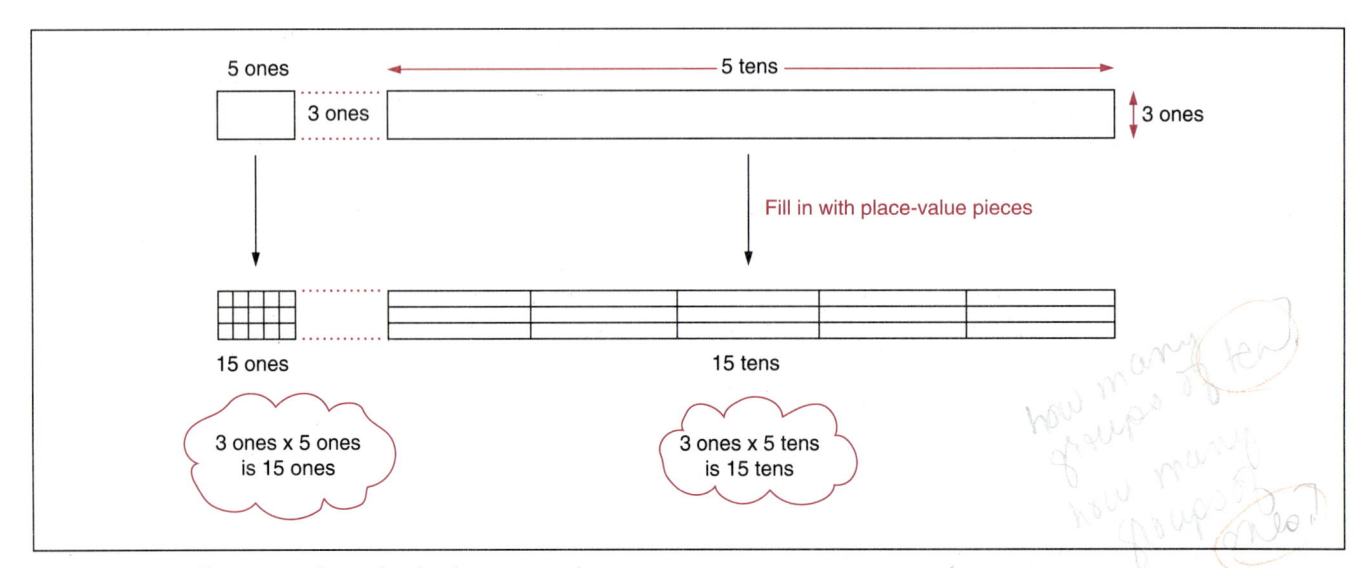

FIGURE 10.9: *Fill in rectangles to develop base ten products.*

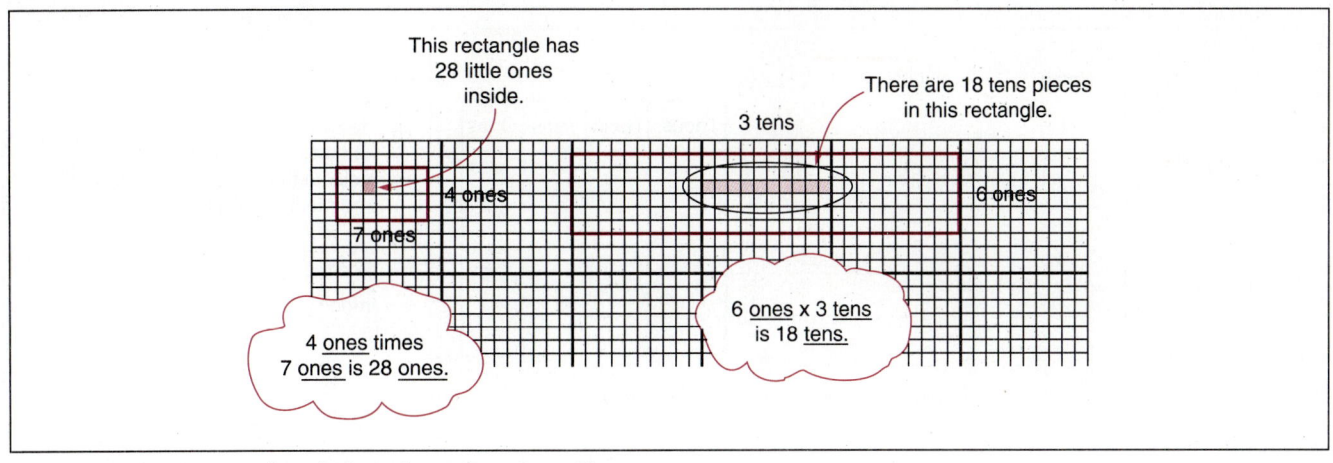

10 / PENCIL-AND-PAPER COMPUTATION WITH WHOLE NUMBERS

FIGURE 10.10: *Draw and read rectangles on base ten grid paper.*

The language of base ten products can be drilled orally. The following series relates basic facts with base ten product language.

Eight times *four* is...................................32
Eight ones times *four* tens is...............32 tens
Eight ones times *four* ones is............32 ones

Recording Base Ten Products

When base ten product language is clearly understood, the next step is to learn to record the products in appropriate place-value columns. Worksheets can be prepared like those in Figure 10.11. The base ten factors are written to the left of the place-value columns, and the products are written in the chart.

FIGURE 10.11: *A place-value record sheet for recording base ten products*

The natural place to write the product of 4 ones times 8 tens is in the tens column. Suggest that children write the two-digit products in columns as in Figure 10.12 (p. 190). "What would you have to do if you had thirty-two tens all in one column?" (Make a trade.) When the class determines that 32 tens will make 3 hundreds and 2 tens, have them cross out the 32 and show the 3 and 2 in the respective columns as shown. Continue with other factors using ones times ones and also ones times tens. Move quickly to recording a product with only one digit in a column and omitting the step of crossing out to show the trade. Students will discover that the last digit of the product will end in the named column.

Empty columns to the right are not filled with zeros. This makes the language agree with what is written. If 32 tens were written with a zero in the ones column, it would be read as "three hundred twenty" and not as "32 tens."

An understanding of the recording of base ten products is the principal conceptual basis for the procedural algorithm. If you yourself do not have a good grasp of the links between filling in the rectangles, how that develops a base ten language, and how these products are recorded in a chart, it would be good to stop here and work through these last two sections again with your own models. Work with a friend, and talk about all of these ideas in your own words. This will also demonstrate to you the value of learning an idea conceptually rather than just learning the rules.

Two-Part Products

Give students a plain rectangle with a long side of 47 cm and a short side of 6 cm. How could this be filled in with base ten pieces?

As shown in Figure 10.13 (p. 190), this rectangle can be "sliced" or separated into two parts so that one part will be 6 ones by 7 ones, or 42 ones, and the other will be 6 ones by 4 tens, or 24 tens. Have children read each part of the rectangle as before, using base ten language. Finally, since they already know how to record these products, show how the recording can be done in two lines beneath the problem as shown. The area of the entire rectangle is easily found by adding the two parts. Since the whole rectangle represents the product of 47 and 6, each section is referred to as a *partial product*.

Now go to the written record, and repeat the problem over orally, pointing to each digit. "Six ones (point to 6)

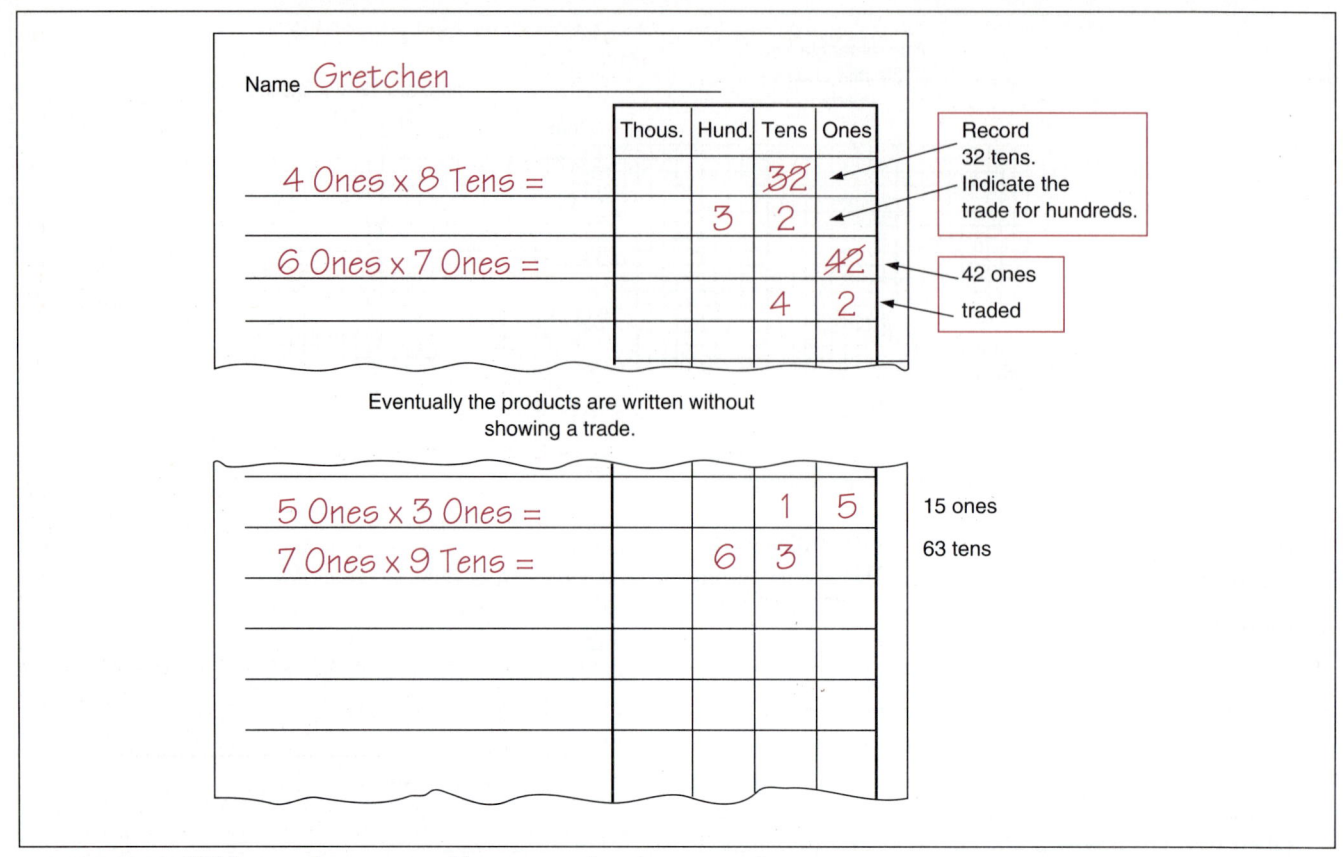

190   10 / PENCIL-AND-PAPER COMPUTATION WITH WHOLE NUMBERS

Name **Gretchen**

	Thous.	Hund.	Tens	Ones
4 Ones × 8 Tens =			3̷2̷	
		3	2	
6 Ones × 7 Ones =				4̷2̷
			4	2

Record 32 tens. Indicate the trade for hundreds.

42 ones traded

Eventually the products are written without showing a trade.

| 5 Ones × 3 Ones = | | | 1 | 5 | 15 ones |
| 7 Ones × 9 Tens = | | 6 | 3 | | 63 tens |

FIGURE 10.12: *Children can learn to record base ten products in correct columns.*

times seven ones (point to 7) is forty-two ones. Six ones (point) times four tens (point) is twenty-four tens." Notice that the base ten value of the factors is determined by which column the digits are in. Students have also learned how to record each partial product with only one digit per column, so they should already know where each product is to be recorded.

It cannot be stressed too strongly that language is the

47
4 tens and 7 ones

6 ones

Fill with base ten pieces.

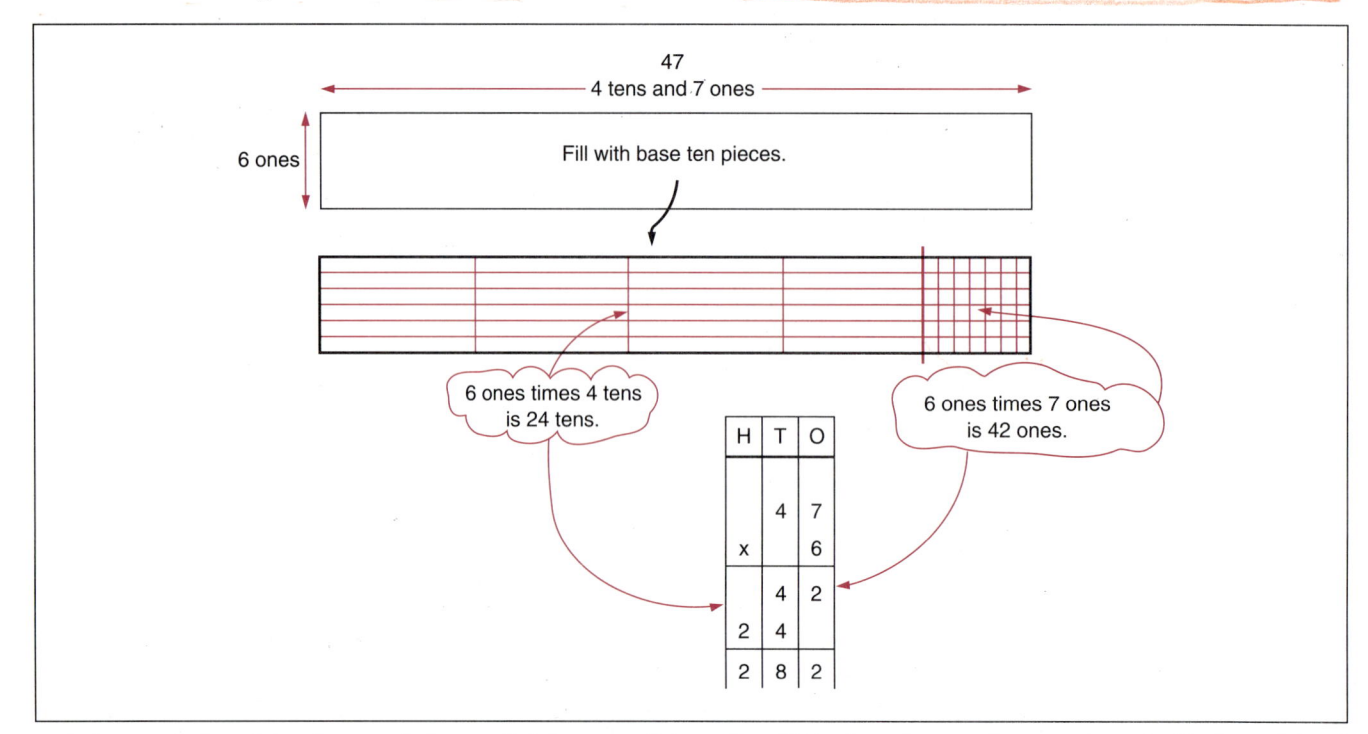

6 ones times 4 tens is 24 tens.

6 ones times 7 ones is 42 ones.

H	T	O
	4	7
×		6
	4	2
2	4	
2	8	2

FIGURE 10.13: *A model and a record of a two-digit by a one-digit product. Notice how the same language is used with the model and the symbolism.*

key key connection between the rectangle model and the written form. Care should be taken to develop the language at each step. Notice that when you read the rectangle you use *exactly the same language* as when you read the written problem.

Students should soon be able to start with a problem such as 39 × 5, draw the appropriate rectangle on base ten grid paper, slice it into two parts, write the two partial products, and explain the connection to the drawing. Stop now, and try this yourself on a sheet of grid paper.

While still connecting the written form with the rectangle model, students write each partial product separately. It can easily be argued that this is good enough for today's needs. For the few times outside the classroom that we actually use this algorithm, we could certainly get by without worrying about a shorter form of the algorithm (with "carrying"). However, if you feel compelled to teach this to children, make the connections shown in Figure 10.14. Many of the errors that children commonly make with multiplication stem from this regrouping procedure. Is it really necessary?

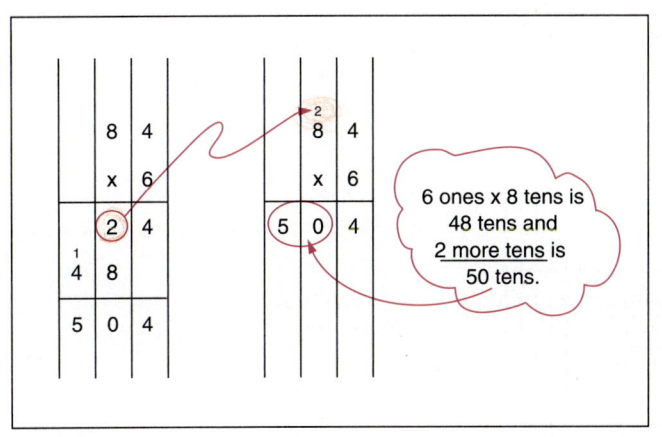

FIGURE 10.14: *Two separate partial products can be added orally and the total written on one line in the standard form.*

With this final symbolic change, the two-digit by one-digit algorithm is complete with no special cases to be considered. Problems with three digits in the top number (e.g., 639 × 7) cannot be modeled with this approach, but the extension is relatively easy. *Ones* times *hundreds* are *hundreds*. The product of 7 *ones* times 6 *hundreds* is 42 *hundreds*. The 2 is written in the hundreds column, the 4 in the thousands column. (Why?) Try the product 639 × 7 yourself. Say each of the three partial products in base ten language, and record each separately.

TWO-DIGIT MULTIPLIERS

If the algorithm for one-digit multipliers is completely understood, the extension to a two-digit multiplier is quite easy. Each of the next three sections is an exact parallel to the preceding development.

Developing Base Ten Products: Tens × Ones and Tens × Tens

On large sheets of paper, provide students with a rectangle with dimensions 40 cm by 60 cm. Have them fill it in with base ten pieces. The obvious choice of pieces is hundreds, as in Figure 10.15. Read the rectangle as before: "4 *tens* by 6 *tens* is 24 *hundreds*." Generalize this new base ten product: *tens times tens are hundreds*. The other new base ten product is the turnaround or commutative partner of the old one: *tens times ones are tens*. Practice these ideas orally with specific numbers as before, and be sure that students can explain them conceptually. For example, "6 *tens* times 8 *ones* is 48 *tens*. 6 *tens* times 4 *tens* is 24 *hundreds*." For tens × tens, help students visualize a large rectangle that clearly will hold hundreds. For tens × ones and ones × tens, the rectangles are always long and skinny. These skinny rectangles hold only tens.

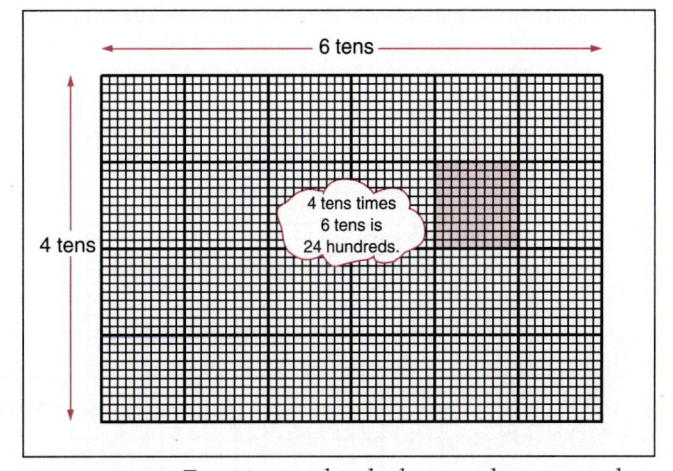

FIGURE 10.15: *Tens × tens = hundreds, a new base ten product*

Recording Base Ten Products

Go through exactly the same process as before in teaching children how to record the new base ten products. That is, permit or even encourage the recording of two digits in one column, and then make a trade symbolically, as in Figure 10.16 (p. 192). It is important to write the verbal form of the two factors to the left of the place-value columns as indicated. Do not move on until children can correctly record any base ten product with only one digit in each column. The step where the whole two-digit product is written in one column is only transitional and should be abandoned before continuing.

Four-Part Products

Give students a rectangle (or have them draw one on base ten paper) with one side of 47 and the other side 36. How could this be filled in with base ten pieces?

While many arrangements are possible, nearly everyone begins with hundreds placed in a corner, with a result similar to Figure 10.17 (p. 192).

When the rectangle is filled in, notice that there are four separate sections; one with hundreds, two with tens pieces,

	Thou.	Hund.	Tens	Ones
4 Tens × 6 Tens =		24		
	2	4		
8 Tens × 5 Ones =			40	
			4	0

Show trades as before.

	Thou.	Hund.	Tens	Ones
3 Tens × 8 Ones =			2	4
6 Tens × 9 Tens =		5	4	
5 Tens × 7 Tens =				
6 Ones × 4 Ones =				

Learn to write in correct column without trades.

Practice all four base ten products.

FIGURE 10.16: *Children learn to record the new base ten products.*

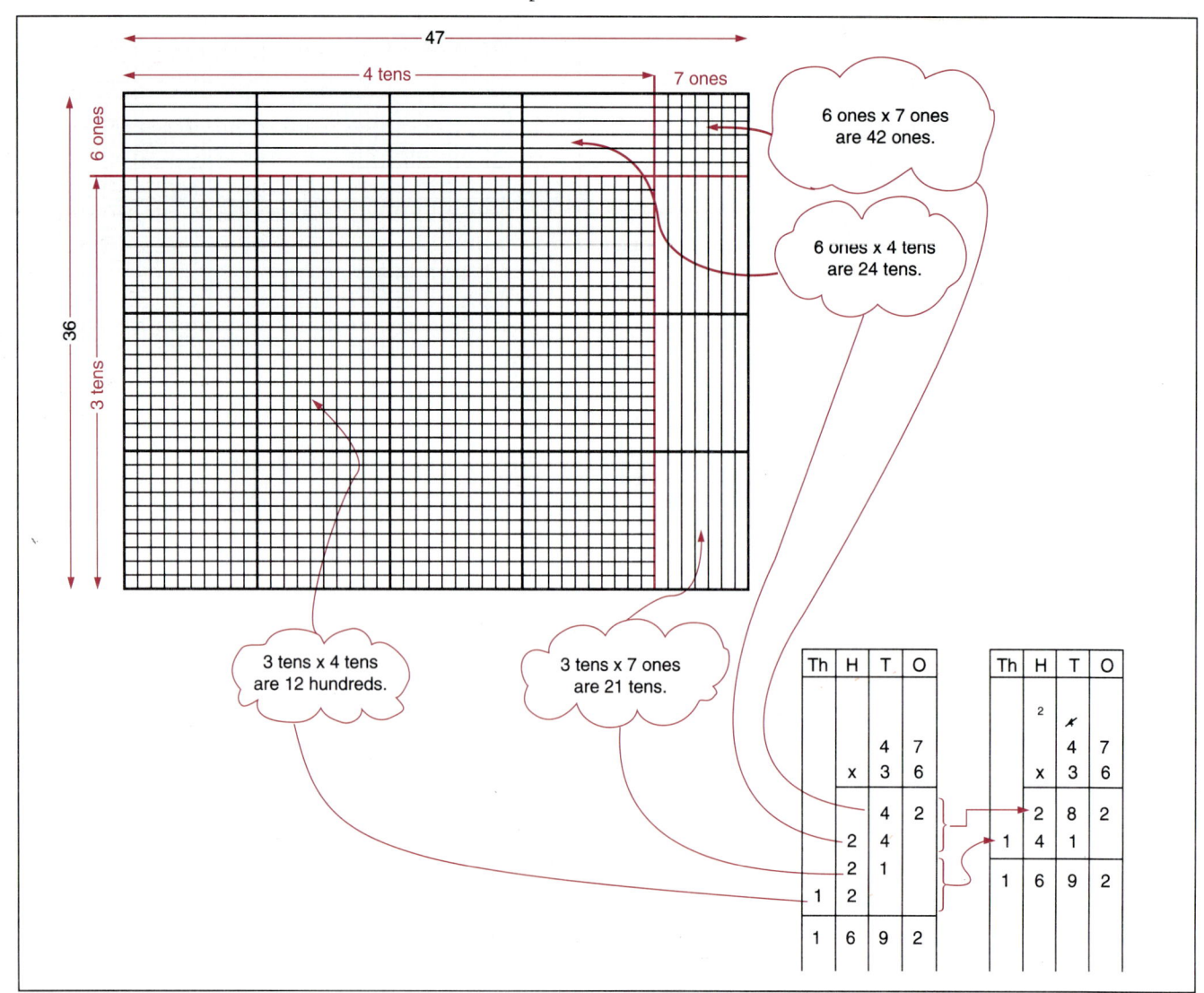

FIGURE 10.17: *A 47 × 36 rectangle filled with base ten pieces. Language connects the four sections of the rectangle to the four partial products.*

192

and one with singles or ones. Have students read each of these sections. As was done for the one-digit multiplier, record each section or partial product in place-value columns under the problem. The order in which the four sections or partial products are recorded is an arbitrary convention of the algorithm. The first two sections are the same as if the tens digit of the multiplier were not there. It is a good idea to record separately all four partial products, as shown in Figure 10.17.

Repeat the language that goes with the problem (reading the rectangles), but this time point to each digit in the problem. As before, the column each digit is in gives its base ten name, and the exact same words can be used with the symbolism, as in the rectangle. Students should be able to explain the connection of each partial product and each factor in the problem with the rectangle model.

You might wish to stop at this point, draw a rectangle that is 56 cm × 34 cm and fill it in with base ten pieces. Then read the four sections as you point to each. Record all four partial products in order, *using the same oral language as you do so.* Try another example drawn on a base ten grid.

A REPEATED ADDITION DEVELOPMENT

The area model for developing the multiplication algorithm, as just discussed, is a nontrivial development requiring time and patience. While it has become more prominent in standard textbooks, adequate pictures of rectangles and explanations require a lot of page space. Quality instruction requires considerable effort beyond what a text can offer.

Why not use a repeated addition model? A repeated addition approach would make a lot of sense if the only goal was to develop a one-digit algorithm. For products such as 37 × 5, it is easy to model five sets of 3 tens and 7 ones. The two products (15 tens and 35 ones) may be computed in either order. The thought process involved transfers nicely to mental computation. (Try 37 × 5 mentally by starting with "five times thirty.")

The difficulty with repeated addition begins with the second digit in the multiplier. It is not easy to model or think about 37 sets of 24. Orally (for this example) when you get to the 2 times 3 part of the computation, it is not correct to say "2 tens times 3 tens" because it is really "20 sets of 3 tens." This is a bit more difficult to handle mentally than "2 tens times 3 tens." Finally, the repeated addition model does not easily carry over to decimal or algebraic computation.

These are simply the arguments on either side of the issue of which approach should be used to develop the algorithm. Figure 10.18 illustrates a typical approach to repeated addition for two-digit multipliers as might be found in a textbook.

DEVELOPING A DIVISION ALGORITHM

Division with one digit is traditionally begun in the third grade, with two-digit divisors introduced in the fourth grade. A two-digit divisor substantially increases the difficulty of the task. Fortunately, there is a movement to totally remove three- and four-digit divisors from the curriculum. For example, the state of Michigan objectives restrict whole-number divisors to two digits. Even these divisors are to be less than 30 except for multiples of 10, which may be as high as 90 (Michigan State Board of Education, 1988). This restriction allows students to focus on the rationale for the algorithm and avoid much of the stress caused by working with larger divisors. It also allows teachers to make better connections with estimation and mental computation methods.

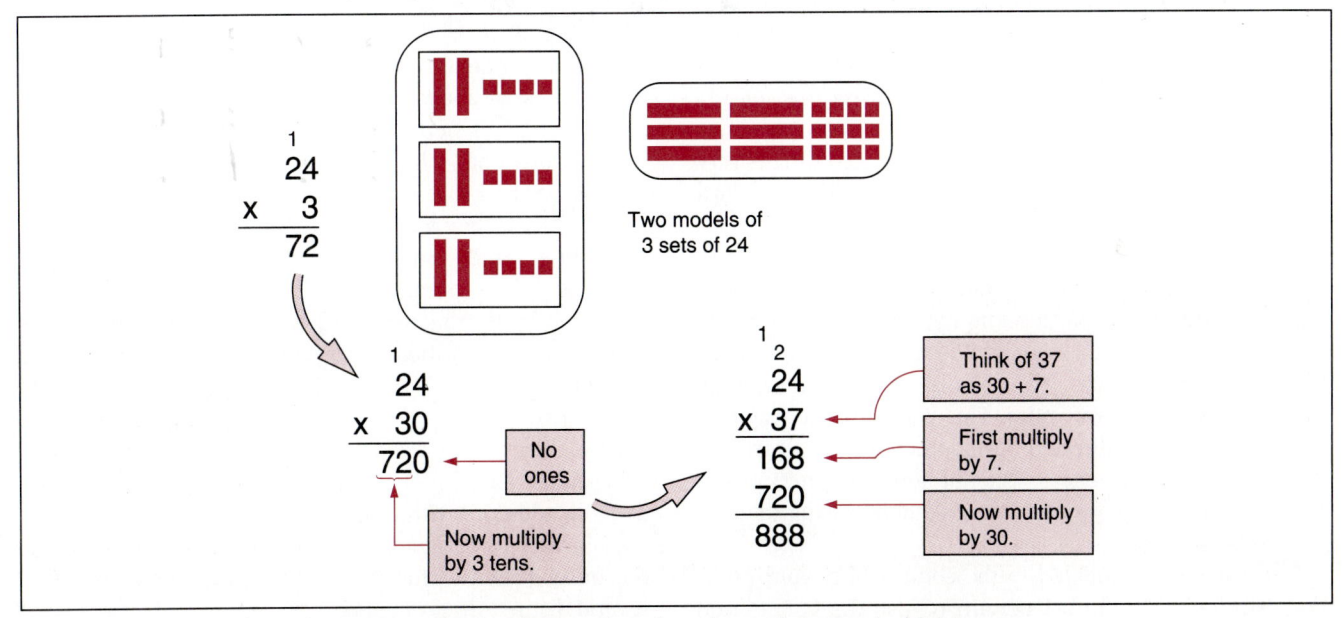

FIGURE 10.18: *A traditional sequence to develop the two-digit algorithm with repeated addition*

EXPLORATIONS AND INVENTED ALGORITHMS

Recall that there are two distinct concepts of division, one involving partitioning or fair shares and the other involving measurement or repeated subtraction. If children are presented division computations without contexts, some children will use a measurement approach, and some will use partition. This provides an excellent opportunity for discussion.

10.4 AN OPEN EXPLORATION OF DIVISION

Have students determine how to solve this exercise in any way they wish:

4) 583

How you present this task and what suggestions you provide will have a significant impact on what methods students use to solve the problem.

One approach is to simply write the division exercise on the board and have students work in pairs to get a solution. As is always the case with an unstructured activity, it is not possible to predict all of the interesting things that students are likely to do. Some may devise a completely mental method. For example: "*One hundred fours is 400, another 25 fours is 500. That leaves 83. Twenty fours is 80. That makes 125 and 20 or 145, and there are three left.*" Other children may do a similar form of thinking but write down the intermediate amounts or show appropriate multiplications. This example is based on measurement: How many fours are in 583?

It is also possible to think about the problem as a partition problem. In that case the reasoning might be like this: "*With 500, that's 125 in each of four sets. The 80 splits up with 20 in each set, so that makes 125 and 20 in each set, or 145 altogether. The three is just left over because you can't separate three into four sets.*" Again, the use of some recording is very likely but in neither case will children use the standard algorithm unless they have been taught.

If you suggest that students use base ten materials to solve a division problem, it is much more likely that they will use a sharing or partitioning approach. For the problem we have been discussing, children set out 5 hundreds, 8 tens, and 3 ones and try to figure out how to separate these materials into four piles. This is exactly the way that is suggested for the directed development of the algorithm. It might be a good idea to get out some base ten pieces right now and see how you would go about using them to solve this problem.

If the exercise is offered in the context of a word problem, students are almost certain to use the conceptual approach suggested by the problem. The use of models remains an option. Consider these two problems, both involving the same division we have been discussing:

> Mr. Martin, who runs a picture-framing shop, has a piece of framing material that is 583 cm long. If he cuts it into four pieces so that he can make a square frame, what is the largest square he can make?
>
> Mrs. Martin buys the frame stock for the framing store. The most popular type of frame material costs $4 per yard. Her most recent bill from the supplier indicates that she spent $583 on this material. How many yards of frame stock did she buy?

Regardless of how you explore division informally with your class, it is almost certain that students will approach the task in a left-hand manner, working with the hundreds before the tens before the ones. That is the way both mental and paper-and-pencil algorithms are done. Therefore, informal explorations of division have a lot of payoff.

DIRECTED DEVELOPMENT OF A DIVISION ALGORITHM

The division algorithm that is most often seen in textbooks is based on the partition or fair-share concept of division. As with addition and subtraction algorithms, the division algorithm can be completely developed conceptually with models before any connection to symbolism is undertaken.

Division with Models Only

Provide students with base ten pieces and six or seven pieces of paper about 6 inches square. (Construction paper cut in fourths is perfect.) Begin with a problem similar to the previous one, 4) 583. Discuss the meaning of the task in base ten language. You have 5 hundreds pieces, 8 tens, and 3 ones to be shared evenly among four sets or piles. Place an emphasis on language that focuses on *8 things, 5 things,* and *3 things* rather that 583 single items. Have students model the quantity 583 with their base ten pieces, and set out four of their paper squares to represent the four sets. Explain that it is a good idea to always begin with the largest pieces, since if they cannot be completely shared evenly, they can be exchanged for smaller pieces. Figure 10.19 illustrates the complete solution to this task as students would do it with models.

It is very important to get students to talk this process out in their own words. Notice that with each place value, several distinct things happen:

4)583

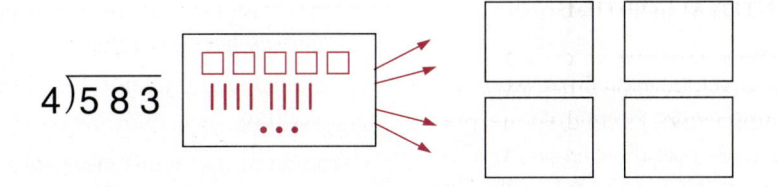

The task is to share these 5 hundreds, 8 tens, and 3 ones among these four sets so that each set gets the same amount.

I'll begin with the hundreds pieces. There are enough so that each set can get 1 hundred. That leaves 1 hundred left that cannot be shared.

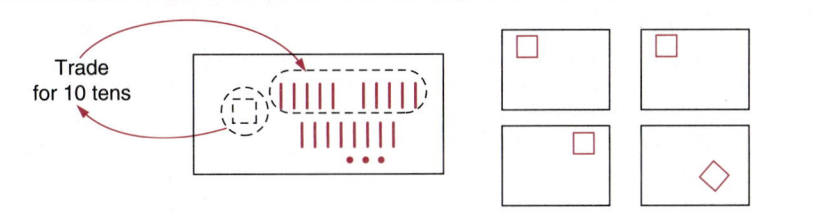

Trade for 10 tens

I can trade the hundred for 10 tens. That gives me a total of 18 tens. With 18 I can put 4 tens in each of the four sets and have 2 tens left. Two is not enough to go around to all four sets.

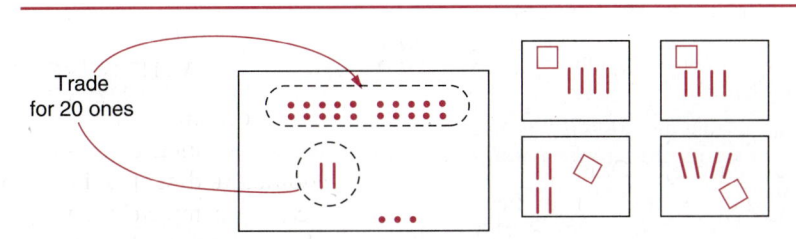

Trade for 20 ones

I can trade the 2 tens for 10 ones each or a total of 20 ones. With the 3 ones I already had, that gives me 23 ones. I can put 5 ones in each of the four sets. That leaves me with only 3 ones left over as a remainder.

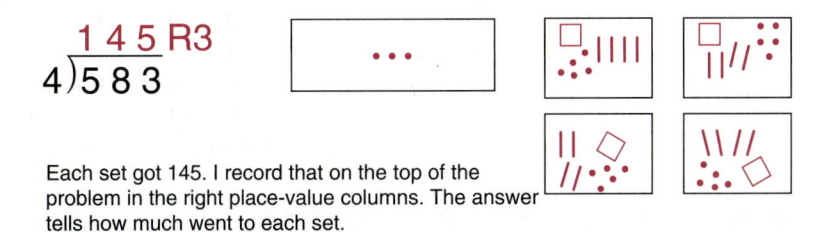

145 R3
4)583

Each set got 145. I record that on the top of the problem in the right place-value columns. The answer tells how much went to each set.

If I added up how much they got altogether, that would be 4 x 145 or 580. the remainder of 3 makes the 583 I started with.

FIGURE 10.19: *Long division at the concept level*

- The number of pieces available is considered, and a decision is made to see if there are enough to be distributed.
- The pieces that each set can get are actually passed out to the sets—that is, placed on the paper mats.
- Any remaining pieces are traded for the next smaller size. The pieces received in trade are added to any that were already there. At this point the process begins over.

The sharing of individual pieces within each place value is the total conceptual basis for the long-division algorithm. Students should do exercises using only models until they can talk through that process (using their own words) in such a way that the sharing of each type of base ten piece and trading to the next size is quite clear.

Notice that in this context the phrase "four *goes into* five" has absolutely no meaning. In fact five hundreds are being *put into* four sets. Teachers especially have difficulty abandoning the "goes into" terminology, because it has become such a strong tradition while doing long division.

In making up the exercises that students should do with models, you will have to keep different problem types and nuances in mind. Figure 10.20 illustrates some of these. If excessive trades are avoided, most exercises with dividends through three digits and divisors of one digit can be easily modeled.

You may wish to stop at this point and try some of these division exercises with your own models. (Use the ones suggested in Figure 10.20.) Pay special attention to the language you are using. Remember you are simply passing pieces out fairly as you would M&Ms to friends. Avoid the "goes into" terminology. Also stress the language that explains the trading.

◆ Connections with Symbolism

The general approach of having students record on paper each manipulative step *as they do it* is the way to connect symbolism to division.

Long division is simply a matter of passing out pieces within each place-value column and then trading any that may be left over for the next smaller size. When recording, however, two in-between steps must also be written down. All four steps are given here. Notice that steps B and C have no corresponding action with models.

- A. *Do*: Share available pieces in the column among the sets.
  Record: The number given to *each* set.
- B. *Record*: The number of pieces given out *in all*. Find this by multiplying the number given to *each* times the number of sets.
- C. *Record*: The number of pieces remaining. Find this by subtracting the total given out from the amount you began with.
- D. *Do*: Trade (if necessary) for equivalent pieces in the next column and combine those with any that may already have been there.
  Record: The total number of pieces now in the next column.

Students must write more than what they do or think about when working manipulatively. In the beginning, students frequently omit steps B and C. The value or necessity of this extra writing may seem obscure to some students while they are still using models.

◆ Two Alternatives for Recording

Along with the traditional method of recording this algorithm, an alternative method is suggested here that helps keep the conceptual meaning more clearly tied to the sym-

---

Two-digit dividends
Not very challenging or interesting

4)67

Be careful of excessive trades.

5)865
3 hundreds for 30 tens (a lot but OK)
1 ten for 10 ones

5)745
2 hundreds for 20 tens
4 tens for 40 ones (excessively tedious)

No hundreds to distribute

4)372

No trades in one or more places

4)852   3)426   3)693

No tens to distribute (zero in an answer)

4)832

Zeros in dividend

3)704

FIGURE 10.20: *Consider a variety of special cases to be worked out using models.*

10 / PENCIL-AND-PAPER COMPUTATION WITH WHOLE NUMBERS   197

Traditional "bring-down" method		Alternative explicit-trade method

(a)
Traditional:
```
    1
5)7 6 3
  5
  ‾
  2
```
A. 1 hundred given to each set. Record in answer space.
B. 5 sets of 1 hundred each is 5 × 1. Record under the 7.
C. 7 − 5 = 2 tells how many hundreds are left.

Alternative:
```
    1
5)7 6 3
  5
  ‾
  2
```

(b)
Traditional:
```
    1
5)7 6 3
  5
  ‾
  2 6
```
D. Trade 2 hundreds for 20 tens plus 6 tens already there is 26 tens. Bring down the 6 to show 26 tens.

OR

Cross out 2 and the 6. Write 26 in tens column.

Alternative:
```
    1
5)7 6̸ 3
  5 26
  ‾
  2̸
```

(c)
Traditional:
```
    1 5
5)7 6 3
  5
  ‾
  2 6
  2 5
  ‾
    1
```
A. Pass out 5 tens to each set. Record in the answer space.
B. 5 sets of 5 each is 5 × 5 = 25 tens. Record the 25.
(Note 2 different recordings.)
C. 26 − 25 = 1 tells how many tens are left.

Alternative:
```
    1 5
5)7 6̸ 3
  5 26
  2̸ 25
       1
```

(d)
Traditional:
```
    1 5 2R3
5)7 6 3
  5
  ‾
  2 6
  2 5
  ‾
    1 3
    1 0
    ‾
      3
```
D. Trade 1 ten for 10 ones plus 3 ones already there is 13 ones. Bring down the 3 to show 13 ones.

OR

Cross out the 1 and 3 and write 13 in the ones column.

A. Pass out 2 ones to each set. Record in the answer space.
B. 5 sets of 2 ones each is 10 ones. Record 10.
C. Subtract 10 from 13. There are 3 ones left.

Alternative:
```
    1 5 2R3
5)7 6̸ 3̸
  5 26 13
  2̸ 25 10
       1̸  3
```

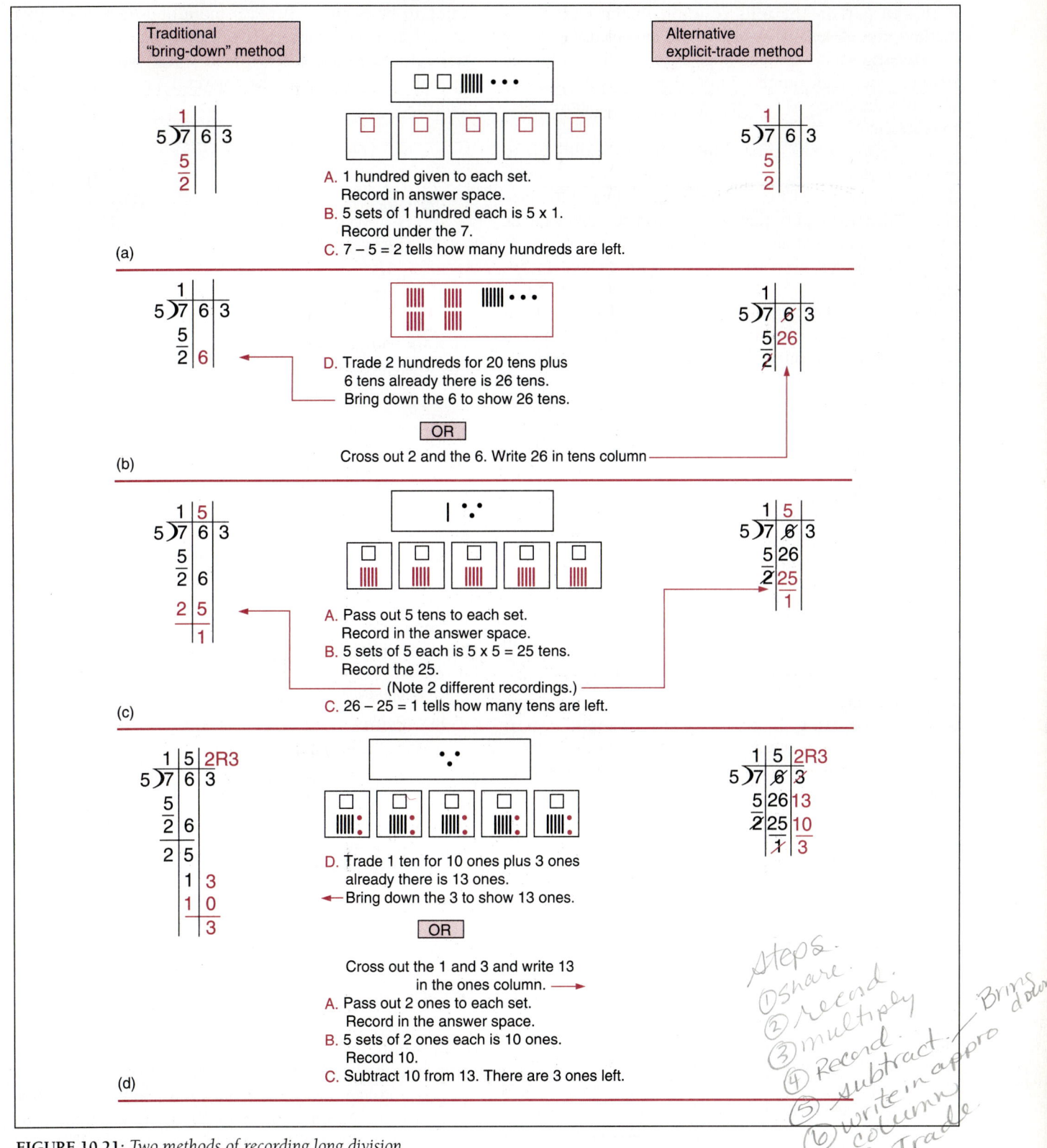

FIGURE 10.21: *Two methods of recording long division*

bolism. Steps A, B, and C are recorded as usual in both methods, as shown in Figure 10.21(a). In the traditional algorithm, the trade is implied when you "bring down" the next digit. Notice that the total is technically in two columns. In the example, the 2 that stood for the remaining hundreds somehow becomes 20 tens when the 6 is brought down [Figure 10.21(b) on the left]. In the alternative explicit-trade version, the trade for tens is made explicit by by crossing out the 2, and the total of 26 tens is written completely in the tens column. Next, multiplication indicates that 25 tens were passed out in all. In the traditional scheme the 25 must be recorded part in the hundreds column and part in the tens column. In the alternative method, the 25 tens is written in the tens column, clearly representing what was done [Figure 10.21(c)]. The idea of writing a two-digit number in one column is not entirely new. In subtracting 637 − 281, we write a 13 above the 3 to represent 13 tens.

The "bring down" procedure seems a bit difficult to explain to children. The explicit-trade method avoids bring downs and is a direct match with the modeling of the procedure. It does require that the digits in the dividend be spaced out more and almost necessitates the use of lines to mark the columns, even as a permanent feature of the algorithm.

The process of doing division problems with models and recording all of the steps is a complex task for young children. Some will get carried away with the materials and forget to record.

In groups of three, one child can be the "doer," one the "recorder," and one the "supervisor." The supervisor's job is to see that all the steps are being written down and to keep the two processes together. All three children should be talking to each other about the problem as it is being solved. Three problems in a period provide each student the chance to take on each task.

◆ Avoid Some Difficulties

Some of the difficulties that young children have with the division algorithm are due to sloppy writing. You can help students by preparing division record sheets. Nine blanks similar to those shown in Figure 10.22 (Black-line Masters) can easily fit on one page. This is especially useful if the alternative recording scheme is used. Similar sheets can be prepared with four columns.

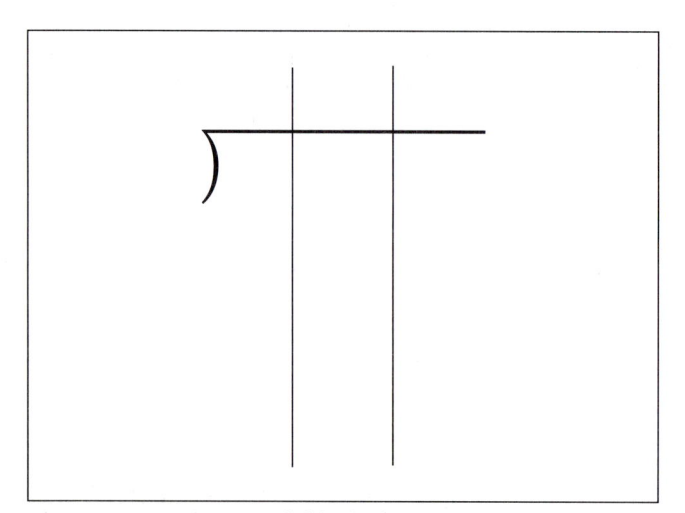

FIGURE 10.22: *Pages with blank charts like this (shown actual size) or with four columns are helpful when children are recording division problems.*

Many children forget to record zeros in the answer. An example is shown in Figure 10.23. The practice of drawing place-value columns or using record sheets as just noted is very helpful since the columns mark a space for each place

value to be filled in. Perhaps more importantly, children should be encouraged to check that their answer makes sense. In the example, 642 things would have at least 100 in each of the six sets. The answer 17 is not even in the ballpark.

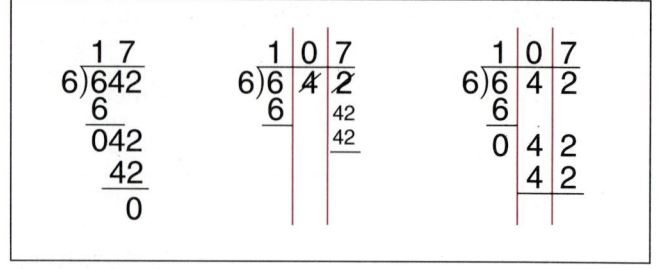

FIGURE 10.23: *Using lines to mark place-value columns can help avoid forgetting to record zero tens.*

◆ TWO-DIGIT DIVISORS

With a two-digit divisor, some form of rounding or similar trick must be used. Not only is rounding complicated, it does not always work. This causes frustration and lots of erasures. Consider the problem in Figure 10.24. The 63 is usually rounded to 60 or, equivalently, 37 divided by 6 is used. Either approach yields an estimate of 6 tens, which turns out to be too much and must be erased. Sometimes we round up instead of down. In these cases it sometimes happens that the partial quotient is too small and must also be erased. Even when the exercises are contrived to make these devices work, children confuse them and/or have difficulty applying them.

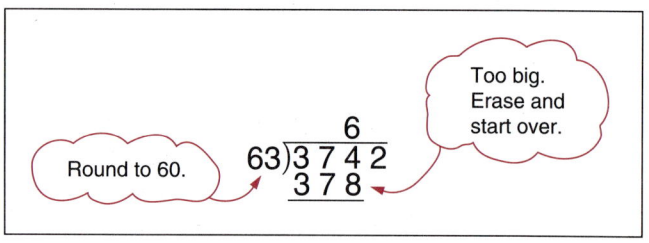

FIGURE 10.24: *Rounding the divisor is not only hard; it does not always work.*

◆ An Intuitive Idea

Suppose that you were sharing a large pile of candies with friends. Instead of passing them out one at a time, you conservatively estimate that each could get at least six pieces. You can always pass out more. So each takes six. What if some are left? If there are enough to go around, you simply pass out more. It would be silly to collect those you already have given out and begin all over. Why not apply this commonsense approach to the sharing of base ten pieces, the conceptual basis for long division?

The principal features of the candy example are first to never overestimate the first time, and second to distribute more if the first attempt leaves enough to pass around. One way to make an easy but always safe estimate the first time is to pretend there are more sets than there really are. For example, if you have 312 pieces for 43 sets, pretend you have 50 sets instead. It is easy to determine that you can distribute 6 to each of 50 sets because 6 × 50 is an easy product. Therefore, you must be able to distribute *at least* 6 to each of the actual 43 sets. The point is, to avoid overestimates, always consider a larger divisor; always round up.

But what if the result shows you could have distributed more? Simply do what you would do with the candy situation—pass out some more.

◆ Apply the Ideas to Long Division

The two preceding ideas can be translated to long division as follows:

1. *Always round the divisor up* to the next multiple of 10 to determine the first estimate.
2. If the first estimate is too small, simply distribute some more.

In Figure 10.25, these ideas are both illustrated, using the same problem that presented difficulties earlier. Both the traditional and the explicit trade methods of recording are illustrated. Children should write the 70 in the little "think bubble" above the divisor. It is easy to run through the multiples of 70 and compare them to 374.

3 × 70 is 210 (too small)

4 × 70 is 280 (too small)

5 × 70 is 350 (close)

6 × 70 is 420 (too big, use 5)

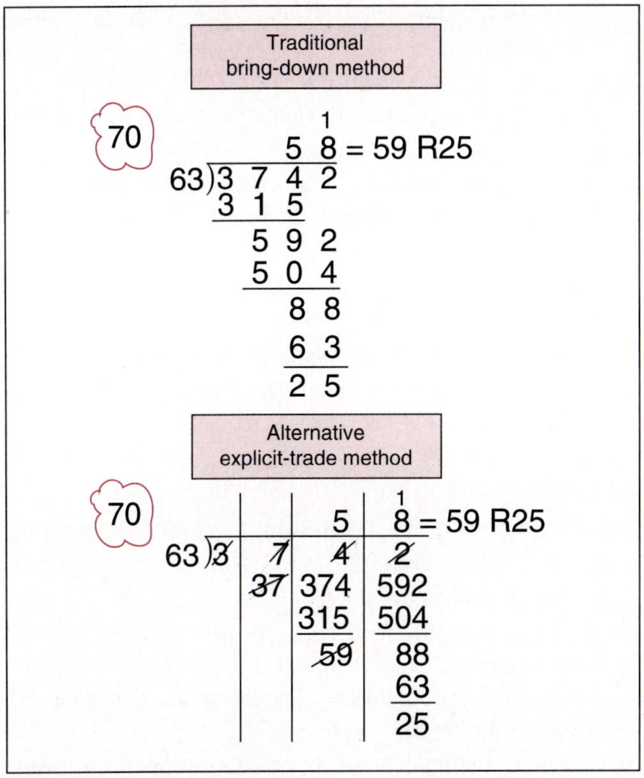

Put 5 tens in each pile. But there are only 63 piles to pass the tens to. Five times 63 tells how many you actually distributed, and subtraction tells how many you have left. In this case, 59 tens is not enough to pass out any more, and so a trade is made for ones. Fifty-nine tens is 590 ones, and 2 ones that were already there is 592 ones. Pretending 70 piles instead of 63 will suggest 8 ones in each pile (8 × 70 is 560, and 9 × 70 is 630). Distributing 8 ones to each of only 63 piles leaves 88. Since that is enough to put one more in each pile, do so. Indicate this by putting a 1 above the 8, and repeat the multiply and subtract steps as always. The result of 5 tens and 9 ones (that is 8 + 1) with 25 left can be written to the right.

This suggestion (always round up and distribute more if needed) is not currently taught in most programs, and it may never be. There is tremendous resistance to change of any sort in the traditions of algorithms. At the same time, the approach has proven to be successful with children in the fourth grade learning division for the first time and with children in the sixth to eighth grades in need of remediation. It reduces the mental strain of making choices and essentially eliminates the need to erase. If an estimate is too low, that's OK. And if you always round up, the estimate will never be too high. Nor is there any reason to ever change to the more familiar approach. It is just as good for adults as for children. The same is true of the explicit-trade notation. It is certainly an idea to consider.

◆◆◆◆◆

REFLECTIONS ON CHAPTER 10: WRITING TO LEARN

1. What is meant by a student-invented algorithm, and why is there any value in having students invent algorithms?
2. What is the most obvious difference between the way students intuitively add two numbers and the usual pencil-and-paper algorithm?
3. Describe how to help children connect what they do with models in the addition algorithm with how they do the algorithm in the usual pencil-and-paper way.
4. When doing a subtraction problem such as 394 − 138 with base ten pieces, why is only the 394 actually modeled?
5. Use the example of 26 × 45 to explain the difference between the area model for multiplication and the repeated-addition model for multiplication.
6. Four different base ten products were described. The first was *ones times ones are ones*. What are the other three? Select any of these other three, and describe how children can be helped to develop these concepts.

FIGURE 10.25: *Think 70 piles instead of 63. In the ones column, pass around 8 to each pile and then pass around one more.*

7. Describe how children learn to record a product such as 3 tens times 7 hundreds.

8. Draw a picture of an area model for 58 × 27 that shows each of the four separate partial products. Your drawing should also indicate the arrangement of the base ten pieces in each of the four sections of the rectangle. Do the computation 58 × 27 showing each of the four partial products separately, and indicate how each corresponds to your drawing.

9. Do the division problem 6⟌748 using both the standard algorithm as you learned it in school and the *explicit-trade* method that was described in this chapter. For the problem as you worked it, answer each of these questions:
   a. How many tens were given to *each* group?
   b. How many ones were given out *in all* (to all the groups combined)?
   c. How many groups are being made?
   d. Before any trades were made, but after all hundreds were shared, how many hundreds were left over?
   e. How many ones were received in trade for the leftover tens?
   f. After trading hundreds for tens, how many tens were available to be shared?

10. Explain why, when dividing by a two-digit divisor, the method of rounding the divisor *up* to the next highest multiple of ten will always assure that the first amount shared with each group in a division problem will be either just right or a little low. If the amount shared is too little, what can be done without doing any erasing? Make up an example to illustrate your answers in this question.

◆◆◆◆◆

FOR DISCUSSION AND EXPLORATION

1. The multiplication algorithm is much more difficult to understand and teach conceptually than any of the other algorithms. With the current deemphasis on pencil-and-paper computation, discuss the proposition that the multiplication algorithm may as well be taught as a rote procedure without any attempt to explain it.

2. Talk with teachers in the upper grades (fifth and above) about how much time they spend teaching algorithms. Do they use base ten models? Would a better development prevent some of the need for remediation? What about the idea of not teaching the algorithms at all until about fourth grade, when students can handle them?

3. What is the educational cost of teaching students to master pencil-and-paper computational algorithms? Here cost means time and effort required over the entire elementary grade span, in contrast with all other topics taught or which could be taught if more time were available. How much about algorithm skill or knowledge do you think is really essential in an age of readily available calculators?

◆◆◆◆◆

SUGGESTED READINGS

Bidwell, J. K. (1987). Using grid arrays to teach long division. *School Science and Mathematics, 87,* 233–238.

Burns, M. (1991). Introducing division through problem-solving experiences. *Arithmetic Teacher, 38*(8), 14–18.

Kamii, C., & Joseph, L. (1988). Teaching place value and double-column addition. *Arithmetic Teacher, 35*(6), 48–52.

Labinowicz, E. (1985). *Learning from children: New beginnings for teaching numerical thinking.* Menlo Park, CA: Addison-Wesley.

Madell, R. (1985). Children's natural processes. *Arithmetic Teacher, 32*(7), 20–22.

Rathmell, E. C., & Trafton, P. R. (1990). Whole number computation. In J. N. Payne (Ed.), *Mathematics for the young child.* Reston, VA: National Council of Teachers of Mathematics.

Robold, A. I. (1983). Grid arrays for multiplication. *Arithmetic Teacher, 30*(5), 14–17.

Schultz, J. E. (1991). Area models—Spanning the mathematics of grades 3–9. *Arithmetic Teacher, 39*(2), 42–46.

Sowder, J. T., & Kelin, J. (1993). Number sense and related topics. In D. T. Owens (Ed.), *Research ideas for the classroom: Middle grades mathematics.* New York: Macmillan Publishing Co.

Stanic, G. M. A., & McKillip, W. D. (1989). Developmental algorithms have a place in elementary school mathematics instruction. *Arithmetic Teacher, 36*(5), 14–16.

Sutton, J. T., & Urbatsch, T. D. (1991). Transition boards: A good idea made better. *Arithmetic Teacher, 38*(5), 4–9.

Thompson, C., & Van de Walle, J. A. (1980). Transition boards: Moving from materials to symbols in addition. *Arithmetic Teacher, 28*(4), 4–8.

Thompson, C., & Van de Walle, J. A. (1981). Transition boards: Moving from materials to symbols in subtraction. *Arithmetic Teacher, 28*(5), 4–9.

Tucker, B. F. (1989). Seeing addition: A diagnosis-remediation case study. *Arithmetic Teacher, 36*(5), 10–11.

Van de Walle, J. A., & Thompson, C. (1985). Partitioning sets for number concepts, place value, and long division. *Arithmetic Teacher, 32*(5), 6–11.

VanLehn, L. (1986). Arithmetic procedures are induced from examples. In J. Hiebert (Ed.), *Conceptual and procedural knowledge: The case of mathematics.* Hillsdale, NJ: Lawrence Erlbaum.

Weiland, L. (1985). Matching instruction to children's thinking about division. *Arithmetic Teacher, 33*(4), 34–45.

11 MENTAL COMPUTATION AND ESTIMATION

ALTERNATIVE FORMS OF COMPUTATION

In our parents' days, to compute meant almost exclusively one thing: a pencil-and-paper computation. Today, a computation by hand is but one of many choices and is usually a last resort.

WHAT ARE THE OPTIONS?

Figure 11.1, from the introduction to the *Standards*, suggests that when a problem situation requiring a computation is met, the first decision to be made concerns the requirements of the solution: Must it be an exact answer or will an approximate answer do? If an approximate result is sufficient, an estimate can usually be arrived at mentally. If an exact answer is required, we frequently make an attempt to work the computation out in our heads—do a mental computation. Mental computation is probably a realistic option a lot more often than you may suspect. However, a mental computation is not always possible due to the particular numbers involved. In those instances where an exact result cannot reasonably be obtained mentally, there are generally three alternatives: Use a calculator, use a computer, or do the computation by hand. The calculator is almost always the method of choice when there are but a few computations to perform. The computer makes sense if there are many similar computations or where computations are related by formulas. Pencil and paper may be the method

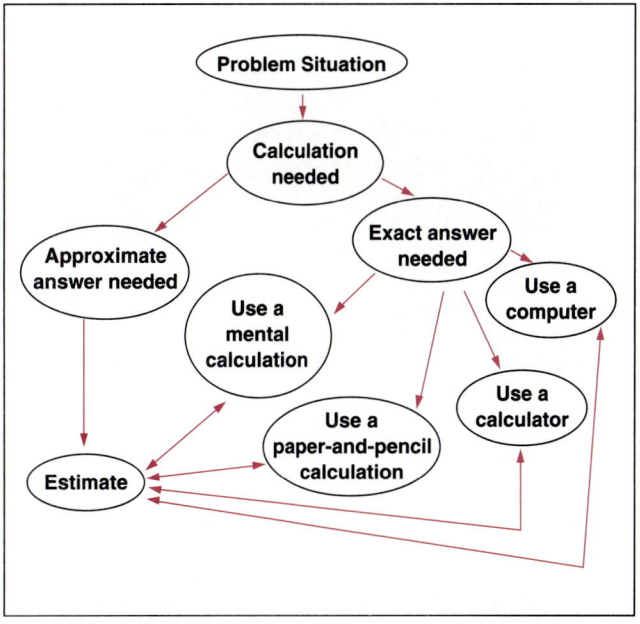

FIGURE 11.1: *Decision about calculation procedures in numerical problems.* SOURCE: NCTM Commission on Standards for School Mathematics (1989). Curriculum and Evaluation Standards for School Mathematics. Reston, VA: National Council of Teachers of Mathematics. Reprinted by permission.

of choice when the computations are not too tedious, when there are not too many of them, and when the calculator is not immediately at hand. (How many of these situations can you think of?)

Notice also how the diagram links estimation with each of the four methods of arriving at an exact answer. In the real world, an estimate may very well be followed up with some form of precise computation. Can you suggest reasons why an estimate might follow an exact computation made by each of the alternative methods?

So the *Standards* has presented a much broader view of computation than was common when you were a child. More importantly, the *Standards* perspective is nothing more than a mirror of the world in which we live. These are the computational choices that should be available to all students who leave our schools. These are the alternatives we must provide for students when they are in our classrooms. There is common agreement that mental methods (computational estimation and mental computation) are much more important than the firmly entrenched traditional pencil-and-paper methods. The focus of computation in schools needs to shift from paper algorithms to mental methods.

REAL-WORLD CHOOSING

Consider each of the following situations. Decide first if an approximate or exact result is appropriate. If an exact result is called for, what is the most reasonable choice of computation? If the numbers were different, would your choice of methods change?

> Melinda walks into a store with $10.00. After picking up 5 packs of paper priced at $1.69 each, she spies a marking pen she has been wanting. It is marked down to $1.35. She needs the 5 packs of paper and does not wish to be embarrassed at the checkout.
>
> Jim works in a restaurant as the cashier. It is his job to figure the total bill when the waiter gives him the prices of the meals purchased. Tax in Jim's city is 4.5%.
>
> In the teacher's lounge, John asks Betty how much she paid for the snacks she bought for her class. Betty had purchased a 19-cent pack of gum for each of her 30 children.
>
> As an automobile salesperson, Ms. Atkins is helping a young couple decide on a car they can afford. They can keep payments down by buying a less expensive car, making a larger down payment, or by opting for one of the special offers of lower interest rates or longer payment periods. Lower rates are tied to higher prices and/or higher down payments. There are a lot of factors to test to see which will suit best their needs and pocketbook.
>
> Professor Able is talking with a student in his office about her grade-point average. Sanya has 78 hours and 216 quality points. She has with her a student teaching application form that requires the average.
>
> Dana makes $3.80 an hour at the day-care center. Last week she worked 20 hours. On her way home she wonders what her paycheck will be (before deductions).

Perhaps what is most interesting when considering real examples of computation is how infrequently pencil and paper is the obvious choice. Mental methods, both for exact and approximate answers, are frequently the most useful of all methods. Would they be the first choice even more often if we had greater mental arithmetic skills? Facility with mental computation is a significant factor in our ability to do computational estimation (Hope, 1986; Leutzinger, Rathmell, & Urbatsch, 1986). Many of us would probably choose to use mental techniques, either exact or approximate, if we only had better skills in mental computation.

MENTAL METHODS IN THE CURRICULUM

Mental methods are discussed in three earlier chapters in this book. Toward the end of Chapter 6, the relationships discussed on small numbers are extended to large numbers. These relationships form a basis for mental mathematics in the early grades. In Chapter 9, on place value, a large section is devoted to number sense. In that context, students are learning to develop a flexibility with numbers that is directly related to mental computation. For example, the idea of selecting an appropriate "nice" number as a substitute is often a first step in estimation. In Chapter 10, students are encouraged to wrestle with computations before being taught the traditional pencil-and-paper algorithms.

In later chapters, the ideas developed with whole number computation will be applied to estimation and mental computations with fractions, decimals, and percents. Mental methods interweave many of the big ideas of numeration throughout the elementary curriculum. Number sense is a complex interaction of number meanings, understanding of operations, and abilities to work mentally with numbers in a flexible manner.

LEARNING AND SELECTING METHODS

The techniques of mental computation and computational estimation are mental algorithms. There are specific methods or procedures that can be taught and practiced the same way that pencil-and-paper algorithms are taught and practiced. You may feel inadequate with mental computation and wonder in awe at people who compute and estimate mentally. Most likely you simply have not been taught. Relatively speaking, mental computation is a very new curriculum area.

In contrast with pencil-and-paper algorithms, there are lots of mental algorithms to learn, not just a few.

Furthermore, it is not always clear-cut which methods to use. Given the same situation, different people will utilize different methods. Consider the following:

48 WIDGETS
13 CENTS EACH
HOW MUCH?

How would *you* estimate the total? Ask some friends how they did it. Are there other ways? Is one method better? What if you wanted an exact answer? Could you do that mentally? How would you approach that task? (Try it before reading on.)

Some possibilities:

Let's see: 48 is about 50; 50 times 10 cents is $5; 50 × 3 cents is $1.50 more; about $6.50.

13 cents is close to 10 cents; 50 at 10 cents is $5.00; Probably more—say $6.00.

Think of 50 times 13; 5 times 13 is 65; must be $6.50, because $65 or 65 cents doesn't seem right.

Instead of 48 and 13, use 50 and 15. Now 5 times 15 is 75. But I made both numbers bigger—$7.50 is too big; probably less than $7.00.

OK, 48 × 10 is $4.80. Then I need 3 times 48 more or about $1.50; $4.80 and $1 is $5.80, and 50 cents is $6.30.

In the classroom you can help children learn a variety of new mental algorithms, and you can provide situations in which they practice newly taught procedures. However, you must also provide opportunity for choice of methods and discussion among class members. Children will learn that different methods are acceptable, and that single correct answers to estimations do not exist. When an exact number is required the choice of method depends a lot on the numbers involved and the skills acquired.

SCOPE AND SEQUENCE

At the present time, there is no standard curriculum for estimation or mental computation. One of the best ways of working on estimation skills seems to be to integrate them with other areas of the mathematics curriculum. In this sense, when there is an opportunity to encourage or discuss estimation of a particular computation, we should capitalize on the opportunity. Textbook authors are now making estimation and mental computation regular features of their books at all grade levels. Generally these features tend to be what publishers refer to as "floating strands," with exercises occurring periodically throughout the texts. The use of these features provides the opportunity to relate estimation and mental computation to the other areas. Few books now have chapters or complete units to teach either mental arithmetic or computation.

Hazekamp (1986) argues that mental multiplication should be taught before children master the pencil-and-paper algorithm. He reasons that once the pencil-and-paper method is mastered, children attempt to use a form of mental blackboard when asked to do mental computation. Try to use the pencil-and-paper algorithm mentally to compute 78 × 6. For most it is very difficult to keep track of trades and adding partial products. Many mental algorithms use a "left-hand" approach. For 78 × 6, think 6 × 70 is 420, and 48 more is 468. Hazekamp's argument is equally applicable to the other operations, with the possible exception of division.

What Hazekamp means is that if we wait to teach mental computational algorithms until after children become proficient with pencil-and-paper methods, we will lose the flexibility that children have before they become completely attached to one procedure. When the one procedure for computation is pencil-and-paper algorithm, then there is a problem. Reconsider the product 78 × 6 for a moment. Another approach is to think of 80 times 6 or 480 and then subtract 12 (for 2 × 6). This approach is like building a rectangle that is 2 too wide and then subtracting off two rows of 6 ones. Both methods can be related directly to the area model presented in the last chapter. The very fact that there is more than one way, and that these methods do not interfere but rather complement instruction in the pencil-and-paper algorithm, suggests at least that these methods of computation might be taught together throughout the curriculum.

LONG-TERM GOALS

Mental algorithms develop and improve in both quality and quantity over years of practice. Mental computation and estimation instruction should begin as early as the first grade and continue beyond the eighth grade. As new concepts and skills are developed, so too can mental algorithms. Mental arithmetic is not a three-week unit but a long-term goal. Each teacher at each grade level must help children develop new ideas and provide children with guided practice.

MENTAL ADDITION AND SUBTRACTION

Each mental algorithm can be presented to your class, discussed, and practiced briefly on different days. As new methods are introduced, some children will select different approaches for the same task. These should certainly be discussed and accepted. Once a method has been taught, be sure to encourage continued practice on a daily basis.

ADDING AND SUBTRACTING TENS AND HUNDREDS

Sums and differences involving multiples of 10 or 100 are easily added mentally, especially if place-value words are used to help keep track of what you are doing.

Example:
300 + 500 + 20
Say: 3 *hundred* and 5 *hundred* is 8 *hundred,* and 20 more is 820.

Use base ten models to help children begin to think in terms of tens and hundreds. Early examples should not include any trades. The exercise 420 plus 300 involves no trades, while 70 plus 80 may be more difficult.

11.1 MOVING MODELS MENTALLY

Display some tens and hundreds base ten models on the overhead. Have students say what number is shown. Next, say how much will be added or subtracted, as in "minus 300," and have students say the result. Finally move to examples with trades. Show 820. "Minus 50." (Think: "Minus 20 is 800, and 30 more is 770.")

11.2 THE CHAIN GAME

Each student has one or more cards with a number written on it. After naming a start number, the teacher calls out for example, "plus 300" or "minus 70." The student holding the result card calls out the result. Then the next change is read by the teacher, and the chain continues. If the cards are prepared so that some numbers occur more than once, then all students must remain alert even after their numbers have been called. Figure 11.2 illustrates the ideas with only seven cards.

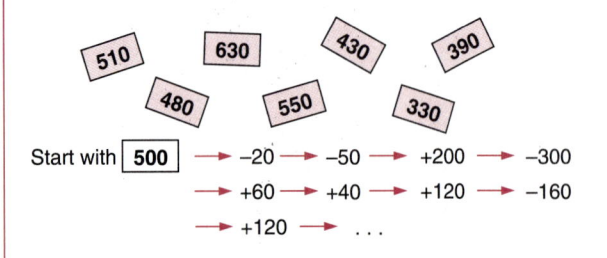

Cards: 510, 630, 430, 390, 480, 550, 330

Start with 500 → −20 → −50 → +200 → −300
→ +60 → +40 → +120 → −160
→ +120 → . . .

FIGURE 11.2: *A chain game*

When working with thousands and hundreds, it is sometimes useful to think of the thousands as hundreds. That is, 3400 can be thought of as 34 hundred. Then 34 hundred and 8 hundred is 42 hundred.

HIGHER-DECADE ADDITION AND SUBTRACTION FACTS

Many children who have mastered their addition and subtraction facts continue to use counting techniques for facts such as 56 + 7 or 73 − 4. These are sometimes referred to as *higher-decade facts.* Present these facts in conjunction with basic facts as in the following examples.

7 + 8 =	12 − 6 =
47 + 8 =	32 − 6 =
77 + 8 =	82 − 6 =
367 + 8 =	552 − 6 =

Of course, facts that do not bridge a decade are easier, such as 48 − 5 or 62 + 7. For those that do cross a decade, the exact strategy may differ with the child or with different numbers. For 44 + 7 you might think of 4 + 7 being 11 and ending with a 1, so the sum must be 51. For 49 + 7 it may be more reasonable to think "One more to 50, and six is 56." Similar differences in thought patterns will appear in subtraction situations. Figure 11.3 shows how a hundreds chart and little base-ten-frames can each be used to help students with these ideas.

The use of higher-decade facts can be extended to adding or subtracting multiples of 10 or 100. To do this, temporarily ignore the other digits to the right.

Example:
374 + 80
Think: 37 + 8 (or 370 + 80) is
45 → 450 → 454
The 4 was temporarily ignored.

FRONT-END APPROACHES

Mental addition and subtraction most frequently begin with the left-hand digits. When working mentally or orally with the number 472, it is easier to think first of 4 hundred, then 70, and finally 2 more, than it is to begin with the 2 and then the 7 tens.

Examples:
46 + 38
Think: 46 and 30 is 76
76 and 8 is 84.

Notice that the first addition converts the problem to a higher-decade fact. Here are two more examples of front-end thinking.

11 / MENTAL COMPUTATION AND ESTIMATION   **205**

---

382 + 75
Think: 380 + 70 (or 38 + 7) is
      450 → 452, + 5 is 457
342 − 85
Think: 340 − 80 (or 34 − 8) is
      260 → 262, − 5 is 257.

11.3 THE OTHER PART OF 100

For example give students the number 28 and have them determine the other part of 100 (72). Discuss their thought patterns. (Can you think of two different approaches?) The other part of 50 is just as easy. Later expand the game to get the other part of 600 or 450 or other numbers that end with 00 or 50.

WORKING WITH NICE NUMBERS

Some numbers are "nice" to work with, like 100 or 700 or 50 or even 450. Try adding nice numbers to not-so-nice numbers. For example, try adding 450 and 27, or add 700 and 248. Subtraction with nice numbers is also easy. For 500 − 73, think 73 and what makes 100. For 650 − 85, first take off 50 then 35 more from 600. Notice the two different ways that nice numbers are used in the last two examples. The next two activities help with this type of thinking.

11.4 50 AND SOME MORE

Say a number and have the students respond with "50 and _____." For 63, the response would be "50 and 13." Use other numbers, such as "450 and some more."

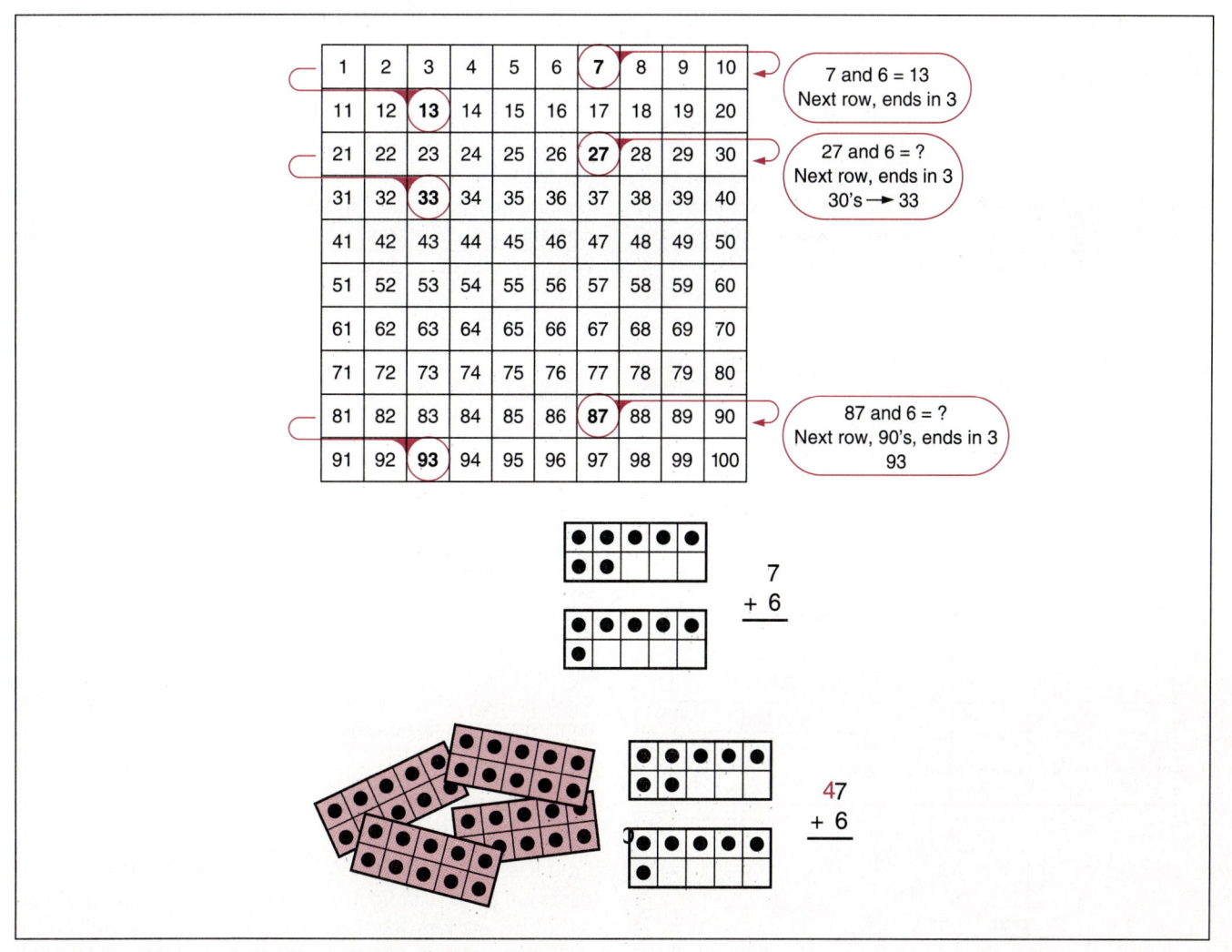

FIGURE 11.3: *Two ways of looking at higher-decade addition*

COMPENSATION STRATEGIES

When students begin to get comfortable with nice numbers, they can use a *compensation* approach when the numbers are almost nice.

Examples:
- 97 + 68 = Think: 100 + 68 is 168; take off 3. 165.
- 126 + 298 = Think: Adding 300 is 426. Then 2 less is 424.
- 473 − 99 = Think: 473 − 100 is 373, so it's 374.
- 452 − 69 = Think: 450 − 50 is 400, less 19 is (other part of 100) 81 → 381. Add back 2 → 383.
- 650 − 85 = Think: 650 − 100 is 550. Then add back 15 → 565.

The last example was used in the previous section and was done differently. Is one method better?

COMPATIBLE NUMBERS

Compatible numbers for addition and subtraction are numbers that go together easily to make nice numbers. Numbers that make tens or hundreds are the most common example. Compatibles also include numbers that end in 5, 25, 50, or 75, since these numbers are also easy to work with. The teaching task is to get students used to looking for combinations that work together and then to look for these combinations in computational situations.

◆◆◆◆◆◆◆◆◆◆◆◆◆◆◆◆◆◆◆◆◆◆◆

11.5 SEARCHING FOR PAIRS

This activity can be used to help children think about "nice combinations" or compatible numbers. In Figure 11.4, see several different searches.

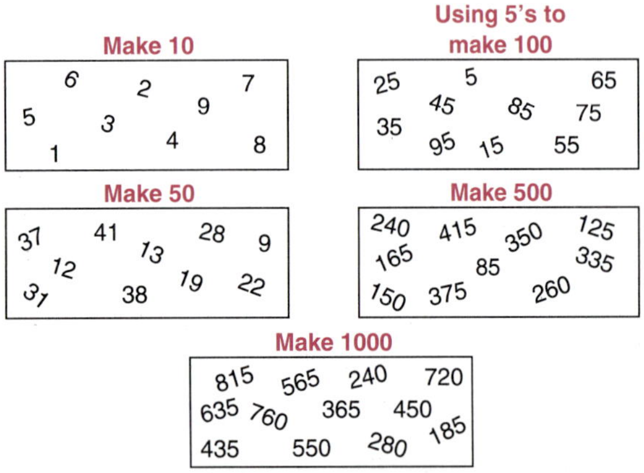

Make 10
6 2 7
5 3 9
1 4 8

Using 5's to make 100
25 5 65
35 45 85 75
95 15 55

Make 50
37 41 28 9
12 13 19 22
37 38

Make 500
240 415 125
165 350 335
150 85 260
375

Make 1000
815 565 240 720
635 760 365 450
435 550 280 185

FIGURE 11.4: *Pair searches*

Pair searches could be worksheets or could be presented on an overhead projector with students writing down or calling out appropriate pairs. As a variation for an independent activity, the numbers could be written on small cards. Students can see how quickly they can pair up the cards.

◆◆◆◆◆◆◆◆◆◆◆◆◆◆◆◆◆◆◆◆◆◆◆

11.6 COMPATIBLE CALCULATIONS FOR +

Present strings of numbers that use compatibles.

30 + 80 + 40 + 50 + 10
25 + 125 + 75 + 250 + 50
95 + 15 + 35 + 5 + 65

Strings such as these can be approached in two ways, and each should be practiced. One way is to search out compatible combinations such as the 5 and 95 in the last example. The other way is to simply add one addend at a time, saying the result as you go. For the first example, that would be "30, 110, 150, 200, 210."

The compensation strategy can now be combined with the compatible numbers approach. "Break off" a small piece of a not-so-nice number to make it into a nice compatible. For example, 47 and 35 is not as easy as 45 and 35. Therefore, break off 2 from 47, and think 45 and 35 is 80 and 2 is 82. To practice this idea, present pairs with a nice number and a nearly nice number, and have students explain their process. This is especially useful with numbers that are close to a 5 (like 137) or close to a 10 (like 62). Alternatively, you can add on a piece. For 37 + 155, add 3 onto 37 to get 40, and 155 is 195. Now take the 3 off again → 192.

MENTAL MULTIPLICATION

Mental multiplication methods are also left-handed or "front-end." Several conceptual ideas may come into play. When multiplication is thought of as repeated addition, the multiplier represents a number of sets. In the computation 562 × 3, it is useful to think of "3 five-hundreds, 3 sixties, and 3 twos." The area model for multiplication can also be useful, especially in the verbal use of base ten products. In multiplying 20 × 67, 2 × 6 is 12, but it must be 12 hundred because tens × tens are hundreds. Clearly place-value concepts are so important.

FACTORS WITH ZEROS

Products such as 400 × 60 that have only a few nonzero digits are prime candidates for mental multiplication. If you write the answers for the exercises below, you will notice that counting the zeros and adding them on the end is easy. If done orally, the zero counting is not as easy. Base ten lan-

guage becomes very important.
   Begin by multiplying by 10, 100, and 1000.

> Examples:
>   7 × 100 = 700
>   6 × 1000 = 6000
>   23 × 100 = 2300 (23 hundred)
>   40 × 100 = 4000 (40 hundred—
>   that's 4 thousand)

When students are proficient with this beginning level, begin to use multiples of 10, 100, and 1000. The nonzero part is done first, and then the zeros are added on or the product adjusted orally.

> Examples:
>   8 × 400 = Think: 4 × 8 is 32; so, 32 hundred.
>   3200.
>   5 × 30 = 150
>   12 × 3000 = Think: 3 × 12 is 36; so, 36 thousand,
>   36,000.

Next, work with examples where both factors involve zeros, as in 40 × 300. In written form this is easy; 4 × 3 is 12, and add 3 zeros. The use of oral base ten products as discussed in Chapter 10 is quite useful.

> Examples:
>   60 × 8 = Think: 6 × 8 is 48; tens × tens are
>   hundreds; so, 48 hundred
>   is 4800.
>   8000 × 20 = Think: 2 × 8 is 16; tens × thousands
>   are ten thousands; so, 16 ten thou-
>   sands, that's 160 thousand
>   is 160,000.
>   Or think: 8 × 20 is 160; That's 160
>   thousands.
>   50 × 600 = Think: 5 × 6 is 30, and tens × hun-
>   dreds are thousands; 30 thousand
>   is 30,000.

Children will need help with dual names for numbers such as 32 hundred for 3 thousand 2 hundred. In the second example, the base ten language produced a nonstandard name, 16 ten thousands. This needs to be discussed with students so that they can use oral alternatives to help their thinking without writing down the numbers. Try writing numbers with zeros, and have children think of at least two ways to read each; then produce two factors that would have that product.

> Example:
>   240,000 Read two ways: 240 thousand, or 24 ten
>   thousands.
>   Factors: 8 × 30,000, or 80 × 3000, or 800
>   × 300.

Again, in written form this is easy. Just count zeros. Without seeing the zeros, base ten language is a key factor.

FRONT-END MULTIPLYING

The product 63 × 8 can easily be thought of as 8 × 60 and 24 more. The idea is to work with the big parts first (front end) and add in the little parts as you work from left to right. The examples below get progressively more difficult.

> Examples:
>   83 × 6 = Think: 80 × 6 is 480, and 18 is 498.
>   370 × 8 = Think: 8 × 300 is 24 hundred, and 560
>   (8 × 70) is 29 hundred 60 or 2960.
>   Or think: 8 × 37. So 8 × 30 is 240, and
>   56 is 296, times 10 is 2960. Here the
>   zero part is used last.
>   42 × 300 = Think: 3 × 40 is 120, and 6 is 126.
>   Then 2 zeros (times hundreds) is 126
>   hundred or 12,600.
>   Or think: 4 × 300 is 12 hundred, so 40
>   × 300 is 12 thousand; 6 hundred more
>   (2 × 300) is 12,600.

In working through examples such as these with your class, write down only intermediate results as they are spoken. Figure 11.5 illustrates this idea.

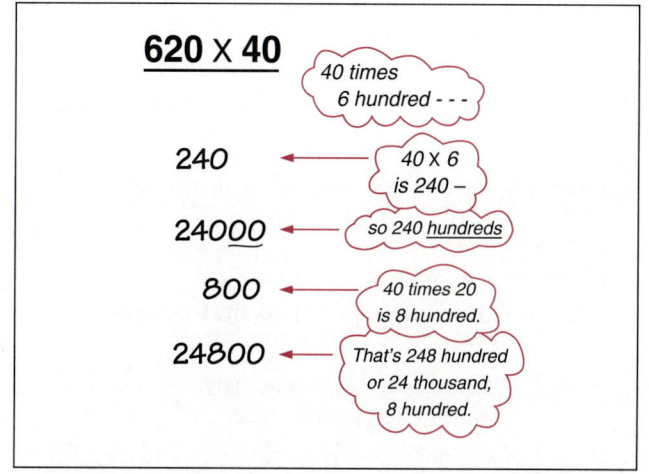

> **620 × 40**
>
> 240   ← 40 times 6 hundred - - -
>   40 × 6 is 240 –
> 24000   ← so 240 hundreds
> 800   ← 40 times 20 is 8 hundred.
> 24800   ← That's 248 hundred or 24 thousand, 8 hundred.

FIGURE 11.5: *Write down intermediate results but not complete calculations.*

Sometimes it is useful to use two-digit parts, especially those that involve 25 or other easy products.

Examples:

3 × 412 = Think: 3 × 4 hundred is 12 hundred, and 36 (3 × 12) is 12 hundred 36, or 1236.

725 × 40 = Think: 4 × 7 hundred is 28 hundred, and 4 × 25 is 1 hundred or 29 hundred, times 10 is 290 hundred, or 29 thousand.

COMPENSATION WITH EIGHTS AND NINES

When one factor ends in 8 or 9, it is useful to use the next higher multiple of 10 and then subtract the difference from the product.

Examples:

7 × 39 = Think: 7 × 40 is 280; but that is 7 too much. So 273.

498 × 6 = Think: 6 × 500 is 30 hundred or 3 thousand. Take back 12; that's 2988.

The use of compensation for eights and nines is more prevalent in money. Prices such as $5.98 or $23.99 abound. Even if your discussion of mental computation does not include decimals, money can easily be dealt with. For example, 7 times $5.98 is 7 × 6 dollars less 7 × 2 cents. Alternatively, the dollar and cents parts can be dealt with separately and then added.

HALVE AND DOUBLE

This approach is based on the idea that if one factor is doubled and the other halved, the product is the same. For example, 8 × 10 is the same as 4 × 20 or 16 × 5. (A really nice exercise for group work is to prepare an explanation for why this is so. Would it work for thirds and triples? for *n*ths and *n* times?) The halve-and-double approach can be applied to any problem with an even factor, but it is most useful with 5, 50, and 500, and also with 25 and 250.

Examples:

682 × 5 = Think: Double 5 is 10 and half of 682 is 341. So 341 times 10 is 3410.

50 × 26 = Think: Double 50 and halve 26 is 100 times 13 or 1300.

60 × 25 = Think: 4 twenty-fives is 100, and a fourth of 60 is 15, 15 hundred.

In the last example, 4 times is the double-and-halve approach applied twice. An alternative thought process might be that each set of 4 twenty-fives is 100. There are 15 fours in 60; so that's 15 hundred. Similar language can be used with 50 or 500. What if the factor to be halved is odd? For 50 × 35, try 34 × 50 and 50 more. Half of 34 is 17, so that's 17 hundred, and one more 50 is 1750. What would you do with 25 × 47 or 500 × 43?

MENTAL DIVISION

Division facts are closely tied to multiplication. For 42 ÷ 7, students are taught to think, "What times 7 is 42?" When long division is introduced, this think-multiplication idea is reduced to a digit-by-digit concept. An advantage here not found with the other operations is that both the pencil-and-paper algorithm and the mental process begin on the left. Here is a case where it would obviously pay to work on both mental and paper algorithms at the same time. For example, 936 ÷ 4 is done by dividing the 9 hundreds and then trading for tens, and so on. If 900 is completely divided up as four sets of 225 and then the 36 is four sets of 9, then the quotient is 225 plus 9 or 234.

ZEROS IN DIVISION

When the dividend has trailing zeros and the divisor is a single digit, a good idea is to emphasize base ten language. In written form, the zeros at the end are just added on.

Examples:

6)24,000 Think: 6 × 4 is 24, so 6 × 4 thousand is 24 thousand; must be 4000.

4500 ÷ 9 = Think: 9 × 5 hundred is 45 hundred. 500.

When both divisor and dividend have zeros, rely on the base ten product language of multiplication.

Examples:

40)3200 Think: 8 × 4 is 32; 40 is 4 tens; the product is hundreds; tens times tens are hundreds; 8 tens. 80.

35,000 ÷ 70 = Think: 7 × 5 is 35; tens times what are thousands? Hundreds—5 hundred. 500.

300)18,000 Think: 6 × 3 is 18; hundreds times tens are thousands; must be 60 (check 60 × 300).

Some suggest that these examples be done by crossing off as many zeros in the dividend as in the divisor. That is, the last example is equivalent to 3)180. This approach is excellent when the numbers are in written form, but it is difficult when the problem is presented orally.

WORKING WITH PART AT A TIME

By thinking missing factor and using base ten language, many division problems can be dealt with mentally.

Examples:
7)637   Think: 637 is 630 and 7; 90 × 7 is 630, and one more is 91.

189 ÷ 3 =   Think: 189 is 180 and 9. What times 3 is 180?—60; and 3 × 3 is 9, so that's 63.

Notice in the last example that the emphasis is on 180 and not 18, as would be done with pencil and paper. Is that necessary? In the following examples, if the base ten language is not used, either a critical zero will be omitted (first example) or an easy-to-use part will not be considered (second example).

Examples:
6)4236   Think: 6 × 700 is 4200, and 6 × 6 is 36. That's 7 hundred and 6 more, 706. (Without the base ten language, an easy error might be 76.)

6)198   Think: 6 × 30 is 180. That leaves 18; so 6 × 3 is 18, or 33 in all.

MISSING-FACTOR PRACTICE

A significant exercise for both mental division and multiplication is to have students find appropriate factors for given products. In Figure 11.6, two exercises are shown that are of a seek-and-find nature with the possible factors given. It is a good idea to include more factors than are needed. Another approach is to give only the products, and have students determine possible factors. This approach will frequently yield more than one pair of factors. For example, 42,000 is 60 × 700, 2100 × 20, or 420 × 100, and many more. Groups might be challenged to find as many factors as possible as long as the products could be computed mentally.

You may have noted that the division examples seem to be contrived. That is, the numbers "work out" nicely. In fact, there are relatively few workable divisions that can be done mentally compared with the other three operations. (Why?) That does not mean that division is less important as a mental computational skill. However, mental division is more of a tool for estimation. Many estimations are done by adjusting numbers to make them workable mentally. So not-nice divisions are sometimes adjusted to those that

Find two factors with this product:

36,000

3 hundred   60   4 thousand   36
    9   6 hundred   120
        1000

Find two factors for each product:

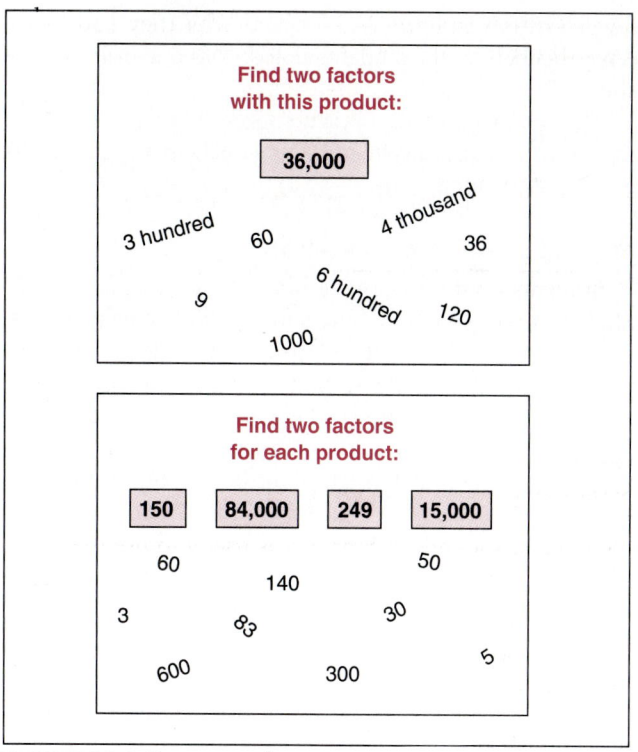

150   84,000   249   15,000

60   140   50
3   83   30
600   300   5

FIGURE 11.6: *Factor searches*

fall into the categories that have been discussed. The missing-factor exercises are useful in helping students learn to think of appropriate nice numbers.

COMPUTATIONAL ESTIMATION

Computational estimation differs from mental computation in that the goal is to quickly determine a result that is adequately accurate for the situation. More often than not, we can be satisfied with an approximate result rather than an exact one. If mental computation was always possible or just as fast as making an estimate, there would be little need for estimation skills. However, many computations do not involve nice numbers, making exact mental computation virtually impossible or difficult. Furthermore, many situations do not require an exact answer, so reaching for a calculator or pencil is not necessary—if one has some good estimation skills. In everyday life, estimation skills prove to be tremendously valuable and time-saving. Teaching these skills to children has become more and more important in recent years.

TEACHING ESTIMATION

Like mental computation, computational estimation can be taught, and children can be encouraged to use their estimation skills regularly. In one fourth-grade class, the teacher suggested to her students that they write down an estimate to each of their computations before they began. Significantly fewer errors resulted. When asked why they had never done this before, the students noted: "No one ever said we should."

The following sections each suggest a general principle that is worth keeping in mind as you help your students develop estimation skills.

◆ Make Estimation a Realistic Topic

Students need to see that estimation is useful in real life. Most exercises should be given in real-life contexts: prices, distances, quantities, and so forth. Have students interview parents and other adults to find out when and where they estimate. Discuss situations where an estimate is satisfactory, and contrast those with situations requiring an exact answer. Curriculum materials are becoming available to help you with this task. Figure 11.7 shows two pages from teacher resource books that illustrate these ideas.

◆ Use the Language of Estimation

The language of estimation includes words and phrases like the following: about, close, just about, a little more (or less) than, and between. Estimation exercises should be posed with this type of terminology. Students should understand that they really are trying to get as close as possible but with quick and easy methods. Estimation does not mean, as some students believe, calculating an exact answer and then rounding it off a bit. Demonstrate that in realistic situations that does not make much sense. For example, when deciding if the $10 you have is enough to buy three drugstore items priced $3.95, $2.57, and three for $6.95, the exact sum is not at all important. Finding that first and then rounding off the answer would be foolish.

◆ Build on Related Skills and Concepts

Estimation skills are highly related to mental computation skills, numeration concepts, and real-world number sense. Most estimation strategies are based on the idea of using nice numbers that are close to the numbers in the computation. The nice numbers lend themselves to mental computation. For example, the cost of 30 soft drinks at 69¢ each can be estimated by using 70 instead of 69. Now 30 times 70 is a mental computation skill. If students have not acquired that skill, the use of 70 is not much help. A good estimation program must have a good mental computation component built in.

Conceptual knowledge of number plays a large part in both estimation and mental computation. Specifically, place-value concepts for both whole numbers and deci-

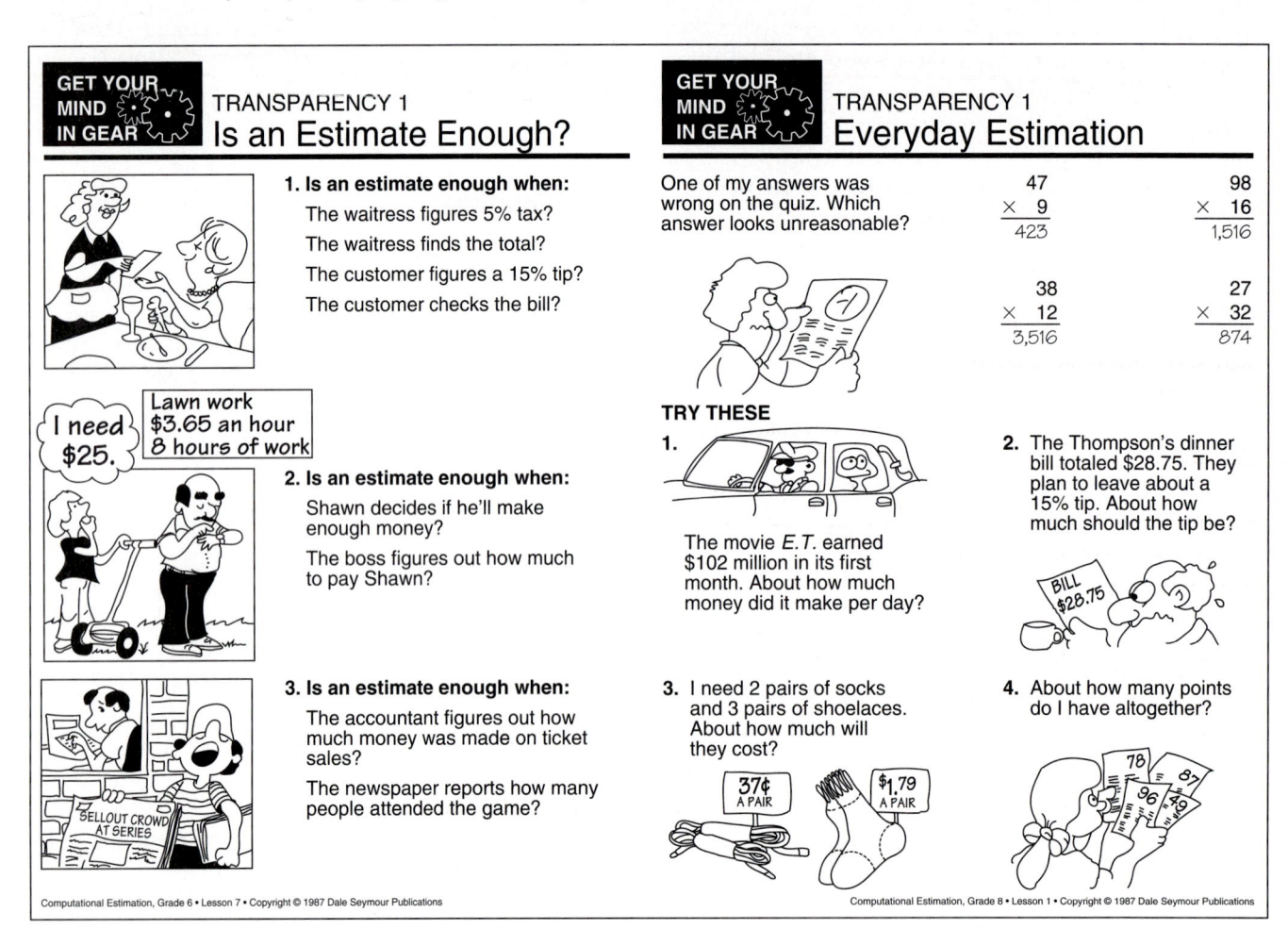

FIGURE 11.7. SOURCE: R. E. Reys, P.R. Trafton, B. Reys, and J. Zawojeski (1987). *Computational Estimation.* Palo Alto, CA: Dale Seymour Publications. Reprinted by permission.

mals, an understanding of fractions, and comfort with percent concepts are critical components of estimation skills.

Finally, what might be called *real-world number sense* is also useful. Is $210, $21, or $2.10 most reasonable for thirty 69¢ soft drinks? Is the cost of a car likely to be $950 or $9500? Could the attendance at the school play be 30 or 300 or 3000? Knowing what is reasonable in familiar situations helps put approximate results in the right ballpark.

◆ Accept a Range of Answers

What estimate would you give for 27 × 325? If you use 20 × 300 you might say 6000. Or you might use 25 for the first factor, divide the second by 4 and get 8100. Or you might use 30 × 300 or 30 × 320 to get 9000 or 9600. Is one of these right and the other wrong?

Too often students believe that for any particular estimation exercise there is one "correct" answer. This is a result of the right-answer syndrome that has pervaded mathematics computation over the years. By listening to the estimates of many students and accepting as "good" estimates those that cover a range of possibilities, students can begin to learn what it means to give an estimate. Two excellent estimators will often produce different estimates for the same computation. Even the same person may give different estimates for the same computation on different days and in different circumstances. The requirements of the situation can change our estimate and our method of producing it. If you want to be sure you have enough money to pay for the five items you've picked up in the drugstore, the accuracy of your estimate depends on how close it appears the total might be to the amount you have in your pocket.

Another way to indicate that estimation is not aimed at one right answer is to avoid the actual estimate altogether. Rather than have students provide an estimate, ask if the result of a particular computation is more or less than a specified value. For example, is 347 + 129 more or less than 500? This is an especially useful technique for the primary grades and when just beginning your estimation program. Obviously the more-or-less-than approach can be used with any operation and any size of numbers.

A similar way to avoid the one-right-answer syndrome is to provide several computations and a single target number. The task is to determine which computation will be closest to the target. (See activity 11.12)

COMPUTATIONAL ESTIMATION STRATEGIES

In the following sections, different algorithms or strategies for the estimation will be explored. When different methods are used on the same computation, different estimates result. Making adjustments after the mental computation is another factor. Experience and conceptual knowledge will eventually help students make better estimates, but it is crucial that children understand that there is no one answer to an estimation.

FRONT-END METHODS

Front-end methods involve the use of the leading or leftmost digits in numbers and ignoring the rest. After an estimate is made based only on these front-end digits, an adjustment can be made by noticing how much has been ignored. In fact, the adjustment may also be a front-end approach.

◆ Front-End Addition and Subtraction

Front end is a very easy estimation strategy for addition or subtraction. This approach is reasonable when all or most of the numbers have the same number of digits. Figure 11.8 illustrates the strategy. Notice that when a number has fewer digits than the rest, that number is ignored completely.

(a) Front-end addition—column form

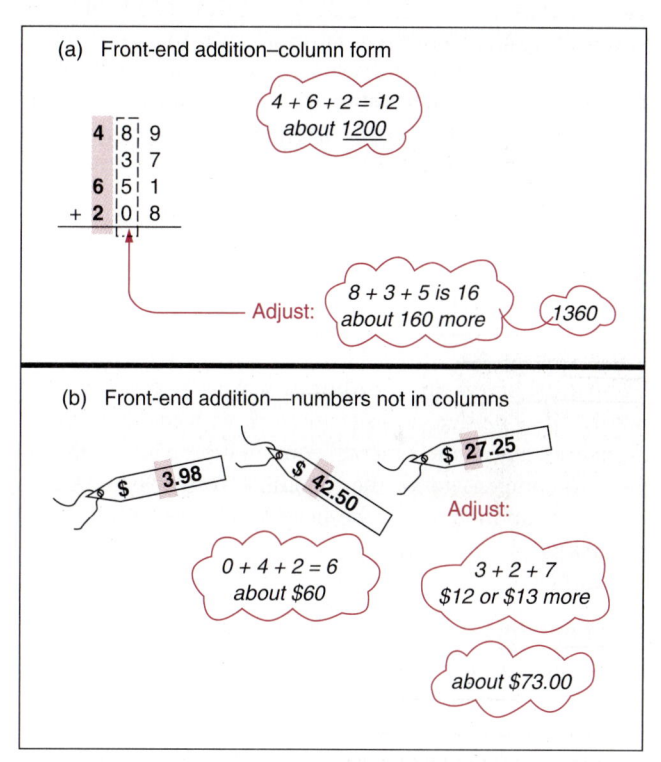

```
  4 | 8 | 9
    | 3 | 7
  6 | 5 | 1
+ 2 | 0 | 8
```

4 + 6 + 2 = 12 about 1200

Adjust: 8 + 3 + 5 is 16 about 160 more → 1360

(b) Front-end addition—numbers not in columns

$ 3.98 $ 42.50 $ 27.25

0 + 4 + 2 = 6 about $60

Adjust: 3 + 2 + 7 $12 or $13 more

about $73.00

FIGURE 11.8

After adding or subtracting the front digits, an adjustment is made to correct for the digits or numbers that were ignored. Making an adjustment is actually a separate skill. For very young children, practice first just using the front digits. Pay special attention to numbers of uneven length when not in a column format.

When teaching this strategy, present additions or subtractions in column form, and cover all but the leading digits. Discuss the sum or difference estimate using these digits. Is it more or less than the actual amount? Is the estimate off by a little or a lot? Later, show numbers in horizontal form or on price tags that are not lined up. What numbers should be added?

The leading-digit strategy is easy to use and easy to teach because it does not require rounding or changing numbers. The numbers used are there and visible, so children can see what they are working with. It is a good first strategy for children as low as the third grade. It is also useful for older children and adults, especially as they learn to make better adjustments in the front-end sum or difference.

◆ Front-End Multiplication and Division

For multiplication and division, the front-end method uses the first digit in each of the two numbers involved. The computation is then done using zeros in the other positions. For example, a front-end estimation of 48 × 7 is 40 times 7 or 280. When both numbers have more than one digit, the front ends of both are used. For 452 × 23, consider 400 × 20 or 8000.

Division with pencil and paper is almost a front-end strategy already. First determine in which column the first digit of the quotient belongs. For 7)3482, the first digit is a 4 and belongs in the hundreds column over the 4. Therefore, the front-end estimate is 400. This method always produces a low estimate, as students will quickly figure out. In this particular example, the answer is clearly much closer to 500, so 480 or 490 is a good adjustment.

ROUNDING METHODS

Rounding is the most familiar form of estimation. Estimation based on rounding is a way of changing the problem to one that is easier to work with mentally. Good estimators follow their mental computation with an adjustment to compensate for the rounding. To be useful in estimation, rounding should be flexible and well understood conceptually.

◆ Rounding in Addition and Subtraction

When a lot of numbers are to be added, it is usually a good idea to round them to the same place value. Keep a running sum as you round each number. In Figure 11.9, the same total is estimated two ways using rounding. A combination of the two is also possible.

In subtraction situations there are only two numbers to deal with. For subtraction and for additions involving only two addends, it is generally necessary to round only one of the two numbers. For subtraction, round only the subtracted number. In 6724 − 1863, round the 1853 to 2000. Then it is easy: 6724 − 2000 is 4624. Now adjust. You took away a bigger number, so the result must be too small. Adjust to about 4800. For 627 + 385, you might round the 627 to 625, because multiples of 25 are almost as easy to work with as multiples of 10 or 100. After substituting 625 for 627, you may or may not want to round 385 to 375 or 400. The point is that there are no right rules. Choices depend on the relationships held by the estimator, on how quickly the estimate is needed, and how accurate an estimate is required. (See "Approximate Numbers and Rounding" in Chapter 9, page 170.)

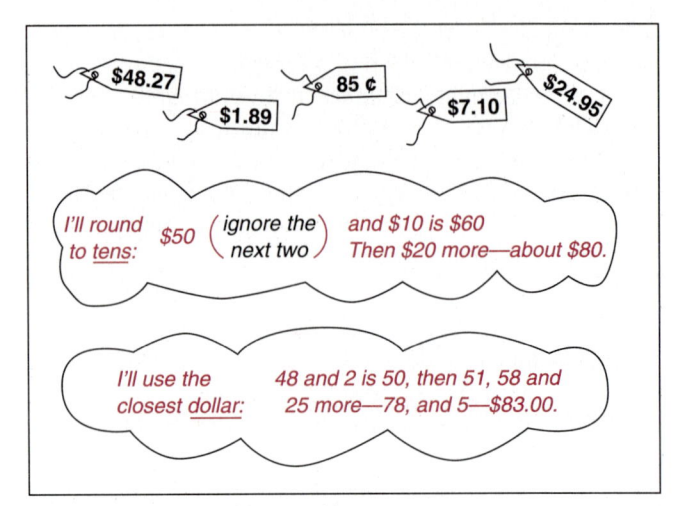

$48.27 $1.89   85¢ $7.10 $24.95

I'll round to tens: $50 (ignore the next two) and $10 is $60. Then $20 more—about $80.

I'll use the closest dollar: 48 and 2 is 50, then 51, 58 and 25 more—78, and 5—$83.00.

FIGURE 11.9: *Rounding in addition*

◆ Rounding in Multiplication and Division

The rounding strategy for multiplication is no different from that for other operations. However, the error involved can be significant, especially when both factors are rounded. In Figure 11.10, several multiplication situations are illustrated, and rounding is used to estimate each.

If one number can be rounded to 10, 100, or 1000, the resulting product is easy to determine without adjusting the other factor.

When one factor is a single digit, examine the other factor. Consider the product 7 × 836. If 836 is rounded to 800, the estimate is relatively easy and is low by 7 × 36. If a more accurate result is required, round 836 to 840, and use a front-end computation. Then the estimate is 5600 plus 280 or 5880 (7 × 800 and 7 × 40). The parts technique relies on the skill of doing the front-end multiplication mentally.

If possible, round only one factor and select the largest one if it is significantly larger. (Why?) For example, in 47 × 7821, 47 × 8000 is 376,000, but 50 × 8000 is 400,000.

Another good rule of thumb with multiplication is to round one factor up and the other down (even if that is not the closest round number). When estimating 86 × 28, the 86 is about in the middle, but 28 is very close to 30. Try rounding 86 down to 80 and 28 to 30. The actual product is 2408, only 8 off from the 80 × 30 estimate. If both numbers were rounded to the nearest ten, the estimate would have been based on 90 × 30 with an error of nearly 300.

With one-digit divisors, it is almost always best to search for a compatible dividend rather than to round off. For example, 7)4325 is best estimated by using the close compatible number, 4200, to yield an estimate of 600. Rounding would suggest a dividend of 4000 or 4300, neither of which is very helpful. (Recall the contrived examples in the "Mental Division" section.)

When the divisor is a two-digit or three-digit number, rounding it to tens or hundreds makes looking for a missing factor much easier. For example, 425)3890, round the divisor to 400. Then think, 400 times what is close to 3890?

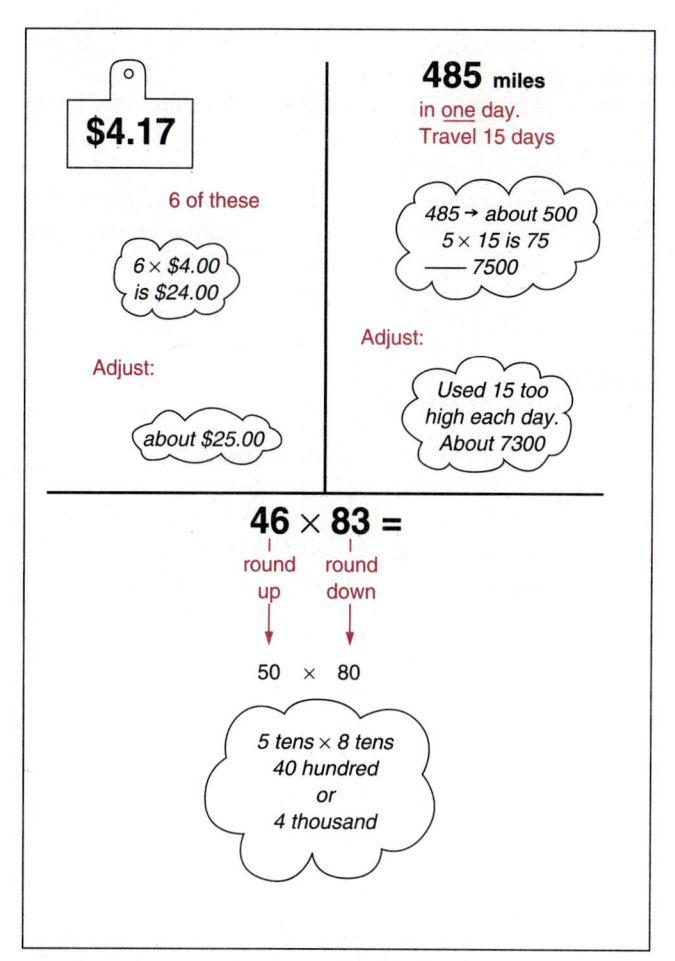

FIGURE 11.10: *Rounding in multiplication*

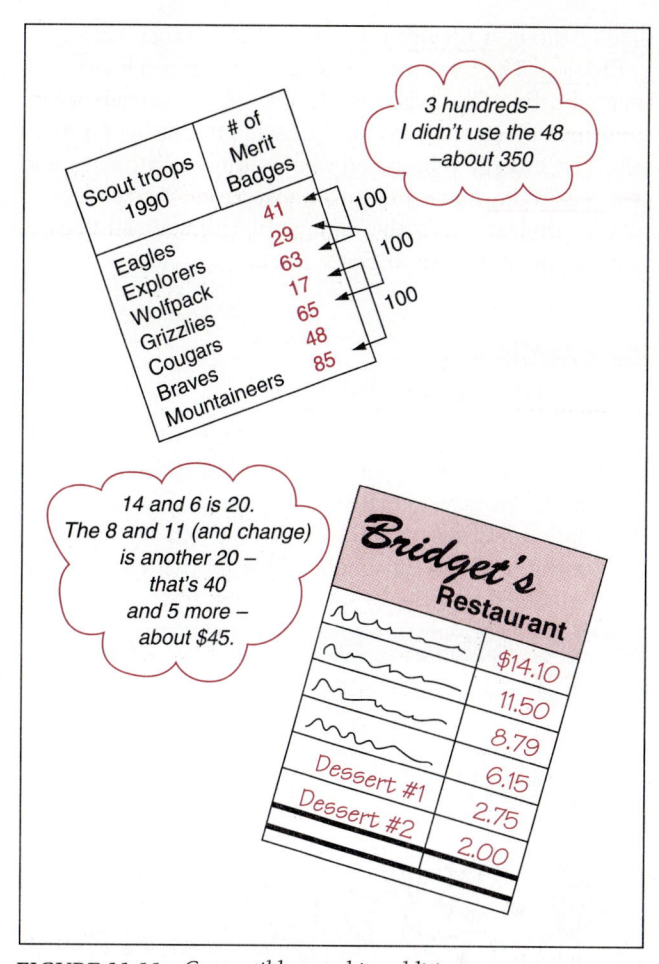

FIGURE 11.11: *Compatibles used in addition*

USING COMPATIBLE NUMBERS

When adding a long list of numbers, it is sometimes useful to look for two or three numbers that can be grouped to make 10 or 100. If numbers in the list can be adjusted slightly to produce these groups, that will make finding an estimate easier. This approach is illustrated in Figure 11.11.

In subtraction, it is frequently easy to adjust only one number to produce an easily observed difference. The thought process may be closer to addition than subtraction, as illustrated in Figure 11.12.

Frequently in the real world, an estimate is needed for a large list of addends that are relatively close. This might happen with a series of prices for similar items, attendance at a series of events in the same arena, cars passing a point on successive days, or other similar data. In these cases, as illustrated in Figure 11.13, a nice number can be selected as representative of each, and multiplication used to determine the total. This is more of an *averaging technique* than a compatible numbers strategy.

FIGURE 11.12: *Compatibles can mean an adjustment that produces an easy difference.*

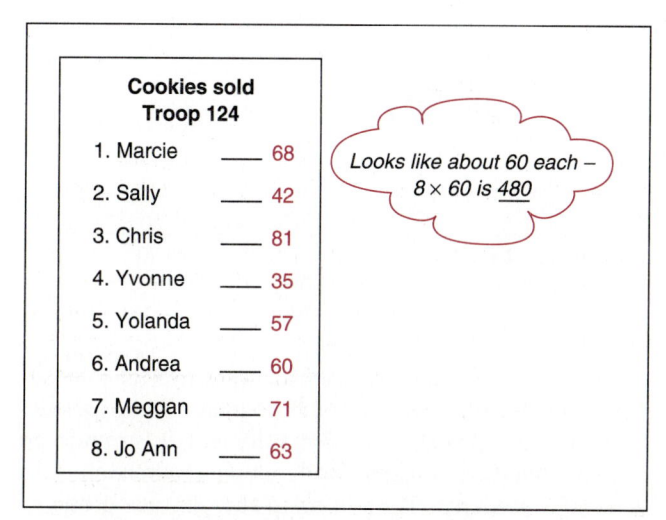

FIGURE 11.13: *Estimating sums using averaging*

One of the best uses of the compatible-number strategy is in division. The two exercises shown in Figure 11.14 illustrate adjusting the divisor and/or dividend to create a division that comes out even and is therefore easy to do mentally. The strategy is based on whole-number arithmetic and is not difficult. Many percent, fraction, and rate situations involve division, and the compatible number strategy is quite useful, as shown in Figure 11.15.

ESTIMATION EXERCISES

The ideas presented so far illustrate some of the types of estimation and thought patterns you want to foster in your classroom. However, making up examples and putting them into a realistic context is not easy and is very time-consuming. Fortunately, good estimation material is now being included in virtually all textbooks. Text material will

FIGURE 11.14. SOURCE: B. J. Reys and R. E. Reys (1983). GUESS (Guide to Using Estimations and Strategies) Box II. Palo Alto, CA: Dale Seymour Publications, Cards 2 and 3. Reprinted by permission.

provide you with hints and periodic exercises for integrating estimation throughout your program. This exposure, however, will probably not be sufficient to provide an ongoing, intensive program for developing estimation skills with your students. Try adding to the text using one or more of the many good teacher resource materials that are available commercially. Some of these are listed at the end of the chapter.

Teacher's guides, resource books, and professional books and articles on estimation offer good activities or activity models that you can easily adapt to your particular needs without additional resources. The examples presented in this section are not designed to *teach* estimation strategies, but offer useful formats to provide your students with practice using skills as they are being developed.

CALCULATOR ACTIVITIES

The calculator is not only a good source of estimation activities, it is also one of the reasons estimation is so important. In the real world, we frequently hit a wrong key, leave off a zero or a decimal, or simply enter numbers incorrectly. An estimate of the expected result is one way we can alert our-

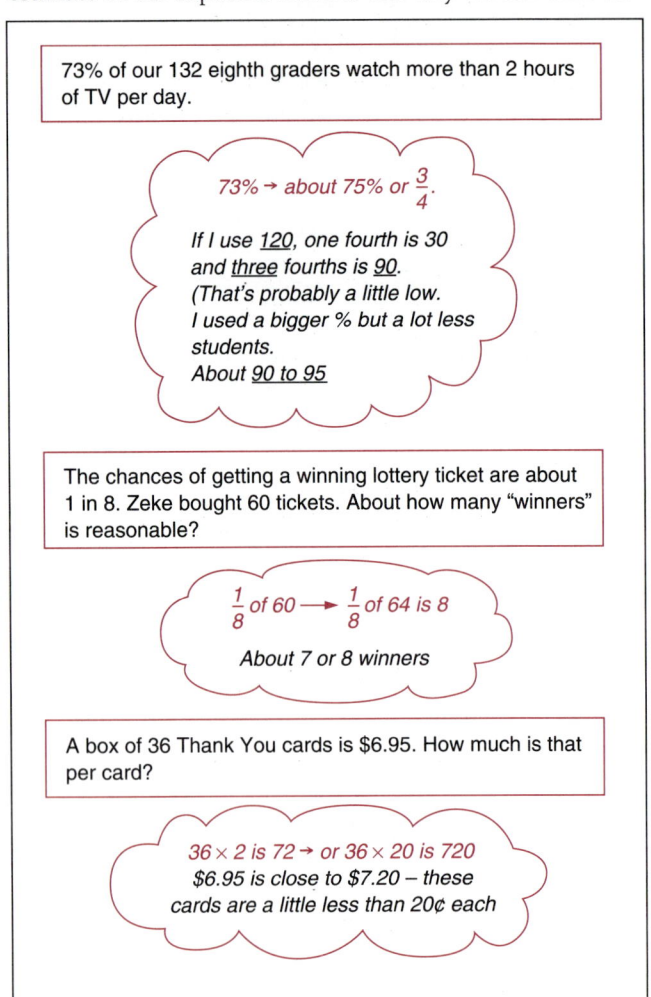

FIGURE 11.15: Using compatible numbers in division

selves to these errors. The calculator as an estimation teaching tool lets students work independently or in pairs in a challenging, fun way without fear of embarrassment. With a calculator for the overhead projector, some of the activities described here work very well with a full class.

11.7 THE RANGE GAME

This is an estimation game for any of the four operations. First pick a start number and an operation. The start number and operation are stored in the calculator. Students then take turns entering a number and pressing ▣ to try to make the result land in the target range. The following example for multiplication illustrates the activity: Suppose a start number of 17 and a range of 800 to 830 is chosen. Press 17 ▣ ▣ to store 17 as a factor. Press a number then ▣. Perhaps you try 25. The result is 425. That is about half the target. Try 50. The result is 850. Maybe 2 or 3 too high. Try 48. Result is 816—in the target! Figure 11.16 gives examples for all four operations. Prepare a list of stat numbers and target ranges. Let students play in pairs and see who can hit the most targets on the list (Wheatley & Hersberger, 1986).

After entering the start # as shown, players take turns pressing a number, then ▣ to try to get a result in the target range.

Addition:

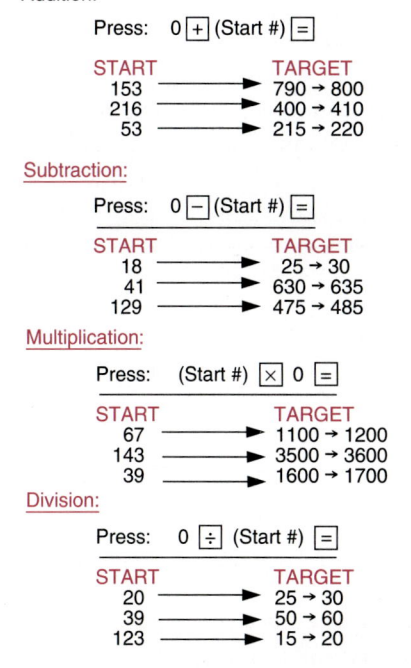

Press: 0 ▣ (Start #) ▣

START	TARGET
153	790 → 800
216	400 → 410
53	215 → 220

Subtraction:

Press: 0 ▣ (Start #) ▣

START	TARGET
18	25 → 30
41	630 → 635
129	475 → 485

Multiplication:

Press: (Start #) ▣ 0 ▣

START	TARGET
67	1100 → 1200
143	3500 → 3600
39	1600 → 1700

Division:

Press: 0 ▣ (Start #) ▣

START	TARGET
20	25 → 30
39	50 → 60
123	15 → 20

FIGURE 11.16: *Calculator Range Game*

The **Range Game** can be played with an overhead calculator with the whole class, or by an individual, or by two or three children with calculators who can race one another. The speed element is important. The width of the range and the type of numbers used can all be adjusted to suit the level of the class.

11.8 SECRET SUM

This calculator activity uses the memory feature. A target number is selected, for example 100. Students take turns entering a number and pressing the ▣ key. Each of the numbers is accumulated in the memory, but the sum is never displayed on the screen. If one player thinks that the other player has made the sum go beyond the target, he or she announces "over," and the memory return key is pressed to check. If a player is able to hit the target exactly, bonus points can be awarded. Interesting strategies quickly develop. (Adapted from an example in *Everybody Counts*, National Research Council, 1989.)

Secret sum can also be played with the ▣ key. First enter a total amount in the memory. Each player's number is followed by a press of ▣ and is subtracted from the memory. Here the first to correctly announce that the other player has made the memory go negative is the winner.

11.9 QUOTIENT ESTIMATION

This calculator activity for estimating quotients is suggested by Coburn (1987). First enter the divisor of a two-digit division problem, and press ▣. Then enter an estimate of the quotient, and press ▣. The result is compared with the dividend, and two subsequent estimates are allowed. There is no need to reenter the divisor.

For example:

$72 \overline{)4197}$
72 ▣ _____ ▣.
_____ ▣.
_____ ▣. _____ ▣.4197 ▣.

Two players can compete by working on the same problem, trying to get the smallest difference in 30 seconds. (Can you see how this game is really the same as the **Range Game**?)

In the **Range Game**, each player benefits from the attempt made by the previous player. In that sense it is an excellent learning technique, and even the weakest children will have an opportunity to succeed. In the variations described next, each turn presents a different estimation task.

11.10 THE RANGE GAME: SEQUENTIAL VERSIONS

Select a target range as before. Next enter the starting number in the calculator, and hand it to the first player. For addition and subtraction, the first player then presses either ∔. or ∸., followed by a number, and then =. The next player begins his or her turn by entering ∔. or ∸. and an appropriate number, operating on the previous result. If the target is 423 to 425, a sequence of turns might go like this:

Start with 119:
⊞ 350 ⊟ → 469 (too high)
⊟ 42 ⊟ → 427 (a little over)
⊟ 3 ⊟ → 424 (success)

For multiplication or division, only one operation is used through the whole game. After the first or second turn, decimal factors are usually required. This variation provides excellent understanding of multiplication or division by decimals. A sequence for a target of 262 to 265 might be like this:

Start with 63:
☒ 5 ⊟ → 315 (too high)
☒ 0.7 ⊟ → 220.5 (too low)
☒ 1.3 ⊟ → 286.65 (too high)
☒ 0.9 ⊟ → 257.98 (too low)
☒ 1.3 ⊟ → 265.72558 (very close!)
(What would you press next?)

Try a target of 76 to 80, begin with 495, and use only division.

◆ Computer Programs

A number of computer programs are available that practice estimation skills. Computer programs can present problems for estimation, control speed, and also compare the result to the actual answer. Most allow the teacher or user to adjust the skill level of the exercises. These programs can be effectively used with a full class using a large monitor or projection system. Students can write down estimates on paper within the allotted time frame. In the MECC package *Estimation,* the "Estimate" program has all of the features mentioned previously (Minnesota Educational Computing Corporation, 1984). While not at all fancy or entertaining, it can be quite effective at a wide range of grade and ability levels.

◆ Activities for the Overhead Projector

The overhead projector offers several advantages. You can prepare the computational exercises ahead of time. You can control how long a particular computation is viewed by the students. There is no need to prepare handouts. Commercial materials such as GUESS cards (Figure 11.14) can be copied onto transparencies for instructional purposes.

11.11 WHAT WAS YOUR METHOD?

Select any single computational estimation problem and put it on the board or overhead. Allow 10 seconds for each class member to get an estimate. Discuss briefly the various estimation techniques that were used. As a variation, prepare a problem with an estimation illustrated. For example, 139 × 43 might be estimated as 6000. Now ask questions concerning this estimate. "How do you think that estimate was arrived at? Was that a good approach? How should it be adjusted? Why might someone select 150 instead of 140 as a substitute for 139?" Almost every estimate can involve different choices and methods. Alternatives make good discussions, help students see different methods, and learn that there is no single correct estimate.

11.12 WHICH ONE'S CLOSEST?

Make a transparency of a page of drill-and-practice computations. Have students focus on a single row or other collection of five to eight problems. Ask them to find the one with an answer that is closest to some round number that you provide (Figure 11.17). One transparency could provide a week's worth of 5-minute drills.

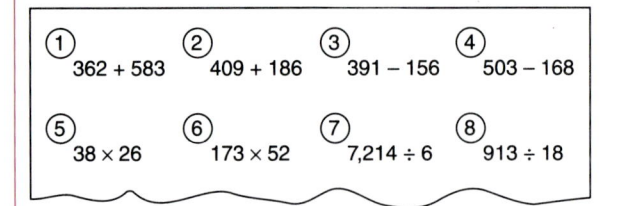

① 362 + 583	② 409 + 186	③ 391 − 156	④ 503 − 168
⑤ 38 × 26	⑥ 173 × 52	⑦ 7,214 ÷ 6	⑧ 913 ÷ 18

Which of these is CLOSEST TO 600? 1000? 10?

FIGURE 11.17: *Use a drill page from your text as an estimation exercise*

11.13 IN THE BALLPARK

Also with a page of problems from a workbook, write in answers to six or seven problems. On one or two problems, make a significant error that can be caught by estimation. For example, write 26)‾5408 = 28 (instead of 208) or 36 × 17 = 342 (instead of 612). Other answers should be correct. Encourage the class to estimate each problem to find those "not even in the ballpark" errors.

11.14 ENOUGH?

Make "Is This Enough?" transparencies. Select an amount such as 10 or 50 (or $10 or $50) and write that at the top of a transparency. Below, put a variety of computations or realistic situation. The task is to decide quickly if the top amount is *clearly enough* (more) or *clearly not enough* (less) or perhaps *too close to call*. An example is shown in Figure 11.18. Show only one example at a time and only for about 10 seconds. This type of estimation is frequently done in real life and does not always call for a very accurate estimate.

$1.50   Clearly enough
        Clearly not enough
        Too close to call

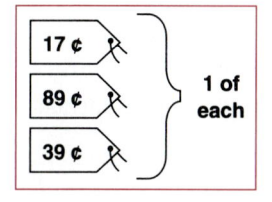

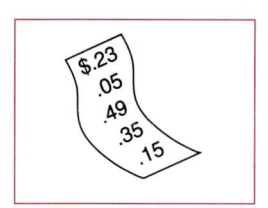

Candy Bar 43¢
5 candy bars

17¢
89¢   1 of each
39¢

$.23
 .05
 .49
 .35
 .15

FIGURE 11.18: *"Is This Enough?"*

ESTIMATING WITH FRACTIONS, DECIMALS, AND PERCENTS

Fractions, decimals, and percents are three different notations for rational numbers. Many real-world situations that call for computational estimation involve the part-to-whole relationships of rational numbers. A few examples are suggested here:

SALE! $51.99. Marked one-fourth off. What was the original price?

About 62% of the 834 students bought their lunch last Wednesday. How many bought lunch?

Tickets sold for $1.25. If attendance was 3124, about how much was the total gate?

I drove 337 miles on 12.35 gallons of gas. How many miles per gallon did my car get?

With the exception of a few examples, this chapter has avoided estimations with fractions, decimals, and percents. To estimate with these numbers first involves an ability to estimate with whole numbers. Beyond this, it involves an understanding of fractions and decimals and what these two types of numbers mean. Calculations with percents are always done as fractions or as decimals. The key is to be able to use an appropriate fraction or decimal equivalent.

In the first of the examples just presented, one approach is based on the realization that if $51.99 (or $52) is the result of one-fourth off, that means $52 is three-fourths of the total. So one-fourth is a third of $52, or a little less than $18. Thus about $52 + $18 = $70 or about $69 seems a fair estimate of the original cost. Notice that the conceptualization of the problem involves an understanding of fractions, but the estimation skill involves only whole numbers. This is also the case for almost all problems involving fractions, decimals, or percents.

From a developmental perspective, it is important to see that the skills of estimation are separate from the conceptual knowledge of rational numbers. It would be a mistake to work on the difficult processes of estimation using fractions or percents if concepts of those numbers were poorly developed. In later chapters, where rational number numeration is discussed, it is shown that an ability to estimate can contribute to an increased flexibility or number sense with fractions, decimals, and percents.

EVALUATING MENTAL COMPUTATION AND ESTIMATION

As with any other part of the mathematics curriculum, proper assessment of mental strategies will help you know how well your students are doing, so you will be able to plan your next lessons accordingly. Assessment is also an important method of communicating to students the values that you do (or do not) attribute to mental forms of computation. If mental computation and estimation are to be the most important forms of computation in your curriculum, your overall assessment and grading scheme should also reflect this value in a proportionate manner.

USE FREQUENT PERIODIC TESTING

Just as development of mental computation and estimation skills is an ongoing matter, so also should evaluation be an ongoing part of your program. Tests or quizzes can be given in as little time as 5 minutes and as often as every week.

Early test results may be very discouraging. Do not despair. And do not hold early low scores against your students. Rather, use these first scores as baseline data against which to measure progress. Children do improve at both mental computation and estimation with instruction and practice. The progress will be reflected in the test scores and will reward both you and your students.

CONTROL TIME

Mental computation and estimation is, by definition, to be done mentally, not with pencil and paper. This is extremely difficult for children to accept. In testing situations (and in many practice activities) it is important that you control the amount of time that is allowed for each exercise. This cannot be done when all of the questions are written on test paper and passed out as in a traditional test. Instead, prepare the test items on cards or on transparencies. Show each problem for a 10-second period, and then immediately go to the next one. If you allow much longer than 10 seconds, students will attempt to use pencil and paper to compute answers.

Another way to emphasize strictly mental methods is to use prepared numbered answer sheets with only a small blank space next to each number. No other markings are to be written on the paper. Some teachers even cut answer sheets into narrow strips so that there is no room left to write anything but the answer.

CONSIDER ANSWERS AND ANSWER FORMATS

For mental computation, you are looking for exact answers. Tests should almost always be open-ended in format. That is, the student should write the answers rather than select from a multiple-choice list. Some care should be taken to look at incorrect answers and determine what caused the error. On quizzes, you may even ask students in upper grades to select one example where they made an error and explain in writing what they did that caused it.

Estimation tests can involve multiple-choice formats as well as open-ended responses. If you use multiple choices, put the answer choices in numeric order and include foils (wrong answers) that are likely to be made. For example, if a good answer is 2400, then 240 and 24,000 may be good foils. Another way to select foils is to use an inappropriate estimation technique. In estimating 585 + 29 + 714 + 693, it would be inappropriate to round to thousands. For skilled estimators who have learned to adjust estimates, a front-end estimate of 1800 is much too low. For open-response estimation tests, you must decide in advance what will be the range of acceptable answers. There are no single correct responses, even if every student used the same approach, which is unlikely.

USE OCCASIONAL INTERVIEWS

It is not always easy to tell what techniques or thought processes are being used by individual students. For example, some students may get overly attached to a particular approach and not be flexible enough to switch to a more efficient method when the numbers call for it. One way to find this out is to conduct a short test on a one-on-one basis. Give perhaps five items for mental computation or estimation. Immediately afterward, return to each answer, and ask the student to explain how the estimate was made. This form of listening to your students will be very valuable in deciding how to pace your program and determining if there are concepts or strategies that require extra emphasis.

◆◆◆◆◆
REFLECTIONS ON CHAPTER 11: WRITING TO LEARN

1. When a real situation or problem requiring a computation confronts us, what is the first decision that is usually made? Give a personal example of each situation.

2. Four alternatives are generally available if an exact answer to a computation is required. Suggest a situation where each may be reasonable.

3. Which should come first: pencil-and-paper computation or mental computation?

4. Why is it not reasonable to select a single method of computing mentally and have students learn and practice it in the same way that pencil-and-paper algorithms are developed?

5. Give an example of how mental computation is an extension of
   a. early number concepts (you may need to return to Chapter 6),
   b. place-value concepts, and
   c. student-invented algorithms.

6. Give an example of an addition and a subtraction mental computation (exact answer) to illustrate each of the following:
   a. higher-decade facts,
   b. front-end approaches,
   c. using nice numbers,
   d. compatible numbers, and
   e. compensation strategies.

7. Illustrate mental multiplication using
   a. front-end digits,
   b. multiples of 10 or 100,
   c. compensation, and
   d. halve and double.

8. Explain why mental division is actually mental multiplication.

9. What are some of the things that are important to communicate to students about computational estimation other than helping them learn various strategies?

10. Describe each of these estimation strategies by using an example and an explanation:

    a. front-end addition or subtraction,

    b. front-end multiplication,

    c. rounding, and

    d. using compatible numbers.

11. How can the calculator be used to practice estimation strategies?

12. What do you see as three of the most important ideas to keep in mind when considering evaluation of mental computation or computational estimation?

◆◆◆◆◆

FOR DISCUSSION AND EXPLORATION

1. What types of mental computation or computational estimation can be taught at the first and second grades?

2. Examine a sixth-, seventh-, or eighth-grade textbook for the total program provided for mental computation and computational estimation. You will want to look at the teacher's edition because many of the best suggestions are found in supplementary activities or pages. Find out how estimation and mental computation is integrated into the rest of the content of the text.

3. Examine one of the following teacher resources:

    Computational Estimation

    Mental Math in the Primary Grades

    Mental Math in the Middle Grades

    Mental Math in the Junior High School

    GUESS, Boxes I and II (Guide to Using Estimation Skills and Strategies) (See references under Suggested Readings.)

    How would the materials you reviewed be used over a one-year period in your classroom?

◆◆◆◆◆

SUGGESTED READINGS

Burns, M. (1992). *About teaching mathematics: A K–8 Resource*. Sausalito, CA: Marilyn Burns Education Associates.

Cobb, P., & Merkel, G. (1989). Thinking strategies: Teaching arithmetic through problem solving. In P. R. Trafton (Ed.), *New directions for elementary school mathematics*. Reston, VA: National Council of Teachers of Mathematics.

Coburn, T. G. (1987). *How to teach mathematics using a calculator*. Reston, VA: National Council of Teachers of Mathematics.

Coburn, T. G. (1989). The role of computation in the changing mathematics curriculum. In P. R. Trafton (Ed.), *New directions for elementary school mathematics*. Reston, VA: National Council of Teachers of Mathematics.

Hazekamp, D. W. (1986). Components of mental multiplying. In H. L. Schoen (Ed.), *Estimation and mental computation*. Reston, VA: National Council of Teachers of Mathematics.

Hope, J. A. (1986). Mental computation: Aliquot parts. *Arithmetic Teacher*, 34(3), 16–17.

Hope, J. A., Leutzinger, L., Reys, B. J., & Reys, R. R. (1988). *Mental math in the primary grades*. Palo Alto, CA: Dale Seymour.

Hope, J. A., Reys, B. J., & Reys, R. (1987). *Mental math in the middle grades*. Palo Alto, CA: Dale Seymour.

Hope, J. A., Reys, B. J., & Reys, R. (1987). *Mental math in the junior high school*. Palo Alto, CA: Dale Seymour.

Hope, J. A., & Sherrill, J. M. (1987). Characteristics of unskilled and skilled mental calculators. *Journal for Research in Mathematics Education*, 18, 98–111.

Lester, F. K. (1989). Mathematical problem solving in and out of school. *Arithmetic Teacher*, 37(3), 33–35.

Leutzinger, L., Rathmell, E. C., & Urbatsch, T. D. (1986). Developing estimation skills in the primary grades. In H. L. Schoen (Ed.), *Estimation and mental computation*. Reston, VA: National Council of Teachers of Mathematics.

McBride, J. W., & Lamb, C. E. (1986). Number sense in the elementary classroom. *School Science and Mathematics*, 86, 100–107.

Reys, B. J. (1986). Teaching computational estimation: Concepts and strategies. In H. L. Schoen (Ed.), *Estimation and mental computation*. Reston, VA: National Council of Teachers of Mathematics.

Reys, B. J. (1988). Estimation. In T. R. Post (Ed.), *Teaching mathematics in grades K–8: Research based methods*. Boston: Allyn and Bacon.

Reys. R., Trafton, P., Reys, B., & Zawojewski, J. (1987). *Computational estimation: (Grades 6, 7, 8)*. Palo Alto, CA: Dale Seymour.

Reys, R. R. (1984). Mental computation and estimation: Past, present, and future. *The Elementary School Journal*, 84, 547–557.

Reys, R. R. (1986). Evaluating computational estimation. In H. L. Schoen (Ed.), *Estimation and mental computation*. Reston, VA: National Council of Teachers of Mathematics.

Sowder, J. (1989). Developing understanding of computational estimation. *Arithmetic Teacher*, 36(5), 25–27.

Sowder, J. T. (1990). Mental computation and number sense. *Arithmetic Teacher*, 37(7), 18–20.

Sowder, J. T., & Kelin, J. (1993). Number sense and related

topics. In D. T. Owens (Ed.), *Research ideas for the classroom: Middle grades mathematics*. New York: Macmillan Publishing Co.

Sowder, J. T., & Wheeler, M. M. (1989). The development of concepts and strategies used in computational estimation. *Journal for Research in Mathematics Education, 20*, 130–146.

Trafton, P. R. (1986). Teaching computational estimation: Establishing an estimation mind-set. In H. L. Schoen (Ed.), *Estimation and mental computation*. Reston, VA: National Council of Teachers of Mathematics.

Van de Walle, J. A. (1991). Redefining computation. *Arithmetic Teacher, 38*(5), 46–51.

Van de Walle, J. A., & Watkins, K. B. (1993). Early development of number sense. In R. J. Jensen (Ed.), *Research ideas for the classroom: Early childhood mathematics*. New York: Macmillan Publishing Co.

12 THE DEVELOPMENT OF FRACTION CONCEPTS

CHILDREN AND FRACTION CONCEPTS

Consider the illustrations of two-thirds in Figure 12.1. Our adult knowledge confirms that each shows two-thirds, yet what is it that these models have in common? They cannot be matched like seven blocks with seven cards or seven fingers. Some are circles, some are dots, and some are lines. Some have many elements, and some have only one. If shown a rectangle, you cannot say what fraction it is. Some other shape or rectangle must also be identified as the unit or whole. Even the symbolism is a problem. The relationship represented by $\frac{2}{3}$ is represented just as well by $\frac{6}{9}$.

The point of the previous discussion is simply to heighten your awareness that fractions are not trivial concepts, even for middle-school children. A fraction is an expression of a relationship between a part and a whole. Helping children construct that relationship and connect it meaningfully to symbolism is the topic of this chapter.*

---

*To be technically correct, we should say that the relationship between a part and a whole is a *rational number*, and that a fraction is one type of symbolism used to represent a rational number. This number versus symbol distinction is not made in this book in the context of rational numbers. The term *fraction* is used in reference to the concept of number as well as the symbolism. The context of the discussion will generally make the intent clear. Furthermore, the distinction is not useful for children, especially not before the seventh or eighth grade.

FRACTIONS IN THE CURRICULUM

It has been traditional to include some minimal exposure to fractions in grades K to 4, with each successive grade spending just a little more time on fraction concepts than the one before. Usually some limited addition and subtraction of fractions is begun in grade 4, with the development of all of the operations for fractions introduced in grade 5. Continued review and reteaching occur in grades 6, 7, and 8. This explosion of procedural knowledge (symbolic rules) at about the fifth grade is generally not supported by strong conceptual knowledge of fraction meanings, because the curriculum simply has not provided time for the complex development that fraction concepts require.

As with many other topics in elementary school mathematics, there is a movement to delay the rush toward symbolism and especially computations with fractions. Research efforts are uncovering more of the difficulties children have with fractions and providing suggestions for how to help children construct these ideas (Behr, Lesh, Post, & Silver, 1983; Post, Wachsmuth, Lesh, & Behr, 1985; Payne, 1976; Pothier, & Sawada, 1983). This delay in symbolism presents some instructional problems that teachers need to confront:

How do you deal with fractions without symbols?
What models should be used? How do you use them?
How can all children have models?

221

222 12 / THE DEVELOPMENT OF FRACTION CONCEPTS

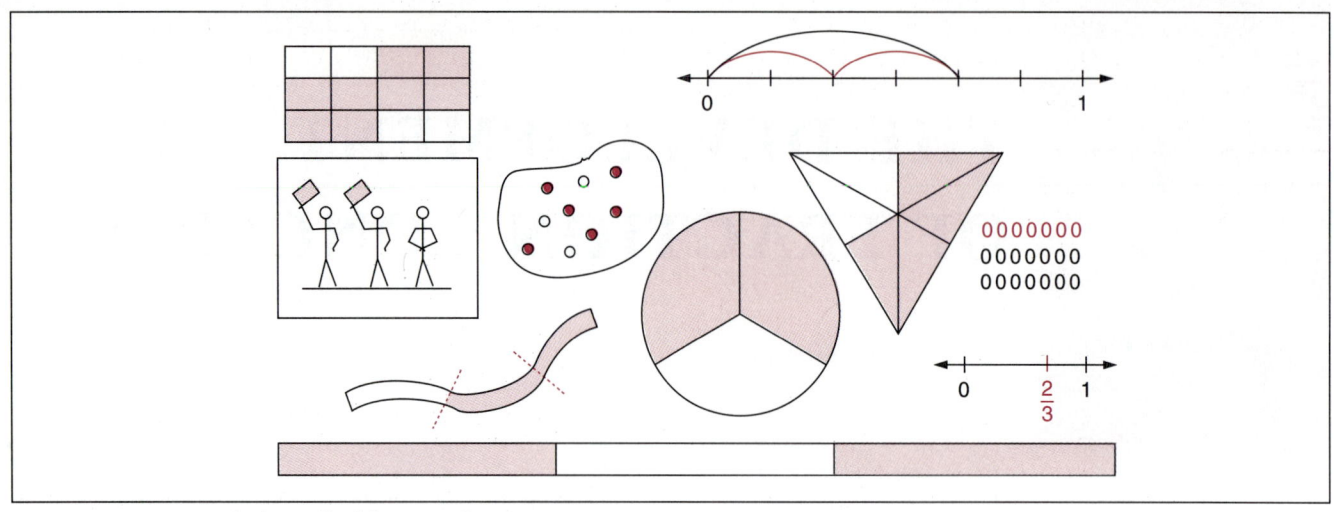

FIGURE 12.1: *Describe how all of these are the same.*

How much can you do with fractions before you add, subtract, multiply, and divide?

How will children learn about fractions without rules?

This entire chapter is about fraction concepts with no mention of adding, subtracting, multiplying, or dividing.

THREE CATEGORIES OF FRACTION MODELS

Models must be used at all grade levels to develop fraction concepts adequately. Further, the variety of representations for fractions suggested by Figure 12.1 indicates that chil-

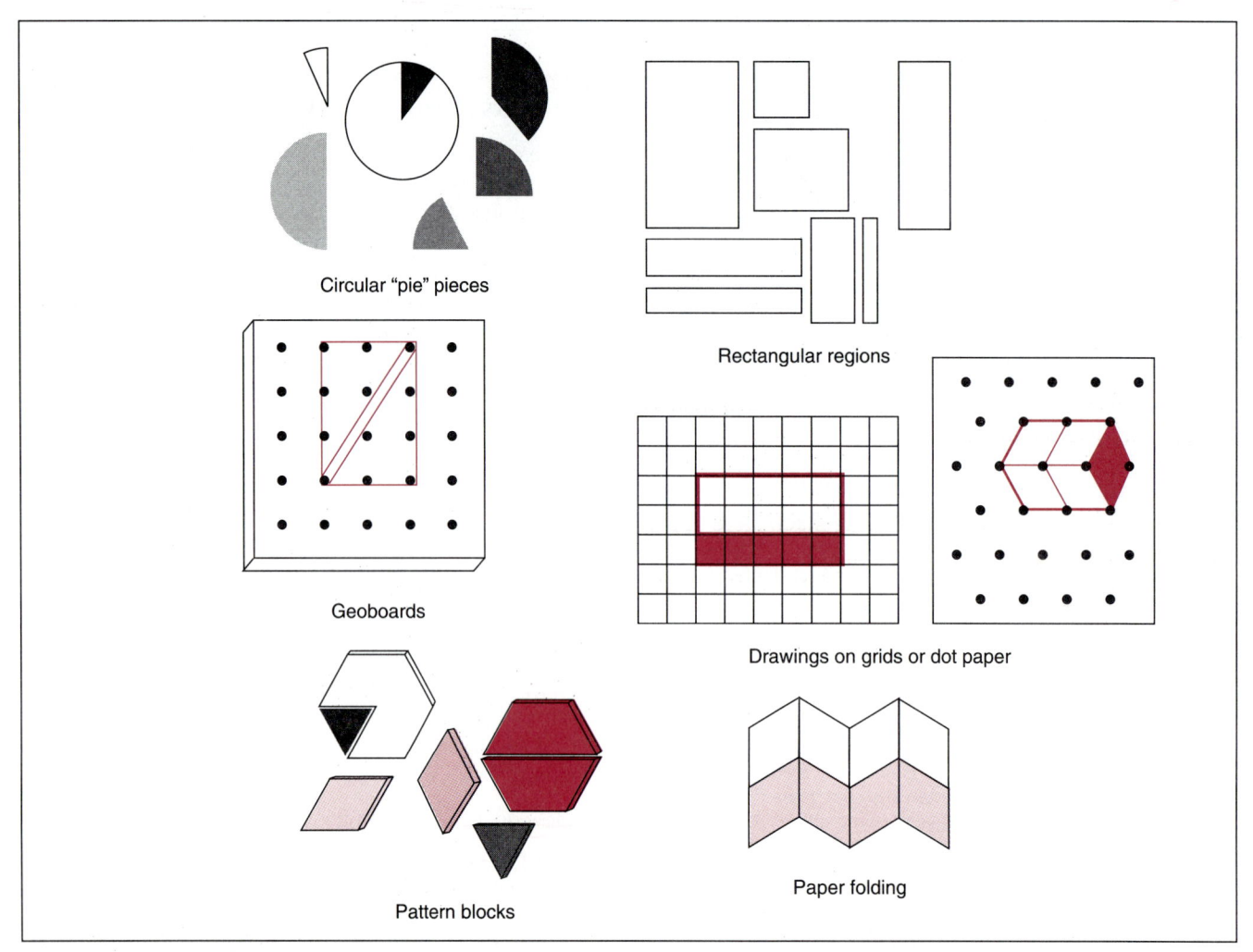

FIGURE 12.2: *Area or region models for fractions*

dren should have experiences with a wide assortment of models. In this section, three categories of models are presented with numerous examples of each. Most of the activities presented in the chapter can be done with all three model types. A change in the model is usually a significant change in the activity from the viewpoint of the children. As you examine the various ideas in the chapter, consider how the same activity could be done with different models and different types of models.

REGION OR AREA MODELS

In region models, a surface or area is subdivided into smaller parts. Each part can be compared with the whole. Almost any shape can be partitioned into smaller pieces. Figure 12.2 shows a variety of models in this category.

Circular regions and rectangle models can be duplicated on tagboard or construction paper, laminated, and cut into fraction kits kept in plastic bags. (Masters for circular models are included in the Black-line Masters and Materials Construction Tips.) Rectangles permit almost any piece to be designated as the whole so that other pieces change in fractional values accordingly. Pattern blocks, geoboards, and grid paper provide the same flexibility. Some commercial models for fractions, such as the rectangles used in "The Fraction Factory" (Holden, 1986), now come in classroom sets. With grid paper or dot paper, children can easily draw pictures to explore fraction ideas. Paper folding is a model available to everyone.

LENGTH OR MEASUREMENT MODELS

Length or measurement models are similar to area models, except the lengths are compared instead of area. Either lines are drawn and subdivided, or physical materials are compared on the basis of length, as shown in Figure 12.3 (p. 224). Manipulative versions provide much more opportunity for trial-and-error and exploration.

Fraction strips are a teacher-made version of Cuisenaire® rods. Both the strips and the rods have pieces that are in lengths of 1 to 10 measured in terms of the smallest strip or rod. Each length is a different color for ease of identification. As an alternative, strips of construction paper or adding-machine tape can be folded to produce equal-sized subparts. Older children can simply draw line segments on paper and subdivide them visually.

The rod or strip model provides the most flexibility while still having separate pieces for comparisons and for trial and error. To make fraction strips, cut 11 different colors of poster board into strips 2 cm wide. Cut the smallest strips into 2-cm squares. Other strips are then 4, 6, 8, ..., 20 cm, producing lengths 1 to 10 in terms of the smallest strip. Cut the last color into strips 24 cm long to produce a 12 strip. If you are using Cuisenaire® rods, tape a red 2 rod to an orange 10 rod to make a 12 rod. In the illustrations for this chapter, the colors of the strips will be the same as the corresponding lengths of the Cuisenaire® rods as given here:

White	1	Purple	4
Red	2	Yellow	5
Light Green	3	Dark Green	6
Black	7	Blue	9
Brown	8	Orange	10

a pink strip or a "rorange" (red-orange) rod is 12

The number line is a significantly more sophisticated measurement model (Bright, Behr, Post, & Wachsmuth, 1988). From a child's vantage, there is a real difference between putting a number on a number line and comparing one length to another. Each number on a line denotes the distance of the labeled point from zero. Place the numbers $\frac{2}{3}$ and $\frac{3}{4}$ on a number line, and consider how a child would think about these numbers in the context of that model.

SET MODELS

In set models the whole is understood to be a set of objects, and subsets of the whole make up fractional parts. For example, 3 objects are one-fourth of a set of 12 objects. The set of 12, in this example, represents the whole or one. It is the idea of referring to sets of counters as single entities that contributes to making set models difficult for primary-age children. However, the set model helps establish important connections with many real-world uses of fractions and with ratio concepts. Sets can profitably be explored by grades 3 or 4. Figure 12.4 (p. 224) illustrates several set models for fractions.

Any type of counter can be used to model fractional parts of sets, including drawing X's and O's. However, if the counters are colored in two colors on opposite sides, then the counters can easily be flipped to change their color to model various fractional parts of a whole set. Two-sided counters can be purchased, or colored picture-mount board can be cut into 2-cm squares. Another simple alternative is to spray-paint lima beans on one side to make two-color beans.

DEVELOPING THE CONCEPT OF FRACTIONAL PARTS

Fractions are the first place in children's experiences where a number represents something other than a count. The

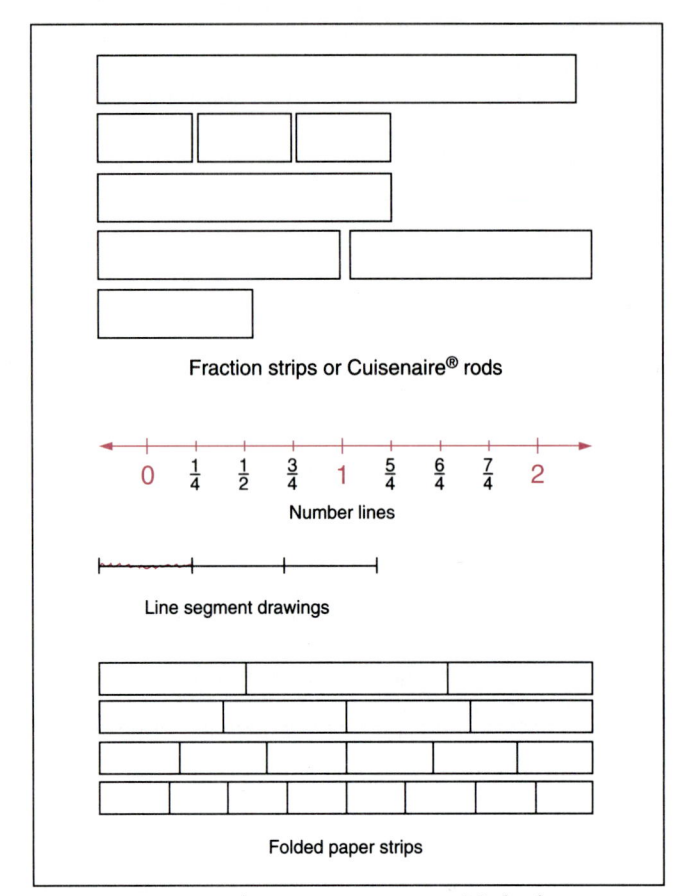

FIGURE 12.3: *Length or measurement models for fractions*

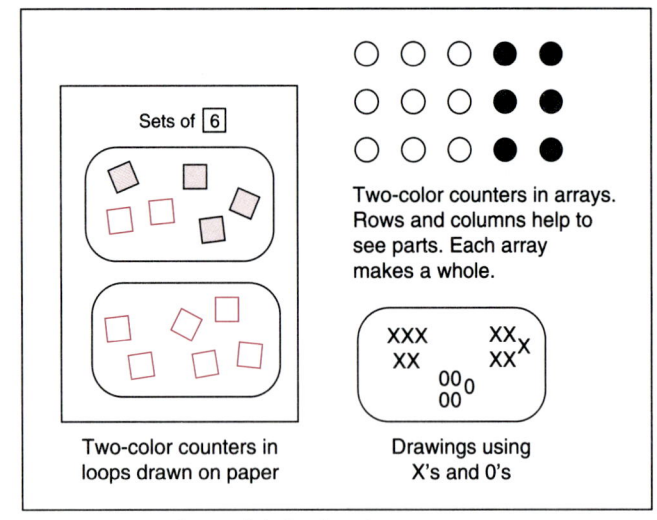

FIGURE 12.4: *Set models for fractions*

notion of a fractional part is completely relative to the whole and, in terms of models, may consist of one piece in some instances and many pieces in others. Helping children develop a firm understanding of fractional parts and of all the related nuances that go with that idea is critical if children are to have any number sense with fractions.

CONSTRUCTING FRACTIONAL PARTS

The one idea that young children do bring to the concept of fractions is the notion of partitioning or sharing. Children seem to understand the idea of separating a quantity into two or more parts to be shared among friends. They have all broken cookies or licorice sticks in half and given one half to a friend. Given a set of candies or a collection of toys, they have engaged in separating these quantities into parts. This understood concept of sharing is therefore a good place to begin the discussion of fractional parts. What must be added or highlighted in the discussion is the element of fair shares—shares that are equal.

Two Requirements for Fractional Parts

In Figure 12.5, some of the regions are divided into fourths, and some are not. Examine those that are not fourths. Are they cut into *equal* or *fair shares*? If they are, then why are they not fourths? (In order to have fourths you need four parts making up the whole.) But other wholes are cut into four parts. Why are these not fourths? (The parts are not fair shares or equal parts.)

A first goal in the development of fractions is to help children construct the idea of *fractional parts of the whole*. A fractional part exists when the whole has been partitioned into *equal-sized parts* or *fair shares*. These latter terms are useful in early discussions of fractions. The *names* of fractional parts are determined by the number of fair shares making up the whole: *halves*, two fair shares; *thirds*, three fair shares; *fourths*, four fair shares; and so on. For any particular fractional part, then, there are two aspects or components on which children need to reflect:

1. There must be the *correct number* of parts or shares making up the whole.
2. Each of the parts must be *equal* or *fair* shares; they must be the same size.

Note that the second requirement does not mean that the parts must be the same shape. Congruence is not necessary to have fractional parts. Nor should the concept of "part" be considered as a singular entity. The term "share" may be more appropriate. For example, a pizza cut into 12 equal slices can be partitioned into four equal shares or parts. Each share would consist of three pieces. The fact that the individual slices are also fractional parts (twelfths) in no way negates the fact that three of them make up a fourth of the pie.

We are in the habit of saying "four equal parts" when talking about fourths. The phrase comes out so quickly that the emphasis on the two requirements of four parts and fair shares may be lost.

Children can generalize the notion of fractional parts right from the beginning. A mistake that is frequently made is to assume that halves, thirds, and fourths are somehow

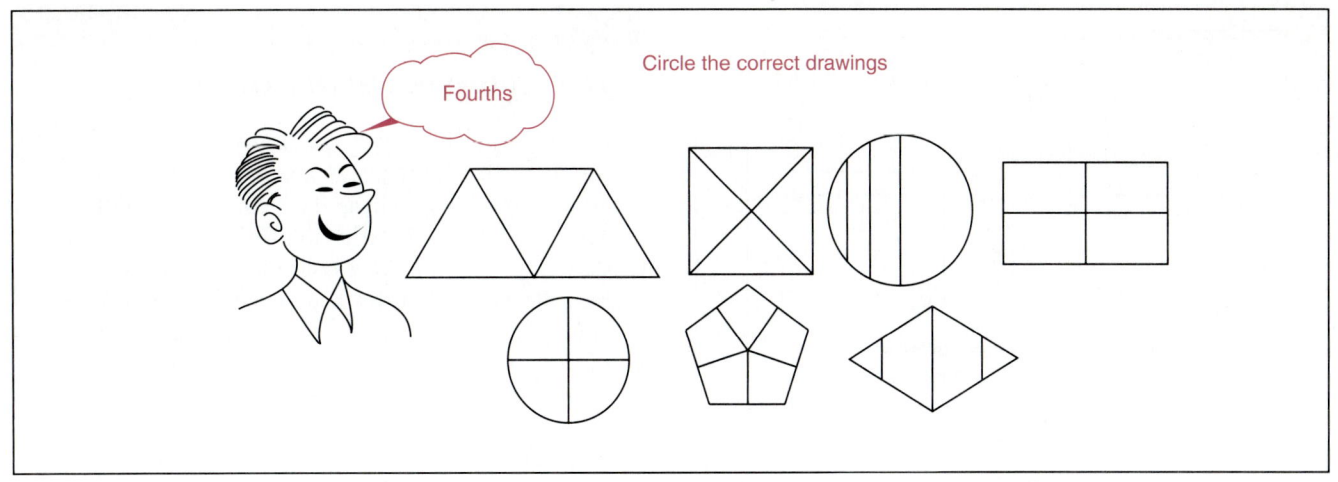

FIGURE 12.5: *Find correct examples of fourths. Why are the nonexamples wrong?*

easier and prior to sixths, eighths, or even twelfths. By introducing the general notion of fractional parts, children have more to explore and discuss. Once this generalization is made, all fractional parts are available. The traditional artificial restriction to one-half, one-third, and one-fourth in the early grades defeats learning more general relationships.

◈ Activities with Fractional Parts and Fraction Words

Fractional parts should be explored using all available models with the possible exception of sets for the very young. The terminology of *the whole* or *one whole* or simply *one* should also be introduced informally at the same time. In all activities, the *fraction words*–halves, thirds, fourths, fifths, and so on–are used orally and written out. These fractional parts or equal shares are the nouns or things or objects of fractions. They are the building blocks of virtually all fraction concepts. The early emphasis is on thirds or sixths, not on one-third or one-sixth.

12.1 CORRECT SHARES

As in Figure 12.5, show examples and non-examples of specified fractional parts. Have students identify the wholes that are correctly divided into the requested fractional parts, and explain why the nonexamples are incorrect. Use length and set models in a similar way.

In the **Correct Shares** activity, the most important part is the discussion of the nonexamples. The wholes are already partitioned either correctly or incorrectly, and the children were not involved in the partitioning. It is also useful for children to create designated equal shares given a whole, as they are asked to do in the next activity.

12.2 FINDING FAIR SHARES

Give students models, and have them find fifths, or eighths, and so on, using the model. (Models should never have fractions written on them.) The activity is especially interesting when different wholes can be designated in the same model. That way, a given fractional part does not get identified with a special shape or color but with the relationship of the part to the designated whole. Some ideas are suggested in Figure 12.6 (p. 226).

Folding paper is another good way to involve children in construction of fractional parts as in the last activity. Folds into halves, fourths, and eighths are clearly the easiest since they are successive halves. With some help, children can fold strips into three parts and from these thirds fold sixths. Having children draw slices to subdivide a circle or even a rectangle is very difficult and is not recommended. On a rectangle, children will frequently draw four lines for fourths and then realize they have made five parts. Others will make three nice equal parts and then realize that the last one is way too large for fourths. These and other difficulties with eye–hand coordination get in the way of the concept development. Children can use models or paper folding and develop fractional part concepts and still not be able to cut a shape into thirds with any accuracy at all. Try to keep the focus on the number and fairness of the shares and less on the ability to draw them.

Notice when partitioning sets that children frequently confuse the number of counters in a share with the name of the share. In the example in Figure 12.6, the 12 counters are partitioned into four sets—fourths. Each share or part has three counters, but it is the number of shares that makes the partition show fourths.

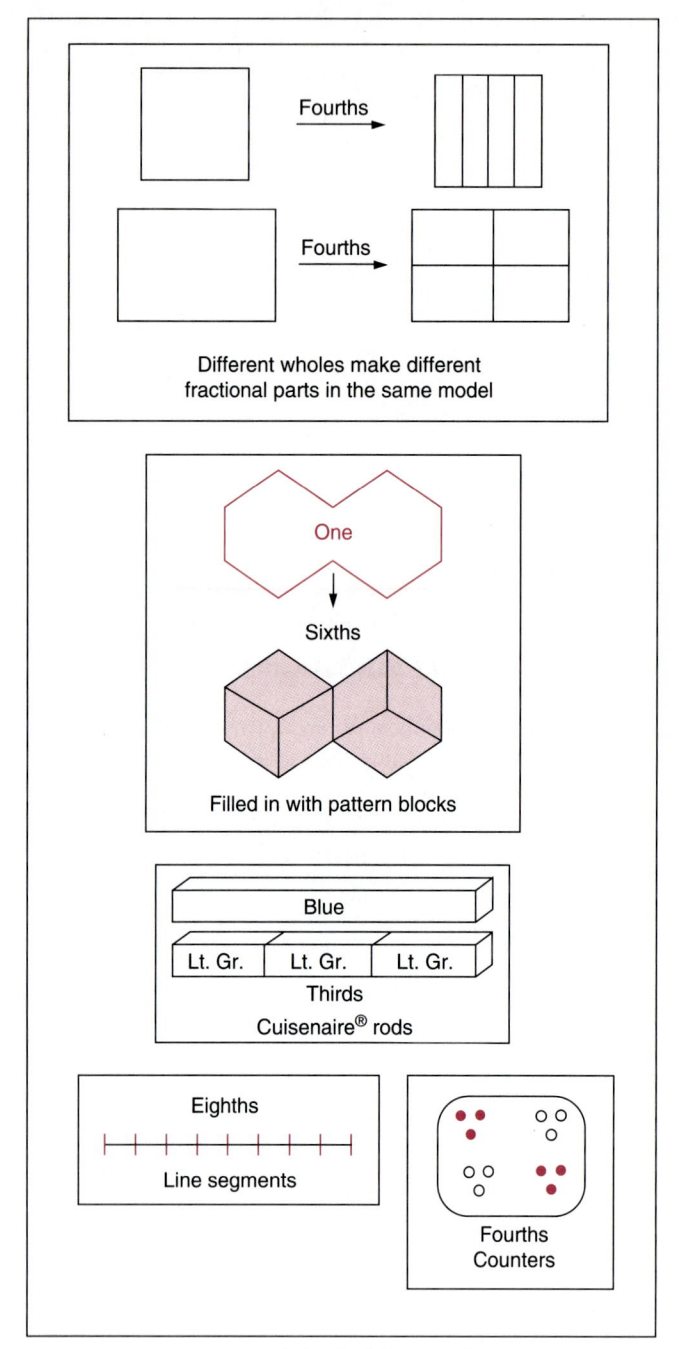

FIGURE 12.6: *Given a whole, find fractional parts.*

◆ Counting Fractional Parts

The importance of fractional parts will become increasingly obvious as you read this chapter. As noted already, fractional parts or equal shares are the objects of fractions. Once these objects are developed as ideas, they can then be called fourths or thirds, or whatever, and counted in the same way one counts apples or other objects. Fractions greater than one whole can similarly be understood this way. Seven-thirds is just seven things called thirds. By counting fractional parts we can help children develop a completely generalized system for naming fractions before they learn about fraction symbolism. The oral names then can be connected easily to the fraction notation.

12.3 COUNTING WITH FRACTIONS

Once students have identified fourths, for example, count fourths. Show five or six fourths on the overhead. "How many fourths? Let's count: one-fourth, two-fourths, three-fourths, four-fourths, five-fourths, six-fourths." Count other collections of fourths. Ask if a collection of fourths that have been counted is more or less than one whole or more or less than two wholes. As shown in Figure 12.7, make informal comparisons among different counts. "Why did we get almost two wholes with seven-fourths, and yet we don't even have one whole with seven-twelfths? What is another way we could say seven-thirds?" (Two wholes and one-third or one whole and four-thirds.)

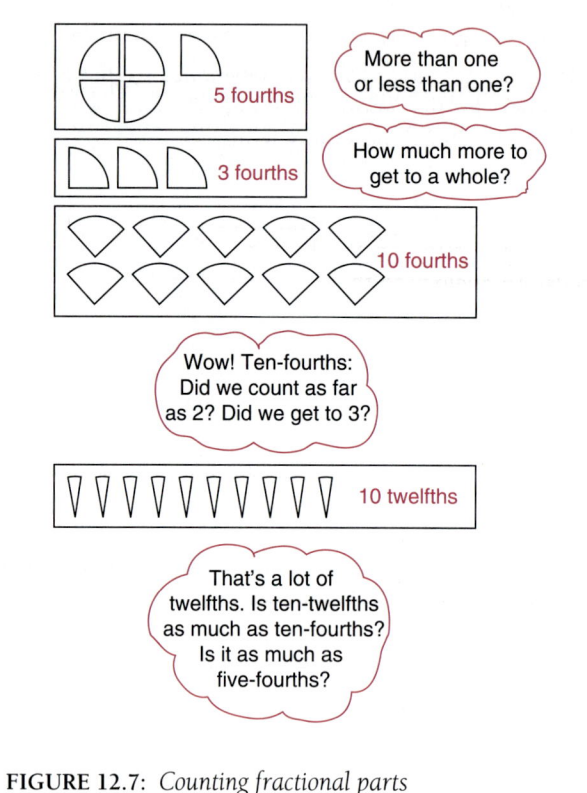

FIGURE 12.7: *Counting fractional parts*

Counting fractional parts will play a major role in much of the development of fractional concepts. In the beginning, counting fractional parts can lay the groundwork for several important ideas. The idea that eighths are smaller than thirds, for example, is an illusive idea. Counting different-sized parts and discussing how many parts it takes to make one whole is one opportunity to begin reflection on this idea. "Count three-fourths and then count out three-twelfths. Which is more? Why?"

Counts should frequently include sets such as 5 thirds, or 11 fourths, or other sets greater than one whole. When

12 / THE DEVELOPMENT OF FRACTION CONCEPTS   227

the count gets to one whole (3 thirds or 4 fourths), stop and discuss other names for that amount. Such counts will dispel the notion that a fraction is something less than one. With older children, counts can also be stopped at places where equivalent fractions are evident such as 6 twelfths or 9 sixths. These discussions will provide nice forerunners for equivalent fractions.

◆ CONNECTING CONCEPTS WITH SYMBOLISM

Fraction symbolism should be delayed as long as possible. The activities in the previous section can all be done orally and with the use of written fraction words such as 7 *fourths* or 3 *eighths*. Eventually the standard symbolic form must be used.

Always write fractions with a horizontal bar, not a slanted one. Write $\frac{3}{4}$, not 3/4.

◆ Meaning of the Top and Bottom Numbers

Consider a grocer who has developed a special way to take orders for his produce. If the order is for "seven apples and three bananas," he simply writes:

$$\frac{7}{A} \quad \frac{3}{B}$$

The number on the top tells him how many items he needs. The letter on the bottom tells him what it is he needs. An analogy to the grocer's idea can be used for fraction symbolism.

Display several collections of fractional parts, and have children count each set as discussed earlier and write the count using fraction words as in "3 fourths." Explain that you are going to show how these fraction words can be written much more easily than writing out the words. For each collection write the standard fraction next to the word. Display some other collections, and ask students if they can tell you how to write the fraction. Rather than an explanation, use the already developed oral language both with models and when writing the fraction symbols.

Next, have children count by fourths as you write the fractions on the board. Repeat for other parts.

$$\frac{1}{4}, \frac{2}{4}, \frac{3}{4}, \frac{4}{4}, \frac{5}{4}, \frac{6}{4}, \frac{7}{4}, \frac{8}{4}, \frac{9}{4}$$

$$\frac{1}{6}, \frac{2}{6}, \frac{3}{6}, \frac{4}{6}, \frac{5}{6}, \frac{6}{6}, \frac{7}{6}, \frac{8}{6}, \frac{9}{6}$$

$$\frac{1}{8}, \frac{2}{8}, \frac{3}{8}, \frac{4}{8}, \frac{5}{8}, \frac{6}{8}, \frac{7}{8}, \frac{8}{8}, \frac{9}{8}$$

Discuss each row. How are they alike? How different? What part of each row is like counting? Why does the bottom number stay the same as you count fourths or sixths? Finally, ask students:

What does the top number in a fraction tell you?

What does the bottom number in a fraction tell you?

Answer these two questions yourself. Try to think in terms of fractional parts and what has been covered up to this point. Write your explanations for top and bottom numbers meanings. Try to use children's language. Explore several different ways of saying the meaning. Your meaning should not be tied to a particular model.

Here are some reasonable explanations for the top and bottom numbers.

> *Top number:* This is the counting number. It tells how many shares or parts we have. It tells how many have been counted. It tells how many parts we are talking about. It counts the parts or shares.

> *Bottom number:* This tells *what* is being counted. It tells what fractional part is being counted. If it is a 4, it means we counted fourths, and if it is a 6, we counted sixths, and so on.

Notice that if the concept of fractional parts is well developed and children can give an explanation for any fraction word, it is not necessary to include that in the meaning of the bottom number. In fact it is clumsy: "It tells how many of the equal parts being counted that it takes to make a whole." Not only is that clumsy, it detracts from the two simple but important ideas:

The top number *counts*.

The bottom number tells *what is counted*.

The *what* of fractions are the fractional parts. They can be counted. Fraction symbols are just a shorthand for saying how many and what.

◆ Numerator and Denominator: A Digression

To count a set is to *enumerate* it. Enumeration is the process of counting. The common name for the top number in a fraction is *numerator*.

A $1 bill, a $5 bill, and a $10 bill are said to be bills of different *denominations*. Similarly, the word *denomination* is used to categorize people by religion (such as Baptists, Presbyterians, Episcopalians, Catholics). A denomination is the name of a class or type of thing. The common name for the bottom number in a fraction is *denominator*.

Up to this point the terms *numerator* and *denominator* have not been used, as will be the case in most of the rest of the chapter. Why? No child in the third grade would mistake the designations top number and bottom number. The words numerator and denominator have no common reference for children. Some may feel it is important that children use these words. Whether used or not, it is clear that the words themselves will not assist young children in understanding the meanings.

Mixed Numbers and Improper Fractions

In the fourth National Assessment of Education Progress, about 80% of seventh graders could change a mixed number to an improper fraction, but fewer than half knew that $5\frac{1}{4}$ was the same as $5 + \frac{1}{4}$ (Kouba et al., 1988a). The result indicates that many children are using a mindless rule that in fact is relatively easy to construct.

12.4 MIXED NUMBER NAMES

Use models to display collections such as 13 sixths or 11 thirds. Have children orally count the displays and give at least two names for each. Then discuss how they could write these different names using numbers. They already know how to write $\frac{13}{6}$ or $\frac{11}{3}$. For two wholes and one-sixth, a variety of alternatives might be suggested: 2 and $\frac{1}{6}$ or 2 wholes and $\frac{1}{6}$, or $2 + \frac{1}{6}$. All are correct. After doing this with other collections, explain that $2 + \frac{1}{6}$ is usually written as $2\frac{1}{6}$ with the "+" being left out or understood.

Now reverse the process. Write mixed numbers on the board and have students make that amount with models, using only one kind of fractional part. When they have done that, they can write the simple fraction that results. (The term *improper* fraction is an unfortunate yet common term for a simple fraction greater than one.)

After much back-and-forth between models and symbols using fractions greater than 1, see if students can figure out a simple fraction for a mixed number and a mixed number for an improper fraction. *Do not provide any rules or procedures.* Let students work this out for themselves. A good student explanation for $3\frac{1}{4}$ might involve the idea that there are 4 fourths in one whole, so there are 8 fourths in two and 12 fourths in three wholes. Since there is one more fourth, that is 13 fourths in all, or $\frac{13}{4}$.

For fractions greater than 2 it is a good exercise to have students find other expressions besides the usual. For example, $4\frac{1}{5}$ is not only $\frac{21}{5}$ but also $3\frac{6}{5}$, $2\frac{11}{5}$, and $1\frac{16}{5}$. This idea is extremely useful later when subtracting fractions, as in $5\frac{1}{8} - 2\frac{3}{8}$.

12.5 CALCULATOR FRACTION COUNTING

Calculators that permit fraction entries and displays are now quite common in schools. The TI-Math Explorer™ and TI-34™ are two examples. If these are available, they can be quite effective in helping children understand fraction symbolism. Counting by fourths, for example, can be done by pressing $+$ 1 $/$ 4 $=$. The display will show $\frac{1}{4}$. Students should have models for fourths to manipulate as they count. When they add successive fourths to the pile, they press $=$ for each piece. At four-fourths the calculator shows 4/4. Continued presses simply add onto the numerator. To see the corresponding mixed number, press the $\boxed{a\,b/c}$ key. The mixed number for $\frac{6}{4}$ is shown as 1 ⌐ 2/4. Since the calculator will not reduce the fraction automatically, the count agrees with the physical models.

Fraction calculators provide an exciting and powerful way to help children develop fractional symbolism. A variation on the last activity is to show children a mixed number such as $3\frac{1}{8}$ and ask how many counts of $\frac{1}{8}$ on the calculator will it take to count that high. The students should try to stop at the correct number ($\frac{25}{8}$) before pressing the $\boxed{a\,b/c}$ key.

PARTS AND WHOLES EXERCISES

The exercises in this section can help children develop their understanding of fractional parts as well as the meanings of the top and bottom numbers in a fraction. Models are used to represent wholes and parts of wholes. Written or oral fraction names represent the relationship between the parts and wholes. Given any two of these—whole, part, and fraction—the students can use their models to determine the third.

Any type of model can be used as long as different sizes can represent the whole. For region and area models it is also necessary that single regions or lengths be used to represent nonunit fractions. Traditional pie pieces do not work since the whole is always the circle, and all the pieces are unit fractions.

Sample Exercises

In Figures 12.8, 12.9, and 12.10 examples of each type of exercise are provided. Each figure includes examples with a region model (freely drawn rectangles), a length model (Cuisenaire® rods or fraction strips), and set models. It would be a good idea to work through these exercises before reading the next section. For the rectangle models, simply sketch a similar rectangle on paper. For the strip or rod models, use Cuisenaire® rods or make some fraction strips. The colors used correspond to the actual rod colors. Lengths are not given in the figures so that you will not be tempted to use an adult-type numeric approach. If you do not have access to rods or strips, just draw lines on paper. The process you use with lines will correspond to what is done with rods.

Answers and explanations are in Figures 12.11, 12.12, and 12.13 (pp. 230–231). The questions that ask for the fraction when given the whole and part require a lot of

12 / THE DEVELOPMENT OF FRACTION CONCEPTS 229

trial-and-error and can frustrate young students. Be sure that appropriate fractional parts are available for the region and length versions of these questions.

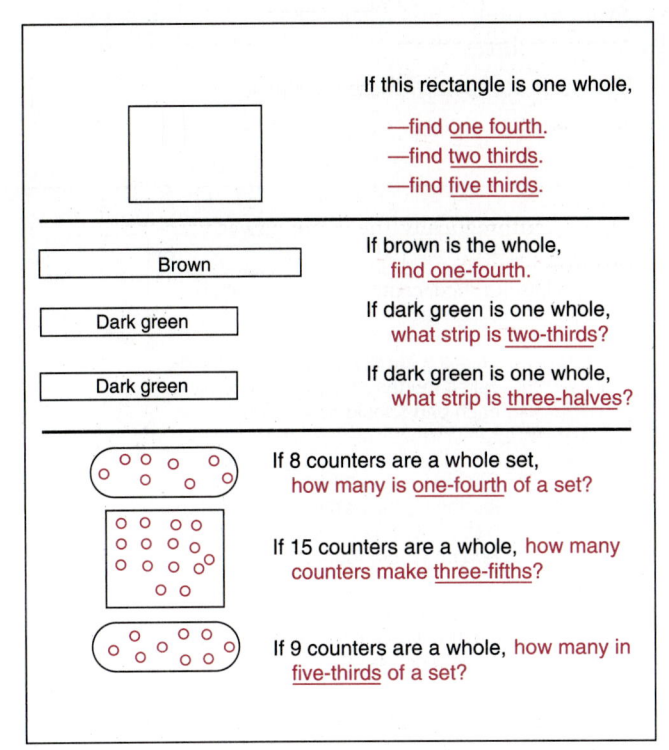

FIGURE 12.8: *Given the whole and fraction, find the part (answers in Figure 12.11).*

- If this rectangle is one whole,
  - —find one fourth.
  - —find two thirds.
  - —find five thirds.
- If brown is the whole, find one-fourth.
- If dark green is one whole, what strip is two-thirds?
- If dark green is one whole, what strip is three-halves?
- If 8 counters are a whole set, how many is one-fourth of a set?
- If 15 counters are a whole, how many counters make three-fifths?
- If 9 counters are a whole, how many in five-thirds of a set?

Part and Whole Exercises in the Classroom

The exercises work much better with physical models as opposed to drawings. The models allow students to use a trial-and-error approach to test their reasoning as they go. Also, younger children simply have limited ability to partition lines or regions into smaller equal parts.

Care must be taken to ask only questions for which there is an answer within the model. For example, if you were using fraction strips, you could ask students "If the blue strip (9) is the whole, what strip is two-thirds?" The answer is the strip that is 6 units long, or dark green. You could not, however, ask students to find three-fourths of the blue strip, since fourths would each be $2\frac{1}{4}$ units long, and no strip has that length. With rectangular pieces of various sizes, you will likewise need to work out your questions in advance and be sure that they are answerable within the set.

Present each exercise to the full class to observe how different children are approaching the task. It may also be good to have students work in groups. After each task, have the students explain or justify their results.

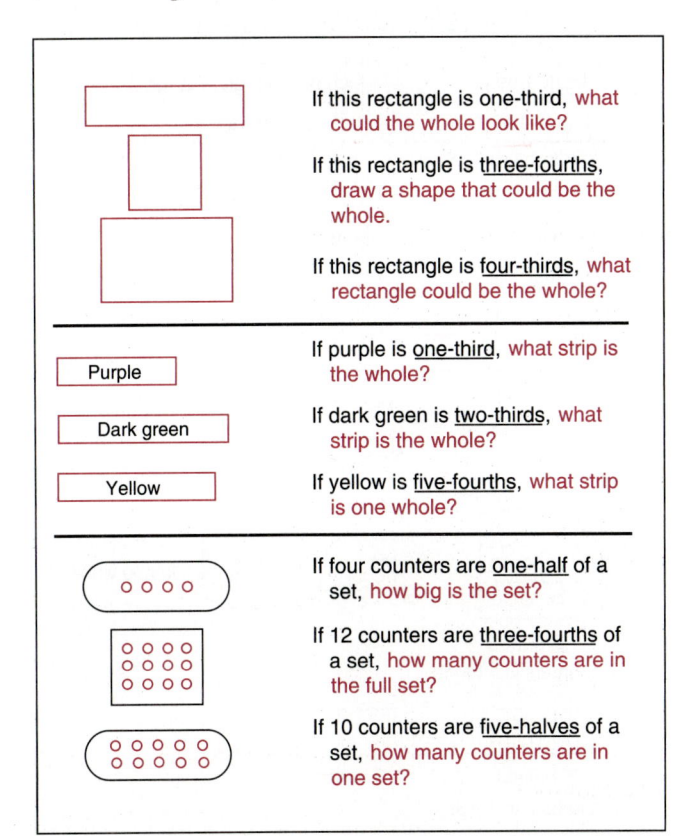

FIGURE 12.9: *Given the part and the fraction, find the whole (answers in Figure 12.12).*

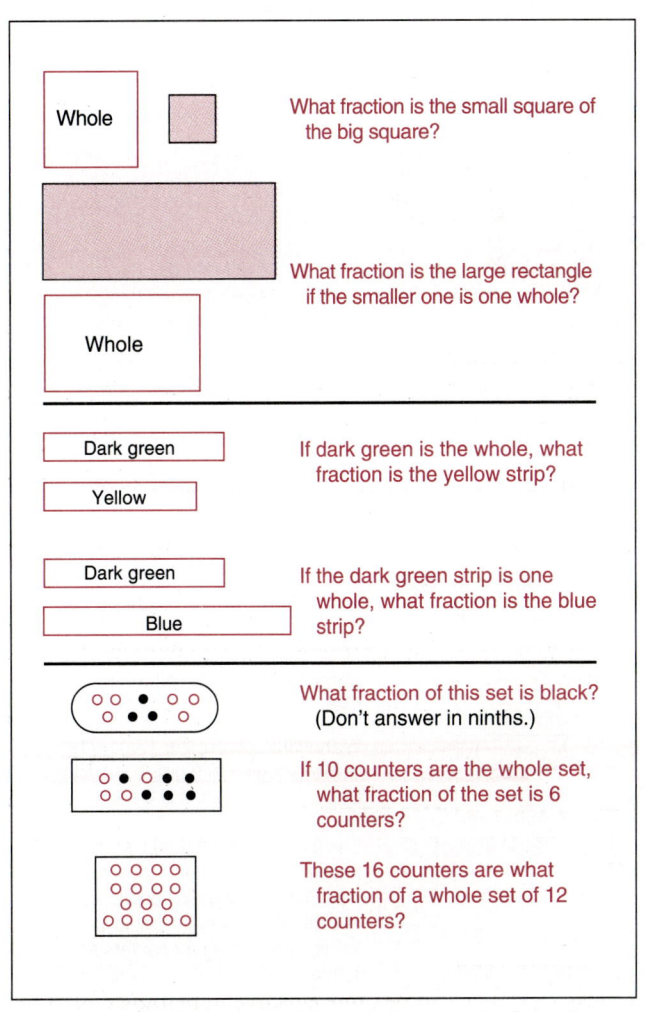

- If this rectangle is one-third, what could the whole look like?
- If this rectangle is three-fourths, draw a shape that could be the whole.
- If this rectangle is four-thirds, what rectangle could be the whole?
- If purple is one-third, what strip is the whole?
- If dark green is two-thirds, what strip is the whole?
- If yellow is five-fourths, what strip is one whole?
- If four counters are one-half of a set, how big is the set?
- If 12 counters are three-fourths of a set, how many counters are in the full set?
- If 10 counters are five-halves of a set, how many counters are in one set?

FIGURE 12.10: *Given the whole and the part, find the fraction (answers in Figure 12.13).*

- What fraction is the small square of the big square?
- What fraction is the large rectangle if the smaller one is one whole?
- If dark green is the whole, what fraction is the yellow strip?
- If the dark green strip is one whole, what fraction is the blue strip?
- What fraction of this set is black? (Don't answer in ninths.)
- If 10 counters are the whole set, what fraction of the set is 6 counters?
- These 16 counters are what fraction of a whole set of 12 counters?

230 12 / THE DEVELOPMENT OF FRACTION CONCEPTS

A *unit fraction* is one that designates a single fractional part. In symbolic form, a unit fraction has a 1 in the numerator. Therefore, $\frac{1}{2}, \frac{1}{6}$ and $\frac{1}{12}$ are unit fractions, and $\frac{3}{4}, \frac{7}{3}$, and $2\frac{1}{6}$ are nonunit fractions. Questions involving unit fractions are generally the easiest. The hardest questions usually involve fractions greater than 1. For example, "If fifteen chips are five-thirds of one whole set, how many chips are in a whole?" The same question can be asked in terms of a mixed number (15 chips are the same as one and two-thirds or $1\frac{2}{3}$). The mixed-number question is more difficult because it must first be translated to an improper fraction.

◆ Exercise Difficulties

Many students find these parts and wholes activities quite challenging and even confusing. There is absolutely no benefit in providing students with rules for solving them or even telling them what to do next. Rather, try to focus on the concepts of unit fractions and fractional parts. By way of example, the following hypothetical interchange illustrates the type of problem students may have. Sketch a rectangle on a piece of paper and follow along.

The student has a small rectangle with the accompanying question: "If this is four-thirds of a box, what might one whole box look like?" The student begins by dividing the box into three parts and then is stumped and does not know what to do.

Teacher: How big is the rectangle?
Student: Four-thirds.
Teacher: Does that mean it is four things or three things
Student: Well, thirds means three. So I divided it into three.
Teacher: So, each of these are thirds?
Student: Yes, thirds.

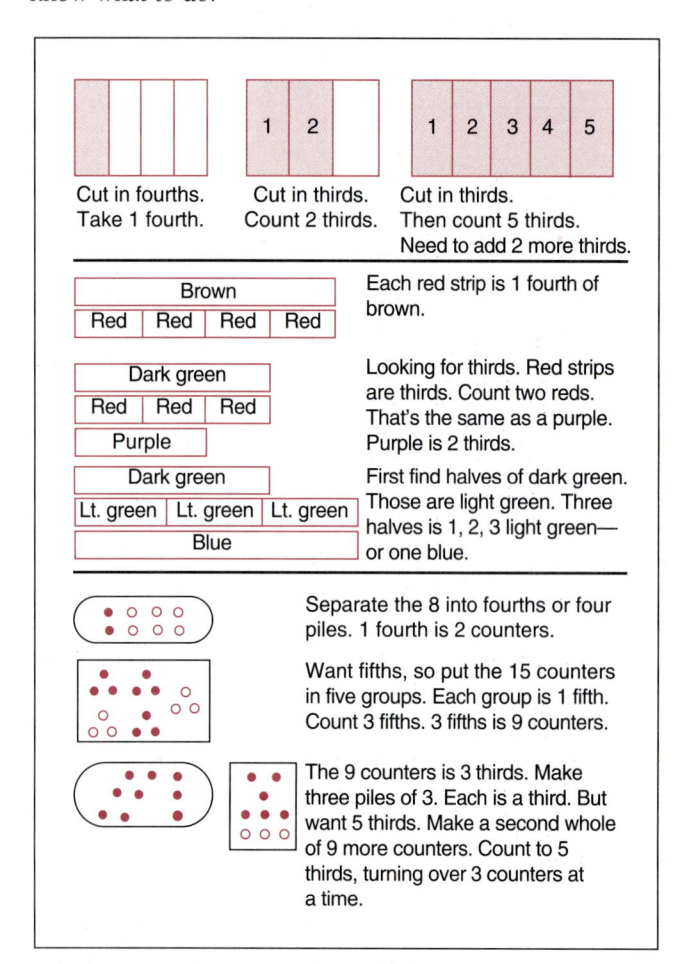

FIGURE 12.11: *Answers to Figure 12.8*

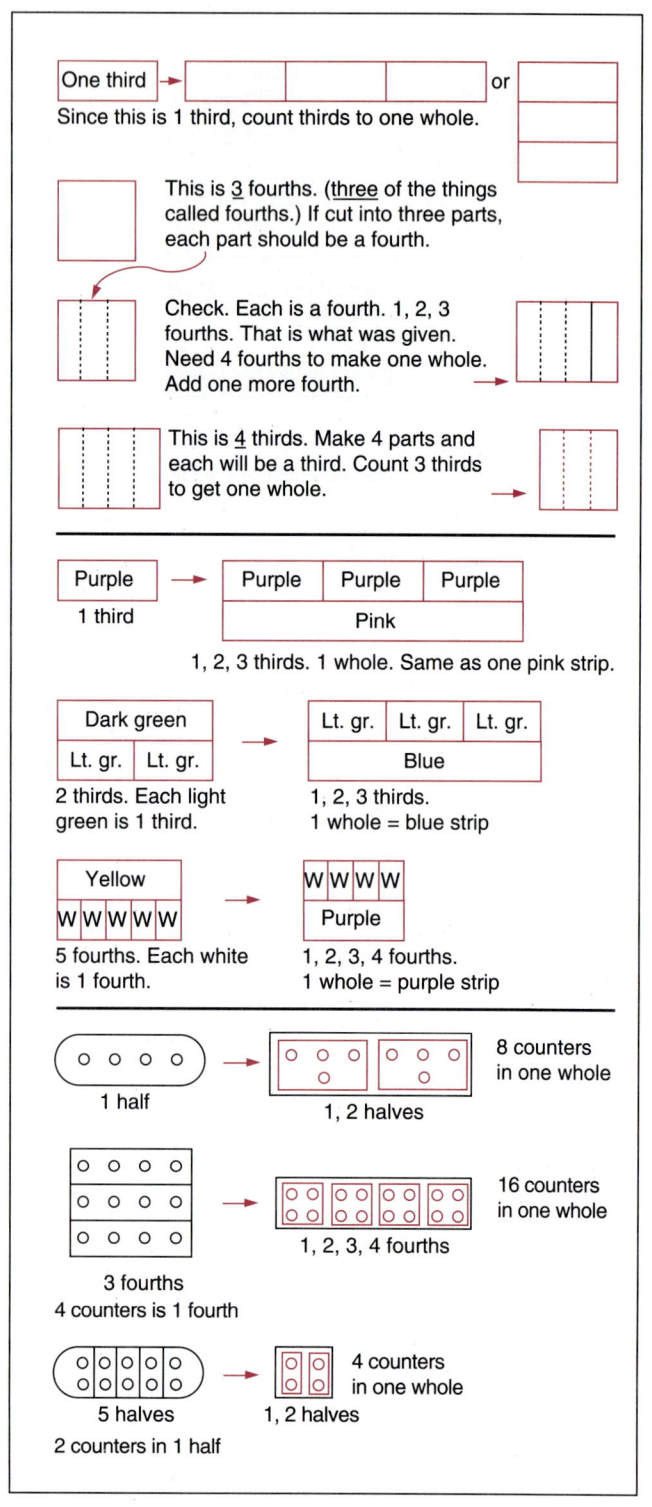

FIGURE 12.12: *Answers to Figure 12.9*

Teacher: Let's count. (together) One-third, two-thirds, three-thirds. How much is three-thirds?

Student: One whole.

Teacher: But the box is not *three*-thirds but *four*-thirds. How many thirds should be in the box you started with?

Student: Four. It's *four*-thirds. I have to make four parts. (Starts over and draws the box divided into four parts.)

Teacher: Now count. These are what kind of pieces?

At this point the teacher wants the child to stick with the idea that the parts are thirds, and he or she has four of them.

In this and every other example, a unit fractional part comes into play. Once this part is identified in the model, students can count unit parts up to one of the pieces (part or whole) that they had to begin with. This confirms they are correct. If the student counts four-thirds, that will agree with what was given, namely that the original box was four-thirds. From that point, counting three-thirds to find the whole is trivial. Again notice the importance of counting parts.

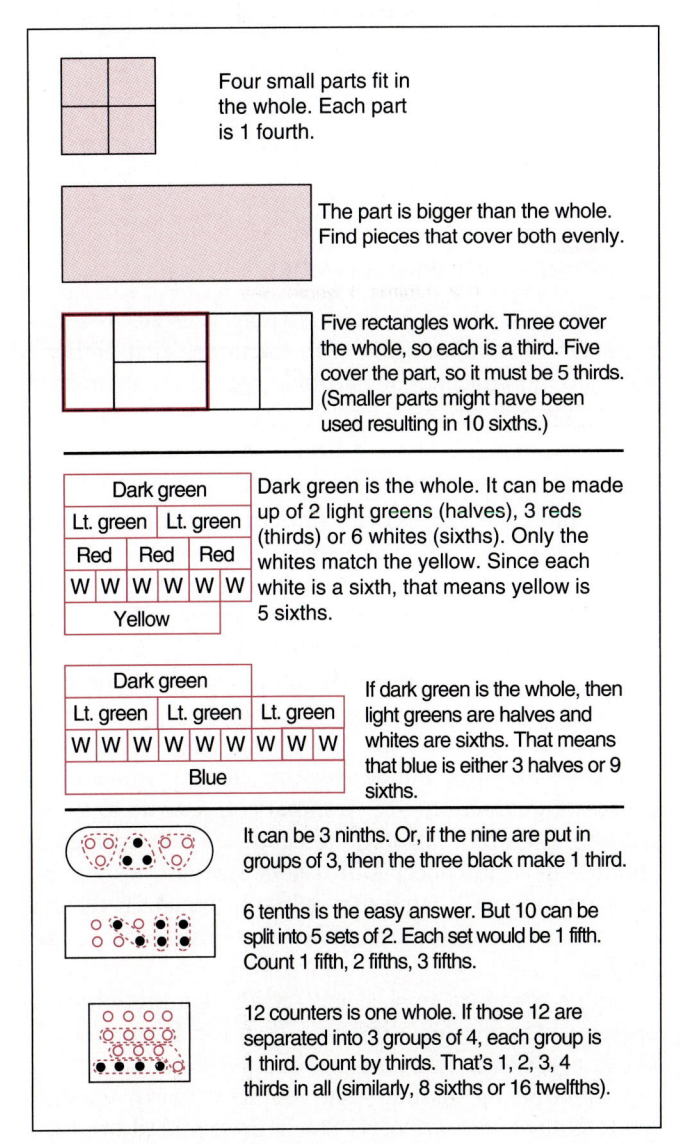

Try to avoid being the answer book for your students. Make students responsible for determining the validity of their own answers. In these exercises, the results can always be confirmed in terms of what is given. Students will learn that they can understand these ideas. There are no obscure rules. It makes sense!

When students do a series of exercises of one type, a pattern will begin to set in because each exercise is similar. When you mix exercises from the first two categories (whole-to-part and part-to-whole), students will experience more difficulty. The goal, however, is not to establish routines but to encourage reflective thought.

FRACTION NUMBER SENSE

The focus on fractional parts in the previous section is an important beginning since it helps students reflect on the meanings of the part-to-whole relationship and the corresponding meanings of the top and bottom numbers of fractions. Number sense with fractions demands more. In particular, a number sense with fractions requires that students have some intuitive feel for fractions. They should know "about" how big a particular fraction is and be able to tell easily which of two fractions is larger. These ideas require building further on the relationships already discussed.

THE RELATIVE SIZE OF FRACTIONAL PARTS

There is a tremendously strong mind-set that children have about numbers that causes them difficulties with the relative size of fractions. In their experience, larger numbers mean more. The tendency is to transfer this whole-number concept to fractions incorrectly: Seven is more than four, so sevenths should be bigger than fourths. The inverse relationship between number of parts and size of parts cannot be told but must be a creation of each student's own thought process.

12.6 ORDERING UNIT FRACTIONS

List a set of unit fractions such as $\frac{1}{3}$, $\frac{1}{8}$, $\frac{1}{5}$, and $\frac{1}{10}$. Ask children to put the fractions in order from least to most. Challenge children to defend the way they ordered the fractions. The first few times you do this activity, have them explain their ideas by using models. Encourage the language of sharing—the more shares in one whole, the smaller each share will be.

FIGURE 12.13: *Answers to Figure 12.10*

Four small parts fit in the whole. Each part is 1 fourth.

The part is bigger than the whole. Find pieces that cover both evenly.

Five rectangles work. Three cover the whole, so each is a third. Five cover the part, so it must be 5 thirds. (Smaller parts might have been used resulting in 10 sixths.)

Dark green is the whole. It can be made up of 2 light greens (halves), 3 reds (thirds) or 6 whites (sixths). Only the whites match the yellow. Since each white is a sixth, that means yellow is 5 sixths.

If dark green is the whole, then light greens are halves and whites are sixths. That means that blue is either 3 halves or 9 sixths.

It can be 3 ninths. Or, if the nine are put in groups of 3, then the three black make 1 third.

6 tenths is the easy answer. But 10 can be split into 5 sets of 2. Each set would be 1 fifth. Count 1 fifth, 2 fifths, 3 fifths.

12 counters is one whole. If those 12 are separated into 3 groups of 4, each group is 1 third. Count by thirds. That's 1, 2, 3, 4 thirds in all (similarly, 8 sixths or 16 twelfths).

This idea is so basic to the understanding of fractions that arbitrary rules ("larger bottom numbers mean smaller fractions") are not only inappropriate but dangerous. Come back to this basic idea periodically. Children will seem to understand one day and revert to their more comfortable ideas about big numbers one or two days later.

CLOSE TO ZERO, HALF, AND ONE

Number sense with whole numbers is partly the ability to understand their relative size. We know that numbers less than 10 are quite a bit smaller than numbers near 100 but that any number less than 100 is relatively small compared to 1000. An analogous familiarity with fractions can be developed by comparing fractions to 0, $\frac{1}{2}$, and 1. For fractions less than one, this gives quite a bit of information. For example $\frac{1}{20}$ is small, close to zero, while $\frac{3}{4}$ is between $\frac{1}{2}$ and 1. The fraction $\frac{9}{10}$ is quite close to 1. Since any fraction greater than one is a whole number plus an amount less than one, this small range of zero to one is quite helpful when thinking about fractions.

12.7 CLOSE-TO FRACTIONS

Have your students name a fraction that is close to 1 but not more than 1. Next have them name another fraction that is even closer to 1 than that. For the second response, they have to explain why they believe the fraction is closer to 1 than the previous fraction. Continue for several fractions in the same manner, each one being closer to one than the previous fraction. Similarly, try close to 0 or close to $\frac{1}{2}$ (either under or over). The first several times you try this activity let the students use models to help with their thinking. Later, see how well their explanations are when they cannot use models or drawings. Focus discussions on the relative size of fractional parts.

Establishing fractions close to 0, $\frac{1}{2}$, and 1 is a good beginning for estimation of fractions. If you see an amount that looks to be a bit more than $\frac{1}{2}$, the type of reasoning in the last activity can help you find a fraction that is a good estimate of the amount you see.

12.8 ABOUT HOW MUCH?

Draw a picture like one of those in Figure 12.14 (or prepare some ahead of time for the overhead). Have each student write down a fraction that he or she thinks is a good estimate of the amount shown (or the indicated mark on the number line). Listen without judgment to the ideas of several students, and

---

and engage them in a discussion of why any particular estimate might be a good one. For these situations there is no correct answer but certainly some estimates will be in the ballpark. If children have difficulty coming up with an estimate, ask if they think the amount is closer to 0, $\frac{1}{2}$, or 1.

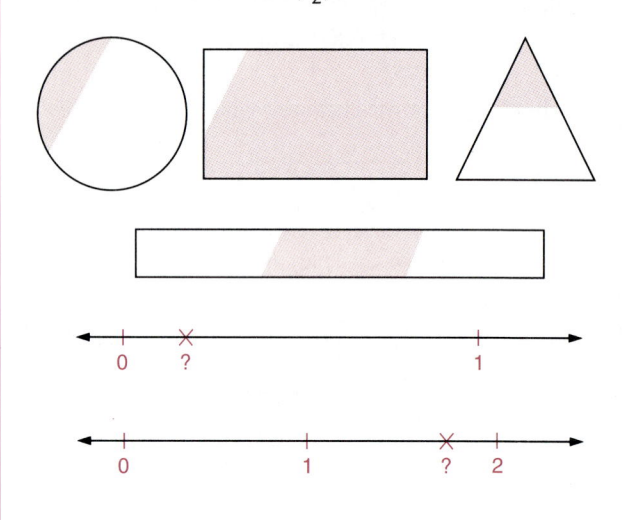

FIGURE 12.14: *About how much? Name a fraction for each.*

THINKING ABOUT WHICH IS MORE

The ability to tell which of two fractions is greater is another aspect of number sense with fractions. That ability is built around concepts of fractions, not on an algorithmic skill or symbolic trick.

◆ Use Concepts Not Rules

You probably have learned rules or algorithms for comparing two fractions. The usual approaches are finding common denominators and cross multiplication. These rules can be effective in getting correct answers but require no thought about the size of the fractions. This is especially true of the cross-multiplication procedure. If children are taught these rules before they have had the opportunity to think about the relative size of various fractions, there is little chance that they will develop any familiarity or number sense about fraction size. Comparison activities (which fraction is more?) can play a significant role in helping children develop concepts of relative size of fractions. But we want to keep in mind that reflective thought is the goal and not an algorithmic method of choosing the correct answer.

Before going on to the next section, try the following exercise. Assume for a moment that you know nothing about equivalent fractions or common denominators or crossmultiplying. Assume that you are a fourth- or fifth-grade student who was never taught these procedures. Now examine the pairs of fractions in Figure 12.15, and select the larger of each pair. Write down or explain one or more reasons for your choice in each case.

Which fraction in each pair is greater?
Give one or more reasons. Try not to use drawings or models.
Do not use common denominators or cross-multiplication.
Rely on concepts.

A. $\frac{4}{5}$ or $\frac{4}{9}$	G. $\frac{7}{12}$ or $\frac{5}{12}$
B. $\frac{4}{7}$ or $\frac{5}{7}$	H. $\frac{3}{5}$ or $\frac{3}{7}$
C. $\frac{3}{8}$ or $\frac{4}{10}$	I. $\frac{3}{8}$ or $\frac{4}{10}$
D. $\frac{5}{3}$ or $\frac{5}{8}$	J. $\frac{9}{8}$ or $\frac{4}{3}$
E. $\frac{3}{4}$ or $\frac{9}{10}$	K. $\frac{4}{6}$ or $\frac{7}{12}$
F. $\frac{3}{8}$ or $\frac{4}{7}$	L. $\frac{8}{9}$ or $\frac{7}{8}$

FIGURE 12.15: *Comparing fractions using concepts*

◆ Conceptual Thought Patterns for Comparison

The first two comparison schemes listed here rely on the meanings of the top and bottom numbers in fractions and on the relative sizes of unit fractional parts. The third and fourth ideas utilize the additional ideas of zero, $\frac{1}{2}$, and one being convenient anchors or benchmarks for thinking about the size of fractions.

More of the same-size parts. To compare $\frac{3}{8}$ and $\frac{5}{8}$, it is easy to think about having three of something and also five of the same thing. When the fractions are given orally, this is almost trivial. When given in written form, it is possible for children to choose $\frac{5}{8}$, simply because five is more than three and the other numbers are the same. Right choice, wrong reason. To compare $\frac{3}{8}$ and $\frac{5}{8}$ should be like comparing three apples and five apples. Pairs B and G in Figure 12.15 can be compared with this idea.

Same number of parts but parts are different sizes. This is the case where the numerators are the same, as in $\frac{3}{4}$ and $\frac{3}{7}$. If a whole is divided into seven parts, they will certainly be smaller than if only divided into four parts. Many children will select $\frac{3}{7}$ as larger, because seven is more than four and the other numbers are the same. That approach yields correct choices when the parts are the same size, but it causes problems in this case. This is like comparing three apples with three melons. You have the same number of things, but melons are larger. Pairs A, D, and H can be compared using this idea.

More and less than one half or one whole. The fraction pairs $\frac{3}{7}$ versus $\frac{5}{8}$ and $\frac{5}{4}$ versus $\frac{7}{8}$ do not lend themselves to either of the previous thought processes. In the first pair, $\frac{3}{7}$ is less than half of the number of sevenths needed to make

a whole, and so $\frac{3}{7}$ is less than a half. Similarly, $\frac{5}{8}$ is more than a half. Therefore, $\frac{5}{8}$ is the larger fraction. The second pair is determined by noting that one fraction is less than 1 and the other is greater than 1. The benchmark numbers of $\frac{1}{2}$ and 1 are useful for making size judgments with fractions. Pairs A, D, F, G, and H can be compared this way.

Closer to one-half or one whole. Why is $\frac{9}{10}$ greater than $\frac{3}{4}$? Not because the 9 and 10 are big numbers, although you will find that to be a common student response. Each is one fractional part away from 1 whole, and tenths are smaller than fourths. Similarly, notice that $\frac{5}{8}$ is smaller than $\frac{4}{6}$, since it is only one-eighth more than a half, while $\frac{4}{6}$ is a sixth more than a half. Pairs C, E, I, K, and L can be compared this way. (Can you use this basic idea to compare $\frac{3}{5}$ and $\frac{5}{9}$? *Hint:* Each is half of a fractional part more than $\frac{1}{2}$.)

How did your reasons for choosing fractions in Figure 12.14 compare to these ideas?

Classroom activities should help children develop informal ideas like those just explained for comparing fractions. However, the ideas should come from student experiences and discussions. To teach "the four ways to compare fractions" would be nearly as defeating as teaching cross-multiplication.

◆◆◆◆◆◆◆◆◆◆◆◆◆◆◆◆◆◆◆◆◆

12.9 COMPARE AND TEST

Provide students with a familiar fraction model to work with. Present a pair of fractions for comparison. (They should be fractions that can be illustrated with the model). Have the students think about which is more (compare), write down their choice, and then test their selection with the models. Be sure they make a commitment first, before they use the models.

When students begin to do well with the **Compare and Test** activity, see if they can begin to give reasons without the use of models. Be sure you include fraction pairs that cover all of the possibilities discussed earlier.

◆◆◆◆◆◆◆◆◆◆◆◆◆◆◆◆◆◆◆◆◆

12.10 WHY IS IT MORE?

Arrange the class in cooperative groups or pairs of students. Give the class a pair of fractions to compare. The task is to find as many good explanations for their choice as possible within an allotted time. Explanations can be written down and then discussed as a full class. The same exercise is a very good homework assignment.

12.11 CATCH MY GOOF

Present a pair of fractions, and make a choice of which is more along with a reason for the choice. For example, "I think that three-fourths is more than ten-twelfths because there is only one more fourth left to get to one." Or, "I think that five-eighths is more than three-sevenths because the five and the eight are both bigger than the three and the seven." The first example used faulty reasoning and produced a wrong result, while the second was a correct result with incorrect reasons. During the game use correct results and reasons as well as incorrect results and/or reasons. The game is a good way to get students to verbalize the conceptual reasoning that has been described and communicates that it is not rules but reasoning that is important. (What do you think you should do in this activity if no one catches an error you make?)

12.12 LINE 'EM UP

Have students put four or five fractions in order from least to most. In this way, a variety of methods for making comparisons can be included within the same exercise. As with all conceptual fraction activities, limit the denominators to reasonable numbers. There are very few reasons to consider fractions with denominators greater than 12.

A word of caution is implicit in the following situation. Mark is offered the choice of a third of a pizza or a half of a pizza. Since he is hungry and likes pizza, he chooses the half. His friend Jane gets a third of a pizza but ends up with more than Mark. How can that be? Figure 12.16 illustrates how Mark got misdirected in his choice. The point of the "pizza fallacy" is that whenever two or more fractions are discussed in the same context, the correct assumption (the one Mark made in choosing a half of the pizza) is that the fractions are all parts of the same whole.

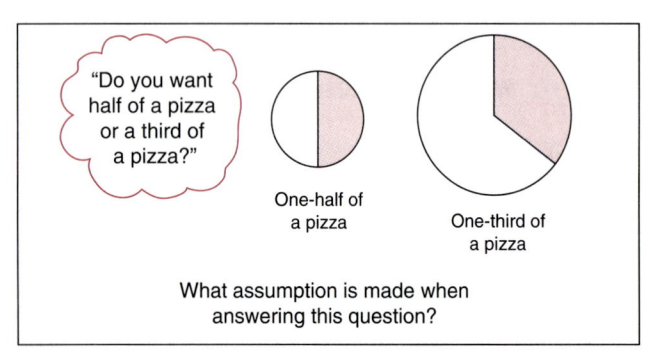

FIGURE 12.16: *The "pizza fallacy"*

Comparisons with any model can only be made if both fractions are parts of the same whole. For example, two-thirds of a light green strip cannot be compared to two-fifths of an orange strip.

OTHER METHODS OF COMPARISON

Certainly not all fractions can be compared by reliance on the conceptual approaches that have been discussed. Most adults would be hard-pressed to compare $\frac{2}{3}$ and $\frac{3}{5}$ without some other methods.

When simple methods fail, a more sophisticated and usually more complex method is required. The most common approach is to convert the fraction to some other form. When the concept and skills related to equivalent fractions are well in place, one or both of the fractions can be rewritten so that both fractions have the same denominator. Another idea is to translate the fraction to a different notation, specifically decimals or percents. Many of these decimal and percent equivalents can and should become second nature and require no computation. In the case of $\frac{2}{3}$ and $\frac{3}{5}$, the equivalents are $66^+\%$ and 60%, respectively. More will be said in later chapters about helping students make meaningful connections between decimals, percents, and fractions.

Even when common denominator methods are developed, they should be a last resort. Mental methods are frequently much quicker, and the symbolic methods detract from thinking in a conceptual way. There is little or no justification for asking students to compare $\frac{7}{17}$ with $\frac{9}{23}$.

EQUIVALENT FRACTION CONCEPTS

CONCEPTS VERSUS RULES

Question: How do you know that $\frac{4}{6} = \frac{2}{3}$? Here are some possible answers:

1. They are the same because you can reduce $\frac{4}{6}$ and get $\frac{2}{3}$.

2. Because if you have a set of 6 things and you take 4 of them, that would be $\frac{4}{6}$. But you can make the 6 into groups of 2. So then there would be 3 groups, and the 4 would be 2 groups out of the 3 groups. That means it's $\frac{2}{3}$.

3. If you start with $\frac{2}{3}$, you can multiply the top and the bottom numbers by 2, and that will give you $\frac{4}{6}$, so they are equal.

4. If you had a square cut into 3 parts and you shaded 2, that would be $\frac{2}{3}$ shaded. If you cut all 3 of these parts in half, that would be 4 parts shaded and 6 parts in all. That's $\frac{4}{6}$, and it would be the same amount.

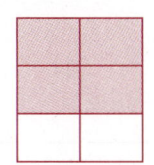

These answers, while all correct, provide clear examples of the distinction between conceptual knowledge and procedural knowledge. Responses (2) and (4) are very conceptual, although not efficient. The procedural responses, (1) and (3), are quite efficient, yet no conceptual knowledge is indicated. All students should eventually know how to write an equivalent fraction for a given fraction. At the same time, the rules should never be taught or used until the students understand what the result means.

The concept: Two fractions are *equivalent* if they are representations for the same amount.

The rule: To get an equivalent fraction, multiply (or divide) the top and bottom numbers by the same nonzero number.

The rule or algorithm for equivalent fractions carries no intuitive connection with the concept. As a result, students can easily learn and use the rule in exercises such as "List the first four equivalent fractions for $\frac{3}{5}$," without any idea of how the fractions in the list are related. It becomes an exercise in multiplication. A developmental approach suggests that students have a firm grasp of the concept and should be led to see that the algorithm is a meaningful and efficient way to find equivalent fractions.

FINDING DIFFERENT NAMES FOR FRACTIONS

The general approach to a conceptual understanding of equivalent fractions is to have students use models to generate different names for models of fractions.

Area or Region Models

Examples of equivalent fraction representations using area models are illustrated in Figure 12.17.

12.13 DIFFERENT FILLERS

Using the same models that students have, draw the outline of several fractions on paper and duplicate them. For example, if the model is rectangles, you might draw an outline (no subdivisions) of a rectangle for $\frac{2}{3}$, $\frac{1}{2}$, and perhaps $\frac{5}{4}$. Have children try to fill the outlines with unit fraction pieces to determine as many simple fraction names for the regions as possible. In class discussion, it may be appropriate to see if students can go beyond the actual models that they have. For example, if the model has no tenths, it would be interesting to ask what other names could be generated if tenths were available. An easier question involves pieces that can be derived from existing pieces. "You found out that five-fourths and

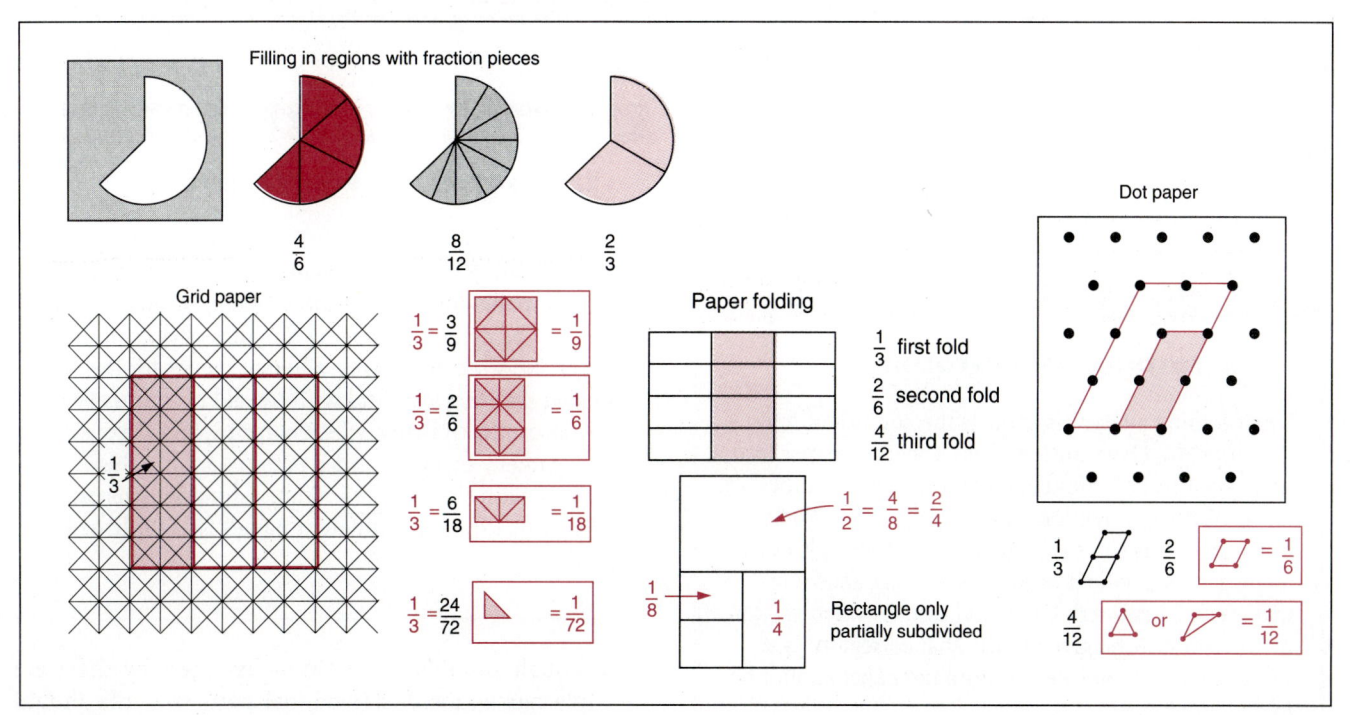

FIGURE 12.17: *Area models for equivalent fractions*

236 12 / THE DEVELOPMENT OF FRACTION CONCEPTS

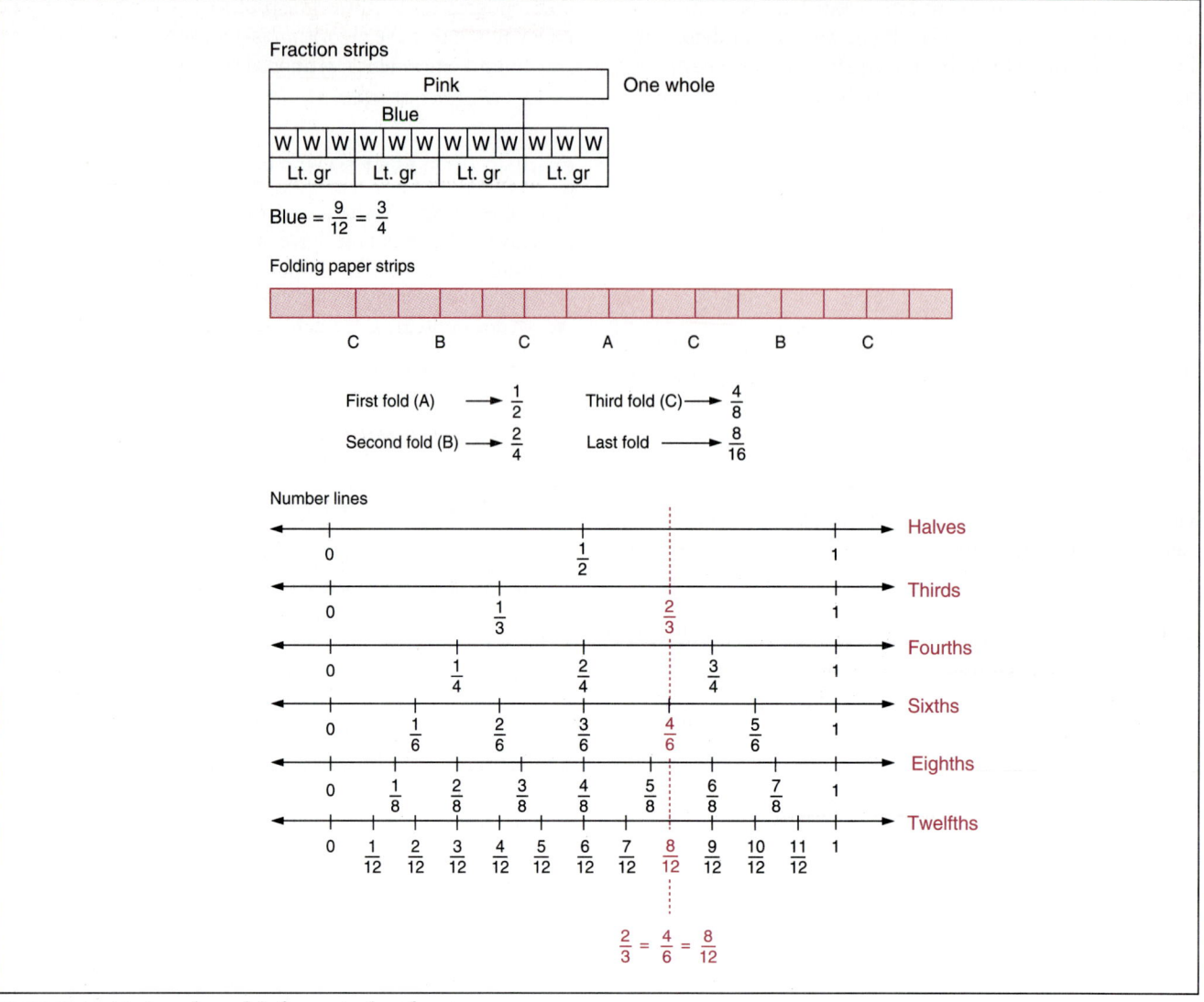

FIGURE 12.18: *Length models for equivalent fractions*

ten-eighths and fifteen-twelfths are all the same. What if we had some sixteenth pieces? Could we cover this same region with those? How many? How can you decide?

12.14 FOLDING AND REFOLDING

Paper folding effectively models the equivalent fraction concept. Have students fold a sheet of paper into halves or thirds. Unfold and color a fraction of the paper. Write the fraction. Now refold and fold one more time. It is fun to discuss, *before opening,* how many sections will be in the whole sheet and how many will be colored. Open and discuss what fraction names can now be given to the shaded region. Is it still the same? Why? Repeat until the paper cannot be folded any longer.

12.15 DOT PAPER EQUIVALENCIES

Use grid paper or dot paper so that regions can easily be subdivided into many smaller parts. Have students draw a model for a whole and shade in some fraction that can be determined using the lines or dots on the paper. Now see how many different names for the shaded part they can find with the aid of the smaller regions in the drawing. (See Figure 12.17.) Make a transparency of a grid or dot pattern, and do this exercise with the full class. Teams or individuals can take turns explaining different names for the shaded part.

Although subdivided regions illustrate why there are multiple names for one fractional part, students should learn that the existence of the subdivisions is not required.

Half of a rectangle is still two-fourths even if no subdivisions are present and even if the other half is divided into three parts. Work toward this understanding by drawing models for fractions and then erasing the subdivision lines. "Is this two-thirds still four-sixths?"

◆ Length Models

Equivalent fractions are modeled with length models in much the same way as area models. One fraction is modeled, and then different lengths are used to determine other fraction names. Some examples are shown in Figure 12.18.

◆ Set Models

The general concept of equivalent fractions is the same with set models as with length and area models, although there are more limitations to how a particular set can be partitioned. For example, if $\frac{2}{3}$ is modeled with 8 out of 12 counters, then that particular representation shows that $\frac{2}{3}$ is also $\frac{4}{6}$ and $\frac{8}{12}$. It cannot be seen as $\frac{6}{9}$ or $\frac{10}{15}$. As shown in Figure 12.19, a given number of counters in two colors can be arranged in different arrays or subgroups to illustrate equivalent fractions.

12.16 ARRANGE THE COUNTERS, FIND THE NAMES

Have students set out a specific number of counters in two colors. For example, 32 counters with 24 red and the rest yellow. This set will be the whole. Then have them find as many names as they can for each color by arranging the counters into different subgroups. Drawings can be made using X's and O's to produce a written record of the activity.

◆ A Transitional Activity

Each of the following four equations is typical of equivalent fraction exercises found in textbooks. Notice the differences among the examples.

$$\frac{5}{3} = \frac{\Box}{6} \qquad \frac{2}{3} = \frac{6}{\Box} \qquad \frac{8}{12} = \frac{\Box}{3} \qquad \frac{12}{8} = \frac{3}{\Box}$$

An excellent exercise is to have students complete equations of this type by using a model to determine the missing number. Some care must be taken to select exercises

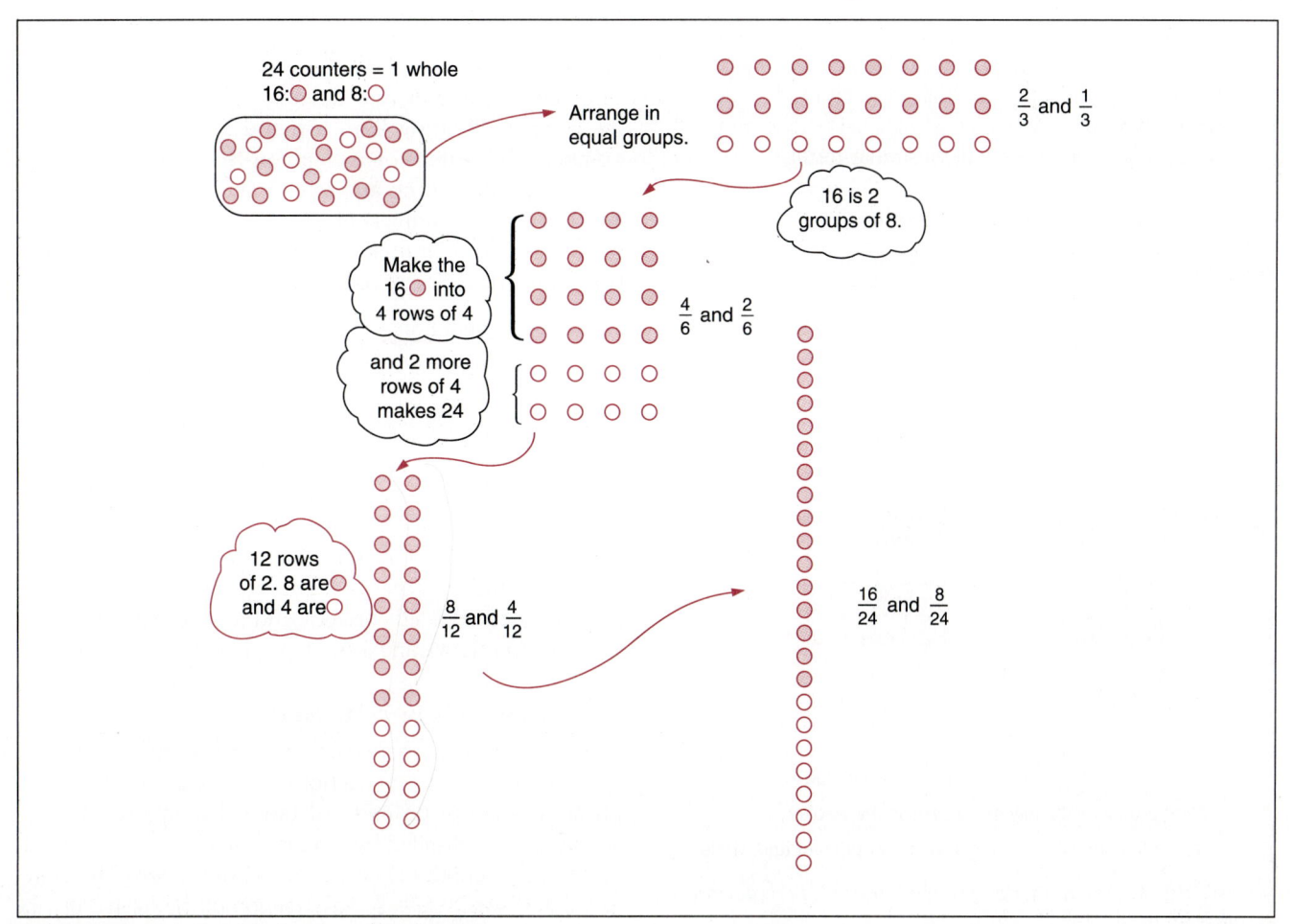

FIGURE 12.19: *Set models for equivalent fractions*

that can be solved with the model being used. For example, most pie-piece sets do not have ninths. Counters or sets are a model that can always be used. Make students responsible for justifying their results with the use of models or drawings. Later, when symbolic rules are developed, these same exercises will be completed in a more algorithmic but probably less meaningful manner.

DEVELOPING AN EQUIVALENT FRACTION ALGORITHM

◆ An Area Model Approach

While there are many possible ways to model the procedure of multiplying top and bottom numbers by the same number, the most commonly used approach is to slice a square in two directions.

12.17 SLICING SQUARES

Give students paper with rows of squares about 3 cm on each side. Have them shade the same fraction in several different squares using vertical subdividing lines. Next, slice each rectangle horizontally into different fractional parts, as shown in Figure 12.20. Help students focus on the products involved by having them write the top and bottom numbers in the fraction as a product. Notice that for each model, the top and bottom numbers will always have a common factor. (Paper folding provides a similar result.)

Start with each square showing $\frac{3}{4}$.

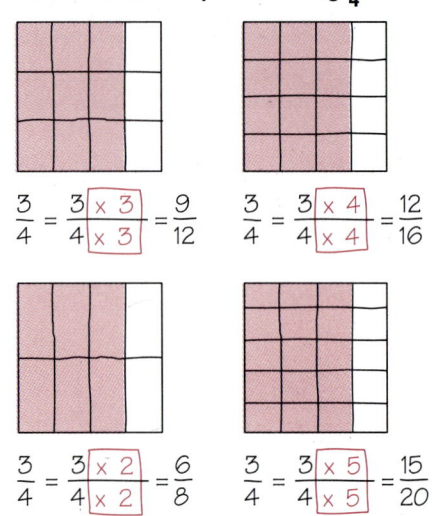

$\frac{3}{4} = \frac{3 \times 3}{4 \times 3} = \frac{9}{12}$ $\frac{3}{4} = \frac{3 \times 4}{4 \times 4} = \frac{12}{16}$

$\frac{3}{4} = \frac{3 \times 2}{4 \times 2} = \frac{6}{8}$ $\frac{3}{4} = \frac{3 \times 5}{4 \times 5} = \frac{15}{20}$

What product tells how many parts are shaded?
What product tells how many parts in the whole?
Notice that the same factor is used for both part and whole.

FIGURE 12.20: *A model for the equivalent fraction algorithm.*

Examine examples of equivalent fractions that have been generated with other models, and see if the rule of multiplying top and bottom numbers by the same number holds there also. If the rule is correct, how can $\frac{6}{8}$ and $\frac{9}{12}$ be equivalent? What about fractions like $2\frac{1}{4}$? How could it be demonstrated that $\frac{9}{4}$ is the same as $\frac{27}{12}$?

◆ Writing Fractions in Simplest Terms

The multiplication rule for equivalent fractions produces fractions with larger denominators. To write a fraction in *simplest terms* means to write it so that numerator and denominator have no common whole number factors. (Some texts use *lowest terms* instead of *simplest terms*.) One meaningful approach to this task of finding simplest terms is to reverse the earlier process, as illustrated in Figure 12.21.

Lower terms →

$\frac{8}{12} = \frac{2 \times 4}{3 \times 4} = \frac{2}{3}$

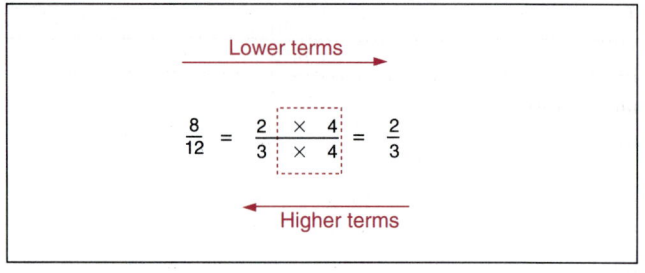

← Higher terms

FIGURE 12.21: *Using the equivalent fraction algorithm to write fractions in simplest terms.*

Of course, finding and eliminating a common factor is the same as dividing both top and bottom by the same number. The search for a common-factor approach keeps the process of writing an equivalent fraction to one rule: Top and bottom numbers of a fraction can be multiplied by the same nonzero number. There is no need for a different rule for rewriting fractions in lowest terms.

Two additional notes:

1. Notice that the phrase "reducing fractions" was not used. This unfortunate terminology implies making a fraction smaller and is rarely used anymore in textbooks.

2. Many teachers seem to believe that fraction answers are incorrect if not in simplest or lowest terms. This total assault on fractions not in simplest terms is also unfortunate. When students add $\frac{1}{6} + \frac{1}{2}$ and get $\frac{4}{6}$, they have added correctly and have found the answer. Rewriting $\frac{4}{6}$ as $\frac{2}{3}$ is a totally separate issue.

◆ The Multiply-by-1 Method

Many junior-high textbooks use a strictly symbolic approach to equivalent fractions. It is based on the multiplication identity property of rational numbers that says any number multiplied by 1 remains unchanged. The number 1 is the identity element for multiplication. Therefore, $\frac{3}{4} = \frac{3}{4} \times 1 = \frac{3}{4} \times \frac{2}{2} = \frac{6}{8}$. Any fraction of the form $\frac{n}{n}$ can be used as the identity element. Furthermore, the numerator and denominator of the identity element can also be fractions. In this way $\frac{6}{12} = \frac{6}{12} \times [\frac{1}{6} \div \frac{1}{6}] = \frac{1}{2}$.

This explanation relies on an understanding of the multiplicative identity property, which most students in grades 4 to 6 do not fully appreciate. It also relies on the procedure for multiplying two fractions. Finally, the argument uses solely deductive reasoning based on an axiom of the rational number system. It does not lend itself to intuitive modeling. A reasonable conclusion is to delay this important explanation until at least seventh or eighth grade in an appropriate prealgebra context and not as a method or a rationale for producing equivalent fractions.

OTHER MEANINGS OF FRACTIONS

In this chapter the only meaning given to fractions is that of an expression of the relationship between a part and a whole. Unit fractional parts are based on the partitioning of the whole into equal-sized parts (denoted by the denominator), and various fractions are taken as multiples of these unit parts (denoted by the numerator). Research generally supports the idea that this is the best way to approach fractions from a developmental perspective. However, the fraction notation does have other meanings, which are introduced to children at about the sixth grade.

FRACTIONS ARE EXPRESSIONS OF DIVISION

If four people were to share 12 candies, the number that each would get can be expressed by 12 ÷ 4. If four people were to share three pizzas, the amount that each would get can be expressed similarly by 3 ÷ 4; that is, three things divided four ways. In the pizza example, each person would receive three-fourths of a pizza. So, $\frac{3}{4}$ and 3 ÷ 4 are both expressions for the same idea: three things divided by four. Similarly, in the candy example, $\frac{12}{4}$ expresses the number each will receive just as well as 12 ÷ 4.

Put simply, a fraction $\frac{a}{b}$ is another way of writing $a \div b$.

Students understandably find this meaning of fractions unusual. First, it is different from the meaning that has been carefully developed. Second, fractions are commonly thought of as amounts of parts of wholes, not as operations. Similarly, expressions such as 7 ÷ 3 are thought of as operations (things to be done), not numbers. In fact, however, 4, 2 + 2, 12 ÷ 3, and $\frac{8}{2}$ are all symbolic expressions for the same number. 12 ÷ 3 is not the question and 4 the answer, but both are expressions for 4. Likewise, 2 ÷ by 3 and $\frac{2}{3}$ are both expressions for the quantity two-thirds. Do you find this a little hard to swallow? So do children. Here for the first time in seven or eight years of school you are telling students that a symbol can represent two different ideas. This is a relatively sophisticated idea. Students should be told quite openly that this is a new and different way to think about a fraction. Real-world problems for division should be used. They should be written in both fraction and whole number division notation and discussed.

FRACTIONS ARE EXPRESSIONS OF RATIOS

Ratios, like part-to-whole fraction concepts, are expressions of a relationship between two quantities. With the part-to-whole concept of fraction, one of those quantities is always fixed and designated as a unit. (Recall the "pizza fallacy.") With ratios there is no fixed unit. If 20 of 30 students in a class are girls, the ratio of boys to girls can be expressed as 10 to 20, as 10:20, or as $\frac{10}{20}$. The colon notation is still used, but is much less common than the fraction notation. The fraction notation is also useful in dealing with equivalent ratios or proportions. Two ratios are proportional if the fractions that express them are equivalent. While 6 of 9 counters and 8 of 12 counters each model two-thirds, they are not the same amounts but equivalent ratios. (Chapter 15 contains a complete discussion of the difficult concepts of ratio and proportion.)

◆◆◆◆◆

REFLECTIONS ON CHAPTER 12: WRITING TO LEARN

1. Give examples of three categories of fraction models. What real models have you used that correspond to each of these?

2. Why are set models more difficult for younger children?

3. Describe fractional parts. What are the two distinct requirements of fractional parts? Explain how children's concepts of partitioning need to be refined to produce a concept of fractional parts.

4. What are children learning in activities where you count fractional parts? How can this type of activity help children learn to write fractions meaningfully?

5. Give a fourth-grade explanation of the meaning of the top number and the bottom number in a fraction.

6. Use a length model and make up part-and-whole questions for each of the following cases:

    Given a part and a nonunit fraction less than one, find the whole.

    Given a part and a nonunit fraction greater than one, find the whole.

    Given a whole and a nonunit fraction less than one, find the part.

    Given a whole and a nonunit fraction greater than one, find the part.

    Try your questions with a friend. Then change each question so that it is in terms of sets. With sets, be sure that a unit fraction is never a single counter. That is, if the question is about fourths, then use whole sets of size 8 or 12 or more.

7. Make up pairs of fractions that can be compared (select the largest) without using an algorithm. See if you can make up examples that utilize the four different ideas for comparing fractions that were suggested.

8. Explain the difference between understanding the concept of equivalent fractions and knowing the algorithm for writing an equivalent fraction.

9. Describe some activities that will help children develop the equivalent-fraction concept.

10. How can you help children develop the algorithm for equivalent fractions?

11. Draw a picture to show that $5 \div 3$ is the same as $\frac{5}{3}$.

◆◆◆◆◆
FOR DISCUSSION AND EXPLORATION

1. A common error that children make is to write $\frac{3}{5}$ for the fraction represented here. Why do you think that they do this? In this chapter, the notation of fractional parts and counting by unit fractions was introduced before any symbols. How could this help avoid the type of thinking that is involved in this common error?

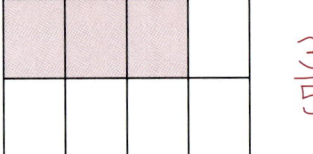

2. Experiment with rods made of connecting cubes (such as Unifix) as a model for fractions. For example, if a bar of 12 cubes is a whole, then bars of three cubes would be fourths. Since the bars can be taken apart and put together, and since the individual cubes can be counted, there are some significant differences between this model and those suggested in the text. Is this a length model or a set model?

3. Experiment with a calculator such as the TI-Math Explorer or TI-34 that shows fractions. Design an activity that takes advantage of the simplify key. What other advantages can you see for a calculator of this type?

4. Read the short article by Steffe and Olive, "The Problem of Fractions in the Elementary School" (see Suggested Readings). Are the children in this article typical of children you have observed in school? Discuss the use of a discomforting situation in helping children learn a new concept.

5. Work with some children. Here are some places to begin to explore their ideas:

   Use a model they have not seen and try some of the part-and-whole questions. Begin with easy examples involving unit fractions.

   See if they can use a model to explain why $2\frac{1}{3}$ is $\frac{7}{3}$. (Or see if they can write $\frac{7}{3}$ as a mixed number and explain their procedure.)

   Try some comparison questions. Encourage children to produce more than one explanation.

   Try to find out if children really believe that $\frac{2}{3}$ and $\frac{8}{12}$ are the same amount. What is their thinking behind this?

   From these initial explorations, you should be able to decide where to proceed for some additional work with your group of children.

6. Read the chapter "Fractions with Cookies" in Marilyn Burns's *A collection of math lessons for grades 3 to 6* (1987). Try these activities with a group of children. While suggested for grade 3, the ideas are easily adapted to grades higher or lower. What is different about this approach to fraction development from the one presented in this chapter?

◆◆◆◆◆
SUGGESTED READINGS

Behr, M. J. Post, T. R., & Wachsmuth, I. (1986). Estimation and children's concept of rational number size. In H. L. Schoen (Ed.), *Estimation and mental computation.* Reston, VA: National Council of Teachers of Mathematics.

Behr, M. J., Wachsmuth, I., & Post, T. R. (1985). Construct a sum: A measure of children's understanding of fraction size. *Journal for Research in Mathematics Education, 16,* 120–131.

Bezuk, N., & Cramer, K. (1989). Teaching about fractions: What, when, and how? In P. R. Trafton (Ed.), *New directions for elementary school mathematics.* Reston, VA: National Council of Teachers of Mathematics.

Bezuk, N. S., & Bieck, M. (1993). Current Research on rational numbers and common fractions: Summary and implications for teachers. In D. T. Owens (Ed.), *Research ideas for the classroom: Middle grades mathematics.* New York: Macmillan Publishing Co.

Delaney, K. (1984). Fraction Games. *Mathematics Teaching, 107,* 8–11.

Ellerbruch, W., & Payne, J. N. (1978). Teaching sequence from initial fraction concepts through the addition of unlike fractions. In M. N. Suydam (Ed.), *Developing computational skills.* Reston, VA: National Council of Teachers of Mathematics.

Holden, L. (1986). *Fraction factory.* Palo Alto, CA: Creative Publications.

Hope, J. A., & Owens, D. T. (1987). An analysis of the difficulty of learning fractions. *Focus on Learning Problems in Mathematics, 9*(4), 25–40.

Kroll, D. L., & Yabe, T. (1987). A Japanese educator's perspective on teaching mathematics in the elementary school. *Arithmetic Teacher, 35*(2), 36–43.

Langford, K., & Sarullo, A. (1993). Introductory common and decimal fraction concepts. In R. J. Jensen (Ed.), *Research ideas for the classroom: Early childhood mathematics.* New York: Macmillan Publishing Co.

Liebeck, P. (1985). Are fractions numbers? *Mathematics Teaching, 111,* 32–34.

National Council of Teachers of Mathematics. (1984). Rational numbers [Focus Issue]. *Arithmetic Teacher, 31*(6).

Payne, J. N., Towsley, A. E., & Huinker, D. M. (1990). Fractions and decimals. In J. N. Payne (Ed.), *Mathematics for the young child.* Reston, VA: National Council of Teachers of Mathematics.

Post, T. R., Behr, M. J., & Lesh, R. (1982). Interpretations of rational number concepts. In L. Silvey (Ed.), *Mathematics for the middle grades (5–9).* Reston, VA: National Council of Teachers of Mathematics.

Post, T., & Cramer, K. (1987). Children's strategies in ordering rational numbers. *Arithmetic Teacher, 35*(2), 33–35.

Pothier, Y., & Sawada, D. (1983). Partitioning: The emergence of rational number ideas in young children. *Journal for Research in Mathematics Education, 14,* 307–317.

Reys, B. J. (1991). *Developing number sense in the middle grades: Addenda series, grades 5–8.* Reston, VA: National Council of Teachers of Mathematics.

Silver, E. A. (1983). Probing young adults' thinking about rational numbers. *Focus on Learning Problems in Mathematics, 5,* 105–117.

Steffe, L. P., & Olive, J. (1991). The problem of fractions in the elementary school. *Arithmetic Teacher, 38*(9), 22–24.

Wearne-Hiebert, D., & Hiebert, J. (1983). Junior high school students' understanding of fractions. *School Science and Mathematics, 83,* 96–106.

13 COMPUTATION WITH FRACTIONS

NUMBER SENSE AND ALGORITHMS

Before beginning the discussion of fraction computation, we need to ask ourselves why we want children to have fraction computation skills in the first place. Pencil-and-paper skills with fractions are almost never used outside of school anymore, and certainly children in the sixth and seventh grades do not need much proficiency with pencil-and-paper computation with fractions. There are, however, some real benefits in having children look at fraction computation. Perhaps most important is the value that can be gained in mental computation and estimation skills. These mental approaches are significantly more important than paper methods. Second, if we can help children approach this problem of fraction computation conceptually and with a problem-solving spirit, there is a significant opportunity for children to begin to really do some mathematics—to figure things out and explain the results. Finally a firm foundation of fraction computation will be very helpful when the same patterns and ideas are translated from numbers to algebraic symbols. The latter reason is not, however, sufficient in itself. Be careful not to tell children, "You're going to need this to do higher math." That is perhaps the weakest reason for learning any mathematics.

THE DANGER OF RULES

In the short term, the rules for fraction computation can be relatively simple to teach. Students can become quite proficient at finding common denominators during a chapter on adding and subtracting simple fractions. Multiplying fractions is such an easy procedure that many experts suggest it should be taught first. The only requirements are third-grade multiplication skills. Division, following the invert-and-multiply rule, is nearly as easy. Fraction rules too can easily become the focus of rote instruction and produce artificial feelings of accomplishment on the quiz at the end of the week.

Focusing attention on fraction rules and answer getting has two significant dangers: First, none of the rules help students think in any way about the meanings of the operations or why they work. Students practicing such rules may very well be doing rote symbol pushing in its purest sense. Second, the mastery observed in the short term is quickly lost. When taken as a group of rules, the procedures governing fraction computation become similar and confusing. "Do I need a common denominator, or can you just add the bottoms?" "Do you invert the second number or the first?"

A NUMBER-SENSE APPROACH

Without a firm understanding of fraction concepts, the development of computational algorithms for fractions can quickly become superficial, rule-oriented, and confusing. That is why this chapter was written separately from the preceding chapter—to place more emphasis on fraction concepts themselves and not to confuse computation with numeration as an objective.

It may be well to begin the development of fraction computation with the following ideas in mind:

1. Connect the meaning of fraction computation with whole-number computation. To think about what $2\frac{1}{2} \times \frac{3}{4}$ might mean, we should ask, "What does 2×3 mean?" The concepts of each operation are the same, and benefits can be had by connecting these ideas.

2. Let estimation play a big role in all phases of the development, especially during initial explorations. "Should $2\frac{1}{2} \times \frac{3}{4}$ be more or less than 1? More or less than 3?" Estimation keeps the focus on the meanings of the numbers and the operations, encourages reflective thinking, and helps build informal number sense with fractions.

3. Explore each of the operations using models. Use a variety of models. Have students defend the solutions to computations using the models. You will find that sometimes it is possible to get answers with models that do not seem to help with pencil and paper approaches. That's fine! The ideas will help children learn to think about the fractions and the operations, contribute to mental methods, and provide a useful background when you eventually do get to the standard algorithms.

In the sections that follow, an informal exploration is encouraged for each operation. There is also a guided development of each traditional algorithm. You are strongly encouraged to use some models and spend a good bit of time with the exploratory sections especially.

ADDITION AND SUBTRACTION

As just suggested, a heavy emphasis should be placed on estimation while developing computation with fractions. Estimation activities can be done at any time; before, during, or after an informal exploration of addition and subtraction.

ESTIMATION

A frequently quoted result from the Second National Assessment (Post, 1981) concerns the following item:

> Estimate the answer to $\frac{12}{13} + \frac{7}{8}$. You will not have time to solve the problem using paper and pencil.

Response choices and how 13-year-olds answered are listed here.

Responses	Percent of 13-year-olds
1	7
2	24
19	28
21	27
I don't know	14

What this result points out all too vividly is that a good concept of fractions is much more significant for estimation purposes than a mastery of the pencil-and-paper procedures. Knowing if a fraction is closer to 0, $\frac{1}{2}$, or 1 proves quite useful. Numbers can be rounded either to the nearest whole or nearest half and then added easily. For example, $2\frac{1}{8} + \frac{4}{9}$ is about the same as $2 + \frac{1}{2}$. A front-end approach is also possible: Deal with the whole numbers, and then look at the fractions using estimates to the nearest half to make an adjustment. The following activities are useful for developing these ideas.

13.1 PICK THE BEST I

Flash fractions between 0 and 1 on an overhead projector. For each fraction, students should record on their answer sheet 0, $\frac{1}{2}$, or 1, depending on which they think the given fraction is closest to.

13.2 PICK THE BEST II

Flash sums or differences of proper fractions. Response options can vary with the age and experience of the students. More than 1 or less than 1 is one alternative. Closer to 0, 1, or 2 is also an easy option. A more sophisticated option is to give the result to the nearest half: 0, $\frac{1}{2}$, 1, $1\frac{1}{2}$, or 2.

13.3 SPEED ESTIMATES

Provide short speed drills for estimating sums and differences with mixed numbers, as illustrated in Figure 13.1 (p. 244).

When more than two addends are involved, an acceptable range may be a bit wider. Encourage students to give whole-number estimates at first and later practice refining estimates to the nearest half. Discuss both front-end and rounding techniques.

244 13 / COMPUTATION WITH FRACTIONS

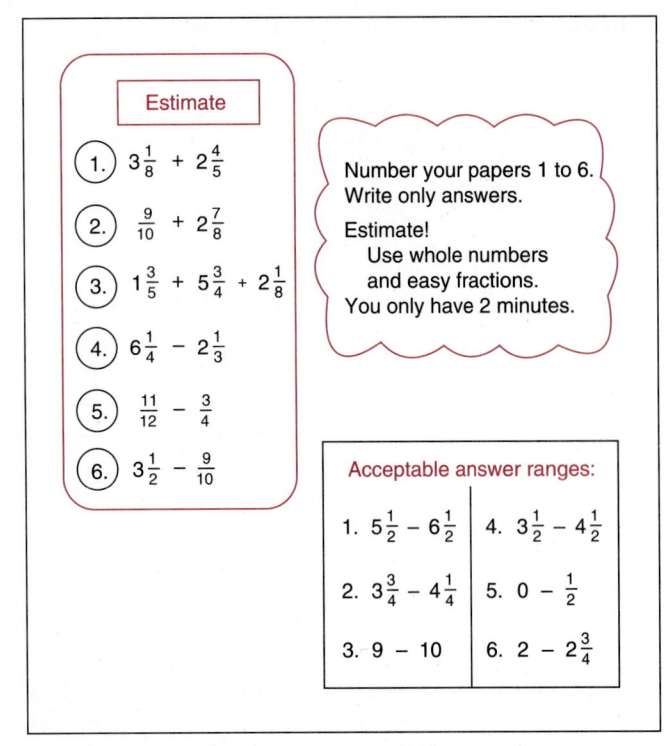

Estimate

1. $3\frac{1}{8} + 2\frac{4}{5}$
2. $\frac{9}{10} + 2\frac{7}{8}$
3. $1\frac{3}{5} + 5\frac{3}{4} + 2\frac{1}{8}$
4. $6\frac{1}{4} - 2\frac{1}{3}$
5. $\frac{11}{12} - \frac{3}{4}$
6. $3\frac{1}{2} - \frac{9}{10}$

Number your papers 1 to 6. Write only answers.
Estimate! Use whole numbers and easy fractions. You only have 2 minutes.

Acceptable answer ranges:
1. $5\frac{1}{2} - 6\frac{1}{2}$ 4. $3\frac{1}{2} - 4\frac{1}{2}$
2. $3\frac{3}{4} - 4\frac{1}{4}$ 5. $0 - \frac{1}{2}$
3. $9 - 10$ 6. $2 - 2\frac{3}{4}$

FIGURE 13.1: *A fraction estimation drill*

▬ **INFORMAL EXPLORATION**

Have students add two fractions using a fraction model. The results should come completely from the use of the model even if some of the students have been exposed to symbolic rules and the idea of a common denominator. For example, suggest that students use their model to find the sum of $\frac{3}{4}$ and $\frac{1}{3}$. Exercises such as this should be explored with area, length, and set models. You must be careful that the problems can actually be worked with the materials the students have available. With circular regions, the assumption is that the circle will always represent the whole. However, as seen in the last chapter, many models allow different representations for the whole. When using these more flexible models (for example, strips or counters) the first thing that must be done is to determine a whole that permits both fractions to be modeled. (Recall the "pizza fallacy" from Figure 12.16.) Figure 13.2 illustrates how beginning addition tasks might be approached using three different models.

Subtraction of two fractions with models is a similar process, as shown in Figure 13.3. Notice that it is sometimes possible to find the sum or difference of two fractions without subdividing either one. Attention is instead focused on the size of a leftover part. When area models are used for addition and subtraction, common denominators are frequently not involved at all. On the other hand, selecting a set size or length that permits modeling of two fractions is mathematically the same as finding a common denominator. (Why?)

$\frac{5}{6} + \frac{1}{2}$ How would you do this with an area model?

Find a strip for a whole that allows both fractions to be modeled.

Dk. green ← Whole
Yellow — $\frac{5}{6}$
Lt. green — $\frac{1}{2}$

Dk. gr.
Yellow | Lt. gr. $\frac{5}{6} + \frac{1}{2}$
         Red

The sum is 1 whole and a red rod more than a whole. A red is $\frac{1}{3}$ of dark green.
So $\frac{5}{6} + \frac{1}{2} = 1\frac{1}{3}$

$\frac{2}{3} + \frac{1}{4}$

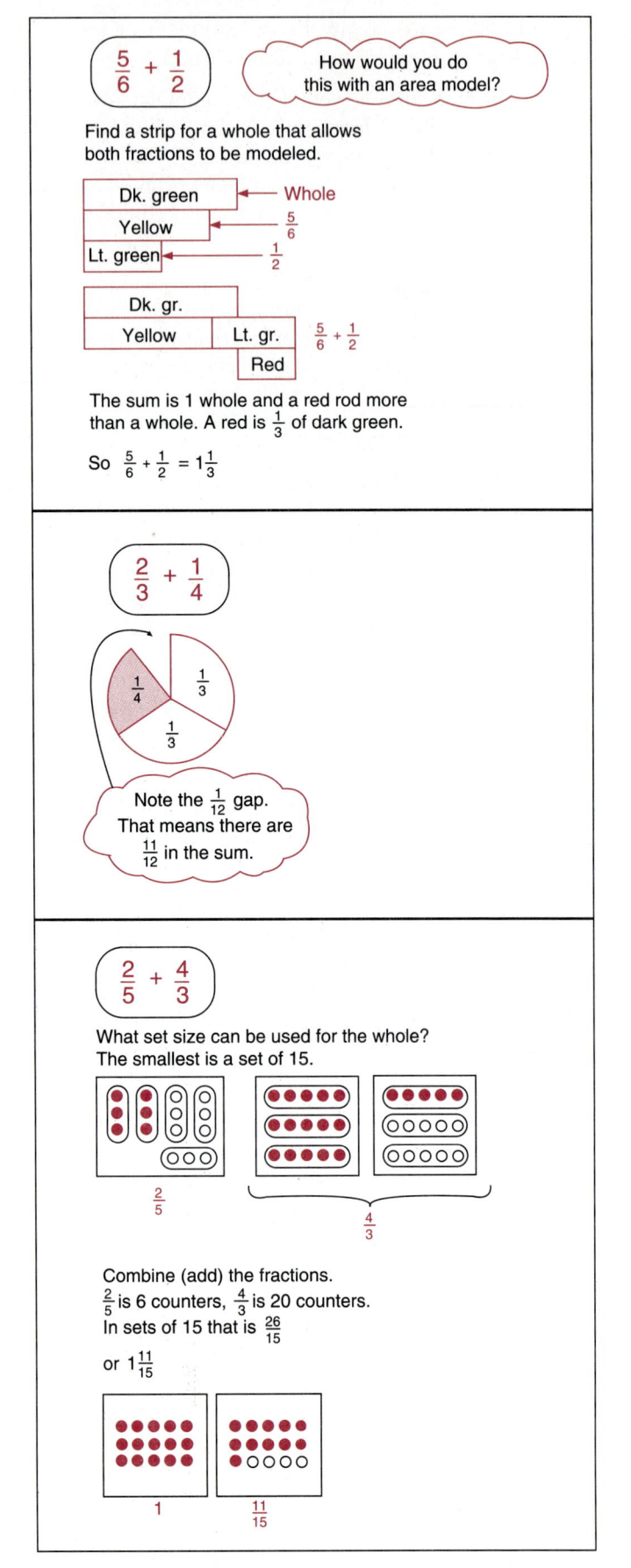

Note the $\frac{1}{12}$ gap. That means there are $\frac{11}{12}$ in the sum.

$\frac{2}{5} + \frac{4}{3}$

What set size can be used for the whole? The smallest is a set of 15.

$\frac{2}{5}$ $\frac{4}{3}$

Combine (add) the fractions.
$\frac{2}{5}$ is 6 counters, $\frac{4}{3}$ is 20 counters.
In sets of 15 that is $\frac{26}{15}$
or $1\frac{11}{15}$

1    $\frac{11}{15}$

FIGURE 13.2: *Using models to add fractions*

13 / COMPUTATION WITH FRACTIONS 245

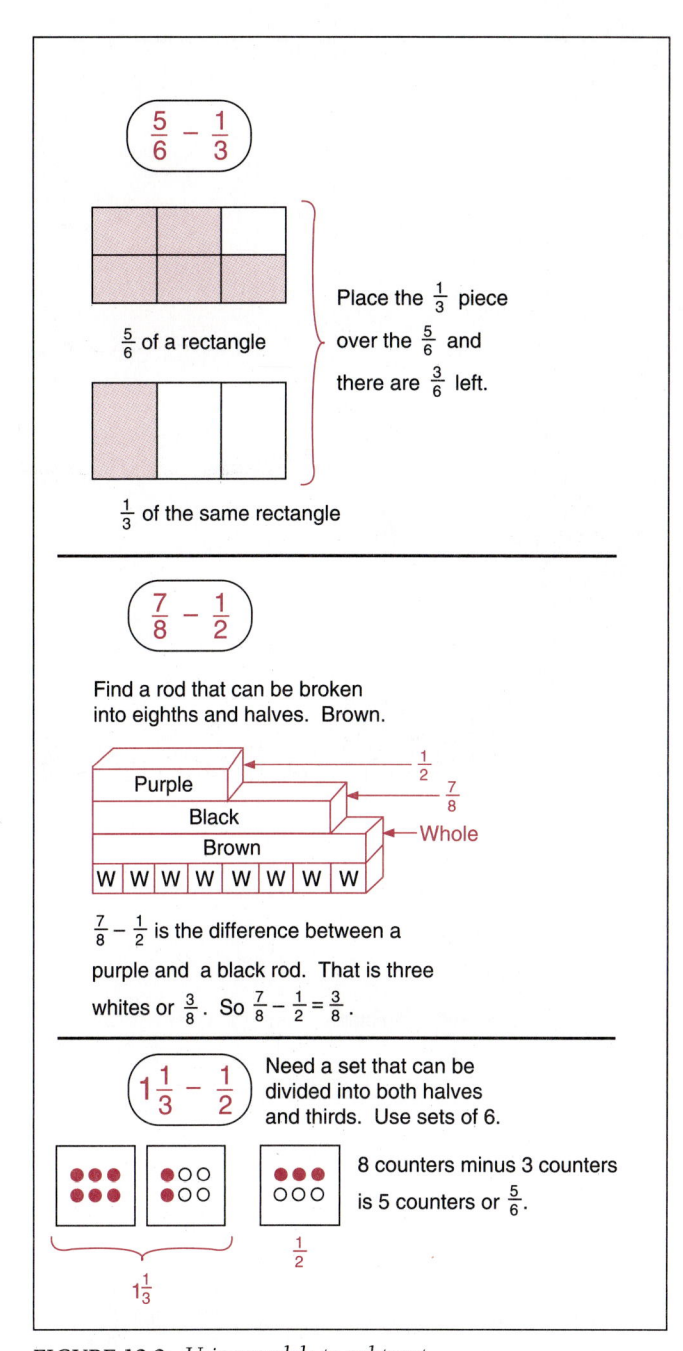

FIGURE 13.3: *Using models to subtract*

When making up addition and subtraction exercises to be done with models, do not be afraid of difficult problems. Include exercises with unlike denominators, fractions greater than one, and mixed numbers. For subtraction, explore problems such as $3\frac{1}{8} - 1\frac{1}{4}$, where a trade of wholes for eighths might occur. At this stage, avoid directing students with a completely formulated method for arriving at a solution. The problem-solving approach along with the use of the models will help students develop relationships. Discuss different solution processes used by different groups. It can be useful to have different groups use different models for their processes.

DEVELOPING THE ALGORITHM

Children are not likely to invent the standard algorithms for addition and subtraction without some guidance. At the same time, they can easily build on the informal explorations and see that the common-denominator approach is quite meaningful.

Like Denominators

If students have worked with models to find sums and differences, exercises with like denominators should be trivial. For example, when adding $\frac{3}{4}$ and $\frac{7}{4}$ with circular pie pieces all that is necessary is to count the fractional parts. In just a few minutes students should be completely comfortable with adding or subtracting any two fractions of the same type (like denominators). There is no need to practice these in pencil-and-paper form. In fact, if you stick with models for a while, the next step in the development will make more sense.

Unlike Denominators

When you use an area model to add or subtract two fractions that are not alike, you are very likely to pay attention to a part left over to make a whole (first example in Figure 13.2) or, in a similar manner, to a part that is a bit more than a whole. While this is excellent thinking, it does not lead to the need for common denominators.

To get students to move to common denominators, consider a task such as $\frac{5}{8} + \frac{2}{4}$. Let students use pie pieces to get the result of $1\frac{1}{8}$ using any approach. Many will note that the models for the two fractions make one whole and there is $\frac{1}{8}$ extra. The key question at this point is, "How can we change this question into one that is just like the easy ones where the parts are the same?" For the current example, it is relatively easy to see that fourths could be changed into eighths. Have students use models to show the original problem and also the converted problem. The main idea is to see that $\frac{5}{8} + \frac{2}{4}$ is exactly the same problem as $\frac{5}{8} + \frac{4}{8}$.

Next try some examples where both fractions need to be changed. For example, $\frac{2}{3} + \frac{1}{4}$. Be careful that the common denominator can actually be modeled with the materials that the students have. Again, focus attention on *rewriting the problem* in a form that is like adding apples and apples, where the parts of both fractions are the same. Students must fully understand that the new form of the problem is actually the same problem. This can and should be demonstrated with models. However, if your students express any doubt about the equivalence of the two problems ("Is $\frac{11}{12}$ really the answer to $\frac{2}{3} + \frac{1}{4}$?"), then that should be a clue that the concept of equivalent fractions is not well understood.

As a result of modeling and rewriting fractions to make the problems easy, students should come to understand that the process of getting a common denominator is really one of looking for a way to change the *statement* of the problem without changing the problem. After getting a

246 13 / COMPUTATION WITH FRACTIONS

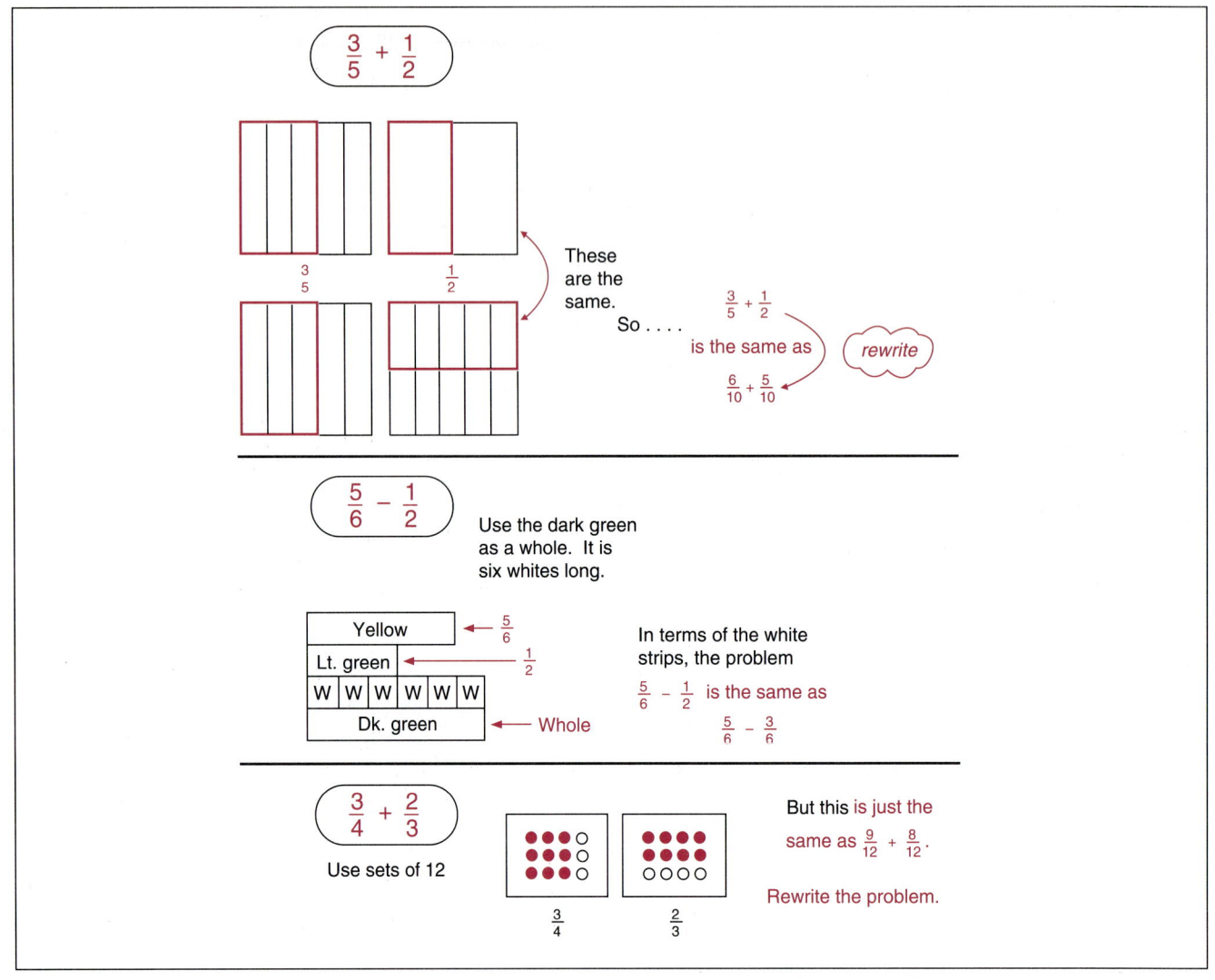

FIGURE 13.4: *Rewriting addition and subtraction problems*

common denominator, it should be immediately obvious that adding the numerators produces the correct answer. These ideas are illustrated in Figure 13.4.

Subtraction of two simple fractions follows exactly the same approach. If the denominators are the same, it is like subtracting apples from apples. When the denominators are different, the problem should be rewritten to make it an easy one.

◆ Common Multiple Practice

Many students have trouble with finding common denominators because they are not able to quickly produce common multiples of the denominators. This is a skill that can be practiced. It also depends on having a good command of the basic facts for multiplication. Here are two activities aimed at the skill of finding least common multiples or common denominators.

13.4 RUNNING THROUGH THE MULTIPLES

For this oral drill, give students a number between 2 and 16 (likely denominators) and have them list the multiples in order. At first, writing the multiples may be helpful. Work toward the skill of doing this exercise orally and quickly. Students should be able to list the multiples to about 50 with ease.

13.5 L.C.M. FLASH CARDS

Make flash cards with pairs of numbers that are potential denominators. Most should be less than 16, as before. For each card, students try to give the least common multiple (Figure 13.5). Be sure to include pairs that are relatively prime, such as 9 and 5, those where one is a multiple of the other, such as 2 and 8, and those with a common multiple, such as 8 and 12.

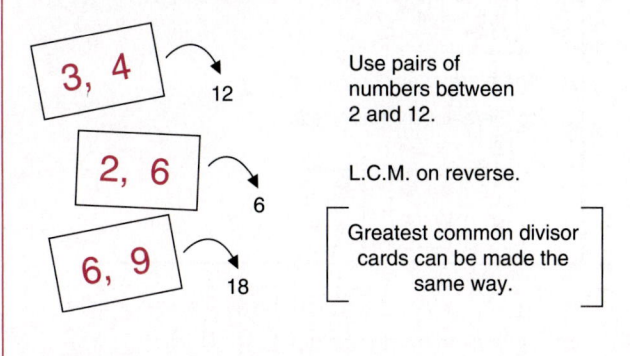

FIGURE 13.5: *Least-common multiple flash cards*

MIXED NUMBERS

When students do addition and subtraction of mixed numbers, they tend to make errors in converting fractions to whole numbers and whole numbers to fractions. This is especially true in subtraction, as illustrated in Figure 13.6.

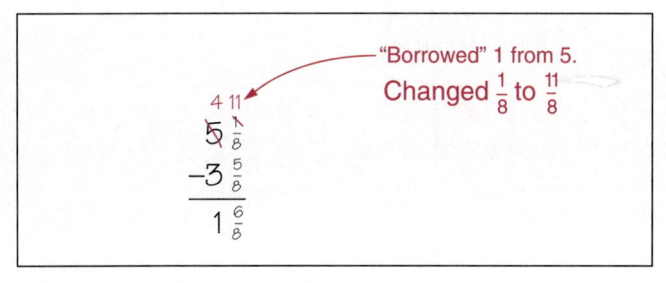

FIGURE 13.6: *A common subtraction error*

Students making this error are using a base ten place-value idea instead of changing a whole for an equivalent set of fractional parts. The following ideas may help with these difficulties.

13.6 FRACTION TRADING

Do fraction-trading activities. Use a two-sided mat and an area model for fractions such as pie pieces. Add fractional parts such as fourths, one at a time, to the right-hand side of the mat. When there are four-fourths, discuss the idea of trading for a whole, as shown in Figure 13.7. In a similar manner, start with a model for three wholes on the left side and begin to take off one-fourth at a time. Help students see how this is both like and different from trading ones for tens and tens for ones. Remember to use the terminology of "trade" and not the old-fashioned "borrow."

Later, allow the numbers and type of part to change at each turn. Roll two dice. Use an ordinary die to tell how many parts to put on or take off the board. A second die indicates halves, thirds, fourths, sixths, eighths, or twelfths. With the two dice, trading games are played the same as with tens and ones. The first player to accumulate three wholes (foward game) or clear the board (backward game) is the winner.

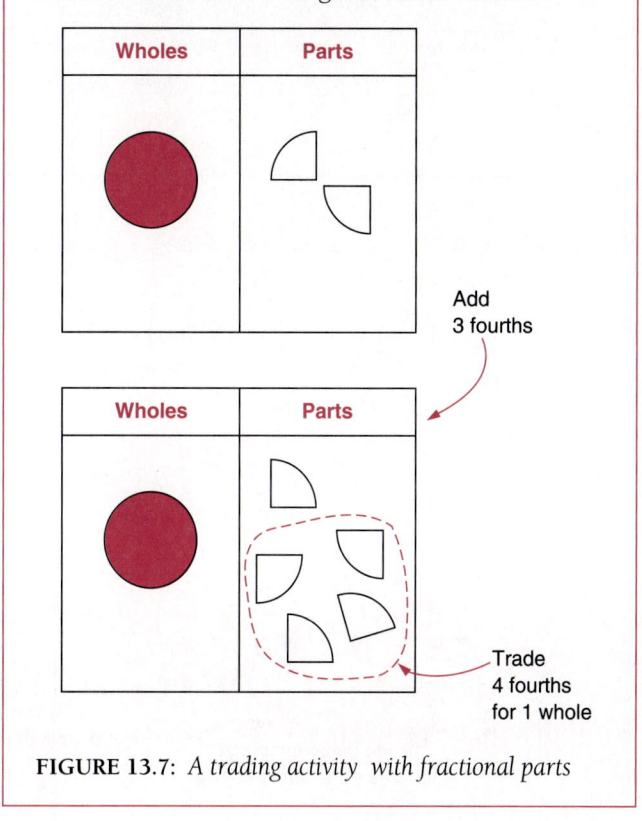

FIGURE 13.7: *A trading activity with fractional parts*

Do not forget to include estimation practice even after you have begun to practice the algorithm. Notice that with mixed numbers the idea of a left-hand or front-end approach is clearly best for addition and subtraction. That is, deal first with the whole-number part. That alone provides a fairly good estimate. If the situation calls for a better estimate, an adjustment can be made by examining the fractional part.

MULTIPLICATION

With multiplication there is a greater need to review the meaning of the operation with whole numbers. Many children have developed the mistaken notion that multiplication always results in a bigger number. When one or both of the factors is a fraction, as in $\frac{3}{4} \times 7$ or $\frac{2}{3} \times \frac{1}{2}$, it is very difficult for children to even make estimates. Therefore, perhaps the first place to begin is with whole numbers and the meaning of multiplication. However, do not forget to make estimates.

248 13 / COMPUTATION WITH FRACTIONS

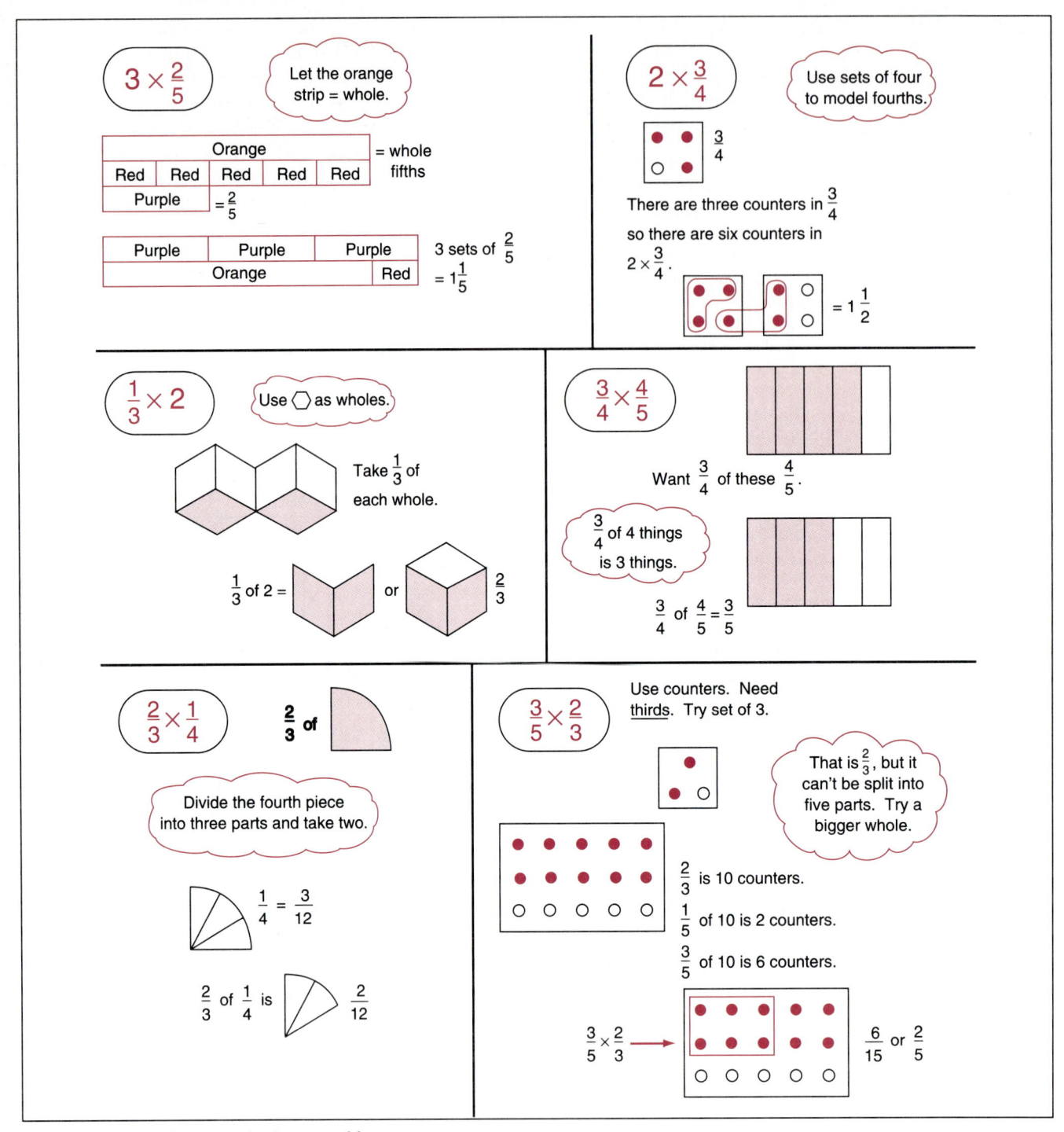

FIGURE 13.8: *Exploring multiplication of fractions*

INFORMAL EXPLORATION

Recall with students that for whole numbers, 3 × 5 means 3 sets with 5 in each set, 3 sets of 5 each. With this in mind, *about* how much do you think $3 \times \frac{4}{5}$ might be? One possible idea is that it will be less than three because it will be three sets with each a bit less than one. Another approach is to use the meanings of the top and bottom number: Three sets of 4 fifths is just three sets of 4 things. There will be 12 things in all—12 fifths. Discuss other examples keeping the first factor a whole number and eliciting a good discussion of techniques.

After a discussion of mental approaches and estimates, have students try to solve some of the same exercises with models. Several examples of how multiplication problems might be modeled are illustrated in Figure 13.8. Work on

13 / COMPUTATION WITH FRACTIONS   249

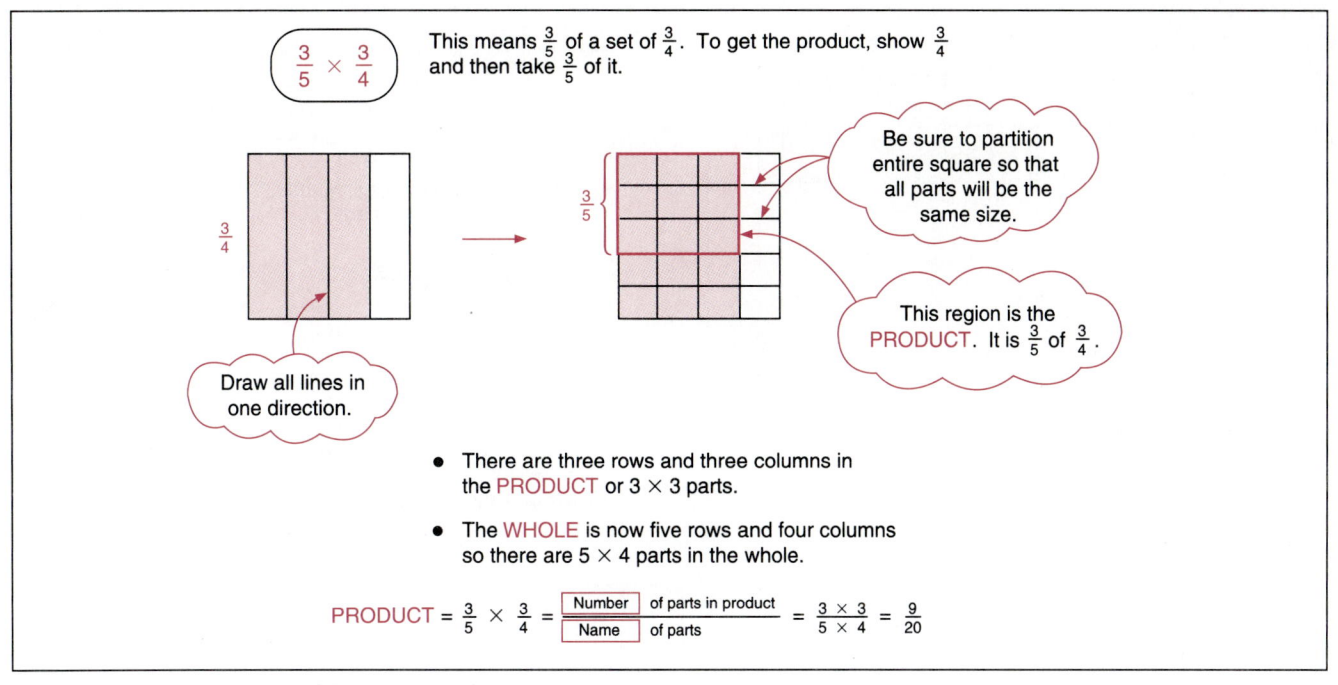

$\boxed{\frac{3}{5} \times \frac{3}{4}}$ This means $\frac{3}{5}$ of a set of $\frac{3}{4}$. To get the product, show $\frac{3}{4}$ and then take $\frac{3}{5}$ of it.

Draw all lines in one direction.

Be sure to partition entire square so that all parts will be the same size.

This region is the PRODUCT. It is $\frac{3}{5}$ of $\frac{3}{4}$.

- There are three rows and three columns in the PRODUCT or 3 × 3 parts.
- The WHOLE is now five rows and four columns so there are 5 × 4 parts in the whole.

$$\text{PRODUCT} = \frac{3}{5} \times \frac{3}{4} = \frac{\boxed{\text{Number}} \text{ of parts in product}}{\boxed{\text{Name}} \text{ of parts}} = \frac{3 \times 3}{5 \times 4} = \frac{9}{20}$$

FIGURE 13.9: *Development of the algorithm for multiplication*

the idea that the first factor tells how many sets and the second factor the size of the sets.

Next, still using models, try examples with a fraction as the first factor and a whole number as the second factor. The meanings remain the same: $\frac{2}{3} \times 4$ means $\frac{2}{3}$ of a set of 4. Try modeling a given product in both orders. For example, try $\frac{1}{6} \times 4$ and $4 \times \frac{1}{6}$. The answers are the same, but the process is quite different.

Finally, see if students can explain the meaning of a fraction times a fraction. For example, what does $\frac{2}{3} \times \frac{3}{4}$ mean? Once again the meanings remain the same: $\frac{2}{3}$ of a set of $\frac{3}{4}$. That is, with models, one would first make a set of three-fourths and then determine what is two-thirds of that set. Notice how easy the last example is compared to $\frac{1}{4} \times \frac{2}{3}$. (Why?) When the first factor is greater than one, children may have extra difficulty. For example try solving $\frac{3}{2} \times \frac{1}{4}$ or $\frac{4}{3} \times 1\frac{2}{3}$ using different models and the basic concepts of multiplication.

Frequently students are perplexed because the answer in multiplication gets smaller instead of larger. When and why a product of two fractions is smaller than either factor is a good question to explore at this preliminary stage.

DEVELOPING THE ALGORITHM

The more careful development of the usual algorithm will build on the informal techniques just discussed.

Both Factors Less than One

While not the only way to generate the multiplication algorithm for fractions, an area model where squares or rectangles are used for the whole is one of the most common. The examples that are easiest to understand are those where both factors are fractions less than one.

Figure 13.9 shows how the square is used first to model the product and then to determine the numerator and denominator of the product in written form. By drawing the lines for each fraction in opposite directions, both the *whole* (the square) and the *product* are arrays made of the same size fractional parts. The numerator of the product is the *number* of parts in the product, the number of rows times the number of columns. The denominator of the product is the *name* of those parts or the number in one whole. Help students see that these two products are also the products of the numerators and the denominators, respectively. After students model a series of fraction products, the rule of multiplying top and bottom numbers will become obvious.

One Factor a Whole Number

For products where the first factor is a whole number, such as $3 \times \frac{4}{5}$, the intuitive meaning of multiplication is quite easy, as noted earlier. Three sets of 4 fifths is 12 fifths: $3 \times \frac{4}{5} = \frac{12}{5}$. Multiply the whole number by the top number. This rule makes sense even without modeling. Since the 3 can be written as 3/1, the rule of multiplying tops and bottoms does apply. Figure 13.10 (p. 250) illustrates how this product would be modeled using squares as already discussed.

For products where the second factor is a whole number, the result is similar, as illustrated in Figure 13.10. The case where the second factor is a whole number can also be looked at by using the "turn-around" property: $\frac{3}{4} \times 2$ is the

250 13 / COMPUTATION WITH FRACTIONS

$3 \times \frac{4}{5}$

The PRODUCT is three sets of $\frac{4}{5}$.

- There are three rows and four columns in the PRODUCT or 3 × 4 parts.
- The WHOLE is still one row and five columns or 1 × 5 parts.

$3 \times \frac{4}{5}$ is the same as $\frac{3}{1} \times \frac{4}{5}$.

$$\text{PRODUCT} = \frac{\boxed{\text{Number}} \text{ of parts}}{\boxed{\text{Name}} \text{ of parts}} = \frac{3 \times 4}{1 \times 5} = \frac{12}{5}$$

Note that the whole is just one square.

$\frac{3}{4} \times 2$

The PRODUCT is $\frac{3}{4}$ of 2 wholes.

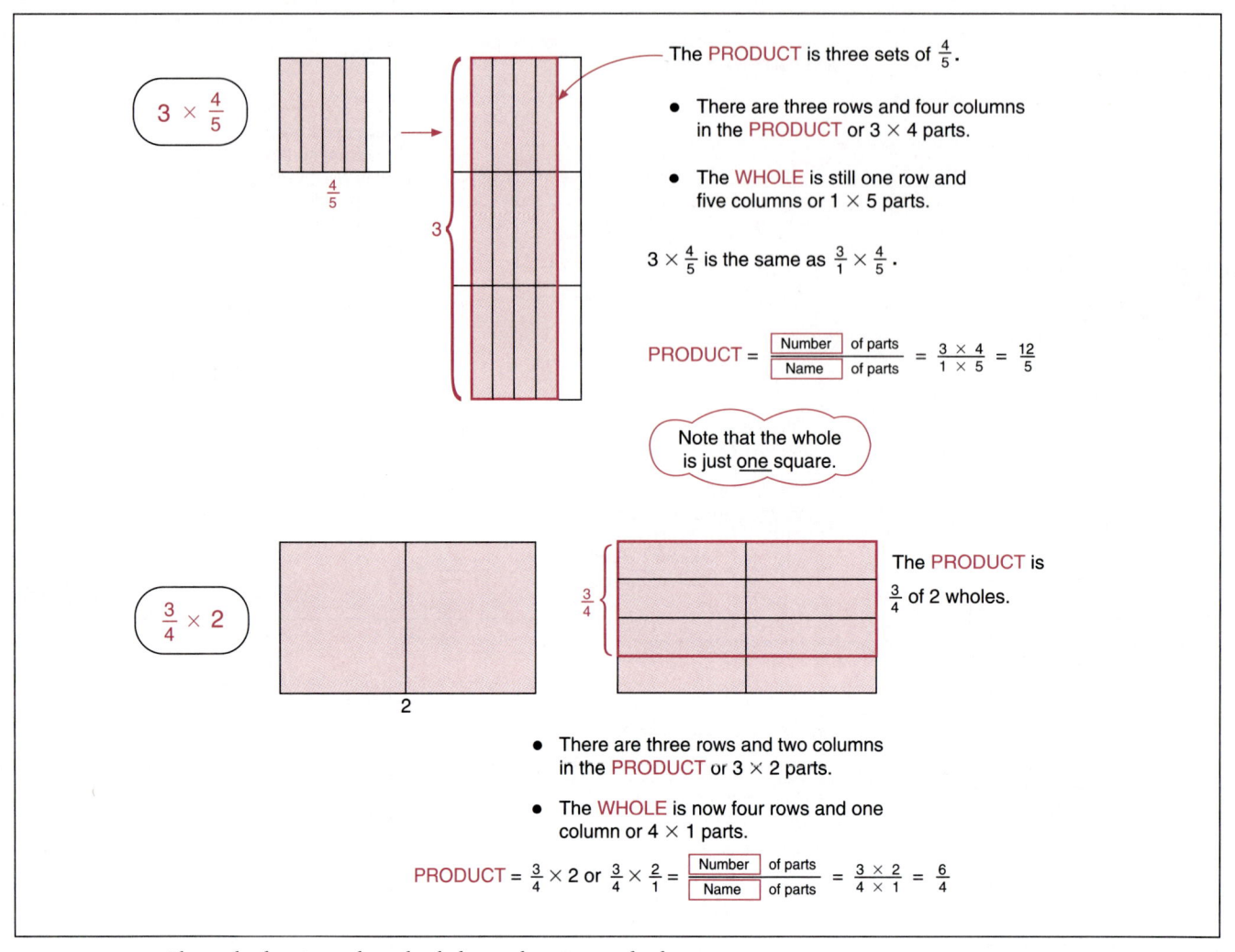

- There are three rows and two columns in the PRODUCT or 3 × 2 parts.
- The WHOLE is now four rows and one column or 4 × 1 parts.

$$\text{PRODUCT} = \frac{3}{4} \times 2 \text{ or } \frac{3}{4} \times \frac{2}{1} = \frac{\boxed{\text{Number}} \text{ of parts}}{\boxed{\text{Name}} \text{ of parts}} = \frac{3 \times 2}{4 \times 1} = \frac{6}{4}$$

FIGURE 13.10: *The multiplication rule with whole numbers is exactly the same.*

---

same as $2 \times \frac{3}{4}$. This is especially useful in mental multiplication. The modeling is done to make the algorithm complete and to see that the rule does not just apply to a special case. Never let mental computation or estimation slip entirely into the background.

◆ Mixed-Number Factors

Recall that with mixed numbers, mental methods suggest that you work with the whole-number parts first. Consider the example of $3\frac{2}{3} \times 2\frac{1}{4}$. To get a first estimate of this product, a good place to begin is with 3 × 2. However, it is not exactly clear what could be done to improve on this estimate. As a problem-solving or performance task, have students work at drawing a picture of the product using the same methods as before. That is, get a picture of $2\frac{1}{4}$ and then show $3\frac{2}{3}$ sets of that $2\frac{1}{4}$. Figure 13.11 shows how this might be done, leading to the use of improper fractions in agreement with the standard algorithm.

This last exercise is a good example of an algorithm getting in the way of mental computation and estimation. The process just does not lend itself well to mental methods. Another idea is to make a connection with whole-number multiplication. Figure 13.12 is very much like Figure 13.11 except that it uses the same length × length = area approach as was developed for the two-digit multiplication algorithm in Chapter 10. Notice that there are four partial products, just as there would be for a product such as 36 × 24. Mentally, work with the larger parts first:

> 3 × 2 is 6. Then there is $3 \times \frac{1}{4}$ or $\frac{3}{4}$ more: $6\frac{3}{4}$. And $2 \times \frac{2}{3}$ is $\frac{4}{3}$, which is $1\frac{1}{3}$. So far that makes a little more than 8 ($6\frac{3}{4}$ and $1\frac{1}{3}$). Then there's a fourth of $\frac{2}{3}$ extra. The answer is probably less than $8\frac{1}{2}$.

While the mental process is difficult to get exactly, the left-hand approach provides a reasonable estimate.

Another significant benefit of the drawing in Figure 13.12 is the connection that is made with something the students have learned earlier, the two-digit products of Chapter 10. Later, a similar drawing can be made for a

13 / COMPUTATION WITH FRACTIONS

$3\frac{2}{3} \times 2\frac{1}{4}$

Whole → 1 $2\frac{1}{4} = \frac{9}{4}$

$3\frac{2}{3}$ or $\frac{11}{3}$

The PRODUCT is $3\frac{2}{3}$ sets of $2\frac{1}{4}$.

There are 11 rows and 9 columns or 11 × 9 parts in the PRODUCT.

The WHOLE now has three rows and four columns or 3 × 4 parts.

$3\frac{2}{3} \times 2\frac{1}{4} = \frac{11}{3} \times \frac{9}{4}$ = PRODUCT =

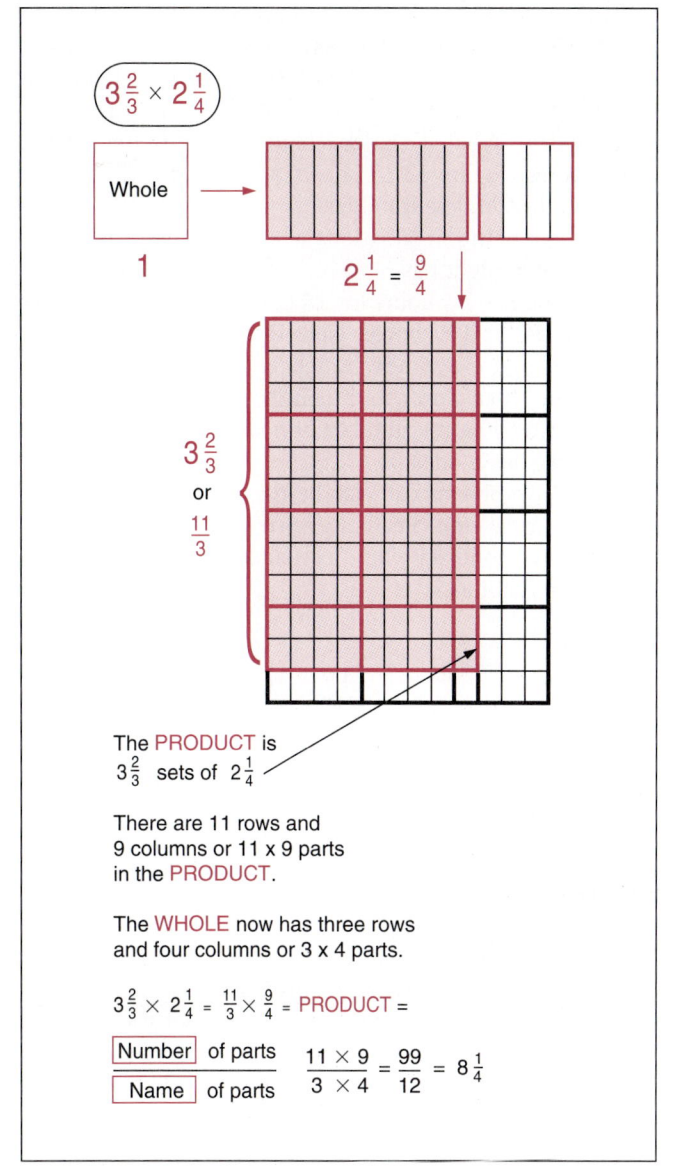

$\boxed{\text{Number}}$ of parts
$\boxed{\text{Name}}$ of parts $\frac{11 \times 9}{3 \times 4} = \frac{99}{12} = 8\frac{1}{4}$

FIGURE 13.11: *The multiplication rule with two mixed numbers.*

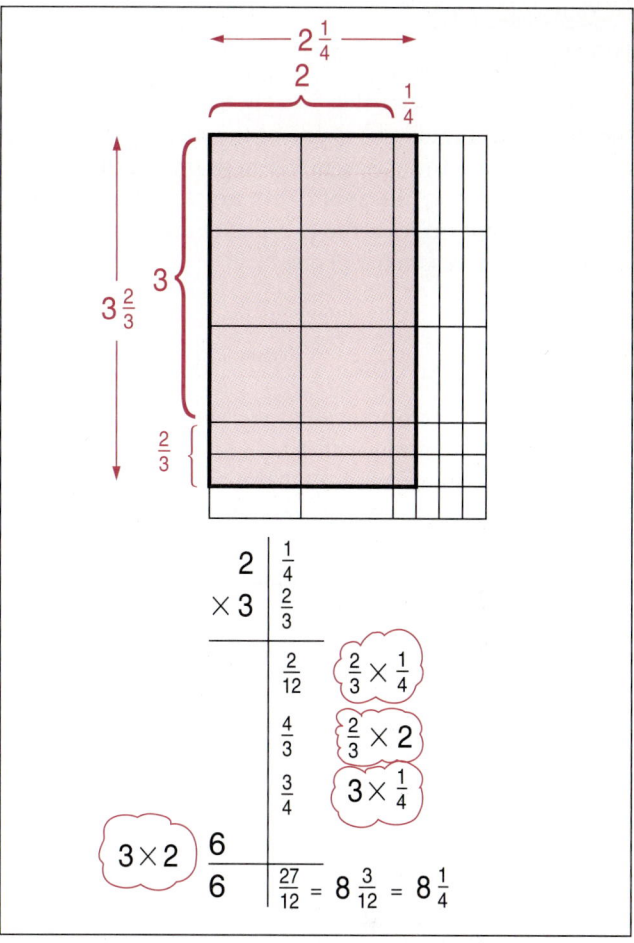

$2\frac{1}{4}$ (2 and $\frac{1}{4}$)

$3\frac{2}{3}$ (3 and $\frac{2}{3}$)

$\begin{array}{r} 2\frac{1}{4} \\ \times\, 3\frac{2}{3} \\ \hline \frac{2}{12} \\ \frac{4}{3} \\ \frac{3}{4} \\ \frac{6}{6} \\ \hline \frac{27}{12} = 8\frac{3}{12} = 8\frac{1}{4} \end{array}$

$\frac{2}{3} \times \frac{1}{4}$
$\frac{2}{3} \times 2$
$3 \times \frac{1}{4}$
3×2

FIGURE 13.12: *The product of mixed numbers as four partial products.*

product with decimal fractions such as 3.4 × 2.6. The same area model is also used in algebra for products such as $(2X + 3)(X + 4)$. It is well worth the time to help students see unifying ideas in mathematics.

◆ More Mental Techniques

In the real world, there are many instances when the product of a whole number times a fraction occurs, and a mental estimate or even an exact answer is quite useful. For example, sale items are frequently listed as one-fourth off, or we read of a one-third increase in the number of registered voters. Fractions are excellent substitutes for percents, as you will see in the next chapter. To get an estimate of 60% of $36.69, it is useful to think of 60% as $\frac{3}{5}$ or as a little less than $\frac{2}{3}$.

These products of fractions with large whole numbers can be done mentally if one utilizes the meanings of the top and bottom numbers. For example, $\frac{3}{5}$ is 3 *one*-fifths. If you want $\frac{3}{5}$ of 350, for example, first think about *one*-fifth of 350, or 70. If *one*-fifth is 70, then *three*-fifths is 3 × 70 or 210. Although this example has very accommodating numbers, it illustrates a process for mentally multiplying a large number by a fraction: First determine the unit fractional part, and then multiply by the number of parts you want. Once again you can see the importance of the fractional-part concept and the meanings of the top and bottom numbers.

When numbers are not so nice, encourage students to use compatible numbers. To estimate $\frac{3}{5}$ of $36.69, a useful compatible is $35. One-fifth of 35 is 7 and so three-fifths is 3 × 7 or 21. Now adjust a bit, perhaps add an additional 50¢ for an estimate of $21.50.

Students should practice estimating fractions times whole numbers in lots of real contexts: $3\frac{1}{4}$ gallons of paint at $14.95 per gallon, or $\frac{7}{8}$ of the 476 students attended Friday's championship football game. When working with decimals and percents, these skills will be revisited and, once again, mathematics will seem more connected than disconnected.

252 13 / COMPUTATION WITH FRACTIONS

DIVISION

The division algorithm that you probably learned (invert the divisor and multiply) is one of the most mysterious to children of all the rules in elementary mathematics. Certainly we want to avoid this mystery at all costs. However, first it makes sense to examine division with fractions from a more familiar perspective.

CONCEPT EXPLORATION

As with the other operations, we want to go back to the meaning of division with whole numbers. Here, recall that there are two meanings of division: partition and measurement. In the following two sections we will review each briefly and look at some real-world situations of the same type that involve fractions. From a number-sense perspective or for the purpose of making estimates, we first have to know what the operation means. (Can you make up a word problem right now that would go with the computation $2\frac{1}{2} \div \frac{1}{4}$?)

◆ Partition Concept

In the partition concept for $12 \div 3$, the task is to separate 12 things into 3 equal parts and determine the amount in each part. Explore this same idea with fractions, first with the divisor a whole number. Consider the following word problem:

> Darlene has $2\frac{1}{4}$ hours to complete 3 household chores. If she divides her time evenly, how many hours can she give to each?

In this problem, the task is to separate or partition the quantity $2\frac{1}{4}$ into three equal parts. Since $2\frac{1}{4}$ is 9 fourths, 3 fourths can be allotted to each task. That is, Darlene can spend $\frac{3}{4}$ of an hour per chore. These numbers turned out rather nicely. What if Darlene had only $1\frac{1}{4}$ hours for the three chores? Figure 13.13 shows how this could be modeled with three different models.

Are there partition situations where the divisor is a fraction? Darlene might have finished half of one of her chores. Suppose she has $2\frac{1}{4}$ hours for $1\frac{1}{2}$ chores? Somewhat harder is $1\frac{1}{4}$ hours for $1\frac{1}{2}$ chores. Note that $1\frac{1}{2}$ is 3 halves. Use models to figure out how much goes in half of a chore (divide into three parts), and then the hours in two of these half-chores will be the number of hours for one chore. (Go ahead, use your models.)

Partition problems can also be made up where the divisor is less than one. Realistic word problems necessarily must involve contexts where partial sets make sense. Here are a few possibilities:

> Dad paid $2.00 for a 3–4 pound box of candy. How much is that per pound? (Divide $2 into 3 parts and then count 4 parts.)
>
> The runner ran $2\frac{1}{2}$ miles in $\frac{3}{4}$ of an hour. At that rate, how many miles could he run in 1 hour? That is, what is his speed in miles per hour? (Divide the $2\frac{1}{2}$ miles into 3 parts. That tells how many miles in each quarter hour; 4 quarters is a full hour.)

Partition division problems with fraction divisors tend to involve rates of time or price rates. Most of these would probably be done using decimals instead of fractions. The ideas and relationships involved in these explorations may, however, have values in themselves.

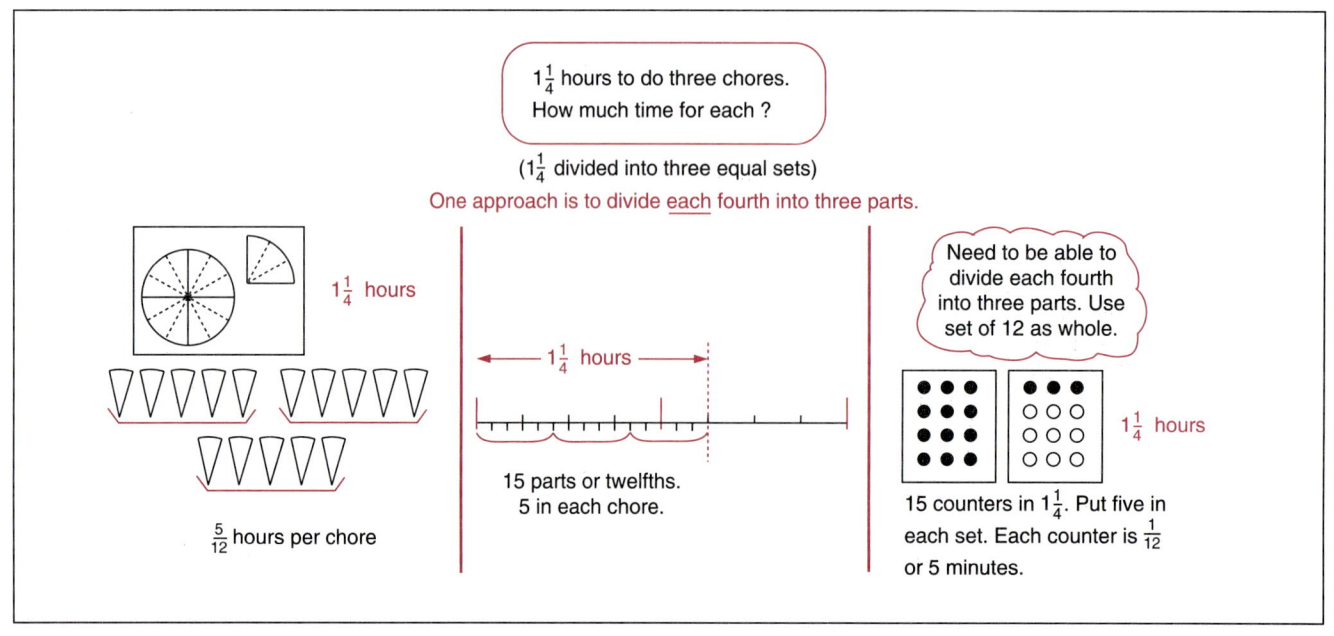

$1\frac{1}{4}$ hours to do three chores. How much time for each?

($1\frac{1}{4}$ divided into three equal sets)

One approach is to divide each fourth into three parts.

$1\frac{1}{4}$ hours

$\frac{5}{12}$ hours per chore

$1\frac{1}{4}$ hours

15 parts or twelfths. 5 in each chore.

Need to be able to divide each fourth into three parts. Use set of 12 as whole.

$1\frac{1}{4}$ hours

15 counters in $1\frac{1}{4}$. Put five in each set. Each counter is $\frac{1}{12}$ or 5 minutes.

FIGURE 13.13: *Partition division: three models*

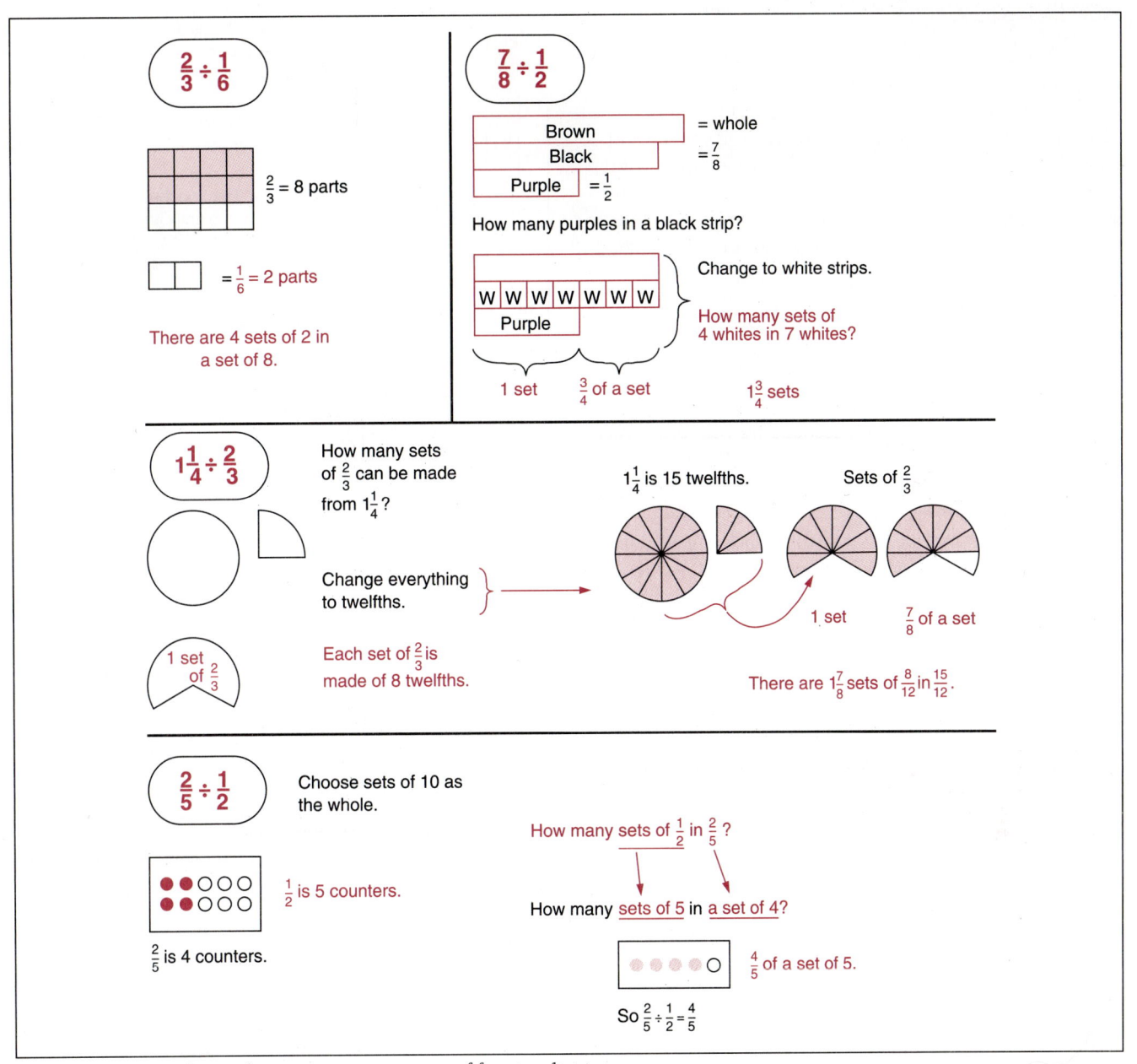

FIGURE 13.14: *Exploring the measurement concept of fraction division*

◆ Measurement Concept

In the measurement concept for 12 ÷ 3, the task is to decide how many sets of 3 things are in a set of 12 things. What would it mean if the problem was $2\frac{1}{4} \div \frac{3}{4}$? The meaning is the same. How many sets of $\frac{3}{4}$ each are in a set of $2\frac{1}{4}$? This might occur in a problem such as the following:

> Farmer Brown measured his remaining insecticide and found that he had $2\frac{1}{4}$ gallons. It takes $\frac{3}{4}$ of a gallon to make a tank of mix. How many tanksful can he make?

In this problem both the total amount and the divisor are in the same type of units or parts, namely fourths. The $2\frac{1}{4}$ is the same as 9 fourths. The question then is, how many sets of 3 fourths are in a set of 9 fourths? The result is 3 sets (not 3 fourths). This is relatively easy to model. Try it.

A similar problem is the following:

> Linda has $4\frac{2}{3}$ yards of material. She is making baby clothes for the bazaar. If each pattern requires $1\frac{1}{6}$ yards of material, how many patterns will she be able to make?

To make this easier, convert the $4\frac{2}{3}$ to sixths also. The problem then becomes, how many sets of 7 sixths are in a set of 28 sixths? Once both quantities are in the same units, it really makes no difference what those units are. The problem $5\frac{3}{5} \div 1\frac{2}{5}$ is identical: How many sets of 7 (fifths) are in a set of 28 (fifths)?

The answers in the foregoing examples were whole numbers. Explore for a minute what $13 \div 3$ might mean. There are 4 complete sets of 3 in a set of 13 with one left. That one constitutes $\frac{1}{3}$ of a set of 3. What if the 13 and 3 were fractional parts, for example, 13 eighths and 3 eighths? There would still be 4 complete sets of 3 eighths and $\frac{1}{3}$ of a set of 3 eighths. The fact that these are eighths is not relevant. Explore this idea using various fraction models and the following exercises:

$$\frac{7}{5} \div \frac{3}{5} = \square \quad \frac{7}{12} \div \frac{3}{12} = \square \quad \frac{3}{4} \div \frac{1}{2} = \square \quad \frac{5}{3} \div \frac{1}{4} = \square$$

As illustrated in Figure 13.14 on page 253, once both the dividend and the divisor are converted to the same amounts, the problem is essentially the same as a whole-number division problem. Pay special attention to a problem such as $\frac{1}{2} \div \frac{3}{4}$, where the answer is less than one.

◆ In the Classroom

Students in your class should explore both measurement and partition problems with fractions. These provide excellent problem-solving situations that can be worked on in groups. Students should be challenged both to come up with an answer and to justify or explain it using a model or a drawing. Even if they have already learned an algorithm for doing the division or are using a fraction calculator to get the results, the explanation of why it makes sense is a terrific performance task. The other important benefit is that these problems focus on the meaning of division and will help students with estimation of fraction division much more than will the algorithm.

DEVELOPING THE ALGORITHM

There are two different algorithms for division of fractions. Methods of teaching both algorithms are discussed here.

◆ The Common-Denominator Algorithm

The common-denominator algorithm relies on the measurement or repeated-subtraction concept of division that was just developed. Consider the problem $\frac{5}{3} \div \frac{1}{2}$. As shown in Figure 13.15, once each number is expressed in terms of the same fractional part, the answer is exactly the same as the whole-number problem $10 \div 3$. The name of the fractional part (the denominators) is no longer important, and the problem is one of dividing the numerators. The resulting rule or algorithm, therefore, is as follows: To divide fractions, first get common denominators, and then divide numerators.

When students learn that a fraction is also a way of indicating division, the rule can be expressed this way: Rewrite the division with common denominators and then $\frac{a}{c} \div \frac{b}{c} = \frac{a}{b}$. Try using pie pieces, fraction strips, and then sets of counters to model $1\frac{2}{3} \div \frac{3}{4}$ and $\frac{5}{8} \div \frac{1}{2}$.

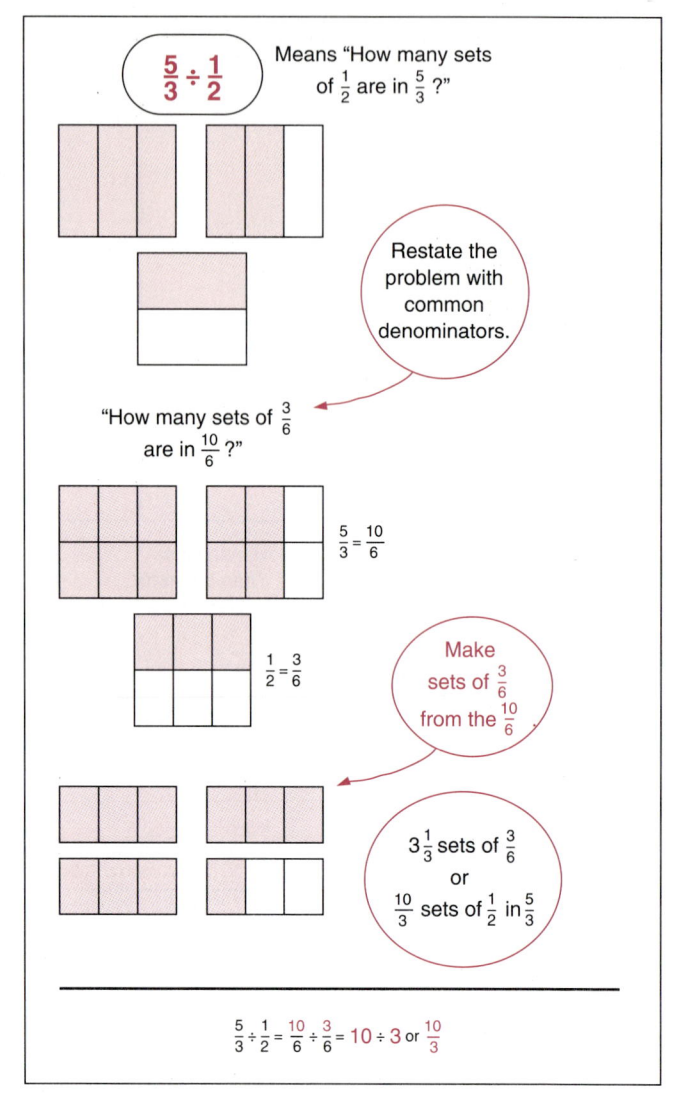

FIGURE 13.15: *Common-denominator method for fraction division.*

◆ The Invert-and-Multiply Algorithm

The more traditional algorithm for division with a fraction is to invert the divisor and multiply. The development of this algorithm relies on a symbolic rationale. One explanation of the invert-and-multiply approach is outlined in Figure 13.16.

◆ Curricular Decisions

Which of these two division algorithm for fractions is "best"? There are values to each. In favor of the common-denominator approach is the fact that it can be explained quite well with models and prior concepts of whole numbers. From a developmental point of view, this is a compelling argument. Very few children below the eighth grade will be able to fully appreciate the explanation for the invert-and-multiply approach. The result may be that they simply learn the rule: Turn the second one upside down

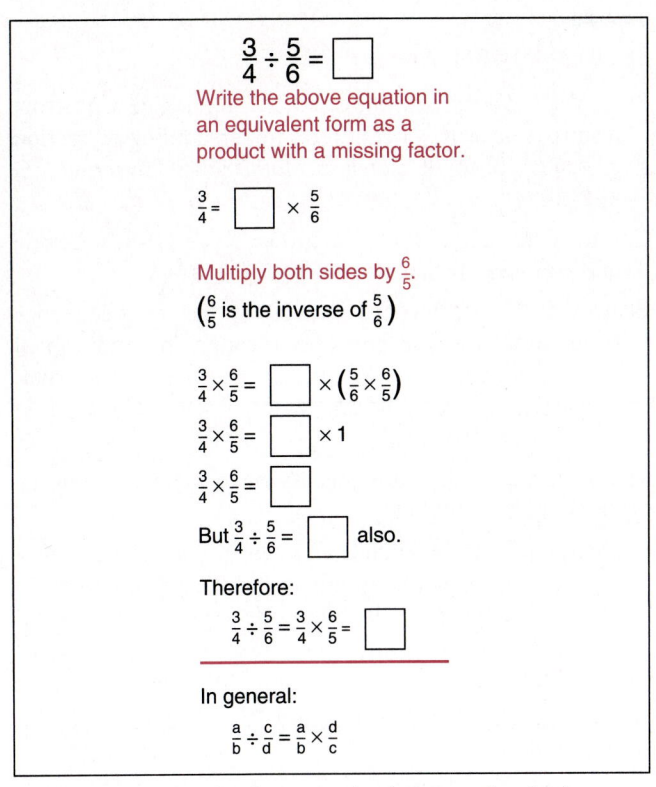

FIGURE 13.16: To divide, invert the divisor and multiply.

$$\frac{3}{4} \div \frac{5}{6} = \Box$$

Write the above equation in an equivalent form as a product with a missing factor.

$$\frac{3}{4} = \Box \times \frac{5}{6}$$

Multiply both sides by $\frac{6}{5}$.

($\frac{6}{5}$ is the inverse of $\frac{5}{6}$)

$$\frac{3}{4} \times \frac{6}{5} = \Box \times \left(\frac{5}{6} \times \frac{6}{5}\right)$$

$$\frac{3}{4} \times \frac{6}{5} = \Box \times 1$$

$$\frac{3}{4} \times \frac{6}{5} = \Box$$

But $\frac{3}{4} \div \frac{5}{6} = \Box$ also.

Therefore:

$$\frac{3}{4} \div \frac{5}{6} = \frac{3}{4} \times \frac{6}{5} = \Box$$

In general:

$$\frac{a}{b} \div \frac{c}{d} = \frac{a}{b} \times \frac{d}{c}$$

and multiply. The idea of "turning a fraction upside down" is antithetical to any meaningful approach to mathematics.

For students who are able to understand the symbolic explanation of the invert-and-multiply approach, it is certainly much more quickly explained with no ambiguities or special cases to consider. For some students in the eighth grade, this may be a very appropriate approach.

Another argument turns on the usefulness of the respective algorithms. Invert and multiply is generally the algorithm taught in algebra (although the common-denominator algorithm works in algebra almost as well). The common-denominator approach and its conceptual rationale may both be more useful for mental estimations of quotients involving fractions.

That brings up an even larger issue. When was the last time you had any need to divide by a fraction where you actually used pencil-and-paper division to obtain an exact result? Perhaps the only real justification for teaching division of fractions at all is as a model for algebra.

◆◆◆◆◆

REFLECTIONS ON CHAPTER 13: WRITING TO LEARN

1. Make up an example and use pie pieces (or draw pictures) to show how two fractions with unlike denominators can be added without first getting a common denominator. In your example, did you get your answer by either looking at the part that was more than a whole or the missing part that was just less than a whole? Explain both of these ideas.

2. Suppose that you got pie pieces out to show $\frac{2}{3} + \frac{1}{2}$ and by some informal means you are now convinced that the sum is $1\frac{1}{6}$. Now suppose that you substitute four $\frac{1}{6}$ pieces for the $\frac{2}{3}$ and two $\frac{1}{4}$ pieces for the $\frac{1}{2}$. What is the sum of these pieces ($\frac{4}{6} + \frac{2}{4}$)? Explain why you would want students to say *immediately* that the sum was $1\frac{1}{6}$. Why is this idea important to the use of common denominators in addition and subtraction?

3. Use the same concept of multiplication as for whole numbers to explain why $\frac{3}{4} \times \frac{8}{5} = \frac{6}{5}$ without using the algorithm and without first getting $\frac{24}{20}$.

4. Describe an estimation activity for addition and subtraction of fractions.

5. Draw pictures of squares for the whole to illustrate these products. Explain each:

   $3 \times \frac{2}{5}$ $\frac{3}{4} \times \frac{2}{3}$ $2\frac{1}{2} \times \frac{2}{3}$

6. Explain at least one mental method (estimation or mental computation) for each of these:

   $\frac{3}{4} \times 5\frac{1}{2}$ $1\frac{1}{8}$ of 679

7. Make up a word problem with a fraction as a divisor. Is your problem a measurement problem or a partition problem? Make up a second word problem with fractions that is of the other type (measurement or partition).

8. Draw pictures to explain each of these divisions using a measurement approach:

   $1\frac{2}{4} \div \frac{1}{4}$ $2\frac{1}{3} \div \frac{2}{3}$ $\frac{3}{4} \div \frac{1}{8}$ $2\frac{3}{4} \div \frac{2}{3}$

   In the second and fourth examples above, the answer is not a whole number. To help you explain the fractional part of the answer, use a set of counters to explain $13 \div 5 = 2\frac{3}{5}$ also using a measurement approach. (That is, how many sets of five are in a set of 13?)

9. Use the same problems you modeled in the previous exercise to explain a common-denominator algorithm for division. Use the same rationale to explain why $\frac{5}{79} \div \frac{3}{79}$ is $\frac{5}{3}$.

10. What is one strong reason for not teaching "invert and multiply"?

FOR DISCUSSION AND EXPLORATION

1. Give two reasons why the following argument is flawed:

    $\frac{1}{2}$ ●○ $\frac{1}{3}$ ●○○

    add

    ●●○○○

    Therefore: $\frac{1}{2} + \frac{1}{3} = \frac{2}{5}$.
    Add tops and bottoms.

2. Discuss the differences and relative difficulty of the following, especially when they are done with models. Each exercise is different in some way from the others.

    $\frac{3}{4} + \frac{3}{4}$ $\frac{2}{3} + \frac{5}{6}$ $\frac{3}{8} + \frac{1}{2}$ $\frac{2}{3} + \frac{5}{6}$

    $1 - \frac{1}{4}$ $\frac{5}{2} - \frac{1}{4}$ $\frac{3}{4} + \frac{2}{3}$ $4\frac{3}{4} - 2\frac{5}{8}$

    $3\frac{3}{8} - 1\frac{3}{4}$

3. Explore a textbook series, and see where each algorithm is first introduced. Does the preparation for the algorithm seem appropriate? Do you think it is useful to spend time on intuitive approaches using models, or is the time required not worth the effort?

4. Several calculators are now available that do computations in fractional form as well as in decimal form. Some of these automatically give results in simplest terms. If you have access to such a calculator, discuss how it might be used in teaching fractions and especially fraction computation. If such calculators become commonplace, should we continue to teach fraction computation?

SUGGESTED READINGS

Behr, M. J., Wachsmuth, I., & Post, T. R. (1985). Construct a sum: A measure of children's understanding of fraction size. *Journal for Research in Mathematics Education, 16,* 120–131.

Curcio, F. R., Sicklick, F., & Turkel, S. B. (1987). Divide and conquer. *Arithmetic Teacher, 35*(4), 6–12.

Ellerbruch, W., & Payne, J. N. (1978). Teaching sequence from initial fraction concepts through the addition of unlike fractions. In M. N. Suydam (Ed.), *Developing computational skills.* Reston, VA: National Council of Teachers of Mathematics.

Feinberg, M. (1980). Is it necessary to invert? *Arithmetic Teacher, 27*(5), 50–52.

Graeber, A. O., & Tanenhaus, E. (1993). Multiplication and division: From whole-numbers to rational numbers. In D. T. Owens (Ed.), *Research ideas for the classroom: Middle grades mathematics.* New York: Macmillan Publishing Co.

Holden, L. (1986). *Fraction factory.* Palo Alto, CA: Creative Publications.

Post, T. R., Wachsmuth, I., Lesh, R., & Behr, M. J. (1985). Order and equivalence of rational numbers: A cognitive analysis. *Journal for Research in Mathematics Education, 16,* 18–36.

Prevost, F. J. (1984). Teaching rational numbers–junior high school. *Arithmetic Teacher, 31*(6), 43–46.

Sweetland, R. D. (1984). Understanding multiplication of fractions. *Arithmetic Teacher, 32*(1), 48–52.

Thompson, C. S. (1979). Teaching division of fractions with understanding. *Arithmetic Teacher, 25*(5), 24–27.

Trafton, P. R., & Zawojewski, J. S. (1984). Teaching rational number division: A special problem. *Arithmetic Teacher, 31*(6), 20–22.

14 DECIMAL AND PERCENT CONCEPTS AND DECIMAL COMPUTATION

CONNECTING FRACTION AND DECIMAL CONCEPTS

There has always been some debate concerning which concepts should be developed first in the curriculum, fractions or decimals. In most U.S. programs, fractions receive the earliest attention. What is abundantly clear is that the ideas of fractions and decimals must be intimately connected. Most of this chapter is devoted to making that connection.

TWO DIFFERENT REPRESENTATIONAL SYSTEMS

The symbols 3.75 and $3\frac{3}{4}$ both represent the same relationship or quantity, yet on the surface the two appear quite different. For children especially, the world of fractions and the world of decimals are very distinct. Even for adults, there is a tendency to think of fractions as sets or regions (three-fourths *of* something), whereas we think of decimals as being more like numbers. The reality is that fractions and decimals are two different systems of representation that have been developed to represent the same ideas. When we tell children that 0.75 is the same as $\frac{3}{4}$, this can be especially confusing. Even though different ways of writing the numbers have been invented, the numbers themselves are not different.

A significant goal of instruction in decimal and fraction numeration should be to help students see how both systems represent the same concepts. A good connection between fractions and decimals is quite useful. Even though we tend to think of 0.5 as a number, it is handy to know that it is a half. For example, in many contexts it is easier to think about $\frac{3}{4}$ than 75 hundredths or 0.75. On the other hand, the decimal system makes it easy to use numbers that are close to $\frac{3}{4}$, such as 0.73 or 0.78. Other conceptual and practical advantages exist for each system. For example, an obvious use of the decimal system is in digital equipment such as calculators, computers, and electronic meters.

To help students see the connection between fractions and decimals we can do three things. First, use familiar fraction concepts and models to explore those rational numbers that are easily represented by decimals: tenths, hundredths, and thousandths. Second, extend the base ten decimal system to include numbers less than one as well as large numbers. Third, help children use models to make meaningful translations between fractions and decimals. If two different symbols describe the same model, they must also represent the same idea. These three components are discussed in the following three sections.

BASE TEN FRACTIONS

Fractions that have denominators of 10, 100, 1000, and so on will be referred to in this chapter as *base ten fractions*. This is simply a convenient label and is not one commonly found in the literature. Fractions such as $\frac{7}{10}$ or $\frac{63}{100}$ are examples of base ten fractions.

◆ Base Ten Fraction Models

Most of the common models for fractions are somewhat limited for the purpose of depicting base ten fractions. Fraction models for tenths do exist, but generally the familiar fraction models cannot show hundredths or thousandths. It is important to provide models for these fractions using the same conceptual approaches that were used for more familiar fractions such as thirds and fourths.

Two very important area or region models can be used to model base ten fractions. First, to model tenths and hundredths, circular disks such as the one shown in Figure 14.1 can be printed on tagboard. (A full-sized master is in the Black-line Masters section.) Each disk is marked with 100 equal intervals around the edge and is cut along one radius. Two disks of different colors, slipped together as shown, can be used to model any fractions less than one. Fractions modeled on this hundredths disk can be read as base ten fractions by noting the spaces around the edge but are still reminiscent of the traditional pie sections model.

The most common model for base ten fractions is a 10 × 10 square. These can be run off on paper for students to shade in various fractions (Figure 14.2 and Blackline Masters). Another important variation is to use base ten place-value strips and squares. As a fraction model, the 10-cm square that was used as the hundred model for whole numbers is taken as the whole or one. Each strip is then 1 tenth, and each small square is 1 hundredth. The *Decimal Squares* materials (Bennett, 1982) include squares in which each hundredth is again partitioned into 10 smaller sections. This permits modeling of thousands by shading in portions of the square. Even one more step is provided by the square in the Black-line Masters section. It is subdivided into 10,000 tiny squares. While too small for paper reproduction, the master can be used to make a transparency. When shown with an overhead projector, individual squares or ten-thousandths can easily be identified and shaded in with a pen on the transparency.

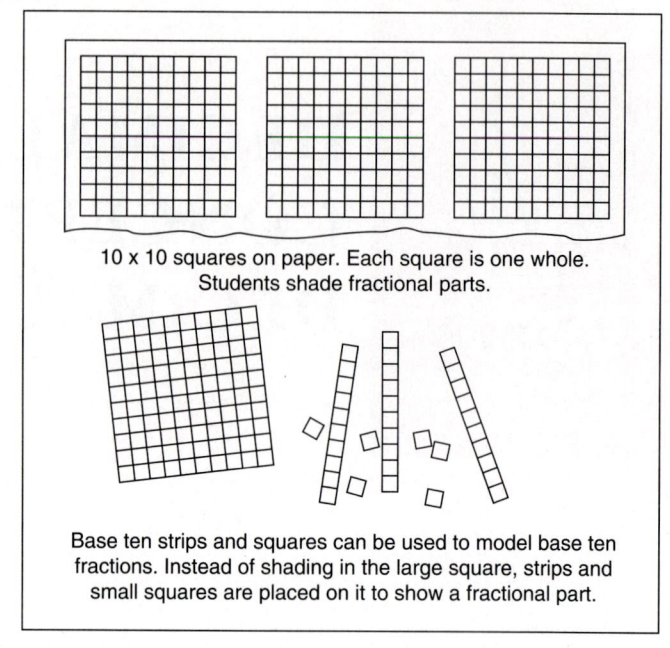

10 x 10 squares on paper. Each square is one whole. Students shade fractional parts.

Base ten strips and squares can be used to model base ten fractions. Instead of shading in the large square, strips and small squares are placed on it to show a fractional part.

FIGURE 14.2: 10 × 10 squares model base ten fractions.

One of the best length models is a meter stick. Each decimeter is one-tenth of the whole stick, each centimeter one-hundredth, and each millimeter is one-thousandth. Any number-line model broken into 100 subparts is likewise a useful model for hundredths.

Many teachers use money as a model for decimals and to some extent this is helpful. However, for children, money is almost exclusively a two-place system. That is, numbers like 3.2 or 12.1389 do not relate to money. Children's initial contact with decimals should be more flexible. Money is certainly an important application of decimal numeration and may be useful in later activities with decimals.

◆ Multiple Names and Formats

Early work with base ten fractions is primarily designed to acquaint students with the models, to help them to begin to think of quantities in terms of tenths and hundredths,

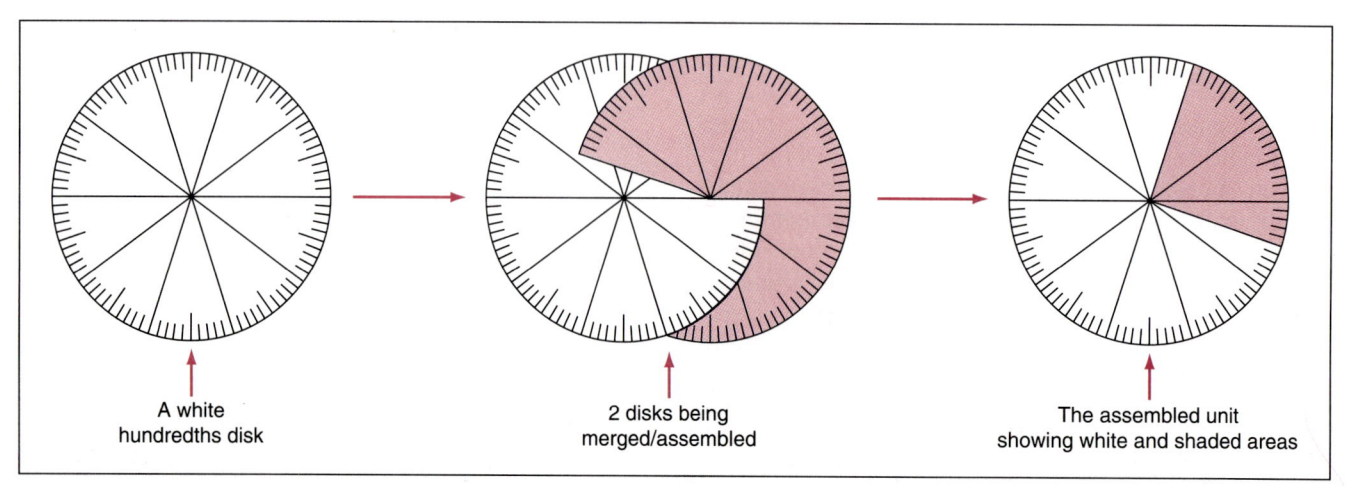

A white hundredths disk → 2 disks being merged/assembled → The assembled unit showing white and shaded areas

FIGURE 14.1: *A hundredths disk for modeling powers-of-ten fractions*

and to learn to read and write base ten fractions in different ways.

Have students show a base ten fraction using any base ten fraction model. Once a fraction, for example $\frac{65}{100}$, is modeled, the following things can be explored:

> Is this fraction more or less than $\frac{1}{2}$? than $\frac{2}{3}$? than $\frac{3}{4}$? That is, some familiarity with these fractions can be developed by comparison with fractions that are easy to think about.
>
> What are some different ways to say this fraction using tenths and hundredths? ("Six-tenths and five-hundredths" or "sixty-five-hundredths") Include thousandths when appropriate.
>
> Show two ways to write this fraction. ($\frac{65}{100}$ or $\frac{6}{10} + \frac{5}{100}$)

The last two questions are very important. When base ten fractions are later written as decimals, they are usually read as a single fraction. That is, 0.65 is read "sixty-five-hundredths." But to understand them in terms of place value, the same number must be thought of as 6 tenths and 5 hundredths. A mixed number such as $5\frac{13}{100}$ is usually read the same way as a decimal: 5.13 is "five and thirteen-hundredths." For purposes of place value, it should also be understood as $5 + \frac{1}{10} + \frac{3}{100}$. Special attention should also be given to numbers such as the following:

$$\frac{30}{1000} = \frac{0}{10} + \frac{3}{100} + \frac{0}{1000}$$

$$\frac{70}{100} = \frac{7}{10} + \frac{0}{100}$$

The expanded forms on the right will be helpful in translating these fractions to decimals. In oral form, fractions or decimals with trailing zeros are sometimes used to indicate a higher level of precision. Seven-tenths is numerically equal to 70 hundredths, but the latter conveys precision to the nearest hundredth.

Exercises at this introductory level should include all possible connections between models, various oral forms, and various written forms. Given a model or written or oral fraction, students should be able to give the other two forms of the fraction, including equivalent forms where appropriate.

EXTENDING THE PLACE-VALUE SYSTEM

The 10-to-1 Relationship Works in Two Directions

Before considering decimal numerals with students, it is advisable to review some ideas of whole-number place value. One of the most basic of these ideas is the 10-to-1

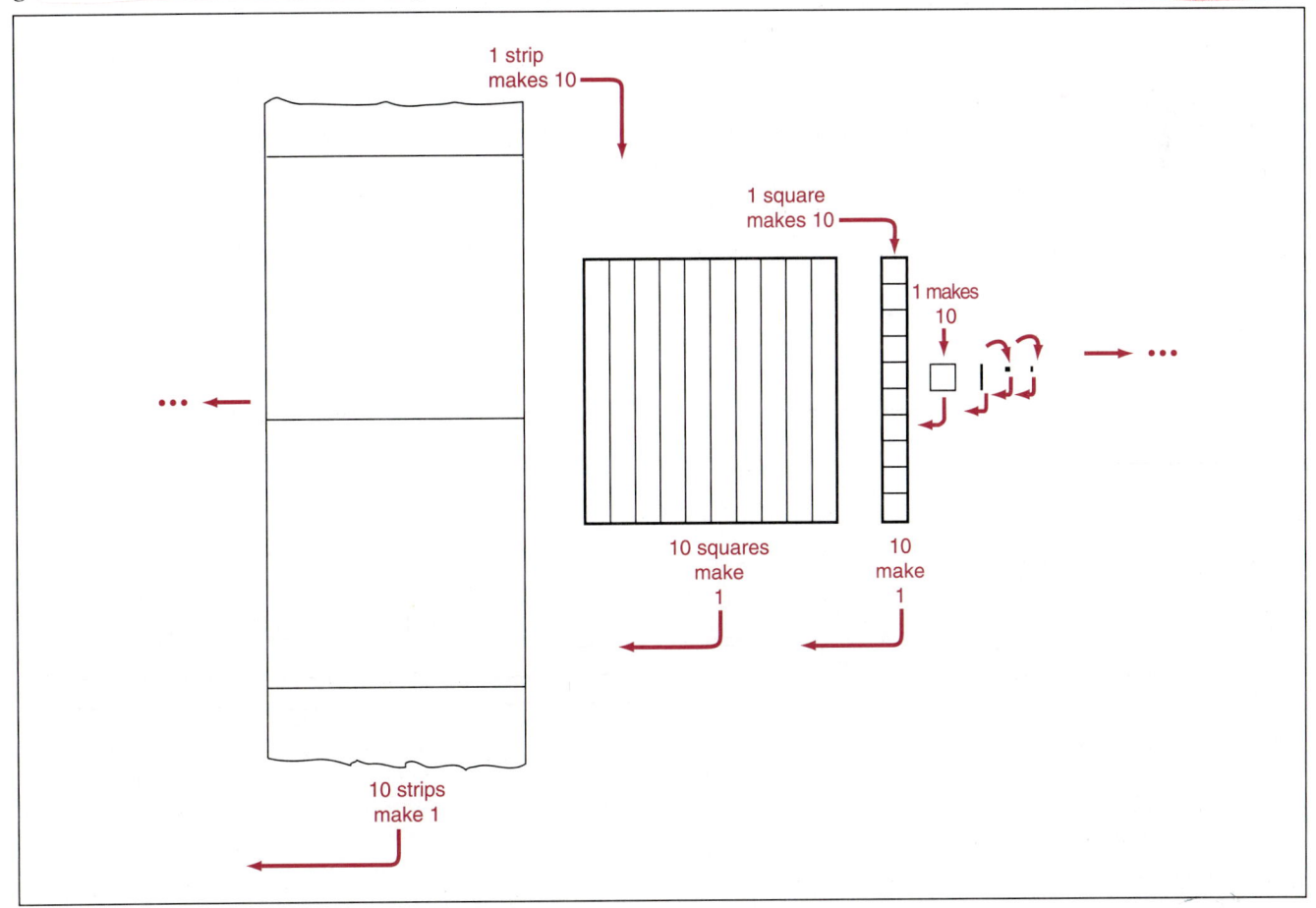

FIGURE 14.3: *The strips and squares extend infinitely in both directions (in the mind's eye).*

relationship between the value of any two adjacent positions. In terms of a base ten model such as strips and squares, 10 of any one piece will make one of the next larger, and vice versa. The 10-makes-1 rule continues indefinitely to larger and larger pieces or positional values. This concept is fun to explore in terms of how large the strips and squares will actually be if you move six or eight places out.

If you are using the strip-and-square model, for example, the strip and square shapes alternate in an infinite progression as they get larger and larger. Having established this idea with your students, focus on the idea that each piece to the right in this string gets smaller by one-tenth. The critical question becomes: "Is there ever a smallest piece?" In the students' experience, the smallest piece is always the one designated as the ones or unit piece. But what is to say that even that piece could not be divided into 10 small strips? And could not these small strips be divided into 10 very small squares, which in turn could be divided into 10 even smaller strips, and so on and so on? In the mind's eye, there is no smallest strip or smallest square.

The goal of this discussion is to help students see that a 10-to-1 relationship can extend *infinitely in two directions, not just one.* There is no smallest piece and there is no largest piece. The relationship between adjacent pieces is the same regardless of which two adjacent pieces are being considered. Figure 14.3 (p.259) illustrates this idea with a strip-and-square model.

◆ Which Is the Unit? The Role of the Decimal Point

An important idea to be realized in this discussion is that there is no built-in reason why any one position should naturally be chosen to be the unit or ones position. In terms of strips and squares, for example, which piece is the ones piece? the small centimeter square? Why? Why not a larger or a smaller square? Why not a strip? *Any piece could effectively be chosen as the ones piece.*

When a number such as 1624 is written, the assumption is that the 4 is in the unit or ones position. But if a position to the left of the 4 were selected as the ones position, some method of designating that position must be devised. Enter the decimal point. As shown in Figure 14.4, the same amount can be written in different ways, depending on the choice of the unit. The decimal point is placed between two positions with the convention that the position to the left of the decimal is the unit or ones position. Thus, *the role of the decimal point* is to designate the unit position, and it does so by sitting just to the right of that position.

A fitting caricature of the decimal is shown in Figure 14.5. The "eyes" of the decimal are always focused up toward the name of the unit or ones. A tagboard disk of this decimal point face can be used between adjacent base ten models or on a place-value chart. If such a decimal point were placed between the squares and strips in Figure 14.4, the squares would then be designated as the units, and the written form 16.24 would be the correct written form for the model.

| Super strips | Squares | Strips | Tinies |

1624 tinies
16.24 squares
1.624 super strips
162.4 strips
0.1624 super squares

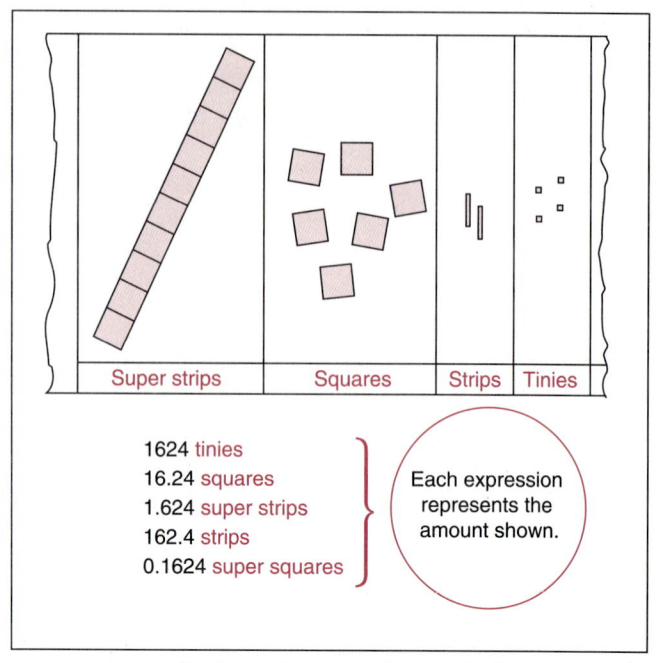

Each expression represents the amount shown.

FIGURE 14.4: *The decimal point indicates which position is the units.*

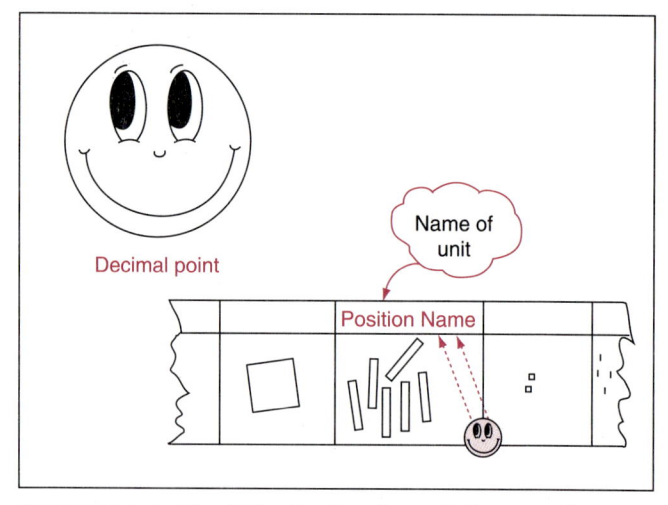

FIGURE 14.5: *The decimal point always looks up at the name of the UNIT position.*

14.1 THE DECIMAL NAMES THE UNIT

Have students display a certain amount of base ten pieces on their desks. For example, put out 3 squares, 7 strips, and 4 tinies. (If you want to have students cut up some one-centimeter-long pieces of flat toothpicks, they can also model a fourth position.) Refer to the pieces as "squares," "strips," and "tinies," and reach an agreement on names for the theoretical pieces both smaller and larger. To the right of tinies can be "tiny strips" and "tiny squares." To the left of squares can be "super strips" and "super squares." Each student should also have a tagboard smiley decimal point (found on the hundredths disk, Appendix B). Now ask students to write and say how many *squares* they have, how many *super strips*, and so on, as in Figure 14.4. The students position their decimal point accordingly and both write and say the amounts.

The value of the last activity is to vividly illustrate how the decimal indicates the named unit, and that unit can change without changing the quantity. A related discussion could involve multiplication or division of the quantity by powers of ten. If the display shows, for example, 3.74 squares, what would 10 times this amount be? Here the decimal remains looking at the squares position since 10 times 3.74 squares will still be squares. However, while the decimal remains fixed, each piece in the display would be enlarged 10 times and thus shift to a position one place to the left. The result would be 37.4 squares with the 3 squares becoming 3 super strips, the 7 strips becoming 7 squares, and the 4 tinies becoming 4 strips.

The notion that the decimal "looks at the units place" is useful in a variety of contexts. For example, in the metric system, seven place values have names. As shown in Figure 14.6, the decimal can be used to designate any of these places as the unit without changing the actual measure. Our monetary system is also a decimal system. In the amount $172.95, the decimal point designates the dollars position as the unit. There are 1 hundreds (of dollars), 7 tens, 2 singles, 9 dimes, and 5 pennies or cents in this amount of money regardless of how it is written. If pennies were the designated unit, the same amount would be written as 17,295 *cents* or 17,295.0 *cents*. It could just as correctly be 0.17295 *thousands of dollars* or 1729.5 *dimes*. The role of the decimal can be explored in this way using money or metric measures as an example. These systems highlight the idea that the choice of a unit is arbitrary but must be designated. In the case of actual measures such as metric lengths or weights, or the U.S. monetary system, the name of the unit is written after the number. You may be 1.62 *meters* tall, but it does not make sense to say you are 1.62 tall. In the paper we may read about Congress spending $7.3 billion. Here the units are billions of dollars, not dollars. A city may have a population of 2.4 million people. That is the same as 2,400,000 single people.

◆ Reading Decimal Numerals

Consider the example of Figure 14.4. What are appropriate ways to read each of the different designations for these amounts? The correct manner is to read them the same way as one would read a mixed fraction. For example, 16.24 squares is read "sixteen *and* twenty-four hundredths squares." Notice that the word *and* is reserved for the decimal point. It could also be read as "sixteen *and* two-tenths and four-hundredths squares." The other important feature is that this language is exactly the same as the language that was used for base ten fractions:

$$16.24 \text{ squares} = 16\tfrac{24}{100} \text{ square} = \tfrac{2}{10} + \tfrac{4}{100} \text{ squares}$$

Help students see that once place-value position is selected as the unit, then the next place to the right is $\tfrac{1}{10}$ or tenths, and the next is $\tfrac{1}{100}$ or hundredths, and so on. As in the previous section on reading fractions, children should become accustomed to reading decimal numerals in two different forms. The oral language then becomes a useful linkage between fraction symbolism and decimal symbols.

◆ Other Decimal Models

Technically, any base ten place-value model can be used as a model for decimals. The strip-and-square model is useful because students can each have a set of these materials at their desks. If the kit includes a tagboard decimal point (Figure 14.5), then no place-value mat is required. Students can arrange their base ten pieces in order and put the decimal point to the right of whatever piece is the designated unit. Usually, the 10-cm square is selected. This allows students to model decimal fractions to hundredths. Some teachers even cut wooden toothpicks into 1-cm lengths to represent the next smaller strip or thousandths. These show up very nicely on an overhead projector.

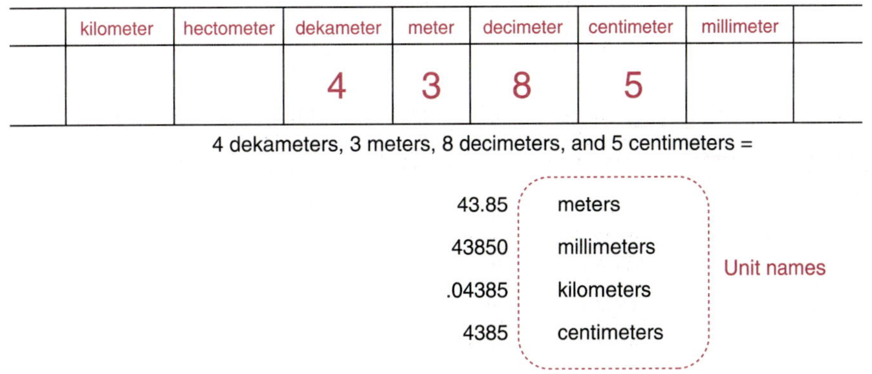

kilometer	hectometer	dekameter	meter	decimeter	centimeter	millimeter
		4	3	8	5	

4 dekameters, 3 meters, 8 decimeters, and 5 centimeters =

43.85 meters
43850 millimeters
.04385 kilometers
4385 centimeters

Unit names

FIGURE 14.6: *In the metric system, each place-value position has a name. The decimal point can be placed to designate which length is the UNIT length.*

The three-dimensional wooden or plastic base ten blocks have four different pieces. If the 10-cm cube or "block" is designated as 1, then the flats, sticks, and small cubes can be used to model decimals to three places.

Any base ten fraction model is also a decimal model. This fact is significant because it points out that decimals and fractions are simply two different symbolisms to represent the same amounts. Three-fourths is shown on the hundredths disk as 7 tenths and 5 hundredths. If these pieces could be cut up and put on a place-value chart, they would be shown as 0.75 with the decimal looking up at the circle's place, as in Figure 14.7.

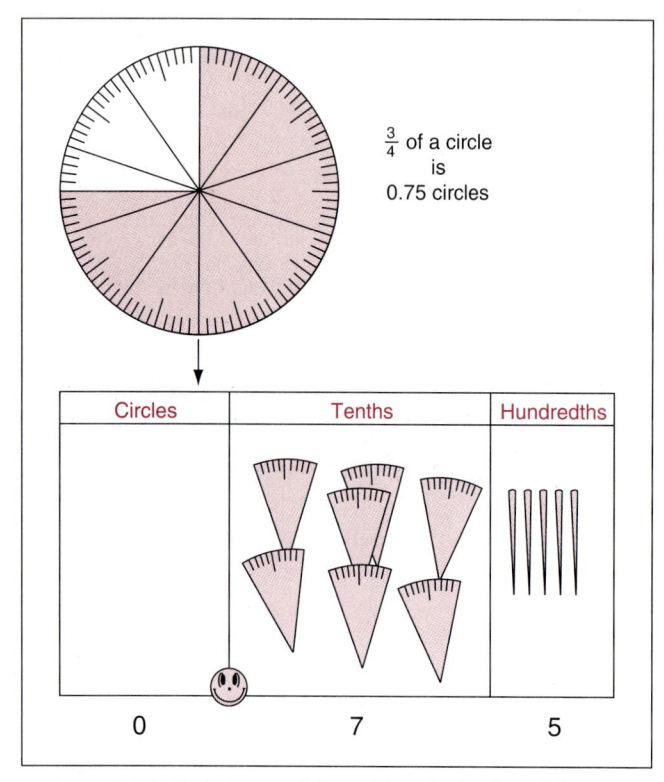

FIGURE 14.7: *Fraction models could be decimal models.*

MAKING THE FRACTION–DECIMAL CONNECTION

To connect the two numeration systems, fractions and decimals, students should make concept-oriented translations from one system to another. The purpose of such activities has less to do with the skill of converting a fraction to a decimal than with construction of the concept that both systems can be used to express the same ideas.

14.2 FRACTIONS TO DECIMALS

Start with a base ten fraction, such as $\frac{35}{100}$ or $\frac{28}{10}$. Have students use at least two different base ten fraction models to show these amounts. Discuss how each model shows the same relationship. Include as one of the models a set of strips and squares with the 10 × 10 square designated as the unit. The strip-and-square model can then be rearranged in standard place-value format indicating how the fraction can be expressed as a decimal (Figure 14.8).

Once a fraction is modeled and written as both a fraction and a decimal, both symbolisms should be read. The oral language for both fraction and decimals will be the same, as has already been discussed.

$\frac{35}{100}$ = 0.35 = "thirty-five-hundredths"

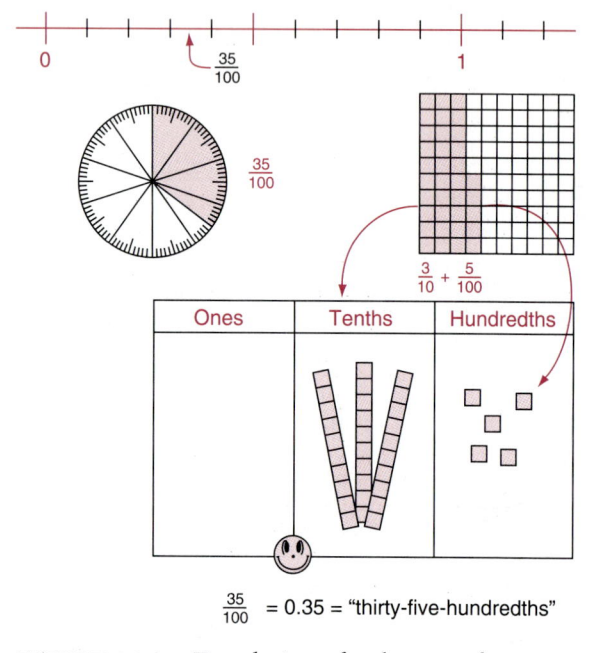

FIGURE 14.8: *Translation of a base ten fraction to a decimal.*

The previous activity suggests starting with a fraction and ending with a decimal. The reverse activity is equally important. That is, begin with a decimal number, and have students use several models to represent it and then read and write the number as a fraction. As long as the fractions are restricted to base ten fractions, the conversions are straightforward.

The calculator can also play a significant role in decimal concept development.

14.3 CALCULATOR DECIMAL COUNTING

Recall how to make the calculator "count" by pressing ⊞ 1 ⊟ ⊟ . . . . Now have students press ⊞ 0.1 ⊟ ⊟ . . . . When the display shows 0.9, stop and discuss what this means and what the display will look like with the next press. Many students will predict 0.10 (thinking that 10 comes after 9). This prediction is even more interesting if, with each press, the students have been accumulating base ten strips as models for tenths. One more press would mean one more strip or ten strips. Why should the calculator not show 0.10? When the tenth press produces a display of 1., the discussion should revolve around trading ten strips for a square. Calculators never display trailing zeros right of the decimal. Continue to count to 4 or 5 by tenths. How many presses to get from one whole number to the next? Try counting by 0.01 or by 0.001. These counts illustrate dramatically how small one hundredth and one thousandth really are. It requires 10 counts by 0.001 to get to 0.01, and 1000 counts will only reach 1.

14.4 DOUBLE CALCULATOR COUNTING

A fascinating variation of the last activity is to use two calculators side by side with the second calculator being one that displays fractions. On the fraction calculator enter [+] 1 [/] 10 [=] and on the standard calculator [+] 0.1 [=]. Continue to press [=] on the two calculators simultaneously. At 10 presses, the standard calculator shows 1. as discussed earlier, but the fraction calculator shows 10/10 and will continue to increment the numerator to 11/10 and so on. If the [a b/c] key is pressed, the mixed number is displayed. The display can be made to toggle back and forth among the mixed number, the improper fraction, and the decimal equivalent. If models are also coordinated with the counting, the activity helps to relate two symbol systems and a conceptual model in a powerful illustration.

DEVELOPING DECIMAL NUMBER SENSE

So far the discussion has revolved around the connection of decimals with base ten fractions because these are the fractions most directly related with the notation of decimals. Number sense implies more. It means having an intuition or friendly understanding of numbers. To this end, it is useful to connect decimals to the fractions with which children have familiarity, to be able to readily and comfortably compare and order decimals, and to approximate decimals with useful familiar numbers.

FAMILIAR FRACTIONS CONNECTED TO DECIMALS

In Chapter 12 on fraction concepts, an emphasis was placed on helping students develop a conceptual familiarity with simple fractions, especially halves, thirds, fourths, fifths, and eighths. We should try to extend this familiarity to the same concepts expressed as decimals. One way to do this is to have students translate these familiar fractions to decimals by means of a base ten model.

14.5 FRIENDLY DECIMAL FRACTIONS

Have students shade in familiar fractions on a 10 × 10 grid. Since the grid effectively translates any fraction to a base ten fraction, these can then easily be written as a decimal. Under each shaded figure, write the fraction in familiar form, then as a base-ten fraction, and finally as a decimal, as shown in Figure 14.9.

$\frac{1}{4} = \frac{25}{100} = .25$

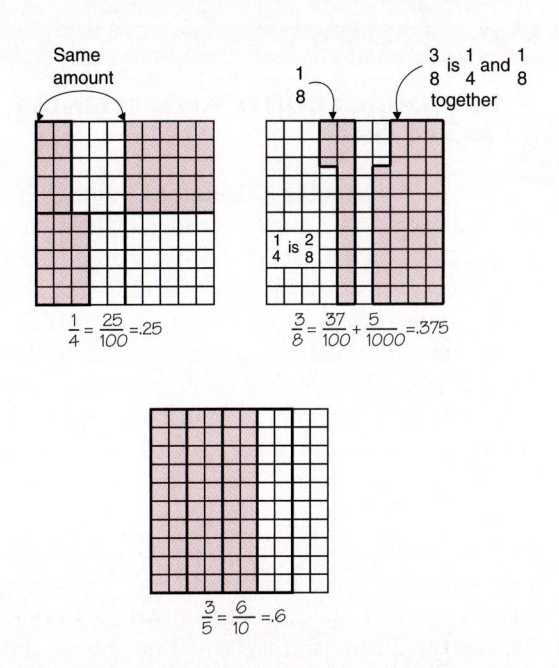

$\frac{3}{8} = \frac{37}{100} + \frac{5}{1000} = .375$

$\frac{3}{5} = \frac{6}{10} = .6$

FIGURE 14.9: *Familiar fractions to decimals using a 10 × 10 square.*

It is good to begin with halves and fifths, since these can be shown with strips of 10 squares. Next explore fourths. Many students will shade one-fourth by shading in a 5 × 5 square of the grid. See if they can then shade the same amount using strips and fewer than 10 small squares. Repeat for other fourths, such as three-fourths or seven-fourths. Eighths present an interesting challenge. One way to find $\frac{1}{8}$ of a 10 × 10 square is to take half as much as is in one-fourth. Since $\frac{1}{4}$ is $\frac{2}{10}$ and $\frac{5}{100}$, $\frac{1}{8}$ is $\frac{2}{10} + \frac{2}{100} + \frac{5}{1000}$. The $\frac{5}{1000}$ part is the same as half of $\frac{1}{100}$. (You should try to explain this fact using a 10 × 10 grid or with the 10,000 grid in the Black-line Masters section.)

The exploration of modeling one-third as a decimal is a good introduction to the concept of an infinitely repeating decimal. Try to partition the whole square into three parts using strips and squares. Each part receives three strips with one left over. To divide the leftover strip, each part gets three small squares with one left over. To divide the small square, each part gets three tiny strips with one left over. (Recall that with base ten pieces, each smaller piece must be a tenth of the preceding size piece.) Each of the three parts will get three tiny strips with one left over. It becomes quite obvious that this process is never-ending. As a result, one-third is the same as 0.333333 . . . or 0.$\overline{3}$. For practical purposes, one-third is about 0.333. In a similar manner, two-thirds is a repeating string of sixes or about 0.667. Later, students will discover that many other fractions cannot be represented by a finite decimal, and this experience is a good background for that idea.

14.6 DUAL COUNTS WITH FRIENDLY FRACTIONS

Repeat 14.4, **Double Calculator Counting**, but this time begin with a familiar or friendly fraction such as $\frac{1}{5}$ or $\frac{1}{4}$ on the fraction calculator and the corresponding decimal on the other calculator. Again, explore counts beyond one or two wholes, and also switch between mixed and improper fractions. Notice also that while the [F↔D] key changes a fraction to a decimal, when going from a decimal to a fraction it uses a base ten fraction. Thus 2.6 is changed to $2\frac{6}{10}$. The [Simp] key can then be used to find the fraction $2\frac{3}{5}$ that you are working with.

It is common in many seventh and eighth grades for students to memorize the decimal equivalents for halves, thirds, fourths, fifths, and eighths. Kouba et al. (1988a) note that in the fourth NAEP, "60 percent of seventh-grade students could express simple fractions as decimals, but only 40 percent could express an improper fraction as a decimal" (p. 16). An explanation for the discrepancy is that the proper fraction equivalents may have been rotely memorized, while to write an improper fraction such as $\frac{7}{5}$ as a decimal would require some conceptual understanding. Even the 60 percent performance speaks to a need for better understanding of these translations. If students know conceptually the decimal translation for the unit fractions $\frac{1}{3}$, $\frac{1}{4}$, $\frac{1}{5}$, and $\frac{1}{8}$, and apply to these a counting-by-unit-fraction concept, the equivalents for all familiar fractions are nearly immediate. For example, $\frac{3}{5}$ is one-fifth three times, or 0.2 + 0.2 + 0.2. Even $\frac{7}{8}$ can be easily thought of as an eighth more than $\frac{3}{4}$ or an eighth less than one. An understanding of mixed fractions and improper fractions can easily be applied to find decimal equivalents for familiar fractions greater than one. For all such exercises, the emphasis should be on conceptual conversions and not on rote memory or other symbolic algorithms.

APPROXIMATION WITH A NICE FRACTION

What do you think of when you see a number such as 7.3962? As is, it is an unruly number to think about, and for many it may even be intimidating. A good approach to these decimal numbers that are not nice is to think of a close nice number. Is 7.3962 closer to 7 or to 8? Is it more or less than $7\frac{1}{2}$? Is 7 or $7\frac{1}{2}$ good enough for your purposes? If more precision is desired, you might look to see if it is close to a nice fraction. In this case, 7.3962 is very close to 7.4, which is $7\frac{2}{5}$. The fraction $7\frac{2}{5}$ is much easier to think with than a four-digit decimal. In the modern world of digital equipment, decimal numbers are frequently produced with much greater accuracy than is necessary. To have good number sense with these decimals implies that you can quickly think of a simple meaningful fraction as a useful approximation for most any decimal number. This facility is analogous to glancing at your digital watch, seeing 8:48' 23", and announcing that it is about 10 'til nine.

To develop this type of familiarity with decimals, children do not need new concepts or skills. They do, however, need the opportunity to apply and discuss the related concepts of fractions, place value, and decimals in exercises and activities such as the following.

14.7 CLOSE TO A FRIENDLY FRACTION

Make a list of decimal numbers that do not have nice fractional equivalents. Have students suggest a decimal number near to the given one that does have a nice equivalent. Try this yourself with this list:

24.8025

6.59

0.9003

124.356

Different students may select different fractions for these numbers. The rationale for their choice presents an excellent opportunity for discussion.

In Figure 14.10, each decimal numeral is to be paired with the fraction expression that is closest to it. The exercise can be made easier by using fractions that are not as close together.

Match each fraction with a decimal number that's close.

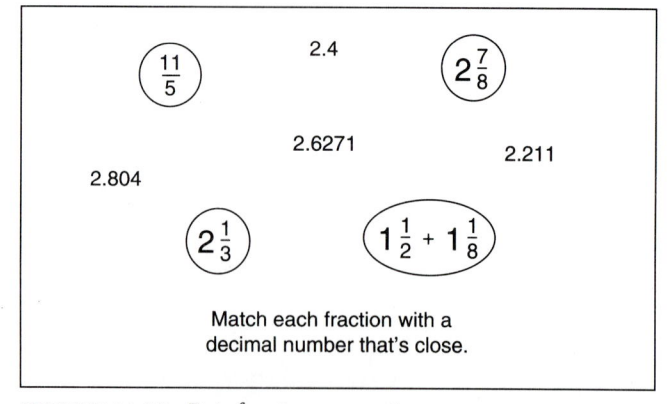

FIGURE 14.10: *Pair fraction expressions*

14.8 FRACTION SIEVES

Place a "fraction sieve" on the overhead, like one of those shown in Figure 14.11. (A master is in the Black-line Masters section.) Above the sieve, write a decimal number between 0 and 1. At each branch students decide if the given decimal is more or less than the fraction at that point. If it is less, they move to the left; if more, to the right. The letters at the bottom of the sieves are there for the purpose of recording answers. For example, if 0.6189 were put in the top left sieve of Figure 14.11, it would come out at Letter F (less than $\frac{3}{4}$ but more than $\frac{3}{5}$).

In these exercises, students should be able to explain and confirm their answer in a meaningful manner. Students who are not able to deal well with these exercises are probably in need of some conceptual remediation.

ORDERING DECIMAL NUMBERS

Putting a list of decimal numbers in order from least to most is a closely related skill to the one just discussed. In the fourth NAEP (Kouba et al., 1988a), only about 50% of seventh graders could identify the largest number in the following list: 0.36, 0.058, 0.375, and 0.4. That result is disturbing. It is unfortunately similar to the results of many other studies. (For example, see Hiebert & Wearne, 1986.) The most common error is to select the number with more digits, which is an incorrect application of whole-number ideas. Some students later pick up the idea that digits way to the right represent very small numbers. They then incorrectly identify numbers with more digits as smaller. Both errors reflect a lack of conceptual understanding of how decimal numbers are constructed. The following activities can help promote useful discussion about the relative size of decimal numbers.

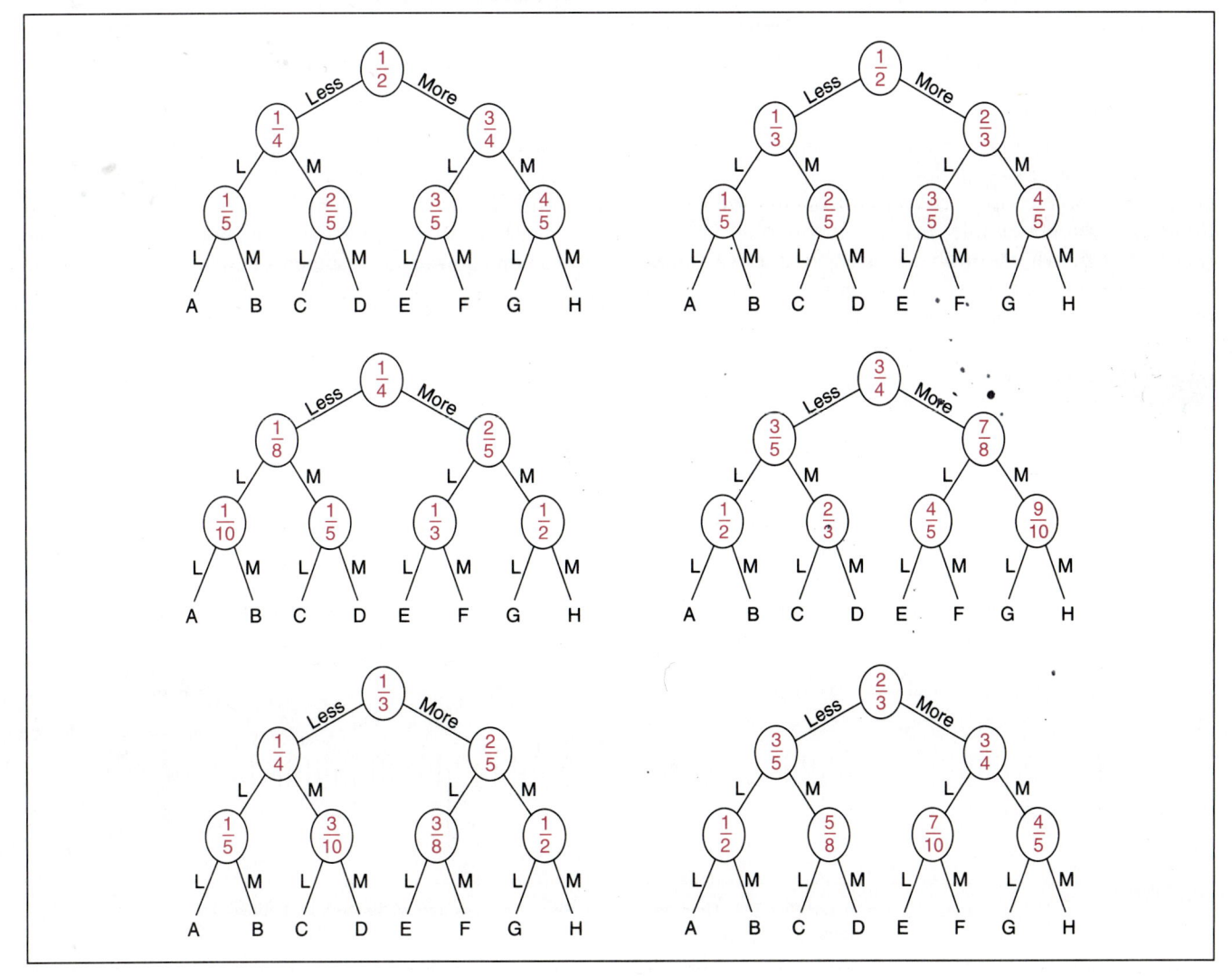

FIGURE 14.11: *Fraction sieves*

14.9 THE LARGER DECIMAL NUMBER

Present two decimal numbers, and have students use models to explain which is larger. A meter stick or a 10 × 10 square are useful for this purpose.

14.10 "ROUNDING" MEANS "CLOSE TO"

Write a four-digit decimal on the board—for example, 3.0917. Start with the whole numbers. Is it closer to 3 or 4? Then go to the tenths. Is it closer to 3.0 or 3.1? Repeat with hundredths and thousandths. At each answer, challenge students to defend their choices with the use of a model or other conceptual explanation. The large number line without numerals described in Figure 14.12 is useful for this activity.

Too often the process of rounding numbers is taught as an algorithm without any reflection on what the purpose is or why the algorithm makes sense. To "round" a number implies that you do something to it or change it in some way. In reality, we round numbers by *substituting* nice numbers as approximations for numbers that are difficult to work with or talk about. In this sense we should also round decimal numbers to *nice fractions* and not just to tenths and hundredths. For example, instead of rounding 6.73 to the nearest tenth, a number-sense perspective might suggest rounding it to the nearest quarter (6.75 or $6\frac{3}{4}$) or to the nearest third (6.67 or $6\frac{2}{3}$).

14.11 LINE 'EM UP

Prepare a list of four or five decimal numbers that students might have difficulty ordering. They should all be between two consecutive whole numbers. Have them first predict the order of the numbers from least to most. Next, have them place each number on a number line with 100 subdivisions, as in Figure 14.12. As an alternative, have students shade in the fractional part of each number on a separate 10 × 10 grid using estimates for the thousandths and ten-thousandths. In either case it quickly becomes obvious which digits contribute the most to the size of a decimal.

OTHER FRACTION/DECIMAL EQUIVALENTS

The only decimal/fraction equivalents that have been discussed so far are those for the nice fractions: halves, thirds, fourths, fifths, and eighths. Also, any base ten fraction is immediately converted to a decimal, and, similarly, simple decimals with two or three decimal places are easily converted to fractions. For the purpose of familiarity or number sense, these fractions and decimal equivalents provide a significant amount of information about decimal numbers. Furthermore, all of this information about decimals can and should be approached conceptually without rules, rote memory, or algorithms. It can be argued that the major focus of decimal instruction should center on these ideas.

At times, however, other fractions must be expressed as a decimal. For example, how do you enter $\frac{5}{6}$ or $\frac{3}{7}$ on a calculator that does not accept fraction notation? The answer, of course, is based on the fact that $\frac{3}{7}$ is also an expression for 3 ÷ 7. If this division is carried out on paper, an infinite but repeating decimal will result. The ninths have very interesting decimal equivalents, and looking for a pattern is a worthwhile activity. (Try $\frac{1}{9}$, $\frac{2}{9}$, . . . on your calculator.) The division process should also be checked out for familiar fraction equivalents. If students have constructed a good understanding of familiar fractions and decimal equivalents, the concept will confirm the division result. If the division 4 ÷ 5 is the only explanation students have for why $\frac{4}{5}$ = 0.8, then a significant lack of a conceptual linkage between fractions and decimals is likely.

Students are frequently shown early in the development of decimal numeration how to use division as a means of converting fractions to decimals. Doing a long division or using a calculator for this purpose does not promote a firsthand familiarity or number sense with fractions and decimals.

In the seventh or eighth grade, students are frequently

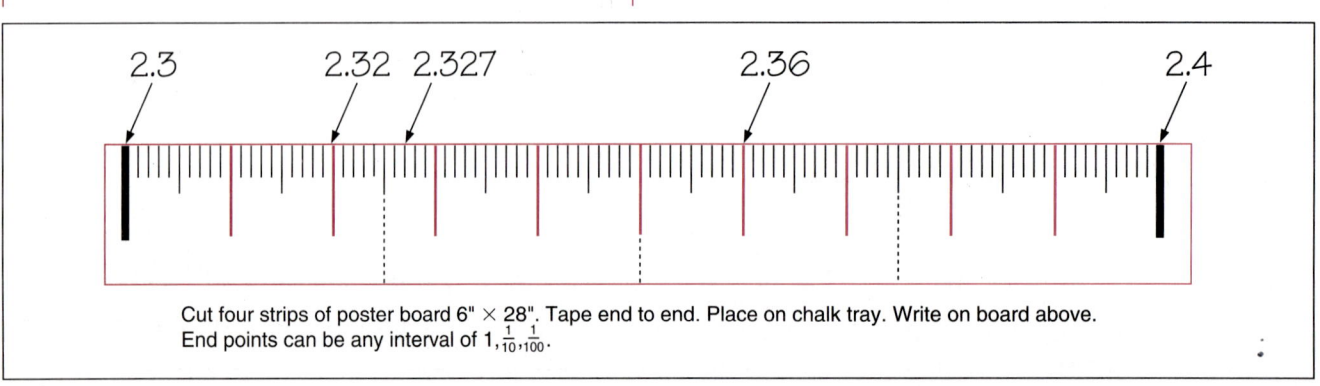

2.3   2.32   2.327   2.36   2.4

Cut four strips of poster board 6" × 28". Tape end to end. Place on chalk tray. Write on board above. End points can be any interval of 1, $\frac{1}{10}$, $\frac{1}{100}$.

FIGURE 14.12: *A decimal number line*

taught to convert a repeating decimal to a fraction. These conversions serve the purpose of demonstrating that every repeating decimal can also be represented as a fraction and therefore as a rational number. This important theoretical result is primarily useful in making a distinction between rational and irrational numbers. The contribution such tedious activities have for number sense is rather minimal.

INTRODUCING PERCENTS

A THIRD OPERATOR SYSTEM

A major goal that has been stressed so far is to help children understand that decimal numerals and fractions are simply two different symbol systems for the same part-to-whole relationships; that is, they are representations for rational numbers. In this sense, both fractions and decimals are interpreted as real numbers. But when we say "three-fifths of a cake" or "three-fifths of the students," the number meaning comes from the cake or the set of students. Certainly $\frac{3}{5}$ of 5 students is different than $\frac{3}{5}$ of 20 students. In fact, it is precisely this operator notion that makes it difficult for children to think of fractions as numbers. They initially learn about them as parts *of* something.

When children have made a strong connection between their concepts of fractions and decimals, the topic of percent can be introduced. Rather than approach percents as a new and strange idea, children should see how percents are nothing more than a different way to write down some of the ideas they have already developed about fractions and decimals. While not a third numeration system, percents are essentially a third symbolism for operators.

◆ Another Name for Hundredths

The term *percent* is simply another name for hundredths. If students can express common fractions and simple decimals as hundredths, then the term *percent* can be substituted for the term *hundredth*. Consider the fraction $\frac{3}{4}$. As a fraction expressed in hundredths, it is $\frac{75}{100}$. When $\frac{3}{4}$ is written in decimal form, it is 0.75. Both 0.75 and $\frac{75}{100}$ are read in exactly the same way, "seventy-five-hundredths." When used as operators, $\frac{3}{4}$ of something is the same as 0.75 or 75% of that same thing. Percent is not a new concept, only a new notation and terminology. Connections with fractions and decimal concepts are developmentally appropriate.

Models provide the principal connecting link between fractions, decimals, and percents, as shown in Figure 14.13. Base ten fraction models are suitable for fractions, decimals, and percents, since they all represent the same idea. (Percent is also a symbol system for ratios, just as fractions are. In this chapter, percents are first connected to fraction and decimal notation and part-to-whole relationships.)

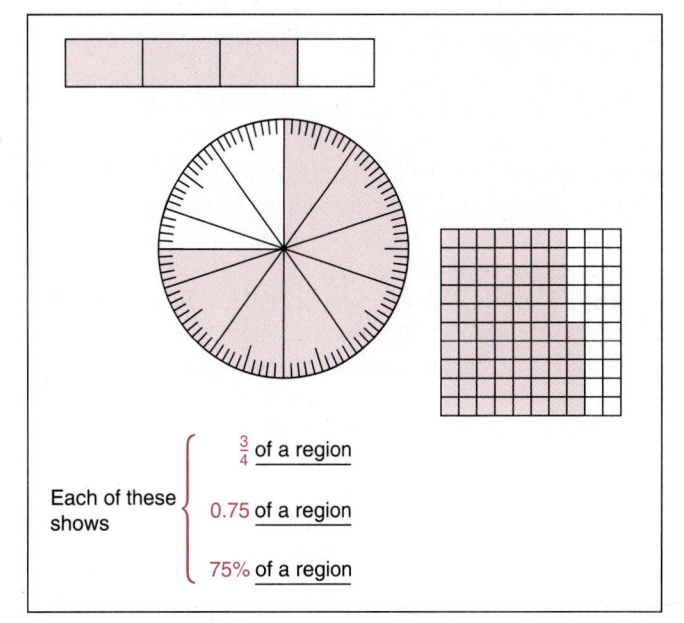

Each of these shows
- $\frac{3}{4}$ of a region
- 0.75 of a region
- 75% of a region

FIGURE 14.13: *Models connect three different notations.*

Another helpful approach to the terminology of percent is through the role of the decimal point. Recall that the decimal identifies the units. When the unit is ones, a number such as 0.659 means a little more than 6 tenths of 1. The word *ones* is understood. But 0.659 is also 6.59 tenths and 65.9 hundredths and 659 thousandths. In each case the name of the unit must be explicitly identified, or else the unit ones would be assumed. In 6.59 tenths, the interpretation is similar to an operator. It is 6 and 59 hundredths *of the things called tenths*. Since percent is another name for hundredths, when the decimal is "identifying" the hundredths position as the units, the word *percent* can be specified as a synonym for hundredths. Thus, 0.659 (of some whole or one) is 65.9 percent or 65.9% of that same whole. As illustrated in Figure 14.14, the notion of placing the decimal point *to identify the percent position* is conceptually more meaningful than the apparently arbitrary rule: "to change a decimal to a percent, move the decimal two places to the right." This rule carries no meaning and is easily confused. "Do I move it right or left?" A better idea is to equate hundredths with percent both orally and in notation.

Ones	Tenths	Percent Hundredths	Thousandths
	3	6	5

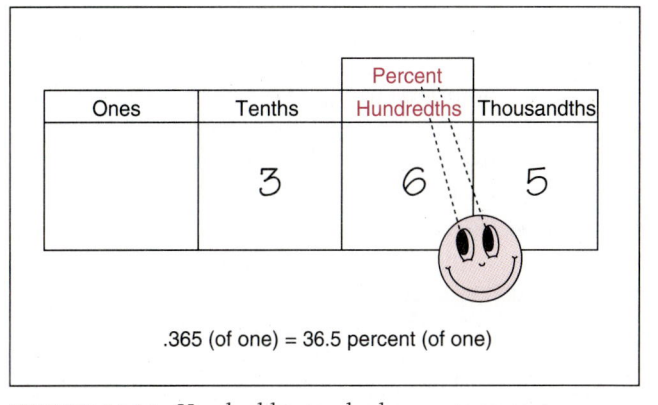

.365 (of one) = 36.5 percent (of one)

FIGURE 14.14: *Hundredths are also known as percents.*

268   14 / DECIMAL AND PERCENT CONCEPTS AND DECIMAL COMPUTATION

◆ Use Percent with Familiar Fractions

Students should use base ten models for percents in much the same way as for decimals. The disk with 100 markings around the edge is now a model for percents as well as a fraction model for hundredths. The same is true of a 10 × 10 square. Each tiny square inside is 1% of the square. Each row or strip of 10 squares is not only a tenth but 10% of the square.

Similarly, the familiar fractions (halves, thirds, fourths, fifths, and eighths) should become familiar in terms of percents as well as decimals. Three fifths, for example, is 60% as well as 0.6. One-third of an amount is frequently expressed as $33\frac{1}{3}$% instead of 33.3333 . . . percent. Likewise, $\frac{1}{8}$ of a quantity is $12\frac{1}{2}$% or 12.5% of the quantity. These ideas should be explored with base ten models and not as rules about moving decimal points.

REALISTIC PERCENT PROBLEMS

◆ The Three Percent Problems

Junior high school teachers talk about "the three percent problems." The sentence " ____ is ____ percent of ____ " has three spaces for numbers; for example, *20 is 25 percent of 80*. The three classic percent problems come from this sterile expression; two of the numbers are given, with the students asked to produce the third. Students learn very quickly that you either multiply or divide the numbers given, and sometimes you have to move a decimal point. This approach to percent is doomed from the start. Students have no way of determining when to do what. As a result, performance on percentage problems is very poor. Furthermore, the major reason or perhaps the only reason for learning about percent is that it is commonly used in our society. Sales, taxes, census data, political information, trends in economics, in farming, in production, and so on, are all full of percent terminology. But in almost none of these is the formula version ____ is ____ percent of ____ utilized. So when asked to solve a realistic percent problem, students are frequently at a loss.

In Chapter 12 three types of exercises with fractions were explored. These involved parts, wholes, and fractions. The three types were determined according to which of the three—part, whole, or fraction—was unknown. Students used models and simple fraction relationships in those exercises. Those three types of exercises are precisely the same as the three percent problems. Developmentally, then, it makes good sense to help students make the connection between the exercises done with fractions and those with percents. How can this be done? Use the same types of models and the same terminology of parts, wholes, and fractions. The only thing that is different is that the word *percent* is used instead of fraction. In Figure 14.15, three exercises from Chapter 12 have been changed to a corresponding percent terminology. A good idea for early work with percent would be to review (or explore the first time) all three types of exercises in terms of percents. The same three types of models can be used. (Refer to Figures 12.8, 12.9, and 12.10.)

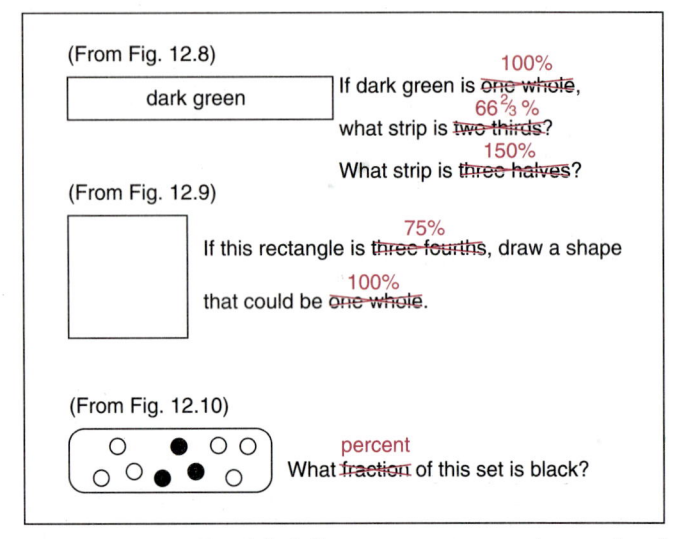

(From Fig. 12.8)
dark green — If dark green is ~~one whole~~ **100%**, what strip is ~~two-thirds~~ **66⅔%**? What strip is ~~three-halves~~ **150%**?

(From Fig. 12.9)
If this rectangle is ~~three-fourths~~ **75%**, draw a shape that could be ~~one whole~~ **100%**.

(From Fig. 12.10)
What ~~fraction~~ **percent** of this set is black?

FIGURE 14.15: *Part/whole/fraction exercises can be translated to percent exercises.*

◆ Realistic Percent Problems and Nice Numbers

While children must have some experience with the noncontextual, straightforward situations in Figure 14.15, it is important to have them explore these relationships in real contexts. Find or make up percent problems and present them in the same way that they appear in newspapers, television, and other real contexts. In addition to realistic problems and formats. follow these maxims for much of your unit on percents:

1. Limit the percents (fractions) to familiar fractions (halves, thirds, fourths, fifths, and eighths) or easy percents ($\frac{1}{10}, \frac{1}{100}$) and use numbers compatible with these fractions. That is, make the computation very easy. The relationships involved are the focus of these exercises. Complex computational skills are not.

2. Do not suggest or provide any rules or procedures for different types of problems. Do not categorize or label problem types.

3. Utilize the terms *part*, *whole*, and *percent* (or fraction). The use of fraction and percent are interchangeable. Help students see these as the same types of exercises they did with simple fractions.

4. Require students to use models or drawings to explain their solutions. It is better to assign three problems requiring a drawing and an explanation rather than 15 problems requiring only computation and answers. Remember that the purpose is the exploration of relationships, not computational skill.

5. Encourage mental arithmetic.

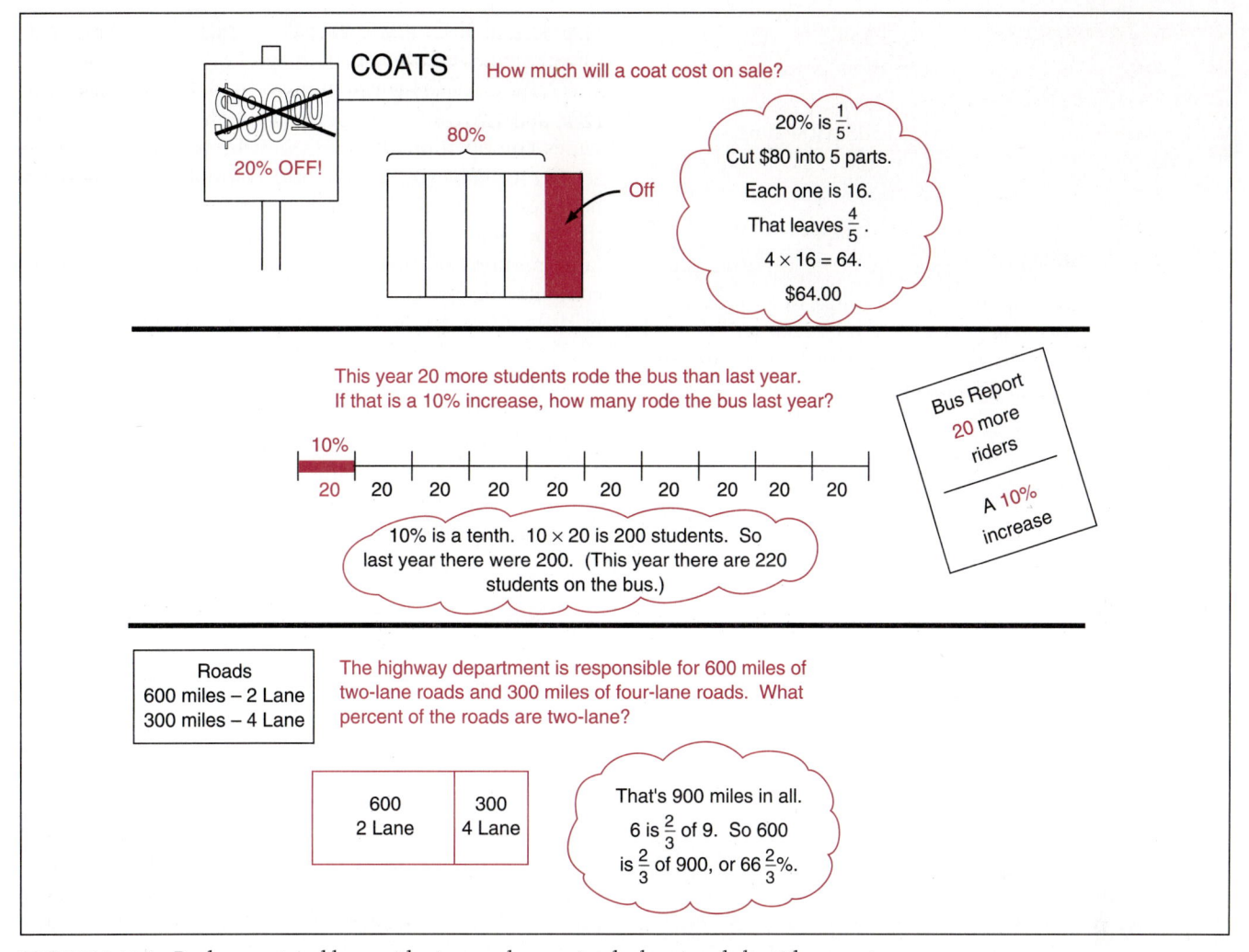

14 / DECIMAL AND PERCENT CONCEPTS AND DECIMAL COMPUTATION   **269**

COATS — How much will a coat cost on sale?

$~~~~$20% OFF!   80%   Off

20% is $\frac{1}{5}$.
Cut $80 into 5 parts.
Each one is 16.
That leaves $\frac{4}{5}$.
$4 \times 16 = 64$.
$64.00

This year 20 more students rode the bus than last year. If that is a 10% increase, how many rode the bus last year?

10% — 20 | 20 | 20 | 20 | 20 | 20 | 20 | 20 | 20 | 20

Bus Report: 20 more riders — A 10% increase

10% is a tenth. 10×20 is 200 students. So last year there were 200. (This year there are 220 students on the bus.)

Roads
600 miles – 2 Lane
300 miles – 4 Lane

The highway department is responsible for 600 miles of two-lane roads and 300 miles of four-lane roads. What percent of the roads are two-lane?

| 600 2 Lane | 300 4 Lane |

That's 900 miles in all. 6 is $\frac{2}{3}$ of 9. So 600 is $\frac{2}{3}$ of 900, or $66\frac{2}{3}\%$.

FIGURE 14.16: *Real percent problems with nice numbers—simple drawings help with reasoning.*

The following example problems meet these criteria for easy fractions and numbers. Try working each problem, identifying each number as a part, a whole, or a fraction. Draw length or area models to explain and/or to work through your thought process. Examples of this informal reasoning are illustrated with additional problems in Figure 14.16.

1. The PTA reported that 75% of the total number of families were represented at the meeting last night. If 320 families are in the school, how many were at the meeting?
2. The baseball team won 80% of the 25 games they played this year. How many games were lost?
3. In Mrs. Carter's class, 20 students or 66 2–3% were on the honor roll. How many students are in her class?
4. George bought his new computer at a 12 1–2% discount. He paid $700. How many dollars did he save by buying it at a discount?
5. If Joyce has read 60 of the 180 pages in her library book, what percent of the book has she read so far?
6. The hardware store bought widgets at 80 cents each and sold them for $1 each. What percent did the store mark up the price of each widget?

◆ Estimation in Percent Problems

Of course, not all real percent problems have nice numbers. Frequently in real life, an approximation or estimate in percent situations is all that is required or is enough to help one think through the situation. Even if a calculator will be used to get an exact answer, an estimate based on an understanding of the relationship can confirm that a correct operation was performed or that the decimal was correctly positioned.

To help students with estimation in percent situations, two ideas that have already been discussed can be applied. First, when the percent (fraction) is not a nice percent, find a close percent or fraction that is nice or easy to work with. Second, in doing the calculation, select numbers that are compatible with the fraction involved to make the calculation easy to do mentally. In essence, convert the not-nice percent problem into one that *is* nice. Here are some examples. Try your hand at estimates of each.

1. The 83,000-seat stadium was 73% full. How many people were at the game?
2. The treasurer reported that 68.3% of the dues had been collected for a total of $385. How much more money could the club expect to collect if all dues are paid?
3. Max McStrike had 217 hits in 842 at-bats. What was his batting average?

Possible estimates:

1. 62,000 (Use $\frac{3}{4}$ and 80,000, then adjust up a bit.)
2. $190 (Use $\frac{2}{3}$ and $380; will collect $\frac{1}{3}$ more.)
3. A bit more than .250 (4 × 217 > 842; $\frac{1}{4}$ is 25% or .250)

The following exercises are also useful in helping students with estimation in percent situations.

> Compare percents instead of decimals to the fractions in the sieves shown in Figure 14.11.
>
> Choose the closest nice fraction. For example, which fraction is closest to 78%: $\frac{1}{2}$, $\frac{2}{3}$, or $\frac{4}{5}$? Multiple-choice exercises such as this one can be done with a full class or can be made into worksheet exercises. Later, students can give a nice fraction that is close to the percent without a multiple choice. Have students justify their answers.
>
> Work with "easy percents," especially 1% and 10%. Begin with exercises where students give 1% and 10% of numbers. Then show them how to use these easy percents to get other percentages, either exact or approximate. For example, to get 15% of $349, think 10% is about $35, and 5% is half of that, or $17.50. So 15% is $35 + $17.50 or $52.50. Similar reasoning can be used to adjust an estimate by 1% or 2%. To find 82% of $400, it is easy to think of $\frac{4}{5}$ (80%) as 4 × $\frac{1}{5}$ of 400 or 4 × $80, which is $320. Since each 1% is $4, add on $8.

Sometimes an exact result, and therefore some calculation, is required. The emphasis on conceptual thinking, nice fractions, and estimation of results will all pave the way for easy work when an exact result is requested. The use of an estimate will determine if the result is in the correct ballpark or if it makes sense. Frequently, the same estimation process will dictate what computation to do so that the problem can be entered on a calculator.

COMPUTATION WITH DECIMALS

Earlier in this text an entire chapter was devoted to the development of computation algorithms for whole-numbers. That level of emphasis is not completely in keeping with the recommendations of the NCTM *Standards*. For the next several years it will probably remain true that pencil-and-paper computation with whole numbers will be a major thrust of the curriculum for grades 2 to 5. For decimals, however, the corresponding emphasis on pencil-and-paper computation simply cannot be justified in light of readily available calculators and computers, the need for better estimation skills, and the strong recommendations of the *Standards*. "It is no longer necessary or useful to devote large portions of instructional time to performing routine computations by hand. Other mathematical experiences for middle school students deserve far more emphasis" (NCTM, p. 94). Readers of this text who have not yet done so should read and reflect on "Standard 7: Computation and Estimation" in the grades 5–8 section of the *Standards*.

The discussion that follows places a heavy emphasis on estimation of decimal computation. As a transitional step before the day when pencil-and-paper computation with decimals is completely gone from the curriculum, some relatively simple alternatives are suggested.

THE ROLE OF ESTIMATION

Students should learn to estimate decimal computations before they learn to compute with pencil and paper. For many decimal computations, rough estimates can be made easily by rounding the numbers to nice whole numbers or simple base ten fractions. In almost all cases a minimum goal for your students should be to have the estimate contain the correct number of digits to the left of the decimal—the whole-number part. Select problems for which estimates are not terribly difficult. Before going on, try making easy whole-number estimates of the following computations. Do not spend time with fine adjustments in your estimates.

a. 4.907 + 123.01 + 56.1234
b. 459.8 − 12.345
c. 24.67 × 1.84
d. 514.67 ÷ 3.59

Your estimates might be similar to the following:

a. between 175 and 200
b. more than 400, or about 425 to 450
c. more than 25, closer to 50 (1.84 is more than 1 and close to 2)
d. more than 125, less than 200 (500 ÷ 4 = 125 and 600 ÷ 3 = 200)

In these examples, an understanding of decimal numeration and some simple whole-number estimation skills can produce rough estimates. When estimating, thinking focuses on the meaning of the numbers and the operations and not on how many decimal places are involved. However, students who are taught to focus primarily on the pencil-and-paper algorithms for decimals may find even simple estimations difficult.

Therefore, a good *place* to begin decimal computation is with estimation. Not only is it a highly practical skill, it helps children look at answers in ballpark terms, can form a check on pencil-and-paper computation, and is one way of placing the decimal in multiplication and division.

A good *time* to begin computation with decimals is well after a firm conceptual background in decimal numeration has been developed. Learning the pencil-and-paper algorithms for decimals will do little or nothing to help students understand decimal numeration.

ADDITION AND SUBTRACTION

From a conceptual standpoint, the addition and subtraction algorithms are nearly identical for decimals and whole numbers. The numbers in like place-value columns are added or subtracted with 10-for-1 trades made when necessary. While this is a relatively straightforward application of place value, many students do have trouble. Errors tend to occur when problems are presented in horizontal format or are in word problems with different numbers of digits to the right of the decimal. For example, students might compute the sum of 3.45 + 12.2 + 0.807 by lining up the right-hand digits 5, 2, and 7 as they would with whole numbers.

To help students with addition and subtraction, have them first estimate the answer as discussed earlier. For the example just given, the sum should be about 16. Next, discuss what the numbers mean in terms of a base ten model. What does each digit represent? How would the problem be written if the numbers were written in the appropriate columns of a place-value chart? (See Figure 14.17.) From such a discussion, it should be reasonable that like place values should be added or subtracted. When students follow this procedure, they should compare the results with their estimate. The estimate should be a confirmation of the computation (and not the reverse). Under no circumstances should students use the purely rote rule of "lining up the decimal points" without being able to justify it.

MULTIPLICATION AND DIVISION

Estimation is also a means of locating the decimal point in multiplication or division. For those two operations, one algorithm that is reasonable is the following: *Ignore the decimal points and do the computation as if all numbers were whole numbers. When finished, place the decimal in the result by estimation.* This approach is illustrated in Figure 14.18. In both examples, notice how a shift of the decimal in either direction just one place would give a result that is not even close to the estimate. While some explanation for the estimation method is required, it is quite useful except for very precise computations such as 0.00987×0.000103. Those seriously needing a precise answer to such computations almost certainly will have a calculator or computer to help them.

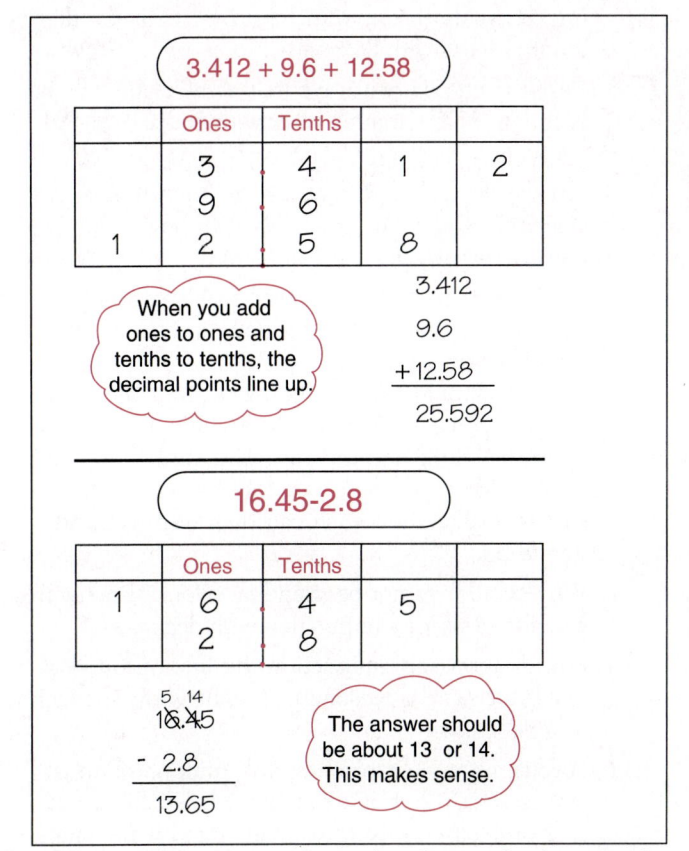

FIGURE 14.17: *Using a place-value chart to develop rules for adding and subtracting decimals.*

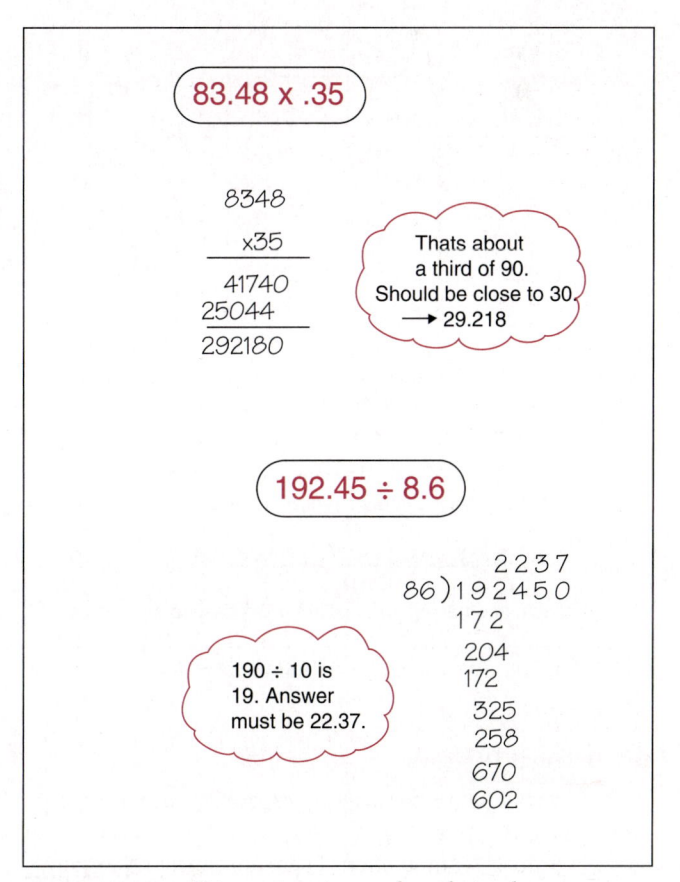

FIGURE 14.18: *Using estimation to place decimals*

Before reading further, estimate each of the following products:

    347   34.7   3.47
  × 2.6   × 2.6   × 2.6

The first is about 900 (2 × 350 is 700 and 6 tenths of 350 is about 200). Similar reasoning would place the second product at about 90 and the third around 9. The whole-number product 347 × 26 is 9022, or about 9000. Each of these products uses exactly the same digits even though the decimal point makes the numbers quite different. This fact can provoke an excellent discussion and provide a useful introduction to decimal multiplication.

◆ Whole-Number Divisors

For whole-number divisors, the whole-number division algorithm is easily extended to decimal dividends and decimal quotients. In each place-value column, a partition is made, and any leftovers are traded to the next column. Since the trade is always 1 for 10 regardless of which two columns are involved, trading in columns to the right of the decimal point is the same as trading to the left of the decimal. Place-value columns in the quotient correspond to those in the dividend. These ideas are illustrated in Figure 14.19.

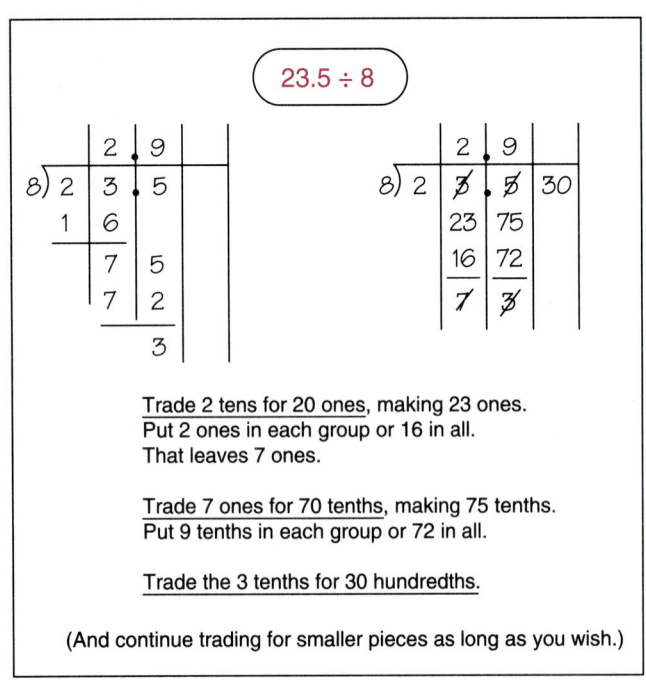

FIGURE 14.19: *Extension of the division algorithm*

◆ Decimal Divisors

Division with a decimal divisor, especially one with two or more digits, is probably a good place to draw the line with paper-and-pencil computation. There are almost no instances outside of classrooms where ordinary people perform divisions with decimal divisors using pencil and paper. This is not to say that such computation cannot be taught meaningfully. It simply means that such computation is not a productive use of school time.

◆◆◆◆◆
REFLECTIONS ON CHAPTER 14: WRITING TO LEARN

1. Describe three different base ten models for fractions and decimals, and use each to illustrate how base ten fractions can easily be represented.

2. Explain how the place-value system extends infinitely in two directions. How can this idea be developed with students in the fifth or sixth grade?

3. Use an example involving base ten pieces to explain the role of the decimal point as identifying the units position. Relate this idea to changing units of measurement such as in money or metric measures.

4. What are the suggested "familiar fractions," and how can these fractions be connected to their decimal equivalents in a conceptual manner?

5. What kinds of things should be emphasized regarding decimals if you want children to have good number sense with decimals?

6. Describe one or two ways that a calculator can be used to develop number ideas with decimals.

7. What does rounding mean? What type of rounding can we do that is different from rounding to the nearest tenth or hundredth? Explain how a number line can be used in rounding.

8. Make up three realistic percent problems where the percents are actually nice fractions and the numbers involved are compatible with the fractions. One problem should ask for the part, given the whole and the percent. One should ask for the percent, given the whole and the part. The third should ask for the whole, given the part and the percent. Model each and show how each can be solved using fraction ideas.

9. Why should we not be spending a lot of time with pencil-and-paper computation with decimals?

10. For addition and subtraction the line-up-the-decimals rule can be reasonably taught without much trouble. Explain.

11. Give an example explaining how multiplication and division with decimals can be replaced for reasonable problems by using estimation and whole-number methods.

FOR DISCUSSION AND EXPLORATION

1. Examine textbooks for one grade level between fifth and eighth for the development of decimal concepts. How would you modify or add to the textbook presentation of this topic if you were teaching at one of these grade levels? Do not forget to check in the teacher's edition for good teaching suggestions.

2. One way to order a series of decimal numbers is to annex zeros to each number so that all numbers have the same number of decimal places. For example, rewrite

    | 0.34 | as | 0.3400 |
    | 0.3004 | as | 0.3004 |
    | 0.059 | as | 0.0590 |

    Now ignore the decimal points and any leading zeros, and order the resulting whole numbers. Discuss the merits of teaching this approach to children.

3. Talk one-on-one with some seventh- or eighth-grade students. First find out if they can do simple fraction exercises that ask for the part, the whole, or the fraction given the other two (as in Chapter 12). Encourage them to make drawings and give explanations. Next ask them to solve simple percent word problems that require the same type of reasoning using simple fractions and compatible numbers. Compare students' abilities with these mathematically identical problems.

SUGGESTED READINGS

Allinger, G. D., & Payne, J. N. (1986). Estimation and mental arithmetic with percent. In H. L. Schoen (Ed.), *Estimation and mental computation*. Reston, VA: National Council of Teachers of Mathematics.

Bennett, A. (1982). *Decimal squares: Step by step teacher's guide, readiness to advanced levels in decimals*. Fort Collins, CO: Scott Resources, Inc.

Boling, B. (1985). A different method for solving percentage problems. *Mathematics Teacher, 78*, 523–524.

Glatzer, D. J. (1984). Teaching percentage: Ideas and suggestions. *Arithmetic Teacher, 31*(6), 24–26.

Grossman, A. S. (1983). Decimal notation: An important research finding. *Arithmetic Teacher, 30*(9), 32–33.

Huinker, D. (1992). Decimals and calculators make sense! In J. T. Fey (Ed.), *Calculators in mathematics education*. Reston, VA: National Council of Teachers of Mathematics.

Langford, K., & Sarullo, A. (1993). Introductory common and decimal fraction concepts. In R. J. Jensen (Ed.), *Research ideas for the classroom: Early childhood mathematics*. New York: Macmillan Publishing Co.

Lichtenberg, B. K., & Lichtenberg, D. R. (1982). Decimals deserve distinction. In L. Silvey (Ed.), *Mathematics for the middle grades (5–9)*. Reston, VA: National Council of Teachers of Mathematics.

Owens, D., & Super, D. B. (1993). Teaching and learning decimal fractions. In D. T. Owens (Ed.), *Research ideas for the classroom: Middle grades mathematics*. New York: Macmillan Publishing Co.

Reys, B. J. (1991). *Developing number sense in the middle grades: Addenda series, grades 5–8*. Reston, VA: National Council of Teachers of Mathematics.

Vance, J. M. (1986a). Estimating decimal products: An instructional sequence. In H. L. Schoen (Ed.), *Estimation and mental computation*. Reston, VA: National Council of Teachers of Mathematics.

Vance J. M. (1986b). Ordering decimals and fractions: A diagnostic study. *Focus on Learning Problems in Mathematics, 8*(2), 51–59.

Zawojewski, J. (1983). Initial decimal concepts: Are they really so easy? *Arithmetic Teacher, 30*(7), 52–56.

15 DEVELOPING THE CONCEPTS OF RATIO AND PROPORTION

PROPORTIONAL REASONING

SOME EXAMPLES FOR EXPLORATION AND DISCUSSION

So that the reading of this chapter is based on some common ground, several exercises are provided here for your exploration. Each embodies the type of relationships that make up the concepts of ratio and proportion. Therefore, rather than begin this topic as many basal textbooks do, with a definition of ratio, you can begin with some intuitive notions. It does not matter what your personal mathematical acquaintance with ratio actually is. Try each of the following. You may wish to discuss and explore the exercises in a group or with a friend.

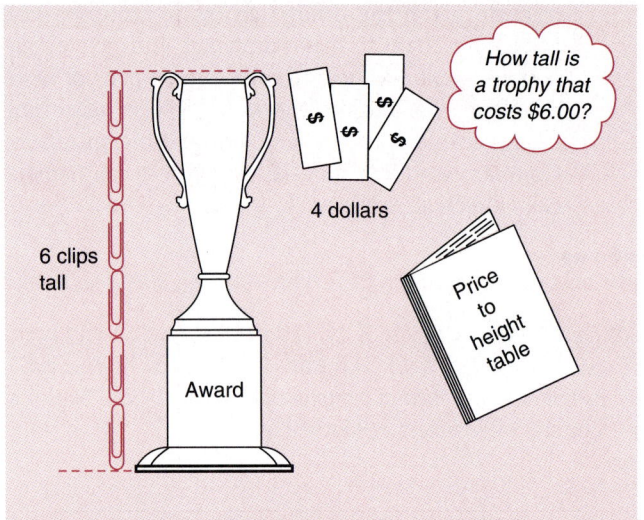

Figure 15.1: *The "Short and Tall" trophy problem*

SHORT AND TALL

Mr. Green sells trophies based on their height. A short trophy measures six paper clips tall, as shown in Figure 15.1. According to his price chart, this short trophy sells for $4.00. Mr. Green has a taller trophy that is worth $6.00 according to the same scale. How many paper clips tall is the $6.00 trophy?

LAPS

Yesterday, Mary counted the number of laps she ran at track practice and recorded the amount of time it took. Today she ran fewer laps in more time than yesterday. Did she run faster, slower, at about the same speed today, or can't you tell? What if she had run more laps in more time?

SIMILAR SHAPES

Place a piece of paper over the dot grid in Figure 15.2. Using the dots as a guide, draw a shape that is *like* the one shown but larger. How many different shapes that are like the given shape can you draw and still stay within the grid provided?

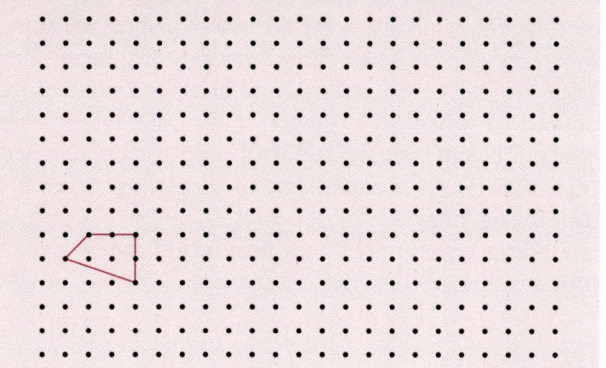

FIGURE 15.2: *Draw some shapes just like this only larger.*

DOT MIXTURES

On a piece of paper, draw a loop about 2 inches across. Fill the loop with black and white dots in a random mixture that appears to be the same mix of black and white dots as in Figure 15.3. Draw your dots about the same size and with the same spacing as the dots in the figure. Do not count the dots in either drawing until you are done. Try to get both mixes of black and white to look the same. When finished, count and compare the number of black and white dots in the book with the number of black and white dots in your drawing. How can these counts help you decide if your drawings have the same mixture? Can you adjust the number of either or both colors so that you could argue that they are in the same mix as those in the book?

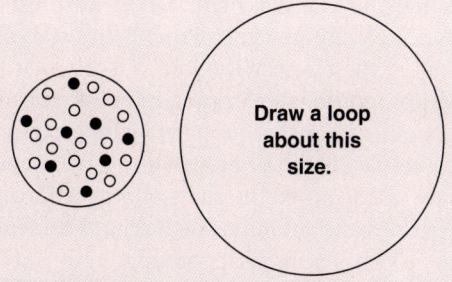

Draw a loop about this size.

Fill a larger loop with black and white dots in about the same mixture. Do not count the dots until you are finished.

FIGURE 15.3: *Dot mixture*

EXAMPLES OF RATIOS IN DIFFERENT CONTEXTS

A *ratio* is an ordered pair of numbers or measurements that are used to express a comparison between the numbers or measures. This succinct definition covers a wide range of situations. Mathematically, all ratios are essentially the same. However, from the student viewpoint, ratios in different settings or contexts may present very different ideas.

◆ All Fractions Are Ratios

Both fractions and ratios are comparisons. Fractions are a means of comparing parts to the whole. The fraction $\frac{3}{4}$ is a comparison of 3 parts to the 4 parts that make up the whole. Both the whole and the part are measured in fourths. Such a comparison of parts to wholes is also a ratio. That is, all fractions are ratios. However, it is not correct to say that all ratios are fractions or that a ratio *is* a fraction. The ratio of 12 to 0, for example, is not a fraction.

◆ Ratios Can Compare Two Parts of the Same Whole

While a fraction is always a comparison of parts to the whole, some ratios are comparisons of one part of a whole to another part. In the "Dot Mixture" example, 8 black dots and 16 white dots are in one whole set of 24 dots. The ratio or fraction of black dots to the whole is 8 to 24, or, one-third of the dots are black. But you could also compare the black to the white dots instead of one color to the total. A comparison of black to white is not a fraction but a ratio of a part to a part. The black and white dots are in the ratio of 8 to 16 or 1 to 2. Other examples of part-to-part ratios are Democrats to Republicans, boys to girls, or sailboats to speedboats.

◆ Ratios Can Be Expressions of Rates

Both part-to-whole and part-to-part ratios are comparisons of two measurements of the same type of thing. The measuring unit is the same for each value. A ratio can also be a rate. A *rate* is a comparison of the measures of two different things or quantities. The measuring unit is different for each value.

In the "Short and Tall" example, the comparison between paper clips and money is an example of ratio that is a rate; in this case, dollars per paper clip. If you figured out the ratio of $2.00 for every 3 clips, then you were able to determine that the tall trophy should be nine clips tall. All prices are rates and are also ratios; 69¢ each, 3 for a dollar, or 12 ounces for $1.39. Each is a ratio of money to a measure of quantity.

The "Laps" example involves comparisons of time to distance, another example of rate. There are no numbers involved in that example, so the reasoning is qualitative rather than quantitative, but the ratio concept is basically the same. All rates of speed are ratios of time to distance: for example, driving at 55 miles per hour, or jogging at 9 minutes per mile.

Other rates that we may not think of as ratios are miles per gallon, square yards of coverage per gallon of paint, inches of tape per roll, passengers per busload, or roses per bouquet. Changes between two units of measure are also rates or ratios. Examples include inches per foot, milliliters per liter, and centimeters per inch. The distance scales on a map are another example of ratio as rate.

Other Examples of Ratio

In geometry, the ratio of the circumference of a circle to the diameter is designated by the Greek letter π (pi). This is one example of a ratio that is not a rational number. Among any set of similar geometric figures, the ratios of corresponding parts are always the same as in the "Similar Shapes" example. The diagonal of a square is always $\sqrt{2}$ times a side. The trigonometric functions can be developed from ratios of sides of right triangles.

The probability of an event occurring is a ratio. The chances of rolling a sum of 7 with two dice are 6 in 36 or 1 in 6. Probabilities are part-to-whole ratios: the number of favorable outcomes compared to the total number of possible outcomes.

In physics, ratios abound. For example, the volume of a fixed amount of gas is determined by the ratio of the temperature to the pressure (Boyle's law). The laws governing the forces of pulleys, levers, and gears all involve ratios.

In music you find a set number of beats to a measure, 8 whole notes or 12 tones to an octave. Pleasant-sounding chords involve tones with wavelengths in special ratios.

In nature the ratio known as the *golden ratio* is found in many spirals from the nautilus shell to the swirls of a pine cone or a pineapple. Artists and architects have utilized this same ratio in creating shapes that are naturally pleasing to the eye.

PROPORTIONS

As students begin to experience and reflect on a variety of examples of ratios, they will also begin to see different comparisons that are in the *same ratio*. In the "Similar Shapes" example, the ratio of any two sides of a small shape is the same as the ratio of the corresponding sides of a larger but similar shape. In "Dot Mixtures," the ratios of black to white dots in two different sets can be seen to be the same even though there are many more dots in one loop than another.

A *proportion* is a statement of equality between two ratios. Different notations for proportions can be used. For example:

$$3:9 = 4:12 \quad \text{or} \quad \frac{3}{9} = \frac{4}{12}$$

These might be read "3 is to 9 as 4 is to 12" or "3 to 9 is in the same ratio as 4 to 12."

A ratio that is a rate usually includes the units of measure when written. For example,

$$\frac{\$12.50}{1 \text{ gallon}} = \frac{\$37.50}{3 \text{ gallons}}$$

12 inches per foot = (is the same as) 36 inches per 3 feet

There is a very real distinction between a proportion and the idea of equivalent fractions. Two equivalent fractions are different symbolisms for the same quantity or amount; they represent the *same rational number* in different forms. If one bag has 3 red and 9 white balls, and another has 4 red and 12 white, the number of red balls in each bag is clearly different, but the ratios of red to white and red to total, that is, the comparisons of these quantities, is the same for both bags.

Finding one number in a proportion when given the other three is called *solving a proportion*.

PROPORTIONAL REASONING AND CHILDREN

Proportional reasoning includes, but is much more than, an understanding of ratios. It involves the ability to compare ratios and to predict or produce equivalent ratios. It requires the ability to compare mentally different pieces of information and to make comparisons, not just of the quantities involved but of the relationships between quantities. As you can see from the examples at the outset of this chapter, proportional reasoning involves quantitative thinking as well as qualitative thinking, yet is not dependent on a skill with a mechanical or algorithmic procedure. (Those interested in Piaget's stages of development theory will note that proportional reasoning has generally been regarded as one of the hallmarks of the formal operational thought stage, acquired at about the time of adolescence.)

A Developmental Perspective

Considerable research has been conducted to determine how children reason in various proportionality tasks and to determine if developmental and/or instructional factors are related to proportional reasoning. (For example, see Karplus, Karplus, & Wollman, 1974; Karplus, Pulos, & Stage, 1983; Noelting, 1980; Post, Behr, & Lesh, 1988.) The results of these research efforts suggest that the development of proportional reasoning in children is a complex issue. It is quite clear that proportional reasoning is not an automatic result of natural growth and development. Many adults have not acquired the skills of proportional thinking, and the performance of junior high school students on proportional tasks has been disappointing.

While perhaps not definitive, research does provide some direction for how to help children develop proportional thought processes. Some of these ideas are outlined here.

1. Provide ratio and proportion tasks in a wide range of settings, since how children (and adults) approach tasks is influenced greatly by context. These contexts might include situations involving measurements, prices, geometric and other visual contexts, and rates of all sorts.
2. Encourage reflective thought, discussion, and experimentation in predicting and comparing ratios.
3. Help children relate proportional reasoning to existing processes. Specifically, the concept of unit fractions is very similar to unit rates. Research indicates that the use of a unit rate for comparing ratios and solving proportions is the most common approach used by junior high students even when cross-product methods have been taught. This approach is explained later.
4. Recognize that the use of symbolic or mechanical methods, such as the cross-product algorithm, for solving proportions does not develop proportional reasoning and should not be introduced until students have had considerable experience with more intuitive and conceptual methods.

◆ Need for Experience and Reflective Thought

Of the foregoing ideas, perhaps the most significant are experience and reflective thought—the opportunity to construct the type of comparative thinking that is fundamental to proportional reasoning. Just as you are able to reflect on and talk informally about the experiences suggested at the outset of this chapter, children must have real experiences to think about.

Each of those four beginning activities involved your active participation. The drawings or models were not answer-getting devices but a means of helping you construct relationships. The example of running laps involved comparisons of relative rates in a qualitative sense, since no numbers were involved. The drawings of similar figures on the dot grid and the visual determination of black and white dot ratios were based initially on intuitive ideas that could then be tested numerically through counting. The exercises provided at least some opportunity for you to construct relationships between quantities or measures. You were then able to think about these relationships and apply them to similar but different situations. You had to construct each ratio mentally before it could be applied proportionally to the next measurement. If you worked together with someone, you were also able to put your proportional thinking into words in order to help clarify it or reflect upon it. Children also require these opportunities if they are to develop their abilities to reason proportionally.

◆ A Contrast with a Mechanical Approach

Suppose that you have been given a procedure for "solving proportions" of the following form:

$\frac{a}{b} = \frac{x}{c}$ Step 1. Set up a proportion. $\frac{a}{b} = \frac{c}{x}$

$a \times c = b \times x$ Step 2. Cross multiply. $a \times x = b \times c$

$x = \frac{a \times c}{b}$ Step 3. Solve for the unknown. $x = \frac{b \times c}{a}$

After exercises involving these procedures, you are given problems like "Short and Tall" and guided to set up the proportion.

$$\frac{4 \text{ dollars}}{6 \text{ clips}} = \frac{6 \text{ dollars}}{x \text{ clips}}$$

Then ignoring the fact that it does not make a lot of sense to multiply clips times dollars, you cross multiply ($4 \times x = 6 \times 6$) and solve for x by division. How would such an activity help you conceptualize ratio in any general way or even help you solve any additional exercises?

Sixth-, seventh-, and eighth-grade curriculums have traditionally focused on the procedural knowledge associated with ratio and proportion. That includes deciding if two ratios given symbolically are equal or unequal (check the cross products) and solving proportions as in the preceding example. This emphasis has been due partly to standardized testing, which has required such skills, and partly to the difficulty of specifying thought processes, such as proportional reasoning, in lists of behavioral objectives that many school systems have developed. It takes considerable class time to provide students with experiences that permit intuitive thought and reflection. It requires that teachers enter into free and open discussions as students begin to discover relationships and verbalize them in their own way. Situations that are very different on the surface must be explored before commonalities show through. The work of the Rational Number Project (Post, Behr, & Lesh, 1988) has indicated that contextual and numerical differences in problems play a significant role in how children approach proportion tasks and how difficult they are to solve. Prescriptions for how to solve proportion problems focus almost entirely on procedures. Such a focus does not help students think about relationships in any global manner.

Time can be found for these exciting activities over the sixth- to eighth-grade years, and procedural knowledge can easily be connected to the resulting concepts. The ultimate responsibility for providing such experiences rests with teachers.

◆ Exploration of Intuitive Methods for Solving Proportions

Other, more meaningful approaches than cross products need to be explored first, with the symbolic method left until much later. To illustrate some of these other approaches, consider the following examples.

1. Tammy bought 3 widgets for $2.40. At the same price, how much would 10 widgets cost?
2. Tammy bought 4 widgets for $3.75. How much would a dozen widgets cost?

Before reading further, it may be interesting to solve these two problems using an approach for each that seems easiest or most reasonable to you.

In the first situation, it is perhaps easiest to determine the cost of one widget, the unit rate or unit price. This can be found by dividing the price of three widgets by 3. Multiplying this unit rate of $0.80 by 10 will produce the answer. This approach can be referred to as the *unit-rate* method of solving proportions.

In the second problem, a unit-rate approach could be used, but the division does not appear to be easy. Since 12 is a multiple of 4, it is easier to notice that the cost of a dozen is 3 times the cost of 4. Using this *factor-of-change* approach on the first problem is possible but awkward. The factor of change between 3 and 10 is $3\frac{1}{3}$. Multiplying the $2.40 by $3\frac{1}{3}$ would produce the correct answer. While the factor-of-change method is a useful way to think about proportions, it is most frequently used when the numbers are compatible. In other situations, students generally utilize a unit-ratio approach (Post, Behr, & Lesh, 1988). Both methods should be explored.

Notice that the unit-rate method is exactly the same as that used in solving many fraction problems. Compare problem 1 with the following fraction exercise: If a given rectangle is $\frac{3}{10}$ of the whole, how big is the whole? Dividing the rectangle into three parts forms a unit fractional part. The unit part can then be "multiplied" to produce any related part, including the whole. (Earlier, we were counting unit fractional parts.) It is developmentally sound to relate new ideas to those that students have already developed.

As illustrated in Figure 15.4, equivalent ratios or rates can be illustrated graphically. Graphs provide another way of thinking about proportions and connect proportional thought to algebraic interpretations. Graphs of equal ratios are a good place for students to see the value of algebraic ideas and to interpret relationships in graphical form. All graphs of equivalent ratios are straight lines through the origin. If the equation of one of these graphs is written in the form $y = mx$, the slope m is always one of the equivalent ratios.

These intuitive approaches to ratio and proportion as well as the procedural algorithm of using cross products should all be explored in the middle or junior high school years. No method should be considered over another, and none should be taught in a rote manner. Rather, activities and guided discussions should help students learn to think proportionally in a variety of ways and contexts in order to make this form of reasoning profitable. Failure to develop an adequate facility with proportional reasoning can preclude study in a variety of disciplines, including algebra, geometry, biology, chemistry, and physics (Hoffer, 1988b).

INFORMAL ACTIVITIES TO DEVELOP PROPORTIONAL REASONING

The activities suggested in the following sections each provide a different opportunity for the development of proportional reasoning. While some are easier than others, they should not be interpreted as being in any definitive sequence. Nor are the activities designed to directly teach specific methods or algorithms for solving proportions. Some activities can be modified or repeated using different numbers to produce more or less difficult situations or to suggest different thought processes. Engaging in these and similar activities over the early adolescent years can help children develop a proportional thought.

SELECTION OF EQUIVALENT RATIOS

The basic approach in the following activities is to present a ratio in some form and have students select an equivalent ratio from among others presented. The focus should be on an intuitive rationale for why the two ratios selected are the same ratio—why they look the same. In some situations,

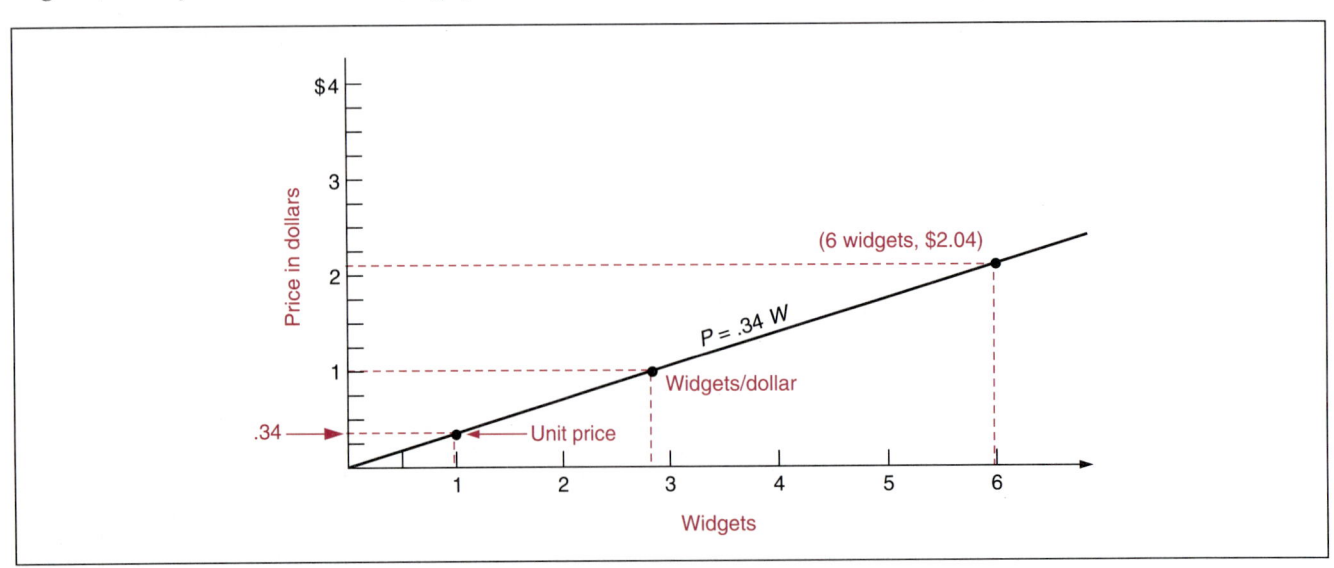

FIGURE 15.4: *Graph of price-to-item ratios*

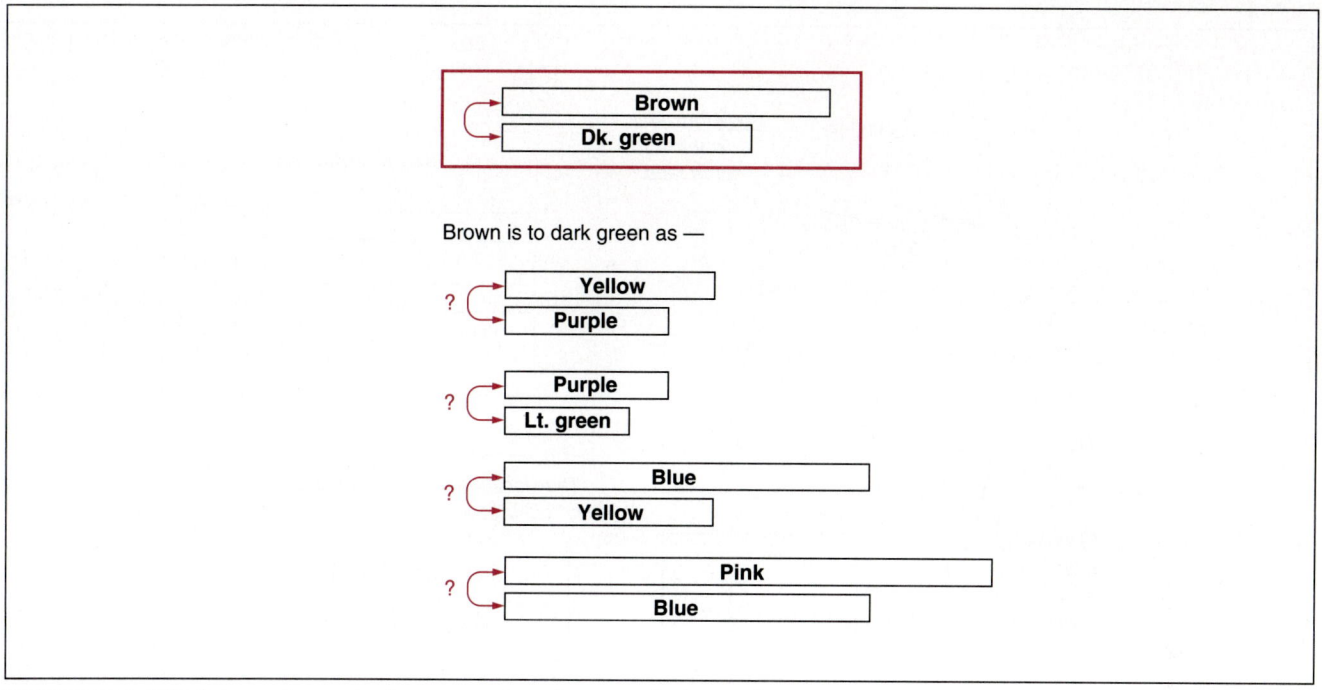

FIGURE 15.5: *Which two pairs are in the same ratio as brown to dark green?*

numerical values will play a part and help students begin to develop numeric methods to validate and explain their reasoning. In later activities, students will be asked to construct or provide an equivalent ratio without choices being provided.

15.1 FIND A PROPORTIONAL PAIR

Choose two fraction strips or Cuisenaire® rods to make a pair in some ratio. Present other pairs of strips, some of which are in the same ratio and some not, as shown in Figure 15.5. Notice that a unit ratio such as 3 to 1 is easier than a nonunit ratio such as 2 to 5. The same activity can be done with any fraction model that allows for units of varying sizes. Simple lines or rectangles can be drawn on paper instead of using a manipulative model, but the opportunity for exploration is reduced.

15.2 LOOK-ALIKE RECTANGLES

Have students classify rectangles. Duplicate a collection of 10 or more different rectangles for students to cut from paper. Prepare the rectangles so that each is similar to two or more other rectangles. Make three or four sets of similar rectangles. (Figure 15.6

illustrates an easy way to draw similar rectangles.) Simply have the students put those rectangles together that are alike. When they have grouped the rectangles, have them make additional observations to justify why they go together; the more obvious the differences in the shapes, the easier the task. Encourage as many ways as possible to describe how the rectangles are alike. Some ideas are suggested in Figure 15.7 (p. 280). The same task can be done with right triangles or with *any* set of triangles, as long as there are pairs that are similar.

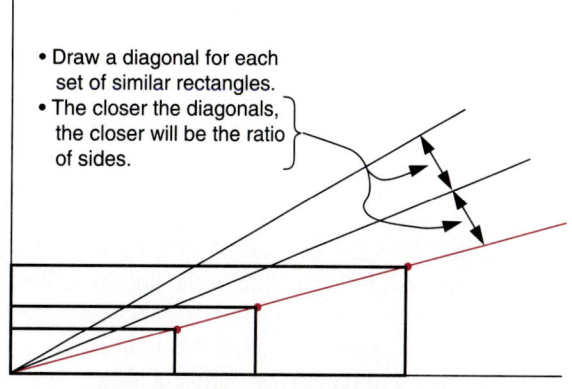

- Draw a diagonal for each set of similar rectangles.
- The closer the diagonals, the closer will be the ratio of sides.

Draw rectangles on each diagonal, trace, and cut out. All rectangles with the same diagonal will be similar.

FIGURE 15.6: *Drawing similar rectangles*

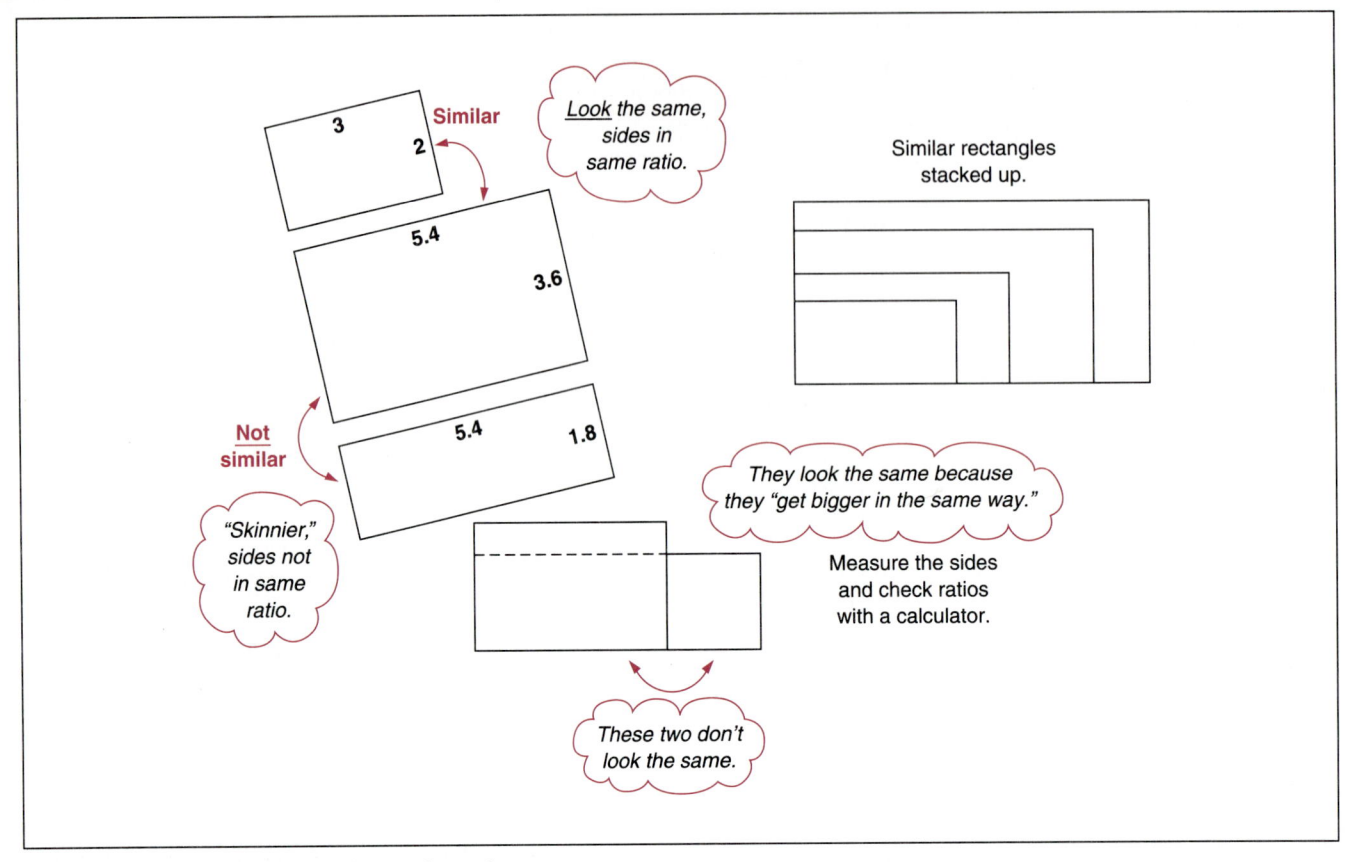

FIGURE 15.7: *Similar rectangles have sides in the same ratio.*

15.3 PROPORTIONAL MIXTURES

Make cards, transparencies, or worksheets with four or five loops of black and white dots. The dots should be in random arrangements, as in the "Dot Mixtures" example. Prepare the dot loops so that at least two loops have dots in the same ratio of black to white but with different total amounts. The task is to select loops of dots where the mixtures look to be about the same. Begin with collections where the ratios are very different, such as 1 to 2, 1 to 4, and 1 to 10. In later exercises narrow the differences to ratios such as 1 to 2, 3 to 5, and 3 to 4. Another variation is to adjust the spread of the dots as well as the numbers. (To avoid having to draw new sets for each exercise, prepare lots of different dot sets on separate paper and record the numbers of each color on the back. Assemble different collections on a photocopier to create different exercises.) Students can count dots within sets they have matched visually and compare part-to-part ratios as well as part-to-whole fractions.

15.4 DIFFERENT OBJECT/SAME RATIOS

In an activity similar to the dot mixtures, prepare cards with two distinctly different objects, as shown in Figure 15.8. Given one card, students are to select a card where the ratio of the two types of objects is the same. This task moves students to a numeric approach rather than a visual one and is an introduction to the notion of rates as ratios. A unit rate is depicted on a card that shows exactly one of either of the two types of objects. In the example shown, the card with three boxes and one truck provides one unit rate. (A unit rate for the other ratio is not shown. What would it be?) The use of objects paired with coins or bills is a good introduction to price as a ratio.

The last activity provides a good example of how numbers can affect thought processes. Different ratios should be used in various examples to encourage both unit rate and factor-of-change approaches. Each of the following pairs of ratios are equivalent, but the numbers in the ratios vary, depending on whether they are multiples of the other number *within* the ratio or *between* the two ratios.

    3 to 3 and 5 to 5 . . . Within and 1 to 1
    3 to 12 and 5 to 20 . . . Within
    2 to 6 and 6 to 24 . . . Between and within
    3 to 7 and 6 to 14 . . . Between
    4 to 6 and 10 to 15 . . . Neither

When there is an even multiple within, the ratio can easily be converted to a unit ratio. For example, 3 to 12 is the same as 1 to 4. Pairs that involve between multiples but not within lend themselves to the factor-of-change approach; 6 to 14 is an even multiple of the 3-to-7 ratio.

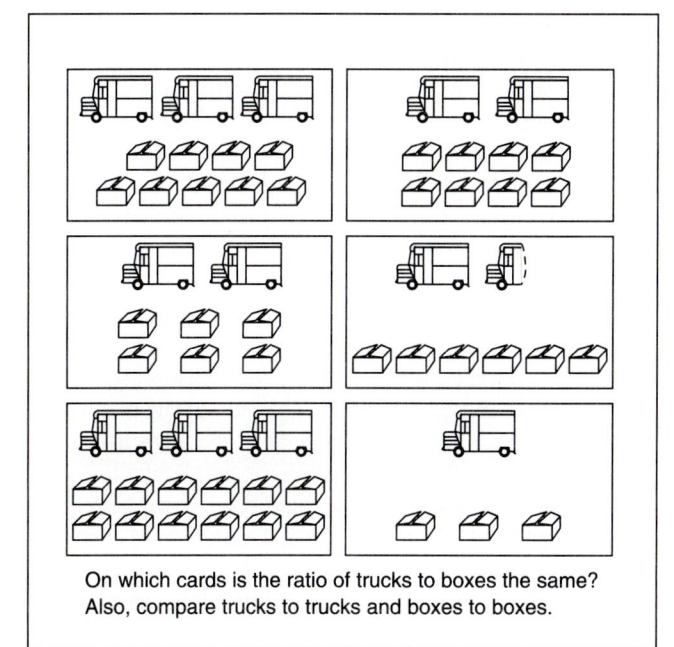

FIGURE 15.8: *Rate cards: Match cards with the same rate of boxes per truck.*

SCALING ACTIVITIES

Scaling activities involve filling in charts where paired entries are related in some way. The format of a chart is not important, and different formats can be used. Scaling up is a matter of providing entries with larger numbers, and scaling down is entering smaller numbers. The following are some examples of scaling up:

    1 foot → 12 inches   1 can → $0.55
    2 feet → 24 inches   2 cans → $1.10
    3 feet → ? inches   3 cans → ?

As illustrated, the items in the charts can be related measures such as time (minutes to hours), money (nickels to quarters), weight (pounds to ounces or pounds to kilograms), or common pairings such as crayons to boxes, hands to fingers, wheels to tricycles, and others. They can also be arbitrary ratios, such as a man who eats three bagels for every two bananas.

In a scaling-down activity, one or two later entries in the list or table are given, and students are asked to provide entries earlier or at the beginning of the list as in the following examples:

    28 days → 4 weeks   12 cups → 24 sugars
    21 days → 3 weeks   11 cups → 22 sugars
    ? days → 2 weeks   10 cups → ? sugars

Be careful not to make these exercises too long or tedious. Allow students to use repeated addition or subtraction as well as multiplication and division. Also, be sure to permit the use of calculators to make computation easy.

Notice that in the examples so far the unit ratio is easy to determine without use of fractions. As long as unit ratios are easily found, scaling activities can be done at an early grade. Later, the use of fractions and decimals can be used in a scaling-type exercise. Students can scale down with one division to get a unit ratio and multiply to get a particular requested entry. This is illustrated in the following example.

    3 boxes for $2.25 → 5 boxes for ?

1 box is $0.75 (unit ratio), so 5 boxes is
5 × $0.75 = $3.75

Suppose that the price of nine boxes or some other multiple of 3 was requested in this example. Then scaling can be done using multiples of the given ratio. Scaling can be done up or down without using a unit ratio.

    Minutes: 5   ?   15 20 25 30 35
    Widgets made: ? 14 21 ?   ?   ?   ?

The next two activities are related to scaling but are less structured.

15.5 WHAT'S IN THE BAG?

This activity involves informal probability concepts as well as ratio. Put colored cubes or other counters of two colors in a bag. For example, use four red and eight blue. Explain that there are cubes of different colors in the bag, but do not tell students the number of cubes or number of colors. Shake the bag and have students draw out a cube, record the color, and return the cube to the bag. After 10 or 15 trials, ask students how many of each color they think might be in the bag and record the guesses. After some more trials, you could ask what is the fewest possible number of cubes that they think could be in the bag. Then you can give them one of the following clues: the total counters that are in the bag or the number of cubes of one of the colors. With this information see if they can predict how many of each color are in the bag. "What if there were more cubes? What are other numbers of each color that might be in the bag?" The discussion is useful even if the students do not guess the correct ratio of colors. You can continue pulling out blocks to see how the ratio stays approximately, but not exactly, the same. Besides changing the ratio of colors on different days, you can also add a third color. After seeing the actual contents of the bag, discuss what other numbers of each color would produce the same result. Groups might explore drawing cubes from bags with equal ratios of colors but different numbers and compare the results.

15.6 GRAPHS SHOWING RATIOS

Have students make a graph of the data from a collection of equal ratios that they have scaled or discussed. Whereas Figure 15.4 is a graph of price-to-items ratios, the graph in Figure 15.9 is of the ratios of two sides of similar rectangles. If only a few ratios have actually been computed, the graph can be drawn carefully and then used to determine other equivalent ratios. This is especially interesting when there is a physical model to coincide with the ratio. In the rectangle example, students can draw rectangles with sides determined by the graphs and compare them to the original rectangles. A unit ratio can be found by locating the point on the line that is directly above or to the right of the number 1 on the graph. (There are actually two unit ratios for every ratio. Why?) Students can then use the unit ratio to scale up to other values and check to see that they are on the graph as well. Note that the slope of any line through the origin is a ratio.

MEASUREMENT ACTIVITIES

In these activities, measurements are made to create physical models of equivalent ratios in order to provide a tangible example of a proportion as well as look at numerical relationships.

15.7 DIFFERENT UNITS/EQUAL RATIOS

Cut strips of adding machine tape all the same length, and give one strip to each group in your class. Each group is to measure the strip using a different unit. Possible units include different Cuisenaire® rods or fractions strips, a piece of chalk, a pencil, the edge of a book or index card, or standard units such as inches or centimeters. When everyone has measured the strip, ask for the measure of one of the groups, and display the unit of measure. Next, hold up the unit of measure used by another group, and have the class compare it with the first unit. See if the class can estimate the measurement that the second group found. The ratio of the measuring units should be the inverse of the measurements made with those units. For example, if two units are in a ratio of 2 to 3, the respective measurements will be in a ratio of 3 to 2. Examine measurements made with other units. Finally, present a unit that no group has used, and see if the class can predict the measurement when made with that unit.

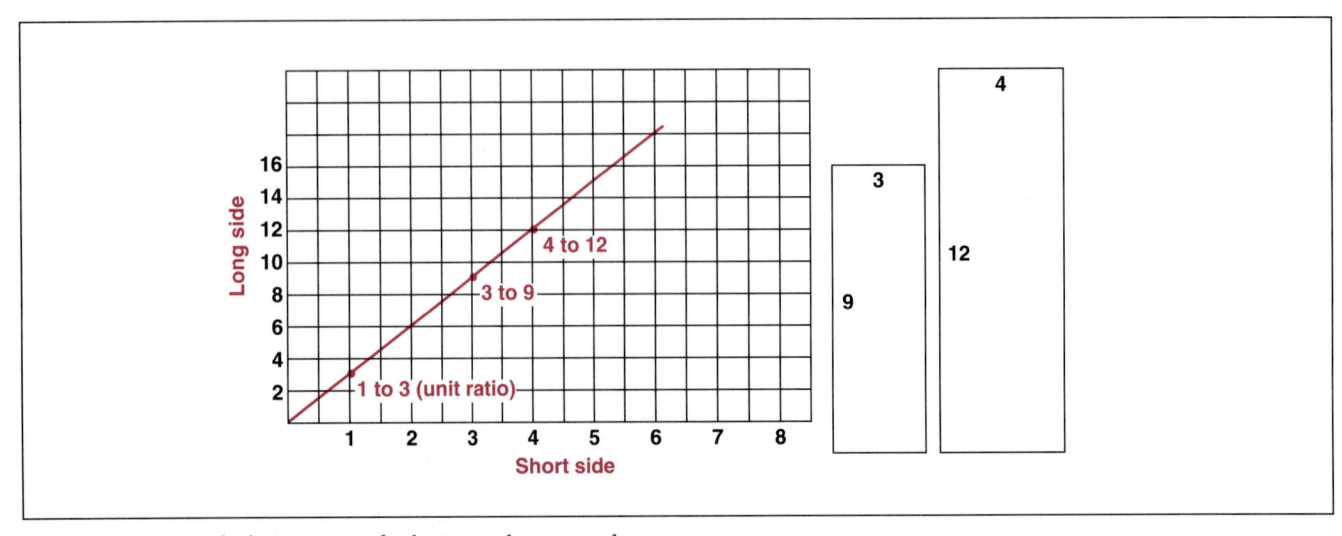

FIGURE 15.9: *Graph shows ratios of sides in similar rectangles.*

15.8 MEASUREMENT, GRAPHS, AND PERCENTAGES

An expansion of the preceding activity is to provide each group with an identical set of four strips of quite different lengths. Good lengths might be 20 cm, 50 cm, 80 cm, and 120 cm. As before, each group measures the strips using a different unit. Next have each group make a bar graph showing their measurements. Have all groups use the same scale for their graphs so that those with short units will have long bars in their graphs and those with long units will have short bars. Before displaying the graphs, have each group also make a circle graph representing the total length of all four strips. This is easily done by adding the measured lengths of the strips and then using a calculator to divide the length of each strip by the total. The results rounded to two decimal places will be the percent of the total contributed by each strip. By tracing a circle around a hundredths disk (Black-line Masters) and using the hundredths marks, you can graph the percent of each strip as a part of a circle. Have all groups make their bar and circle graphs with the strips in the same order and color them with the same colors. All of the bar graphs will be different heights due to the use of different units, but the pie graphs will all be nearly identical. Now you can discuss a variety of different ratios and proportions as suggested in Figure 15.10. The fact that certain ratios are the same and that the pie graphs are all the same provides vivid examples of proportionality.

15.9 DENSITY RATIOS

An activity involving weight and density can be conducted in a similar manner to the previous one. Instead of four different length strips, provide each group with four small containers of different sizes. The four containers must be the same for each group. Have each group fill their containers with a different "filler." Select fillers that vary greatly in density. For example, use dry oatmeal, rice, sand, and small metal washers. Each group weighs the contents of their containers (not including the containers) to the nearest gram and makes a bar graph of the results. A pie graph is made of the total weight of all containers. The results will be similar to the length experiment. Here, instead of different units of measure, the different densities of the fillers will produce different weights. Since the volumes are the same, the ratios of weights for each group will be the same, and each group should get about the same pie graph.

Circle graph computation for unit A.

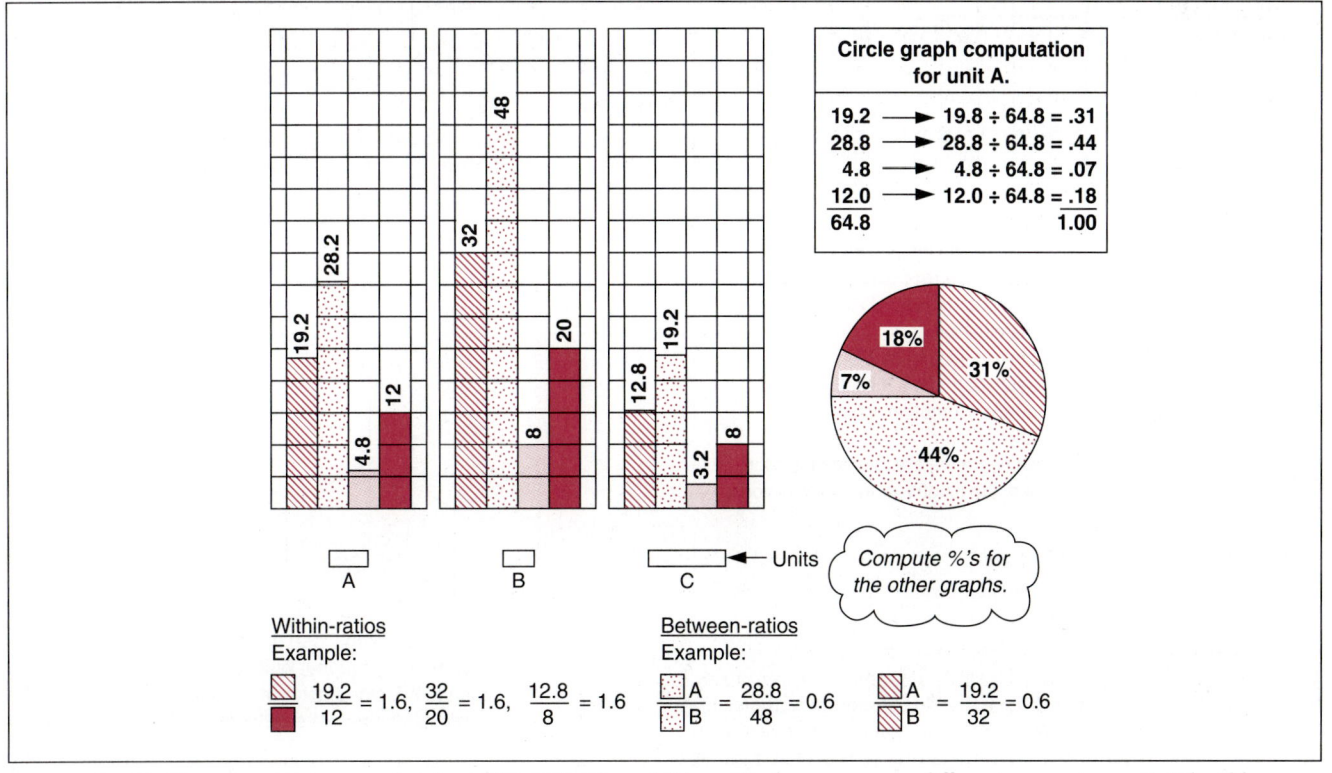

19.2	→	19.8 ÷ 64.8 = .31
28.8	→	28.8 ÷ 64.8 = .44
4.8	→	4.8 ÷ 64.8 = .07
12.0	→	12.0 ÷ 64.8 = .18
64.8		1.00

Within-ratios
Example: $\frac{19.2}{12} = 1.6$, $\frac{32}{20} = 1.6$, $\frac{12.8}{8} = 1.6$

Between-ratios
Example: $\frac{A}{B} = \frac{28.8}{48} = 0.6$ $\frac{A}{B} = \frac{19.2}{32} = 0.6$

Compute %'s for the other graphs.

FIGURE 15.10: Three groups measure 4 strips of 80, 120, 20, and 50 cm. Each group uses a different measuring unit. What like ratios can be observed?

15.10 SCALE DRAWINGS

On dot paper (Black-line Masters), have students draw a simple shape using straight lines with vertices on the dots. After one shape is complete, have them draw a larger or a smaller shape that is the same or similar to the first. This can be done either on the same size or on a different size grid, as in Figure 15.11. After completing two or three pictures of different sizes, the ratios of the lengths of different sides can be compared. Corresponding sides from one figure to the next should all be in the same ratio. The ratio of two sides within one figure should be the same as the ratio of the corresponding two sides in another figure. This activity connects the geometric idea of similarity with the numerical concept of ratio.*

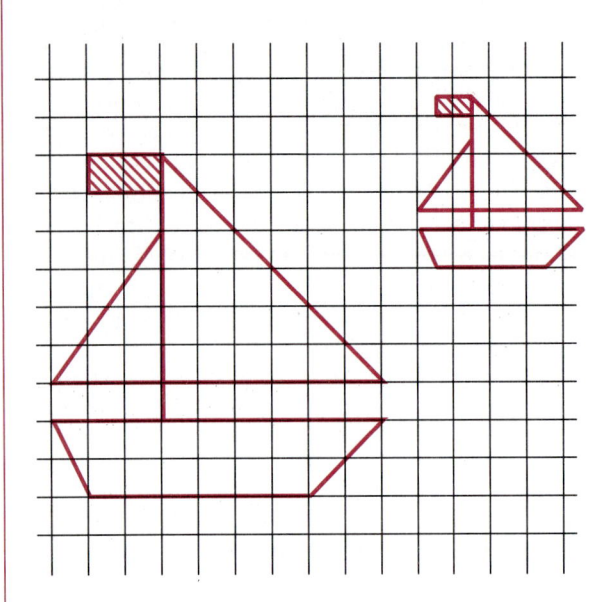

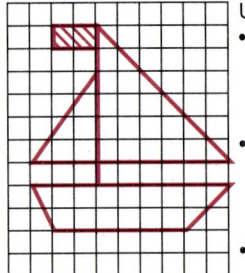

Use a metric ruler
- Choose two lengths on one boat and form a ratio (use a calculator!). Compare to ratio of same parts of the other boats.
- Choose two boats. Measure the same part of each boat and form a ratio. Compare with ratios of another part.
- Compare the areas of the big sails with the lengths of the bottom sides.

FIGURE 15.11: *Comparing similar figures drawn on grids*

---

* If the area of the figures can be easily determined by counting squares or half-squares, then it is interesting to compare the ratios of areas with the ratios of lengths. Areas of plane figures vary with the *square* of the sides and are not, therefore, in the same ratio as the sides. If the ratio of corresponding sides is 1 to 3, the areas will be in the ratio of 1 to 9. (What if the sides are in the ratio of 2 to 3?)

15.11 LENGTH, SURFACE, AND VOLUME RATIOS

A three-dimensional version of the last activity can be done with blocks, as shown in Figure 15.12. Using 1-inch or 2-cm wooden cubes, make a simple "building." Then make a similar but larger building, and compare measures. A different size can also be made using different-size blocks. To measure buildings made with different blocks, use a common unit such as centimeters. (Notice that volumes and surface areas do not vary proportionally with the edges of solids. However, these are relationships that are interesting to observe.)

Similar "buildings" can be made by changing the number of blocks in each dimension (factor of change) or by using different size blocks.

FIGURE 15.12: *Similar constructions*

In all these activities, measurements have been made to observe equal ratios or proportions. In the following activity, perception is used without measurement to create a proportionate length or shape. Measurement follows the perceptual judgment to see how good the estimate is.

15.12 STOP WHEN THEY'RE THE SAME

On the board, draw two lines labeled A and B as in Figure 15.13. Draw a third line, C, that is significantly different from A. Begin drawing a fourth line, D, under C. Have students tell you when to stop drawing so that the ratio of C to D is the same as the ratio of A to B. Measure all four lines and compare the ratios with a calculator. (Notice that here a single decimal number represents a ratio. How do you explain that?)

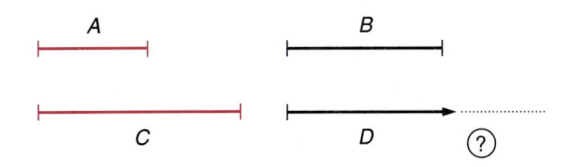

FIGURE 15.13: *Where should line D end so that A:B as C:D?*

Other activities that have been explored can also be transformed into estimation activities. For example, prepare a figure on plain paper using graph paper placed underneath as a guide. The figure can be as simple as a right triangle or more complicated, such as a drawing of a truck or a house. Make a transparency of the drawing. Students will see a very large figure with no measurements or grid. On paper with a very small grid, perhaps 0.5 cm, students attempt to draw a figure similar to the one on the overhead. While a perfectly similar figure may be difficult, there may be many that are close or even exact.

SOLVING PROPORTIONS

Notice that everything that has been discussed so far has been aimed at helping students develop an intuitive concept of ratio and proportion, to help in the development of proportional reasoning.

One practical value of proportional reasoning is to use observed proportions to find unknown values. Given that two ratios are equivalent, it is frequently the case that knowledge of one ratio is used to find a value in the other. Comparison pricing, using scales on maps, and solving percentage problems are just a few examples where proportions are involved and solving proportions is common. Students need to learn to set up proportions symbolically and to solve them.

THE CROSS-PRODUCT ALGORITHM

When a proportion cannot be solved mentally (unit-rate or factor-of-change methods) and an exact answer is desired, some sort of algorithm is useful. Usually this means you set up the known proportion including a single unknown quantity, cross multiply, and solve for the unknown. As with every other algorithm in mathematics, it should make sense and not just be a mechanical method of solving problems.

Investigate Known Proportions

Before jumping in with the magic and simplicity of the cross-product algorithm, we can help students examine some situations in which they have already established equal ratios. Data might come from a scaling experience or perhaps from a measuring activity such as one of those discussed in the previous section. For example, suppose that the class had completed the following chart for prices of tennis balls:

Tennis balls: 3   6   9   12   15
Price: $2.49 $4.98 $7.47 $9.96 $12.45

Have students select any two ratios from the chart that they think should be equal and write them in an equation as fractions. One possibility is the following, although any pair of ratios from the table could be used.

$$\frac{3 \text{ balls}}{\$2.49} = \frac{9 \text{ balls}}{\$7.47}$$

Now write the cross product, which for this example is

$$3 \text{ balls} \times \$7.47 = 9 \text{ balls} \times \$2.49$$

The numerical part of this equation is correct (3 × 7.47 = 9 × 2.49). Will the numbers always work out that way in a proportion? The answer is clearly "yes," but students should investigate this with real data and be personally convinced. Students could quickly set up other proportions from the table and check the cross products with a calculator.

But what about the units? Balls times dollars does not make much sense unless you want to invent a unit called "ball-dollars." But the equation in that form is not very useful anyway. A more likely situation occurs when one of the four factors in the equation is unknown. The fact that the cross products are equal can be used to solve for the missing value. Pretend that one of the four factors in the equation is unknown and, with the class, write a corresponding question or problem. For example, if the $7.47 factor was unknown, a problem might be: "If three balls cost $2.49, how much are nine balls?" For this problem, the equation would be

$$\$7.47 = \frac{9 \text{ balls} \times \$2.49}{3 \text{ balls}}$$

There are now two meaningful ways to interpret this equation. First, if written in the form

$$\$7.47 = 9 \text{ balls} \times \frac{\$2.49}{3 \text{ balls}} = 9 \text{ balls} \times \frac{\$0.83}{1 \text{ ball}}$$

then the unit rate is apparent. The fraction $\frac{\$2.49}{3 \text{ balls}}$ is equivalent to the fraction $\frac{\$0.83}{1 \text{ ball}}$, which is the cost per ball. It is not really necessary to compute this unit rate before multiplying by 9 to get the cost of nine balls, but students can see that the cross product can make sense in terms of unit rates.

If the equation is grouped in the following way, the factor of change is apparent:

$$\$7.47 = \frac{9 \text{ balls}}{3 \text{ balls}} \times \$2.49 = 3 \times \$2.49$$

The fraction $\frac{9 \text{ balls}}{3 \text{ balls}}$ tells how many sets of 3 balls are in a set of 9 balls. (As noted earlier, ratios of like items are usually written without units.) Three is a factor of change that would probably have been the intuitive choice for solving this particular problem.

Return to the original equation, and select any other factor as the unknown. A similar analysis can be made. After solving for one factor, we can always group the other side of the equation to show a unit rate times a quantity, or we can group it to illustrate a factor of change times a quantity. If both unit rates and factors of change have been used intuitively in previous investigations, then students can begin to make connections between the procedural approach of the cross-product algorithm and their conceptual understanding. (It would be a useful exercise for you to solve this tennis ball/price proportion for each of the other three values. Pretending that one value is an unknown, first make up a word problem, and interpret the equation as a unit-ratio approach and then as a factor-of-change approach.)

As an aside, it is easy to see mathematically why the cross product of two equal ratios produces an equality. By definition, two ratios are equal if one is a multiple of the other. Therefore, all equal ratios are of the form

$\frac{a}{b} = \frac{ac}{bc}$ where c is a nonzero rational number.

The cross product of this expression is clearly an equality.

◆ Find New Values

Students should also investigate the cross-product method within a familiar context to find values that are not part of their original data. Using the same tennis-ball example, suppose that balls could be purchased individually. What would the cost of 10 balls be? Help students see that any other known ratio of balls to price can be used along with the incomplete ratio of 10 balls to the unknown price. For example,

$$\frac{6 \text{ balls}}{\$4.98} = \frac{10 \text{ balls}}{x \text{ dollars}}$$

Using cross products and solving for the price of 10 balls, we find

$$x \text{ dollars} = \frac{\$4.98 \times 10 \text{ balls}}{6 \text{ balls}} = \$8.30.$$

In this example, the fraction $\frac{\$4.98}{6 \text{ balls}}$ is the unit ratio or unit price, which is multiplied times 10 balls, or the factor of change $\frac{10}{6}$ is multiplied times $4.98, since 10 balls would cost $\frac{10}{6}$ times as much as the price of 6 balls. Check to see that the result of $8.30 is a reasonable fit in the original chart. (Compared with the costs of 9 and 12 balls, does this result seem reasonable?) Again, the cross-product approach not only works mathematically, but it can be interpreted in terms of the intuitive experiences of the students. Try solving for other unknown values. Consider the question, "How many balls can be purchased for $12.00?" The cross-product algorithm produces an answer of 14.4578 balls. Does this answer make sense? Situations like this should be discussed with your students to help them see that any results must be interpreted in real contexts.

WRITING PROPORTIONS FROM A MODEL

Given a ratio word problem, students have great difficulty writing ratios in a correct proportion. The task is to write an equation of two ratios, one of which includes the missing value. Since many apparently different equations can be written for the same proportion, some students become even more confused.

◆ Draw a Simple Model

Rather than drill and drill in hopes that they will somehow eventually get it, show students how to sketch a simple model or picture that will help them determine what parts are related. In Figure 15.14, a simple model is drawn for a typical rate or price problem. If a known price-to-weight ratio is used as one of the ratios, then a ratio with the unknown price can easily be determined from the

---

Apples are 3 lb. for 89¢. How much should you pay for 5 lb.?

3 lb. → 89 ¢

5 lb. → n ¢

Within-ratios or **Between-ratios**

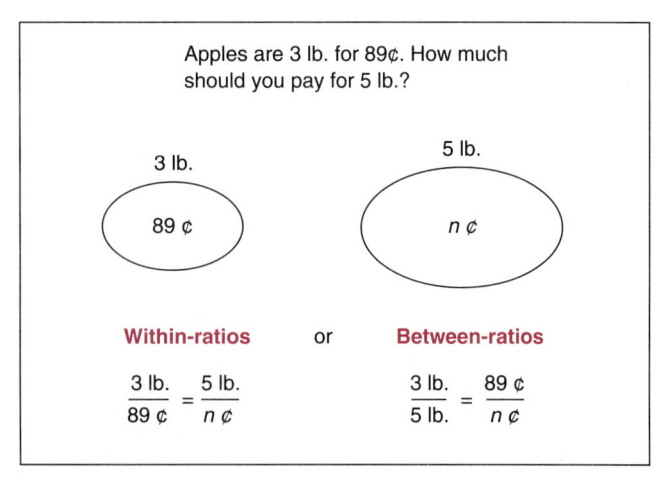

$$\frac{3 \text{ lb.}}{89 \text{ ¢}} = \frac{5 \text{ lb.}}{n \text{ ¢}} \qquad \frac{3 \text{ lb.}}{5 \text{ lb.}} = \frac{89 \text{ ¢}}{n \text{ ¢}}$$

FIGURE 15.14: *A simple drawing helps in a price-ratio problem.*

model by choosing the corresponding elements in the same order. These are within ratios. That is, the ratio of two numbers within one situation is equated to the corresponding ratio in the other. (Notice that the reciprocal ratios, weight to price, produce exactly the same cross product.)

It is just as easy and just as reasonable to equate the between ratios. For the same example, a ratio of price to price can be equated to the ratio of weight to weight. That is, if two situations are in proportion, then the ratios of corresponding proportional elements can be equated. Notice that the cross product is the same as for the within ratios and that the reciprocal ratios are also equivalent to the other forms.

In Figure 15.15, a problem involving rates of speed is modeled with a simple line representing the two distances. The distance and the time for each run is modeled with the same line. You cannot see time, but it fits into the distance covered. All equal rates of speed problems can be modeled this way. There really is no significant difference from the drawing used for the apples. Again, it is just as acceptable to write between ratios as within ratios, and students need not worry about which one goes on top as long as the ratios are written in the same order. The model helps with this difficulty.

Jack can run an 8-km race in 37 minutes. If he runs at the same rate, how long should it take him to run a 5-km race?

8 km	5 km
37 min.	x min.

Within-ratios

$$\frac{8 \text{ km}}{37 \text{ min.}} = \frac{5 \text{ km}}{x \text{ min.}}$$

Between-ratios

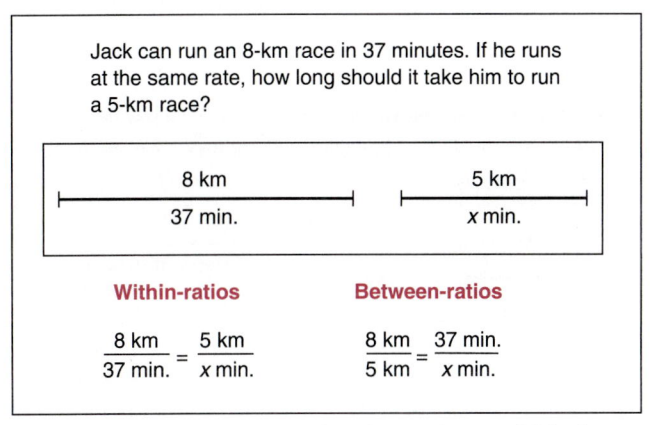

$$\frac{8 \text{ km}}{5 \text{ km}} = \frac{37 \text{ min.}}{x \text{ min.}}$$

FIGURE 15.15: *Line segments can be used to model both time and distance.*

[Figure 15.16: scale drawing with trapezoid inside rectangle, dimensions 28, 19, 44; Enlargement with 30, How long? X, How long? Y]

Within-ratios

$$\frac{28}{19} = \frac{X}{30} \qquad \frac{19}{44} = \frac{30}{Y}$$

Between-ratios

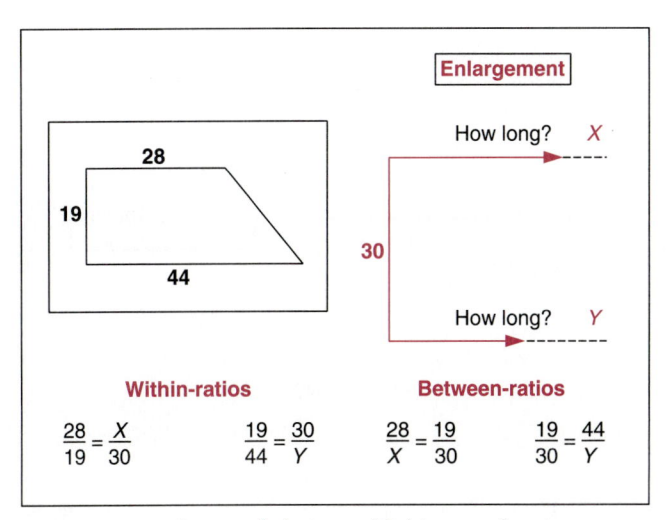

$$\frac{28}{X} = \frac{19}{30} \qquad \frac{19}{30} = \frac{44}{Y}$$

FIGURE 15.16: *Pictures help in establishing equal ratios.*

In Figure 15.16 a scale drawing is being made. As before, with the use of the simple sketch, students can easily find two like ratios. The drawing provides security without being a meaningless trick. It also helps to illustrate that there is not just one correct way to set up the equation.

PERCENT PROBLEMS AS PROPORTIONS

In most sixth- through eighth-grade textbooks, you will find a chapter on ratio, proportion, and percent. Percent has traditionally been included as a topic with ratio and proportion because percent is one form of ratio, a part-to-whole ratio. In many older programs, the unit on ratio and proportion focused a lot of time on solving percent problems as a proportion and relatively little time developing proportional thought. In the last chapter, it was shown that the solution to percent problems can be connected to concepts of fractions. In this section, the same part-to-whole fraction concept of percent is extended to ratio and proportion concepts. Ideally, all of these ideas (fractions, decimals, ratio, proportion, and percent) should be conceptually integrated. The better that students connect these ideas, the more flexible and useful their reasoning and problem-solving skill will be.

Equivalent Fractions as Proportions

Before considering percents specifically, consider first how equivalent fractions can be interpreted as a proportion using the same simple models already suggested. In Figure 15.17 a line segment or bar is partitioned in two different ways: in fourths on one side and in twelfths on the other. In the previous examples, proportions were established based on two amounts of apples, two different distances or runs, and two different sizes of drawings. Here there is only one thing measured—the part of a whole—but it is measured or partitioned two ways: in fourths and in twelfths.

[Figure showing bar partitioned with labels 4, 3, 9, 12 — Fourths and Twelfths]

Within ratios

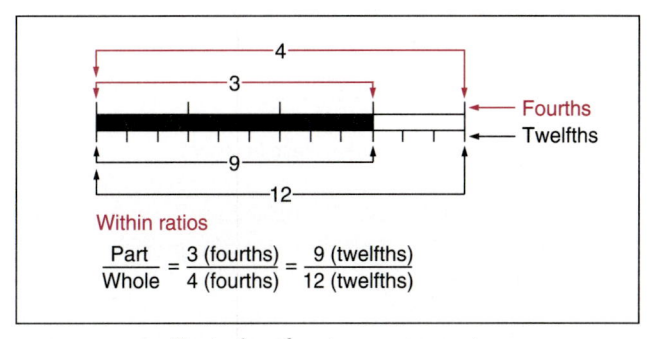

$$\frac{\text{Part}}{\text{Whole}} = \frac{3 \text{ (fourths)}}{4 \text{ (fourths)}} = \frac{9 \text{ (twelfths)}}{12 \text{ (twelfths)}}$$

FIGURE 15.17: *Equivalent fractions as proportions*

The within ratios are ratios of part to whole within each measurement. Within ratios result in the usual equivalent fraction equation, $\frac{3}{4} = \frac{9}{12}$ (3 fourths are to 4 fourths as 9 twelfths are to 12 twelfths). The between proportion equates a part-to-part ratio with a whole-to-whole ratio, or $\frac{3}{9} = \frac{4}{12}$ (3 fourths are to 9 twelfths as 4 fourths are to 12 twelfths). The between ratios here might be confusing to children. They illustrate, however, how this drawing is like those of Figures 15.14 through 15.16.

A simple line segment drawing similar to the one in Figure 15.17 could be drawn to set up a proportion to solve any equivalent fraction problem, even those that do not result in whole number numerators or denominators. Two examples are shown in Figure 15.18.

involve a part and whole measured in some unit and the same part and whole measured in hundredths—that is, in percents. A simple line segment drawing can be used for each of the three types of percent problems. Using this model as a guide, a proportion can then be written and solved by the cross-product algorithm. Examples of each type of problem are illustrated in Figure 15.19.

It is tempting to teach all percent problems in this one way. That is, whenever there is a percent problem, make a

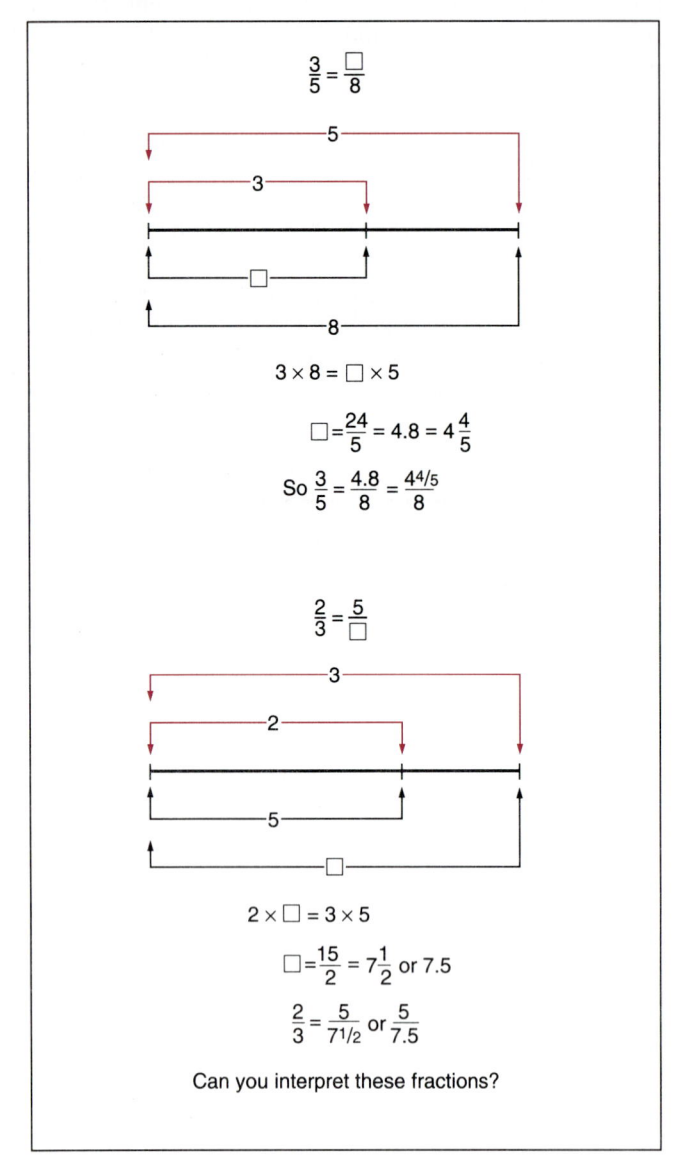

FIGURE 15.18: *Solving equivalent fraction problems by the cross-product algorithm.*

Percent Problems

The equivalent fraction examples illustrate how any fraction can be sketched easily on a simple line segment showing the part and the whole measured or partitioned two different ways. All percent problems are exactly the same as this. They

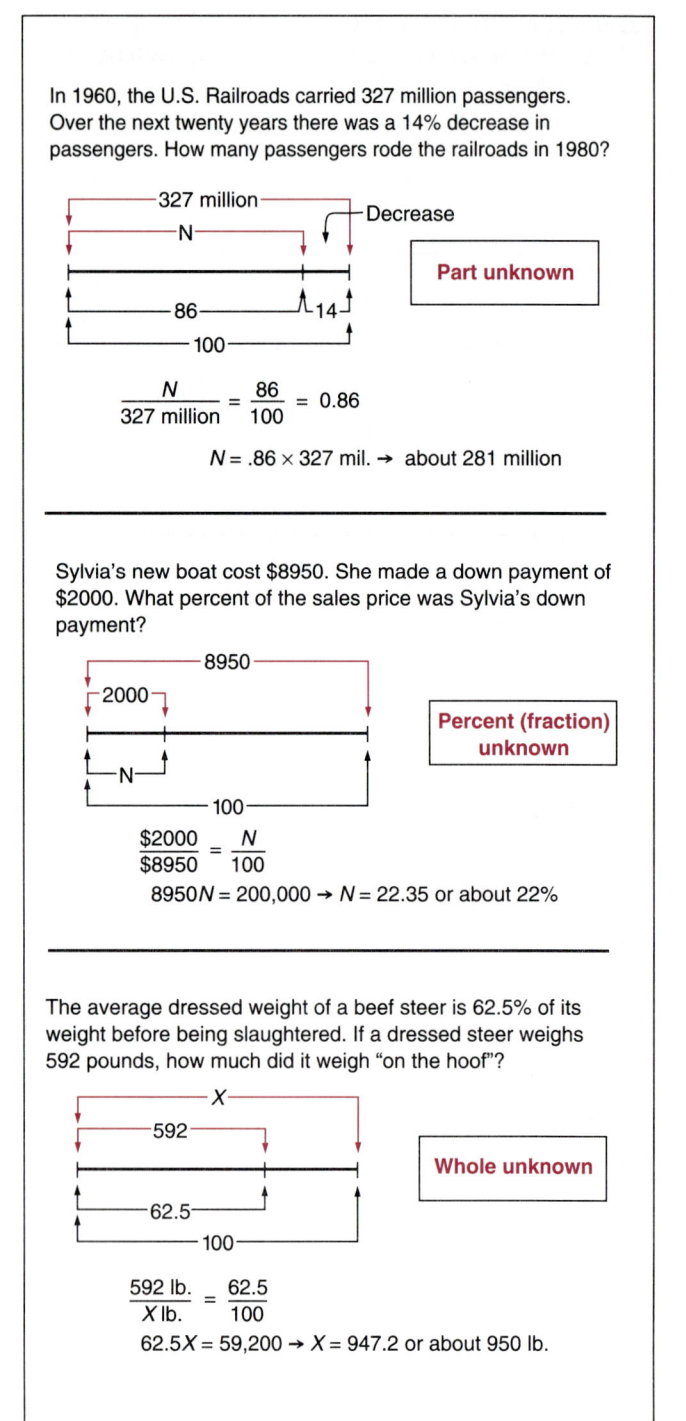

FIGURE 15.19: *The three percent problems solved by setting up a proportion using a simple line segment model.*

little drawing, set up a proportion, and solve by cross products. Developmentally, such an approach is not recommended. First, and most important, even though the approach is conceptual it does not translate easily to intuitive ideas, mental arithmetic, or estimation as discussed in the last chapter. Second, research does not seem to support the notion of focusing on a single algorithmic approach to solving percent problems (Callahan & Glennon, 1975). The modeling and proportion approach of Figure 15.19 is suggested only as a way to help students connect percent concepts with fraction and proportion ideas and analyze problems that may verbally present some difficulty. The approach of the last chapter, which relates percent to part-whole fraction concepts, should probably receive the major emphasis in teaching percent.

◆◆◆◆◆
REFLECTIONS ON CHAPTER 15: WRITING TO LEARN

1. Describe the idea of *ratio* in your own words. Explain how your idea fits with each of the following:
   a. A fraction is a ratio.
   b. Ratios can compare things that are not at all alike.
   c. Ratios can compare two parts of the same whole.
   d. Rates such as prices or speeds are ratios.

2. What is a proportion? For each of the situations in Question 1, give an example of a proportion.

3. Make up a realistic proportional situation that can be solved
   a. by a factor-of-change approach, and
   b. by a unit-rate approach.
   Explain each.

4. Much of this chapter is about providing students with activities that help them observe ratios and develop proportional reasoning abilities. These activities have been placed in three groups:
   a. selecting equivalent ratios,
   b. scaling activities, and
   c. measurement activities.
   Pick one activity from each of these sections that you did not do as part of class experiences. Do the activity, and describe briefly how you think the activity would contribute to students' proportional reasoning.

5. Consider the problem: If 50 gallons of fuel oil costs $56.95, how much can be purchased for $100? Draw a sketch to illustrate the proportion, and set up the equation in two different ways. One equation should equate within ratios and the other between ratios.

6. Make up a realistic percent problem, and set up a proportion. Draw a model to help explain why the proportion makes sense. Illustrate how this method could be used for any of the three types of percentage problems.

◆◆◆◆◆
FOR DISCUSSION AND EXPLORATION

1. "Proportional thinking is quite different from being able to solve a proportion." Is this statement true? If so, is proportional thinking really important? Why?

2. Examine a teacher's edition of a basal textbook for the sixth, seventh, or eighth grade. How is the topic of ratio developed? What is the emphasis? Select one lesson, and write a lesson plan that extends the ideas found on the student pages and actively involves the students.

3. In the last chapter the three percent problems were developed around the theme of which element was missing—the part, the whole, or the fraction that related the two. In this chapter, percent is related to proportions, an equality of two ratios with one of these ratios a comparison to 100. How are these two approaches alike? Do you have a preference?

4. Get some percent problems from any basal textbook, and solve them by using the line segment method of setting up a proportion. In solving percent problems by proportions, are the ratios that are used generally within or between ratios?

◆◆◆◆◆
SUGGESTED READINGS

Cramer, K., Post, T., & Currier, S. (1993). Learning and teaching ratio and proportion: Research implications. In D. T. Owens (Ed.), *Research ideas for the classroom: Middle grades mathematics*. New York: Macmillan Publishing Co.

Dewer, J. M. (1984). Another look at the teaching of percent. *Arithmetic Teacher, 31*(7), 48–49.

Haubner, M. A. (1992). Percents: Developing meaning through models. *Arithmetic Teacher, 40,* 232–234.

Hoffer, A. R. (1988). Ratios and proportional thinking. In T. R. Post (Ed.), *Teaching mathematics in grades K–8: Research based methods*. Boston: Allyn & Bacon.

Lamon, S. J. (1993). Ratio and proportion: Connecting content and children's thinking. *Journal for Research in Mathematics Education, 24,* 41–61.

Lappan, G., Fitzgerald, W., Winter, M. J., & Phillips, E. (1986). *Middle grades mathematics project: Similarity and equivalent fractions.* Menlo Park, CA: Addison-Wesley.

Quintero, A. H. (1987). Helping children understand ratios. *Arithmetic Teacher, 34*(9), 17–21.

Strickland, J. F., & Denitto, J. F. (1989). The power of proportions in problems solving. *Mathematics Teacher, 82,* 11–13.

Vance, J. H. (1982). An opinion poll: A percent activity for all students. In L. Silvey (Ed.), *Mathematics for the middle grades (5–9).* Reston, VA: National Council of Teachers of Mathematics.

Wiebe, J. H. (1986). Manipulating percentages. *Mathematics Teacher, 79*(21), 23–26.

16 DEVELOPING MEASUREMENT CONCEPTS

THE MEANING AND PROCESS OF MEASURING

CONCEPTUAL AND PROCEDURAL KNOWLEDGE OF MEASUREMENT

Measuring would appear to be a very procedural activity. The tendency is to teach children "how to measure" rather than teaching "what it means to measure." True, it is very difficult to separate the procedural activity of measurement from the concept. The concept of measurement is best embodied by the process. A typical group of first graders measures the length of their classroom by laying strips 1 meter long end to end. But the strips sometimes overlap, and the line weaves in a snakelike fashion around the desks. Which do these children not understand: *how to measure* or the *meaning* of measurement? In the fourth NAEP (Kouba et al., 1988b), most seventh-grade students could read a ruler to the nearest quarter inch. However, if a line segment was not aligned with the ruler, as in Figure 16.1, fewer than half of the seventh-grade students and very few third-grade students could determine its length. These results point to a similar difference between *using* a measuring device and *understanding* how and why it works.

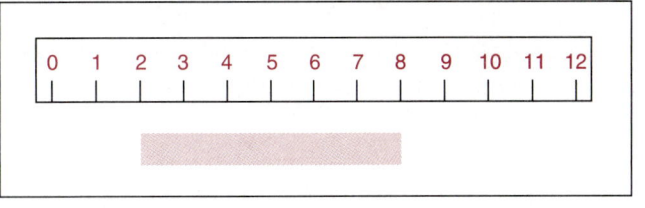

FIGURE 16.1: *"How long is this segment?"*

BASIC MEASUREMENT CONCEPTS

Suppose you were asked to measure an empty box. The first thing you would need to know or decide is what about the box is to be measured. Various lengths could be measured such as width, height, or distance around. The area of one or all sides could be determined. The box also has a volume and a weight. Each of these *things that can be measured* is an *attribute* of the box.

Once the attribute to be measured is determined, a unit of measure can be chosen. The unit must have the attribute that is being measured. Length is measured with units that have length, volume with units that have volume, and so on. Technically, a *measurement* is a number that indicates a comparison between the attribute of the object being measured and the same attribute of a unit of measure. A gross comparison might indicate that the measured attribute was more or less, longer or shorter, heavier or lighter, and so on, than the unit of measure. It is much more common to use small units of measure and to determine in some way a numeric relationship (the measurement) between what is measured and the unit. These comparisons are done in many different ways depending on what is being measured. For example, to measure a length, the comparison can be done by lining up copies of the unit directly against the length being measured. Notice that this is what you do when you use a ruler or tape measure. To measure weight, which is a pull of gravity or a force, the weight of the object might first be applied to a spring. Then the comparison is made by finding out how many units of weight produce the same effect on the spring. In either case, the number of units is the measure of the object.

For most of the attributes that are measured in schools, we can say that *to measure* means that the attribute being measured is "filled" or "covered" or "matched" with a unit of measure that has the same attribute (as illustrated in Figure 16.2). This concept of measurement will adequately serve the purposes of this chapter and is a good way to discuss with children what a measurement is. It is appropriate with this understanding, then, to say that the *measure of an attribute* is a count of how many units are needed to fill, cover, or match the attribute of the object being measured.

In summary, to measure something one must

1. decide on the attribute to be measured;
2. select a unit that has that attribute; and
3. compare the units by filling, covering, matching, or some other method, with the attribute of the object being measured.

Measuring instruments such as rulers, scales, protractors, or clocks are devices that make the filling or covering process easier. A ruler lines up the units of length and numbers them. The protractor lines up the unit angles and numbers them. A clock can be thought of as lining up units of time and marking them off.

DEVELOPING MEASUREMENT CONCEPTS AND SKILLS

Return to the children measuring the length of the classroom. Did they understand the concept of length as an attribute of the classroom? Did they understand that each strip of 1 meter had this attribute of length? Did they understand that their task was to fill smaller units of length into the longer one? What they most likely understood was that they were supposed to be making a line of strips stretching from wall to wall (and from their vantage they were doing quite well). They were performing a procedure instrumentally, without a conceptual basis.

A GENERAL PLAN OF INSTRUCTION

A basic understanding of measurement suggests how to help children develop conceptual knowledge of measuring.

Conceptual Knowledge to Develop	Type of Activity to Use
1. Understand the attribute being measured.	1. Make comparisons based on that attribute.
2. Understand how filling, covering, or matching an attribute with units produces a measure.	2. Use physical models of measuring units to actually fill, cover, or match the attribute.
3. Understand the way measuring instruments work.	3. Make measuring instruments, and use along with unit models.

Attribute: Length
units: rods, straws, toothpicks
How many units of length cover the height of the box?

Attribute: Volume
units: cubes, balls, cups of rice
How many units of volume fill the box?

Attribute: Area
units: index cards, squares of paper, triangles, tiles
How many units of area cover the surface of the box?

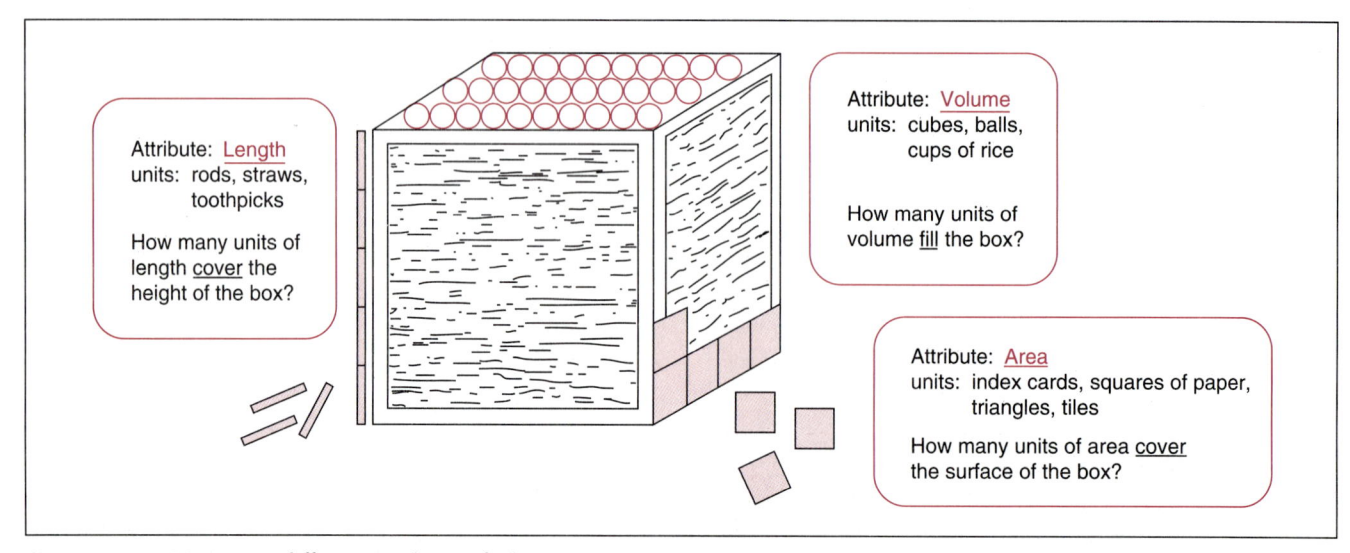

FIGURE 16.2: *Measuring different attributes of a box*

The different types of instructional activities are discussed briefly in the following sections.

◆ Making Comparisons

When students compare objects on the basis of some measurable attribute, that attribute becomes the focus of the activity. For example, is the capacity of one box more than, less than, or about the same as the capacity of another? No measurement is required, but some manner of comparing one volume to the other must be devised. The attribute of "capacity" (how much a container can hold) is inescapable.

Many attributes can be compared *directly,* such as placing one length directly in line with another. In the case of volume or capacity, some *indirect* method is probably required, such as filling one box with beans and then pouring the beans into the other box. Using a string to compare the height of a waste can to the distance around is another example of an indirect comparison. The string is the intermediary. It is impossible to compare these two lengths directly.

Constructing or making something that is the same in terms of a measurable attribute is another type of comparison activity. For example, cut the straw to be just as long as this piece of chalk, or draw a rectangle that is about the same size (area) as this triangle.

◆ Using Models of Units

For most attributes that are measured in elementary schools, it is possible to have physical models of the units of measure. Time and temperature are exceptions. (Many other attributes not commonly measured in school also do not have physical units of measure. Light intensity, speed, loudness, viscosity, and radioactivity are just a few examples.) Unit models can be found for both informal units and standard units. For length, for example, drinking straws (informal) or tagboard strips 1 foot long (standard) might be used as units.

Unit models can be used in two ways. The most basic and easily understood method is to actually use as many copies of the unit as are needed to fill or match the attribute measured. To measure the area of the desktop with an index card unit, you can literally cover the entire desk with index cards. Somewhat more difficult, especially for younger children, is to use a single copy of the unit with an iteration process. The same desktop area can be measured with a single index card by moving it from position to position and keeping track at which area the card has covered.

Another type of activity that helps children focus on measurement in a conceptual way is to make or construct objects with the same measure as a given object. Making (drawing, building, finding) an object with the same measure as a given one is quite different from simply measuring a series of objects and writing the results. For example, to cut a piece of paper that has just as much area as the surface of a can is an effective way to focus on the attribute of area as well as on the meaning of area units of measure. It is also possible when using units to talk about a numeric difference in the attributes of two objects instead of just making more, less, and same comparisons. "The capacity of the bucket is $2\frac{1}{2}$ liters larger than the capacity of this box."

It is a good idea to measure the same object with different- sized units and discuss the results. This will help students understand that the unit used is as important as the attribute being measured. It makes no sense to say, "The book weighs 23," unless you say 23 of what unit. The fact that smaller units produce larger numerical measures, and vice-versa, is very hard for young children to understand. This inverse relationship can only be constructed through experience with varying sized units.

◆ Making and Using Measuring Instruments

If students actually make simple measuring instruments using the same unit models that they have already measured with, it is more likely that they will understand how an instrument actually measures. A ruler is a good example. If students line up physical units along a strip of tagboard and mark them off, they can see that it is the spaces on rulers and not the marks or numbers that are important. It is quite important the measurement with actual unit models be compared with the measurement using an instrument. Without this comparison, students may not understand that these two methods are really two means to the same end.

A discussion of student-made measuring instruments for each attribute is provided in the sections that follow. Of course, children should also use standard, ready-made instruments such as rulers and scales. The use of these devices should still be compared directly with the use of the corresponding unit models.

INFORMAL UNITS AND STANDARDS UNITS: REASONS FOR USING EACH

It is common in primary grades to use nonstandard or informal units to measure lengths and sometimes area. Unfortunately, measurement activities in the upper grades, where other attributes are measured, frequently do not begin with informal units. There are a number of values in using informal units for beginning measurement activities at all grade levels.

1. Informal units make it easier to focus directly on the attribute being measured. For example, instead of using square inches to measure area, an assortment of different units, some of which are not square, can be

used to help understand the essential features of area and units of area.

2. By selecting units carefully, the size of the numbers in early measurements can be kept reasonable. The measures of length for first-grade students can be kept less than 20 even when measuring long distances. An angle unit much larger than a degree can be significantly easier for a sixth-grader to use, since a degree is very, very small and it is not reasonable to have individual copies.

3. The use of informal units can avoid conflicting objectives in the same beginning lesson. Is your lesson about what it means to measure area or understanding square centimeters? Learning to measure is different from learning about the standard units used to measure.

4. Informal units provide a good rationale for standard units. A discussion of the need for a standard unit can be quite meaningful after each of the groups in your class measured the same objects with their own units.

5. Informal units can be fun.

The use of standard units is also important in your measurement program at any grade level.

1. Students must eventually develop a familiarity with the most common standard units. That is, knowledge of standard units is a valid objective of a measurement program and must be addressed. Students must not only develop a familiarity with standard units, but they must also learn appropriate relationships between them.

2. Once a measuring concept is fairly well developed, it is frequently just as easy or even easier to use standard units. If there is no good instructional reason for using informal units, why not use standard units and provide the exposure?

There is no simple rule for when or where to use standard or informal units. Children's initial measurement of any particular attribute should probably begin with informal units and progress over time to the use of standard units and standard measuring tools. How much time should be spent using informal unit models varies with the age of the children and the attributes being measured. For example, first-grade children need a lot of experience with a variety of informal units of length, weight, and capacity. Informal units might be used at this level all year. On the other hand, the benefits of nonstandard measuring units may be diminished in two or three days for measurements of mass or capacity at the middle school level.

THE ROLE OF ESTIMATION WHILE LEARNING MEASUREMENT

It is very important to have students estimate a measurement before they make it. This is true with both informal and standard units. There are three reasons for including estimation in measurement activities:

1. Estimation helps students focus on the attribute being measured and the measuring process. Think how you would estimate the area of the front of this book with standard playing cards as the unit. To do so, you have to think about what area is and how the units might be fitted into the book cover.

2. Estimation provides intrinsic motivation. It adds fun and interest to measurement activities. It is fun to see how close you can come to estimating a measurement or if your team can make a better estimate than the other teams in the room.

3. When standard units are used, estimation helps to develop a familiarity with the unit. If you estimate the height of the door in meters before measuring, you have to devise some way to think about the size of a meter.

HELPING CHILDREN WITH MEASUREMENT ESTIMATES

Having said that estimating measures is very important and should occur with almost all measurement activities, there remains the problem of how to go about this. Later we will discuss teaching children specific estimating strategies that can be used throughout life. However, until these strategies are developed along with familiarity with standard units, children will have a great deal of difficulty making estimates. Here are three ways to ask for estimates that avoid asking children to come up with an actual number or measure estimate before they are capable.

1. Ask for a comparison estimate rather than a measure. For example, is the teacher taller, shorter, or about the same as 2 meters? Does the book weigh more, less, or about the same as 1 pound? The same can be done with informal units. Is the area of the desktop more, less, or about the same as 20 index cards?

2. Ask which of two or more suggested measures the actual measure is closer to. For example, is the angle measure closer to 15°, 60°, or 90°? Is the capacity of the box closer to 1 liter or 3 liters?

3. Provide an actual unit or set of units as a comparison. The children can make an initial estimate privately on paper. Then show the unit or set of 10 or 100 units, and let them adjust their first estimate accordingly. The use of the provided set of units allows them to mentally mark off or make a comparison in some way without having to make a blind guess. Making the

first estimate adds a bit of interest. For example, if estimating the length of the chalkboard in Unifix cube units, the number may be rather large and quite difficult for children. After seeing 10 cubes placed at one end of the chalkboard, the children can count by tens as they visually mark off these bars along the board. The use of 10 or 100 is suggested for place-value reasons. For large units such as a meter or a kilogram, one copy of the unit can be placed directly alongside the object being estimated. Of course with an attribute such as weight, the children should be able to handle the unit(s) as well as the object.

Notice that there is a progression to these three ideas in terms of how specific an estimate is requested. Use this progression as you develop estimation skills with children. These techniques can be applied to almost any measurement activities.

EMPHASIZE THE APPROXIMATE NATURE OF MEASUREMENT

In all measuring activities, emphasize the use of approximate language. The desk is *about* 15 orange rods long. The chair is *a little less than* four straws high. The use of approximate language is very useful with younger children using large units because many measurements to not come out even. Older children will begin to search for smaller units or will use fractional units to try to measure exactly. Here is an opportunity to develop the idea that all measurements include some error. Each smaller unit or subdivison does produce a greater degree of precision. For example, a length measure can never be more than a half unit in error. And yet, since there is mathematically no "smallest unit," there is always some error involved.

The notion of precision related to the size of the unit is an important idea in all measuring tasks. There are times when precision is not required, and a larger unit is much easier to deal with. At other times, the need for precision is significant, and smaller units become important. For example, measuring a pane of glass for a window requires a different precision than measuring a wall to decide how many 4-ft. × 8-ft. panels are needed to cover it. An awareness of precision due to unit size and the need for precision in different situations is an important aspect of measurement, especially at the upper grades.

MEASURING ACTIVITIES

For each attribute that we want children to measure, we can identify the three types of activities that have been discussed. Comparison activities should generally precede the use of units and measuring instruments should be dealt with last. Within each of these categories there is also a rough guideline of progression that can be considered as shown in the following chart.

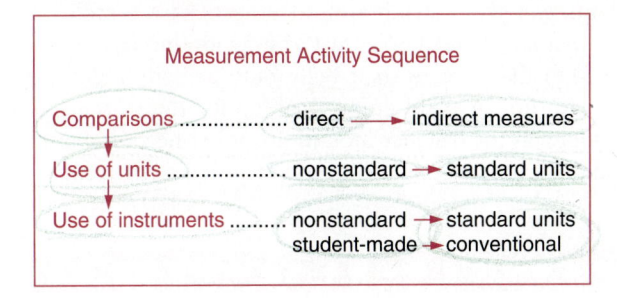

```
Measurement Activity Sequence

Comparisons .............. direct ——→ indirect measures

Use of units .............. nonstandard ——→ standard units

Use of instruments ...... nonstandard ——→ standard units
                          student-made ——→ conventional
```

Almost *all* activities should include an estimation component. Familiarity with standard units is a separate objective related to understanding the measurement process.

Notice especially in the chart that estimation and standard unit familiarity are important considerations. In the sections that follow, the focus is on activities for comparison, use of units, and instruments for each attribute. Separate sections on standard units and estimation follow, pointing to the importance of these objectives as well.

MEASURING LENGTH

◆ Comparison Activities

At the kindergarten level, children should begin with direct comparisons of two or more lengths.

16.1 LONGER, SHORTER, SAME

Make a sorting-by-length station at which students sort objects as longer, shorter, or about the same as a specified object. It is easy to have several such stations in your room. The reference object can be changed occasionally to produce different sorts. A similar task is to put objects in order from shortest to longest.

16.2 LENGTH HUNT

Go on a length hunt. Give groups of two a strip of tagboard, a stick, a length of rope, or other object with an obvious length dimension. The task on one day might be to find five things in the room that are shorter than (or longer than, or about the same as) their object. They can draw pictures or write the name of the things they find. (Label things in the room to help.)

16 / DEVELOPING MEASUREMENT CONCEPTS

It is important to compare lengths that are not in straight lines. One way to do this is with string or rope. Students can wrap string around objects in a search for things that are, for example, just as long around as the distance from the floor to their belly button. Body measures are always fun. A child enjoys looking for things that are just as long as the distance around his or her head or waist.

Indirect comparisons are a next step in length comparisons.

16.3 CROOKED PATHS

Make some crooked or curvy paths on the floor with masking tape. Provide students with pieces of rope longer than the total path. Assign teams of two students a starting mark (tape) on the floor and a direction. Have them place a second piece of tape on the floor so that the distance between the marks is just as long as one of the crooked paths. Have students explain to the class how they solved the problem and to demonstrate why they think their straight path is just as long as the crooked one. (This is a good outdoor activity, also.)

The crooked path activity can also be done with the small distances at students' desks. A simple worksheet might be prepared like the one in Figure 16.3. Instead of crooked paths, students can make straight paths as long as the distance around simple shapes (perimeter).

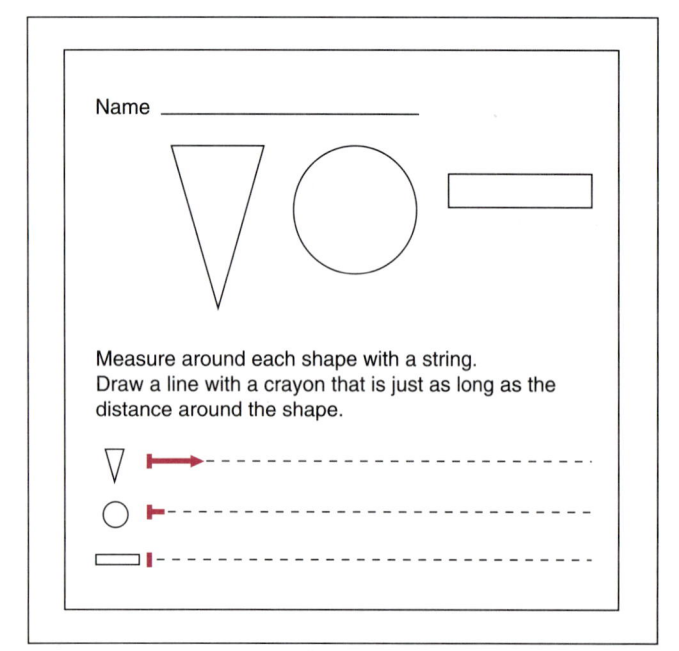

FIGURE 16.3: *Making a straight path just as long as a crooked path.*

Using Units of Length

Students can use a wide variety of informal units to begin measuring length. Some examples of units of different lengths are suggested here:

Giant Footprints: Make about 20 copies of a large footprint about $1\frac{1}{2}$ to 2 feet long cut out of poster board.

Measuring Strips: Cut strips of poster board about 5 cm wide. Several sets can be made to provide different units. Some can be the long dimension of the poster board, some the short and a third set about 1 foot long. Make each set of a different color.

Measuring Ropes: Cut cotton clothesline into lengths of 1 meter. These are useful for measuring curved lines and around large objects such as the teacher's desk.

Plastic Straws: Drinking straws are inexpensive and provide large quantities of a useful unit. Straws are easily cut into smaller units. A good idea is to string straws together with a long string. The string of straws is an excellent bridge to a ruler or measuring tape.

Short Units: Toothpicks, Unifix cubes, strips of tagboard, wooden cubes, and paper clips are all useful as units for measuring shorter lengths. Cuisenaire® rods are one of the nicest sets of units since they come in 10 different lengths, are easily placed end to end, and can be related to each other. Paper clips can easily be made into chains of about 20 clips for easier use.

For young children, initial measurements should be along lines or edges. If different teams of students measure the same distances and get different results, discuss why they may have gotten these differences. The discussion can help focus on the reason why units need to be lined up end to end and in a straight line and why units such as ropes must be stretched to their full length.

16.4 GUESS AND MEASURE

Make lists of things to measure around the room. For younger children, run a piece of masking tape along the dimension of each object to be measured. On the list, designate the units to be used. Do not forget to include curves or other distances that are not straight lines. Distances around desks, doors, or balls are some examples (Figure 16.4). Include estimates before the measures. Young children will not be very good at estimating distances at first.

16 / DEVELOPING MEASUREMENT CONCEPTS

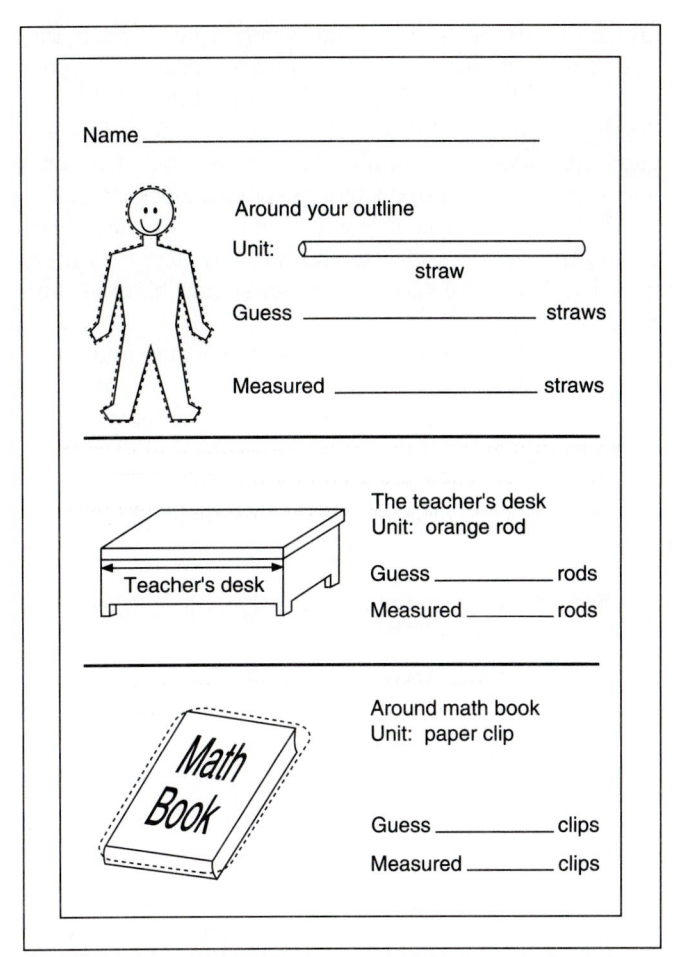

FIGURE 16.4: *Record sheet for measuring with informal length units.*

16.5 CHANGING UNITS

Have students measure a distance with one unit, then provide them with a different unit, and see if they can predict the measure with the new unit before actually doing the measurement. Students should write down their predictions and an explanation of why they arrived at it. These predictions and explanations when shared and discussed with the class will be the most important part of the activity. The first few times you do this activity, the larger unit should be a simple multiple of the smaller unit. If the two units are not related by a whole-number multiple, the task becomes difficult numerically and can be frustrating.

As children begin to develop a need for more precision, two units can be used at the same time. The second unit should be a smaller subunit of the first. For example, with Unifix cubes, the first unit can be bars of 10 cubes and the second, individual cubes. With measuring strips, make subunits that are one-fourth or one-tenth as long as the longer strip. Cut plastic straws so that an even number of paper clips is equal to one straw. Cuisenaire® rods allow for a variety of possibilities. For example, four reds make a brown, or ten whites make an orange. Have children measure with the larger unit until no more of that unit will fit and then add on sufficient smaller units to fill up the distance (Figure 16.5). Report measures in two parts: 8 straws and 3 clips long. For older students, smaller units can simply be fractional parts of longer units.

After a measurement with two related units has been made and recorded, have students figure out how to report the same measurement in terms of either unit. For example, $5\frac{2}{3}$ blue rods or 17 light green rods (3 light green = 1 blue). This provides a readiness exercise for standard units. For example, a measurement of 4 feet 3 inches is sometimes reported as 51 inches or as $4\frac{1}{4}$ feet. The use of two units is also good readiness for subdivision marks on a ruler.

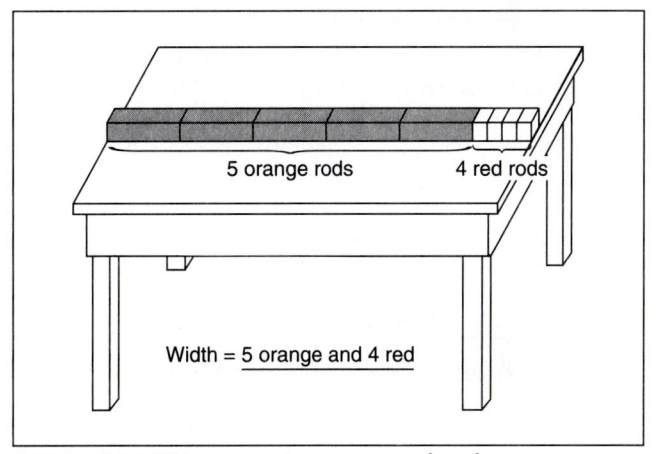

FIGURE 16.5: *Using two units to measure length*

◆ Making and Using Rulers

Rulers or tape measures can be made for almost any unit of measure that students have used. It is important that students have used the actual unit models before making the rulers.

16.6 RULER CONSTRUCTION

After doing some measuring with orange Cuisenaire® rods (or any unit not shorter than about 5 cm), give students narrow strips of construction paper in two contrasting colors. Have students use a unit model as a guide and cut the strips into lengths as long as the unit. Discuss how the paper strips could be used for measuring instead of the actual units. Have students paste the paper unit strips end to end along the edge of a long tagboard strip about 5 cm wide, as shown in Figure 16.6.

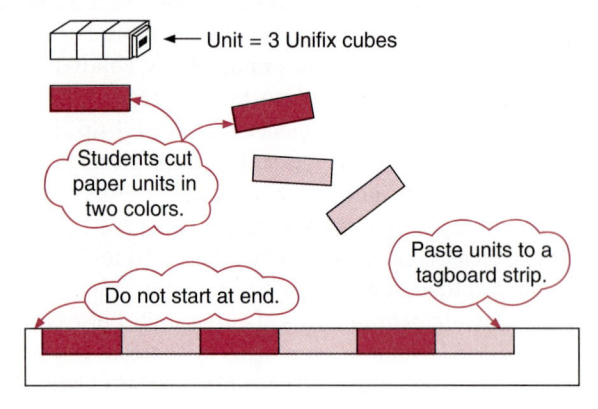

FIGURE 16.6: *Making a simple ruler*

Pasting down copies of the units on a ruler maximizes the connection between the spaces on a ruler and the actual units. Older children can make rulers by using a real unit to make marks along a tagboard strip and then coloring in the spaces. Rulers made with very small units are more difficult for students to make simply because they require better fine-motor skills. If the first unit on a ruler does not coincide with the end of the ruler, the student is forced to attend to aligning the units on the ruler with the object measured. Children should not be encouraged to use the end of a ruler as a starting point because many real rulers are not made that way.

Students should eventually put numbers on their homemade rulers, as shown in Figure 16.7. For young children, numbers should be written in the center of each unit to make it clear that the numbers are a way of precounting the units. When numbers are written in the standard way, at the ends of the units, the ruler becomes a number line. This format is more sophisticated and should be carefully discussed with children. (Number lines are generally a poor number model for children below the third grade. The development of a ruler in the manner just described is a good introduction to number lines in general. A number line is, in fact, a measurement model for numbers.)

After students make rulers, it is important to use them. In addition to the estimate-and-measure activities mentioned earlier, some special activities should be done with rulers.

Have teams measure items once with a ruler and a second time with actual unit models. While the results supposedly should be the same, inaccuracies or incorrect use of the ruler may produce differences that are important to discuss.

Use the ruler to measure lengths that are longer than the ruler.

16.7 MORE THAN ONE WAY

Challenge students to find different ways to measure the same length with one ruler. (Start from either end; start at a point not at the end; measure different parts of the object and add the results.)

Tape measures, especially for measuring around objects, can be made in a variety of ways using the same approach as rulers. With long units such as meters, a clothesline can be marked at the end of each meter with a piece of masking tape or a marking pen. Grosgrain ribbon is easily marked with a ballpoint pen. Even adding-machine tape can be used to make temporary tape measures.

After working with simple rulers and tapes, have students make rulers with subunits or fractional units. This should follow the use of two units for measuring as described in the previous section.

Much of the value of using student-made rulers can be lost if careful attention is not given to transfer of this knowledge to standard rulers. Give children a standard ruler and discuss how it is alike and how it is different from the ones they have made. What are the units? Could you make a ruler with paper units the same as this? Could you make some cardboard units and measure the same way as with the ruler? What do the numbers mean? What are the other marks for? Where do the units begin?

MEASURING AREA

Comparison Activities

When comparing two areas, there is the added consideration of shape. One of the purposes of early comparison activities with areas is to help students distinguish between size (or area) and shape, lengths, and other dimensions. A

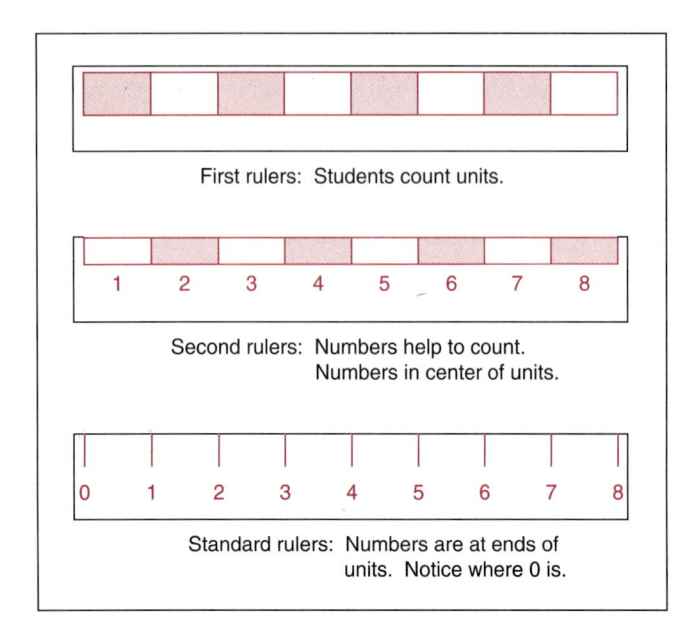

FIGURE 16.7: *Give meaning to numbers on rulers.*

long skinny rectangle may have less area than a triangle with shorter sides. This is an especially difficult concept for young children to understand. (Piagetian experiments with conservation of area indicate that many children eight or nine years old do not understand that rearranging areas into different shapes does not affect the amount of area.)

Direct comparison of two areas is nearly always impossible except when the shapes involved have some common dimension or property that makes it possible. For example, two rectangles with the same width can be compared directly as can any two circles. Comparison of these shapes, however, fails to deal with the attribute of area. Instead of comparison activities for area, activities in which one area is rearranged are suggested. By cutting a shape in two parts and reassembling it into a different shape, the intent is that students will understand that the before and after shapes have the same area, even though they are different shapes. While obvious to adults, the idea is not at all obvious to children in the K-to-2 grade range.

16.8 TWO-PIECE SHAPES

Give children a rectangle of construction paper and have them fold and cut it on the diagonal, making two identical triangles. Next have them rearrange the triangles into different shapes, including the original rectangle. The rule is that only sides of the same length can be matched up and must be matched exactly. Have each group find all the shapes that can be made this way, pasting the triangles on paper as a record of each shape (Figure 16.8). Discuss the size and shape of the different results. Is one shape bigger than the rest? How is it bigger? Did one take more paper to make, or do they all have the same amount of paper? Help children conclude that while each figure is a different shape, they all have the same *area*. ("Size" in this context is a useful substitute for area with very young children, although it does not mean exactly the same thing.)

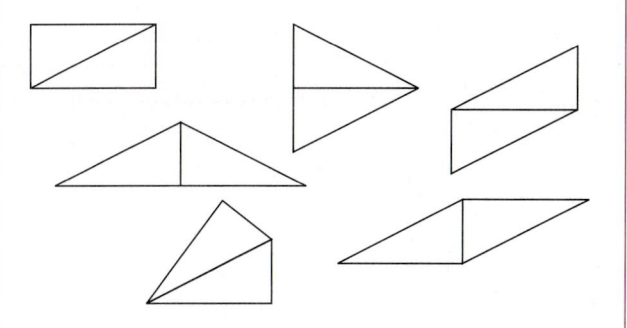

FIGURE 16.8: *Different shapes, same size*

The preceding activity can be extended to three or four triangles to produce even more shapes. [If two each of two colors are used, it is also exciting to find all the different color patterns for each shape (Burns & Tank, 1988). Tangrams, a very old and popular set of puzzle shapes, can be used in a similar way. The standard set of seven tangram pieces is cut from a square, as shown in Figure 16.9. The two small triangles can be used to make the parallelogram, the square, and the medium triangle. Four small triangles will make the large triangle. This permits a similar discussion about the pieces being the same size (area) but different shape (Seymour, 1971). A black-line master for tangram pieces is in the last section of the text.

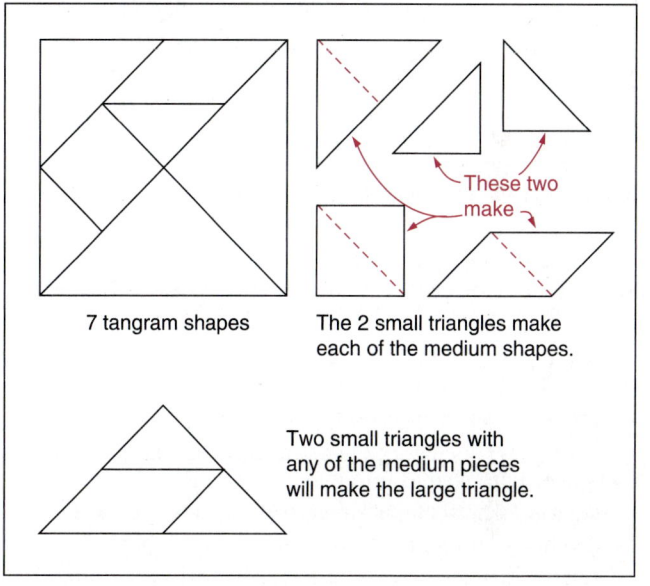

FIGURE 16.9: *Tangrams provide a nice opportunity to investigate size and shape concepts.*

In the following activities, two different methods are used for comparing areas without measuring.

16.9 TANGRAM AREAS

Draw the outline of several shapes made with tangram pieces, as in Figure 16.10 (p. 300). Let students use tangrams to decide which shapes are the same size, which are larger, and which smaller. Shapes can be duplicated on paper, and children can work in groups. Let students explain how they came to their conclusions. There are several different approaches to this task, and it is best if students determine their own solutions rather than blindly follow your directions. You might stop here, get a set of tangrams, and make the area comparisons suggested in Figure 16.10.

300 16 / DEVELOPING MEASUREMENT CONCEPTS

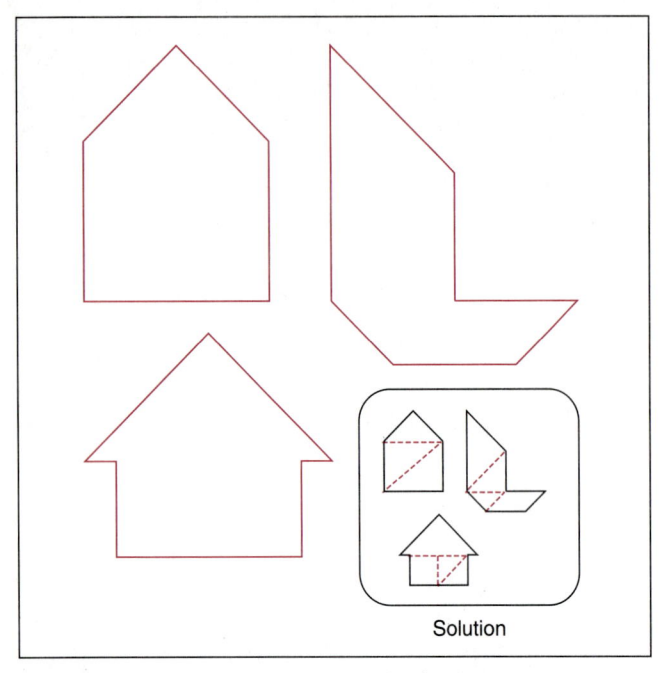

FIGURE 16.10: *Compare shapes made of tangram pieces.*

16.10 CUT TO COMPARE

Duplicate two simple shapes on a piece of paper. Have students cut them out. Suggest that they cut one or both shapes into two pieces so that they can decide if one shape is larger or if they are the same. As shown in Figure 16.11, make the two shapes related in some way so that the comparison requires no measurement.

◆ Using Units of Area

While squares are very nice units of area (and are far and away the most commonly used), area units need not be squares. Any tile can be used that conveniently fills up a plane region. When children are first learning about measuring areas by actually filling up the region with copies of the unit, it may be preferable if the tiles that are used fit together without leaving any gaps. However, even filling a region with uniform circles provides a useful idea of what it means to measure areas. Here are some suggestions for area units that are easy to gather or make in the large quantities you will need.

- Cut squares or triangles (diagonals of squares or rectangles) from corrugated cardboard. (Use a paper cutter.) Large squares or triangles (about 20 cm on a side) are good for large areas. Smaller units should be about 5 to 10 cm on a side.
- Sheets of newspaper make excellent units for very large areas.
- Poster board can easily be cut into large quantities of congruent tiles for smaller units. Include 2 by 1 rectangles, equilateral triangles, and right triangles as well as squares. These tiles can be from 2 to 5 cm on a side.
- Pattern blocks provide six different units. The hexagon, trapezoid, large diamond, and triangle can be related to each other in a manner similar to the tangrams.
- Playing cards, index cards, or old business cards make good medium-sized units.
- Round plastic chips, pennies, or even lima beans can be used. It is not necessary at a beginning stage that the area units fit with no gaps.

Children can use units to measure surfaces in the room such as desktops, bulletin boards, or books. Large regions can be outlined with masking tape on the floor. Use the gym

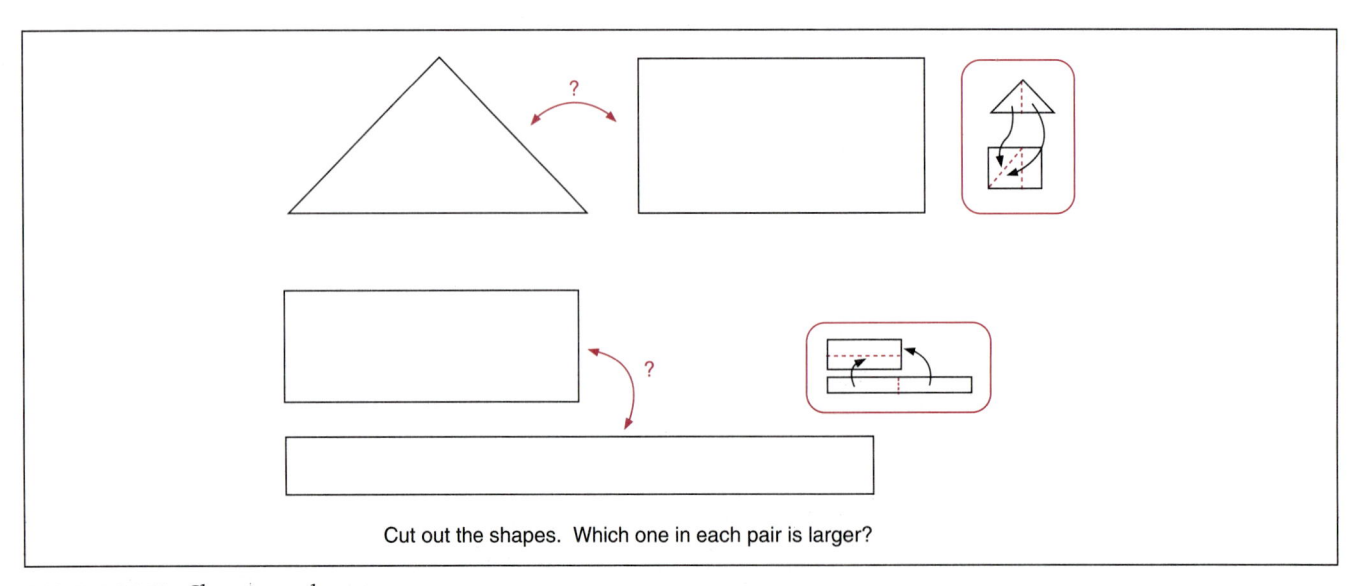

Cut out the shapes. Which one in each pair is larger?

FIGURE 16.11: *Changing a shape to compare area*

or hallway for very large areas. Small regions can be duplicated on paper so that students can work at their desks. Odd shapes and curved surfaces provide more challenge and interest. The surfaces of a watermelon or of the side of the wastebasket are quite difficult but fun to explore.

For length measurements, it is only the last unit that may not completely fit, but in area measurements there may be lots of units that only partially fit. You may wish to begin with shapes you have designed so the units fit. That is, build a shape with units, and draw the outline. By third or fourth grade, students should begin to wrestle with counting partial units and mentally putting together two or more partial units to count as one (Figure 16.12).

Your objective in the beginning is to develop the idea that area is a *measure of covering*. Do not introduce formulas for areas. Simply have the students fill the shapes and count the units. Be sure to include estimation before measuring (this is significantly more difficult than for length), use approximate language, and relate precision to the size of the units in the same manner as with length. When two or more groups measure the same region, different measures are very likely. Discuss these differences with the children, and point to the difficulties involved in making estimates around the edges. Avoid the idea that there is some precise right answer that everyone should have gotten. The following activities are additional suggestions for area measurement.

16.11 SMALLER TO LARGER

Present a series of quite different shapes with only slightly different areas. Let students predict the order of the shapes from small to large and then use units to determine the correct order. Students or groups should each decide on a predicted order of the shapes by size and record this. Then they should be challenged to determine the correct order using any methods and units they wish. The determination of the "correct" order is left to the students, and a profitable class discussion can be held concerning different approaches to determining the order of the shapes.

16.12 SAME-SIZE RECTANGLE

Have students make a rectangle that has the same size as another shape that you have given them. The given shape may be a blob or other irregular shape, or it may be a triangle or even a rectangle. The activity can be done with small shapes that can be drawn on paper or very large shapes drawn on the floor with tape or chalk. This activity works best when children work in groups. All groups in the class should be given the same shape to work with but no directions as to how to design their same-size rectangle. Almost certainly each group will come up with a rectangle of different dimensions, even if all have very good solutions. Groups should also be asked to explain why they believe their rectangle has the same area as the given shape. They should be allowed to use whatever materials they wish to prove the two shapes have the same area. This activity can be done before the area formula for a rectangle has been introduced. In fact, the area formula may not be useful at all.

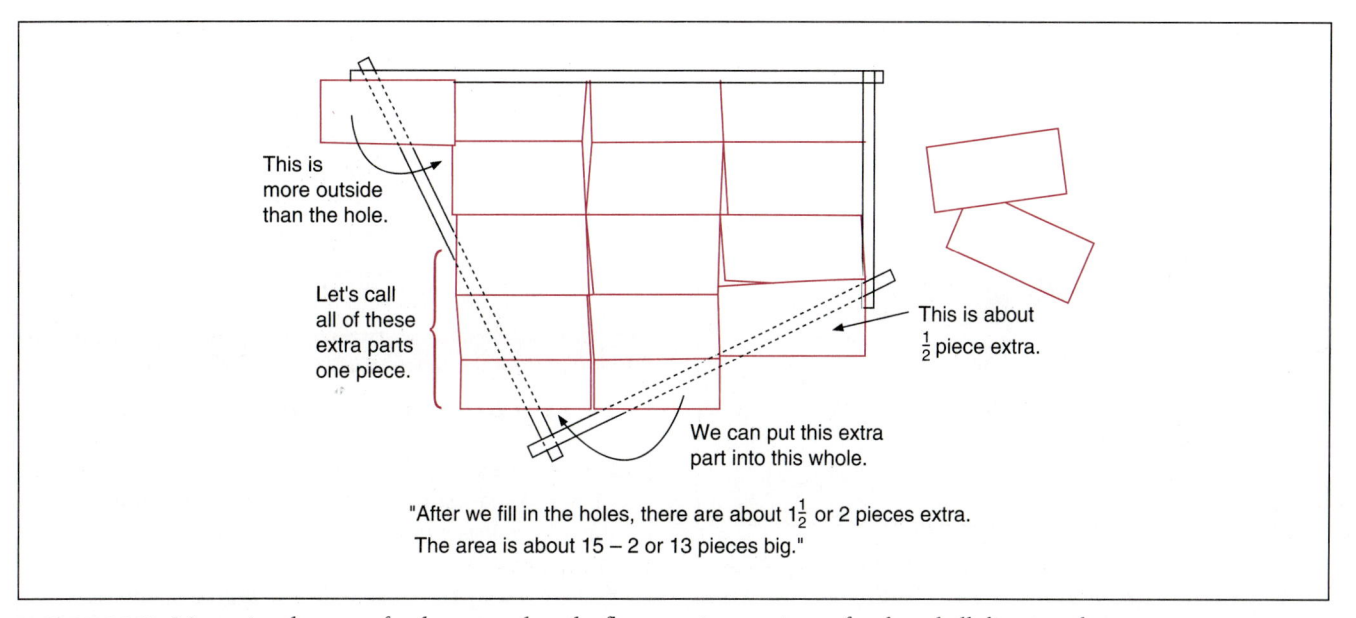

This is more outside than the hole.

Let's call all of these extra parts one piece.

This is about $\frac{1}{2}$ piece extra.

We can put this extra part into this whole.

"After we fill in the holes, there are about $1\frac{1}{2}$ or 2 pieces extra. The area is about 15 − 2 or 13 pieces big."

FIGURE 16.12: *Measuring the area of a shape taped on the floor—units are pieces of tagboard all the same shape.*

16.13 TWO AREA UNITS

Measure the same shape with two different but related units. As with length, have students predict the second measure after making the first. Do not forget to have them write an explanation of how they made their prediction as well as how they made the actual units.

Using Grids

With the exception of professional drafting equipment and computer methods, there really are no instruments designed for measuring areas. However, grids of various types can be thought of as a form of "area ruler." A grid of squares or triangles does exactly what a ruler does: It lays out the units for you. Square grids, isometric grids (equilateral triangles), and square grids with diagonal lines making smaller triangles are all available in Appendix B. All can be used for measuring areas with different units.

16.14 TRANSPARENT GRIDS

Make transparencies of any grid paper. Have students place the grid over a region to be measured and count the units inside. An alternative method is to trace around a region on a paper grid.

16.15 FLOOR GRIDS

For larger regions on the floor, make two lines with tape in an **L**, as in Figure 16.13. On the tape mark off appropriate units of length. Discuss how the two lines are the edges of a grid and can be used to help visualize imaginary squares without drawing them. Large shapes cut from butcher paper can be positioned into the corners of an **L** that has been drawn on the floor.

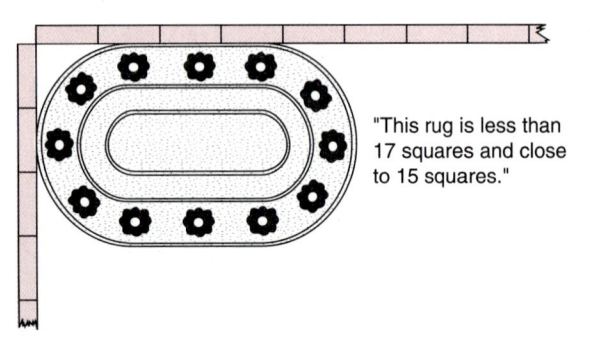

"This rug is less than 17 squares and close to 15 squares."

FIGURE 16.13: *Two tape rulers on the floor make an imaginary grid to "measure" areas of big shapes.*

Of course, the most useful grid is one of squares, since most standard units of area are squares derived from standard units of length (square inches, square centimeters, and the like). As children use square grids to determine areas of rectangular shapes, many will begin to see that the product of the numbers of rows times the numbers of columns is an easy way to count the squares. The **L** approach (which can also be used for small shapes on paper) provides a hint at using lengths of sides to determine areas.

MEASURING VOLUME AND CAPACITY

Volume and *capacity* are both terms for measures of the "size" of three-dimensional regions. Standard units of volume are expressed in terms of length units, such as cubic inches or cubic centimeters. Capacity units are generally applied to liquids or the containers that hold liquids. Standard capacity units include quarts and gallons, liters and milliliters.

Comparison Activities

Most solid shapes and containers must be compared indirectly. By far the easiest comparisons are made between containers that can be filled with something that is then poured into another container.

Young children should have lots of direct experiences with comparing the capacities of different containers. To this end, collect a large assortment of cans, small boxes, plastic jars, and other containers. Try to gather as many different shapes as possible. Also gather some plastic scoops and funnels. Rice or dried beans are good fillers to use. Sand and water are both considerably messier.

16.16 CAPACITY SORT

Sort containers by volume. Select one container as the standard, and, by pouring fillers from this to other containers, determine which hold less, more, and about the same as the standard. For each container, have students mark a slip of paper *more, less,* or *about the same* before they test it and then mark the slip again with the actual result.

16.17 CAPACITY LINE-UP

Given a series of five or six containers of different sizes and shapes, have students attempt to order them from least volume to most. This can be quite challenging. Again, do not provide answers. Let students work in groups to come up with a solution and also explain how they arrived at it. If there are differences between groups, this provides a good opportunity for discussion.

Solids such as a rock, a ball, a block, an apple, or an eggplant can also be compared in terms of their volumes, but it is a bit more difficult. Some method of displacement must be used. One of several approaches is to use a container that will hold each of the items to be compared. Place an object in the container and fill with rice to a level that will be above all of the objects. Mark this level on the container. Then remove the object and mark the new level. The difference is equal to the volume of the object. Each new object should be buried in the container and the rice filled to the same initial level. If students are challenged to devise their own methods, the problem-solving nature of the activity is enhanced and discussions around solution methods will be interesting and profitable.

◆ Using Units of Volume

There are two types of units that can be used to measure volume and capacity: solid units and containers. Solid units are things like wooden cubes or old tennis balls that can actually be used to fill the container being measured. The other type of unit model is a small container that is filled and poured repeatedly into the container being measured. The following are a few examples of units that you might want to collect.

- Thimbles, plastic caps, and liquid medicine cups are all good for very small units.
- Plastic jars and containers of almost any size can serve as a unit.
- Wooden cubic blocks or blocks of any shape can be units as long as you have a lot of the same size.
- Styrofoam packing peanuts, walnuts, or even marbles can be used, even though they do not pack perfectly. They still produce conceptual measures of volume.
- For large containers, borrow some big cardboard or wooden blocks from the kindergarten room.
- A large sack of old tennis balls might be collected from parents or from tennis pro shops.

Measuring activities for volume and capacity are similar to those for length and area. Estimation of volumes is a lot more fun as a class activity because it is much more difficult. Finding ways to measure large containers such as a large cardboard carton in terms of a relatively small container-type unit can be an excellent challenge to present to groups of fourth or fifth graders. This can be done long before volume formulas are developed and can even be done with unusual containers such as a wastebasket with slanted sides.

When measuring with solid units such as cubes or balls, it is very difficult to use only one and iterate it, as can be done with length or volume. However, a worthwhile challenge is to determine the volumes of containers given only enough units to cover the bottom. Do not forget very large volumes such as your room. How many cubic meters or how many basketballs will the room hold? Be sure to make these group projects where children devise their methods and provide written explanations and drawings.

◆ Making and Using Measuring Cups

Instruments for measuring capacity are generally used for small amounts of liquids or pourable materials such as rice or sand. These tools are commonly found in kitchens and laboratories. As with other instruments, if children work at making their own, they are likely to develop a better understanding of the units and the approach to the measuring process.

16.18 MAKE A MEASURING CUP

A measuring cup can be made easily by using a small container as a unit. Select a large transparent container for the cup and a small container for a unit. Fill the unit with beans or rice, empty it into the large container, and make a mark indicating the level. Repeat until the cup is nearly full. If the unit is quite small, marks may only be necessary after every 5 or 10 units. Numbers need not be written on the container for every marking. Students frequently have difficulties reading scales in which not every mark is labeled. This is an opportunity to help them understand how to interpret lines on a real measuring cup that are not labeled.

Students should use their measuring cup and compare the measures with those made by directly filling the container from the unit. The cup is very likely to produce errors due to inaccurate markings. This is a good opportunity to point out that measuring instruments themselves can be a source of error in measurement. The more accurately made the instrument is, and the finer the calibration, the less error will be attributed to that source.

MEASURING MASS AND WEIGHT

Weight is a measure of the pull or force of gravity on an object. *Mass* is the amount of matter in an object and technically a measure of the force needed to accelerate it. On the moon, where gravity is much less than on earth, an object has a smaller weight but the identical mass as it has on earth. For practical purposes, on the surface of the earth, the measures of mass and weight will be about the same. In this chapter the terms *weight* and *mass* will be used interchangeably.

◆ Making Comparisons

The most conceptual way to compare the weights of two objects is to hold one in each hand, extend your arms, and experience the relative downward pull on each. This is an effective way of communicating to a very young child what "heavier" or "weighs more" means. This personal experience can then be transferred to one of two basic types of scales which can then be used to make other comparisons.

Children should first use their hands to decide which of two objects is heavier. When they then place these in the two pans of a balance, the pan that goes down can be understood to hold the heavier object. Even a relatively simple balance will detect very small differences. If two objects are placed one at a time in a spring scale, the heavier object pulls the pan down further. Both balances and spring scales have real value in the classroom. Figure 16.14 shows a homemade version of each. Simple scales of each type are available through school-supply catalogs. (Technically, spring scales measure weight, and balance scales measure mass. Why?)

With either scale, sorting and ordering tasks are possible with very young children. For older children, comparison activities for weight are not necessary. (Why?)

◆ Using Units of Weight or Mass

Any collection of uniform objects with the same mass can serve as weight units. For very light objects, wooden or plastic cubes are quite reasonable. Very large metal washers found in hardware stores are effective for weighing slightly heavier objects. You will need to rely on standard weights to weigh things as heavy as a kilogram or more.

16.19 SCALES WITH UNITS

In a balance scale, the object is placed on one pan and weights in the other until the two pans are balanced. In a spring scale, first place the object in and mark the position of the pan on a piece of paper taped behind the pan. Remove the object and place just enough weights in the pan to pull it down to the same level. Discuss with the children how equal weights will pull the spring or rubber band with the same force.

While the concept of heavier and lighter is learned rather early, the notion of units of weight or mass is a bit more mysterious. At any grade level even a brief experience with informal unit weights is a good preparation for a discussion of standard units and scales.

◆ Making and Using a Scale

Most scales that we use in our daily lives produce a number when an object is placed on or in it. There are no visible unit weights. How does the scale come up with the right number? By making a scale that can give a numeric result without recourse to any units, children can see that scales are not all that mysterious.

16.20 SPRING SCALE CALIBRATION

Students can use informal weight units and calibrate a simple rubber-band scale like the one in Figure 16.14 to help understand how dial scales work. Mount the scale with a piece of paper behind it, and place weights in the pan. After every five weights make a mark on the paper. The resulting marks correspond to the markings around the dial of the standard scale. The pan serves as the pointer. In the dial

Two-pan balance

Rubber-band scales

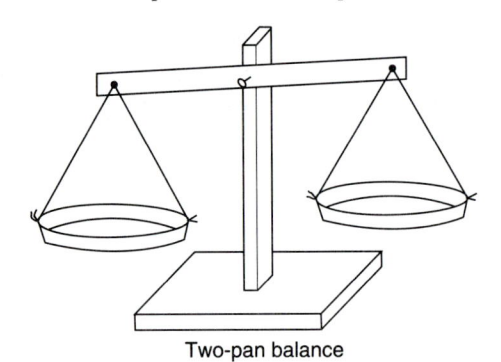

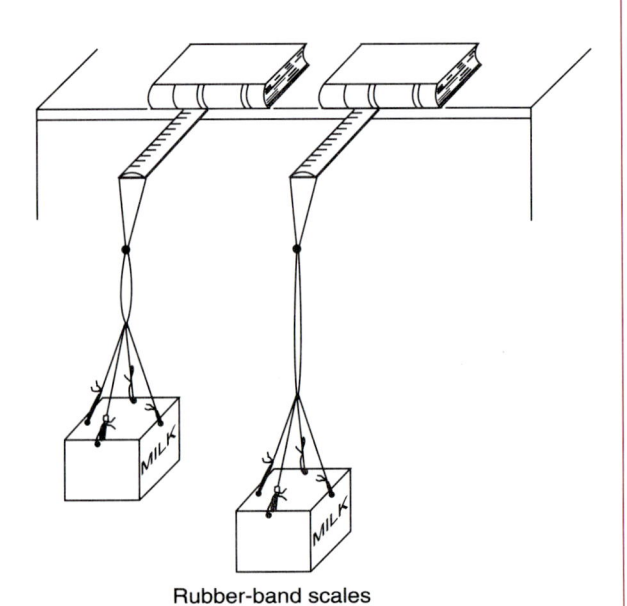

FIGURE 16.14: *Two types of scales*

scale, the downward movement of the pan mechanically causes the dial to turn.

After making the markings on the rubber-band scale, use it to measure objects. Then measure them using the same units, but on a two-pan balance. The pan balance is likely to be more sensitive.

The principal value of the last activity is seeing how scales are made. Even digital readout scales are based on the same basic principle.

MEASURING ANGLES

Comparing Angles

The attribute of an angle that is measured or compared might be called the "spread of the angle's rays," although this is somewhat ambiguous. Angles are composed of two rays that are infinite in length. The only difference in their shape is how widely or narrowly the two rays are spread apart. (There is no word like *length* or *area* for angular measure.) The relatively simple exercise of making a direct comparison of two angles can help children conceptualize this attribute. Two angles can be compared by tracing one and placing it over the other, as in Figure 16.15. Be sure to have students compare angles where the angle sides are of different lengths. A wide angle with short sides may seem smaller than a narrow angle with long sides. This is a common misconception among students.

It is not necessary to spend a lot of time with these activities. As soon as students can tell the difference between a large angle and a small one, regardless of the length of the sides, you can move on to measuring angles.

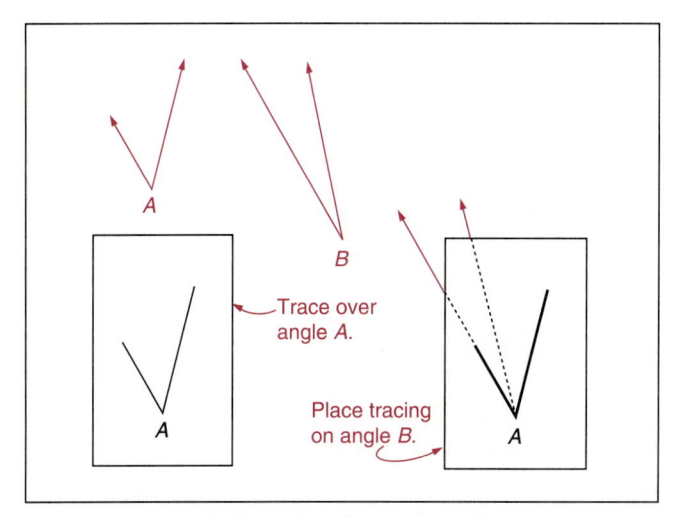

FIGURE 16.15: *Which angle is largest? Smallest? Use tracings to compare.*

Using Units of Angular Measure

A unit for measuring an angle must be an angle. Nothing else has the same attribute of "spread" that we want to measure. (Contrary to popular opinion, it is not necessary to use degrees to measure angles.)

16.21 A UNIT ANGLE

Give each student an index card or small piece of tagboard. Have them draw a narrow angle on the tagboard, using a straightedge, and then cut it out. The resulting wedge can then be used as a unit of angular measure by counting the number that will fit in a given angle. Pass out a worksheet with assorted angles on it and have them use their unit to measure them. Since everyone makes their own unit angles, the results will be different and can be discussed in terms of the size of their units.

The last activity illustrates that measuring an angle is the same as measuring length or area. Unit angles are used to fill or cover the spread of an angle just as unit lengths fill or cover a length. Once this concept is well understood, it is reasonable to move on to measuring instruments.

Making a Protractor

The protractor is one of the most poorly understood measuring instruments found in schools, and yet it is commonly and frequently used. Part of the difficulty arises because the units (degrees) are so very small. It would be physically impossible to cut out a single degree and use it as in activity 16.21. Another problem is due to the fact that there are no visible angles showing; there are only little marks around the outside edge of the protractor. Finally, the numbering that appears on most protractors is double, running both clockwise and counterclockwise along the marked edges—"Which numbers do I use?" By making a protractor with a large unit angle, all of these mysterious features can be understood. Then, a careful comparison with a standard protractor will permit that instrument to be used with understanding.

16.22 THE WAXED-PAPER PROTRACTOR

Tear off about a foot of ordinary waxed paper for each student. Have the students fold the paper in half and crease the fold tightly. Fold in half again so that the folded edges match. Repeat this two more times, each time bringing the folded edges together and creasing tightly. Cut or tear off the resulting wedge shape about 4 or 5 inches from the vertex and

306 16 / DEVELOPING MEASUREMENT CONCEPTS

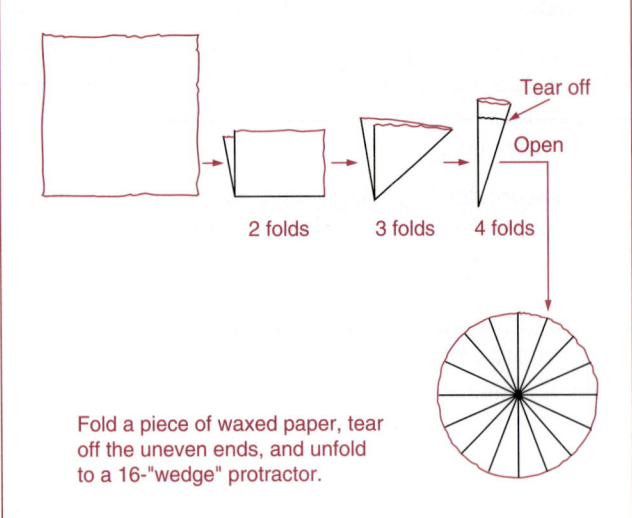

Fold a piece of waxed paper, tear off the uneven ends, and unfold to a 16-"wedge" protractor.

2 folds → 3 folds → 4 folds → Tear off / Open

FIGURE 16.16: *Making a waxed-paper protractor*

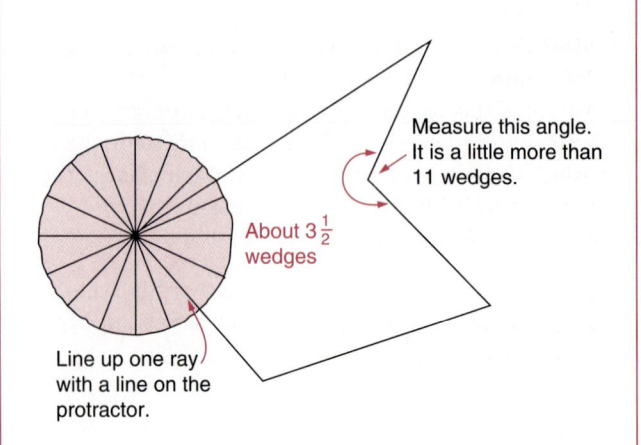

Measure this angle. It is a little more than 11 wedges.

About $3\frac{1}{2}$ wedges

Line up one ray with a line on the protractor.

FIGURE 16.17: *Measuring inside angles in a polygon using a waxed-paper protractor.*

unfold. If done correctly, there will be 16 angles surrounding the center, as in Figure 16.16. This serves as an excellent protractor with a unit angle that is one-eighth of a straight angle. It is sufficiently transparent that it can be placed over an angle on paper, on the blackboard, or on the overhead projector to measure angles, as shown in Figure 16.17. Reasonable estimates of angle measures can be made with the waxed-paper protractor on drawings as small as the one in Figure 16.17. In that figure, one angle is measured for you. Use a waxed-paper protractor and measure the other four angles as carefully as possible. Use fractional estimates. Your sum for all five interior angles should be very close to 24 wedges. There are two possible ways to get the measure of the angle indicated with the arrow. How would you measure that angle if your protractor was only a half circle instead of a full circle?

waxed-paper wedges

degrees

tagboard protractor

standard protractor

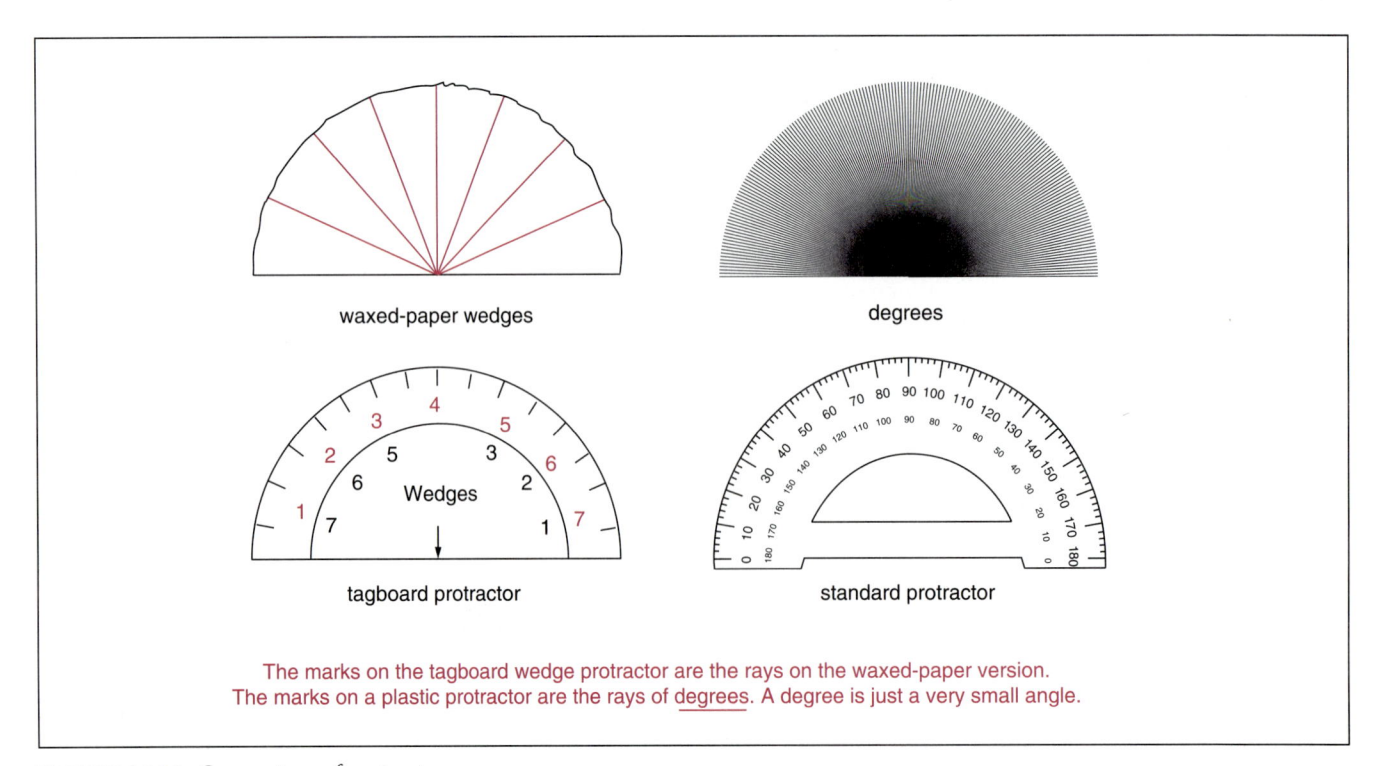

The marks on the tagboard wedge protractor are the rays on the waxed-paper version. The marks on a plastic protractor are the rays of <u>degrees</u>. A degree is just a very small angle.

FIGURE 16.18: *Comparison of protractors*

The waxed-paper protractor makes it quite clear how a protractor fits unit angles into an angle for measurement. When measuring angles, students can easily estimate halves, thirds, or fourths of a "wedge," a possible name for this informal unit angle. This is sufficiently accurate to measure, for example, the interior angles of a polygon and discover the usual relationship between number of sides and sum of the interior angles. For a triangle, the sum is 8 wedges or 8^w. For a quadrilateral, the sum is 16^w. And in general, the sum for an n-sided polygon is $(n - 2) \times 8^w$. The superscript w is a forerunner of the degree symbol.

Figure 16.18 (see p. 306) illustrates how a tagboard semicircle can be made into a protractor to measure angles in wedges. This tagboard version is a bit closer to a standard protractor, since the rays do not extend down to the vertex and the markings are numbered in two directions. Both of these features are confusing to students who begin angle measurement with small plastic protractors. The only difference between this protractor and a standard one is the size of the unit angle. The standard unit angle is the *degree*. The degree is simply a very small angle (Figure 16.18). A standard protractor is not very helpful in teaching the meaning of a degree. But an analogy between wedges and degrees and between these two protractors is a very useful approach.

INTRODUCING STANDARD UNITS

As was pointed out earlier in the chapter, there are a number of reasons for using nonstandard units while teaching measurement. However, standard units of measure remain very important for children to know about. Measurement sense demands that children are familiar with the more common measurement units, that they can make estimates in terms of these units and can meaningfully interpret measures depicted with standard units.

Perhaps the biggest error that is made with respect to measurement instruction is the failure to recognize and separate two types of objectives: first, understanding the meaning and technique of measuring a particular attribute, and second, learning about the standard units that are commonly used to measure that attribute. It is clear that these two objectives can be developed separately, that measurement understanding is necessary to understand the use of any units including standard units, and that when both objectives are attempted together, confusion is very likely. Reread the list of reasons for using nonstandard units, and be very clear about why you would use them. It is only when students are comfortable with measurement of an attribute that they can focus on things like cups and quarts or the number of inches in a foot or feet in a yard or have a feel for grams and kilograms. While both domains of knowledge are important, it may be useful to keep the objectives separate when planning your lessons.

INSTRUCTIONAL GOALS

Three broad goals relative to standard units of measure can be identified:

1. **Familiarity with the unit.** Familiarity means that when a commonly used unit is encountered there is a basic idea of its size and what it measures. Without this familiarity, measurement sense is impossible. It is more important to know about how much one liter of water is or to be able to estimate a shelf as five feet long than to have the ability to measure either of these accurately.

2. **Ability to select an appropriate unit.** Related to unit familiarity is knowing what is a reasonable unit of measure in a given situation. The selection of an appropriate unit is also a matter of required precision. (Would you measure your lawn to purchase grass seed with the same precision as you would use in measuring a window to buy a pane of glass?) Students need practice in using common sense and personal judgment in the selection of appropriate standard units.

3. **Knowledge of a *few important* relationships between units.** The emphasis should be kept to those relationships that are commonly used, such as inches, feet, and yards or milliliters and liters. Tedious conversion exercises do little to enhance measurement sense. The goal of unit relationships is the least important of all measurement objectives.

Developing Unit Familiarity

The importance of familiarity with standard units of measure is highlighted by the reluctance (even fear) on the part of a majority of U.S. citizens to adopt the metric system. The average adult has developed a "feel" for, a familiarity with, those units he or she has used throughout life: inch, foot, mile, pound, cup, gallon. These *customary* units are certainly not easier to use in any way. They simply are known, while *metric* units are not. Helping students develop that same sense of familiarity with the most frequently used units of measure, from *both* the metric and the customary systems, is an important goal of instruction.

Two simple types of activities can be utilized to help develop familiarity with most standard unit:

1. comparison activities with a focus on a single unit, and

2. development of familiar and personal referents or benchmarks for single units or easy multiples of units.

16.23 ABOUT ONE UNIT

Give students a model of a standard unit, and have them search for things that measure about the same as that one unit. For example, to develop familiarity with the meter, give students a piece of rope 1 meter long. Have them make lists of things that are about 1 meter. Things that are a little less (or more) or twice as long (or half as long) should be noted in separate lists. Encourage students to find familiar items that are in their daily lives. In the case of lengths, be sure to include circular lengths. Later, students can try to predict if a given object is more, less, or close to 1 meter.

The same activity can be done with other unit lengths. Parents can be enlisted to help students find familiar distances that are about 1 mile or about 1 kilometer. Suggest in a letter that they check the distances around the neighborhood, to the school or shopping center, or along other frequently traveled paths.

For capacities, students need a container that holds or has a marking for a single unit. They should then find other containers that hold about as much, more, and less as in the length example. Remember that the shapes of containers can be every deceptive when estimating their capacity. Have students look for examples at home as well as at school.

For the standard weights of gram, kilogram, ounce, and pound, students should have ample opportunity to compare objects on a two-pan balance with single copies of these units. It may be more effective to work with 10 grams or 5 ounces. Students can be encouraged to bring in familiar objects from home to compare on the classroom scale.

Standard areas are in terms of lengths such as square inches or square feet so that familiarity with lengths is more important. Familiarity with a single degree is not as important as some idea of 30, 45, 60, and 90 degrees. The turtle graphics of Logo provide an excellent avenue for degree familiarity. (Logo is a very accessible computer language with a simple graphics capability that even first-grade children can control. This language is discussed more fully in Chapter 21.)

A second and somewhat different approach to unit familiarity is to begin with very familiar items and use their measures as references or benchmarks. A doorway is a bit more than two meters. A bag of flour is a good reference for 5 pounds. Your bedroom may be about 10 feet long. A paper clip is about a gram and about 1 centimeter wide. A gallon of milk weighs a little less than 4 kilograms.

16.24 FAMILIAR REFERENCES

For each unit of measure you wish to focus on, have students make a list of at least five things that they are very familiar with and have them measure them using that unit. For lengths, encourage them to include long and short things; for weight, to find both light and heavy things; and so on. The measures should be rounded off to nice whole numbers. Discuss lists in class so that different ideas are shared.

Of special interest for length are benchmarks found on our own bodies or using our bodies. These become quite familiar over time and can be used as approximate versions of units or rulers in many situations. Perhaps you know some personal reference that you use for an inch, a foot, or a yard. Even though young children grow quite rapidly, it is useful for them to know the approximate lengths that they carry around with them.

16.25 PERSONAL BENCHMARKS

Measure your body. About how long is your foot, your stride, your hand span (stretched and with fingers together); the width of a finger; your arm span (finger to finger and finger to nose); your height to waist, to shoulder, and to head; the distance around your wrist and around your waist? Perhaps you cannot remember all of these, but some may prove to be very useful benchmarks, and some may be excellent models for single units. (The average fingernail width is about 1 cm, and most people can find a 10-cm width in some way with their hands.)

To help remember these references, they must be used in activities where lengths, volumes, and so on are compared to the benchmarks in order to estimate measurements.

◆ What Unit Is Appropriate

Should the room be measured in feet or inches? Should the concrete blocks be weighed in grams or in kilograms? The answers to questions such as these involve more than simply knowing how big the units are, although that is certainly required. Another consideration involves the need for accuracy. If you were measuring your wall in order to cut a piece of molding or woodwork to fit, you would need to measure it very accurately. The smallest unit would be an inch or a centimeter, and you would also use small fractional parts. But if you were determining how many 8-foot molding strips to buy, the nearest foot would probably be sufficient.

16.26 GUESS THE UNIT

Find examples of measurements of all types in newspapers, signs, or other everyday occurrences. Present the context and situation to students and see if they can predict what units of measure are used. There may well be disagreement. Have students discuss their choices.

Developing Relationships between Units

The number of inches in a foot or yard or the number of cups in a quart is the type of information that must eventually be committed to memory. Practice with conversions is something that lends itself well to pencil-and-paper work. This procedural aspect of unit familiarity is another example of objectives that have, in the past, been overworked in the curriculum largely due to the ease of testing rather than the need to know. Conversions between one unit and another should not be overemphasized.

Many students may know that there are 16 ounces in 1 pound, but when they try to determine how many pounds are in 90 ounces they get confused over which operation to use. This is partly a matter of understanding the meanings of operations. However, simple common sense can help a lot also. In this example, since pounds are bigger or heavier than ounces, it is reasonable to end up with fewer pounds than ounces.

The customary system involves an unfortunate variety of conversion factors, and as long as the United States continues to use it, teachers will have to deal with helping children commit the most commonly used factors to memory.

Exact conversions between the metric and the customary system should never be done and are almost nonexistent in textbooks today. From the standpoint of familiarity with these systems, "soft" or "friendly" conversions may even be useful as long as we live in a country that seems bent on having two systems of measurement. For example, a liter is a "gulp more" than a quart, and a meter is a "bit longer" than a yard. The same is true for benchmarks. One-hundred meters is about 1 football field plus 1 end zone, or about 110 yards.

IMPORTANT STANDARD UNITS

Both the customary and the metric system include many units that are not used for everyday living. Table 16.1 lists those units that are most common.

The Metric System

Unit familiarity with the popularly used units should be the principal focus of almost all instruction with standard units.

Before students have developed a full understanding of decimal notation, there is very little advantage in teaching students all of the very nice relationships in the metric system. While the customary system frequently mixes units (3 pounds, 6 ounces or 6 ft., 2 in.) the standard version of the metric system insists on the use of a single unit for each measure. To adhere to this rubric of the metric system means that a primary-grade child should report a measure in the smallest unit used since decimal notation is not meaningful. For example, a length would be recorded as 235 cm instead of 2.35 m or 2 m and 35 cm. How important this rule is for the second grade may be open to debate.

	METRIC SYSTEM	CUSTOMARY SYSTEM
LENGTH	millimeter	inch
	centimeter	foot
	meter	yard
	kilometer	mile
AREA	square centimeter	square inch
	square meter	square foot
		square yard
VOLUME	cubic centimeter	cubic inch
	cubic meter	cubic foot
		cubic yard
CAPACITY	milliliter	ounce*
	liter	cup
		quart
		gallon
WEIGHT	gram	ounce*
	kilogram	pound
	metric ton	ton

*In the U.S. Customary System, the term *ounce* refers to a weight or *avoirdupois* unit, 16 of which make a pound, and also a volume or capacity unit, 8 of which make a cup. While the two units have the same name, they are not related.

TABLE 16.1

Perhaps one of the worst errors in metric measurement curriculum is having students "move decimal points" to convert from one metric unit to another prior to a complete development of decimal notation. Faced with confusion, children memorize rules about moving decimals so many places this way or that, and the focus becomes rules and right answers.

As children begin to appreciate the structure of decimal notation, the metric system can and should be developed with all seven places: three prefixes for smaller units (deci-, centi-, milli-) and three for larger units (deka-, hecto-, kilo-). With decimal knowledge and familiarity with the basic and popularly used units, the complete decimal system is easy to learn.

Making conversions within the metric system can be approached in two related ways. As we saw earlier in Figure 14.6, a place-value chart is used to give a metric name to each of seven consecutive places. If it is understood that the decimal point is always identifies the unit position, then given any metric measurement, each digit is in a position with a metric name. The decimal point can be repositioned if the name of the unit is changed accordingly. For example, in the measure 17.238 kg, the decimal indicates that the 7 is in the unit position. The label "kg" indicates that the name of the position is kilograms. Therefore, the 2 is in the hectogram position, the 3 in the dekagram position, and the 8 in the gram position. Repositioning the decimal to indicate grams as the unit makes the same measure read 17,238 g or 17,238.0 g.

An alternate rationale is to think of decimal point shifting as multiplying or dividing by powers of 10. In the example above, since there are 1000 grams in a kilogram, change to grams by multiplying by 1000 or shift the decimal three places to the right.

It should be emphasized once again that unit conversion is perhaps the least important part of learning the metric system or any standard system. It simply is a skill that is not used that frequently.

◆ The Customary System

The "familiar" system of units is technically known as the U.S. Customary System. It is difficult because it lacks any common structure or common conversion ratios. After an attempt by schools during the 1970s to go completely metric, most schools and textbooks have resigned themselves to teaching both metric and customary systems.

Conversions of units within the customary system is difficult for children for two reasons. There are more conversion ratios to memorize, and they are not conveniently related to the decimal system. For example, many children will add 3 feet 8 inches to 5 feet 6 inches and get 9 feet 4 inches instead of 9 feet 2 inches. (Why?) Similar difficulties occur with all conversions in the customary system. Once again, while it is important to know how many inches in a foot and a few other relationships that we use regularly, an overemphasis on the type of conversions indicated by this last example is unwarranted.

ESTIMATING MEASURES

Measurement estimation is the process of using mental and visual information to measure or make comparisons without the use of measuring instruments. About how long is the fence? Find a one gallon container for the juice. Will this paper cover the box? Each of these involves estimation.

Measurement estimation is a very practical skill. Almost every day we make estimates of measures. Do I have enough sugar to make the cookies? Will the car fit in that space? Can you throw the ball 50 feet? Is this suitcase over the weight limit?

Besides its value outside the classroom, it was pointed out earlier that estimation in measurement activities helps students focus on the attribute being measured, adds intrinsic motivation, and helps develop familiarity with standard units. Therefore, measurement estimation both helps our measurement instruction and provides students with a valuable like skill.

TECHNIQUES OF MEASUREMENT ESTIMATION

Like computational estimation, specific strategies or approaches exist for making measurement estimations. These strategies can be identified and taught specifically.

1. Develop and use benchmarks or referents for important units. (This strategy was also mentioned as a good way to help students develop a familiarity with units.) Students should have a good referent for single units and also useful multiples of standard units. Referents or benchmarks for 1, 5, 10, and perhaps, 100 pounds (or something near these amounts) might be useful, but there is little value in a referent for 500 pounds. On the other hand, a referent for 500 milliliters is very useful. These benchmarks can then be compared mentally to objects being estimated. That tree is about as tall as four doorways or between 8 and 9 meters.

2. Utilize a "chunking" procedure when appropriate. Figure 16.19 is a common example. It may be easier to estimate the shorter chunks along the length of the wall than to estimate the whole length as one. The weight of a stack of books is easier if some estimate is given to an "average" book.

3. Use subdivisions. This is a similar strategy to chunking. However, here the chunks are imposed on the object by the estimator. For example, if the wall length to be estimated has no useful chunks, it can be mentally divided in half and then in fourths or even eighths by repeated halving until a more manageable length is arrived at. Length, volume, and area measurements all lend themselves to this technique.

4. Iterate a unit either mentally or physically. For length, area, and volume, it is sometimes easy to visually mark off single units. You might use your hands, make markings or folds, or use other methods of keeping track as you go. For length, it is especially useful to use a body measure as a unit and iterate with that. If you know, for example, that your stride is about $\frac{3}{4}$ meter, you can walk off a length and then multiply to get an estimate. Hand and finger widths are useful for shorter measures.

MEASUREMENT ESTIMATION ACTIVITIES

Tips for Teaching Estimation

Each of the strategies listed in the last section should be taught directly and discussed with students. But the best approach to improving their estimation skills is to have them do a lot of estimating. As you conduct estimation activities, keep the following tips in mind:

1. Help students learn strategies by having them use a specified approach. Later activities should permit students to choose whatever techniques they wish.

2. Periodically discuss how different students made their estimates. This will help students understand that there is no single right way to estimate and also remind them of different approaches that are useful.

3. Accept a range of estimates. Think in relative terms about what is a good estimate. Within 10% for length is quite good. Even 30% off may be reasonable for weights or volumes.

4. Sometimes have students give a range of measures that they believe includes the actual measure. This not only is a practical approach in real life but also helps focus on the approximate nature of estimation.

5. Let students measure to check estimates. However, it is only necessary that one or two students or one team do the measurement if the focus of the activity is on estimation. If all students are required to follow their estimates with a measure, they may correctly wonder why they bothered estimating.

6. Make measurement estimation an ongoing activity during the year. A daily measurement to estimate can be posted on the bulletin board. Students can turn in their estimates on paper and discuss them quickly in a five-minute period. Older students can even be given the task of making up the things to estimate with a team assigned this task each week.

7. Make an effort to include estimations of all attributes. It is easy to get carried away with length and forget about area, volume, weight, and angles.

Activities

Estimation activities need not be elaborate or involve a lot of effort on your part. Any measurement activity can have an "estimate first" component. For more emphasis on the process of estimation itself, simply think of things that can be estimated, and have students estimate. Here are a few suggestions to think about.

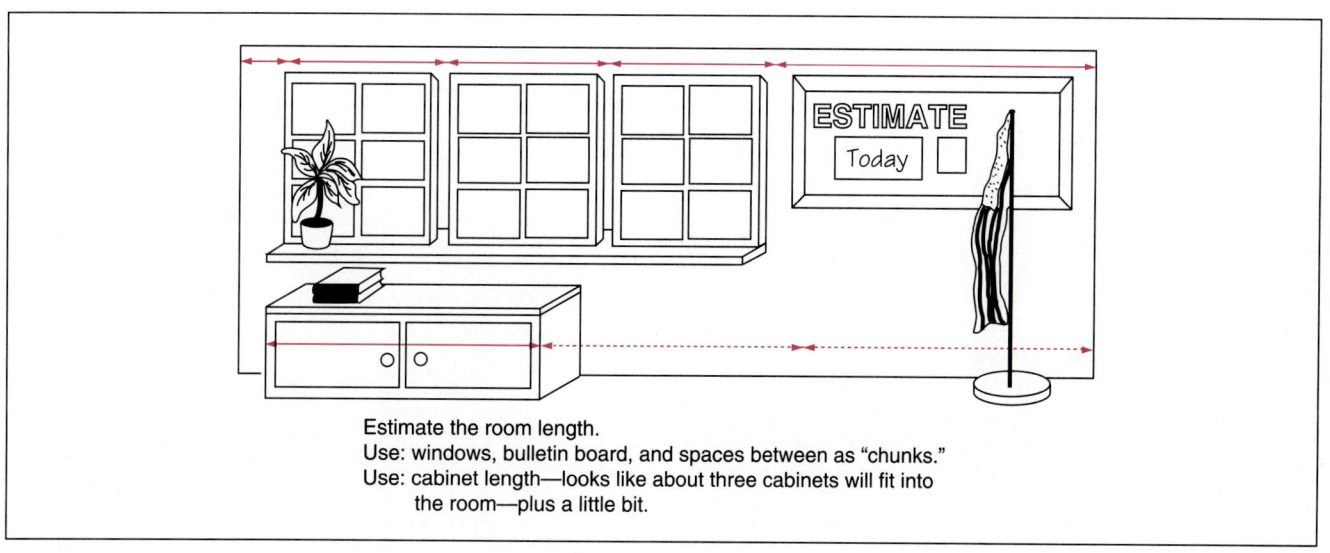

Estimate the room length.
Use: windows, bulletin board, and spaces between as "chunks."
Use: cabinet length—looks like about three cabinets will fit into the room—plus a little bit.

FIGURE 16.19: *Measuring by chunking*

312 16 / DEVELOPING MEASUREMENT CONCEPTS

16.27 ESTIMATION QUICKIE

Select a single object such as a box, a watermelon, a jar, or even the principal. Each day select a different attribute or dimension to estimate. For a watermelon, for example, students can estimate its length, girth, weight, volume, and even its surface area.

16.28 ESTIMATION SCAVENGER HUNT

Conduct measurement scavenger hunts. Give teams a list of measurements, and have them find things that are close to having those measurements. Permit no measuring instruments. For example, a list might include

a length of 3.5 meters

something that weighs more than 1 kg but less than 2 kg

a container that holds about 200 milliliters

Let students suggest how to judge results in terms of accuracy.

16.29 E-M-E SEQUENCES

Use estimate-measure-estimate sequences (Lindquist, 1987). Select pairs of objects to estimate that are somehow related or close in measure but not the same. Have students estimate the measure of the first and check by measuring. Then have them estimate the second. Some examples of pairs are

width of a window, width of wall
volume of coffee mug, volume of a pitcher
distance between eyes, width of head
weight of handful of marbles, weight of bag of marbles

This type of activity is a good way to help students understand how benchmarks are used in estimation.

DEVELOPING FORMULAS

Frequently measurement and geometry are presented in the same chapters of a textbook. The relationship between these two areas of mathematics is most evident in the development of formulas for measures of geometric figures. Formulas help us use easily made measures to determine indirectly some other measure that is not so easily found.

For example, it is easy to measure the three dimensions of a box with a ruler, but it is not easy to directly measure the volume of the same box. By use of a formula, the volume can be determined from the length measures.

Children should never use formulas without participating in the development of those formulas. Formulas for area and volume can and should all be developed by children. Developing the formulas and seeing how they are connected and interrelated is significantly more important than having children apply formulas blindly to a series of figures. Plugging numbers in formulas is little more than computational tedium and has little to do with measurement sense or geometry.

COMMON DIFFICULTIES

Many children become so encumbered with the use of formulas and rules that an understanding of what these formulas are all about becomes completely lost.

◆ Overemphasizing Formulas

Figure 16.20 shows results from the fourth NAEP (Kouba et al., 1988b) concerning area and perimeter. Even the seventh-grade results are disappointing. The most common errors involved computing area for perimeter and vice-versa. Notice the difference between questions C and D. The results

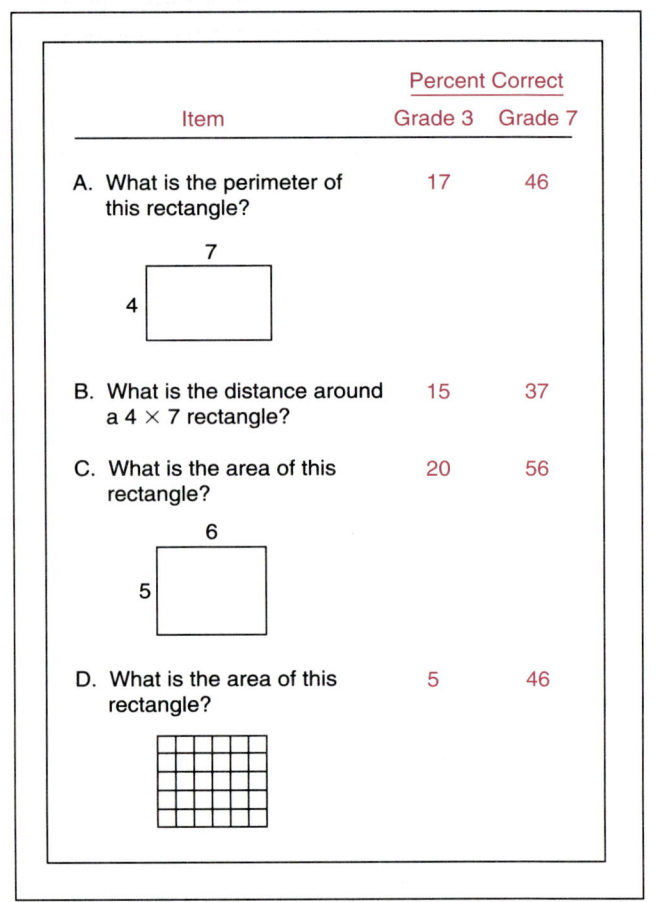

Item	Percent Correct Grade 3	Grade 7
A. What is the perimeter of this rectangle? (7 × 4)	17	46
B. What is the distance around a 4 × 7 rectangle?	15	37
C. What is the area of this rectangle? (6 × 5)	20	56
D. What is the area of this rectangle? (grid)	5	46

FIGURE 16.20. SOURCE: Data from the Fourth National Assessment of Educational Progress (NAEP), 1988.

suggest that students rely on formulas much more than on concepts. In item A, if measures for all four sides are given, the results improve dramatically. However, if all four sides are given in item C, the results go down even further. Can you explain this effect in terms of over-reliance on formulas? Premature use of formulas can also lead to the use of words without meaning. Notice the difference between items A and B in the figure and the corresponding results. Is "perimeter" just a formula? If not, why would the results be lower for item B? It is not uncommon even for adults to define area as length times width rather than as a measure of covering or the space inside. The tasks in Figure 16.21 become quite impossible with this formula definition of area.

◆ Height or Side

Another error that is commonly made when students use formulas is due to not conceptualizing the meaning of *height* in geometric figures, both two- and three-dimensional. The shapes in Figure 16.22 each have a slanted side and a height given. Students tend to confuse these two. *Any* side or flat surface of a figure can be called a *base* of the figure. For each base that a figure has, there is a corresponding height. If the figure were to slide into a room on its base, the height would be the height of the shortest door it could pass through without bending over: the perpendicular distance to the base. Children have a lot of early experiences with $L \times W$ formula for rectangles, in which the height is exactly the same as the length of a side. Perhaps this is the source of the confusion.

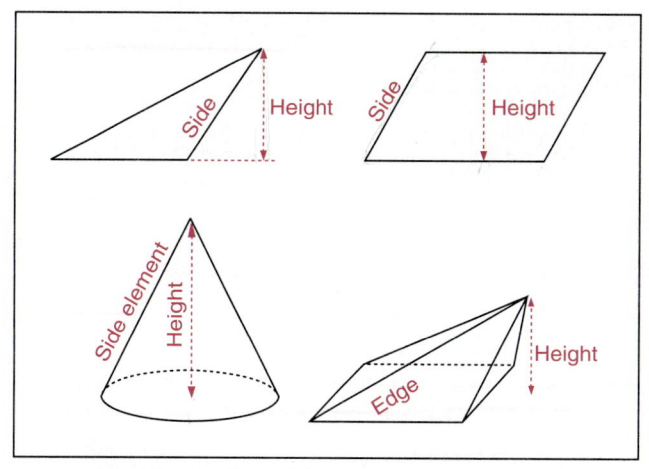

FIGURE 16.22: *Heights of figures are not always measured along an edge or surface.*

AREAS OF SIMPLE PLANE SHAPES

The development of area formulas is a fantastic opportunity to follow the spirit of the NCTM *Standards*. First, a problem-solving approach can meaningfully involve students and help them see that mathematics is a sense-making endeavor. Second, the connectedness of mathematics can be clearly seen. In this section you will see how all of the standard area formulas are intimately related and can be developed and learned as an integrated whole rather than a collection of isolated facts. An integrated approach to these formulas highlights mathematics as a science of pattern and order.

◆ Rectangles to Trapezoids

The formula for the area of a rectangle is one of the first that is developed and is usually given as $A = L \times W$ or *length times width*. Looking forward to other area formulas, an equivalent but more unifying idea might be $A = b \times h$ or *base times height*. The base-times-height formulation can be generalized to all parallelograms and is useful in developing the area formulas for triangles and trapezoids. Furthermore, the same approach can be extended to three dimensions where volumes of cylinders are given in terms of the *area of the base* times the height. Base times height, then, helps to connect a large family of formulas that otherwise must be mastered independently.

The following sequence of exercises to develop the area formula for a rectangle is illustrated in Figure 16.23 (p. 314).

Approach each step of this sequence with a problem-solving spirit: How can we figure this out?

1. Have students determine the areas of rectangles drawn on square grids or geoboards. Alternatively, have students draw rectangles that have specified

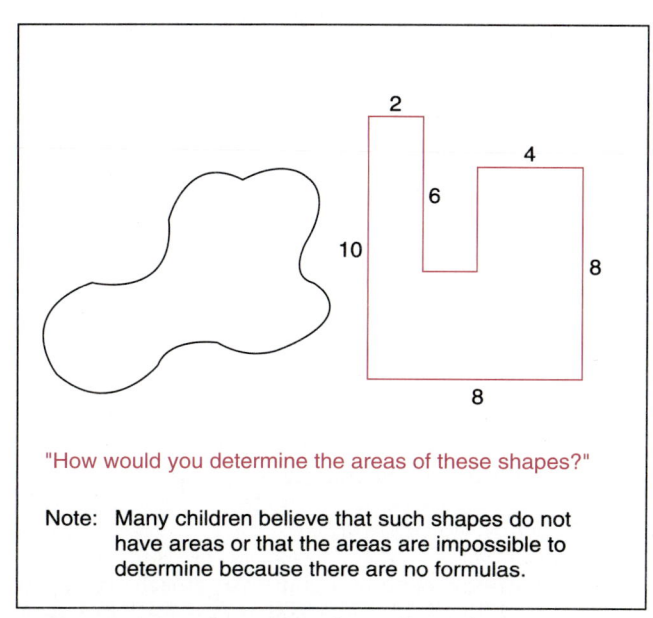

FIGURE 16.21: *Understanding the attribute of area*

Before formulas involving heights are discussed, children should discuss the meaning of height of a geometric figure. They should be able to identify where a height could be measured for any base that a figure has.

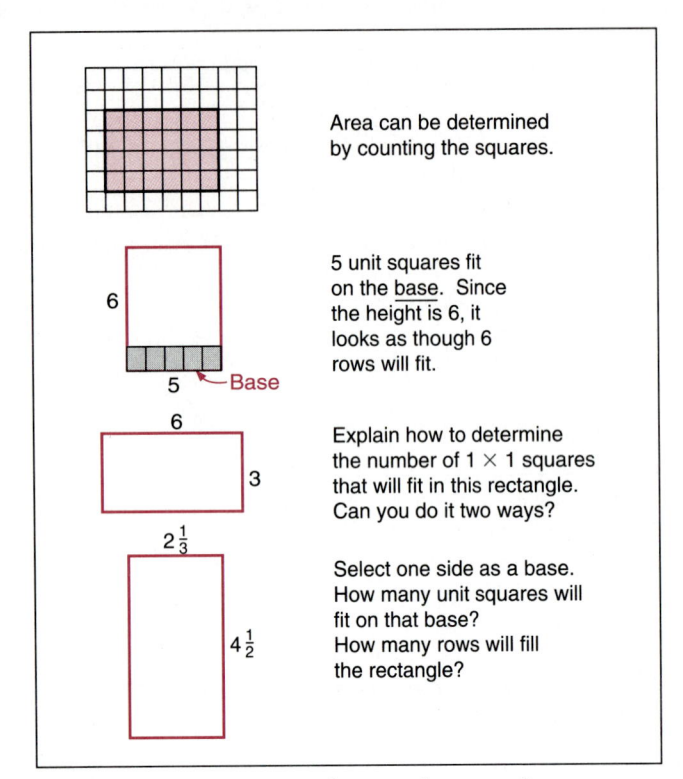

FIGURE 16.23: *Determining the area of a rectangle*

areas (but do not give the dimensions). Some may count every square while others may multiply to find the total number of squares.

2. Examine rectangles, not on a grid but with whole-number dimensions. Designate one side as the base, and line unit squares along this side. How many such rows can fit into the rectangle? On the same rectangle repeat this approach using the other side as the base.

3. Give students rectangles with only the dimensions provided and have them determine the area. Require them to justify their results. Encourage approaches similar to those in step 2.

4. Examine rectangles with dimensions that are not whole numbers. If the base is $4\frac{1}{2}$ units, then $4\frac{1}{2}$ unit squares will fit along the base. If the height is $2\frac{1}{3}$ units, then there are $2\frac{1}{3}$ rows with $4\frac{1}{2}$ squares in each, or $2\frac{1}{3}$ sets of $4\frac{1}{2}$.

As illustrated in Figure 16.24, parallelograms that are not rectangles can be transformed into rectangles having the same area. The new rectangle has the same height and two sides the same as in the original parallelogram. Students should explore these relationships on grid paper, on geoboards, or by cutting paper models, and should be quite convinced that the areas are the same and that it can always be done. As a result, the area of a parallelogram is base times height, just as for rectangles.

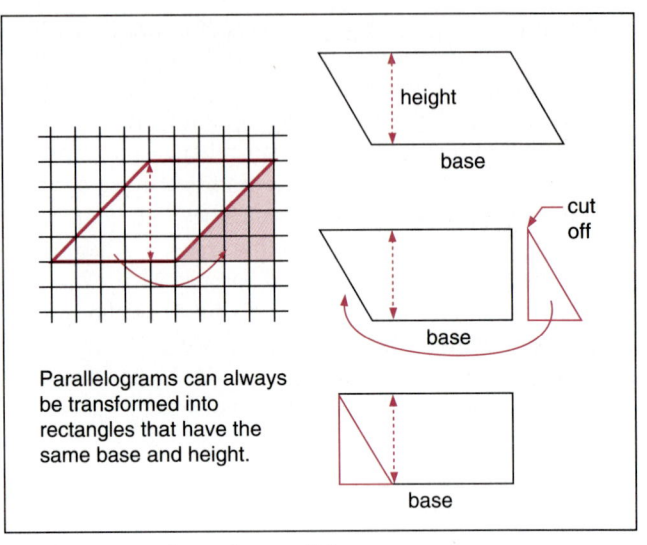

FIGURE 16.24: *Area of a parallelogram*

The area of triangles can easily be determined by showing that two identical triangles can always be arranged to form a parallelogram, as shown in Figure 16.25.

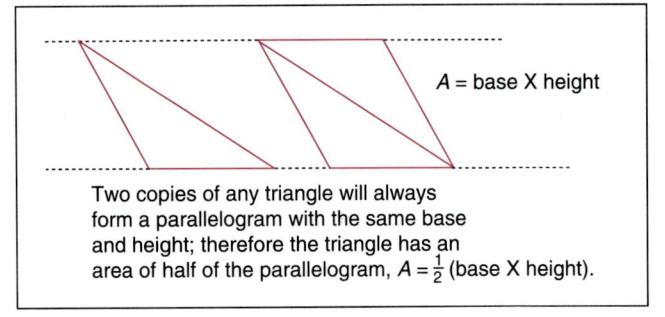

FIGURE 16.25: *Parallelograms of triangles*

◆◆◆◆◆◆◆◆◆◆◆◆◆◆◆◆◆◆◆

16.30 THE TRIANGLE CONNECTION

Have students draw any triangle on a piece of paper and cut out two identical copies. These should then be rearranged into a parallelogram. (If all sides are different, three different parallelograms are possible.) For each parallelogram, have students identify a base and a height. Are these also the base and height of the triangle? The area of the parallelogram is still $b = h$. What is the area of the triangle?

For trapezoids, explore a similar approach. Have students cut out any two identical trapezoids, put them together to make a parallelogram, and relate the area of that to the area of the trapezoid as in Figure 16.26. Students should be able to explain how the areas of triangles, rectangles, parallelograms, and trapezoids are all related. If they cannot, it is a good possibility that they do not understand the development of these formulas.

(Do you think that a special formula for squares should be taught? What about formulas for perimeter?)

---

base 2

base 1

base = base 1 + base 2
A = height × (base 1 + base 2)

Two trapezoids always make a parallelogram with the same height and a base equal to the sum in the trapezoid. Therefore,

$A = \frac{1}{2} \times height \times (base\ 1 + base\ 2)$

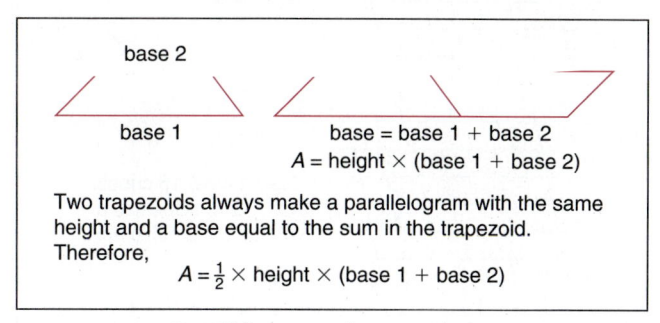

FIGURE 16.26: *Parallelograms of trapezoids*

◆ Circle Formulas

One of the most interesting relationships that children can discover in geometry is that between the *circumference* of a circle (the distance around) and the length of the *diameter* (a line through the center joining two points on the circle). In Chapter 17, students measure diameters and circumferences of circles of all sizes and use calculators to compare these two measures. The circumference of every circle is about 3.14 times as long as the diameter. That exact ratio is an irrational number close to 3.14 and is represented by the Greek letter π. That is, $\pi = \frac{C}{D}$, the circumference divided by the diameter. In a slightly different form, $C = \pi D$ or $2\pi r$, where r is the length of the radius.

In Figure 16.27, an argument is presented for the area formula $A = \pi r^2$. This development is one commonly found in textbooks. Another informal proof is based on the notion that the area of a polygon inscribed in a circle gets closer and closer to the area of the circle as the number of sides increases.*

VOLUMES OF COMMON SOLID SHAPES

The relationships between the various formulas for volume are completely analogous to those for area. As you read through this section, notice the similarities between rectangles and prisms, between parallelograms and "sheered" or oblique prisms, between triangles and pyramids. Not only are the formulas related in a similar manner, the process for development of the formulas can be seen as similar.

---

*The relationship between regular polygons and the circle is well known to those who have tried to draw circles with Logo. For them, the latter approach to the area of a circle may be especially appealing.

$C = 2\pi r$

The circle and each shape made from sectors all have the same area.

8 sectors can be arranged in a "near parallelogram."

24 sectors is even closer to a parallelogram.

As the number of sectors gets larger, the figure becomes closer and closer to a rectangle (a special parallelogram).

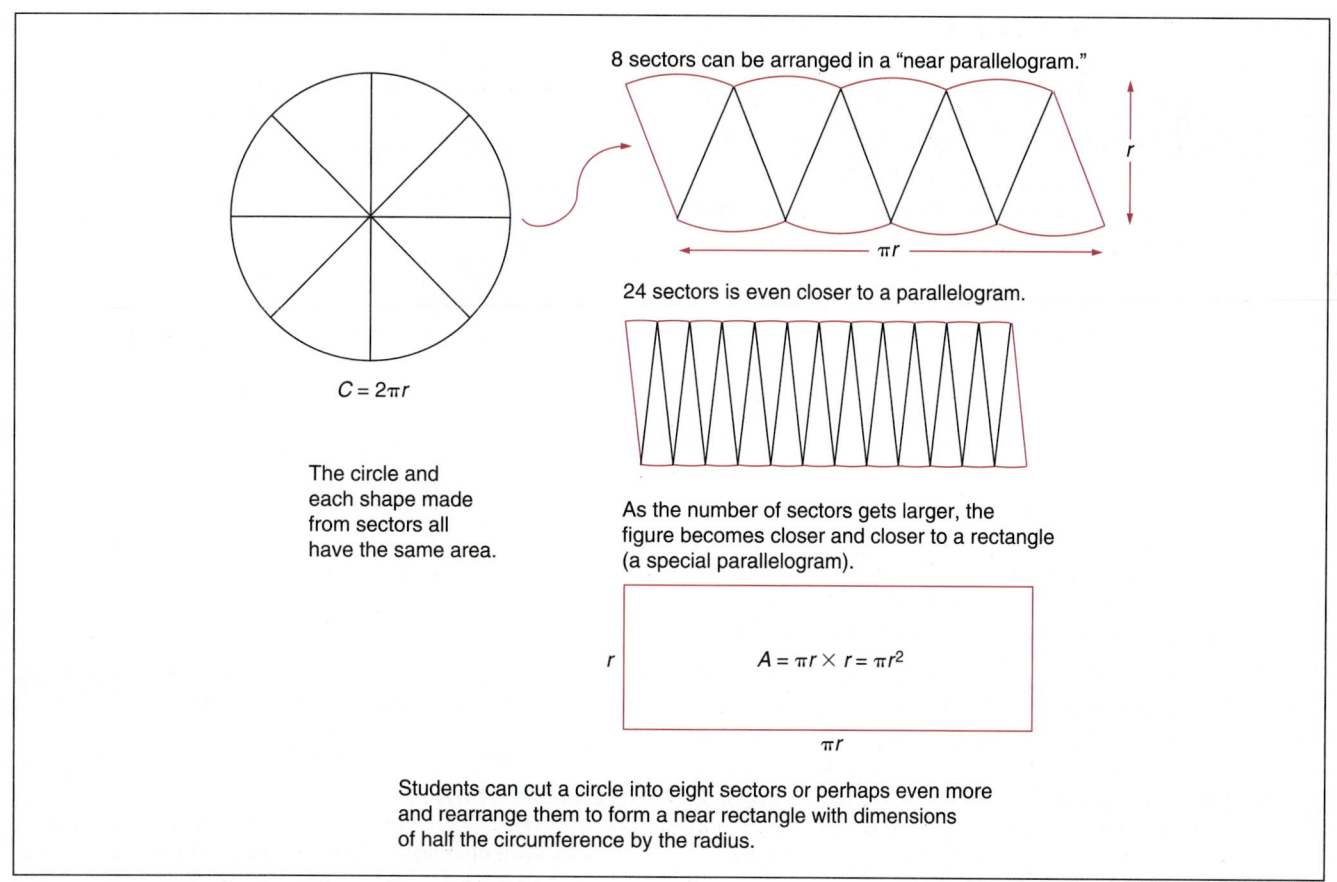

$A = \pi r \times r = \pi r^2$

Students can cut a circle into eight sectors or perhaps even more and rearrange them to form a near rectangle with dimensions of half the circumference by the radius.

FIGURE 16.27: *Development of a circle area formula*

316 16 / DEVELOPING MEASUREMENT CONCEPTS

◆ Volumes of Cylinders

A *cylinder* is a solid with two congruent parallel bases and sides with parallel elements. There are several special classes of cylinders such as *prisms* (polygons for bases), right prisms, *rectangular prisms,* and *cubes* (see Chapter 17, p. 343). Interestingly, all of these solids have the same volume formula, and that one formula is analogous to the area formula for parallelograms.

16.31 BUILDING THE VOLUME

Provide students with some wooden cubes and square grid paper that matches the face of the cubes. Have students draw a 3 × 5 rectangle on paper, and place 15 cubes on the rectangle. This makes a box with a height of one unit. The volume of this box is one times the area of the base. Now place a second layer of cubes on the first. What is the height? the volume? Continue to add layers up to five or six (Figure 16.28). For each new layer notice that the total number of cubes is the number of cubes on the bottom layer times the number of layers. The number on the bottom layer (or any layer) is the area of the base. The number of layers is the height. Therefore, the volume of the solid is $V = A \times h$, the *area of the base times the height.*

Recall how the area formula for rectangles was developed (Figure 16.23), and notice how that development is like the one just presented for volume. The following activity extends the same idea to other cylinders.

16.32 SHEERED PRISMS

Make a stack of three or four decks of playing cards (or a stack of any cards or paper). When the cards are stacked straight, they form a rectangular solid. The volume is $V = A \times h$. The area of the base is the area of one play card. Now if the stack is slanted to one side (sheered), a different cylinder is formed as in Figure 16.29. But the new cylinder (or prism) has the same base, the same height, and obviously the same volume. (Why?) Therefore, the volume formula remains $V = A \times h$.

What if the cards in this activity were some other shape—any shape? If they were circular, the volume would still be the area of the base times the height; if they were triangular, still the same. The conclusion is that the volume of any cylinder is equal to the area of the base times the height.

◆ Volumes of Cones and Pyramids

Just as there is a nice relationship between the areas of parallelograms and triangles, there is a similar relationship between the volumes of cylinders and cones.

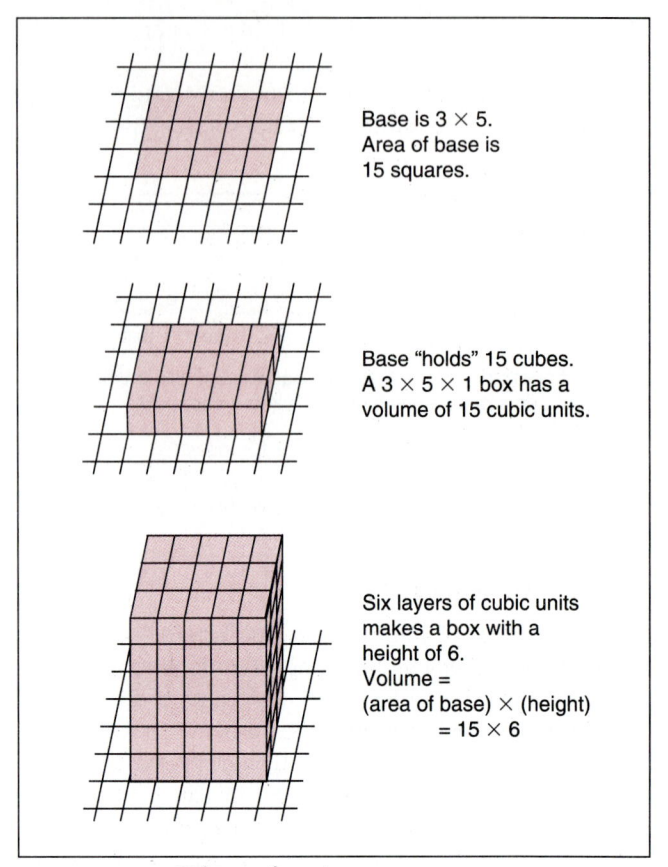

Base is 3 × 5.
Area of base is 15 squares.

Base "holds" 15 cubes. A 3 × 5 × 1 box has a volume of 15 cubic units.

Six layers of cubic units makes a box with a height of 6.
Volume = (area of base) × (height) = 15 × 6

FIGURE 16.28: *Volume of a prism*

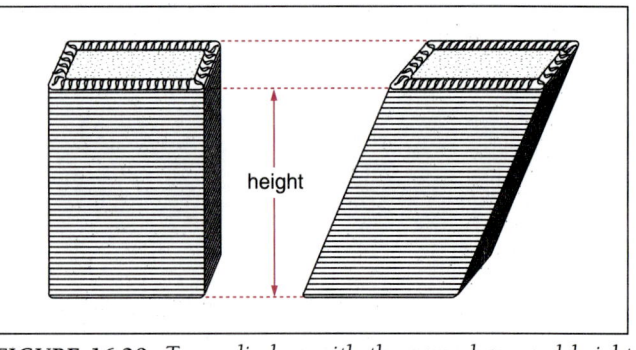

FIGURE 16.29: *Two cylinders with the same base and height have the same volume.*

16.33 THE PYRAMID/CONE CONNECTIONS

Help students make a poster board prism and a pyramid with the same height and base. (Plastic models of these related shapes can be purchased.) Leave the base of each open. Score fold lines for accurate folds. Dimensions for a square pyramid and prism are given in Figure 16.30. Have students estimate the number of times the pyramid will fit into the prism. Then have them test their prediction by filling the pyramid with beans or rice and emptying it into the prism. They will discover that exactly three pyramids will fill a prism with the same base and height. (An alternative method is to carefully make models from clay and compare the weights.)

The 3-to-1 ratio of volumes is true of all cylinders and cones with the same base and height regardless of the shape of the base or the position of the vertex. That is, for any cone or pyramid, $V = \frac{1}{3}(A \times h)$. Recall that for parallelograms and triangles that have the same base and height, the areas are in a 2-to-1 ratio. Area is to two-dimensional figures what volume is to three-dimensional figures. Further, triangles are to parallelograms as pyramids are to prisms.

The volume of a sphere is a bit more difficult to observe experimentally. The resulting formula $V = (\frac{4}{3}) \pi r^3$ can be demonstrated using an intuitive argument based on knowledge of the volume of a pyramid. For this development and for truly fascinating explorations of shapes and their measurement, the reader is urged to consider the book *Experiencing Geometry* by James Bruni (1977). Bruni's book is aimed at helping the middle school student explore geometric relationships in an intuitive, hands-on manner.

TIME AND CLOCK READING

MEASURING TIME

Time is measured in the same way that other attributes are measured: A unit of time is selected and used to "fill" the time to be measured. Time can be thought of as the duration of an event from its beginning to its end. An informal unit of time might be the duration of a pendulum swing, the steady drip of a water faucet, or the movement of the sun's shadow between two fixed points (as on a sundial). To measure time, the units of time are started at the same time as the activity being measured ("timed") and counted until the activity is finished. Thus, the pendulum swings, for example, are "fit into" the duration of time that it takes the child to print his or her name. Young children enjoy timing events with informal units. Older children can appreciate the measurement of time as a process similar to the measurement of other attributes and thus see commonality in all measuring.

CLOCK READING

Telling time has very little to do with measurement of time conceptually. The skills of clock reading are related to the skills of reading any meter that uses pointers or hands on a numbered scale. Clock reading is a difficult skill to teach in the first and second grades, and yet nearly everyone learns to tell time by middle school.

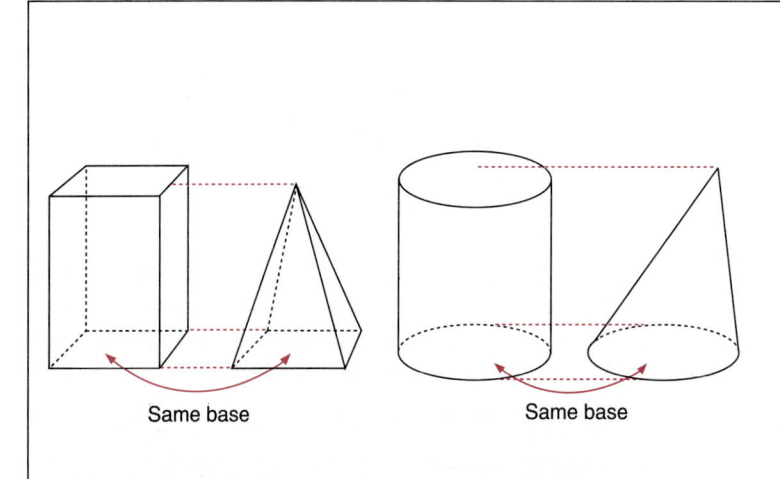

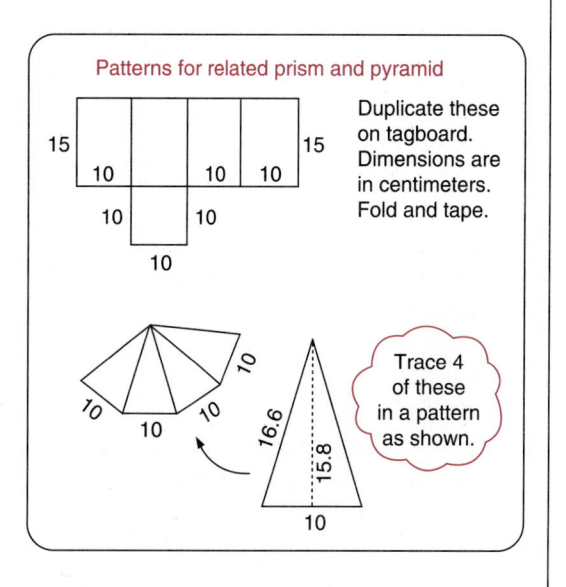

FIGURE 16.30: *Comparing volumes*

318   16 / DEVELOPING MEASUREMENT CONCEPTS

◆ Some Difficulties

Young children's problems with clock reading may be due to the curriculum. Children are usually taught first to read clocks to the hour, then the half and quarter hours, and finally to five- and one-minute intervals. In the early stages of this sequence, children are shown clocks set exactly to the hour or half hour. Many children who can read a clock at 7:00 or 2:30 have no idea what time it is at 6:58 or 2:33.

Digital clocks permit students to read times easily but do not relate times very well. To know that a digital reading of 7:58 is nearly eight o'clock, the child must know that there are 60 minutes in an hour, that 58 is close to 60, and that two minutes is not a very long time. These concepts have not been developed by most first-grade children and not all second-grade children. The analog clock (with hands) can show "close-to" times without the need for understanding big numbers or even how many minutes in an hour.

The standard approach to clock reading ignores the distinctly different actions and functions of the two hands. The little hand indicates broad, approximate time (nearest hour), and the long hand indicates time (minutes) after and until an hour. When we look at the hour hand, we focus on where it is pointing. With the minute hand, the focus is on the amount that has gone around the clock or the amount yet to go to get back to the top.

◆ A Suggested Approach

The following suggestions can be successfully used to help students understand analog clocks and be able to read them.

1. Begin with a one-handed clock. A clock with only an hour hand can be read with reasonable accuracy. Read clocks with only an hour hand, and use lots of approximate language. It is about seven o'clock. It is a little past nine o'clock. It is halfway between two o'clock and three o'clock (Figure 16.31).

2. Discuss what happens to the big hand as the little hand goes from one hour to the next. When the big hand is at 12, the hour hand is pointing exactly to a number. If the hour hand is about halfway between numbers, about where would the minute hand be? If the hour hand is a little past or before an hour (10 to 15 minutes), about where would the minute hand be?

3. Use two real clocks, one with only an hour hand and one with two hands. (Break off the minute hand from an old clock.) Cover the two-handed clock. Periodically during the day, direct attention to the one-handed clock. Discuss the time in approximate language. Have students predict where the minute hand should be. Uncover the other clock and check.

4. Teach time after the hour in five-minute intervals. After step 3 has begun, count by fives going around the clock. Instead of predicting that the minute hand is pointing at the 4, encourage students to say it is about 20 minutes after. As skills develop, suggest that students always look first at the little or hour hand to learn *about* what time it is and then focus on the minute hand for precision.

5. Predict the reading on a digital clock when shown an analog clock, and vice-versa; set an analog clock when shown a digital clock. This can be done with both one-handed and two-handed clocks.

"About 7 o'clock"

"Halfway between 2 o'clock and 3 o'clock"

"A little bit past 9 o'clock"

FIGURE 16.31: *Approximate time with one-handed clocks*

◆ A Teaching Clock

A simple clock for use on the overhead projector can easily be made with three sheets of acetate. On one, draw a simple clock face but with no hands. Cut a small hole in the center, and insert a brass fastener from the back side. Leave the fastener stick straight up; do not spread the prongs. Cut two circles from the other sheets. The radius of the circles should be the same as the minute and hour hands of the clock. Cut small holes in the center of these circles and draw a clock hand on each that extends from the center to the edge of the circle. If you slip just the smaller hour hand over the brass fastener, you have a very nice one-handed clock. For a clock with two hands, place the larger circle on first.

Use your overhead clock to practice one-handed clock reading emphasizing approximate language. Set the little hand at various places on the clock, and have children suggest how to read it. Let students set the hour hand at times like half-past five or a little before eight o'clock.

If an overhead projector is not convenient, make a large laminated clock face with detachable hands. Use masking tape on the backs of the hands to hold them temporarily in place.

Related Concepts

Students also need to learn about seconds, minutes, and hours and to develop some concept of how long these units are. If you make a conscious effort to note the duration of short and long events during the day, this will help. Timing small events of $\frac{1}{2}$ minute to 2 minutes is fun and useful. TV shows and commercials are a good standard. Have students time familiar events in their daily lives: brushing teeth, eating dinner, riding to school, time in the reading group.

As students learn more about two-digit numbers, the time after the hours can also be related to the time left before the hour. This is not only useful for telling time but helpful for number sense as well. Note that in the sequence suggested, time after the hour is stressed almost exclusively. Time before or 'til the hour can come later.

Problem-solving exercises such as, "If it was 7:30 when Bill left home and the trip took eight hours, what time did he arrive?" are important. Even middle-school students have difficulty with these ideas. Adding and subtracting time involves understanding the relationships between minutes and hours and also the two cycles of 12 hours in the day.

REFLECTIONS ON CHAPTER 16: WRITING TO LEARN

1. Explain what it means to measure something.
2. A general instructional plan for measurement has three parts. Explain how the type of activity to use with each part accomplishes the instructional goal.
3. Five reasons were offered for using informal units instead of standard units in instructional activities. Which of these seem most important to you, and why?
4. For each of the following attributes, describe a comparison activity, one or two possible informal units, and a group activity that includes an estimation component:
   a. length
   b. area
   c. volume
   d. weight
   e. capacity
5. With a straightedge, draw a triangle, a quadrilateral, and a five-sided figure. Make each about as large as a sheet of notebook paper. Make a waxed paper protractor, and measure each interior angle. Did the sum of the angles for each figure come close to what is predicted?
6. What is a degree? How would you help children learn what a degree is?
7. What do students need to know about standard units? Of these, which is the most and least important?
8. Develop in a connected way the area formulas for rectangles, parallelograms, triangles, and trapezoids. Draw pictures, and provide explanations.
9. Explain how the volume formula for a right rectangular prism can be developed in a completely analogous manner to the area formula for a rectangle.
10. Explain how the area of a circle can be determined using the basic formula for the area of a parallelogram. (If you have a set of fraction "pie pieces," these can be used as sectors of a circle.)
11. Describe the differences between the typical approach or sequence for teaching clock reading and the one-handed approach described here.

FOR DISCUSSION AND EXPLORATION

1. Measure a length with an informal unit model. Select your unit and object to measure so that the measurement is at least 10 units. Do the measurement in two ways. First use enough copies of the unit that they can be lined up end to end. Second, measure the same thing using only one copy of the unit. Observe how much more difficult the second method is. Try the same activity with area.
2. Make your own measuring instrument for an informal unit of measure. Select one attribute or make one for each attribute. Use your instrument to measure with and then make the same measurement directly with a unit model. What are some of the values and limitations of each method? Can you see the importance of children doing this both ways?
3. Set up an estimation activity for metric measurement and have the class or your friends try it out. Make it for length, capacity, or weight.
4. Find three good metric benchmarks on your body. Use them to measure some things. Include long distances such as the length of the hall and short distances that you can measure on your desktop. Also use your body benchmarks to estimate measures of distances around objects. Check your estimates made this way with measures made with rulers or tape measures.
5. Get a teacher's edition of a basal textbook for any grade level and look at the chapter(s) on measurement. How would you expand on the activities in the text? What special ideas in the teacher's edition do you like?
6. Read "Chapter 10: Foot Activities" in *A Collection of Math Lessons from Grades 3 to 6* (Burns, 1987). Identify two good ideas in the sequence of lessons. Modify the activities to suit your own needs, and try them out with a class of children.

SUGGESTED READINGS

Clopton, E. L. (1991). Area and perimeter are independent. *Mathematics Teacher, 84,* 33–35.

Coburn, T. G., & Shulte, A. P. (1986). Estimation in measurement. In H. L. Schoen (Ed.), *Estimation and mental computation.* Reston, VA: National Council of Teachers of Mathematics.

Corwin, R. B., & Russell, S. J. (1990). Measuring: From paces to feet [grades 3–4]. From *Used numbers: Real data in the classroom.* Palo Alto, CA: Dale Seymour Publications.

Gerver, R. (1990). Discovering *pi*—Two approaches. *Arithmetic Teacher, 37*(8), 18–22.

Hart, K. (1984). Which comes first—length, area, or volume? *Arithmetic Teacher, 31*(9), 16–18, 26–27.

Hiebert, J. (1984). Why do some children have trouble learning measurement concepts? *Arithmetic Teacher, 31*(7), 19–24.

Liedtke, W. W. (1990). Measurement. In J. N. Payne (Ed.), *Mathematics for the young child.* Reston, VA: National Council of Teachers of Mathematics.

Lindquist, M. M., & Dana, M. E. (1977). The neglected decimeter. *Arithmetic Teacher, 25*(1), 10–17.

Shaw, J. M. (1983). Student-made measuring tools. *Arithmetic Teacher, 31*(3), 12–15.

Shaw, J. M., & Cliatt, M. J. P. (1989). Developing measurement sense. In P. R. Trafton (Ed.), *New directions for elementary school mathematics.* Reston, VA: National Council of Teachers of Mathematics.

Szetela, W., & Owens, D. T. (1986). Finding the area of a circle: Use a cake pan and leave out the pi. *Arithmetic Teacher, 33*(9), 12–18.

Thompson, C. S., & Van de Walle, J. A. (1981). A single handed approach to telling time. *Arithmetic Teacher, 28*(8), 4–9.

Thompson, C. S., & Van de Walle, J. A. (1985). Learning about rulers and measuring. *Arithmetic Teacher, 32*(8), 8–12.

Van de Walle, J. A., & Thompson, C. S. (1985). Estimate how much. *Arithmetic Teacher, 32*(9), 4–8.

Wilson, P. S. (1990). Understanding angles: Wedges to degrees. *Mathematics Teacher, 83,* 294–300.

Wilson, P. S., & Osborne, A. (1992). Foundational ideas in teaching about measure. In T. R. Post (Ed.), *Teaching mathematics in grades K–8: Research-based methods* (2nd edition). Boston, MA: Allyn & Bacon.

Wilson, P. S., & Rowland, R. E. (1993). Teaching measurement. In R. J. Jensen (Ed.), *Research ideas for the classroom: Early childhood mathematics.* New York: Macmillan Publishing Co.

Zweng, M. J. (1986). Introducing angle measurement through estimation. In H. L. Schoen (Ed.), *Estimation and mental computation.* Reston, VA: National Council of Teachers of Mathematics.

17

GEOMETRIC THINKING AND GEOMETRIC CONCEPTS

EXPLORATIONS FOR YOUR REFLECTION

The geometry curriculum in grades K–8 should provide an opportunity to experience and reflect on shapes in as many different forms as possible. These should include shapes built with blocks, sticks, or tiles; shapes drawn on paper or with a computer; shapes observed in art, nature, and architecture. Hands-on, highly reflective and interactive experiences are at the heart of good geometry activities at the elementary and middle-school levels.

THREE ACTIVITIES FOR YOUR EXPLORATION

It is very likely that most of you have quite different ideas about geometry in the elementary school. In order to provide some common view of the nature of elementary school geometry and how young children approach geometric concepts, three simple activities are offered here for you to do and experience. The activities will provide some idea of the spirit of informal, elementary school geometry as well as background for a discussion of children's geometric thinking. All you will need is a pencil, several pieces of paper, scissors, and 15 to 20 minutes.

Different Triangles

Draw a series of at least five triangles. After the first triangle, each new one should be different in some way from those already drawn. Write down why you think each is different.

Shapes with Triangles

Place a piece of paper over the dot grid in Figure 17.1 (p. 322) and draw three or four different figures by connecting adjacent dots. Each figure should have an area of 10 small triangles. A few simple rules for drawing your shapes are explained in Figure 17.2 (p. 322). Count to find the distance around each figure or the perimeter, and record this next to each drawing. You may want to explore any observations you may have by drawing some more shapes.

A Tiling Pattern

First make at least eight copies of any shape in Figure 17.3 (p. 322). An easy way to do this is to fold a piece of paper so there are eight thicknesses. Trace the shape on an outside section and cut through all eight thicknesses at once. You may want more copies of these shapes.

Think of the shapes you cut out as tiles. The task is to use the tiles to make a regular tiling pattern. A tiling pattern made with one shape has two basic properties. First, there are no holes or gaps. The tiles must fit together without overlapping and without leaving any spaces. Second, the tiles must be arranged in a repeating pattern that could be extended indefinitely. That is, if you were to tile an endless floor with your pattern, the design in one section of the floor would be the same as that found in any other section. Experiment with the paper tiles to decide on a pattern that you like. There are several different tiling patterns possible for each of the three tiles.

FIGURE 17.1: *Dot grid*

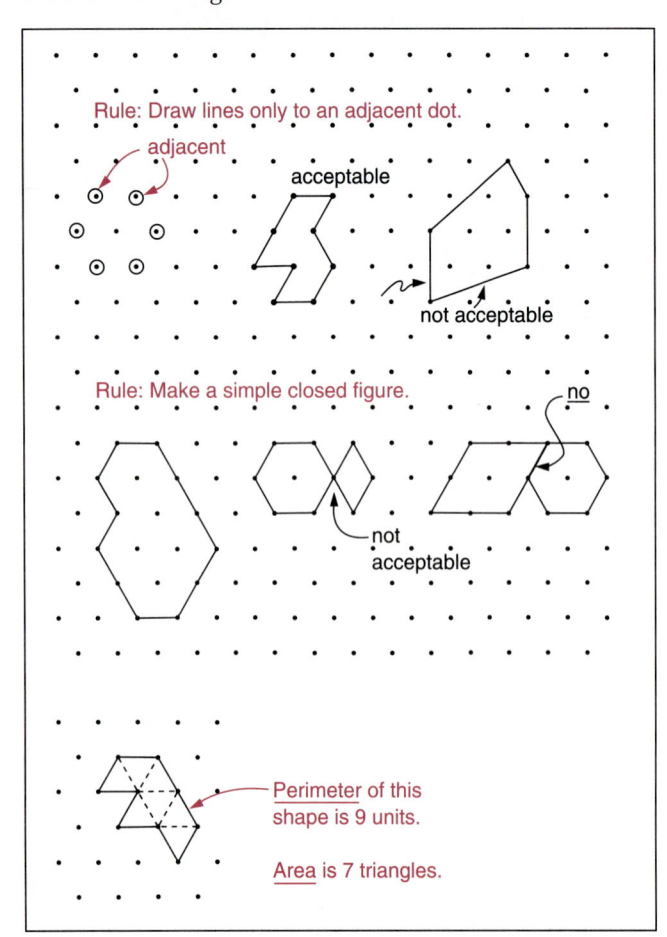

FIGURE 17.2: *Rules for making shapes with triangles on the dot grid.*

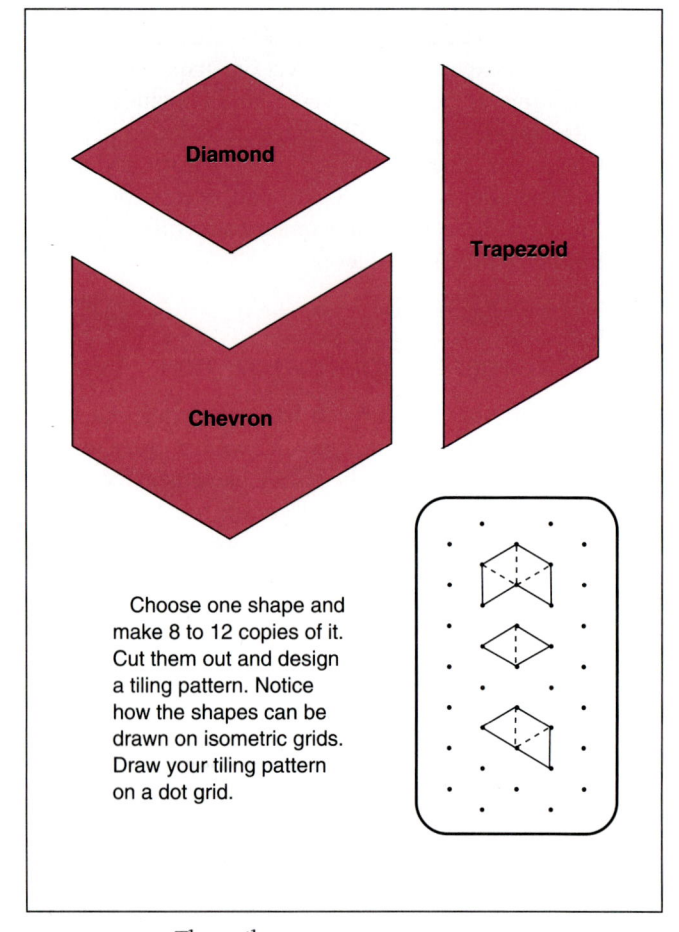

Rule: Draw lines only to an adjacent dot.
adjacent
acceptable
not acceptable

Rule: Make a simple closed figure.
no
not acceptable

Perimeter of this shape is 9 units.
Area is 7 triangles.

FIGURE 17.3: *Three tile patterns*

Diamond

Trapezoid

Chevron

Choose one shape and make 8 to 12 copies of it. Cut them out and design a tiling pattern. Notice how the shapes can be drawn on isometric grids. Draw your tiling pattern on a dot grid.

Notice that each of the tile shapes is made up of triangles and can be drawn on the dot grid of Figure 17.1. When you have decided on a tiling pattern, place a piece of paper over the dot grid, and draw your tiling pattern with small versions of the tile shape using the dots as a guide. Cover most of the grid with your pattern. Finally, suppose your tiles came in two colors, and you want to add a color pattern to your tile pattern. With a pen or pencil, shade in some of the tiles to make a regular pattern in two colors.

SOME INITIAL OBSERVATIONS ABOUT THESE ACTIVITIES

Rather than discuss these three activities in detail, the following are some observations on which you might reflect. These comments are about geometry in the elementary school in general as well as reflections or observations about the activities themselves.

Different People Think about Geometric Ideas in Different Ways

Compare your response to the activities on p. 321 with those of your peers. Are there qualitative differences as well as objective differences? How would primary children approach these activities compared to an eighth grader? Figure 17.4 shows how two students, one in the fifth grade and one in the eighth grade, responded to the triangle task.

Research indicates that age is not the major criterion for how students think geometrically. The kinds of experiences a child has may be a more significant factor. What were your grade school experiences with geometry?

Explorations Can Help Develop Relationships

The more that you play around with and think about the ideas in these activities, the more there is to think about. A good teacher might be able to extend each of these activities to develop the ideas beyond the obvious. For example:

> For "Different Triangles": How many different ways can two triangles be different? Could you draw five or more *quadrilaterals* that were each different?

> For "Shapes with Triangles": What did you notice about the shapes that had smaller perimeters as opposed to those with the larger perimeters? If you tried the same activity with rectangles on square grid paper, what would the shapes with the largest/smallest perimeters look like? What about three-dimensional boxes? If you built different boxes with the same number of cubes, what could you say about the surface areas?

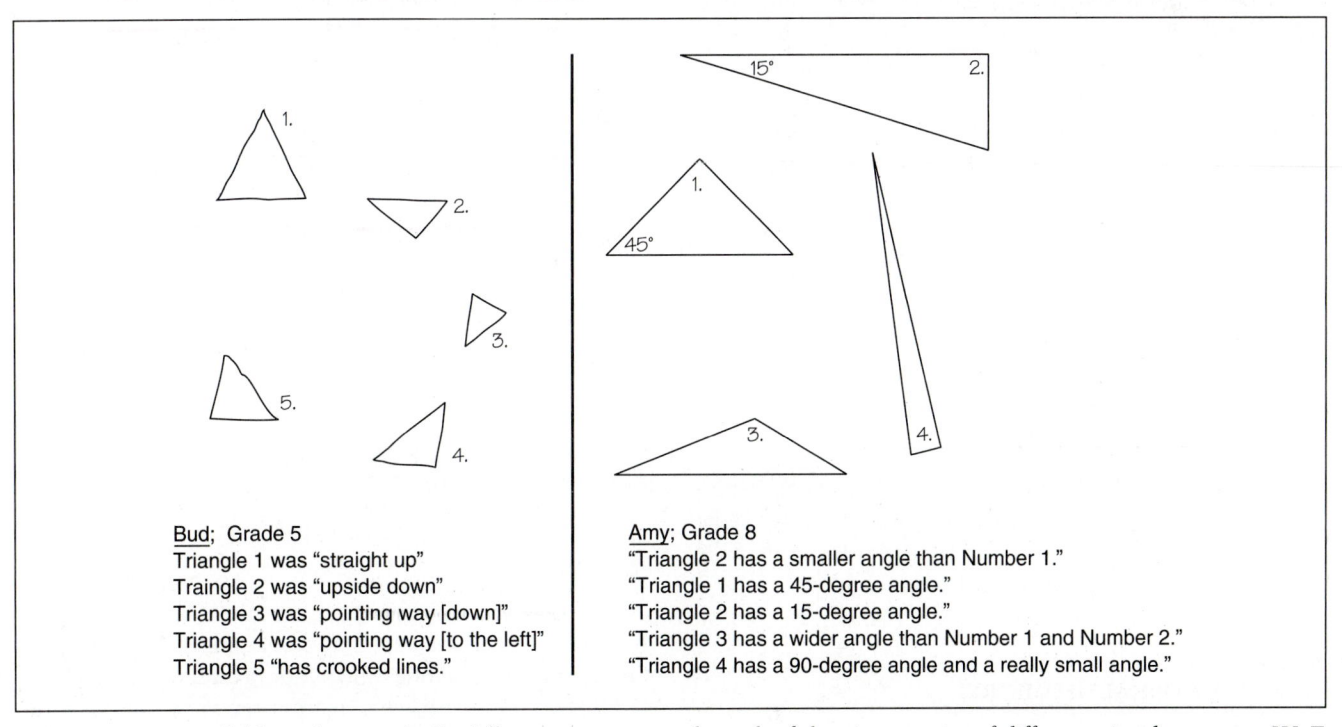

Bud; Grade 5
Triangle 1 was "straight up"
Traingle 2 was "upside down"
Triangle 3 was "pointing way [down]"
Triangle 4 was "pointing way [to the left]"
Triangle 5 "has crooked lines."

Amy; Grade 8
"Triangle 2 has a smaller angle than Number 1."
"Triangle 1 has a 45-degree angle."
"Triangle 2 has a 15-degree angle."
"Triangle 3 has a wider angle than Number 1 and Number 2."
"Triangle 4 has a 90-degree angle and a really small angle."

FIGURE 17.4: *Two children show markedly different responses to the task of drawing a series of different triangles.* SOURCE: W. F. Burger and J. M. Shaughnessy (1986). Characterizing the van Hiele levels of development in geometry. *Journal for Research in Mathematics Education, 17.*

> For "A Tiling Pattern": How many different tiling patterns are there for this shape? Can any shape be used to tile with? Can you see any bigger shapes within your pattern?

Notice that it takes more than just doing an activity in order to learn or create a new idea. The greatest learning occurs when you stop and reflect on what you did and begin to ask questions or make observations.

GEOMETRY ACTIVITIES ARE ALSO PROBLEM-SOLVING ACTIVITIES

Good geometry activities almost always have a spirit of inquiry or problem solving. Many of the goals of problem solving are also the goals of geometry. Reconsider each of the three activities as examples of problem solving. Can you identify at least one problem in each that needs to be solved? In the context of those problems, consider the goals of problem solving, especially perseverance, willingness to take risks, understanding the problem, evaluating the results, and going beyond the initial problem.

GOOD GEOMETRY ACTIVITIES INVOLVE HANDS-ON MATERIALS

Even the simple paper tiles used in "A Tiling Pattern" gave you the opportunity to explore spatial relationships and search for patterns much more easily than without them. Activities on paper such as the dot grid in "Shapes with Triangles" are a second-best alternative to real physical objects. The same area and perimeter activity is much more effective with a collection of cardboard triangles that can be rearranged to form different shapes. The first activity is the least enticing of the three, but at least you could freely draw pictures. Virtually every activity that is appropriate for K–8 geometry should involve some form of hands-on materials, models, or at least paper such as graph paper or dot paper that lends itself to easy spatial explorations.

INFORMAL GEOMETRY: WHAT AND WHY

The activities suggested in the last section are simple examples of informal geometry activities. What are some of the characteristics of these and similar activities that make them important enough for us to do with children?

WHAT IS INFORMAL GEOMETRY?

Most if not all the geometry that is taught in grades K to 8 can be referred to as *informal geometry*. This term suggests more about the nature of geometry activities than it does about the goals or the content of the geometry curriculum. Good informal geometry activities are:

- **Experiential and Exploratory:** Informal geometry activities provide children with the opportunity to explore, to feel and see, to build and take apart, to make observations about shape in the world around them as well as the world they can create with drawings and models. Activities involve constructing, measuring, visualizing, comparing, transforming, and classifying geometric figures. As we will see, the experiences and explorations can take place at different levels: from shapes and their appearances, to properties of shapes, to relationships between and among these properties. Good activities will avoid teacher telling and student memorizing.

- **Hands-On:** Virtually every good activity at any level of informal geometry involves some form of materials. These materials may be tiles, sticks, blocks, paper with line or dot grids, computer programs, rulers, mirrors, clay, and so on. The list of materials, some very specialized, some with many different uses, is nearly endless. Activities that focus on notation or emphasize words and definitions in preference to student activity and student descriptions should be discouraged.

Frequently, informal geometry will involve:

- **A problem-solving approach.** Students will figure out how to solve a puzzle, discover why shapes are alike and different, find ways to construct shapes with certain characteristics, or discover that certain combinations of properties are impossible.

- **An artistic component.** Many excellent geometry activities involve the construction of an artwork. Color and pattern are combined with shape and form at various levels of sophistication to let students at any level see from a personal perspective a distinct connection between art and mathematics.

Regardless of the specifics, informal geometry can be just plain fun. Tedious exercises of memorizing definitions or trying to remember the correct symbol for congruence can be boring if not disheartening. In contrast, informal geometry activities should always be enjoyable for teacher and students alike.

WHY STUDY GEOMETRY?

Informal geometry is aimed at development of *spatial sense*, "an intuitive feel for one's surroundings and objects in them" (NCTM, 1989, p. 49). This spatial sense grows and develops over the entire time children are in school. Students should have several in-depth geometric experiences every year. As you will learn in the section to follow, rich geometric experiences are the most important factor in the development of children's spatial thinking and reasoning.

Unfortunately, geometry frequently takes a back seat to other topics in school mathematics, and so, "Why study geometry?" is a fair question. In fact, if you were to decide that you really enjoyed doing informal geometry activities with your students (some of the most fun that you can have), you may feel the need to justify your actions. The following list of reasons for studying geometry is not intended to be complete or definitive but it is a good place to begin.

1. Geometry helps people have more complete appreciation of the world in which they live. Geometry can be found in the structure of the solar system, in geological formations, in rocks and crystals, in plants and flowers, and even animals. It is also a major part of our synthetic universe: Art, architecture, cars, machines, and virtually everything that humans create has an element of geometric form.

2. Geometric explorations can develop problem-solving skills. Spatial reasoning is an important form of problem solving, and problem solving is one of the major reasons for studying mathematics.

3. Geometry plays a major role in the study of other areas of mathematics. By way of example: Fraction concepts are related to geometric part-to-whole constructs. Ratio and proportion are directly related to the geometric concept of similarity. Measurement and geometry are clearly related topics, each adding to the understanding of the other.

4. Geometry is used by many people daily in their professional as well as their everyday lives. Scientists of all sorts, architects and artists, engineers, and land developers are just a few of the professions that use geometry regularly. In the home, geometry helps build a fence, design a dog house, plan a garden, or even decorate a living room.

5. Geometry is fun and enjoyable. While fun may not be a reason in and of itself, if geometry is a way to entice students into a little more love of mathematics in general, then that makes much of the effort worthwhile.

THE DEVELOPMENT OF GEOMETRIC THINKING

Until recently, the geometry curriculum in the United States has been very poorly defined. You could count on children in the early grades learning to identify a few basic shapes. Beyond this primitive bit of knowledge there has not been much to guide teachers in terms of what is important. The work of two Dutch educators, Pierre van Hiele and Dina van Hiele-Geldof, is beginning to have an impact on the design of geometry instruction and curriculum.

THE VAN HIELE LEVELS OF GEOMETRIC THOUGHT

The van Hieles' work began in 1959 and immediately attracted a lot of attention in what was then the Soviet Union. For nearly two decades, the work of the van Hieles received little attention in this country (Hoffer, 1983; Hoffer & Hoffer, 1992). While research continues, the van Hiele theory or some variant of it has become the most influential factor forming our geometry curriculum.

The most prominent feature of the model is a five-level hierarchy of ways of understanding spatial ideas. Each of the five levels is descriptive of the thinking processes that one uses in geometric contexts. The levels describe how one thinks and what types of geometric ideas one thinks about rather than how much knowledge a person has. What follows is a very brief description of the five van Hiele levels of geometric thinking.

Level 0: Visualization

Students recognize and name figures based on the global, visual characteristics of the figure; a gestalt-like approach to shape. Students operating at this level are able to make measurements and even talk about properties of shapes, but these properties are incidental to the shape. It is the appearance of the shape that defines it for the student. A square is a square because "it looks like a square." Because appearance is dominant in the student's thinking at this basic level, appearances can actually overpower properties of a shape. For example, a square that has been rotated so that all sides are at a 45° angle to the vertical may not appear to be a square for a level 0 thinker. Students at this level will sort and classify shapes based on their appearances—"I put these together because they all look sort of alike." The level 0 thinker is explicitly aware of the appearances of shapes. Using these appearances to form classifications suggests that geometric properties or defining characteristics of shapes are present implicitly or under the surface.

Level 1: Analysis

Students at the analysis level of thinking tend to pay more attention to the properties of a shape than to its appearance. By focusing more on what makes a rectangle a rectangle (four sides, opposite sides parallel, opposite sides same length, four right angles, congruent diagonals, . . .) the irrelevant features fade into the background. At this level students begin to appreciate that the reason a collection of shapes goes together has something to do with properties. It makes sense to define and sort shapes by these properties rather than by appearances. Students use properties at this level to define classes of shapes and understand that if a shape belongs to a particular class then it has the corresponding properties of that class. "All cubes have six congruent faces, and each of these faces is a square." These properties, only implicit or subsurface for the level 0 thinker, become the objects of thought at level 1. Implicit or subsurface for the level 1 thinker are the relationships among the properties and among the classifications of shapes made by the properties. Students operating at level 1 may be able to list all properties of squares, rectangles, and parallelograms but not see that these are subclasses of one another, that all squares are rectangles and all rectangles are parallelograms. In defining a shape, level 1 thinkers are likely to list as many properties of a shape as they know. The explicit objects of thought are the properties, but not the relationships among the properties. Deductive arguments that create relationships between different classes are not spontaneously made at this level.

Level 2: Informal Deduction

The relationships between classes, only implicit in the thinking of the previous level, are the objects of thought for the thinker at level 2. With greater attention to "if-then" reasoning, shapes can be classified by using only minimum characteristics. For example, four congruent sides and at least one right angle can be sufficient to define a square. Rectangles are parallelograms with a right angle. Observations go beyond properties themselves and begin to focus on logical arguments *about* the properties—"If the figure is a parallelogram, then it must follow that the opposite angles are congruent." Students at level 2 will be able to follow and appreciate an informal deductive argument. Proof may be more intuitive than deductive. Informal logical reasoning is explicit at this level. The appreciation of the axiomatic structure of a formal deductive system remains, however, under the surface. Thus, the actual construction of clear, tight deductive proof is not likely for students at level 2.

Level 3: Deduction

The structure of the axiomatic system complete with axioms, definitions, theorems, corollaries, and postulates, implicit at level 2, now becomes the explicit object of thought for the level 3 thinker. At this level, students can appreciate the need for a system of logic that rests on a minimum set of assumptions and from which other truths can be derived. The student at this level is able to work with abstract statements about geometric properties and make conclusions based more on logic than intuition. This is the level of the traditional tenth-grade geometry course. A student in such a course who can clearly see that the diagonals of a rectangle bisect each other and yet can still appreciate the need to prove this from a series of deductive arguments is operating at level 3. The level 2 thinker, by contrast, follows the argument but fails to appreciate the need. The result or relationship is the focus of the level 2 thinker while the logical argument is the object of thought for the level 3 thinker.

Level 4: Rigor

At this fifth and highest level of the van Hiele hierarchy, the object of attention is now axiomatic systems themselves, not just the deductions within a system. There is an appreciation of the distinctions and relationships between different axiomatic systems. This is generally the level of a mathematician who is studying geometry as a branch of mathematical science.

CHARACTERISTICS OF THE LEVELS OF THOUGHT

As you look back over the descriptions of the five levels of the van Hiele theory, you will notice that at each successive level what was only implicit at the previous level becomes the explicit object of thought at the next level. It is as if objects (ideas) must be created at each level so that relationships on these objects can become the focus of the next level. In addition to this key concept of the van Hiele theory, four related characteristics of the levels of thought merit special attention.

1. The levels are sequential. That is, to arrive at any level above level 0, the basic level, students must move through all prior levels. To move through a level means that one has experienced geometric thinking about the type of objects that is explicit at that level and is ready to shift attention to the objects or relationships that are only implicit or in the background at that level. To miss a level would be to break the chain. This rarely occurs.

2. The levels are not age-dependent (as, for example, are the well-known developmental stages of Piaget). A third grader as well as a high school student may be at level 0. Many high school students and adults remain at the basic level 0, and a significant number of adults have not reached level 2. At the same time, age is certainly related to the amount and types of geometric experiences that we have. Therefore, it is reasonable for all children in the K – 2 range to be at level 0 as are the large majority of children in grades 3 and 4.

3. Geometric experience is the largest single factor influencing advancement through the levels. Activities that permit children to explore, talk about, and interact with content at the next level, while increasing their experiences at their current level, have the best chance of advancing the level of thought for those children.

4. When instruction or language is at a level higher than that of the student, real learning may not occur. Students required to wrestle with ideas above their level may be forced to rote learning and achieve only temporary and superficial gain. A student can, for example, memorize that all squares are rectangles without having constructed that relationship. At another level, a student may memorize a geometric proof but fail to create the steps or understand the rationale involved (Crowley, 1987; Fuys, Geddes, & Tischler, 1988).

IMPLICATIONS FOR INSTRUCTION

The van Hiele theory provides the thoughtful teacher with a framework within which to conduct geometric activities. The theory does not specify content or curriculum but can be applied to most activities. While most activities can be designed to begin with the assumption of a particular level, most can also be raised or lowered in terms of the types of questioning and guidance provided by the teacher.

Instructional Goals: Content and Levels of Thought

If you examine just the bold-print portion of the K–4 and 5–8 geometry standards (in the NCTM *Standards*), you will find the following 16 verbs used:

describe	model	develop (an appreciation)
classify	investigate	draw
relate	recognize (geometry in the world)	predict
identify	compare	visualize
represent	explore	understand
apply		

Only one of these (identify) suggests the age-old "knowing the names of the basic shapes." The *Standards* describes an active, process-oriented view of what children should be doing and learning in geometry. The goal is much broader than a more traditional collection of facts and bits of knowledge about various geometric ideas. It is best summed up as *spatial sense*.

The van Hiele theory fits very nicely with a *Standards*-view of geometry. It focuses our attention on how students think in geometric contexts and the object of their thinking: shape → properties → informal logic → deductive principles. If the van Hiele theory is correct, and there is much evidence today to support it, then a major goal of the K–8 curriculum must be to advance students' level of geometric thought. If students are to be adequately prepared for the deductive geometry curriculum of high school, then their thinking should have advanced to at least level 2. (Level 3 is assumed in the traditional tenth-grade course.)

This is not to say that content knowledge is no longer appropriate. Spatial sense is clearly enhanced by an understanding of shapes, what they look like, and even what they are named. The concepts of symmetry, congruence and similarity play a major role in understanding our geometric world. And the interaction with measurement that allows us to analyze angle measure and relationships between geometric entities is equally important. But these must all be seen and developed not in the context of "things to master" but rather as ways of knowing and understanding the geometric world in which we live.

Therefore, as we design geometry lessons for children, it is good to build them around identifiable content but with a view toward the thought levels of our children. Each lesson should be seen as an opportunity to challenge students to begin to think at the next level: Those at level 0 can be encouraged to consider properties of shapes; those at level 1 can be challenged to formulate and engage in informal arguments or deductions.

Teach at the Student's Level of Thought

A developmental approach to instruction demands that we listen to children and begin where we find them. The van Hiele theory highlights the necessity of teaching at the child's level. Almost any activity can be modified to span two levels of thinking, even within the same classroom. This can be done by respecting the types of responses and observations made by children that suggest a lower level of thought while encouraging and challenging children to operate at the next level of thought. Remember that it is the type of thinking that children are required to do that makes a difference in learning, not the specific content.

The following are some suggestive features of instruction appropriate for each of the first three van Hiele levels.

Level 0 activities should

- use lots of physical models that can be manipulated by the students;
- include many different and varied examples of shapes so that irrelevant features do not become important (many students, for example, believe that only equilateral triangles are really triangles or that squares turned 45° are no longer squares);
- involve lots of sorting, identifying, and describing of various shapes; and
- provide opportunities to build, make, draw, put together, and take apart shapes.

Level 1 activities should

- continue to use models as with level 0, but include models that permit the exploration of various properties of figures;

- begin to focus more on properties of figures rather than on simple identification. Define, measure, observe, and change properties with the use of models;
- classify figures based on properties of shapes as well as by names of shapes. For example, find different properties of triangles that make some alike and others different; and
- use problem-solving contexts in which properties of shapes are important components.

Level 2 activities should

- continue to use models with a focus on defining properties. Make property lists and discuss which properties are necessary and which are sufficient conditions for a specific shape or concept;
- include language of an informal deductive nature: all, some, none, if-then, what if, and so forth;
- investigate the converse of certain relationships for validity: for example, the converse of "if it is a square, it must have four right angles" is "if it has four right angles, it must be a square";
- using models and drawings as tools to think with, begin to look for generalizations and counterexamples; and
- encourage hypothesis making and hypothesis testing.

Most of the content of the elementary school curriculum can be adapted to any of the three levels. An exception may be an inappropriate attention to abstract concepts such as point, line, ray, and plane as basic elements of geometric forms. These abstract ideas are not even appropriate at level 2.

Listen to your children during a geometry activity. Compare the types of comments and observations that they make with the descriptions of the first two van Hiele levels. Be sure that the activities you plan do not require students to reason above their level of thought.

The activities suggested in the remainder of this chapter have been grouped according to the first three van Hiele levels. This is just a suggestion for getting started. Each activity has the potential of being addressed at a slightly lower or higher level.

The activity descriptions are generally rather brief. There are so many excellent geometry activities in the literature that this entire book could easily be filled with them. The intent here is to get you started and to illustrate the wide variety of things that can be done. Find ideas that you like, and develop them fully. Search out additional resource books to help you.

INFORMAL GEOMETRY ACTIVITIES: LEVEL 0

The emphasis at level 0 is on shape and form experiences. While properties of figures are included, they are only explored informally. Remember that a level 0 activity does not necessarily mean a primary-grade activity. Not all students in the upper grades have had sufficient opportunity to experience ideas and develop their thinking beyond a beginning level.

Any categorization of informal geometry activities according to van Hiele levels is likely to be a bit fuzzy. You may think some activities represent level 1 better, and many can be extended easily to include that level.

EXPLORING SHAPES AND PROPERTIES

In this section the activities all begin with shapes already made or drawn, and students work with them in various ways. The shapes may be two- or three-dimensional. Some activities can be done with either a set of flat two-dimensional shapes or equally well with a set of solid shapes.

Sorting and Classifying Shapes

Sorting or classifying shapes using models is a good way to introduce geometric ideas. Names of shapes and properties can be provided as students begin to recognize and discuss them in their own words.

17.1 SHAPE SORT I

Make collections of posterboard shapes similar to those in Figure 17.5. You will want many more shapes than those shown. Have groups of children find sets of shapes that are alike in some way. If you prepare the pieces so that the ideas you want children to learn are represented by at least four or five different examples, it is likely that students will notice that concept. When students have sorted out some shapes indicating that they recognize some idea that is common, you have an opportunity to label the concept or provide the proper name of a shape without trying to define it formally. Figure 17.6 suggests some concepts that can be explored in this way. Notice the use of varied examples of each idea.

When students omit shapes from a category they have identified or fail to create a category you hoped they would discover, it is a clue to their perceptual thinking. Interact informally by selecting an appropriate shape and have children discuss why they think it may be different or the same as other shapes. Avoid definitions and right or wrong answers at this level.

Shape sorts can be done with three-dimensional shapes. Wooden or plastic collections are available as one option. Another is to make some solids from tagboard or modeling clay. Real objects such as cans, boxes, balls, or Styrofoam shapes are another source of three-dimensional models. Figure 17.7 (p. 330) illustrates some classifications of solids.

17 / GEOMETRIC THINKING AND GEOMETRIC CONCEPTS 329

FIGURE 17.5: *An assortment of shapes for sorting*

Matching Shapes

17.2 FEEL-IT MATCH

Prepare two identical collections of shapes. One set is placed in view. Without children seeing, place a shape from the other set in a box or bag. Children reach in the box, feel the shape without looking, and attempt to find the matching shape from those that are displayed. The activity can be done with either two-dimensional tagboard shapes or with solids. The shapes in the collection determine what ideas will be focused on and how difficult the activity is.

- All shapes have different numbers of sides or faces.
- All shapes are different but belong to one category, such as quadrilaterals, triangles, pyramids, prisms, curved surfaces, curved edges.

Shapes with curved edges

Opposite sides "go the same way" —parallelograms.

Three sides —triangles

Shapes with a "square corner" —right angle

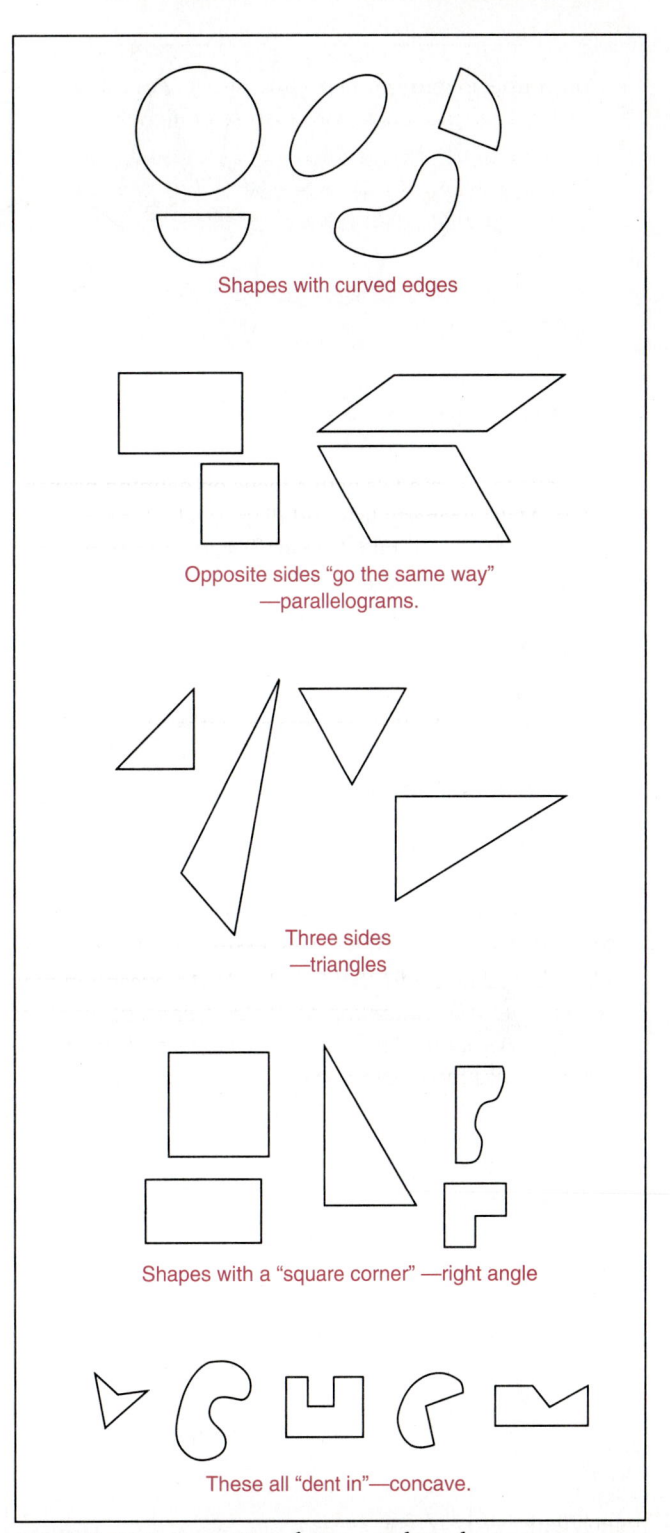

These all "dent in"—concave.

FIGURE 17.6: *By sorting shapes, students begin to recognize properties.*

A variation is to have the hidden shapes be small versions of the shapes the students see. Matching a small shape with a larger one provides an informal introduction to the notion of similarity: same shape, different size.

The following matching activity is similar but can be done with the entire class and provides more opportunity for informal discussion of how shapes are alike and what their properties are.

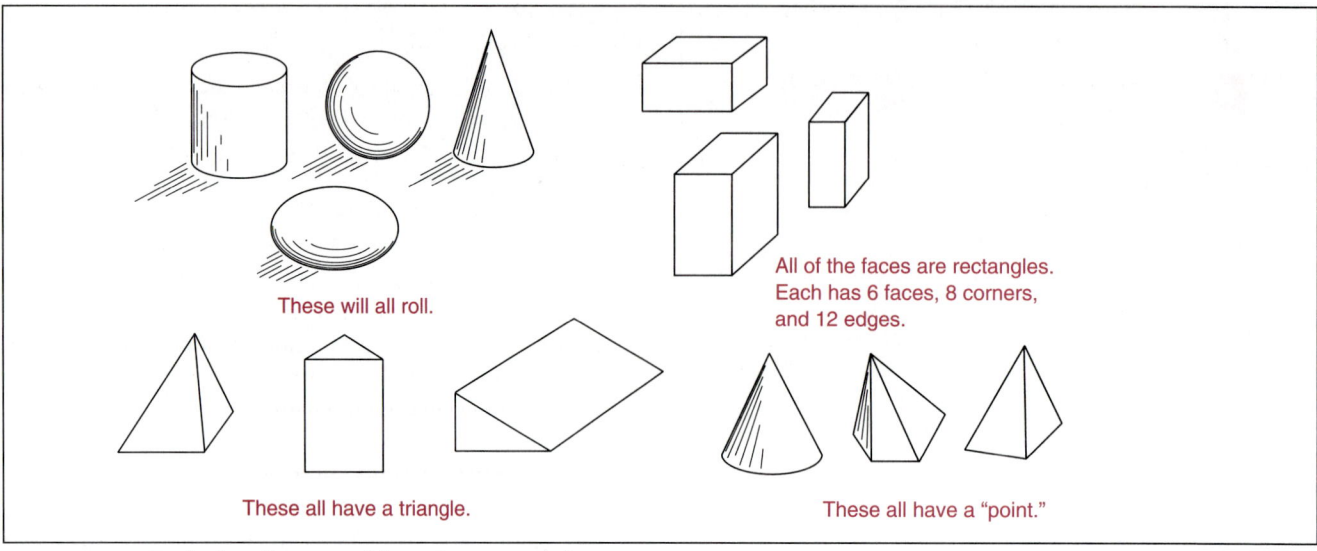

330 17 / GEOMETRIC THINKING AND GEOMETRIC CONCEPTS

These will all roll.

All of the faces are rectangles. Each has 6 faces, 8 corners, and 12 edges.

These all have a triangle.

These all have a "point."

FIGURE 17.7: *Early classifications of three-dimensional shapes*

17.3 WHAT'S LIKE THIS?

Display a collection of shapes for all to see (either two- or three-dimensional). Show the class a shape that has something in common with one or more of the shapes in the collection. Students are to select the shape that is like your shape in some way and explain their choice. The target shape may be another example of a particular shape or it may be entirely different with only some property in common with another shape. There may be excellent choices and reasons that you did not even think about. Figure 17.8 illustrates only a few of the many possibilities for this activity.

Another type of matching activity matches solid shapes with copies of the faces. "Face cards" can be made by tracing around the different faces of a solid shape, as in Figure 17.9.

17.4 FACE MATCH

Two different matching activities can be done with face cards: Given a face card, find the solid; or, given a solid, find the face card. If cards are made with only one face per card, students can select from a larger collection those faces that belong with a particular shape. Another variation is to show only one face at a time as clues.

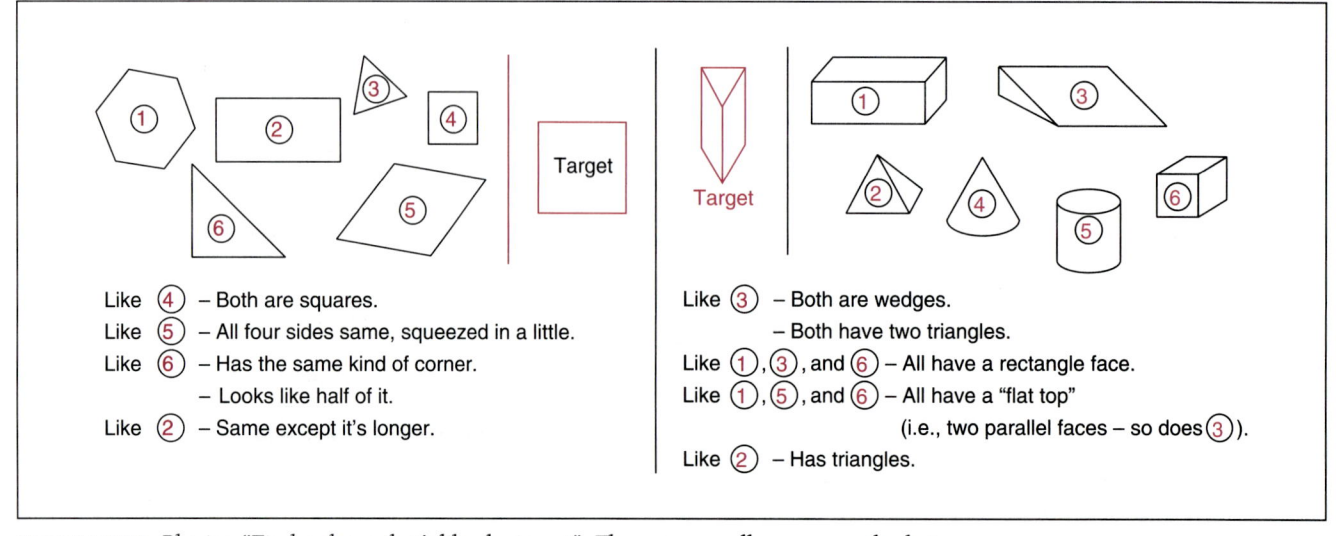

Target

Like ④ – Both are squares.
Like ⑤ – All four sides same, squeezed in a little.
Like ⑥ – Has the same kind of corner.
  – Looks like half of it.
Like ② – Same except it's longer.

Target

Like ③ – Both are wedges.
  – Both have two triangles.
Like ①, ③, and ⑥ – All have a rectangle face.
Like ①, ⑤, and ⑥ – All have a "flat top"
  (i.e., two parallel faces – so does ③).
Like ② – Has triangles.

FIGURE 17.8: *Playing "Find a shape that's like the target": There are usually many good solutions.*

Going on a "shape hunt" is a well-worn but still worthwhile activity.

17.5 SHAPE HUNT

Have students search not just for triangles, circles, squares, and rectangles, but also for properties of shapes. (What you have students look for is a way to make this either a level 0 activity or a level 1 activity.) A shape hunt will be much more successful if you let students look for either one thing or a specific list. Different groups can hunt for different things. Some examples of things to search for are

parallel lines (lines "going in the same direction")

right angles ("square corners")

curved surfaces or curved lines

two or more shapes that make another shape

circle inside each other (concentric)

shapes with "dents" (concave) or without dents (convex)

shapes used over and over in a pattern (brick wall, chain-link fence)

solids that are somehow *like* a box, a cylinder (tube), pyramid, a cone

five shapes that are alike somehow (specify solid or flat)

shapes that are symmetrical

BUILD, DRAW, MAKE

In this section the activities all have children creating shapes in some way. With a new set of materials to work with, it is good to begin by letting students freely make whatever shapes or designs they wish. This permits children to experience the new materials (D-stix, clay, straws, tiles, ...) and to do the construction they want to. Prepared activities can then go beyond this free-play level and challenge children to build shapes having a particular property or feature. These challenges promote reflective thinking about the properties involved and are a good way to encourage level 1 thinking without pushing children too hard.

◆ Using Tiles to Make Shapes

A good way to explore shapes at level 0 is to use smaller shapes or tiles to create larger shapes. Different criteria or directions can provide the intended focus to the activity. Among the best materials for this purpose are pattern blocks, but many teacher-made materials can be used. In Figure 17.10 (p. 332) a variety of different shapes are suggested. Class sets can be cut from poster board and placed in plastic bags for individual or group use. Some of the activities that follow can be repeated using a different set of tiles to provide not only variety but a different perspective.

17.6 TILE SHAPES

Make a specific type of shape using one or two different types of tiles. Exactly what shapes can be suggested is determined by the tiles being used. Try some of

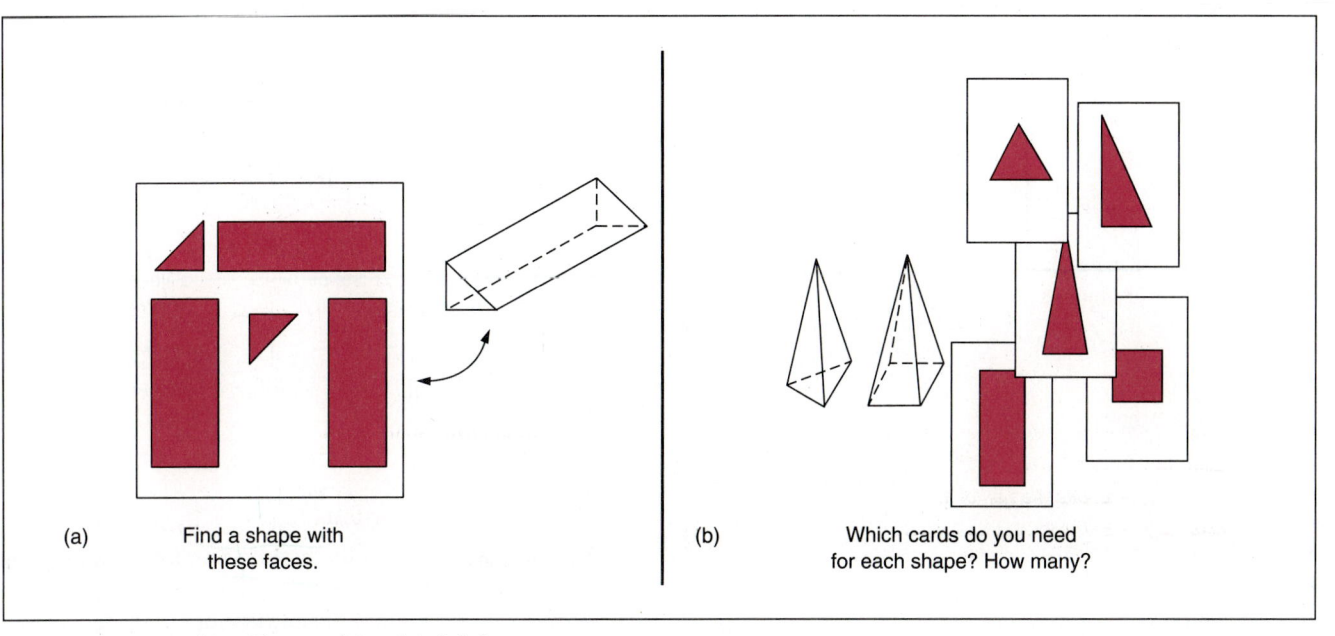

(a) Find a shape with these faces.

(b) Which cards do you need for each shape? How many?

FIGURE 17.9: *Matching "face cards" with solid shapes*

332 17 / GEOMETRIC THINKING AND GEOMETRIC CONCEPTS

these ideas with different tiles, and make up more of your own.

- Make some different triangles. Make big ones and little ones. What do you notice?
- Make some rectangles. How many different ones can you make?
- Make some shapes with four sides. (Try other numbers.)
- Make some parallelograms (or trapezoids, rhombuses, squares, or hexagons). How are your shapes alike? How are they different?

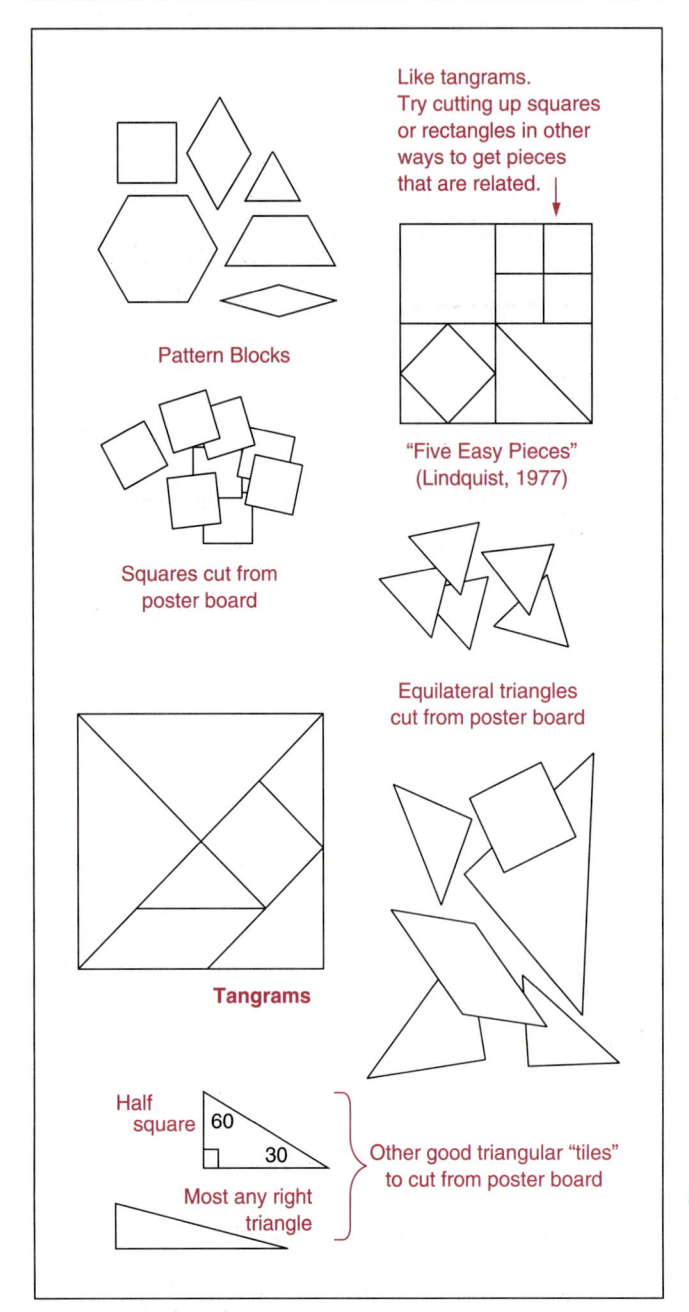

FIGURE 17.10: *Collections of tiles can include an assortment of shapes or can be all the same shape.*

17.7 TILE CHALLENGES

Make some shapes that _____. Fill this in with different properties. Again, what is possible will vary with the tiles being used.

- Make a shape that has parallel sides (or that has no parallel sides, or two, or even three sets of parallel sides) [Figure 17.11(a)].
- Make some shapes that have square corners. Make some with two right angles, three, four, . . . [Figure 17.11(b)].
- Make a shape that has a line of symmetry. Check it by placing a mirror on the shape [Figure 17.11(c)].

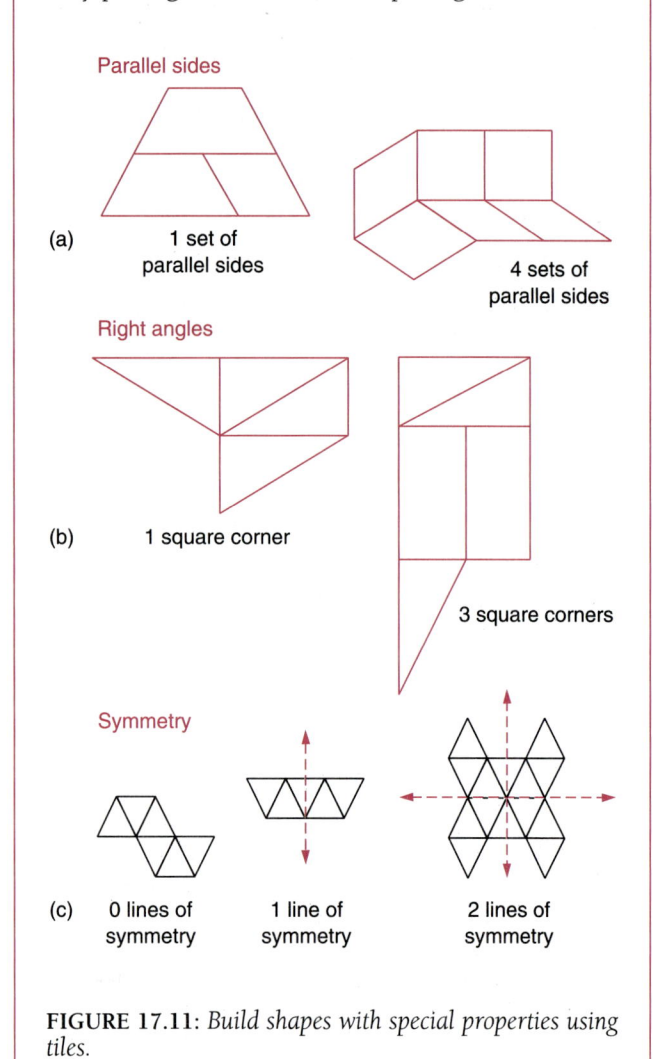

FIGURE 17.11: *Build shapes with special properties using tiles.*

◆ Geoboard Explorations

The geoboard is one of the best devices for drawing two-dimensional shapes. Literally hundreds of activities, task cards, and worksheets have been developed for geoboards. Here are just a few activities for a beginning level.

17 / GEOMETRIC THINKING AND GEOMETRIC CONCEPTS 333

17.8 GEOBOARD COPY

Copy shapes, designs, and patterns from prepared cards as in Figure 17.12. Begin with designs shown with dots as on a geoboard, and later have students copy designs drawn without dots.

Have children copy shapes from pattern cards onto a geoboard.

Besides pattern cards with and without dots, have children copy <u>real</u> shapes–tables, houses, letters of the alphabet, etc.

FIGURE 17.12: *Shapes on geoboards*

17.9 GEOBOARD TILES

Challenge the students to make shapes on the geoboard using combinations of only one smaller shape as if the smaller shape were a tile (Figure 17.13). Cards with the smaller shapes can be prepared to direct the activity.

Geoboard Notes

Two practical comments should be made about geoboards. First, have lots of them available in the classroom. It is better for two or three children to have 10 to 12 boards at a station

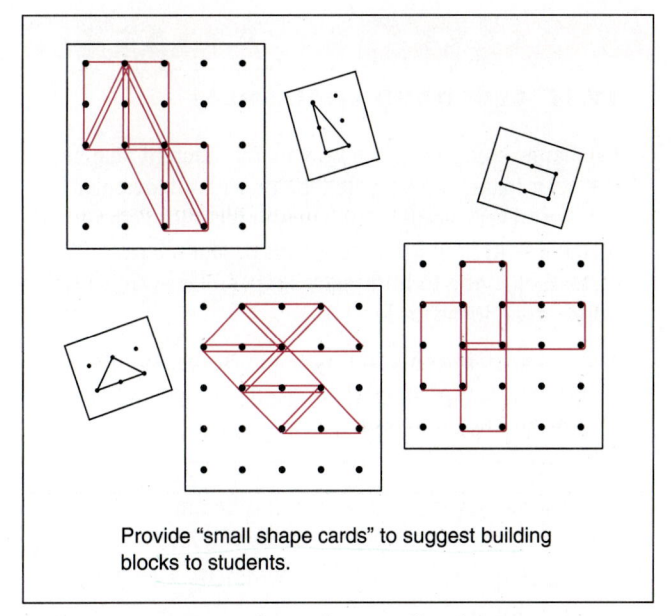

Provide "small shape cards" to suggest building blocks to students.

FIGURE 17.13: *Bigger shapes from smaller shapes*

17.10 CONGRUENT PARTS

Copy a shape from a card, and then have students subdivide or cut it into smaller shapes. Specify the number of smaller shapes. Also specify whether they are all to be congruent or simply of the same type as shown in Figure 17.14. Depending on the shapes involved, this activity can be made quite easy or relatively challenging.

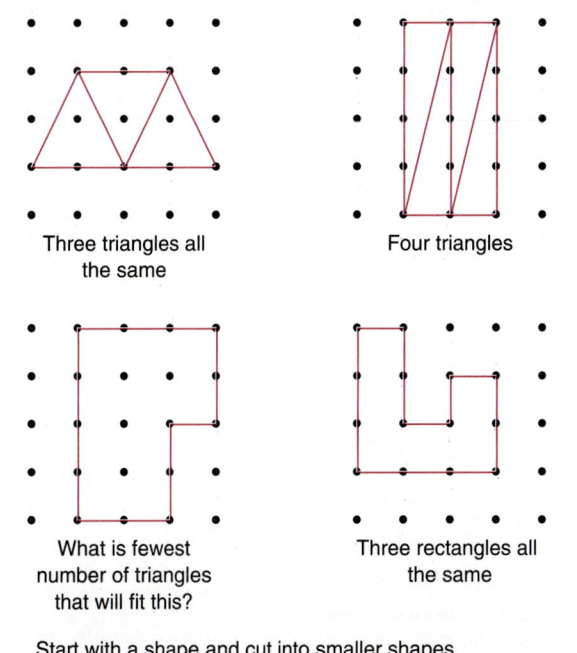

Three triangles all the same

Four triangles

What is fewest number of triangles that will fit this?

Three rectangles all the same

Start with a shape and cut into smaller shapes. Add special conditions to make the activity challenging.

FIGURE 17.14: *Subdividing shapes*

17.11 GEOBOARD CHALLENGES

Challenge students to see how many different shapes they can make of a specific type or with a particular property. (Very young children will feel more comfortable searching for *three* shapes or *four* shapes instead of trying to find many shapes.) Here are some appropriate ideas for level 0.

Make shapes with five sides (or some other number).

Make shapes with all square corners. Can you make one with three sides? four sides? six, seven, eight sides?

Make some trapezoids that are all different (or any other shape that students can identify).

Make a shape that has a line of symmetry. Check it with a mirror.

---

than for each to only have one. This way a variety of shapes can be made and compared before they are changed.

That leads to the second point. Teach students from the very beginning to copy their geoboard designs onto paper. Paper copies permit students to create complete sets of drawings that fulfill a particular task. If a student wants to make a series of different six-sided shapes but has only one geoboard, the paper permits him or her to copy the entire series. Drawings can be placed on the bulletin board for classification and discussion, made into booklets illustrating a new idea that is being discussed, and sent home to show Mom and Dad what is happening in geometry.

Younger students can use paper with a single geoboard on each sheet. At the kindergarten or first-grade level, the paper geoboard can be the same size as the real board. Later, a paper board about 10 cm square is adequate. Older children can use centimeter dot paper or a page of small boards, each about 5 cm square. (Black-line Masters).

In the very early grades, children will have some difficulty copying geoboard designs onto paper, especially designs with slanted lines. To help, suggest that they first mark the dots for the corners of their shape. (Example: "second row, end peg.") With the corners identified it is much easier for them to draw lines to make the shape.

◆ Dot and Grid Paper Explorations

Geoboards are excellent due to the ease with which drawings can be made and changed. They do have limitations in terms of size, arrangement, and number of pegs. Assorted dot and grid papers provide an alternative way to make drawings and explore shapes. Virtually all of the activities suggested for tiles and geoboards can also be done on dot or grid paper. Changing the type of paper changes the activity and provides new opportunity for insight and discovery. Figure 17.15 shows several possibilities for dot and grid papers, included in the Black-line Masters.

Square grids

Isometric grids

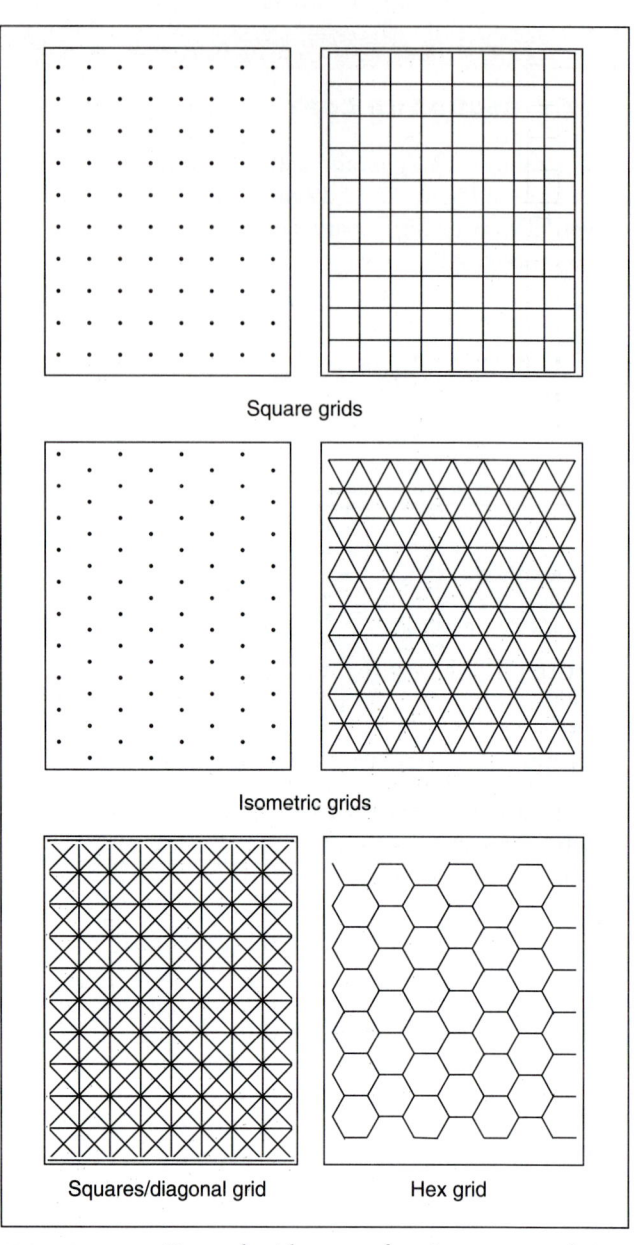

Squares/diagonal grid Hex grid

FIGURE 17.15: *Dot and grid paper of various types and sizes can be used for many geometric explorations.*

Besides the geoboard and tiles activities mentioned so far, here are some additional ideas that lend themselves particularly well to dot and grid paper.

17.12 THREE-DIMENSIONAL DRAWINGS

Isometric dot paper is an effective way to draw solid shapes that are built with cubes. Square dot paper can be used to draw side and top views, as shown in Figure 17.16. Building a simple shape with cubes and then drawing plan and perspective views is an excellent activity for middle-grade students to help with perspective. (See Winter, Lappan, Phillips, & Fitzgerald, 1986, for a series of activities.)

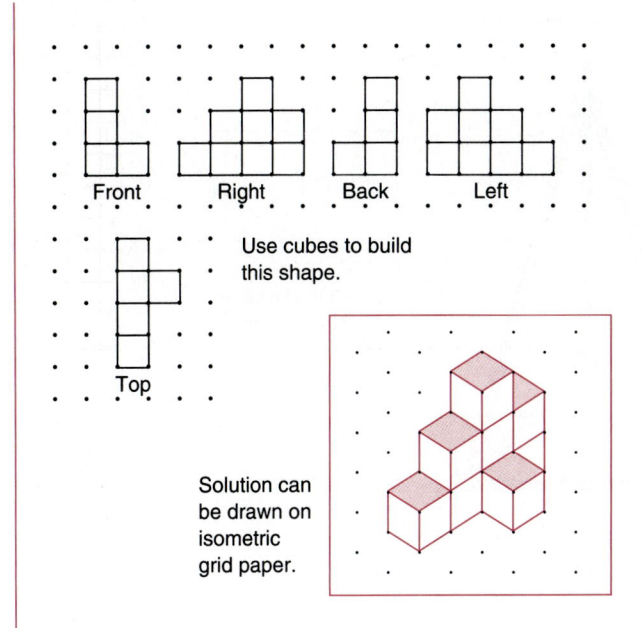

FIGURE 17.16: *Develop perspective and visual perception with cubes and plan views. Draw block "buildings" on isometric dot grids.*

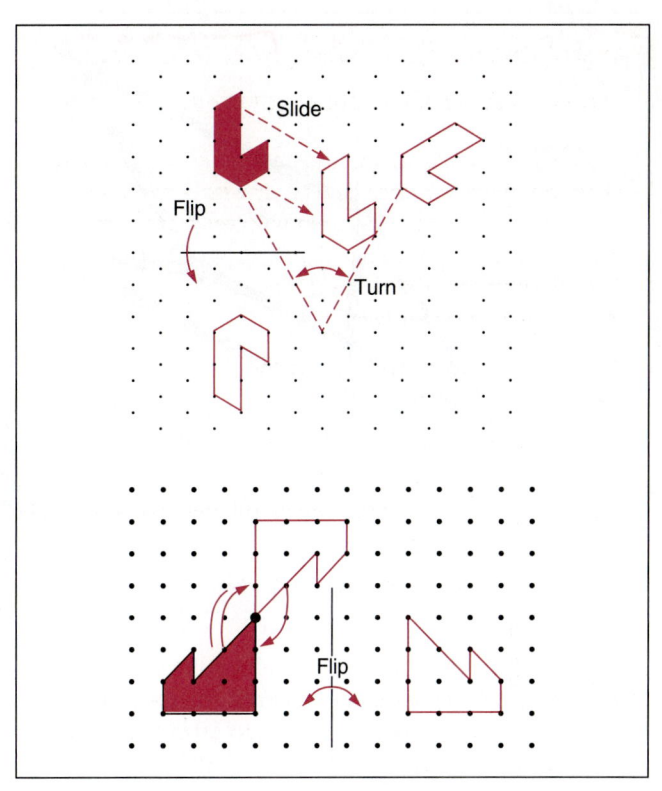

FIGURE 17.17: *Slides, flips, and turns can be explored on most any type of grid paper.*

A slightly different idea is to essentially reverse the last activity. Provide students with a perspective drawing that you have made, and have them build the structure with blocks. Drawings can be prepared on worksheets or you can begin to develop a collection of cards to be used at work stations. While an answer key is possible in the form of a top view with the number of cubes indicated in each square, it is best to let students work together to decide if they have made an accurate construction.

17.13 SLIDES, FLIPS, AND TURNS

Slides, flips, and turns can be investigated on any grid. Start with a simple shape. Draw the same shape flipped over, or turned, or placed in a different orientation. Trace the original shape, and cut it out. This copy can be flipped or reoriented to test the drawings that are made (Figure 17.17).

Building Solids

Building solid or three-dimensional shapes presents a little more difficulty than two-dimensional shapes, but is perhaps even more important. Building a model of a three-dimensional shape is an informal way to get to know and understand the shape intuitively in terms of its component parts.

Skeleton models are three-dimensional "solids" that are built using sticks of some sort. There are many ways that this can be done. In addition to commercial materials such as D-stix, the following two ideas are highly recommended.

Plastic drinking straws with flexible joints: Cut the straws lengthwise with scissors from the top down to the flexible joint. These slit ends can then be inserted into the uncut bottom ends of other straws making a strong but flexible joint. Three or more straws are joined in this fashion to form two-dimensional polygons. To make skeletal solids, use tape or wire twist ties to join two polygons side to side. When more and more polygons are connected, a three-dimensional solid can be formed (Prentice, 1989).

Rolled newspaper rods: Fantastic super-large skeletons can be built using newspaper and masking tape. Roll three large sheets of newspaper on the diagonal to form a rod approximately one meter long. The more tightly the paper is rolled, the less likely the rod is to bend. Secure the roll at the center with a bit of tape. The ends of the rods are thin and flexible for about six inches where there is less paper. Connect rods by bunching this thin part together and fastening with tape. Use the tape freely, wrapping it several times around each joint. Additional rods can be joined after two or three are already taped. Structures will very soon become quite large (Figure 17.18, p. 336).

Regardless of the method used, students should first experience the rigidity of a triangle and compare it with the lack of rigidity of polygons with more than three sides. Point out how the rigidity of the triangle is used in many bridges, in the long booms of construction cranes, in gates and screen doors,

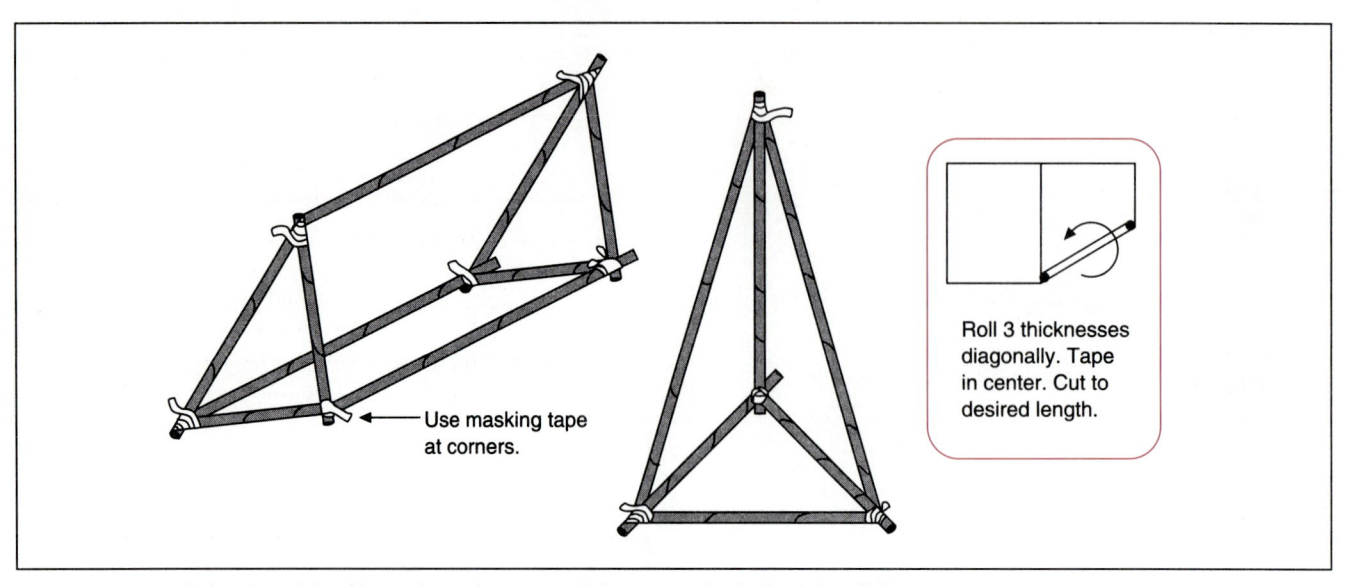

FIGURE 17.18: *Skeletal models of large three-dimensional shapes can be built with rolled-up newspaper.*

and in the structural parts of many buildings. As children begin to build large skeleton structures, they will find that they need to add diagonal members to form triangles. The more triangles, the less likely their structure will collapse.

17.14 UNFOLDED SOLIDS

Have students design and test nets for various solids. (A *net* for a solid is a flat shape that will fold up to make that solid.) If a square centimeter grid is used, parallel lines and angles can be drawn without tedious measuring (Figure 17.19). Paper nets can be traced or pasted to tagboard for a sturdier solid. Circular cones are easily made by cutting a sector from a circle. Experiment with different circle sizes and different sectors. Much of the value in folding up a solid from tagboard is planning what shape the faces will be and where faces can be connected. Encourage groups to solve these problems themselves; provide only as much help as necessary.

Solids can also be built from blocks. Either small wooden cubes (1 in. or 2 cm) or plastic connecting cubes are ideal. The following tasks can make such activities challenging and valuable.

17.15 BLOCK BOXES

In groups, have students see how many different rectangular solids can be made using just 12 cubes in each. (A *rectangular solid* has six faces, and each face is a rectangle.) Try other numbers of cubes. When are

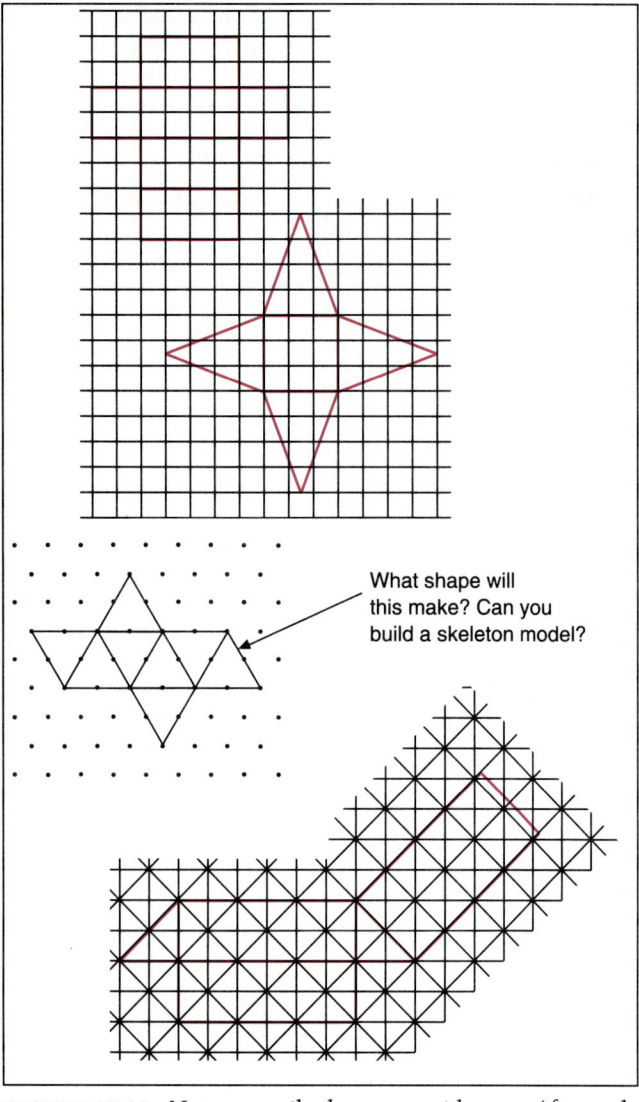

FIGURE 17.19: *Nets are easily drawn on grid paper. After making one net for a solid, try to fiind others that will fold up to make the same solid.*

two rectangular solids congruent (exactly the same)? How would you have to turn one solid to get it in the same orientation as another that is the same shape?

17.16 BUILDING PLANS

From a simple building that you have built, make a task card with five views of the building, as in Figure 17.20. Students use trial-and-error to try to build the same building from the five views. Slightly harder is to omit the top view but give the total number of cubes. This is a good exercise in spatial perception.

GEOMETRIC PROBLEM-SOLVING ACTIVITIES

Many excellent geometry activities are essentially spatial process problems. Content of a traditional nature may be minimal while the problem-solving value is significant. These geometric problem-solving activities are just as important as verbal problem solving and also provide opportunity for growth in geometric thinking and spatial sense.

At level 0, geometric problems involve the manipulation, drawing, and creation of shapes of all types. The tasks revolve around global features of shapes rather than any analysis of properties or relationships between classes or figures. Many activities already suggested have a problem-solving orientation. The following are additional suggestions.

◆ **Geometric Puzzles**

17.17 TANGRAM PUZZLES

Tangrams have been a popular geometric puzzle for years. Figure 17.21 (p. 338) shows tangram puzzles of different difficulties. Easy ones are appropriate even for preschool children. Several good books have a wide assortment of tangram puzzles, for example, Fair (1987) and Seymour (1971).

17.18 PENTOMINOES

Pentominoes are the set of all shapes that can be made from five squares connected as if cut from a square grid. Each square must have at least one side in common with another square. It is an excellent geometric puzzle: students find the complete set of 12 different pentominoes shown in Figure 17.22 (p. 338). (Do not tell the students how many different shapes there are. Good discussions can come from deciding if some shapes are really different and if all shapes have in fact been found.)

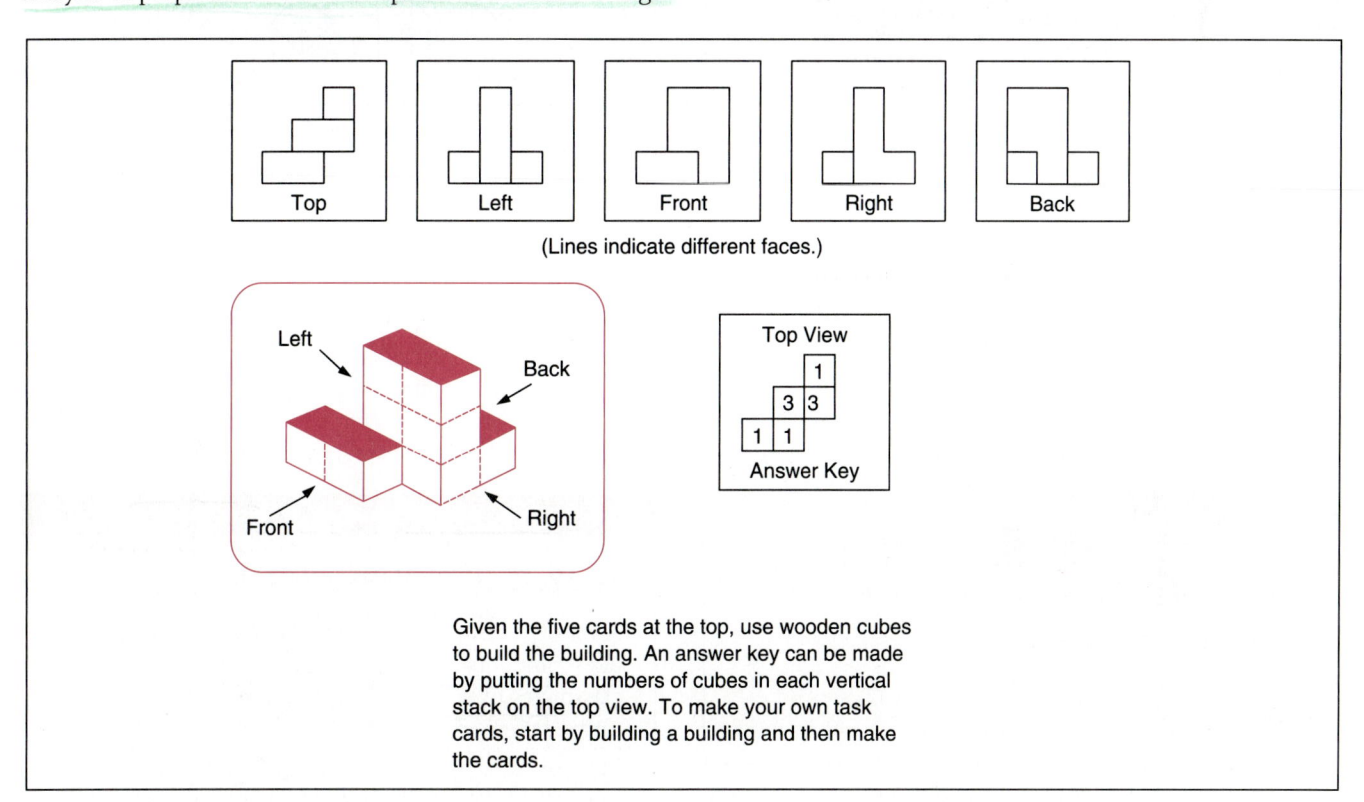

Given the five cards at the top, use wooden cubes to build the building. An answer key can be made by putting the numbers of cubes in each vertical stack on the top view. To make your own task cards, start by building a building and then make the cards.

FIGURE 17.20: *Views of a simple building*

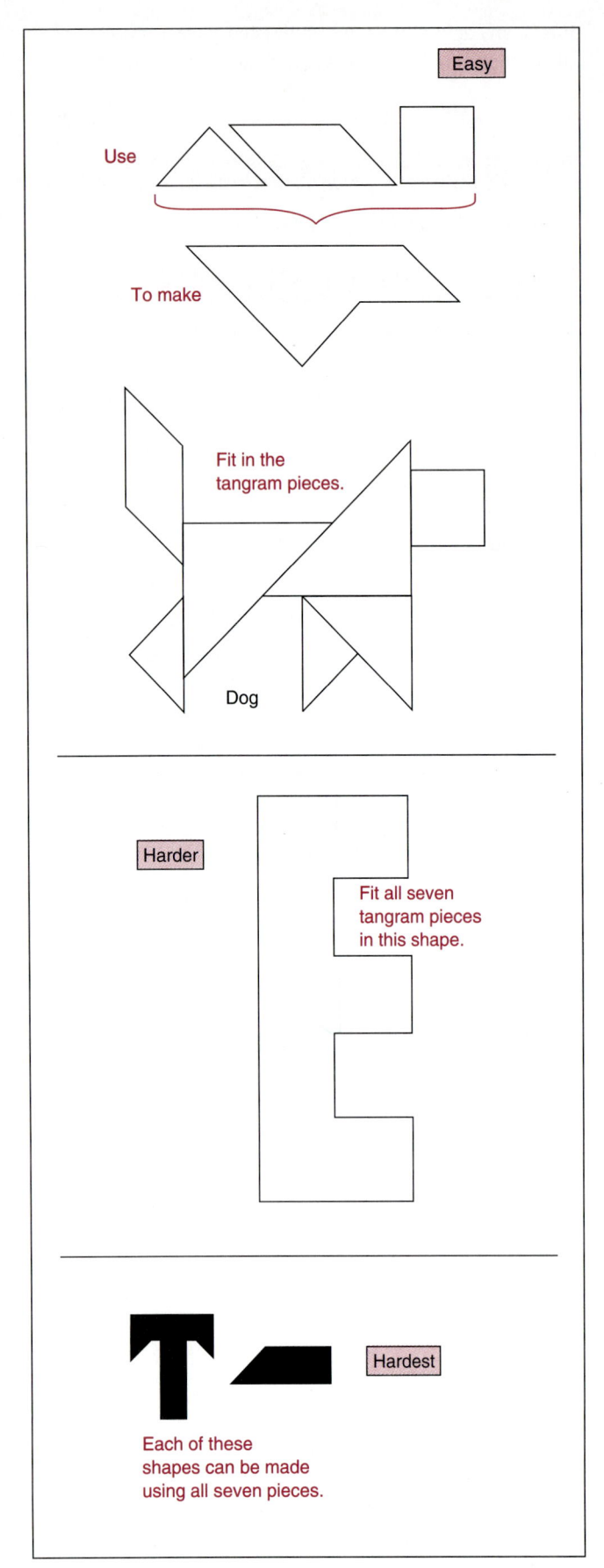

338   17 / GEOMETRIC THINKING AND GEOMETRIC CONCEPTS

Easy

Use [three tangram pieces] To make [larger triangle]

Fit in the tangram pieces.

Dog

Harder

Fit all seven tangram pieces in this shape.

Hardest

Each of these shapes can be made using all seven pieces.

FIGURE 17.21: *Four different types of tangram puzzles illustrate a range of difficulty levels.*

There are 12 pentominoes.

Finding all possible shapes made with five squares, or six squares (called "hexominoes"), or six equilateral triangles, etc., is a good excercise in spatial problem solving.

Four of the different shapes that six equilateral triangles make.

Four of the different shapes that four "half-square" triangles will make.

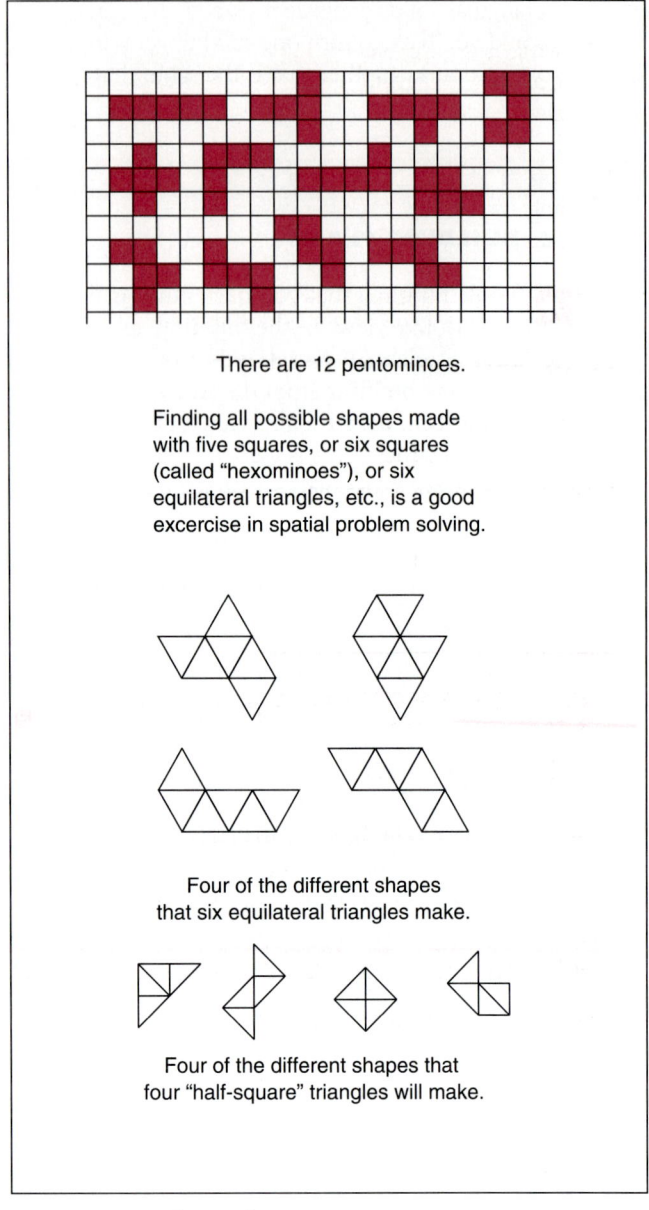

FIGURE 17.22: *Pentominoes*

Once students have decided that there are just 12 pentominoes, the 12 pieces can then be used in a variety of activities. Paste the grids with the children's pentominoes onto tagboard and let them cut out the 12 shapes. These can be used in the next two activities.

17.19 PENTOMINOE PUZZLES

A variety of puzzles can be made from the pentominoes. Students can make puzzles to challenge their classmates by selecting four or more pieces and putting them together into a single shape. The outline of this shape is then drawn on grid paper. The challenge is to find a way to create the shape without

17 / GEOMETRIC THINKING AND GEOMETRIC CONCEPTS   339

knowing which pieces were used. The more pieces there are, the more difficult will be the puzzle. Many puzzles will have more than one solution. Much harder is to use all 12 pieces to construct a rectangle. It is possible to make a 6 × 10, 5 × 12, and a 4 × 15 rectangle using the 12 pentominoes.

17.20 PENTOMINOE SQUEEZE

Play **Pentominoe Squeeze** on an 8 × 8 grid of squares. Pieces are dealt out randomly to two players. The player with the cross places it on the board. In turn players place pieces on the board without overlapping any piece that is already placed. The last player to be able to play is the winner. Smaller boards can be used to make the game more difficult.

17.21 HEXOMINOES

Find out how many hexominoes there are (made with six squares). This is quite a challenge.

17.22 TRIANGLE PATTERNS

Use six equilateral triangles to make shapes in the same way as pentominoes. An even more challenging activity is to see how many shapes can be made from five 45° right triangles (halves of squares). Sides that touch must be the same length. (There are 14 shapes when only four right triangles are used.)

◆ "How Many Ways" Problems

Many of the tasks with geoboards, tiles, and other materials involve seeing how many different shapes can be made with a particular property. Similar challenges can be posed that do not involve a standard geometric concept.

17.23 HOW MANY WAYS/GEOBOARDS

On a geoboard, how many shapes can you make:

that have no pegs inside (or exactly one peg; two pegs)?

that will fit around this shape (any cardboard shape)?

that touch exactly five pegs (or some other number)?

that can be cut into two identical parts with one line on the geoboard?

17.24 HOW MANY BUILDINGS?

With cubes, how many different buildings can you make with exactly five cubes? Touching cubes must have a complete face in common. This is a three-dimensional version of pentominoes. (With plastic interlocking cubes, the number of possibilities increases. Why?) How many ways can you build a building using eight cubes if they must all touch on a whole face and no more than three cubes can be in any one row? (The numbers can be changed to change the problem.)

TESSELLATIONS

A *tessellation* is a tiling of the plane using one or more shapes in a repeated pattern with no holes or gaps. The "Tiling Pattern" activity at the beginning of the chapter was a tessellation activity. Making tessellations, or tiling patterns, is a good way for level 0 students from first grade to eighth grade to engage in geometric problem solving. Not only is there a nonanalytic interaction with geometric form, there is also challenge and artistic interest as well. One-shape or two-shape tessellation activities can vary considerably in relative difficulty and still remain level 0 activities.

Some shapes are easier to tessellate with than others, as illustrated in Figure 17.23 (p. 340). When the shapes can be put together in more than one pattern, both the problem-solving level and the creativity increase. Literally hundreds of shapes can be used as tiles for tessellations. Every one of the 12 pentominoes will tessellate. It is fun to create shapes on various grid papers and then test them to see if they can be used to tessellate.

Most children will benefit from using actual tiles with which to create patterns. Tiles can be cut from construction paper. Simple tiles can be cut quickly on a paper cutter. Other tiles can be traced onto construction paper and several thicknesses cut at once with scissors. Older children may be able to use various dot or line grids and plan their tessellations with pencil and paper. Spend one period with tiles or grids, letting children experiment with various tiling patterns. To plan a tessellation, use only one color so that the focus is on the spatial relationships. To complete an artistic-looking tessellation, add a color design. Use only two colors with younger children and never more than four. Color designs should also be repeated regularly all over the tessellation.

Tessellations can be made by gluing construction paper tiles to large sheets of paper, by drawing them on dot or line grids, or by tracing around a poster board tile. Do not worry about the edges of tessellations. Work from the center out, leaving ragged edges to indicate that the pattern should go on and on.

INFORMAL GEOMETRY ACTIVITIES: LEVEL 1

The activities at level 1 begin to focus more on the properties of shapes including some analysis of those properties. For example, at level 0, triangles might have been sorted by "big" and "little," "pointy" and "not as pointy," or even "with square corners" and "without square corners." At level 1, the same set of triangles can be sorted by the relative sizes of the angles or the relative lengths of the sides. Combinations of these categories produce even more relationships. Most of the materials in the suggested activities are the same as those used at level 0. It is quite reasonable to have several similar investigations proceeding within one classroom with different groups pursuing tasks at different van Hiele levels.

CLASSIFYING SHAPES BY PROPERTIES

Sorting activities are grouped here in a similar manner as was done for level 0 activities. To promote the kind of thinking that is appropriate for level 1, shapes are presented so that specific properties and categorizations are clearly evident. For example, the shapes used might be quadrilaterals or even a collection of two or three types of quadrilaterals. You may have to be more direct in pointing out particular properties or categories that students do not notice. When a classification of shapes has been made and discussed and is well understood, the appropriate name for the classification can be supplied.

Special Categories of Two-Dimensional Shapes

Listed below are some important categorizations (or definitions) of two-dimensional shapes. Examples of these shapes can be found in Figure 17.24.

Simple Closed Curves

- **Concave or convex.** An intuitive definition of concave might be a "shape with a dent in it." If a simple closed curve is not concave, it is convex. A more precise definition of concave may be interesting to explore with older students.

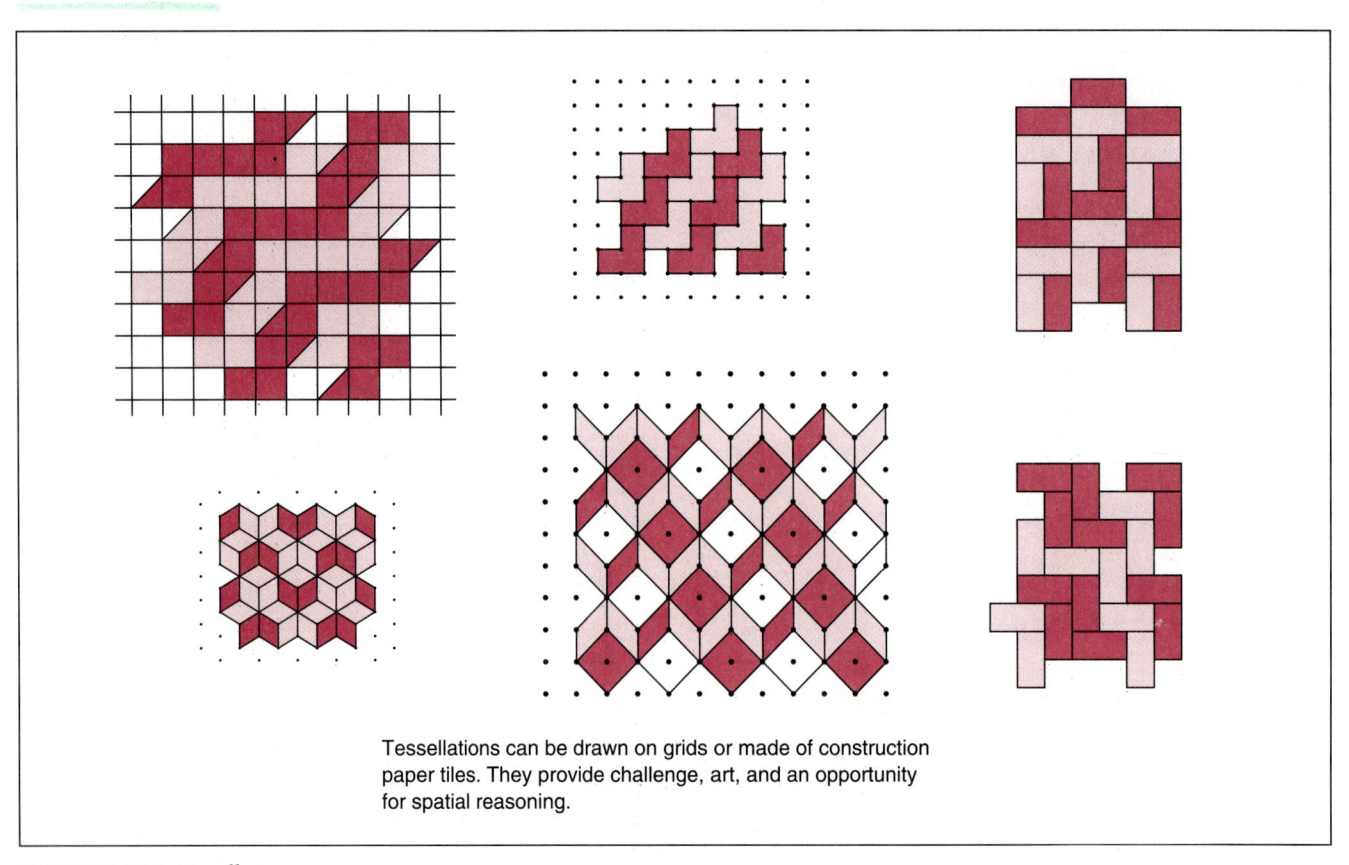

Tessellations can be drawn on grids or made of construction paper tiles. They provide challenge, art, and an opportunity for spatial reasoning.

FIGURE 17.23: *Tessellations*

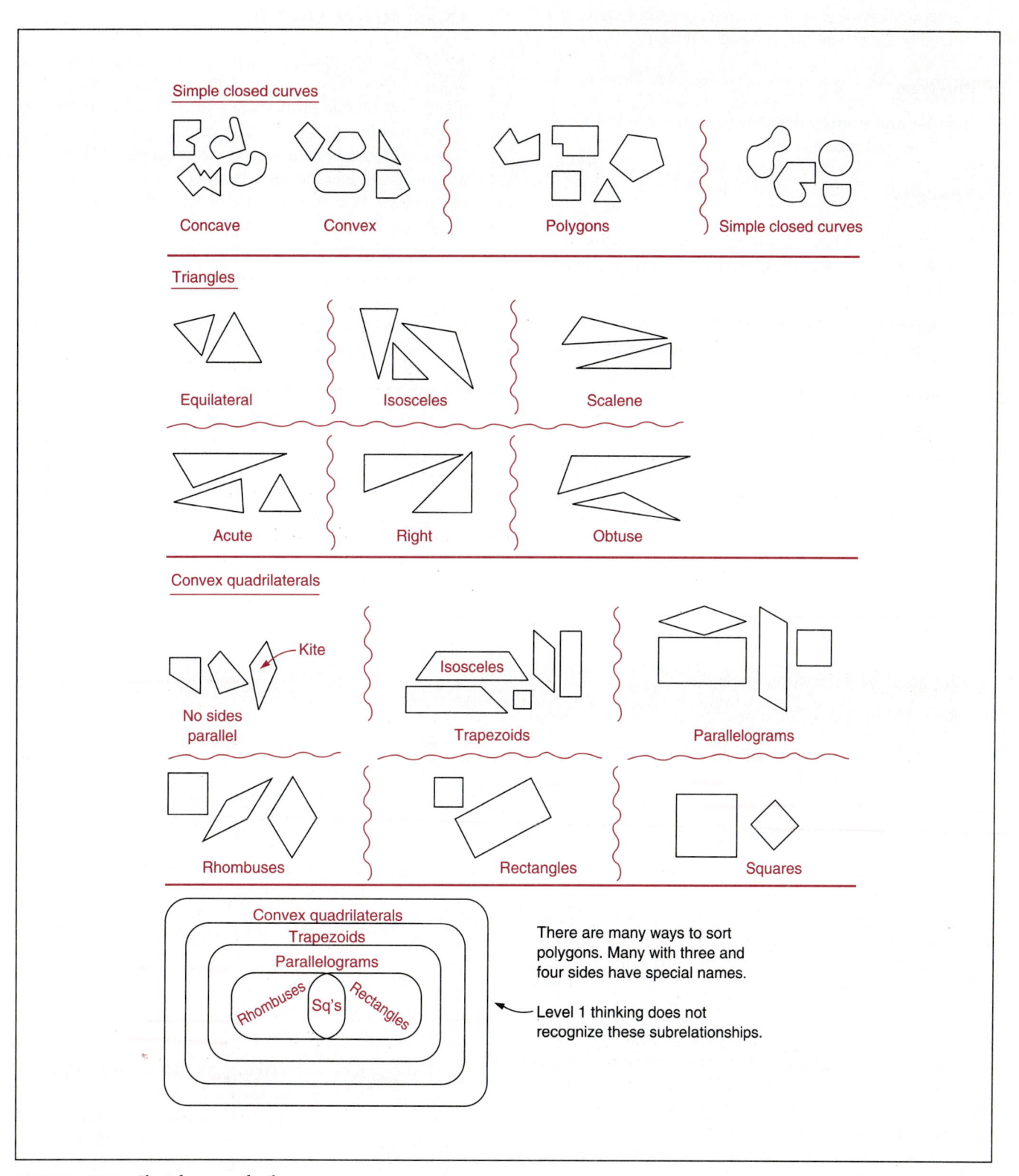

FIGURE 17.24: *Classification of polygons*

- **Symmetrical or nonsymmetrical.** Shapes can have one or more lines of symmetry and may or may not have rotational symmetry. These concepts will require more detailed investigation, as discussed later.

Polygons

Regular and nonregular. A regular polygon has all sides and all angles congruent.

Triangles

Classified by sides. Equilateral (all sides congruent), isosceles (only two sides congruent), scalene (no sides congruent).

Classified by angles. Right, acute (all smaller than right angles), obtuse (one angle greater than a right angle).

Quadrilaterals

Classified by the number of parallel sides. No sides parallel (no standard name by this scheme), trapezoid (at least one pair of parallel sides), parallelogram (two pairs of parallel sides).

Trapezoids

Isosceles or not isosceles.

Parallelograms

Classified by angles. Rectangle (has a right angle).

Classified by sides. Rhombus (has all sides congruent).

Square. Has a right angle and all sides congruent.

In the classification of quadrilaterals and parallelograms, the subsets are not all disjoint. For example, a square is a rectangle that is a rhombus. All parallelograms are trapezoids, but not all trapezoids are parallelograms.* Children at level 1 have difficulty seeing this type of subrelationship. They may quite correctly list all the properties of a square, a rhombus, and a rectangle and still identify a square as a "nonrhombus" or a "nonrectangle." Is it wrong for students to refer to subgroups as disjoint sets? By fourth or fifth grade, it is only wrong to promote or encourage such thinking. Burger (1985) points out that upper elementary students correctly use such classification schemes in other contexts. For example, individual students in a class can belong to more than one club. A square is an example of a quadrilateral that belongs to two other clubs.

Several specific approaches to sorting activities can help students grow in their understanding of how shapes are related to each other.

---

*Some definitions of trapezoid specify only one pair of parallel sides, in which case parallelograms would not be trapezoids.

17.25 SHAPE SORT II

Sort shapes by naming properties and not by names of the shapes. When two or more properties are combined, sort by one property at a time. "Find all of the shapes that have opposite sides parallel." (Find these.) "Now find those that *also* have a right angle." (This group should include squares as well as nonsquare rectangles.) After sorting, discuss what the name of the shapes is. Also try sorting by the same combinations of properties but in a different order.

Use loops of string to keep track of shapes as you sort them. Have students put all of the shapes with four congruent sides in one string and those with a right angle in the other group. Where do squares go? Let them wrestle with this dilemma until they can see that the two loops must be overlapped with the squares belonging to the intersection or overlapping region.

17.26 MYSTERY DEFINITION

Use an "all of these, none of these" type of activity as in Figure 17.25.

◆ Special Categories of Three-Dimensional Shapes

There is a wide variety of important and interesting shapes and relationships in three dimensions. Some classifications of solids are given below. At least an example or two of each of these can be made with clay or with tagboard. Other suggestions for making solids are given later.

All Solids

Sorted by edges and vertices (corners). No edges and no vertices, edges but no vertices, edges and vertices. Students can find real-world examples as well as use clay to make unusual examples.

Sorted by faces and surfaces. (A *face* is a flat surface of a solid.) Solids can be sorted by various combinations of faces and curved surfaces. Some have all faces, all curved surfaces, some of each, with and without edges, with and without vertices.

Parallel faces. Find, sort, or make solids with one or more pairs of parallel faces. Since faces are two-dimensional, students can refer to the shapes of the faces; for example, solids with two square faces that are parallel and two pairs of rectangle faces that are parallel.

17 / GEOMETRIC THINKING AND GEOMETRIC CONCEPTS   343

All of these have something in common.

None of these has it.

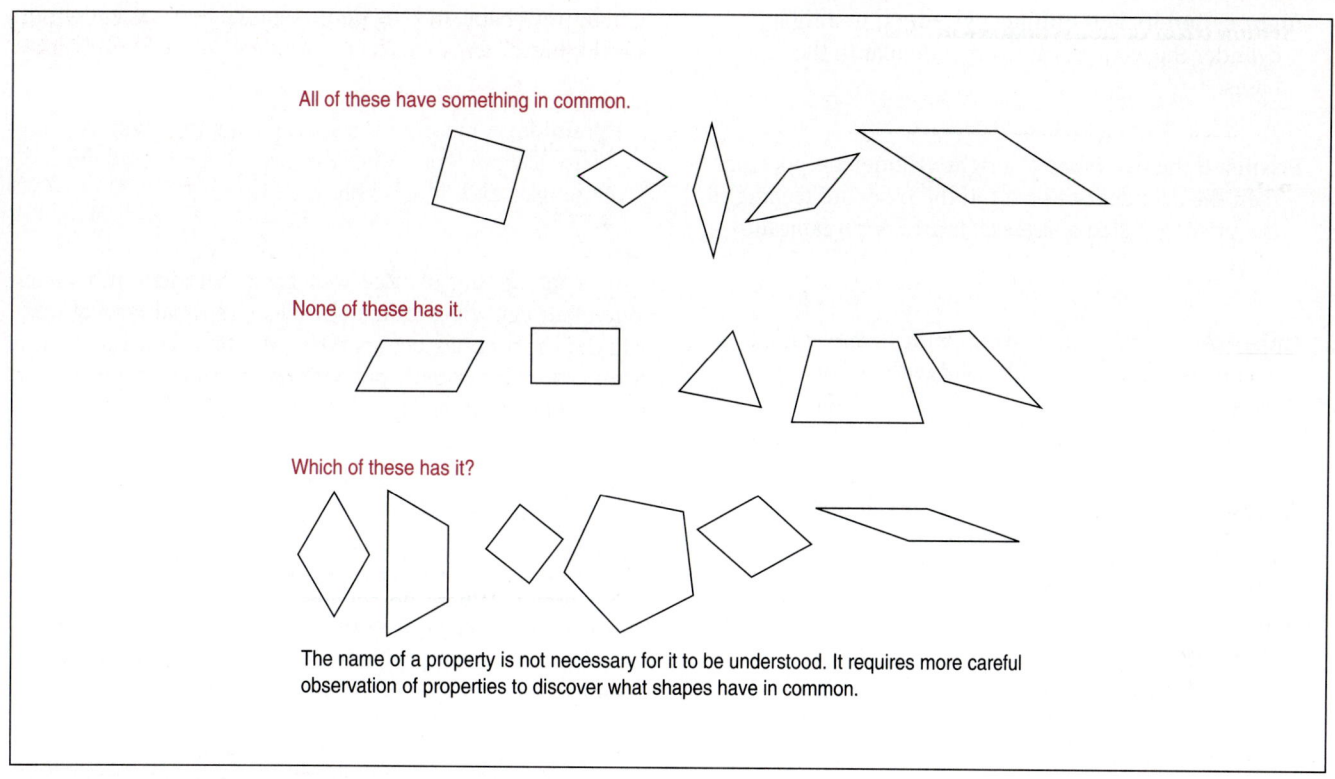

Which of these has it?

The name of a property is not necessary for it to be understood. It requires more careful observation of properties to discover what shapes have in common.

FIGURE 17.25: *All of these/none of these*

Cylinders

Examples of cylinders are shown in Figure 17.26. Two properties separate cylinders from other solids. First, *cylinders* have two congruent faces called *bases*. The bases are in parallel planes. Second, all lines joining corresponding points on the bases are parallel. These parallel lines are called *elements*. In verbal form this description is quite difficult. Models are very useful for discussion.

Cylinders

Not cylinders

Special cylinders

Prisms   Right prisms   Right cylinders (not prisms)

Cylinders have two parallel faces, and parallel lines join corresponding points on these faces. If the parallel faces are polygons, the cylinder can be called a prism.

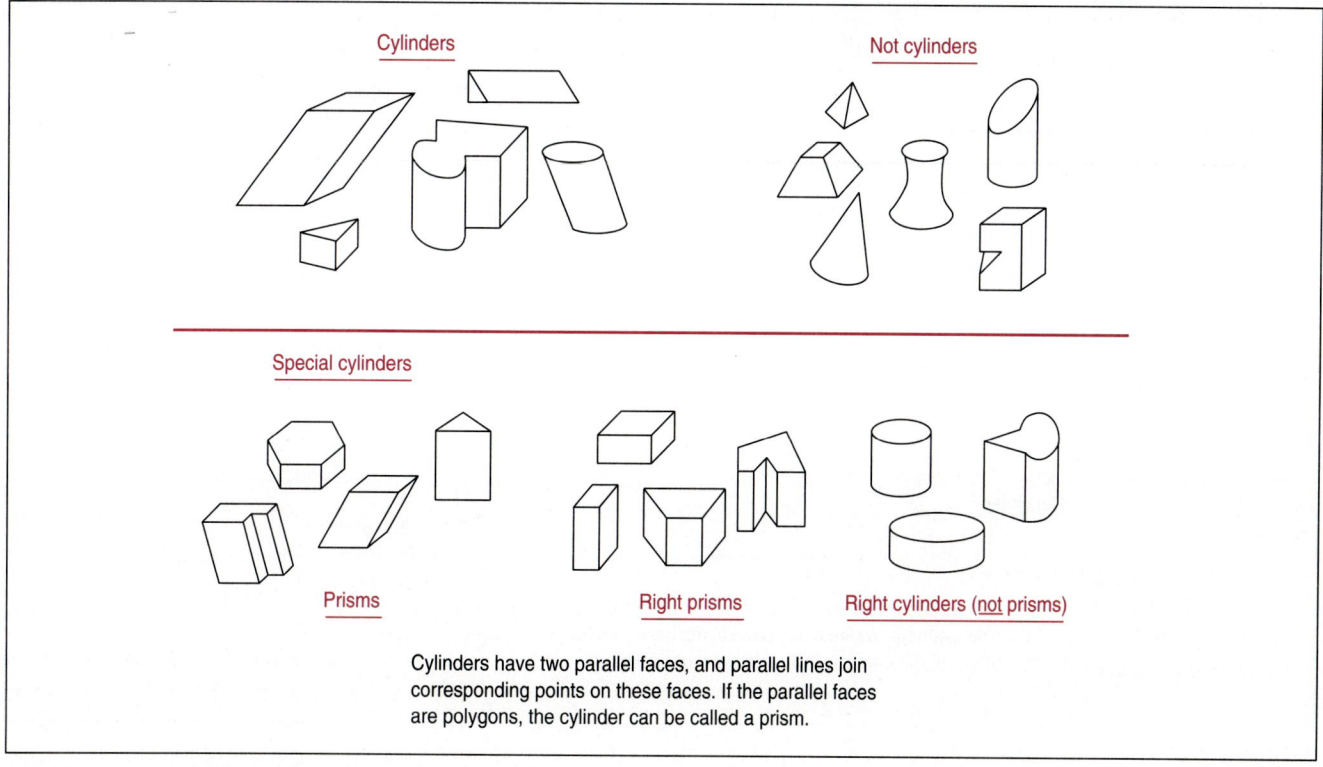

FIGURE 17.26: *Cylinders and prisms*

344   17 / GEOMETRIC THINKING AND GEOMETRIC CONCEPTS

Right cylinders and oblique cylinders. In a right cylinder the elements are perpendicular to the bases.

Prisms. If the two bases of a right cylinder are polygons, then the cylinder is a prism. If the bases are rectangles, the prism is called a *rectangular prism* or *rectangular solid*.

Cubes. A cube is a square prism with square sides. A cube is the only possible solid with all square faces.

Cones

A *cone* is a solid with at least one face called the base and a vertex that is not on the face. It is possible to draw a straight line (element) from any point on the edge of the base to the vertex.

Sorted by the shape of the base. If the base is a circle, the cone is a circular cone, which is the type most people associate with the word *cone*. But the base can be any shape and the figure is still a cone, as shown in Figure 17.27.

Pyramids. A pyramid is a special cone in which the base is a polygon. All of the faces of a pyramid are triangles except, possibly, the base.

It is interesting to note that both cylinders and cones contain straight lines called *elements*. A special type of both cylinders and cones occurs when the base is a polygon: a *prism* is a cylinder with polygon bases; a *pyramid* is a cone with a polygon base.

BUILD, DRAW, MAKE, AND MEASURE

Constructing shapes in both two and three dimensions is still one of the most profitable types of activities that can be done in geometry. As children begin to demonstrate level 1 thinking, the tasks for construction activities can be posed in terms of the properties of shapes rather than how they look.

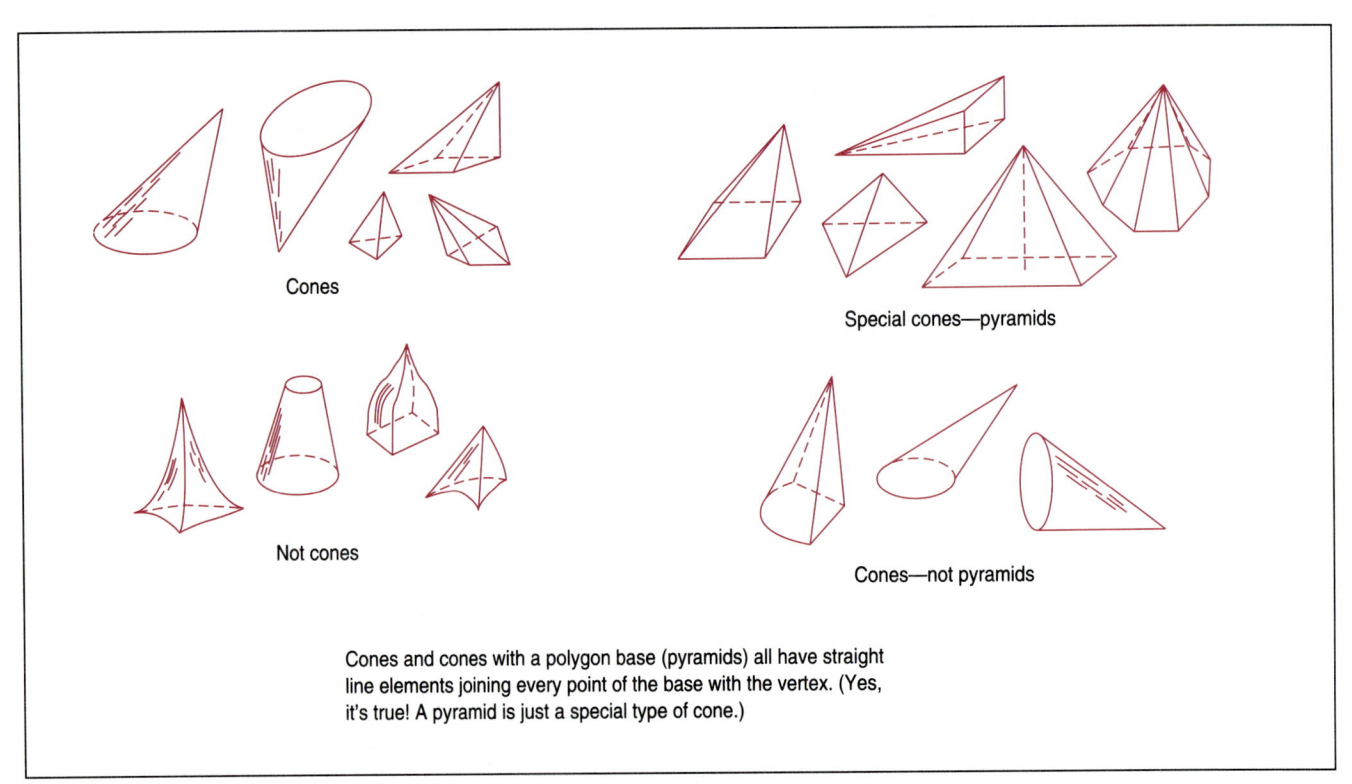

Cones

Special cones—pyramids

Not cones

Cones—not pyramids

Cones and cones with a polygon base (pyramids) all have straight line elements joining every point of the base with the vertex. (Yes, it's true! A pyramid is just a special type of cone.)

FIGURE 17.27: *Cones and pyramids*

In addition to geometric properties, measurements of shapes can help students develop even more relationships. Specifically, area, perimeter, surface area, volume, angles, radii, and circumferences are examples of things that can be measured on various shapes. For example, students can measure interior angles of various polygons and discover that when the number of sides of two polygons is the same, so is the sum of the measures of the interior angles. Direct comparisons, informal units, and simple student-made measuring devices are sufficient for exploring almost all interesting relationships involving measures. Do not let sophisticated measurements interfere with the activities.

◆ Two-Dimensional Constructions

Tiles, geoboards, and dot and grid paper continue to be the best construction materials for two-dimensional shapes. However, in the first activity here (adapted from the 5–8 section of the NCTM *Standards*), ordinary flat toothpicks are used to draw lines.

17.27 TOOTHPICK TRIANGLES

Provide children with a supply of toothpicks. They are to arrange the toothpicks in straight lines to make triangles. Begin with three picks, then four, then five, and so on. For each number of picks they first decide if any triangle is possible. If so, they sketch it (showing each pick), and then see if there are other possible triangles that can be made with the same number of picks. For example, only one triangle is possible with three picks, none are possible with four picks, and two different triangles can be made with seven picks.

Toothpick Triangles is a good example of an opportunity to extend the reasoning to level 2. If students are asked to explain why certain triangles are impossible ("Why is only one triangle possible with 8 picks?"), students may get beyond just looking at properties of triangles and begin to do some informal reasoning. The sum of any two sides of a triangle must be greater than the third side. (Why?)

17.28 PROPERTY CHALLENGE

This activity can be done with almost any materials that allow students to easily draw or construct shapes. Give students properties or relationships, and have them construct as many shapes as possible that have these properties. Compare shapes made by different groups. For example:

Make a four-sided shape with two opposite sides the same length but not parallel.

Make some six-sided shapes. Make some with one, two, and then three pairs of parallel sides, and some others with no parallel sides.

Make shapes with all square corners. Can you make one with three sides? four sides? five sides? six, seven, or eight sides?

Make some six-sided shapes with all square corners. Count how many squares are inside each. What is the distance around each?

Make five different triangles. How are they different? (This is also good for four-, five-, or six-sided shapes.)

Make some triangles with two sides equal (congruent).

Make some four-sided shapes with three congruent sides.

Try five-sided shapes with four congruent sides.

Make some quadrilaterals with all sides equal (or with two pairs of equal sides).

Make a shape with one or more lines of symmetry or that has rotational symmetry. (A longer discussion of symmetry is provided later in the chapter.)

It is quite easy (and fun) to make up property challenges. Explore some of these, and make up others that suit your needs. One idea is to use combinations of previously explored concepts.

Can you make a triangle with two right angles?

Can you make a rhombus that has a right angle? a parallelogram with only two equal sides?

Make a chart like the one shown here:

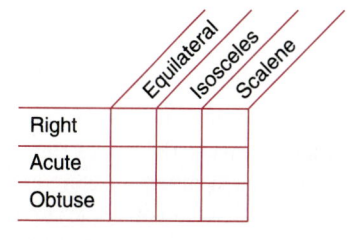

	Equilateral	Isosceles	Scalene
Right			
Acute			
Obtuse			

Challenge children to draw or construct triangles that fit into each of the nine cells. Of the nine, two are impossible. (Can you tell which ones?)

Combination challenges can also include the notions of perpendicular, angle measurement, area, perimeter, similarity, concave and convex, regularity, and symmetry, just to name a few. Using the materials that your students will use, prepare a series of combination challenges. Also encourage your students to come up with some of their own. Notice that some of the preceding challenges are not possible. Discovering that some combination of relationships is not possible and why is just as valuable as learning the relationships themselves.

17.29 MEASURE INVESTIGATION

Make shapes according to special measurement requirements.

Make at least five different shapes with an area of _____ (appropriate number for your materials). What is the perimeter of each?

Make shapes with a fixed perimeter and examine the areas.

Try to make the shape with the largest area for a given perimeter or the smallest perimeter for a given area. (For polygons, the largest area for a fixed perimeter is always regular. Try it.)

Angle and side specifications provide lots of opportunities to examine properties. Angles can usually be kept to multiples of 30° and 45°, or informal units can be used. Length measurements can be made using grid dimensions if diagonal segments are avoided.

Make several different triangles that all have one angle the same. Next make some different triangles with two angles the same (for example, 30° and 45°). What did you notice?

Can you make a parallelogram with a 60° angle? Make several. Are they all alike? How? How are they different?

Make some parallelograms with a side of 5 and a side of 10. Are they all the same?

Draw some polygons with 4, 5, 6, 7, and 8 sides. Divide them all up into triangles, but do not let any lines cross or triangles overlap. What do you discover? Measure the angles inside each polygon.

Draw an assortment of rectangles, and draw the diagonals in each. Measure the angles that the diagonals make with each other. Measure each part of each diagonal. What do you observe? Try this exercise with squares, rhombuses, other parallelograms, and kites (quadrilaterals with 2 pairs of adjacent sides congruent).

◆ **Three-Dimensional Constructions**

Shapes are so difficult to visualize in drawings that the only way to experience some of the truly interesting relationships in three dimensions is to build models. The activities below suggest different ways to construct solids and potential explorations to go with the constructions.

17.30 STRING CYLINDERS AND CONES

An easily made model for exploring cylinders, cones (and prisms and pyramids) is a string model with poster board bases. Students can help you make your initial collection of models, and soon you will have a wide assortment. Directions are in Figure 17.28. When both bases are held together, the vertex can be adjusted up, down, or sideways to produce a family of cones or pyramids with the same base (Figure 17.29). By moving one base up to the knot and adjusting the other base, you can model a family of cylinders or prisms. The bases can also be tilted (not kept parallel) and/or twisted so that noncylindrical shapes are formed. Be sure that students notice that for cylinders and prisms, the elements (strings) remain parallel regardless of the position of the bases. The angle that each makes with the base is the same for all elements. For cones and pyramids, examine how the angles change relative to each other as the vertex is moved. String models like these are well worth the small cost and effort simply for the opportunity to explore how these related solids change as the bases or vertices are moved.

Materials: Soft cotton string or embroidery yarn, metal washers about $\frac{3}{4}$ inch in diameter (about 15 per model), poster board, and a hole punch.

Directions:
1. Cut two identical models of the base from poster board. There are no restrictions on the shape. The size should be roughly 4 to 6 inches across.
2. Place the two bases together and punch an <u>even number</u> of holes around the edges, punching both pieces at the same time so that the holes line up. The holes should be about 1 cm apart.
3. Cut pieces of string about 5 inches long. You will need half as many pieces as holes.
4. Run each piece of string through a washer and thread the two ends up through two adjacent holes of both bases. Pull all the ends together directly above the base and tie in a knot.

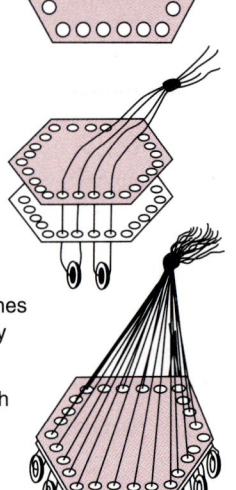

FIGURE 17.28: *A model for cylinders and cones*

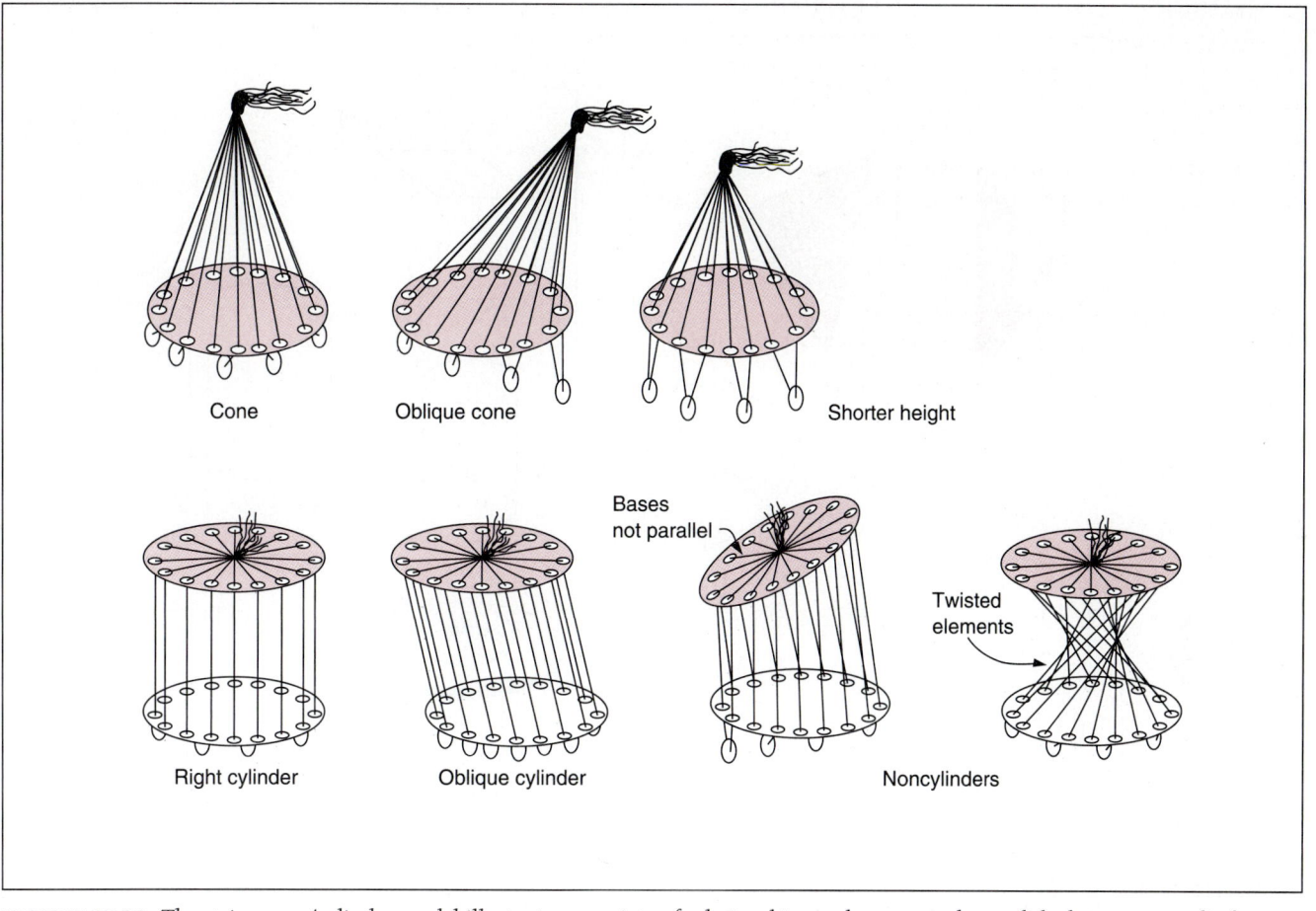

FIGURE 17.29: *The string cone/cylinder model illustrates a variety of relationships and can even show solids that are noncylinders.*

17.31 CLAY SHAPES AND SLICES

Modeling clay is a useful way to make almost any shape. Use an oil-based craft clay that will not dry out. Perhaps the best reason to go to this trouble is that clay models can be sliced to investigate the resulting faces on the slices. An inexpensive tool called a "piano wire," a wire with two handles, is used to cut the clay. This works much better than a knife and can be purchased in an art supply store. Have students make cubes, cones, prisms, cylinders, a torus (doughnut shape), and other shapes. Precision is not important as long as the essential features are there. There are two types of challenges. First, suggest where to make a slice, and see if students can predict what the face of the slice will be. Second, given a solid, ask students to find a way to slice the shape to produce a slice-face with a particular shape. For example, how could you slice a cube so that the slice-face is a trapezoid? Will it be isosceles? Always? Figure 17.30 (p. 348) suggests a few ways that different solids can be cut (Carroll, 1988).

17.32 GENERATING SOLIDS

Students can "generate" imaginary solids by sliding or revolving a plane surface, as shown in Figure 17.31 (p. 348). The resulting shapes can be made with clay. Notice that *all* cylinders can be generated by sliding a base. Of all cones, only a right circular cone can be spun from a right triangle, but many other shapes can be made by spinning a shape about an edge. How would you generate a torus (doughnut shape) or other shapes with holes in them? What shapes around the room could be generated by rotating or sliding a surface? Which ones cannot? How are shapes that can be generated in this manner different from those that cannot? From one triangle or rectangle, how many different shapes can be generated by sliding? how many by rotation? Are the answers the same for all triangles? all rectangles?

Using cubes to build shapes was a suggestion for level 0. At level 1 it is interesting to examine rectangular prisms (box shapes) using cubes because surface area and volume

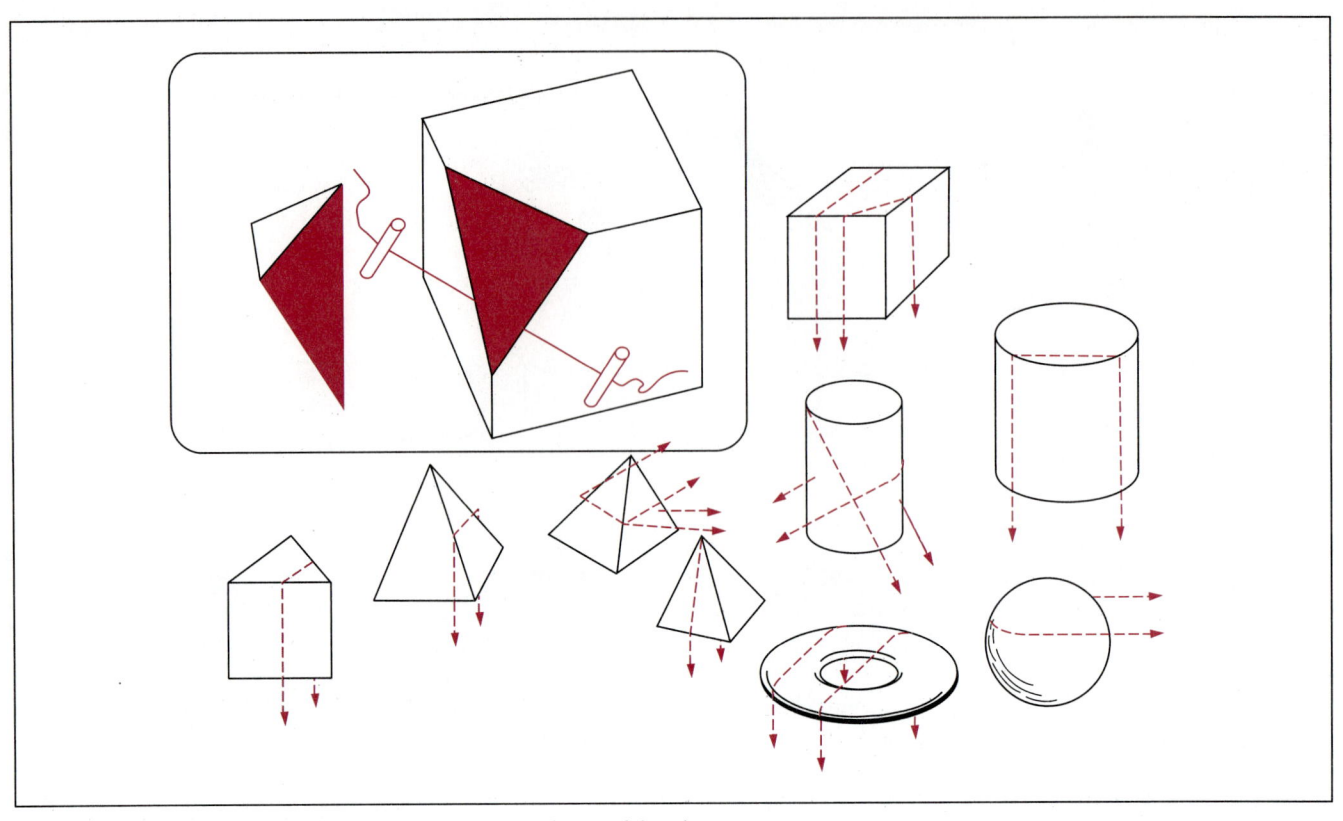

FIGURE 17.30: *Predict the slice face before you cut a clay model with a piano wire.*

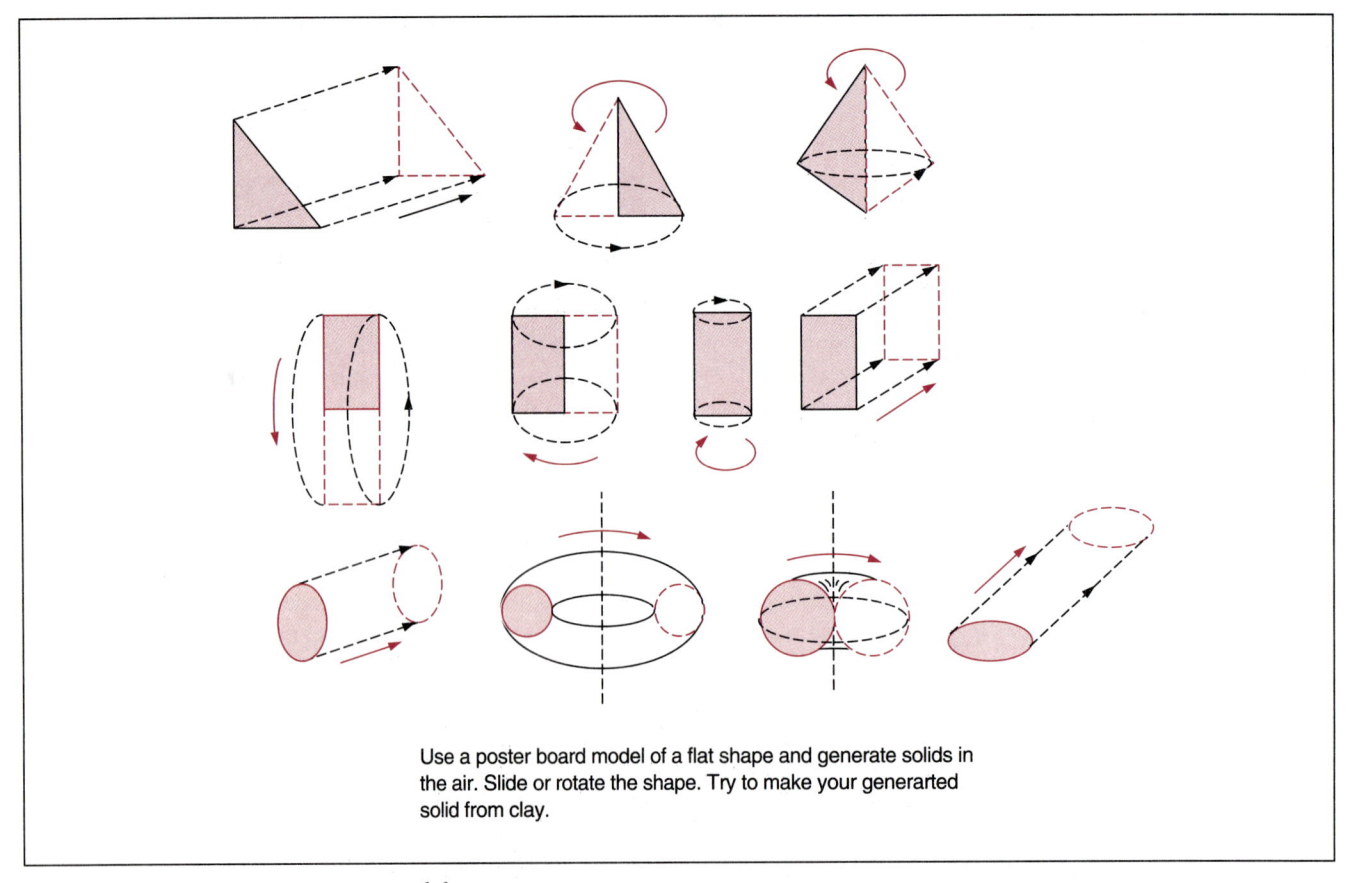

Use a poster board model of a flat shape and generate solids in the air. Slide or rotate the shape. Try to make your generarted solid from clay.

FIGURE 17.31: *Generating imaginary solids*

are so easily determined. Tasks involving surface area and volume are analogous to tasks in two dimensions that involve perimeter and area.

17.33 SURFACE/VOLUME INQUIRY

Working with 36 cubes, have students figure out how many boxes (rectangular prisms) they can make that use all 36 cubes. For each, they should list the dimensions and the total surface area. Clearly all boxes do not have the same surface area. What can you say about boxes with larger surface areas? smaller surface areas? Try 60 cubes and then 64 cubes. (Good numbers to use will have several factorizations into three factors.) Later, reverse the problem. Have students construct boxes that have a given surface area. Can boxes with the same surface area have different volumes? Try surface areas of 24, 54, or 96 squares.

One general conclusion from the **Surface/Volume Inquiry** is that boxes that are the closest to cubical will have the smallest surface areas. (Recall that rectangles with the smallest perimeters were squares.) If the experiment could be carried out for boxes with more and more sides, you would see that shapes with the most volume and least surface area tend to get closer and closer to the shape of a sphere. (A cube is sort of a six-sided "sphere.") This, then, explains why soap bubbles are spherical. The volume of air is fixed (trapped inside), and the soap film tends to contract. When the surface is minimal for the amount of air, the bubble is a sphere.

EXPLORATION OF SPECIAL PROPERTIES AND RELATIONSHIPS

Some geometric properties of shape are worthy of special attention. In this section you will recognize some of the more traditional content of geometry. The activities are generally adaptable to several levels of thought in the van Hiele scheme. Since the activities are designed to investigate properties, they are suitable for level 1 thinking. Those students still at level 0 will be able to work at the activities but may not arrive at general principles or be able to recognize the same ideas when they appear in different contexts. Those at or ready for level 2 thinking can be challenged to see how properties are related or what conditions give rise to particular properties.

Line Symmetry

Line symmetry (bilateral symmetry or mirror symmetry) is fun and challenging to explore using a variety of different materials.

17.34 SYMMETRY ON A GRID

On a geoboard or on any dot or grid paper, draw a line to be a mirror line. Next draw a shape on one side that touches the line as in Figure 17.32. Try to draw the mirror or symmetric image of the shape on the other side of the line. Try starting with two intersecting lines of symmetry. Draw the beginning shape between two of the rays formed by these lines. Reflect the shape across each line, a step at a time, until further reflections produce no new lines on your drawing. Mirrors can be used to test the results.

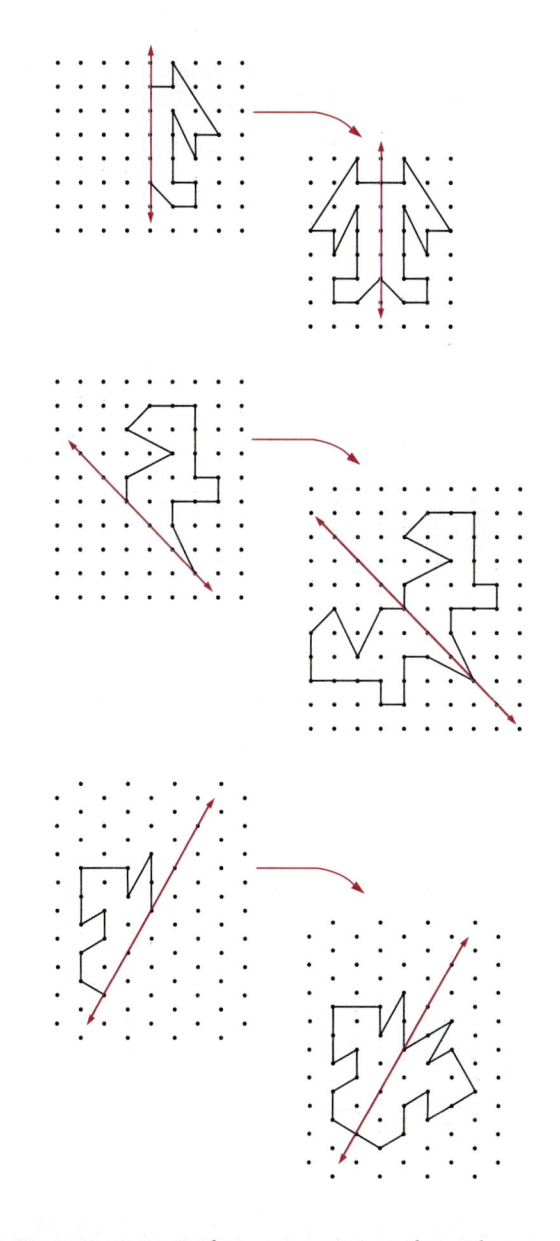

FIGURE 17.32: *Exploring symmetry on dot grids*

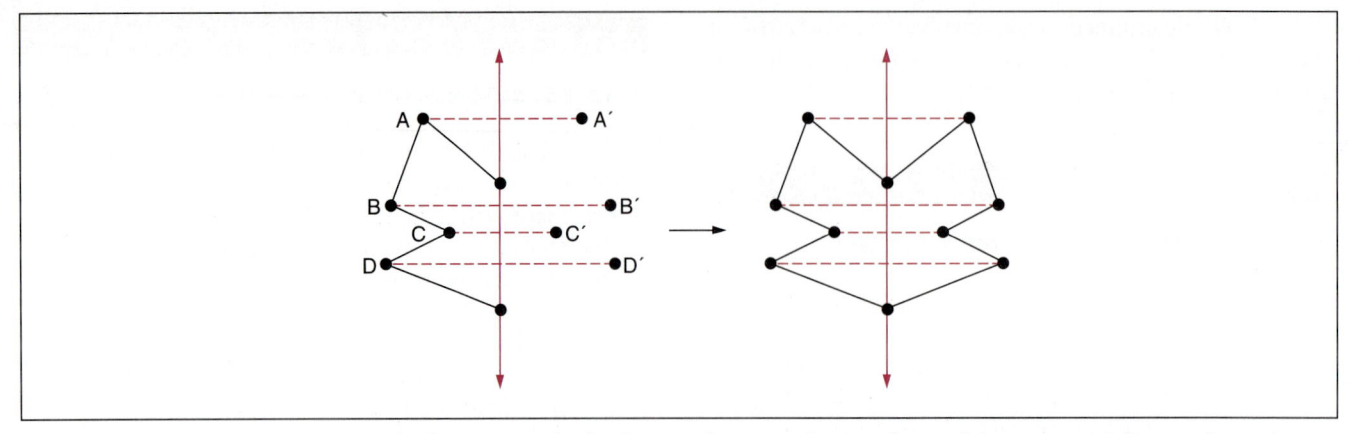

FIGURE 17.33: *From points on one side, draw perpendicular lines to the mirror line, and extend them an equal distance beyond.*

17.35 LINE SYMMETRY ANALYSIS

On a piece of dot paper use the technique of the previous exercise to create a symmetric drawing. Fold the paper on the mirror line, and notice how corresponding points on each side of the line match up. Open the paper and connect several corresponding points with a straight line. Notice that the mirror line is a perpendicular bisector of the lines joining the points. Use this property to create a symmetric drawing on plane paper. Draw the line and half of the figure as before. From several critical points, draw perpendicular lines to the mirror line and extend them an equal distance beyond (Figure 17.33).

A *flip* or *reflection* of a shape through a line is very similar to the concept of symmetry. If you plan to investigate transformations (slides, flips, and turns), the last few activities are good beginnings.

A very useful device for studying symmetry and transformations is the Mira, a piece of red plexiglass (Figure 17.34), that stands perpendicular to the table surface. The Mira is essentially a see-through mirror. You can reach behind the Mira and draw the image that you see in the mirror. Since you can see the image behind the Mira as well as the reflected image, it is possible to match images with reflections and draw the mirror line at the base of the Mira. This feature of the Mira allows one to draw perpendicular bisectors and angle bisectors, reflect images, and to do a variety of constructions more easily than with a compass and straight edge and in a very conceptual manner.

The following activity offers a slightly different view of symmetry and provides a good preparation for rotational symmetry.

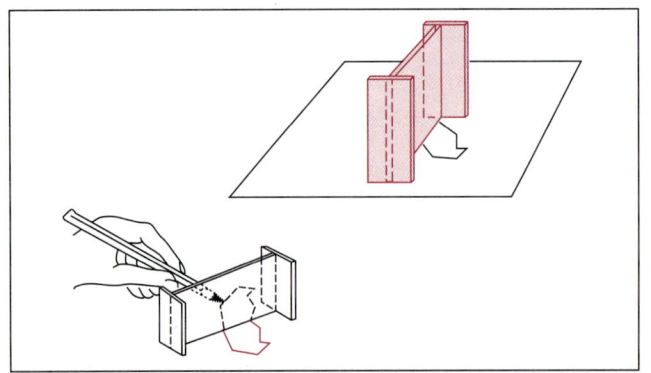

FIGURE 17.34: *The Mira is a "see-through mirror" that allows you to draw the image that you see in the glass.*

17.36 FLIP INTO THE BOX

Cut out a small rectangle from paper or cardboard. Color one side and label the corners A, B, C, and D, so that they have the same label on both sides. Place the rectangle on a sheet of paper, and trace around it. Refer to the traced rectangle as a "box" for the cut-out rectangle. The question is, "How many different ways can you flip the rectangle over so that it fits in the box?" Before each flip, place the rectangle in the box in the initial orientation. As shown in Figure 17.35, each flip into the box is a flip through a line, and these lines are also lines of symmetry. Students can discover that for a plane shape there are as many lines of symmetry as there are ways to flip a figure over and still have it fit into its box. Try with other figures: a square, nonsquare rhombus, a kite, a parallelogram with unequal sides and angles, triangles, regular pentagons, and others.

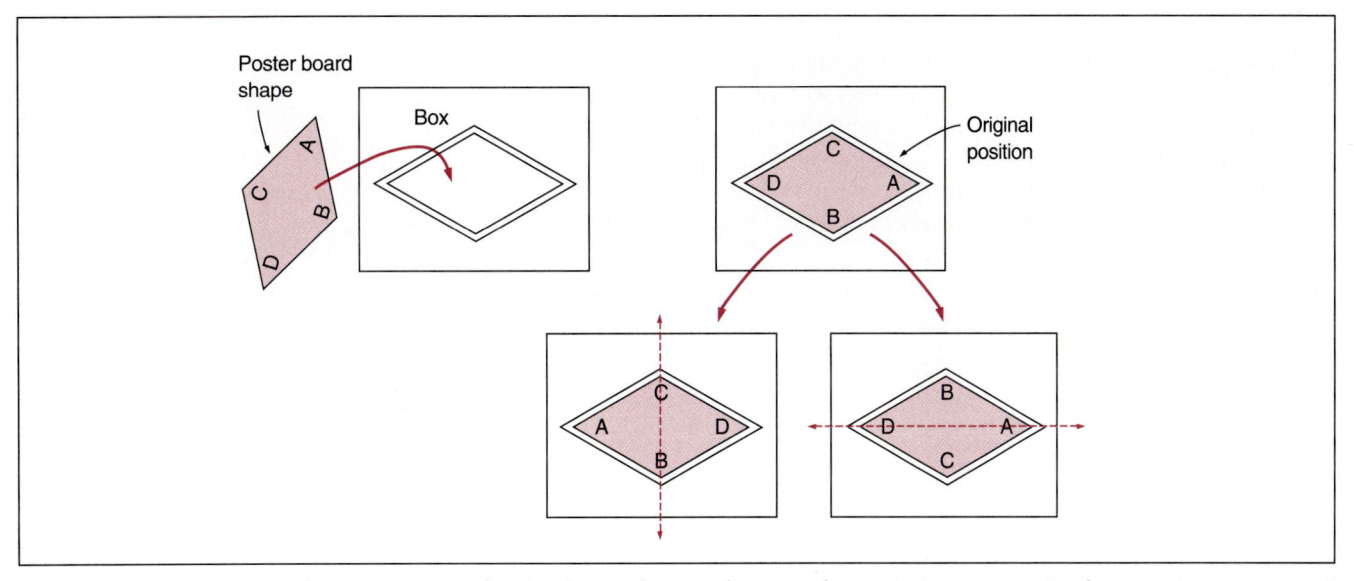

FIGURE 17.35: *There are at least two ways to flip this diamond into its box. Are there more? Cut one out and try it!*

Rotational Symmetry

One of the easiest introductions to rotational symmetry is to create a box for a shape by tracing around it as in the last activity. If a shape will fit into its box in more than one way without flipping it over, it has *rotational symmetry*. The *order of rotational symmetry* is the number of different ways it can fit into the box. Thus a square has rotational symmetry of order 4 as well as four flip lines or lines of symmetry. A parallelogram with unequal sides and angles has rotational symmetry of order 2, but no lines of symmetry (Figure 17.36).

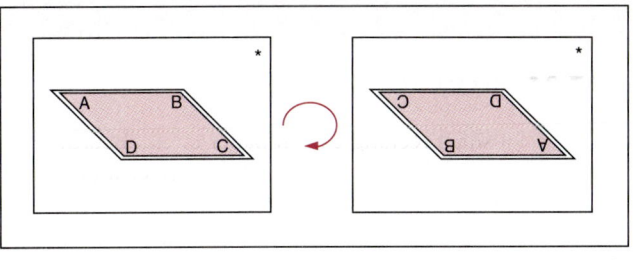

FIGURE 17.36: *This parallelogram fits in its box two ways without flipping it over. Therefore, it has rotational symmetry of order 2.*

17.37 BUILDING "ROTATABLE" SHAPES

Use tiles, geoboards, or dot or grid paper to draw a shape that has rotational symmetry of a given order. Except for regular polygons, this can be quite challenging. To test a result, trace around it, and cut out a copy of the shape. Try to rotate it on the drawing.

17.38 DRAWING "ROTATABLE" SHAPES

On a piece of paper lightly draw three rays from one point separated by 120°. Cut out any shape from a piece of tagboard. Draw a line through it anywhere. Push a pin or compass point through the shape at any point along the line. Place the point at the center of the three rays so that the line on the shape coincides with one of the rays. Trace around the shape. Rotate the shape to each of the other two lines and repeat. By ignoring the interior lines, the three tracings will form a shape with rotational symmetry of order 3. The center point is called the *center of rotation*. Connect any three corresponding points from the three tracings, and observe an equilateral triangle (Figure 17.37, p. 352). The same approach can be used for rotational symmetries of any order.

Symmetries in Three-Dimensional Figures

A plane of symmetry in three dimensions is analogous to a line of symmetry in two dimensions. Each point in a symmetric solid corresponds to a point on the other side. The plane of symmetry is the perpendicular bisector of the line segments joining each pair of points. Figure 17.38 (p. 352) illustrates a shape built with cubes that has a plane of symmetry.

17.39 PLANE-SYMMETRY BUILDINGS

With cubes, build a building that has a plane of symmetry. If the plane of symmetry goes between cubes, the shape can be sliced by separating the building into two symmetrical parts. Try making buildings with two or more planes of symmetry. Examine various prisms that you can build. Do not forget that a plane can slice diagonally through the blocks. Try using clay to build solids with planes of symmetry and slice through these planes with a piano wire.

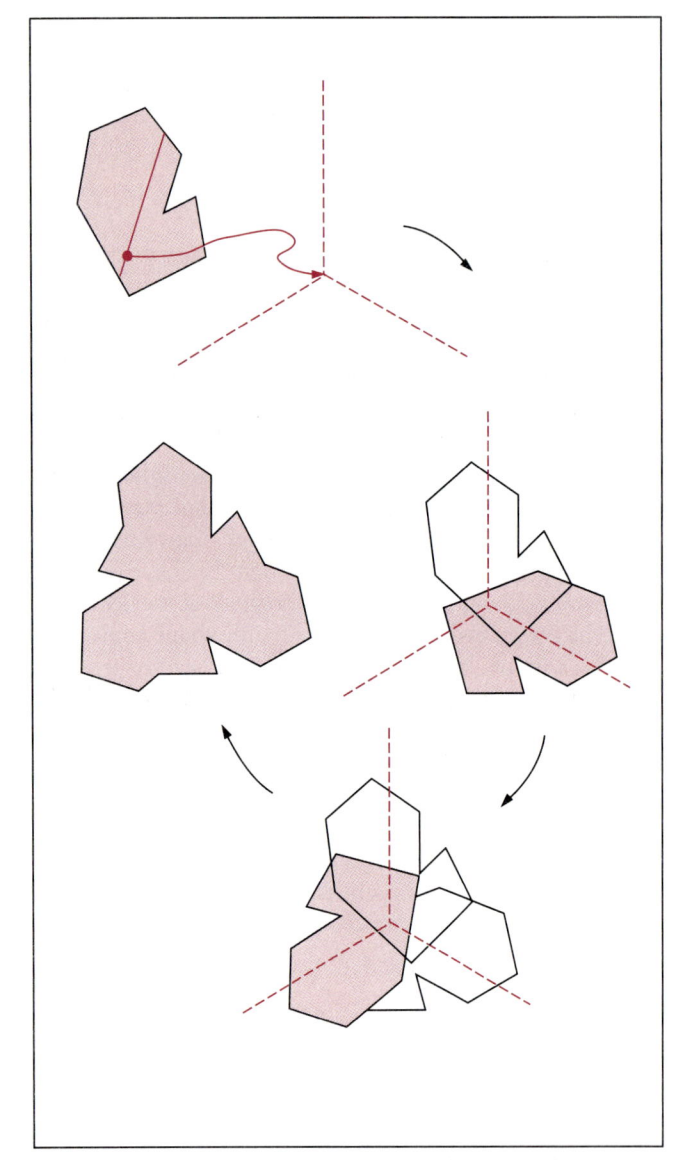

FIGURE 17.37: *How to build a shape with rotational symmetry of order 3. How would you build one with order 2 or 5?*

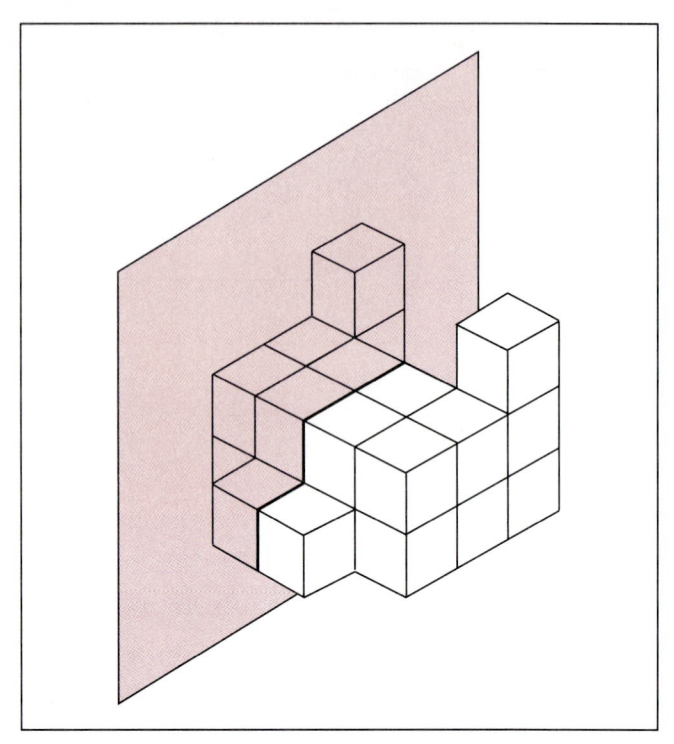

FIGURE 17.38: *A block building with one plane of symmetry*

rotation. For each axis of symmetry there is a corresponding order of rotational symmetry. A regular square pyramid has only 1 axis of symmetry that runs through the vertex and the center of the square. A cube, on the other hand, has a total of 13 axes of symmetry: 3 (through opposite faces) of order 4, 4 (through diagonally opposite vertices) of order 3, and 6 (through midpoints of diagonally opposite edges) of order 2.

17.40 SPINNING ON AN AXIS

To investigate an axis of symmetry, use a small tagboard model of the solid, and insert a long pin or wire through an axis of symmetry. Color or label each face of the solid to help keep track of the different positions. Hold the axis (pin) vertically, and rotate the solid slowly, observing how many times it fills the same space as it did in the original position as in Figure 17.39 (Bruni, 1977).

Diagonals of Quadrilaterals

The usual approach to quadrilaterals is in terms of the sides (parallel or not, congruent or not) and angles. Another way to analyze quadrilaterals is in terms of the diagonals. The diagonals of quadrilaterals provide a wealth of interesting relationships to observe. Consider the following:

Length: either equal in length or not (two possibilities)

Angle of intersection: either right angle or not (two possibilities)

Rotational symmetry in the plane also has an analogous counterpart in three dimensions. Whereas a figure in a plane is rotated about a point, a three-dimensional figure is rotated about a line. Such a line is called an *axis of symmetry*. As a solid with rotational symmetry revolves around an axis of symmetry, it will occupy the same position in space (its box), but in different orientations. While plane figures have only one center of rotation, a solid can have more than one axis of

17 / GEOMETRIC THINKING AND GEOMETRIC CONCEPTS   353

With A on top, the cube fits in its "box" four ways. Through this axis the order of rotational symmetry is order 4.

These two axes also have rotational symmetry of order 4.

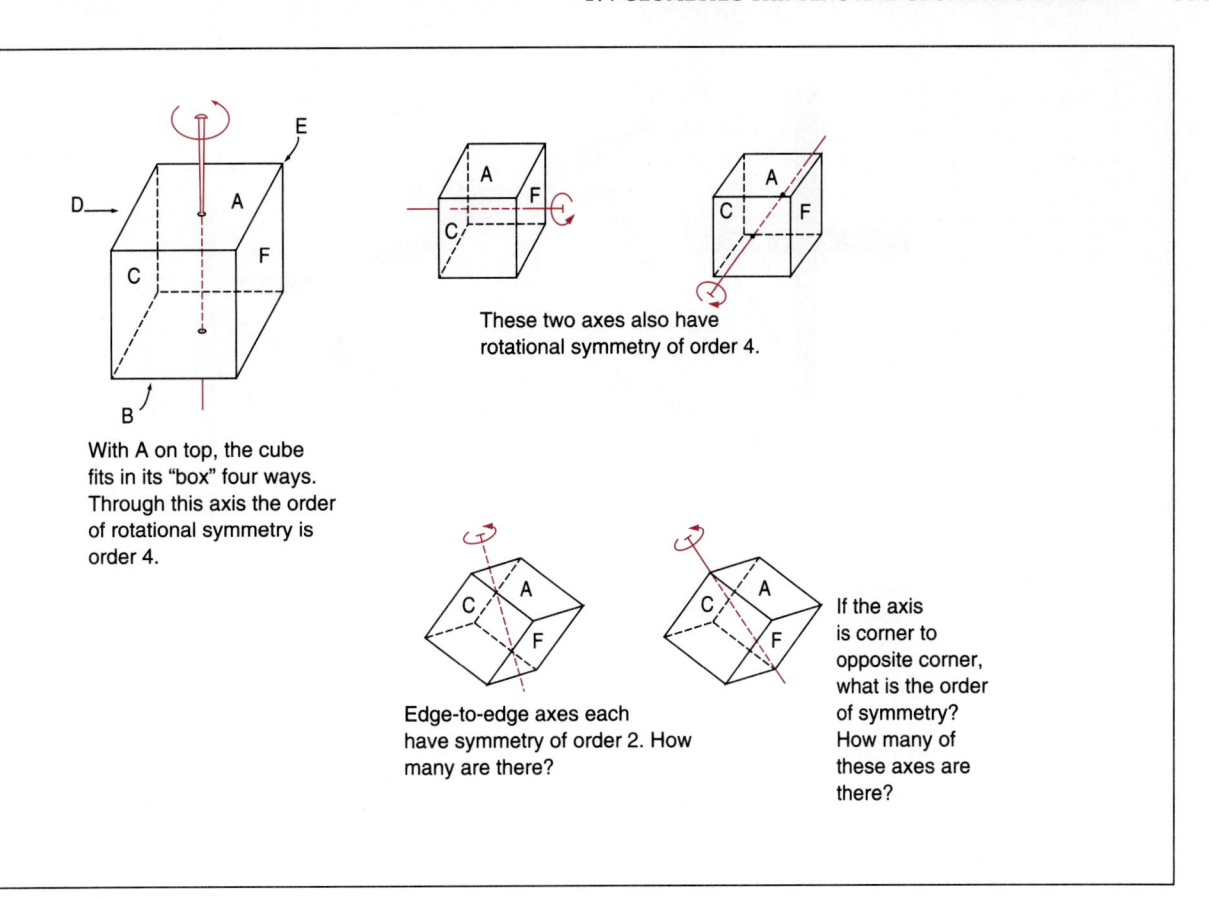

Edge-to-edge axes each have symmetry of order 2. How many are there?

If the axis is corner to opposite corner, what is the order of symmetry? How many of these axes are there?

FIGURE 17.39: *Rotations of a cube*

Ratio of parts (from the corners to the intersection): one diagonal is bisected, both are bisected, neither bisected but parts proportional, or none of these (4 possibilities)

17.41 DIAGONAL INVESTIGATION

Have students select a particular type of quadrilateral (say, parallelogram) and draw several different examples of it. Use dot or grid paper to make drawing parallel lines and congruent angles easy. For each example they should draw the diagonals and make whatever measurements and calculations they desire. The goal is to discover what properties the diagonals of that particular type of quadrilateral might have. Different groups of students can investigate different quadrilaterals, and information can be shared and discussed as a class.

In the preceding activity the students begin with a quadrilateral and examine properties of the diagonals. A similar activity begins with the diagonals.

17.42 DIAGONAL STRIPS

For this activity students need three strips of tagboard about 2 cm wide. Two should be the same length (about 30 cm) and the third somewhat shorter (about 20 cm). Punch nine holes equally spaced along the strip. (Punch a hole near each end. Divide the distance between the holes by 8. This will be the distance between the remaining holes.) Use a brass fastener to join two strips. A quadrilateral is formed by joining the four end holes as shown in Figure 17.40 (p. 354). Provide students with the list of possible relationships for angles, lengths, and ratios of parts, and see if they can use the strips to determine the properties of diagonals that will produce different diagonals. The strips are there to help in the exploration. Students may want to make additional drawings on dot grids to test their various hypotheses. (*Challenge*: What properties will produce a nonisosceles trapezoid?)

Every type of quadrilateral can be uniquely described in terms of its diagonals using only the conditions listed earlier.

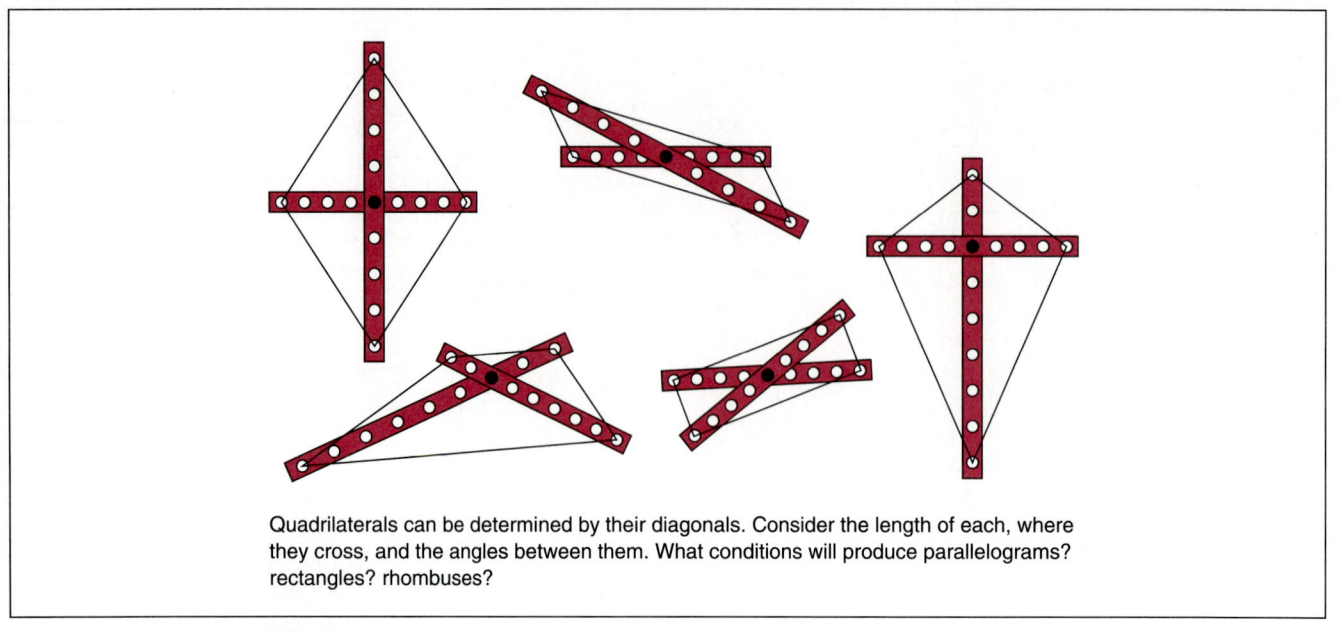

FIGURE 17.40: *Diagonals of quadrilaterals*

Quadrilaterals can be determined by their diagonals. Consider the length of each, where they cross, and the angles between them. What conditions will produce parallelograms? rectangles? rhombuses?

◆ Angles, Lines, and Planes

The relationships between angles within a figure can be explored quite well by tracing angles for comparisons, by comparing angles to a square corner, and by using informal units. With only such simple techniques, students can begin to look at relationships such as the following:

- The angles made by intersecting lines or by lines crossing two parallel lines. Which angles are equal? Which add up to a straight angle?
- The sum of the interior angles of polygons of different types. Is there a relationship in the number of sides and the sum of the angles? What if the shape is concave?
- The exterior angles of polygons. Extend each side in the same direction (for example, clockwise) and observe the sum of these angles. How is the exterior angle related to the interior angle?

In three dimensions, students can begin to observe the angles formed by intersecting planes and between lines and planes. How can two lines not intersect but not be parallel? Can a line and a plane be parallel? How could you describe a line that is perpendicular to a plane? Where are there some examples of these relationships in the real world?

◆ Similar Figures

In both two and three dimensions, two figures can be the same shape but be different sizes. At level 0, students can sort out shapes that look alike. At that level the concept of similar is strictly visual and not likely to be precise. At level 1, students can begin to measure angles, lengths of sides, areas, and volumes (for solids) of shapes that are similar. By investigation, relationships between similar shapes can be observed. For example, students will find that all corresponding angles must be congruent, but other measures vary proportionately. If one side of a larger yet similar figure is triple that of the smaller figure, so will all linear dimensions be triple those of the smaller figure. If the ratio of corresponding lengths is 1 to n, the ratio of areas will be 1 to n^2, and the ratio of volumes will be 1 to n^3.

17.43 BUILDING SIMILAR FIGURES

Have students build or draw a series of similar shapes. Start with one shape, and pick one side (or for solids, one edge) to be the control side. Make the corresponding side of the similar figures in easy multiples of the control side. For example, make shapes with a corresponding side two, three, and four times as long as the control (Figure 17.41). Make measures of lengths, areas, and volumes (for solids), and compare ratios of corresponding parts. If the control sides are in the ratio 1 to 3, then all linear dimensions will be in the same ratio, areas in the ratio 1 to 9, and volume in the ratio 1 to 27.

◆ Circles

The circle, an apparently simple shape, is also very important. Consider the world around you. What are some of the ways the circle is used? Why do you think it is so frequently used instead of other shapes? Many interesting relationships can be observed between measures of different parts of the circle (see Bruni, 1977). Among the most astounding and important is the ratio between measures of the circumference and the diameter.

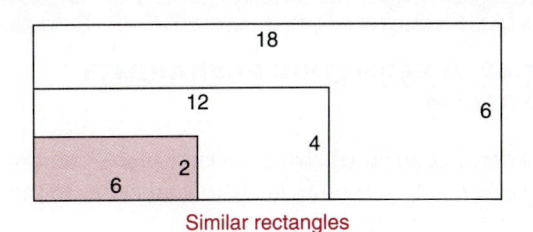

17 / GEOMETRIC THINKING AND GEOMETRIC CONCEPTS 355

Similar rectangles

Compare the ratios of lengths of sides and ratios of areas.

Example: Small to Large
Length   2 to   6   (1 to 3)
Area   12 to 108   (1 to 9)

Similar shapes have corresponding dimensions in predictable ratios. What measures stay the same? How do lengths, areas, and volumes change? Is that true of all shapes?

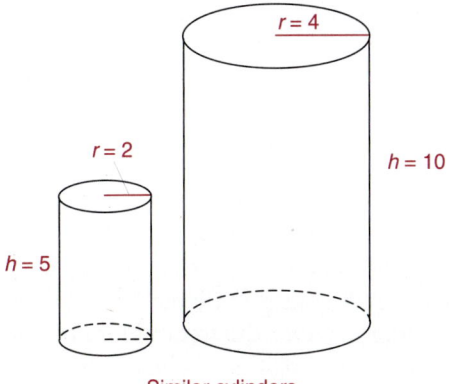

Similar cylinders

Height, radius, and circumference are all in ratio 1 to 2.

Surface Areas:
Sides 20π to 80π } 1 to 4
Tops 4π to 16π
Volumes: 20π to 160π or 1 to 8

FIGURE 17.41: *Similar solids*

◆◆◆◆◆◆◆◆◆◆◆◆◆◆◆◆◆◆◆◆◆

17.44 DISCOVERING π

Have groups of students carefully measure the circumference and diameters of many different circles. Each group can be responsible for a different collection of circles.

Carefully wrap string around jar lids, tubes, cans, and similar items, and mark the circumference. Measure the string and the diameters to the nearest millimeter.

Draw larger circles using a string, as shown in Figure 17.42. String can be used to measure the circumference.

Measure the large circles found on gym floors and playgrounds. Use a trundle wheel or rope to measure the circumference.

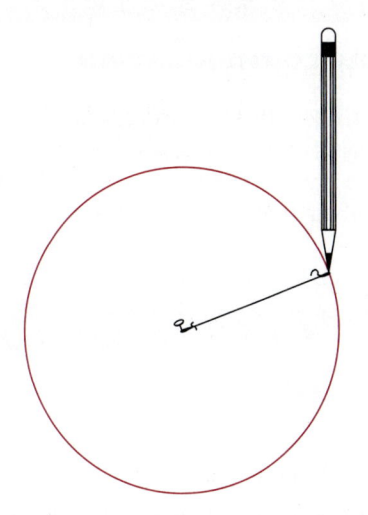

A string with two loops can be used to draw very large circles that are larger than those drawn with a compass. Two students working together will have better success than just one.

FIGURE 17.42: *Drawing a circle*

Using a calculator, students should record the ratio of the two measures for each circle. With careful measurement, the results should be close to 3.1 or 3.2. The exact ratio is an irrational number that is represented by the Greek letter π, which is about 3.14159.

What is most important in the above activity is that students develop a clear understanding of π as the ratio of circumference to diameter in any circle. The quantity π is not some strange number that appears in math formulas.

There are many other explorations that can be done with circles. For example, a short article by Gerver (1990) describes some interesting area activities that also get at the value of π in an intuitive manner. Drawing circles with compasses and then drawing arcs with the same radius through the center has delighted students for ages. By making three folds of a circular disk into the center, we form a very nice equilateral triangle, complete with flaps on each edge. These can then be glued together to make three of the five Platonic solids (see p. 359). It is interesting to notice that all triangles drawn with two vertices on opposite ends of a diameter and the third at some other point on the circle have something in common. These and many other explorations suggest that the circle is a good place to look for geometric experiences.

GEOMETRIC PROBLEM-SOLVING ACTIVITIES

Problem solving continues to be an important feature of geometry activities at level 1. Many activities already suggested are explorations posed as problems. A few additional explorations of a problem-solving nature are suggested here.

17.45 TRIANGLE COMMUNICATION

Draw any triangle. Choose three measurements of either angles or lengths of sides, and use only these to tell a partner how to draw a triangle that is congruent to yours. What combinations of angles and sides will work?

17.46 AREA PROBLEMS

Determine the areas of odd shapes and surfaces such as those in Figure 17.43 for which there are no formulas or for which dimensions are not provided.

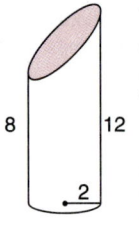

What is the lateral surface area (not including the two faces)?

Hint: Recall how the area of a triangle was derived from the area of a parallelogram.

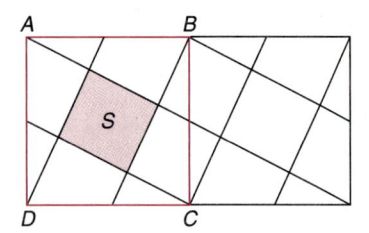

If the big square (*A, B, C, D*) is 1 unit in area, what is the area of *S*?

45	25
X	15

The rectangle is divided into four smaller rectangles with areas 45, 25, 15, and *X*. What is the area of *X*?

FIGURE 17.43: *Examples of area problems that require some analysis.* SOURCE: Problems adapted from Milauskas, 1987.

17.47 A GEOMETRIC PROBABILITY PROBLEM

If a dart has an equal chance to land at any point on a circular target, is it more likely to land closer to the center or closer to the edge?

TESSELLATIONS REVISITED

Earlier you saw that a tessellation was a regular tiling pattern that covers the plane. Students at level 1 can begin to create more interesting and/or complex tessellations with attention being given to certain geometric properties.

The Dutch artist M. C. Escher is well known for his tessellations, where the tiles are very intricate and usually take the shape of things like birds, horses, angels, or lizards. What Escher did was take a simple shape such as a triangle, parallelogram, or hexagon and perform transformations on the sides. For example, a curve drawn along one side might be translated (a slide) to the opposite side. Another idea was to draw a curve from the midpoint of a side to the adjoining vertex. This curve was then rotated about the midpoint to form a totally new side of the tile. These two ideas are illustrated in Figure 17.44(a). Dot paper is used to help draw the lines. Escher-type tessellations, as these have come to be called, are quite popular projects for students in the sixth to eighth grades. Once the tile has been designed, it can be cut from two different colors of construction paper instead of drawing the tessellation on a dot grid.

A *regular* tessellation is made of a single tile that is a regular polygon (all sides and angles congruent). Each vertex of a regular tessellation has the same number of tiles meeting at that point. A checkerboard is a simple example of a regular tessellation. A *semiregular* tessellation is made of two or more tiles, each of which is a regular polygon. At each vertex of a semiregular tessellation the same collection of regular polygons comes together in the same order. A vertex (and therefore the complete semiregular tessellation) can be described by the series of shapes meeting at a vertex. Under each example of these tessellations in Figure 17.44(b), the vertex numbers are given. Students can figure out what polygons are possible at a vertex and design their own semiregular tessellations.

Books by Ranucci and Teeters (1977) and Bezuszka, Kenney, and Silvey (1977) are among a number of excellent resources that can be used to explore this fascinating topic further with junior high students.

INFORMAL GEOMETRY ACTIVITIES: LEVEL 2

At the van Hiele level 2 of geometric thinking, students begin to use informal deductive reasoning. That is, they can

(1) Start with a simple shape. (2) Draw the same curve on two opposite sides. This tile will stack up in columns. (3) Rotate a curve on the midpoint of one side. (4) Rotate a curve on the midpoint of the other side. Use this tile for tesselation (below, left).

(a)

A column of this tile will now match a like column that is rotated one complete turn. Find these rotated columns in the tessellation below.

(a)

4-3-4-6

3-4-3-3-4

8-8-4

3-3-3-3-6

(b)

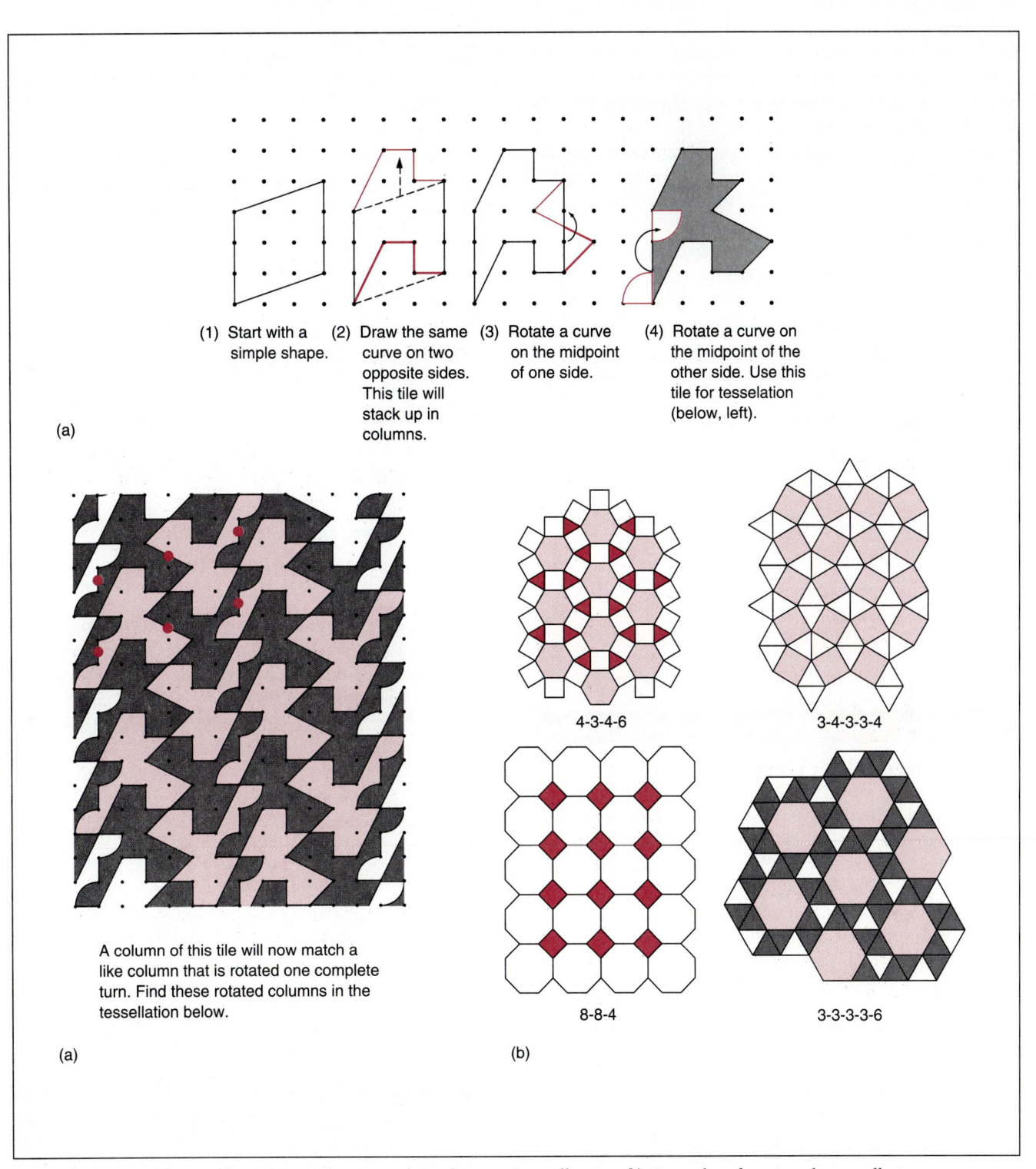

FIGURES 17.44: (*a*) *One of many ways to create an "Escher-type" tessellation.* (*b*) *Examples of semiregular tessellations.*

357

follow and utilize logical arguments, although they will have a difficult time constructing a proof of their own as in tenth-grade geometry. Physical models and drawings are still important but for different reasons. At level 1 students' explorations lead to inductive conclusions about shapes. For example, the diagonals of parallelograms bisect each other. Students at level 1 are satisfied that such results are so because it seems to happen that way when they try it. At level 2, students can use a drawing to help them follow a deductive argument supplied by the teacher. They may also use models to test conjectures or to find counterexamples. Models become more of a tool for thinking and verification than one of exploration.

Many topics in the seventh- and eighth-grade curriculum lend themselves to projects that promote level 2 thinking. It is important that students be challenged to reason through these topics and not just memorize formulas and procedures by rote. Eighth-grade geometry may be one of the last opportunities for good activities that prepare students for the level 3 thinking required in a high school geometry course.

DEFINITIONS AND PROPERTIES

Classification activities at level 2 can begin to focus on the definitions of shapes and how different classes of shapes are related one to another.

17.48 PROPERTY LISTS

Consider a particular type of shape, for example a rectangle, and list all of the properties of that shape that students can think of. One possible list is suggested here. What is the shortest list that can be chosen from this one that will still determine a rectangle? Are there different sublists that will work? Which property or properties are necessary if the rectangle is to be nonsquare?

A. Four sides.
B. Four right angles.
C. Opposite sides parallel.
D. Diagonals bisect each other.
E. Adjacent sides perpendicular.
F. Opposite sides equal in length.
G. Diagonals congruent.
H. Only two lines of symmetry.
I. Diagonals not perpendicular.

17.49 IF/THEN—TRUE OR FALSE

Explore statements in these forms:

If it is a _____, then it is also a _____.
All _____ s are _____.

Fill the blanks with names of shapes or statements of properties and relationships. Let students use drawings or models to decide if the statements are true or to find counterexamples. A few examples are suggested here.

If it is a cylinder, then it is a prism.
If it is a prism, then it is a cylinder.
If it is a square, then it is a rhombus.
All parallelograms have congruent diagonals.
All quadrilaterals with congruent diagonals are parallelograms.
If two rectangles have the same area, then they are congruent.
All prisms have a plane of symmetry.
All right prisms have a plane of symmetry.
If a prism has a plane of symmetry, it is a right prism.

Obviously not all of the above statements are true. For those that you think may be false, find a counter example. You and your students can make up similar puzzles to stump the class.

INFORMAL PROOFS

The Pythagorean Theorem

An area interpretation of the Pythagorean theorem states that if a square is made on each side of a right triangle, the areas of the two smaller squares will together be equal to the area of the square on the longest side or hypotenuse. Figure 17.45 illustrates two proofs that students can follow and begin to appreciate.

As a variation to the Pythagorean theorem, have students explore what happens if the squares formed on each side of a right triangle are replaced by triangles or semicircles. Will the same relationship between the three areas hold? What about other shapes (pentagons, hexagons, . . .)?

Area and Volume Formulas

In Chapter 16, suggestions are given for helping students develop formulas for areas and volumes of common shapes. Formula development is a major connection

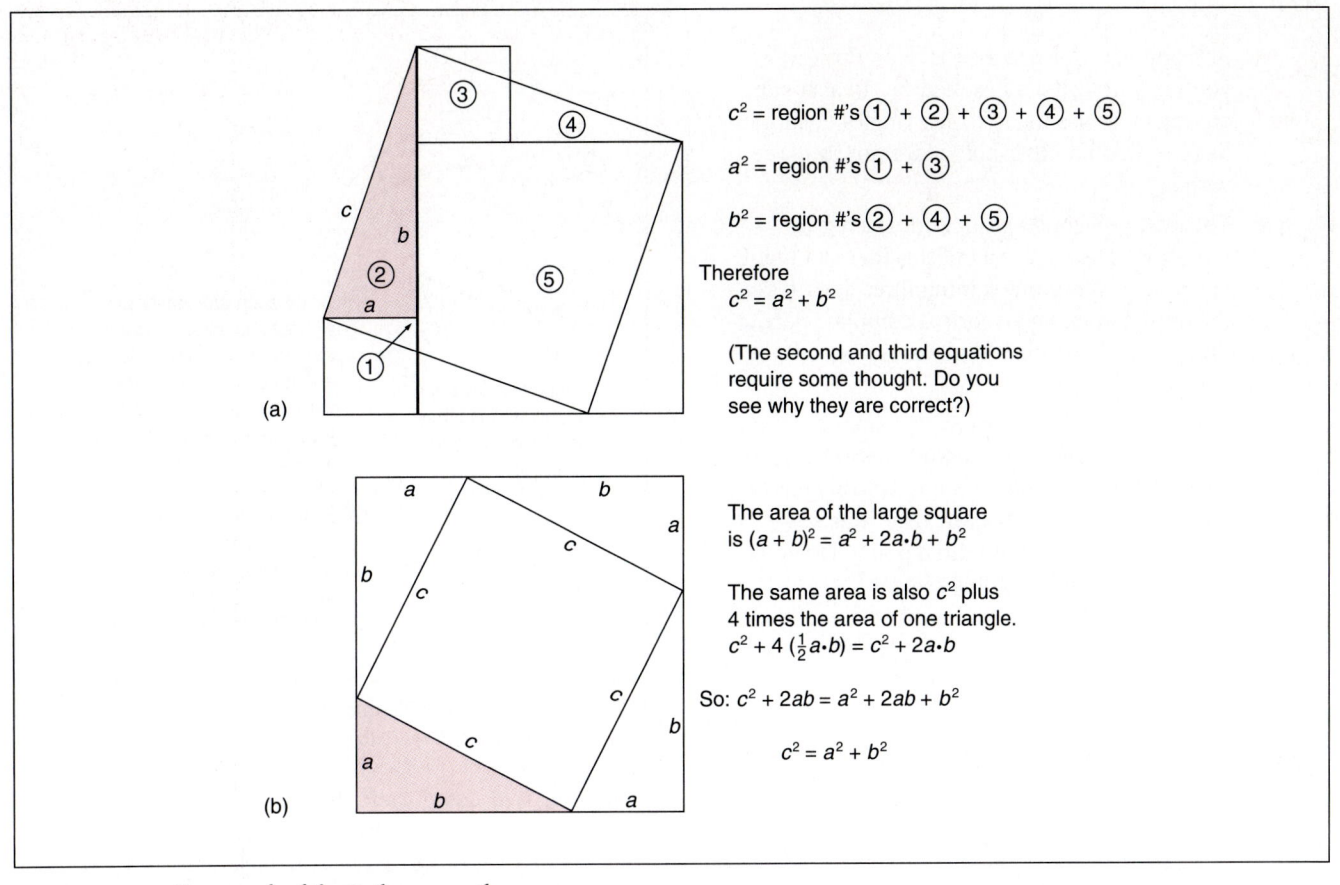

FIGURE 17.45: *Two proofs of the Pythagorean theorem*

between measurement and geometry and is also a good bridge between level 1 and level 2 of geometric thinking.

◆ The Platonic Solids

The *Platonic* solids are *completely regular* polyhedra. By completely regular it is meant that each face is a regular polygon, and every vertex has exactly the same number of faces joining at that point. It is quite reasonable for level 2 students to work through an informal argument that demonstrates there can be only five such completely regular polyhedra and also allows the excitement of constructing them. To examine these solids, students need a supply of congruent equilateral triangles cut from tagboard, some tagboard squares, tagboard regular pentagons, and a few regular hexagons. One possible approach to the Platonic solids might go as follows:

> Since the polyhedra must have regular faces, begin with the simplest possible face, an equilateral triangle, and see what polyhedra can be constructed.
> 1. Put three triangles together at a point. Tape all edges together. This will leave a space for one more triangle to be inserted forming a *tetrahedron* (tetra = four). Observe that this is a completely regular polyhedron. (Why is a six-sided polyhedron made of two tetrahedra placed face to face not completely regular?)
> 2. Start over but this time with four triangles coming together at a point. Taping four together forms a square pyramid. Tape another triangle to each edge of the square and bring these four to a point. This will result in a completely regular eight-sided polyhedron or *octahedron* (octa = eight).
> 3. Begin again with triangles. Tape five together at a point forming a pentagonal pyramid. Continue to add triangles on open edges until there are five at each vertex. It is very exciting to see and feel this form into a ball-like structure of 20 triangles, an *icosahedron* (icosa = twenty).
> 4. When you try six triangles at a point they make a flat surface, not a solid. Clearly no more than six will work either. Therefore, there are no more regular polyhedra possible with triangular faces. The next regular polygon to try is a square. Put three together at a point and at each of the other vertices, and you will

find you have formed a cube. Now, however, you know to call it a hexahedron. In a similar argument as was made with triangles, it should be clear that no other polyhedra can be made with squares.

5. The next polygon to work with is a regular pentagon. Three can be put together at a point, and you can continue joining three at each vertex until 12 pentagons form a complete polyhedron—a *dodecahedron* (dodeca = two + ten, or twelve).

6. Finally, conclude by experimentation that there are no other polyhedra possible with pentagons or with hexagons. After seeing that three hexagons lay flat, it is fairly clear that three septagons would not fit around a point. Thus, there are no other possible regular polyhedra, and you have found all five of the Platonic solids.

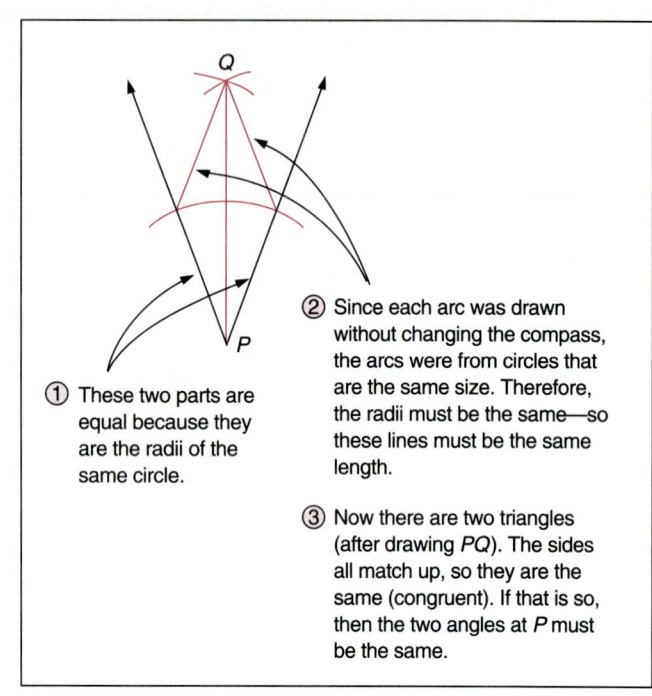

① These two parts are equal because they are the radii of the same circle.

② Since each arc was drawn without changing the compass, the arcs were from circles that are the same size. Therefore, the radii must be the same—so these lines must be the same length.

③ Now there are two triangles (after drawing *PQ*). The sides all match up, so they are the same (congruent). If that is so, then the two angles at *P* must be the same.

FIGURE 17.46: *A possible informal explanation for bisecting an angle with compass and straightedge.*

An absolutely fantastic skeletal icosahedron can be built out of the newspaper rods described earlier in this chapter. Start with five at a point and join the other ends with five rods to form a pentagonal pyramid. Add two more rods at each vertex of the base and continue from there. If you simply keep in mind that you want five rods at each vertex and that all faces make triangles, the result is guaranteed. The icosahedron will be about four feet in diameter and will be amazingly rigid.

Another way to make all of the Platonic solids in skeletal form is to use the bendable drinking straws described earlier. There is enough rigidity that even the cube and dodecahedron will turn out all right.

◆ Constructions

Too frequently students are taught to perform constructions with a straightedge and compass but have no idea why the constructions work. In most instances, the constructions represent simple theorems that students can follow and provide reasons for the steps. An example of bisecting an angle is illustrated in Figure 17.46.

The Mira offers a different approach to doing many constructions that have traditionally been done with compass and straightedge. With either set of tools, determining on their own how to do these constructions is also a valuable form of deductive reasoning.

◆ INFORMAL GEOMETRY ON THE COMPUTER

More and more the computer offers interesting and powerful methods of exploring geometric ideas. Some of the most interesting software for mathematics is aimed at the development of spatial sense. While there are many programs available, two categories of geometric software deserve some attention in this chapter. First are programs that allow drawings to be easily made and measured (length, area, angles, ratios) in a way that invites conjecturing and exploration of geometric properties. Second is the computer-programming language called Logo. While Logo is a regular computer-programming language, here we will examine briefly only the graphics feature of the language as it relates to geometry.

GEOMETRY COMPUTER TOOLS

A "tool" in this context refers to a piece of software that is analogous to a word processor or a spreadsheet. It has capabilities to do things, but it only does what the user asks of it. In that sense, it does not "teach" but allows the user to explore and measure shapes in two dimensions. The two software packages discussed here are not the only such tools available, but at the present time they adequately represent the range of available tools of this genre.

◆ The Geometric *Supposers*

A family of computer programs called *Geometric Supposers* from Sunburst were the first tool programs in this category. They are now available in all popular computer formats but have changed little from the original versions. The first *Supposer* (Schwartz & Yerushalmy, 1985) explores triangles. Later came similar programs for investigating quadrilaterals and circles, and a fourth program called the *preSupposer* that includes most of the drawing capabilities of the other three.

The *Geometric Supposer: Triangles* starts by drawing a triangle. Users can select different types of triangles or can draw their own choosing various combinations of sides and angles. The *Supposer* generates its shapes with random dimensions so that each time an isosceles triangle is selected the program will draw a different isosceles triangle and in a different orientation on the screen. Once a shape is on the screen, the user selects additional constructions or measurements from menus. Angles can be bisected; circles can be inscribed or circumscribed; line segments can be partitioned into equal segments; lines can be extended, reflected, or drawn parallel or perpendicular to other lines; and points can be joined. Lengths, angles, and areas can all be measured, either in terms of a fixed unit or in terms of some other line segment or area. As a result, it is easy to make comparisons or observe ratios of distances or areas. Figure 17.47 shows the *Geometric Supposer: Triangles* after several constructions and measurements have been made.

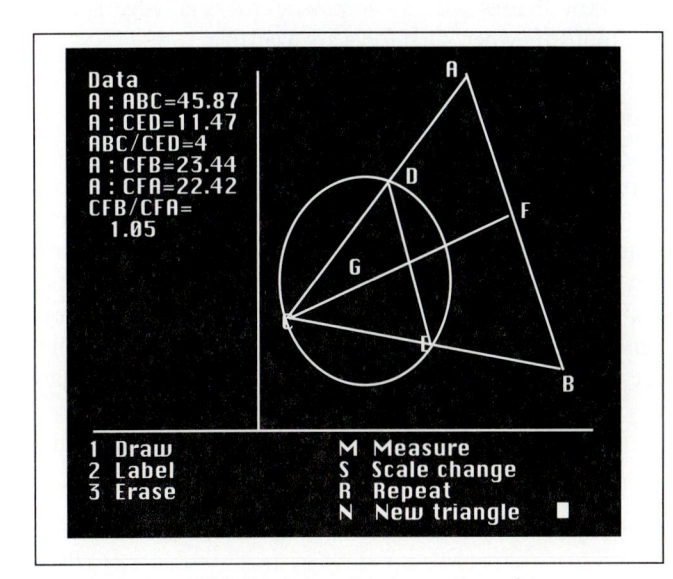

FIGURE 17.47: *The Geometric Supposer: Triangles*

One of the most interesting features of the *Supposers* is their ability to repeat a series of constructions on previous figures or on new figures. When students observe a relationship on one figure, they are encouraged to test it immediately on a whole series of figures. With the *Geometric Supposer: Quadrilaterals,* for example, the teacher might suggest joining the midpoints of the sides of quadrilaterals. The first experiment on a random parallelogram produces another parallelogram. Was that just a coincidence? Try it quickly on another parallelogram. What about on trapezoids? How about rectangles, kites, rhombuses, squares, or even a random quadrilateral? Since the *Supposer* draws these shapes quickly and accurately, and because measurements on any shape are easily make by the computer, the student can focus attention on relationships and make observations. Drawing and measuring skills are not required.

The *Geometer's Sketchpad*

The *Geometer's Sketchpad* (Key Curriculum Press, 1992) is designed to provide the same conjecturing and exploration needs originally envisioned for the *Supposers* but is much, much more powerful. The *Sketchpad* offers a pallet of drawing tools (points, circles, segments/rays/lines) that are used with a mouse to draw almost any geometric objects. Objects can be drawn in a freehand manner or with a specified relationships to existing objects. Once drawn, any object (point, line, circle, etc.) can be moved or dragged with the mouse, making the drawing dynamic or manipulatable. When objects are drawn freely, they can be moved anywhere on the screen. This allows for polygons to be stretched and positioned in almost any imaginable manner. When objects are created in relationship to other objects, those relationships remain even though objects are moved. Parallel lines remain parallel, and when a figure is changed so does any transformation of that figure that may also have been constructed. Consider the following very simplified example:

Follow these steps in Figure 17.48:

1. Points A, B, and C were placed freely, and a segment was drawn between A and B.
2. A line was drawn through C parallel to AB.
3. A point D was placed on the line through C, and segments AD and DB were drawn.

(At this time, the line through C can be moved, but it will remain parallel to AB. If A or B is moved, C remains fixed, but the line through C will rotate so that it remains parallel to AB. Through all of this, the point D will remain on the line, and triangle ABD will adjust accordingly.)

4. A segment is constructed (some steps are omitted here) so that it runs from D to a hidden line through AB and is always perpendicular to AB. This line is clearly the height of triangle ABD. The *Sketchpad* permits labeling any way you wish.
5. Measures are made of AB, the height, and the area of ABD. The formula AB × height/2 is also calculated.

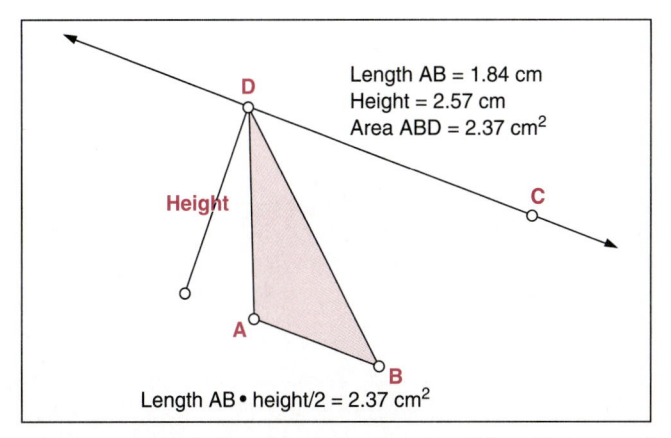

FIGURE 17.48: *A* Sketchpad *construction. When D is moved along the line DC, all measures are updated.*

Now D can be dragged along line DC, and the height of ABD will remain constant, as will the length of AB. It can clearly be seen that the area of the triangle remains fixed as long as the base and height remain fixed.

In Figure 17.49, the midpoints of a freely drawn quadrilateral ABCD have been joined. The diagonals of the resulting quadrilateral (EFGH) are also drawn and measured. No matter how the points A, B, C, and D are dragged around the screen, even inverting the quadrilateral, the other lines will maintain the same relationships (joining midpoints and diagonals), and the measurements will be instantly updated on the screen.

These two examples can only provide a small hint at the dynamic features of the *Sketchpad* without your actually experiencing it. In addition, the *Sketchpad* permits drawings to be saved and programs that produce drawings to be created and saved. Among other capabilities is the ability to examine slides, flips, rotations, and dilations—all dynamically—and to define combinations of these transformations.

◆ **In the Classroom**

Tools such as the *Sketchpad* and the *Supposers* can be used by students in a laboratory setting where all students or students working in pairs have access to their own computers. In this mode, the class should have discussed a potential exploration or defined a problem to solve. For example, on the *Quadrilateral Supposer,* students might explore what relationships they can discover about the diagonals of different types of quadrilaterals. They can then test their ideas by using the *Supposer* to draw a specific quadrilateral beginning with the diagonals. This is essentially the same exploration found in activities 17.41 and 17.42. Worksheets on which students are guided as to what types of constructions to make and what measurements to look for are very helpful. Middle-school students will need careful guidance when working with these tools in a lab setting. There is an excellent book of guided explorations available for the *Sketchpad* (Bennett, 1992), and there are a number of books that have been written to support the *Supposer.* (Refer to a *Sunburst* catalog.)

These geometry tools are also very effective under teacher direction with just a single computer and an overhead liquid crystal display panel. In this mode, the programs become high-tech electronic blackboards. Students can interact in groups and/or with the full class in making conjectures. Their ideas can be tested or explored on the computer. The verbal interaction between students and with the teacher is so powerful in this mode that it is difficult to argue that the lab setting is better.

Many of the activities suggested for level 1 and level 2 in this chapter can be augmented by or completely implemented with these or similar programs. Tedious drawing of geometric figures with a compass and straightedge is difficult to justify when such tools are available.

LOGO AND INFORMAL GEOMETRY

Turtle graphics is only one aspect of Logo, a highly accessible yet powerful computer language that has many uses other then geometry. However, the ease with which Logo can be used to draw pictures on the monitor makes it a natural vehicle for geometric explorations. Since Logo is both very easy to use and very powerful, it can be used by children as young as kindergarten and also be a challenging medium for college students. Very briefly, in Logo graphics, a tiny triangle or picture of a turtle is controlled with sim-

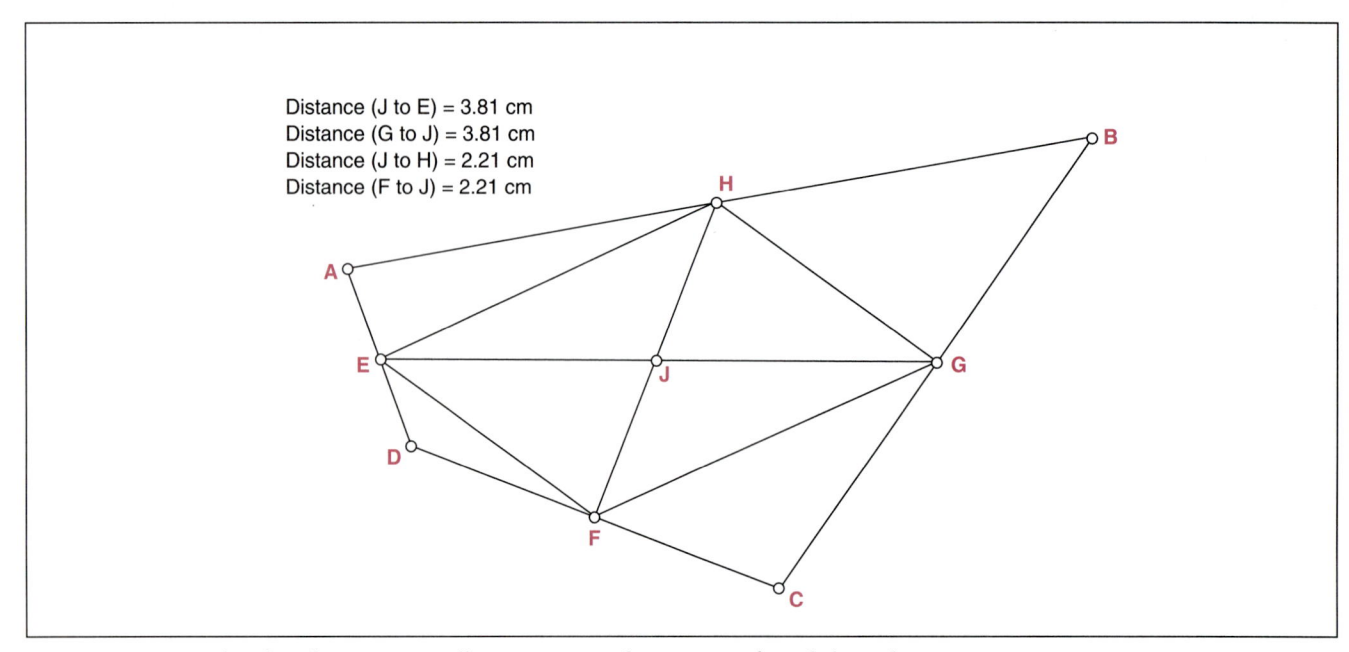

FIGURE 17.49: *A Sketchpad construction illustrating several properties of quadrilaterals*

ple keyboard commands. It can be made to go forward or backward any distance. A FORWARD 50 command makes the turtle move forward about 5 cm and draw a line in its path. The turtle can also be turned in its place. The command RIGHT 90 turns the turtle 90 degrees to its right. With these and other commands coupled with the ability to put lists of commands into procedures (little programs), children have both power and flexibility with which they can draw on the computer screen.

The purpose of the sections that is only to illustrate some of the ways that Logo can be used in an informal geometry program. No attempt is made to explain the details of the Logo programming involved. (For a more detailed explanation of the Logo language, see Chapter 21.)

◆ Shapes in Logo

Children can learn to use regular Logo commands and Logo programming as early as the first grade. Second and third graders can learn to define procedures to create shapes of their own. Within the simplest turtle-graphics programming, children develop intuitive understandings of angles and distances.

Consider a child who is trying to write a procedure to draw a rectangle. He or she has a visual image of "rectangle" and may even have one drawn on paper. But the turtle must be "taught" to draw each part separately, and the relationship between the parts must be clearly understood. The resulting procedure is a form of definition of a rectangle developed by the child. Further, it is a definition based on component parts and relationships rather than a gestalt visual image.

A more profound difference between visualizing shapes and defining procedures to draw shapes can be observed when watching children who want the turtle to draw a circle. The turtle moves only in straight lines, not curves. A common technique for helping children translate shapes into turtle actions is to have them "play turtle" or walk the shape as if they were the turtle on the screen. How can you walk in a circle if you can only move forward? Children quickly conclude that they should go forward just a little bit, and then turn a little bit, and repeat this over and over. Logo includes a REPEAT function, which makes it easy to do a series of actions over and over. A common "first circle" is produced by the line,

REPEAT 360 [FD 1 RT 1]

Compare the following Logo procedures:

TO SQUARE
　REPEAT 4 [FD 50 RT 90]
END

TO EIGHTSIDES
　REPEAT 8 [FD 10 RT 45]
END

TO 36 SIDES
　REPEAT 36 [FD 2 RT 10]
END

Each procedure draws a regular polygon. The 36-sided polygon looks exactly like a circle, and yet it was conceived as a polygon. "Circles" of different sizes can be made by changing the amount of the turn and/or the forward step. This example illustrates a completely different form of thinking and analysis of shape than is provided in other mediums. How do you know how much to turn? What if you wanted to make a triangle or a hexagon? How is a circle like an octagon or even a square? How do you make different triangles? Can you make a curve that is not a circle?

Activities in the regular Logo environment can be designed for almost any level of geometric thought. Most of these will be at level 1 or higher.

◆ "Instant" Logo

Programs in Logo can be written by teachers that permit very young children to press only a single key, possibly followed by RETURN, to control the turtle. Most of these so-called "instant" Logo routines have a key for each of the following actions: FORWARD 10, BACK 10, RIGHT 30, and LEFT 30.

In Figure 17.50, these actions are caused by pressing F, B, R, and L. The keys S, T, and X make the turtle draw a square, triangle, and a rectangle (box). The D key clears the screen. The N and P keys are for PENUP and PENDOWN. Keys can be labeled with stickers to correspond to the turtle action.

FIGURE 17.50: *An "instant" Logo program allows young children to experiment with lines, shapes, and angles on the computer.*

With an instant Logo program children can explore drawing shapes and putting simple shapes together on the screen. There is a significant difference between this activity and using tiles or a geoboard. With tiles and similar materials, children visualize shapes as whole entities. Shapes are easily twisted, turned, and shoved together. When drawing with the Logo turtle, the interaction with shape and position is quite different. Even with an instant program, children must consider how shapes are formed, at what corner of a shape the turtle starts, in what direction it turns, and what direction the turtle is facing when it begins. This is not to say that instant Logo is better than tiles or geoboards, only that it provides a significantly different perspective of shape and orientation.

All of the major versions of Logo include a simple version of an instant program in the manuals. A complete listing and description of an excellent version is included in Campbell's article, "Microcomputers in the Primary Mathematics Classroom" (1988), or in *Learning with Logo* (Watt, 1983).

Instant Logo activities are appropriate for level 0 thinking and provide a nice bridge to level 1 activities and to regular versions of Logo.

◆ More Ideas with Logo

The few ideas that follow are only intended to give some flavor for what can be done with Logo and geometry.

Variables can be used in procedures to create many shapes with the same characteristics. For example, the following procedure will draw rectangles of any dimensions. Executing RECTANGLE 20 20 will produce a square. How are RECTANGLE 20 60 and RECTANGLE 60 20 alike? How are they different?

```
TO RECTANGLE :L :W
    REPEAT 2 [FD :L RT 90 FD :W]
END
```

Not only can forward distances be variable, but angles can be, too. How would you write a procedure to draw any parallelogram?

Symmetry can be explored in Logo to provide a different perspective. A mirror image of a shape (reflection) is achieved by changing all right turns to left, and vice-versa. It is also easy to create shapes, images, and rotational symmetry.

Try writing a procedure with no variables to create a small shape or drawing. Next write a new procedure from the original in which all distances are multiplied by the same variable. By changing the value of the variable, proportional drawings (similar figures) are produced.

REFLECTIONS ON CHAPTER 17: WRITING TO LEARN

1. Describe what is meant by informal geometry. How is it different from the geometry usually taught in high school? How is it different from what you remember from school?

2. What do you think are the two best reasons for studying geometry? Explain.

3. Describe in your own words the first three van Hiele levels of geometric thought (levels 0, 1, and 2). Note in your description how the ideas implicit or under the surface at each level become the explicit objects of thought for the subsequent levels.

4. Describe the four characteristics of the van Hiele levels of thought. For each, reflect on why each characteristic might be important for teachers.

5. How would activities aimed at levels 0, 1, and 2 be different?

6. Select an activity from each of the levels, and do it (perhaps with a friend). Describe your experiences.

Note: There are so many activities in this chapter that you are probably overwhelmed. The best way to get excited and comfortable with informal geometry is to start doing activities, the same ones that are designed for students. Do not worry about being "right." The only thing that is right is to *do something,* to think about what you are doing, and have fun. If you get interested, you will begin to search for ideas and relationships, and you may even decide you want to do some further reading. Good geometry resource books for teachers are quite plentiful, and articles in the *Arithmetic Teacher* are also quite common. You could not go wrong by simply exploring activities from this chapter, especially those designed for levels 0 and 1. A possible limitation may be materials. An effort has been made to include many activities that use only teacher-made materials.

FOR DISCUSSION AND EXPLORATION

1. Examine the teacher's edition of a basal textbook at any grade level. Select any lesson on geometry. Remember that the authors of the pupil's book are restricted to the printed page by the very nature of books. Teachers are not so restricted. How would you teach this lesson so that it was a *good* informal geometry lesson? Your lesson should include a hands-on activity and have a problem-solving spirit.

2. Read the Geometry Standard for both K–4 and 5–8 levels of the NCTM *Standards* book. What do you think the *Standards* authors mean by "spatial sense"? What message is in the *Standards* that is most important to you?
3. Get a copy of the book *Tangramath* (Seymour, 1971) or *Tangram Treasury*, Book A, B, or C (Fair, 1987), or any other book of tangram activities. Explore one idea that you like, and develop a lesson for children around that activity.
4. Get one of the *Geometric Supposers,* the *Geometer's Sketchpad,* or a similar computer program, and play around with it. There are suggested explorations in the manuals, or you might pursue one of the activity booklets that has been designed to accompany the *Supposers.*

◆◆◆◆◆

SUGGESTED READINGS

Barson, A. (1971). *Geoboard activity cards: (Primary, intermediate).* Fort Collins, CO: Scott Resources, Inc.

Beaumont, V., Curtis, R., & Smart, J. (1986). *How to teach perimeter, area, and volume.* Reston, VA: National Council of Teachers of Mathematics.

Bennett, D. (1992). *Exploring geometry with the Geometer's Sketchpad.* Berkeley, CA: Key Curriculum Press.

Bezuszka, S., Kenney, M., & Silvey, L. (1977). *Tessellations: The geometry of patterns.* Palo Alto, CA: Creative Publications.

Bidwell, J. K. (1987). Using reflections to find symmetric and asymmetric patterns. *Arithmetic Teacher, 34*(7), 10–15.

Bruni, J. V., & Seidenstein, R. B. (1990). Geometric concepts and spatial sense. In J. N. Payne (Ed.), *Mathematics for the young children.* Reston, VA: National Council of Teachers of Mathematics.

Burger, W. F. (1982). Graph paper geometry. In L. Silvey (Ed.), *Mathematics for the middle grades (5–9).* Reston, VA: National Council of Teachers of Mathematics.

Carroll, W. M. (1988). Cross sections for clay solids. *Arithmetic Teacher, 35*(7), 6–11.

Chazan, D. (1989). *How to use conjecturing and microcomputers to teach geometry.* Reston, VA: National Council of Teachers of Mathematics.

Creative Publications. (1986). *Hands on pattern blocks, Books 1, 2, & 3.* Palo Alto, CA: Creative Publications.

Creative Publications. (1986). *Hands on geoboards, Books 1, 2, & 3.* Palo Alto, CA: Creative Publications.

Crowley, M. L. (1987). The van Hiele model of the development of geometric thought. In M. M. Lindquist (Ed.), *Learning and teaching geometry, K–12.* Reston, VA: National Council of Teachers of Mathematics.

Dana, M. E. (1987). Geometry—a square deal for elementary teachers. In M. M. Lindquist (Ed.), *Learning and teaching geometry, K–12.* Reston, VA: National Council of Teachers of Mathematics.

Dana, M. E., & Lindquist, M. M. (1978). Let's try triangles. *Arithmetic Teacher, 26*(1), 2–9.

Del Grande, J. (1993). *Geometry and spatial sense: Addenda series, grades K–6.* Reston, VA: National Council of Teachers of Mathematics.

Fuys, D. J., & Liebov, A. K. (1993). Geometry and spatial sense. In R. J. Jensen (Ed.), *Research ideas for the classroom: Early childhood mathematics.* New York: Macmillan Publishing Co.

Geddes, D. (1992). *Geometry in the middle grades: Addenda series, grade 5–8.* Reston, VA: National Council of Teachers of Mathematics.

Geddes, D., & Fortunato, I. (1993). Geometry: Research and classroom activities. In D. T. Owens (Ed.), *Research ideas for the classroom: Middle grades mathematics.* New York: Macmillan Publishing Co.

Giganti, P., Jr., & Cittadino, M. J. (1990). The art of tessellation. *Arithmetic Teacher, 37*(7), 6–16.

Hill, J. M. (Ed.). (1987). *Geometry for grades K–6: Readings from the Arithmetic Teacher.* Reston, VA: National Council of Teachers of Mathematics.

Hoffer, A. R. (1988). Geometry and visual thinking. In T. R. Post (Ed.), *Teaching mathematics in grades K–8: Research based methods.* Boston: Allyn & Bacon.

Kaiser, B. (1988). Explorations with tessellating polygons. *Arithmetic Teacher, 36*(4), 19–24.

Lappan, G., & Even, R. (1998). Scale drawings. *Arithmetic Teacher, 35*(9), 32–35.

National Council of Teachers of Mathematics. (1990). Spatial sense [Focus issue]. *Arithmetic Teacher, 37*(6).

Ranucci, E. R., & Teeters, J. L. (1977). *Creating Escher-type drawings.* Palo Alto, CA: Creative Publications.

Senk, S. L., & Hirschhorn, D. B. (1990). Multiple approaches to geometry: Teaching similarity. *Mathematics Teacher, 83,* 274–280.

Serra, M. (1993). *Discovering geometry: An inductive approach.* Berkeley, CA: Key Curriculum Press.

Shaw, J. M. (1983). Exploring perimeter and area using centimeter squared paper. *Arithmetic Teacher, 31*(4), 4–11.

Shroyer, J., & Fitzgerald, W. (1986). *Mouse and elephant: Measuring growth.* Menlo Park, CA: Addison–Wesley.

Skinner, J. (1987). Extracts from a teacher's diary. *Mathematics Teaching, 121,* 23–26.

Teppo, A. (1991). Van Hiele levels of geometric thought revisited. *Mathematics Teacher, 84,* 210–221.

Van de Walle, J. A., & Thompson, C. S. (1980). Concepts, art, and fun from simple tiling patterns. *Arithmetic Teacher, 28*(3), 4–8.

Van de Walle, J. A., & Thompson, C. S. (1981). A triangle treasury. *Arithmetic Teacher, 28*(6), 6–11.

Van de Walle, J. A., & Thompson, C. S. (1984). Cut and paste for geometric thinking. *Arithmetic Teacher, 32*(1), 8–13.

Walter, M. J. (1970). *Boxes, squares and other things.* Reston, VA: National Council of Teachers of Mathematics.

Willcutt, B. (1987). Triangular tiles for your patio. *Arithmetic Teacher, 34*(9), 43–45.

Winter, J. J., Lappan, G., Phillips, E., & Fitzgerald, W. (1986). *Middle grades mathematics project: Spatial visualization.* Menlo Park, CA: Addison-Wesley.

18 LOGICAL REASONING: ATTRIBUTE AND PATTERN ACTIVITIES

◆ IN CHAPTER 2, MATHEMATICS WAS DESCRIBED AS THE SCIENCE of pattern and order. The search for and analysis of pattern and order are an integral part of really doing mathematics, whether it is developing a computational algorithm, exploring properties of shapes, or figuring the solution to a probability problem. Children can learn the processes of doing mathematics as they learn mathematics content. However, the ability to reason is so important in this science of pattern and order that attention should be given explicitly to helping children develop their reasoning skills.

Before you read any further, one important note: The activities in this chapter are *not just for early childhood* classrooms. There has been a tradition that work with patterns and attribute materials is done only in K–1 classrooms. Put that notion aside. There are activities in this chapter for all students K–8. The objectives of these activities are integral to *all* mathematics learning.

OBJECTIVES

Again turning your attention to an earlier chapter, we saw in Chapter 4 that in addition to the knowledge base of mathematics, a good problem-solving program has three additional sets of goals: (1) *strategy goals*, (2) *metacognitive goals*, and (3) *affective goals*. The objectives of attribute and patterning activities are nearly identical with those found in those three categories.

STRATEGY GOALS

The activities you will find throughout this chapter are clearly aimed at higher-order thinking skills. Logical reasoning of some form is important in nearly every activity. Students are encouraged to formulate and check hypotheses. Because many of these logic problems are posed with manipulative materials or calculators, the strategy of try-and-adjust or guess-and-check is encouraged. Attribute activities get specifically at the skills of classification (observing likenesses and differences). Patterning activities develop directly a sense of pattern and regularity, and practice the skills of searching for pattern, extending patterns, and making pattern generalizations.

Another feature of all of these activities is that correctness can always be determined by the students. Certainly students will ask, "Is this right?" but you can always respond with, "How can you tell? Does what you did fit the pattern?" The ability to self-assess is absolutely critical in all of mathematics. The habit and the process of self-assessment are developed in these activities along with the expectation that it can be done.

METACOGNITIVE GOALS

In all of these tasks there is a clear sense of working toward a goal—figuring out the classification scheme, discerning the pattern, extending the pattern. Thus it is quite natural to ask students the three questions suggested in Chapter 4: *What* are you doing? *Why* are you doing it? *How* does it help you? If you keep the habits of self-monitoring in mind, you can get as much mileage out of these activities as any problems in the curriculum. Students learn from these experiences to reflect on their own activity, to be aware of their progress or lack of progress, and to change their strategies or approaches as need dictates.

AFFECTIVE GOALS

Affective goals, you will recall, fall under two headings: attitudes and beliefs. Under attitudes we find *willingness* to try, *enjoyment* in solving problems, and *perseverance* when faced with difficulty. Most attribute or pattern problems can be adjusted so that there is just sufficient difficulty to be challenging and interesting without being disheartening. When the tasks are set at the right level, children love to work on them. They are fun things to do in groups. There is a lot of natural give-and-take between children as they work. As a result, children learn that they really can solve tough problems and enjoy doing it. They learn that occasionally it takes time, but the work pays off.

Under the heading of beliefs we find *self-confidence*, belief in *personal abilities*, and the belief that *methods can be discovered*. Again, the good-spirited, fun approach, coupled with the use of manipulatives, make attribute and pattern activities ideally suited to address these goals.

ATTRIBUTE MATERIALS AND ACTIVITIES

Attribute activities help students reason about likenesses and differences. Classifications or groupings are discovered. Schemes for presenting these groupings are developed. The problems and materials help children understand that things can be looked at in a number of ways. (An object has several attributes. It may be like another object when thought about one way but different when considered in another way.) Classification skills are also science skills. Many good science programs, especially those that are process-oriented, include a heavy emphasis on classification.

In the 1960s and early 1970s, classification activities became very popular as readiness activities for number. They were viewed as "prenumber" activities. The view adopted in this book is that there is no demonstrated relationship between classification activities and number-concept development. The activities presented here should not be considered prenumber experiences.

ATTRIBUTE MATERIALS

Attribute materials are sets of objects that lend themselves to being sorted and classified in different ways. Natural or *unstructured* attribute materials include such things as sea shells, leaves, the children themselves, or the set of the children's shoes. The *attributes* are the ways that the materials can be sorted. For example, hair color, height, and gender are attributes of children. Each attribute has a number of different *values*: for example, blond, brown, or red (for the attribute of hair color); tall or short (for height); male or female (for gender).

A *structured* set of attribute pieces has exactly one piece for every possible combination of values for each attribute. For example, several commercial sets of plastic attribute materials have four attributes: color (red, yellow, blue), shape (circle, triangle, rectangle, square, hexagon), size (big, little), and thickness (thick, thin). In the set just described there is exactly one large, red, thin, triangle as well as one each of all other combinations. The specific values, number of values, or number of attributes that a set may have is not important.

In Figure 18.1 three teacher-made sets of attribute pieces are illustrated. The attribute shapes are easily made in nice large sizes out of poster board and laminated for durability. A Woozle Cards master is in the Black-line Masters. These can be duplicated on tagboard and quickly colored in two colors before laminating. The ice-cream cones can be made from construction paper and laminated or could be duplicated on paper and cut out by the students.

The value of using structured attribute materials (instead of unstructured materials) is that the attributes and values are very clearly identified and easily articulated to students. There is no confusion or argument concerning what values a particular piece possesses. In this way we can focus our attention in the activities on the reasoning skills that the materials and activities are meant to serve. Even though a nice set of attribute pieces may contain geometric shapes of different colors and sizes, they are not very good materials for teaching shape, color, or size. A set of attribute shapes does not provide enough variability in any of the shape attributes to help students develop anything but very limited geometric ideas. In fact, simple shapes, primary colors, and two sizes are chosen because they are most easily discriminated and identified by even the youngest of students.

ACTIVITIES WITH ATTRIBUTE MATERIALS

Most attribute activities are best done in a teacher-directed format. Young children can sit on the floor in a large circle where all can see and have access to the materials. Older children can work in groups of four to six students, each group with its own set of materials. In that format, problems can be addressed to the full class, and groups can explore them independently. All activities should be conducted in an easygoing manner that encourages risks, good thinking, attentiveness, and discussion of ideas. The atmosphere should be nonthreatening, nonpunitive, and nonevaluative.

Attribute Shapes

Attributes (values): shape (circle, square, triangle, diamond, rectangle)   face (happy, sad)
60 pieces   color (red, gray, white)   size (big, little)

Woozle Cards
(Black-line Masters)

Attributes (values): shape (rounded, straight)
16 pieces   color (red, gray)
                  dots (one, two)
                  hair (bald, fuzzy)

Ice Cream Cones

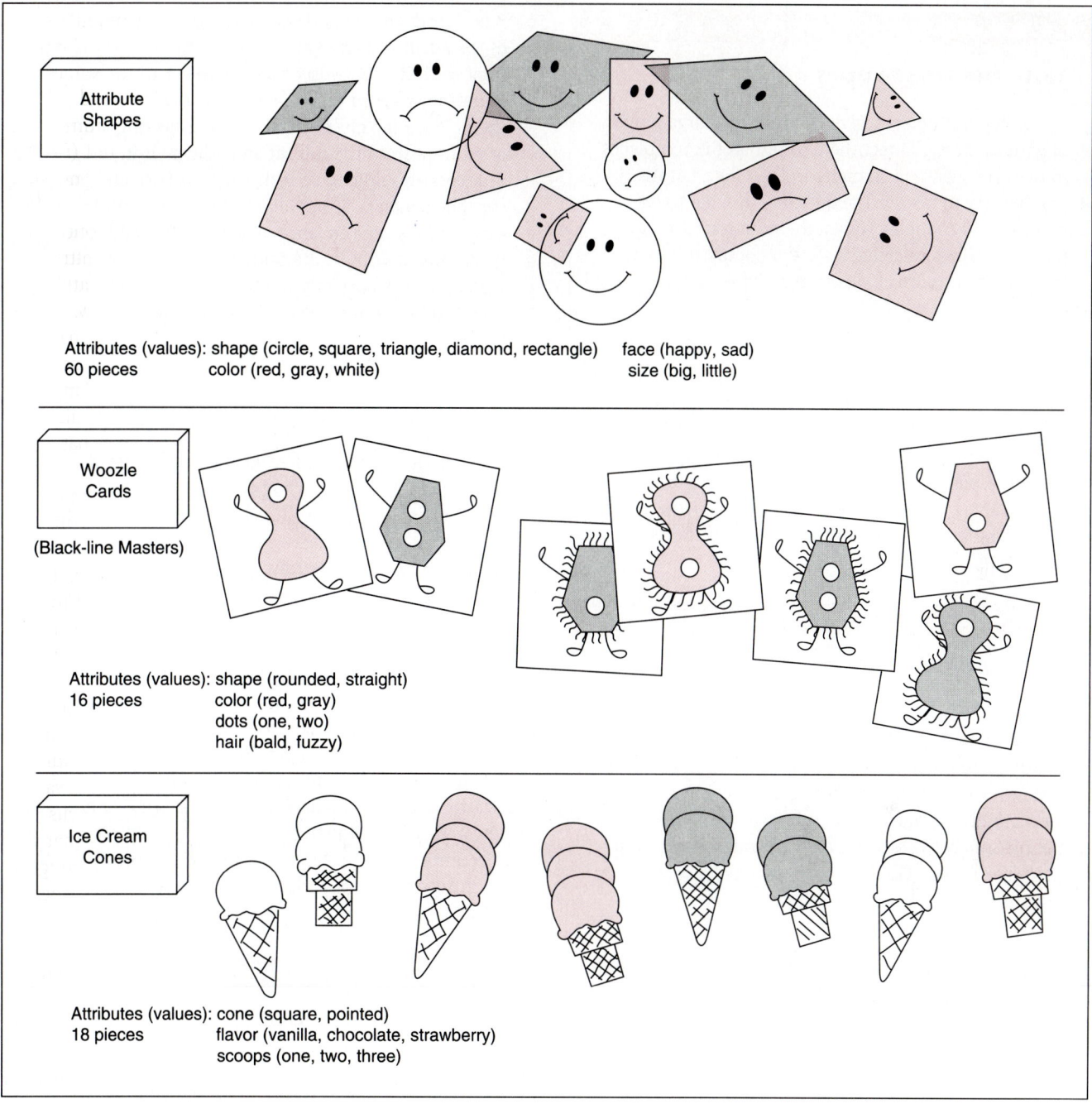

Attributes (values): cone (square, pointed)
18 pieces   flavor (vanilla, chocolate, strawberry)
                  scoops (one, two, three)

FIGURE 18.1: *Three teacher-made attribute sets*

Most of the activities in this section will be described using the geometric shapes in Figure 18.1. However, each could be done with any structured set, and some could be done with nonstructured materials.

◆ Learning Classification Schemes

Several attribute activities involve using overlapping loops, each containing a designated class of materials. Loops are made of yarn or are drawn on paper to hold a designated class of pieces such as "red" or "not square." When two loops overlap, the section that is inside both loops is for the pieces that have both properties. Children as young as kindergarten can have fun with simple loop activities. With the use of words such as *and, or,* and *not,* the loop activities become challenging for children in the upper grades as well.

Before children can use these loops in a problem-solving activity, the scheme itself must first be understood. A good way to accomplish this is to simply do a few activities that involve the loops. Children find these interesting and fun. After several days of working with these initial activities, you will be able to move to the logic problems described in the next section that involve the same formats. Children will be able to attend to the problems because they are familiar with the way the loops are used.

18.1 THE FIRST LOOPS

At the beginning level, give children two large loops of yarn or string. Direct them to put all the red pieces in one string and all triangles in the other. Let the children try to resolve the difficulty of what to do with the red triangles. When the notion of overlapping the strings to create a section common to both loops is clear, more challenging activities can be explored.

Once the idea of how a loop can be used to hold a particular type of piece, "strings" or loops can be drawn on poster board or on large sheets of paper. If you happen to have a magnetic blackboard, try using small magnets on the backs of the pieces and conduct full-class activities with the pieces on the board. Students can come to the board to place or arrange pieces in strings drawn on the board with colored chalk.

18.2 LABELED LOOPS

Label the overlapping loops with cards indicating values of different attributes. Let children take turns randomly selecting a piece from the pile and deciding in which region it belongs. Pieces belonging in neither loop are placed outside. Let other students decide if the placement is correct, and occasionally have someone else explain. Do this even when the choice of regions is correct.

A significant variation of the last activity is to introduce negative attributes such as "not red" or "not small." Other labels that are important include the use of the *and* and *or* connectives as in "red and square," "big or happy." The use of *and*, *or*, and *not* significantly widens children's classification schemes. It also makes these activities quite difficult for very young children. In Figure 18.2, three loops are used in a string game illustrating some of these ideas.

In addition to using loops to arrange or display overlapping classes of objects, a matrix or grid can also be used as shown in Figure 18.3. Generally the values for one attribute are listed across the top and for another attribute down the side. Each region in the grid holds all objects that share the two corresponding values.

18.3 ATTRIBUTE GRIDS

Draw a large grid on a poster board or on a large sheet of butcher paper. If you draw a 3 × 3 or a 3 × 4 grid, the same drawing can be used regardless of the attributes and values selected. Use cards as was done with the loops to designate the attribute values. Place the value cards face up so the students can see them. Have the children take turns selecting a piece at random from the full set and placing it in the grid. Students should decide if the piece is correctly placed (not the teacher). If not all values of an attribute are included among the labels placed on the grid, some pieces may not fit in the grid at all. Instead of simply placing pieces correctly in the grids, students can be asked to find a piece that fits in a particular cell of the grid.

◆ Solving Logic Problems

The activities described so far have students attempting to classify materials according to our schemes. That is, the teacher creates a classification, and the children fit pieces into it. While this is important for the purpose of understanding classifications by more than one attribute (the overlapping loops), the activities make relatively few cognitive demands. All that is required to do these activities is an understanding of the loop method of classification and the ability to discriminate the attributes. When the words *and*, *or*, and *not* are used, children gain experience with those logical connectives. However, very limited logical reasoning or problem solving is actually going on. A much more significant mental activity is to infer how things have been classified when the scheme is not clearly articulated. The following activities are examples of those that require students to make and test conjectures about how things are being classified. These activities move classification squarely into the domain of problem solving.

18.4 GUESS MY RULE

For this activity try using students instead of shapes as attribute "pieces." Decide on an attribute of your students such as "wearing blue jeans" or "has stripes on clothing," but do not tell your rule to the class. Silently look at one child at a time and move him or her to the left or right according to this attribute rule. After a number of students have been sorted, have the next child come up and ask students to predict which group he or she belongs in. Before the rule is articulated, continue the activity for a while so that others in the class will have an opportunity to determine the rule. This same activity can be done with virtually any materials that can be sorted. When unstructured materials such as students, students' shoes, shells, or buttons are used, the classifications may be quite obscure, providing an interesting challenge.

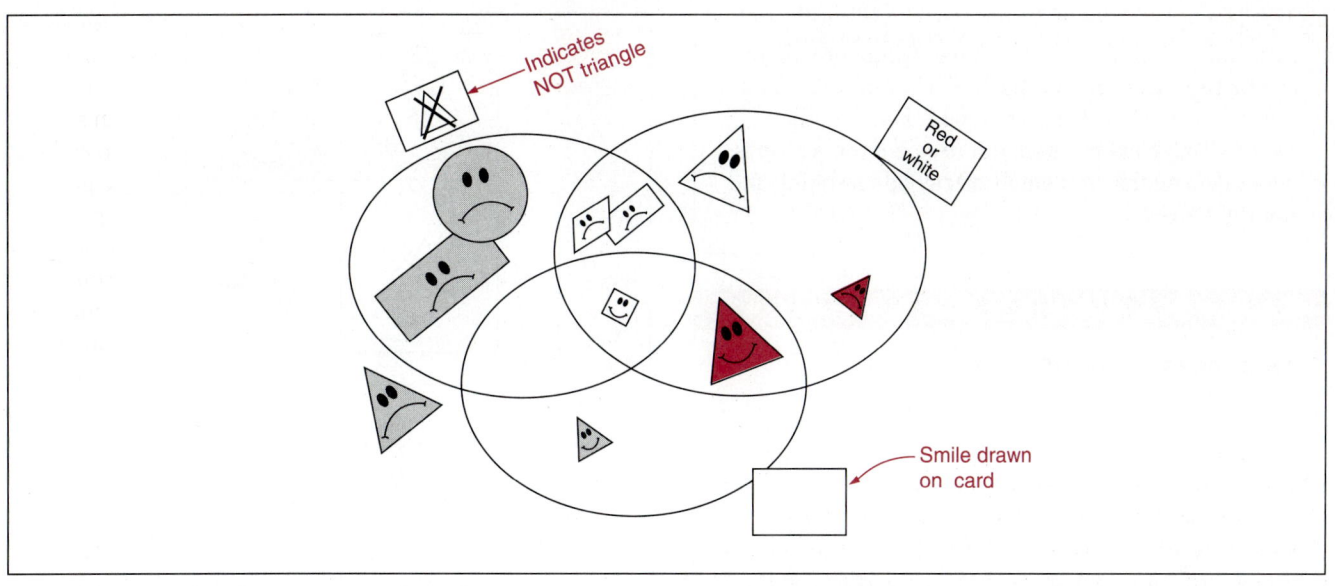

FIGURE 18.2: *A three-loop activity with attribute pieces*

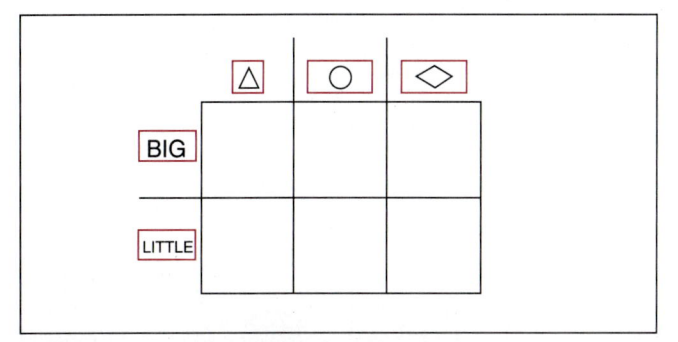

FIGURE 18.3: *A two-way classification grid*

18.5 HIDDEN LABELS

The same inference approach can be applied to the string game. Select label cards for the strings and place them face down. Begin to sort pieces according to the turned-down labels. As you sort, have students try to determine what the labels are for each of the loops. Let students who think they have guessed the labels try to place a piece in the string, but avoid having them guess the labels aloud. Students who think they know the labels can be asked to "play teacher" and respond to the guesses of the others. Point out that one way to test an idea about the labels is to select pieces that you think might go in a particular section. Do not turn the cards up until most students have figured out the rule. Notice that some rules or labels are equivalent: "not large" is the same as "small." With the use of three loops and logical connectives, this activity can become quite challenging even for middle-school students. With simple, one-value labels and only two loops it can easily be played in kindergarten.

18.6 SECRET GRIDS

The **Attribute Grids** activity can be converted to one that requires inference as was done with the loops. Select labels for each row and column, but turn the cards face down. Place pieces on the grid where they belong. The object is to determine what the attribute values (on the cards) are for each row and column. Encourage students to use trial and error to test what they think the cards say.

18.7 WHICH ONE DOESN'T BELONG?

The "Sesame Street" game "One of these things is not like the others" is easily conducted with any attribute set. Select four pieces so that three of the pieces have some feature in common that is not a feature of the fourth. The students try to decide which piece is different. In Figure 18.4, there are two pieces that are

FIGURE 18.4: *"One of these things is not like the others"*

18 / LOGICAL REASONING: ATTRIBUTE AND PATTERN ACTIVITIES

each different from the other three. Frequently there can be three or even four possible choices, each for a different reason. The students should explain their reasons, and classmates should decide if the reason is good. Be sure that you emphasize good reasoning and not right answers.

18.8 SETS OF FOUR

Choose one value of any two attributes, such as happy triangles or large red pieces. Select any four pieces that share both values and arrange them in a 2 × 2 array. The challenge is to make more arrays simi-lar to or like the original one. To make new arrays, you have to first decide how the pieces in the original array are alike, and use those same attributes but with different values to make the other arrays. Look at the example in Figure 18.5. The original array was all happy triangles: attributes of face and shape. The values of color and size are mixed. Within each new array, face and shape values are the same. Corresponding pieces each match the original array in color and size. If this activity is done with sets such as the Woozle Cards or the ESS People Pieces (Elementary Science Study, 1966), where each of the attributes has exactly two values, four arrays will use up all of the pieces.

Be prepared to adjust the difficulty of these and similar activities according to the skill and interest of your children. Remember that children like to be challenged, but an activity that is either too easy or too difficult is likely to result in restless children and discipline problems.

◆ Difference Games

As an introduction to the activities in this section, let each child select an attribute piece. Then you hold up one piece for all to see and ask questions such as:

Who has one that is like mine? How is it like mine?

Who has one that is different? Explain how it is different.

Look at your neighbor's piece. Tell how yours is like his or hers. Let your partner explain how his or hers is different from your piece.

Students will soon find that sometimes a piece differs in three or four ways, and other pieces will differ in perhaps one or two ways. To focus attention on this, ask, "Who has a piece that is different in *just exactly one way* from mine? Who has a piece that is different in *exactly two ways*?" Notice that for a set with four attributes, a piece differing in three ways is alike in one way. It is usual to limit attention to either one difference or two differences.

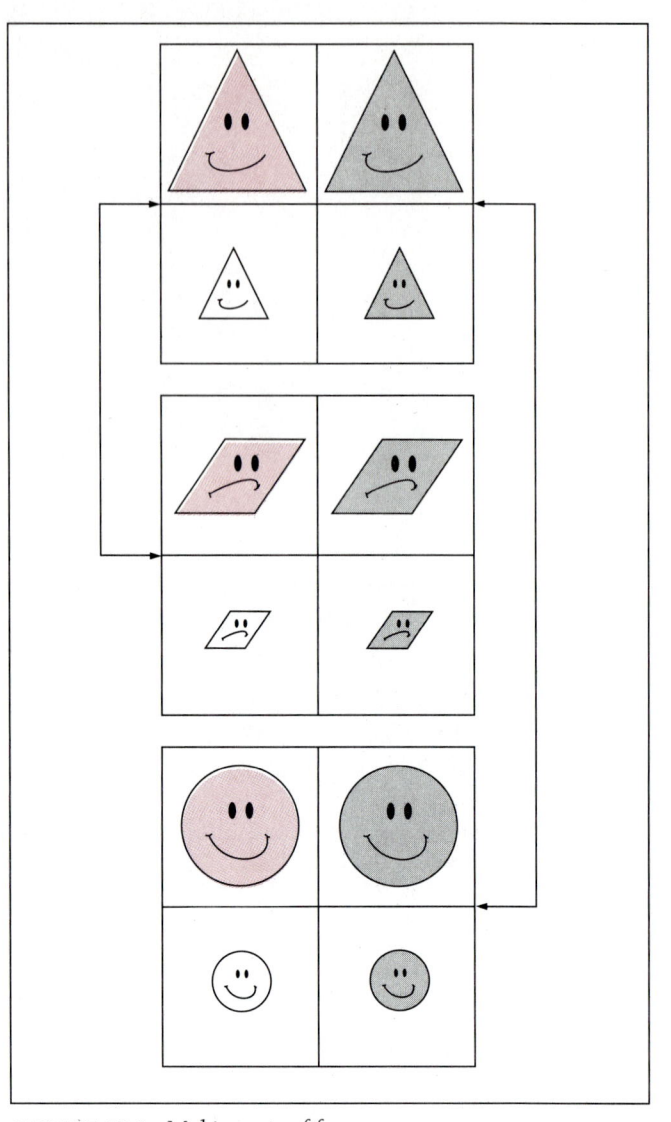

FIGURE 18.5: *Making sets of four*

18.9 DIFFERENCE TRAINS AND LOOPS

Place an attribute piece in the center of the group. The first student finds a piece that is different from this piece in exactly one way. Let students take turns finding a piece that differs from the *preceding* piece in just one way, creating a "one-difference train." The train can be made as long as the students wish or until no more pieces will fit the one-difference rule.

As a variation, draw a circular track on a piece of paper or poster board with 6 to 10 sections on it. Place the first piece in one of the sections. Subsequent pieces can be placed to the right or the left around the loop, but must differ in one way from the adjacent piece. Placing the last piece may be very difficult or even impossible; it must differ in exactly one way from the piece on either side. A sample is shown in Figure 18.6.

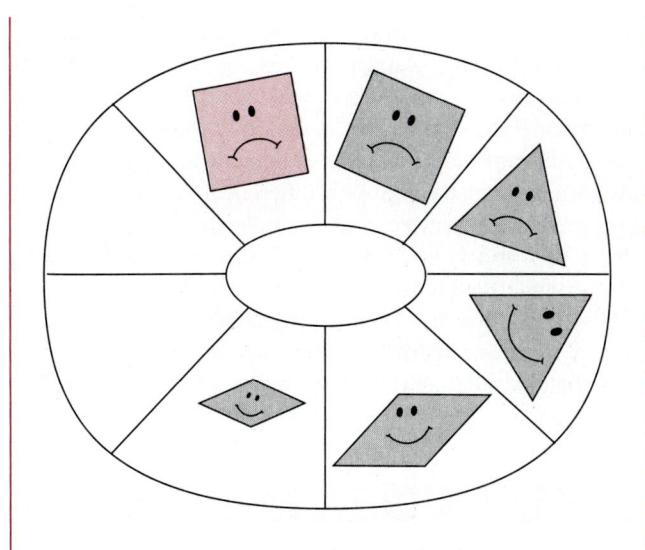

FIGURE 18.6: *Can you finish this one-difference track?*

Difference train and loop games can be two-difference games as well as one-difference games. These are not significantly more difficult but add variety.

18.10 DIFFERENCE GRIDS

Draw a 4 × 5 grid on a poster board so that each space will hold an attribute piece. Select one piece, and place it anywhere on the board. In turn students try to put pieces on the board in any space above, below, to the left, or right of a space that is filled. The rule is that adjacent pieces up and down must differ in two ways; to the left and right they must differ in one way. It frequently will be impossible to complete the grid but establishing that fact is a significant challenge. It is important for students to see that not every problem is solvable.

WORKING WITH PATTERNS

Pattern is so pervasive in all of mathematics that most any attempt to cover pattern in grades K–8 will be incomplete. Repeating patterns are appropriate for the early grades, especially grades K to 3. Growing patterns can begin as early as grade 3 or 4 but extend even to high school. These two categories are discussed first. Finally, a collection of other patterns and pattern activities is discussed including special attention to the use of calculators.

REPEATING PATTERNS

Identifying and extending patterns is an important process in mathematical thinking. Simple repetitive patterns can be explored as early as kindergarten. Young children seem to love to extend patterns, such as those made with colored blocks, Unifix cubes, buttons, and the like, across an entire room. The internal positive feedback they receive from knowing they are right and successful is significant.

Using Materials in Patterning

Almost all patterning activities should involve some form of physical materials to make up the pattern. This is especially true of repeating patterns in grades K to 4, but it is also true of virtually all patterning activities, even at the junior-high level. When patterns are built with materials, children are able to test the extension of a pattern and make changes as they discover errors without fear of being wrong. The materials permit experimentation or trial-and-error approaches that are almost impossible without them.

Many kindergarten and first-grade textbooks have pages where students are given a pattern such as a string of colored beads. The task may be to color the last bead or two in the string. The difference between this and the same activity done with actual materials is twofold. First, by coloring or marking a space on the page the activity takes on a significant aura of right versus wrong. There is clearly a correct way to finish the pattern. If a mistake is made, correction on the page is difficult and can cause feelings of inadequacy. With materials, a trial-and-error approach can be used. Second, pattern activities on worksheets prevent children from extending patterns beyond the few spaces provided by the page. Most children enjoy using materials such as colored blocks, buttons, Unifix cubes, and so on, to extend their patterns well beyond what could possibly be provided by a printed page. Children are frequently observed continuing a pattern with materials half-way across the classroom floor. In doing so, children receive a great deal of satisfaction and positive feedback from the activity itself. "Hey, I understand this! I can do it really well. I feel good about how I solved my pattern problem."

The same benefit of using materials can be built into patterning activities at the upper grades. There, the satisfaction comes not so much from extending a repeating pattern as it does from seeing how an observed relationship actually exists in a particular design or arrangement of materials. Not only is patterning a form of problem solving and logical reasoning, it can be very satisfying and self-rewarding. It is very important that students connect such positive feelings with mathematical thinking.

Repeating Pattern Activities

The concept of a repeating pattern and how a pattern is extended or continued can be introduced to the full class in several ways. One possibility is to draw simple shape patterns on the board and extend them in a class discussion. Oral patterns can be joined in by all children. For example, "do, mi, mi, do, mi, mi, . . ." is a simple singing pattern. Up, down, and sideways arm positions provide three elements with which to make patterns: up, side, side, down, up, side,

374   18 / LOGICAL REASONING: ATTRIBUTE AND PATTERN ACTIVITIES

side, down, . . . Boy, girl patterns or stand, sit, and squat-down patterns are also fun. From these ideas the youngest children learn quickly the concept of patterns. Students can begin to work more profitably in small groups or even independently once a general notion of patterns is developed.

18.11 PATTERN STRIPS

Students can work independently or in groups of two or three to extend patterns made from simple materials: buttons, colored blocks, Unifix cubes, toothpicks, geometric shapes, and a wide variety of other materials, most of which you can easily gather. For each set of materials, draw two or three complete repetitions of a pattern on strips of tagboard about 5 cm by 30 cm. The students, using actual materials, copy the pattern shown and extend it as far as they wish. Discussions with students can help them verbalize the patterns, uncover errors, and encourage students to make up patterns on their own. Figure 18.7 illustrates some possible patterns for a variety of materials. It is not necessary to have class-sized sets of materials. Make 10 to 15 different pattern strips for each set of materials. With six to eight sets, your entire class can work at the same time, with small groups working with different patterns and different materials.

The *core* of a repeating pattern is the shortest string of elements that repeats. Notice in the example pattern cards that the core is always fully repeated and never only partially shown. If the core of a pattern was –oo, then a card might have –oo–oo (two repetitions of the core), but it would be ambiguous if the card showed –oo–oo– or –oo–.

A significant step forward mathematically is to see how two patterns constructed with different materials can actually be the same pattern. For example the first patterns in Figure 18.7 and Figure 18.8 (p. 375) can be "read" A-B-B-A-B-B-, while the second patterns in each figure are A-B-C-C-A-B-C-C- patterns. Challenging students to translate a pattern from one medium to another or to find two patterns that are alike, even though made with different materials, is an important activity that helps students focus on the relationships that are the essence of repeated patterns.

18.12 PATTERN MATCH

On a chalkboard or overhead show six or seven patterns with different materials or pictures. Teach students to use an A, B, C method of reading a pattern. Half of the class can close their eyes while the other half uses the A, B, C scheme to read a pattern that you point to. After hearing the pattern, the students who had their eyes closed examine the patterns and try to decide which pattern was read. If two of the patterns in the list have the same structure, the discussion can be very interesting.

The following independent activity involves translation of a pattern from one medium to another, which is another

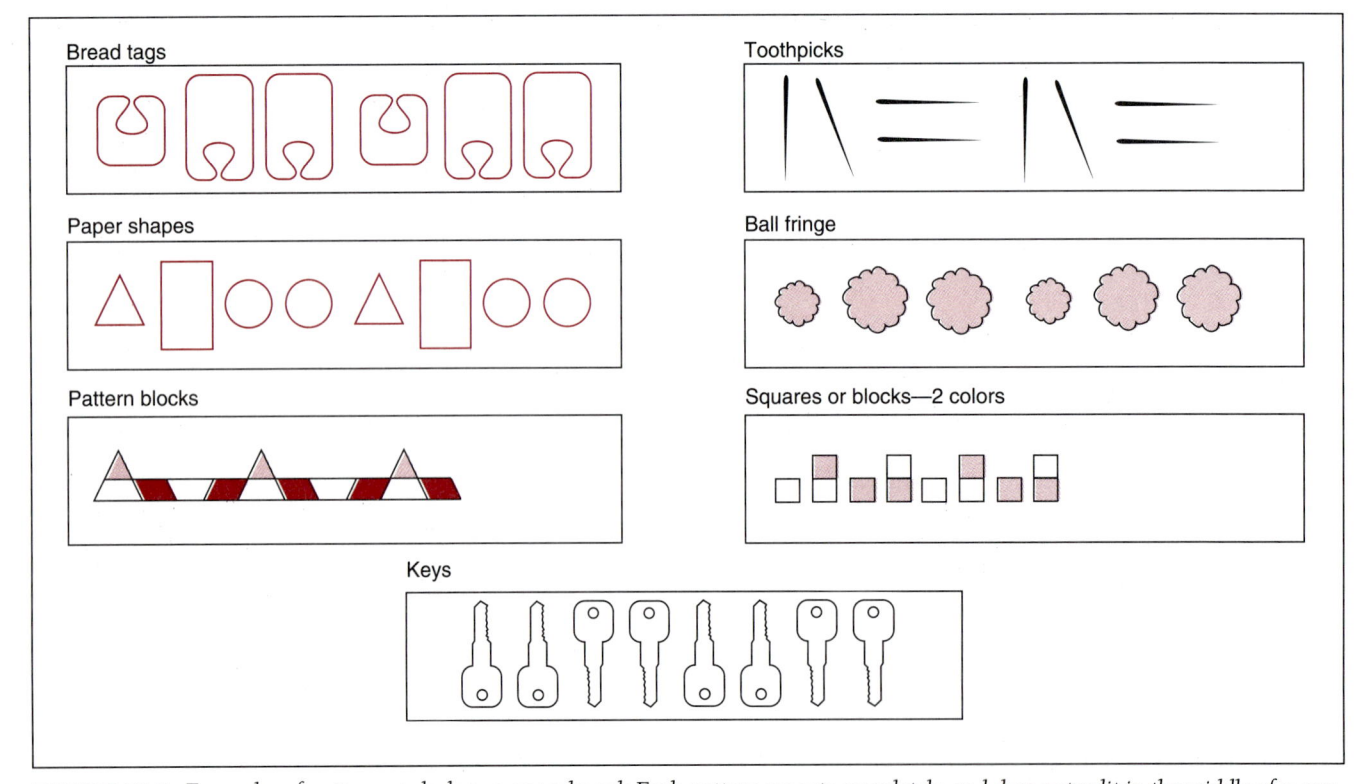

FIGURE 18.7: *Examples of pattern cards drawn on tagboard: Each pattern repeats completely and does not split in the middle of a core.*

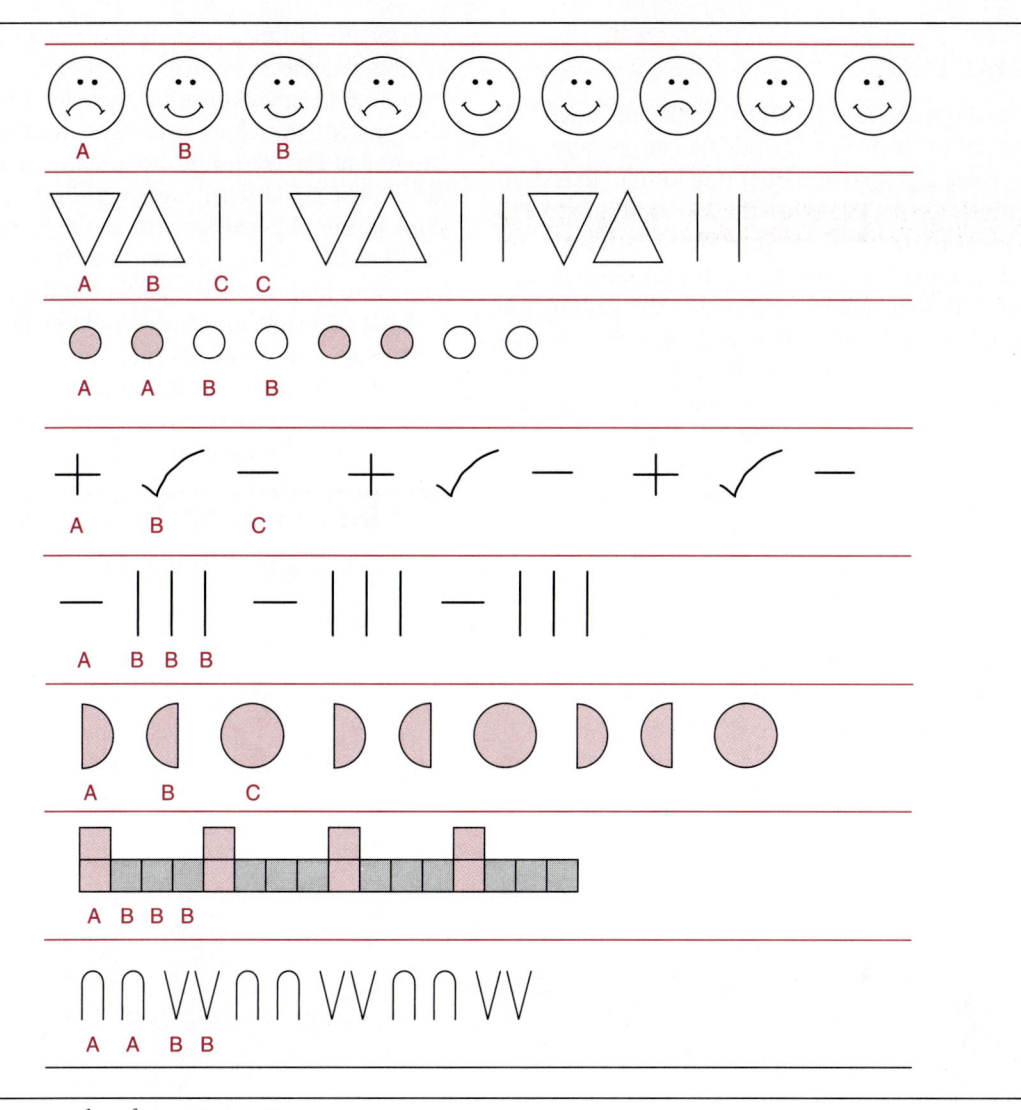

FIGURE 18.8: *More examples of repeating patterns*

18.13 SAME PATTERN/DIFFERENT STUFF

Have students make a pattern with one set of materials given a pattern strip showing a different set. This activity can easily be set up by simply switching the pattern strips from one set of materials to another. A similar idea is to mix up the pattern strips for four or five different materials and have students find strips that have the same pattern. To test if two patterns are the same, children can translate each of the strips into a third set of materials or can write down the A, B, C pattern for each.

◆ Two-Dimensional Patterns

Figure 18.9 illustrates how patterns can be developed on a grid instead of a straight line. Children at the primary level find completion of these patterns quite challenging. Pattern cards can be made by coloring or drawing on a piece of grid paper. If blank grids the same width as the pattern cards are laminated, then students can use colored blocks or colored squares of construction paper on the blank grids. This provides the same trial-and-error potential that was noted with repeating patterns.

Extend these patterns.

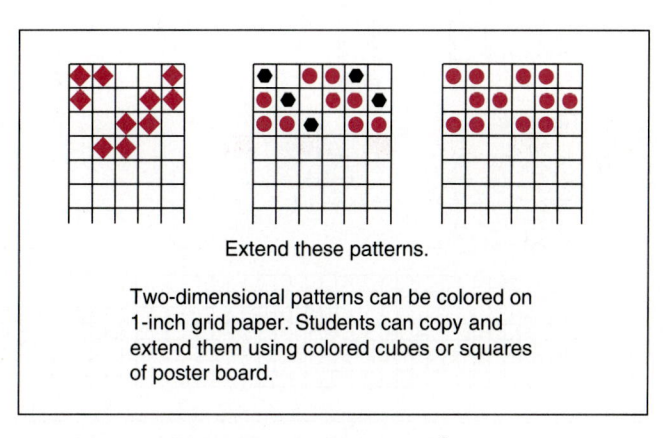

Two-dimensional patterns can be colored on 1-inch grid paper. Students can copy and extend them using colored cubes or squares of poster board.

FIGURE 18.9: *Repeating patterns on a grid*

18 / LOGICAL REASONING: ATTRIBUTE AND PATTERN ACTIVITIES

◆ GROWING PATTERNS

Beginning at about the fourth or fifth grade and extending through the junior-high years, students can explore patterns that involve a progression from step to step. In technical terms these are called *sequences*. We will simply call them *growing patterns*. With these patterns, students not only extend patterns but look for a generalization or algebraic relationship that will tell them what the pattern will be at any point along the way. Therefore, these activities not only develop the mathematical processes noted at the outset of the chapter, they are also an excellent example of the concept of function and can be used as one way to provide meaningful experiences with this very important mathematical idea.

◆ Materials, Frames, and Charts

Like repeating patterns, growing patterns are best developed with materials or, at the very least, drawings. Each element of the pattern is a progression of some sort from the previous element of the pattern. Several examples of this type of pattern are shown in Figure 18.10. Students should discuss how each frame or group in the pattern is different from the preceding group. If each new frame can be built by adding on to the previous frame, then the discussion should include how this can be done. For example, each stair step in Figure 18.10 can be made by adding a column of blocks to the preceding stair steps.

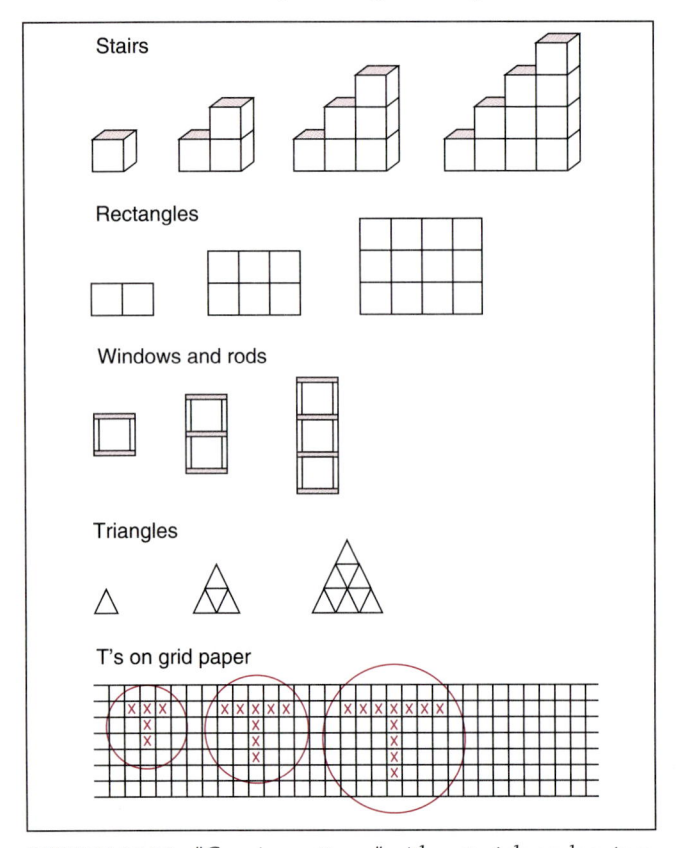

FIGURE 18.10: *"Growing patterns" with materials or drawings*

Growing patterns also have a numeric component, the number of objects or elements in each frame. As shown in Figure 18.11, a chart can be made for any growing pattern. One row of the chart is always the frame number, and another is for recording how many elements are in that frame. The initial challenge is to discover spatial relationships in the physical patterns as they change or grow from frame to frame. The next and much more significant challenge is to find a numeric pattern in the chart. Frequently it is only reasonable to actually build or draw the first five or six frames. The numbers often get large and materials or space or time become barriers to extending the patterns. A good question then becomes, "How many items will be in the next frame? How many in the tenth frame? or the twentieth?"

◆ Searching for Numeric Relationships in the Patterns

There are two different places in the chart where you and your students can look for numeric relationships: first, in the progression from one frame to the next; and second, in the relationship between the frame number and the number of objects in that frame. With the help of the clues provided, see what numeric relationships you can find in Figure 18.11. You should also make charts for the patterns in Figure 18.10 and examine these for relationships.

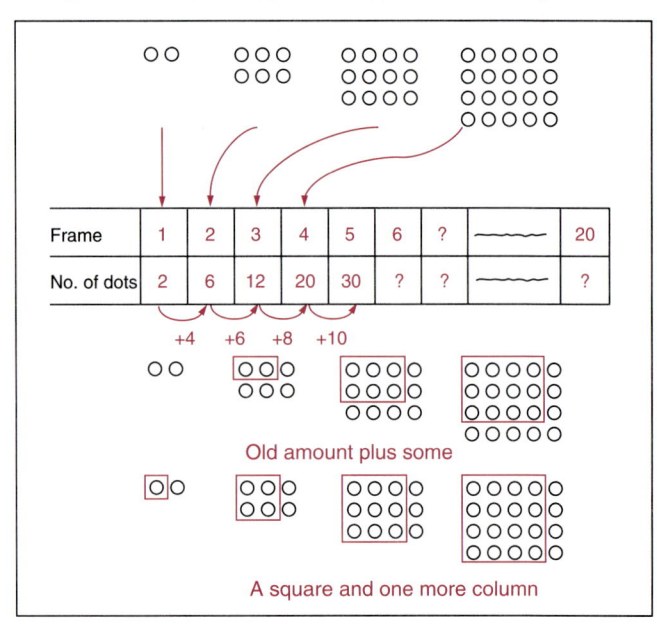

FIGURE 18.11: *Two different relationships in a visual pattern*

As students first begin to explore or build growing patterns they should be asked to predict what the tenth or perhaps twentieth frame will be like or how many elements it will have. Since the actual construction may not be reasonable, this challenge encourages a search for useful patterns and relationships. A relationship from one frame to the next is often easy to see. For example, with the triangles in Figure 18.10, the value of each successive frame can be found by adding successive odd numbers.

A numeric relationship between the frame number and the number of objects in the frame is the most powerful. If a rule can be discovered that can give the value of any frame in terms of the frame number, the number of objects in any frame can be determined without building or calculating all of the intermediate frames in the pattern. Have students develop numeric expressions for each frame using the frame numbers in the expressions. Work toward finding similar expressions for each frame with the frame number being the only part of the expression that changes. It may take much searching and experimenting for students to come up with an expression that is similar for each frame. Sometimes clues to this relationship can be found in the drawings more easily than in the charts. This search can be an exciting class or group activity. Do not be nervous or upset if students have difficulty. Encourage the search for relationships to continue, even if it takes more than one day. The search for and discovery of relationships is the most significant portion of these activities.

When students have discovered expressions for each frame that are the same except for frame numbers, write them with brackets around the frame numbers as shown in Figure 18.12. Notice that the bracketed numbers vary from one expression to the next, while the other numbers in the expressions remain the same. Now, the bracketed numbers can be replaced by a letter or *variable* resulting in a general formula. The formula then defines a *functional relationship* between frame numbers and frame values. Even if these terms are not used at the time, the activity provides an excellent early example of function and variable for students to use later when those ideas are encountered.

It would be a good idea for you to explore each of the patterns in Figures 18.10, 18.11, and 18.12 to see if you can find a formula (functional relationship) for the value of each frame in terms of the frame numbers.

The calculator should be used at all times to promote discovery of relationships in growing patterns. For example, if the frame-to-frame relationship is observed, the number of items for a large frame number is easily calculated. The result can then be verified by building or drawing that particular frame.

OTHER PATTERNS TO EXPLORE

Patterns pervade all of mathematics and much of nature. It would certainly be in error to leave the impression that repeat patterns and growing patterns as discussed so far are all that children need to know about patterns. In this section you will be exposed to some other patterns or approaches to looking at patterns. But there really is no end to the variation. Mathematicians continue to search for and discover new patterns and relationships. Applications of mathematical patterns have lead to solutions of real-world problems that previously were thought to be unsolvable. Patterns are powerful ideas.

The Fibonacci Sequence

For a growing pattern that is just a little bit different, see Figure 18.13 (p. 378). It begins with a little square. Each successive frame is formed by building a new larger square onto the previous design. (Can you see how to continue drawing this pattern?) If the side of the first two little squares is 1 each, then the sides of each new square are the numbers of most interest in this pattern. For those squares shown in the figure, the sides are 1, 1, 2, 3, 5, 8, and 13. What would the side of the next square be? This series of numbers, known as the Fibonacci sequence, is named for an Italian mathematician, Fibonacci (ca. 1180–1250). The sequence occurs in a variety of living things. For example, if you count the sets of spirals that go in opposite directions on a pineapple or the seeds of a sunflower, the two numbers will be adjacent numbers in the Fibonacci sequence; usually 8 and 13 for a pineapple and 55 and 89 for sunflowers.

Frame	1	2	3	4
Dots	6	10	14	

Notice: Each long side has one more dot than the short side. Take these away and ×4 helps tell how many dots.

[1] ×4 +2
[2] ×4 +2
[3] ×4 +2 — General formula
[4] ×4 +2 — $(n \times 4) + 2$

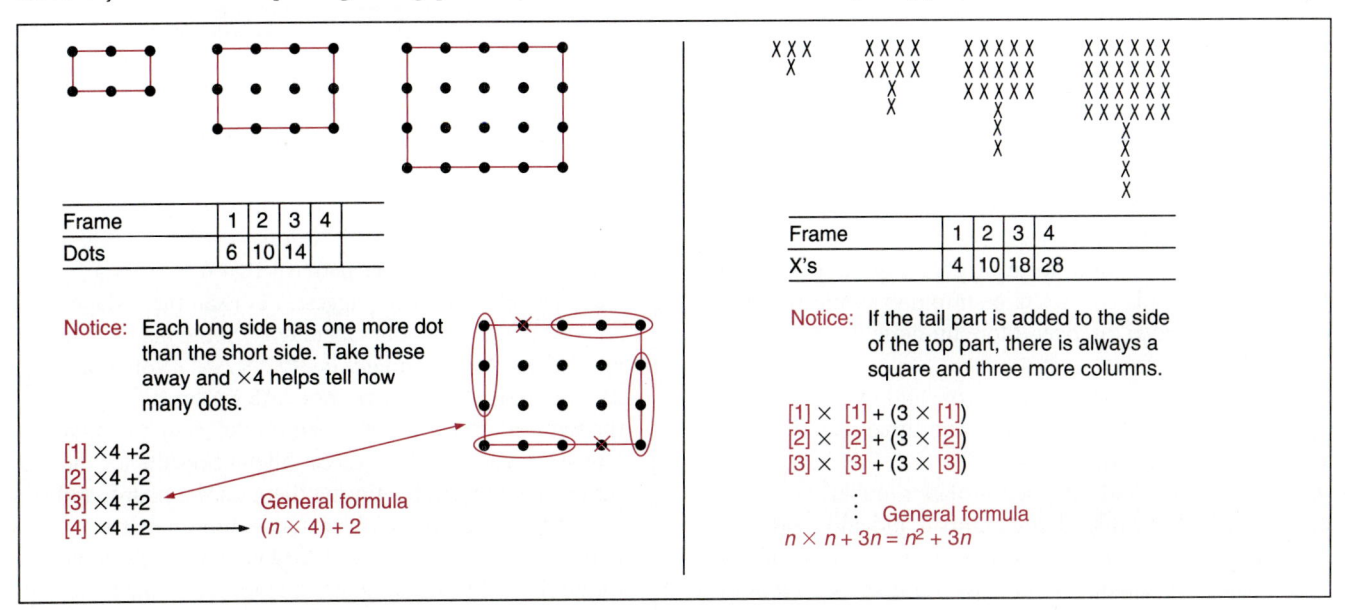

Frame	1	2	3	4
X's	4	10	18	28

Notice: If the tail part is added to the side of the top part, there is always a square and three more columns.

[1] × [1] + (3 × [1])
[2] × [2] + (3 × [2])
[3] × [3] + (3 × [3])
⋮   General formula
$n \times n + 3n = n^2 + 3n$

FIGURE 18.12: *Generalizing relationships*

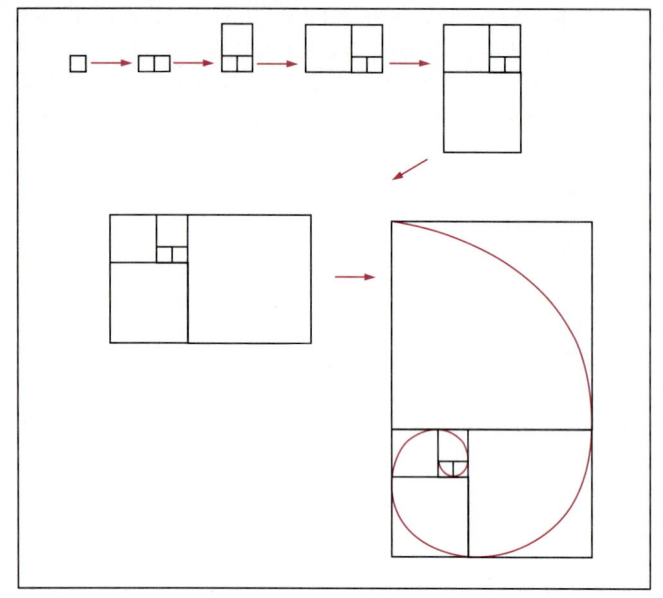

FIGURE 18.13: *A growing pattern of squares: Each new rectangle is a little closer to a "golden rectangle."*

Another interesting fact about the Fibonacci sequence is that the ratio of adjacent numbers in the sequence gets closer and closer to a single fixed number known as the "golden ratio," a number very close to 1.618. Each larger rectangle in Figure 18.13 has sides in ratio a little closer to the golden ratio. A rectangle in that ratio is called a *golden rectangle*, examples of which can be found in most of the prominent examples of ancient Greek architecture as well as in much art and architecture through the ages. The spiral that is drawn in the last rectangle shown (made from quarter circles drawn in each square) is the same spiral found in the chambered nautilus shell.

◆ Numeric Patterns

Drawings and manipulative objects should be a principal feature of patterning programs because they permit experimentation and trial-and-error approaches without threat. Students can validate their observations and conjectures without recourse to the teacher or answer book. However, many worthwhile patterns can be observed with numbers alone. These can be simple repeating patterns such as 1, 2, 1, 2, ... Even very young children can use numbers in patterns like these. Generally, however, numeric patterns involve some form of progression. The pattern 1, 2, 1, 3, 1, 4, 1, 5, . . . is a simple example that even young students can discover. More numeric patterns are:

2, 4, 6, 8, 10, . . . (even numbers, or "add two each time"?)
1, 4, 7, 10, 13, . . .(start with one; add three each time)
1, 4, 9, 16, ...(squares: 1 × 1, 2 × 2, 3 × 3, etc.)
0, 1, 5, 14, 30, . . .(add the next square number)
2, 5, 11, 23, ...(double the number, and add one)
2, 6, 12, 20, 30, . .(products of successive pairs of numbers)
3, 3, 6, 9,15, 24, . .(add the last two numbers—a Fibonacci sequence)

The challenges in these patterns or sequences of numbers are not only to find and extend the pattern but to try to determine a general rule to produce the *n*th number in the sequence. Informal or exploratory approaches are similar to those described for growing patterns in the preceding section.

◆ Patterns with the Calculator

The calculator provides a powerful approach to patterns. Listed here are a few examples, but there are many more.

18.14 SKIP COUNT PATTERNS

Choose a start number between 0 and 9 and add a constant repeatedly to that number. Remember to use the automatic constant feature. For example, to start with 7 and add 4 repeatedly, you press 7 ⊞ 4 ⊟ ⊟ ⊟. . . . What digits appear in the ones place? (1, 5, 9, 3, 7, 1, 5, 9, 3, . . .) How long is the pattern before it repeats? Are all patterns the same length? Are there shorter ones? Can you find one that is length six? Why not? How does this change when the start number changes? How does it change when the skip number changes?

As an addition to the last activity, supply children with 100s charts printed on paper. (Six charts conveniently fit on one page.) For each pattern (start number and constant number) have students color in all of the results. Visual patterns will appear on the charts as well as in the numbers. Also, do not forget to look for other patterns in these series of numbers. There are more patterns than those you may find in the ones digits.

18.15 CREATING MATCHING PATTERNS

In **Pattern Match** (activity 18.12), children learned that patterns in different forms can be coded using letters. They learn to talk about an AAB pattern, or an ABBC pattern. Young children can use the calculator to create number patterns that match these letter patterns that you suggest. For example, "Make an AAB pattern on your calculator." Some may key in 44844844. Another might be 99299299. (Note that 299299 is an ABB pattern, not AAB.) After 8 digits the pattern cannot be extended on the 8-digit display. However, you can share these different results for the same pattern form by writing them on the board, and asking how they should be extended and how the patterns are alike. As an added twist, you might look for all the patterns that match the form and have the same "core sum," the sum of the digits making up the core. The core sums for the two examples here are 16 and 20, respectively.

18.16 SECRET FUNCTION I

This game can be modified for any grade level K–8. It can be introduced using an overhead calculator with the full class, and then pairs of children can play independently. Without the class looking, store a one-step operation (secret function) in the calculator. Here is an example for each operation.

Addition: Secret = +8 . . . press 0 [+] 8 [=] [±] [=]
Subtraction: Secret = −4 . . . press 0 [−] 4 [=] [±] [=]
Multiplication: Secret = ×6 . . press 6 [×] 0 [=]
Division: Secret = ÷3 . . . press 0 [÷] 3 [=]

The display will show 0 after the function has been stored. After a secret function has been entered, students try to guess the secret rule or function. They can get up to three clues. To get the first clue, they enter any number and press [=] to see what the secret function does to the entered number. They can get a second clue to the secret by pressing [=] again, and the third clue by pressing [=] a third time. After each clue they should try to guess the secret function and predict the display after the next [=] press.

Notice that **Secret Function I** can be played at any grade with addition and subtraction. Be prepared to talk about numbers less than zero. When division is used, decimals will certainly appear. At the upper grade levels, the multiplication and division functions may be decimals or fractions. Consider the discussion when the rule is "× 0.5" and a student guesses "÷ 2." As a variation, instead of pressing [=] successively, a different number can be entered each time before pressing [=]. If any of the operation keys are pressed, the function will no longer be stored.

Your calculator must have a sign-change key [±] to enter addition or subtraction functions. Some calculators store the second rather than the first factor for multiplication. In that case, reverse the order in the hidden function.

18.17 SECRET FUNCTION II

A graphing calculator allows for a different version of **Secret Function I**. Compound functions such as $(2x + 1) / 2$ or $x^2 - 4x$ can be stored. Students attempt to guess the rule by entering different numbers and seeing the output. The goal is to try to guess the function using as few input trials as possible. On the TI-81, the following simple program will take an entry, display the output, and continue to accept inputs until you press [QUIT].

    Prgm1: GUESS
    :Lbl 1
    :Disp "ENTER X"
    :Input X
    :2X+5→A (This line stores the secret function.)
    :Disp A
    :Goto 1

Students using the program will see only:

    Prgm1: GUESS
    ENTER X
    ?3
              11
    ENTER X
    ?5
              15

The function line of the program is easily changed.

18.18 AMAZING DIGITS

Enter 9 [×] n, where n is any number 1 to 9. Press [=]. Now press other numbers followed by [=]. Even if students know their nines facts, this step will serve to clarify the process and illustrate that each new press of [=] multiplies the display by 9. Now enter 99 [×] n [=]. Try other values of n followed by [=]. What is the pattern? Try 999 [×] and even 9999 [×] or 99999 [×]. Students should play with and explore this idea as long as they wish. Next try using repeating digits for n (3333 or 66). Instead of using nines for the multiplier, try using 0.009 or 99.9. Also experiment with other repeat-digit multipliers. If students are interested, the patterns with nines might be analyzed by looking at 999 as $1000 - 1$ or as 9×111.

18.19 MORE AMAZING DIGITS

Especially if there has been interest in **Amazing Digits**, try division by 9. Begin by just dividing single digit numbers by 9. (The calculator remembers the last divisor, so after the first division just enter the new number and press [=].) After this there are all sorts of variations to try:

(a) Divide by 99, 999, . . .
(b) Divide by 0.9, 0.09, 0.009, . . .

(c) Divide by 9, but use two-digit dividends. Can you predict the results?
(d) Try three-digit dividends and divisor of 9.
(e) Combine (a) or (b) with (c) or (d).

The patterns in **More Amazing Digits** are spectacular and interesting by themselves, but you may want to also try using a digit other than 9 in that exercise.

18.20 CONSECUTIVE ODD NUMBERS

Before doing this activity you will need to explain what consecutive odd numbers are and how to add them on the calculator. Consecutive odd numbers are odd numbers that come together in the counting order. Thus, 1, 3, 5, 7, . . . are consecutive odd numbers as are 27, 29, 31, ... To add a string of these numbers, some students may benefit by writing the numbers down before they begin to add them on the calculator. Young students are confused when adding more than two numbers in a row. When they press ➕ after the second addend, the display shows the sum as if ＝ had been pressed.

Part 1: Have students use their calculators to make a chart that lists the sums of the first *n* consecutive odd numbers.

1 1
1 + 3 4
1 + 3 + 5 9
1 + 3 + 5 + 7 ... 16
etc.

The sums should look familiar. If students (about grades 5 to 8) do not recognize these as squares, suggest that press the √ key after they get the sum. It would be a good idea for them to express the sums as squares ($1^2, 2^2, 3^2, 4^2, \ldots$). Can they predict the sum of the first 20 or first 50 consecutive odd numbers?

Part 2: If the consecutive odd numbers are written in a triangular list as shown here, and the sum of each horizontal row is recorded, there is an interesting result.

```
            1
          3   5
        7   9   11
      13  15  17  19
           etc.
```

As follow-up to the odd sum activity, students may want to examine consecutive even numbers or simply consecutive numbers. For example, what can you tell about the sum of any 4 consecutive numbers? any 5? Pick a number less than 100. Can you find a string of consecutive numbers that add up to your number? Can you find two different consecutive strings of numbers that have the same sum? (3 + 4 + 5 + 6 = 5 + 6 + 7). What numbers have only one sum?

As you can see, the calculator not only can be used to create patterns, it also can hide patterns for discovery, and it can help with the tedium of uncovering patterns that otherwise students would be unwilling to explore. The extent to which the patterns are explored depends on the interest and the age level of the students. Encourage students to explore as much as they are able. Within these patterns are some very nice mathematical ideas.*

REFLECTIONS ON CHAPTER 18: WRITING TO LEARN

1. Explain how attribute activities and patterning activities meet the goals of problem solving in each of these categories: (1) strategic, (2) metacognitive, and (3) affective.

2. What is the difference between a structured and an unstructured set of attribute materials? Give an example of each to support your distinctions.

3. Loop activities in which children attribute place pieces into loops can be done in two different ways. First they are done with the label cards showing, and later they are done with the label cards face down—the teacher deciding on the correct placement of each piece. Explain why the latter activity is a logic activity while the first really is not.

4. How is a two-difference game played on an oval track? Draw a picture showing a two-difference solution for a track with seven spaces.

5. Make up three pattern strips showing repeating patterns for some common objects that might be found in the classroom. Label each using an A-B-C scheme. No two schemes should be alike but all should use the same materials. What is the core of each?

6. The following growing pattern consists of square borders. The elements of each frame are the number of squares shaded in the border. Draw the fifth frame and the tenth frame. Make and complete a chart showing the number of elements in the first ten frames. What patterns can you discover? Try to use the pictures you have drawn to help you find a general formula for the number of squares in the *n*th frame.

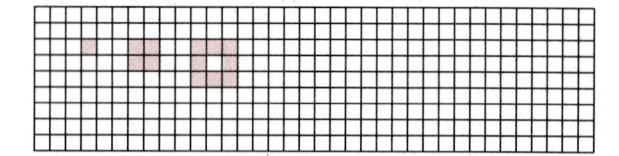

---

*Many of the ideas found in the section on calculator patterns were adapted from *CAMP-LA (Calculators and Mathematics Project, Los Angeles)* (Pagni, 1991). Each of four books in this series for grades K–8 has a chapter on patterns and functions and is complete with lesson plans and worksheets. The books are highly recommended.

7. Make your calculator store each of these secret functions: +3 and ×5. Explain how to do this and how to **Secret Function I.**

8. Explore one other of the suggested ideas for calculators, and explain your results briefly.

◆◆◆◆◆
FOR DISCUSSION AND EXPLORATION

1. Get a copy of the *Teacher's Guide for Attribute Games and Problems* (Elementary Science Study, 1966). Although quite old, this book still contains some of the best attribute activities. In this book you will find activities for "people pieces," which are a set of 16 tiles after which the Woozle Cards were patterned. Any activity that can be done with people pieces can be done with Woozle Cards.

2. Invent a growing pattern. Use blocks, pattern blocks, or grid or dot paper. See how many patterns or relationships you can find in your pattern. Can you generalize any of these numeric patterns? If you are having trouble with ideas, check the *Pattern Factory* (Holden & Roper, 1980).

3. In the NCTM *Standards*, the third standard in both the K–4 and 5–8 sections is "Mathematics as Reasoning." Read these two standards. Discuss briefly how the activities of this chapter fit those goals. What else is meant by "mathematics as reasoning"?

4. Look up additional information about the Fibonacci sequence. The classic Disney film, *Donald in Mathmagic Land* (now available in video), also includes information about this sequence.

◆◆◆◆◆
SUGGESTED READINGS

Baratta-Lorton, M. (1976). *Mathematics their way.* Menlo Park, CA: Addison-Wesley.

Bezuszka, S. J., & Kenney, M. (1982). *Number treasury: A sourcebook of problems for calculators and computers.* Menlo Park, CA: Dale Seymour.

Burk, D., Snider, A., & Symonds, P. (1988). *Box it or bag it mathematics: Teachers resource guide* [Kindergarten, First–Second]. Salem, OR: The Math Learning Center.

Burns, M. (1992). *About teaching mathematics: A K–8 resource.* Sausalito, CA: Marilyn Burns Education Associates.

Creative Publications. (1986). *Hands on attribute blocks.* Palo Alto, CA: Creative Publications.

Johnson, J. J. (1987). Do you think you might be wrong? Confirmation bias in problem solving. *Arithmetic Teacher, 34*(9), 13–16.

Masalski, W. J. (1975). *Color cube activities.* Fort Collins, CO: Scott Resources.

Perl, T. (1974). *Relationshapes activity cards.* White Plains, NY: Cuisenaire Corporation of America.

Phillips, E. (1991). *Patterns and functions: Addenda series, grades 5–8.* Reston, VA: National Council of Teachers of Mathematics.

Thompson, A. G. (1985). On patterns, conjectures, and proof: Developing students' mathematical thinking. *Arithmetic Teacher, 33*(1), 20–23.

Thompson, C. S., & Van de Walle, J. A. (1985). Patterns and geometry with Logo. *Arithmetic Teacher, 32*(7), 6–13.

Trotter, T., Jr., & Myers, M. D. (1980). Number bracelets: A study in patterns. *Arithmetic Teacher, 27*(9), 14–17.

Van de Walle, J. A. (1988). Hands-on thinking activities for young children. *Arithmetic Teacher, 35*(6), 62–63.

Van de Walle, J. A., & Holbrook, H. (1987). Patterns, thinking, and problem solving. *Arithmetic Teacher, 34*(8), 6–12.

Van de Walle, J. A., & Thompson, C. S. (1985). Promoting mathematical thinking. *Arithmetic Teacher, 32*(6), 7–13.

19 EXPLORING BEGINNING CONCEPTS OF PROBABILITY AND STATISTICS

PROBABILITY AND STATISTICS IN ELEMENTARY SCHOOLS

The related topics of probability and statistics represent two of the most prominent uses of mathematics in our everyday lives. We hear about the possibility of contracting a particular disease, having twins, winning the lottery, or living to be 100. Simulations of complex situations are frequently based on simple probabilities used in the design of highways, storm sewers, medical treatments, sales promotions, and spacecraft. Graphs and statistics bombard the public in advertising, opinion polls, reliability estimates, population trends, health risks, and the progress of students in schools and schools in school systems, to name only a few areas.

In order to deal with this information, students should have ample opportunity throughout the school years to have informal yet meaningful experiences with the basic concepts involved. The emphasis from the primary level into high school should be placed on activities leading to intuitive understanding and conceptual knowledge rather than computations and formulas.

Organizations such as the Joint Committee on the Curriculum in Statistics and Probability of the American Statistical Association and the NCTM have promoted attention to the topics of probability and statistics in schools. In recent years, a number of factors can be identified that indicate an increase in the quantity and quality of probability and statistics instruction in the elementary school:

- An increased awareness of the importance of probability and statistics concepts and methods.
- An emphasis on experimental or simulation approaches to probability (instead of rules and formulas).
- The use of simplified yet powerful plotting techniques to describe data visually without complicated procedures.
- The use of readily available calculators and computers, especially graphing calculators, to (a) conduct thousands of random trials of experiments from flipping coins to simulating baseball batting performances, (b) to do the tedious work of constructing graphs, and (c) to almost instantly perform computations on large sets of numbers.

With new approaches and attitudes toward the development of the conceptual knowledge of probability and statistics, it is almost certain that even more emphasis will be placed on these topics in the near future.

AN INTRODUCTION TO PROBABILITY

Probability may be a foreign or difficult-sounding idea for elementary teachers to consider teaching. With an informal approach, however, the ideas can be quite simple and interesting.

TWO EXPERIMENTS

Consider answering the following two questions by actually performing the experiments enough times to make a reasonable guess at the results.

1. **Tossing a cup**. Toss a paper or Styrofoam cup once or twice, letting it land on the floor. Notice that there are three possible ways for the cup to land: upside down, right side up, or on its side. If the cup were tossed this way 100 times, about how many times do you think it will land in each position?

2. **Flipping two coins**. If you were to flip one coin 100 times, you would expect that it would come up heads about as many times as tails. If two coins were tossed 100 times, about how many times do you think that they will both come up heads?

A quick way to conduct these experiments is to work in groups. If 10 people each do 10 trials and pool their data, the time needed for 100 trials is not very long. Even if you do not actually do the experiments, jot down your predictions now before reading on.

THEORETICAL VERSUS EXPERIMENTAL PROBABILITY

In the cup toss, there is no practical way to determine the results before you start. However, once you had results for 100 flips, you would undoubtedly feel more confident in predicting the results of the next 100 flips. If you gathered data on that same cup for 1000 trials, you would feel even more confident. Suppose for example that your cup lands on its side 78 times out of 100. You might choose a round figure of 75 or 80 for the 100 tosses. If, after 200 flips, there were 163 sideways landings, you would feel even more confident of the 4-out-of-5 ratio and predict about 800 sideways landings for 1000 tosses. The more flips that are made, the more confident you become. You have determined an *experimental probability* of $\frac{4}{5}$ or 80% for the cup to land on its side. It is experimental because it is based on the results of an experiment rather than a theoretical analysis of the cup.

In a one-coin toss, the best prediction for 100 flips would be 50 heads, although you would not be too surprised if actual results were between 45 and 55 heads. The prediction of 50% heads could confidently be made before you flipped the coin, based on your understanding of a fair coin. Your prediction for 2 heads in the two-coin version may be more difficult. It is quite common for people to observe that there are three types of outcomes: both heads, both tails, and one of each. Based on this analysis they predict that 2 heads will come up about one-third of the time. (What did you predict?) The prediction is based on their analysis of the experiment, not on experimental results. When they conduct the experiment, however, they are surprised to find that two heads come up only about one-fourth of the time. With this experiential base they might return to their original analysis and look for an error in their thinking.

There is only one way for 2 heads to occur and one way for 2 tails to occur. However, there are two ways that a head and a tail could result: Either the first coin is heads and the second tails, or vice versa. As shown in Figure 19.1, that makes a total of four different outcomes, not three. The assumption that each outcome is equally likely was correct. Therefore the correct probability of 2 heads is 1 out of 4 or $\frac{1}{4}$, not $\frac{1}{3}$. This *theoretical probability* is based on a logical analysis of the experiment, not on experimental results.

First coin	Second coin
Head	Head
Head	Tail
Tail	Head
Tail	Tail

FIGURE 19.1: *Four possible outcomes for two coins*

When we talk about probabilities we are assigning some measure of chance to an experiment. An *experiment* is any activity that has two or more clearly discernible results or *outcomes*. Both tossing the cup and tossing the two coins were experiments. Observing tomorrow's weather and shooting 10 free throws on the basketball court are also experiments. The collection of all outcomes is generally referred to as the *sample space*. As you have already seen, the toss of two coins has four outcomes in the sample space. Tomorrow's weather can be described in many ways: precipitation or no precipitation, or dry, rain, sleet, or snow. An *event* is any subset of the outcomes or any subset of the sample space. For the two-coin experiment, the event we were concerned with was getting two heads. For the free-throw shooting, we might be interested in the event of getting 5 or more out of the 10.

When all possible outcomes of a simple experiment are equally likely, the *theoretical probability* of an event is

$$\frac{\text{Number of outcomes in the event}}{\text{Number of possible outcomes}}$$

In real-world situations, outcomes frequently cannot be determined to be equally likely as they are for coin flips or dice rolls. The cup-tossing experiment, while not a practical situation, is "muddy" and real. The outcomes are difficult to predict, and they are not equally likely. In situations like these, we can determine the observed relative frequency of the event by performing the experiment a lot of times. The *relative frequency* of an event is

$$\frac{\text{Number of observed occurrences of the event}}{\text{Total number of trials}}$$

It should be clear that the relative frequency is not a good predictor of the chance of the event happening unless the number of trials is very large. The *experimental probability* of an event is the ratio that the relative frequency gets closer and closer to as the number of trials gets infinitely large. Because it is impossible to perform an infinite number of trials, we must be satisfied with some large number of trials. The more trials, the more confident we might be that the experimental probability is close to the actual probability.

IMPLICATIONS FOR INSTRUCTION

There are many reasons why an experimental approach to probability, actually conducting experiments and examining outcomes, is important in the classroom. An experimental approach

- Is significantly more intuitive and conceptual. Results begin to make sense and certainly do not result from some abstract rule.
- Eliminates guessing at probabilities and wondering, "Did I get it right?" Counting or trying to determine the number of elements in a sample space can be very difficult without some intuitive background information.
- Provides an experiential background for examining the theoretical model. When you begin to sense that the probability of two heads is $\frac{1}{4}$ instead of $\frac{1}{3}$, the analysis in Figure 19.1 seems more reasonable.
- Helps students see how the ratio of a particular outcome to the total number of trials begins to converge to (get closer and closer to) a fixed number. For an infinite number of trials, the relative frequency and theoretical probability would be the same.
- Develops an appreciation for a simulation approach to solving problems. Many real-world problems are actually solved by conducting experiments or simulations.
- Is a lot more fun and interesting! It even makes searching for a correct explanation in the theoretical model more interesting.

Whenever possible, then, we should try to use an experimental approach in the classroom. If a theoretical analysis (such as with the two-coin experiment) is possible, then it should also be examined and results compared. Rather than correcting a student error in an initial analysis, we can let experimental results guide and correct student thinking.

Sometimes it is possible to develop theoretical explanations from results of experiments. For example, the results of the cup toss might be compared with the ratio of the height of the cup to the diameter of the opening. The cup can then be cut to different heights and the experiment repeated. A reasonable connection can be made between the ratio of height to top opening and the probability of a side landing. (For a better-controlled experiment, try a variety of open-ended cylinders such as paper tubes and tin cans.)

DEVELOPING CONCEPTS OF PROBABILITY

It is no longer reasonable to wait until high school to begin helping children develop informal ideas about probability. Informal ideas developed early provide necessary background for concepts to be constructed at the middle and secondary levels.

EARLY CONCEPTS OF CHANCE

Children in kindergarten and primary grades need to develop an intuitive concept of chance: the idea that some events, when compared with others, have a better or worse chance or an approximately equal chance of happening.

Many young children believe that an event will happen because "it's my favorite color" or "because it's lucky" or "because it did it that way last time." Many games such as *Candy Land* or *Chutes and Ladders* are very exciting for young children who do not comprehend that the outcomes are entirely random chance. When they finally learn that they have no control over the outcome, children begin to look for other games where there is some element of player determination.

◆◆◆◆◆◆◆◆◆◆◆◆◆◆◆◆◆◆◆◆◆◆◆

19.1 IS IT LIKELY?

Ask students to judge events as *certain, impossible,* or *maybe.* For example, consider these:

It will rain tomorrow.

Drop a rock in water, and it will sink.

Trees will talk to us in the afternoon.

The sun will rise tomorrow morning.

Three students will be absent tomorrow.

George will go to bed before 8:30 tonight.

You will have two birthdays this year.

◆◆◆◆◆◆◆◆◆◆◆◆◆◆◆◆◆◆◆◆◆◆◆

19.2 WHO WILL WIN?

Play simple games where the chance of one side winning can be controlled. Before playing the games, have students predict who will win and why. Afterwards, discuss why they think it happened that way. For example, the hockey game in Figure 19.2 starts with a counter in the center. Two players take turns spinning a spinner. The counter moves one space toward the goal that comes up on the spinner. Play the game with different spinners. As a variation, let students choose a spinner on each turn. Ask them to explain their choices.

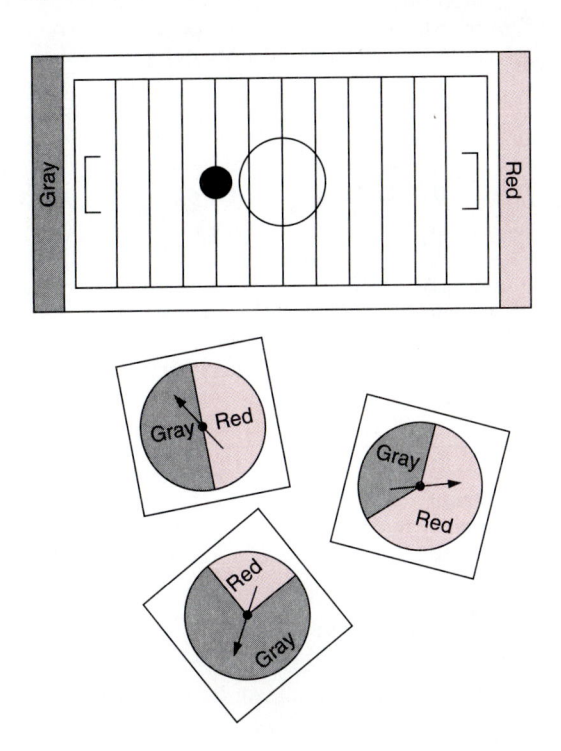

FIGURE 19.2: *A simple game of chance played with different spinners helps young children with basic concepts of chance.*

19.3 PREDICTIONS

Have students make predictions about the outcomes of simple experiments using the terms *more, less, all,* and *none*. For example, show children how many each of red and yellow cubes you have in a bag. You will let children draw cubes one at a time and put them back each time. "If we do this 10 times, will there be more reds, less reds, all reds, or no reds?" Change the number of each color cube and repeat. The same activity can be done with spinners, rolling dice, drawing cards, or any random device that you can adjust. Include situations that are certain, such as using all yellow cubes in the bag.

DETERMINING PROBABILITIES FOR SIMPLE EVENTS

From a basic understanding that one event can be more or less likely than another, students can begin to predict specific ratios of outcomes of simple events. Before students have worked with part-to-whole ratios, use language such as "65 out of 100" instead of using fractional probabilities. A discussion of reasons for their predictions is always important. The experiment should then be conducted and results compared with expected outcomes.

Figure 19.3 illustrates a number of simple random devices that can be used for experiments. To get large numbers, let groups of students conduct the same experiment and tally their results. Group results can quickly be combined to get larger numbers. Ask students to notice how the results for smaller numbers of trials are all different and how they frequently are quite different from what might be expected.

RANDOM NUMBERS AND ELECTRONIC DEVICES

If a coin is tossed repeatedly, the long run will produce a ratio very close to one-half heads and one-half tails, but what happens in the short run? Do heads and tails alternate? If heads come up six times in a row, what is the probability that tails will come up next? The answers to questions such as these reflect an understanding of randomness and are worth discussing.

Occasionally have students list the outcomes of their experiments in a row, to show the order of the outcomes. Interestingly, truly random events do not alternate. They frequently appear in clusters or runs. If eight odd numbers in a row come up on a die roll, the chance remains exactly one-half that the next roll will be odd. The die has no memory. The previous roll of a die cannot affect the next roll. It is very unlikely that even numbers will come up seven times in a row (one chance in 128). However, if that does happen, the chance for an even number on the next roll is still one-half.

Hands-on random devices such as those in Figure 19.3 (p. 386) provide an intuitive feel for randomness. It is important to conduct experiments using these because students can believe in the unbiased outcomes. The downside is that the use of these devices requires a lot of time to produce a large number of trials and keep track of the results. This is where the computer, and more recently the programmable calculator (usually the graphing calculator), can help out enormously. These electronic devices are designed to produce random numbers. Usually the random numbers are in the form of decimal numbers between 0 and 1. Students who are going to use these random number generators should understand what these numbers look like and how they can use them. One idea is simply to make the computer or calculator produce some of these one at a time. The outcome may look like this:

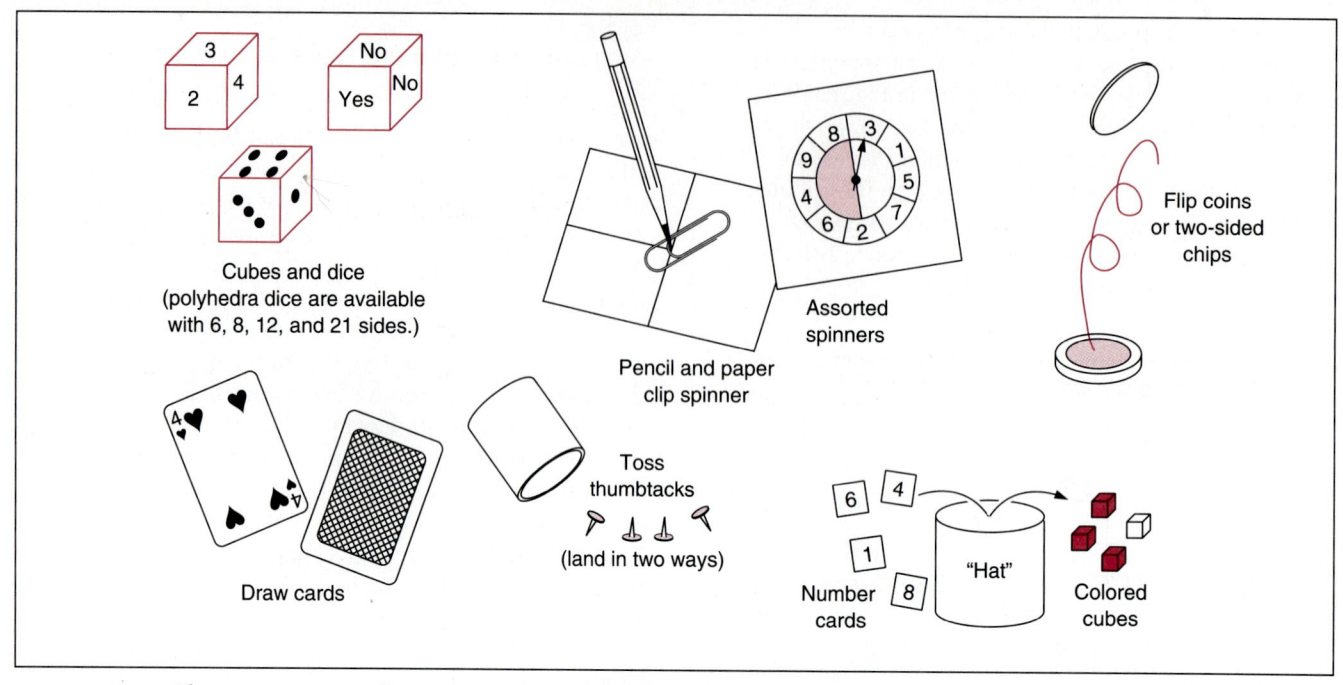

386 19 / EXPLORING BEGINNING CONCEPTS OF PROBABILITY AND STATISTICS

FIGURE 19.3: *There are many simple ways to produce random outcomes.*

0.0232028877
0.8904433368
0.1693227117
0.1841957303
0.5523325715

How could a list of decimals like this replace flipping a coin or spinning a spinner? If you multiplied each of these numbers by 2, they would then be numbers between 0 and 2. If you then ignored the decimal part of the number you would get a series of 0's and 1's. These might stand for heads and tails or boys and girls, or true or false or any other pair of outcomes that are equally likely. If we wanted to get outcomes that were the same as a $\frac{1}{4}$ versus $\frac{3}{4}$ spinner, we could multiply the random numbers by 4, throw out the decimal parts (producing 0's, 1's, 2's and 3's). We could assign one of the digits to the $\frac{1}{4}$ color on the spinner and the other three to the other color. How could you use random digits like these to simulate dice throwing?

Students who are going to use an electronic random number device (calculator or computer) should understand how the numbers can be used to simulate a physical device. It may even be a good idea to compare the outcomes of the two devices for a small run before using the electronic version for a large run.

The program in Figure 19.4 is for a TI-81 calculator. It will simulate rolling a die as many times as you want. The tenth line (:IPart 6 Rand+1 =>X) multiplies the random number by 6, takes the integer part (discarding the decimal), adds 1 (to give a number between 1 and 6) and stores it in the X box of the memory. The result of each roll is stored in a 6 × 1 matrix, which the calculator displays when the program is finished. This program can "roll a die" 1000 times in about 80 seconds.

```
:6=>Arow
:1=>Acol
:6=>Brow
:1=>Bcol
:0=>[B]
:Disp "NO OF ROLLS?"
:Input N
:1=>J
:Lbl 1
:IPart 6 Rand+1=>X
:0=>[A]
:1=>[A] (X,1)
:[A]+[B]=>[B]
:IS>(J,N)
:Goto 1
:Disp [B]
:End
```

FIGURE 19.4: *This TI-81 program can be used to simulate thousands of dice rolls and accumulate the results.*

EXPERIMENTS WITH TWO OR MORE INDEPENDENT EVENTS

Flipping two coins and observing the result of each is an example of an experiment with two independent events. The flip of one coin has no effect on the other. The events are *independent*. Another example is that of drawing a card and spinning a spinner. Many interesting experiments involve two or more separate, independent events.

Determining the experimental probability of compound events is no different than for simple events. The experiment is performed numerous times, and the number of favorable results is compared to the total trials as before. The challenges come in trying to reconcile experimental results with theoretical ones.

To illustrate, suppose that a class is conducting an experiment to determine the probability of rolling a seven with two dice. They might tally their results in a chart showing each sum from 2 to 12 as a single event, as in Figure 19.5(a).

The results of their experiment will show clearly that these events are not equally likely, and in fact the sum of 7 has the best chance of occurring. To explain this, they might look for the combinations that make 7: 1 and 6, 2 and 5, and 3 and 4. But there are also three combinations for 8. It seems as though 8 should be just as likely as 7 and yet it is not.

Now suppose that the experiment is repeated. This time, for the sake of clarity, suggest that students roll two different colored dice and that they keep the tallies in a chart like the one in Figure 19.5(b). A TI-81 program very similar to the one in Figure 19.4 can roll two dice and record the outcomes in a similar manner.

The results of a large number of dice rolls indicate what one would expect, namely, that all 36 cells of this chart are equally likely. But there are more cells with a sum of 7 than any other number. Therefore they were really looking for the event that consists of any of the 6 ways, not 3, that two dice can make a 7. There are 6 outcomes in the desired event out of a total of 36 for a probability of $\frac{6}{36}$ or $\frac{1}{6}$.

When investigating the theoretical probability of a compound event, it is useful to use a chart or diagram that keeps the two independent events separate and illustrates the combinations. The matrix in Figure 19.5(b) is one good suggestion when there are only two events. A tree diagram is another method that can be used with any number of events (Figure 19.6, p. 388).

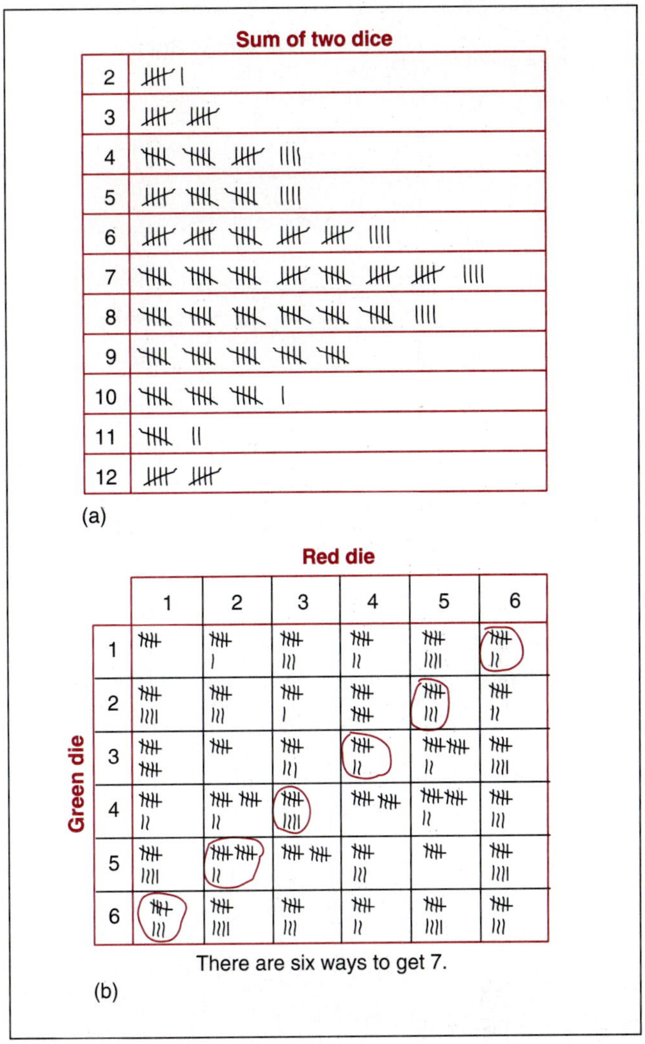

FIGURES 19.5: (a) Tallies that account only for the total. (b) Tallies that keep track of the individual dice.

19.4 COMPOUND EXPERIMENTS

The following are examples of compound experiments with independent events. Determine the probability of

rolling an even sum with two dice

spinning blue and flipping a cup on end

getting two blues out of three spins (depends on spinner)

having a tack or a cup land up if each is tossed once

getting *at least* two heads from a toss of four coins

Words and phrases such as *and*, *or*, *at least*, and *no more than* can also cause children some trouble. Of special note is the word *or* since its everyday usage is generally not the same as the strict logical use found in mathematics. In mathematics, *or* includes the case of *both*. So, in the tack-and-cup example, the event includes tack up, cup up, and *both* tack *and* cup up.

THEORETICAL PROBABILITIES WITH AN AREA MODEL

The method just suggested for determining the theoretical probability of a compound event is to list all possible outcomes and count those that are favorable; that is, those that make up the event. This is very useful and intuitive as a first approach. However, it has some limitations. First, what if the events are not all equally likely and are not made up of smaller events? An example is the cup toss. Second, it is difficult to move from that approach to even slightly more sophisticated methods. An area model approach has been used successfully with fifth-grade students. It is quite useful for some reasonably difficult problems (Armstrong, 1981). The following example will illustrate how an area model works.

Suppose that after many experiences, you have decided that your cup lands on its side 82% of the time. The experiment is to toss the cup and then draw a card from a deck. What is the probability that the cup will land on the side *and* you will draw a spade? Draw a square to represent one whole. First partition the square to represent the cup toss,

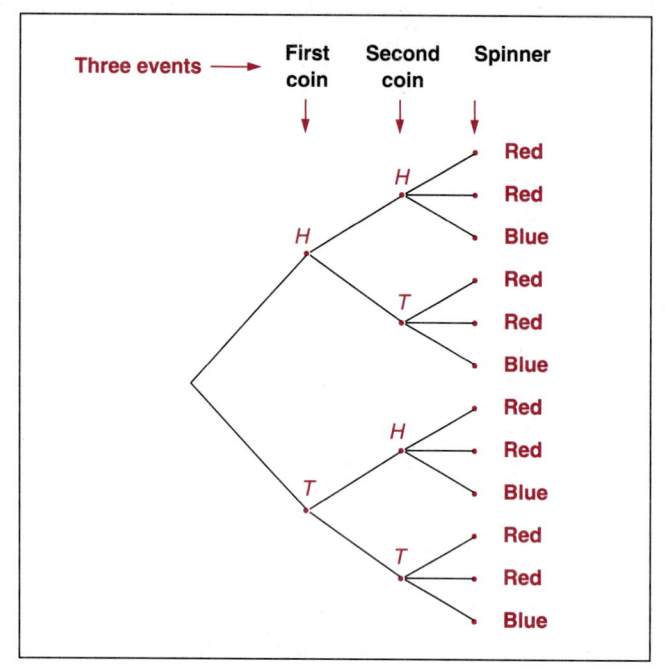

FIGURE 19.6: *A tree diagram showing all possible outcomes for two coins and a spinner that is $\frac{2}{3}$ red.*

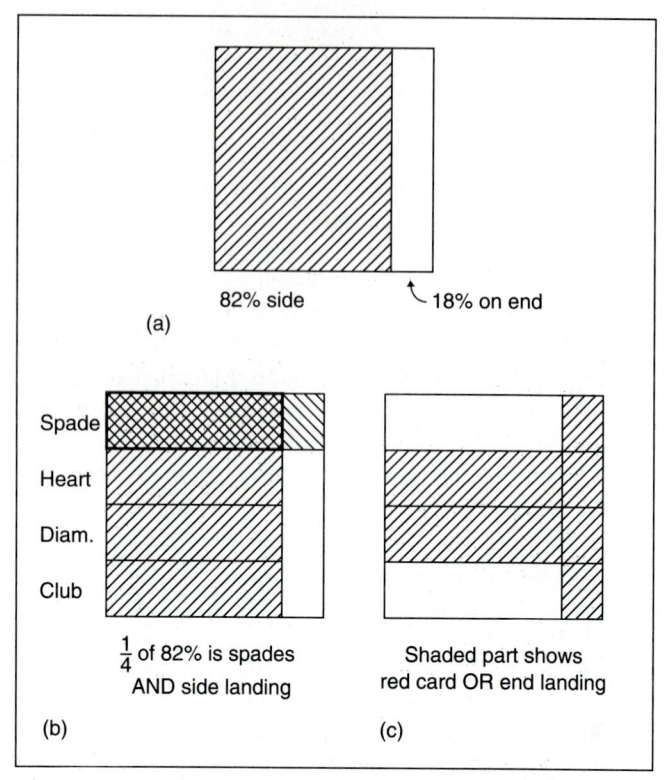

FIGURE 19.7: *An area model for determining probabilities*

82% and 18% as in Figure 19.7(a). Now partition the square in the other direction to represent the four equal card suits and shade $\frac{1}{4}$ for spades [Figure 19.7(b)]. The overlapping region is the proportion of time that both events, sideways and spades, happen. The area of this region is $\frac{1}{4}$ of 82% or 20.5%.

You can use the same drawing to determine the probability of other events in the same experiment. For example, what is the probability of the cup landing on either end *or* drawing a red card? As shown in Figure 19.7(c), half of the area of the square corresponds to drawing a red card. This section includes the case of drawing a red card *and* an end landing. The other half of the 18% end landings happen when a red card is not drawn. Half of 18% is 9% of the area. The total area for a red card *or* an end landing is 59%.

The area approach is easy for students to use and understand for experiments involving two independent events where the probability of each is known. For more than two independent events, further subdivision of each region is required but is still quite reasonable. The use of *and* and *or* connectives is easily dealt with. It is quite clear, without memorization of formulas, how probabilities should be combined.

EXPLORING DEPENDENT EVENTS

The next level of difficulty occurs when the probability of one event depends on the result of the first. For example, suppose there are two identical boxes. In one box is a dollar bill and two counterfeit bills. In the other box is one of each. You may choose one box and from that box select one bill without looking. What are your chances of getting a genuine dollar? Here there are two events: selecting a box and selecting a bill. The probability of getting a dollar in the second event depends on which box is chosen in the first event. These events are *dependent*, not independent.

As another example, suppose that you are a prisoner in a faraway land. The king has pity on you and gives you a chance to leave. He shows you the maze in Figure 19.8. At the start and at each fork in the path, you must spin the spinner and follow the path that it points to. You may request the key to freedom be placed in one of the two rooms. In which room should you place the key to have the best chance of freedom? Notice that the probability of ending the maze in any one room is dependent on the result of the first spin.

Either of these two problems could be solved with an experimental approach, a simulation. A second approach to both problems is to use the area model to determine the theoretical probabilities. An area-model solution to the prisoner problem is shown in Figure 19.9.

It would be good to stop at this point and try the area approach for the problem of the counterfeit bills. (The chance of getting a dollar is $\frac{5}{12}$.)

The area model will not solve all probability problems. However, it fits very well into a developmental approach to the subject because it is conceptual, it is based on existing knowledge of fractions, and because more symbolic approaches can be derived from it. Figure 19.10 shows a tree diagram for the same problem with the probabilities of each path of the tree written in. After some experience with probability situations, the tree diagram model is probably

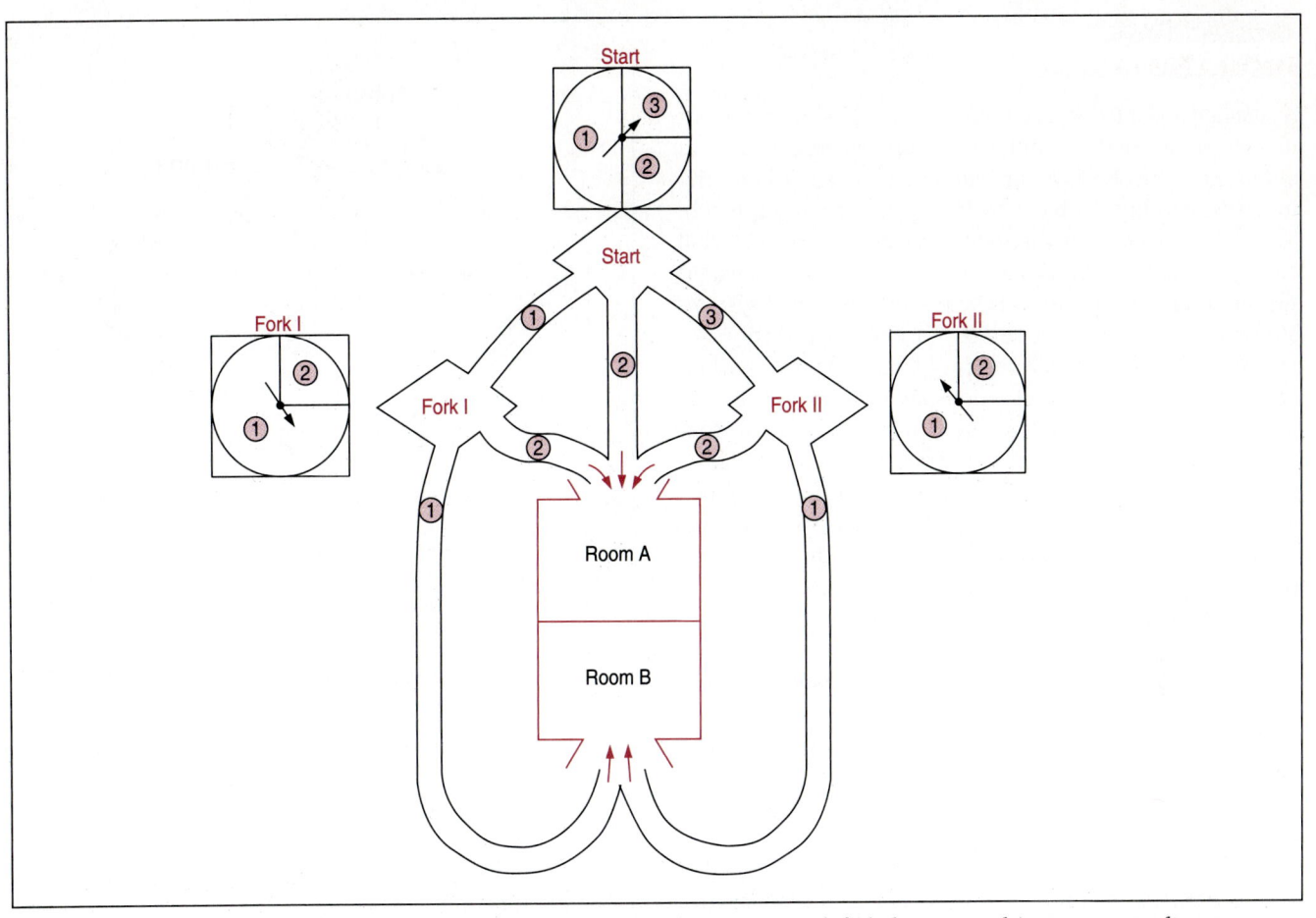

FIGURE 19.8: *Should you place your key to freedom in Room A or Room B? At each fork the spinner determines your direction.*

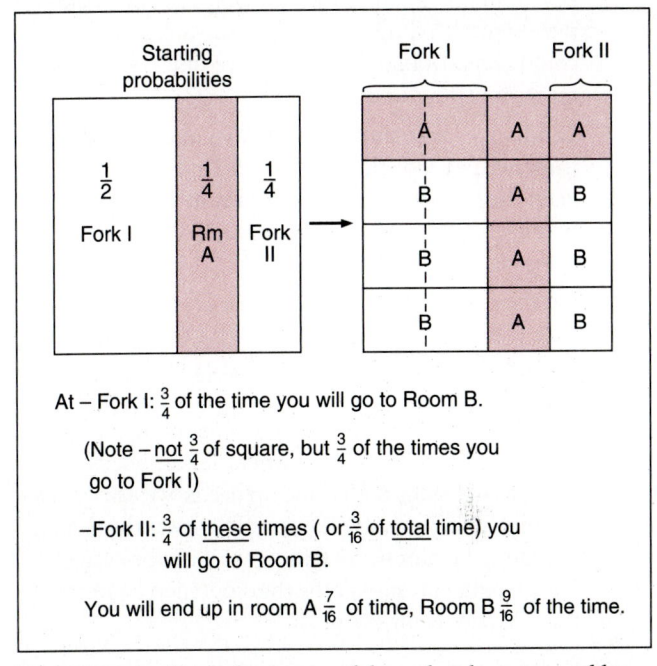

At – Fork I: $\frac{3}{4}$ of the time you will go to Room B.

(Note – <u>not</u> $\frac{3}{4}$ of square, but $\frac{3}{4}$ of the times you go to Fork I)

–Fork II: $\frac{3}{4}$ of <u>these</u> times (or $\frac{3}{16}$ of <u>total</u> time) you will go to Room B.

You will end up in room A $\frac{7}{16}$ of time, Room B $\frac{9}{16}$ of the time.

FIGURE 19.9: *Using the area model to solve the maze problem*

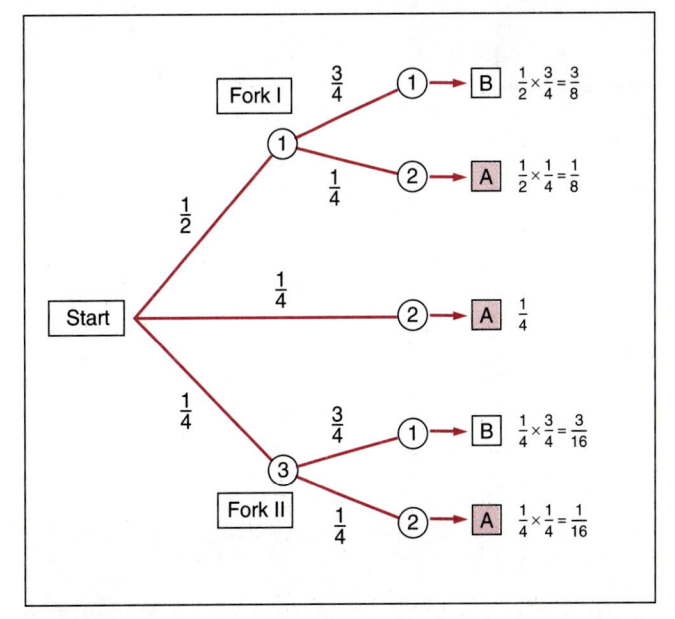

FIGURE 19.10: *A tree diagram is another way to model the outcomes of two or more dependent events.*

easier to use and adapts to a wider range of situations. You should be able to match up each branch of the tree diagram in Figure 19.10 with a section of the square in Figure 19.9.

Use the area model to explain why the probability for each complete branch of the tree is determined by multiplying the probabilities along the branch.

SIMULATIONS

A *simulation* is a technique used for answering questions or making decisions in complex situations where an element of chance is involved. A simulation is very much like solving a probability problem by an experimental approach. The only difference is that one must design a model that has the same probabilities as the real situation. For example, in designing a rocket, a large number of related systems all have some chance of failure. Various combinations of failures might cause serious problems with the rocket. Knowing the probability of serious failures will help decide if redesign or backup systems are required. It is not reasonable to make repeated tests of the actual rocket. Instead, a model that simulates all of the chance situations is designed and run repeatedly, most likely with the help of a computer. The computer model can simulate hundreds or even thousands of flights, and an estimate of the chance of failure can be made.

Many real-world situations lend themselves to simulation analysis. In a business venture, the probability of selling a product might change depending on a variety of chance factors, some of which can be controlled or changed and others not. Will advertising help? How much chance is there that a competitor will enter the market? Should high-cost materials be used? What location provides the best chance of sales? If a reasonable model can be set up that simulates these factors, then an experiment can be run before actually entering into the venture to determine the best choices.

A MODEL FOR CONDUCTING SIMULATIONS

The following problem and model are adapted from the excellent materials developed by the Quantitative Literacy Project (Gnanadesikan, Schaeffer, & Swift, 1987). In Figure 19.11, a diagram shows water pipes for a pumping system connecting A to B. The five pumps are aging, and it is estimated that at any given time the probability of pump failure is $\frac{1}{2}$. If a pump fails, water cannot pass that station. For example, if pumps 1, 2, and 5 fail, water can flow through 4 and 3. Consider the following questions that might well be asked about such a system:

What is the probability that water will flow at any time?

On the average, about how many stations need repair at any time?

What is the probability that the 1-2 path is working at any time?

For any simulation, a series of steps or a model can serve as a useful guide.

1. **Identify key components and assumptions of the problem.** The key component in the water problem is the condition of a pump. Each pump is either

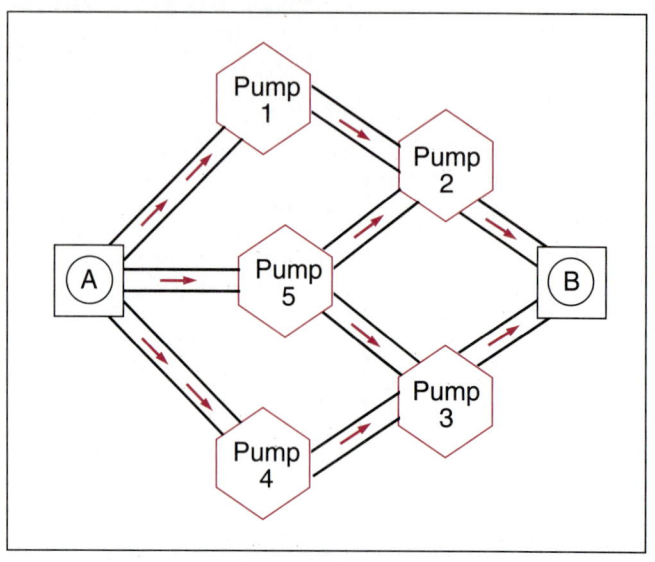

FIGURE 19.11: *Five pumps, each with a 50% chance of failure. What is the probability that some path from A to B is working?*

working or not. The assumption is that the probability that a pump is working is $\frac{1}{2}$.

2. **Select a random device for the key components.** Any random device can be selected that has outcomes with the same probability as the key component, in this case the pumps. Here a simple choice might be tossing a coin with heads representing a working pump.

3. **Define a trial.** A trial consists of simulating a series of key components until the situation has been completely modeled one time. In this problem, a trial could consist of tossing a coin five times, each toss representing a different pump.

4. **Conduct a large number of trials, and record the information.** For this problem it would be good to keep the record of heads and tails in groups of 5 because each set of 5 is one trial and represents all of the pumps.

5. **Use the data to draw conclusions.** There are four possible paths for the water, each flowing through two of the five pumps. As they are numbered in the drawing, if any of the pairs 1–2, 5–2, 5–3, or 4–3 are open, it makes no difference whether the other pumps are working. By counting those trials where at least one of these four pairs of coins both came up heads, we can estimate the probability of water flowing. To answer the second question, the number of tails per trial can be averaged. How would you answer the third question concerning the 1–2 path being open?

Steps 4 and 5 are the same as solving a probability problem by experimental means. The problem-solving and interesting aspects of simulation activities are in the first three steps, where the real-world situation is translated into a model. Translation of real-world information into models is the essence of applied mathematics.

Data Management

Intro.   predicting, surveying, modelling.
Investigate   posing question for a survey.
Culmination   making a graph
Reflection   analyzing results.

Math Language

label	data	most	bar graph
tally	collect	least	pictograph
graph	survey	difference	circle graph
scale	predict	equal	Venn Diagram
title	estimate	model	
	more	organize	

Introduction: Activities: whole class
- survey
- concrete graphs < blocks / pictures
- analyze & interpret
- journal entry

see Pizza Party in Linking Ax Binder.

Getting to Know You (Surveys)
- Favourites
- Family
- Measurements
- Birthdays
- Animals

Here are a few more examples of problems that can be solved by simulation and are easy enough to be tackled by middle-school students.

> In a true-false test, what is the probability of getting 7 out of 10 questions correct by guessing alone? (Key component: answering a question. Assumption: chance is $\frac{1}{2}$ of getting it correct.) What if the test were multiple choice with 4 choices?

> In a group of 5 people, what is the chance that 2 were born in the same month? (Key component: month of birth. Assumption: all 12 months are equally likely.)

> Casey's batting average is .350. What is the chance he will go hitless in a complete 9-inning game? (Key component: getting a hit. Assumptions: probability of a hit for each at-bat is .35. Casey will get to bat 4 times in the average game.)

> Krunch-a-Munch cereal packs 1 of 3 games in each box. About how many boxes should you expect to buy before you get a complete set? (Key component: getting one game. Assumption: each game has a $\frac{1}{3}$ chance. Trial: Use a $\frac{1}{3}$ random device repeatedly until all 3 outcomes appear; the average length of a trial answers the question.) Answer this question: What is the chance of getting a set in 6 or fewer boxes?

GATHERING AND MAKING SENSE OF DATA

The NCTM *Standards* includes a standard for statistics at both the K–4 and 5–8 levels. This does not mean that elementary children should be engaged in using complex formulas. It does mean that students should be

- involved in the collection and description of data;
- constructing and interpreting charts and graphs;
- making inferences and arguments based on their analysis of data; and
- examining arguments based on data that others have analyzed.

For children in school, these processes are not only relevant, interesting, and important for daily living, they constitute a real form of problem solving. A variety of simple techniques for graphing and making sense of data are quite simple and accessible to elementary school children. These same techniques can be directly applied to the real world.

COLLECTING DATA

One of the most important rules to follow in conducting graphing and statistics activities is to let students gather their own data. Tables of numbers produced in textbooks tend to be sterile and uninteresting. Real data, gathered by children, are almost always more interesting. For example, one class of students gathered data concerning which cafeteria foods were most often thrown in the garbage. As a result of these efforts, certain items were removed from the regular menu. The activity illustrated to students the power of organized data, and it helped them get food that they liked better.

SOURCES OF DATA

There are all sorts of ways for students to gather data that may be of interest to them. Some examples are suggested here, but these represent only a few ideas. Use your imagination, the interests of your students, and special events and activities in your class, school, and community.

Classroom Surveys

One of the easiest ways to get data in the primary grades is to get one piece of information from each student in your class. The resulting information will have manageable numbers and everyone will be interested. Here are some ideas:

Favorites. TV shows, fruit, season of the year, color, football team, pet, ice cream. When there are lots of possibilities, restrict the number of choices.

Numbers. Bus number, number of pets, sisters, hours of sleep, birthday (month or day in month), time to go to bed.

Measures. Height, weight, and other body measures, long-jump distances, length of name, time to button sweater, number of beans in a "handful," seeds in a slice of watermelon, weight of a potato (or measures of any object that students could bring from home).

School or Grade-Level Surveys

The news media frequently uses the phrase "A survey shows that the typical _____ . . . " when describing families, businesses, teenagers, drug addicts, or others. Two things in these stories can be used to attract the attention of your students. First, how did they survey everyone? Of course they only sampled a small percentage and used that data to describe the whole group. Second, what does "typical" mean? Are all people in the group typical? Is anyone in the group typical? Do the students in your class believe they are typical of children at that grade level? An excellent activity is to get children involved in describing the typical student at their grade level. What questions should they ask to decide? How many questions? Should they use multiple-choice answers, or if not, how will they group the responses? Will they survey everyone in the class or in the grade level? Is your class typical of other classes? These and many related decisions are an integral part of using statistics in real situations.

Gathering school data can involve sampling techniques such as randomly selecting 10 students per grade level instead of surveying everyone. Consider school issues such as cafeteria likes and dislikes, preferred lunch order, or use of the playground or gym. Political or social issues that all students may know about are also useful and allow you to integrate other subject areas.

Besides surveys, the school has a wealth of interesting data. Attendance by day of week, by grade level, and by month is one example. What materials are used and how much, how many tests are given in what grades and in what subjects, how many parents attend PTA are a few additional ideas. Some "people data" can be compared to similar statistics for the population at large. Examples include left-handed people, eye color, race, average family size, or years living at present address.

Consumer Data

The ingredients, prices, weights, and volumes of popular grocery items such as cereal, candy bars, paper towels, or laundry soap provide all sorts of interesting data. Catalogs and menus are another method of getting consumer data in the classroom.

Other Sources

Count various things in the newspaper (number of letters in headlines, number of vowels in 100 words, number of common words such as "and" used on a page). Almanacs, sports records, and assorted government publications can provide a wide variety of interesting statistics.

GRAPHICAL REPRESENTATIONS

Once data are gathered, what are you going to do with them? Students should be involved in the decisions that go into answering this question. To whom do you want to communicate the information? What ideas in the data are most important, and what are some ways to best show these off?

Children with no experiences with various ways of picturing data will not even be aware of the many options that are available to them. Sometimes you can suggest a new way of displaying data and have children learn to construct that type of graph or chart. Once they have made the display they can discuss its value. Did this graph (or chart or picture) tell about our data in a clear way? Compared to other ways of displaying data, how is this better?

The emphasis or goal of our instruction should be to help children see how graphs and charts tell about information; that different types of representations tell different things about the same data. Constructing and using graphs to tell about real data that they themselves have been involved in will help them interpret other graphs and charts that they see in newspapers and on TV.

What we should not get overly anxious about is the tedious details of graph construction. The message is more important then the technique! Fortunately, technology has provided us with many tools for constructing simple yet powerful representations. With the help of computers or graphing calculators, it is possible to actually construct a lot of pictures of the same data with very little effort. Then the discussion can be placed where it belongs: What form of data representation is the most appropriate for our purposes? What does this picture of the data show?

BAR GRAPHS

Bar graphs are one of the first ways to group and present data and are especially useful in grades K to 3. At this early level, bar graphs should be made so that each bar is made of countable parts such as squares, objects, or pictures of objects. No numeric scale is necessary. Graphs should be simple and quickly constructed. As shown in Figure 19.12, simple bar graphs can be real graphs, picture graphs, or symbolic graphs.

Real graphs use the actual objects being graphed. Examples include types of shoes, sea shells, or books. When making a real graph, each item should be placed in a square so that comparisons and counts are easily made.

Picture graphs use a drawing of some sort that represents what is being graphed. Students can make their own drawings or you can duplicate drawings to be colored or cut out to suit particular needs.

Symbolic graphs use something like squares, blocks, or X's to represent the things being graphed.

It is easier for young students to understand real and picture graphs, but exposure to all different types is important. To quickly make a graph of class data, follow these steps:

1. Decide on what groups of data will make the different bars. It is good to have from two to six different bars in a graph.

2. Everyone should decide on or prepare their contribution to the graph before you begin. For real or picture graphs, the object or picture should be ready to be placed on the graph. For symbolic graphs, students should write down or mark their choice.

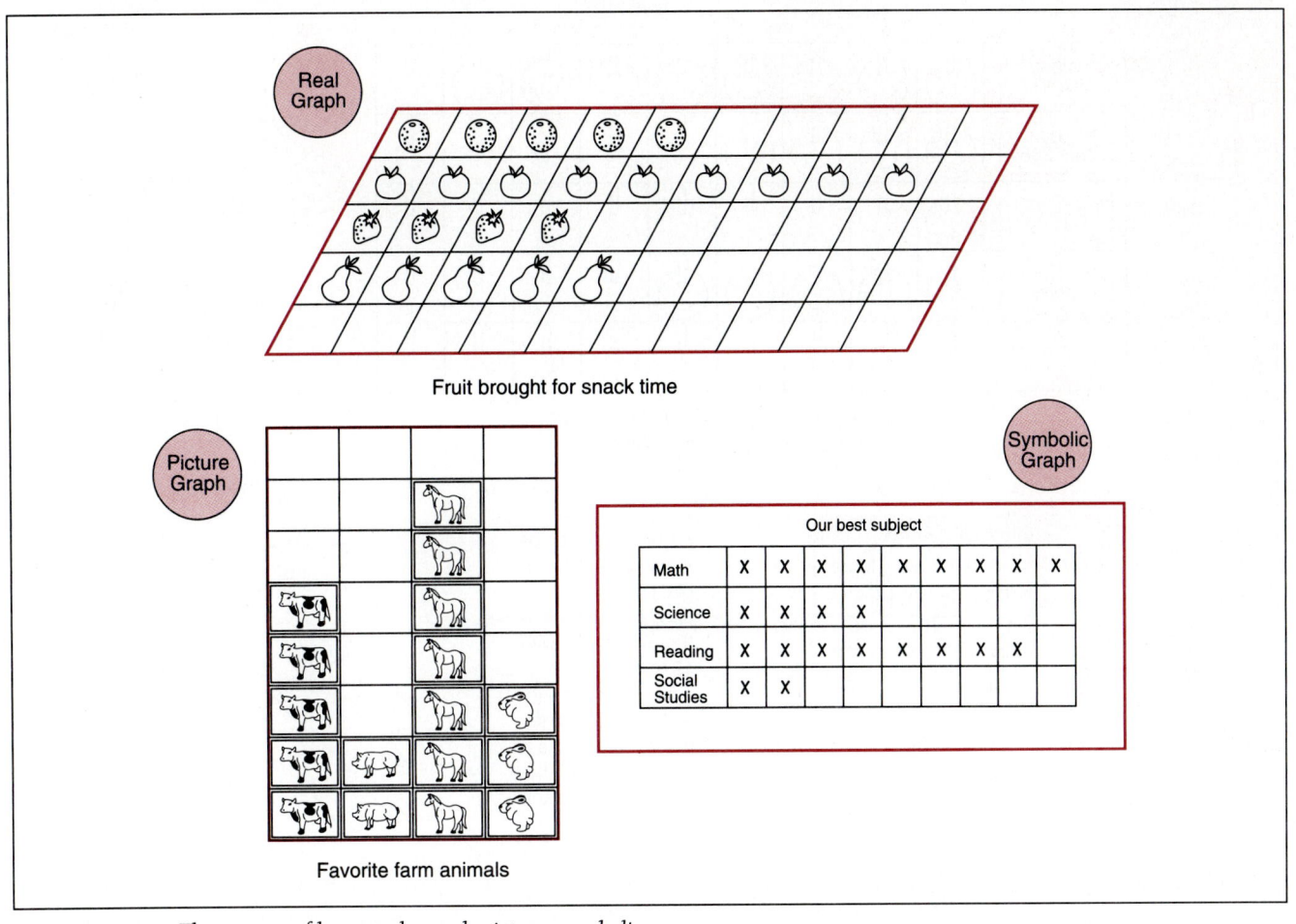

FIGURE 19.12: *Three types of bar graphs: real, picture, symbolic*

3. In small groups have students quickly place or mark their entry on the graph. A graph mat can be placed on the floor, or a chart prepared on the wall or blackboard (Figure 19.13, p. 394). If tape or pins are to be used, have these ready.

By following these steps, a class of 25 to 30 students can make a graph in less than 10 minutes, leaving ample time to use it for questions and observations.

Questions like the following should be considered for every graph that is made in the early grades:

Which has the most? least?
Are there more _____ or more _____?
Are there less _____ or less _____?
Are there any the same?
How many _____ and _____ are there together?
How many more (or less) _____ are there than _____?

Children in the intermediate grades can also profit from making and using bar graphs. Individual groups can graph data that they gather themselves rather than one element per person. As part of their task, let students decide what the categories should be, what scales should be used, and whether to use pictures, squares, or continuous bars. Single pictures or squares on the graph might represent 5, 10, or 100 things, rather than 1.

STEM-AND-LEAF PLOTS

Stem-and-leaf plots are a newer form of bar graph where numeric data is plotted by using the actual numerals in the data to form the graph. By way of example, suppose that the American League baseball teams had posted the following win records over the past season.

Baltimore	45	Milwaukee	91
Boston	94	Minnesota	98
California	85	New York	100
Chicago	72	Oakland	101
Cleveland	91	Seattle	48
Detroit	102	Toronto	64
Kansas City	96	Texas	65

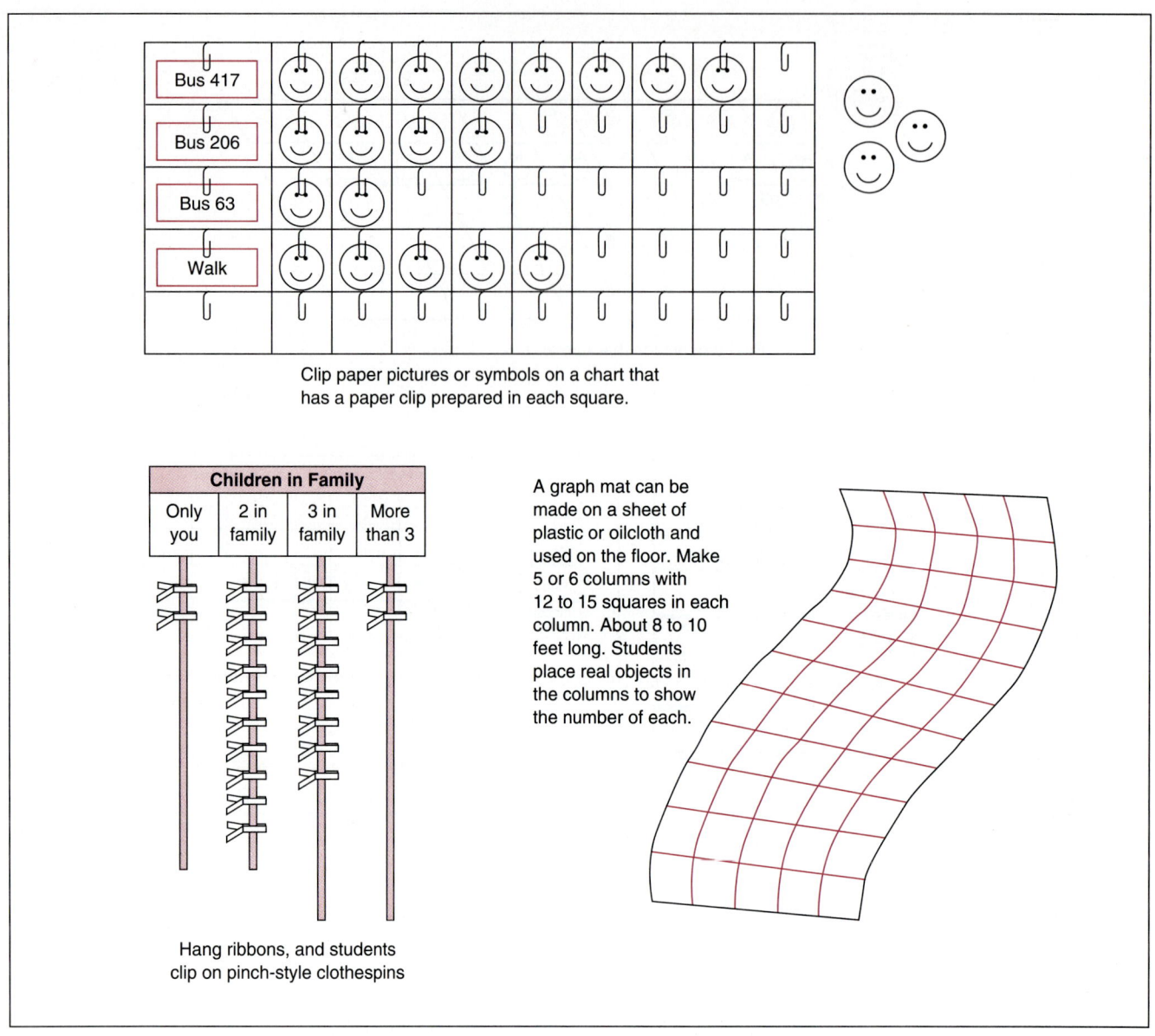

FIGURE 19.13: *Some ideas for quick graphs that can be used over and over*

If the data are to be grouped by tens, then list the tens digits in order and draw a line to the right as in Figure 19.14(a). These form the "stem" of the graph. Next go through the list of scores and write the ones digits next to the appropriate tens digit as in Figure 19.14(b). These are the "leaves." There is no need to count or group the data ahead of time. The process of making the graph does it all for you. Furthermore, every piece of data can be retrieved from the graph. (Notice that stem-and-leaf plots are best made on graph paper so that each digit takes up the same amount of space.)

To provide more information, the graph can be quickly rewritten, ordering each leaf from least to most [Figure 19.14(c)]. In this form it may be useful to identify the number that belongs to a particular team, indicating its relative place within the grouped listing.

Stem-and-leaf graphs are not limited to two-digit data. For example, if the data ranged from 600 to 1300, the stem could be the numerals from 6 to 13 and the leaves made of two-digit numbers separated by commas. If the ones digit is not important, round the data to the nearest ten, and use only the tens digit in the leaves. Figure 19.15 shows the same data in two different stem-and-leaf plots.

Figure 19.16 illustrates two additional variations. When two sets of data are to be compared, the leaves can extend in opposite directions from the same stem. In the same example, notice that the data are grouped by fives instead of tens. When plotting 42, the 2 is written next to the 4; for 47, the 7 is written next to the dot below the 4.

Stem-and-leaf plots are significantly easier for students to make than bar graphs, all of the data are maintained, they provide an efficient method of ordering data, and indi-

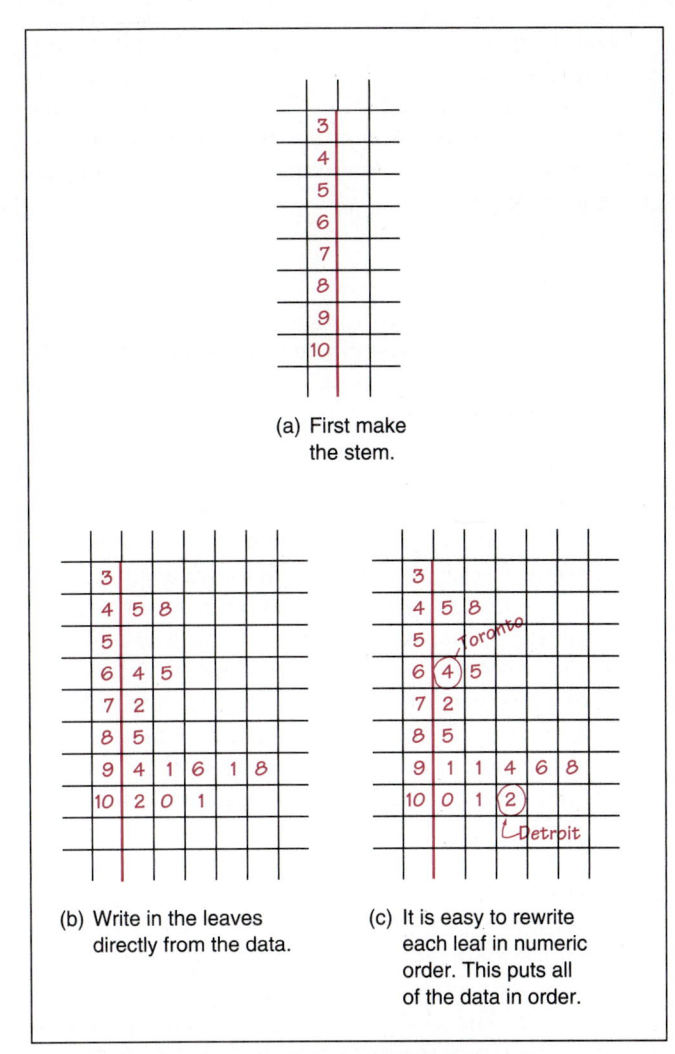

19 / EXPLORING BEGINNING CONCEPTS OF PROBABILITY AND STATISTICS   395

FIGURE 19.14: *Making a stem-and-leaf plot*

FIGURE 19.15: *In the bottom plot the data from above are rounded to the nearest ten.*

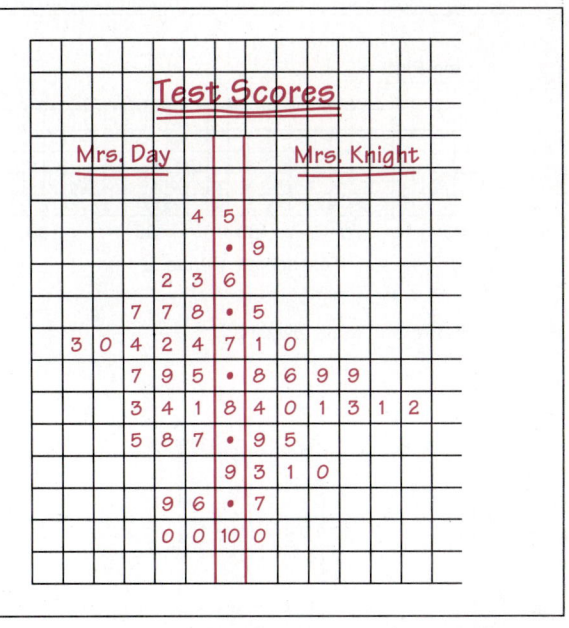

FIGURE 19.16: *Stem-and-leaf plots can be used to compare two sets of data.*

vidual elements of data can be identified. For these reasons, stem-and-leaf plots are considered by many to be preferable to bar graphs (Landwehr, Swift, & Watkins, 1987).

CONTINUOUS DATA GRAPHS

The bar graphs just discussed are useful for illustrating categories of data that have no numeric ordering; for example, colors or TV shows. When data are grouped along a continuous scale, it make sense to represent the data in that order and perhaps to show progressions from one point in the scale to the next. Examples of such information include temperatures that occur over time, height or weight over age, and percents of test takers scoring in different intervals along the scale of possible scores.

◆ Line Plots

Line plots are useful *counts* of things along a numeric scale. To make a line plot, a number line is drawn and an X is made above the corresponding value on the line for every corresponding data element. One advantage of a line plot is that every piece of data is shown on the graph. It is also a very easy type of graph for students to make. It is essentially a bar graph with a potential bar for every value possible. A simple example is shown in Figure 19.17 (p. 396). (A stem-and-leaf plot is a lot like a line plot except that the data are grouped.)

◆ Histograms

A *histogram* is a form of bar graph where the categories are consecutive intervals along a numeric scale. The inter-

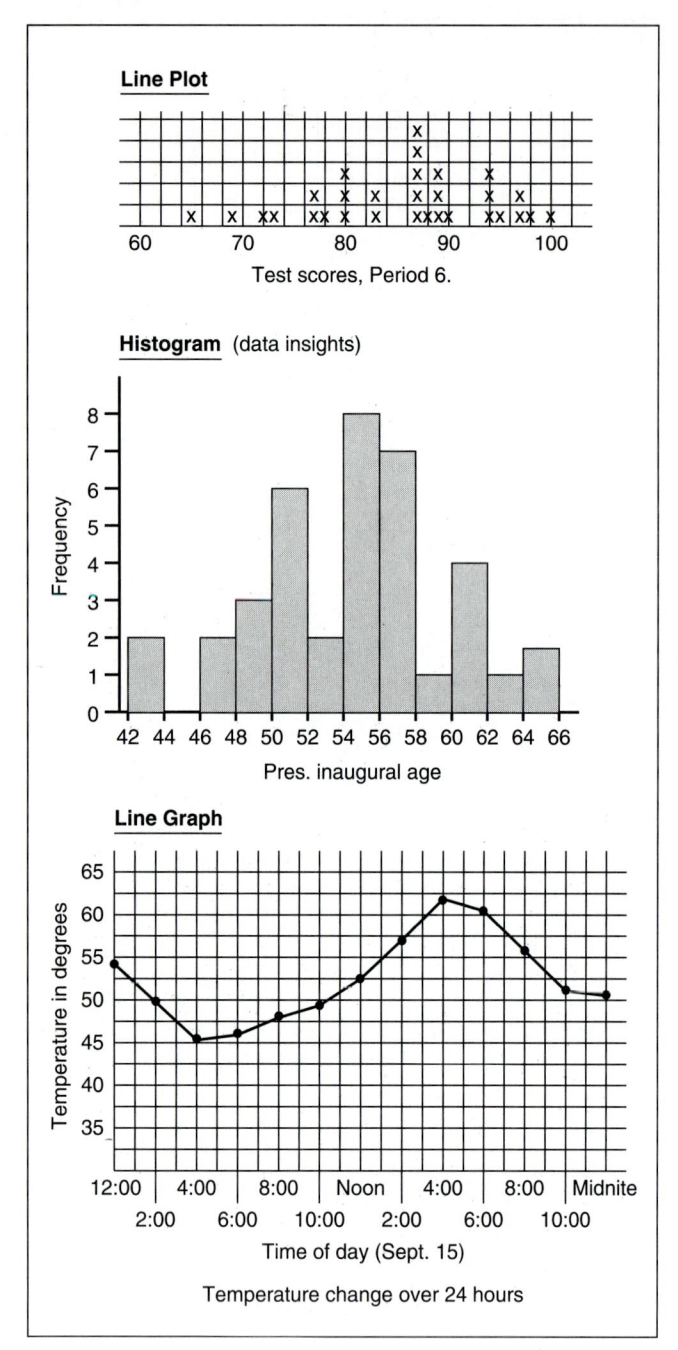

FIGURE 19.17: *Three forms of graphs of data over continuous intervals. Notice that the horizontal scale must show some progression and is not just a grouping as in a bar graph.*

vals should always be the same size with no gaps between. The height or length of each bar is determined by the number of data elements falling into that particular interval. Histograms are not difficult in concept but can cause students difficulty in constructing them. What is the appropriate interval to use for the bar width? What is a good scale to use for the length of the bars? All of the data must be grouped and counted within each interval causing another difficulty. Technology can help us with all of these decisions, allowing children to focus on the graph and its message. *Data Insights* (Edwards & Keogh, 1990) is just one example of an easy-to-use program that quickly allows students to display a histogram of their own data, selects "optimum" scales, and permits adjustments. The graphing calculator will also produce histograms without much difficulty. Figure 19.17 shows an example of a histogram produced with *Data Insights*.

◆ Line Graphs

A *line graph* is used when there is a numeric value associated with equally spaced points along a continuous number scale. Points are plotted to represent two related pieces of data, and a line is drawn to connect the points. For example, a line graph might be used to show how the length of the flagpole shadow changed from one hour to the next during the day. The horizontal scale would be time and the vertical scale, the length of the shadow. If measurements were taken every hour or half hour during the day, these discrete points would be plotted and a straight line drawn to connect the points. There is always an assumption with a line graph that even though the only real data points are those that are plotted, the line between points represents implied data. In the example of the shadow, a shadow did exist at all times, and its length did not jump or drop from one plotted value to the other. It changed continuously as suggested by the graph. See the example in Figure 19.17.

A histogram is a lot like a crude line graph. Imagine the intervals in the histogram getting narrower and narrower. The tops of the bars would come very close to forming a smooth curve. The TI-81 graphing calculator will draw a line graph on top of the histogram. It uses the left edge of each bar as the plotted point (the center of the bar would be more accurate).

CIRCLE GRAPHS

A *circle* or *pie graph* is used when a total amount has been partitioned into parts and interest is in the ratio of each part to the whole and not so much in the particular quantities. In Figure 19.18, two graphs each show the percentages of students with different numbers of siblings. One graph is based on classroom data and the other on school data.
Since the pie graphs display ratios not quantities, the small class data can be compared to the large school data. That could not be done with bar graphs.

◆ Easily Made Circle Graphs

There are a variety of ways that circle graphs can be made easily. Circle graphs of the students in your room can be made quickly and quite dramatically. Suppose, for example, that each student picked his or her favorite bas-

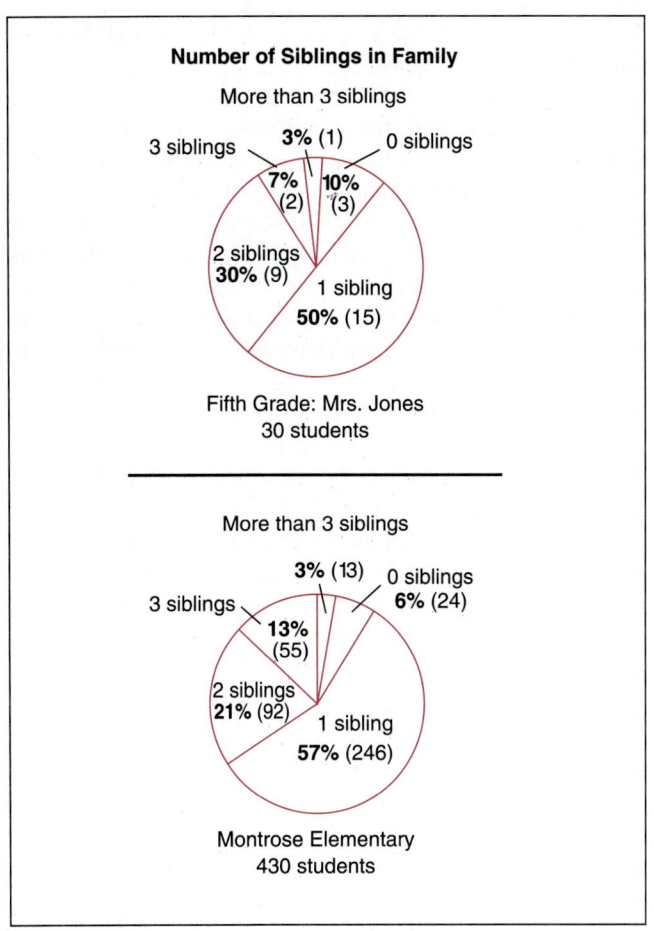

FIGURE 19.18: *Circle graphs show ratios of part to whole and can be used to compare ratios.*

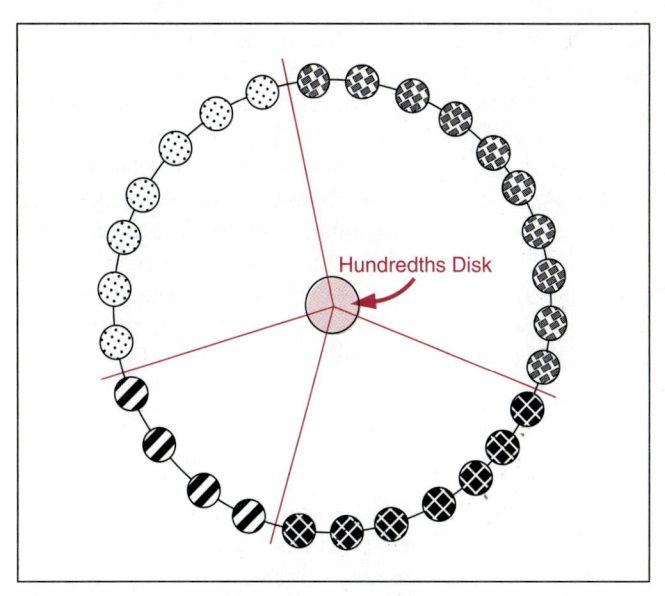

FIGURE 19.19: *A human pie graph: Students arranged in a circle with string stretched to show the divisions.*

ketball team in the NCAA tournament's "Final Four." Line up all of the students in the room so that students favoring the same team are together. Now form the entire group into a circle of students. Tape the ends of four long strings to the floor in the center of the circle and extend them to the circle at each point where the teams change. Voilà! A very nice pie graph with no measuring and no percentages. If you copy and cut out a hundredths disk (Black-line Masters) and place it on the center of the circle, the strings will show approximate percentages for each part of your graph (Figure 19.19.) By the way, if the students in the graph rearrange themselves into four rows, they can also be a human bar graph.

Another easy approach to circle graphs is similar to the human pie graph. Begin by having students make a bar graph of the data. Once complete, cut out the bars themselves, and tape them together end to end. Each bar should be a different color. Next, tape the two ends together to form a circle. Estimate where the center of the circle is, draw lines to the points where different bars meet, and trace around the full loop. The result is, again, a very meaningful circle graph. You can determine the percentages using the hundredths disk as before.

From Percents to Pie Graphs

Pie graphs can also be made from computed percentages in the traditional way. If students have experienced either of the two methods described in the previous section, the calculations will make more sense. The numbers in each category are added to form the total or whole. (That's the same as taping all of the strips together or lining up the students.) By dividing each of the parts by the whole with a calculator, numbers between zero and one result—fractional parts of the whole. If rounded to hundredths, these numbers are now percentages of the whole. (Check that the total is 1 whole, or 100%. Rounding may cause some error.) With a copy of the hundredths disk, they can easily make a pie chart and never have to mess with degrees and protractors. Trace around the disk to make the outline of the pie. Mark the center through a small hole in the disk, and draw a line to the circle. Start from that point, and use the disk to measure hundredths around the outside.

Again, it is well worth noting that computer software is now available to produce a variety of graphs. Circle graphs are included in almost every package you might consider. *Data Insights* is one of the best examples of student-friendly software to produce graphs, including pie graphs. Most all business-oriented spreadsheet programs (e.g., *Excel* or *Lotus 1,2,3*) will make fantastic pie graphs and permit them to be custom-labeled in any way.

DESCRIPTIVE STATISTICS

Pictures of data (graphs) are excellent ways to describe data. However, it is common to measure the data in some way to produce a number that describes the data. These numbers are *statistics*, measures of the data that quantify some attribute of the data. A graph and a statistic are analogous to a picture of a person and a measure of his or her height. One shows something about the person, the other is a number that describes some feature of the person. The things that are most often described numerically about a set of data are the range (distance between the upper and lower data values), some measure of where the center of the data is (an average), and how dispersed the data are within the range. School children can get an idea of the importance of these statistics by exploring the ideas informally.

AVERAGES

The term *average* is heard quite frequently in everyday usage. Sometimes it refers to an exact arithmetic average, as in, "the average daily rainfall." Sometimes it is used quite loosely, as in, "she is about average in height." In either situation, an *average* is a single number or measure that is descriptive of a larger collection of numbers. If your test average is 92, it is assumed that somehow all of your test scores are reflected by this number. The *mean*, *median*, and *mode* are each specific types of averages or *measures of central tendency*. While other averages exist, these three are the ones that are generally taught in the elementary school.

The *mode* is that value or values that occur most frequently in the data set. Of these three statistics, the mode is the least useful and could perhaps be ignored completely. Consider the following set of numbers:

1, 1, 3, 5, 6, 7, 8, 9

The mode of this set is 1. In this example, the mode is not a very good description of the set, and that is often the case. If the 8 in this string of numbers were a 9, a change of only one, there would be two modes. If one of the 1's were changed to a 2, there would be no mode at all. That is, the mode is a statistic that does not always exist, does not necessarily reflect the center of the data, and can be "unstable" or changeable with very small changes in the data.

The *mean* is computed by adding all of the numbers in the set and dividing the sum by the number of elements added. This is the statistic that is sometimes referred to as the average. The mean of the above set of 5 (40 divided by 8). A bit later in the chapter, the mean is discussed in more detail.

The *median* is the middle value in an ordered set of data. Half of all values lie at or above the median and half below. For the eight numbers listed above, the median is between 5 and 6 or 5.5. The Quantitative Literacy Project, in its book *Exploring Data* (Landwehr & Watkins, 1987), places a heavy emphasis on the median in preference to the mean or mode. They note that the median is easier to understand, easier to compute, and is not affected by one or two extremely large or extremely small values outside the range of most of the data as is the mean.

BOX-AND-WHISKER PLOTS

Box-and-whisker plots (or just *box plots*) are an easy method for visually displaying not only the median statistic but also information about the range and distribution of data. In Figure 19.20, the ages in months for 27 sixth-grade students are listed along with stem-and-leaf plots for the full class and the boys and girls separately. Box-and-whisker plots are shown in Figure 19.21.

The following numbers represent the ages in months of a class of sixth-grade students. The numbers 1 to 14 are the boys, and 15 to 27 the girls.

1.	132	8.	122	15.	140	22.	131
2.	140	9.	130	16.	129	23.	128
3.	133	10.	134	17.	141	24.	131
4.	142	11.	125	18.	134	25.	132
5.	134 *Joe B.*	12.	147	19.	124 *Whitney*	26.	130
6.	(137)	13.	131	20.	129	27.	127
7.	139	14.	129	21.	(125)		

All students

```
12 | 2, 4
 • | 5, 5, 7, 8, (9,) 9, 9
13 | 0, 0, 1, 1, (1,) 2, 2, 3, 4, 4, 4
 • | (7,) 9
14 | 0, 0, 1, 2
 • | 7
```

Boys

```
12 | 2
 • | 5, 9
13 | (0,) 1, 2, 3, |4, 4
 • | 7, (9)
14 | 0, 2
 • | 7
```

Girls

```
12 | 4
 • | 5, 7, |8, 9, 9
13 | (0,) 1, 1, 2, |4
 • |
14 | 0, 1
 • |
```

FIGURE 19.20: *Ordered stem-and-leaf plots grouped by 5. Medians, upper, and lower quartiles are found on the stem-and-leaf plots. Medians and quartiles are circled if data elements, represented by a bar (|), fall between two elements.*

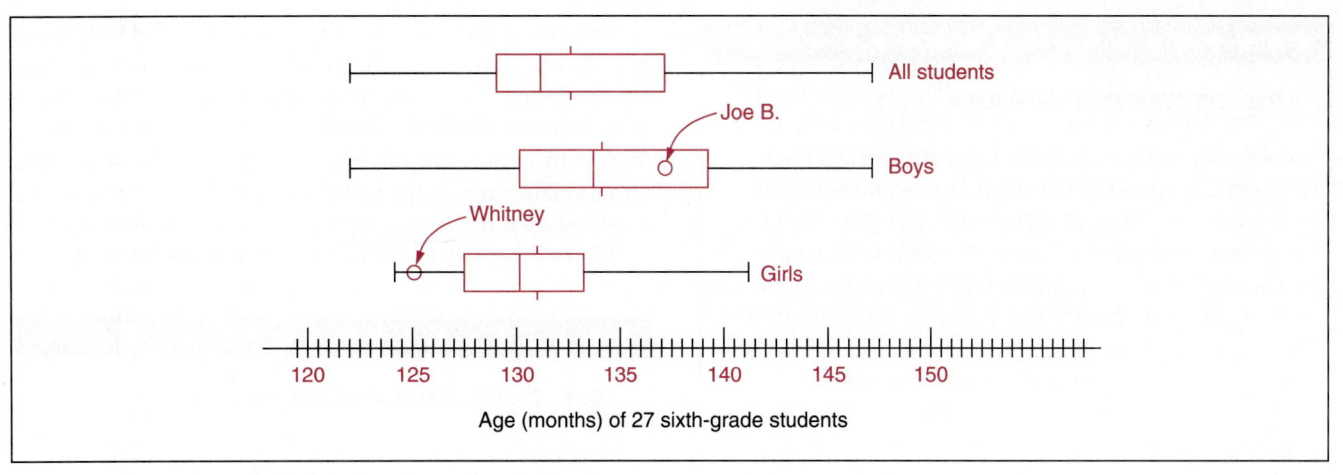

FIGURE 19.21: *Box-and-whisker plots show a lot of information.*

Each box-and-whisker plot has these three features:

A box which contains the "middle half" of the data, one fourth to the left and right of the median. The ends of the box are at the *lower quartile*, the median of the lower half of data, and the *upper quartile*, the median of the upper half of the data.

A line inside the box at the median of the data.

A line extending from the end of each box to the *lower extreme* and *upper extreme* of the data. Therefore each line covers the upper and lower fourths of the data.

Look at the information these box plots provide at a glance! The box, in comparison to the lengths of the lines, provides a quick indication of how the data are spread out or bunched together. Since the median is shown, this spreading or bunching can be determined for each quarter of the data. The entire class in this example is much more spread out in the upper half than the lower half. The girls are much more closely grouped in age than are either the boys or the class as a whole. It is immediately obvious that at least three-fourths of the girls are younger than the median age of the boys. The *range* of the data (difference between upper and lower extremes) is represented by the length of the plot, and the extreme values can be read directly. It is easy to mark and label entries of particular interest. For example, Joe B. and Whitney might be the class officers.

Making box-and-whisker plots is quite simple. First, put the data in order. An easy and valuable method is to make a stem-and-leaf plot and order the leaves, providing another visual image as well. Next, find the median. If the data is listed in a single row or column, young students can put their fingers on the ends and move each toward the middle one step at a time. Older children will simply count the number of values and determine the middle one. This can be done directly on the stem-and-leaf plots as was done in Figure 19.20. To find the two quartiles, ignore the median itself, and find the medians of the upper and lower halves of the data. Again, this can be done on the stem-and-leaf plots. Mark the two extremes, the two quartiles and the median above an appropriate number line. Draw the box and the lines. Box plots can also be drawn vertically.

Note that the means for the data in the above example are each just slightly higher than the medians (class = 132.4; boys = 133.9; girls = 130.8). For this example, the means themselves do not provide nearly as much information as the box plots.

Box-and-whisker plots are graphical representations, and yet they are pictures of statistics perhaps more than pictures of data. They show the range and the median and pictorially indicate a sense of the spread of data. The more traditional measures of spread, the *variance* and *standard deviation*, are not necessary to consider at the elementary level. However, once a concept of "spread" or dispersion of data is developed informally with box-and-whisker plots, the standard deviation may have more meaning. This statistic is rather tedious to compute but is available on most scientific calculators at the press of a button.

UNDERSTANDING THE MEAN

Due to ease of computation and stability, the median when compared to the mean has some advantages as a practical average. However, the mean will continue to be used in popular media and in books, frequently along with the median. For smaller sets of data such as your test scores, the mean is perhaps a more meaningful statistic. Finally, the mean is used in the computation of other statistics such as the standard deviation. Therefore, it remains important that students have a good concept of what the mean tells them about a set of numbers. How do you describe the mean other than how to compute it? The two activities in this section will help students construct intuitive ideas about the mean.

19.5 LEVELING THE BARS

Make a bar graph using wooden cubes for the bars. Choose a graph with five or six bars with lengths of no more than 10 to 12. (The graph in Figure 19.22 represents the prices of six toys chosen so that the total of the prices is a multiple of 6.) Sketch a picture of the graph on the board as a record. Have students compute the mean using the usual numeric methods. Discuss with them what they think this number called the mean tells them about the data. After listening to their ideas, compare the process of adding up the numbers to the process of piling all the cubes into one stack. Dividing the sum by the number of values is the same as separating the one big stack of block into equal stacks. Rearrange the cubes into bars of equal length. The mean, then, is the number you get if all of the values are leveled off. Make different graphs with cubes, and let students find the mean by leveling out the stacks rather than using any computation.

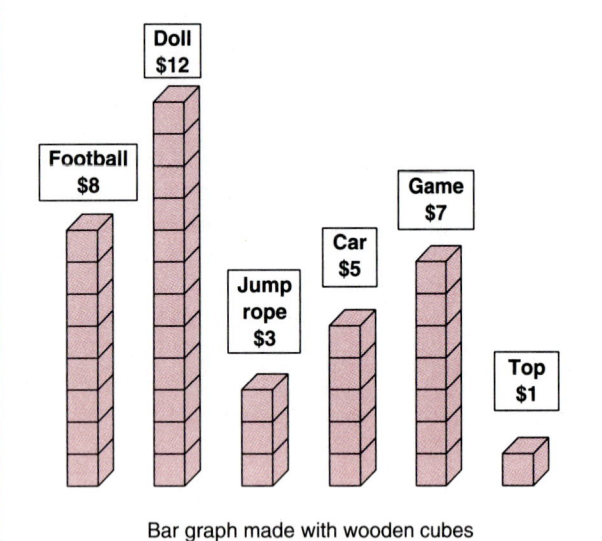

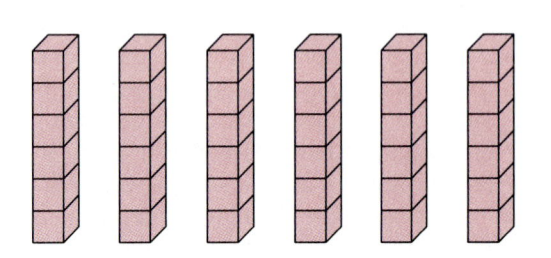

Bar graph made with wooden cubes

The same cubes rearranged into equal stacks. Height is the mean value of the bars above.

FIGURE 19.22: *Understanding the mean*

This bar graph/mean activity can also be done by making the original bar graph on paper. Numbers of any size can be used. After computing the mean, cut out all the bars, and tape them together end to end. The length of the total strip is the sum of the numbers. Fold the long strip into as many equal parts as there are bars in the graph. The length of each part is the mean.

The next activity is a good follow-up to the last one.

19.6 OVER AND UNDER THE MEAN

Duplicate a simple bar graph on graph paper, and pass it out to the class. *Without doing any computation*, ask each student to draw a line across the graph where they estimate the mean to be. Next, using the scale on the graph, have them add up the pieces of the bars that are above their estimated mean line and record this number. Similarly, for bars that are below their estimated mean line, add up the spaces between the tops of the bars and the mean line (Figure 19.23). Compare the two sums. Should they be the same, different, or is it impossible to know? Now have them compute the actual mean with a calculator, draw a new line, and check the totals of the bars above and below as before. The two sums should be exactly the same when the mean line is drawn accurately. Notice that the pieces that stick up above the mean could be cut off and fitted into the spaces below the line to make the bars all equal (as with the blocks). Visually estimating the mean of a bar graph is a fun challenge.

Technology provides teachers with an effective method of examining the mean with their classes. Spreadsheets are unbelievably easy to use for things such as adding lists of numbers, ordering them, and computing the mean. In the example in Figure 19.24, data values can be changed as quickly as they can be typed, and all the sums and means will be recomputed immediately. With this tool it is easy to add one "far out" number to the data, delete numbers, or change the data in any way at all to observe the effect on the mean. The program *Data Insights* has a spreadsheet function built right into it that requires no programming and allows for this same sort of exploration. (It also draws box-and-whisker plots.) Also, do not forget the graphing calculator. Data sets can be entered in much the same way as on a spreadsheet.*

---

*The graphing calculator and *Data Insights* will both produce scatter plots and best-fit lines. These ideas have not been discussed in this chapter, but both are accessible to students in the seventh and eighth grades. The use of technology is extremely beneficial in helping students explore scatter plots and best-fit lines.

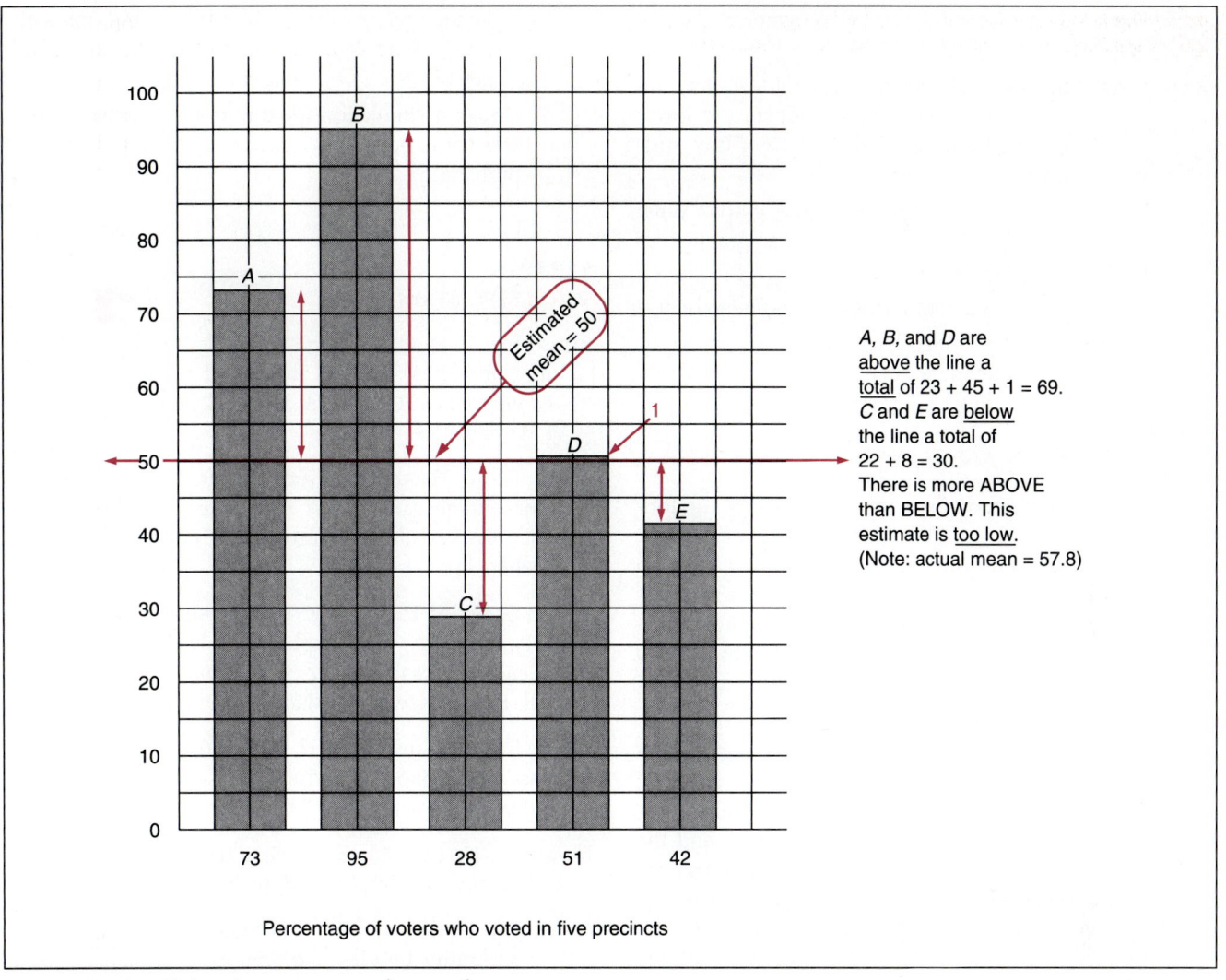

[Bar graph showing percentages of voters who voted in five precincts: A=73, B=95, C=28, D=51, E=42, with estimated mean line at 50.]

A, B, and D are above the line a total of 23 + 45 + 1 = 69. C and E are below the line a total of 22 + 8 = 30. There is more ABOVE than BELOW. This estimate is too low. (Note: actual mean = 57.8).

FIGURE 19.23: *Estimating a mean on a bar graph*

```
File: Mean exp.    REVIEW/ADD/CHANGE    Escape: Main Menu
======A======B======C======D======E======F======G======H===
  10|
  11|                   500                34
  12|                    54                89
  13|                    90                67
  14|                   129               120
  15|                   301                23
  16|                   346                44
  17|                   665
  18|                    44
  19|                   765
  20|                   973
  21|                  1000
  22|
  23|     Sum:        4867    Sum:        377
  24|     Mean:    442.4545   Mean:   62.83333
  25|
  26|     Combined Mean of Both:  308.4705
  27|
-----------------------------------------------------------
F24: (Value) @AVG(F11...F21)
```

FIGURE 19.24: *Screen display from AppleWorks Spreadsheet. In this example, column sums and column means are computed instantly for any data in those two columns.*

♦♦♦♦♦
REFLECTIONS ON CHAPTER 19: WRITING TO LEARN

1. Describe the difference between theoretical probability, relative frequency, and experimental probability.

2. Why do you think it is a good idea to have students conduct experiments before trying to figure out probabilities?

3. What are the first ideas that young students should develop about the concept of chance? How can this be done?

4. Describe how you could use a random-number generator on a computer or calculator to produce a simulation of a three-part spinner with all parts equal?

5. What is a simulation?

6. What questions should young children learn to answer about a bar graph they have just made?

7. Draw a picture of a line plot, a histogram, and a line graph. How are all of these alike?

8. Put at least 30 numbers in a stem-and-leaf plot, and use it to determine the median, upper, and lower quartiles and the range and to draw a box-and-whisker plot.

9. What are three ways to make a circle graph? What does a circle graph tell you that a bar graph does not? What does it not tell?

10. What are three different forms of averages? Which is the most "stable"? Explain.

11. What is meant by the mean?

◆◆◆◆◆

FOR DISCUSSION AND EXPLORATION

1. Set up a simulation experiment for each of the four examples on p. 391. If you do not use a computer to generate random numbers, you will need to decide on a random device such as coins or spinners or drawing cards from a hat. A computer is not necessary for any of these exercises. Run at least 10 trials of your experiment. Was the outcome about what you expected?

2. Make up two lists of 30 numbers each ranging from 50 to 100. Make one list with most of the numbers clustering in the 60s and a median of about 65. Make the second list with a median of 75 and the numbers somewhat evenly distributed across the range. Make box-and-whisker plots of both sets of numbers. Explain how the box plots describe your two sets of scores. How would you go about adding 40 more scores to one of the sets of data without having to change the box plot?

3. Examine one of six books from the *Used Numbers Project*.* These books have some of the best statistics activities for elementary students that have been developed. You might: (a) share an activity or idea in the book that excited you, (b) react to a lesson described in the book, or (c) share an idea in the book with a teacher and get his or her reaction.

4. Use an area model and a tree diagram to determine the theoretical probability for the following experiment:

   Dad puts a $5 bill and three $1 bills in the first box. In a second box he puts another $5 bill but just one $1 bill. Junior, for washing the car, gets to take one bill from the first box without looking and put it in the second box. After these have been well mixed, he then gets to take one from the second box without looking. What is the probability that he will get $5?

5. Design a simulation for the problem in Exercise 4, and try it out. Does it agree with your theoretical probability?

◆◆◆◆◆

SUGGESTED READINGS

Armstrong, R. D. (1981). An area model for solving probability problems. In A. P. Shulte (Ed.), *Teaching statistics and probability*. Reston, VA: National Council of Teachers of Mathematics.

Bright, G. W., Harvey, J. G., & Wheeler, M.M. (1981). Fair games, unfair games. In A. O. Shulte (Ed.), *Teaching statistics and probability*. Reston, VA: National Council of Teachers of Mathematics.

Bright, G. W., & Hoeffner, K. (1993). Measurement probability, statistics, and graphing. In D. T. Owens (Ed.), *Research ideas for the classroom: Middle grades mathematics*. New York: Macmillan Publishing Co.

Bruni, J. V., & Silverman, H. J. (1986). Developing concepts in probability and statistics—and much more. *Arithmetic Teacher, 33*(6), 34–37.

Bryan, E. H. (1988). Exploring data with box plots. *Mathematics Teacher, 81*, 658–663.

Corwin, R. B., & Friel, S. N. (1990). Statistics: Prediction and sampling [grades 5–6]. From *Used numbers: Real data in the classroom*. Palo Alto, CA: Dale Seymour Publications.

Corwin, R. B., & Russell, S. J. (1990). Measuring: From paces to feet [grades 3–5]. From *Used numbers: Real data in the classroom*. Palo Alto, CA: Dale Seymour Publications.

Fair, J., & Melvin, M. (1986). *Kids are consumers, too! Real-world mathematics for today's classroom*. Menlo Park, CA: Addison-Wesley.

Friel, S. N., Mokros, J. R., & Russell, S. J. (1992). Middles, means, and in-between [grades 5–6]. From *Used numbers: Real data in the classroom*. Palo Alto, CA: Dale Seymour Publications.

Gnanadesikan, M., Schaeffer, R., & Swift, J. (1987). *The art and techniques of simulation: Quantitative literacy series*. Palo Alto, CA: Dale Seymour Publications.

Goldman, P. H. (1990). Teaching arithmetic averaging: An activity approach. *Arithmetic Teacher, 37*(7), 38–43.

Kelly, I. W., & Bany, B. (1984). Probability: Developing student intuition with a 20 × 20 array. *School Science and Mathematics, 84*, 598–604.

---

*Used Numbers is a set of six books, each covering two or three grades between K and 6. The books are a result of an NSF-supported project at Technical Education Research Centers (TERC) and Lesley College. The books are an excellent resource for elementary statistics activities. See the suggested readings.

Landwehr, J. M., Swift, J., & Watkins, A. E. (1987). *Exploring surveys and information from samples: Quantitative literacy series.* Palo Alto, CA: Dale Seymour Publications.

Landwehr, J. M., & Watkins, A. E. (1987). *Exploring data: Quantitative literacy series.* Palo Alto, CA: Dale Seymour.

Leutzinger, L. P. (1990). Graphical representation and probability. In J. N. Payne (Ed.), *Mathematics for the young child.* Reston, VA: National Council of Teachers of Mathematics.

Lindquist, M. M. (1992). *Making sense of data: Addenda series, grades K–6.* Reston, VA: National Council of Teachers of Mathematics.

National Council of Teachers of Mathematics. (1990). Data Analysis [Minifocus Issue]. *Mathematics Teacher, 83*(2).

Newman, C. M., Obremski, T. E., & Schaeffer, R. L. (1987). *Exploring probability: Quantitative literacy series.* Palo Alto, CA: Dale Seymour Publications.

Russell, S. J., & Corwin, R. B. (1989). Statistics: The shape of the data [grades 4–6]. From *Used numbers: Real data in the classroom.* Palo Alto, CA: Dale Seymour Publications.

Russell, S. J., & Corwin, R. B. (1990). Sorting: Groups and graphs [grades 2–3]. From *Used Numbers: Real Data in the Classroom.* Palo Alto, CA: Dale Seymour Publications.

Russell, S. J., & Friel, S. N. (1989). Collecting and analyzing real data in the elementary classroom. In P. R. Trafton (Ed.), *New directions for elementary school mathematics.* Reston, VA: National Council of Teachers of Mathematics.

Russell, S. J., & Stone, A. (1990). Counting: Ourselves and our families [grades K–1]. From *Used numbers: Real data in the classroom.* Palo Alto. CA: Dale Seymour Publications.

Schultz, H. S., & Leonard, B. (1989). Probability and intuition. *Mathematics Teacher, 82,* 52–53.

Shaughnessy, J. M., & Dick, T. (1991). Monty's dilemma: Should you stick or switch? *Mathematics Teacher, 84,* 252–256.

Shulte, A. P., & Choate, S. S. (1977). *What are my chances? (Books A & B).* Palo Alto, CA: Creative Publications.

Watkins, A. E. (1981). Monte Carlo simulation: Probability the easy way. In A. P. Shulte (Ed.), *Teaching statistics and probability.* Reston, VA: National Council of Teachers of Mathematics.

Zawojewski, J. S. (1991). *Dealing with data and chance: Addenda series, grades 5–8.* Reston, VA: National Council of Teachers of Mathematics.

20 PREPARING FOR ALGEBRA

WHAT IS PREALGEBRA?

For many years, the content of middle-grade mathematics courses has been largely review. Flanders (1987) examined three popular textbook series and concluded that only about half of the material in the sixth grade could be considered "new," and even less was new in the eighth grade. In the report of the Second International Mathematics Study, *The Underachieving Curriculum* (McKnight et al., 1987), the U.S. curriculum is criticized for spending too much time on previously taught, lower-level skills with little intense effort on new topics, especially those related to algebra. For example, U.S. eighth-grade teachers estimate spending 18% of their time on fractions and 14% of the time on ratio, proportion, and percent—nearly one-third of the year. Japanese teachers report spending only 6% of the time on these two topics together. In contrast, U.S. teachers reported teaching algebraic topics 20% of the time, compared with 37% in Japan. A more recent trend in middle-school mathematics curricula is to increase the emphasis on problem solving, geometry, probability, statistics, and measurement as well as to develop the important foundational concepts of algebra. (Refer to the summary of emphasis changes suggested by the NCTM *Standards* for grades 5–8 in the Appendix.)

All of the topics in this chapter—primes and factors, integers, variables, and function—will be found in most middle-grade curriculums. Some of the number-theory concepts, such as odd and even numbers and primes, are commonly explored in lower grades as well. The middle-school curriculum should begin to prepare students for the more abstract curriculum of the high school. This chapter focuses on those topics that are most important for the study of algebra. The objective is to show how students can develop and explore the basic building blocks of algebra using intuitive approaches connected to their understanding of arithmetic, so that algebra can be a natural progression rather than an abrupt change.

EXPLORING TOPICS IN NUMBER THEORY

Number theory is the study of relationships found among the natural numbers. At the elementary level, number theory includes the concepts of prime number, odd and even numbers, and the related notions of factor, multiple, and divisibility. Prime factorization is frequently connected with finding common denominators and reducing fractions. More importantly, the concepts of prime, factor, and multiple are also used in algebraic expressions. Students should develop an intuitive understanding of these topics with numbers so that the algebraic generalizations can be built upon them. Number-theory topics also provide an opportunity for problem solving and for student discovery of many fascinating relationships.

PRIMES AND FACTORIZATION

◆ Discovering Primes and Other Factors

Prime numbers can be viewed as fundamental building blocks of the other natural numbers. Simply defining a prime number and searching for primes can be a rather dull experience. The activities described here are examples of interesting things we can explore with children and still have fun while developing basic concepts of prime, factor, and multiple.

20.1 LOOKING FOR RECTANGLES

Have students work in groups using square tiles or cubes or just square grid paper. Begin with the number 12 or 16. Have students find as many different rectangles as possible made up of that many squares. Share results. Students should agree that 1×12 and 1×16 rectangles should be included and that a square (4×4) is also a special kind of rectangle. When the process is understood completely, have them try to build as many rectangles as possible for each number up to 100. Give different groups different numbers to work on. Draw pictures on graph paper for each number, and make a display. Students will discover that some numbers (primes) have only one rectangle. The dimensions of each rectangle are two factors of that number. Use this idea to develop a definition of factor and of prime number. (A number other than 1 is a *prime* number if its only factors are itself and 1.) Help students use the idea of rectangular dimensions to develop definitions of prime numbers and factors. Rectangles are a way to think of numbers as a product of two factors. Rectangular solids are a model for a number expressed as a product of three factors.

20.2 LOOKING FOR BOXES

With cubes, find ways to build boxes for a number. For example, 12 cubes can be arranged into a $2 \times 2 \times 3$ box (Figure 20.1). What can you say about a number for which that many cubes can be used to

Build only one box?

Build only two different boxes?

Build only one box that has no edges of length one?

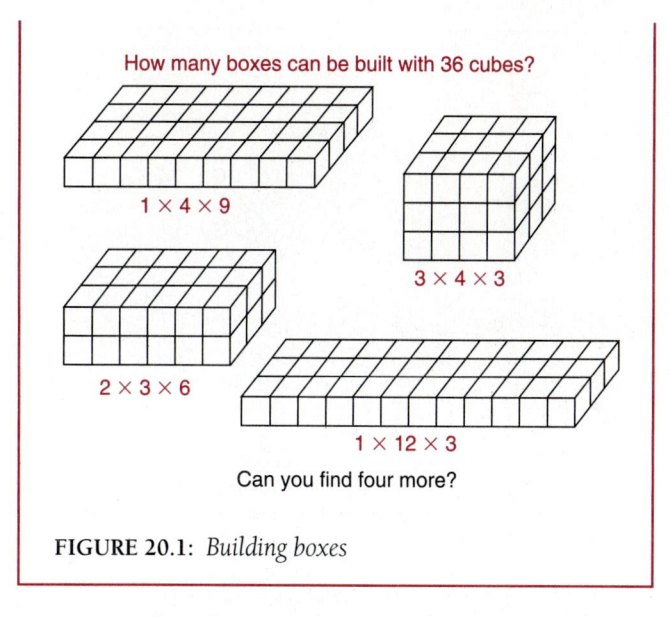

How many boxes can be built with 36 cubes?

$1 \times 4 \times 9$

$3 \times 4 \times 3$

$2 \times 3 \times 6$

$1 \times 12 \times 3$

Can you find four more?

FIGURE 20.1: *Building boxes*

Once students have an idea of what a prime number and a factor are, they can explore a variety of other relationships involving primes. Calculators should be available for all explorations.

- Given a number, is it a prime or not a prime? (Nonprime numbers other than 1 are called *composite numbers*.) What are some ways to test if a number is a prime?

- Explore Goldbach's conjecture: Any even number greater than 2 can be written as the sum of two prime numbers. For example: $38 = 31 + 7$. Goldbach's conjecture has never been proved or disproved. (Goldbach lived between 1690 and 1764.)

- Have students make up and explore different conjectures of their own. They might consider the difference of primes, or the sum of three primes. Can they find a formula that will always produce primes? For example, $2^n - 1$ works for quite a few values of n. The formula $n^2 - n + 41$ yields prime numbers for any value of n between 1 and 40 inclusive. Let students try to find other "prime generators."

- Have students write down all of the prime factors of a number, including repeats. Write the number as a product of these prime factors placing them in order from least to most. For example, $360 = 2 \times 2 \times 2 \times 3 \times 3 \times 5$. Before comparing results from different groups, discuss the various strategies or approaches that were used. If a different approach is used, will it still result in the same factorization? (The answer, of course, is "yes." This result is known as the Fundamental Theorem of Arithmetic.) Some different ways to find the prime factorization of a number are shown in Figure 20.2 (p. 406).

- The terms *divisor* and *factor* are synonyms. For a nonprime number, find all of the *proper divisors* (divisors other than the number itself) and add them up. (See which groups can come up with the most clever ways of finding all of the factors of a number.) If the sum is

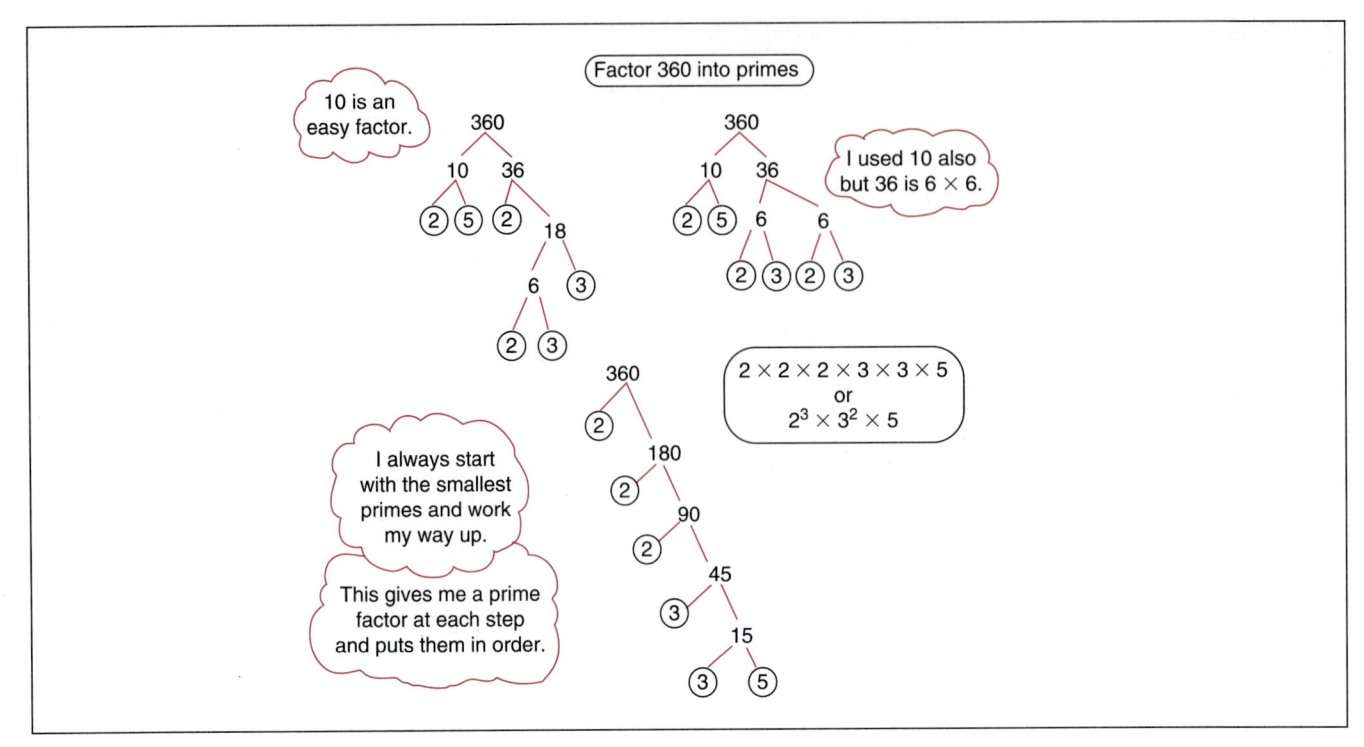

FIGURE 20.2: *Several different routes to the same factors*

less than the number, it is called *deficient,* if more than the number, it is called *abundant;* if it is equal to the sum of its proper divisors, it is called *perfect.* For example, 6 is perfect, 8 is deficient, and 12 is abundant. Those are the easy ones. What are the next three of each type?

Students can also explore looking for twin primes, those that are two apart, as are 11 and 13.

PATTERNS ON A HUNDREDS CHART

One of the truly fascinating aspects of mathematics is the way that numbers tend to appear in patterns. In Chapter 18, the calculator was suggested as a good way to explore some interesting numeric patterns. The Fibonacci sequence was also discussed as an intriguing source of patterns and relationships. In the following activities students can use a hundreds chart to discover other interesting phenomena of numbers. As you explore these ideas, consider how prime numbers and factors play a role in the results. How are these patterns like the calculator patterns in Chapter 18? Encourage students to explore some of these ideas and make their own observations. A self-discovered pattern is much more exciting than one that we point out to students.

20.3 HUNDREDS CHART PATTERNS

Duplicate lots of small copies of a hundreds chart (or 0-to-99 chart). Start with a number less than 10 and have students color in that number and all multiples of that number. For every number, some pattern will emerge (Figure 20.3). Which numbers make diagonal patterns, and which only produce columns or parts of columns? Notice that the pattern for sixes is made up of numbers that are also in the threes pattern. How can you tell if one pattern will be a part of another? Do they ever overlap, or is one always completely inside of another? What happens if you start a jump-by-threes pattern at two instead of at zero?

Circle the threes.
Color the sixes.
What columns are these in?

FIGURE 20.3: *Patterns of multiples*

20.4 CHANGING THE CHARTS

Change the dimensions of the chart in the last activity by changing the lengths of the rows (Figure 20.4). If the rows have six squares, which numbers will make patterns in rows and which in columns? If there are seven squares in the row, only one number will make a column pattern, the multiples of 7. Why? Try other row lengths including rows with more than 10 squares. All patterns seem to be either columns or diagonals. What does the row length have to do with the patterns involved? Consider looking at primes and factors.

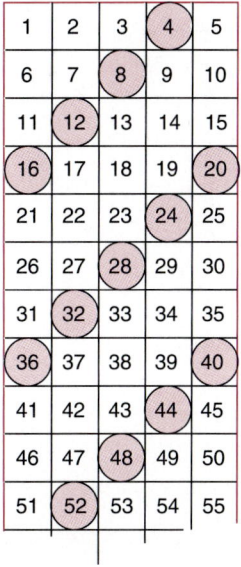

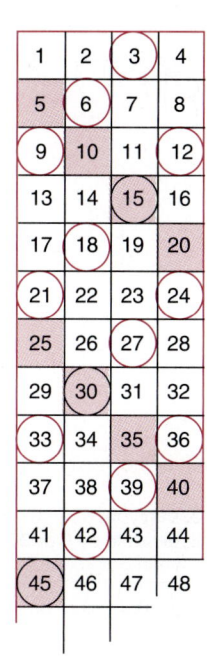

Fours on a 5-wide chart

Threes and fives on a 4-wide chart

FIGURE 20.4: *Looking for patterns on different hundreds charts.*

TESTS FOR DIVISIBILITY

It is easy to tell whether or not a number is divisible by 10. For example, 198,456 is not, but 650,270 is. The test for divisibility by 10 is to look at the last digit, the digit in the ones position. You probably know this already. Experience indicates that products with a factor of 10 end in 0. Divisibility by 5 is similar: If the last digit is either 5 or 0, then the number is divisible by 5.

There are similar tests for divisibility by other numbers.

A number is divisible by

3 if the sum of the digits is divisible by 3

9 if the sum of the digits is divisible by 9

2 if the last digit is even (or divisible by 2)

4 if the last two digits are divisible by 4

8 if the last three digits are divisible by 8

6 if it is divisible by 3 and 2

Many students find these intriguing. The limited practical value of such tests indicates that they should be investigated as a problem-solving task rather than a topic of mastery. The more intriguing and more valuable question is, why do these tests work? Since the answer lies in the base ten place-value representation of the numbers, divisibility rules also provide an opportunity to review place-value concepts (Figure 20.5, p. 408). One approach is to give students the divisibility rule and suggest that they use base ten materials or drawings to figure out why it works. Having discovered one rule, the students can then be asked to find a rule that is similar. For example, the argument for divisibility by 3 is identical to that for 9 and the arguments for 4 and 8 are similar to that for 2.

EXPONENTS

In algebra classes students get very confused trying to remember the rules of exponents. For example, when you raise numbers to powers, do you add or multiply the exponents? Here is an excellent example of procedural knowledge that is frequently learned without supporting conceptual knowledge. Before algebra, students should have ample opportunity to explore working with exponents on whole numbers rather than with letters or variables. By doing so, they are able to deal directly with the concept and actually generate the rules themselves.

A *whole-number exponent* is simply a shorthand symbolism for repeated multiplication of a number times itself; for example, $3^4 = 3 \times 3 \times 3 \times 3$. That is the only conceptual understanding that is required.

There are also conventions of symbolism that must be learned. As conventions, they are arbitrary rules with no conceptual basis. The first is that an exponent applies to its immediate base. For example, in the expression $2 + 5^3$, the exponent 3 only applies to the 5, so the expression is equal to $2 + (5 \times 5 \times 5)$. However, in the expressions $(2 + 5)^3$, the 3 is an exponent of the quantity $2 + 5$ and is evaluated as $(2 + 5) \times (2 + 5) \times (2 + 5)$ or $7 \times 7 \times 7$.

The other convention involves the order of operations; multiplication and division are always done before addition and subtraction. Notice that since exponentiation is repeated multiplication, it also is done before addition and subtraction. In the expression $5 + 4 \times 2 - 6 \div 3$, 4×2 and $6 \div 3$ are done first. Therefore the expression is evaluated as $5 + 8 - 2 = 13 - 2 = 11$. If done in left-to-right order the result would be 4. Parentheses are used to group operations that are to be done first. Therefore, in $(5 + 4) \times 2 - 6 \div 3$, the addition can be done inside the parentheses first, or the distributive property can be used, and the final result is 16. The phrase "please excuse my dear Aunt Sally"

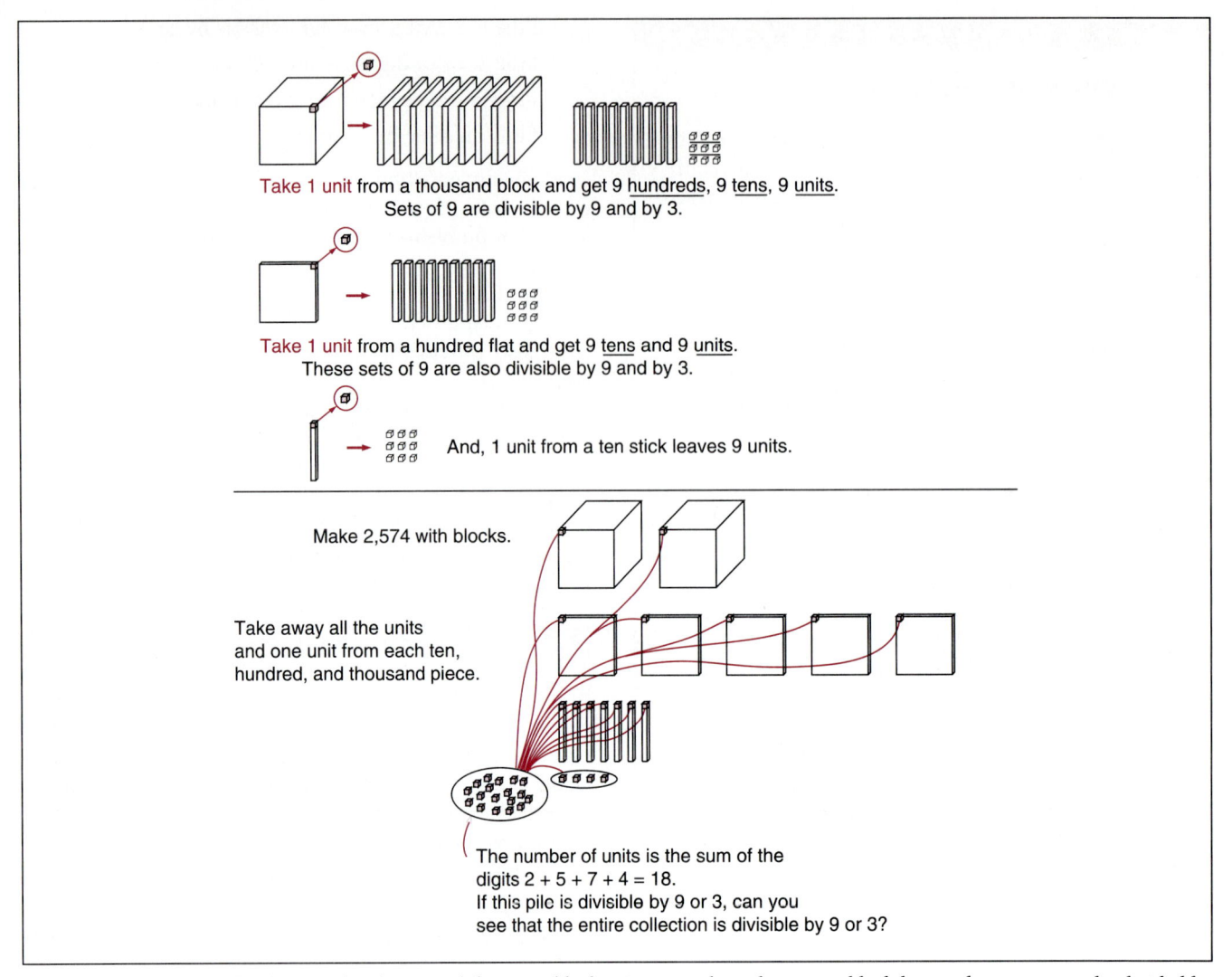

FIGURE 20.5: *Divisibility by 9 and 3 shown with base ten blocks. Can you show that every block larger than a ten can be divided by 4? Use this fact to establish divisibility by 4.*

is sometimes used to help students recall that operations inside *parentheses* are done first, then *exponentiation* and then *multiplication* and *division* are done before *addition* and *subtraction*.

◆ Using Calculators

Most scientific calculators, such as those now popularly used in high school, employ "algebraic logic" that will evaluate expressions correctly and also allow grouping with parentheses. However, the simple four-function calculators generally available and commonly used in the elementary schools do not use algebraic logic. Operations are processed as they are entered. For these calculators, the following two keying sequences produce the same results:

Key: → 3 (+) 2 (×) 7 (=)
Display → 3   2 5 7 35

Key: → 3 (+) 2 (=) (×) 7 (=)
Display → 3   2 5   7 35

Whenever an operation sign is pressed, the effect is the same as pressing (=), and then the operation. Of course, neither result is correct for the expressions 3 + 2 × 7, which should be evaluated as 3 + 14, or 17. In recent years the TI-MathMate™ and the TI-Explorer™ have appeared, and these do use algebraic logic and include parentheses keys so that both 3 + 2 × 7 and (3 + 2) × 7 can be keyed in the order that the symbols appear. With these and also with scientific calculators, the display only shows one number at a time as shown here.

Key: → 3 (+) 2 (×) 7 (=)
Display → 3   2   7 17

Notice that the display does not change when (×) is pressed.

Key: → (() 3 (+) 2 ()) (×) 7 (=)
Display → [3   2 5   7 35

The right parenthesis is not displayed, because the operation 3 + 2 could be completed.

20 / PREPARING FOR ALGEBRA   409

The graphing calculator offers the best solution to these problems and at the same time provides other advantages far beyond the scope of this chapter. When the expression $3 + 2 \times (6^2 - 4)$ is keyed in, the display shows the full expression in the display screen. Nothing is evaluated until you press ENTER or EXE. Then the result appears on the next line to the right of the screen.

```
┌─────────────────────────────┐
│  3 + 2 * (6² – 4)        67 │
└─────────────────────────────┘
```

Moreover, the last expression entered can be recalled and edited so that students can see how different expressions are evaluated. Only minimum key presses are required.

```
┌─────────────────────────────┐
│  3 + 2 * (6² – 4)        67 │
│  (3 + 2) * (6² – 4)     160 │
│  (3 + 2) * 6² – 4       176 │
│  3 + 2 * 6² – 4          71 │
└─────────────────────────────┘
```

Returning to the simple four-function calculator—it remains a powerful tool regardless of its limitations. For example, to evaluate 3^8, press 3 × = = = = = = = . (The first press of = will result in 9, or 3 × 3.) Students will be fascinated with how quickly numbers grow. Enter a number, press ×, and then continue to press =. Try two-digit numbers. Try it with 0.2.

◆ **Explorations**

Students should have ample opportunity to explore expressions involving mixed operations and exponents with only the conventions and the meaning of exponents as their guides. No rules for exponents should be promoted. When experience has provided students with a firm background, the rules of exponents will make good sense and should not require rote memorization.

◆◆◆◆◆◆◆◆◆◆◆◆◆◆◆◆◆◆

20.5 WHAT'S IN AN EXPRESSION?

Provide students with numeric expressions to evaluate with their calculators. Some examples of the type of expressions that can be valuable are

$3 + 4 \times 8$	$3^6 + 2^6$	$3^4 \times 7 - 5^2$	$3^4 \times 5^2$
$4 \times 8 + 3$	$(3 + 2)^6$	$(3 \times 7)^4 - 5 \times 2$	$(3 \times 5)^6$

$\dfrac{5^3 \times 5^2}{5^6}$	$4 \times 3 - 2^3 \times 5 + 23 \times 9$	$\dfrac{4 \times 3^5}{2}$ $4 + \dfrac{3^5}{2}$

When experiencing difficulty, students should write equivalent expressions without exponents and/or include parentheses to indicate explicit groupings. For example,

$$(7 \times 2^3 - 5)^3 = (7 \times (2 \times 2 \times 2) - 5)$$
$$\times (7 \times (2 \times 2 \times 2) - 5)$$
$$\times (7 \times (2 \times 2 \times 2) - 5)$$
$$= ((7 \times 8) - 5) \times ((7 \times 8) - 5)$$
$$\times ((7 \times 8) - 5)$$
$$= (56 - 5) \times (56 - 5) \times (56 - 5)$$
$$= 51 \times 51 \times 51$$

When discussing results, place all of the emphasis on the procedures rather than the answer. The fact that two groups got the same result does not help a group that got a different result. For many expressions there is more than one way to proceed, and one may be easier to do or to understand than another.

Of course calculators with algebraic logic and graphing calculators will automatically produce correct results. It remains important for students to know the correct order of operations. The calculator should not replace an understanding of the rules. The order-of-operation rules apply to symbolic manipulation in algebra and also must be understood if a calculator without algebraic logic is to be used.

The following activity involves an understanding of exponents and estimation and is also a good problem-solving task.

◆◆◆◆◆◆◆◆◆◆◆◆◆◆◆◆◆◆

20.6 THE MYSTERY NUMBER

Before class, write down any simple expression involving exponents, and evaluate it. Give the expression to the class with at least two of the numbers in the expression replaced by stars. Give students the value of the expression, and see which group, using their calculators, can find correct values for the stars first. For example:

$$(* + 7*)^3 = 41,063,625$$

The size of the result and the fact that it ends in 25 are clues that may help. There is also more than one solution to this one, so do not be too quick to tell students their answers are not correct.

ROOTS AND IRRATIONALS

INTRODUCING THE CONCEPT OF ROOTS

How could you use a calculator to estimate the side of the square or the edge of the cube (assume there is no square-root key on the calculator) in Figure 20.6? These are excellent challenges for students and provide a good introduction to the process of finding the root of a number. The solutions will satisfy these equations:

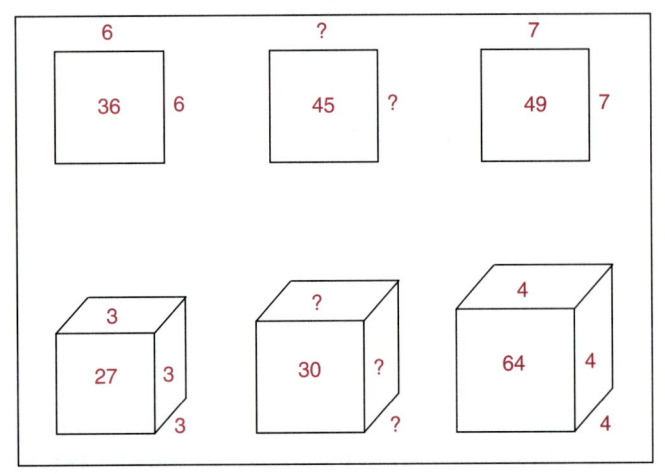

□ × □ = 45 or □² = 45 and
□ × □ × □ = 30 or □³ = 30

FIGURE 20.6: *A geometric interpretation of square roots and cube roots.*

The calculator permits students to guess at a solution and quickly test to see if it is too big or too small. For example, to solve the cube problem, you might start with 3.5 and find that 3.5^3 is 42.875, much too large. Quickly you will find that the solution is between 3.1 and 3.2. But where? Again, try halfway: 3.15^3 is 31.255+. The next try should be lower. By continued trial and error, a simple calculator can get the result correct to six decimal places.

From a simple introduction such as this, students can be challenged to find solutions to equations such as $□^6 = 8$. These students are now prepared to understand the general definition of the *nth root* of a number *N* as that number that when raised to the *n*th power equals *N*. The square and cube roots are simply other names for the second and third roots. With this approach, the concept of the *n*th root is developed in general. The notation involving the radical sign can come last in this development. It should then be clear that $\sqrt{6}$ is a number and not an exercise to be done. The cube root of eight *is the same as* $\sqrt[3]{8}$, which is the same as two.

REAL NUMBERS

While it may be open to argument, eighth-grade students probably do not need to have a very sophisticated knowledge of the real number system. There are a few powerful ideas, however, that can be explored or discussed informally.

◆ Some Numbers Are Not Rational

There are two characterizations of a rational number. One is that it is a number that can be written as a fraction of two whole numbers. Alternatively, it is true that a number is rational if and only if, when written as a decimal, the decimal part is either finite or repeats infinitely. Thus 3.45 and 87.19363636. . . are rational numbers and can each be converted to their fractional forms. But what about a decimal number that just goes on and on and on, with no repetition? Or, what about the number 3.101001000100001000001 . . . ? These never repeat and are not finite and therefore are not rational. A number that is not rational is called *irrational*.

The numbers π and $\sqrt{2}$ are both irrational numbers. The number π is a ratio of two measures in one of the most common shapes we know, the circle. While it is not possible to prove the irrationality of p at this level, the fact that it is irrational implies that it is impossible to have a circle with the lengths of both the circumference and the diameter rational. (Why?) A proof that $\sqrt{2}$ is irrational is generally explored at the junior-high level. The usual argument assumes that $\sqrt{2}$ is rational, which then leads to a contradiction.

◆ Rational Numbers Are Dense

The property of being *dense* means that between any two numbers in the system, there is another one. It is impossible to find two rational numbers that are "next" to each other because there are always some more in between. The denseness of the rationals is fun and easy to demonstrate and is a powerful concept for students to experience. In fraction form, consider $\frac{7}{15}$ and $\frac{8}{15}$. They seem relatively close. But there are clearly at least nine fractions between $\frac{70}{150}$ and $\frac{80}{150}$, the same two numbers. And for any two of those, say $\frac{73}{150}$ and $\frac{74}{150}$, the same process can be used again. It never ends. There are always more in between.

While the density of the rationals is a fairly impressive idea in itself, what is even more astounding is that the irrationals are also dense. And the irrationals and the rationals are all mixed up together. The density of the irrationals is not as easy to demonstrate and is not within the scope of the elementary school.

◆ Not All Roots Are Irrational

One reason it is difficult to comprehend irrational numbers is that we have very little firsthand experience with them. The irrational numbers we are most familiar with are roots of numbers. For example, it has already been noted that students are frequently shown a proof for the irrationality of $\sqrt{2}$. An intuitive notion of that fact is difficult to come by. Most calculators will only show eight digits, requiring at least some leap of faith to accept that the decimal representation is infinite and nonrepeating.

What can unfortunately happen is that whenever students see the radical sign, they think the number is irrational. A possible approach is to consider the concept of roots from the opposite direction. Rather than ask what is the square root of 64 or the cube root of 27, we might suggest that every number is the square root, the cube root, the fourth root, ..., of some number. (For example, 3 is the square root of 9, the third root of 27, etc.) From this vantage point, students can see that "square root" is just a way of indicating a relationship between two numbers. That the cube root of 27 is 3 indicates a special relationship between 3 and 27.

INTEGER CONCEPTS

Negative numbers are an important set of numbers. They can and should be explored before they are encountered in algebra. In fact, students almost every day either have some interaction with negative numbers or experience a phenomenon that negative numbers can model. Some examples:

A loss of money is a negative cash flow.

Slowing down the car is negative acceleration, and driving in reverse is negative velocity.

Time *before* "blast off" is negative time.

Below zero temperature and below ground level are negatives relative to a scale.

In fact, almost any concept that is quantified and has direction probably has both a positive and a negative value.

Generally, negative values are introduced with integers (the whole numbers and their negatives or opposites) instead of with fractions or decimals. One reason for this is so that the ideas can be modeled before they are generalized to all real numbers.

INTUITIVE EXAMPLES OF SIGNED QUANTITIES

As with any new types of numbers that students encounter, some form of real model or examples must be used. Negative numbers or situations that model negative numbers do exist that students can relate to. It is a good idea to discuss some of these with your class before jumping directly into computation with signed numbers.

Bills and Cash *or* Debits and Credits

Suppose you are the bookkeeper for a small business. At any time your records show how many dollars the company has in its account. There are always so many dollars in cash (credits or receipts) and so many dollars in bills (debits). The difference between the debit and credit totals can tell the value of the account. If there is more cash than bills, the account is positive or "in the black." If there are more bills, the account is in debt, or shows a negative cash value, or is "in the red." Suppose further that all transactions are handled by mail. The mail carrier can bring mail, a positive action, or take mail, a negative action.

With this scenario, it is easy to discuss addition and subtraction of signed quantities. An example is illustrated in Figure 20.7.

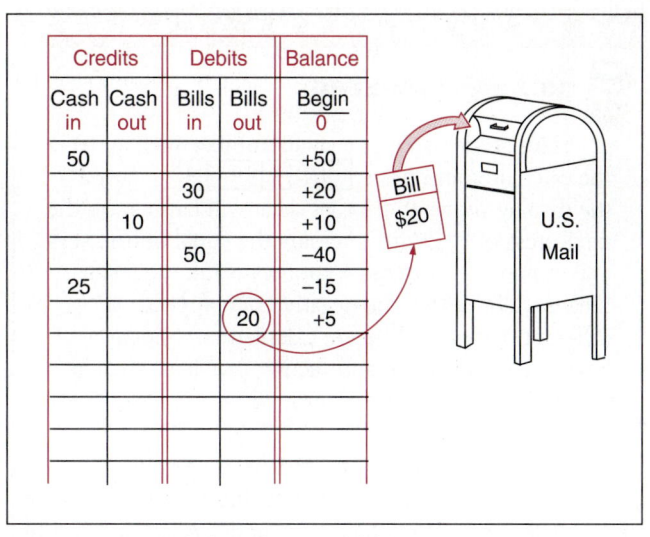

Credits		Debits		Balance
Cash in	Cash out	Bills in	Bills out	Begin 0
50				+50
		30		+20
		10		+10
		50		−40
25				−15
			20	+5

FIGURE 20.7: *A ledger sheet model for integers*

Football

The standard football field has a 50-yard line in the center extending to 0 at each goal. A mathematical football field might have the center be the 0-yard line with one goal being the +50 goal and the other being the −50 goal. Any position on the field is determined by a signed number between +50 and −50. Gains or losses are like positive and negative quantities. A positive team moves toward the positive goal, and a negative team toward the negative goal. If the negative team starts on the −15-yard line and has a loss of 20 yards, they will be on the +5-yard line. (You should sketch a picture of this example, and convince yourself that it is numerically the same example as the one in Figure 20.7.)

Contrived situations such as mailing debits and credits and the football field are suggested as introductory discussion models. They can help students to think intuitively about what happens to quantities when an action causes them to be less than zero. They also provide an example of a joining or positive action and a removal or negative action of both positive and negative quantities. With these models there are two specific types of questions we can pose for students.

1. Give students a beginning and an end value, and have them describe different ways that the change might have occurred. For example, how could the ball get from the +40-yard line to the −10-yard line if the negative team has the ball?

2. Give students either a beginning or an ending value and one or more actions, and have them determine the value not given. For example, if the company received $20 in cash and $35 in bills, resulting in a balance of negative $5, what did they have to begin with?

The calculator is another model that might be explored early in the discussion of signed numbers. It gives correct and immediate results that students seem to believe. The major drawback is that no rationale for the result is provided.

20.7 BEYOND ZERO

Have students enter a small number such as 10 in the calculator and press $-$ 1 $=$ $=$ $=$ ... until the display shows 0. Discuss what will happen on the next press of $=$. After pressing the equal button several more times, discuss what they think the result means. (Notice that the negative sign appears in different places on different calculators.) Similarly, have students explore subtraction problems such as $5 - 8 = ?$, and discuss the results.

A negative number is entered on a calculator by first pressing the number and then pressing the change-of-sign key (usually $+/-$). Thus to add -3 and -8, press 3 $+/-$ $+$ 8 $+/-$ $=$. Students can benefit by using the calculator along with the intuitive examples mentioned earlier, or they can use the calculator to answer questions of both types previously mentioned. For example, how can you get from -5 to -17 by addition? or 13 minus *what* is 15?

MATHEMATICS DEFINITION OF NEGATIVE NUMBERS

The mathematician defines a negative number in terms of whole numbers. Therefore, the definition of negative 3 is the solution to the equation $3 + \square = 0$. In general, the *opposite of n* is the solution to $n + \square = 0$. If n is a positive number, then the opposite of n is a *negative number*. The set of *integers*, therefore, consists of the positive whole numbers, the opposites of the whole numbers, or negative numbers, and zero, which is neither positive nor negative. This is the definition that you may have learned in a college mathematics class. It is also found in student textbooks. Like many things in mathematics, abstract or symbolic definitions are best when there is some intuitive or conceptual framework with which to link the idea. For one who has never thought about the concept of a negative quantity, the symbolic definition of "the opposite of a number" is relatively vague if not totally strange. Virtually all new ideas in mathematics develop originally from playing around with familiar ideas and relationships, frequently aided by real models. It is a major error to let students believe otherwise.

OPERATIONS WITH THE INTEGERS

Up until students encounter the integers, the plus and minus signs are used only for the operations of addition and subtraction. Notation for signed numbers represents a real problem for many students. For example, the sum of 3 and negative 7 can be written as $3 + (-7)$ or as $3 + {}^{-}7$. The latter form might be clear in a printed book, but may be obscure in handwritten form. The use of parentheses is awkward, especially in expressions already involving parentheses.*

MODELS FOR THE OPERATIONS

Two models are popular for helping students understand how the four operations ($+$, $-$, $\times$, and $\div$) work with the integers. One model consists of counters in two different colors, one for positive counts and one for negative counts. Two counters, one of each type, cancel each other out. Thus if blues are positive and reds are negative, five blues and seven reds is the same as two reds, each representing negative two (Figure 20.8). It is important with this model that students understand that it is always possible to add to or remove from a pile a pair consisting of one positive and one negative counter without changing the value of the pile. (Intuitively, this is like adding equal quantities of debits and credits.) The actions of addition and subtraction are the same as for whole numbers; addition is joining or adding counters, and subtraction is removing or taking away counters.

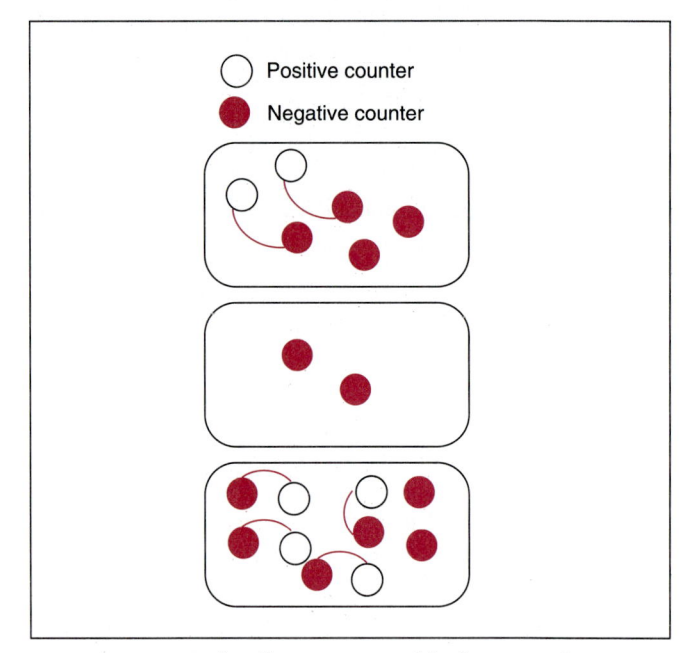

FIGURE 20.8: *Each collection is a model of negative 2.*

The other commonly used model is the number line. It is a bit more traditional and mathematical, and yet many find it somewhat confusing. The football-field model provides an intuitive background. Positive and negative numbers are measured distances to the right and left of zero. It is important to remember that signed values are *directed distances* and not points on a line. The points on the number

---

*The Comprehensive School Mathematics Program (CSMP) uses a chevron over a numeral to represent a negative number. Thus negative four is written $\hat{4}$, and $3 - (-7)$ becomes $3 - \hat{7}$. This notation avoids much of the confusion that students have, especially with the notation for subtraction. While you could easily adopt this notation, it is unlikely that it will ever be commonly used.

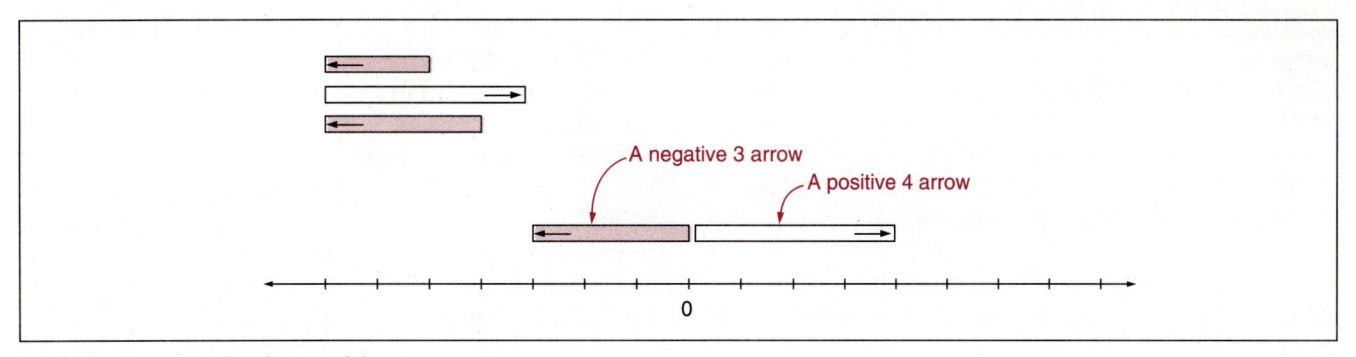

FIGURE 20.9: *Number line model*

line are not models of integers; the directed distances are. To emphasize this for students, represent all integers with arrows, and avoid referring to the coordinates on the number line as numbers. Poster board arrows of different whole-number lengths can be made in two colors, blue pointing to the right for positive quantities and red to the left for negative quantities (Figure 20.9). The physical arrows help students think of integer quantities as directed

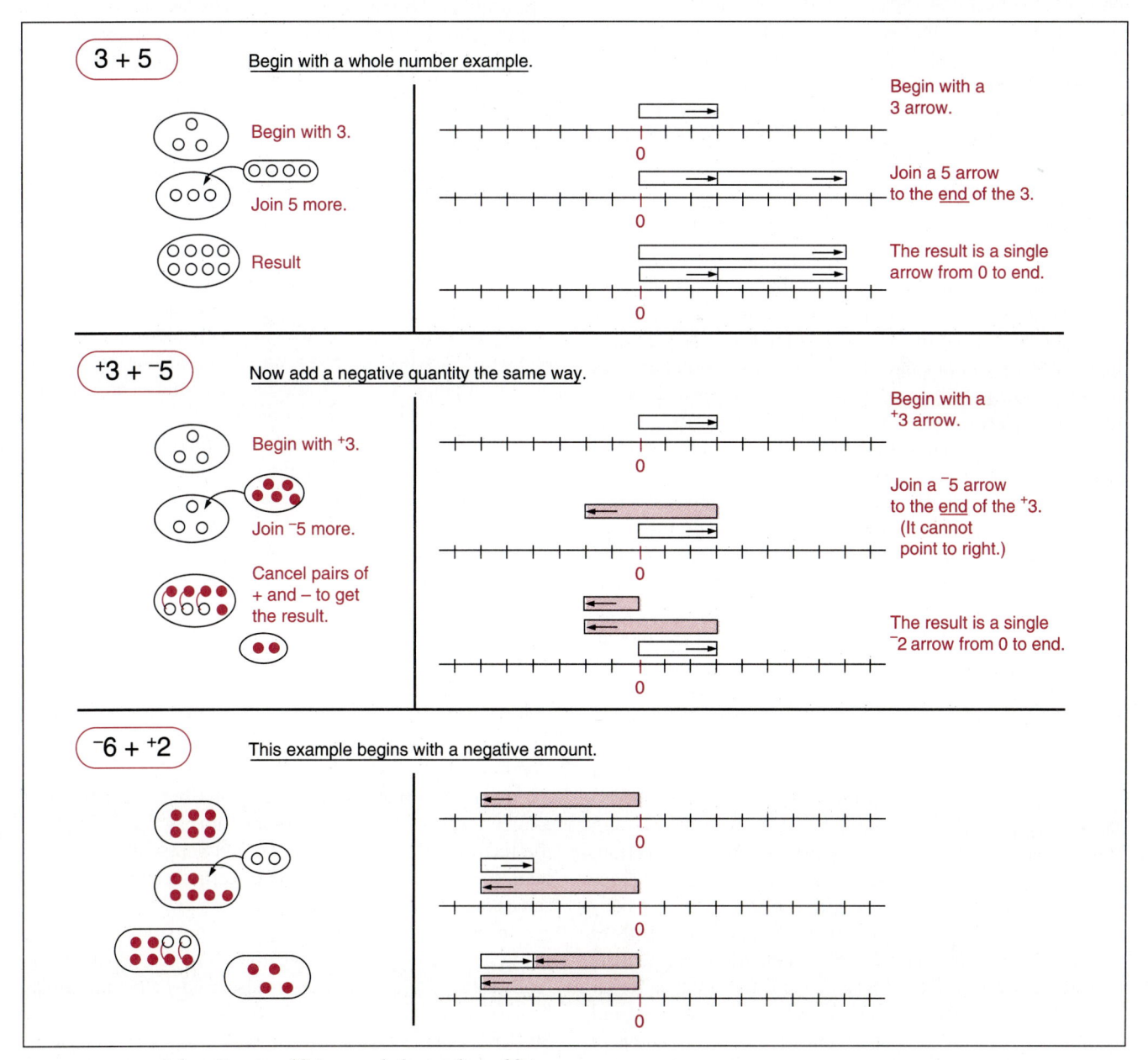

FIGURE 20.10: *Relate integer addition to whole-number addition.*

distances. A positive arrow never points left, and a negative arrow is never pointing to the right. Furthermore, each arrow is a quantity with both length (magnitude or absolute value) and direction (sign). These properties remain for each arrow regardless of its position on the number line.

Before discussing how these models are used to explain the operations, it is worth pointing out that the rules (procedural knowledge) for operations on the integers are generally easier to simply give to students than are explanations with a model. The conceptual explanations do not make the rules easier to use, and it is never intended that students continue to think in terms of these models as they practice integer arithmetic. Rather, it is important that students do not view the procedural rules for manipulating integers as being arbitrary and mysterious. Here then is a case where we must remember to make students responsible for the conceptual knowledge. If we emphasize only the procedural rules, there is little reason for students to attend to the conceptual justifications. In your discussions with students, do not be content to get right answers, but demand explanations. You might even try *giving* students the answers and having them explain with a model why they must be so.

ADDITION AND SUBTRACTION

Adding or subtracting integers with the models is straightforward and analogous to the corresponding debit/credit model or the football model. Since middle-school or junior-high students will not have used counters or number lines for some time, it would be good to begin work with either of these models using positive whole numbers. After a few examples to help them get familiar with the model for addition or subtraction with whole numbers, have them work through an example with integers using exactly the same reasoning. Remember, the emphasis should be on the rationale and not how quickly students can get correct answers.

In Figure 20.10 (see p. 413) several examples of addition are modeled, each in two ways: with positive and negative counters and with the number line and arrow model. First examine the counter model. After the two quantities are joined, any pairs of positive and negative counters cancel each other out, and students can remove these making it easier to see the result.

To add using the arrow model, note that each added arrow begins at the point end of the previous arrow. If you help students with the analogy between these arrows and the football situation, when the arrows change direction, that is like the ball changing teams. Addition is the advance of a team from the previous position. In the $^+3 + {}^-5$ example, the positive arrow (+ team) starts at zero and ends at positive 3. From that point, the negative arrow begins (the − team takes over and *advances* toward the negative direction). The result, then, is an arrow beginning at zero and ending where the second arrow ended. The same change of direction (change of ball possession) takes place in the $^-6 + {}^+2$ example. If a negative were added to a negative, the arrows would each go toward the left or negative (no change of possession) just as there was no change of direction for the 3 + 5 example.

Subtraction is interpreted as "remove" in terms of the counter model and "back up" in terms of the arrow model. In Figure 20.11, for $^-5 - {}^+2$, both models begin with a representation of $^-5$. In order to remove two positive counters from a set that has none, a different representation of $^-5$ must first be made. Since any number of neutral pairs (one positive, one negative) can be added without changing the value of the set, two pairs are added so that two positive counters can be removed. The net effect is to have more negative counters. This is like removing credits from your ledger if you are already in debt. The result is to leave you further in debt. A similar change in the representation of the beginning amount is always necessary when you need to subtract a quantity of a different sign.

With the number line and arrow model, subtraction means to back up or to move in the opposite direction. Using the football-field analogy, either team moves backwards when they are penalized or lose yardage. They move in the opposite direction from their own goal. In the example of $^-5 - {}^+2$, the first arrow ends at $^-5$. If the $^+2$ were to be added, it would be in the dotted position, ending at $^-3$. But it backs up instead. The result of the operation is an arrow from zero to the back end of the $^+2$ arrow. In the football analogy, the ball is first at the $^-5$-yard line, changes hands to the positive team that proceeds to lose two yards, leaving the ball at the $^-7$ yard line. In the second example ($^-4 - {}^-7$) the ball does not change hands, but the negative team loses yardage.

You want your students to draw pictures to accompany computations that they do with integers. Set pictures are easy enough. They may consist of X's and O's, for example. For the arrow model, there is no need for anything elaborate either. Figure 20.12 (p. 416) illustrates how a student might draw arrows for simple addition and subtraction exercises without even sketching the number line. Directions are shown by the arrows, and magnitudes are written on the arrows. For your initial modeling, however, the poster board arrows in two colors will help them see that negative arrows always point left, and that addition is a forward movement for either type of arrow while subtraction is a backward movement.

An effort is usually made to see that $^+3 + {}^-5$ is the same as $3 - {}^+5$ and that $^+2 - {}^-6$ is the same as $^+2 + {}^+6$. With the method of modeling addition and subtraction described here, these expressions are quite discernible and yet have the same result as they should have.

On graphing calculators, these expressions are entered using a separate key for "negative" and the usual key for "subtraction." The difference is also evident in the display. The redundant upper-case plus signs are not shown. Students can see that $3 + {}^-5$ and $3 - 5$ each result in $^-2$, and $3 - {}^-5$ and $3 + 5$ are also alike.

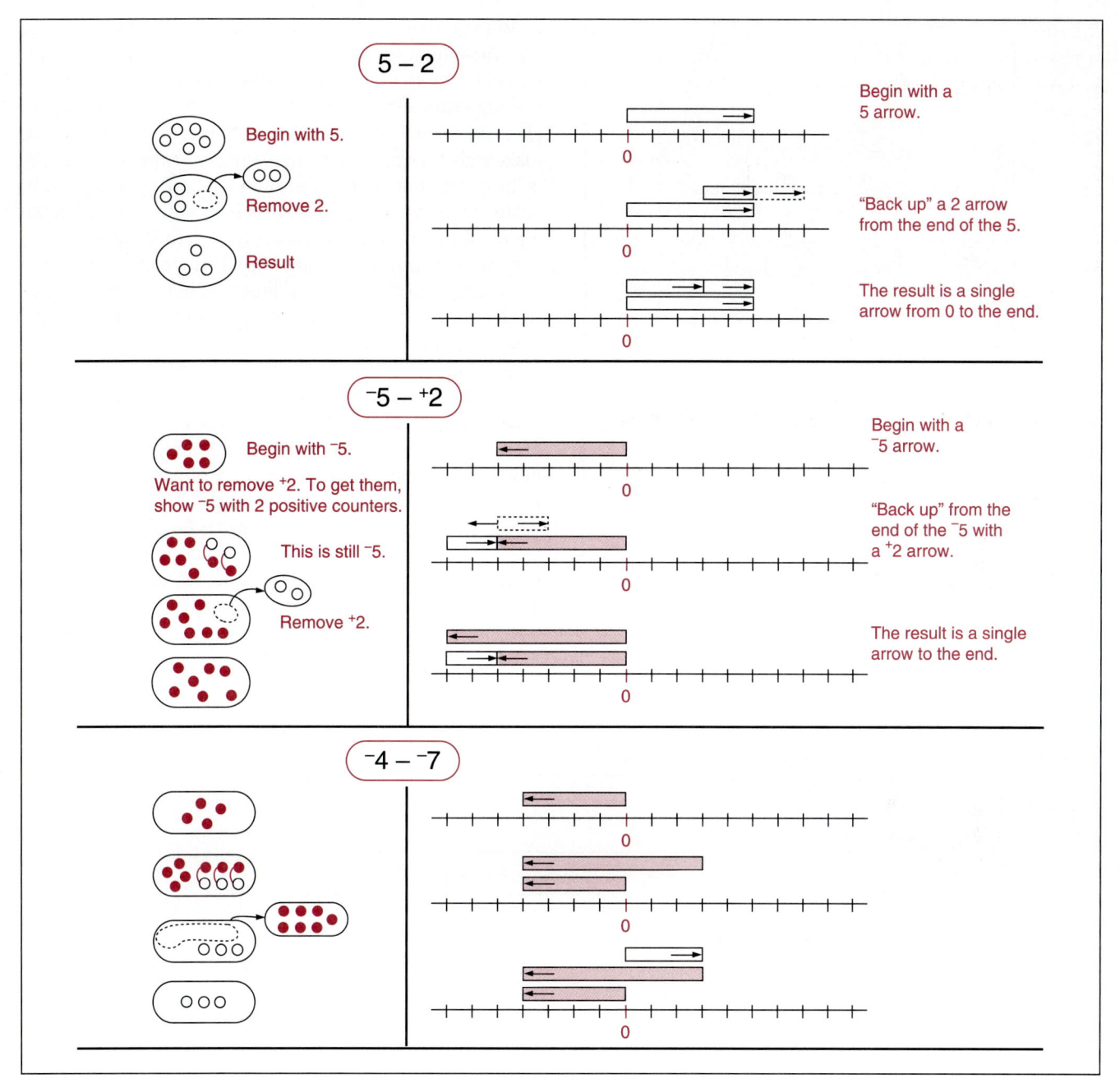

FIGURE 20.11: *Integer subtraction is also related to whole numbers.*

MULTIPLICATION AND DIVISION

Multiplication of integers should be a direct extension of multiplication for whole numbers, just as addition and subtraction were connected to whole-number concepts. We frequently refer to whole-number multiplication as repeated addition. The first factor tells how many sets there are or how many are added in all, beginning with zero. This translates to integer multiplication quite readily when the first factor is positive, regardless of the sign of the second factor. The first example in Figure 20.13 (p. 416) illustrates a positive first factor and a negative second factor.

What could the meaning be when the first factor is negative, as in ⁻2 × ⁻3? If the first factor positive means repeated addition (how many times added to 0), the first factor negative should mean repeated subtraction (how many times subtracted from 0). The second example in Figure 20.13 illustrates how multiplication with the first factor negative can be modeled. The success of your students in understanding integer multiplication depends on how well they understood integer addition and subtraction. Notice that there really are no new ideas, only an application of addition and subtraction to multiplication.

The deceptively simply rules of "like signs yield positive products" and "unlike signs negative products" are quickly established. However, one more time, it is not as important that your students be able to produce answers correctly and skillfully but that they be able to supply a rationale.

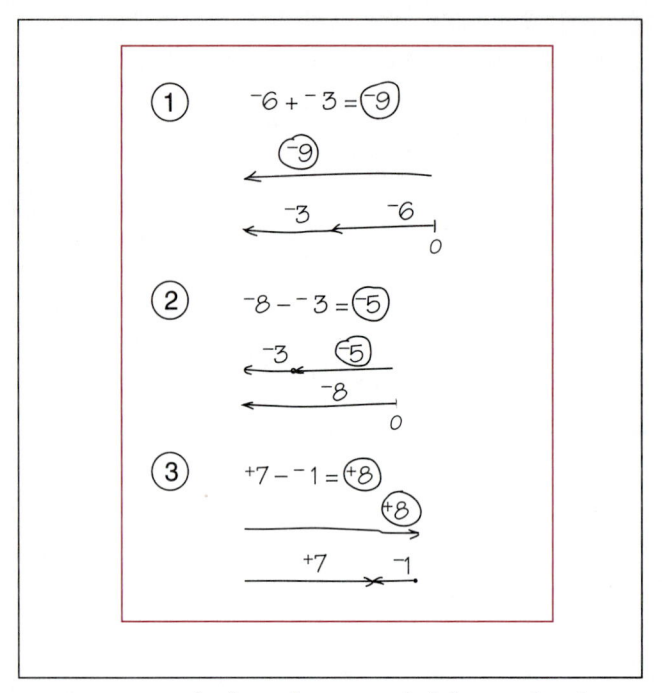

FIGURE 20.12: *Students do not need elaborate drawings to think through the number-line model.*

With division of integers it is again a good idea to explore the whole-number case first. Recall that 8 ÷ 4 with whole numbers has two possible meanings corresponding to two missing-factor expressions: 4 × ☐ = 8 asks "Four sets of ? make eight?" while ☐ × 4 = 8 asks "How many fours make eight?" Generally, the measurement approach (☐ × 4) is the one used with integers, although both concepts can be exhibited with either model. It is helpful to think of building the dividend with the divisor from zero in the same way that we fill units into an amount when we measure it. The first example in Figure 20.14 illustrates how the two models work for whole numbers. Following that is an example where the divisor is positive, but the dividend is negative. How many sets of $^+2$ will make $^-8$ or ? × $^+2$ = $^-8$. With the set model, if we try to add positive counters, the result will be positive, not negative. The only way to use sets of $^+2$ to make $^-8$ is to remove them from 0. This means we must first change the representation of zero as illustrated. For the arrow model, consider how $^+2$ arrows can be placed end to end to result in a distance eight to the left of zero. Starting at zero, the arrows must be backed up. In both models, the arrows are being repeatedly subtracted or added a negative number of times. Try now to model

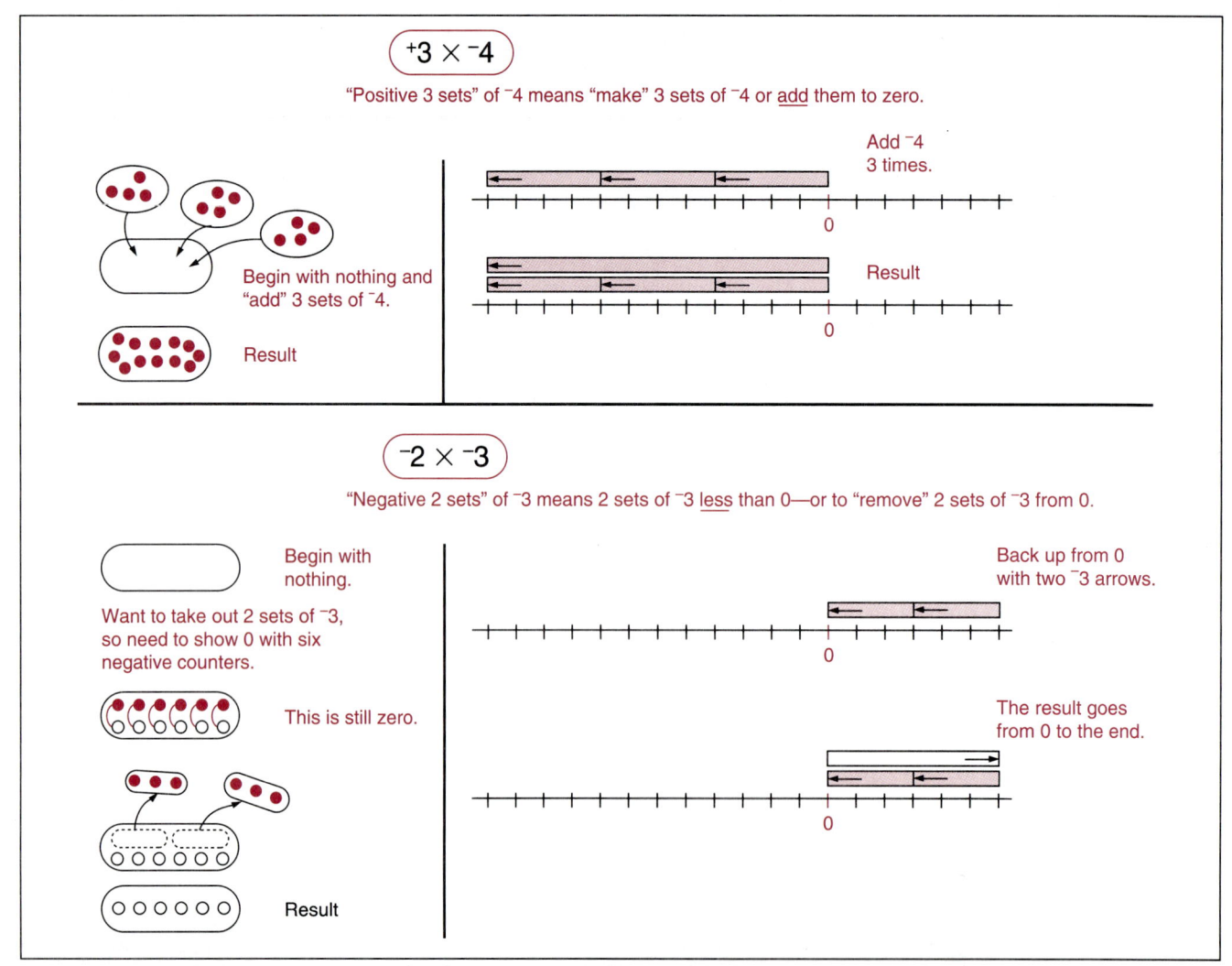

FIGURE 20.13: *Multiplication by a positive first factor is repeated addition. Multiplication by a negative first factor is repeated subtraction.*

$^+9 \div {^-3}$ using both models. The approach is very similar to this example. Then try a negative divided by a negative. That case is much easier to understand. However, the entire understanding of integer division rests on a good concept of a negative first factor for multiplication and a knowledge of the relationship between multiplication and division.

There is no need to rush your students on to some mastery of use of the models. It is much better that they first think about how to model the whole-number situation and then figure out, with some guidance from you, how to deal with integers.

DEVELOPING CONCEPTS OF VARIABLE

If mathematics is the science of pattern and order, one of the most powerful ways we have of expressing that pattern and order is with variables. With variables we can begin to use mathematical symbolism as a tool to think with as well as physical objects and drawings. But if variables are to be included with "thinker toys" and are to become powerful means of expressing ideas, then we must help children construct clear ideas of what variables are all about.

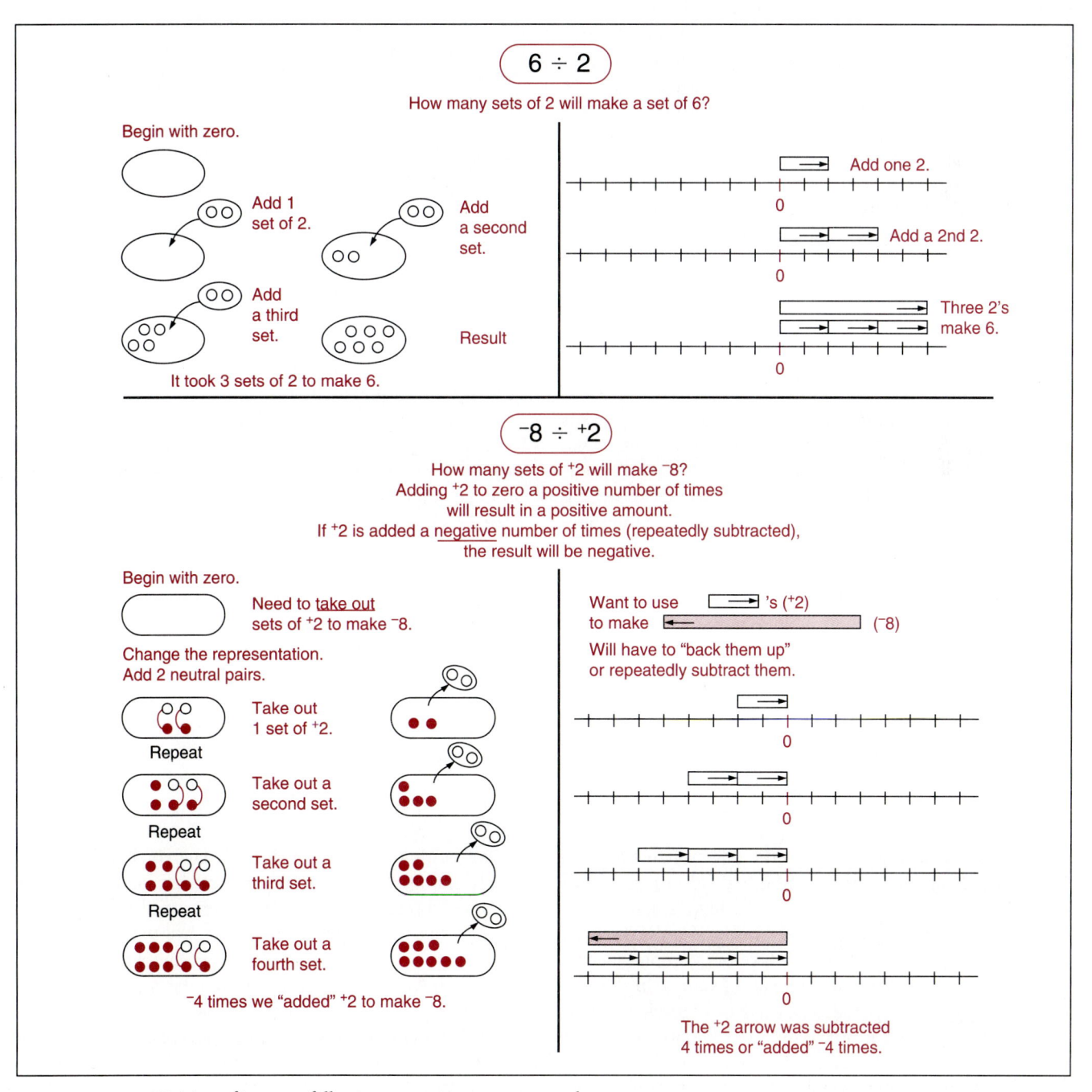

FIGURE 20.14: *Division of integers following a measurement approach*

MISUNDERSTANDINGS

Even though students are exposed throughout the elementary years to boxes and letters in arithmetic expressions, studies indicate that most children have a very narrow understanding of variable (for example, see Booth, 1988; Chalouh & Herscovics, 1988). The most common interpretation is that a letter stands for a particular number. Wagner (1981) found many 12- to 17-year-old students believe that $7w + 22 = 109$ and $7n + 22 = 109$ will have different solutions. Some students might accept L for length but would have difficulty using the letter r instead. Some assume that the letter represents an object rather than a numeric value. For example, one child explained that $8y$ could possibly mean "8 yachts or 8 yams" (Booth, 1988). Notation used with variables only compounds the difficulties. For example, in algebra, $3n$ means "three times the value of n," but in arithmetic 37 means "three tens and seven ones." So in the expression $5n$, if n is 2, many students will interpret the result as 52.

That these and other difficulties with variables exist suggests that we should take extra care to help students develop appropriate meanings for variable as a readiness for algebra and the many places in mathematics where variables are commonly used.

MEANINGS AND USES OF VARIABLES

Some of the difficulties with the notion of variable stems from the fact that variables are used in different ways. Meanings of variables vary with the way they are used. Usiskin (1988) identified four uses and corresponding meanings for variable. Of these, three meanings commonly occur in middle-school mathematics.

1. **A specific yet unknown quantity**

    Example: If $3x + 2 = 4x - 1$, solve for x.

2. **A representative from a range of values on which other values depend**

    Example: $y = 3x - 4$ or $p = \frac{1}{t}$.

3. **As a pattern generalizer**

    Example: For all real numbers, $a \times b = b \times a$ or $N + 0 = N$.

The first meaning is closest to the one students tend to develop during the early elementary school years. In expressions such as $3 + \square = 8$, the $\square$ has a single correct value. When moving to the other two meanings, the variables are not unknowns but may take on or represent any value from among a specified domain of values. In the second meaning, the variables are used to express relationships. In the equation $A = L \times W$, there are no numbers at all, but we can talk about what happens to the area for example, as the length doubles, then triples, and so on.

Here, as Usiskin puts it, the "variables vary." In the third meaning, the letters again represent any value in a domain, but the purpose is to illustrate a property or pattern. Things that are known from arithmetic are generalized.

Students need to have experiences with these different meanings and uses of variables and become comfortable with them in different contexts. It is neither necessary nor appropriate to expect students to articulate different meanings or definitions of "variable." At the same time, it is also not appropriate to offer definitions that are inaccurate or incomplete. For example, it is overly narrow to say "a variable is an unknown quantity" or "a variable stands for a number." These "definitions" relate almost exclusively to the first use of a variable as representing an unknown quantity, but hinder the understanding of the variable as used in the second meaning. In the equation $y = f(x) = 3x + 2$, the y and the x represent numbers, but not specific ones, and there is no unknown to be found.

EXPRESSIONS AS QUANTITIES

Consider the following expressions involving variables:

$$3B + 7 \quad \text{and} \quad 3B + 7 = B - C$$

The first is an expression of a quantity. It may be a specific quantity, or it may be a whole range of quantities related to a range of values of B. In the second expression, the same quantity is set equal to another quantity involving variables. The equal sign means that the quantity on the left is the same as the quantity on the right, even though they are expressed with different letters and different operations. In order to interpret these expressions in this way, students must also interpret similar arithmetic expressions with only numbers in the same way. Expressions such as $3 + 5$ or 4×87 must be understood as single quantities, and the equal sign ($=$) must be understood as a means of showing that two quantities represented are the same.

Unfortunately, while we use both of these ideas and assume all through elementary school that students understand them, a large majority of students do not completely comprehend them (Herscovics & Kieran, 1980; Kieran, 1988; Behr, Erlwanger, & Nichols, 1976). Students tend to look on expressions such as $3 + 5$ and 4×87 as commands or things to do. The $+$ tells you to add, and students think of adding as a verb. The idea that $5 + 2$ is another way of writing 7 is not considered. The equal sign is commonly thought of as an operator button, something like pressing $=$ on a calculator. As students read left to right in an equation, the $=$ tells them, "Now give the answer." In a similar sense, students think of $=$ as a symbol that is used to separate the question or problem from the answer.

The following simple activities are suggested as ways to help students with these basic concepts. They each initially use only numbers but can easily be extended to include variables.

20.8 NAMES FOR NUMBERS

Challenge students to find different ways to express a particular number, for example 10. At first suggest that they use only + or only − expressions, such as 5 + 5 or 12 − 2. Later, devise more challenging tasks such as "How many names for eight can you find using only numbers less than 10 and using at least three different operations?" One solution is (5 × 6 + 2) ÷ 4. Make it a point to emphasize that each expression is a way of representing a number. Restrictions on which operations and how many and which numbers can be used can produce real challenges. Groups can compete with each other for the most expressions in a specified time.

20.9 TILT OR BALANCE

On the board draw a simple two-pan balance. In each pan write a numeric expression and ask which pan will go down, or whether it will balance [Figure 20.15(a)]. Challenge students to write expressions for each side of the scale that will make it balance. For each, write a corresponding equation or inequality. Soon the scale feature can be abandoned, and students can simply write equations according to the directions given. For example, use at least three operations on each side and do not use any number twice.

The last two activities are clearly related. To understand the balance or equation idea requires an understanding that the expressions in each pan or side of the equation represent quantities.

After a short time with these two activities, variables can be added to each. Instead of names for numbers, have students write expressions for quantities such as the following:

two more than three times a number (2 + 3 □
   or 2 + 3 × N)

any odd number (2N + 1)

the average age in this class; use S for the sum of our ages $\left(\frac{S}{23}\right)$

the average age of the students in any class $\left(\frac{S}{n}\right)$

a number cubed less another number squared $(A^3 − B^2)$

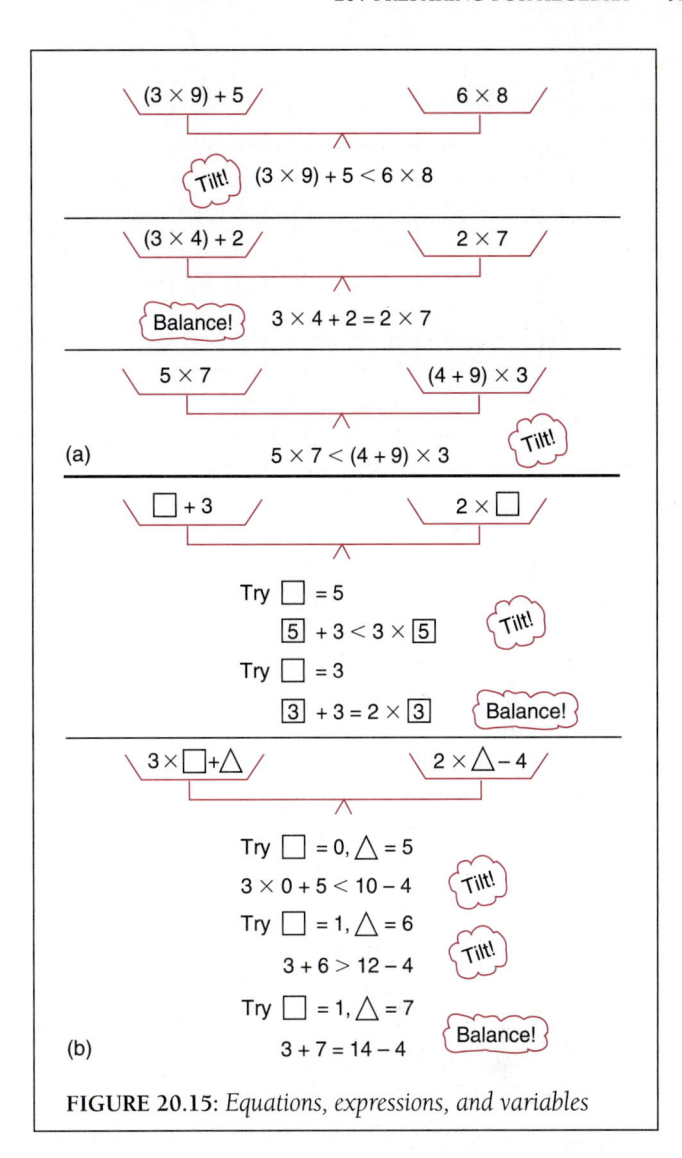

Notice in these examples that variables are used as both specific unknowns and as unspecified quantities.

FIGURE 20.15: *Equations, expressions, and variables*

(a)
- (3 × 9) + 5   6 × 8   Tilt!   (3 × 9) + 5 < 6 × 8
- (3 × 4) + 2   2 × 7   Balance!   3 × 4 + 2 = 2 × 7
- 5 × 7   (4 + 9) × 3   5 × 7 < (4 + 9) × 3   Tilt!

(b)
- □ + 3   2 × □
- Try □ = 5
- 5 + 3 < 3 × 5   Tilt!
- Try □ = 3
- 3 + 3 = 2 × 3   Balance!
- 3 × □ + △   2 × △ − 4
- Try □ = 0, △ = 5
- 3 × 0 + 5 < 10 − 4   Tilt!
- Try □ = 1, △ = 6
- 3 + 6 > 12 − 4   Tilt!
- Try □ = 1, △ = 7
- 3 + 7 = 14 − 4   Balance!

20.10 VARIABLES IN THE BALANCE

In the two-pan balance, have students write expressions in either or both pans that include variables. Use directions similar to the ones that were just listed. Students may then use a trial-and-error approach to find numbers that can be substituted for the variables to make the equation balance (true equation) or tilt (a false equation). Use calculators for tedious calculations [Figure 20.15(b)].

MORE EARLY EXPERIENCES WITH VARIABLES

Evaluating Expressions and Formulas

In Figure 20.16 (p. 420), examples of charts are shown. Each chart has a separate column for the variable and the

n	(n+2)	3(n+2)	3n+6
0			
2			
−1			
−2			

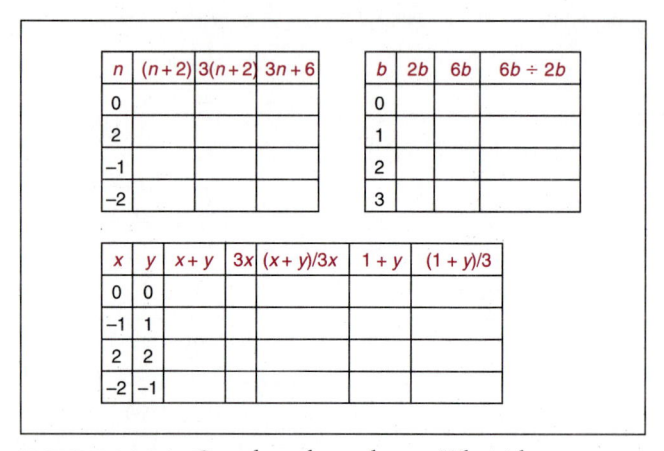

b	2b	6b	6b ÷ 2b
0			
1			
2			
3			

x	y	x+y	3x	(x+y)/3x	1+y	(1+y)/3
0	0					
−1	1					
2	2					
−2	−1					

FIGURE 20.16: *Complete these charts. What ideas can you discover?*

variable expressions across the top of the chart are all related. Students can evaluate the expressions and fill in the charts. In this type of activity, several important ideas about notation and properties of the number systems can be explored informally. For example, $4n$ and $4 \times n$ can be placed in separate columns to help students see this convention for expressing multiplication. Other ideas include

$8x - 3x$ with $5x$ (many students think $8x - 3x$ is 5)

$b(3 + b)$ with $3b$, b^2, and $3b + b^2$

$(3 + 2x) - x$ with $3 + x$

$\frac{y}{5}$ with $y \div 5$ (students may not realize these mean the same thing)

$-x$ and $(-1)x$

$3k + \frac{1}{k}$ with $(3k + 1)$ and $\frac{3k+1}{k}$

The entries in the charts can also be expressions that appear in geometric formulas or formulas from other areas such as science. As an example, while reviewing volume and surface area formulas, you might have students help you build a chart for a right circular cylinder with the following expressions:

$$r \quad h \quad \pi \quad 2\pi r^2 \quad \pi r^2 h \quad 2h\pi r$$

With two variables, you may want to have students hold one variable fixed for several values of the other. In the preceding example, observe the values of the various columns as r changes from 1 to 5 and h remains fixed at 10. Then change h to 20 and repeat.

Another variation with these tables is to enter values in the columns and have students determine the value of the variable. Notice with this variation two different uses of variable are now included in the same example: letting the value of the variable vary and also solving for a fixed value.

The graphing calculator will store values "in" a variable. For example, if you enter $6 \rightarrow X$ (store 6 in X), and then enter $3(X + 2)$, the result will be 24. If you have also stored 4 in A, then $5X - 2A$ will be evaluated as 22. The values stored in variables remain until they are changed.

◆ Solving Equations

Students can learn appropriate techniques for solving simple linear equations before they reach a formal algebra class. A good technique is to capitalize on the two-pan balance approach that was suggested earlier. With the balance, the idea of doing to one side what you do to the other can be developed meaningfully.

In Figure 20.17, a balance scale is set up with two numeric expressions. A box is drawn around both of the 5s. The task is to manipulate the numbers so that there is only one boxed 5 on the left. You can operate on numbers or on boxed 5s, but you may not combine the 5 in the box with any other numbers. Obviously, in order to keep the scale balanced, you must do the same things to both sides of the balance. At each step along the way, the balance can be confirmed.

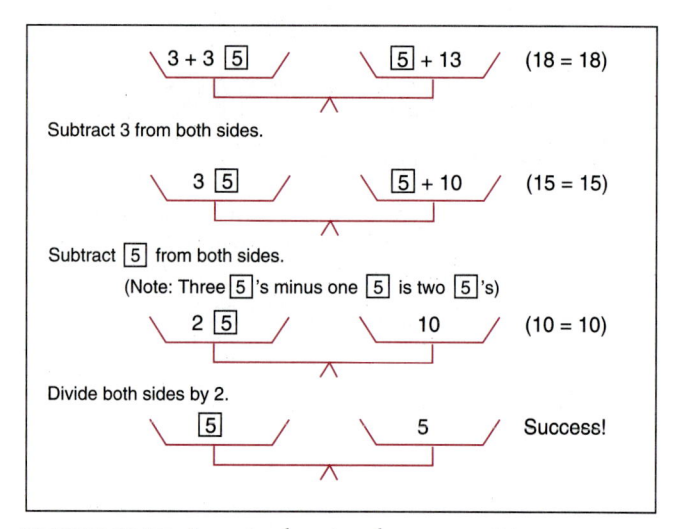

$3 + 3\boxed{5} \quad\quad \boxed{5} + 13 \quad (18 = 18)$

Subtract 3 from both sides.

$3\boxed{5} \quad\quad \boxed{5} + 10 \quad (15 = 15)$

Subtract $\boxed{5}$ from both sides.
(Note: Three $\boxed{5}$'s minus one $\boxed{5}$ is two $\boxed{5}$'s)

$2\boxed{5} \quad\quad 10 \quad (10 = 10)$

Divide both sides by 2.

$\boxed{5} \quad\quad 5 \quad$ Success!

FIGURE 20.17: *Learning how to solve an equation*

After one or two similar examples, put variable expressions in the balance pans and have students decide how to isolate the variable on one side. The other side should all be numeric.

Even after you have stopped drawing the balance, it is a good idea to refer to the scale or pan-balance concept of equation and the idea of keeping the two sides balanced. Operations on both sides of the equation should be done as illustrated in Figure 20.18, underneath the last equation. Division of both sides by a constant is best done with fraction notation.

A goal in these activities is to promote the use of inverse operations to solve an equation. Teachers are frequently frustrated when students prefer to solve simple equations such as $3N + 2 = 11$ by inspection or trial and error. One effective tactic is to use large numbers and avoid nice whole-number answers. For all computations, have students use a calculator to avoid tedium. For example, the solution process for $3.68N + 47.5 = 16$ is the same as for $3N + 2 = 11$.

20 / PREPARING FOR ALGEBRA   421

(a)

$4 - 6x \quad = \quad 3(1 + x)$

Subtract 4 from both sides and multiply right-hand expression.

$-6x \quad = \quad 3 + 3x - 4$

Subtract $3x$ from both sides.

$-9x \quad = \quad -1$

Divide both sides by -9.

$x \quad = \quad \frac{1}{9}$

Check:

$4 - \frac{6}{9} \quad = \quad 3(1 + \frac{1}{9})$

Both sides $= 3\frac{1}{3}$.

(b)

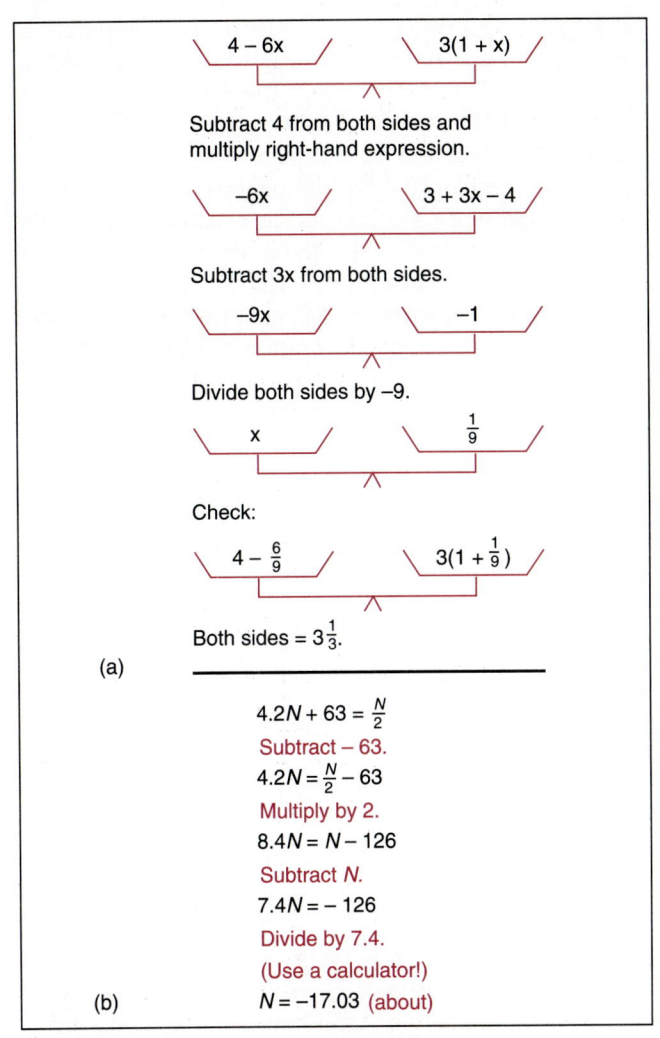

$4.2N + 63 = \frac{N}{2}$
Subtract -63.
$4.2N = \frac{N}{2} - 63$
Multiply by 2.
$8.4N = N - 126$
Subtract N.
$7.4N = -126$
Divide by 7.4.
(Use a calculator!)
$N = -17.03$ (about)

FIGURE 20.18: *Scale balance with variable*

GRAPHS AND RELATIONSHIPS

In the past, a significant amount of the time spent in algebra has been devoted to learning how to graph equations. Graphs approached in this manner require that students learn a lot of rules and memorize equation forms that go with certain types of graphs. In fact, many algebra curricula continue with this approach. There is compelling evidence that students do not have a clear understanding of how equations, graphs, and function charts all are related. If so, what good was the graph? Graphs are pictures of relationships. Sometimes a graph is a graph of a function relationship. But before worrying about equations and functions and how to graph assorted classes of equations, students should develop some basic ideas about how graphs illustrate relationships.

GRAPHS OF REAL SITUATIONS

One approach that can be taken to develop an idea of graphs as pictures of relationships is to begin with real-world, meaningful relationships.

◆ **Cutting Lawns**

Consider the following situation.

> Yolanda's mother agrees to loan her the price of a new lawn mower so that Yolanda can earn money cutting grass over the summer. The lawn mower costs $230. Yolanda has lined up some jobs and has figured that she can average $18 per lawn after paying for gasoline. She thinks she will be able to cut about 5 lawns per week. When will Yolanda be able to pay her mother back? How much profit can she make over the summer?

One place to begin with this is to construct a chart that shows Yolanda's profit in terms of the number of lawns she cuts. The chart might begin with the beginning—0 lawns cut—and go in increments of 5 lawns at a time. Note that Yolanda's profit at the outset is a negative number: -230 dollars. Students should determine about 4 or 5 entries. Next consider a way to show these values on a graph. The graph needs two axes, one for the number of lawns and one for profit. It is reasonable that the horizontal axis be used to represent the number of lawns and the vertical axis represent profit. Where on the graph would you show that Yolanda had cut 5 lawns resulting in a profit of -140 dollars? Solicit ideas. The idea of plotting points using coordinates is not difficult. The important thing in this example is that students see that the point $(5, -140)$ represents five lawns cut resulting in a profit of negative $140. Plot other points in the same manner (See Figure 20.19).

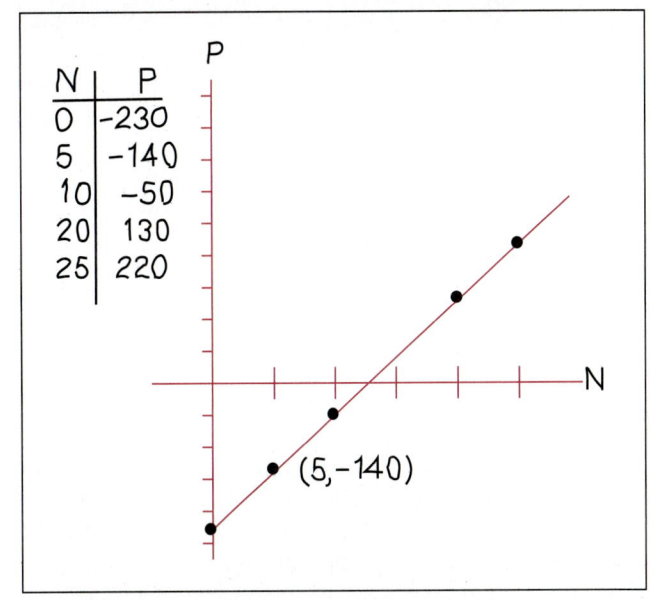

N	P
0	-230
5	-140
10	-50
20	130
25	220

FIGURE 20.19: *Beginning to show the information in the lawn-cutting situation.*

The next step is to see if students can develop a general rule for Yolanda's profit in terms of the grass cut. Examine the computations students used to fill in the chart. Write them in equation form. For example:

14 lawns: 15 × 18 − 230 = 40

20 lawns: 20 × 18 − 230 = 130

What about a value that is not yet in the table; perhaps 13?

13 lawns: 13 × 18 − 230 = 4

Notice that the last equation means that Yolanda will be $4 ahead after 13 lawns. But more importantly, what is the pattern in the equations? What expression will give us Yolanda's profit in terms of lawns cut? A reasonable ideas is $N \times 18 - 230$. Put some values in for N, and get other values for profit (P). Soon the graph will have lots of points, and they will all be on a straight line. Does it make sense to connect these points? How can you use the graph to determine some values that are not yet in the chart? Do these values satisfy the equation $P = N \times 18 - 230$? How can you use the graph to figure out how many lawns Yolanda will have to cut to make $200? How can you use the equation to find out how many lawns must be cut to make $300? (A try-and-adjust strategy with the calculator is one approach. Others may want to solve the equation for N.) Does it make sense for the line to extend to the left of the vertical axis? What would that mean? Should the line extend indefinitely? (Yolanda will probably work for only about 10 or 12 weeks.) What can you say about Yolanda's profit in terms of the lawns she cuts? (The more lawns she cuts, the more profit she makes.) What happens to the graph as the number of lawns increases?

Notice in all of the discussion above that all questions are posed in terms of the lawn-cutting scenario. Questions can be asked about the number chart, the equation, and the graph, all in terms of lawns and profit. This is designed to reinforce the notion that the graph, the equation, and the chart are all nothing more than different representations for the same relationship. The graph is a picture of that relationship.

What would happen to the graph if the points on the N-axis or horizontal axis represented the number of weeks that Yolanda worked? If she cuts exactly five lawns per week, the graph remains the same. What would the graph look like if Yolanda took a two-week vacation with her parents between the fifth and seventh weeks?

Now try to draw the following picture: Show Yolanda's profit in terms of the number of weeks she works. For the first 5 weeks she cuts 5 lawns per week. Then she goes on vacation for 2 weeks. Then she returns to cutting lawns, but adds 2 more lawns each week. If summer lasts 13 weeks, will Yolanda be able to make up for missing the two weeks by finishing out the summer with 7 lawns each week? Can you use the graph to answer your question?

◆ A General Scheme

The suggestion is that a situation be presented and discussed with the class to get some general idea of what is going on. Next, students develop a chart to begin to get some data and a sense of a pattern to the computation. Points are plotted, and a curve or line is drawn connecting the points (if that makes sense). Use letters that the students suggest. Do not be a slave to the use of X and Y. Return to the chart and the computations to see if an equation can be developed from the pattern that is observed. Use the graph to get solutions to the equation, and use the equation to get points on the graph. Be sure to work from each of the variables to the other. (In the case of the lawns, this was number of lawns used to get profit and profit used to determine the corresponding number of lawns.)

If it is reasonable or it makes sense, discuss modifications in the situation. These may result in other graphs with or without the benefit of new equations. (In the lawn mowing case, the number of lawns cut was changed to number of weeks. The vacation and the increased rate of cutting was introduced without benefit of equations or charts).

Here are a few suggestions for situations.

- Mr. Calloway wants to build a fenced pen against one side of his shed. The shed is 15 feet long, and he wants to use the full side of the shed. The pen is to be in the shape of a rectangle with two sides being 15 feet long. How much fence will he have to buy if he knows how long the other two sides of the pen will be (side vs. fence length)? Add in a gate that is 3 feet long and costs $32. If fencing is $4.25 per foot, rethink the problem in terms of side versus cost. You can also discuss area and side length.

- Mark is an avid cyclist. He can average 17 miles per hour for about 4 hours. He leaves home and travels for $2\frac{1}{2}$ hours at this rate, rests for lunch for a $\frac{1}{2}$ hour, and then starts home. What is his distance from home at any given time? Suppose he goes faster for 1 hour and slower for 2 more hours and then returns home. Suppose he has a flat tire and has to stop for 15 minutes to repair it. How fast will he have to go in order to return home at the same time as scheduled including the same lunch break? Does it make any difference where the breakdown occurs?

- Pleasant's Hardware buys widgets for $4.17 each, marks them up 35% over wholesale, and sells them at that price. Relate widgets sold to total income. Consider profit instead of income. Add in a sale using a reduced price.

GRAPHS WITHOUT EQUATIONS

In the last section, the idea was to begin to think about a graph as an expression of a relationship. The relationship was developed from a situation that had numbers, and eventually an equation and a chart were developed. It is also useful to sketch graphs that show a relationship but without using any numbers. For example, try to sketch a graph for each of the following situations. No numbers are to be used and certainly no formulas.

a. The temperature of a frozen dinner from 30 minutes before it's removed from the freezer until it is removed from the microwave and placed on the table. Consider time = 0 to be the moment the dinner is removed from the freezer.

b. The value of a 1970 VW Beetle from the time it was purchased to the present. It was kept by a loving owner and is in top condition.

c. The level of water in the bathtub from the time you begin to fill it to the time it is completely empty after your bath.

d. Profit in terms of number of items sold.

e. The height of a baseball in terms of time from when it is thrown straight up to the time it hits the ground.

f. The speed of the baseball in the previous situation.

In Figure 20.20 are six graphs that match these situations but not in order. Can you find your sketches in Figure 20.20? Or, can you match the graphs in the figure with the situations above?*

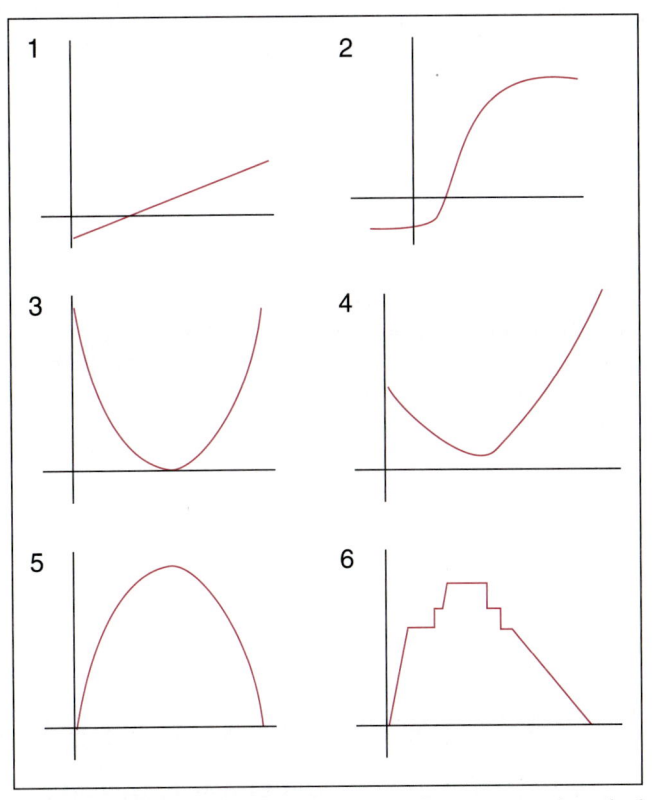

FIGURE 20.20: *Match each graph with the situations described. Talk about what is happening in each case.*

BEGINNING FUNCTIONAL IDEAS

For each of the situations found in the last section, it is possible to discuss the relationships observed using functional language and even functional notation without ever having an equation. For example, it is reasonable to say, "Profit is a function of sales." We can write this as follows: $P = F(S)$. Here P stands for profit and S for sales. The letter F is the name of the function or the relationship between profit and sales. Furthermore, we can describe this functional relationship in words: "As sales increase, profit increases. With no sales, profit is negative or there is a loss. There is a break-even point, and then profit continues. The gain in profit is steady or uniform."

Try a similar description and symbol definition for the other relationships.

MORE ABOUT FUNCTIONS

The concept of function is one of the big ideas or common threads that pervade a large portion of mathematics. While the concept of function is critical to the study of mathematics, it is not an easy concept for students to understand completely, even in the ninth-grade algebra course (Markovits, Eylon, & Bruckheimer, 1988). The informal experiences discussed so far are designed to help students construct informal ideas beginning well before a regular algebra course. The intent is to develop a background of knowledge about relationships and how these can be represented both symbolically and graphically before definitions and formal ideas are introduced. Formal function concepts can be added to or built on these beginnings.

FUNCTION DEFINITION

A *function* is a rule of correspondence between two sets of elements such that each element in the first set corresponds to one and only one element in the second. The first set is called the *domain* of the function and the second set is referred to as the *range*. In the lawn-mowing example, profit is a function of the number of lawns being cut. Here the domain is the number of lawns that can be cut, and the range is the profit that is accumulated for that number of lawns. Thus there is one profit number associated with each possible number of lawns.

For the frozen-food example, if you tell me the time relative to taking the food from the freezer, the graph can tell the corresponding temperature. This can be done by drawing a vertical line from the specific time on the time-axis until it hits the graph. That point on the graph has two coordinates, (time, temperature). The second coordinate is the temperature, the range element associated uniquely with that particular time. The times are the domain elements, and the temperatures are the range elements. Consider how an arrow could be drawn from a point on the horizontal axis, up or down until it hits the graph, and then horizontally over from that point to the vertical axis. The arrow would be connecting two points, one from each axis. The graph would be the rule that determines what

---

*The graphs match the situations in the text as follows: 1–d, 2–a, 3–f, 4–b, 5–e, 6–c.

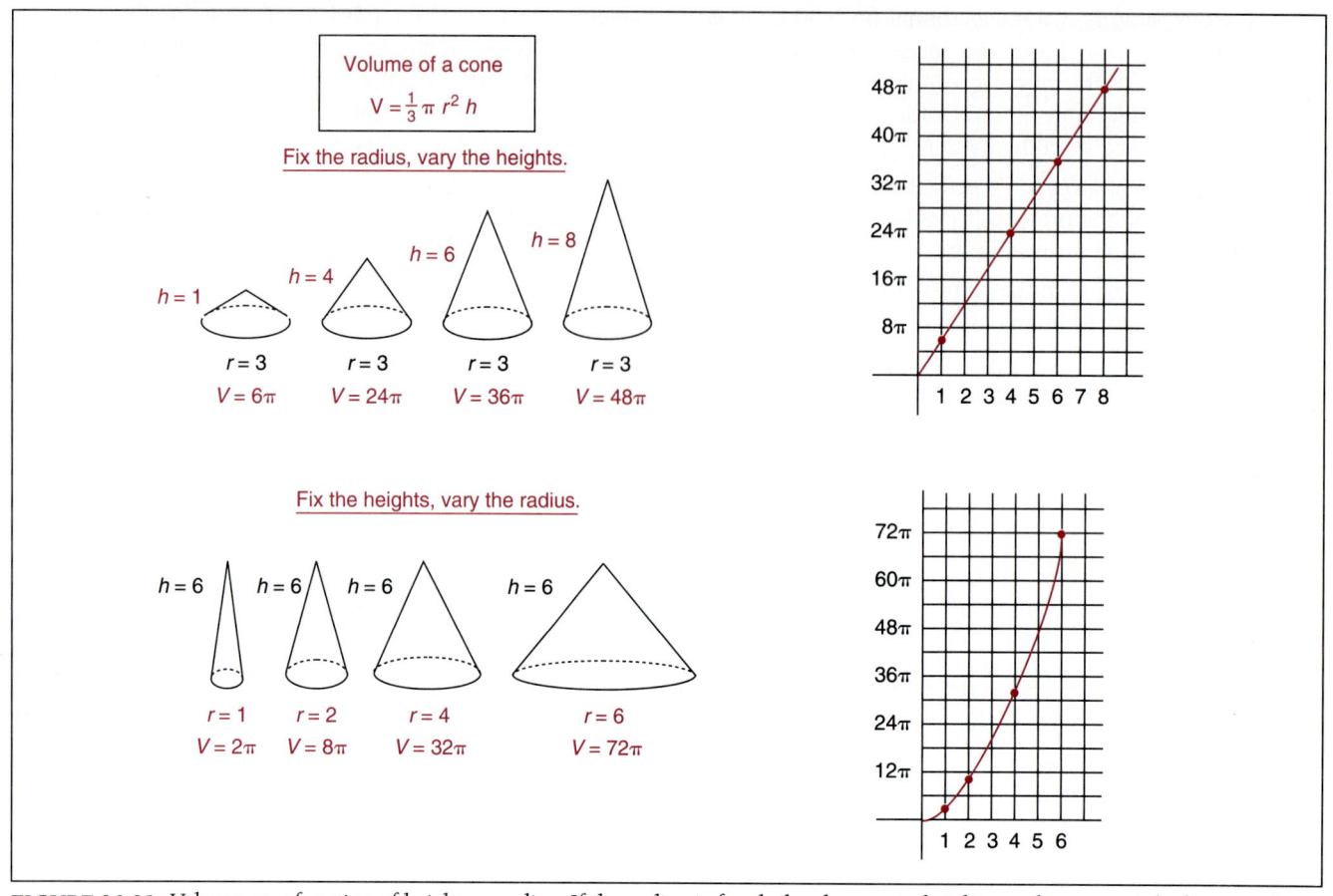

FIGURE 20.21: *Volume as a function of height or radius. If the radius is fixed, the changes in heights produce a straight-line graph, but if the height is fixed, the radius changes produce a curved line.*

points are connected. The graph is the relationship, the rule or function—in picture form.

An equation such as $Y = 3X + 4$ can define a function as we saw in the lawn example. Each value that is given to X determines a corresponding value of Y. Thus, the equation is a "rule of correspondence" that associates a value for Y with each value of X.

Recall the chart from the lawn-mowing example. The two columns of the chart create a set of pairs of numbers. Each number on the left side of the chart is associated by the rule of correspondence with one number on the right.

While there are other ways to think about and represent functions, those that have been presented here are the ones that are most common in algebra: graphs, charts, and equations. The important thing is for students to see how these three ideas are simply representations for this one idea. It is equally important for students to recognize functional relationships in real-world contexts and to be able to discuss these relationships as functions.

MORE EXAMPLES OF FUNCTIONS

Functions exist in all sorts of situations. In this section we will look at a few more examples of functions that students can explore.

Functions in Geometry and Measurement

In most geometric figures, in either two or three dimensions, a change in one dimension affects the measurement of another. This provides a good example of a functional relationship.

Consider any formula for measuring a geometric shape with which students are familiar. For example, $V = \left(\frac{1}{3}\right)\pi r^2 h$ is the formula for the volume of a circular cone. Figure 20.21 shows the graphs of volume as a function of height for a fixed value of the radius and as a function of the radius for a fixed value of height.

◆◆◆◆◆◆◆◆◆◆◆◆◆◆◆◆◆◆◆◆◆◆

20.11 FUNCTIONS FROM SHAPE MEASURES

Show the class a particular shape, either flat or solid. Let different groups select any two measures that are not directly related by a familiar formula. By systematically letting one of these measures change by making different drawings and measuring with rulers and protractors, they can observe the relationships between the two measures. Suggest that they make large drawings. The results can be graphed even if a formula cannot be found (Figure 20.22).

Draw a whole family of isosceles triangles on centimeter graph paper. Keep the height fixed. Let the base (between the equal sides) vary from one to 20. How will the angle at the top change?

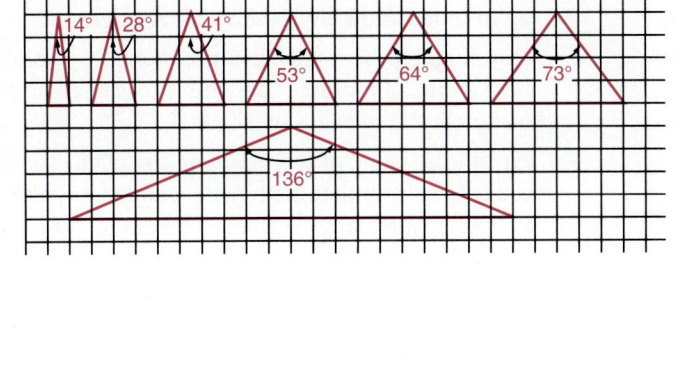

If the heights of these triangles are at least 10 cm, the angles can be measured suprisingly accurately. No formulas are required, and the graph paper makes the triangles easier to draw.

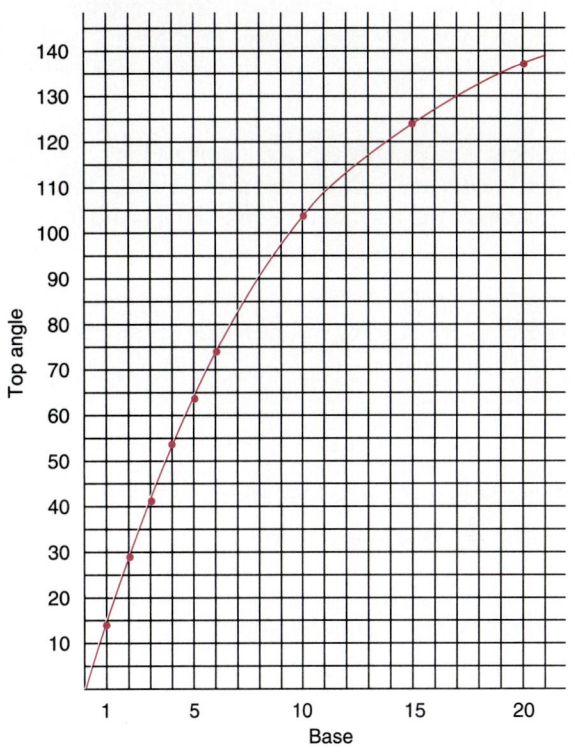

FIGURE 20.22: *One angle of an isosceles triangle as a function of the base*

Length of string	Number of swings in 10 seconds
2'	6
4'	$4\frac{1}{2}$
6'	4
1'	8
7'	$3\frac{1}{2}$
3'	5
$\frac{1}{2}$'	22

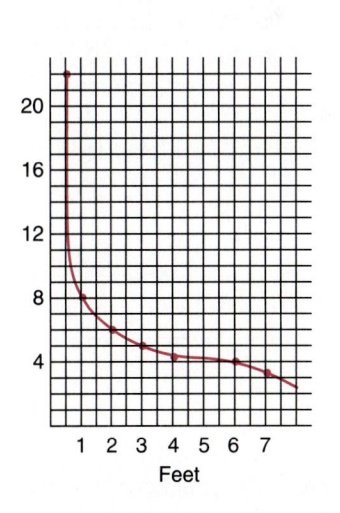

FIGURE 20.23: *Swings of a pendulum as a function of the length of the string*

◆ Functions from Real Data

Students can take measurements from an experiment and plot the data for two related measures on a graph. From this plot of points, they can draw a smooth line that approximates the points. This line can then be used to predict the results for other trials. It is not necessary that an equation for the curve be determined.

One example of an experiment that involves functional relationships is determining the number of swings of a pendulum in 10 seconds compared with the length of the string (Figure 20.23). Variations would be:

> The height of students in the school compared with the height of their head (or arm span). Get data from students in different grades to find different lengths. When two or more students have the same height, use their average head height.

> The time it takes a toy car to roll down a long inclined plane as compared to the angle of inclination or the height of the end of the board.

> The length of a shadow cast by a pole as compared to different times of day.

◆ Functions as General Solutions to Word Problems

Consider the following problem:

> Two out of every 3 students who eat in the cafeteria drink a pint of white milk. If 450 students eat in the cafeteria, how many gallons of milk are consumed?

In this problem there is a fixed number of students and a single answer to the problem. A more practical problem is to write an equation that will give the number of gallons of milk *as a function of* the number of students. For this problem the equation $G = (\frac{2}{3})S \div 8$ is a solution. (There are 8 pints in a gallon, and two-thirds of the students drink milk.)

The following problems lend themselves to similar solutions:

> Mr. Schultz pays $4.00 for a box of 12 candy bars, which he sells for 45 cents each. How much profit will he make on n boxes of candy bars?

> If each recipe of lemonade will serve 20 people, how many recipes are needed to serve n people? If it takes 3 cans of concentrate to make one recipe, how many cans should be purchased to serve n people?

To answer the second question, write two functions, one that outputs recipes for n people and another that outputs cans for recipes. Now substitute one function in place of the variable in the other.

◆◆◆◆◆ REFLECTIONS ON CHAPTER 20: WRITING TO LEARN

1. Distinguish between a prime and a composite number in terms of the dimensions of rectangles.
2. Why might it be valuable for students to examine the patterns of multiples on a hundreds chart and/or to explore the rules for divisibility?
3. Explain the value of the graphing calculator's ability to display a complete arithmetic expression (with or without a variable) on a single line and to evaluate it on another line.
4. How would you explain the difference between a rational and an irrational number to a middle-school student?
5. Use both the arrow model and the counter model to demonstrate the following:

   $^-10 + {^+}13 = {^+}3 \quad ^-4 - {^-}9 = {^+}5 \quad ^+6 - {^-}7 = {^+}13$

   $^-4 \times {^-}3 = {^+}12 \quad ^+15 \div {^-}5 = {^-}3 \quad ^-12 \div {^-}3 = {^+}4$

6. Complete the tables in Figure 20.16. Explain what students are supposed to be learning by doing this exercise.
7. Explain how to solve the equation $4X + 3 = X + 12$, using a pan balance. To understand this idea, students need to be able to think of expressions such as $3 + 12$ or $X + 3$ as objects instead of operations or something to do. Explain why this understanding is important to the notion of solving equations.
8. Make up a real-world situation that defines a functional relationship. Use your example to do the following:
   a. Sketch a graph illustrating the relationship.
   b. State the relationship using the language of functions.
   c. Build a chart with numbers that might go with your relationships.
   d. Explain how the graph and the chart are both ways of illustrating your function; that is, how are these both the same information in different forms.
   e. Explain how your example meets the formal definition of a function.

◆◆◆◆◆ FOR DISCUSSION AND EXPLORATION

1. Read the article, "Why Elementary Algebra Can, Should, and Must Be an Eighth-Grade Course for Average Students" (Usiskin, 1987). In this article Usiskin argues for moving the curriculum forward

to algebra in the eighth grade for all students. His position is somewhat different from that of NCTM *Standards*. What is your view?

2. Calculator exercises
   a. Use a calculator to estimate the cube root of 10 to five decimal places.
   b. What do you think will happen if you enter 1000 in your calculator and then press ÷ 2 = = = ... ? Try it.
   c. What do you think will happen if you enter 1000 in your calculator, and then continually press the square root key? Before you try it, try to explain why you think it will happen.
   d. With the help of your calculator, see if you can find a prime number that is between 500 and 1000, and prove that it is prime.

3. Think up an experiment that produces a set of data like the drawings of triangles or the pendulum-swing experiment. Can you find a functional relationship in your experiment? If so, make a graph of your function. What is the domain variable? What is the range variable?

4. Examine the contents of a seventh- and eighth-grade textbook for the topics of integers, function, and variables. How much emphasis are these topics given? Describe the best idea you can find for each of these topics. Teacher's editions will be best.

5. Examine a textbook labeled "prealgebra" for the same three topics: integers, functions, and variables. How much emphasis are these topics given? Make a comparison between a regular eighth-grade book and a prealgebra book. Describe one good idea you found for teaching each of these topics. Use a teacher's edition.

6. If your instructor or library has a copy of the NCTM videotape *Algebra for Everyone* (Mathematics Education Trust Committee, 1991), view this 19-minute tape. The tape demonstrates useful classroom techniques for helping students develop abstract algebraic ideas. Discuss your reaction to the video with a group, or write a short reaction paper.

◆◆◆◆◆

SUGGESTED READINGS

Battista, M. T. (1983). A complete model for operations on integers. *Arithmetic Teacher, 30*(9), 26–31.

Bennett, A., Jr. (1988). Visual thinking and number relationships. *Mathematics Teacher, 81,* 267–272.

Blubaugh, W. L. (1988). Why cancel? *Mathematics Teacher, 81,* 300–302.

Coxford, A. F. (Ed.) (1988). *The ideas of algebra.* Reston, VA: National Council of Teachers of Mathematics.

Curcio, F. R. (1989). *Developing graph comprehension: Elementary and middle school activities.* Reston, VA: National Council of Teachers of Mathematics.

Edwards, F. M. (1987). Geometric figures make the LCM obvious. *Arithmetic Teacher, 34*(7), 17–18.

Fitzgerald, W., Winter, M. J., Lappan, G., & Phillips, E. (1986). *Middle grades mathematics project: Factors and multiples.* Menlo Park, CA: Addison-Wesley.

Geer, C. P. (1992). Exploring patterns, relations, and functions. *Arithmetic Teacher, 39*(9), 19–21.

Hector, J. H. (1992). Graphical insight into elementary functions. In J. T. Fey (Ed.), *Calculators in Mathematics Education.* Reston, VA: National Council of Teachers of Mathematics.

Heid, M. K., Sheets, C., & Matras, M. A. (1990). Computer-enhanced algebra: New roles and challenges for teachers and students. In T. J. Cooney (Ed.), *Teaching and learning mathematics in the 1990s.* Reston, VA: National Council of Teachers of Mathematics.

Herscovics, N., & Kiernan, C. (1980). Constructing meaning for the concept of equation. *Mathematics Teacher, 73,* 572–580.

Kiernan, C. (1991). Helping to make the transition to algebra. *Arithmetic Teacher, 38*(7), 49–51.

Kiernan, C., & Chalouh, L. (1993). Prealgebra: The transition from arithmetic to algebra. In D. T. Owens (Ed.), *Research ideas for the classroom: Middle grades mathematics.* New York: Macmillan Publishing Co.

Loewen, A. C. (1991). Lima beans, paper cups, and algebra. *Arithmetic Teacher, 38*(8), 34–37.

Martinez, J. G. R. (1988). Helping students understand factors and terms. *Mathematics Teacher, 81,* 747–751.

Morelli, L. (1992). A visual approach to algebra concepts. *Mathematics Teacher, 85,* 434–437.

Osborne, A., & Wilson, P. S. (1988). Moving to algebraic thought. In T. R. Post (Ed.), *Teaching mathematics in grades K–8: Research based methods.* Boston: Allyn & Bacon.

Schoenfeld, A. H., & Arcavi, A. (1988). On the meaning of variable. *Mathematics Teacher, 81,* 420–427.

Sobel, M. A., & Maletsky, E. M. (1988). *Teaching mathematics: A sourcebook of aids, activities, and strategies* (2nd ed.). Englewood Cliffs, NJ: Prentice Hall.

Whitman, B. S. (1982). Intuitive equation-solving skills. In L. Silvey (Ed.), *Mathematics for the middle grades (5–9).* Reston, VA: National Council of Teachers of Mathematics.

21 TECHNOLOGY AND ELEMENTARY SCHOOL MATHEMATICS

◆ THE NATIONAL COUNCIL OF TEACHERS OF MATHEMATICS HAS for many years been very clear in its support of technology in the mathematics classroom. Considerable attention is given to this issue in the introduction to the *Standards* as well as throughout that document. Two statements in the introduction are reflective of the theme of this chapter.

> The new technology not only has made calculations and graphing easier, it has changed the very nature of problems important to mathematics and the methods mathematicians use to investigate them. (p. 8)

> Access to this technology is no guarantee that any student will become mathematically literate. Calculators and computers for users of mathematics, like word processors for writers, are tools that simplify, but do not accomplish, the work at hand. Thus, our vision of school mathematics is based on the fundamental mathematics students will need, not just on the technological training that will facilitate the use of that mathematics. (p. 8)

Technology has permitted the mathematical world to investigate ideas that were never before possible. In the same spirit, technology in schools permits students to investigate and learn mathematics that was inaccessible without it. Not only does technology affect *what* we teach in schools, it likewise affects *how* we teach.

CALCULATORS IN THE CLASSROOM

In 1976, NCTM published a special issue of the *Arithmetic Teacher* on the use of calculators. Since that time numerous professional groups, including NCTM, have published recommendations encouraging a full integration of calculators into the mathematics classroom at all levels. In 1986, NCTM issued a position statement on calculators that recommended "the integration of the calculator into the school mathematics program at all grade levels in classwork, homework, and evaluation." In February of 1991 a new and expanded position statement was adopted by the Board. This position (see box) not only advocates the regular use of calculators by all students, it strongly points to the use of calculators in evaluation and testing instruments, suggests calculator usage be promoted by teachers at every level, and directs school divisions, authors, educators, and teachers to understand and promote the proper use of calculators in the curriculum. Coupled with the *Standards,* this NCTM position on calculators is difficult for a thoughtful professional to ignore or dispute.

In 1987, eleven years after the first *Arithmetic Teacher* focus issue on calculators, a second such issue was published. In 1992, NCTM devoted an entire yearbook to the subject of calculators.

Since 1976, the quality of calculators has increased, prices have gone down, batteries have been made unnecessary, and virtually every business, industry, and home relies heavily on their use.

CALCULATORS AND THE EDUCATION OF YOUTH

The following statement is an official NCTM position.

Calculators are widely used at home and in the workplace, Increased use of calculators in school will ensure that students' experiences in mathematics will match the realities of everyday life, develop their reasoning skills, and promote the understanding and application of mathematics. The National Council of Teachers of Mathematics therefore recommends the integration of the calculator into the school mathematics program at all grade levels in classwork, homework, and evaluation.

Instruction with calculators will extend the understanding of mathematics and will allow all students access to rich, problem-solving experiences. This instruction must develop students' ability to know now and when to use a calculator. Skill in estimation and the ability to decide if the solution to a problem is reasonable are essential adjuncts to the effective use of the calculator.

Evaluation must be in alignment with normal everyday use of calculators in the classroom. Testing instruments that measure students' understanding of mathematics and its applications must include calculator use. As the availability of calculators increases and the technology improves, testing instruments and evaluation practices must be continually upgraded to reflect these changes.

The National Council of Teachers of Mathematics recommends that all students use calculators to:

- explore and experiment with mathematical ideas such as patterns, numerical and algebraic properties, and functions;
- develop and reinforce skills such as estimation, computation, graphing, and analyzing data;
- focus on problem-solving processes rather than the computations associated with problems;
- perform the tedious computations that often develop when working with real data in problem situations;
- gain access to mathematical ideas and experiences that go beyond those levels limited by traditional paper-and-pencil computation.

The National Council of Teachers of Mathematics also recommends that every mathematics teacher at every level promote the use of calculators to enhance mathematics instruction by:

- modeling the use of calculators in a variety of situations;
- using calculators in computation, problem solving, concept development, pattern recognition, data analysis, and graphing;
- incorporating the use of calculators in testing mathematical skills and concepts;
- keeping current with the state-of-the-art technology appropriate for the grade level being taught;
- exploring and developing new ways to use calculators to support instruction and assessment.

The National Council of Teachers of Mathematics further recommends that:

- school districts conduct staff development programs that enhance teachers' understanding of the use of appropriate state-of-the-art calculators in the classroom;
- teacher preparation institutions develop preservice and in-service programs that use a variety of calculators, including graphing calculators, at all levels of the curriculum;
- educators responsible for selecting curriculum materials make choices that reflect and support the use of calculators in the classroom;
- publishers, authors, and test and competition writers integrate the use of calculators at all levels of mathematics;
- mathematics educators inform students, parents, administrators, and school boards about the research that shows the advantages of including calculators as an everyday tool for the student of mathematics.

Research and experience have clearly demonstrated the potential of calculators to enhance students' learning in mathematics. The cognitive gain in number sense, conceptual development, and visualization can empower and motivate students to engage in true mathematical problem solving at a level previously denied to all but the most talented. The calculator is an essential tool for all students of mathematics.

(February 1991)

The popular everyday use of calculators in society, along with the lengthy history of support for calculators in schools, has had less than spectacular impact on the mathematics classroom, especially at the elementary level. It is finally safe to say that resistance to the use of calculators is beginning to diminish. The public at large is slowly beginning to understand and accept that mathematics is more than computation and that thinking is perhaps more important than long division. It remains difficult for many, however, to accept that the calculator is not going to prevent Johnny from learning something basic about mathematics. Nothing could be further from the truth.

> **Don't be an adding machine -- Buy one!**
>
> Burroughs Corporation, early 1920s

REASONS FOR USING CALCULATORS

Perhaps when the next edition of this book is published, this section will be unnecessary. Perhaps for you, it is already. If you need some ammunition to talk to parents or principals, or just to help convince yourself, here are some reasons why calculators should be an everyday part of every classroom at every level.

COMMON USAGE IN SOCIETY

The fact is, almost everyone uses calculators in almost every facet of life that involves any sort of exact computation—except school children. The traditional reasons for teaching pencil-and-paper computation, especially with numbers involving more than two or three digits, have all but evaporated. It is more than a little hypocritical to forbid the use of calculators.

It also makes good sense that students should know now to use this popular tool effectively. Many adults have not learned how to use the memory keys, how to do a chain of mixed operations, how to utilize the automatic constant feature, or how to quickly judge if a gross error has been made. These are important practical skills that can easily be learned over the school years if the calculator is simply there for open, everyday use.

AFFECTIVE AND INDIRECT BENEFITS

The overwhelming conclusions of numerous research studies have found that students' attitudes toward mathematics are better in classrooms where calculators are used than where they are not (Hembree & Dessart, 1986; Reys & Reys, 1987). Students using calculators tend to be enthusiastic and are more confident and persistent in solving problems.

In addition to positive affective results, students using calculators discover a wide variety of interesting ideas that might otherwise remain unnoticed. For example, decimal numbers and negative numbers are almost inescapable, and children learn to explore these ideas at an early age. The number of digits that result in some computations is very different than in others. For example, 2 ÷ 7 fills up the display but 1245 ÷ 5 has only three digits. Students quickly find that it is easy to make errors on a calculator and develop a real appreciation for estimation. Students also learn that it is frequently easier to do a mental computation than to search for a calculator and press buttons. All of these outcomes are positive benefits that can result without any direct instruction. They can happen just by having calculators in every school, on every desk, every day, all of the time, at every grade.

INSTRUCTION ENHANCED

Not only does the calculator make suspect the value of spending significant portions of the school curriculum teaching pencil-and-paper computation, the calculator is as much a teaching tool as the chalkboard or overhead projector. That is, the calculator not only replaces much of the tedious computation required in school, it can be used to help develop concepts and skills in other areas of mathematics. While activities for teaching with the calculator have been suggested throughout the book, a few additional activities are presented here as examples of teaching concepts.

21.1 AND THE REMAINDER IS . . .

Have students find a method for determining the whole number remainder in a division problem using only the calculator. Suggest that they might begin with divisors that end in 2 or 5 so that the complete result will be displayed.

21.2 FOLD AND DIVIDE

Give students a strip of paper exactly 22 cm long and a centimeter ruler. Have them measure and record the length. Next, carefully fold the paper in half, and measure the result as accurately as possible. On the calculator, divide 22 by 2, and compare this result with the measurement. Fold the paper in half one more time, and again measure carefully. Divide the last result on the calculator by 2, and compare. Do this fold, measure, and divide sequence several more times. How do the results of the measurement compare with the results of dividing by 2 on the calculator? Try this with other lengths. (This activity, suggested by a second-grade teacher, is reported by Shumway, 1988).

In the last activity, there is a vivid contrast between symbolic manipulation and physical reality. What other ideas might develop at least informally from that exploration? What happens if after the first division by 2, you continue to press = over and over?

Almost any problem-solving activity that involves computation is enhanced by letting students use a calculator. Besides encouraging trial and error or exploration of different approaches to a solution, students also learn the value of recording intermediate results as they go along. Many interesting process problems arise from the use of the calculator itself. Following are examples of such problems.

21.3 TOO HARD FOR THE CALCULATOR?

Find a way to use the calculator to compute the product of two numbers such as 3456 and 88,888. (The standard method of entering this product on the calculator causes an overload.)

21.4 KEYPAD PARTNER NUMBERS

For the purpose of this activity (there really is no other purpose), define a *keypad partner number* (KPN) as a two-digit number that can be entered on a calculator by pressing two adjacent keys, either vertically, horizontally, or diagonally in any direction. Examples of KPNs are 48, 63, 12, and 21. The numbers 73 and 28 are not KPNs. The initial task is to find two different pairs of KPNs that have the same sum. (For example, 32 + 65 = 62 + 35.) Once the idea of pairs of KPNs with the same sum is established, try to find some sort of pattern or generalization about KPNs with like sums. When a pattern has been found that seems exciting, challenge students to figure out an explanation for why the pattern works. You might also suggest looking at differences of KPNs. Patterns also exist in products of KPNs, but these are a bit more difficult to come by.

The KPN activity is an interesting exercise in searching for patterns and reasons. It is surprising how many different ideas students can discover, especially if they work in groups to share ideas. Ideas that rely on place value include looking at the sum of the digits or examination of the addition algorithm. Others have drawn pictures of the keypad and discovered geometric patterns of various sorts. These then have interesting descriptions as well as explanations. Is the activity valuable? This is a good example of doing mathematics; searching for patterns, communicating ideas, justifying results.

The following activity is adapted from an article by Goldenberg (1991). It is useful for children in grades 2 to 5 who have limited concepts of decimals.

21.5 NUMBERS IN BETWEEN

First examine the idea that a number can be quickly multiplied by itself by pressing the number followed by ⊠ ⊟. That is, 4 ⊠ ⊟ produces 16. What number can be multiplied by itself to get 43? After some discussion around 6 being too little and 7 too large, introduce the idea that there are numbers *between* 6 and 7. (Children do not think of fractions as numbers at this age.) List in a column the numbers 6.1, 6.2, . . . 6.9 and explain simply that these are numbers that are more than 6 and less than 7. Have students use their calculators to multiply each of these by itself. This will show that 6.5 is too small, and 6.6 is too large, to produce 43. Next, suggest that there are numbers between 6.5 and 6.6 and list these in a column: 6.51, 6.52, . . . 6.59. Repeat the exercise of squaring each of these, and list all of the results. The process can be repeated again if there is interest.

The **Numbers in Between** activity is not the only way to discuss decimals and not even the best way. But it does introduce ideas in a meaningful way that adds to student understanding of decimal numbers in a powerful manner.

Many books and resources for calculator activities are available with excellent ideas. The *CAMP-LA* books, for example, have literally hundreds of excellent activities organized by content and grade level (see Suggested Readings). Almost every chapter in this text has included calculator activities as well.

NOT USING CALCULATORS WASTES TIME

Pencil-and-paper computation is time-consuming, especially for young students who have not developed a high degree of mastery. Why should time be wasted having students add numbers to find the perimeter of a polygon when the lesson is on geometry? Why compute averages, find percents, convert fractions to decimals, or solve problems of any sort with pencil-and-paper methods when pencil-and-paper skills are not the objective of the lesson? Defending laborious and time-consuming computations in noncomputational lessons is indeed difficult.

It really comes down to this:

Why teach children to do inefficiently what a $5.00 machine can do efficiently?

WHEN AND WHERE TO USE CALCULATORS

Calculators should be in or on students' desks at *all times* at *all grade levels*. This position may seem quite radical to some. Here are a few arguments to support immediate availability at all times:

The need to make a special effort to use calculators for any activity is diminished. Throughout this text, activities have been interspersed that utilize calculators. If we have to stop the flow of our lessons to pass out calculators, we are likely not to take the time. Instead we will save calculators for special "calculator lessons," or we may not even take the time at all. Many excellent calculator explorations will happen spontaneously and/or will take up only a few minutes of class time. These activities simply will not get done if we must stop to distribute and collect calculators.

Ready availability of calculators allows students on their own to choose when it seems appropriate to use calculators. There are many times when it is much easier or quicker to use a mental computation or estimation or even to use the pencil that happens to be in our hands rather than to reach for the calculator. How can students ever learn to make these choices if we decide for them when to use calculators by keeping them out of sight unless otherwise directed?

It simply does no harm to have calculators available! This is very difficult for many teachers, prospective teachers, and parents to accept. But the fact of the matter is that there is virtually no research to suggest that students fail to develop basic skills when taught in the presence of a calculator. Even basic facts and pencil-and-paper skills will still be learned with the calculator on the desk.

Calculators have been promoted by mathematics educators for over twenty years. They are inexpensive. There are excellent resource materials directing their use. There is no evidence that they do any harm. There is lots of evidence that they can enhance learning in many ways. They are still not accepted. Perhaps the reason is that not enough teachers have simply put them on the students' desks.

PRACTICAL CONSIDERATIONS CONCERNING CALCULATORS

Like any other tools in the classroom, there are things to consider after you have decided you want them. There are different types of calculators with different features. Which ones should I use? How can all of my students have them?

CALCULATOR TYPES AND FEATURES

Automatic Constant Feature

Simple four-function calculators are all that are necessary for most of the activities that have been suggested in this book. The one important feature that is not on all simple calculators is the automatic constant for addition and subtraction. (That is the one that allows you to enter ⊞ 1 ⊟ ⊟ ⊟ ... and have the calculator count.) Almost all calculators have this feature for multiplication and division. It is especially important to note that automatic constant features do not operate exactly the same on all calculators.

Algebraic Logic

Demana and Osborne (1988) make a strong case for the use of calculators that include algebraic logic. In essence, that means that expressions are evaluated according to the correct order-of-operations rules rather than evaluated as they are entered. On a standard four-function calculator, if 5 ⊞ 3 ⊠ 4 ⊟ is keyed in, the result will be 32. The rules concerning order of operations dictates that multiplication and division precede addition and subtraction. Thus the correct result of $5 + 3 \times 4$ is 17. At the very least you should be aware of these differences. Texas Instruments introduced MathMate™ in 1990, a four-function calculator for young children that has algebraic logic and keys for left and right parentheses. The TI Math Explorer also has these features. Scientific calculators also are designed with algebraic logic. (For a more detailed discussion of algebraic logic, see p. 408, Chapter 20.)

Fraction Capabilities

A few calculators now allow for common fractions to be entered and operation results displayed in fraction form. In addition to being able to enter and operate on fractions, it is possible to simplify a fraction using any appropriate factor (e.g., $\frac{12}{16}$ can be simplified to $\frac{6}{8}$ or $\frac{3}{4}$). Displays can be switched back and forth between fraction and decimal forms and between mixed and improper forms. Several activities have been suggested for this feature in Chapters 12, 13, and 14. The Math Explorer and TI-34 calculators are among those that have these features. (The Math Explorer also has a few other unique features that make it an interesting tool for the middle grades.)

Overhead Projector Versions

Several suppliers now make transparent versions of their calculators that can be placed on the overhead projector. These allow you to show students exactly what keys you are pressing for demonstrating calculator usage. The overhead calculator is especially useful for showing young children how to use the calculator or for illustrating new features for older children. Overhead calculators permit many good activities to be conducted with the full class instead of having every child pressing his or her own buttons. There are many times when this approach has definite advantages.

GRAPHING CALCULATORS

The graphing calculator is another level of sophistication from the calculators discussed so far. Teachers who are using them at the high-school level are very excited about the potential that they have for exploring all sorts of mathematics. They literally are changing what can be taught at the high school level. It is a mistake, however, to disregard the idea of using these amazing machines with students beginning in about the sixth grade. Vonder Embse (1992), makes a compelling case for their inclusion at the middle grade level. He suggests the following values:

- The large-screen display permits compound expressions such as $3 + 4(5 - 6/7)$ to be completely displayed before being evaluated. Further, the expressions are easily modified (e.g., adding or removing parentheses) with corresponding results displayed. This can aid in understanding notation and order of operations.
- The large screen allows students to construct tables of values by inserting different numbers into formulas or equations. The table of values that a student might develop this way can also be stored in the calculator for further analysis.
- The ability to enter different values in a formula or expression is a way to get at the idea of a variable as "something that varies."
- Students can plot points on a coordinate screen and begin to appreciate the relationships between tables of values, equations, and graphs.

These four points are all included in the discussion of function and variable found in Chapter 20. Perhaps the argument for the graphing calculator at the middle school level is actually stronger than that. The ideas mentioned so far are the ones that are most obvious to anyone who first picks up one of these tools and plays around with it. But these calculators have many more features, some of which could be quite useful to middle grade students. Here are just a few more points that should be considered:

- The built-in statistical functions allow students to examine mean, median, and standard deviation of reasonably large sets of realistic numbers. There is no reason not to use real data with large numbers for any exploration. You would not feel as comfortable doing this using any other calculator because the lists of numbers you would be working with cannot be stored, ordered, added to, or changed. Other calculators do not record your entries or store lists of data as does a graphing calculator.
- Scatter plots of ordered pairs of data allow students to examine trends in data and to begin to understand fairly sophisticated ideas of statistical inference without requiring fancy symbolic mathematics.
- Random-number generators allow for the simulation of a wide variety of probability experiments that would never be possible without this device. A few simple examples are included in Chapter 18.
- The graphing calculator is programmable. Programs are very easily written and can be understood at the middle-grade level. For example, a program to compute the quadratic formula adds another dimension to the discussion of the Pythagorean theorem.

Besides all of these specific reasons, there is the overriding notion of opening up real mathematics to young children, a way to excite them and provide them with opportunities to explore well beyond the lessons you may have planned.

It is quite clear that there are considerable advantages to the use of the graphing calculator beginning at about the sixth grade. What are the disadvantages? Cost is the obvious issue to be raised. In response, the cost of these "computer/calculators" continues to drop. At the time this is being written, at least three different brands of graphing calculators can be purchased for well under the cost of even a moderately priced pair of "must-have" sneakers or about four CDs. It is quite amazing how almost all kids seem to have these items even though the values are fleeting. The calculator purchased in the sixth grade is the last calculator that will be needed through high school and college. The cost is very cheap from this perspective.

Most of the other arguments are similar to those against any calculator: "Will it keep my students from learning the basics?" Or there is the not-uncommon response found at most any grade level: "It is OK for students to use one of those things *next* year, but this year we are going to learn the *real* way to do math." These arguments are hollow and cannot be supported with reality.

The graphing calculator is not yet part of the standard set of tools for the middle grades. It is probably only a matter of time. The time could be now in your classroom. It should be seriously considered.

AVAILABILITY

Calculators should be available for use at *all* times to all students in the classroom at *every* grade level. The price of a calculator is now so low that it is entirely reasonable to require that each of your students provide his or her own calculator at the start of the year the same way that pencils, notebooks, and other supplies are required. Be sure to provide parents with a short list of particular calculators that you recommend. Include both brand and model number when appropriate. In this way you can be assured that every student has the features you desire and that all of the calculators in the room operate the same way. Nothing is more frustrating than having to stop in the middle of an activity to provide special directions to one or two students with a nonconforming calculator. Another good idea is to have a few extra calculators available to fill in when one breaks or is left at home.

A possible exception to the suggestion for uniformity in the classroom is in order. By the time students reach the fourth or fifth grade, they should be aware that all calculators are not alike and do not even use the same logic. Having some calculators around that perform differently may provide some interesting opportunities for learning. For example, if some calculators employed algebraic logic as discussed earlier, the results on that calculator could be contrasted with the results on a simple four-function calculator. The need for a rule concerning order of operations will become apparent in this context.

In some school districts, classroom sets of calculators are being purchased by the schools. Some book publishers have also provided calculators at reduced prices as part of their sales promotions. Major manufacturers have designed calculators specifically for this market and sell them at reasonable prices in large quantities. There is no need, however, to wait for someone else to supply your room with calculators.

COMPUTERS AND MATHEMATICS EDUCATION

The age of computers in schools is over two decades. The excitement of simply having students be on a computer has just about faded. Nearly all schools have some form of computer capability. It is quite reasonable to assume that computers are here to stay.

The pace at which technology improves has shown no signs of slowing down. Improved technology means more capability for programmers to create quality software. In fact, the improvement in technology also has a down side. The Apple IIe, the backbone machine for many schools throughout the nation, is extremely limited in comparison to current DOS-based or Macintosh computers. The best programs today simply will not run on these older machines, and yet that is the hardware that represents a substantial investment already made by schools. Interactive laser disc and CD-ROM capabilities bring the real world into the classroom in fantastic ways that could never be imagined only 10 years ago. Quality graphics, so very important in mathematics programs, demand the speed of the more advanced machines.

At the same time, another advance in technology may be both the short-term and even the long-term solution to high-cost and rapidly changing hardware. Liquid-crystal-display (LCD) panels, connected to the video output of a computer, sit directly on an ordinary overhead projector stage and permit the full computer image to be sharply projected onto a large screen. Most newer versions even permit color displays. While not inexpensive, one LCD panel and one newer generation computer are considerably less expensive than a complete lab of computers with limited capability. Some excellent programs are quite effective in the hands of the teacher interacting with the full class. The benefits of computer technology do not always depend on students actually pressing the keys.

The NCTM *Standards* notes in the introduction that "a computer should be available in every classroom for demonstration purposes; every student should have access to a computer for individual and group work; and students should learn to use the computer as a tool for processing information and performing calculations to investigate and solve problems" (p. 8). This position highlights the appropriate use of the computer as a tool for learning rather than an object to learn about. It is worth noting that there is no standard in the document that speaks to learning about computers or learning how to program. Computer literacy is not a topic of mathematics education.

In the remainder of this chapter, the different ways that computers can be used in learning mathematics will be discussed. The purpose is to help you understand the different uses of this instructional tool so that you can begin to make reasoned rather than emotional decisions about how to include computers in your instructional program.

If you have never used a computer for one of the purposes described, the best recommendation is to find some software in that category, and try it out. Be careful not to make quick judgments based on limited experiences. Talk to friends, teachers, and other educational specialists who have experience with computers. Ask for their opinions and ideas about particular software programs to try. Find out how they are best used with children, what works well and what does not. If possible, try different software and different computer uses with children. At the end of each section a few popular software packages are listed. These listings are only to provide examples. They are by no means recommendations of the best there is. Be aware that new software is constantly being generated.

INSTRUCTIONAL SOFTWARE

Several terms with similar meanings are commonly used when talking about computers and instruction. *Computer-assisted instruction* (CAI) refers to a method of teaching that uses the computer to present instructional material. Students using a CAI program interact with the information on the screen. The learning situation might be drill and practice, tutorial (where concepts or skills are developed), a simulation, or an instructional game or problem-solving situation. The terms *computer-assisted learning, computer-based instruction,* or other variants are frequently used instead of CAI.

The programs that enable the computer to present instructional information are examples of *software* or *courseware*. Generally, software refers to commercially produced programs, although that is not a requirement. *Software* is a broad term that encompasses not only CAI programs but also tool programs such as spreadsheets, graphing programs, and programming languages such as Logo or Pascal.

DRILL-AND-PRACTICE SOFTWARE

Drill-and-practice programs do exactly that; they provide students practice with skills or concepts that are assumed to be learned elsewhere. More drill-and-practice programs exist than any other type of software for elementary mathematics. In general, a drill-and-practice program poses ques-

THE USE OF COMPUTERS IN THE LEARNING AND TEACHING OF MATHEMATICS

The following statement is an official NCTM position.

Computer technology is changing the ways we use mathematics; consequently, the content of mathematics programs and the methods by which mathematics is taught are changing. Students must continue to study appropriate mathematics content, and they also must be able to recognize when and how to use computers effectively when doing mathematics. Teachers must know how and when to use the tools of computer technology to develop and expand their students' understandings of mathematics.

It is the position of the National Council of Teachers of Mathematics that the computer is an appropriate tool that can be used in a variety of ways for the enhancement of mathematics learning, teaching, and evaluation. Changes are therefore needed in mathematics curricula, instructional methods, access to computer hardware and software, and teacher education:

- The content of school mathematics courses must be modified to reflect the changes brought about by computer technology. Curricular revisions should provide for the deletion of topics that are no longer useful, the addition of topics that have acquired new importance, and the retention of topics that remain important. In implementing revised curricula, educators must ensure that the time and emphasis allocated to topics are consistent with their importance in an age of increased access to technology. Instructional materials that capitalize on the power of computers must continue to be developed for students at all levels.

- Teachers should use computers as tools to assist students with the exploration and discovery of concepts, with the transition from concrete experiences to abstract mathematical ideas, with the practice of skills, and with the process of problem solving. In mathematics education computers must be instructional aids, not the object of instruction. Similarly, computer programming activities in mathematics classes should be used to support mathematics instruction; they should not be the focus of instruction. The amount of classroom time spent by mathematics students in learning a programming language must be consistent with the expected gains in mathematical understanding.

- Schools should be equipped with computers, peripherals, and courseware in sufficient quantity and quality for them to be used consistently in the teaching and learning of mathematics. Every classroom in which mathematics is taught should be equipped with computing hardware that includes a large-screen display device. Computer laboratories should be available to all students on a regular basis for the extended exploration of mathematical topics by individuals or groups. School systems must budget for the ongoing acquisition, maintenance, and upgrading of hardware and courseware for use in classrooms and computer laboratories at all grade levels.

- All preservice and in-service teachers of mathematics should be educated on the use of computers in the teaching of mathematics and in examining curricula for technology-related modifications. Teachers should be prepared to design computer-integrated classroom and laboratory lessons that promote interaction among the students, the computers, and the teacher. Mathematics teachers should be able to select and use electronic courseware for a variety of activities such as simulation, generation and analysis of data, problem solving, graphical analysis, and practice.

Mathematics teachers should be able to appropriately use a variety of computer tools such as programming languages and spreadsheets in the mathematics classroom. For example, teachers should be able to identify topics for which expressing an algorithm as a computer program will deepen student insight, and they should be able to develop or modify programs to fit the needs of classes or individuals. Keeping pace with advances in technology will enable mathematics teachers to use the most efficient and effective tools available.

Changes in curricula and in the availability of hardware and courseware are not sufficient to guarantee that teachers use computers appropriately. Ongoing in-service programs must be readily available to help teachers take full advantage of the unique power of the computer as a tool for teaching and learning mathematics.

(April 1987)

---

tions that the user answers either by directly entering the response or by selecting from options. The program evaluates the response and reacts accordingly.

Computers have proven to be an effective but not necessarily superior method of providing drill and practice (see, for example, Carrier, Post, & Heck, 1985; Fuson & Brinko, 1985; Suydam, 1984). Some advantage may exist in the provision of motivation and change of pace provided by attractive graphics and action, the ability of the computers to provide immediate feedback, and the possibility of selecting drills specifically to meet individual needs. Another positive feature offered by many programs is the ability to keep records of individual student's progress. At the present time, the significant disadvantages of cost and limited access to computers in the classroom must be kept in mind when considering these values.

Besides basic fact practice, one of the best uses of the computer for drill and practice is for computational estimation and mental arithmetic. Here the computer can control response time and immediately evaluate responses in terms of percent of error, and students can respond without a pencil, thus eliminating the temptation to do the computations by hand instead of mentally.

How the program responds or gives feedback is one of the most important features to consider when selecting a drill program. The most sophisticated drill programs are coordinated with conceptual tutorials. If a student misses a certain percentage of questions, the program suggests stopping the drill and moving into a different program that provides conceptual or developmental assistance. These latter programs are few in number at the present time. the desired type of feedback will vary with different types of drills.

Examples of software in this category:

Addition Logician, MECC, grades 2–4

Teasers by Tobbs, Sunburst, grades 3–6

Math Blaster, Davidson, grades 1–7

Math Strategies—Estimation, SRA, grades 5–7

Fraction Munchers, MECC, grades 4–9

TUTORIAL SOFTWARE

Tutorial software is designed to teach rather than to simply drill. Teaching or conceptual development of almost any topic is significantly more complex than providing drill. To design good tutorial programs requires not only good pedagogical understanding and anticipation of wide varieties of student responses but also a high level of programming skill. The enormous amounts of time required to design and program good tutorial programs has left quality software of this type quite scarce. Richard Shumway, a firm believer in the use of computers at every grade level, notes that "the use of a computer to simulate teachers is not yet realized and very difficult" (1988, p. 339). Since conceptual development in mathematics is highly dependent on models, tutorial software has its greatest potential in providing graphics to stand in place of physical models. Many programs provide static pictures of counters, fraction pieces, base ten place-value models, or geometric figures. Some better programs show movement, such as the formation of a ten-stick from 10 single squares. The best of these programs allow students to freely manipulate the computer model with keystrokes or a joystick, thus simulating actual manipulation of physical materials. For example, students may be able to place as many ones, tens, and hundreds pieces on the screen as they wish. Once there, the pieces might be moved, grouped, or separated. The student can then use this manipulation to respond to a question posed by the program.

Some of the most promising programs to date are designed to run only on larger, more sophisticated computers with faster processors, more memory, and better graphics capabilities.

Examples of software in this category:

Balance!, HRM, grades 8–10

Decimal Squares Computer Programs, Scott Resources, grades 5–8

Learning Place Value, Mindscape, grades 1–5

IBM Math Concepts, IBM, grades K–8

PROBLEM-SOLVING SOFTWARE

Programs in which games or activities require the use of logical reasoning and spatial or geometric reasoning have been among some of the most innovative software packages yet developed. These differ from tutorial software in that they present visual, numeric, or logical challenges or problems rather than drill or instruction. Activities requiring identifying and extending patterns and assorted logic activities with attribute pieces can be found in several programs. Other programs involve students with more sophisticated if-then reasoning and the use of *and* and *or* connectives. Numeric challenges involving number theory and rules of inference are also part of this genre of software.

The argument can be made that most of these activities can be conducted without a computer. Especially for primary children, patterning and attribute activities should certainly be done with actual physical materials before moving students to the computer, and some teachers have found ways to combine physical activities directly with the computer programs.

On the other hand, the computer provides an element of motivation, especially with interesting graphics displays. It also allows students to work independently or in groups of two or three on one computer without constant teacher intervention. This sort of independent involvement is difficult to provide in most noncomputer logic activities.

A different sort of problem-solving software is designed to teach specific problem-solving heuristics such as guess-and-check or making an organized list. Other programs are designed to help children solve translation problems. These higher-order skills are difficult to develop outside of the rich environment of a class discussion led by a skilled teacher. Software of this type may best be used with an entire class rather than by individual students.

Examples of software in this category:

Problem Solving Strategies, MECC, grades 4–9

The Factory, Sunburst, grades K–2

Heath Math Worlds: Strategies, Data Analysis, grades 4–6, 6–9

King's Rule, Sunburst, grades 4–10

Pond, Sunburst, grades 2–6

Gertrude's Secrets, The Learning Company, grades K–4

COMPUTER TOOLS

More and more what seem to be the best computer materials in mathematics education are those that can be placed under the general category of tool software. These programs do not teach, drill, or convey information. Like calculators, they dramatically expand the user's abilities to do things. Most of these tools are aimed at the middle- and high-school levels with some promising exceptions for younger students in the area of geometry.

SPREADSHEETS

Spreadsheets are programs that can manipulate rows and columns of numeric data. Formulas can be entered easily so that, for example, the average, highest, and lowest values from each of three different columns of numbers is automatically entered into another position. Every time a value is changed, all other values that depend on that one are recalculated. Words can also be written into a spreadsheet, so that, for example, a class of students' names could be entered and a column for each of their test grades listed as in a grade book. Besides determining both row and column averages or other statistics that may be of interest, the spreadsheet could be used to order any row or column either numerically or alphabetically. *Spreadsheet* is a generic term, with many different companies producing spreadsheet software. Many exist that are simple to use in the classroom.

In addition to their obvious use as an aid in statistics investigations, spreadsheets are a powerful way to investigate functions numerically. Figure 21.1 was generated by the *Appleworks* spreadsheet and shows comparative values of several simple and not-so-simple mathematics functions. The left-hand column contains a list of values for which each function was evaluated.

The spreadsheet is an indispensable tool in the business world. As a result, very sophisticated spreadsheets such as *Lotus 1,2,3* and *Excel* include very nice graphics capabilities that would be useful for students. For the things that young students would use a spreadsheet for, these high-end versions are actually easier to use than the *Appleworks* spreadsheet. Here is an interesting example of useful technology that is generally not available to students due to cost and hardware requirements.

```
File: CH20 FIG 1    REVIEW/ADD/CHANGE    Escape: Main Menu
======B======C======D======E======F======G======H======I===
    4 |
    5 |    X      X^2      X^3    X^3-X^2    (X^3+2X)/(3+X)
    6 |=======================================================
    7 |   -5       25     -125     -150        67.50
    8 |   -4       16      -64      -80        72.00
    9 |   -3        9      -27      -36        ERROR
   10 |   -2        4       -8      -12       -12.00
   11 |   -1        1       -1       -2        -1.50
   12 |    0        0        0        0         0.00
   13 |    1        1        1        0          .75
   14 |    2        4        8        4         2.40
   15 |    3        9       27       18         5.50
   16 |    4       16       64       48        10.29
   17 |    5       25      125      100        16.88
   18 |    6       36      216      180        25.33
   19 |    7       49      343      294        35.70
   20 |    8       64      512      448        48.00
   21 |
       -------------------------------------------------------
G20: (Value) ((D20)+(2*(B20)))/(3+(B20))
```

FIGURE 21.1: *Spreadsheet*

GRAPHING PACKAGES

Software that quickly converts numeric data into graphs is one of the most powerful tool uses of the computer in the mathematics classroom. Statistical graphing packages typically produce pie charts, bar graphs, and line graphs from any sets of data that are entered. *Graphing* (MECC), *Exploring Tables and Graphs* (Weekly Reader Family Software) and *Data Insights* (Sunburst) are three popular examples. Algebraic graphing packages graph the curves of functions and equations that are entered by the user. *Green Globs* (Sunburst) was one of the first algebraic graphing packages, but there are many from which to select that take advantage of today's more powerful computers. Figure 21.2 was generated by *Green Globs*.

Both types of graphing programs are powerful teaching tools because they demonstrate changes in the graphs very

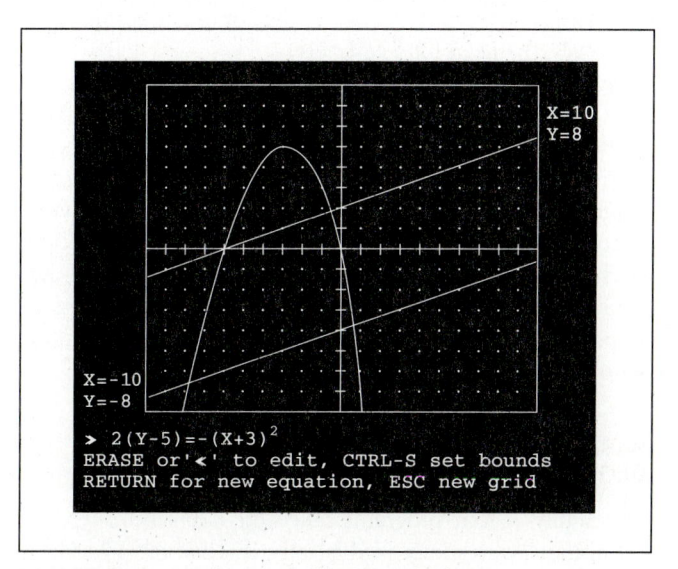

FIGURE 21.2: *Graphs generated by* Green Globs.

quickly with different changes in the data or functions to be plotted. Students' attention can be focused on understanding and interpretation of graphs and relationships without the tedium of plotting points, determining percentages, measuring bars, and so on. Most programs allow several line graphs and curves to be plotted on the same axis in order to make contrasts and comparisons.

SYMBOL MANIPULATORS

Graphing utilities have been built into another category of tool that is very popular in high-school mathematics—symbol manipulators. The two most widely used packages in schools at this writing are the *Mathematics Exploration Tool Kit* (IBM) and *Derive* (Soft Warehouse, Inc.). Similar but more sophisticated and more powerful packages are frequently used by scientists and mathematicians. These programs require a DOS-type computer and a fair amount of memory. But what they can do is quite amazing. The graphing capabilities of these machines are considerably better than *Green Globs*, which runs on an Apple IIe. In addition, they can perform algebraic manipulations such as simplifying very complex expressions, solving equations, factoring or expanding expressions, doing some operations in calculus, and much more. Functions can be defined and graphed. Programs can be written.

These packages have begun to change the curriculum at the high-school level because they make ideas and skills accessible that simply could not be done with pencil and paper, although the concepts were well within reach. At the same time, as the calculator has done with pencil-and-paper arithmetic, these programs have made virtually obsolete much of what has traditionally been taught in algebra courses.

Interestingly enough, the graphing calculator does many of the things that the symbol manipulator does with the significant exception of solving and simplifying algebraic expressions. There are advantages and disadvantages of each tool. The main drawback to the computer approach is, obviously, the need for a computer.

While these relatively fancy tools are not at the present time having an impact on the elementary or middle-school program, their effect on the high-school curriculum speaks loudly concerning what should be happening in the middle grades. Very simply, tedious skills such as equation solving, factoring, simplifying expressions, and function graphing are not nearly as important as the ability to reason, to evaluate, to solve problems, to discern patterns and relationships. These represent the directions in which the middle school must be headed.

GEOMETRY UTILITIES

The family of four programs called the *Geometric Supposers* (Sunburst) were discussed briefly in Chapter 17. These programs permit a shape to be quickly produced on the screen and any series of constructions or measurements performed on that shape. For example, the side of a triangle can be subdivided into as many as nine equal parts and lines drawn from the opposite vertex to each subdividing point. Angles, lengths, and areas can then be measured and compared. Furthermore, the same series of constructions can then be repeated on a different triangle for the purpose of making comparisons and conjectures. The programs encourage students to explore their own ideas about geometric relationships and actually invent theorems of their own. This type of activity is much closer to the spirit of real mathematics. Students create ideas rather than demonstrate proofs of ideas that were constructed by someone else.

The *Geometer's Sketchpad* (Key Curriculum Press) represents a prime example of how more powerful computers have made a good idea much better. The *Supposers* have a track record of excellence because they promote the type of interaction and thinking that is desired in all of mathematics, not just in geometry. The *Sketchpad* took the same philosophy around which the *Supposers* were designed and adapted the considerable capabilities of the Macintosh computer to create a program that is vastly more capable and more attractive to the user. (A brief discussion of the *Sketchpad* can be found in Chapter 17.)

A simpler program, *Elastic Lines* (Sunburst), is essentially an electronic geoboard that allows shapes to be drawn on a 10×10 dot grid, produces flips and rotations, and makes some measurements.

The number of programs that permit shapes to be easily drawn, manipulated, and measured will certainly increase in the near future. This is especially true for computers at the level of the MacIntosh, the IBM, or the IBM-compatible. The greater speed and superior graphics of more powerful machines will give teachers blackboards to draw and move shapes in ways they never could have imagined.

LEARNING THROUGH PROGRAMMING

When computers first became popular in schools, there was a heavy emphasis on teaching programming. Most often this was (and is) done under the title of "computer literacy." Computer literacy refers to practical knowledge about computers that every person in our society should know. As such, it is not a topic of mathematics as much as it is social studies. Today, the emphasis has fortunately shifted from teaching students to use computers to using computers for learning. From a practical standpoint very few people in society need to know how to program a computer. Is there any value, then, in teaching grade-school children to program? How much programming knowledge do children need to have? What can be learned by learning to program other than a social utility knowledge of how computers work? These and similar questions are still being researched and debated.

LEARNING MATHEMATICS THROUGH PROGRAMMING

Many mathematics educators believe that students can learn mathematics through the process of programming (Shumway, 1987, 1988; Smith, 1984). This might happen in two different but related ways. First, if students, even very young students, are given a computer program to explore, some good mathematics can be learned in the process. Second, if students are asked to program a computer to perform some task, they (not the computer) must understand any mathematics that might be involved.

Exploring a Simple Program

By way of example, consider the following two programs written in BASIC.

```
10 FOR N = 1 to 12          10 GR
20 PRINT N, N+N, N+N+N+N    20 COLOR = 7
30 NEXT N                   30 FOR N = 0 to 39
                            40 PLOT N, N
                            50 NEXT N
```

The first program (on the left) will print 3 columns of 12 numbers very quickly. The second (written for an Apple computer) will graph 40 large "points" in a diagonal line down the screen from the upper left corner. The screens produced by these two programs are shown in Figures 21.3 and 21.4, respectively.

The first point to make is that even first graders can develop some understanding of how these programs work, run them, examine their output, make changes, and compare results. Any programming knowledge that is required can be taught as new examples are presented. It is not necessary for children to first learn to program.

Can you think of any mathematics that might be learned from either of these programs? Here are some possibilities: For the left program, in Figure 21.3, students might:

Notice that all of the numbers in the second and third columns are even. What other formulas could be put in line 20 to produce even numbers? What about odd numbers?

Notice that some of the numbers in the second column are also in the third but much higher up. What is the pattern? How often does that happen? What if line 20 were changed?

For line 20, try

20 PRINT N, 2 * N, N+N

or

20 PRINT N, 4 * N, N+N+N+N

Second graders could begin to explore multiplication this way.

Try

20 PRINT N, 1/N, 1(N * N)

and begin to look at the limiting process in the seventh or eighth grades. Change line 10 to allow for longer lists and/or different steps. For example, try

10 FOR N = 0 TO 20 STEP .1

For the right-hand program, in Figure 21.4 students might wonder what caused the program to print the dots where it did. To check, they might change line 30, line 40, or both to see if they can figure out how this works. That is a mathematical investigation.

Change line 40 to

40 PLOT N, N/2,

```
]LIST
10   FOR N = 1 TO 12
20   PRINT N,N + N,N + N + N + N
30   NEXT N

]RUN
1          2          4
2          4          8
3          6         12
4          8         16
5         10         20
6         12         24
7         14         28
8         16         32
9         18         36
10        20         40
11        22         44
12        24         48

]
```

FIGURE 21.3

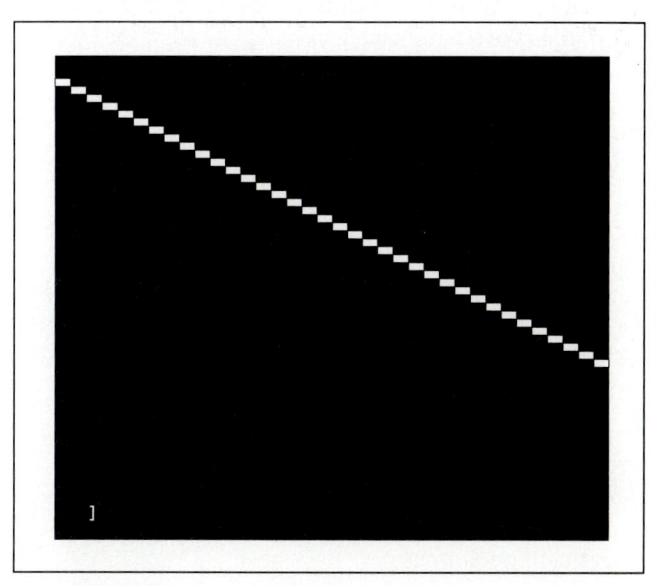

FIGURE 21.4

or

40 PLOT N, N/3.

This provides an informal exploration into the notion of slope of a line. Why do the little rectangles get longer? Does "PLOT" round off the numbers or throw away the decimal parts?

Add a line 45.

45 PLOT N, N+3

Compare the results. Parallel lines have the same slopes. What could be done to make them intersect?

More significantly, in both programs students are using variables in a meaningful manner. In the graphics program the relationships among variables, real numbers, and coordinates on the screen are powerful ideas. The computer provides a medium within which these ideas are explored at an informal, experimental level.

The notion that students can be encouraged to be more reflective about specific mathematical concepts through programming exercises has a lot of appeal. While there remain some advocates of this idea, there is currently little research evidence to support it, and very few schools have pursued this particular use of computers.

PROGRAMMING AND PROBLEM-SOLVING SKILLS

Another possible benefit of students programming computers involves the mental processes that are involved rather than the mathematics content that is addressed. The issue is one of transfer of these problem-solving processes to environments other than the computer.

Consider the things you might have to do and to think about in order to program a computer to perform one of the following tasks:

> Print in order all of the even numbers from 1 to 100. To make it a bit harder, also include all of the numbers ending in 5.
>
> Play "Guess My Number." The computer selects a number between 1 and 1000, and the player tries to guess it. Each guess should get an appropriate and helpful response from the computer.

Now compare the different stages and thought processes you might have to go through with the following list of generic problem-solving skills:

understanding or analyzing the problem

breaking the problem down into smaller parts

selecting and planning a solution strategy

monitoring process toward the solution

evaluating solutions

extending solutions

When students are taught how to approach a programming task in a thoughtful and well-planned manner, each of the foregoing skills can generally be identified as part of that process. Many educators believe that if students are engaged in programming experiences these higher-order process skills will transfer from the specific domain of computer programming to the broader area of solving nonroutine problems (for example, Blume & Schoen, 1988; Dalbey & Linn, 1985; Wells, 1981). The computer, in essence, provides a whole class of extremely rich and interesting process problems within which to develop the skills of problem solving.

Research evidence supporting transfer of computer problem-solving skills is at this time inconclusive and sketchy. The methods of instruction used in teaching programming, the amount of time required to develop transferable skills (months or years?), and the ability to detect effects in controlled research settings are all factors that make it difficult to be definitive about the possible effects of programming.

Problem Solving and the Logo Language

The most vocal proponents of the programming-for-problem-solving position are the advocates of Logo (for example, Au, Horton, & Ryba, 1987; Campbell, & Clements, 1990; Clements, 1985a, b, c; 1986; Papert, 1980). Unlike BASIC, Logo permits students to easily design relatively powerful programs using user-defined words. The turtle-graphics feature, only one aspect of the Logo language, can be used to create complex designs, draw pictures, and produce geometric shapes. The power of Logo is accessible to very young children, is captivating in its graphics capabilities, and yet is powerful enough to be used in high school and college.

Originally developed at MIT by Seymour Papert, Logo has been the object of numerous research efforts. Much of this interest is due to the claims of Papert (1980) and his followers for the cognitive gains and problem-solving capabilities of children who have had experience with Logo programming. Papert's basic thesis is that the computer is a tool to think with. By exploring or analyzing an idea on the computer, a learner has an object that can be manipulated, viewed, modified, combined, expanded, and, in a sense, played with. In fact, Papert's original explorations during the development of Logo were with a robot turtle that moved around on the floor and could draw pictures with a marker pointing down from the bottom of the turtle. Several robot floor turtles are now available that are controlled with Logo. The screen version of Logo uses a small triangular shape about 1 cm tall that is called a "turtle," or else a small picture of a turtle. It is controlled the same way that the floor turtle would be. The computer thus adds a personal reality to a child's experimentation and should, according to Papert, enhance logical reasoning abilities.

Papert's arguments are so inviting and the Logo language so captivating and powerful that troops of teachers and mathematics educators have been on a headlong and enthusiastic pursuit of Logo. What have been the results? In a review of Logo-related research, Clements (1985c) included these observations:

> Programming appears to facilitate the development of specific problem-solving behaviors.
>
> Younger students may benefit more than older students.
>
> Logo may enhance social interaction, positive self-images, positive attitudes toward learning, and independent work habits.

Researchers are very cautious about making definitive claims for Logo. Exactly what the teacher and students are doing in a Logo experience is not clear in many of the studies that have been conducted. The transfer of thinking processes to situations that do not involve Logo is not always well supported. After nearly two decades of Logo in the classroom, much of the interest in Logo emanates from Logo enthusiasts who are convinced of the values of Logo more by their intuition than by careful research. For example, Clements says, "I believe that, over several years, working with Logo has changed the way I think. Years, not weeks. I have every respect for the students who made the significant leap from their Logo work to the non-Logo problem-solving tasks they were given. I have every respect for their teachers who, unlike too many researchers, are trying to find better ways to work with Logo. I have fewer illusions, but . . . I have every hope for the future" (1986, p. 25).

A BRIEF LOOK AT LOGO

It is difficult to appreciate the enthusiasm that Logo tends to generate without some firsthand experience. If you have never played around with Logo, you are *strongly encouraged to get some firsthand experience*. Even if you have had a negative experience with programming in another language such as BASIC, you owe it to yourself to at least try Logo. It is a truly different way to work with computers, and almost everyone has fun.

One minimal approach is to work through the examples and explorations provided in the following section while actually sitting at a computer. A better idea is to get one of the many books about Logo and teaching Logo and work through some of it on your own or with a friend. A few such books are suggested at the end of the chapter.

Some Logo Turtle-Graphics Experiences

This section is strictly for those who have never experienced Logo.

To begin you will need a Logo language disk. This section will not provide you with all of the details of getting the language booted up, correcting typos, or other nuances. A good idea is to find someone who knows a little about Logo to help you get started. If you are using MIT Logo (or Terrapin Logo), you will be able to follow the directions given here exactly. If you are using Logo Writer or some other dialect of Logo, there are a few minor differences. Where these differences occur you will be directed to check a manual.

Simple Turtle Commands. To get Logo to show a screen where the turtle draws pictures, type DRAW and press RETURN. (This command varies with different Logo versions.) There are a few commands the turtle understands without being taught. Here are some of them.

FD: Short for FORWARD. You must also say how far forward you want the turtle to go. For example, FD 100 or FD 30 will make the turtle go forward. Notice the space between FD and 100. Spaces are important in Logo.

BK: Short for BACK. BK works like FD and needs an "input" or a number that tells how far, like BK 140.

RT: Short for RIGHT. The turtle can turn, but you have to tell it how much, like RT 30 or RT 90 or RT 200. You will soon discover that those turn numbers are degrees.

LT: LEFT. Works the same as RT.

Try these commands, and make the turtle scribble around. You have to press the RETURN key (‹R›) before it will do anything. When you want to start all over with a clean screen, type DRAW. (Check this command in other versions of Logo.) The turtle has a "pen" that is usually down causing the line to be drawn when it moves. If you want the turtle to move without drawing a line, type PU for PENUP (and press ‹R›). To draw again, type PD for PENDOWN.

Before going any further, play with these commands. Try to draw something: your initials, a rectangle, or a triangle. What happens if you send the turtle really far, like FD 2000? Try that when the turtle is turned just a little from "north." Can you figure out what happens when the turtle goes off the side of the screen?

There are four lines of text at the bottom of the screen. Sometimes the turtle gets "under" these lines and cannot be seen. Don't worry. It is still there.

Before you go any further, try to draw a simple house like the one on Figure 21.5. All of the distances should be

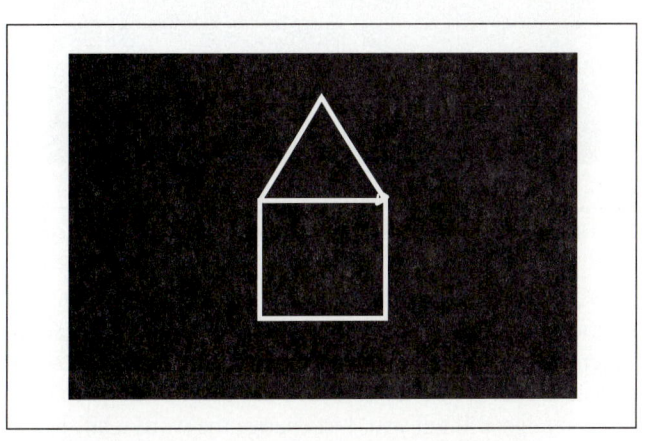

FIGURE 21.5: *A typical house drawn with Logo*

the same. Plan on paper the list of turtle commands you want to use. Test out your list and revise it if necessary.

Teaching the Turtle. One of the nice things about Logo is that there are very few words to remember. You can define the language as you go along, using your own words for whatever you want them to mean. This is called *defining procedures*. Young children might call it "teaching the turtle."

By way of illustration, teach the turtle how to do some simple little squiggle: for example, FD 60 LT 45 BK 20. Since this is just two lines at an angle, you can call it anything you want, like GEORGE, CHECK, SQUIGGLE, or even YZR or S3.

Suppose you decide on SQUIGGLE. Type TO SQUIGGLE and press ‹R›. (Look up "defining procedures" in other versions.) The word TO tells Logo you want to define the word that follows it. The screen now looks different. This is called the *edit mode* where procedures are defined. Now type each of the three commands FD 60, LT 45, and BK 20. It is a good idea to press ‹R› after each. Then type END for the end of the procedure. If you made any mistakes, you can use the arrow keys and the delete key to make changes. When you are done, press Ctrl-C. (Hold the control key down and press C.) Logo responds, SQUIGGLE DEFINED. Now the turtle knows the word SQUIGGLE.

Now try out the new word. Type SQUIGGLE. Try it again. Use SQUIGGLE along with the other commands. Figure 21.6 shows some experiments with SQUIGGLE.

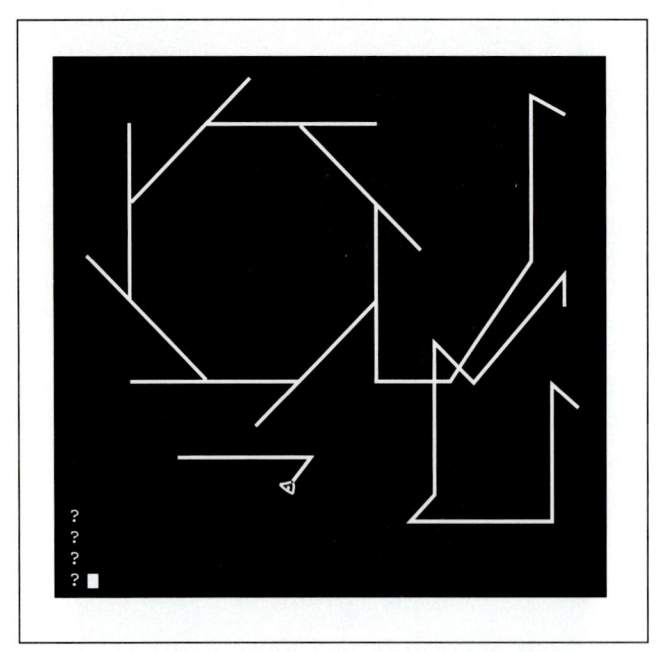

FIGURE 21.6: *Playing with Logo and the SQUIGGLE procedure*

SQUIGGLE is an example of a procedure that you defined. It is actually a little Logo program. Define some other words. Define your initial or a dashed line or some simple shape. If you try to define a word that Logo already knows, Logo will tell you. Just change the word a bit. If you want to change SQUIGGLE, just type TO SQUIGGLE, and then use the arrow keys to add to it or make changes. Remember Ctrl-C to finish or "define the procedure."

Before you go any further, try to define two more procedures. Call them BOX and TRI. Define BOX to be a square and TRI to be a triangle so that all of the sides of BOX and TRI are the same length.

Putting Procedures Together. Once you have defined a procedure, that word can be used just like any other Logo word. That means it can be used inside another procedure. For example, here are four procedures. Each of the first two are used in the third, and the third is in the fourth. Use a pencil and paper, pretend you are the turtle, and see if you can predict what each will do. Then define these procedures on the computer and try them out.

TO CANE	TO RECTS	TO DOWNTOWN
FD 60	CANE	RECTS
RT 90	LT 90	BK 50
FD 20	CORNER	RT 90
END	RT 90	RECTS
	CORNER	BK 50
TO CORNER	LT 90	RT 90
FD 30	CANE	CANE
RT 90	END	RT 90
FD 30		CANE
END		END

If you were successful, your DOWNTOWN procedure should draw what looks a little like a building. Try putting the building in different places on the screen. Use CANE, CORNER, and RECTS to make other procedures. Add some ideas of your own.

Finally, before you continue, can you use your BOX and TRI procedures to make HOUSE as in Figure 21.5? Try to make a procedure that draws three houses.

This was a very limited exposure to Logo. You should not generalize or make major conclusions about Logo from this experience. If you are at all intrigued by this little exercise, or if you just had fun, there is much, much more that Logo has to offer you. Your next explorations should involve the REPEAT command and also variables. Get a book and a friend. Take a course or go to a workshop.

◆◆◆◆◆

REFLECTIONS ON CHAPTER 21: WRITING TO LEARN

1. Review the arguments for using calculators in elementary school. Select two that are most appealing to you, and describe them as if you were arguing for the inclusion of calculators in schools.

2. Explain briefly these features of calculators and why they may be important:
   a. automatic constant feature for +,
   b. algebraic logic,
   c. fraction capability.

3. What are two reasons for promoting the use of graphing calculators in grades 6 to 8? Describe and defend your own opinion about this.

4. Describe briefly these classifications of software:
   a. drill and practice,
   b. tutorial,
   c. problem solving.

5. Four types of computer tools were described. What are these and what do they each do?

6. Explain the argument that students can learn mathematics content by working on certain programming problems.

7. Another argument suggests that students may learn problem-solving processes that will transfer to problem-solving activities in noncomputer environments. What type of thinking do students do while working on a program that the proponents of this position believe will transfer?

◆◆◆◆◆
FOR DISCUSSION AND EXPLORATION

1. The February 1987 issue of *Arithmetic Teacher*, 34(6), is all about teaching mathematics using calculators. Find at least one article of interest to you in that issue and read it.

2. Since 1987, most issues of *Arithmetic Teacher* have a two-page feature called "Teaching Mathematics with Technology." Look at some of these articles and find an idea of particular interest to you. Prepare a lesson or sequence of lessons using the idea(s) you find.

3. Examine one or more of the resource books for calculator activities in the following list. Try the ideas you find with a calculator. Frequently when we read a calculator activity its value escapes us until we actually try it. Choose several of your favorite ideas, and plan a lesson around calculators.

   The *Keystroke* series (Reys et al., 1980).

   How to Teach Mathematics Using a Calculator (Coburn, 1987).

   CAMP-LA (Calculators and Mathematics Project, Los Angeles) [four books: K–2, 3–4, 5–6, & 7–8], (Pagni, ed., 1991).

4. Talk with some teachers about their use or lack of use of calculators in the classroom. How do they use them if they do use them? What are their main reasons for not using them if they do not? Neither the *Standards*, the NCTM position statement on calculators, nor this book offer much room for hedging on the issue of calculators in schools. If you or one of the teachers you talk with do not accept this position fully, how would you and/or the teacher argue against these various authors?

5. The following are three good resources for the use of spreadsheets. Examine one and make a short report.

   Spreadsheet Activities in Middle School Mathematics (Russell, 1992).

   Spreadsheets in Mathematics and Science Teaching (Whitmer, 1992).

   How to Use the Spreadsheet as a Tool in the Secondary School Classroom (Masalski, 1990).

6. The following two books are excellent ways to begin an exploration of the Supposers or the Geometer's Sketchpad. The first is applicable to both tools while the second is directly related to the Sketchpad.

   How to Use Conjecturing and Microcomputers to Teach Geometry, (Chazan & Houde, 1989).

   Exploring Geometry with the Geometer's Sketchpad, (Bennett, 1992).

   Use one of these books and try out the software. Report on your reactions.

7.
   a. If you have never ever tried to play around within the Logo language, get a Logo language disk and at the very least work through the introductory exercises beginning on p. 441.

   b. If you do know something about Logo, think of some concept in mathematics that possibly could be explored through a Logo activity. Sketch out some of your ideas, and share them with someone else who knows about Logo.

◆◆◆◆◆
SUGGESTED READINGS

Aieta, J. F. (1985). Microworlds: Options for learning and teaching geometry. *Mathematics Teacher, 78,* 473–480.

Barrett, G., & Goebel, J. (1990). The impact of graphing calculators on the teaching and learning of mathematics. In T. J. Cooney (Ed.), *Teaching and learning mathematics in the 1990s*. Reston, VA: National Council of Teachers of Mathematics.

Battista, M. T., & Clements, D. H. (1988). A case for a Logo-based elementary school geometry curriculum. *Arithmetic Teacher, 36*(3), 11-17.

Calculator-enhanced Mathematics Instruction Steering Committee. (1992). *Calculators for classrooms* [video and guidebook]. Reston, VA: National Council of Teachers of Mathematics.

Campbell, P., & Clements, D. H. (1990). Using microcomputers for mathematics learning. In J. N. Payne (Ed.), *Mathematics for the young child*. Reston, VA: National Council of Teachers of Mathematics.

Campbell, P., & Stewart, E. L. (1993). Calculators and computers. In R. J. Jensen (Ed.), *Research ideas for the classroom: Early childhood mathematics*. New York: Macmillan Publishing Co.

Clements, D. H. (1989). *Computers in elementary mathematics education*. Englewood Cliffs, NJ: Prentice-Hall.

Coburn, T. (1987). *How to teach mathematics using a calculator*. Reston, VA: National Council of Teachers of Mathematics.

Demana, F., & Waits, B. K. (1990). Enhancing mathematics teaching and learning through technology. In T. J. Cooney (Ed.), *Teaching and learning mathematics in the 1990s*. Reston, VA: National Council of Teachers of Mathematics.

Fey, J. T. (Ed.). (1992). *Calculators in mathematics education*. Reston, VA: National Council of Teachers of Mathematics.

Hembre, R., & Dessart, D. J. (1992). Research on calculators in mathematics education. In J. T. Fey (Ed.), *Calculators in mathematics education*. Reston, VA: National Council of Teachers of Mathematics.

Jensen, R. J., & Williams, B. S. (1993). Technology: Implications for the middle grades. In D. T. Owens (Ed.), *Research ideas for the classroom: Middle grades mathematics*. New York: Macmillan Publishing Co.

Lilly, M. W. (Ed.). (1987). Calculators. Special issue of *Arithmetic Teacher, 34*(6).

Maddux, C. D. (Ed.). (1985). Logo in the schools. Special issue of *Computers in the Schools, 2* (2 & 3).

McDonald, J. L. (1988). Integrating spreadsheets into the mathematics classroom. *Mathematics Teacher, 81*, 615–622.

Pagni, D. (Ed.). (1991). *CAMP-LA* (Calculators and Mathematics Project, Los Angeles), [four books: K–2, 3–4, 5–6, & 7–8]. Fullerton, CA: Cal State Fullerton Press.

Papert, S. (1980). *Mindstorms: Children, computers, and powerful ideas*. New York: Basic Books.

Reys, B. (1989). The calculator as a tool for instruction and learning. In P. R. Trafton (Ed.), *New directions for elementary school mathematics*. Reston, VA: National Council of Teachers of Mathematics.

Reys, R. R., et al. (1980). *Calculator activities for young students*. Palo Alto, CA: Creative Publications.

Riedesel, C. A., & Clements, D. H. (1985). *Coping with computers in the elementary and middle schools*. Englewood Cliffs, NJ: Prentice-Hall.

Russell, J. C. (1992). *Spreadsheet activities in middle school mathematics*. Reston, VA: National Council of Teachers of Mathematics.

Shumway, R. J. (1992). Calculators and computers. In T. R. Post (Ed.), *Teaching mathematics in grades K–8: Research based methods* (Second Ed.). Boston, MA: Allyn & Bacon.

Shumway, R. J. (1984). Young children, programming, and mathematical thinking. In V. P. Hansen (Ed.), *Computers in mathematics education*. Reston, VA: National Council of Teachers of Mathematics.

Vonder Embse, C. (1992). Concept development and problem solving using graphing calculators in the middle school. In J. T. Fey (Ed.), *Calculators in mathematics education*. Reston, VA: National Council of Teachers of Mathematics.

Wheatley, G. H., & Shumway, R. (1992). The potential for calculators to transform elementary school mathematics. In J. T. Fey (Ed.), *Calculators in mathematics education*. Reston, VA: National Council of Teachers of Mathematics.

Whitmer, J. C. (1992). *Spreadsheets in mathematics and science teaching*. School Science and Mathematics Association, Classroom Activities Series, No. 3.

22 PLANNING FOR DEVELOPMENTAL INSTRUCTION

◆ TEACHING IS BOTH A SCIENCE AND AN ART. THE SCIENCE PORtion comes from knowledge of subject matter, theories of learning and development, and knowledge of instructional materials and activities that research and practice have found effective. The artistic portion comes from adding to this scientific base the human qualities of both the teacher and the students in the classroom and the design of specific lessons and interactions that will promote effective and pleasurable learning.

Recognition of the artistic nature of teaching suggests that there is no single formula or plan that can be set down and followed mechanically. Different content requires different strategies. Young children require different approaches from those for older children. Review is distinct from initial development. And each teacher and class is a unique combination of human factors. The suggestions in this chapter are not rigid lesson plan formulas. Rather, they point toward broad guidelines and practical ideas that you might consider while maintaining an individual style.

CONCEPTUAL AND PROCEDURAL KNOWLEDGE: BALANCE AND PACE

Recall that knowledge of mathematics includes both conceptual knowledge and procedural knowledge. Furthermore, relational understanding is characterized by the degree to which knowledge of all types is connected or integrated with other ideas that we may have. A heavy emphasis on procedural knowledge (symbolic skills and procedures) can hinder the development of connections with concepts or with other procedures. In other words, instruction that focuses largely on procedural outcomes is likely to sacrifice understanding. Pressures from standardized tests that have also been very skill- and procedure-oriented, plus a commonly accepted value system that prized computational skills, have influenced many classroom teachers. As a result of these influences, teachers have tended to spend a majority of their instructional efforts on procedural aspects of mathematics at the expense of well-integrated conceptual development. Fortunately we are moving toward a better balance. The classroom teacher, however, remains the one person who is ultimately in control of how instruction will be focused in his or her classroom.

USE A CHAPTER/UNIT PERSPECTIVE

To develop relational understanding children need time for reflective thought. They need to construct new ideas and relationships, test them, integrate them with existing knowledge, verbalize them, and make them their own. They need to make mistakes and have the opportunity to learn from them. While an individual lesson may have a specific goal, it is important that each lesson be seen as contributing to the larger, more complex set of concepts and skills. Understanding is a unit objective, not a lesson objective (Burns, 1987; Trafton, 1984).

BALANCE

In Chapter 3 it was suggested that 50% to 60% of your time should be devoted to conceptual development and/or connection of concepts to procedural knowledge. This maxim of 50% applies to the unit goal, not to each lesson. That means that roughly half of the lessons in a unit might well be conceptually oriented (not half of each lesson). To significantly alter this ratio in favor of procedural knowledge is to risk teaching rules and procedures without concepts.

The 50% rule is a departure from what is found in most traditional classrooms. It is not an easy rule to follow, and it is almost completely violated by following page by page through a basal textbook. Many teachers fear that time spent on conceptual development is simply not available. There is, however, evidence to support the notion that if up-front time is given to development of concepts, and explicit efforts are made to make connections with procedural skills, the time required at the procedural level will be significantly reduced (Suydam & Higgins, 1977; Madell, 1985; Baroody, 1987; Hiebert & Lefevre, 1986; Good & Grouws, 1979). Meaningful skills simply take less time to master than those not conceptually supported. Even more time is made up by reducing the need for remediation. The investment in conceptual development will, in the long term, more than make up for the time spent.

PACE

A common lament of student teachers is, "I had to cut the lesson short because the children were getting restless" (or they "were tired," or they "were misbehaving," or they "were distracted"). There are times when children really are tired or distracted, and we have to adjust for that. But it is interesting that children do not seem to get tired or distracted as often for the good, experienced teachers. Children and adults enjoy learning. Children and adults also get uncomfortable with boring activity and with activities they cannot or do not understand.

Knowing when to move forward requires real skill at listening to children. The interest and enthusiasm of students is one indicator that can be generalized across content. Children who are asked to do activities requiring no new conceptual growth soon get bored and restless. Children who are asked to do activities that are too difficult or not well connected to conceptual knowledge will likewise soon become fidgety and disinterested.

New relationships can only be constructed in terms of existing ones. If concepts are moved too quickly, required conceptual anchors to current knowledge will not be available. The artistic trick is to keep children moving forward without getting beyond them but not so slowly as to bore them.

PACING OF PROCESS OBJECTIVES

Many unit objectives culminate in an observable skill, the ability to do something. Changes in emphasis within the curriculum have created a focus on problem solving, estimation, number sense, and mental computation. These topics cannot and should not be dealt with entirely in units or chapters. We never really master problem solving. We can only improve at it over a lifetime. The real objectives of problem solving and other process-oriented skills are mental processes or complex interactions or mental processes and concepts.

These process objectives require that we attend to them on a continual basis virtually over the entire school year and at every grade level. It is best that some specific activities related to problem solving and estimation/mental computation are included every week if not every day. Process skills develop slowly and continuously over time. Current abilities and progress can be assessed periodically even though a mastery-level approach is inappropriate.

GUIDELINES FOR DEVELOPMENTAL LESSONS

A constructivist view of learning suggests that students should be actively engaged in the learning process. This implies more than simply doing work. Mechanistic or routine activity, even with physical models and diligently performed, does not promote reflective thought or growth in conceptual knowledge. Effective teachers tend to use a significant portion of their class time in some way that encourages this interactive involvement and development of ideas and relationships (Good, Grouws, & Ebmeier, 1983).

A GENERAL MODEL TO WORK FROM

In their model for fourth-grade teachers (Figure 22.1), Good and his colleagues suggest that at least 20 minutes of a lesson be devoted to "development" and 15 minutes to "seatwork." The first 8 minutes of their model include a brief review, checking homework, and a mental computation activity. A maximum of two minutes is allotted for concluding the lesson and assigning homework.

We should not take the weekly plan or the lesson outline as rigid models. Many other models for instructional planning have merit. Good, Grouws, and Ebmeier designed theirs for fourth- through eighth-grade classrooms. However, the success of their model and its emphasis on active student involvement suggest that its basic features be given serious attention, especially the use of a development and consolidation (seatwork) portion.

	Monday	Tuesday	Wednesday	Thursday	Friday
	Weekly Review (20 Min.)	Homework, Review, Mental Computation (8 Min.)	Homework, Review, Mental Computation (8 Min.)	Homework, Review, Mental Computation (8 Min.)	Homework, Review, Mental Computation (8 Min.)
		Developmental (20 Min.)	Developmental (20 Min.)	Developmental (20 Min.)	Developmental (20 Min.)
	Development (10 Min.)				
	Seat Work (10 Min.)	Seat Work (15 Min.)	Seat Work (15 Min.)	Seat Work (15 Min.)	Seat Work (15 Min.)
	Lesson Conclusion & Homework Assign. (2 Min. Max)	Lesson Conclusion & Homework Assign. (2 Min. Max)	Lesson Conclusion & Homework Assign. (2 Min. Max)	Lesson Conclusion & Homework Assign. (2 Min. Max)	Lesson Conclusion & Homework Assign. (2 Min. Max)

FIGURE 22.1: *A general model for weekly lesson guide. Fourth grade.* SOURCE: *T. L. Good, D. A. Grouws, and H. Ebmeier (1985), Active Mathematics Teaching. New York: Longman. Reprinted by permission.*

◆ Development and Consolidation

The major distinction between development and seatwork is the degree of teacher involvement with the full class. The *development portion* of a lesson is interactive, involving students and teacher. During the development portion the teacher also prepares the way for the seatwork portion. The seatwork portion may be labeled more broadly as a *consolidation period*. During this consolidation time students independently explore the concepts and relationships or practice the procedures that have just been taught. The teacher is active, working with small groups or individuals, diagnosing, receiving feedback, and determining the direction of subsequent lessons.

The next two sections offer guidelines for development and consolidation activities.

▶ DEVELOPMENT PORTION OF A LESSON

The teaching objective for the development portion of the lesson is always the exploration of a new or developing idea or skill. The content may include conceptual and/or procedural knowledge. A problem-solving flavor or approach is appropriate most of the time.

◆ Getting Started

"Today we are going to talk about fractions. Who can tell me something about fractions?" This opening may appear to be reasonable, but it is certainly not one that would cause students to get excited, care, or begin to really think about the topic of the day. Many will yawn and leave the response to the classroom whiz kid. Good lessons need good beginnings, something that causes students to sit up and take notice and entices them to become involved. A number of possibilities exist.

Start with an Interesting Question or Problem. Figure 22.2 shows two examples of questions that could be posed as a lesson beginning. In the primary grades the cake prob-

How many different ways can you slice this cake in half? What if you don't stay on the lines?

(a)

Use four of these numbers in the boxes. How close can you get the sum to 1?

① ③ ④ ⑤ ⑥ ⑦

Perhaps:
(Also try it using ⑪ but no ①.)

(b)

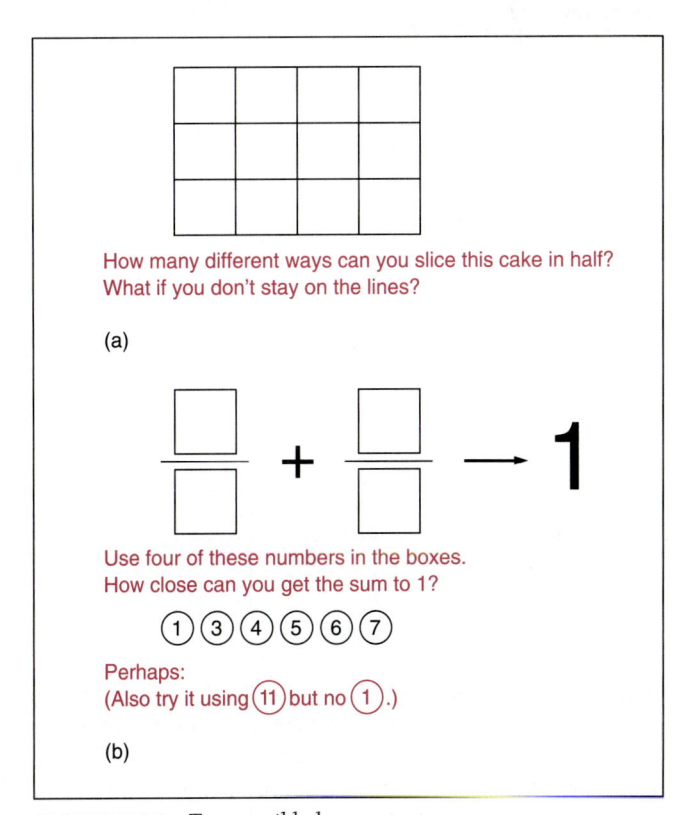

FIGURE 22.2: *Two possible lesson starters*

lem could lead to a discussion of what "half" means, or it might begin an area lesson in measurement. The fraction challenge could be used as an introduction to estimation with fractions or common denominators and computation.

Begin with New or Exciting Materials. As an example of this beginning, a new set of attribute pieces with different properties might be kept in a bag and pulled out one at a time. Likenesses and differences could be discussed as you proceed. Consideration of what the rest of the bag might hold is both profitable as well as captivating at the K–2 level. Distribution of base ten pieces, geoboards, or fraction strips, along with a quick and easy challenge, is frequently a good mind capture. Challenges with most any of the materials in this book can provide interesting lesson starters.

Start with a Model or Picture on the Board or Overhead Projector. If you are not ready for your students to actually have materials in their hands, they can still be presented to the class. Many materials show up nicely when placed on the overhead projector. Fraction strips, counters, base ten pieces, and geometric shapes or drawings are just a few examples. Computer software is another alternative that can be used with an overhead display screen. As an example of starting with materials, the dot grid and drawing in Figure 22.3 could be prepared on a transparency. Students could have geoboards at their seats to explore the question posed. Many teachers make poster board models of their materials and attach small pieces of magnetic tape to the backs. These are easily presented and manipulated on a magnetic blackboard.

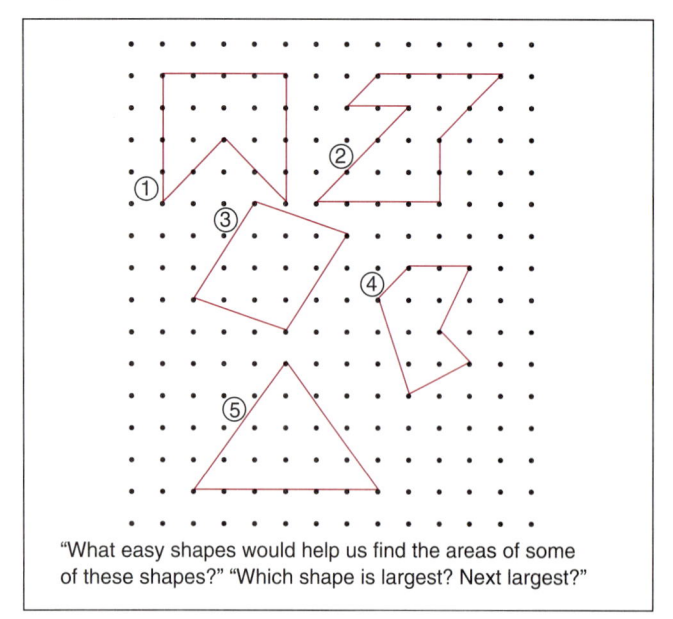

"What easy shapes would help us find the areas of some of these shapes?" "Which shape is largest? Next largest?"

FIGURE 22.3: *Geoboard shapes*

Begin with a Quickly Played Game. The game should relate to the lesson to follow. Max/Min is a good example of a game with a lot of potential and is easily played with several variations. All students have a drawing of one of the boards in Figure 22.4. Number cards with the digits 0 through 9 are drawn at random. As each card is drawn students must immediately write the number in one of the blanks. Which blank they select is their choice, but once written, no changes can be made. Suppose that the game was "Max Sum." After 7 numbers are drawn (1 for the discard), each student will have 2 three-digit numbers on his or her paper. These are added, and the student with the greatest sum is the winner. The game can be played to try to get the smallest ("Min") result or can be played as "Target" where each student tries to make their result closest to some designated number. These games might be used to begin lessons on computation, estimation, probability (What if the numbers are not replaced, or there are twice as many small numbers?), or numeration. There are other examples of games throughout the book that provide good lesson starters.

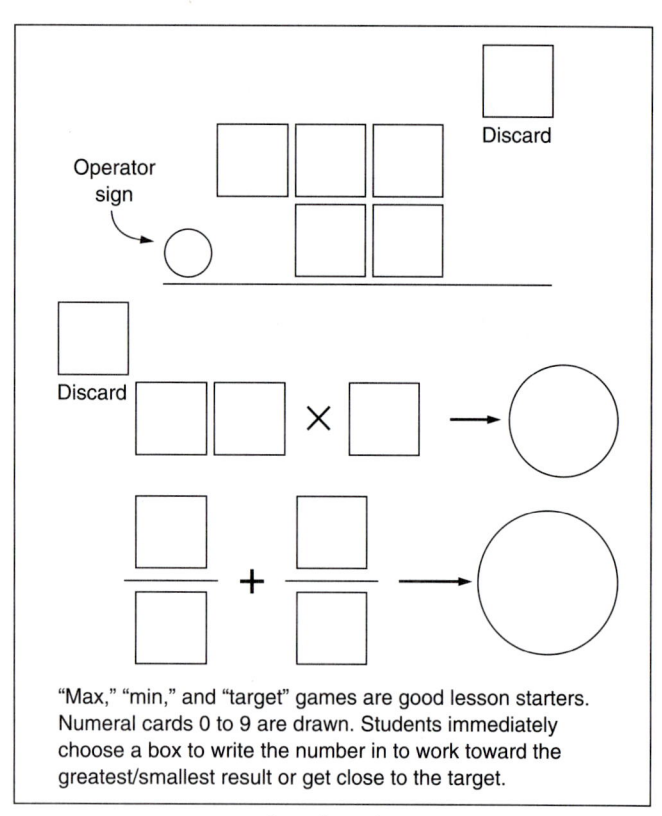

"Max," "min," and "target" games are good lesson starters. Numeral cards 0 to 9 are drawn. Students immediately choose a box to write the number in to work toward the greatest/smallest result or get close to the target.

FIGURE 22.4: *Max/Min boards and target games*

◆ Maintaining Involvement

Once you have students involved with a good lesson beginning, the next trick is to keep that level of interest and enthusiasm throughout the lesson. In planning a lesson, the key is to consider the students' perspective. While planning your lesson the question, "What will the students be doing?" is much more important than "What will I do next?" If your plans revolve around phrases such as "I will say . . ." or "I will do . . . ," then the students in your lesson are very likely to be passive watchers or listeners. Most children do not do either very well and certainly not for very long. The key is to get the students *doing* something as quickly as possible.

Whenever possible, think about how to get manipulative materials into students' hands early in the lesson. If distributed to individuals or groups of students at their desks, a directed activity can and should begin immediately. Students should be given a problem to solve or a concept to demonstrate with the materials as soon as they get them. If the materials are on the desks, that is also where the students' attention will be. Do not pass out materials until you are ready to use them. Activities involving such things as counters, base ten materials, number lines, fraction models, calculators, measurement devices, or geometric materials can be conducted by modeling step-by-step procedures as the students work along with you. Each new example is then planned with increased student independence.

Another approach with younger children is to use a single set of materials that all can see when seated on the floor in a large circle. Each child should have some of the materials in his or her hands. Questions posed or activities conducted should permit all children to be involved. For example, if each child has one piece of a large set of attribute materials, the question can be: "Think how your piece is different from mine. Who has a piece that is different in just one way? Who has one different in two ways?"

◆ Student Responses

How you get responses from individuals in a full class or large group is very important. Children frequently follow the leader without thinking. Others are shy or insecure with their thoughts and prefer no response to potential embarrassment. A lesson that involves a series of questions from the teacher to the full class or group can be conducted effectively if thought is given to how children will respond. Several possibilities can be considered.

Accept Many Responses. After a question is posed to the class, students should respond only after being called upon. Allow sufficient time for all students to have a chance to think about their answers. Avoid calling only the first hand raised or only the children you are confident will have the correct answer. Develop the habit or responding to the first one or two answers in a neutral or noncommittal manner: "That's a good idea. Who else has an idea?" Accept student answers that have already been offered. These repetitious answers may well be personal, genuinely independent thoughts, and not just the result of following the leader. Slower students frequently are thinking very hard while the quicker children are answering. They deserve to have their ideas heard as well.

Observe Responses with Physical Materials. If the children can answer your questions by selecting an object from those on their desk or by doing something with objects on their desks, you will be able to observe most of the students on each question simply by moving about the class. Using base ten materials to show the number you say is a good example.

Use Response Cards. Whenever you plan a series of questions with simple quick responses, some method in which every child can respond by holding up the answer works well. If the answers to a series of questions are to be numbers, students can have numeral cards arranged on their desks to be held up simultaneously on your signal. This way, each child must respond, and you can check on the response of all members of the class. Two sets of cards from 0 to 9 will allow responses up to 99. Yes/no, true/false, greater/less than, and other paired responses can all be coded by using two colors of construction paper, each color assigned to one of the responses. For example, give each child a square of red and a square of blue construction paper. On the board make it clear that red stands for "yes," and blue for "no" (or whatever pair of responses). At your signal, after asking a question, all students hold up their answer card.

An advantage of this method of getting responses over oral answers is that every child must respond to every question and therefore must attend to every question.

Use Written Responses. Have all students write their responses on paper as you work through the lesson. These can be observed by walking around the room. Collecting and grading is not important. The requirement to write a response suggests to each child that he or she must think about the question. It permits shy children to respond without public scrutiny and allows the teacher to reinforce correct, thoughtful, or creative responses. Not all responses are numbers. They can be drawings, words, or sentences.

◆ CONSOLIDATION PORTION OF A LESSON

Most lessons should involve at least a short consolidation time in which students have the opportunity to work through and think about the ideas that were developed in the developmental portion. If the focus of the lesson was on concept development, the consolidation period should have students continuing to do activities that promote and develop those concepts. Even if this second phase of a concept lesson begins to connect procedural knowledge to the concepts, the models and/or verbal aspect of the lesson should not be abandoned. If the first phase of the lesson was directed at connecting conceptual and procedural knowledge, then the consolidation should also give students the opportunity to make these connections. Consolidation should only be completely drill or practice when drill, practice, or review was the intent of the first half of the lesson.

The activity in this part of your lesson should be a familiar one to the students or a direct follow-up to the development portion. If pie pieces were the model for the fraction activity conducted with the full class, then pie pieces should probably be used in the consolidation part of the lesson. If you choose to change models, say to fractions strips, that model should at least be familiar. Different groups or individual students may work with different but

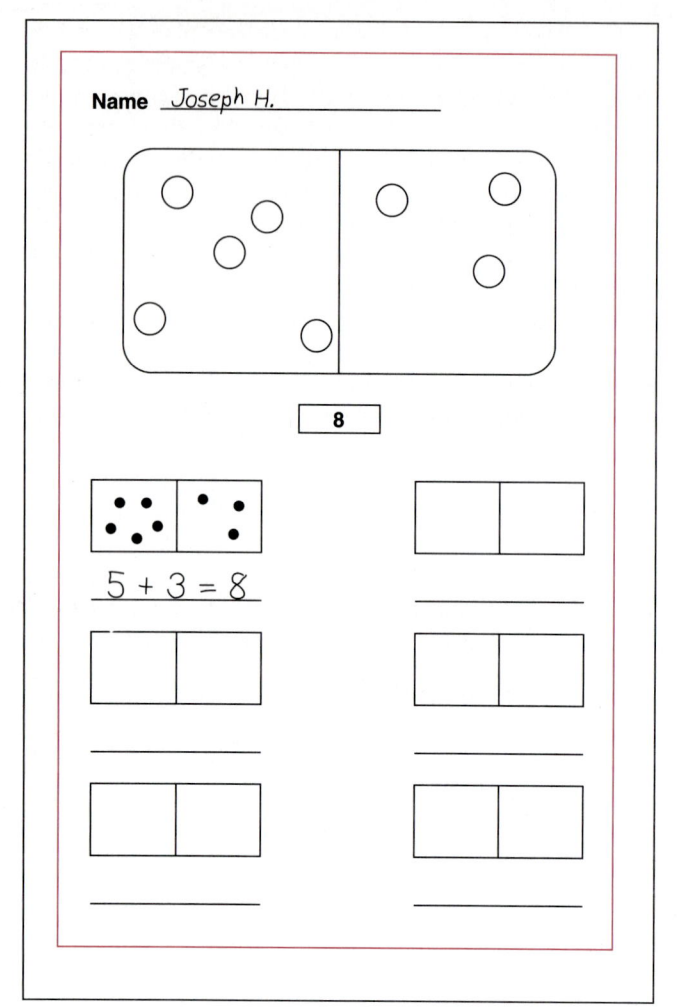

FIGURE 22.5: *A worksheet with counters*

previously introduced models, games, or activities. The content of the lesson remains the same.

Too frequently a lesson begins with an excellent discussion of a concept or relationship and then shifts dramatically to "here's how you do these problems." Students develop the idea that their real task in mathematics class is to learn how to do the exercises, how to "do the page." Older students even tell teachers, "Don't bother with all that explanation. Just show us how to do the problems." If, however, the consolidation portion of each lesson is at the same level as the development, then the goal of the lesson will be clear in the development. The practice time is a time to work on what was introduced. Students can and should be made responsible for demonstration of conceptual knowledge as well as procedural knowledge.

Specific strategies for consolidation activities are suggested in the following sections.

◆ Worksheets with Manipulatives

A common misconception concerning use of manipulative materials is that any activity done with them must be a group or class activity directed by the teacher. This simply is not the case. One of the best ways to approach the consolidation portion of a concept or connecting level lesson is to design a worksheet that requires the use of physical models. Such worksheets guide the specifics of a manipulative activity to be done at a desk or table. The students draw simple pictures to show what they have done and/or record numeric results of a manipulative activity. A few examples are provided here to suggest the idea.

Figure 22.5 is a worksheet for number combinations that might be used in first grade. Counters are placed in the area at the top of the page and separated into two parts, and dots are drawn in the small versions at the bottom. Textbooks are beginning to incorporate physical models into first- and second-grade lessons in this way.

Many activities on worksheets can involve base ten materials with the recording being done as a drawing or with numerals. The worksheet in Figure 22.6 is designed to follow a full-class activity involving trading with base ten

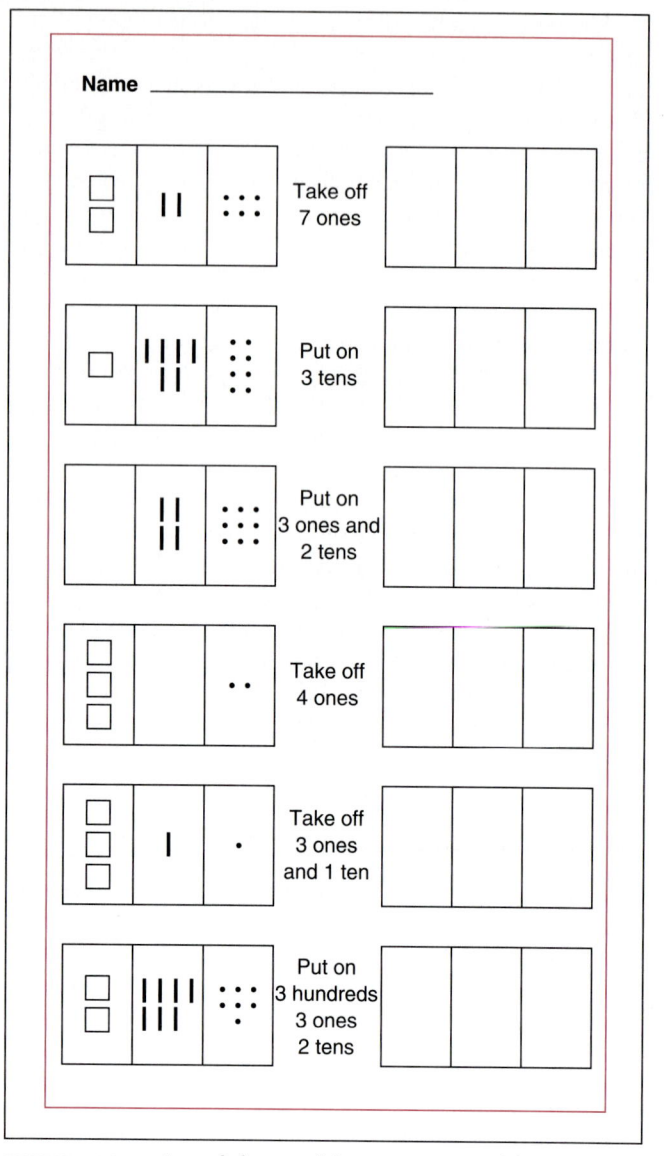

FIGURE 22.6: *A worksheet with base ten materials*

models. Students draw small dots, sticks, and squares to represent the base ten materials that they worked with to do the exercise.

The seventh-grade example in Figure 22.7 involves percentages approximated by familiar fractions. The students could use pie pieces and the hundredths disk (see earlier Figure 14.1), to determine the approximate equivalences without doing any computation. Several examples of the worksheet activity could be explored as part of the developmental portion of the lesson.

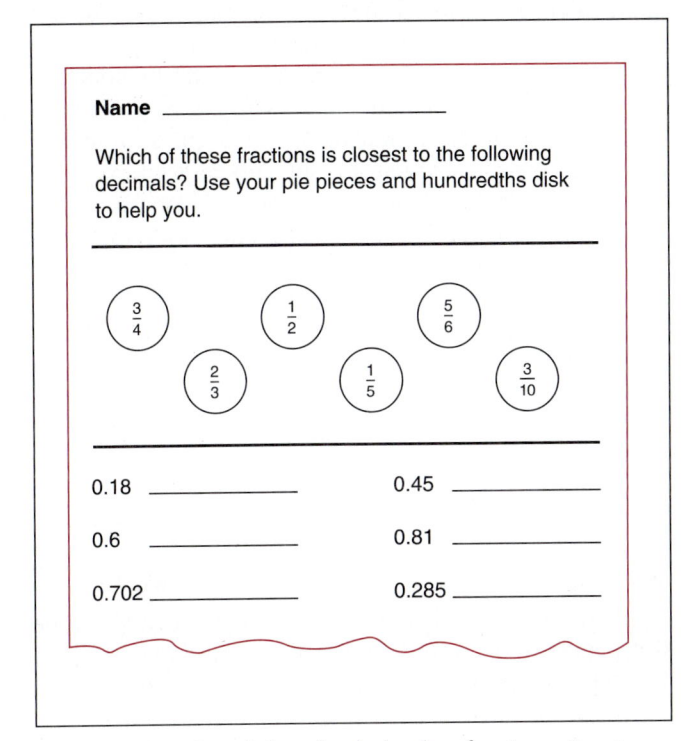

Name _____

Which of these fractions is closest to the following decimals? Use your pie pieces and hundredths disk to help you.

$\frac{3}{4}$ $\frac{1}{2}$ $\frac{5}{6}$

$\frac{2}{3}$ $\frac{1}{5}$ $\frac{3}{10}$

0.18 _____   0.45 _____
0.6 _____   0.81 _____
0.702 _____   0.285 _____

FIGURE 22.7: *A worksheet for decimal-to-fraction estimation*

◆ Workstations

A workstation approach is a profitable way to get children in Grades K through 2 actively involved. In this approach, materials that go with a particular activity are placed in separate tubs (plastic dishpans) or boxes. These materials might include special manipulatives, work cards to help guide the activity, and, if required, such things as scissors, paper, or paste. Each activity is taught and practiced in groups or with the full class for several days before being assigned to a tub. In this way, when children select an activity, they already know exactly how to do it and what is expected. For a given topic you might prepare as many as 8 or 10 related activities to be placed in workstations. These are carefully introduced over several days. Some activities may be games to be played by two or three children. Activities may be duplicated at more than one station.

The same containers for the activities are used throughout the year. Each tub is labeled and assigned to an area of the room: a space on a carpet, a table, or collection of desks shoved together. A sign at each area indicates which tub belongs there. Helpers can set out the tubs at the appropriate places in the room. Students can then freely select a station or they can be assigned. This procedure can remain the same throughout the year with the activities within the tubs changing as they are taught to the class.

Several additional comments about the workstation approach are appropriate. It is important that all students go to the workstations at once or that any group not going to a workstation have a similar activity to do at their seats. In this way you are available to interact with and monitor the activity at the stations.

Early in the year, station activities should be so easy that all children will have no difficulty. During this early period you can focus attention on the form of acceptable behavior at the stations, how to get them out, and how to put them away.

Some form of recording of the activity should be included when possible. This may involve drawing pictures, pasting down paper counters, or recording symbolically what was done manipulatively. The recording serves several purposes. Most significantly it represents accountability and thereby lends an air of responsibility and importance to the activity. Second, recording is a way to connect symbolic activity to conceptual activity. Finally, the records are lasting. They can be displayed on bulletin boards, sent home for parents, or placed in students' portfolios.*

◆ Workstations in the Upper Grades

Variations on the workstation approach are useful even in middle school. The idea is simply to get all students involved independently or in small groups with familiar activities. For example, during a unit on percentage in the seventh grade, a number of different models and conceptual activities may be introduced during the first week. In the second week, a short portion of each period can be devoted to working on activities involving these models. If not enough materials are available, different materials can serve different activities. Written directions can be prepared so that class time is not required to explain such activity. Students can get materials and return with them to their seats without the need for tubs or stations. The teacher is free to interact, assist, diagnose, and remediate.

Many teachers in the sixth, seventh, and eighth grades permit students to work on homework during the last 15 or 20 minutes of class. More challenging activities can be used in a workstation approach, because the teacher is there to interact and materials can be provided.

---

*The workstation/tubbing approach was adapted from the *Mathematics Their Way Newsletter V* (Center for Innovation in Education, 1977). This method has been used effectively around the country in kindergartens and primary classrooms as a way to get students actively involved with materials. It provides choice and variety with many materials without the need for classroom sets of each type.

◆ Writing an Explanation

At the upper grades, students are able to express their ideas in writing. Rather than finish a conceptual lesson with the usual symbolic exercises in the text, select only two or three to be done. However, require that they be done with manipulatives or with the aid of drawings. On paper, the students' task is to write out a conceptual explanation of why the exercise was done the way it was or why the answer makes sense. Encourage the use of drawings in the explanation. To emphasize to the students that the responsibility is for the conceptual explanation and not for the answer, sometimes provide answers along with the exercises.

DEVELOPING STUDENT RESPONSIBILITY FOR CONCEPTS

Today, conceptual knowledge and process skills are the focus of school mathematics, not just a means of improving skills. As teachers, we must convey this value system to our students, or our efforts at promoting conceptual understanding may fail.

How do students know what is important and what is not? The answer lies in those things for which we hold them accountable. In school, accountability is determined by quizzes, worksheets, tests, and other forms of assessment. It is the results of these things that go home to Mom and Dad.

Most children are not able to see beyond the short-term goal of surviving Friday's test. They do not appreciate the value of understanding when all we evaluate is "how to do it." If the development of conceptual knowledge is as important as we think it is, we had better let students know it, and we had better back that up with what we evaluate. Students know that procedural knowledge is important because we constantly test and grade it. By not doing the same with conceptual knowledge, we deliver a clear, albeit unintended, message that it is not important. By fourth or fifth grade, students actually resist conceptual activities because they are not seen as important.

SUGGESTIONS

The following are ways we can tell our students, "It is important to understand this!"

◆ Be Interested and Enthusiastic About Conceptual Activities

Spend entire class periods on developmental activities, not just the first 10 minutes. Interact with student discussions and discoveries. Praise students verbally and often for "making a connection" or discovering a relationships. Let students know how hard you have worked to make materials.

◆ Keep Checklists of Student Involvement

During developmental activities, keep a grade book handy, and give students credit for their efforts (not just right answers). Let students know that a part of their grade is determined by their cooperation with activities and their participation in class discussions.

◆ Use Portfolios

Many developmental activities have products: something made, a recording worksheet, a drawing, a record of a game, and so on. If you have students do these things, then let them know they are important. Give credit for completing activities, even when they involve errors. Reward efforts. Make helpful comments on these products before returning. Provide a system for students to keep evaluated work, especially that showing conceptual understanding, in a working portfolio. Then, in consultation with you, each student contributes material from the working portfolio to an assessment portfolio at the end of the unit or grading period. The portfolio should then be used as an integral component of your evaluation program. At the same time, it will send a strong message to students about what you value.

An extensive discussion of portfolios and performance assessment can be found in Chapter 5.

◆ Assign Conceptual Activities for Homework

Almost any type of conceptual activity that can be done independently at a student's desk can be done at home. Instead of assigning exercises "1 to 39, odd" for homework, select two or three to explain with models. (Homework is discussed in greater detail later in the chapter.)

◆ Include Conceptual Activities on Tests

Older children can be required to write explanations and draw pictures of what they have thought through with models. The models can and should be available during the test or quiz. Younger students can also be taught simple ways to make responses that demonstrate understanding. If tests or quizzes are administered orally, children can use materials on their desks in response. With groups of 8 to 10, you can see what everyone is doing. Such quizzes need not be lengthy, and you can quickly evaluate the entire class.

◆ Share Conceptual Activities in Your Class with Parents

Let your room visually reflect a commitment to concepts. Models should be visible, and student work that shows conceptual knowledge should be prominently displayed. Keep examples of students' concept activities, and share them with parents during conferences.

◆ Make Your Grading Scheme Reflect Conceptual Understanding

Let students and parents know how the various evaluations suggested here will be used to determine grades. We are fortunately in a time when almost all parents will be highly supportive of this approach.

COOPERATIVE LEARNING GROUPS

For both developmental and consolidation activities, organizing the class into small cooperative learning groups is an excellent idea for maximizing student involvement and interaction. Groups have been suggested for many activities throughout the book. In this section some suggestions for setting up and working with groups are offered.

SIZE AND COMPOSITION

There is no magic number for the size of a group. For example, Johnson and Johnson (1990) lean toward groups of three for mathematics, while Burns (1990) uses groups of four. Johnson, Johnson, and Holubec (1986) point out that the more members there are, the more possible interactions need to be accommodated. At some point, this causes difficulty. If students have not worked in groups before, it may well be good to begin with groups of two. In pairs, students can attend more easily to this new idea of working with someone. Later you will probably want to shift to groups of three or four.

"Think-Pair-Share" is a simple alternative to the more structured groups that many teachers use and that may well be a good beginning. In this method, a question or problem is posed to the full class, and individual students first think about how they would solve it or what ideas they might have. Next, they pair up with a partner next to them and share ideas.

Students within a group obviously need to work together, so some accommodation in the physical arrangement of the room is usually required. Desks can be shoved together or tables and chairs arranged to allow groups of students to work together. These arrangements can be either permanent or temporary. It is not necessary that the students remain in their group arrangement throughout the day or period.

A good case can be made for forming groups heterogeneously; that is, with a range of ability levels within the groups. This permits less able students the chance to hear ideas expressed by others and thus learn from them. At the same time, more capable students can benefit from explaining their ideas to the group. Ability grouping will quickly create a feeling of inferiority for those in the slower groups. For the purposes of interactive group work there is no benefit to homogeneous grouping.

Random assignment of students to groups on a daily or weekly basis is one method of easily obtaining heterogeneous groups. If group composition changes daily or at least weekly, quarrels over who gets to be with whom can be minimized. Tomorrow it will be different. Burns suggests dealing out playing cards, using four each of as many numbers ace, two, three, and so on, as are required to give one card to each class member. The ace group, the two group, and so on, are thereby quickly and randomly established.

For groups to be truly effective, students must develop interpersonal skills of cooperation, sharing, listening and responsibility. The group must view itself as a unit, not as individuals sitting at the same table. These skills may best be developed by having groups remain intact for at least several weeks or even several months. If you use a fixed-group approach, spend considerable time composing groups. Work on more than balance in abilities. Try also to mix boys and girls, to spread out both the leaders and the followers, to blend troublemakers with those who are cooperative. Do not put your problems all in one group.

RESPONSIBILITIES OF GROUP MEMBERS

Working cooperatively and productively in groups is not something that children do automatically or easily. Students need direction and practice with working together. Three simple but important rules for groups have been suggested in a number of places and are offered here, since they have proven to be quite effective:

1. Each student in the group is responsible for his or her own work and his or her own behavior.
2. Each student must be willing to help any group member who asks for help.
3. The only time during group work that a student may ask the teacher for help is when every student in the group has the same question.

Each of these rules is designed to help both students and teacher. Each requires discussion and periodic review with the class. The first is a basic rule of individual responsibility. It prevents students from complaining that they are being penalized for behavior of others and therefore encourages students to monitor both their own work and the group work. The rule does *not* mean that a student should work independently inside the group. Rather, it means each child has a responsibility to see that the group works effectively. Answer questions when asked; ask questions when you need help; listen to the ideas of others during discussions. If there is individual work to be done, that work is *yours*. It may not be borrowed. But group projects are group responsibilities, and each group member is responsible for seeing that the group does its best.

The second rule not only cuts down on the bothersome procedural questions, but lets each student know that there are helpers near by. It also is a way of telling each student to help others. Helping is part of everyone's responsibilities.

The last rule is the most significant. It not only cuts

down on procedural questions but promotes interaction within the group. This rule will be the most difficult of the three for *you* to keep. As teachers we all want to be responsive and helpful. If a child asks you a question during group work, you must remind him or her to check with the group. In the long run you are helping children to become more independent and responsible. The fact that the group members are there to offer help gives children a means to get assistance without needing the guidance of an adult.

When groups are given a task, the group is responsible for the task. One very effective way to ensure that the group works together is to allow only one worksheet or one record per group. The group can decide on a recorder to write down results. A presenter can be appointed to share results of the group work with the class, but each student in the group should be prepared to contribute. Where appropriate, other tasks such as gathering data or taking measurements can be assigned within each group. When every group member has a specific job or responsibility, the group has a better chance of working together. Allow the groups to make these decisions after offering suggestions. Group work does not mean four students working independently in a circle.

These general rules are a principal basis from which to develop effective cooperative group work. The more you regiment groups and make decisions for them, the more time you will spend on management. Invest efforts in helping children understand the rules of working in groups and the advantages that groups have for them. Expect cooperation from them and help them work together without your constant intervention.

During a lesson you can shift from full-class discussion to group work several times. Just because the groups have been formed at the beginning of a lesson does not mean that all work must be done by groups. Individual tasks can still be assigned, and questions can be posed to the whole class. At the same time, when an activity would benefit from reflection or interaction, the group structure is there to support it.

The way that groups function and the way that individuals function within the groups will require a lot of attention. Do not assume that you can tell students how to work in groups, and everything will then work perfectly. The skills of groups will grow as do other skills. As with any other objective, you need to periodically assess the performance of your groups and involve them in that assessment. In Figure 22.8, items for a self-rating scale are suggested (Collison, 1992). Collison suggests that the completed checklist be shared with the other members of the group, who are given the opportunity to disagree with that member's self-evaluation. After hearing the group's response, changes can be made if desired and comments added. The important idea is to actively involve students in reflecting on and evaluating their own and their group's behavior and effectiveness. Pride in the quality functioning of a group can easily result.

HOMEWORK

We have traditionally used homework as a means of providing extra drill and practice on the procedures that have been taught that day. Drill-oriented homework is an obvious extension of lessons that end with a procedural "how to do it" recipe. In a developmental approach to instruction, many lessons do not have this mechanical goal. Rather, the instruction requires reflection and students validating their results through the use of models or other methods. Drill-oriented homework is not an appropriate follow-up to these lessons.

If symbolic drill and practice is not appropriate, does it follow that no homework can be assigned? Certainly not! Homework can serve the same purpose as the consolidation portion of a lesson. It can be a time when students are required to reflect on the ideas that have been explored during the lesson.

Homework is also an effective way to communicate to both students and parents the importance of conceptual understanding. If only drill lessons are followed by homework, students may well infer that drill is more important than understanding. When students take work home that involves models or requires writing, developing oral explanations, or gathering data, parents can see that their children are being required to do more than drill. Homework is a parent's window to your classroom. Students may discuss homework with their parents, and most parents will help see that it is done. If developmental lessons are not followed by developmental homework assignments, parents will not see this portion of your instruction.

Homework also builds self-reliance as students are forced to grapple with ideas apart from the teacher's guidance. In this sense, homework is a way of communicating to students, "I know you are able to do this on your own."

Finally, practice *is* an important part of learning and requires time. When you have activities that require practice, from basic-fact drill to solving verbal percent problems, homework is an effective way to provide that practice.

SUGGESTIONS FOR HOMEWORK ASSIGNMENTS

The design of good homework follows much the same approach as the consolidation portion of the lesson described earlier. Students can use manipulatives, draw pictures, write explanations, find and correct errors, or prepare explanations for class. Return to the ideas presented for consolidation activities, and modify them for homework.

The manipulative worksheet approach is the easiest to adapt for home use. Many times the only materials that are required are simple counters. Children can use lima beans, buttons, or pennies as counters. If materials such as base ten pieces or fraction pieces are required, paper versions that students can cut out and take home are usually ade-

GROUP-PERFORMANCE RATING FORM

A. Group Participation
1. Participated in group discussion without prompting
2. Did his or her fair share of the work
3. Tried to dominate the group, interrupted others, spoke too much
4. Participated in the group's activities

B. Staying on the Topic
5. Paid attention, listened to what was being said and done
6. Made comments aimed at getting the group back on the topic
7. Got off the topic or changed the subject
8. Stayed on the topic

C. Offering Useful Ideas
9. Gave ideas and suggestions that helped the group
10. Offered helpful criticism and comments
11. Influenced the group's decisions and plans
12. Offered useful ideas

D. Consideration
13. Made positive, encouraging remarks about group members and their ideas
14. Gave recognition and credit to others for their ideas
15. Made inconsiderate or hostile comments about a group member
16. Was considerate of others

E. Involving Others
17. Got others involved by asking questions, requesting input, or challenging others
18. Tried to get the group working together to reach group agreements
19. Seriously considered the ideas of others
20. Involved others

F. Communicating
21. Spoke clearly, was easy to hear and understand
22. Expressed ideas clearly and effectively
23. Communicated clearly

G. Overall Experience
24. This group helped me to improve my understanding of the problems and the ways of solving them more than if I had worked alone.
25. Working with the group was an enjoyable experience.

FIGURE 22.8: *A group-performance rating scale. Students respond to each item with* Almost always, Often, Sometimes, *or* Rarely *(Collison, 1992).*

quate. For younger children, a periodic letter to parents may be required to explain how to help with cutting out models or providing manipulatives. Take the opportunity to tell the parent what your students are learning and enlist their support.

Drawing pictures and/or writing out explanations for three or four problems is a much better homework assignment for a connecting-level lesson than having students do 10 or 20 symbolic exercises. There is no law that says that all of the exercise problems in the text must be worked. Select a few and require students not only to produce the answers but also (and more importantly) the explanations. The responsibility to understand is a significant feature of such an assignment.

Of course there are times when the lesson is at the symbolic level, and a strictly procedural homework assignment is appropriate. For such assignments it is important to keep the purpose of the assignment in perspective. Provide sufficient practice to meet the needs of the students, but avoid creating a tedious burden just to show that you are giving homework.

USING HOMEWORK IN THE CLASSROOM

How you attend or do not attend to what was done for homework sends a clear value message about that homework to your students. However, homework checking can easily become a major time-consuming burden to your plans. It is important to find ways of dealing with homework that are quick yet effective.

Be sure that it was done. This does not mean it is necessary to take all homework up and grade or evaluate it. Simply checking that the homework assigned was completed is a minimal activity. This can usually be accomplished by simply requiring that it be placed out on the desks for you to see. Make it clear that you do keep a daily record of homework done and not done.

If an assignment consists of practice exercises, it is rarely necessary to go over each individual item. Try randomly selecting one or two exercises for quick review. For those assignments that required explanations or other conceptual work, a few students can be selected to share their ideas with the class. Not every student needs to share every day as long as you acknowledge that their work was done. Models, drawings, and written explanations should be out on desks while others are making presentations. Instead of having all students explain or present all of their work, ask students to compare what they did with what was shown first. "Martha, I see a different drawing on your paper. Can you tell us about that part?" Or, "Pete, it looks like you may have done about the same thing. Is that right, or did you do something differently?"

When homework is conceptual, it is not necessary to provide answer keys or solutions to every assignment. When students begin to appreciate that they are responsible for their understanding and that it is important both to you and to their grades, they will soon learn to check their own ideas against those shared in class.

THE BASAL TEXTBOOK: AN OVERVIEW

The basal textbook has been by far the most significant factor influencing instruction in the elementary classroom. In order to make decisions about the use of the textbook, it is good to have an objective view of textbooks and the role they can serve in instruction.

HOW ARE TEXTBOOKS DEVELOPED?

It is worthwhile to remember that publishing textbooks is a business. If the very best ideas from mathematics educators were incorporated into a textbook that did not sell, those excellent books sitting in warehouses would be of no value to students and would cost the publisher millions of dollars.

Most publishers enlist as authors mathematics educators and teachers who are quite knowledgeable about teaching mathematics. They also do extensive market research to determine what will sell and what teachers think they want in a book. There is frequently a significant gap between what the authors think would be good and what the publisher thinks will sell. Compromise between author and market becomes the rule. Rather than be forward with new ideas for content or pedagogy, publishers make a serious attempt to offend no one. As a consequence, there is frequently a significant time lapse between the state-of-the-art in mathematics education and what appears in textbooks.

Experience has demonstrated to publishers that if the text requires the teacher to provide additional materials or models, then there is an increased likelihood that the text will not sell. As a result, authors are somewhat limited in writing textbooks that incorporate all the principles of learning mathematics they know to be effective. Since 1990, there has finally been enough demand for models to be used that publishers have tried to respond. Punch-out cardboard models come with many books. Kits of materials, a very high-cost item for publishers, are made available as a purchase option.

Textbook authors are also limited by other factors. The textbook is a printed medium. Pictures can be put on pages, but manipulative models cannot. To convincingly illustrate movement and relationships, such as grouping 10 ones to make a single ten requires a lot of page space. Young children have a great deal of difficulty following time-lapse-sequence drawings that show materials in several stages of an activity. To use more pages simply creates a book that is too expensive and too heavy. The result is frequently a simple illustration followed by symbolic exercises that require less space.

THE TEACHER'S EDITIONS

The teacher's editions provide considerably more author freedom. Some publishers have provided one or two extra pages for each lesson as well as additional pages with ideas, explanations, and so on, in the front and back of each chapter. Frequently, the teacher's editions suggest excellent ways to teach the content of the lesson completely apart from doing the activity presented on the student pages. Teachers should take advantage of this information, especially the suggestions for activities. The package of off-page activities and student page exercises is meant by the authors as the suggested manner of teaching the concepts or skills of the lesson or unit. Too many teachers interpret the textbook curriculum as "getting students through the student pages" when, in fact, the real objectives require the much broader scope presented in the teacher editions. Pupil pages are just one tool for instruction. Pupil pages are not the objective, nor are they the curriculum.

THE TWO-PAGE LESSON FORMAT

The usual textbook lesson in the pupil book is presented on two pages. An observable pattern to these lessons can be seen in almost all popular textbook series. A portion of the first page consists of pictures and illustrations that depict the concepts for that lesson. The teacher is to use this section of the page to discuss the concepts with the students. Next are well-explained examples for the students to follow or an exercise guided by the text. Finally, the lesson has a series of exercises or practice activities. Thus, many lessons move from conceptual development to symbolic or procedural activities rather than a unit or chapter moving gradually over a period of days from concepts to procedures.

This three-part characterization of a textbook lesson is unfairly oversimplified. At the K–2 level, where the children write directly in consumable workbooks, what the student writes for all or most of the two pages may be closely tied to meaningful pictures or even simple hands-on models. The clear adherence to the two-page lesson is not always evident in seventh- and eighth-grade texts. As noted earlier, there is a definite movement in the textbook industry to develop pupil pages that require some form of hands-on models. When this is done, authors can write activities that are much like the consolidation portion of a lesson, activities that really are connected to conceptual development.

The traditional two-page format sends a clear message to students that the pictures, concepts, and discussion part of a lesson can be ignored. They begin to tune out until we begin to explain how to do the exercises. Following page-by-page and assigning procedural exercises from every lesson can easily negate all other efforts to communicate the importance of conceptual understanding.

SUGGESTIONS FOR TEXTBOOK USE

Perhaps the best maxim for avoiding the textbook focus on procedures is *teach the content, not the page*. Our task as teachers is to help children construct relationships and ideas, not to get them to do pages. We should look on the textbook simply as one of a variety of teaching tools available in the classroom, not the object of instruction.

If one considers the limitations of the printed medium and understands that the authors and publishers had to make compromises, then the textbook can be a source of ideas for designing lessons rather than dictate what the lessons will be. Here are some suggestions:

- Consider chapter objectives rather than lesson activities. The chapter or unit viewpoint will help focus on the learning objectives rather than the activity required to complete a page.

- Let the pace of your lessons through a unit be determined by student performance and understanding rather than the artificial norm of two pages per day.

- Use the ideas in the teacher's edition.

- Consider the conceptual portions of lessons as ideas or inspirations for planning more manipulative, interactive, and reflective activities. Think about how these practice exercises could be modified if students were to use models, write explanations, or discuss outcomes or approaches. The exercises are there for our *use*. We should use them wisely, not let them rule how we teach.

- Remember that there is no law saying every page must be done or every exercise completed. Select activities that suit your instructional goals rather than designing instruction to suit the text.

- Choose a few selected exercises to be done with manipulatives, drawings, discussions, or written explanations, rather than requiring students to complete all exercises.

- Feel free to omit pages and activities you believe to be inappropriate for the needs of your students and your instructional goals, assuming of course that the change is consistent with your district objectives.

- If the general approach in the text for a particular unit is not the same as the approach you prefer, omit its use for that unit altogether, or select only exercises that provide appropriate practice after you have developed the concepts with your method.

The text is usually a good guide for scope and sequence, especially for computational objectives. There is no reason that you as a teacher should be required to be a curriculum designer. If exercises are selected from pages covering the objective we are teaching, we can be reasonably sure that they have been designed to work well for that objective. It is not easy to make up good exercises for all activities. Take advantage of the text.

When drill and practice at the symbolic level are desired, look to the textbook as one source of such activity.

Problem solving, estimation, and mental arithmetic are areas that are becoming more and more visible in textbook series. These are difficult topics to teach, and we need all the help we can get. There is no need to do these activities only when they appear in the book. Good problems and activities for these areas are difficult to come by. Pull them from all over your book as suits the needs of your class.

Textbooks and other supplementary materials (ancillaries) supplied by publishers usually include evaluation instruments that may be of use for diagnosis, for guiding the pace of your instruction, or for evaluation. At the present time such tests are more likely to assess computational skills rather than conceptual understanding, problem solving, or other process skills. However, there is a definite effort to develop better testing methods for these higher-order skills. The availability of good tests provided by a publisher would be of significant value to the classroom teacher.

◆◆◆◆◆
REFLECTIONS ON CHAPTER 22: WRITING TO LEARN

1. It was suggested that over a given unit of instruction you should attend to balance and pace with respect to conceptual and procedural knowledge. What is meant in this context by "balance," and what is meant by "pace"?

2. In general, what is the purpose of the development portion of a lesson, and what happens in the consolidation portion of a lesson?

3. What are some ideas for getting a lesson started?

4. What are three ways that you can use to get feedback from students during a lesson?

5. Describe workstations as they would work for either primary grades or for upper-level classrooms.

6. Describe at least three things that can be done to help students focus on and take responsibility for concept learning.

7. Explain each of the three rules for cooperative groups that were suggested.

8. What are two valid reasons for assigning homework? Would the nature of the assignment be the same for each of these reasons? Explain.

9. Describe briefly what is meant by the "two-page lesson format" that is often adhered to in traditional basal textbooks. Why is this format generally not in keeping with a chapter or unit perspective for a lesson? What can the teacher do if using a standard text of this sort?

◆◆◆◆◆
FOR DISCUSSION AND EXPLORATION

1. Examine a textbook for any grade level. Look at a topic for a whole chapter, and decide on one or two major objectives that the chapter addresses. Then look at each individual lesson. How do lesson objectives relate to the unit objectives? How are students guided to work on conceptual development and connection of concepts to skills? Are any lessons predominantly conceptual? Are any clearly focused at connections?

2. Select a lesson from a textbook, and try to structure a developmental activity for the lesson based on ideas found either within the lesson itself or on the pages of the teacher's edition. What materials would you need? How could you get the lesson started with a problem-solving emphasis? See how you can use the exercises in the textbook lesson to your advantage. Remember it is not necessary to assign every exercise.

3. For the same lesson chosen in Exercise 2 or for another lesson, design either a consolidation activity or a homework assignment that is not purely symbolic or procedural.

4. Look through some back issues of *Arithmetic Teacher*, and try to find activities that would be good lesson starters. A file of these activities is a good resource to start and build on throughout your career. When planning a lesson it is not always easy to remember all of those neat games, problems, and other ideas you have seen in the past.

◆◆◆◆◆
SUGGESTED READINGS

Artzt, A. F., & Newman, C. M. (1990). *How to use cooperative learning in the mathematics class.* Reston, VA: National Council of Teachers of Mathematics.

Azzolino, A. (1990). Writing as a tool for teaching mathematics: The silent revolution. In T. J. Cooney (Ed.), *Teaching and learning mathematics in the 1990s.* Reston, VA: National Council of Teachers of Mathematics.

Behounek, K. J., Rosenbaum, L. J., Brown, L., & Burcalow, J. V. (1988). Our class has twenty-five teachers. *Arithmetic Teacher, 36*(4), 10–13.

Bell, A. (1986). Diagnostic teaching: 2 developing conflict-discussion lessons. *Mathematics Teaching, 116,* 26–29.

Davidson, N. (Ed.). (1990). *Cooperative learning in mathematics: A handbook for teachers.* Menlo Park, CA: Addison-Wesley Publishing Co.

Davidson, N. (1990). Small-group cooperative learning in mathematics. In T. J. Cooney (Ed.), *Teaching and learning mathematics in the 1990s.* Reston, VA: National Council of Teachers of Mathematics.

Ellis, A. K. (1988). Planning for mathematics instruction. In T. R. Post (Ed.), *Teaching mathematics in grades K–8: Research based methods.* Boston, MA: Allyn & Bacon.

Fitzgerald, W. M., & Bouck, M. K. (1993). Models of instruction. In D. T. Owens (Ed.), *Research ideas for the classroom: Middle grades mathematics.* New York: Macmillan Publishing Co.

Johnson, D. W., & Johnson, R. T. (1989). Cooperative learning in mathematics education. In P. R. Trafton (Ed.), *New directions for elementary school mathematics.* Reston, VA: National Council of Teachers of Mathematics.

Johnson, D. W., Johnson, R. T., Holubec, E. J., & Roy, P. (1984). *Circles of learning: Cooperation in the classroom.* Alexandria, VA: Association for Supervision and Curriculum Development.

Keedy, M. L. (1989). Textbooks and curriculum—Whose dilemma? *Arithmetic Teacher, 36*(7), 6.

Leinhardt, G. (1989). Math lessons: A contrast of novice and expert competence. *Journal for Research in Mathematics Education, 20,* 52–75.

Madsen, A. L., & Baker, K. (1993). Planning and organizing the middle grades mathematics curriculum. In D. T. Owens (Ed.), *Research ideas for the classroom: Middle grades mathematics.* New York: Macmillan Publishing Co.

Rowan, T. E., & Cetorelli, N. D. (1990). An eclectic model for teaching elementary school mathematics. In T. J. Cooney (Ed.), *Teaching and learning mathematics in the 1990s.* Reston, VA: National Council of Teachers of Mathematics.

Slavin, R. E. (1987). Cooperative learning and individualized instruction. *Arithmetic Teacher, 35*(3), 14–16.

Sutton, G. O. (1992). Cooperative learning works in mathematics. *Mathematics Teacher, 85,* 63–66.

Suydam, M. N. (1990). Planning for mathematics instruction. In J. N. Payne (Ed.), *Mathematics for the young child.* Reston, VA: National Council of Teachers of Mathematics.

Thornton, C. A., & Wilson, S. J. (1993). Classroom organization and models of instruction. In R. J. Jensen (Ed.), *Research ideas for the classroom: Early childhood mathematics.* New York: Macmillan Publishing Co.

Trafton, P. R. (1984). Toward more effective, efficient instruction in mathematics. *The Elementary School Journal, 84,* 514–528.

Walberb, H. J. (1985). Homework's powerful effects on learning. *Educational Leadership, 42*(7), 76–79.

Weber, L. J., & Todd, R. M. (1984). On homework. *Arithmetic Teacher, 31*(5), 40–41.

23 MATHEMATICS AND CHILDREN WITH SPECIAL NEEDS

EXCEPTIONAL CHILDREN

The Education for All Handicapped Children Act, or Public Law 94-142, was passed in 1975. The law mandates the most appropriate education for all handicapped children in the "least restrictive environment." Since the passage of P.L. 94-142, more and more children with some form of handicapping condition are being placed in regular education classrooms for all or part of the school day. The trend to accommodate these exceptional children within regular educational settings appears to be increasing.

Another group of exceptional children are the talented and gifted. These students also require special consideration. They represent a significant national resource and should be stimulated and provided with opportunities to reach their full potential.

Both regular and special education teachers should understand that the basic principles, strategies, and materials appropriate for any sound developmental instruction are also the principles, strategies, and materials appropriate for exceptional children.

CHILDREN WITH PERCEPTUAL AND COGNITIVE PROCESSING DEFICITS

CHARACTERISTICS OF SPECIFIC PROCESSING DEFICITS

A variety of terms are commonly used to label children who have difficulties processing information. Frequently these special children are referred to as *learning disabled* (LD) or *specific learning disabled* (SLD). Others are said to have *attention deficit disorders* (ADD). According to Glennon and Cruickshank (1981), *perceptual and cognitive processing deficit* would be a more accurate term than learning disability. These processing deficits are thought to be the result of some neurological dysfunction. Children with a processing deficit may be quite intelligent and by most definitions are not retarded. A boxed list of specific perceptual and cognitive processing deficits has been adapted from one developed by Bley and Thornton (1981).

When teaching mathematics to a child with a processing

PERCEPTUAL AND COGNITIVE PROCESSING DEFICITS	
PERCEPTUAL DEFICITS Figure-ground Discrimination Spatial organization	**INTEGRATIVE DEFICITS** Closure Receptive language Expressive language Abstract reasoning
MEMORY DEFICITS Short-term memory Long-term memory Sequential memory	**ATTENTION DISORDERS** Attention Impulsivity Compulsiveness

deficit, your focus should be on the child's pattern of learning strengths and weaknesses and not exclusively on failures in mathematics. Processing deficits span content areas. They are not unique to particular disciplines.

It is important that the teacher have as much detailed information as possible about children with cognitive or perceptual deficits so that instruction can be modified in the most effective manner. The school psychologist is the principal source of help to the teacher in identifying a child's specific learning problem. The teacher can work with the psychologist to aid in an accurate diagnosis. The psychologist can help the teacher better understand the nature of the dysfunction and how it affects learning. Both can work together to design the most effective instructional plan. Without a clear diagnosis, there can be a tendency to treat all children with "learning disabilities" the same. This global approach fails to capitalize on individual strengths and deal directly with specific weaknesses.

◆ Perceptual Deficits

Perceptual deficits may be visual or auditory. Children with *figure-ground* problems have difficulty sorting out or recognizing component parts of what they see or hear. Separate steps of computational algorithms may be confused with others. For example, the multiplication and subtraction procedures within long division may be confused or missed completely. When listening to directions with several parts, these children may have difficulty understanding the complete procedure. An entire page of exercises can be an incomprehensible maze.

Discrimination difficulties refer to the inability to discern differences in things that are seen or heard. Numbers or problems on the page become confused. The word "forty" may be heard as "fourteen." Numbers are frequently written backward. Coins are frequently confused.

Spatial organization deficits are observed in children who cannot organize their work on paper, cannot understand pictures of three-dimensional solids, or are confused with orientations of up, down, left, right, under, and so on. These children frequently have difficulty writing numbers or drawing figures.

◆ Memory Deficits

Memory deficits can also be specifically visual or auditory. Some children may have more difficulty recalling things seen than things heard, or vice versa. Children with *short-term memory* deficits can have trouble recalling things for even a few seconds. This causes difficulty copying from the board or the book or in following directions. From the time the child's eyes leave the board to the time he or she begins to write, the idea may already be forgotten. When he or she reads or listens to a verbal problem, the first parts of the problem may not be retained. After the reading, the entire problem may be mysterious.

Long-term memory deficits are manifested quite differently. Children with this problem will show less trouble with new material on the day presented but will have difficulty retaining material over days or weeks. Mastery of basic facts is a hallmark problem for children with long-term memory problems.

Sequential memory problems refer to the inability to retain an order to a sequence of events, or a series of steps in a procedure. These children frequently will ask, "What do I do next?" when doing a computational procedure such as adding fractions with unlike denominators. They may know how to do the steps but they may not know which comes next.

◆ Integrative Disorders

Integrative disorders refer to some dysfunction that makes it difficult for children to make connections or associations between ideas, to see similarities, to build one idea upon another.

Children with *closure* difficulties do not see or hear things in the same obvious groupings or clusters that they are presented in. They fail to see simple repeating patterns, to group the digits in large numbers that they are reading in a logical order, or to make connections between two similar problems or procedures. Counting on may be difficult. To solve word problems that require identifying key numbers, relationships, or actions and then seeing how these are the same as one of the four basic operations may be exceedingly difficult for these children.

A *receptive language* deficit is exhibited by the child who frequently asks, "What do you mean?" or gives a blank expression when a question is asked. If a problem is explained a bit differently than what was presented earlier, the child may not connect this different approach with the previous learning. The same can be true in visual presentations. If the model or diagram for a concept is different superficially although still representing the same idea, these children may not see the common relationships.

Expressive language deficits refer to the inability to verbalize what is known. The difficulty is in retrieving the specific words or symbols that are used to express their idea. These children would be able to identify a procedure that was done incorrectly, but they would be unable to explain the error.

Children with *abstract reasoning* deficits have much more difficulty constructing relationships and connecting abstract ideas with symbolism than do other children. Their reliance on models is much greater than most children. Memory, language, and proficiency with the models may not present a difficulty. However, since it is difficult for these children to connect ideas, to integrate new thoughts with existing ones, it is also difficult for them to abstract a relationship from their work with materials.

◆ Attention Disorders

Many children with processing deficits are very easily distracted, unable to focus for any length of time on one task. They are distracted by the slightest sounds, touches, visual stimuli, smells, hunger, and so forth, that would likely go unnoticed by the average student. Another feature of these *attention disturbances* is an overattention to unimportant details such as a variation in color of two essentially alike base ten models, or a wrinkle in a sheet of paper. They may repeatedly erase numbers or drawings to fix unimportant features.

Impulsivity is observed in children who tend to respond almost immediately to questions but give little thought to their answers. Consequently they frequently make what seem to be strange or even totally unrelated responses either orally or in written format.

Compulsiveness refers to a tendency to continue on with a train of thought without attending to changes in the task or without sensing when a task is completed. On a page of exercises with mixed operations these children may do all problems using the same operation as the first.

ADAPTING INSTRUCTION FOR CHILDREN WITH PROCESSING DEFICITS

Instructional methods for children with perceptual and cognitive processing deficits must be designed by taking into consideration the specific deficit or deficits involved. Children will not outgrow or overcome their problems with learning, but they can be helped to cope with them. Three general principles can be kept in mind:

1. Make an effort to add structure to your instruction, avoiding free and open discovery approaches.
2. Rely more heavily on models, spending a longer time and with less variation.
3. Teach to strengths while avoiding weaknesses.

◆ Teaching to Perceptual Deficits

With perceptual deficits the general rule is to focus on the mode of learning that is a strength for the child and avoid or provide special helping techniques for the mode of learning that causes difficulties.

The following specific suggestions may be useful:

Seat the child near you and/or near the chalkboard.

Keep the child's desk or workspace clear of clutter, focusing on one task at a time.

Provide directions in written form for children with auditory problems and in verbal form for those with visual difficulties.

Keep your voice moderate and at an even level. Repeat main ideas.

Structure the page for the child. Use a marker to block out all but one row of exercises or cut out a tagboard template to mask all but one problem at a time (Figure 23.1). Cut up worksheets into single problems or rows of exercises, or prepare separate exercise sheets with tasks arranged in separate boxes to avoid confusion.

Help the child with methods of organizing written work. Have computations done on centimeter grid paper with one numeral per square. Provide paper with columns drawn or templates in which to record computations.

Use a tape recorder (and headphones) with instruction on the tape to help students who have difficulty learning from visual materials.

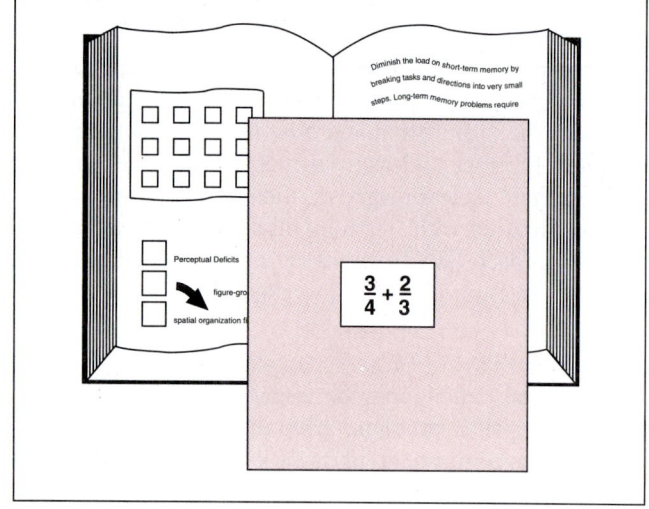

FIGURE 23.1: *A template frames the exercise to decrease perceptual problems.*

◆ Teaching to Memory Deficits

Diminish the load on short-term memory by breaking tasks and directions into very small steps. Long-term memory problems require overlearning, frequent practice, and as many associations with other ideas as possible. The following more specific suggestions may useful:

For short-term auditory memory problems, make an effort to give only one instruction at a time rather than a series of steps. If possible, write the steps of a task down on paper.

Provide a model exercise. Make each step clear (Figure 23.2).

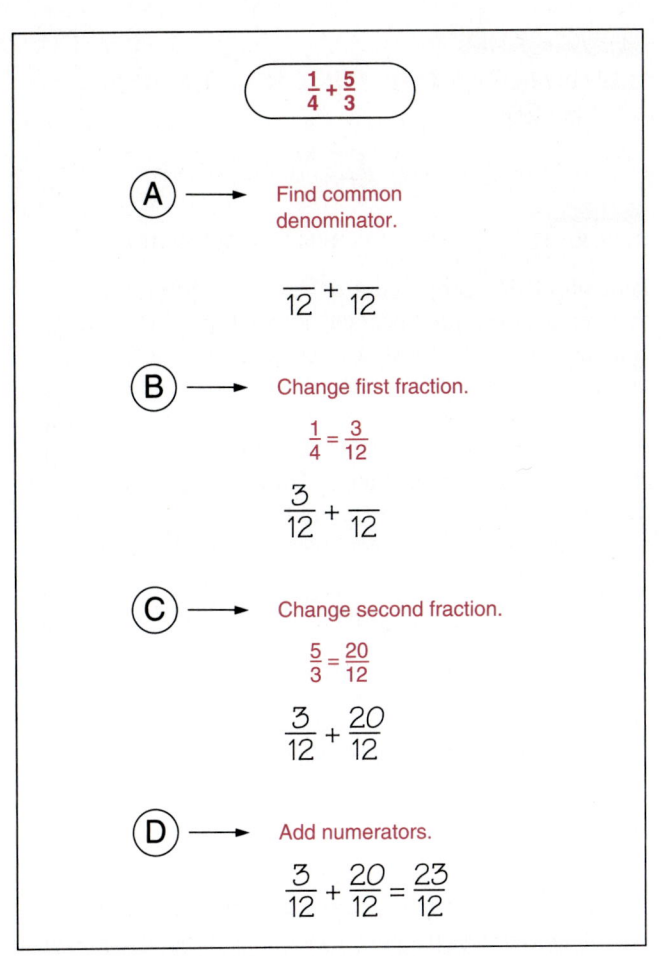

FIGURE 23.2: *An example problem in step-by-step form. Each step can be put on a separate card.*

For basic facts use strategies and relationships as in Chapter 8. Provide fact charts (Figure 23.3), and cross out facts that are mastered.

Permit the regular use of calculators. (Good for all students!)

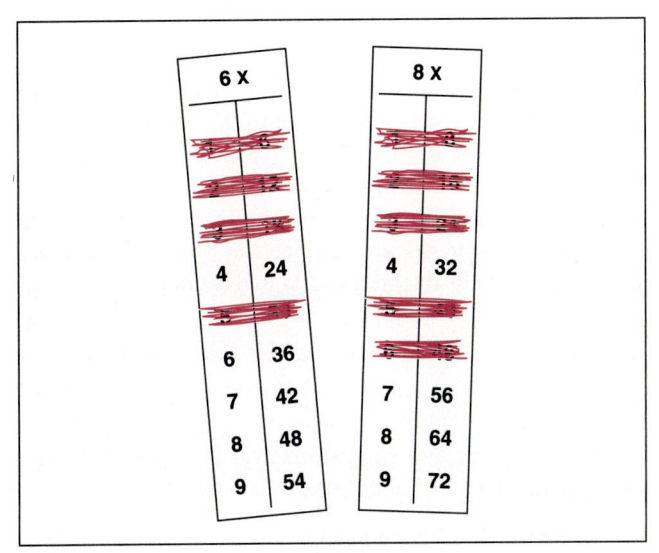

FIGURE 23.3: *Permit the use of fact charts, but encourage memorization.*

The ideas in Figures 23.1–23.3 have been adapted from Bley and Thornton, *Teaching Mathematics to the Learning Disabled* (1981).

◆ Teaching to Integrative Disorders

The best general principle for helping children with integrative disorders is to emphasize the same methods of a conceptual, developmental approach to learning.

Stay with the use of models much longer than usual.

Simplify instruction. Use familiar models with simple directions.

Avoid any symbolism that is unfamiliar and not useful to the child.

Provide repetitive practice of a newly learned idea. Children will feel comfortable with the success provided by doing a mastered task, and the overlearning will help make connections.

◆ Teaching to Attention Disorders

For children with attention problems, try to diminish the sources of distraction and decrease the demand for long-term attention.

Seat the child facing a bare wall. Isolate him or her from potential disturbances in the room.

Make assignments very short. Provide only one or two exercises to be done, and then give one or two more when those are complete.

Have the child work with a buddy.

Play a blank tape on which a tone or bell sounds intermittently every 2 to 5 minutes. The tone reminds the child to refocus his attention to the task. Similarly, set a timer to buzz in 3 minutes. After each buzz, the child resets the timer and returns to work.

SLOW LEARNERS AND THE MILDLY MENTALLY HANDICAPPED

CHARACTERISTICS OF SLOW LEARNERS AND MILDLY MENTALLY HANDICAPPED CHILDREN

The term *mentally handicapped* is used to describe those children with general intellectual functioning that is significantly below average. Frequently the term *educable mentally retarded* (EMR) is used instead. These children have measured intelligence scores (IQ scores) between 50 and 70 or between two and three standard deviations below the mean. The *slow learner* also exhibits general below-average intelligence, usually having an IQ score between 70 and 90 (Callahan & MacMillan, 1981). These children are usually well below their expected grade level in all areas of the curriculum.

Many mildly handicapped children will spend all or part of their school day in a regular education classroom. Most slow learners do not qualify for special education classes and are almost always taught by the regular education teacher.

MODIFICATIONS IN INSTRUCTION

Callahan and MacMillan (1981) contend that mildly mentally handicapped children "seem to learn in the very same fashion as their nonretarded peers do, albeit a little less efficiently" (p. 156). The most significant difference in their learning is in the time that is required. The implication of this conclusion, they explain, is that there is no need for some special set of materials or techniques for these children.

There are some additional suggestions that teachers will find useful. First, in selecting models and materials, simplicity and clarity of purpose should be the major criterion. Use the one best model over and over, rather than attempt to use a variety of models for the same concept. For example, these children may not be able to see how pie sections, fraction strips, and sets of counters all model the same idea since they are very slow to construct relationships. Second, make your instruction much more intense and explicit than is necessary with other children. Repeat and model introductory ideas several times. Provide very simple tasks using models that the students can easily repeat until they become comfortable with the idea. Third, since many of these special children have difficulty attending to task, instruction should be in small increments with no more than one new idea at a time.

Care should be taken that instruction in the classroom and instruction provided by any supplementary program be carefully coordinated. Vocabulary, methods, and materials should all match as closely as possible. Without this coordination, the added diversity of two different programs can inadvertently cause confusion rather than help.

MODIFICATIONS IN CURRICULUM

Since mentally handicapped children learn much more slowly than other children, it follows that less content can be learned during the years they are in school. It makes sense to focus the instructional time on those areas that are going to be of the most value to these students as adults. For example, while computational skills take up much of the regular curriculum, they are very complex and difficult for the mentally handicapped. These skills are accessible to them with the use of the calculator. At the same time, it is important that these students develop a conceptual knowledge of numbers, especially in terms of real-world referents. Therefore, the curriculum should be strong in the development of number concepts and should assist the student in learning numbers in the context of money and familiar measurements.

MATHEMATICS FOR THE GIFTED AND TALENTED

DESCRIPTION OF THE MATHEMATICALLY GIFTED

Renzulli (1978) proposed a definition of *gifted and talented* that requires the presence and interaction of three distinct characteristics: (1) above-average general intelligence (not necessarily superior), (2) task commitment, and (3) creativity. In addition to the traits described by the Renzulli definition, the *mathematically gifted* student will exhibit specific characteristics in mathematics. These will include a clearly demonstrated interest in things mathematical, mastery of mathematical skills at an early age, an ability to reason analytically, and an ability to perceive mathematical patterns and generalizations (Ridge & Renzulli, 1981; House, 1987).

INSTRUCTION FOR THE MATHEMATICALLY GIFTED

◆ Acceleration or Enrichment

Debate is common over the relative merits of vertical acceleration versus horizontal acceleration or enrichment. *Vertical acceleration* refers to movement through the regular curriculum at a more rapid pace. *Enrichment* is the expansion of the regular curriculum to include topics not generally encountered and the study of many standard topics in greater depth. The NCTM position on acceleration was first articulated in *An Agenda for Action,* which noted that "programs for the gifted student should be based on a sequential program of enrichment through ingenious problem solving opportunities rather than acceleration alone" (NCTM, 1980). In 1983 the council issued a position statement on vertical acceleration, noting that "vertical acceleration should be implemented *only* for the extremely talented and productive student" and then only in consultation with the student, parents, teachers, and counselors. Later, in 1986, the council issued a position on providing for the mathematically talented in which a preference for "a broad and enriched view of mathematics in a context of higher expectation" was clearly articulated. (See the box on "Provisions for Mathematically Talented and Gifted Students.")

◆ Gifted Students in the Regular Classroom

The classroom teacher who has only one or two gifted students has a difficult struggle meeting their needs. To accommodate various extensions and projects, the classroom teacher has several options to consider:

1. Allow advanced students to work independently or in a small group while directing the rest of the class as a group.

PROVISIONS FOR MATHEMATICALLY TALENTED AND GIFTED STUDENTS

The following statement is an official NCTM position.

All students deserve the opportunity to achieve their full potential; talented and gifted students in mathematics deserve no less. It is a fundamental responsibility of all school districts to identify mathematically talented and gifted students and to design and implement programs that meet their needs. Further, it is the responsibility of mathematics educators to provide appropriate instruction for such students.

The identification of mathematically talented and gifted students should be based on multiple assessment measures and should involve teachers, counselors, administrators, and other professional staff. In determining admission to talented and gifted programs, the evaluators must consider the student's total educational development as well as his or her mathematical ability, achievement, and aspirations. Eligible students and their parents should fully understand the nature and demands of the program before making a commitment to participate. Unqualified students should not be admitted for any reason.

The needs of mathematically talented and gifted students cannot be met by programs of study that only accelerate these students through the standard school curriculum, nor can they be met by programs that allow students to terminate their study of mathematics before their graduation from high school. The curriculum should provide for all mathematically talented and gifted students every year they are in school. These students need enriched and expanded curricula that emphasize higher-order thinking skills, nontraditional topics, and applications of skills and concepts in a variety of contexts.

Therefore, the National Council of Teachers of Mathematics recommends that all mathematically talented and gifted students should be enrolled in a program that provides a broad and enriched view of mathematics in a context of higher expectation. Acceleration within such a program is recommended only for those students whose interests, attitudes, and participation clearly reflect the ability to persevere and excel throughout the entire program.

(October 1986)

2. Group students by ability within the class, providing some instruction with each group as is frequently done in reading.
3. Group gifted students from several classrooms or grade levels to create a larger group of students who can work together and be provided with planned, directed instruction.

Each of these options has advantages and disadvantages. Approaches that permit talented students to interact or work together provide a more productive environment for enrichment but require extra work and preparation. Separation and special classes also raise difficult questions: How will the students be selected? What will the program content be? Will these students miss material in the regular class?

◆ A Triad Model for Planning Activities

Too frequently the highly capable student is provided with clever puzzles, computer games or explorations, or a variety of independent study projects that fail to fit together in any cohesive manner to create a program. While certainly better than nothing, these projects can become superficial pastimes or busy work and even turn students off of the idea of a serious investigation.

Renzulli (1977) suggested a model for the design of enrichment programs consisting of three types of activities. Type I activities are *general explorations to stimulate interest.* Type II activities are called *group training activities,* in which processes and skills related to a particular interest are developed and refined. Type III activities are *investigations of real problems.*

The following example illustrates the Renzulli model. A brief introduction to the Logo computer language is provided. Students are given an assortment of task cards from which to select and work on. The open exploration of Logo might go on for several sessions with challenges coming from the teacher, from books or task cards, or from the students themselves. These Type I activities are designed to explore and develop interest. If interest is there, the students are in need of Type II activities to develop their problem-solving skills and their knowledge of some of the procedural aspects of the Logo language. The teacher can plan and assist these activities by providing guided exploration with any number of available books. Students need not become Logo experts. However, they should realize that there is a knowledge base that they can develop and apply to their explorations. A Type III activity might then consist of a real project that will sustain their interest and draw on the processes of the Type II experiences. This might be a Logo contest to be entered at the school or system level, the design or modification of a game that other students in the class can then play, or perhaps the application of Logo to assist the teacher with planning or grading. An important feature of the Type III activity is the sharing of the results of the efforts with other students, teachers, or groups outside the school. This exposition removes the project from the realm of superficiality and provides a sense of real worth.

◆ Ideas and Beginnings

The start of good explorations for the gifted is not quite as difficult as it may seem on the surface. Type I activities are not all high-tech or even highly advanced. Many common materials and teacher resource books provide good beginnings from which further study into Type II and III activities can easily develop.

- Geometric models such as pattern blocks, Mira, kaleidoscopes, string designs, construction of geometric solids from sticks or poster board, the design of unusual tessellations, geoboards, and grid paper explorations all can be used to pose interesting tasks and explorations. Many commercial geometric materials come with resource books to assist teachers.

- Calculators, even simple ones, can provide interesting challenges. For example, how can the calculator be used to multiply two large numbers if the product is more than eight digits causing an overload? How many times do you have to multiply 0.99 by itself before the result is less than 0.01? Graphics calculators open up an even larger door of explorations.

- Probability problems can be explored at a wide range of levels. Computers or tables of random numbers can be used to create simulations of interesting phenomena such as batting averages, rainfall, lottery chances, traffic patterns, or even airplane crashes. Theoretical probabilities can be compared with actual experiments and discrepancies in the results can be discussed.

- Many computer software packages provide opportunities for exploration without special computer knowledge. Examples include *Rockey's Boots* (Learning Company), *Factory* and *Super Factory* (Sunburst), and more open-ended explorations such as those provided by the *Geometric Supposer* packages (Sunburst).

- Most any collection of problem-solving task cards, strategy games or puzzles, or number games offer good and challenging explorations.

Every teacher should make it a habit to collect a resource of ideas, materials, and activities to provide challenges for able students and to enrich their own knowledge of mathematics as well. There is no need to wait until a gifted child appears in the classroom. Examine commercial materials that are available, read the *Student Math Notes* provided quarterly to NCTM members, and read *Arithmetic Teacher* and/or *Mathematics Teacher* regularly. With a little effort it is easy to get hooked on activities that go beyond the regular curriculum. You will find yourself using many ideas with your entire class.

◆◆◆◆◆

REFLECTIONS ON CHAPTER 23: WRITING TO LEARN

1. Describe briefly in your own words, each of the following:
   a. perceptual deficit,
   b. memory deficit,
   c. integrative disorder,
   d. attention disorder.

2. Describe one or two adaptive teaching strategies for each of the above deficits or disorders.

3. According to Callahan and MacMillan, what is the main learning difference between regular education children and those identified as mildly mentally handicapped?

4. Describe one additional instructional change that is appropriate for mentally handicapped students.

5. What is the position of the NCTM relative to the issue of acceleration versus achievement for mathematically gifted children?

6. Describe briefly the three types of activities in the Renzulli model for programs for the mathematically gifted.

◆◆◆◆◆

FOR DISCUSSION AND EXPLORATION

1. Select a basal textbook for a grade level of your interest and a chapter on any topic within the book. Consider an introductory lesson in the chapter, and discuss how the lesson would be modified for:

   a. A student with a visual perceptual processing deficit.

   b. A student with an integrative processing deficit.

   Or, select any particular processing disorder of your interest.

2. What physical models would you select for teaching early basic number concepts to slow learners? Design a series of lessons for these children to develop number concepts. How do your lessons differ from what you might have planned for other students?

3. Repeat question two for one of the following topics:

   base ten numeration

   fraction concepts

   decimal and/or percent concepts

4. What broad topics in the fourth- to eighth-grade curriculum do you think are the most important for the mildly mentally handicapped? Which topics do you think could be eliminated? Defend your selections and compare them with others.

5. Refer to one of the resources at the end of this chapter or to one suggested by your instructor. Find an activity that could serve as a Type I exploration for talented and gifted students in a grade level of your choice. Explore the activity yourself. What additional related study would be appropriate as a Type II activity? Can you design a "real problem" or exploration that would be a Type III activity?

SUGGESTED READINGS

Bley, N. S., & Thornton, C. A. (1982). Help for learning disabled students in the mainstream. In L. Silvey (Ed.), *Mathematics for the middle grades (5–9)*. Reston, VA: National Council of Teachers of Mathematics.

Bley, N. S., & Thornton, C. A. (1989). *Teaching mathematics to the learning disabled* (2nd ed.). Austin, TX: PRO-ED.

Brown, S. I., & Walter, M. I. (1983). *The art of problem posing*. Philadelphia: Franklin Institute Press.

Cawley, J. F. (1985). *Cognitive strategies and mathematics for the learning disabled*. Austin, TX: PRO-ED.

Glennon, V. J. (Ed.). (1981). *The mathematical education of exceptional children and youth: An interdisciplinary approach*. Reston, VA: National Council of Teachers of Mathematics.

Haag, V., Kaufman, B., Martin, E., & Rising, G. (1986). *Challenge: A program for the mathematically talented*. Menlo Park, CA: Addison–Wesley.

House, P. A. (Ed.). (1987). *Providing opportunities for the mathematically gifted, K–12*. Reston, VA: National Council of Teachers of Mathematics.

Johnson, S. W. (1979). *Arithmetic and learning disabilities: Guidelines for identification and remediation*. Boston: Allyn & Bacon.

Jones, S. M. (1982). Don't forget math for special students: Activities to identify and use modality strengths of learning disabled children. *School Science and Mathematics, 82*, 118–126.

Lamon, W. E. (Ed.). (1984 Summer). Educating mathematically gifted and talented children. Special issue of *Focus on Learning Problems in Mathematics, 6*.

Meyers, J. M., & Burton, G. (1989). Yes you can . . . plan appropriate instruction for learning disabled students. *Arithmetic Teacher, 36*(7), 46–50.

Milgram, R. M. (Ed.). (1989). *Teaching gifted and talented learners in regular classrooms*. Springfield, IL: Charles C. Thomas.

Moyer, M. B., & Moyer, J. C. (1985). Ensuring that practice makes perfect: Implications for children with learning disabilities. *Arithmetic Teacher, 33*(1), 40–42.

Thornton, C. A. (Ed.). (in press). *Windows of opportunity: Mathematics for children with special needs*. Reston, VA: National Council of Teachers of Mathematics.

Vance, J. H. (1986). The low achiever in mathematics: Readings from the *Arithmetic Teacher*. *Arithmetic Teacher, 33*(5), 20–23.

Wilmot, B., & Thornton, C. A. (1989). Mathematics teaching and learning: Meeting the needs of special learners. In P. R. Trafton (Ed.), *New directions for elementary school mathematics*. Reston, VA: National Council of Teachers of Mathematics.

APPENDIX: SUMMARIES OF CHANGES IN CONTENT AND EMPHASIS

CURRICULUM AND EVALUATION STANDARDS FOR SCHOOL MATHEMATICS*

K–4 MATHEMATICS

INCREASED ATTENTION

◆ **NUMBER**

Number sense
Place-value concepts
Meaning of fractions and decimals
Estimation of quantities

◆ **OPERATIONS AND COMPUTATION**

Meaning of operations
Operation sense
Mental computation
Estimation and the reasonableness of answers
Selection of an appropriate computational method
Use of calculators for complex computation
Thinking strategies for basic facts

DECREASED ATTENTION

◆ **NUMBER**

Early attention to reading, writing, and ordering numbers symbolically

◆ **OPERATIONS AND COMPUTATION**

Complex paper-and-pencil computations
Isolated treatment of paper-and-pencil computations
Addition and subtraction without renaming
Isolated treatment of division facts
Long division
Long division without remainders
Paper-and-pencil fraction computation
Use of rounding to estimate

---

*NCTM Commission on Standards for School Mathematics (1989). *Curriculum and Evaluation Standards for School Mathematics.* Reston, VA: National Council of Teachers of Mathematics, pages 18, 74, and 75. Reprinted by permission.

INCREASED ATTENTION

GEOMETRY AND MEASUREMENT

Properties of geometric figures
Geometric relationships
Spatial sense
Process of measuring
Concepts related to units of measurement
Actual measuring
Estimation of measurements
Use of measurement and geometry ideas throughout the curriculum

PROBABILITY AND STATISTICS

Collection and organization of data
Exploration of chance

PATTERNS AND RELATIONSHIPS

Pattern recognition and description
Use of variables to express relationships

PROBLEM SOLVING

Word problems with a variety of structures
Use of everyday problems
Applications
Study of patterns and relationships
Problem-solving strategies

INSTRUCTIONAL PRACTICES

Use of manipulative materials
Cooperative work
Discussion of mathematics
Questioning
Justification of thinking
Writing about mathematics
Problem-solving approach to instruction
Content integration
Use of calculators and computers

DECREASED ATTENTION

GEOMETRY AND MEASUREMENT

Primary focus on naming geometric figures
Memorization of equivalencies between units of measurement

PROBLEM SOLVING

Use of clue words to determine which operation to use

INSTRUCTIONAL PRACTICES

Rote practice
Rote memorization of rules
One answer and one method
Use of worksheets
Written practice
Teaching by telling

5–8 MATHEMATICS

INCREASED ATTENTION

◆ PROBLEM SOLVING

Pursuing open-ended problems and extended problem-solving projects

Investigating and formulating questions from problem situations

Representing situations verbally, numerically, graphically, geometrically, or symbolically

◆ COMMUNICATION

Discussing, writing, reading, and listening to mathematical ideas

◆ REASONING

Reasoning in spatial contexts

Reasoning with proportions

Reasoning from graphs

Reasoning inductively and deductively

◆ CONNECTIONS

Connecting mathematics to other subjects and to the world outside the classroom

Connecting topics within mathematics

Applying mathematics

◆ NUMBER/OPERATIONS/COMPUTATION

Developing number sense

Developing operation sense

Creating algorithms and procedures

Using estimation both in solving problems and in checking the reasonableness of results

Exploring relationships among representations of, and operations on, whole number, fractions, decimals, integers, and rational numbers

Developing an understanding of ratio, proportion, and percent

DECREASED ATTENTION

◆ PROBLEM SOLVING

Practicing routine, one-step problems

Practicing problems categorized by types (e.g., coin problems, age problems)

◆ COMMUNICATION

Doing fill-in-the-blank worksheets

Answering questions that require only yes, no, or a number as responses

◆ REASONING

Relying on outside authority (teacher or an answer key)

◆ CONNECTIONS

Learning isolated topics

Developing skills out of context

◆ NUMBER/OPERATIONS/COMPUTATIONS

Memorizing rules and algorithms

Practicing tedious paper-and-pencil computations

Finding exact forms of answers

Memorizing procedures, such as cross multiplication, without understanding

Practicing rounding numbers out of context

INCREASED ATTENTION

PATTERNS AND FUNCTIONS

Identifying and using functional relationships

Developing and using tables, graphs, and rules to describe situations

Interpreting among different mathematical representations

ALGEBRA

Developing an understanding of variables, expressions, and equations

Using a variety of methods to solve linear equations and informally investigate inequalities and nonlinear equations

STATISTICS

Using statistical methods to describe, analyze, evaluate, and make a decision

PROBABILITY

Creating experimental and theoretical modes of situations involving probabilities

GEOMETRY

Developing an understanding of geometric objects and relationships

Using geometry in solving problems

MEASUREMENT

Estimating and using measurement to solve problems

INSTRUCTIONAL PRACTICES

Actively involving students individually and in groups in exploring, conjecturing, analyzing, and applying mathematics in both a mathematical and a real-world context

Using appropriate technology for computation and exploration

Using concrete materials

Being a facilitator of learning

Assessing learning as an integral part of instruction

DECREASED ATTENTION

PATTERNS AND FUNCTIONS

Topics seldom in the current curriculum

ALGEBRA

Manipulating symbols

Memorizing procedures and drilling on equation solving

STATISTICS

Memorizing formulas

PROBABILITY

Memorizing formulas

GEOMETRY

Memorizing geometric vocabulary

Memorizing facts and relationships

MEASUREMENT

Memorizing and manipulating formulas

Converting within and between measurement systems

INSTRUCTIONAL PRACTICES

Teaching computations out of context

Drilling on paper-and-pencil algorithms

Teaching topics in isolation

Stressing memorization

Being the dispenser of knowledge

Testing for the sole purpose of assigning grades

REFERENCES

Armstrong, R. D. (1981). An area model for solving probability problems. In A. P. Schulte (Ed.), *Teaching probability and statistics*. Reston, VA: National Council of Teachers of Mathematics.

Au, W. K., Horton, J., & Ryba, K. (1987). Logo, teacher intervention, and the development of thinking skills. *The Computing Teacher, 15*(3), 12–16.

Azzolino, A. (1990). Writing as a tool for teaching mathematics: The silent revolution. In T. J. Cooney (Ed.), *Teaching and learning mathematics in the 1990s* (pp. 92–100). Reston, VA: National Council of Teachers of Mathematics.

Backhouse, J., Haggarty, L., Pirie, S., & Stratton, J. (1992). *Improving the learning of mathematics*. Portsmouth, NH: Heinemann.

Baker, J., & Baker, A. (1990). *Mathematics in process*. Portsmouth, NH: Heinemann.

Baratta-Lorton, M. (1976). *Mathematics their way*. Menlo Park, CA: Addison-Wesley.

Baratta-Lorton, M. (1979). *Work jobs II*. Menlo Park, CA: Addison-Wesley.

Baroody, A. J. (1985). Mastery of the basic number combinations: Internalization of relationships or facts? *Journal for Research in Mathematics Education, 16*, 38–98.

Baroody, A. J. (1987). *Children's mathematical thinking: A developmental framework for preschool, primary, and special education teachers*. New York: Teachers College Press.

Behr, M. J., Erlwanger, S., & Nichols, E. (1976). How children view equality sentences. *PMDC Technical Report No. 3*. Tallahassee: Florida State University.

Behr, M. J., Lesh, R., Post, T. R., & Silver, E. A. (1983). Rational-number concepts. In R. Lesh & M. Landau (Eds.), *Acquisition of mathematics concepts and processes*. New York: Academic Press.

Bennett, A. (1982). *Decimal squares: Step by step teachers guide, Readiness to advanced levels in decimals*. Fort Collins, CO: Scott Scientific, Inc.

Bennett, D. (1992). *Exploring geometry with the Geometer's Sketchpad*. Berkeley, CA: Key Curriculum Press.

Bezuszka, S., Kenney, M., & Silvey, L. (1977). *Tessellations: The geometry of patterns*. Palo Alto, CA: Creative Publications.

Bley, N. S., & Thornton, C. A. (1981). *Teaching mathematics to the learning disabled*. Rockville, MD: Aspen Systems Corp.

Blume, G. W., & Schoen, H. L. (1988). Mathematical problem-solving performance of eighth-grade programmers and nonprogrammers. *Journal for Research in Mathematics Education, 19*, 142–156.

Bolster, L. C., et al. (1988). *Invitation to mathematics: 2*. Glenview, IL: Scott, Foresman.

Booth, L. R. (1988). Children's difficulties in beginning algebra. In A. F. Coxford (Ed.), *The ideas of algebra, K–12*. Reston, VA: National Council of Teachers of Mathematics.

Borasi, R. (1990). The invisible hand operating in mathematics instruction: Students' conceptions and expectations. In T. J. Cooney (Ed.), *Teaching and learning mathematics in the 1990s* (pp. 174–182). Reston, VA: National Council of Teachers of Mathematics.

Bright, G. W., Behr, M. J., Post, T. R., & Wachsmuth, I. (1988). Identifying fractions on number lines. *Journal for Research in Mathematics Education, 19*, 215–232.

Brownell, W., & Chazal, C. (1935). The effects of premature drill in third grade arithmetic. *Journal of Educational Research, 29*, 17–28.

Bruner, J. S. (1960). *The process of education*. Cambridge, MA: Harvard University Press.

Bruner, J. S. (1963). *The process of education*. New York: Vintage Books.

Bruni, J. V. (1977). *Experiencing geometry*. Belmont, CA: Wadsworth.

Burger, W. F. (1985). Geometry. *Arithmetic Teacher, 32*(6), 52–56.

Burger, W. F., & Shaughnessy, J. M. (1986). Characterizing the van Hiele levels of development in geometry. *Journal for Research in Mathematics Education, 17*, 13–48.

Burk, D., Snider, A., & Symonds, P. (1988). *Box it or bag it mathematics: Teachers resource guide* [Kindergarten, First-Second]. Salem, OR: The Math Learning Center.

Burns, M. (1987). *A collection of math lessons from grades 3 through 6*. (The Math Solutions Publications), White Plains, NY: Cuisenaire Company of America (distributor).

Burns, M. (1989). *The Math Solutions newsletter,* Spring issue.

Burns, M. (1990a). *A collection of math lessons from grades 6 through 8*. New Rochelle, NY: Cuisenaire.

Burns, M. (1990b). The math solution: Using groups of four. In N. Davidson (Ed.), *Cooperative learning in mathematics: A handbook for teachers* (pp. 21–46). Menlo Park, CA: Addison-Wesley Pub. Co.

Burns, M. (1992). *About teaching mathematics: A K–8 resource*. Sausalito, CA: Marilyn Burns Education Associates.

Burns, M., & Tank, B. (1988). *A collection of math lessons from grades 1 through 3*. (The Math Solutions Publications), White Plains, NY: Cuisenaire Company of America (distributor).

Callahan, L. G., & Glennon, V. (1975). *Elementary school mathematics: A guide to current research* (4th ed.). Washington, DC: Association for Supervision and Curriculum Development.

Callahan, L. G., & MacMillan, D. L. (1981). Teaching mathematics to slow-learning and mentally retarded children. In V. J. Glennon (Ed.), *The mathematical education of exceptional children and youth: An interdisciplinary approach*. Reston, VA: National Council of Teachers of Mathematics.

Campbell, P. F. (1988). Microcomputers in the primary mathematics classroom. *Arithmetic Teacher, 35*(6), 22–30.

Campbell, P. F., & Clements, D. H. (1990). Using microcomputers for mathematics learning. In J. N. Payne (Ed.), *Mathematics for the young child*. Reston, VA: National Council of Teachers of Mathematics.

Campione, J. C., Brown, A. L., & Connell, M. L. (1989). Metacognition: On the importance of understanding what you are doing. In R. I. Charles & E. A. Silver (Eds.), *The teaching and assessing of mathematical problem solving* (pp. 93–114). Reston, VA: National Council of Teachers of Mathematics.

Carey, D. A. (1991). Number sentences: Linking addition and subtraction word problems and symbols. *Journal for Research in Mathematics Education, 22*, 266–280.

Carrier, C., Post, T. R., & Heck, W. (1985). Using microcomputers with fourth-grade students to reinforce arithmetic skills. *Journal for Research in Mathematics Education, 16*, 45–51.

Carpenter, T. P. (1986). Conceptual knowledge as a foundation for procedural knowledge: Implications from research on the initial learning. In J. Hiebert (Ed.), *Conceptual and procedural knowledge: The case of mathematics*. Hillsdale, NJ: Lawrence Erlbaum.

Carpenter, T. P., & Moser, J. M. (1983). The acquisition of addition and subtraction concepts. In R. Lesh & M. Landau (Eds.), *Acquisition of mathematics concepts and processes*. New York: Academic Press.

Carroll, W. M. (1988). Cross sections of clay solids. *Arithmetic Teacher, 35*(7), 6–11.

Cauley, K. M. (1988). Construction of logical knowledge: Study of borrowing in subtraction. *Journal of Educational Psychology, 80*, 202–205.

Center for Innovation in Education. (1977). *Mathematics their way newsletter V*. Saratoga, CA.

Chalouh, L., & Herscovics, N. (1988). Teaching algebraic expressions in a meaningful way. In A. F. Coxford (Ed.), *The ideas of algebra, K–12*. Reston, VA: National Council of Teachers of Mathematics.

Charles, R., et al. (1985). *Problem-solving experiences in mathematics (grades 1 to 8)*. Menlo Park, CA: Addison-Wesley.

Charles., R., & Lester, F. (1982). *Teaching problem solving: What, why & how*. Palo Alto, CA: Dale Seymour.

Charles, R., Lester, F., & O'Daffer, P. (1987). *How to evaluate progress in problem solving*. Reston, VA: National Council of Teachers of Mathematics.

Chazan, D., & Houde, R. (1989). *How to use conjecturing and microcomputers to teach geometry*. Reston, VA: National Council of Teachers of Mathematics.

Clements, D. H. (1985a). *Computers in early and primary education*. Englewood Cliffs, NJ: Prentice-Hall.

Clements, D. H. (1985b). Logo programming: Can it change how children think? *Electronic Learning, 4*(4), 28, 74–75.

Clements, D. H. (1985c). Research on Logo in education: Is the turtle slow but steady, or not even in the race? *Logo in the Schools, 2*(2/3), 55–71.

Clements, D. H. (1986). Early studies on Logo and problem solving. *Logo Exchange, 5*(2), 23–25.

Cobb, P., & Merkel, B. (1989). Thinking strategies: Teaching arithmetic through problem solving. In P. R. Trafton (Ed.), *New directions for elementary school mathematics* (pp. 70–84). Reston, VA: National Council of Teachers of Mathematics.

Coburn, T. (1987). *How to teach mathematics using a calculator*. Reston, VA: National Council of Teachers of Mathematics.

Collison, J. (1992). Using performance assessment to determine mathematical dispositions. *Arithmetic Teacher, 39*(6), 40–47.

Coombs, B., & Harcourt, L. (1986). *Explorations 1*. Don Mills, Ontario: Addison-Wesley.

Countryman, J. (1992). *Writing to learn mathematics: Strategies that work, K–12*. Portsmouth, NH: Heinemann.

Crowley, M. L. (1987). The van Hiele model of the development of geometric thought. In M. M. Lindquist (Ed.), *Learning and teaching geometry, K–12*. Reston, VA: National Council of Teachers of Mathematics.

Dalbey, J., & Linn, M. C. (1985). The demands and requirements of computer programming: A literature review. *Journal of Educational Computing Research, 1*, 253–274.

Davidson, P. S. (1975). *Chip trading activities: Teacher's guide*. Fort Collins, CO: Scott Scientific, Inc.

Davis, E. J. (1978). Suggestions for teaching the basic facts of arithmetic. In M. N. Suydam (Ed.)., *Developing computational skills*. Reston, VA: National Council of Teachers of Mathematics.

Davis, R. B. (1986). *Learning mathematics: The cognitive science approach to mathematics education*. Norwood, NJ: Ablex.

Demana, F., & Osborne, A. (1988). Choosing a calculator: Four-function foul-ups. *Arithmetic Teacher, 35*(7), 2–3.

Dienes, Z. P. (1960). *Building up mathematics*. London: Hutchinson Educational Ltd.

Dossey, J. A., Mullis, I. V. S., Lindquist, M. M., & Chambers, D. L. (1988). *The mathematics report card: Are we measuring up?* Princeton, NJ: Educational Testing Service.

Elementary Science Study. (1966). *Teacher's guide for attribute games and problems*. Nashua, NH: Delta Education.

Erlwanger, S. (1975). Case studies of children's conceptions of mathematics: 1. *Journal of Children's Mathematical Behavior, 1*, 157–183.

Fair, J. (1987). *Tangram treasury (Books A, B, and C)*. White Plains, NY: Cuisenaire Company of America.

Fischer, F. E. (1990). A part-part-whole curriculum for teaching number in the kindergarten. *Journal for Research in Mathematics Education, 21*, 207–215.

Fitch, D. M. (1987). *Logo data tool kit*. Cambridge, MA: Terrapin, Inc.

Flanders, J. R. (1987). How much of the content in mathematics textbooks is new? *Arithmetic Teacher, 35*(1), 18–23.

Fortunato, I., Hecht, D., Tittle, C. K., & Alvarez, L. (1991). Metacognition and problem solving. *Arithmetic Teacher, 39*(4), 38–40.

Fuson, K. C. (1984). More complexities in subtraction. *Journal for Research in Mathematics Education, 15*, 214–225.

Fuson, K. C. (1992). Research on whole number addition and subtraction. In D. A. Grouws (Ed.), *Handbook of research on teaching and learning* (pp. 243–275). New York: Macmillan.

Fuson, K. C., & Brinko, K. T. (1985). The comparative effectiveness of microcomputers and flash cards in the drill and practice of basic mathematics facts. *Journal for Research in Mathematics Education, 16*, 225–232.

Fuson, K. C., & Hall, J. W. (1983). The acquisition of early number word meanings: A conceptual analysis and review. In H. P. Ginsburg (Ed.), *The development of mathematical thinking*. New York: Academic Press.

Fuson, K. C., Secada, W. G., 7 Hall, J.W. (1983). Matching, counting, and conservation of numerical equivalence. *Child Development, 54*, 91–97.

Fuys, D., Geddes, D., & Tischler, R. (1988). The van Hiele model of thinking in geometry among adolescents. *Journal for Research in Mathematics Education Monograph, 3*.

Garofalo, J. (1987). Metacognition and school mathematics. *Arithmetic Teacher, 34*(9), 22–23.

Gelman, R., & Gallistel, C. R. (1978). *The child's understanding of number*. Cambridge, MA: Harvard University Press.

Gelman, R., & Meck, E. (1986). The notion of principle: The case of counting. In J. Hiebert (Ed.), *Conceptual and procedural knowledge: The case of mathematics*. Hillsdale, NJ: Lawrence Erlbaum.

Gerver, R. (1990). Discovering pi—Two approaches. *Arithmetic Teacher, 37*(8), 18–22.

Ginsburg, H. P. (1977). *Children's arithmetic: The learning process*. New York: Van Nostrand.

Glennon, V. J., & Cruickshank, W. M. (1981). Teaching mathematics to children and youth with perceptual and cognitive processing deficits. In V. J. Glennon (Ed.), *The mathematical education of exceptional children and youth: An interdisciplinary approach*. Reston, VA: National Council of Teachers of Mathematics.

Gnanadesikan, M., Schaeffer, R., & Swift, J. (1987). *The art and techniques of simulation: Quantitative literacy series*. Palo Alto, CA: Dale Seymour.

Goldenberg, E. P. (1991). A mathematical conversation with fourth graders. *Arithmetic Teacher, 38*(8), 38–43.

Good, T. L., & Grouws, D. A. (1979). The Missouri mathematics effectiveness project: An experimental study in fourth-grade classrooms. *Journal of Educational Psychology, 71*, 355-362.

Good, T. L., Grouws, D. A., & Ebmeier, H. (1983). *Active mathematics teaching*. New York: Longman.

Goodnow, J., Hoogeboom, S., Moretti, G., Stephens, M., & Scanlin, A. (1987). *The problem solver: Activities for learning problem solving strategies*. Palo Alto, CA: Creative Publications.

Hatano, G. (1982). Learning to add and subtract: A Japanese perspective. In T. P. Carpenter (Ed.), *Addition and subtraction: A cognitive perspective*. Hillsdale, NJ: Lawrence Erlbaum.

Hazekamp, D. W. (1986). Components of mental multiplying. In H. Schoen (Ed.), *Estimation and mental computation*. Reston, VA: National Council of Teachers of Mathematics.

Hembree, R., & Dessart, D. D. (1986). Effects of hand-held calculators in precollege mathematics education: A meta-analysis. *Journal for Research in Mathematics Education, 17*, 83–99.

Hendrickson, A. D. (1986). Verbal multiplication and division problems: Some difficulties and some solutions. *Arithmetic Teacher, 33*(8), 26–33.

Herscovics, N., & Kieran, C. (1980). Constructing meaning for the concept of equation. *Mathematics Teacher, 73*, 573–580.

Hiebert, J., & Carpenter, T. (1992). Learning and teaching with understanding. In D. A. Grouws (Ed.), *Handbook of research on teaching and learning* (pp. 65–100). New York: Macmillan.

Hiebert, J., & Lefevre, P. (1986). Conceptual and procedural knowledge in mathematics: An introductory analysis. In J. Hiebert (Ed.), *Conceptual and procedural knowledge: The case of mathematics*. Hillsdale, NJ: Lawrence Erlbaum.

Hiebert, J., & Lindquist, M. M. (1990). Developing mathematical knowledge in the young child. In J. Payne (Ed.), *Mathematics for the young child* (pp. 17–36). Reston, VA: National Council of Teachers of Mathematics.

Hiebert, J., & Wearne, D. (1986). Procedures over concepts. The acquisition of decimal number knowledge. In J. Hiebert (Ed.), *Conceptual and procedural knowledge: The case of mathematics*. Hillsdale, NJ: Lawrence Erlbaum.

Hoffer, A. (1983). Van Hiele-based research. In R. Lesh & M. Landau (Eds.), *Acquisition of mathematics concepts and processes*. New York: Academic Press.

Hoffer, A. (1988). Ratios and proportional thinking. In T. R. Post (Ed.), *Teaching mathematics in grades K–8: Research based methods*. Boston: Allyn & Bacon.

Hoffer, A. R., & Hoffer, S. A. K. (1992). Geometry and visual thinking. In T. R. Post (Ed.), *Teaching mathematics in grades K–8: Research based methods* (pp. 249–277). Boston: Allyn & Bacon.

Holden, L. (1986). *Fraction factory*. Palo Alto, CA: Creative Publications.

Holden, L., & Roper, A. (1980). *The pattern factory*. Palo Alto, CA: Creative Publications.

Hope, J. A. (1986). Mental calculation: Anachronism or basic skill. In H. Schoen (Ed.), *Estimation and mental computation*. Reston, VA: National Council of Teachers of Mathematics.

Hope, J. A., Leutzinger, L., Reys, B. J., & Reys, R. R. (1988). *Mental math in the primary grades*. Palo Alto, CA: Dale Seymour.

Hope, J. A., Reys, B. J., & Reys, R. (1987). *Mental math in the middle grades*. Palo Alto, CA: Dale Seymour.

Hope, J. A., Reys, B. J., & Reys, R. (1988). *Mental math in the junior high school*. Palo Alto, CA: Dale Seymour.

Houde, R., & Yerushalmy, M. (1988). *Geometry problems and projects: Blackline masters for use with the Geometric Supposers*. Pleasantville, NY: Sunburst Communications.

House, P. A. (Ed.) (1987). *Providing opportunities for the mathematically gifted, K–12*. Reston, VA: National Council of Teachers of Mathematics.

Howden, H. (1989). Teaching number sense. *Arithmetic Teacher, 36*(6), 6–11.

Hyde, A. A., & Hyde, P. R. (1991). *Mathwise: Teaching mathematical thinking and problem solving*. Portsmouth, NH: Heinemann.

Institute for Research on Teaching. (1988). Reinventing the meaning of "knowing" in mathematics. *Communication Quarterly, 11*(1).

Janvier, M. (Ed.) (1987). *Problems of representation in the teaching and learning of mathematics*. Hillsdale, NJ: Lawrence Erlbaum.

Johnson, D. W., & Johnson, R. (1990). Using cooperative learning in math. In N. Davidson (Ed.), *Cooperative learning in mathematics: A handbook for teachers* (pp. 103–125). Menlo Park, CA: Addison-Wesley Pub. Co.

Johnson, D. W., Johnson, R., & Holubec, E. (1986). *Circles of learning* (rev. ed.). Edina, MN: Interaction Book Co.

Kamii, C. K. (1985). *Young children reinvent arithmetic*. New York: Teachers College Press.

Kamii, C. (1989). *Young children continue to reinvent arithmetic, 2nd grade*. New York: Teachers College Press.

Kamii, C., & Joseph, L. (1988). Teaching place value and double-column addition. *Arithmetic Teacher, 36*(6), 48–52.

Karplus, E. F., Karplus, R., & Wollman W. (1974). Intellectual development beyond elementary school (Vol. IV): Ratio, the influence of cognitive style. *School Science and Mathematics, 74,* 476–482.

Karplus, R., Pulos, S., & Stage, E. K. (1983). Proportional reasoning of early adolescents. In R. Lesh & M. Landau (Eds.), *Acquisition of mathematics concepts and processes*. New York: Academic Press.

Kieran, C. (1988). Two different approaches among algebra learners. In A. F. Coxford & A. P. Shulte (Eds.), *The ideas of algebra, K–12*. Reston, VA: National Council of Teachers of Mathematics.

Kouba, V. L., Brown, C. A., Carpenter, T. P., Lindquist, M. M., Silver, E. A., & Swafford, J. O. (1988a). Results of the fourth NAEP assessment of mathematics: Number operations, and word problems. *Arithmetic Teacher, 35*(8), 14–19.

Kouba, V. L., Brown, C. A., Carpenter, T. P., Lindquist, M. M., Silver, E. A., & Swafford, J. O. (1988b). Results of the fourth NAEP assessment of mathematics: Measurement, geometry, data interpretation, attitudes, and other topics. *Arithmetic Teacher, 35*(9), 10–16.

Labinowicz, E. (1980). *Piaget primer: Thinking, learning, teaching*. Menlo Park, CA: Addison-Wesley.

Labinowicz, E. (1985). *Learning from children: New beginnings for teaching numerical thinking*. Menlo Park, CA: Addison-Wesley.

Labinowicz, E. (1987). Assessing for learning: The interview method. *Arithmetic Teacher, 35*(3), 22–24.

Lampert, M. (1990). When the problem is not the question and the solution is not the answer: Mathematical knowing and teaching. *American Educational Research Journal, 27,* 29–63.

Landwehr, J. M., Swift, J., & Watkins, A. E. (1987). *Exploring surveys and information from samples: Quantitative literacy series*. Palo Alto, CA: Dale Seymour.

Landwehr, J. M., & Watkins, A. E. (1987). *Exploring data: Quantitative literacy series*. Palo Alto, CA: Dale Seymour.

Lappan, G., & Schram, P. W. (1989). Communication and reasoning: Critical dimensions of sense making in mathematics. In P. R. Trafton (Ed.), *New directions for elementary school mathematics* (pp. 14–30). Reston, VA: National Council of Teachers of Mathematics.

Lesh, R., Post, T., & Behr, M. (1987). Representations and translations among representations in mathematics learning and problem solving. In C. Janvier (Ed.), *Problems of representation in the teaching and learning of mathematics* (pp. 33–40). Hillsdale, NJ: Lawrence Erlbaum.

Lesh, R., & Zawojeski, J. S. (1992). Problem solving. In T. R. Post (Ed.), *Teaching mathematics in grades K-8: Research-based methods* (2nd ed.) (pp. 49–88). Boston: Allyn & Bacon.

Lester, F. (1985). Methodological considerations in research on mathematical problem-solving instruction. In E. A. Silver (Ed.), *Teaching and learning mathematical problem solving: Multiple research perspectives* (pp. 41–70). Hillsdale, NJ: Lawrence Erlbaum.

Lester, F. (1989). Reflections about mathematical problem-solving research. In R. I. Charles & E. A. Silver (Eds.), *The teaching and assessing of mathematical problem solving* (pp. 115–124). Reston, VA: National Council of Teachers of Mathematics.

Leutzinger, L. P., Rathmell, E. C., & Urbatsch, T. D. (1986). Developing estimation skills in the primary grades. In H. Schoen (Ed.), *Estimation and mental computation*. Reston, VA: National Council of Teachers of Mathematics.

Liedtke, W. (1988). Diagnosis in mathematics: The advantages of an interview. *Arithmetic Teacher, 36*(3), 26–29.

Lindquist, M. M. (1987). Estimation and mental computation: Measurement. *Arithmetic Teacher, 34*(5), 16–17.

Lindquist, M. M. (1989). It's time to change. In P. R. Trafton (Ed.), *New directions for elementary school mathematics* (pp. 1–13). Reston, VA: National Council of Teachers of Mathematics.

Madell, R. 1985). Children's natural processes. *Arithmetic Teacher, 32*(7), 20–22.

Madell, R., & Larkin, E. (1977). *Picturing numeration*. Palo Alto, CA: Creative Publications.

Markovits, Z., Eylon, B. S., & Bruckheimer, M. (1988). Difficulties students have with the function concept. In A. F. Coxford (Ed.), *The ideas of algebra, K–12*. Reston, VA: National Council of Teachers of Mathematics.

Masalski, W. (1990). *How to use the spreadsheet as a tool in the secondary school classroom*. Reston, VA: National Council of Teachers of Mathematics.

Mason, J., Burton, L., & Stacey, K. (1982). *Thinking mathematically*. London: Addison-Wesley.

McKnight, C. C., Crosswhite, F. J., Dossey, J. A., Kifer, E., Swafford, J. O., Travers, K. J., & Cooney, T. J. (1987). *The underachieving curriculum: Assessing U.S. school mathematics from an international perspective*. Champaign, IL: Stipes.

Michigan State Board of Education. (1899). *Michigan essential goals and objectives for mathematics education*. Lansing, MI: Michigan Department of Education.

Milauskas, G. A. (1987). Creative geometry problems can lead to creative problem solvers. In M. M. Lindquist (Ed.), *Learning and teaching geometry, K–12*. Reston, VA: National Council of Teachers of Mathematics.

Minnesota Educational Computing Consortium. (1984). *Estimation* (computer software). St. Paul, MN: MECC.

Morris, J. (1981). *How to develop problem solving using a calculator.* Reston, VA: National Council of Teachers of Mathematics.

National Council of Teachers of Mathematics. (1980). *An agenda for action: Recommendations for school mathematics of the 1980s.* Reston, VA: The Council.

National Council of Teachers of Mathematics. (1983). *A position statement: Vertical accelerations.* Reston, VA: The Council.

National Council of Teachers of Mathematics. (1986). *A position statement: Provisions for mathematically talented and gifted students.* Reston, VA: The Council.

National Council of Teachers of Mathematics: Commission on Standards for School Mathematics. (1989). *Curriculum and evaluation standards for school mathematics.* Reston, VA: The Council.

National Council of Teachers of Mathematics: Commission on Teaching Standards for School Mathematics. (1991). *Professional standards for teaching mathematics.* Reston, VA: The Council.

National Research Council. (1989). *Everybody counts: A report to the nation on the future of mathematics education.* Washington, DC: National Academy of Sciences.

National Research Council. (1990). *Reshaping school mathematics: A philosophy and framework for curriculum.* Washington, DC: National Academy of Sciences.

Noelting, G. (1980). The development of proportional reasoning and the ratio concept: 1. Differentiation of stages. *Educational Studies in Mathematics, 11,* 217–253.

Pagni, D. (Ed.). (1991). *Calculators and mathematics, Los Angeles (CAMP-LA)* (Book 1, Grades K–2; Book 2, Grades 3–4; Book 3, Grades 5–6; Book 4, Grades 7–8). Fullerton, CA: Cal State Fullerton Press.

Papert, S. (1980). *Mindstorms: Children, computers, and powerful ideas.* New York: Basic Books.

Payne, J. N. (1976). Review of research on fractions. In R. A. Lesh & D. A. Bradbard (Eds.), *Number and measurement: Papers from a research workshop* (pp. 145–187). Columbus, OH: ERIC/SMEAC.

Payne, J. N., & Rathmell, E. C. (1975). Number and numeration. In J. N. Payne (Ed.), *Mathematics learning in early childhood.* Reston, VA: National Council of Teachers of Mathematics.

Polya, G. (1957). *How to solve it: A new aspect of mathematical method.* Princeton, NJ: Princeton University Press.

Post, T. (1981). Fractions: Results and implications from the national assessment. *Arithmetic Teacher, 28*(9), 26–31.

Post, T. R., Behr, M. J., & Lesh, R. (1988). Proportionality and development of prealgebra understandings. In A. Coxford (Ed.), *The ideas of algebra, K–12.* Reston, VA: National Council of Teachers of Mathematics.

Post, T. R., Wachsmuth, I., Lesh, R., & Behr, M. J. (1985). Order and equivalence of rational numbers: A cognitive analysis. *Journal for Research in Mathematics Education, 16,* 18–36.

Pothier, Y., & Sawada, D. (1983). Partitioning: The emergence of rational number ideas in young children. *Journal for Research in Mathematics Education, 14,* 307–317.

Prentice, G. (1989). Flexible straws. *Arithmetic Teacher, 37*(3), 4–5.

Quintero, A. H. (1986). Children's conceptual understanding of situations involving multiplication. *Arithmetic Teacher, 33*(5), 34–37.

Ranucci, E. R., & Teeters, J. L. (1977). *Creating Escher-type drawings.* Palo Alto, CA: Creative Publications.

Rathmell, E. C. (1978). Using thinking strategies to teach the basic facts. In M. N. Suydam (Ed.), *Developing computational skills.* Reston, VA: National Council of Teachers of Mathematics.

Renzulli, J. S. (1977). *The enrichment triad model: A guide for developing defensible programs for the gifted and talented.* Wethersfield, CT: Creative Learning Press.

Renzulli, J. S. (1978). What makes giftedness? Reexamining a definition. *Phi Delta Kappan, 60,* 180–184, 261.

Resnick, L. (1983). A developmental theory of number understanding. In H. P. Ginsburg (Ed.), *The development of mathematical thinking.* New York: Academic Press.

Resnick, L. B. (1988). Treating mathematics as an ill-structured discipline. In R. I. Charles & E. A. Silver (Eds.), *The teaching and assessing of mathematical problem solving* (pp. 32–60). Reston, VA: National Council of Teachers of Mathematics.

Reys, R., Bestgen, B., Coburn, T., Schoen, H., Shumway, R., Wheatley, C., Wheatley, G., & White, A. (1979). *Keystrokes: Calculator activities for young students* (four booklets). Palo Alto, CA: Creative Publications.

Reys, R. E., & Reys, B. J. (1983). *Guide to using estimation skills and strategies (GUESS) Box I & II.* Palo Alto, CA: Dale Seymour.

Reys, R. E., & Reys, B. J. (1987). Calculators in the classroom: How can we make it happen? *Arithmetic Teacher, 34*(6), 12–14.

Reys, R., Trafton, P., Reys, B., & Zawojewski, J. (1987). *Computational estimation: (Grades 6, 7, 8).* Palo Alto, CA: Dale Seymour.

Ridge, H., & Renzulli, J. S. (1981). Teaching mathematics to the talented and gifted. In V. J. Glennon (Ed.)., *The mathematical education of exceptional children and youth: An interdisciplinary approach.* Reston, VA: National Council of Teachers of Mathematics.

Romberg, T. (1992). Perspectives on scholarship and research methods. In D. A. Grouws (Ed.), *Handbook of research on teaching and learning* (pp. 49–64). New York: Macmillan.

Ross, S. H. (1986). The development of children's place-value numeration concepts in grades two through five. Paper presented at the annual meeting of the American

Educational Research Association, San Francisco. *ERIC Document Reproduction Service no. ED 2773 482.*

Ross, S. H. (1989). Parts, wholes, and place value: A developmental perspective. *Arithmetic Teacher, 36*(6), 47–51.

Russell, J. C. (1992). *Spreadsheet activities in middle school mathematics.* Reston, VA: National Council of Teachers of Mathematics.

Scheer, J. K. (1980). The etiquette of diagnosis. *Arithmetic Teacher, 27*(9), 18–19.

Schoenfeld, A. H. (1985). *Mathematical problem solving.* New York: Academic Press.

Schoenfeld, A. H. (1988). What's all the fuss about metacognition? In A. H. Schoenfeld (Ed.), *Cognitive science and mathematics education.* Hillsdale, NJ: Lawrence Erlbaum.

Schoenfeld, A. H. (1989). Problem solving in context(s). In R. I. Charles & E. A. Silver (Eds.), *The teaching and assessing of mathematical problem solving* (pp. 82–92). Reston, VA: National Council of Teachers of Mathematics.

Schoenfeld, A. H. (1992). Learning to think mathematically: Problem solving, metacognition, and sense making in mathematics. In D. A. Grouws (Ed.), *Handbook of research on teaching and learning* (pp. 49–64). New York: Macmillan.

Schroeder, T. L., & Lester, F. K., Jr. (1989). Developing understanding in mathematics via problem solving. In P. R. Trafton (Ed.), *New directions for elementary school mathematics* (pp. 31–42). Reston, VA: National Council of Teachers of Mathematics.

Schwartz, J., & Yerushalmy, M. (1985). *The Geometric Supposers.* Cambridge, MA: Educational Development Center. (Available from Sunburst Communications.)

Seymour, D. (1971). *Tangramath.* Palo Alto, CA: Creative Publications.

Shumway, R. J. (1987). *101 ways to learn mathematics using BASIC (K–8).* Englewood Cliffs, NJ: Prentice-Hall.

Shumway, R. J. (1988). Calculators and computers. In T. R. Post (Ed.), *Teaching mathematics in grades K–8.* Boston: Allyn & Bacon.

Silver, E. A. (1986). Using conceptual and procedural knowledge: A focus on relationships. In J. Hiebert (Ed.), *Conceptual and procedural knowledge: The case of mathematics.* Hillsdale, NJ: Lawrence Erlbaum.

Skemp, R. (1978). Relational understanding and instrumental understanding. *Arithmetic Teacher, 26*(3), 9–15.

Smith, S. (1984). Microcomputers in the middle school. In V. P. Hansen & M. J. Zweng (Eds.), *Computers in mathematics education.* Reston, VA: National Council of Teachers of Mathematics.

Steinberg, R. M. (1985). Instruction on derived facts strategies in addition and subtraction. *Journal for Research in Mathematics Education, 16,* 337–355.

Stenmark, J. K. (1989). *Assessment alternatives in mathematics: An overview of assessment techniques that promote learning.* Berkeley, CA: EQUALS, University of California.

Stenmark, J. K. (Ed). (1991). *Mathematics assessment: Myths, models, good questions, and practical suggestions.* Reston, VA: National Council of Teachers of Mathematics.

Suydam, M. (1987). Indications from research on problem solving. In Curcio, F. R. (Ed.), *Teaching and learning: A problem-solving focus* (pp. 99–114). Reston, VA: National Council of Teachers of Mathematics.

Suydam, M., & Higgins, J. L. (1977). *Activity based learning in elementary school mathematics: Recommendations from research.* Columbus, OH: ERIC/SMEAC.

Suydam, M. N. (1984). Microcomputers in mathematics instruction. *Arithmetic Teacher, 32*(2), 35.

Thompson, C. S. (1990). Place value and larger numbers. In J. N. Payne (Ed.), *Mathematics for the young child.* Reston, VA: National Council of Teachers of Mathematics.

Thompson, C. S., & Hendrickson, A. D. (1986). Verbal addition and subtraction problems: Some difficulties and some solutions. *Arithmetic Teacher, 33*(7), 21–25.

Thompson, C. S., & Van de Walle, J. A. (1984a). Modeling subtraction situations. *Arithmetic Teacher, 32*(2), 8–12.

Thompson, C. S., & Van de Walle, J. A. (1984b). The power of 10. *Arithmetic Teacher, 32*(3), 6–11.

Thornton, C. A. (1982). Doubles up–Easy! *Arithmetic Teacher, 29*(8), 20.

Thornton, C. A., & Noxon, C. (1977). *Look into the facts: (Addition, subtraction, multiplication, division).* (Palo Alto, CA: Creative Publications.

Thornton, C. A., & Toohey, M. A. (1984). *A matter of facts: (Addition, subtraction, multiplication, division).* Palo Alto, CA: Creative Publications.

Trafton, Paul R. (1984). Toward more effective, efficient instruction in mathematics. *The Elementary School Journal, 84,* 514–528.

Usiskin, Z. (1987). Why elementary algebra can, should and must be an eighth-grade course for average students. *Mathematics Teacher, 80,* 428–438.

Usiskin, Z. (1988). Conceptions of school algebra and uses of variables. In A. F. Coxford (Ed.), *The ideas of algebra, K–12.* Reston, VA: National Council of Teachers of Mathematics.

Vonder Embse, C. (1992). Concept development and problem solving using graphing calculators in the middle school. In J. T. Fey (Ed.), *Calculators in Mathematics Education.* Reston, VA: National Council of Teachers of Mathematics.

Wagner, S. (1981). Conservation of equation and function under transformation of variable. *Journal for Research in Mathematics Education, 12,* 107–118.

Watt, D. (1983). *Learning with Logo.* New York: McGraw-Hill.

Wells, G. W. (1981). The relationship between the processes involved in problem solving and the processes involved in computer programming. *Dissertation Abstracts International*, 42, 2009A–2010A. (University Microfilms No. 81-23, 791).

Wheatley, G. H., & Hersberger, J. (1986). A calculator estimation activity. In H. Schoen (Ed.), *Estimation and mental computation.* Reston, VA: National Council of Teachers of Mathematics.

Whitin, D. J. (1989). The power of mathematical investigations. In P. R. Trafton (Ed.), *New directions for elementary school mathematics* (pp. 183–190). Reston, VA: National Council of Teachers of Mathematics.

Whitmer, J. C. (1992). *Spreadsheets in mathematics and science teaching.* School Science and Mathematics Association, Classroom Activities Series, No. 3.

Willoughby, Stephen S. (1990). *Mathematics education for a changing world.* Alexandria, VA: Association for Supervision and Curriculum Development.

Winter, M. J., Lappan, G., Phillips, E., & Fitzgerald, W. (1986). *Middle grades mathematics project: Spatial visualization.* Menlo Park, CA: Addison-Wesley.

Yackel, E., Cobb, P., Wood, T., Wheatley, G., & Merkel, G. (1990). The importance of social interaction in children's construction of mathematical knowledge. In T. J. Cooney (Ed.), *Teaching and learning mathematics in the 1990s* (pp. 12–21). Reston, VA: National Council of Teachers of Mathematics.

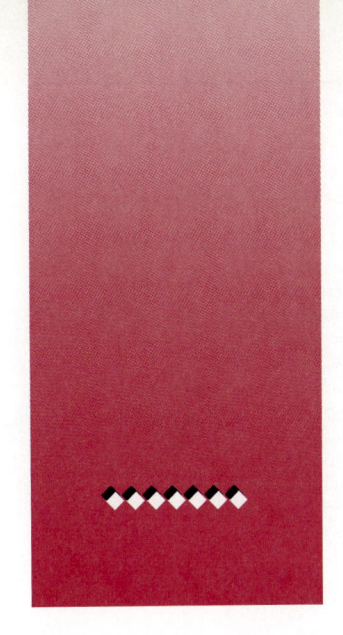

INDEX

Note: Reference to black-line masters is by title and number ("BLM," followed by the number of the individual black-line master).

Addition
  basic fact mastery for, 136–141
  comparison concept of, 113–115
  with fractions, 243–247
  front-end estimation strategy for, 211–212
  integer, 414
  order property in, 115
  part-part-whole concept of, 110–113
  repeated. See Repeated addition
  rounding in, 212
  using mental computation, 203–206
  word problems for, 115–118
  zero in, 115, 116
Addition algorithms. See also Algorithms
  directed development of, 182–183
  explorations and invented, 181–182
Addition and subtraction record blanks, BLM 16
Add one more set, 149
Affective goals
  assessment and, 62–63, 78
  for problem solving, 44, 60–61, 367, 368
Algorithms
  addition, 181–183, 245–247, 271
  cross-product, 285–286
  division, 192–199, 254–255, 271–272
  equivalent fraction, 238–239
  explanation of, 179
  mental. See Estimation; Mental computation
  methods of computing, 179
  multiplication, 185–192, 249–251, 271–272
  prerequisites for, 181
  standard pencil-and-paper, 180–181
  student-invented, 180
  subtraction, 183–185, 245–247, 271
American Statistical Association, 382
Analytic scoring scale, 72
Angles
  measurement of, 305–307
  relationships between, 352

Anxiety, mathematics, 28
Area
  activity to find, 13, 15–16
  comparison activities with, 298–300
  concept for multiplication, 121–122
  formulas for, 313–315, 358–359
  grids to measure, 302
  use of units of, 300–302
Area models
  for fractions, 222, 223, 235–238
  for multiplication algorithms, 186–187
Arrays
  drawing square, 149
  used to illustrate order property in multiplication, 122
Assessment. See also Grading; Performance-assessment tasks; Tests
  advantages of, 66
  approaches to, 74–79
  Curriculum and Evaluation Standards for School Mathematics and, 65–66
  grading and, 83–84
  interviews used for, 81–83
  of mental computation and estimation skills, 217–218
  performance, 68–74. See also Performance assessment; Performance-assessment tasks
  portfolio, 79–81, 452
  of problem-solving ability, 62–63
  purpose of, 66
  what to look for during, 66–68
Attention disorders
  teaching to, 463
  types of, 462
Attitudes
  assessment of, 67
  development of positive, 60–61
  regarding mathematics, 42

Attribute activities, 368
  loop activities, 367–371
Attribute materials
  activities using, 368–373
  description of, 368
Averages
  explanation of, 398
  techniques used with, 213

Backward trading game, 166–168
Bar graphs
  comparisons in, 105, 106
  description of, 392–394
  examination of mean by using, 400, 401
  stem-and-leaf, 393–395, 398
Baroody, A. J., 89, 90
Basal textbooks. See Textbooks
Base ten fractions
  explanation of, 257–258
  models for, 258, 267
Base ten grid paper, BLM 15
Base ten models
  counting with, 164
  groupable–pregrouped, 157–159
  used to learn algorithms, 181–182, 188–194
Base ten riddles, 164
Basic fact mastery
  for addition, 136–141
  for division, 150–151
  explanation of, 133
  for multiplication, 146–149
  remediation with upper-grade students, 151–152
  for subtraction, 141–146
  three-step approach to, 133–136
Basic facts, 133
Bean sticks, 98

I-1

I-2 INDEX

Beliefs
  assessment of, 67
  regarding mathematics, 42
Black-line masters. See individual titles
Blank hundreds chart (10 x 10 square), BLM 18
Box-and-wisker plots, 398–399
Bruner, Jerome, 28
Build-down through ten strategy, 143–144
Build-up through ten strategy, 143
Burns, Marilyn, 8

CAI See Computer-assisted instruction
Calculator activities
  adding five, 141
  connecting symbols to sequential grouping and trading, 168
  for decimal concept development, 262–263
  involving estimation, 214–216
  missing parts, 101
  for numeral writing, 90–91
  two-more-than, 96
Calculators
  computation and, 180–181
  fraction, 228
  graphing, 420, 432–433
  histograms produced by, 396
  modeling through use of, 32
  multiplication on, 120, 146
  NCTM position on, 428, 429
  patterns with, 378–380
  problem solving with, 12–13, 430–431
  reasons for using, 2, 430–431
  scientific, 408–409
  types and features of, 432
  used during assessment, 68, 78
  when and where to use, 431–434
Callahan, L. G., 464
CAMP-LA books, 431
Capacity, 302–303
Cardinality rule, 88
Cartesian products
  as multiplication concept, 120–121
  word problems with, 123
Charles, R., 57
Children
  exceptional, 460
  gifted and talented, 464–466
  mathematics from perspective of, 9–10
  with perceptual and cognitive processing deficits, 460–463
  slow learners and mildly mentally handicapped, 463–464
Circle
  formulas for, 315
  relationships between different parts of, 354–355
Circle graphs, 396–397
Circular fraction pieces, BLM 19–22
Classification
  emphasis on, 368
  learning schemes for, 369–370
  of shapes, 341–344
Classroom environment
  conducive to problem solving, 10–11
  requirements for, 17
  shifts in, 4
  teachers' role in, 17, 62
Clock reading, 317, 318
Closed questioning, 74
Cognitive schemas, 29
Combinations, 120–121
Combining Piles: Computation, 12, 15
Common-denominator algorithm, 254, 255
Communicative property, 122

Comparison
  addition concepts, 113–114
  with fractions, 232–234
  of measurable attributes, 293, 295–296, 304
  subtraction concepts, 114, 115
  word problems with, 117, 118
Compatible numbers
  for addition and subtraction, 206
  computational estimation using, 213–214
Compensation strategies
  for mental addition and subtraction, 206
  for mental multiplication, 208
Computation. See also Calculators; Computers; Mental computation
  calculators and, 180–181
  choices regarding method of, 202
  deemphasis of, 180
  estimation vs., 180
  fraction, 242
  options for, 201
  time involved in, 179
Computational estimation. See Estimation
Computer-assisted instruction (CAI), 434
Computer programming
  learning mathematics through, 438–440
  Logo and, 441–442. See also Logo
  problem-solving skills and, 440–441
Computers
  function of, 23
  impact on school mathematics, 2
  role in mathematics education, 434, 435
  used during assessment, 68
Computer software
  drill-and-practice, 434–436
  for estimation skills, 216
  for geometry, 360–364
  problem-solving, 436–437
  tutorial, 436
Computer tools
  geometry utilities, 438
  graphing packages, 437–438
  spreadsheets, 400, 401, 437
  symbol manipulators, 438
Concave, 340
Conceptual knowledge
  assessment of, 67
  connecting procedural to, 32–35, 445, 446
  explanation of, 22–24
  of measurement, 291, 292
  methods of promoting, 452–453
  of probability and statistics, 382
Cones
  classification of, 343, 344
  volume of, 316–317
Construction tips. See final section of text
Constructivist approach
  description and examples of, 28–29, 446
  methods of, 29
  reflective thought and, 29–30
Conventional knowledge, 22
Convex, 340
Cooperative learning groups
  group member responsibilities in, 453–455
  reflective thought in, 30
  size and composition of, 453
Counting
  development of skill in, 88
  fractional parts, 226–227
  meaning attached to, 88–89
  oral, 91
  of sets, 91–93
Counting-on strategy, 140
Countryman, J., 30
Crazy Mixed-Up Numbers, 97, 103
Cross-product algorithm, 285–286

Cubes
  explanation of, 343
Cuisenaire® rods, 222–223
Curriculum and Evaluation Standards for School Mathematics (NCTM)
  on assessment, 65–66
  on computation, 154, 201, 202, 270
  on geometry, 327
  impact of, 2–3
  on middle school instruction, 404
  publication of, 1
  on statistics, 391
  summary of, A-1—A-4
  on use of technology, 428, 434
  vision of, 3–4
Customary system, 310
Cylinders
  classification of, 343, 344
  volume of, 316

Data
  functions from real, 425
  graphical representations of, 392–397
  sources of, 391–392
Data Insights, 396, 397, 400
Decimal/fraction sieve, BLM 26
Decimals
  connections between fractions and, 257–262
  equivalents with fractions, 262, 266–267
  estimation with, 217, 270–272
  listed in order, 265–266
  number sense with, 263–267
  place value and, 259–262
  problem-solving explorations using, 15
  random numbers in form of, 385, 386
Denominators
  common, 234, 245, 254
  exercises using, 246–247
  unlike, 245–246
  use of term, 227
Derive (Soft Warehouse, Inc.), 438
Descriptive statistics
  averages in, 398
  box-and-wisker plots used in, 398–399
  explanation of, 397–398
Developmental instruction
  approaches to, 35–36
  balance and pace in, 445–446
  basal textbooks and, 456–458
  cooperative learning groups used for, 30, 453–454
  lesson development guidelines for, 446–452
  role of homework in, 454–456
  student responsibilities regarding, 452–453
Dienes, Zoltan, 28
Difference games, 372–373
Distributive property, 122
Dividend, 126
Divisibility tests, 407
Division
  activities involving, 126–127
  basic fact mastery for, 150–151
  classroom use of word problems using, 128
  at concept level, 195
  curriculum trends regarding, 192, 194
  fair sharing or partition concept of, 124–126
  with fractions, 242, 252–255
  fractions as expressions of, 239
  integer, 415–417
  measurement problems using, 128
  using mental computation, 208–209
  methods of recording, 196–198
  notation, language, and remainders in, 125–126, 129
  partition problems using, 128

Division *(continued)*
  repeated subtraction or measurement concept of, 125, 126
  rounding in, 212
  by zero, 127
Division algorithms
  curriculum trends and, 192, 194
  directed development of, 194–198
  explorations and invented, 194
  for two-digit divisors, 198–199
Divisors
  decimal, 272
  use of term, 126
  whole-number, 272
Dot cards, BLM 1–6
  activities using, 101–102
Dot paper
  activities using, 334–335
  problem-solving explorations using, 13
Double relationships, 103
Doubles and double again approach, 148, 149
Doubles and one more set, 148, 149
Doubles facts
  addition, 137, 138
  multiplication, 146
  subtraction, 143
Doubles plus two facts, 140
Draw-a-picture strategy, 48
Drill-and-practice software, 434–436
Drills
  features of effective, 151
  on homework, 454
  role in fact mastery, 134–135

Education for All Handicapped Children Act (Public Law 94–142), 460
Elastic Lines (Sunburst), 438
Enrichment, 464
Equations
  graphs and, 421–422
  graphs without, 422–423
  learning to solve, 420, 421
  with number patterns, 111
  problem-solving explorations using, 13, 16–17
Equivalent fractions
  concepts vs. rules for, 234–235
  development of algorithm for, 238–239
  models used to understand, 235–238
  as proportions, 287–288
Equivalent ratios, 278–281
Equivalent representations, 162–163
Erlwanger, S., 25, 26
Escher, M. C., 356
Estimation
  assessment of students using, 217–218
  using compatible numbers, 213–214
  with decimals and percents, 217, 269–272
  exercises using, 104, 214–217
  with fractions, 217, 242–244, 251, 264
  front-end methods for, 211–212
  of measurement, 104, 294–295, 310–312
  mental computation vs., 209
  rounding methods for, 212
  teaching methods for, 209–211
Evaluation standards, 4
Events
  dependent, 388–389
  independent, 386–387
  simple, 385
  theoretical probability of, 383
Everybody Counts: A Report to the Nation on the Future of School Mathematics (National Research Council), 1–2, 8
Experimental probability
  of simple and random events, 386–387

  theoretical probability vs., 383–384. *See also* Probability
Exponents
  explorations with, 409
  rules of, 407–408
  using scientific calculators, 408–409

Fact mastery. *See* Basic fact mastery
Factorization, 405–406
Factor-of-change approach, 278
Factors
  less than one, 249
  mixed-number, 250–251
  use of term, 126
  whole number as one, 249–250
Fair sharing, 124–125
Fibonacci sequence, 377–378, 406
Finding Areas: Making Connections within Mathematics, 13, 15–16
Find the Same Amount, 88, 89
5-cm triangle tiles, BLM 38
Fives facts, 146–147
Formulas
  area, 313–315, 358–359
  circle, 315
  errors resulting from use of, 312–313
  participation in development of, 312
  for volume, 315–317, 358–359
Forward trading game, 166–168
Fractional parts
  construction of, 224–227
  exercises using, 228–231
  relative size of, 231–232
  symbolism and, 227–228
  understanding of, 223, 224, 228
Fractions
  addition and subtraction with, 243–247
  categories of models, 222–223
  connections between decimals and, 257, 262
  in curriculum, 221–222
  division with, 242, 252–255
  equivalent, 234–239
  equivalents with decimals, 262, 266–267
  estimation with, 217, 242–244, 251, 264
  explanation of, 221
  as expressions of division, 239
  as expressions of ratios, 239
  multiplication with, 242, 247–251
  number sense with, 231–234, 242–243
  in simplest terms, 238
  unit, 230
Fraction strips, 222–223
Fraction words, 225
Front-end estimation
  for addition and subtraction, 211–212
  with fractions, 243, 247
Functions
  concept of, 423
  definition of, 423–424
  in geometry and measurement, 424, 425
  from real data, 425
  as solutions to word problems, 425–426

Geoboard pattern, BLM 35
Geoboard recording sheets, BLM 36
Geoboards
  problem-solving explorations using, 13
  used to draw shapes, 332–334
Geometer's Sketchpad (Key Curriculum Press), 361–362, 438
Geometric puzzles, 337–339
Geometric Supposers (Sunburst), 360–362, 438
Geometric thinking
  activities as approach to, 321–323
  characteristics of, 326–327

  and implications for instruction, 327–328
  levels, 325–326
Geometry
  approaches to, 323–324
  computer software for, 360–364
  functions in, 424, 425
  informal, 324
  reasons to study, 324–325
Geometry activities
  as approach to geometric thinking, 321–323
  hands-on material for, 324
  level 0, 327–340
  level 1, 327–328, 340–356
  level 2, 328, 356–360
  as problem-solving activities, 324
Gerver, R., 355
Gifted and talented children
  description of, 464
  instruction for, 464–466
  provisions for, 465
Golden ratio, 276
Grading. *See also* Assessment; Scoring; Tests
  on conceptual understanding, 453
  myths of, 83–84
  scales used for, 84
Graphical representations
  bar graphs as, 105, 106, 392–394
  box-and-wisker plots as, 398–399
  circle graphs as, 396–397
  using computers, 437–438
  connection of real world with, 105, 106
  continuous data graphs as, 395–396
  without equations, 422–423
  equations and, 421–422
  stem-and-leaf plots as, 393–395, 398
Grid paper, 334–335
Grids, 302, 370, 371
Grouping, 161–164. *See also* Trading
Group work. *See also* Cooperative learning groups
  assessment during, 68
  problem-solving environment for, 11

Half-centimeter square grid, BLM 29
Halve-and-double approach, 149, 208
Height, 313
Higher-decade facts, 204, 205
Histograms, 395–396
Holistic scoring, 72–73
Homework
  conceptual activities for, 452
  function of, 454
  suggestions for, 454–456
  used in classroom, 456
Howden, H., 87
Hundreds chart patterns, 406, 407
Hundreds master for bean stick base ten pieces, BLM 11
Hundredths, 267
Hundredths disk, BLM 24

If-then strategy, 51–52
In-Between Numbers: Development of a Concept, 12–13, 15
Informal geometry. *See* Geometry
Informal units of measurement
  benefits of using, 293–294
  explanation of, 293
Instrumental understanding
  explanation of, 24
  procedural knowledge and, 25
Integers
  adding and subtracting, 414, 415
  models for operations with, 412–414
  multiplying and dividing, 415–417
  negative, 411–412

INDEX

Integrative disorders
  teaching to, 463
  types of, 461–462
Interviews
  to assess mental computation and estimation skills, 218
  assessment through, 81–82
  planning for, 82
  tips for effective, 83
Invert-and-multiply algorithm, 254–255
Irrational numbers, 410–411

Japan, mathematics instruction in, 13, 98, 404
Joining
  activities involving, 111–112
  separating and, 115–117

Kamii, C., 182, 183
Knowledge
  assessment of, 67
  conceptual and procedural, 22–23. *See also* Conceptual knowledge; Procedural knowledge
  explanation of, 21
  individuality of, 35
  types of, 21–22

Lampert, M., 17
Length
  comparison activities with, 295–296
  measurement of, 111
  models for, 222–224, 237
  use of units of, 296–298
"Less" concept, 89–90
Lesson development
  consolidation and, 449–452
  elements of, 447–449
  general model for, 446–447
Lester, F., 57
Line graphs, 396
Line plots, 395, 396
Line symmetry, 349–350
Listening
  importance of, 10, 30–31, 36
  methods of, 26
List making, as problem-solving strategy, 50–51
Little base-ten-frames, BLM 12–13
Logical reasoning. *See also* Patterns
  problems involving, 370–372
  strategy for, 51–52
Logico-mathematical knowledge, 21–22
Logo
  explanation of, 362–363
  for gifted and talented students, 465
  ideas with, 364
  "instant," 363–364
  programming with, 441–442
  shapes in, 363

MacMillan, D. L., 464
Madell, R., 181–182
Make-a-table strategy, 50
Make-ten extended, 140
Make-ten facts, 139
Mass
  explanation of, 303
  measurement activities using, 304–305
Mathematics
  anxiety associated with, 28
  benefits of relational understanding of, 26–28
  conceptual knowledge and, 24. *See also* Conceptual knowledge
  connections within, 3–4, 13, 23–24
  giftedness in, 464
  negative attitudes regarding, 42
  relational vs. instrumental understanding of, 24–26
  revolution in school, 1–2
  as science of patterns and order, 7–8, 367, 417
Mathematics Assessment: Myths, Models, Good Questions, and Practical Suggestions (NCTM), 79, 81
Mathematics Exploration Tool Kit (IBM), 438
Mathematics instruction. *See* Developmental instruction
Mean
  compared to median, 399
  examination of, 399–401
  explanation of, 398
Measurement
  approximate nature of, 295
  concept of division, 125, 126, 253–254
  conceptual and procedural knowledge of, 291, 292
  development of formulas for, 312–317
  estimation of, 104, 294–295, 310–312
  functions in, 424, 425
  standard units of, 307–310
  of time, 317–319
  word problems using, 128
Measurement activities
  using angles, 305–307
  using area, 298–302
  to develop proportional reasoning, 282–285
  using length, 295–298
  using mass and weight, 303–305
  using models of units, 293
  using volume and capacity, 302–303
Measurement concepts
  development of, 292–295
  discussion of, 291–292
Measurement models, 222–224
Measuring cups, 303
Measuring instruments, student-made, 293
Median
  compared to mean, 399
  explanation of, 398
Memory, effect of relational understanding on, 27
Memory deficits
  teaching to, 462–463
  types of, 461
Mental computation. *See also* Computation
  for addition and subtraction, 203–206
  assessment of students using, 217–218
  computational estimation vs., 209. *See also* Estimation
  in curriculum, 202–203
  for division, 208–209
  extensions to early, 105–107
  with fractions, 242
  methods involved in, 180, 181
  for multiplication, 203, 206–208
  as option, 201
  using tens and hundreds, 172–173
Metacognition
  assessment of, 67
  development of, 59–60
Metacognitive goals, 40–41, 44, 62, 367–368
Meter sticks, 127
Metric system, 309–310
Mildly mentally handicapped children, 463–464
Mira, 350, 360
Missing-part activities
  importance of, 98–99
  types of, 100–101
Missing-part blanks, BLM 8
Mixed numbers
  addition and subtraction of, 247
  as factors, 250–251
  improper fractions and, 228

Mode, 398
Models
  assessment of use of, 71
  basic meanings developed with, 109–110
  benefits of using, 10–11
  examples of, 31–32
  fraction, 222–223, 258
  as links, 33, 34
  for operations, 412–414
  for place value, 157–160
  promoting reflective thought through use of, 30
  for repeated addition, 118, 119
  as separate language, 110
  translations involving, 34–35
  unit, 293
  uses of, 32, 48
"More" concept, 89–90
Multiplicand, 126
Multiplication
  area concept of, 121–122
  basic fact mastery for, 146–149
  Cartesian products concept of, 120–121
  distributive property in, 122
  with fractions, 242, 247–251
  front-end estimation strategy for, 212
  integer, 415–416
  using mental computation, 203, 206–208
  order property in, 122
  repeated addition concept of, 118–120
  role of zero and one in, 122
  rounding in, 212
  word problems for, 122–124
Multiplication algorithms
  directed development of, 187–188
  explorations and invented, 185–187
  for one-digit multipliers, 188–191
  repeated addition approach to, 192, 193
  for two-digit multipliers, 191–193
Multiplication and division record blanks, BLM 17
Multiplier, 126

National Council of Teachers of Mathematics (NCTM). *See also Curriculum and Evaluation Standards for School Mathematics* (NCTM); *Professional Standards for Teaching Mathematics* (NCTM)
  on acceleration, 464, 465
  position on calculator use, 428, 429
  position on computer use, 435
Near-double relationships, 103
Near-doubles facts
  addition, 137–139
  subtraction, 143
Negative numbers
  exploration of, 411
  mathematics definition of, 412
Nifty nines facts, 147–148
Nonproportional materials, 158–160
Nonroutine problems
  examples of, 46–47
  explanation of, 45
Number concepts
  development of, 87–93
  fact mastery and, 134
Number lines
  to assist with near and nice numbers, 172
  to examine relative magnitude, 169–170
  as models for multiplication, 118
  for primary students, 111
  used with integers, 412–413
Number relationships
  one to ten, 93–102
  ten to twenty, 102–103
Numbers, 410–411
  approximate, 170–171

beyond 1000, 173–176
compatible, 206, 213–214
irrational, 410–411
negative, 411, 412
production of random, 385–386
rational, 410
Number sense
with algorithms, 242–243
components of, 87
development of, 169–173
expansion of early, 104–106
explanation of, 106–107, 154
with fractions, 231–234
fractions and, 242–243
with whole numbers, 154, 232
Number theory
divisibility tests, 407
explanation of, 404
exponents, 407–409
patterns on hundreds chart, 406, 407
primes and factorization, 405–406
Number tiles, 98
Numeral recognition, 90–91
Numeral writing, 90
Numerator, 227

Observations
recording, 75–77
scoring, 73–74
1-cm hex grid, BLM 34
1-cm isometric dot grid, BLM 32
1-cm square/diagonal grid, BLM 33
1-cm square dot grid, BLM 30
1-cm square grid, BLM 28
One Equation: Consolidation of Simple Concepts, 13, 16–17
One-less-than facts, 142
One-more-than facts, 136
Ones, in multiplication, 122, 147
Open questioning, 74–75
Open sentences strategy, 53–54
Operations
addition and subtraction, 110–118. *See also* Addition; Subtraction
division, 124–129. *See also* Division
fact mastery and, 134
multiplication, 118–124. *See also* Multiplication
sources of meaning of, 109–110
Order property
in addition, 115
in multiplication, 122
Overhead projector, 216

Pace of instruction, 446
Parallelograms, 314, 315, 342
Partition
division problems using, 124–126, 128
with fractions, 252
Part-part-whole relationships
activities with, 99–100
addition, 110–112, 114
early mental math and, 106
elements of, 98–99
missing-part activities as variation of, 98–101
subtraction, 112–114
word problems with, 117
Patterns
using calculators, 378–380
Fibonacci sequence, 377–378, 406
growing, 376–377
on hundreds chart, 406, 407
mathematics as science of, 7–8, 367, 417
numeric, 378
problem solving by searching for, 12, 14–15, 49
recognition of, 93–95
repetitive, 373–375

Percents
approaches to, 267–268
estimation with, 217, 269–270
exercises with, 268–269
as proportions, 287–288
Perceptual deficits
teaching to, 462
types of, 461
Performance-assessment tasks. *See also* Assessment
creation of, 71–72
examples of, 68–71
scoring of, 72–74
Physical knowledge, 21
Piaget, Jean, 21, 28, 29
Picture graphs, 392, 393
Pigs-and-Chickens problem, 40, 41, 46
Place value
basic ideas of, 155–158
explanation of, 154
models for, 157–160
and numbers beyond 1000, 173–176
number sense development and, 169–173
relationship with 10, 102–103
Place-value concepts
algorithms and, 181
decimals and, 259–262
diagnosis of, 176–177
grouping activities to develop, 161–164
oral names for numbers to develop, 164–165
trading activities and written names to develop, 165–169
Place-value mat (with ten-frames), BLM 14
Platonic solids, 355, 359–360
0.5-cm square grid, BLM 29
Polya, G., 40, 42
Polygons, 341, 342
Portfolios
assessment of, 79–81, 452
content suggestions for, 79, 80
evaluation of, 81
explanation of, 79
problem-solving, 63
Prealgebra
explanation of, 404
function concepts and, 423–426
graphs and relationships and, 421–423
integer concepts and, 411–417
number theory and, 404–409
roots and irrationals and, 409–411
variable concepts and, 417–421
Prime numbers, 405–406
Prisms, 343, 344
Probability. *See also* Statistics
developing concepts of, 384–389
in elementary schools, 382
experimental approach to, 384
as ratio, 276
theoretical vs. experimental, 383–384
use of simulations in, 390–391
Problems, 39–40. *See also* Word problems
Problem solving. *See also* Word problems
affective goals for, 60–62, 367
assessment of skills in, 62–63
attitudes and beliefs regarding, 42
with calculators, 12–13, 430–431
computer programming and, 440
effect of relational understanding on, 27
environments conducive to, 10–11, 30
as focus of curriculum, 3, 8
four-step approach to, 42–44
goals for programs in, 367
instructional goals for, 44
metacognitive processes and, 40–41, 44, 59–60
for nonroutine problems, 45–47
in primary levels, 56–57

for routine problems, 45
software for, 436–437
teaching children about, 57–59, 62
in upper grades, 129–131
Problem-solving explorations
Combining Piles: Computation, 12, 15
discussion of, 11–12, 14
Finding Areas: Making Connections within Mathematics, 13, 15–16
In-Between Numbers: Development of a Concept, 12–13, 15
One Equation: Consolidation of Simple Concepts, 13, 16–17
Start and Jump: Searching for Patterns, 12, 14–15
Problem-solving portfolios, 63
Problem-solving strategies
assessment of, 67
draw a picture as, 48–49
explanation of, 40
list making as, 50–51
look for patterns as, 49
make table or chart as, 50
try-and-adjust as, 40, 47–48, 53
try simpler problem as, 52–53
used in combination, 54–56
work backward as, 51–52
write equation or open sentence as, 53–54
Procedural knowledge
assessment of, 67
connecting conceptual to, 32–35, 445, 446
in dealing with ratio and proportion, 277
explanation of, 23, 24
of measurement, 291
of numbers, 90
understanding of, 24–26
Procedures, 23
Processing deficits
instruction for children with, 462–463
types of, 460–462
Process problems, 46
Product, 126
Professional Standards for Teaching Mathematics (NCTM)
abbreviated list of standards for teaching, 18–19
classroom environment shifts listed in, 4, 17
statements endorsed by, 4–5
teaching standards listed in, 5
Proportional reasoning
in children, 276–278
equivalent ratios to develop, 278–281
measurement activities to develop, 282–285
scaling activities to develop, 281–282
Proportions
approaches to, 276–277
cross-product algorithm to solve, 285–286
exercises that explore, 274
explanation of, 276
intuitive methods for solving, 277–278
percent problems as, 287–288
written from models, 286–287
Protractors, 305–307
Pyramids
classification of, 343, 344
volume of, 316–317
Pythagorean theorem, 358, 359

Quadrilaterals
classification of, 342
diagonals of, 352–354
Quadrilateral Supposer, 362
Quantitative Literacy Project, 390
Quantities
multiples of, 123, 124
understanding expressions as, 418–419

Questioning
  closed vs. open, 74–75
  development of good, 75
Quotient, 126

Random numbers, 385–386
Rate
  ratios as expression of, 275–276
  times quantity, 123
Rational Number Project, 277
Rational numbers, 410
Ratios
  associations with concept of, 24, 25
  examples of, 276
  exercises that explore, 274
  as expressions of rates, 275–276
  fractions as, 239, 275
  selection of equivalent, 278–281
Real graphs, 392, 393
Reasoning. See also Proportional reasoning
  assessment of, 67
  mathematics as, 3, 8
  strategy of logical, 51–52, 370–372
Recording strips, 168, 169
Rectangles, 313–314
Reflective thought
  in dealing with proportional reasoning, 277
  explanation of, 29, 36, 445
  importance of, 35
  ways to promote, 30–31, 60
Region models, 222, 223, 235–237
Regrouping. See also Trading
  errors in, 191
  term of, 181
Relational understanding
  approaches to development of, 28–32
  benefits of, 26–28
  explanation of, 24
Relative magnitude, 169–170
Remainders
  in division, 126
  in word problems, 129
Remediation, 151–152
Renzulli, J. S., 464, 465
Repeated addition
  models for, 118, 119
  with multiplication algorithms, 187, 192, 193
  as multiplication concept, 118
Repeated subtraction, 125
Resnick, L., 17, 98
Right-triangle tiles, BLM 37
Risk taking
  encouragement of, 10
  within groups, 11, 30
Romberg, T., 9, 17
Roots, 409–410
Rotational symmetry, 350–352
Rounding
  in addition and subtraction, 212
  development of ability for, 171–172
  in multiplication and division, 212
Rulers, 297–298

"Same" concept, 89–90
Say It/Press It, 168
Scales, 304–305
Scaling activities, 281–282
Schoenfeld, A. H., 17
Scientific calculators, 408–409. See also Calculators
Scoring
  holistic, 72–73
  through observation, 73–74
Self-assessment, 77

Self-validation
  encouragement of, 11
  promoting reflective thought through use of, 30
Sequences, 376. See also Patterns
Set models, 223, 224, 237
Shapes
  building, drawing, and making, 331–337
  classified by properties, 340–343
  using Logo, 363
  matching, 329–331
  sorting and classifying, 328, 329
Simulations, 390–391
Skemp, R., 24, 27
Slow learners, 463–464
Software. See Computer software
Solids, 342–343. See also Three-dimensional shapes
Sort-backward strategy, 51
Speed tests, 79
Spreadsheets, 400, 401, 437
Squares on the checkerboard problem, 43, 46
Standards. See Curriculum and Evaluation Standards for School Mathematics (NCTM); Professional Standards for Teaching Mathematics (NCTM)
Standard units
  benefits of using, 294, 307
  customary system, 310
  explanation of, 293
  familiarity with, 307–308
  metric system, 309–310
  relationships between, 309
  selection of appropriate, 308
Start and Jump: Searching for Patterns, 12, 14–15
Statistics. See also Probability
  collecting data for, 391–392
  descriptive, 397–401
  in elementary schools, 382
  use of graphical representations in, 392–397
Stem-and-leaf plots, 393–395, 398
Stenmark, J. K., 77, 81
Strategies. See also Problem-solving strategies
  for addition facts, 136–141
  approach of using, 135
  development of efficient, 134
  for division facts, 150–151
  for multiplication facts, 146–149
  selection or retrieval of, 134–135
  for subtraction facts, 141–146
Strategy goals
  assessment of, 62
  for problem solving, 40, 367
Subtraction
  basic fact mastery for, 141–146
  comparison concept of, 113–115
  with fractions, 243–247
  front-end estimation strategy for, 211–212
  integer, 414, 415
  using mental computation, 203–206
  part-part-whole concept of, 110–113
  rounding in, 212
  word problems for, 115–118
  zero in, 115, 117
Subtraction algorithms
  directed development of, 183–185
  explorations and invented, 183
Symbolic graphs, 392, 393
Symbolism
  explanation of, 23
  fraction, 227–228
  ideas represented through, 33
  and math curriculum, 221–222
  translations involving, 35
Symbol manipulators, 438
Symbols
  connections with, 168–169
  as separate language, 110

Symmetry
  axis of, 352
  exploration of line, 349–350
  rotational, 350, 351
  in three-dimensional figures, 351–352

Tangrams, BLM 39
Tape measures, 298
Teachers
  listing of professional standards for, 18–19
  problem-solving climate set by, 62
  role in creation of classroom environment, 17, 62
Teaching
  as child-centered activity, 35–36
  developmental. See Developmental instruction
  as science and art, 445
Technology. See Calculators; Computers
Ten-frame facts
  addition, 139–141
  subtraction, 142, 143
Ten-frames
  benefit of, 98, 105–106
  showing numbers on, 96, 97
  10,000 grid, BLM 25
  10 x 10 bean chart, BLM 10
  10 x 10 grids, BLM 23
  10 x 10 multiplication array, BLM 7
Tessellations
  complex, 356, 357
  construction of, 339–340
  explanation of, 339
Tests. See also Assessment
  to assess mental computation and estimation skills, 217–218
  conceptual activities on, 452
  construction of, 78–79
  explanation of, 78
  timed, 78–79
Textbooks
  development of, 456
  lesson format in, 457
  suggestions for use of, 457–458
  teacher's edition of, 456–457
Theoretical probability. See also Probability
  with area model, 387–388
  experimental probability vs., 383–384
Think-addition
  extended use of, 144–145
  subtraction as, 141–142
Thinker toys, 32, 68
Think-multiplication, 144
Think-Pair-Share, 453
Thornton, C. A., 103, 149
Three-digit number names, 165
Three-dimensional shapes
  categories or, 342–343
  construction of, 335–337, 346–349
  symmetries in, 351–352
TI-Explorer calculator, 408
TI-MathMate calculator, 408
Time
  concepts related to, 319
  measurement of, 317
  telling, 317–318
Toohey, M. A., 149
Trading, 165–169, 181
Translation activities
  explanation of, 45
  with patterns, 374–375
  results of, 110
  solving modified, 45–46
  in upper grades, 129–131
  use of, 34–35

Trapezoids
  classification of, 342
  formula for, 314
Triangles
  area of, 314
  classification of, 341, 342
  construction of, 345
Try-and-adjust strategy, 40, 47–48, 53
Tutorial software, 436
Two-apart facts, 140
2-cm isometric grid, BLM 31
2-cm square grid, BLM 27
Two-digit number names, 164–165
Two-dimensional patterns, 375
Two-dimensional shapes
  categories of, 340–342
  construction of, 343, 345–346
Two-less-than facts, 142
Two-more-than facts, 136

Unit fraction, 230
Unit-rate method, 278
U.S. Customary System, 310

Van Allen, R., 33
Van Hiele levels of geometric thought, 325–326
  characteristics of, 326–327
  implications of, 327–328
Variables
  early experiences with, 419–421
  expressions as quantities, 418–419
  meanings and uses of, 418
  misunderstandings regarding, 418
  pattern and order expressed with, 417
Vertical acceleration, 464
Volume
  formulas for, 315–317, 358–359
  measurement activities using, 302–303
Vonder Embose, C., 433

Weight
  explanation of, 303
  measurement activities using, 304–305
Whole-number exponents, 407
Wirtz, R., 96, 98
Woozle cards, BLM 40

Word problems. *See also* Problem solving
  using addition and subtraction, 118
  using division, 128–129
  functions as solutions to, 425–426
  join and separate, 115–117
  measurement, 128
  using multiplication, 122–124
  operation meanings from, 110
  partition, 128
  part-part-whole, 117
  ratio, 286–287
  as separate language, 110
  in upper grades, 129–131
Worksheets
  information provided from student, 75
  with manipulators, 450–451
Workstations, 451

Zero
  in addition and subtraction, 115, 116, 185
  in division, 127, 208–209
  facts with, 136, 137, 142
  in multiplication, 122, 147, 206–207

BLACK-LINE MASTERS AND MATERIALS CONSTRUCTION TIPS

BLACK-LINE MASTERS

Permission is given to reproduce any of the black-line masters for classroom use. Pages are perforated.

Dot cards   1–6
10 × 10 multiplication array   7
Missing-part blanks   8
Base ten materials grid   9
10 × 10 bean chart   10
Hundreds master for bean stick base ten pieces   11
Little base-ten-frames   12–13
Place-value mat (with ten-frames)   14
Base ten grid paper   15
Addition and subtraction record blanks   16
Multiplication and division record blanks   17
Blank hundreds chart (10 × 10 square)   18
Circular fraction pieces   19–22
10 × 10 grids   23
Hundredths disk   24
10,000 grid   25
Decimal/fraction sieve   26
2-cm square grid   27
1-cm square grid   28
0.5-cm square grid   29
1-cm square dot grid   30
2-cm isometric grid   31
1-cm isometric dot grid   32
1-cm square/diagonal grid   33
1-cm hex grid   34
Geoboard pattern   35
Geoboard recording sheets   36
Right-triangle tiles   37
5-cm triangle tiles   38
Tangrams   39
Woozle cards   40

SUGGESTIONS FOR USE AND CONSTRUCTION OF MATERIALS

CARD STOCK MATERIALS

A good way to have many materials made quickly and easily for students is to have them duplicated on *card stock* at a photocopy store. Card stock is a heavy paper, not quite as heavy as tagboard, that comes in a variety of colors. It is also called *cover stock* or *index stock*. The copy stores use it for report covers, and it can be printed on, just as paper can be. The price is about twice that of paper.

Card stock can be laminated and then cut into smaller pieces, if desired, and the laminate adheres very well.

BLACK-LINE MASTERS AND MATERIALS CONTRUCTION TIPS

Laminate first, and then cut into pieces afterwards. Otherwise you will need to cut each piece two times.

Materials are best kept in plastic bags with zip-type closures. Freezer bags are recommended for durability. Punch a hole near the top of the bag so that you do not store air. Lots of small bags can be stuffed into the largest bags. You can always see what you have stored in the bags.

The following list is a suggestion for materials than can be made from card stock using the masters in this appendix. Quantity suggestions are also given.

◆ Dot Cards

First make a duplicate of the first page. This is so that there will be adequate cards with 1, 2, and 3 dots. One complete set of cards will serve four to six children. Duplicate each set in a different color, so that mixed sets can be separated easily. Laminate and then cut with a paper cutter.

◆ 10 × 10 Multiplication Array

Make one per student in any color. Lamination is suggested. Provide each student with an L-shaped piece of tagboard.

◆ Base Ten Pieces (Centimeter Grid)

Use the grid (number 9), and make a master as directed. Run copies on white card stock. One sheet will make 4 hundreds and 10 tens or 4 hundreds and a lot of ones. Mount the printed card stock onto white poster board using either a dry-mount press or permanent spray adhesive. (Spray adhesive can be purchased in art-supply stores. It is very effective, but very messy to handle.) Cut into pieces with a paper cutter. For the tens and ones pieces, it is recommended that you mount the index stock onto *mount board* or *illustration board,* also available in art-supply stores. This material is thicker and will make the pieces easier to handle. It is recommended that you *not* laminate the base ten pieces. A kit consisting of 10 hundreds, 30 tens, and 30 ones is adequate for each student or pair of students.

◆ Bean Stick Base Ten Pieces

Use either great northern or pinto beans to make bean sticks. Craft sticks can be purchased in craft stores in boxes of 500. Use white glue (such as Elmer's) or a glue gun. Also dribble a row of glue over the beans to keep them from splitting off the sticks. (The white glue dries clear.) For hundreds, use the master mounted onto poster board as described for the grids above.

◆ Little Base-Ten-Frames

There are two masters for these materials. One has full ten-frames and the other has from 1 to 9 dots including two with 5 dots. Copy the 1 to 9 master on one color of card stock and the full ten-frames on another. Cut off most of the excess stock (do not trim) and then laminate. Cut into little ten-frames. Each set consists of 20 pieces: 10 full ten-frames and 10 of the 1 to 9 pieces, including 2 fives. Make a set for each child.

◆ Place-Value Mat (with Ten-Frames)

The mats can be duplicated on any pastel card stock. It is recommended that you not laminate these because they tend to curl and counters like beans slide around too much. Make one for every child.

One way to make a three-place place-value mat is to simply tape a half sheet of blank card stock to the left edge of a two-place mat. Use strapping tape (filament tape used for packages). The tape will act as a hinge and permit the extra piece to be folded under for storage.

◆ Circular Fraction Pieces

First make three copies of each page of the master. Cut the disks apart and tape onto blank pages with three of the same type on a page. You will then have a separate master for each size with three full circles per master. Duplicate each master on a different color card stock. Laminate and then cut the circles out. A kit for one or two students should have two circles of each size piece.

◆ Hundredths Disk

These can be made on paper but are much more satisfying on card stock. Duplicate the master on two contrasting colors. Laminate and cut the circles and also the slot on the dotted line. The smiley face is used as a decimal point on the desk top. Make a set for each student. It's easy and worth it.

◆ Tiles

Included here are two masters for triangular tiles. You could easily draw other masters for large squares or 2 × 1 rectangles or the set of "five easy pieces" shown in Figure 17.10. If tiles are duplicated on card stock, laminated, and cut out on a paper cutter, they make quite nice geometric manipulatives. If you take the extra effort to first mount them onto poster board, they will be even better. A small set would consist of about three sheets of tiles, and you will want about one set for every pair of students.

◆ Tangrams

Tangrams should be copied on card stock and then laminated. Especially for younger children, the card stock should first be mounted onto poster board to make the pieces a bit thicker and easier to put together in puzzles. You will want one set per student. Keep individual sets in plastic bags.

BLACK-LINE MASTERS AND MATERIALS CONSTRUCION TIPS

◆ Woozle Cards

Copy the woozle card master on white or off-white card stock. You need two copies per set. Before laminating, color one set blue and the other red. (Any two colors will be fine.) An easy way to color the cards is to simply make one pass around the inside of each woozle leaving the rest of the creature white. If you color the entire woozle, the dots may not show up. Make one set for every four students.

TRANSPARENCIES

A copy of any page can be made into a transparency with thermal transparency masters and a Thermo-fax machine. Masters come in various colors on a clear background and black line on various colored backgrounds. Follow the directions for making the masters on the box. (Check with your media specialist.)

Some masters are useful to have as transparency mats to use for demonstration purposes on the overhead. The 10 × 10 array and the large geoboard are examples. The place-value mat can be used with strips and squares or with counters and cups, directly on the overhead. The 10 × 10 bean chart is a useful hundreds board as is the blank hundreds board. The missing part blank, the record blanks for the four algorithms, and the decimal/fraction sieve are examples of pages that you may wish to use as write-on transparencies. Of course you will want to simply copy these and many other pages on paper for your students to write on.

A transparency of the 10,000 grid is the easiest way there is to show 10,000 or to model four-place decimal numbers. You will need to be careful in making the transparency. If too dark, the squares run together, and if too light, you will find that some squares do not reproduce. It can be done! If you pull the overhead away from the screen until the square is as large as possible, each tiny square can be seen across the average room and individual squares or strips of squares can be colored with a pen.

All of the line and dot grids are useful to have available as transparencies. You may find it a good idea to make several copies of each and keep them in a folder where you can get to them easily.

For the Woozle cards and dot cards, make a reduction of the master on a photocopy machine. Then make transparencies of the small cards, cut them apart, and use them on the overhead. The dot cards are best on a colored transparency. The Woozle cards are best on a clear transparency. Color them with a permanent transparency marker.

Tiles can also be cut out of transparencies, but they are very difficult to handle.

MAKING GEOBOARDS

It is possible to mass-produce geoboards so that large numbers of them can be made quickly and quite clearly. The master (number 35) is for a $7\frac{1}{2}$-inch board. Seventy-two boards this size can be made for about $15.00. Get one or two other teachers to do this with you so that the mass production technique and cost savings can best be appreciated.

Use $\frac{5}{8}$-inch particle board. This can be purchased at lumber stores or home supply stores and is very inexpensive, especially if purchased in a 4-foot by 8-foot sheet. Have someone with a table saw cut the board into $7\frac{1}{2}$-inch squares. These squares should be cut fairly accurately. (Go to the shop teacher at the junior or senior high school for help.) Purchase 1-inch #16 wire brads. These have no heads. The #16 refers to the thickness. You want the nails as thick as possible so that they will not bend. Buy nails in bulk at a hardware store. You will want about a pound of nails to make 25 boards.

Make a "geoboard maker" out of a piece of $\frac{1}{2}$-inch-thick plywood or other half-inch-thick lumber. You will need about a 1-foot square. Tape a copy of the geoboard master in the center of the board. Nail strips of wood (about one inch by six inches) around all four sides of the master. These strips should be about $\frac{1}{16}$ of an inch outside of the outline of the geoboard. This will allow the squares of particle board to fit snugly but allow for minor errors in cutting them out. Now drill small holes through each dot on the master. The holes should be just barely large enough for one of the brads to be pulled through. Be certain that the holes are drilled perpendicular to the board.

Now you are ready to make geoboards. Put the geoboard maker over one of the squares of particle board. Place a nail in each hole and hammer it flush. Then with a screwdriver, gently pry the geoboard maker off of the board. The nails will each be a uniform half-inch height and all perfectly arranged. Try not to bang on the geoboard maker because it eventually suffers from overuse. However, one maker should serve the production of several hundred geoboards.

You may want to paint your geoboards a dark color before nailing in the nails (or spray paint them afterward). This makes it easier to see brown rubber bands on the boards.

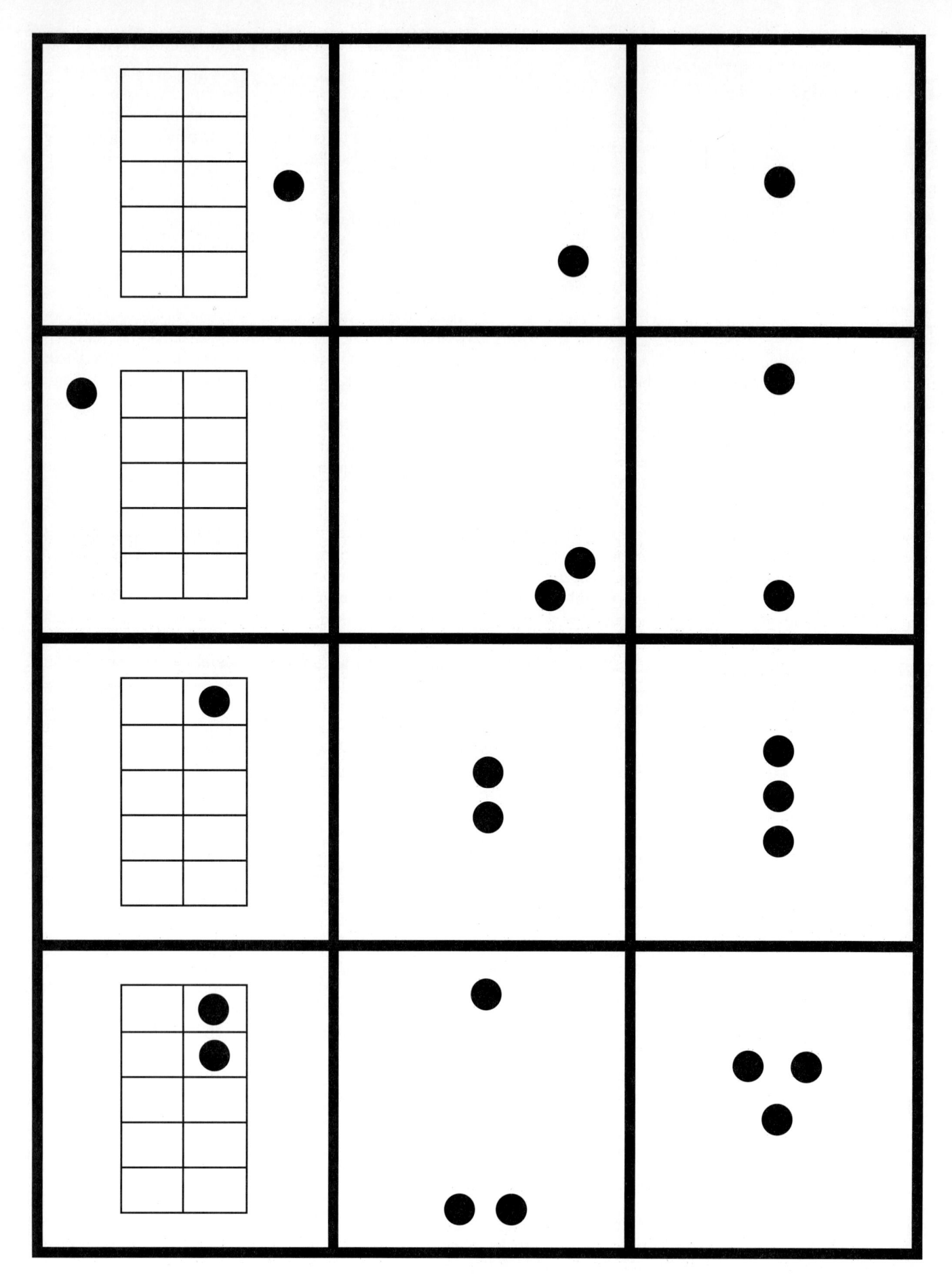

Dot cards—1

Copyright © 1994 by Longman Publishing Group.

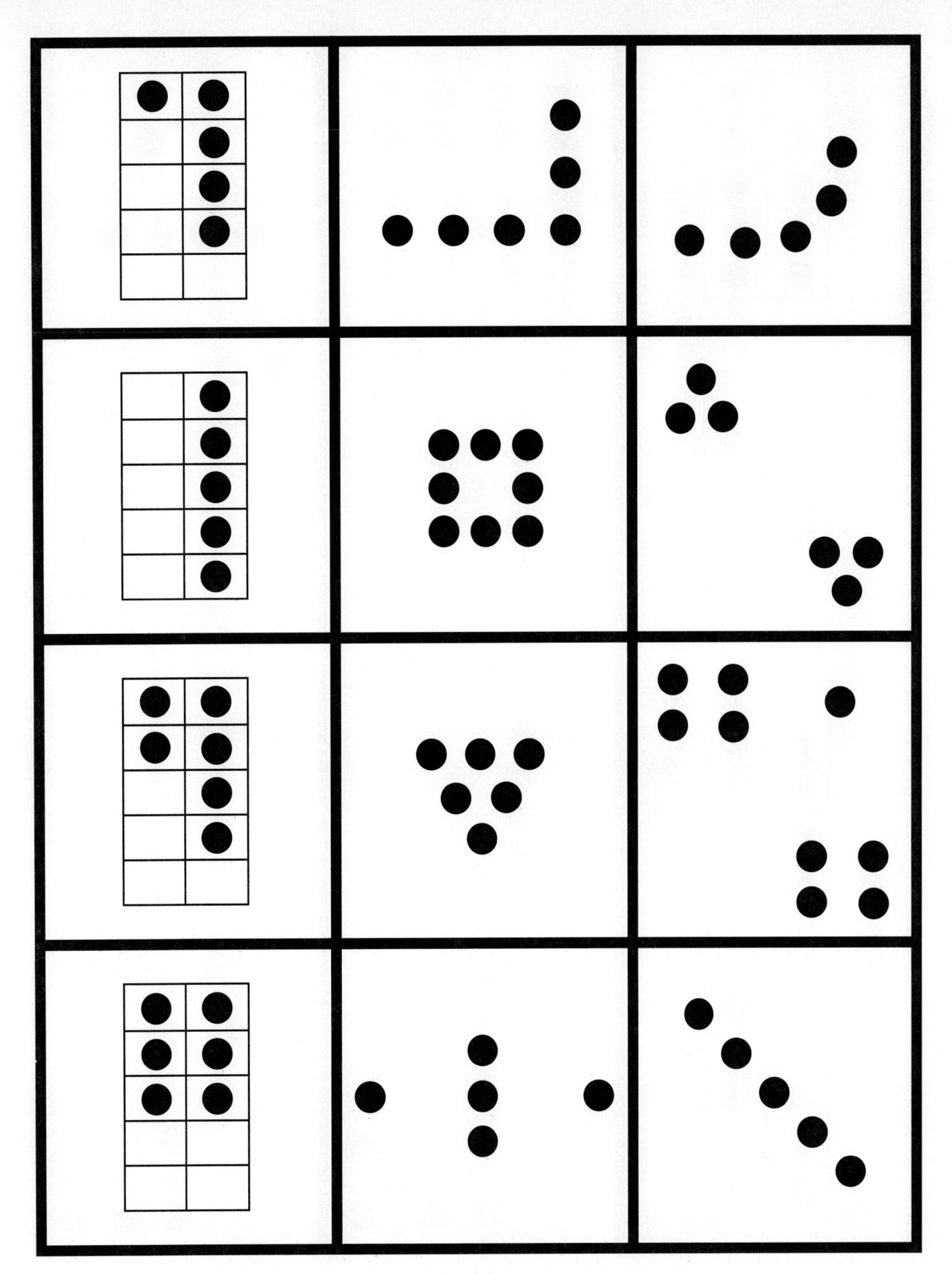

Dot cards—2

Copyright © 1994 by Longman Publishing Group.

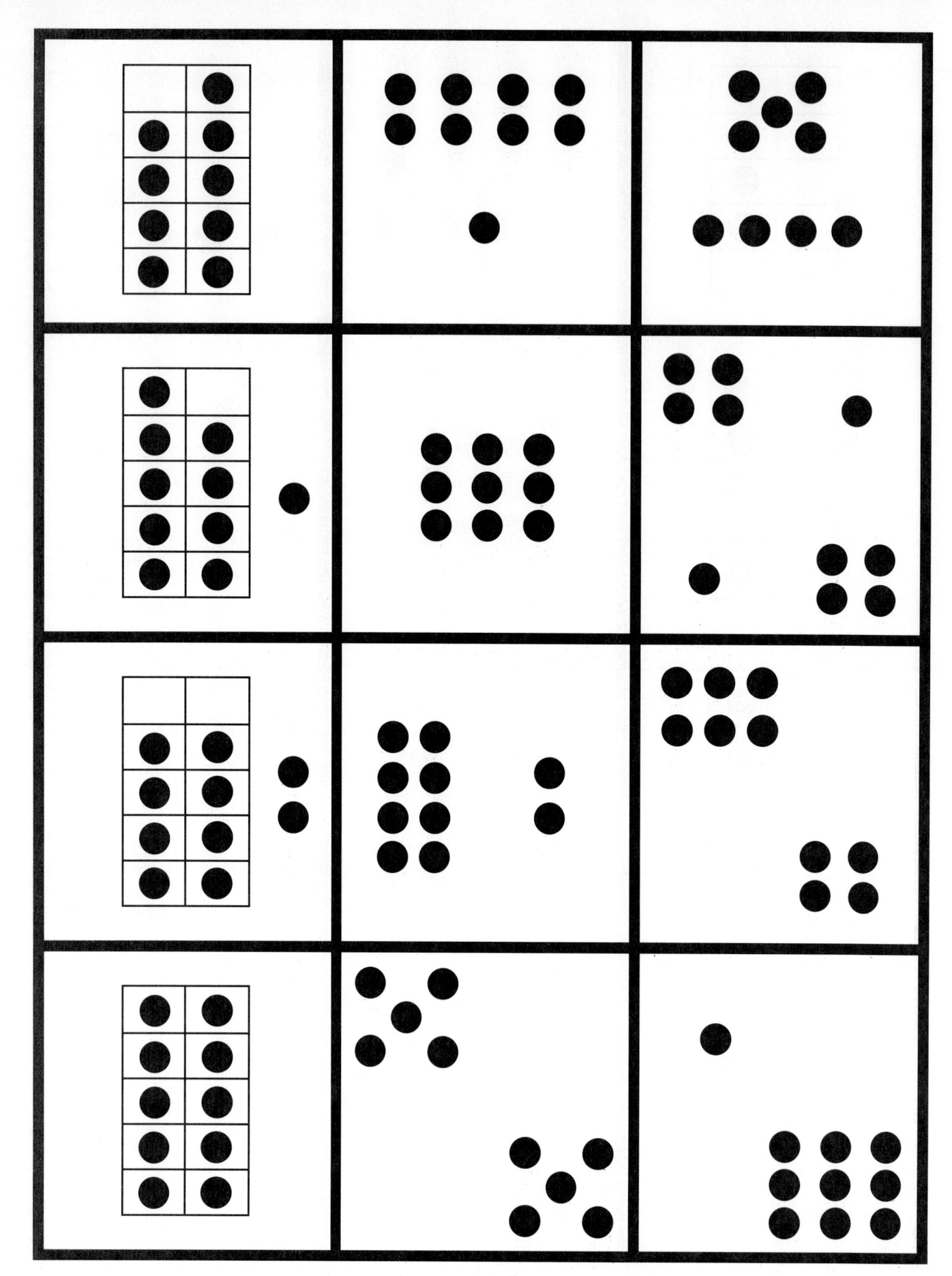

Dot cards—3

Copyright © 1994 by Longman Publishing Group.

Dot cards—4

Copyright © 1994 by Longman Publishing Group.

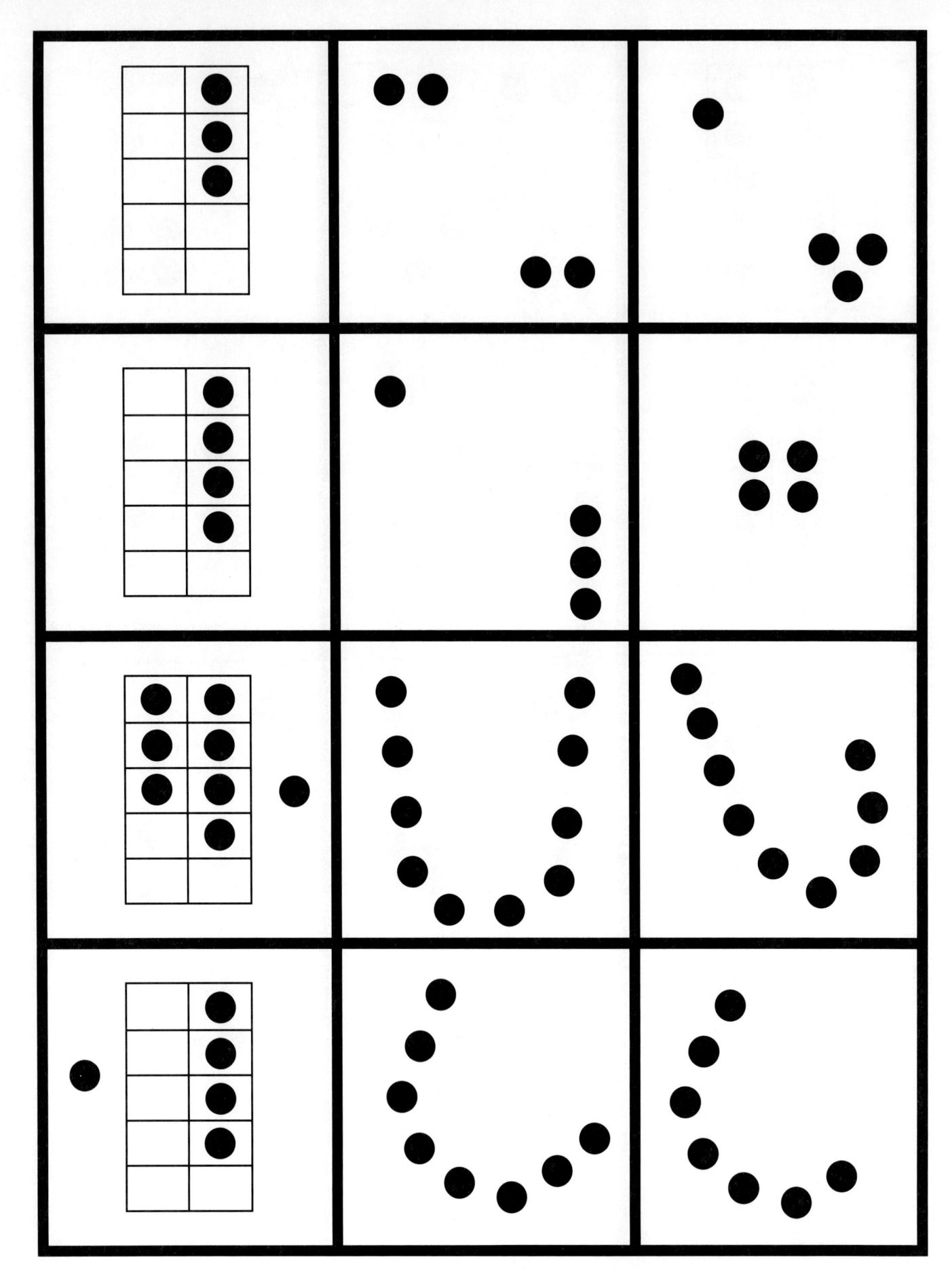

Dot cards—5

Copyright © 1994 by Longman Publishing Group.

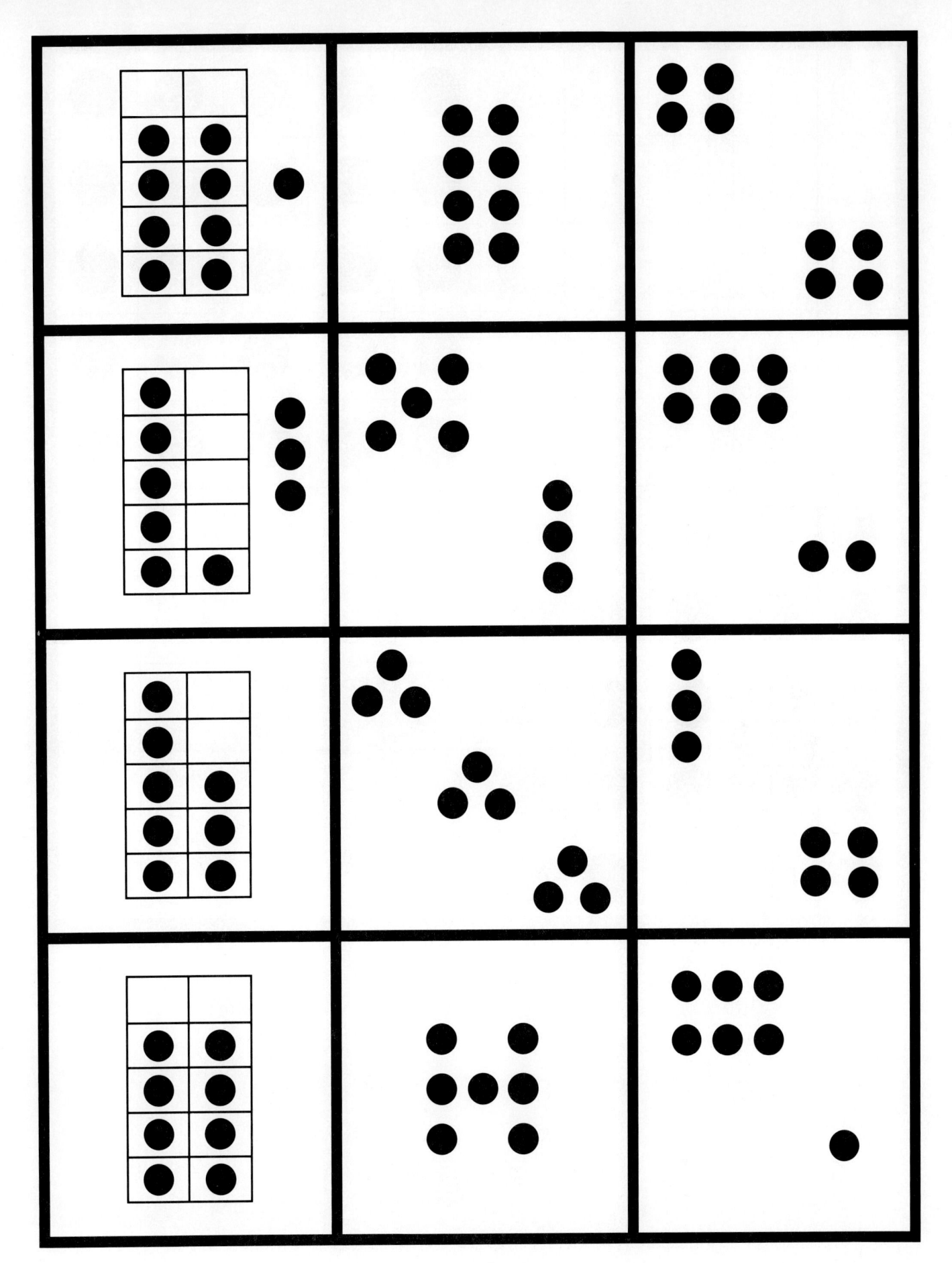

Dot cards—6

Copyright © 1994 by Longman Publishing Group.

10 × 10 multiplication array—7

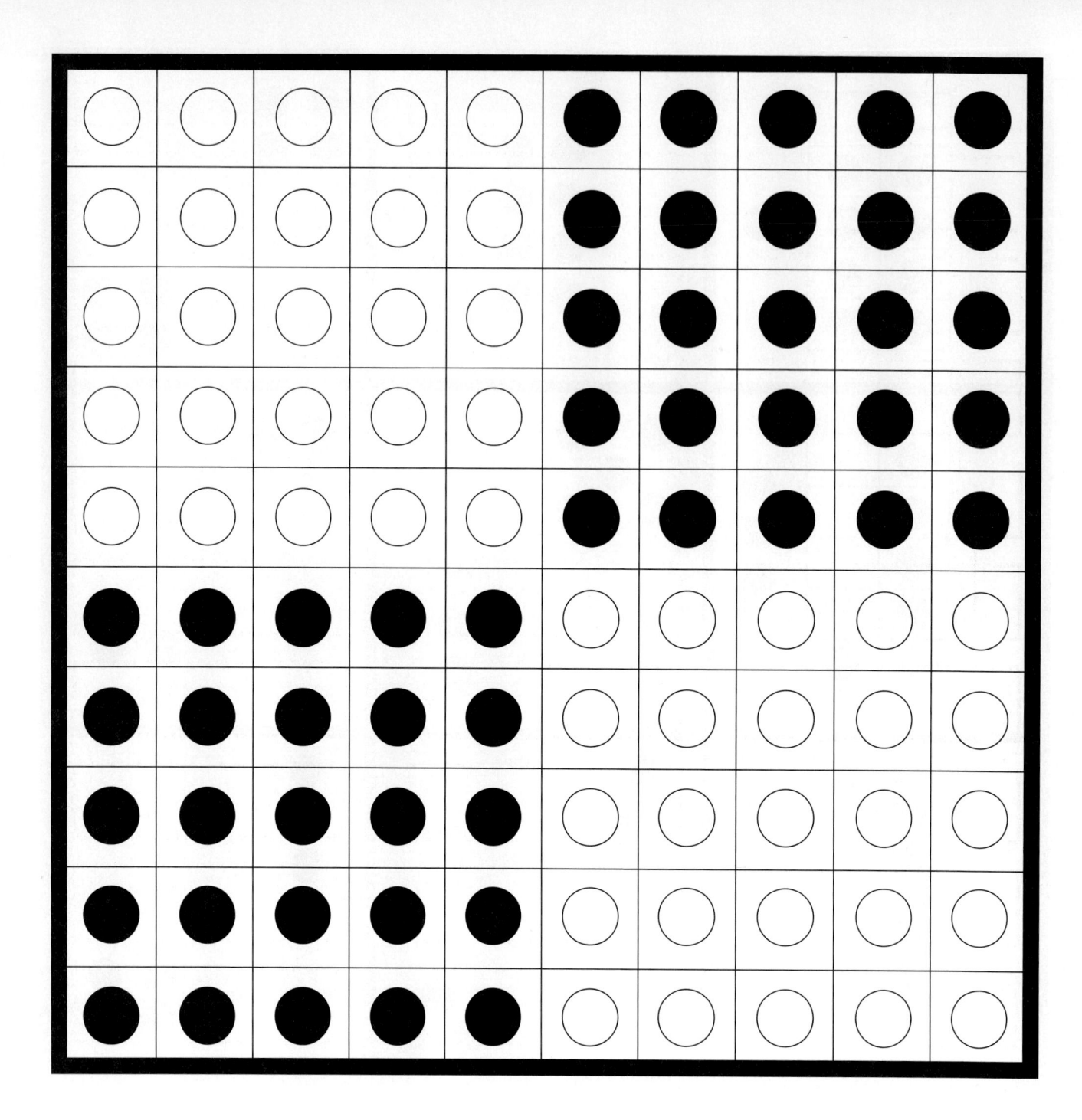

Copyright © 1994 by Longman Publishing Group.

Missing-part blanks—8

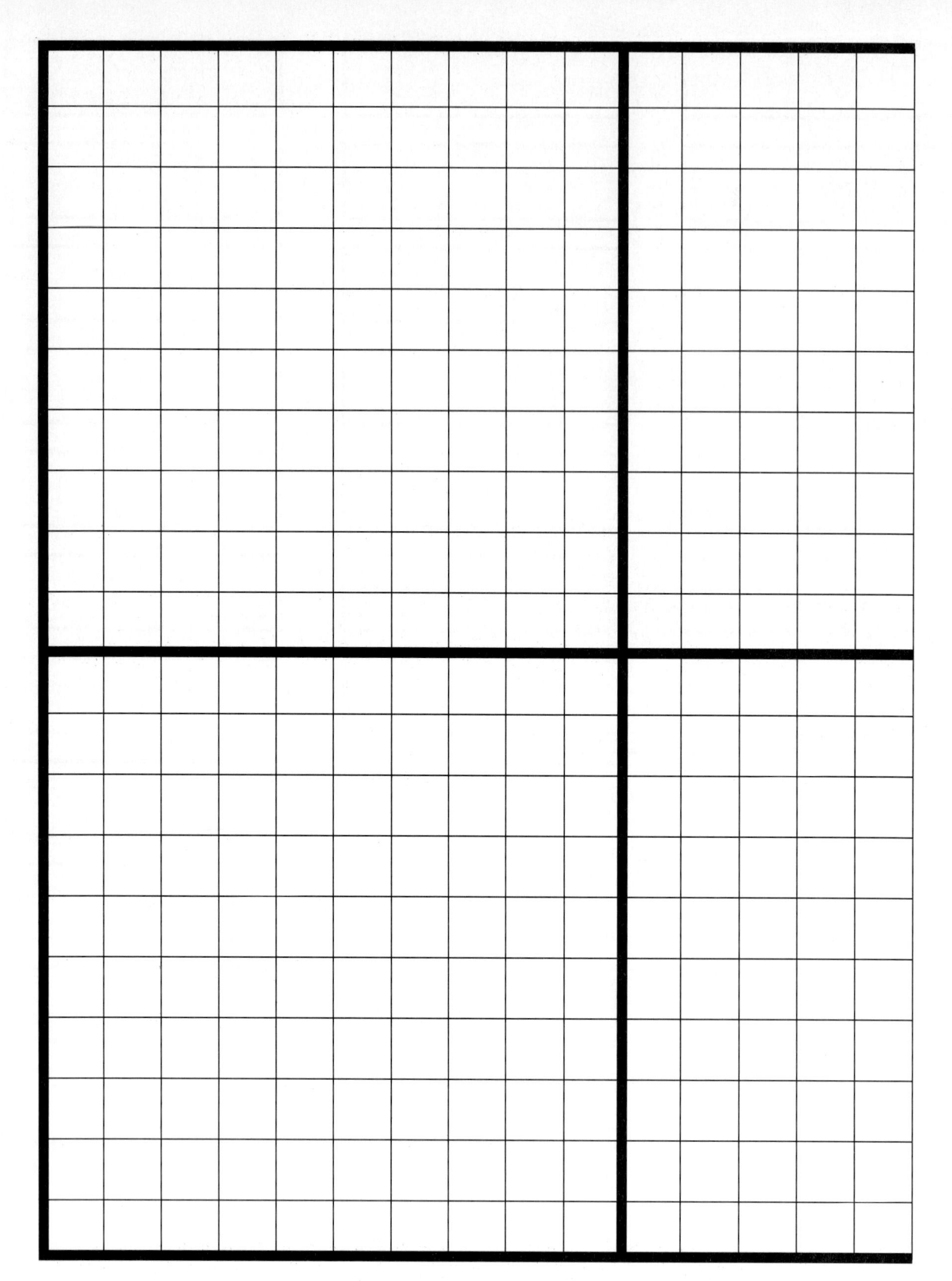

1. Make two copies of this page. Cut out the grid from each copy.
2. Overlap the two grids and tape onto a blank sheet to form a 20 by 25 centimeter grid with 4 complete hundreds squares and 2 rows of 5 tens each.
3. Use this as a master to make copies on card stock.

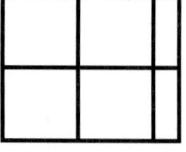

Base ten materials grid—9

Copyright © 1994 by Longman Publishing Group.

10 × 10 bean chart—10

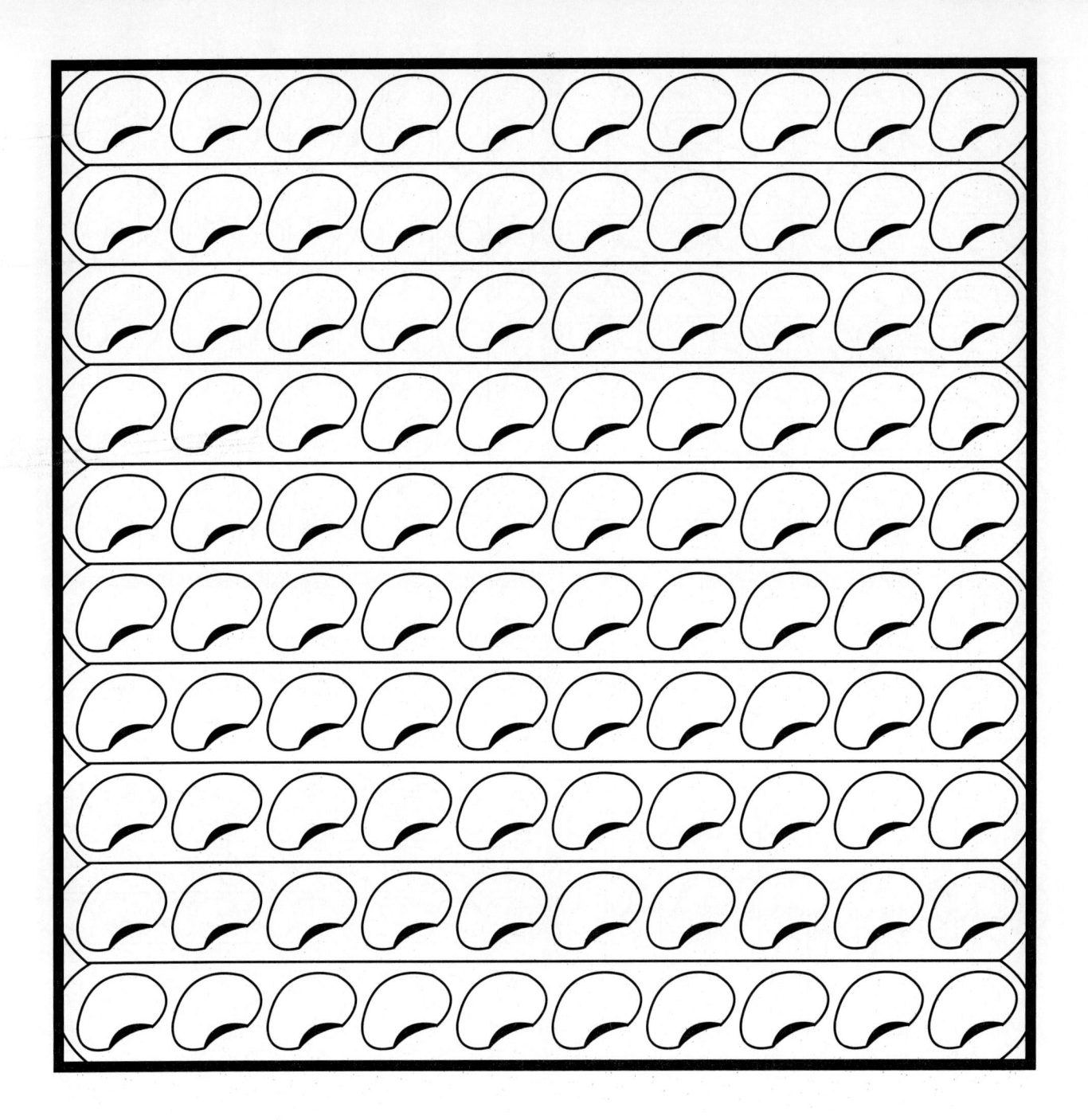

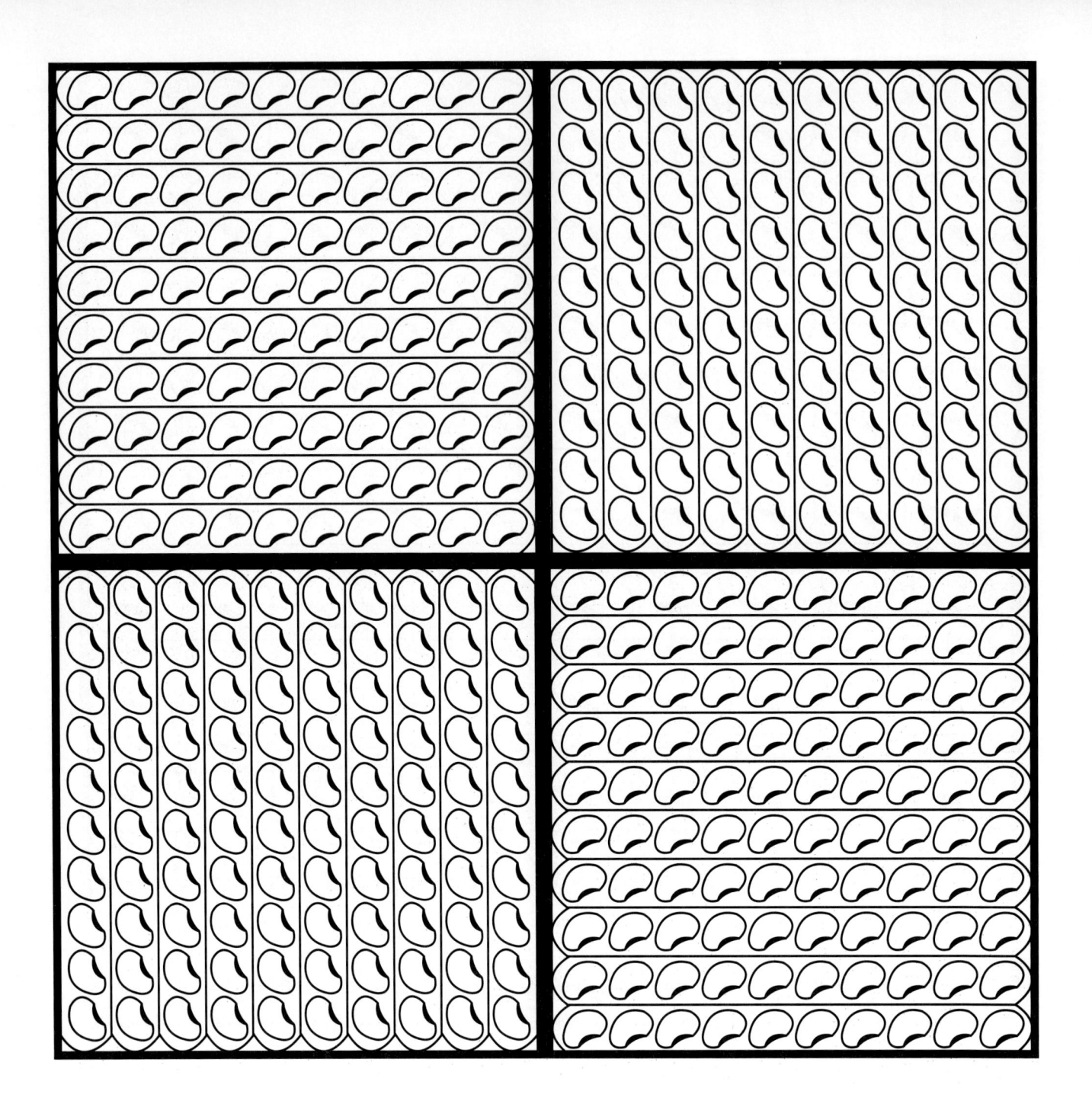

Hundreds master for bean stick base ten pieces—11

Copyright © 1994 by Longman Publishing Group.

Duplicate this sheet on index or card stock, available at most copy stores. Be sure to use a different color for the tens than you use for the sheet with 1 to 9 dots in each frame. Cut off the excess paper and laminate. Then cut into individual ten-frames. One kit has ten of these and ten of the 1 to 9 frames.

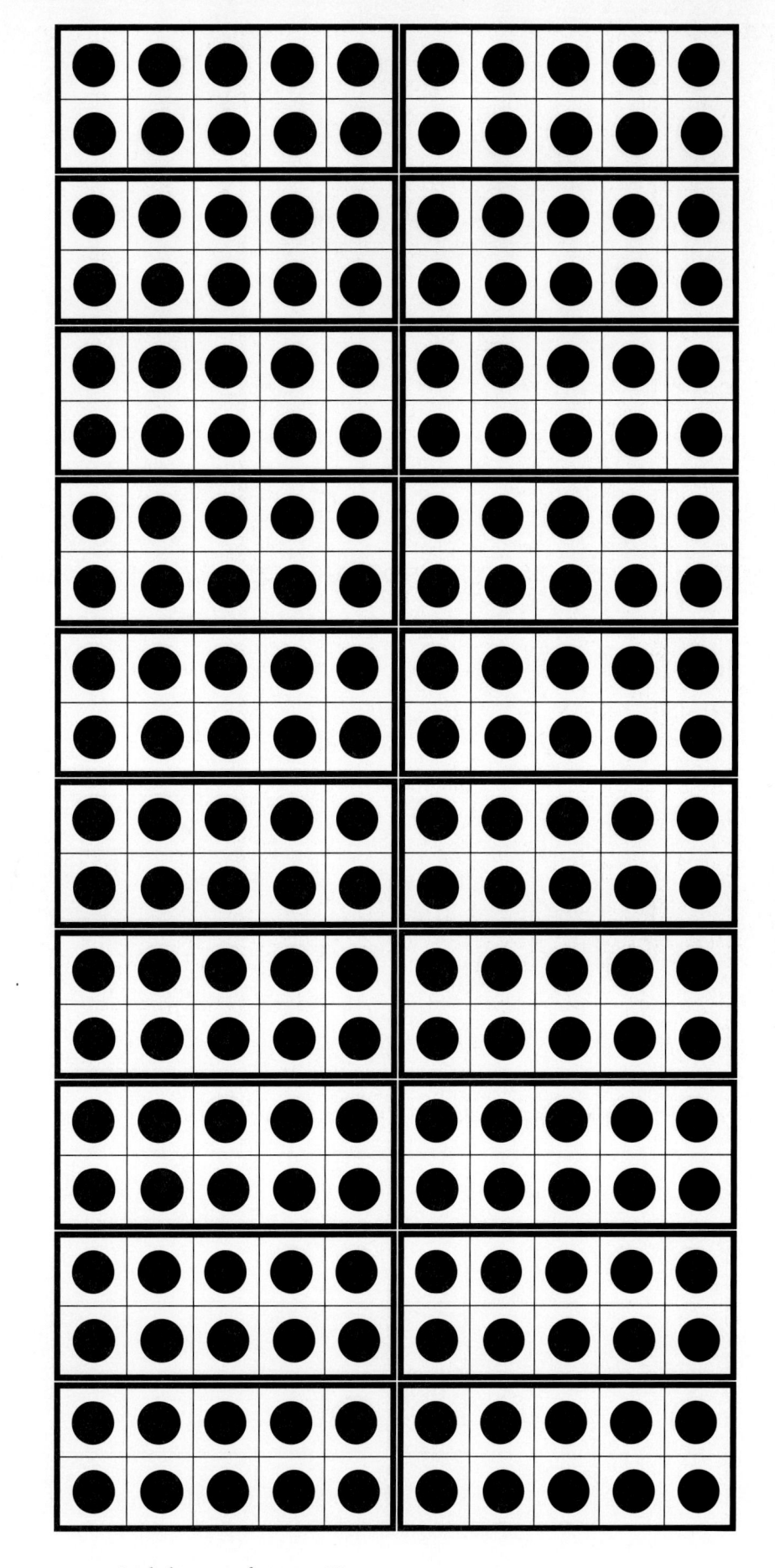

Little base-ten-frames—12

Copyright © 1994 by Longman Publishing Group.

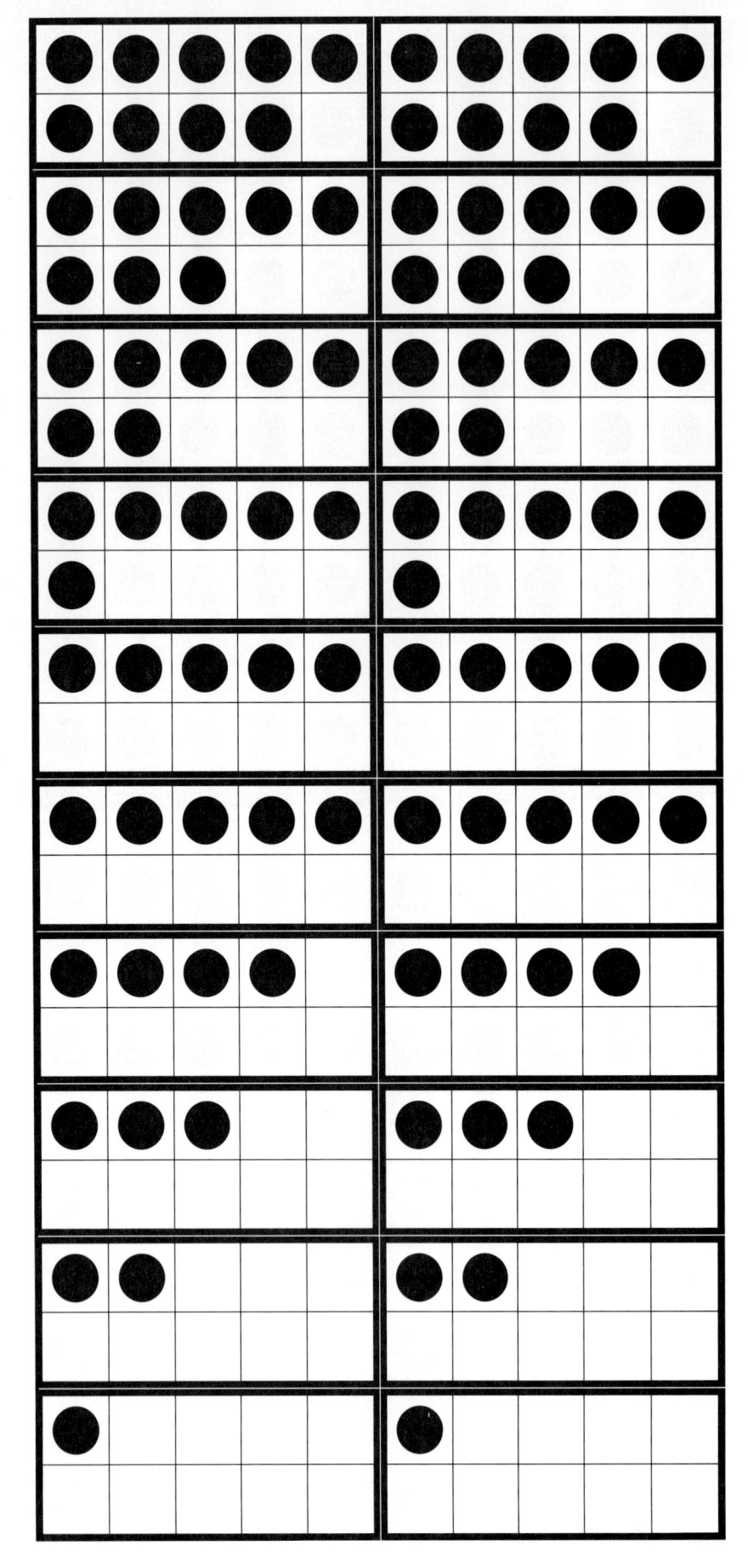

Duplicate this sheet on index or card stock, available at most copy stores. Be sure to use a different color for these pieces than for the sheet with full ten-frames. After cutting off the excess paper, laminate and cut into individual ten-fames. Each kit gets one of each number plus an extra five.

Little base-ten-frames—13

Copyright © 1994 by Longman Publishing Group.

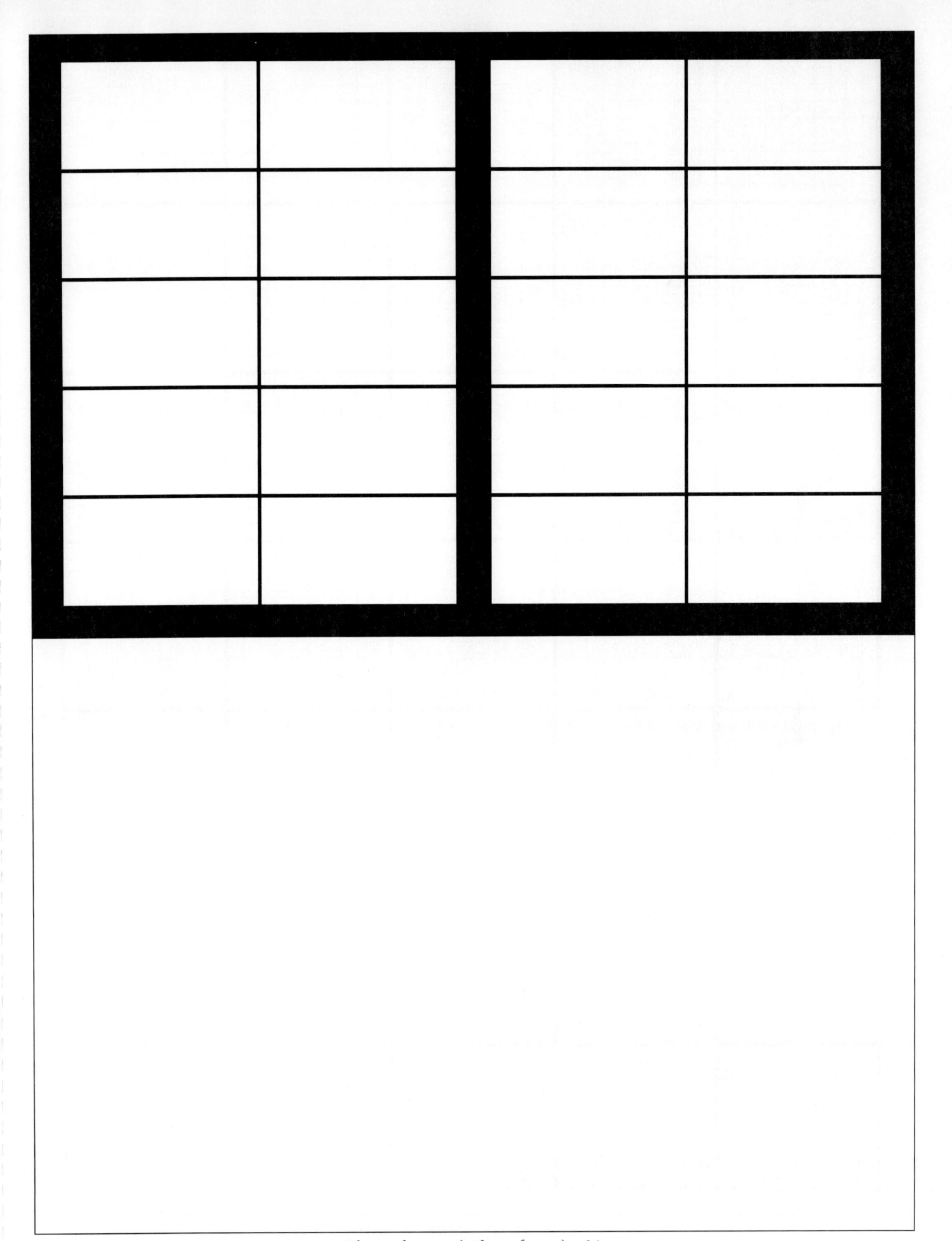

Place-value mat (with ten-frames)—14

Copyright © 1994 by Longman Publishing Group.

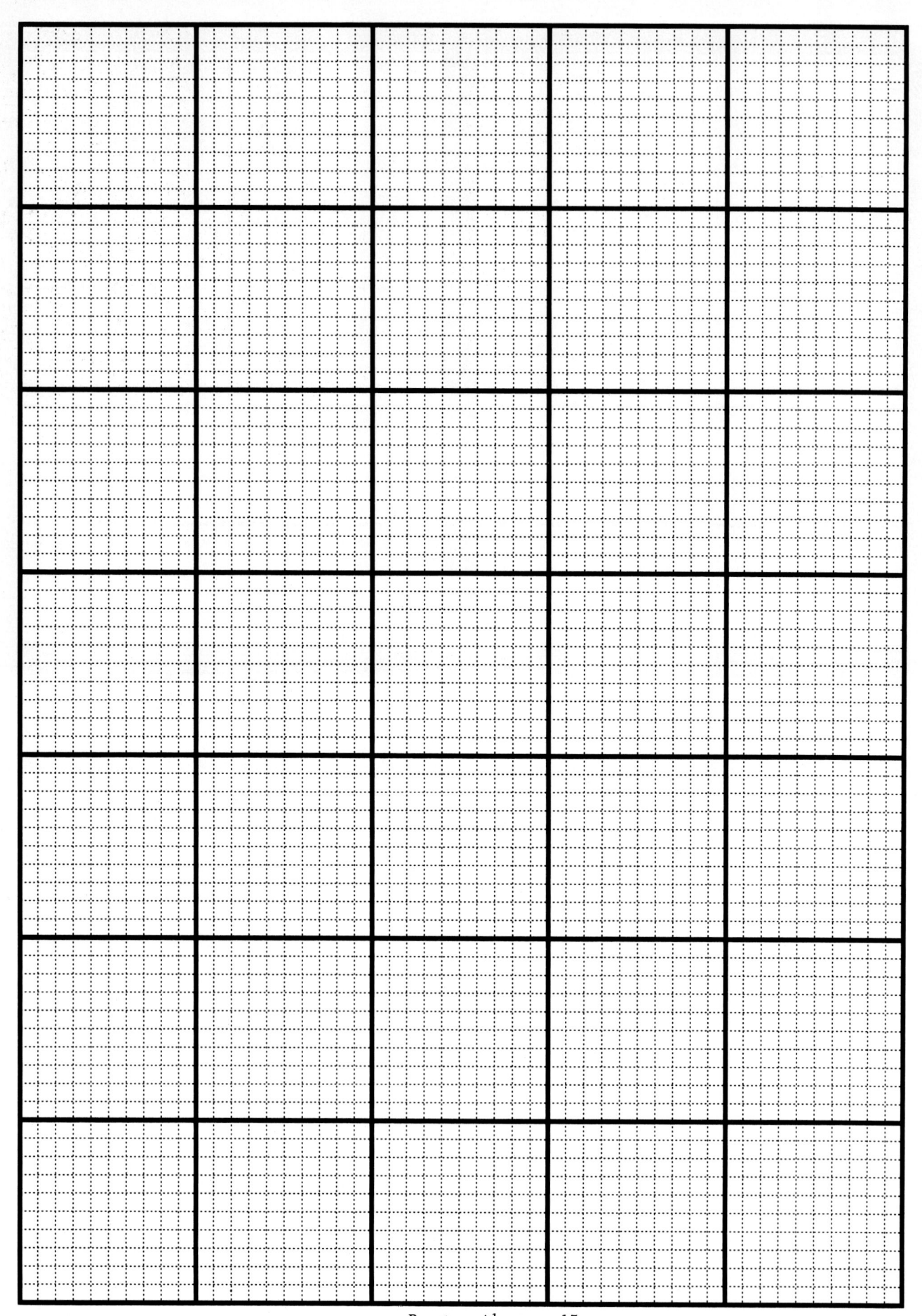

Base ten grid paper—15

Addition and subtraction record blanks—16

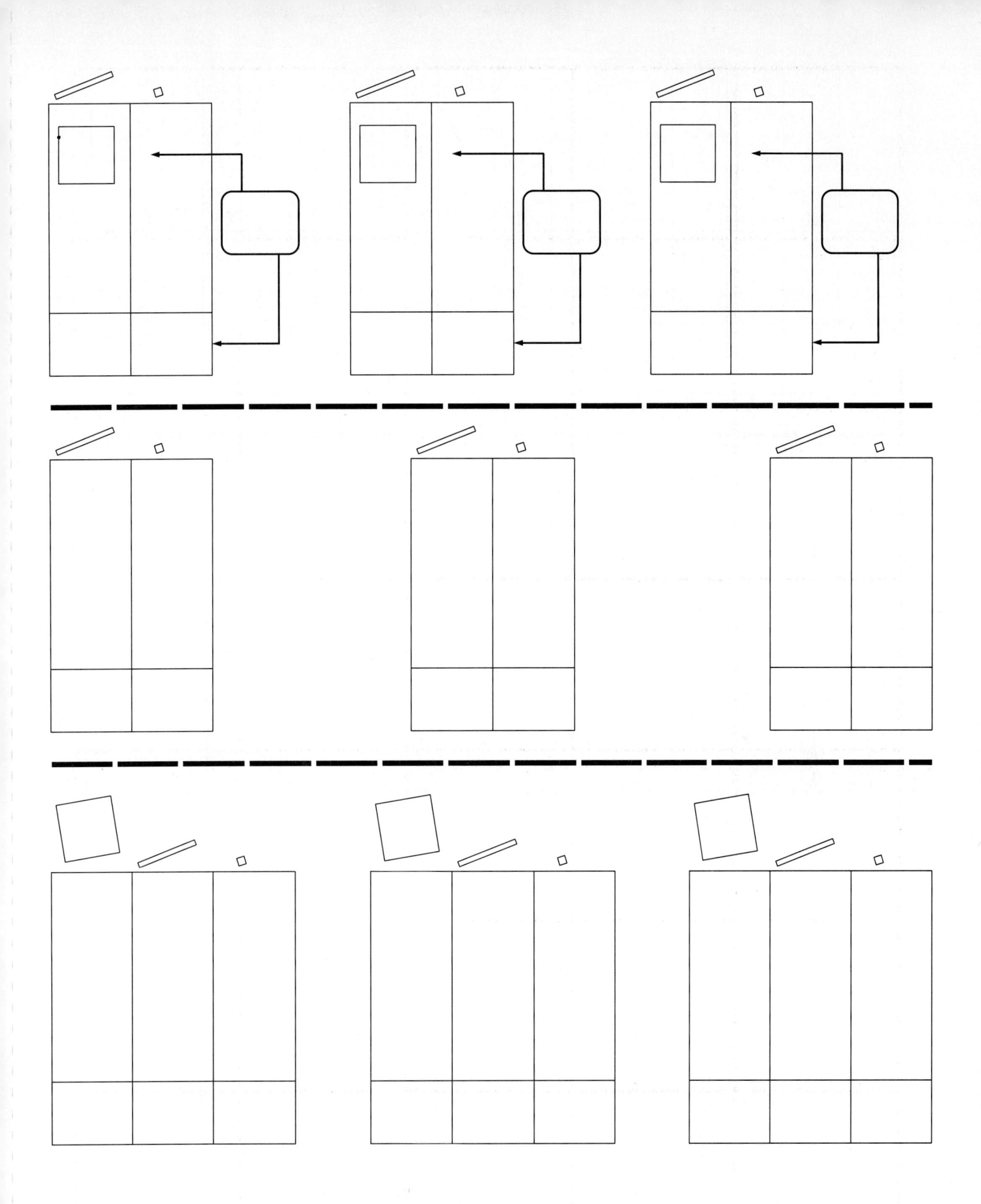

Copyright © 1994 by Longman Publishing Group.

Multiplication and division record blanks—17

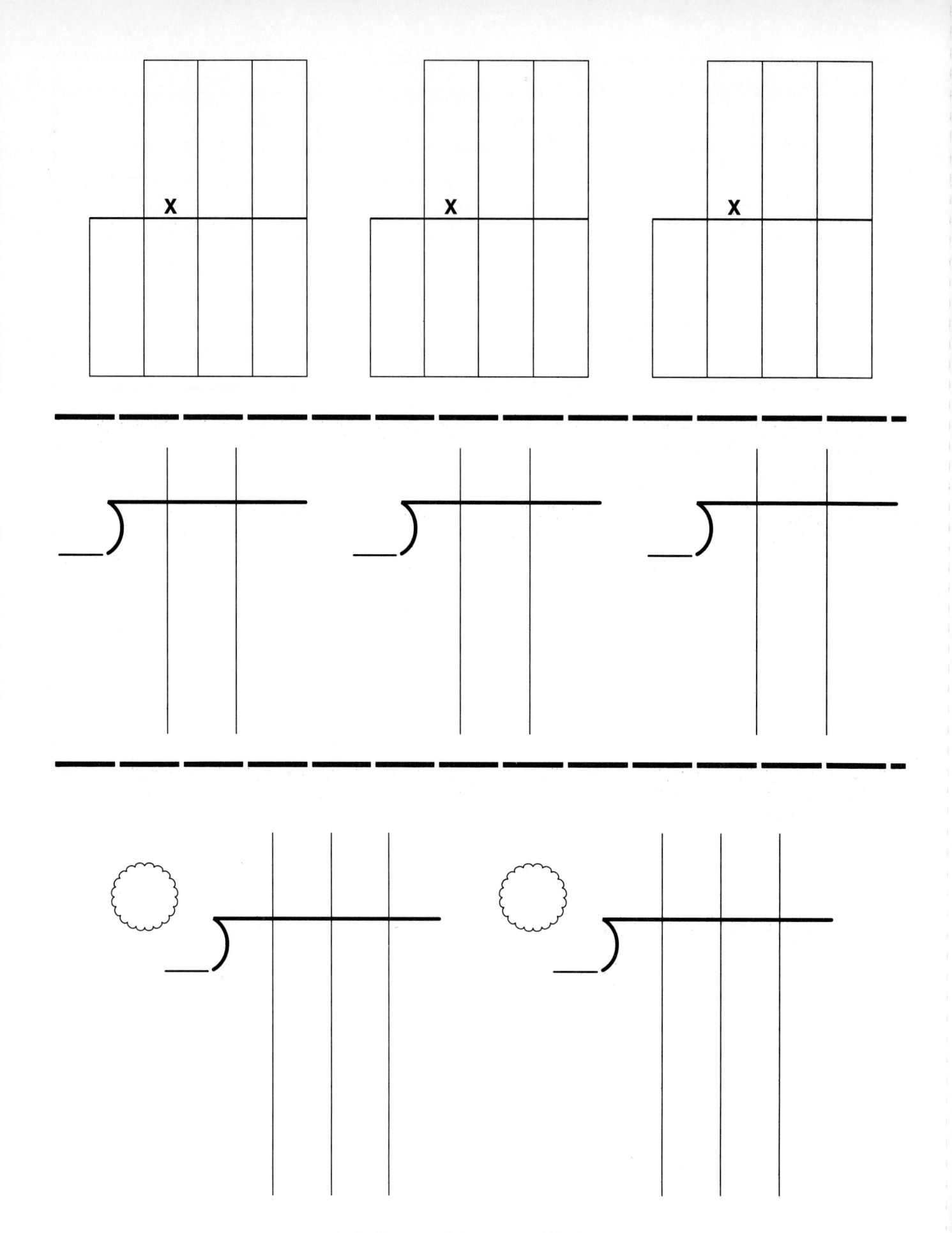

Blank hundreds chart (10 × 10 square)—18

Copyright © 1994 by Longman Publishing Group.

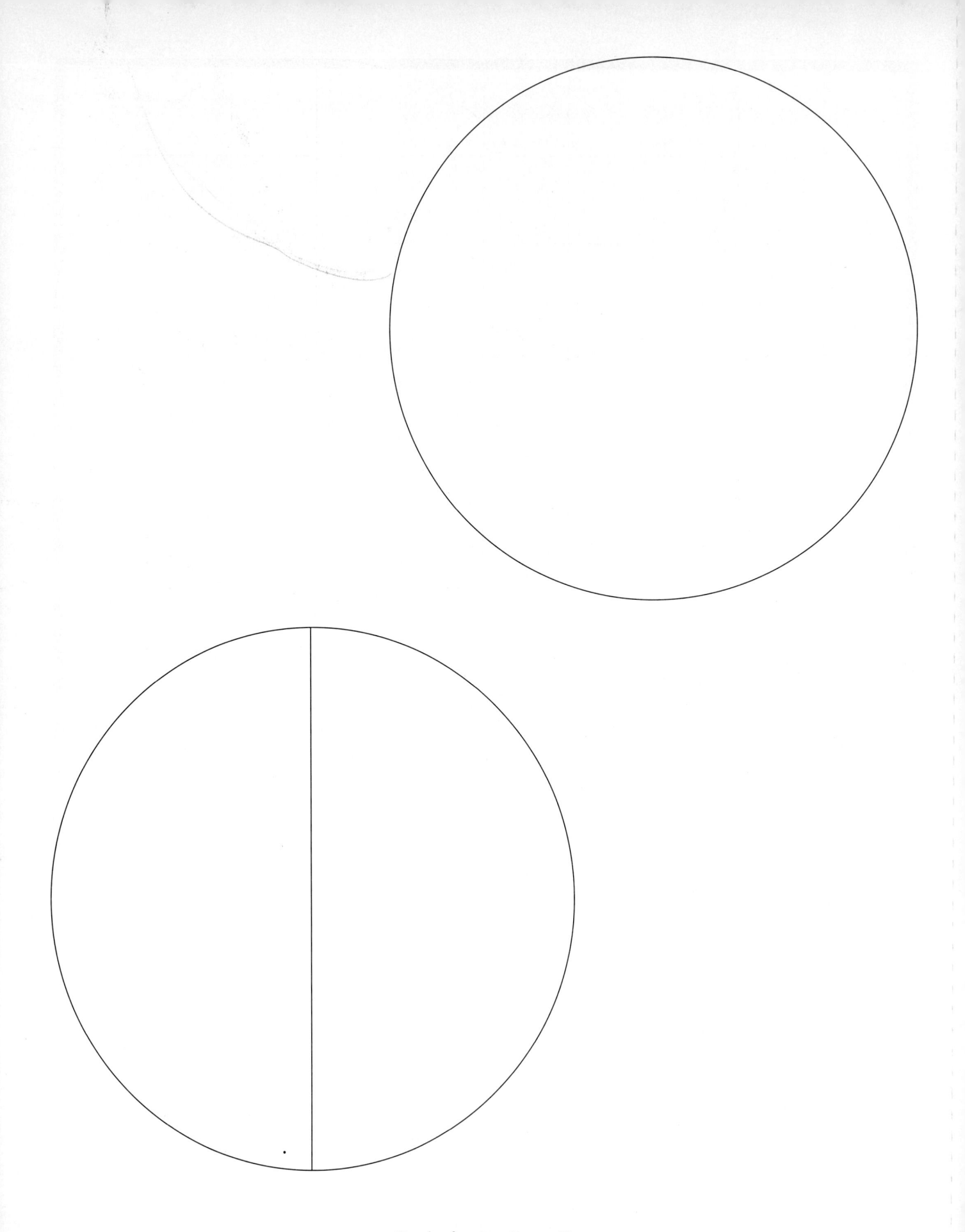

Circular fraction pieces—19

Copyright © 1994 by Longman Publishing Group.

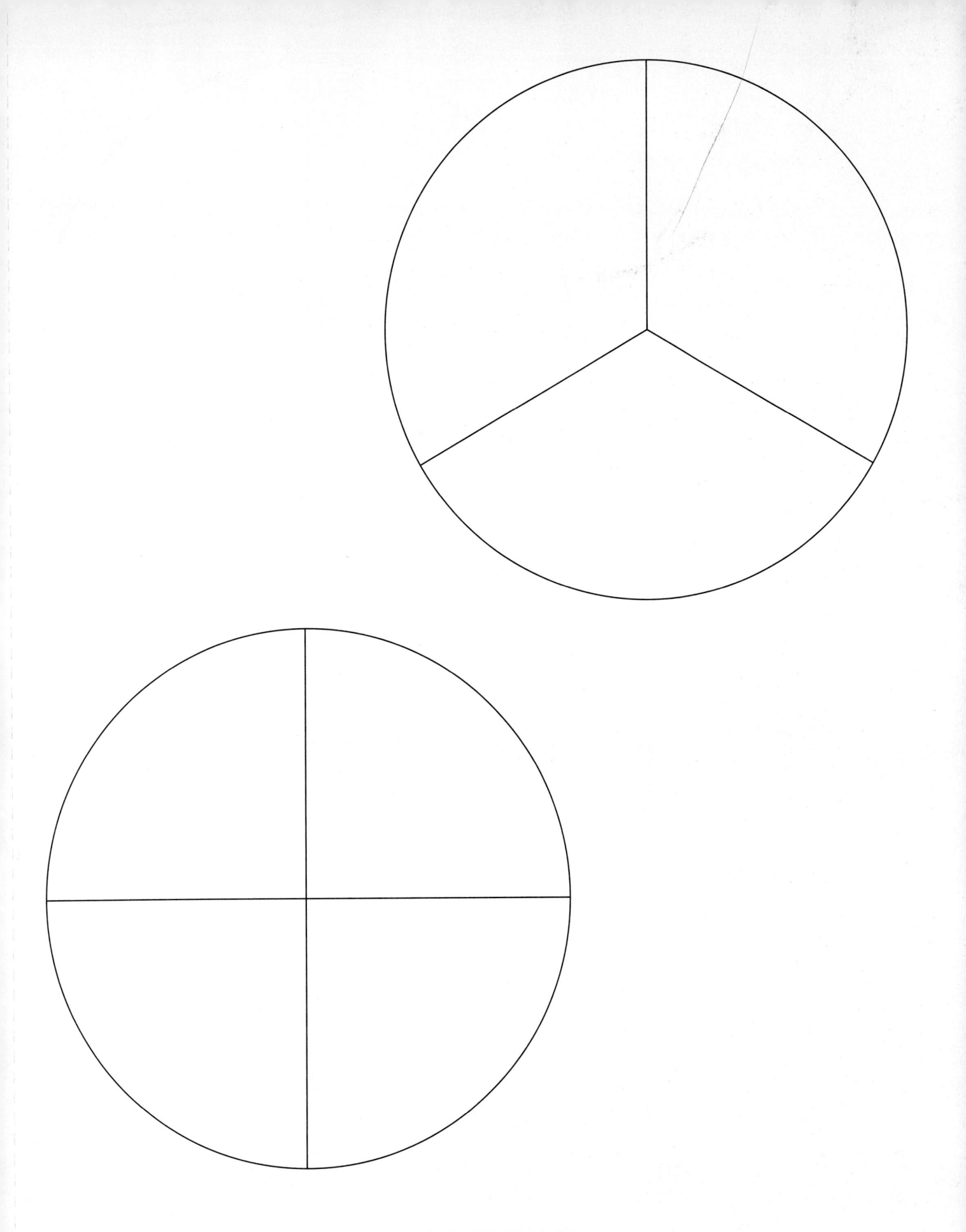

Circular fraction pieces—20

Copyright © 1994 by Longman Publishing Group.

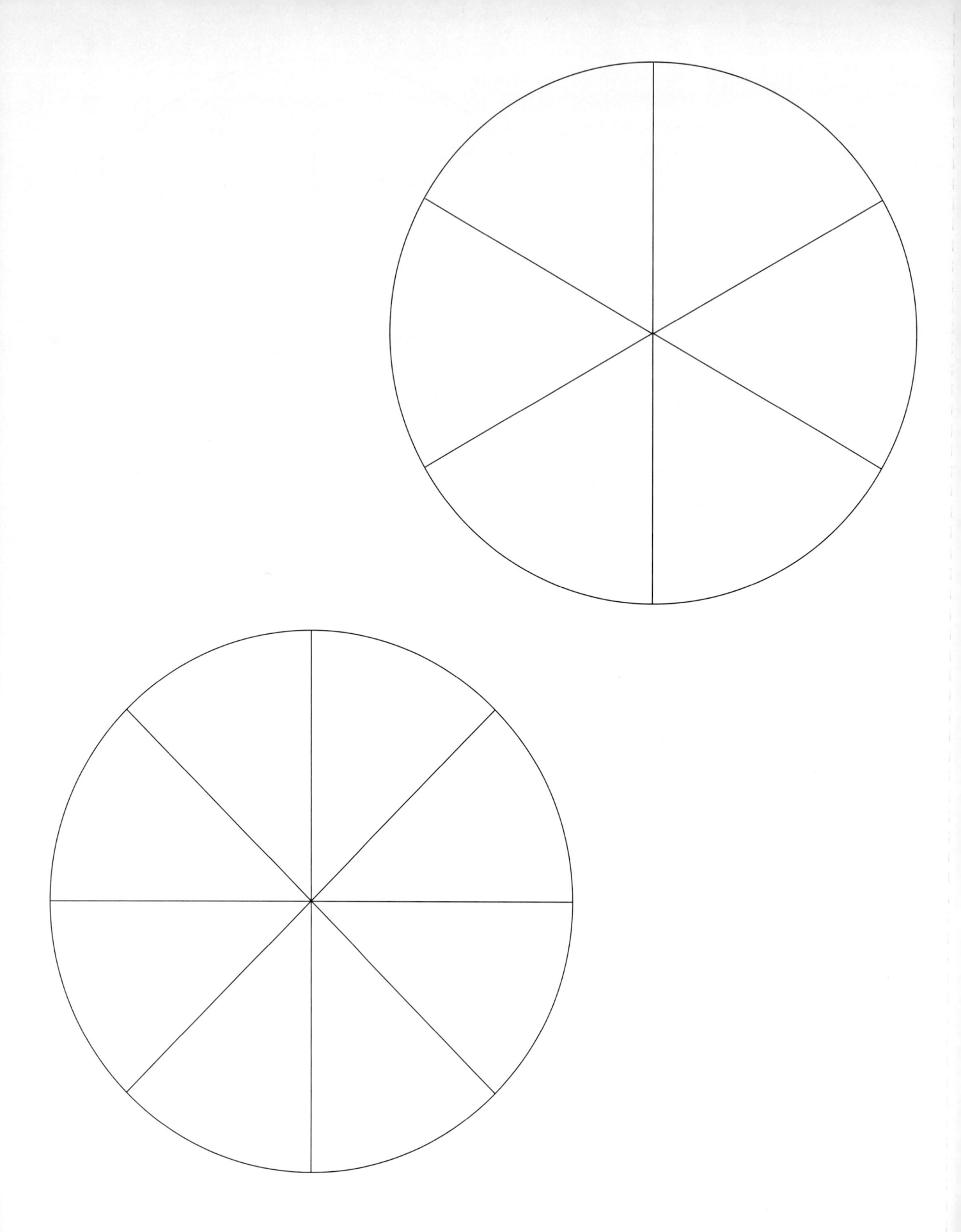

Circular fraction pieces—21

Copyright © 1994 by Longman Publishing Group.

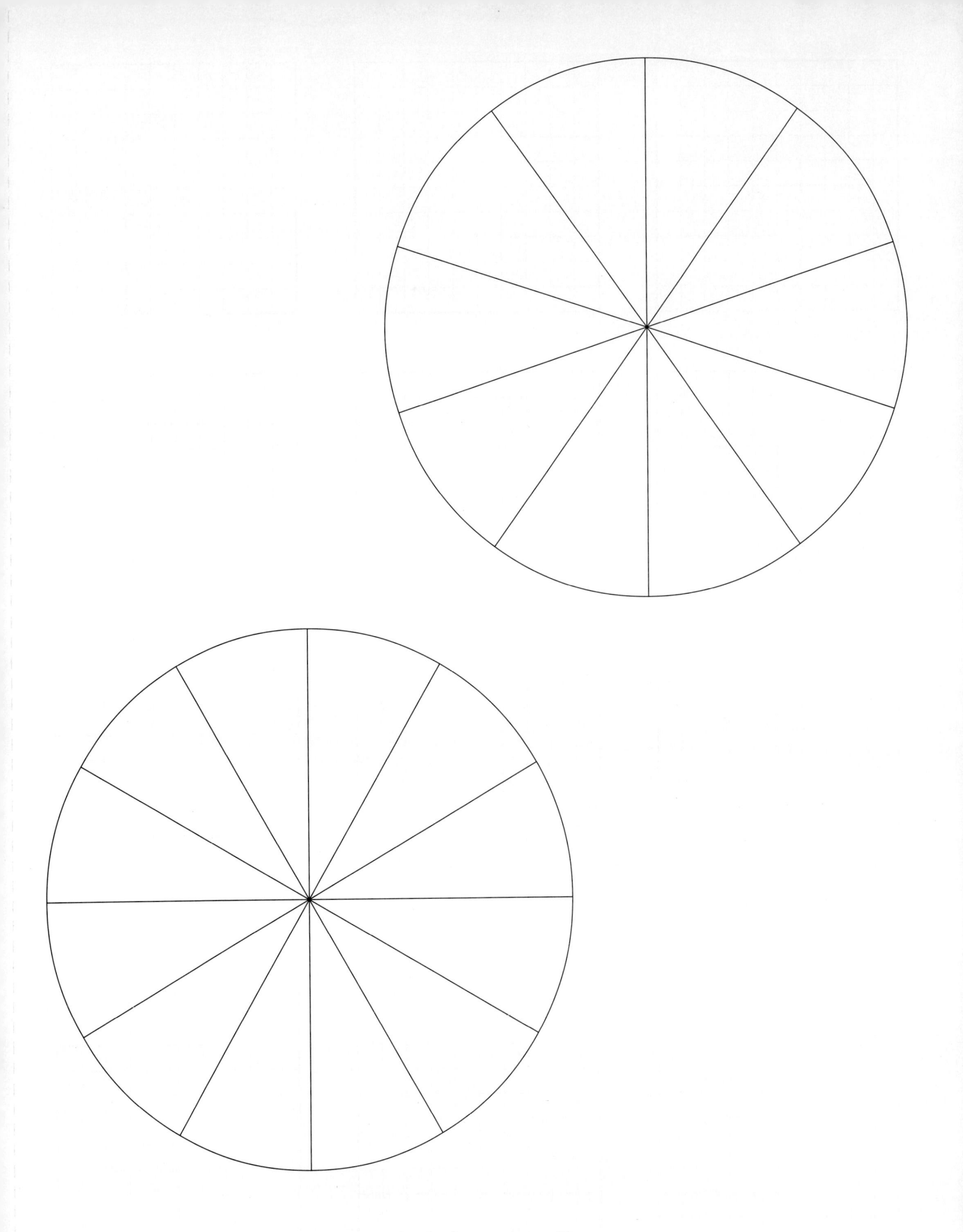

Circular fraction pieces—22

Copyright © 1994 by Longman Publishing Group.

10 × 10 grids—23

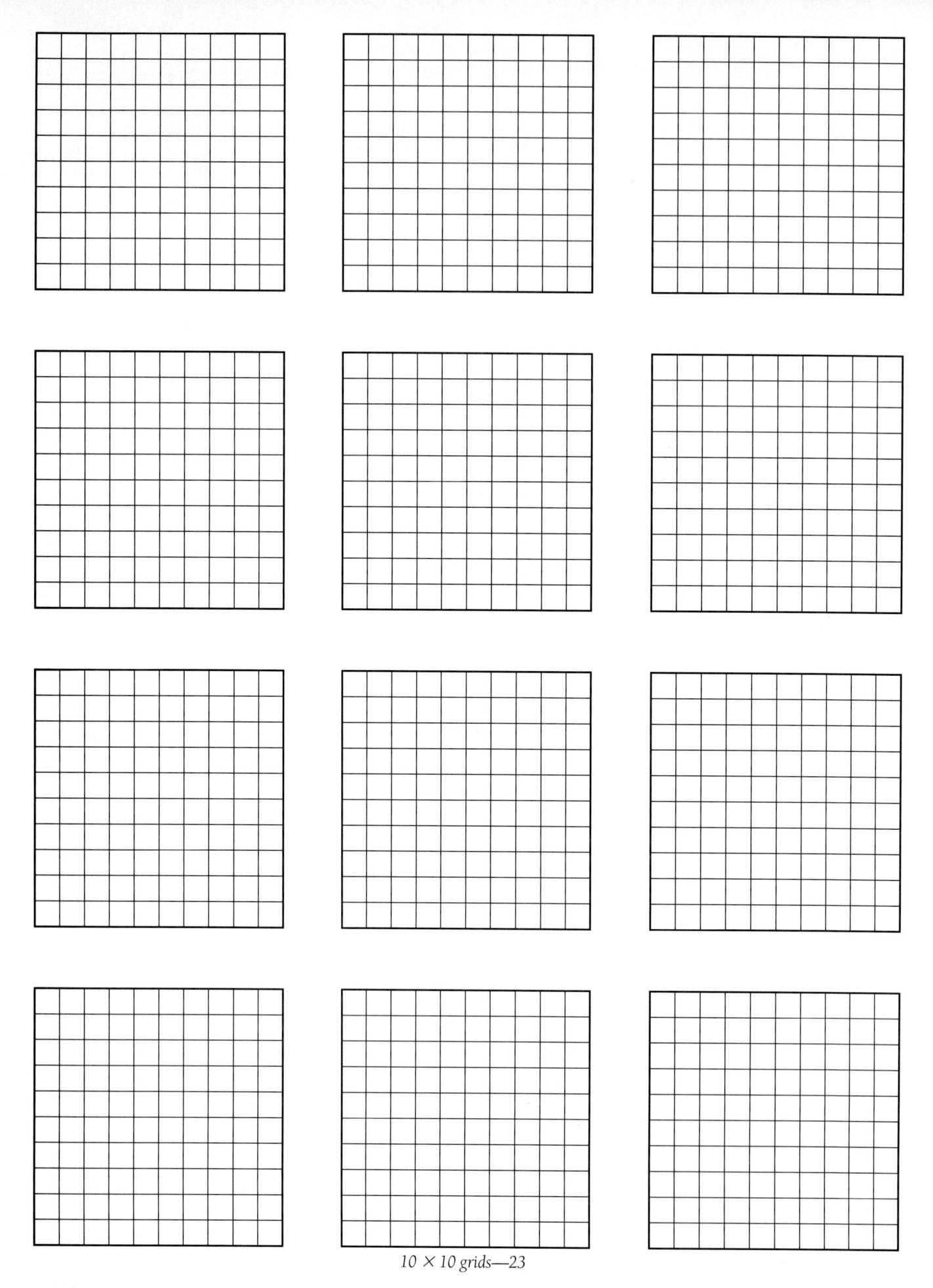

Copyright © 1994 by Longman Publishing Group.

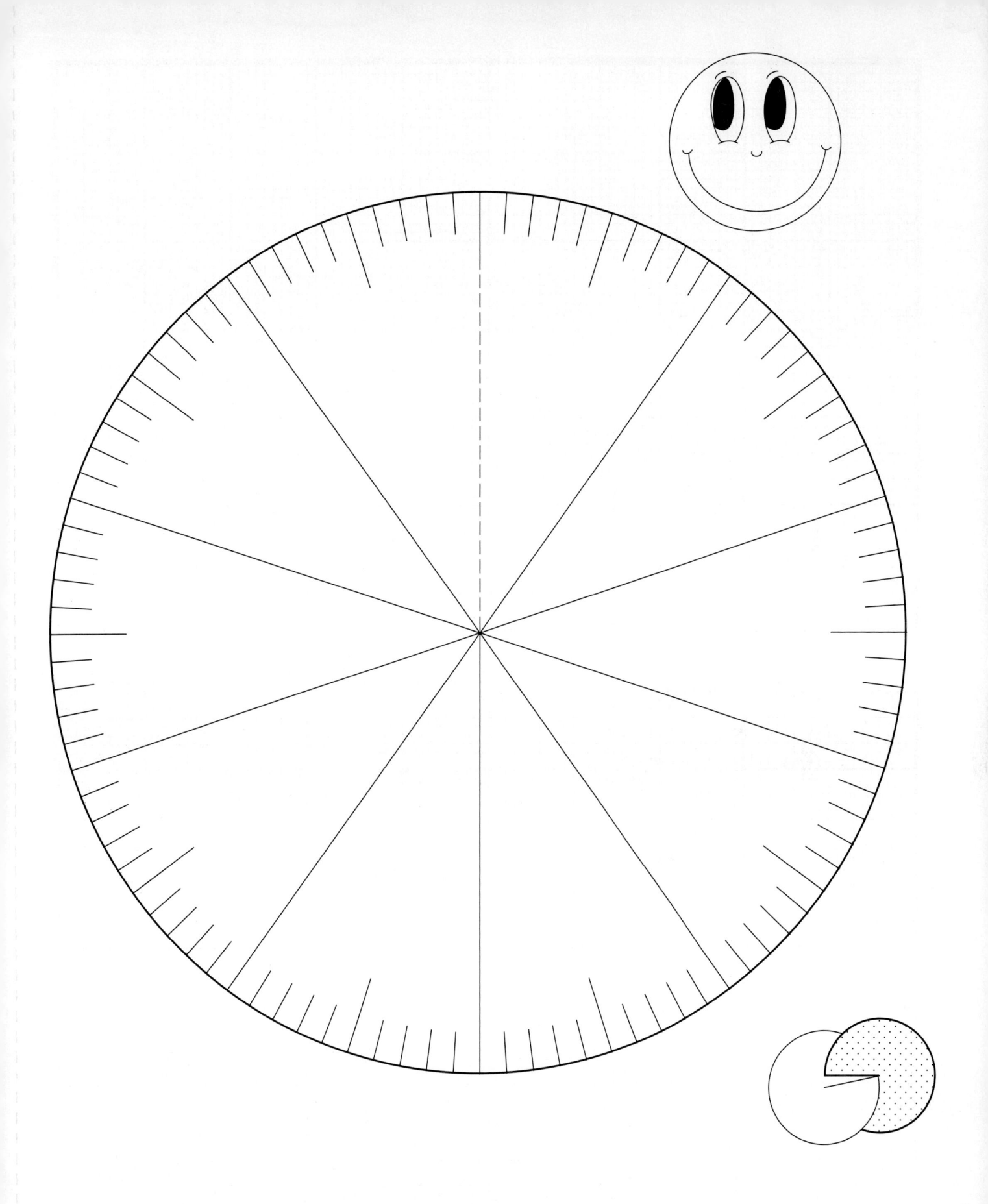

Hundredths disk—24

Copyright © 1994 by Longman Publishing Group.

10,000 grid—25

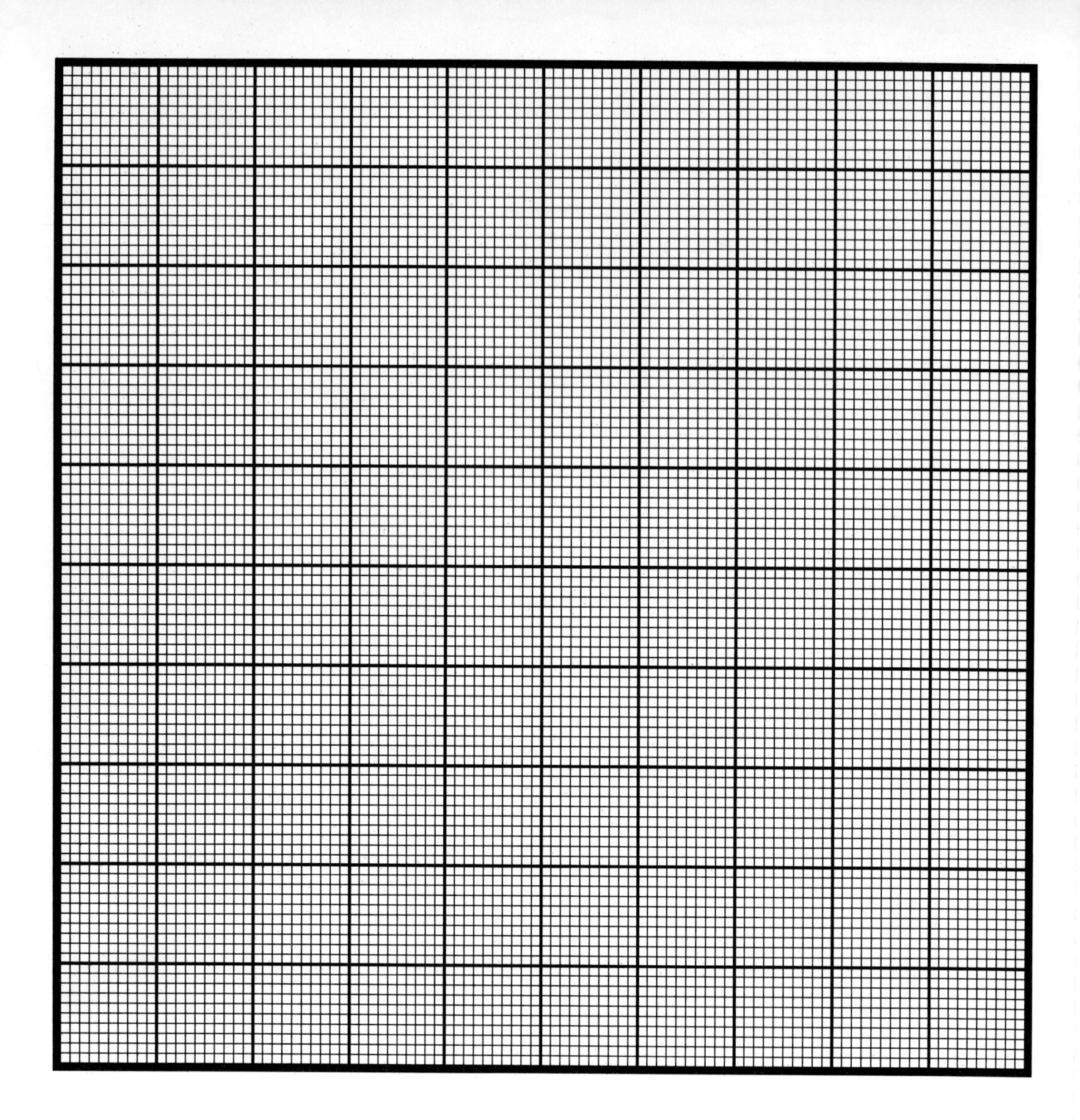

Decimal/fraction sieve—26

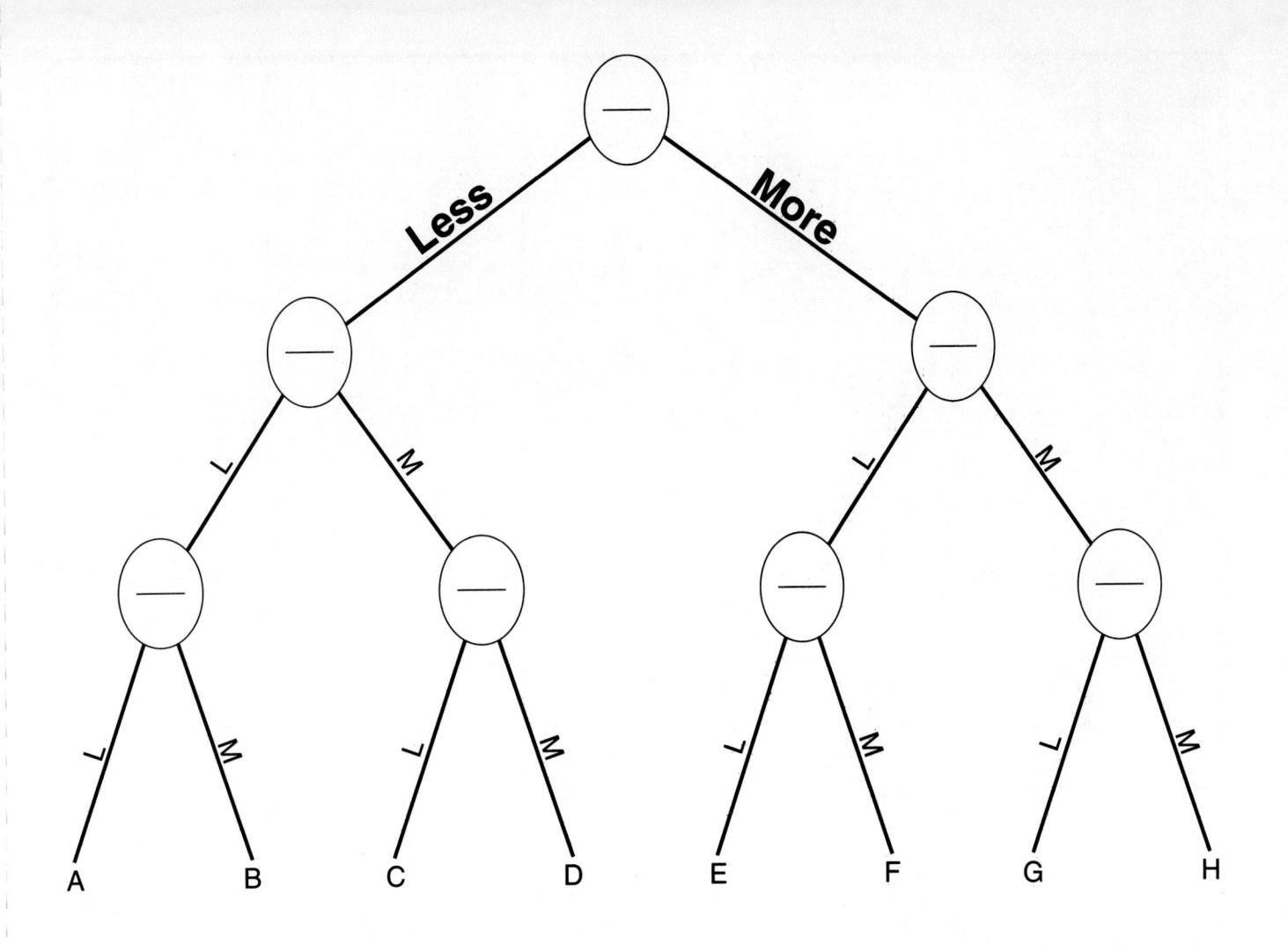

2-cm square grid—27

1-cm square grid—28

0.5-cm square grid—29

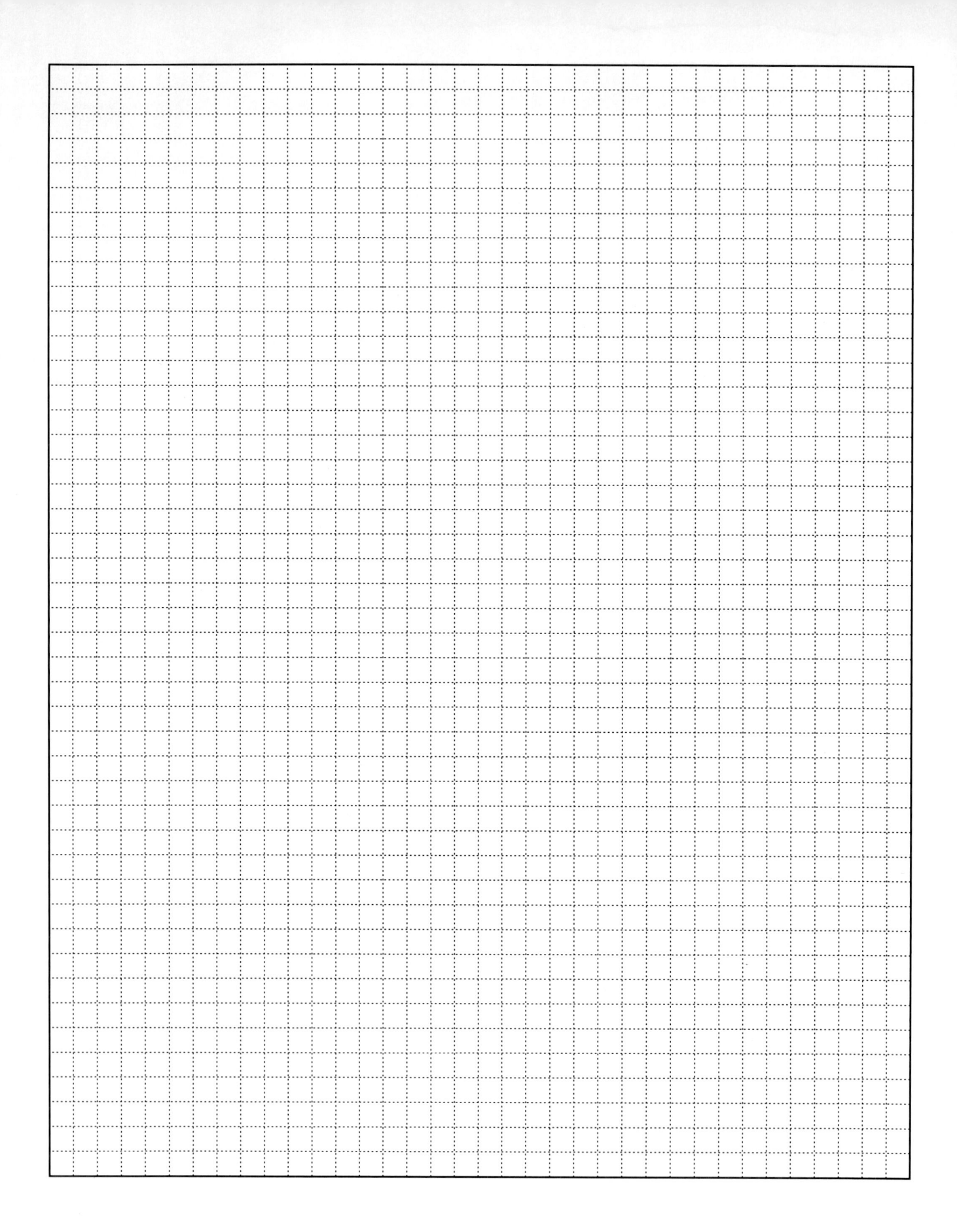

1-cm square dot grid—30

2-cm isometric grid—31

Copyright © 1994 by Longman Publishing Group.

1-cm isometric dot grid—32

Copyright © 1994 by Longman Publishing Group.

1-cm square/diagonal grid—33

Copyright © 1994 by Longman Publishing Group.

1-cm hex grid—34

Copyright © 1994 by Longman Publishing Group.

Geoboard pattern—35

Copyright © 1994 by Longman Publishing Group.

Geoboard recording sheets—36

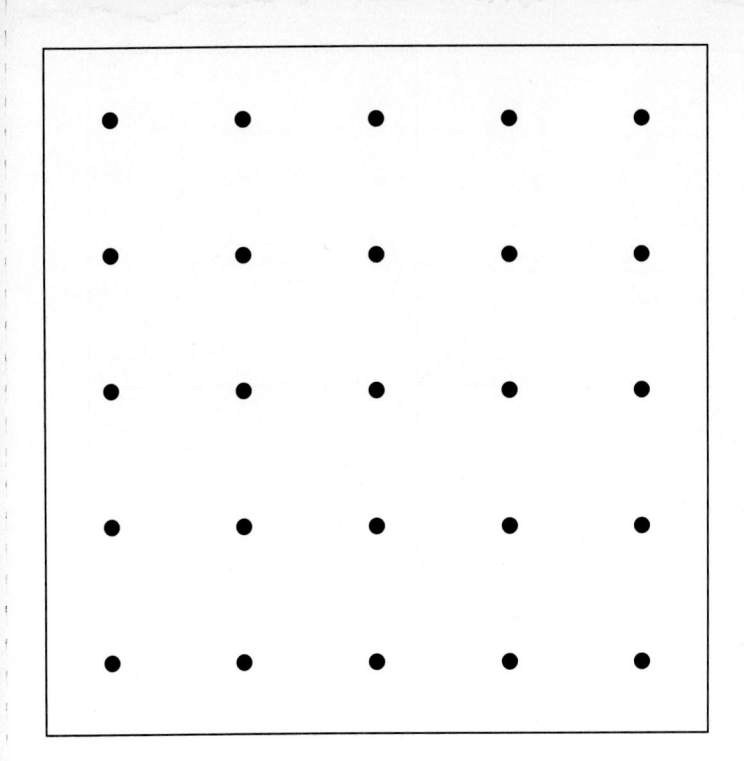

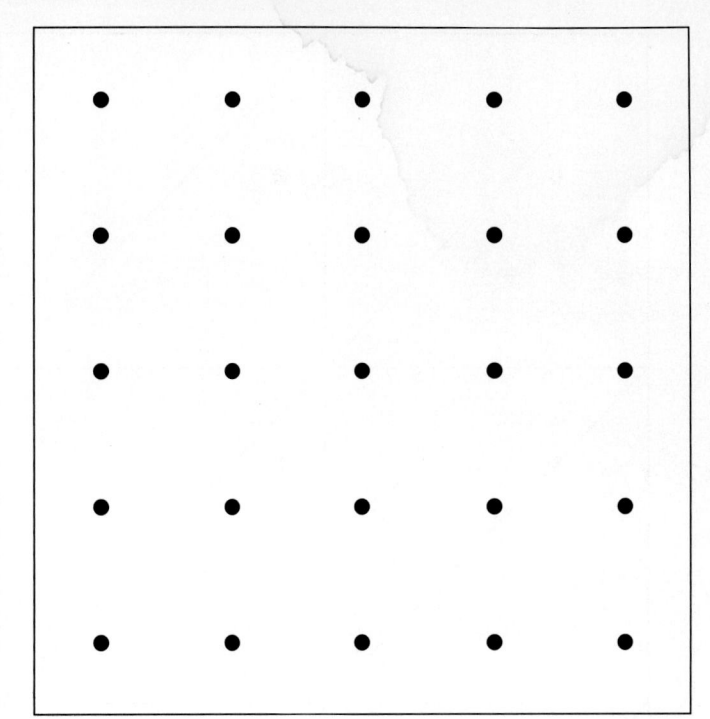

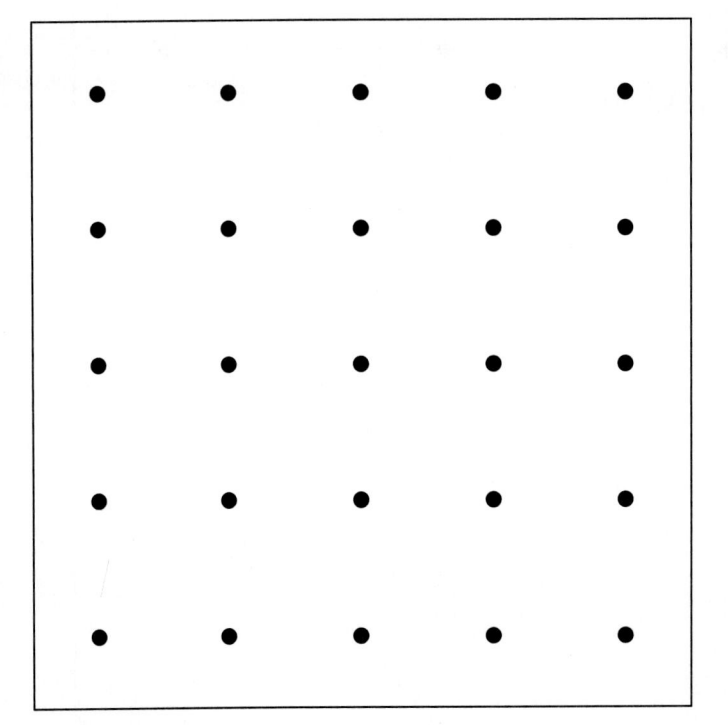

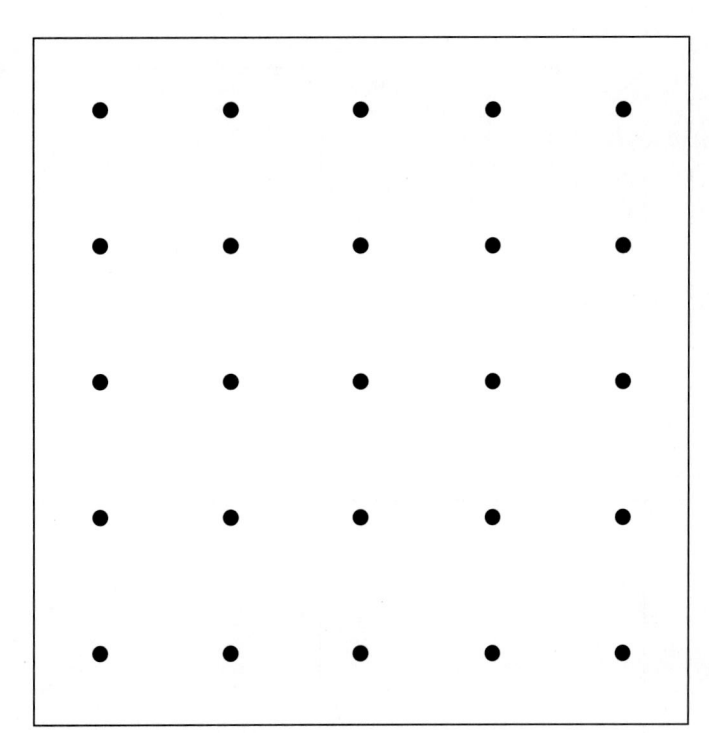

Copyright © 1994 by Longman Publishing Group.

Right-triangle tiles—37

Copyright © 1994 by Longman Publishing Group.

5-cm triangle tiles—38   Copyright © 1994 by Longman Publishing Group.

Tangrams—39

Woozle cards—40

Copyright © 1994 by Longman Publishing Group.